MW00995487

COLLEGE PHYSICS

COLLEGE PHYSICS

Putting It All Together

Ron Hellings
Montana State University

Jeff Adams
The Pennsylvania State University

Greg Francis
Montana State University

UNIVERSITY SCIENCE BOOKS
MILL VALLEY, CALIFORNIA

University Science Books
Mill Valley, California
www.uscibooks.com

Editor: *Jane Ellis*
Production manager: *Julianna Scott Fein*
Manuscript editor: *Lee Young*
Proofreader: *MaryEllen Oliver*
Illustrator: *Laurel Muller*
Text design: *Yvonne Tsang at Wilsted & Taylor*
Cover design: *Yvonne Tsang at Wilsted & Taylor*
Compositor: *Laurel Muller*
Printer and binder: *Edwards Brothers Malloy*

This book is printed on acid-free paper.

Print ISBN 978-1-938787-93-5
eBook ISBN 978-1-938787-95-9

Library of Congress Cataloging-in-Publication Data

Names: Hellings, Ronald W., author. | Adams, Jeffrey P., author. | Francis, Gregory E., author.
Title: College physics : putting it all together / Ron Hellings, Jeff Adams, Greg Francis.
Description: Mill Valley, California : University Science Books, [2017] | Includes index.
Identifiers: LCCN 2017010603| ISBN 9781938787935 (alk. paper) | ISBN 1938787935 (alk. paper) | ISBN 9781938787959 (eBook) | ISBN 1938787951 (eBook)
Subjects: LCSH: Physics--Textbooks.
Classification: LCC QC21.3 .H45 2017 | DDC 530--dc23
LC record available at https://lccn.loc.gov/2017010603

Printed in the United States of America
10 9 8 7 6 5 4 3 2

To our wives and families for their continual support
and to all of our former students who taught us so much.

BRIEF CONTENTS

CONTENTS

PREFACE

What should a student expect to find in a physics textbook? What should be included that the reader will find useful and worth his or her effort? And what things that are commonly included in physics textbooks might effectively be left out? These are the questions that motivated us as we produced this book. We would like to preface our book by stating how it all worked out. We would like to explain what the reader may expect to find and not to find in these pages.

First, most students in general college physics classes will depend on lectures by the professor for the introduction of new material, for explanations of the principles that describe how the universe behaves. But new material will often not be grasped at the first presentation. An important goal of a textbook should be to provide a second look at the principles the student has learned in class. We want to be as clear as we can; we want to get quickly to the point of the principles being taught, without getting off on tangents; but we also want to provide enough background information that the situations where the principles apply are clearly understood.

Second, with an eye to learning where the principles apply, we provide many worked examples in which we use the principles to solve sample problems. But we do not feel that our examples need to explain how to solve all physics problems that the student may ever encounter. On the contrary, the goal of the worked examples is simply to help the reader understand the principles by showing typical ways in which they apply. In particular, students should not expect to be able to read a problem from the back of the chapter and then browse through the worked examples to find one that tells them exactly how to solve their problem. That would mean that we were teaching procedures, not principles.

Third, the path the student should follow through the textbook should be clear. Many textbooks attach boxes and sidebars and web links, places where the reader will find all kinds of supplemental material to their narrative. These might provide things such as biographies of famous physicists, block diagrams showing how to solve certain kinds of problems, summaries of reasoning strategies, concept checks, etc. This often leaves the reader not knowing whether to stop and follow a parallel path or jump forward to stay with the text. We provide a single string to follow. Even when we break the narrative to display a worked example, the example is meant to be read in sequence. It will usually be introduced in the main text and then discussed in the text afterwards.

Fourth, the text should be readable. We find that a more informal style is easier to follow and more interesting. We talk with the reader about the physics. If we find something funny about our subject, we will share it with the reader. If we know of something that is difficult to understand, we will try to summarize and restate it in clear, short sentences.

Fifth, the entire text should be short and to the point. Most textbooks spend pages and pages explaining interesting ways in which the principles of the chapter can be used to understand everyday things as diverse as the tides in the Bay of Fundy or the iridescence of dragonfly wings. We understand why books do this. We, ourselves, are fascinated by these applications and would dearly love to tell you all about them. But it all takes time to read and, truth be told, it does not help the student learn the physics any better or any more quickly. You will find a few such applications in our pages, but only a few.

In summary, let us just say that we have tried to always keep in mind that this book is for you, not for us.

COLLEGE PHYSICS

1

INTRODUCTION

You may have heard that physics is hard. It isn't. So why does it have this reputation? We think it's because your physics class is fundamentally different from most of your other classes. In some of your classes, you find hundreds of facts that you need to learn. You have probably figured out how to do well in these classes. You read the textbook, highlighting the significant facts in yellow; you commit these facts to memory just before the test; you find that the test asks you for the facts you have memorized; and so you get your "A." In other kinds of classes, your textbook will show you how to solve a particular kind of problem; the exercises at the back of the chapter will give you practice solving the same kind of problems in the same way; and then, on the test, you will be given the same kinds of problems. In physics, we do not really try to teach facts *or* techniques; we teach principles. Each chapter will explain some new principles and then follow the explanations with worked examples. The goal of the examples is *not* just to show you how to solve certain kinds of problems, but to make it clear what the principle means and how it applies. Then, when you go to the problems in the back of the chapter, you will find some that are *not* exactly like the examples, though they may be solved using the same principles.

As you might guess, this process often leaves you with a little uncertainty in your mind about whether you are doing things in the right way. You will have come up with a path to the solution on your own; the book did not show it to you, and so you may worry whether you have solved the problem correctly. This is what we think makes physics seem hard. But once you see how all the principles fit together you will be amazed at how simple it really is.

That's the good news. The bad news is that, if you are like most of us, you have already taught yourself a lot of physics while you were growing up. You tried to make sense of the physical world around you and you invented rules that the world appears to follow. You learned to put your car in second gear and ease on the gas slowly when stuck on mud or ice. In fact, the gear you choose neither helps nor hurts you in getting unstuck. We call these kinds of things "misconceptions." Much of the learning of physics that you will do in this class will actually be unlearning misconceptions. We have spent much of our careers studying the most common of these misconceptions and looking for ways to help students unlearn them. We have used what we have learned from our research in physics education to inform this book.

Your authors have taught together at Montana State University in Bozeman, Montana. Between us, we have taught over 25,000 students in more than 300 classes. (Yes, we are getting old.) We discovered during this experience that most of our students were not finding the traditional textbook to be a useful resource. Even the students who were reading the text faithfully were usually just letting the words pass in front of their eyeballs, searching for the end of the chapter. Many only skimmed the text looking for examples that might help them solve the homework assignments. We also noticed that before every exam we would have students visit our office with the same disturbing message. "I have read the textbook and done all the homework. But I still don't get it." We wanted to write a book for those students, for you. With each major topic we have tried to picture a frustrated student in our office, and we have attempted to paint the big picture in a way that the students leave the office with a big smile, asking "Oh, is that all there is to it?" We hope we have created a useful tool that will actually help you.

To the end of actually helping you, we have been careful to discuss only those things that you will truly need to know to be able to solve your physics problems. This keeps your reading time to a minimum. We will occasionally derive the formulas that express a principle, because this is often necessary if you are to understand how to apply it. And, if you compare our textbook with others, you might see that we often spend a little more time explaining a principle, rather than just telling you what it is. This we do for the same reason. But we generally try to avoid unnecessary topics or non-essential applications, even when we ourselves find these fascinating. After all, this book is for you, not for us.

Our advice is to read each chapter carefully. Pay special attention to the worked examples and go through them in detail, making sure that you understand not only what was done, but *why* it was done. Then, before you go to the homework problems, ask yourself the following questions—What important principles have I learned? To what situations would they apply? What behavior do they predict? Then go do the assigned homework problems, remembering that whatever struggle they require as you work through them will repay you liberally with knowledge and insight. It is as you work through the problems that you will come to understand the principles. Once you have that, you can go on to take any professor's exam with confidence.

1.1 Measurement

The basic goal of physics is to be able to make a set of measurements of the properties of some physical system at some initial time (balls on a pool table, a pot of water on a stove, a battery hooked up to a light bulb, etc.), and then to use the laws and principles of physics to correctly predict what the measurements of the properties of the system will be at some future time. So let's talk about measurement.

Measurement of a physical property requires two things. First, we must have a *unit* for the measurement. A unit is some physical object that serves as the standard for that type of measurement. The standard for a distance measurement is a meter stick, a ruler that everyone agrees to use by international agreement. Second, we must have a *process*, or an operation, that compares the property of the thing being measured with the unit. For example, we can measure the distance between two points by placing meter sticks end-to-end, beginning at the first point and ending at the second. The distance is the number of meter sticks required. Note that we have required both a unit and a process to *define* what we mean by "distance."

For all of the physics we study in this book, there are only four fundamental measurements that are required. These are time, distance, mass, and electrical charge. Our discussion of electrical charge will be put off until Chapter 16. Here we discuss the three remaining basic measurements we will need—time, distance, and mass. For most of what we do, we will use the International System of units (or SI, for *Système international*), a system based on the meter for distance, the kilogram for mass, and the second for time. They are also known as *MKS* units, the initials standing for the three basic units used.

Time. The SI unit of time is the second (s). It was originally defined as a fraction of the solar day. Since 1967, the SI definition has been: "The second is the duration of 9,192,631,770 periods of the radiation corresponding to the transition between the two hyperfine levels of the ground state of the cesium 133 atom." In short, time is defined by counting cycles from a highly stable cesium oscillator. These are the "ticks" of the clock. Because the unit is based on a fundamental property of matter, anyone with a good laboratory apparatus can create the fundamental time unit in their lab. The measurement of a time interval is accomplished by counting how many times the clock ticks between the beginning and the end of the interval.

Distance. The SI unit of distance is the meter (m). In 1785, a brass bar was constructed in Paris with a length that was one ten-millionth (10^{-7}) of the estimated distance from the North Pole to the equator, along a meridian passing through Paris. This became the international unit of length. In 1874, the brass bar was replaced by a platinum-iridium bar, kept in the newly created International Office of Weights and Measures (*Bureau international de poids et mesures*, or BIPM) in Sèvres, on the outskirts of Paris. In 1960, the definition was changed to be a number of atomic wavelengths. Finally, in 1983, the SI meter was again redefined: "The meter is the distance traveled by light in vacuum during a time interval of one 299,792,458th of a second." Thus, the meter is now based on the definition of the second, in such a way as to make the speed of light exactly 299,792,458 m/s, by definition. Because of this new definition, it is possible to create an exact fundamental meter unit in any laboratory that has an accurate fundamental clock. The measurement of distance is defined by placing calibrated meter sticks end-to-end.

Mass. The SI unit of mass is the kilogram (kg). It is the only unit, as of this writing in 2017, that is still based on a particular artifact rather than on any fundamental property of matter. Since 1889, the international standard kilogram has been a particular cylinder of platinum-iridium housed in the BIPM (see Figure 1.1). Because of its definition, there is only one standard kilogram. Any laboratory that needs a precise unit of mass must create it by comparing with the kilogram in Paris. At present, several alternative definitions of the kilogram are being considered that rely on fundamental properties of matter, rather than on the arbitrary properties of this particular lump of metal. But, for the present, this is all we have. The process used to define mass is also interesting, and perhaps a little surprising. As we will see in Chapter 5, "mass" is a measure of the inertia of a body—its resistance to forces applied to it. However, the process used to measure mass does not employ the inertial property of an object, but its weight—the force that gravity exerts on it. The procedure involves placing an object whose mass is to be measured on one side of a balance beam and putting the kilogram, or carefully verified fractions of the kilogram, on the other. The *mass* of the object is the number of kilograms that it takes to balance the apparatus.

Figure 1.1 The prototype kilogram, housed under 2 bell jars in Sèvres at the BIPM.

The standard units of the SI system are not always of appropriate size for all scales of objects. For atomic-scale physics, we would want a unit of length that is very much smaller than a meter and for astronomical objects we would want a unit that is very much bigger. Starting with the fundamental SI units, other units have been defined by adding prefixes to the name of the unit, as displayed in Table 1.1.

TABLE 1.1 Metric Prefixes

Bigger			Smaller		
Prefix	Abbreviation	Meaning	Prefix	Abbreviation	Meaning
			centi	c	$\times 10^{-2}$
kilo	k	$\times 10^{3}$	milli	m	$\times 10^{-3}$
mega	M	$\times 10^{6}$	micro	μ	$\times 10^{-6}$
giga	G	$\times 10^{9}$	nano	n	$\times 10^{-9}$
tera	T	$\times 10^{12}$	pico	p	$\times 10^{-12}$
peta	P	$\times 10^{15}$	femto	f	$\times 10^{-15}$

Thus, a picosecond (ps) would be 1 trillionth (10^{-12}) of a second and 3 megagrams (Mg) would be 3×10^{6} grams, or 3×10^{3} kg.

Before leaving the subject of units, we should mention the two lesser-used systems. The *cgs* system is based on centimeters, grams, and seconds. Its units for length and mass are smaller than the corresponding *MKS* units, though (go figure!) this system is often used by astronomers. (We suspect that the reason is that because astronomical numbers are so . . . astronomical, that a few more powers of 10 don't really matter.) The English system, also called the *fps* system after its fundamental units, is still commonly used in the U.S. and partially in the U.K. English units use the second for time and the foot for distance, but the third fundamental unit in the English system is not a unit of mass, but of weight, or force. The English unit of weight is the pound (lb, for the Latin *libra*). Because of our familiarity with these units in the U.S., we will occasionally use them in this text in the situations where they are most familiar to us.

1.2 Arithmetic

Yes. We know that you already know how to do arithmetic. That's just the point. Because you already know how to do arithmetic, you already know several important things about how physics works. So let us remind you of some of the things you already know.

Multiplication. You already know that 3×2 means $2 + 2 + 2$. The 2 is the thing and the 3 is the number of times you add it to itself. Similarly, if we say that something is 3 meters long, it means that the length is the same as if you added meter sticks, one after the other, 3 times. So a length of 3 meters can be thought of as 3 times one meter, or $3 \times$ meter. Units can be thought of as things that you multiply some number of times, thereby turning a pure number (3) into a measured quantity (3 meters).

Addition and Subtraction. You already know that you can't add apples and oranges. Addition and subtraction just don't make any sense unless you are adding the same kinds of things together. So if you come to a place in a calculation where you are adding a mass to a length, you know you are making a mistake. Either you have misinterpreted a formula, or you have made an algebra mistake, or something. On the other hand, there is nothing wrong with adding 1 km to 1 m, as long as you don't expect to get 2 of anything. Both m and km are lengths, and, as we just saw, 1 km means $1 \times$ km, or $1 \times (1000 \times \text{m})$, so 1 km + 1 m = 1000 m + 1 m = 1001 m.

Division. You know that dividing something by 2 means to distribute whatever you have into 2 equal parts. Thus 6 dollars ÷ 2 = 3 dollars in each part. But consider a case where you go to a bakery with your 6 dollars and use it to buy three baguettes. What is the price of the bread? What you do, of course, is take the 6 dollars and divide them evenly over the 3 loaves, as shown in Figure 1.2. The unit of the bread, the loaf, becomes the way to distribute the dollars. You have divided the 6 dollars not by a number, but by a quantity of bread. Six dollars divided by three loaves equals two "dollars per loaf," written dollars/loaf. That is the unit of price. The "per" tells us that, to define the quantity called price, we have divided one quantity that has fundamental units by another quantity that has fundamental units. The unit of price is a *derived unit*. Dollars may be measured by counting; loaves of bread are measured by counting; but price is defined only by dividing the one quantity by the other.

Figure 1.2 Dividing 6 dollars by 3 loaves gives a quotient of 2 dollars/loaf.

Multiplication Again. Anyone who has ever tried to use a wrench to unscrew a rusty nut from a bolt knows what is required. If the nut isn't turning, you can do one of two things. You can either apply more force to the wrench or you can get more leverage by applying your force further out along the handle of the wrench. Your ability to force the nut to turn is called the torque, a quantity we will discuss properly in Chapter 9. But let us just tell you here that it turns out that the ability to force the nut to turn is exactly proportional to both things. If you double the force, you double the torque and if you double the lever arm, you double the torque. If you doubled both, you would get four times as much torque. The torque is therefore equal to the product of the force and the lever arm (Figure 1.3). But what does this mean? If I measure force in pounds and lever arm in feet, how do I multiply pounds times feet? I know that 4 × 3 ft = 12 ft, because four times anything means adding it to itself four times. But how can I add three feet to itself "four pounds times"? The answer is that, to define the torque, we will have to slightly redefine what we mean by multiplication. We will *define* 3 feet times 4 pounds to mean

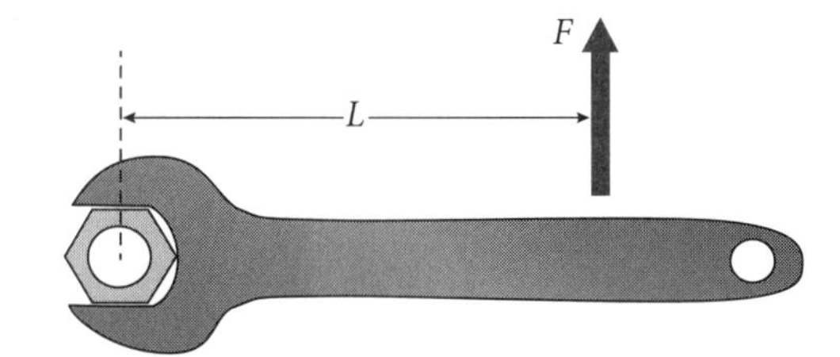

Figure 1.3 The torque is the lever arm times the force.

$$3 \text{ ft} \times 4 \text{ lbs} = 3 \times 4 \text{ ft} \cdot \text{lbs} = 12 \text{ ft} \cdot \text{lbs}.$$

That is, we agree to do the multiplication of the integers in the usual way and be left with a new unit, a ft · lb. This new unit for torque is another derived unit. There is no physical object called a ft · lb, nothing I can count, nothing that I can picture. This unit arose only as we defined a useful new quantity by multiplying together two other quantities. We will see a lot of this kind of thing as we go through this text, and every time we multiply or divide two physical quantities to create a new quantity, we create a new derived unit.

Naming Units. As we said, the three fundamental units are the second, the meter, and the kilogram. Any other units, until we get to electrical charge in Chapter 16, are derived units. They come from defining new quantities by multiplying or dividing fundamental measured quantities. Sometimes, the new unit is given a name, typically the name of some scientist or other eminent person (generally deceased) who helped advance that particular field. For example, someday scientists might decide to name the ft·lb of torque the "chubbychecker," after the rock and roll star who contributed so much to the twist. Future Ford repair manuals might then require tightening flywheel bolts to 150 chubbycheckers.

1.3 Working with Units

Thinking of units as multiplicative factors, as we did in the last section, helps us see how to treat them when we are calculating things. We all know that if we travel at a speed of 80 miles per hour (mi/hr) for 2½ hours, we will travel a distance of 200 miles. The formula we used to do the calculation in our heads was $d = vt$, with d for distance, v for velocity (or speed), and t for time. The calculation we performed was

$$d = vt = 80\frac{\text{mi}}{\text{hr}} \cdot 2\frac{1}{2}\,\text{hr} = 200\frac{\text{mi}}{\cancel{\text{hr}}} \cdot \cancel{\text{hr}} = 200\ \text{mi}.$$

Note how we treated the hr units in the third step, cancelling them like we would any other algebraic factor.

As a more complicated example, let's look at a more complicated formula. This one is a formula for calculating the constant acceleration of an object that started from rest and attained a velocity v as it moved a distance D. What the formula is for doesn't really matter here (we talk about it properly in the next chapter); the point is to watch how the units are combined during the calculation. The formula for the acceleration is

$$a = \frac{v^2}{2D}.$$

The units of D are meters (m) and the units of v are meters per second (m/s). So the units of acceleration must be found from:

$$\frac{v^2}{2D} \Rightarrow \frac{(\text{m/s})^2}{\text{m}} = \frac{\text{m}^2}{\text{s}^2}\frac{1}{\text{m}} = \frac{\text{m}}{\text{s}^2}.$$

Note how the 2 is a unitless number that does not affect the final units at all. Note also that dividing by (m) is the same as multiplying by (1/m). Finally, note how the units are cancelled to arrive at the derived units for the calculated quantity, the acceleration.

Often we are given some quantity in one set of units, but we need it in another. When this happens, it is important to know how to convert units. This process is often confusing to students, who sometimes multiply when they should be dividing or divide when they should be multiplying. There is, however, one sure-fire way to convert units that is easy to remember and never goes wrong. It is this. Suppose you are given a length in furlongs (a quaint British unit, equal to 1/8 mile, based on the length of a furrow that an ox could plow before it had to rest) and that you need to convert it to meters. Say that the length is

$$L = 3.5\ \text{furlongs}.$$

Then that is the correct length, and we should do nothing to change the length, only the units. There are two things you need to remember if you want to change the units. First, multiplying something by 1 does not change its value. If we multiply the length by 1, it will still be the same length. Second, anything divided by itself is 1. Since 1 mile = 8 furlongs, it is true that

$$\left(\frac{1 \text{ mi}}{8 \text{ furlongs}}\right) = 1.$$

We also know that 1 mile is about 1.6 km and that 1 km is 1000 m. Now the way to transform units is to multiply the length L by several cleverly chosen ones, chosen in such a way that we cancel the units we don't want and end up with the units we do want. We do it like this:

$$L = (3.5 \text{ furlongs})(1)(1)(1) = (3.5 \cancel{\text{furlongs}})\left(\frac{\cancel{1 \text{ mi}}}{8 \cancel{\text{furlongs}}}\right)\left(\frac{1.6 \cancel{\text{km}}}{\cancel{1 \text{ mi}}}\right)\left(\frac{1000 \text{ m}}{\cancel{1 \text{ km}}}\right) = 700 \text{ m}$$

The length has not changed; L is still 3.5 furlongs. But we now know that it is also 700 m. Note that in each set of parentheses, we have put one quantity in the numerator and put an equal quantity in the denominator. The choice of which goes where is determined by what unit we want to cancel algebraically at each step.

As another example, let's convert a speed of 180 km/hr into m/s.

$$v = 180 \frac{\text{km}}{\text{hr}}\left(\frac{1000 \text{ m}}{1 \text{ km}}\right)\left(\frac{1 \text{ hr}}{3600 \text{ s}}\right) = 50 \frac{\text{m}}{\text{s}}$$

Note that, in the second set of parentheses, the 1 hr was put in the numerator so that it would cancel the hr we wanted to eliminate in the denominator of the km/hr.

As a help, a number of equivalences between units are given in Appendix C. And, while we're at it, let us also direct your attention to Appendix B where you will find values for several fundamental constants and several useful physical constants.

1.4 Math Requirements

One last thing. The mathematics background you need in order to work through this book includes high-school algebra and trigonometry, but not calculus. We promise. You will find a short review of the trigonometry we use in Chapter 3. If even your algebra is a little bit rusty, you might also want to take a look at Appendix A.

1.5 Summary

From this chapter, you should have learned the following:

- You should understand that we will use three fundamental quantities—time, distance, and mass. All other quantities are defined quantities, formed by multiplying and dividing these fundamental quantities by each other. SI units for the fundamental quantities are the second for time, the meter for length, and the kilogram for mass. The units of defined quantities are derived units, formed by multiplying or dividing fundamental units by other fundamental units.
- You should be able to calculate things while carrying the units along in the calculation, cancelling units like you would factors in an algebraic expression.
- You should be able to convert units for fundamental quantities or defined quantities by sequentially multiplying the quantity by cleverly chosen forms of the number 1.

PROBLEMS

In the rest of this book, we will classify problems by their difficulty. A 1-star problem represents a simple application of a principle, or may be thought of as an exercise that helps clarify a single concept or a quantity. A medium-difficulty problem gets 2 stars. And, as you might suspect, 3-star problems are more difficult, either because they require you to merge together multiple concepts or because they require a more complicated path to a solution or more complicated algebra techniques to use in your solution. We thought we would be nice and not put any 3-star problems in this first chapter.

You may notice, as you read through the problems, that 1-star and 2-star problems are mixed up. This is because the order of the problems was set up to reflect the order of topics in the chapter, not the difficulty of the problems. To help you connect the problems to the readings, we have placed section numbers next to some of the problem numbers. These indicate the section you should have previously read in order to solve that problem and any subsequent problems in the chapter.

1.1 * 1. Legend suggests that the English unit of length, the yard, was officially defined in the twelfth century as the distance from the tip of Henry I's nose to the end of his outstretched thumb. If you were king or queen of England, how many inches would there be in a yard based on your nose and thumb? [Yes, you just go measure it.]

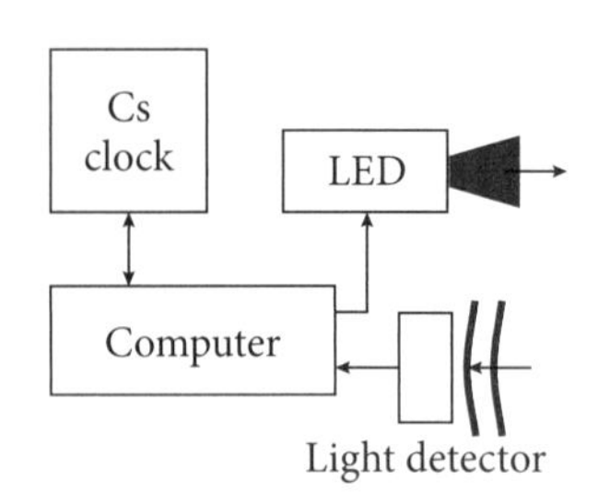

** 2. A computer is hooked up to an LED flash and a light detector. If the computer's clock speed is controlled by a laboratory cesium standard, the time between a flash emitted from the LED and the received pulse of reflected light in the detector can be accurately measured by the computer. Explain how this apparatus could be used in a laboratory to produce an accurate length that is one meter long. [Hint: There are several valid ways you might set this up.]

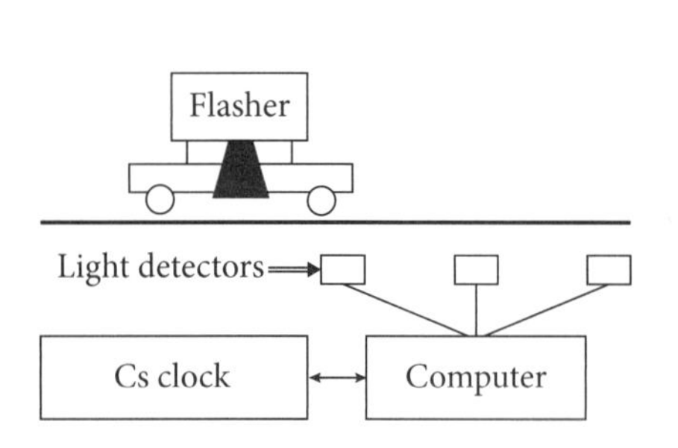

** 3. You have constructed a flasher clock that will produce a flash of light once every second. You can test the accuracy of that clock by detecting the flashes in a light detector hooked up to a computer that is controlled from a cesium standard clock. The computer-measured interval between the flashes it detects will tell you if the flashes from your flasher clock are exactly one second apart or not. But you wonder if your flasher clock will work properly if it is moving. So you set your flasher clock on a cart in your laboratory and start it down a track at 1 m/s. Devise a precise method to measure the accuracy of the moving flasher clock. [Hint: You can have as many light detectors connected to your computer as you want, placed wherever you want to put them.]

* 4. What is the mass in kg of a milligram of silver?

* 5. Data storage in a computer is measured in quantities called bytes. How many megabytes are there in a terabyte?

In the next five problems (6–10), we give formulas for calculating various quantities and ask you to analyze the units and do the calculations when asked. You are not expected to remember these formulas or to understand what the quantities represent.

1.3 *6. In industry, the effort required to produce a product is measured by multiplying the number of employees working on the project (N) by the average number of hours worked by each employee (T). The formula is $E = NT$. If a project requires 7 employees, each working for 12 hours, what is the effort required? What are the units of this "effort"?

**7. The distance traveled by a body with initial velocity v_0 and constant acceleration a is given by the formula:

$$x = v_0 t + \frac{1}{2}at^2.$$

If $a = 5\ \text{m/s}^2$ and $v_0 = 7\ \text{m/s}$, find the distance traveled after time $t = 2\,\text{s}$, taking care to carry and cancel all units. Does your result have units of distance? When you did the addition, were you adding things with the same units as each other?

**8. In the following three formulas, x is a distance in meters, v is a velocity in m/s, and a is an acceleration in m/s^2. Which of the formulas cannot be correct because they would involve adding together unlike things? Can we conclude that the formulas properly adding like things are actually correct?

a) $v = \dfrac{x - \frac{1}{2}at^2}{t}$ b) $a + \dfrac{v}{t} = 2\dfrac{x}{t^2}$ c) $x^2 + v^2 t = a^2 t^4$

**9. The thermal resistance of a piece of thermal insulation is defined by the formula:

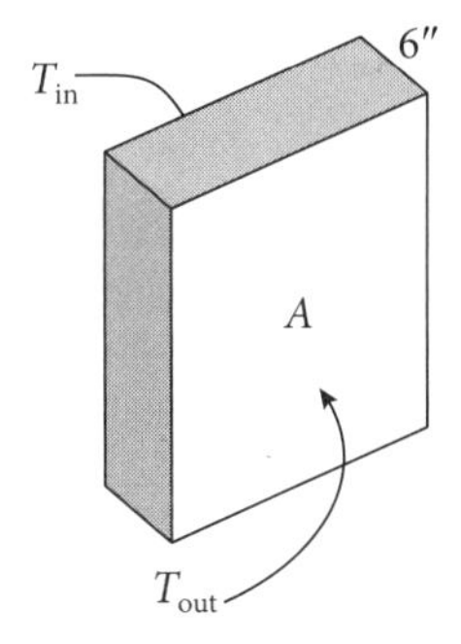

$$R = \frac{A(T_{in} - T_{out})}{Q}t$$

where Q is the heat in BTUs (British Thermal Units) transferred during a time t in seconds, A is the area of the wall in ft^2, and T_{in} and T_{out} are inside and outside temperatures in degrees Fahrenheit. In these English units, a common 6-inch-thick slab of fiberglass wall insulation has an R-value of 19. What are the units of this thermal resistance R?

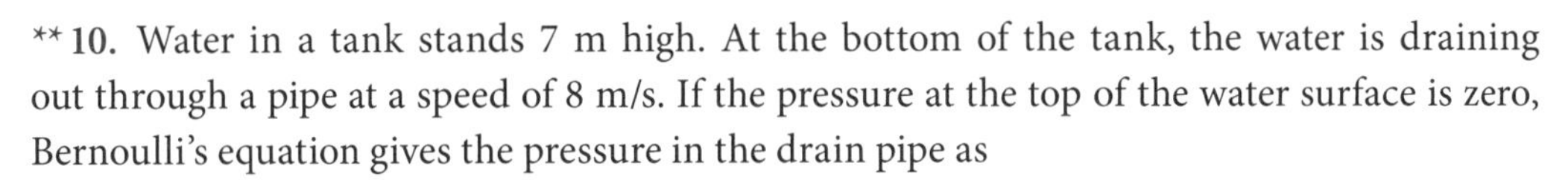

**10. Water in a tank stands 7 m high. At the bottom of the tank, the water is draining out through a pipe at a speed of 8 m/s. If the pressure at the top of the water surface is zero, Bernoulli's equation gives the pressure in the drain pipe as

$$P = \rho h g - \frac{1}{2}\rho v^2$$

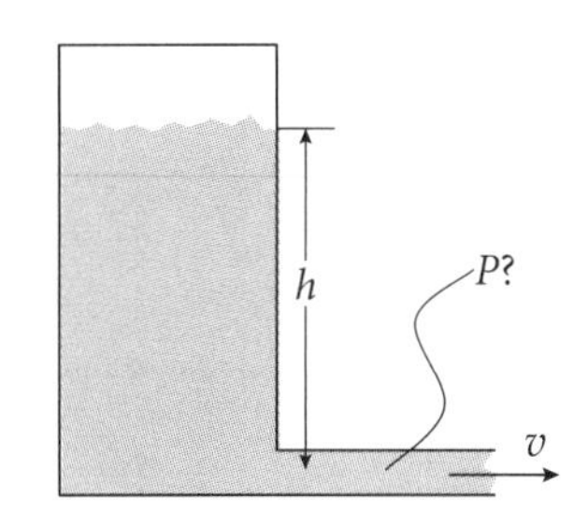

where ρ is the density of water (10^3 kg/m^3), h is the height of the water in the tank, g is the gravitational field strength (about 10 m/s^2), and v is the speed of the fluid in the pipe. [Note: Remember, we are not asking you to understand or remember this formula. We will not use it again.]

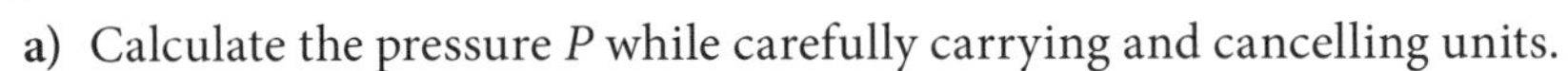

a) Calculate the pressure P while carefully carrying and cancelling units.

b) What are the units of pressure in this calculation? (This unit is named the Pascal, after Blaise Pascal, the 17th-century French scientist and philosopher who made important contributions to the physics of fluids.)

*11. Find the number of seconds in a day by multiplying $t = 1$ day by appropriately chosen forms of the number 1.

** **12.** Convert a speed of 3 cm/s into units of furlongs per fortnight. You will need to know that 1 fortnight = 2 weeks. [You will also need to know the number of days in a week, etc. But we expect we don't have to tell you *everything.*]

** **13.** The exact connection between the English standard of length and the SI unit of length is made by defining 1 inch = 2.54 cm exactly. Use this relationship, along with others you already know, to exactly convert 5 ft to meters.

** **14.** A light year is defined as the distance that a beam of light travels in one year. Convert a length of one light year to a length in kilometers. [Hint: Remember that the speed of light was given in terms of the defined meters and seconds on page 3. It is also found in Appendix B.]

** **15.** Floor space in an apartment is found by multiplying the length of a room by the width of the room, giving an area in ft^2. In Europe, floor space is measured in m^2.

a) Find the number of ft^2 in 1 m^2.

b) A 2-bedroom apartment in Paris advertises 42 m^2 of floor space. How many ft^2 is this?

** **16.** Figure out how to convert a volume in cubic meters (m^3) to a volume in cubic feet (ft^3) by finding the number of ft^3 in 1 m^3.

** **17.** A U.S. gallon is defined to be exactly 231 cubic inches (in^3). Find the number of cubic meters in a 5-gallon drum.

** **18.** The formula for the volume of a rectangular slab is $V = lwt$, where l is the length of the slab, w is its width, and t is its thickness. You have a garden plot that is 36 ft long by 24 ft wide, and you want to cover it in 3 inches of mulch. How many cubic yards of mulch will you need? (In case you come from someplace where English units are not used, let us explain that there are 12 inches in a foot and 3 feet in a yard.)

Note to the student: Problem statements throughout this book conform to the usual significant-figure conventions, with one exception. When we display a number without a decimal point, it should be taken to be a theoretically exact number. Our reason for this is that most of our assigned problems are set up to illustrate a theoretical principle, and we generally do not wish to add to a problem the complexity of worrying about its measurement accuracy. In these theoretically exact cases, other non-exact numbers in the problem, or a criterion of reasonable accuracy, demonstrating that you have done the problem correctly, would dictate the number of significant figures you would want to display.

2

ONE-DIMENSIONAL KINEMATICS

As we said in our Introduction, physics is mostly about predicting the future. We start with the description of some physical system at the present time and predict what its state will be at some future time. Most of the first half of this book is simply about figuring out where something is and where it is going to be. It is about motion. The description of the motion, without concern for the cause of the motion, is known as *kinematics*. This is what we do in this chapter. In Chapter 5, we will turn our attention to *mechanics*, the study of the *causes* of motion. But, for the moment, we just want to discuss how to characterize where something is and how it is moving.

To make things easy, we begin our study of kinematics by limiting ourselves to motion in one dimension. One-dimensional motion is motion along a straight line. It is called one-dimensional because a single number (dimension) is sufficient to specify the position of an object on the line.

2.1 The Coordinate System

Our first task is to define where something is. In one dimension, we can do this by imagining that we lay a ruler along the straight path so that, at any time, we can check the position of the object against the ruler and note the reading.

If we wanted to characterize an entire line, we would need a ruler that extended off to infinity in both directions. In this case, the zero of the ruler, the point we will call the *origin*, could be chosen to lie near the place where the motion takes place, but there is no reason that this has to be one place rather than another. Also, because the zero cannot be at the end of an endless ruler, the number scale for an entire line must be positive to one side of zero and negative to the other. But we are still free to choose which direction is to be positive and which is to be negative, a choice we will call the *orientation* of the ruler.

The combination of an origin and an orientation is called a *coordinate system*, and it may be defined however we please. The freedom to choose a coordinate system is one of our inalienable mathematical rights. In fact, we think that there ought to be a Physics Constitution or something that says, "Congress shall make no law establishing the origin of a coordinate system or restricting the free orientation thereof."

The coordinate system is a way of assigning an address to each point along the one-dimensional path. Giving the address defines the position of an object. A common convention is to use the variable x to stand for position. In Figure 2.1 (next page) we show a straight section of highway with mile markers set up as coordinates. A car at marker 14 along the interstate would be said to be at position $x = 14$ mi. That is its address.

Figure 2.1 The mile-marker number line (car and markers not to scale).

2.2 Displacement

As an object changes position, we say that it is undergoing *translation*. The amount by which the object translates is called the *displacement*. We generally use a letter D for displacement. The idea of displacement involves not only how far the object has moved, but also in which direction. Since we limit ourselves here to motion in one dimension, the direction of the displacement can be toward more positive numbers on our ruler (a positive displacement) or toward more negative numbers (a negative displacement). Thus, if a car moved from an initial position at mile marker 14 to a final position at mile marker 7, we could write this as $x_{\text{initial}} = 14$ mi and $x_{\text{final}} = 7$ mi. The car has moved a distance of 7 miles, but it moved in the negative direction. We therefore say that the displacement of the car was $D = -7$ mi.

As we talk about displacement, and other quantities in this chapter, we will take advantage of a commonly used convention of using the Greek capital letter delta (Δ) to indicate a change. Displacement is the change in position, so we often write $D = \Delta x$ (pronounced "delta-x"). To understand the sign of the change in a quantity, let us think about a more familiar situation. If you want to find out how much your bank account balance changed today, you take the balance at the end of the day and subtract the balance at the beginning of the day. If the number is positive, you made money; if the number is negative, you lost money. In the same way, to find your displacement between an initial time and a final time, you start with your final position and subtract your initial position from it. We write this symbolically as

$$\Delta x = x_{\text{final}} - x_{\text{initial}}. \tag{2.1}$$

This convention—that the change is the final minus the initial—is something that we will use consistently throughout this book. In the case of our car, the displacement can be written following this same convention:

$$D = \Delta x = x_{\text{final}} - x_{\text{initial}} = 7 \text{ miles} - 14 \text{ miles} = -7 \text{ miles}.$$

Finally, let us define *distance* as the absolute value of the displacement. It is how far it is between your initial and final points, regardless of direction, and it has no sign.

2.3 Velocity

Most everyone is familiar with the two terms that are used for how fast something is going—speed and velocity. Most people use them interchangeably. In physics, however, we will want to carefully discriminate between the two. The term *speed* is used for the idea of how fast we are going, regardless of direction, while the term *velocity* is used to describe how fast *and* in what direction we are going. For one-dimensional motion, the direction can be completely specified by making the velocity either positive or negative. Because this is so simple, the distinction between speed and velocity in one dimension is often overlooked. However, when we consider two-dimensional motion, in the next chapter, the distinction will be much more important.

Everyone who drives a car knows what speed is. It is what your speedometer reads; it determines how quickly you can get from one place to another; it is the rate at which you are covering distance. Once you know your speed, it is a simple thing to convert it into a velocity. If you are going west with a speedometer reading 80 mi/hr, and if you decide to orient your ruler so it is positive toward the east, then your velocity is –80 mi/hr. The velocity is negative because you are going west while the coordinate system is oriented toward the east. A negative velocity does not mean that the car is driving backwards.

There are actually two kinds of velocity. One is called instantaneous velocity and the other is called average velocity. *Instantaneous velocity* is what a speedometer measures, supplemented by a sign to specify the direction. We will give a slightly more careful definition below, but, for now, let's just say that instantaneous velocity is based on what a speedometer measures. Before we go on to define average velocity, let's consider the special case where the two kinds of velocity end up being identical. This is the case of uniform motion.

Uniform motion is a situation where your instantaneous velocity does not change. If your speedometer registers an even 60 mi/hr over some extended period of time, your velocity is constant and we say your motion is uniform during that time. As a concrete example, consider a case where you are driving your car down a straight road with the cruise control set at 60 mi/hr (one mile per minute) while the numbers on the mile markers are going down. Then your velocity is a constant –60 mi/hr. If we started at mile marker 97, we could get to mile marker 82 (a distance of 15 miles) in 15 minutes. The relationship between uniform velocity, the distance traveled, and the time elapsed is a simple formula that you probably already know. It is

$$v_{\text{uniform}} = \frac{\Delta x}{\Delta t}. \tag{2.2}$$

So, without ever looking at the speedometer, we could find our uniform speed from

$$\begin{aligned} v_{\text{uniform}} &= \frac{\Delta x}{\Delta t} = \frac{x_{\text{final}} - x_{\text{initial}}}{t_{\text{final}} - t_{\text{initial}}} = \frac{82 \text{ miles} - 97 \text{ miles}}{\frac{1}{4} \text{ hour}} \\ &= \frac{-15 \text{ miles}}{\frac{1}{4} \text{ hour}} = -60 \frac{\text{mi}}{\text{hr}}. \end{aligned}$$

As we have just seen, the units for velocity are a distance divided by a time. The standard SI units of velocity are meters per second (m/s), but we have used, and will routinely use, other common units like miles per hour.

We are now ready to define what we will mean by *average velocity.* Let us first relax our uniform motion restriction and allow the velocity to change. To be specific, let's consider a case where a car is traveling at a velocity of +40 mi/hr for the first 15 minutes of a trip and then travels at a velocity of +60 mi/hr for the last 15 minutes of the trip. Because the car was going +40 mi/hr for exactly as long as it was going +60 mi/hr, its *average velocity* is halfway between the two, at +50 mi/hr. This result was found by a true process of averaging, just like test scores are averaged for a class. It is what we call an *average over time,* because equal times at a certain speed are given equal weight in determining the average, just like equal numbers of students with a certain test score are given equal weight in the process of computing the class average, or an "average over the class."

But there is also a simpler way to find the average velocity. During the first 15 minutes, the velocity was uniform, so we can calculate the distance the car traveled by using Equation 2.2. The distance is

$$\Delta x = v_{\text{uniform}} \Delta t = (40 \text{ mi/hr})(\tfrac{1}{4} \text{ hr}) = 10 \text{ mi}.$$

And, during the second 15 minutes, the car traveled

$$\Delta x = v_{\text{uniform}} \Delta t = (60 \text{ mi/hr})(\tfrac{1}{4} \text{ hr}) = 15 \text{ mi}.$$

The total distance traveled during the ½ hour is 25 miles. Now note this interesting fact: The average velocity can be calculated, not just by averaging the velocity of the car over time, but also by considering the distance traveled during the entire trip and the total time it took to make the trip. The formula is

$$v_{\text{average}} = \frac{\Delta x}{\Delta t} = \frac{x_{\text{final}} - x_{\text{initial}}}{t_{\text{final}} - t_{\text{initial}}}. \quad (2.3)$$

In our example, this gives

$$v_{\text{average}} = \frac{\Delta x}{\Delta t} = \frac{25 \text{ mi}}{0.5 \text{ hr}} = 50 \frac{\text{mi}}{\text{hr}}.$$

And this always works. The average velocity you get by averaging your instantaneous velocity over time is the same as the average velocity you get by dividing your total displacement during a time interval by the elapsed time in the interval. When we speak of "average velocity," we will always mean an average over time, so Equation 2.3 will always give the average velocity.

EXAMPLE 2.1 Average Speed

A car travels down a straight road 100 miles at 50 mi/hr and then travels another 100 miles at 80 mi/hr. What is its average velocity?

ANSWER It is tempting to think that, because the car traveled equal distances at the two velocities, the average should be halfway between them, or 65 mi/hr. However, such a calculation would be an "average over distance," in which we would give equal weight to equal distances in computing the average. But we have agreed that "average velocity" will always mean average over time, and this definition is consistent with Equation 2.3. So, to find the average velocity, we must use Equation 2.3, which means that we must know the total distance covered and the total time. The first hundred miles will require a time of (100 mi)/(50 mi/hr) = 2 hours to cover and the second hundred will take (100 mi)/(80 mi/hr) = 1¼ hrs. Therefore, the total time required to cover the entire 200 miles is 3¼ hr, giving an average velocity of

$$v_{\text{average}} = \frac{\Delta x}{\Delta t} = \frac{200 \text{ mi}}{3\tfrac{1}{4} \text{ hr}} = 61\tfrac{1}{2} \frac{\text{mi}}{\text{hr}}.$$

Working with Equation 2.3 is the correct way to compute average velocities.

Now let's consider what the average velocity can and cannot tell us. Imagine that you are driving down a straight highway with passengers who like to stop for restroom breaks and snacks all the time. At 9:00 am you start driving east along the highway, and at 3:00 pm you stop the car, having covered a total distance of 240 miles. You traveled 240 miles east in

6 hours, so your average velocity was +40 mi/hr eastward. So here's a question. What was your instantaneous velocity at 1:00 pm? The answer is that there is no way to know. Unless you know that the motion was uniform, the average velocity for the trip cannot tell you the instantaneous velocity at any particular time. For all we know the car was stopped at a rest stop at 1:00 pm.

But there is still a relationship between the instantaneous velocity of an object and the distance it covers in a time interval. Suppose that the car in the last paragraph had an instantaneous velocity of +30 m/s (a little over 60 mi/hr) at 1:00 pm. Then it is a pretty safe bet that, one second after 1:00 pm, it would still be traveling 30 m/s, because cars do not change speed by much in one second. In that one second, it would have traveled 30 meters. On the other hand, knowing the speed at 1:00 pm does not tell us much about the speed at 1:05 pm because a lot can happen in five minutes. So we cannot say much about the distance traveled during those five minutes if all we know is the instantaneous velocity at 1:00 pm. If we want to be able to relate the instantaneous velocity at 1:00 pm to a distance traveled, we must choose a very short time interval. Let us use a small Greek letter delta (δ) to refer to a very small change in something. Then we can define instantaneous velocity to be the ratio

$$v_{\text{instantaneous}} = \frac{\delta x}{\delta t}, \tag{2.4}$$

where the time interval, δt, is sufficiently small.

"Sufficiently small." That's pretty vague, isn't it? Let's try to be more precise. If you imagine that an object, a car or a space shuttle, has a speedometer attached to it, and that the speedometer does not change over some time interval, then you could divide the distance traveled during that interval by the elapsed time to find the speed. If you took an interval half as long, the object would travel half as far, so the ratio of the distance to the time would be the same. If you took a time interval 1/100th of that, both the time and the distance would be tiny, but the ratio would still be the same, so the velocity would still be the same. In the limit that δt goes right down to zero, δx will also go to zero; but the ratio of the distance and the time will remain unchanged and well-defined. It is the instantaneous velocity at the chosen time.

You may know that taking this kind of limit is the basis for differential calculus, and we promised not to use calculus in this book. But we haven't really broken our promise. We have only presented this discussion to help us explain the concept of instantaneous velocity, not as instructions on how to calculate it. You will never be asked to determine the instantaneous velocity of something by estimating its displacement over a time interval that goes to zero.

2.4 Straight-Line Acceleration: Speeding Up and Slowing Down

If all motion were uniform, and all velocities constant, the physics goal of predicting the future would be a breeze. All you would have to do is multiply whatever velocity an object has now by how far into the future you want to predict its motion, and you would be able to calculate what the exact displacement of the object would be, relative to where it started. Unfortunately for this program, objects can change their velocity. They can accelerate.

Everyone who drives a car knows what acceleration is. It is what the accelerator pedal produces. But how do you measure acceleration? One method that is often used to describe an automobile's ability to accelerate is to specify the amount of time it takes the car to go from 0 to 60 mi/hr. But this is a little cumbersome. Let's see if we can find a way to specify the acceleration with a single number. Suppose you had a car that could go from 0 to 60 mi/hr in 10 seconds. Cars do not accelerate uniformly in time, but, if they did, you would be able to put together a table of speed versus time. A table giving the speed of a car that is accelerating uniformly from 0 to 60 mi/hr in 10 s would look like this:

TABLE 2.1 Speed as a Function of Time for Uniformly Accelerated Motion

Time (s)	0	1	2	3	4	5	6	7	8	9	10
Speed (mi/hr)	0	6	12	18	24	30	36	42	48	54	60

Is there a way we could characterize this performance with a single number? A couple of moments looking over the table will probably be enough for you to notice that, at each second, the speed is 6 mi/hr faster than it was the second before. Let us use *this* number as the measure of acceleration. We can say that the acceleration of the car in Table 2.1 is 6 "miles per hour per second," or 6 mi/hr/s.

Let us write this prescription in a formal definition. (At this point, we will begin a convention of using a subscript i for initial value and a subscript f for a final value.) If the speed of something at an initial time t_i is v_i, and if at the final time t_f it ends up at a speed v_f, then we will define acceleration, in this case where the acceleration is constant, to be the change in speed divided by the elapsed time:

$$a_{\text{uniform}} = \frac{v_f - v_i}{t_f - t_i} = \frac{\Delta v}{\Delta t}. \tag{2.5}$$

As we can see, the units of acceleration are the units of velocity divided by time, such as the mi/hr/s as in our example from Table 2.1. The standard SI unit for velocity is m/s, so the standard unit for acceleration is m/s/s, which we should read as "meters per second per second." In many ways, it is unfortunate that the common convention is to write this as m/s^2. It still means "meters per second per second," although it is often read as "meters per second squared." While this isn't wrong, it is worth noting that there is no such thing as a square second.

Look at Equation 2.5 and consider a case where v_f and v_i are both positive. Suppose, too, that the velocity has decreased, so that the final velocity v_f is *less* than the initial velocity v_i. Then the numerator will be negative and the acceleration will be negative. Because we have this possibility of negative numbers for acceleration, we see that we can speak of "acceleration" whether something is speeding up or slowing down. This means that your brake pedal is as much an "accelerator" as your accelerator pedal. The term "deceleration" is commonly used when something is slowing down, but the term is unnecessary, since, if we use both positive and negative numbers, acceleration can refer to both cases. To reduce confusion, we will avoid the use of the term "deceleration" altogether and talk only about acceleration.

Unfortunately, the possibility of positive and negative signs for acceleration can create a confusion of its own. Let's take a concrete example. Imagine a bicycle rolling toward the east at a speed of +6 mi/hr, as in Figure 2.2a. The plus sign means that we have taken positive eastward. At the same time, the bicyclist is accelerating at +2 mi/hr/s, meaning that the velocity changes by +2 mi/hr each second. If we start our timer at $t = 0$ s with the bike going +6 mi/hr, then, at $t = 1.0$ s, the bike will be going +8 mi/hr. At $t = 2.0$ s, it will be going +10 mi/hr, and so on. Note that when we ask how fast the bike is "going" we mean its instantaneous velocity, not its acceleration. The acceleration tells us about the *change* in the velocity.

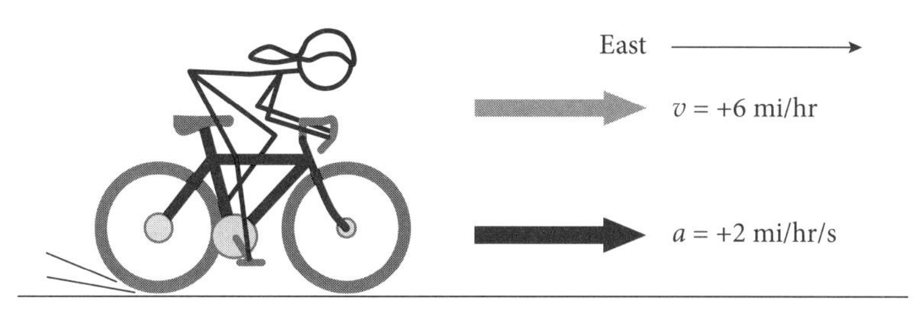

Figure 2.2a Velocity and acceleration are both positive. The bicycle's speed increases.

Now, if we imagine the bicycle moving at –6 mi/hr at time t = 0 s, with the same +2 mi/hr/s acceleration, the pattern is quite different. First, we remember that the negative sign for the velocity does not mean that the bicycle is going backwards, but only that the bicycle is going west while we chose to orient our coordinates toward the east. This situation is shown in Figure 2.2b, with the negative velocity being a velocity toward the west, while the acceleration is positive, so toward the east. Note that, in Figure 2.2a, the +2 mi/hr/s acceleration described a bike that was speeding up, the bicyclist leaning on the pedals and kicking up dirt from the back wheel. However, in the second scenario, Figure 2.2b, the *same* acceleration describes a bicycle that is slowing down as the cyclist hits the brakes, skidding and kicking dirt forward from the front wheel. At t = 0, the bike starts at –6 mi/hr; at t = 1.0 s it is going –4 mi/hr; and at t = 2.0 s it is going –2 mi/hr. In both cases, we add +2 mi/hr to the velocity every second, but for a positive velocity this increases the speed, while for a negative velocity the speed decreases.

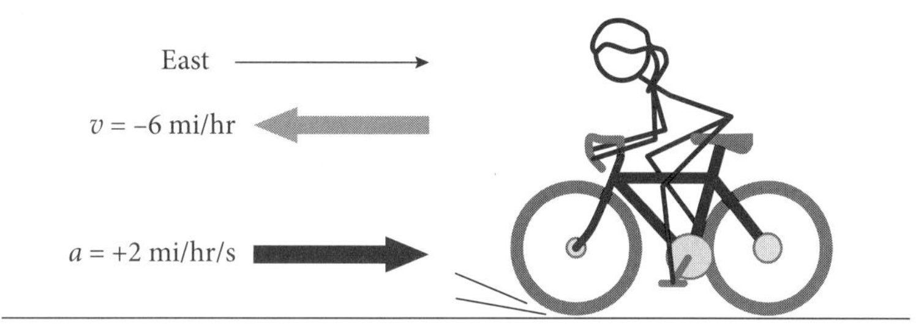

Figure 2.2b Velocity is negative, not because the bike is going backwards (careful cyclists don't ride backwards), but because it is going west. Acceleration is positive, and the speed is reduced.

What this emphasizes is that it is wrong to assume that positive acceleration always means speeding up while negative always means slowing down. What matter are the relative directions of the velocity and the acceleration. When the velocity and acceleration point in the same direction, the object speeds up, and when they point in opposite directions, the object slows down.

Now, please don't think that this is some trick that physicists have played on the world to make things difficult. As long as we need to be able to describe things that move north or south, or up or down, or left or right, this kind of thing is necessary. We must have positive and negative positions, which means we must have positive and negative velocities, which means that there must be positive and negative accelerations. And this is the only consistent way to define them.

Let us look at a couple of examples that illustrate what we mean.

EXAMPLE 2.2 The Acceleration Direction—I

A water balloon has an acceleration directed upward and the balloon is slowing down. In what direction is it going?

ANSWER If the balloon is slowing down, its velocity and acceleration must be in opposite directions. If the acceleration is up, the velocity must be down. The answer to the question, "In what direction is it going?" always means the direction of the velocity, not the acceleration. So the balloon is going down. By the way, one way to produce this kind of motion would be to catch a falling balloon and bring it slowly to rest. After your hands first contact the balloon, the balloon continues to move down as it slows—an upward acceleration on a downward-moving balloon.

EXAMPLE 2.3 The Acceleration Direction—II

Describe the directions of the velocity and acceleration in Example 2.2 using a coordinate system in which "up" is taken as positive. How does your answer change if the coordinate system is reversed?

ANSWER If up is positive, the upward acceleration would be positive and the initial downward velocity would be negative. If we reverse the coordinate system, both answers change sign. As simple as this example appears, it emphasizes the critical point that the motion does not depend on the coordinate system used to analyze it. The coordinate system is just a helpful way to systematically keep track of things.

In Equation 2.5, we gave a definition for acceleration in the case of uniform, or constant, acceleration. However, it is also possible to consider situations in which the acceleration changes. In this case we will have to be careful, just as we had to be careful in how we defined velocity in a situation where the velocity could change. We will define the average acceleration over some time between t_i and t_f as

$$a_{\text{average}} = \frac{v_f - v_i}{t_f - t_i} = \frac{\Delta v}{\Delta t}, \tag{2.6}$$

in analogy with how we defined average velocity in Equation 2.3. If the acceleration is changing, we should recognize that the average acceleration might be different for different choices of t_i and t_f. We can also define the instantaneous acceleration in a way that is analogous to instantaneous velocity. It is

$$a_{\text{instantaneous}} = \frac{\delta v}{\delta t}, \tag{2.7}$$

where we again use δ to refer to a tiny change in velocity and a tiny change in time, in the limit that both differences go to zero together and preserve a well-defined ratio.

However, for the rest of this chapter, we will limit ourselves to the case of constant acceleration. In this case there is no distinction between the average and the instantaneous acceleration, and Equations 2.5, 2.6, or 2.7 are all sufficient definitions of the term "acceleration."

Also, from here on, we will use a simple a to mean instantaneous acceleration, without any subscript. And, while we're at it, let's drop the subscript from $v_{\text{instantaneous}}$ as well. When we mean instantaneous velocity we will just use the letter v.

2.5 Graphing the Motion

One thing that helps in understanding one-dimensional motion is to graph the position or the velocity of an object. Let us begin with a simple case of a truck going down a highway at a constant 60 mi/hr, passing mile markers along the way. If the driver is going at constant velocity he will cover equal distances in equal time intervals. Note how this produces a straight line graph in Figure 2.3.

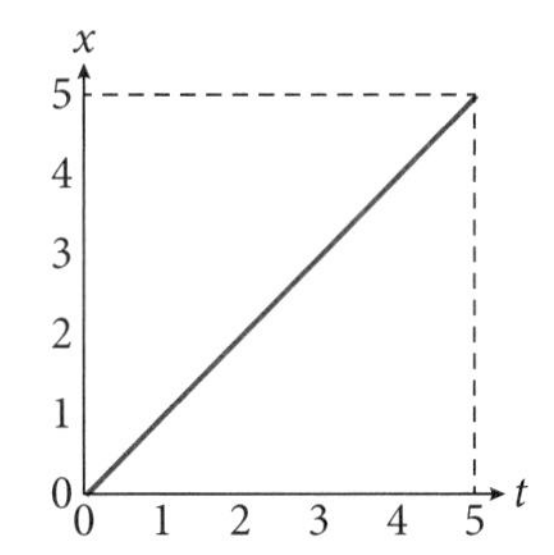

Figure 2.3 A graph of x-position in miles versus time in minutes for a truck going 60 mi/hr.

A more complicated graph of position versus time is shown in Figure 2.4. This represents the position of a toy radio-controlled car moving along a straight strip of sidewalk. Look first at the time between $t = 0$ and $t = 1$. Here, the graph is a straight line, covering 1 meter in that 1 second. We conclude that the car starts out at a constant speed of 1 m/s. There is another straight-line section between $t = 3$ s and $t = 4$ s, so here, again, the velocity is constant. This time it is 3 m/s. Between $t = 1$ s and $t = 3$ s, the graph is not straight, but is curving upward. The velocity during these 2 seconds is increasing from the initial 1 m/s to the final 3 m/s. This positive acceleration shows up in the position graph as a section of the line that is increasing its slope. If you think of it as the shape of a hill, the hill is getting steeper as you go from left to right. The portion of the graph between $t = 4$ s and $t = 6$ s, with its decreasing slope, must then represent a decreasing velocity. The car is slowing down.

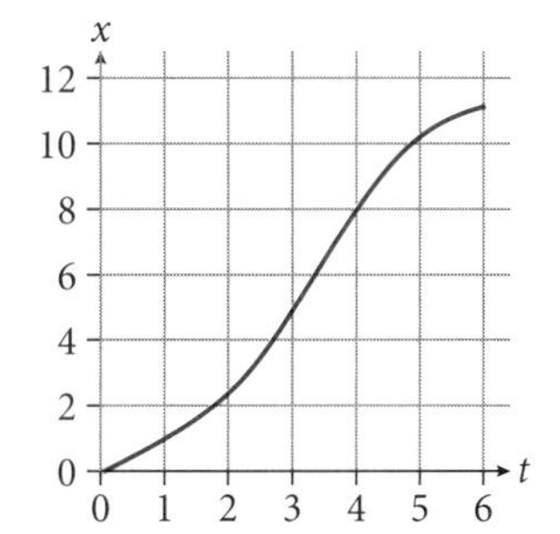

Figure 2.4 The position (in meters) of a toy car versus time (in seconds).

Let us look more closely at the section of the graph in Figure 2.4 between $t = 4$ s and $t = 6$ s. Since the car covers 3 m in those 2 seconds (from $x = 8$ m to $x = 11$ m), the average speed is $v_{\text{average}} = 3\,\text{m}/2\,\text{s} = 1\frac{1}{2}$ m/s. This average is depicted by the dashed line in Figure 2.5. But, since the slope of the line is decreasing during this time, this is not the instantaneous velocity at all points in the interval. Let us, for example, consider the instantaneous velocity at the time $t = 4\frac{1}{2}$ s. A comparison of the slope of the red line at this point with the slope of the dashed black average-velocity line makes it clear that the red line is steeper than the dashed line. We can actually measure how steep the curve is at this point in time, and thereby measure the instantaneous velocity, by drawing a straight line that just touches the curve at $t = 4\frac{1}{2}$ s and is tangent to the curve at that point. This line is thus a measure of how steep the hill is there. This tangent line is the solid black line in Figure 2.5. Since the tangent line is straight, it is easy to use it to measure how many meters the car would move if this were its constant velocity, instead of its instantaneous velocity. If you measure the slope of the tangent line carefully (we measured it *very* carefully), you will see that it crosses the $t = 4$ s grid line at $8\frac{3}{16}$ m and that it hits the $t = 5$ s grid line at $10\frac{7}{16}$ m. Therefore, if the car had had a constant velocity equal to its instantaneous velocity at $t = 4\frac{1}{2}$ s, the distance the car would have traveled in the 1 second between $t = 4$ s and $t = 5$ s is $2\frac{1}{4}$ m. So, at $t = 4\frac{1}{2}$ s, the car's instantaneous velocity is $2\frac{1}{4}$ m/s.

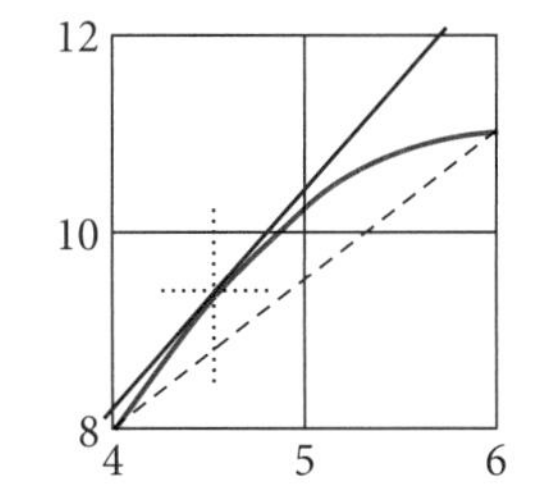

Figure 2.5 The last 2 seconds of Figure 2.4.

Finally, let us see what the instantaneous velocity is at $t = 6$ s. The solid black line in Figure 2.6 is tangent to the curve at $t = 6$ s, and it is completely horizontal. At exactly $t = 6$ s, the car is neither moving to higher or lower values of x. It has stopped, and its instantaneous velocity is zero.

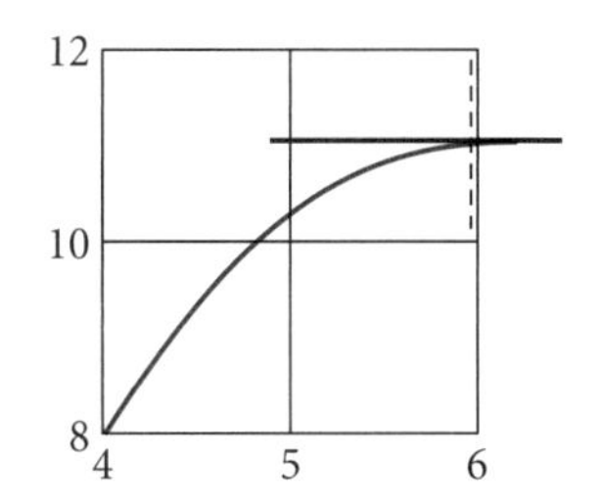

Figure 2.6 The tangent line at $t = 6$ s.

In addition to graphing the position of an object as a function of time, it is also instructive to plot the velocity as a function of time. Seeing how the velocity varies helps us to identify what the accelerations are. For example, if we plotted the velocity of the toy car in

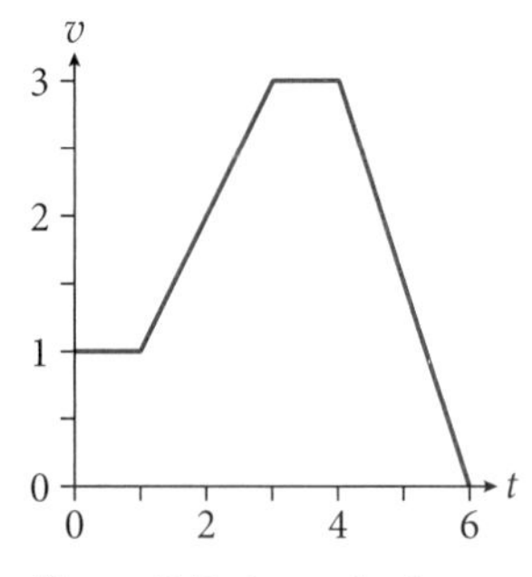

Figure 2.7 A graph of velocity versus time for the motion shown in Figure 2.4.

Figure 2.4, as a function of time, we would get the graph shown in Figure 2.7. Here, you can see the initial constant 1 m/s velocity for the first second. During the next 2 seconds, the straight-line rise from 1 m/s to 3 m/s is characteristic of constant acceleration, in which the ratio of the change in velocity to the increase in time is the same all along the line. Then there is another flat segment with constant velocity (and zero acceleration) at 3 m/s. Finally, the car has a negative acceleration for 2 seconds, ending at rest with $v = 0$. The final acceleration drops the speed from 3 m/s to zero, at a constant (straight-line) acceleration of 1½ m/s². Note that the steeper the slope of the velocity graph, the greater the magnitude of the acceleration, the sharp drop for the final 1½ m/s² being steeper downward than the climb in the 1 m/s² upward slope between $t = 1$ s and $t = 3$ s.

As further examples of how graphing helps us to characterize the motion, let's take a look at the following.

EXAMPLE 2.4 Mary Jane and the Evil Professor

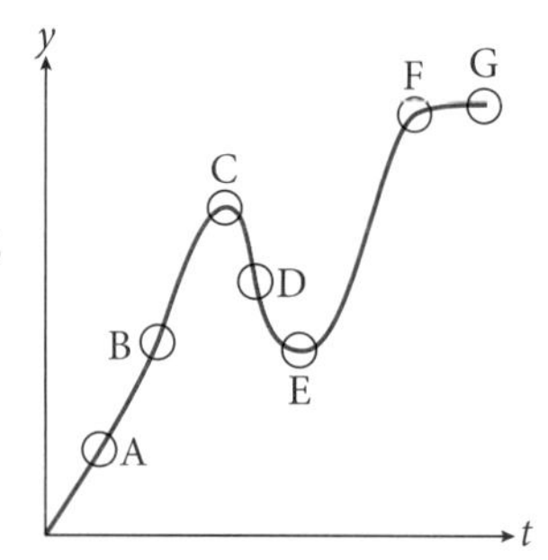

Mary Jane is in an elevator heading to the 10th floor, when the evil Professor Ron speeds the elevator up, faster and faster until the cable breaks. Fortunately, Spider-Man catches the elevator with a strand of spider web and pulls it up to the 10th floor, all according to the graph at right. Describe what is happening in each of the little circles in the figure.

ANSWER

A. The red line is nearly straight, so this is where the elevator is moving upward at constant speed.

B. The slope starts to get steeper. This is Professor Ron accelerating the elevator upward.

C. The elevator has reached its highest point before falling back down the shaft. If we are careful, we see that the cable must have broken just before C, at the point where the slope started to flatten out. The elevator was momentarily stopped at C. After C, the slope gets steeper and steeper downward.

D. The slope of the line changes from getting steeper (speeding up) to getting shallower (slowing down). Point D must be the point where the web caught the elevator.

E. The elevator has stopped its fall (the line is horizontal) and started to accelerate upward again.

F. The elevator's speed decreases so that it can stop smoothly at the 10th floor.

G. The elevator has reached its highest point and stopped.

EXAMPLE 2.5 The Lamborghini Giulia Acceleration Curve

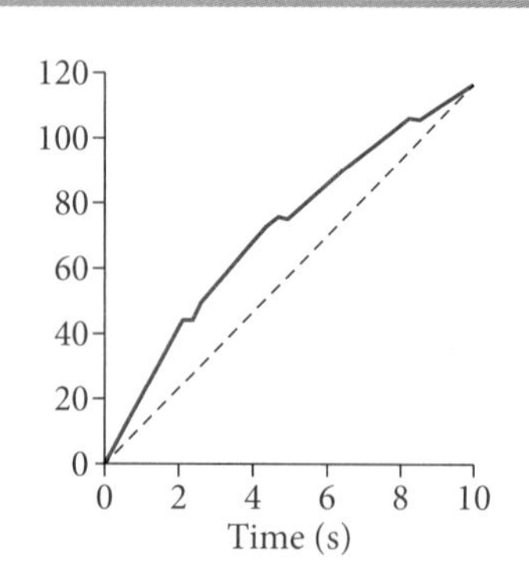

The red curve at right shows the speed in mi/hr of an accelerating Lamborghini Giulia with a 6-speed transmission. a) Find the average acceleration over the 10 seconds. b) Identify the time of the greatest acceleration. c) Find the instantaneous acceleration at 1 s.

ANSWER

a) The dashed line connects the initial zero velocity with the final 115 mi/hr velocity, giving an average acceleration of

$$a_{\text{average}} = \frac{\Delta v}{\Delta t} = \frac{115 \text{ mph}}{10 \text{ s}} = 11.5 \text{ mi/hr/s}.$$

b) We expect that everyone has identified the little glitches in the curve as the places where the driver is changing gears. The steepest slope is when the car is in first gear between zero and a little after 2 s.

c) At $t = 1$ s, the curve is nearly straight, so the tangent would lie along the line from $t = 0$ to $t = 2$ s, for an instantaneous acceleration of $a = (40 \text{ mi/hr})/(2 \text{ s}) = 20 \text{ mi/hr/s}$.

2.6 Equations Governing Uniformly Accelerated Motion

In this section, we develop a set of equations that work when the acceleration is constant. These equations will enable us to connect acceleration, position, time, and velocities, *as long as this constant-acceleration condition is realized.* Because we will see many examples of constant acceleration over the next few chapters, it is important to become familiar with the use of these equations. We begin with a numerical example of a constant-acceleration problem, one that asks us to find the acceleration using the basic definition from Equation 2.5.

EXAMPLE 2.6 A Ball Rolling Down a Ramp

A ball rolling down a ramp changes speed at constant acceleration from 3 m/s to 7 m/s during a time of 2 seconds. What is the ball's acceleration during this interval?

ANSWER In later chapters, we will learn how we know that a ball rolling down a ramp will do so with constant acceleration. For now, we will just accept this because the problem said the acceleration was constant. We see that the ball is speeding up, so we know that the acceleration must point in the same direction as the velocity, down the slope. We know that the velocity changes by 4 m/s in a time of 2 s, which gives an acceleration $a = (4 \text{ m/s})/(2 \text{ s}) = 2 \text{ m/s/s}$ or 2 m/s^2. We report our answer as 2 m/s^2 down the slope.

An alternative solution involves defining a coordinate system. We arbitrarily choose down the slope as positive. Consistent with this, the data given in the problem are $v_i = +3$ m/s, $v_f = +7$ m/s, and $\Delta t = 2$ s.

Using the definition of acceleration in Equation 2.5, we find

$$a = \frac{v_f - v_i}{\Delta t} = \frac{(7 \text{ m/s}) - (3 \text{ m/s})}{2 \text{ s}} = \frac{4 \text{ m/s}}{2 \text{ s}} = 2\frac{\text{m}}{\text{s}^2}.$$

The positive sign of the result means that the acceleration is down the slope.

Now let us go on from there. Let us assume that the acceleration of the ball in Example 2.6 continues at a constant rate, and let us find the velocity 5 seconds after it was first observed. We know that the speed changes by 2 m/s per second, and that the acceleration is down the slope. In five seconds, the speed will change by (2 m/s²)(5 s) = 10 m/s, and the direction of the change will be down the ramp. Since it began with a velocity of 3 m/s down the ramp, then its velocity after 5 seconds will be 3 m/s + 10 m/s, or 13 m/s, still down the ramp. For procedures that we may want to repeat over and over like this, it is helpful to summarize things in the form of a new formula. It is easy to confirm that what we have just been doing is captured by the formula

$$v_\mathrm{f} = v_\mathrm{i} + a\Delta t. \tag{2.8}$$

In our case, it gives $v_\mathrm{f} = 3\ \mathrm{m/s} + (2\ \mathrm{m/s^2})(5\ \mathrm{s}) = 3\ \mathrm{m/s} + 10\ \mathrm{m/s} = 13\ \mathrm{m/s}$.

Here's another question we might like to answer. How far did the ball roll during the 5 seconds? Let's think about this. If it had continued at its initial uniform speed of 3 m/s for the whole 5 seconds, it would have traveled 15 m. If it had instead been moving uniformly with its final speed of 13 m/s for the 5 seconds, it would have traveled 65 m. It is tempting to just average these values and suppose that the ball moved 40 m during this time. Well, guess what. That would be exactly right.

Back in Equation 2.3, we saw that there was a simple relationship between the average velocity and the displacement. If we multiply both sides of that equation by Δt, we can get a formula for the distance rolled,

$$\Delta x = v_\mathrm{average}\Delta t.$$

It only remains to find the average velocity of the rolling ball over the 5 seconds. In Figure 2.8, we graph velocity as a function of time for this case of constant acceleration. This is a graph of Equation 2.8 with values $v_\mathrm{i} = 3$ m/s and $a = 2$ m/s² (note how the velocity increases by 2 m/s every second). We claim that the horizontal line on the graph is the average velocity. Let's see why. We remember that average velocity means averaged over time. For the velocity to average out at the horizontal line, the line will have to be placed so that, for every instant that the velocity is some amount above the line (like the dashed line on the right), there is another instant where the velocity is the same amount below the average (like the dashed line on the

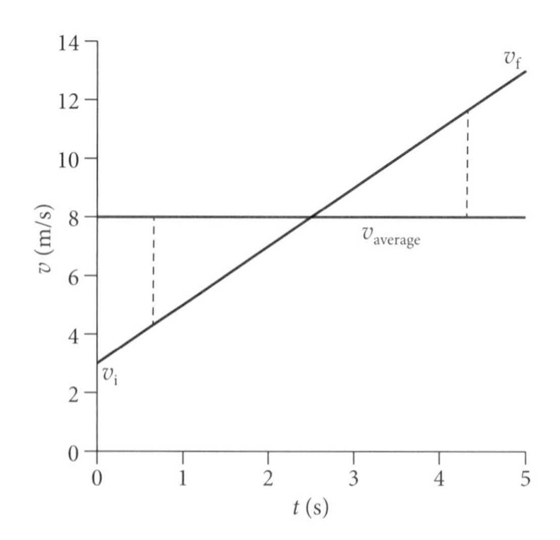

Figure 2.8 A graph of v versus t (the sloping line), also showing v_average (the horizontal line).

left). The horizontal line that satisfies this requirement will be halfway between the initial and the final velocity:

$$v_{\text{average}} = \frac{1}{2}(v_i + v_f).$$

We can thus write down a general formula capturing what we have discovered about the distance that an object will travel when its velocity is changing due to a uniform acceleration. The formula is

$$\Delta x = v_{\text{average}}\Delta t = \frac{1}{2}(v_i + v_f)\Delta t. \tag{2.9}$$

Equations 2.8 and 2.9 contain a total of 5 variables: Δx, Δt, v_i, v_f, and a. With a little algebra, we can combine them to create two additional formulas that are very useful. If we use Equation 2.8 for the final velocity, and substitute it into Equation 2.9, we get

$$\Delta x = v_i\Delta t + \frac{1}{2}a\Delta t^2. \tag{2.10}$$

And, if we solve Equation 2.9 for Δt and substitute it into Equation 2.10, we get

$$v_f^2 = v_i^2 + 2a\Delta x. \tag{2.11}$$

It is common practice to define the initial time to be zero and for the initial position to be set as zero. In this case, these equations can be written with the Δ's removed.

Let us gather together Equations 2.8, 2.9, 2.10 and 2.11 in Table 2.2. Next to each equation, we indicate which variables appear in that equation and which do not.

TABLE 2.2 Equations Relating Kinematic Quantities in Cases with Uniform Acceleration

No.	Equation	Δx	Δt	v_i	v_f	a
2.8	$v_f = v_i + a\Delta t$	✗	✓	✓	✓	✓
2.9	$\Delta x = \frac{1}{2}(v_i + v_f)\Delta t$	✓	✓	✓	✓	✗
2.10	$\Delta x = v_i\Delta t + \frac{1}{2}a\Delta t^2$	✓	✓	✓	✗	✓
2.11	$v_f^2 = v_i^2 + 2a\Delta x$	✓	✗	✓	✓	✓

The equations in Table 2.2 may be used as part of a systematic way to solve kinematics problems for one-dimensional uniformly accelerated motion. We are now going to explain this system to you. Of course, we're sure you remember what we said at the beginning of Chapter 1, about how the goal of a physics class is to teach principles, not to simply develop techniques. And so you might be wondering why we are, in fact, showing you a turn-the-crank technique for solving a particular kind of problem. There are two reasons for this. First, you are probably new at physics, and we find that it helps most people to start off with a procedure where they feel confident that they are doing things the right way. But, second, and most importantly, we are not *trying* to make physics hard. If there is a clear, well-defined procedure that works for a large class of problems, we would be remiss not to teach it to you.

Don't worry. There will be plenty of problems to come where there is no general technique to use and you will have to be creative with the deep principles you understand.

The process boils down to the following basic steps:

1. Identify explicitly the instant in the problem that you will consider the initial time and the instant that you will consider the final time.
2. Define the coordinate system you will use throughout the problem.
3. Identify as many of the five variables, Δx, Δt, v_i, v_f, and a, as you can from the problem, and identify what variable you are trying to find—the unknown. For a problem to be solvable you must know at least three of the five.
4. Find the equation in Table 2.2 that has checks for the variable you are trying to find and for the variables you already know. (Or, you can find the one that has an ✗ for the variable that does not appear in the problem.)
5. Solve for the variable you want to find and substitute the values from the problem into the equation.

Let's see how this process works.

EXAMPLE 2.7 Solving for the Distance

A car, initially driving at 15 m/s, doubles its speed while maintaining a constant acceleration of 3 m/s^2. How far does the car travel while it is speeding up?

ANSWER For this solution, we will go through the steps listed above in order. We don't normally have to be this explicit, but sometimes this attention to detail can help organize the solution to a more complicated problem.

1. We take the initial instant as the point where the car is traveling 15 m/s and the final instant as the point where it is traveling twice as fast, or 30 m/s.
2. We take the initial direction of motion as positive. (We could call this negative, and we would get the correct answer, but there would be a lot of extra negative signs running around, waiting to be missed.)
3. We know $v_i = 15$ m/s; $v_f = 30$ m/s; $a = 3$ m/s^2, and we want to find the displacement, Δx.
4. The equation that contains v_i, v_f, a, and Δx is $v_f^2 = v_i^2 + 2a\Delta x$, Equation 2.11 in Table 2.2.
5. Solving for Δx and substituting in the values of the variables gives

$$\Delta x = \frac{v_f^2 - v_i^2}{2a} = \frac{(30 \text{ m/s})^2 - (15 \text{ m/s})^2}{2(3 \text{ m/s}^2)} = \frac{900 - 225}{6} \frac{\text{m}^2}{\text{m/s}^2 \ \text{s}^2} = 113 \text{ m}.$$

2.7 Free Fall

Objects that fall near earth's surface are found to all change their velocities in a downward direction by about 10 m/s each second, provided that air resistance acting on the object is small enough to be ignored. Therefore, an object thrown upward with a speed of 40 m/s will be traveling upward at 30 m/s after 1 s, will momentarily reach a speed of zero after 4 seconds, and will be traveling downward at 10 m/s after 5 seconds. The magnitude of this acceleration due to gravity is denoted by the letter g.

We will generally use this value, $g = 10\ \text{m/s}^2$, in our problems and examples, because this will allow you to do many calculations in your head. In some cases, where the number we are finding represents an important physical property of something, we will use the more accurate $g = 9.8\ \text{m/s}^2$, or sometimes the even more accurate $g = 9.81\ \text{m/s}^2$. But, for most of our applications, we will simply pretend that $g = 10\ \text{m/s}^2$ exactly, without worrying about the fact that we are making a 2% error in the value of g. When you see how many quick calculations this allows you to do in your head, you will thank us. However, we don't want anyone going around saying that Hellings, Adams, and Francis said that the acceleration due to gravity is $10\ \text{m/s}^2$, not $9.8\ \text{m/s}^2$.

EXAMPLE 2.8 The Maximum Height of a Ball Thrown Upward

You throw a ball straight up in the air with an initial speed of 40 m/s. Find the maximum height the ball reaches above the release point.

ANSWER From the first paragraph in this section, we know that at time $t = 4$ seconds, the ball momentarily has zero velocity. Just before this instant, the ball was still going upward, and, just after this instant, the ball will be going downward. So this instant at $t = 4\ s$ represents the top of the ball's motion. During the 4-second interval in which it was traveling upward, the ball's average upward speed was $\frac{1}{2}(40\ \text{m/s} + 0\ \text{m/s}) = 20\ \text{m/s}$. If its average speed is 20 m/s for the 4 seconds, it must have traveled upward a distance of $(20\ \text{m/s})(4\ \text{s}) = 80\ \text{m}$.

You may have noticed that we did not follow the formal steps of our 5-step program this time. There is nothing wrong with using a simpler method, if you are sure of it. But you can always count on the 5-step method for security. We encourage you to redo this example using the 5 steps and confirm that you get the same answer.

Example 2.8 raises an issue that is often confusing to students. It is the question of the values of the velocity and the acceleration at the highest point in the motion. We have already established that the ball momentarily has zero velocity at this point, but what about the acceleration? It is tempting to say that, since there is no motion at that point, the acceleration is also zero. However, we have to remember that acceleration does not tell us how a thing is moving, but how its motion is *changing*. If, at some point, an object with zero velocity also had zero acceleration, then it would remain at zero velocity forever. But a ball does not rise to some height and then just hang there. So the acceleration at the top *cannot* be zero. Rather, what has happened is that the ball has stopped; its velocity is zero. But it is, at that instant, still changing its velocity downward by $10\ \text{m/s}^2$, so it will be moving downward a split-second later.

2.8 How to Solve Physics Problems

You have now read 25 pages of your physics textbook and, we hope, have already worked several of the problems at the ends of the chapters. You probably think it is a little late for us to be getting around to explaining how to do physics problems. But we did this for a reason. Our experience has been that the suggestions we are about to make will make much more sense after you have worked a few problems already. So let's begin.

There are four steps in the solution of most physics problems.

1. **English.** The first step is to carefully read the problem to pick out what information is given and what you are supposed to find. While not critical, it is often helpful to start by writing down the word "Given:" and then combing through the problem for the given quantities, writing them down as you go, with appropriate memory-jogging letters. For example, for the question in Example 2.7, you might write down "Given: $v_i = 15$ m/s." Calling the initial velocity v_i will remind you of all the formulas where the initial velocity is called v_i. Sometimes the information is not explicitly given as a number. For example, the final velocity in Example 2.7 was not stated explicitly, but you were told that the car doubled its speed. So you would do this in your head and write down "$v_f = 30$ m/s." In Example 2.8, you were not told the final velocity explicitly, but the fact that the ball has reached the top of its trajectory lets you know that you should write down "$v_f = 0$." Finally, it often helps to write down the word "Find:" and to then write down a letter that represents the thing you need to find. In all of this process, do not underestimate the value of a picture in helping you to see the problem clearly and identify things like the initial and final points.
2. **Physics.** We admit it. This is usually the hardest step in solving a problem. The goal here is to take what you know about the problem and figure out a path to get to what you want to find. In the process, you must decide which principles apply to this situation and which equations or formulas will determine the unknown, based on what is known. For example, the question in Example 2.7 states that the acceleration is constant, letting you know that the formulas in Table 2.2 will all apply in solving this problem. If you had not been told explicitly that the acceleration is constant, you would either have to decide for yourself if it is constant (you know that something falling near the earth's surface would have a constant acceleration) or you would have to conclude that the acceleration might *not* be constant, so that the equations in Table 2.2 could not be used. In this case, you would have to do something else (which, admittedly, we haven't discussed yet). In all cases, a well-done English step helps a lot with the physics. It is much easier to stare at a list of "Givens" and a "Find" than it is to keep re-reading some complicated paragraph. If you can't find a single equation that gives your unknown in terms of a bunch of knowns, as we did in Example 2.7, then you will have to be more creative. You might find an equation giving the unknown in terms of things that are not yet known, and then ask if you can figure out a way to find these missing quantities. Or you might ask, given what you do know, what things can you calculate that might have some bearing on the problem? There is no single path to solve all physics problems, so the only general help we can offer is this: As you work problems, you will get better at it.
3. **Algebra.** Once you have an equation or set of equations that allow you to determine your unknown in terms of your knowns, your next step should almost always be algebra. It is usually best not to put in numbers until you have a single formula with the quantity you want on the left and things you know on the right. The reason for this is to avoid errors and save effort. Often, as you derive your formulas, you will be able to cancel terms algebraically, and so not have to calculate these at all. And, most importantly, as long as you keep the symbols you keep a memory of what is going into your

calculation. This is a good hedge against careless mistakes that arise by using a number without remembering where it came from. Finally, let us just point out that, if your algebra is a little rusty, you might benefit from the Algebra Review in Appendix A.

4. **Arithmetic.** Only at the very end should you put numbers into your formula and calculate the final answer, keeping careful track of units. This lets you concentrate on all of the units in one place and not have to carry them along through half-calculated intermediate steps.

Let's see how this works by doing an example.

EXAMPLE 2.9 The BB Gun and the Birds

A kid shoots a BB gun directly upward at a flock of birds. The muzzle velocity of the BB is 21 m/s. If a BB hits a bird at a speed ≤ 1 m/s, it will not harm the bird. What is the minimum height at which the birds may safely fly?

ANSWER We define the initial time as the time when the BB leaves the gun. The criterion for safety is that the speed of the bullet is less than 1 m/s, so we define the final time to be when the velocity drops to 1 m/s. We choose positive upward.

Given: $v_i = 21$ m/s $v_f = 1$ m/s $a = -g = -10$ m/s^2

Find: Δx

English

Since the acceleration is constant, the equations in Table 2.2 apply. Equation 2.11, in particular, connects Δx to v_i and v_f. It is

$$v_f^2 = v_i^2 + 2a\Delta x.$$

Physics

Solving for Δx, we find

$$2a\Delta x = v_f^2 - v_i^2$$

$$\Delta x = \frac{v_f^2 - v_i^2}{2a}.$$

Algebra

Substituting in the given values, we get

$$\Delta x = \frac{v_f^2 - v_i^2}{2a} = \frac{(1 \text{ m/s})^2 - (21 \text{ m/s})^2}{2(-10 \text{ m/s}^2)} = \frac{1 \text{ m}^2/\text{s}^2 - 441 \text{ m}^2/\text{s}^2}{-20 \text{ m/s}^2} = \frac{-440}{-20} \text{ m} = 22 \text{ m}.$$

Arithmetic

We conclude that the birds will be safe if they are more than 22 m off the ground.

With the English step written out, as in the previous example, the process of finding the right equation to use to find the unknown Δx—step 4 of the 5-step program on page 27—is made even a little more straightforward.

Our next example is one that points out the importance of doing the English step carefully and translating the paragraph into the right set of "givens."

EXAMPLE 2.10 The Catapult

A catapult shoots a rock directly upward at an initial speed of 30 m/s. The rock rises to a height of 45 m and then falls directly back to the ground. If the catapult operator wants to save his machinery from getting smashed when the rock comes back down, how long after the rock is shot upward does he have to move his catapult?

ANSWER We begin by writing

Given: (taking positive upward)

$v_i = 30$ m/s
$\Delta x = 45$ m
$a = -10$ m/s^2

Find: Δt

We use Equation 2.10, $\Delta x = v_i \Delta t + \frac{1}{2} a \Delta t^2$, rewritten as $\frac{1}{2} a \Delta t^2 + v_i \Delta t - \Delta x = 0$. This is a quadratic equation for Δt. (If you're asking yourself, "What is a quadratic equation again?" check out the Algebra Review in Appendix A.) Using the quadratic formula, with $A = \frac{1}{2}a$, $B = v_i$, and $C = -\Delta x$, we solve for the elapsed time. We first substitute the "givens" for the A, B, and C of the quadratic equation. This gives

$$\Delta t = \frac{-B \pm \sqrt{B^2 - 4AC}}{2A} = \frac{-v_i \pm \sqrt{v_i^2 - 4(-\Delta x)\left(\frac{1}{2}a\right)}}{a}.$$

We then substitute the values of the givens to get

$$\Delta t = \frac{-30 \text{ m/s} \pm \sqrt{(30 \text{ m/s})^2 - 4(-45 \text{ m})(-5 \text{ m/s}^2)}}{-10 \text{ m/s}^2}.$$

If we carry out the calculations, we get

$$\Delta t = \frac{-30 \text{ m/s} \pm \sqrt{900 \text{ m}^2/\text{s}^2 - 900 \text{ m}^2/\text{s}^2}}{-10 \text{ m/s}^2} = \frac{-30 \text{ m/s} \pm 0}{-10 \text{ m/s}^2} = 3 \text{ s}.$$

So the rock spends 3 s in the air. Right? . . . WRONG! Do you see what we did wrong? First, it's easy to show that this answer *must* be wrong by calculating the velocity the rock has after 3 seconds, the time when it supposedly hits the ground. If we use Equation 2.8 to find the velocity after 3 s, we get

$$v_f = v_i + a\Delta t = 30 \text{ m/s} + (-10 \text{ m/s}^2)(3 \text{ s}) = 0.$$

The rock hit the ground at zero velocity? That can't be right. So what was our mistake?

Well, the problem is that we were careless in defining what the final point was in this problem. We wanted to find the time for the entire flight of the rock, so we should take the initial point to be when it left the catapult, as we did, and the final point should be when it hit the ground. But what is Δx for these two points? Remember that $\Delta x = x_f - x_i$, and that, at the final point, the rock is back where it started. Therefore, the correct value is $\Delta x = 0$, not $\Delta x = 45$ m.

The information that the rock flew 45 m into the air before falling back to earth is true, but it is actually irrelevant. It was put into the problem by malevolent authors who wanted to lead you astray (all right, it was us). Remember that Equation 2.10 relates elapsed time to *change in position*, not to distance traveled. If the final position is the same as the initial position, then the change is zero. With the correct $\Delta x = 0$ in the quadratic formula, we get

$$\Delta t = \frac{-v_i \pm \sqrt{v_i^2 + \Delta x a}}{a} = \frac{-30\ \text{m/s} \pm \sqrt{(30\ \text{m/s})^2 + 0}}{-10\ \text{m/s}^2} = \begin{Bmatrix} 0 \\ 6\ \text{s} \end{Bmatrix}.$$

There are two answers to the quadratic formula, 0 s and 6 s. The reason for the spurious $\Delta t = 0$ solution is that the formula is not very discriminating. It tells us the times when $\Delta x = 0$, including both the time when it left the catapult ($\Delta t = 0$) and the time when it came back to earth ($\Delta t = 6$ s).

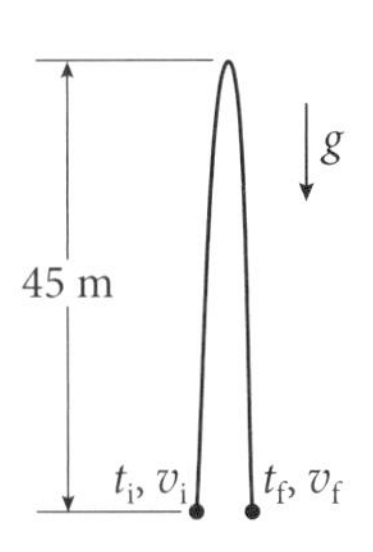

Our point in this example is to remind you that, in setting up a problem, you cannot just mine the paragraph for numbers that you can assign a letter to. You must read the problem for comprehension of what has actually been given and what, in fact, you are being asked to calculate. Finally, please note how drawing a diagram like the one at right could have helped you avoid the mistake we just made.

2.9 Summary

From this chapter, you should have learned the following things:

- You should understand the definitions of position, displacement, velocity, and acceleration, and you should feel comfortable with what it means when these have positive and negative signs, when they have the same signs as each other, and when they have the opposite signs as each other.
- You should be familiar with the use of the equations in Table 2.2, remembering that they are valid only in the case of constant acceleration and knowing exactly what the variables in the equation mean. These formulas are especially important, since they will continue to be used in future chapters.
- You should be able to be systematic in your solution of physics problems, either employing our 4 steps or something like them.

CHAPTER FORMULAS

(A box like this one will be placed after the chapter summary for the rest of the chapters in this book. You still have to read the text to know what a formula means, but having all the new formulas from the chapter in one place like this will help you to find the correct form without having to page through the whole chapter.)

For constant velocity: $\Delta x = v\Delta t$

For constant acceleration: $v_f = v_i + a\Delta t$ $\quad \Delta x = \frac{1}{2}(v_i + v_f)\Delta t$

$\Delta x = v_i\Delta t + \frac{1}{2}a\Delta t^2 \quad v_f^2 = v_i^2 + 2a\Delta x$

PROBLEMS

2.3 * **1.** You are driving along the highway at a constant 60 mi/hr. You look down to adjust your in-car entertainment center for 1 second. How many feet did you travel during that dangerous second when you were not watching the road?

* **2.** In a process of calibrating a car's speedometer, 5 measured miles are covered in 4 minutes and 26 seconds. What is the average speed of the car?

** **3.** Rita gets out of class and wants to drive from the parking lot to a coffee shop on Main St. Unfortunately, the most direct route, the one that is only 1.2 km long, will take her through a 0.2-km-long school zone, with its speed limit of 15 mi/hr (6.6 m/s). Of course, she *could* take a route that avoids the school zone, but it is actually 1.4 km long. The speed limit in town is 25 mi/hr (11.1 m/s) everywhere but in the school zone (we are very careful here in Bozeman). Assuming she follows the speed limits, how long would each route take her?

** **4.** Two cars are traveling from point A to point C, as shown in the figure below. Each car moves with constant velocity throughout the entire trip. Point B is 25 km from point A, and point C is 40 km beyond point B, as shown.

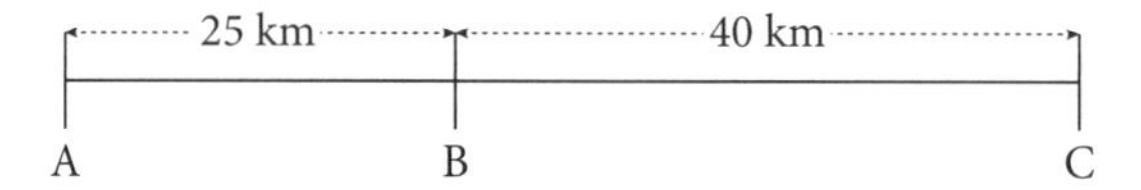

Car 1 leaves point A at 12:00 and reaches point B at 12:20. Car 2 leaves point A at 12:10 and reaches point B at 12:27. Which car will arrive at point C first? Explain your reasoning.

** **5.** A car travels 30 mi/hr for 10 minutes. It then goes 40 mi/hr for the next 10 minutes. Finally it travels for the last 10 minutes at 50 mi/hr.

a) What is its average speed over the entire ½ hour?

b) What was its average speed for the first 20 minutes?

c) How far will it have traveled after 20 minutes?

d) How far will it have traveled after ½ hour?

2.4 * **6.** A car going initially 60 mi/hr hits the brakes and skids to a stop in 4 seconds. What is the acceleration in mi/hr/sec?

* **7.** You are driving westward on I-90, from Bozeman toward Butte, at 60 mi/hr. You then accelerate in such a way that your acceleration points east with a magnitude of 4 mi/hr/second. How fast will you be going in 8 seconds?

* **8.** You see two cars side-by-side as they exit a tunnel on the freeway. The red car has a velocity of 40 m/s and is accelerating at 2 m/s^2. The blue car has a velocity of 20 m/s and an acceleration of 5 m/s^2. Which car is passing the other the instant they leave the tunnel? Explain your reasoning.

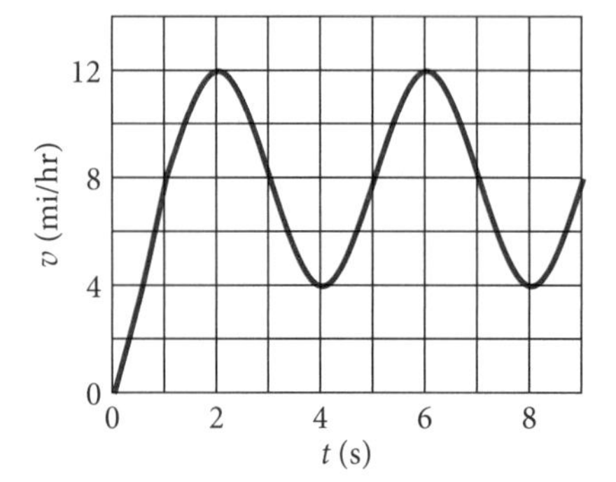

2.5 ** **9.** A student driver is having a hard time figuring out the clutch on a car with a manual transmission. As a result, the speed of the car as a function of time is as shown in the figure at the left.

a) Consider the portion of the graph between 1 s and 9 s and determine the average speed during this 8-second interval. [Hint: See the discussion on page 22 of how the average velocity was found for Figure 2.8.]

b) Is the average speed over the whole 9 s shown in the graph more or less than this?

c) What is the average acceleration of the car between $t = 0$ and $t = 2$ s?

d) What is the average acceleration between $t = 0$ and $t = 4$ s?

** **10.** The graph below shows the speed as a function of time for an annoying driver who is trying to make his passengers carsick.

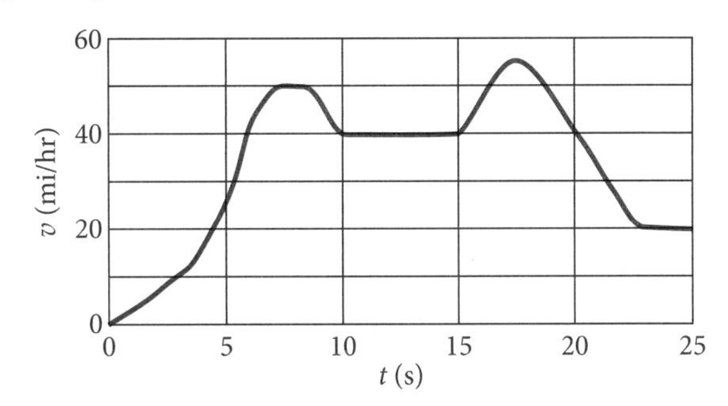

a) At what approximate time was the car's speed the greatest?

b) At what approximate time is his acceleration the greatest?

c) Was the acceleration ever zero during the motion? If so, where?

d) Estimate the instantaneous acceleration at $t = 2$ s.

** **11.** A car goes from a dead stop to 12 mi/hr in 1.2 s, following the graph shown below.

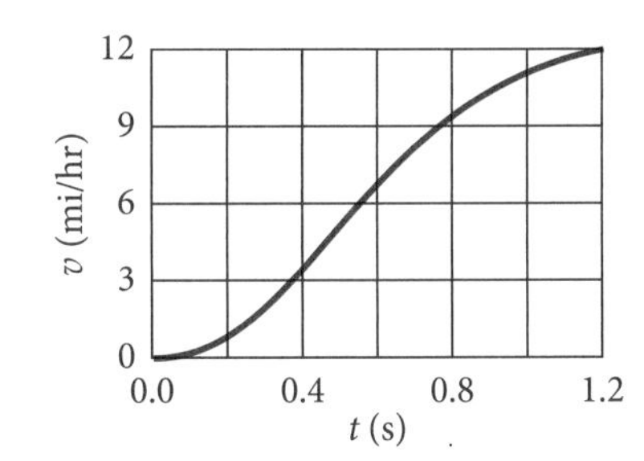

a) What is the average acceleration of the car?

b) Estimating from the graph, at approximately what point in time do you think the acceleration is the greatest?

** **12.** The positions of objects A and B are graphed at right. Answer the following questions and explain.

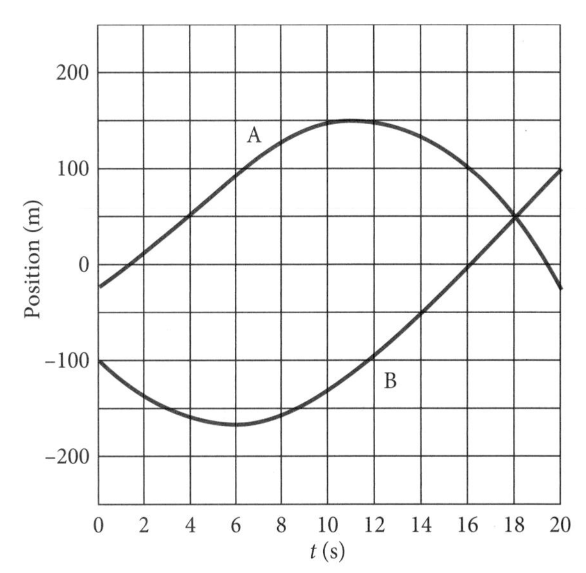

a) Is the acceleration of object A positive, negative, or zero at $t = 2$ s?

b) At $t = 2$ s, is object B speeding up, slowing down, or moving with constant speed?

c) Is the acceleration of object B positive, negative, or zero at $t = 2$ s?

d) Do the two objects ever have the same position? If so, when?

e) Which object has the greater speed at $t = 11$ s?

f) Find the average velocity of object B for the 20-second interval shown.

g) Estimate the instantaneous velocity of object A at $t = 15$ s.

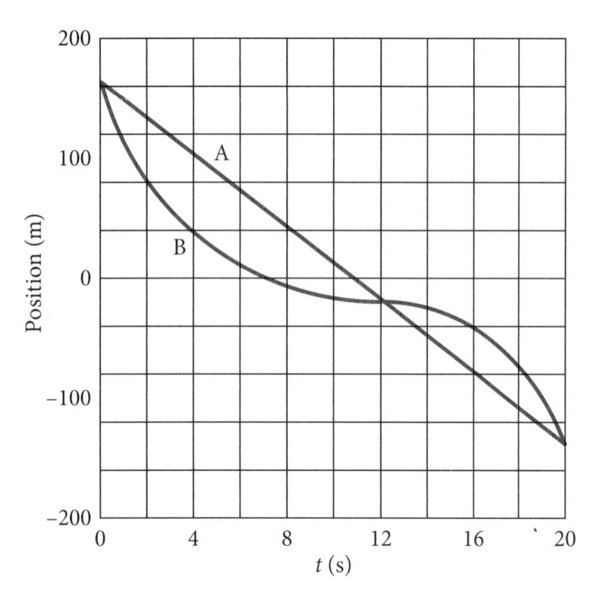

** 13. The positions of two cars (labeled A and B) on a straight driveway are graphed at left. Answer the following questions and explain each answer.

a) Consider the speed of car B. At $t = 16$ s, is it speeding up, slowing down, or moving with constant speed?

b) Is the acceleration of car B positive, negative, or zero at $t = 16$ s?

c) Is the acceleration of car A positive, negative, or zero at $t = 4$ s?

d) Which car is passing the other at $t = 12$ s?

e) During the time interval shown ($t = 0$ s to $t = 20$ s), is the average velocity of car B greater than, less than, or equal to the average velocity of car A?

f) Estimate the instantaneous velocity of car B at $t = 2$ s.

** 14. A car accelerates from rest for 5 seconds at 8 mi/hr/s, heading west. It then continues without acceleration for 10 seconds. The brakes then slow the car to half its previous speed in 4 seconds and then to a stop in 3 more seconds. Graph the speed of the car as a function of time.

** 15. An otter runs horizontally along the ground at a constant speed of 10 m/s for 5 seconds before he comes to the river bank. At the river bank, he throws himself down and skids along the surface of the water, slowing down with a constant acceleration of magnitude 2 m/s^2 until he comes to a stop.

a) Graph the speed of the otter as a function of time.

b) Graph the position of the otter as a function of time.

** 16. A car starts at $x = 0$ and accelerates from rest at a rate of 5 m/s^2, for 5 seconds. It then continues at constant speed for 5 more seconds.

a) Draw a careful graph of the velocity of the car as a function of time, with enough numerical detail to enable the velocities to be read from your graph.

b) Draw a careful graph of the position of the car as a function of time, with enough numerical detail to enable the positions to be read from your graph.

** 17. The graph below shows the position of a car moving along a highway.

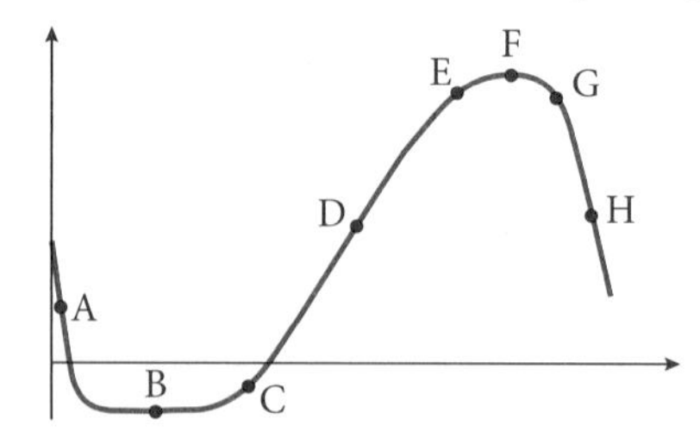

Identify the labeled points where

a) the velocity is zero;

b) the velocity is positive;

c) the velocity is negative;

d) the acceleration is zero;

e) the acceleration is positive;

f) the acceleration is negative.

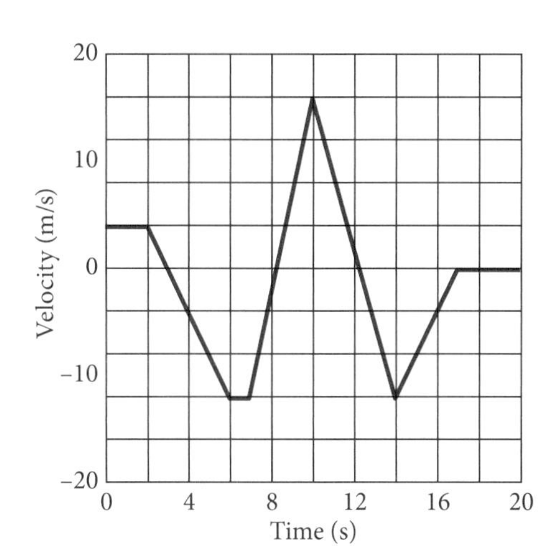

*** 18. The figure at right shows the velocity of an object as a function of time.

a) Sketch a graph for the acceleration of the object as a function of time.

b) Assuming that the object starts its motion from position $x = 0$, sketch a graph for the position of the object as a function of time.

** 19. Starting from rest, a car experiences constant acceleration over a distance of 20 meters and reaches a final velocity of 10 m/s. What would be the final velocity of the car if it had experienced this same constant acceleration over a distance of 80 meters?

2.6 ** 20. A marble is released from rest on an inclined track. It travels a distance of 4 m down the track in a time of 1.5 s. What is the magnitude of the marble's acceleration after it is released?

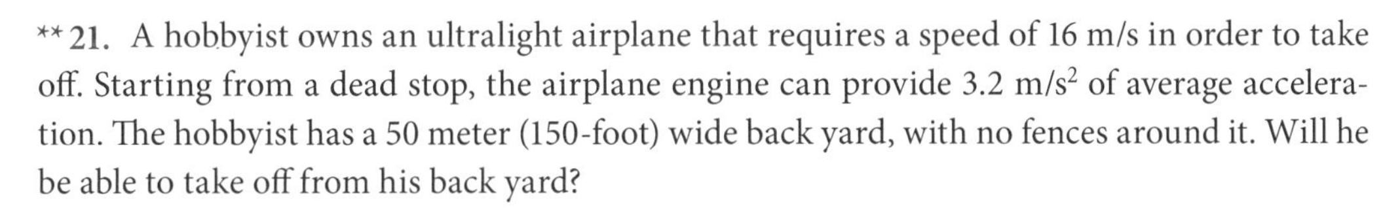

** 21. A hobbyist owns an ultralight airplane that requires a speed of 16 m/s in order to take off. Starting from a dead stop, the airplane engine can provide 3.2 m/s^2 of average acceleration. The hobbyist has a 50 meter (150-foot) wide back yard, with no fences around it. Will he be able to take off from his back yard?

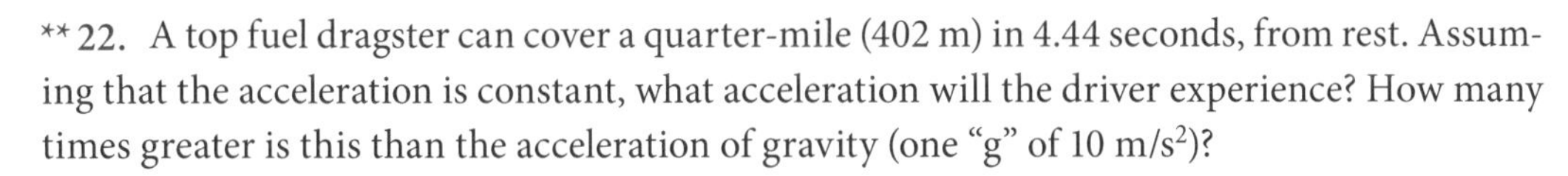

** 22. A top fuel dragster can cover a quarter-mile (402 m) in 4.44 seconds, from rest. Assuming that the acceleration is constant, what acceleration will the driver experience? How many times greater is this than the acceleration of gravity (one "g" of 10 m/s^2)?

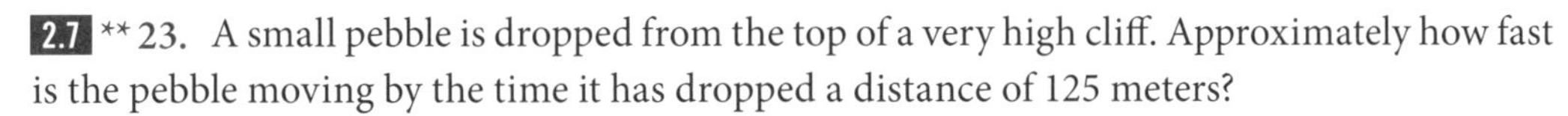

2.7 ** 23. A small pebble is dropped from the top of a very high cliff. Approximately how fast is the pebble moving by the time it has dropped a distance of 125 meters?

** 24. The picture at right shows a 150-meter-high cliff from which a water balloon is dropped from rest. Neglect air resistance.

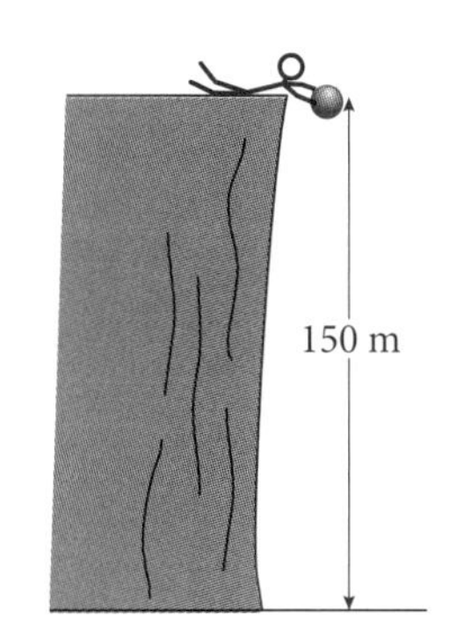

a) Sketch your own picture of the cliff and carefully draw a series of large dots on the sketch, showing the position of the water balloon after 1 s, 2 s, 3 s, 4 s, and 5 s.

b) Next to each dot, indicate the speed of the water balloon at that point.

** 25. Ron finds that he is holding a ticking stink-bomb. Panicking, he throws the stink-bomb straight up (good choice) with an initial velocity of 40 m/s (great arm). The bomb explodes six seconds after it leaves his hands.

a) How fast is the stink-bomb moving when it explodes?

b) How far is the stink-bomb from Ron when it explodes?

** 26. Butch and Sundance are standing on the edge of a cliff, arguing about whether or not they should jump into the water below. Butch drops a small rock and is able to count to 7-Mississippi (i.e., 7 seconds) before it hits the water. Assume that $g = 10$ m/s^2. Neglect any air resistance.

a) If they were to jump, how fast (in m/s) would they be going just before they hit the water?

b) How high is the cliff in meters?

c) Transform the velocity in (a) to mi/hr and the height in (b) to feet. Should they jump? Would you?

** 27. You go to the edge of the roof of the physics building and throw one of those *other* physics textbooks straight up with a speed of 15 m/s. How fast will the book be traveling just before it hits the ground, 12 m below you?

** 28. A delivery boy wants to throw a small package up to a young woman on a balcony 10 m above him. If he wants to make it as easy as he can for the woman to catch it, with what speed should it leave his hand?

** 29. A car traveling at 20 m/s (about 45 mi/hr) hits a brick wall. From what height would the car have to fall to hit the ground at that same speed? How many stories high would that be? [Suggestion: Estimate one story as about 3 m.]

** 30. The gravitational acceleration on the surface of the moon is 1/6 what it is on earth. If a man could jump straight up 0.7 m (about 2 feet) on the earth, how high could he jump on the moon?

** 31. If a bullet accelerates to a muzzle velocity of 900 m/s as it moves through a barrel of length 81 cm, find the average acceleration the bullet experiences.

** 32. A basketball center holds a basketball straight out, 2.0 m above the floor, and releases it. It bounces off the floor and rises to a height of 1.5 m.

a) What is the ball's velocity just before it hits the floor?

b) What is the ball's velocity just after it leaves the floor?

c) If the ball is in contact with the floor for 0.02 seconds, what are the magnitude and direction of the ball's average acceleration while in contact with the floor?

** 33. A brick is held over the edge of a tall building and dropped from rest.

a) What is its instantaneous speed at the end of 2 s of fall?

b) What is its instantaneous speed at the end of 3 s of fall?

c) What was the average speed during that third second?

d) How far did the brick travel during the third second?

** 34. A wrench slips off the floor and falls from an open helicopter that is climbing directly upward at a uniform speed of 2 m/s.

a) What is the velocity of the wrench 3 seconds after it leaves the helicopter?

b) How far below the helicopter is the wrench at that time?

*** 35. It is Monday morning and Asha is running a little late. She sees her commuter train start to accelerate away from her at 1 m/s^2 while she is still 10 m behind it. She can run 5 m/s (she is wearing business slacks and comfortable shoes). How many seconds will she have to run before she catches her train? Can you explain why there are two answers to the algebra of this problem?

*** 36. A resident in a high-rise apartment building sees a falling flower pot pass her 2-meter-tall window in $1/10$ s. She wonders whose flower pot it was. She knows it was someone directly above her, but which floor? She makes the reasonable assumption that the pot fell from rest and finds that she can calculate how high above the top of her window the pot was when it began to fall. How far was it?

*** 37. A certain car can accelerate at a rate of 4 m/s^2 and can brake at a rate -6 m/s^2. If the distance between two stop signs is 120 m and the speed limit is 24 m/s (about 55 mi/hr), what is the shortest time it can take the driver to legally cover the distance between the two stop signs?

*** 38. On a training exercise, a Navy SEAL rides in a helicopter that is moving directly upward at 10 m/s for 30 s. He then steps out of the rising helicopter and waits for 8 seconds before pulling the ripcord on his parachute, after which the parachute immediately deploys. Before he pulled the ripcord, his downward acceleration was 10 m/s^2. Once the parachute deploys, his acceleration is reduced to 3 m/s^2.

a) Sketch a graph of the Navy SEAL's height as a function of time. [Hint: The instant he steps out of the helicopter, he is still rising at 10 m/s.]

b) What will be the greatest height above the ground to which he rises?

c) How long after the helicopter initially lifted off the ground will he be back on the ground again?

d) How fast will he be moving when he hits the ground?

[Warning: Do not try this at home. There are many unrealistic assumptions in this problem.]

3

VECTORS

3.1 Superposition

In 1602, Galileo was doing research on projectile motion and discovered that the complicated parabolic trajectory of a projectile could easily be understood as a combination of two simpler motions if he required what he called the *principle of superposition*:

> If a body is subject to two or more separate influences, each producing a characteristic type of motion, it responds to each without modifying its response to the other.

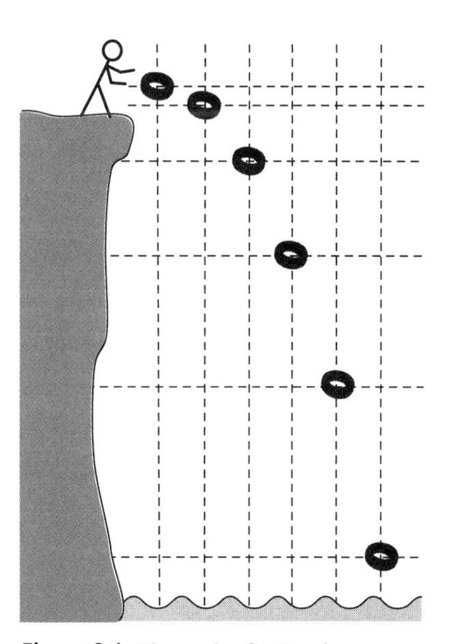

Figure 3.1 The path of a tire thrown from a cliff.

Consider the diagram in Figure 3.1, where an old tire is thrown from a cliff into the ocean (an environmentally irresponsible thing to do). There are two influences at work here. Horizontally, there is no acceleration, so the tire will travel at constant velocity in the horizontal direction (as long as air resistance can be neglected). In the vertical direction, gravity accelerates the tire downward, so the tire drops with increasing velocity. Superposition requires that the vertical motion be unchanged by the horizontal motion, and vice versa. The tire simply does both things at the same time.

In two or more dimensions, mathematical objects called *vectors* were expressly created to obey the principle of superposition. Since we will go beyond one-dimensional motion in most of the rest of this book, vectors will have to become second nature to you. But don't worry. They will.

3.2 Vector and Scalars

Vector quantities are those that have both magnitude and direction. Consider a desert island with a single tree (Figure 3.2a). A pirate wishes to bury his treasure on the island and then write down instructions for finding it. How can he do this? He cannot just say that the treasure is 50 paces from the tree, because that could as well place him in the ocean, 50 paces south of the tree, as it would place him at the treasure. However, he could wait until dark, sight toward the North Star to establish a north–south line, and then write these instructions: "Turn 37° east of north and walk 50 paces." This displacement is an example of a vector quantity. It is not completely specified until both the *magnitude* (50

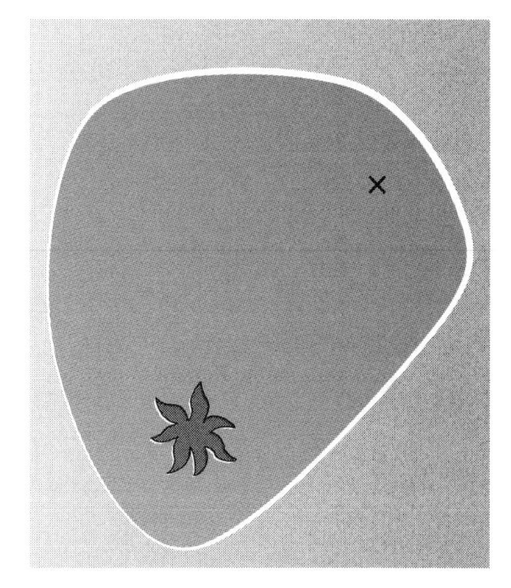

Figure 3.2a *The problem*: how to exactly describe the point where the treasure is buried.

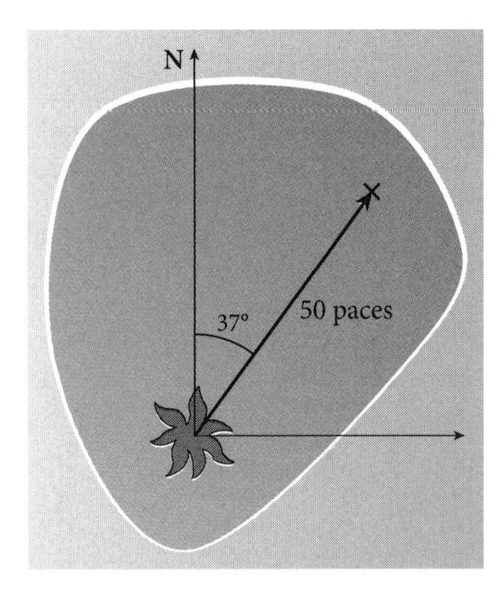

Figure 3.2b *The solution*: draw a vector—an arrow from the origin at the tree to the location of the treasure, giving the displacement magnitude and direction.

paces) and *direction* (37° east of north) are given. The arrow in Figure 3.2b is a graphical representation of the pirate's displacement vector. The length of the arrow on the page is scaled to represent the 50-pace distance and the angle on the diagram is drawn to represent the vector's direction east of north on the island.

Velocity is another example of a vector quantity. As there is a big difference between the instruction "walk 50 paces from the tree" and "walk 50 paces from the tree in a direction 37° east of north," there is also a big difference between driving 60 mi/hr and driving 60 mi/hr east. If going east will get you home, then the simple instruction to drive 60 mi/hr will not give you enough information to find the right place. We discuss velocity much more carefully in the next chapter. Other examples of vector quantities are force and momentum (concepts we will see in later chapters).

Things that do *not* have direction, things that are completely described with a single number, are called *scalars*. Temperature is an example of a scalar. If we say that the temperature in the room is 68 degrees, then that is all we can say about it. It is meaningless to say that the temperature is 68 degrees heading southeast.

In the remainder of our book, we will use a special notation for vectors. We will denote a vector by writing a letter with an arrow over it, like $\vec{D}$. The same letter, with no arrow above it, will represent the magnitude of the vector. When we specify the particular value of a vector, we will enclose its magnitude and direction inside curly brackets. So we would say that $\vec{D} = \{50 \text{ paces}, 37°\}$ is the vector from the tree to the treasure. The magnitude of the vector would be $D = 50$ paces. Another way to denote the value of a vector is to give its components along two perpendicular directions, as we will see later in this chapter.

3.3 Vector Addition

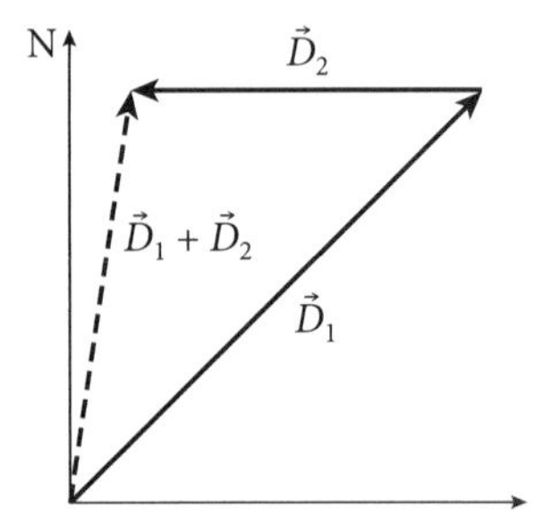

Figure 3.3 The addition of two displacement vectors. We denote vectors by placing a small arrow over the letter.

The mathematics of vectors is the mathematics of arrows (arrows drawn on a page, not bow-and-arrow arrows). We define addition of vectors, as we do with most mathematical concepts, in such a way as to make the concept useful. For instance, if I go 50 paces northeast, I end up 50 paces northeast of where I started. The tip of the displacement vector arrow represents where I am. If I go 50 paces northeast and then go 30 paces west, where do I end up? That is the question that mathematicians have defined vector addition to answer. The answer may be found by representing each displacement as an arrow, adding the arrows tail-to-head, and drawing another arrow with its tail at the tail of the first and its head at the head of the second, as in Figure 3.3. This vector, $\vec{D}_1 + \vec{D}_2$, is called the *resultant*.

You should especially note how different this adding of vectors is from adding scalars together. A vector of length 4 added to a vector of length 3 will *not* produce a resultant vector of length 7, unless the two vectors are pointing in the same direction. The actual sum of the two vectors can be anywhere between 1 and 7, depending on the directions in which the vectors point.

A little experience will convince you that the order of vector addition does not matter. Mathematicians call this freedom of order in a mathematical operation the "commutative" property. We demonstrate the commutativity of vector addition in the next example.

EXAMPLE 3.1 Vector Addition of Two Vectors

Consider the two vectors at right, and show that $\vec{B}$ added to $\vec{A}$ is the same as $\vec{A}$ added to $\vec{B}$.

ANSWER We find the sum of two vectors by placing the tail of the second vector on the head of the first vector. Note that we may move a vector to add it, but we may not change the two features of the vector that are important, its magnitude (length of the arrow) and direction. Below we show two ways to add the vectors and note that the results are identical. When we add $\vec{B}$ to $\vec{A}$, we get the picture below left. When we add $\vec{A}$ to $\vec{B}$, we get the picture below right. The dashed arrow representing the sum, $\vec{C}$, is the same in both cases.

When we add many vectors we stack them—tail to head—one after the other. The resultant vector is the one that starts at the tail of the first vector and ends at the head of the last vector. Let's do an example that shows how this is done.

EXAMPLE 3.2 Vector Addition of Many Vectors

Find the resultant vector produced by the addition of the four vectors shown.

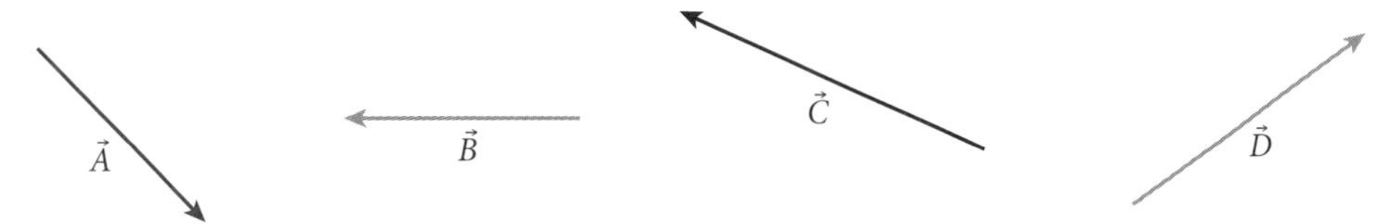

ANSWER We find the sum of the vectors by graphically stacking them tail to head, one after another. We actually show two ways to add the vectors and note that the results are again identical. The resultant vectors are the dashed arrows, labeled $\vec{E}$.

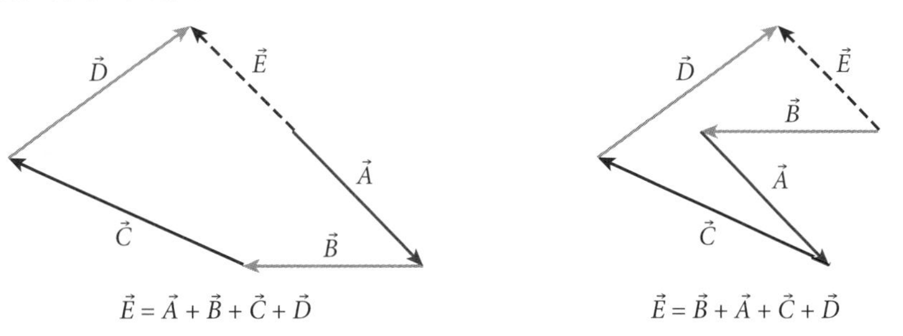

Before leaving this introduction to vector addition, let us discuss what it means to multiply a vector by a number. As $3 \times 2 = 2 + 2 + 2$, so $3 \times \vec{A} = \vec{A} + \vec{A} + \vec{A}$. We know how to add the three vectors, so, if you think about it, you will see that multiplying a vector $\vec{A}$ by 3 will produce a vector 3 times as long as $\vec{A}$ that points in the same direction as $\vec{A}$. The 3 multiplies the magnitude, not the direction. It is also possible to define multiplication of a vector by a scalar quantity that has its own units, one that is not just a pure number like 3. For example, if the velocity of some object is $\vec{v}$ = {3 m/s, north}, then we can *define* the product of $\vec{v}$ and a mass m = 2 kg to be a vector pointing in the same direction as $\vec{v}$, but with a magnitude of 6 kg·m/s. The resulting vector will be a completely different kind of quantity than velocity, with different units, so it may not be added to a velocity vector. In fact, its magnitude cannot even be compared to a velocity vector's magnitude. That would be like asking which is bigger, a mile or a kilogram? The new vector may be represented by an arrow, just like any vector, but there is no reason that the scale used for drawing $m\vec{v}$ arrows on a page should be related in any way to the scale used for $\vec{v}$ arrows.

3.4 Vector Subtraction

Having defined vector addition, it is easy to extend this to define vector subtraction, the difference between two vectors. There are two ways to think of this.

First, if we are interested in the difference in height between two children, we would stand them up, back to back, and compare their heights. They start at the same place, heel to heel. The same is true when we are subtracting vectors to see how different they are. Suppose we have one vector that describes a car's initial velocity $\vec{v}_i$ and another that represents its final velocity $\vec{v}_f$. The difference between the two is

$$\Delta\vec{v} = \vec{v}_f - \vec{v}_i. \tag{3.1}$$

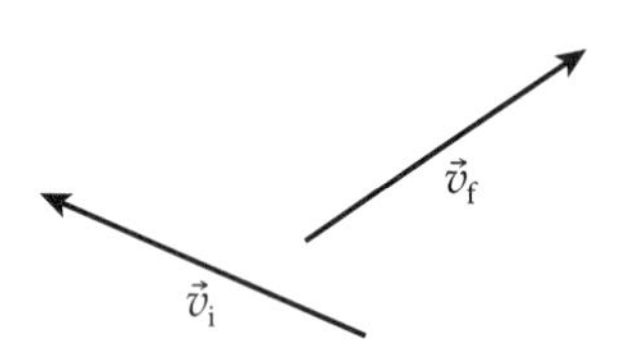

Figure 3.4a The vectors.

If we start with the two vectors in Figure 3.4a, we can find $\Delta\vec{v}$ graphically by moving one of the vectors (careful not to change its magnitude or direction) so that it is heel to heel with the other vector, and we then compare the two. To see what we mean by "compare," let us re-write Equation 3.1 by adding $\vec{v}_i$ to both sides to produce

$$\vec{v}_i + \Delta\vec{v} = \vec{v}_f.$$

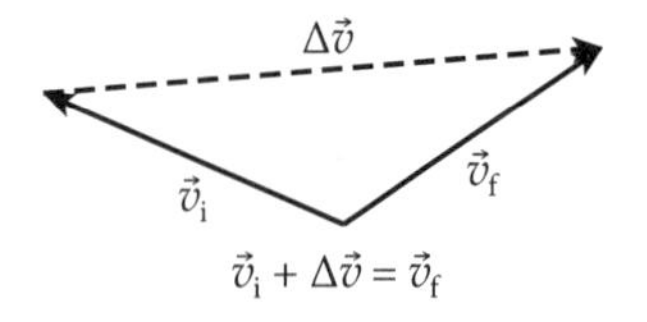

Figure 3.4b Subtracting vectors tail-to-tail.

So the way to think of the difference vector $\Delta\vec{v}$ is that it is the vector that must be added to the initial velocity vector to turn it into the final velocity vector. This is shown in Figure 3.4b.

Another way to think of vector subtraction is to define the inverse of a vector $\vec{v}$ as a vector $-\vec{v}$ that is the same length as $\vec{v}$, but that points in the opposite direction. This allows us to define a second method for finding $\Delta\vec{v} = \vec{v}_f - \vec{v}_i$. In this method, we subtract by adding the inverse vector:

$$\Delta\vec{v} = \vec{v}_f + (-\vec{v}_i).$$

This method is illustrated in Figure 3.4c. Note that the resultant $\Delta\vec{v}$ vector is identical in both methods.

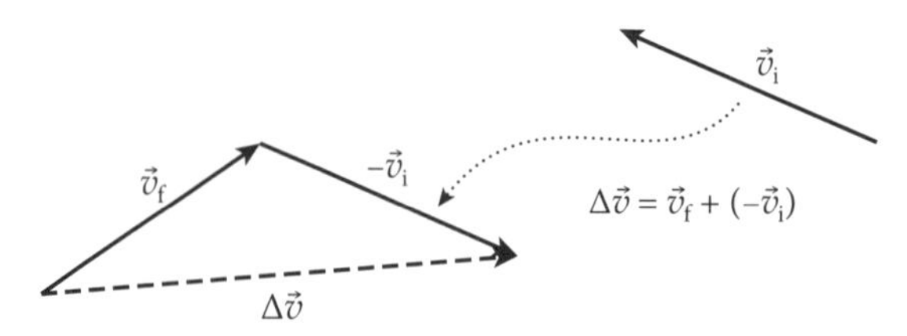

Figure 3.4c Subtracting by adding the inverse vector.

3.5 Trigonometry Review

In the last section, we defined addition of vectors by a process of drawing vectors on a piece of paper and measuring them to find their magnitudes and directions. This is the fundamental basis for the addition of vectors and, though it seems that the procedure is always an approximate one, the sum of two vectors may be found to any precision desired by use of this definition. If one wants to find the sum of two vectors with a precision of four digits, one can draw the arrows on a sheet of paper 10 m square and measure lengths to the nearest millimeter. Of course, this is getting a little silly, so you will be happy to know that there exists a way to add vectors numerically, rather than graphically, a method that can produce whatever accuracy you desire. This will, however, require familiarity with trigonometry applied to right triangles. The purpose of this section is to serve as a review of the relevant trigonometry.

By "right triangles" we mean triangles that have one 90° angle in them. In such a triangle, if you know one more angle, you know all the angles, because the two non-right angles must add up to 90°. Therefore, any right triangle with a 23° angle in it must have 67° for its last angle. So all right triangles with a 23° angle in them have all of their angles the same, which means that they all look alike. In geometry, we say that triangles with the same angles in them are "similar." Similar triangles can be big or small, but the big ones will always look exactly like the small ones, except that they have been "blown up," like blowing up a copy on a copy machine. All the sides of the triangle get bigger but the angles remain the same.

Let's look at the example in Figure 3.5, where a 37° right triangle has been blown up by a factor of 2. The triangle at left is the original and the one at right is the enlarged copy. The figure shows what we mean when we say that they look the same. One of the characteristics of the picture on the right is that the length of each side is exactly twice as big as in the picture on the left. Because of this strict proportionality, the ratios of corresponding sides are the same in

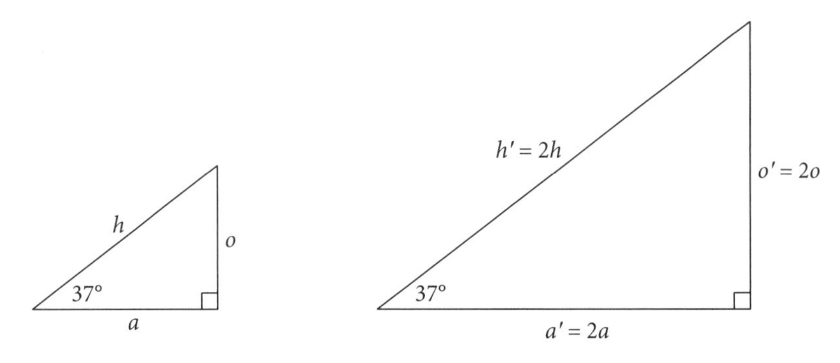

Figure 3.5 Characteristics of similar 37° right triangles.

both pictures. Let us first concentrate on the side opposite the 37° angle (the one not touching the 37° angle). Although the hypotenuse h' of the large triangle is not the same length as the hypotenuse h of the small triangle, the ratio of the opposite side to the hypotenuse is the same for both triangles (and for all similar triangles). This ratio is given the name "sine of 37°," or sin(37°). For 37°, this ratio is always about 0.6. So, for either triangle, we have

$$\frac{o}{h} = \frac{o'}{h'} \equiv \sin(37°) = 0.6. \tag{3.2}$$

And this ratio will always be the same no matter how big or small the triangle gets. Likewise, we define the cosine of 37°, or cos(37°), to be the ratio of the adjacent side (the side touching the 37° angle) to the hypotenuse.[1] For the 37° triangle, this ratio is

$$\frac{a}{h} = \frac{a'}{h'} \equiv \cos(37°) = 0.8. \tag{3.3}$$

Finally, we define the tangent of 37°, or tan(37°), to be the ratio of the opposite side to the adjacent side for all right triangles having a 37° angle. This is

$$\frac{o}{a} = \frac{o'}{a'} \equiv \tan(37°) = 0.75. \tag{3.4}$$

All that remains, if we want to know how the sides of all right triangles are related to each other, is to figure out what these ratios are for every possible angle. We need values for all of these sines and cosines and tangents. Fortunately, these ratios have been known for hundreds of years. When your authors were students, we had to look these things up in a trig table. Today, all that you need to know about getting these trig functions is that they are a button on your scientific calculator. You type in the angle on your calculator (making sure, if you are inputting angles in degrees, that your calculator is set to accept angles in degrees and not in radians or something else). You then hit the "SIN" button, and the sine of the angle will automatically appear.[2]

We will do a lot of trigonometry in this book, but we will always be solving one or the other of the same two basic problems. The first problem arises whenever we know the magnitude (length) of the hypotenuse and the angle, and we want to find the lengths of the other two sides. The usefulness of doing this will be apparent in the next section (Section 3.6). Let's start by doing an example of this first procedure.

EXAMPLE 3.3 Finding the Sides of a Triangle

A right triangle is laid out on a football field with a hypotenuse that is 35 m long and with an angle of 23°, as shown. Find the other two sides of the triangle.

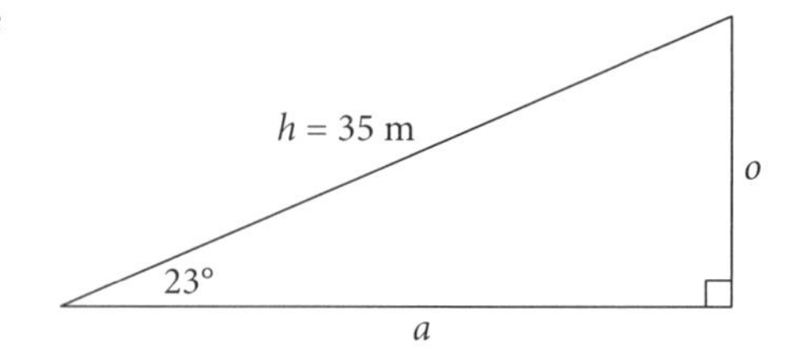

ANSWER The ratios of the sides in any right triangle with a 23° angle in it will always be the same. The ratio of the side opposite the 23° angle to the hypotenuse will always be

$$\frac{o}{h} = \sin(23°).$$

1. Some people who have trouble remembering which ratio is the sine and which is the cosine are helped by remembering that the cosine is the one that uses the side that is "cozy" with the angle, the side that is next to the angle.

2. If you don't currently have a scientific calculator, you may want to purchase one. In the meantime, we have given an abbreviated trig table (just a few useful angles) in Appendix B.

We look up the sine of 23° in a trig table, or use our calculator, to find that this ratio is $\sin(23°) = 0.39$. We can therefore write

$$o = h\sin(23°) = (35\text{ m})(0.39) = 14\text{ m}.$$

For the adjacent side, we begin by writing

$$\cos(23°) = \frac{a}{h}.$$

If we again consult our calculator, we find $\cos(23°) = 0.92$. We then solve this equation for a and find that

$$a = h\cos(23°) = (35\text{ m})(0.92) = 32\text{ m}.$$

Look back over these calculations and note how we started with the hypotenuse and the angle, and used trigonometry to find the two remaining sides.

The truth is that you will soon get to the point where you look at a right triangle with hypotenuse F and angle θ, and you will just automatically write down the lengths of the opposite and adjacent sides. Take a look at the triangle in Figure 3.6 and think about what the lengths of the opposite and adjacent sides will be. Did you have to think about the ratios defined in Equations 3.2, 3.3, and 3.4, or did you just see the sides as $F\sin\theta$ across from θ and $F\cos\theta$ for the side adjacent? Don't worry if you didn't. You will.

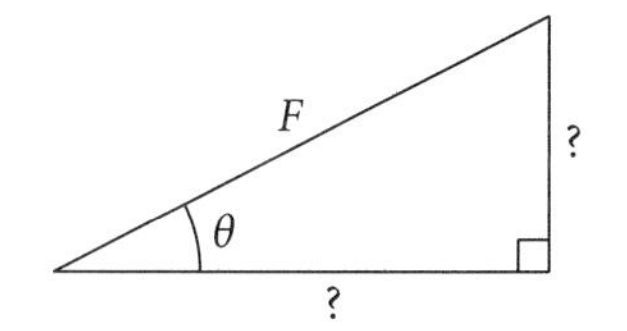

Figure 3.6 The sides of a right triangle with angle θ.

The second trig problem that arises over and over again is just the opposite of the first. This is the case where we know the opposite and adjacent sides of a right triangle, and need to use this information to find the hypotenuse of the triangle and one of the non-right angles. As we promised for the previous procedure, the value of this operation for vectors will be apparent when we get to Section 3.6.

EXAMPLE 3.4 Finding the Hypotenuse and the Angle

A right triangle has the side opposite to an angle θ that is 3 m long and an adjacent side of length 4 m, as shown. Find the magnitude of the hypotenuse, h, and find the angle θ.

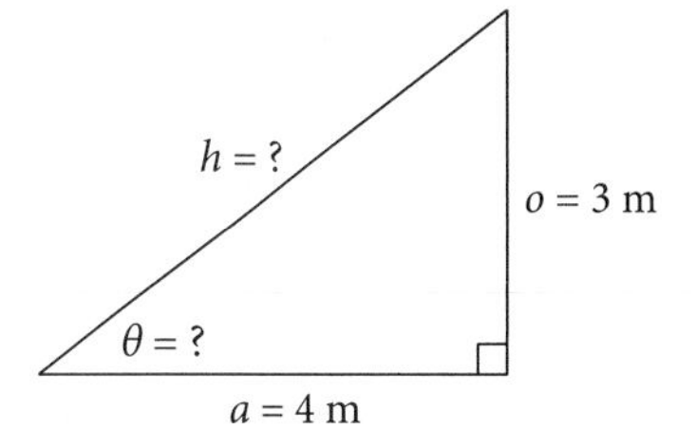

ANSWER We can easily find the magnitude of the hypotenuse by using the Pythagorean Theorem. It is

$$h^2 = a^2 + o^2 = (4\text{ m})^2 + (3\text{ m})^2 = 25\text{ m}^2, \quad \text{or} \quad h = 5\text{ m}.$$

Notice that the answer, $h = 5$ m, is an integer number of meters. This particular triangle is famous. It is called a 3-4-5 triangle and is sometimes used by carpenters as a way to create a right angle. Three pieces of string, 3 ft, 4 ft, and 5 ft in length, will create a triangle with a 90° angle.

The angle θ is found using the trig function we haven't used yet, the tangent function. Remember that the tangent of the angle θ is defined in Equation 3.4 as the ratio of the opposite side to the adjacent side, which in this case is (3 m)/(4 m) = 0.75. We need only find the angle whose tangent is equal to 0.75. The inverse tangent function, or $\tan^{-1}$, is just another button on your calculator. Simply input 0.75 on your calculator, hit the $\tan^{-1}$ button, and you should get the answer 37°. (If you got 0.644, your calculator is set up wrong. It is set up for angles in radians.)

The right triangle with a 37° angle (and a 53° angle) in it is the 3-4-5 right triangle. It will also be very common in homework assignments because they are triangles with integer sides and trig functions you consequently know (sin 37° = 0.6, cos 37° = 0.8, and tan 37° = 0.75). This way, you can concentrate on other parts of the problem, without worrying about going to the calculator all the time. However, not all of the triangles you will see will be 3-4-5 triangles. You will ultimately need to be able to use your calculator for doing trigonometry problems.

3.6 Vector Components

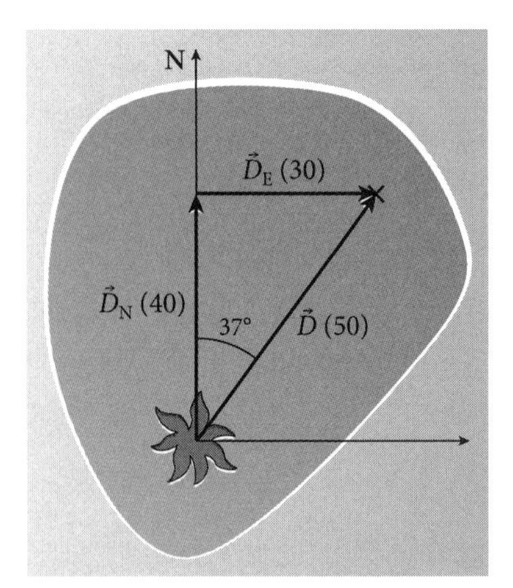

Figure 3.7 The north and east components of the displacement.

As we said in the last section, the reason for this digression into trigonometry was to enable us to add vectors numerically, instead of graphically. The key to being able to do this is the ability to take any vector and resolve it into what are called a vector's *components*. Let us see what we mean by this.

If we return to our example of the pirate and his buried treasure, we saw that a single vector, with its 37° direction (there's that 37° angle already) and its 50-pace magnitude, was sufficient to uniquely determine the position of the treasure. On the other hand, the same result could have been achieved if, instead of giving magnitude and direction instructions, the pirate had given the following instructions: "First walk 40 paces north, then walk 30 paces east." This set of instructions is diagrammed in Figure 3.7. The point that we want to make here is that these two vectors, $\vec{D}_N$ = {40 paces, north} and $\vec{D}_E$ = {30 paces, east}, add up to the same position as $\vec{D}$ = {50 paces, 37° east of north}. The two vectors, $\vec{D}_E$ and $\vec{D}_N$, are called *components* of the vector $\vec{D}$, because the vector sum of the two components takes you to exactly the same place. We can say that $\vec{D}_E + \vec{D}_N$ is equivalent to $\vec{D}$, or that $\vec{D}$ is actually replaced by the vector sum of its components.

The upshot of this is that we can now write down another way to specify a particular vector. We first choose two mutually perpendicular directions. Next, we determine the components of the vector along those two directions. Finally, we put the magnitudes of the two components inside parentheses. If, in the buried treasure example, we agree to put the eastward component first and the northward component second, we can write the vector from the origin at the tree out to the location of the treasure as

$$\vec{D} = (30\text{ paces}, 40\text{ paces}).$$

In summary, there are two ways of writing a vector. We can specify a single direction (like north) and agree to measure angles from that direction (like eastward from north), and we can then write the magnitude and the angle in curly brackets. Equivalently, we can choose two mutually perpendicular directions (like east and north) and write the components of the vector in these directions inside parentheses. For the component notation inside the parentheses, we will generally follow the common convention of calling one of the directions the x-axis and the perpendicular direction the y-axis, and we will agree to always write the x-component first inside the parentheses and the y-component second.

Now look at Figure 3.8. Here, we display a single vector $\vec{A}$ referred to two systems of x and y axes having different orientations. The point of the figure is that the magnitude of a vector does not change if we rotate the coordinates by some fixed amount, but that the angle used to define the direction of the vector does change when there is a change in the choice of the reference directions. Also, the components of a vector both change when there is such a change in the coordinates. A vector does not have an absolute direction angle or an absolute set of components. It acquires a direction angle and components only *after* a choice of the orientation of the axes has been made.

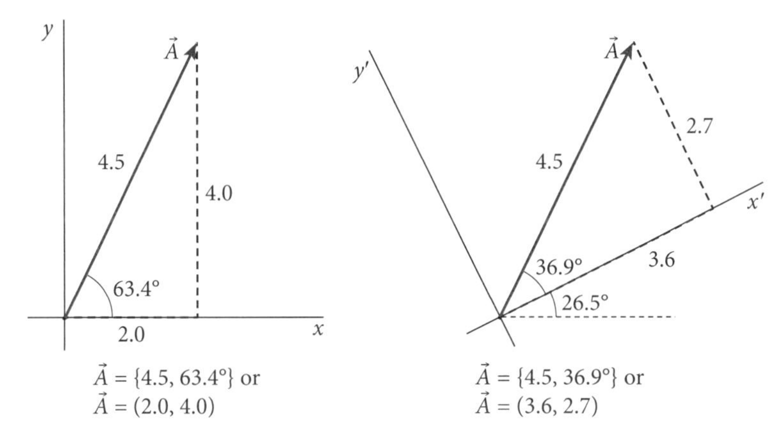

Figure 3.8 The vector $\vec{A}$ referred to two different systems of axes. The primed system is rotated 26.5° relative to the unprimed system. Angles and components differ between the two systems, even though the vector $\vec{A}$ is the same.

Please be clear that the vector is the same in both pictures but that its description depends on the direction of the axes chosen. If we change the axes, the direction angle inside the curly brackets changes. If we change the axes, the components inside the parentheses change. But the vector stays the same.

3.7 Vector Addition (Revisited)

There are two keys to the component method for adding vectors. The first is the fact that vectors may be resolved into components and that the components replace the vector in all ways, as discussed in the last section. The second key is the simplicity of adding vectors that point in the same direction. If two vectors point in different directions, vector addition is defined graphically, by drawing 2-dimensional arrows to scale. But vectors that point in the same direction can be added numerically with the rules of ordinary addition. The sum of a vector 4 m long pointing east and a vector 2 m long pointing east is a vector 6 m long pointing east. The sum of a vector 6 m long pointing north and a vector 5 m long pointing south (so it is a −5 m vector north) is a vector 1 m long pointing north. Isn't that easy?

So, here's the plan. To add vectors numerically by the component method, we pick two perpendicular directions and then we replace each vector by its components along the two directions. When we are done, we will have component vectors pointing either in one of our two directions or in the other, never in-between. These can then be added numerically, since the vectors in each set point in the same direction as each other. We then end up with two sums, one for the x-direction and one for the y-direction. The results may be combined, as in the trig procedure we demonstrated in Example 3.4, to find the magnitude and direction of the resultant vector. The easiest way to explain this whole process is probably just to do a couple of examples.

EXAMPLE 3.5 Adding Vector Displacements

You first walk 10 miles in the direction 35° east of north. You then walk 5 miles due east. Where are you (magnitude and direction) relative to your initial point?

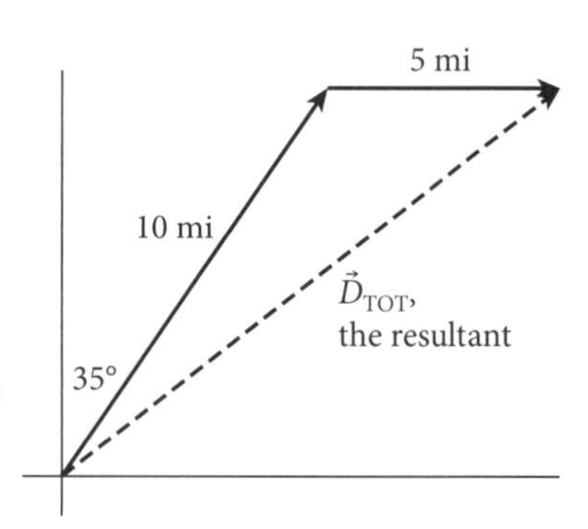

ANSWER It is always best to start by drawing a diagram as shown at right. Even if you are going to solve the problem numerically, as we are here, it helps by giving you a sense of what the answer will be and what calculations you will need to do. Of course, the diagram does not have to be drawn exactly, because you are never going to measure anything on it to get your answer, but some care in getting it close will often help.

The next step is to choose a set of mutually perpendicular directions along which your vector components will lie. Sometimes the problem will ask for some particular directions, but most often you will be free to choose them. The best choice is generally one in which fewer vectors will have to be resolved into components, thus reducing the amount of work you will have to do. In this problem, one of the vectors already points due east, so it is already resolved into components if we pick east and north (90° away from east) as our component directions. We call the eastward direction the x-axis and the northward direction the y-axis.

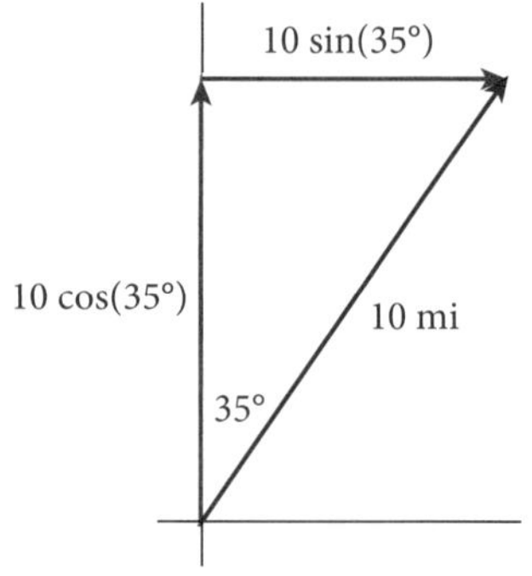

The 10-mile vector does not lie along either of the axes, so it will have to be resolved into components. In the process shown in the diagram at right, the two components (the red arrows) form the sides of a right triangle. So we can use the trigonometry we have just reviewed to find the lengths of the two sides. The northward vector is 10 cos(35°) = 8.2 miles, and the eastward vector is 10 sin(35°) = 5.7 miles. By this procedure, the 10-mile vector is divided into components as shown in the diagram. (By the way, some students carry the notion, probably remembered from some high school math class, that the x-component always uses the cosine, but this is not a safe bet. It depends on how the axes are chosen and on what angle of the triangle is known.)

With the triangle correctly analyzed, we write the two vectors in miles as:

$$\vec{D}_{10} = (5.7, 8.2) \text{ miles} \quad \text{and} \quad \vec{D}_5 = (5, 0) \text{ miles.}$$

Once all vectors are resolved into components, we can add all x-components to get the x-component of the resultant vector and similarly all the y-components. In our case, this gives us the components of the total displacement in miles as

$$D_x = 5 + 10\sin(35°) = 10.7$$
$$D_y = 0 + 10\cos(35°) = 8.2$$

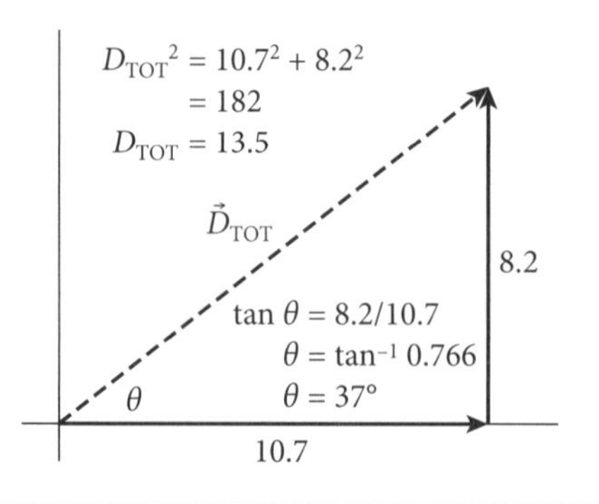

The last step is to combine the x- and y-components to find the magnitude and direction of the final displacement, the sum of the original 5-mile and 10-mile vectors. Here, we know the 8.2 and 10.7 sides of a right triangle, and we need to get the hypotenuse and the angle. The calculation of these two quantities is shown in the diagram at right. The result is $\vec{D}_{TOT} = (13.5 \text{ mi}, 37°)$.

EXAMPLE 3.6 The Desperate Duck

A duck leaves Amy Pond and flies 4 mi northeast to Veronica Lake. Hunters on the shore of the lake scare the duck, who then flies 3 miles southeast to Jonathan Creek. But the current here is too swift to land, so he continues on, flying 2 miles directly south to Gerald Ford. What is the duck's final displacement relative to Amy Pond?

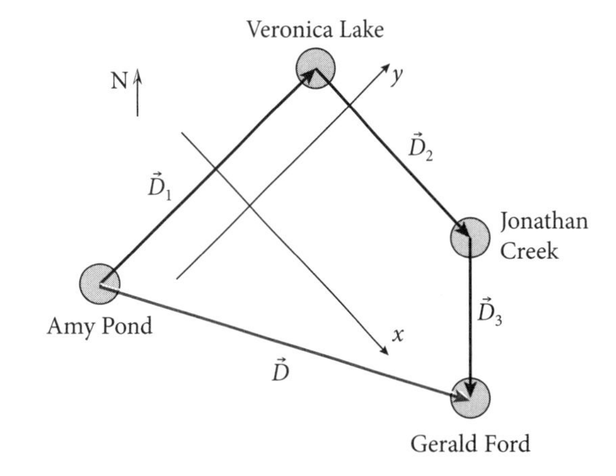

ANSWER Let's begin with a diagram of the legs of the flight, as in the figure above. Now the natural choice of map axes would probably be east and north. However, since two of the vectors lie along lines that are oriented 45° to these directions, it may save time to choose axes as shown in the diagram, x to the southeast and y to the northeast. We label the displacements $\vec{D}_1$, $\vec{D}_2$, and $\vec{D}_3$ and we may easily write (remembering that the x-component is written first inside the parentheses)

$$\vec{D}_1 = (0, 4\text{ mi}) \qquad \vec{D}_2 = (3\text{ mi}, 0),$$

leaving only $\vec{D}_3$ to resolve into components. From the diagram at right, we see that $\vec{D}_3$ has a positive component in the x-direction and a negative component in the y-direction. It is therefore

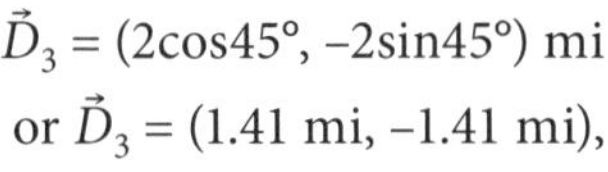

$$\vec{D}_3 = (2\cos 45°, -2\sin 45°)\text{ mi}$$
$$\text{or } \vec{D}_3 = (1.41\text{ mi}, -1.41\text{ mi}),$$

and we are ready to add $\vec{D}_1 + \vec{D}_2 + \vec{D}_3 = \vec{D}$. First, we find the x-components,

$$D_x = D_{1x} + D_{2x} + D_{3x} = 0 + 3\text{ mi} + 1.41\text{ mi} = 4.414\text{ mi},$$

and then we add up the y-components,

$$D_y = D_{1y} + D_{2y} + D_{3y} = 4\text{ mi} + 0 - 1.41\text{ mi} = 2.59\text{ mi}.$$

The duck's displacement vector can therefore be written

$$\vec{D} = (4.41, 2.59)\text{ mi}.$$

It has a magnitude in miles of

$$\sqrt{4.41^2 + 2.59^2} = \sqrt{26.2} = 5.12\text{ mi}$$

and a direction determined from the triangle at right

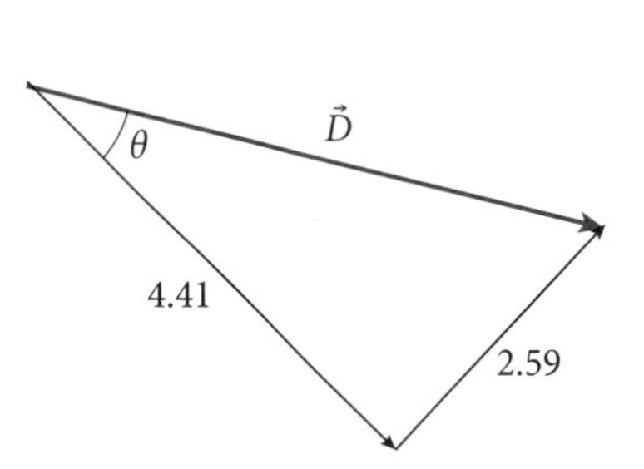

$$\theta = \tan^{-1}\left(\frac{2.59}{4.41}\right) = \tan^{-1}(0.59) \approx 30°.$$

Since the direction is 30° above the x-axis, which is itself 45° south of east, the cardinal direction is 45° − 30° = 15° south of east.

There is sometimes confusion about the best way to specify the direction of a vector. The direction of the velocity in the above example could be described as:

- 15° south of east,
- 30° north of southeast,
- 345° counterclockwise from the east.

All of these work. What matters is to be clear what direction you mean. Often the best way is to label a diagram, leaving no room for misunderstanding.

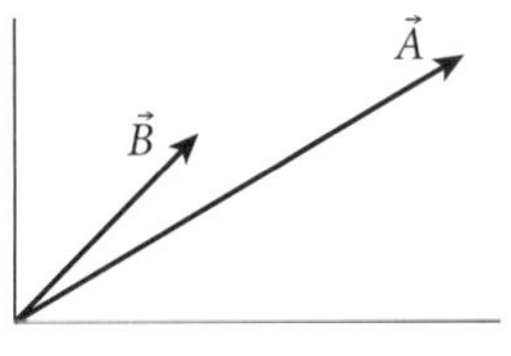

Figure 3.9

There is one last consideration we want to be very clear about. Most of you will find this obvious, but some do not. Suppose we want to add two vectors; the first is vector $\vec{A}$ which is 20 m long at an angle of 30°, and the second is vector $\vec{B}$ which is 10 m long, and at an angle of 45° (Figure 3.9). We now have two ways of expressing the vectors. Using curly brackets, we would write

$$\vec{A} = \{20\text{ m}, 30°\} \quad \text{and} \quad \vec{B} = \{10\text{ m}, 45°\},$$

or, using components placed inside parentheses, we would have

$$\vec{A} = (17.32\text{ m}, 10\text{ m}) \quad \text{and} \quad \vec{B} = (7.07\text{ m}, 7.07\text{ m}).$$

We have seen that there are two ways to add the vectors. We may draw them, using the magnitude–angle (curly-bracket) descriptions to draw the arrows on the paper and then adding them graphically with the help of a ruler and a protractor. Or we may use the component notation with parentheses and add like components—the x-component to the x-component and the y-component to the y-component. But one thing that we absolutely cannot do, because it just doesn't have the same meaning as adding vectors tail to head, is to add the entries inside the curly bracket notations. The sum of $\vec{A} = \{20\text{ m}, 30°\}$ and $\vec{B} = \{10\text{ m}, 45°\}$ is simply **not** $\{30\text{ m}, 75°\}$.

The mathematical addition of vectors will prove to be one of the more important skills you will need for future work in this book. You will see it again and again and again. You need to learn it. But the process is straightforward—break each vector into components, add up all the components in each of the two directions to get the two components of the resultant, combine the two components to get a magnitude and direction, if required.

You do not have to be a clever physicist to add vectors—you only have to be a careful bookkeeper. Clever physicists are always happy to do simple bookkeeping whenever they can. It just feels so good to know you are doing something right. So work slowly, follow the program, and don't make mistakes.

3.8 Summary

From this chapter, you should have learned the following things:

- You should know how to take a vector whose magnitude and direction are known and break it into components.
- You should know how to begin with two known components and combine them to calculate the magnitude and direction of the vector.
- You should be able to do the vector component method of vector addition and subtraction like a good bookkeeper, slowly, methodically, and without mistakes.

CHAPTER FORMULAS

Formulas for a right triangle are:

$h^2 = o^2 + a^2$ and

$\sin\theta = \frac{o}{h}$ $\cos\theta = \frac{a}{h}$ $\tan\theta = \frac{o}{a}$.

Thus, $o = h\sin\theta$ $a = h\cos\theta$ $o = a\tan\theta$

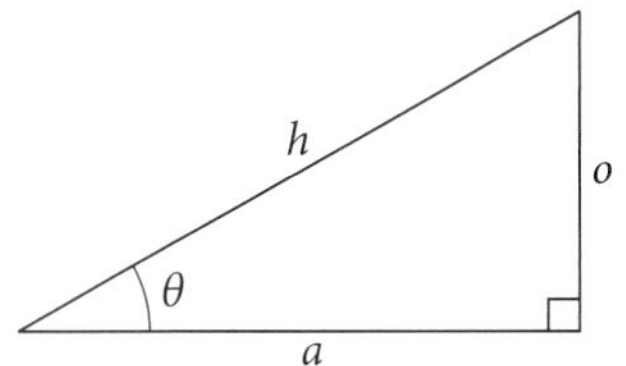

PROBLEMS

3.3 *1. Using a ruler and a protractor, choose a scale and draw arrows representing a vector 50 m west and a vector 70 m south. Add the two vectors graphically with an accuracy of the nearest meter.

*2. Using a ruler and a protractor, choose a scale and draw arrows representing a vector 10 ft due east and a vector 12 ft 30° west of north. Add the two vectors graphically with an accuracy of a tenth of a foot.

*3. A knight on a chessboard moves two squares forward and one square to the right. If each square is 4 cm on a side, draw the displacement vectors on a page and measure the total displacement (magnitude and direction) for this move. [Assume that the knight begins and ends in the center of its square.]

3.4 *4. Bob and Irene start off together. Bob walks 500 m north and Irene walks 700 m west. Find the distance between them and determine what direction Bob would have to walk to get to Irene, assuming she waits where she is.

**5. For each pair of velocity vectors shown, graphically estimate the change in velocity $\Delta\vec{v} = \vec{v}_f - \vec{v}_i$ by copying the $\vec{v}_f$ and $\vec{v}_i$ vectors on a page and drawing the $\Delta\vec{v}$ vector for each case.

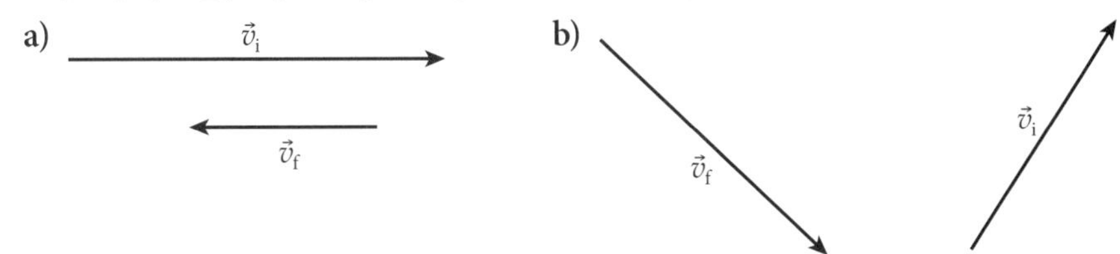

**6. For each pair of velocity vectors in problem 5, find $\Delta\vec{v} = \vec{v}_f - \vec{v}_i$ by drawing $\vec{v}_f$, drawing a vector representing $-\vec{v}_i$, and then adding $\vec{v}_f + (-\vec{v}_i)$.

3.5 **7. For each of the following right triangles, the magnitudes (lengths) of the two sides have been given. Find the magnitude of the hypotenuse and find the angle labeled θ in each drawing.

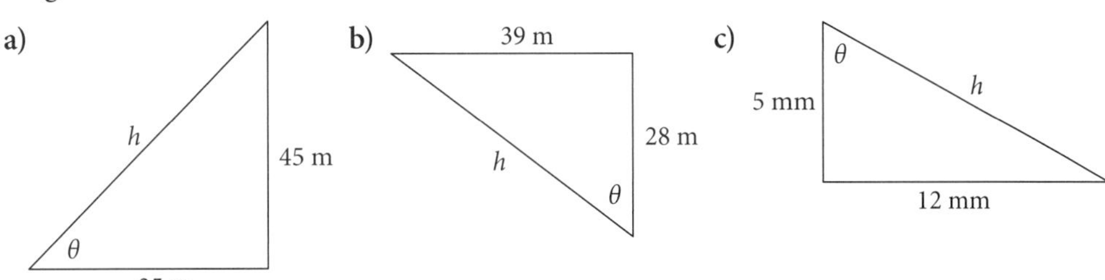

** 8. For each of the following right triangles, the magnitude of the hypotenuse and one of the angles are given. Find the magnitudes (lengths) of the other two sides.

a) 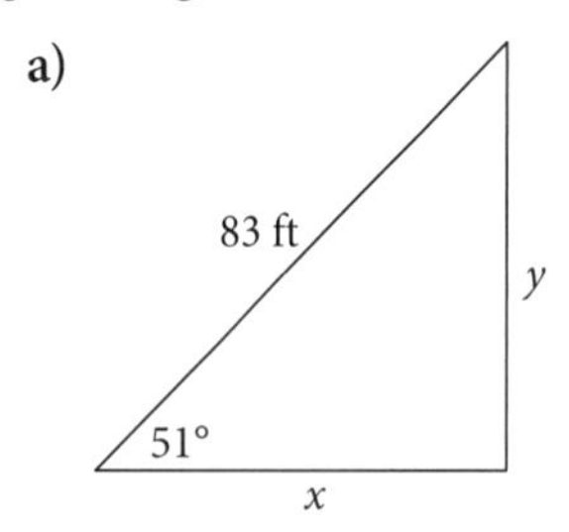

b) 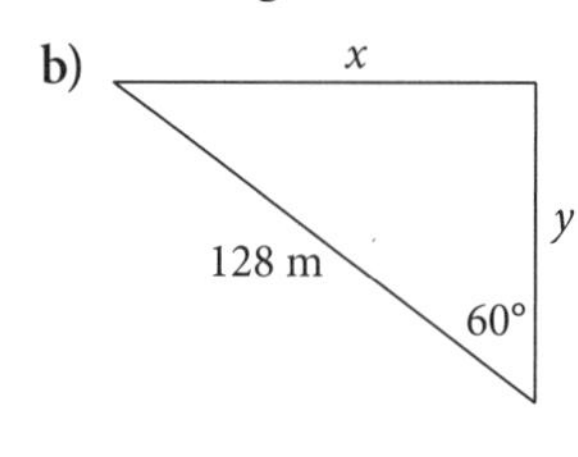

c) 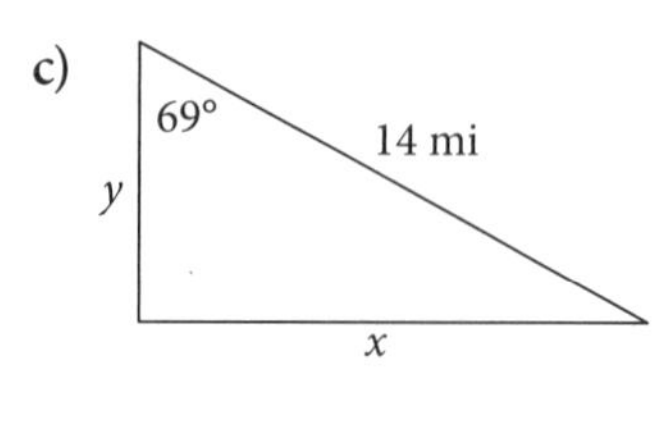

** 9. The magnitudes and directions of three velocity vectors are indicated in the drawings. Using the coordinate directions shown, find their x- and y-components.

a) 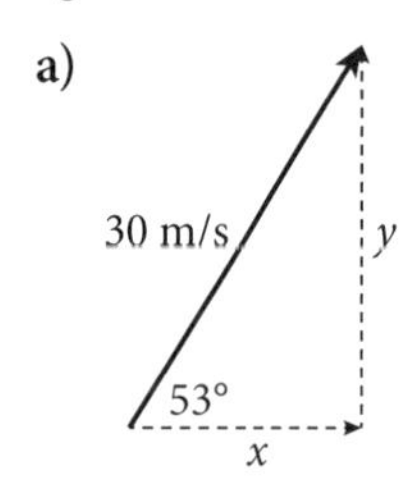

b) 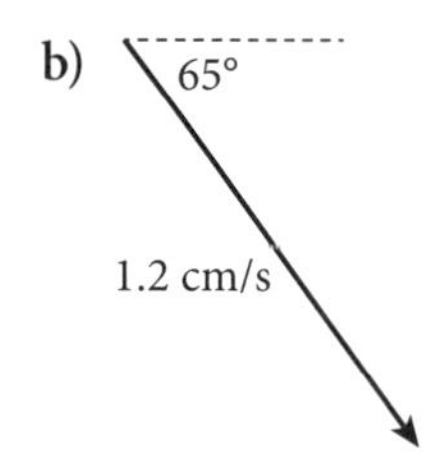

c) 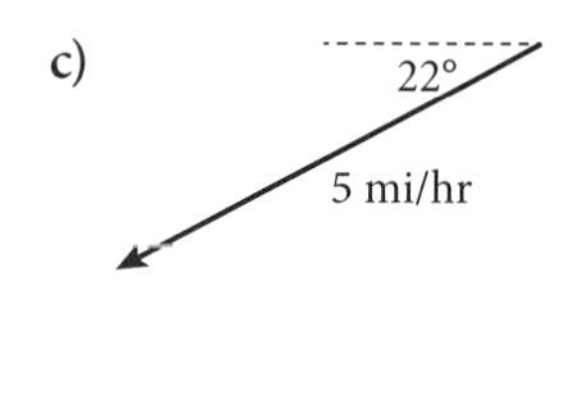

** 10. For each of the following velocity vectors, the x- and y-components of the vector are indicated. Find magnitude and θ-direction of each vector.

a) 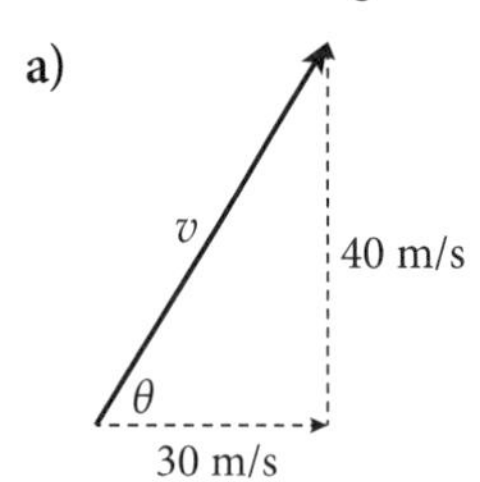

b) 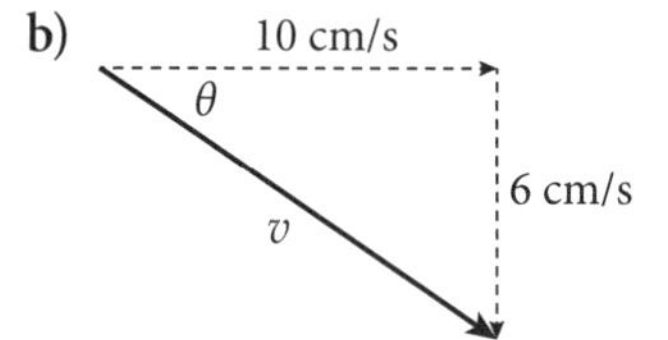

c) 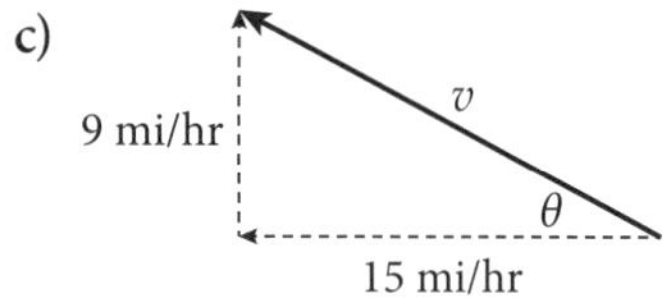

** 11. Vector $\vec{A}$ is 5 ft long and points east. Vector $\vec{B}$ is 3 ft long. Find $\vec{A} + \vec{B}$ if $\vec{B}$

a) points east b) points north c) points west.

3.7 ** 12. We want to add four position vectors together. The vectors in component form are $\vec{A}$ = (10 m, 10 m), $\vec{B}$ = (−4 m, 1 m), $\vec{C}$ = (5 m, −6 m), and $\vec{D}$ = (−8 m, −2 m). Practice your vector bookkeeping skills by drawing out a ledger like the one below and carefully finding the components of the resultant $\vec{A} + \vec{B} + \vec{C} + \vec{D}$.

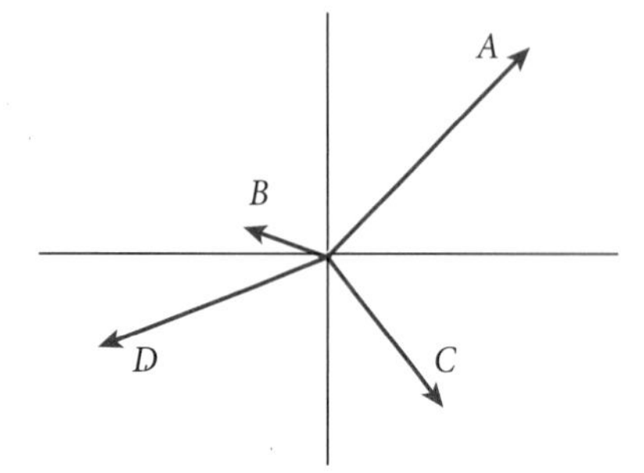

Vector	*x*-component	*y*-component
A		
B		
C		
D		
Resultant		

** 13. Forces are vectors. In the English system of units, they are measured in pounds (see page 4). Add the four forces drawn at right (and given in the ledger below). Practice your vector bookkeeping skills by setting up and completing the ledger, including finding the magnitude and angle for the vector $\vec{E} = \vec{A} + \vec{B} + \vec{C} + \vec{D}$.

Vector	Magnitude	Angle from *x*-axis	*x*-component	*y*-component
A	14 lbs	10°		
B	10 lbs	220°		
C	6 lbs	45°		
D	8 lbs	280°		
E				

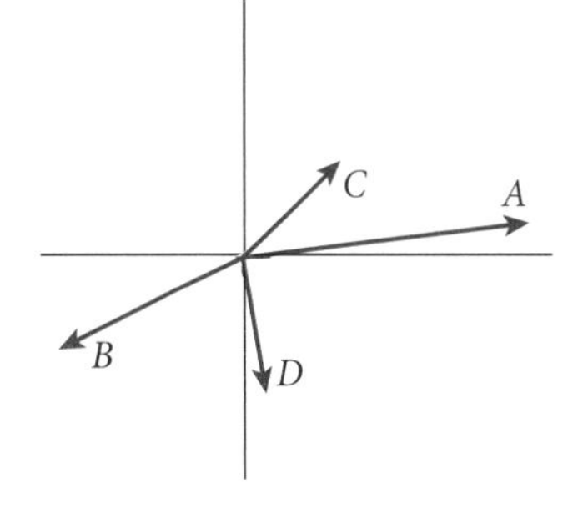

** 14. While playing ultimate Frisbee, you run north for 10 m, then east for 30 m, and finally 34 m in a direction 37° south of east (as shown). What is your total displacement (magnitude and direction) during this run?

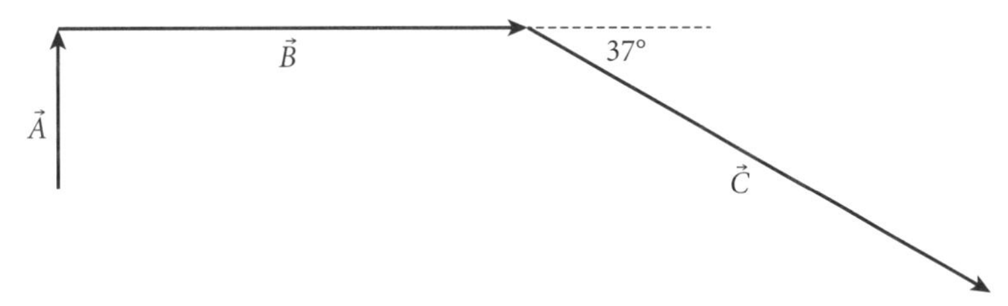

** 15. Scoutmaster Greg and his scout troop are working on their orienteering merit badge. They start at an old stump in the center of a large open meadow. They first march 400 m west. Then they march 1000 m in a direction 37° south of east. They then march 1500 m in a direction 53° north of east. Finally, they march 500 m west. Find their total displacement (magnitude and direction) from the stump.

** 16. Masahisa's ball is on a miniature golf green, 3 m west of the hole, as shown below. Unfortunately, Masahisa hits the ball in the wrong direction, 18° south of east. The ball travels 5 m then hits the curb on the edge of the green and rolls north for 2.5 m before stopping. Find the ball's final position relative to the hole (magnitude R and direction α).

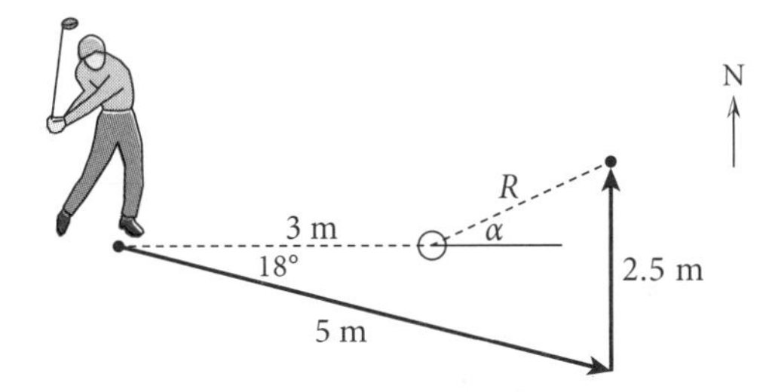

Note: This is a schematic diagram only. It is not drawn to scale, so you cannot measure on the picture to get an answer.

** 17. Christina is an avid cross-country skier. On the first day of her trip she travels 25 km in a direction 37° south of east and sets camp for the night. The following day she travels another 25 km in a direction 53° south of west and sets camp. In the final day of her trip she skis 30 km due north. What is her total displacement (direction and magnitude) for the 3-day period?

** 18. An Apollo astronaut lands on the moon and exits the Lunar Excursion Module (LEM). Starting at the base of the LEM, he drives the lunar rover 17 km at an angle of 28° south of lunar east. At this point, the rover breaks down, so he steps out of the rover and walks 12 km directly south and plants a flag pole. What is the displacement of the flag from the LEM?

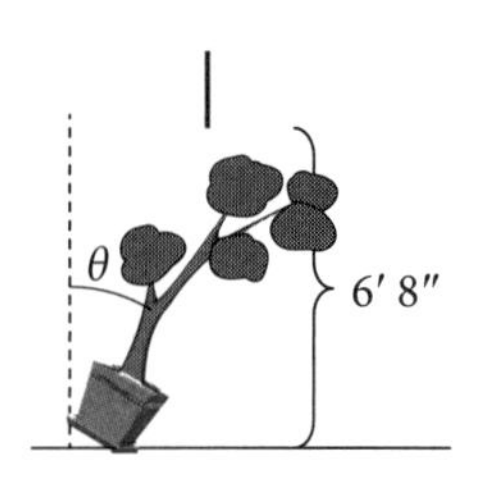

** 19. Llewellyn needs to get her 8-foot-tall potted plant through a door that is 6 feet 8 inches tall. She is worried that, if she tips the plant too far, some of the dirt will fall out onto her clean floor. At what angle to the vertical must she tip her plant to just barely get it through the door?

** 20. A glider out over the ocean dives at a speed 80 m/s at an angle 37° below the horizontal, as shown. If the sun is directly overhead, how fast is the plane's shadow moving along the surface?

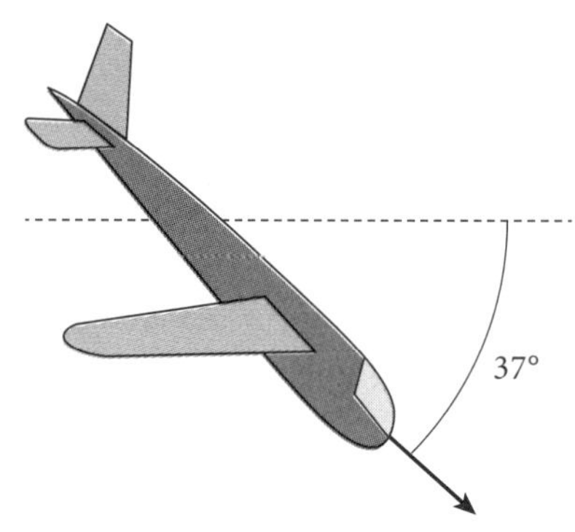

** 21. A speedboat travels on a calm lake at a speed of 25 mi/hr, on a heading 68° east of north. It leaves the south shore of the lake (whose shoreline is uniformly straight along an east-to-west line) at 10:00 am. How far from the south shore will the boat be at 10:10 am?

** 22. At noon, on December 22 in Bozeman, Montana, the sun shines from a height that is only 21° above the horizon. How long a shadow will the 40-foot flagpole at the fairgrounds cast on the ground at noon during the Bozeman Winter Fair?

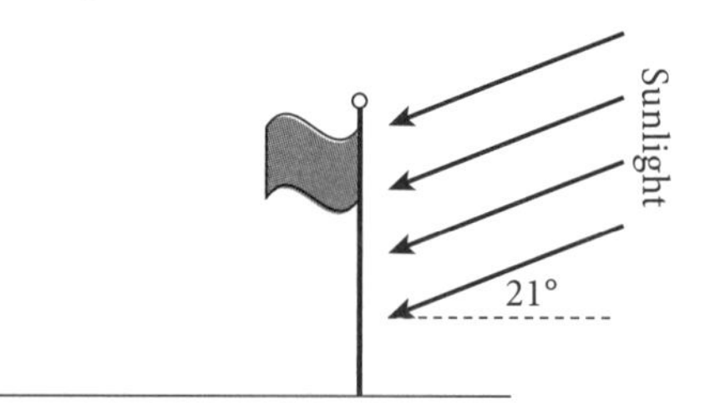

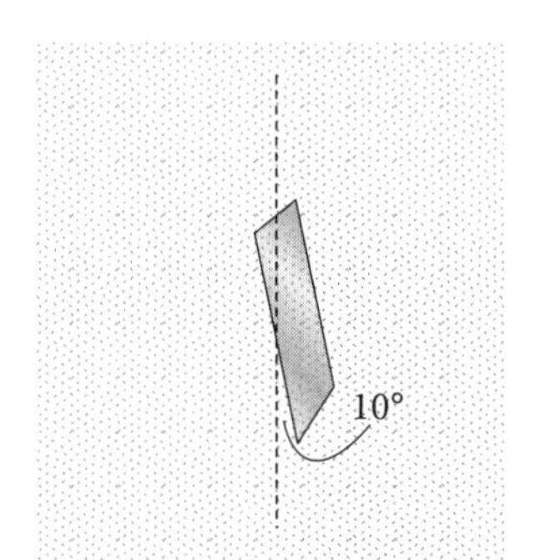

*** 23. You are carrying a pane of glass from your truck to your garage, when a summer storm hits. Hailstones are falling straight down at a rate of 10 stones/sec for each square foot of ground area. Your windowpane has an area of 8 ft^2, so that if the pane were to be held horizontally, it would be hit by 80 hailstones every second. But you decide to tip the pane upward, so that it makes an angle of 10° with the vertical, as shown at left. How many hailstones will now hit it each second?

Formula-Finding Problems: In all the problems you have done so far, the goal of the problem has been to use a formula or an equation to find a numerical answer to a numerical question. Often, however, physicists need to work out a *new* formula for something, a formula that will work regardless of the values chosen. This is like doing the first three steps of our four steps in solving physics problems (see page 26), but not doing the last step—the arithmetic. From time to time, we will assign a problem in which you need to find, not a numerical answer, but a new formula for the quantity requested. Instead of looking for values in the problem statement, you should look for variable names and find your formula in terms of these variables. Problems like this will be indicated with a superscript f, as in the next three problems.

** 24^{f}. Consider a right triangle and use the definitions of sine and cosine, along with the Pythagorean Theorem, to prove that, for angles θ less than 90°,

$$\cos\theta = \sqrt{1 - \sin^2\theta}.$$

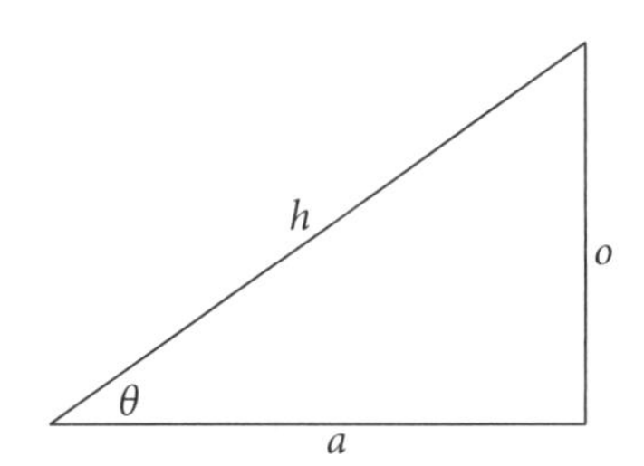

*** 25^{f}. A lens is constructed of glass that is flat on one side and has a section of a sphere as the other side, as in the figure below. If the radius of the sphere whose shape gives the curved side of the lens is R, and if the total width of the lens is $2h$, as shown, find a formula for the thickness of the lens, x.

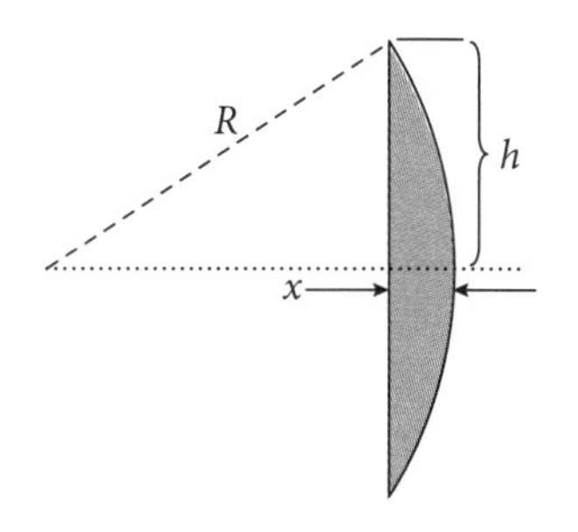

*** 26^{f}. A series of three vectors, each of magnitude A, are directed in such a way that each one points at angle γ relative to the vector before it. Show that the sum of the three vectors is a vector with magnitude $A(1 + 2\cos\gamma)$, pointing at an angle γ relative to the first vector. [Hint: Think of the dashed line's length as being divided into segments, each segment being either a side of a right triangle or of a rectangle, with each one of the three vectors being one of the other sides in each geometrical figure.]

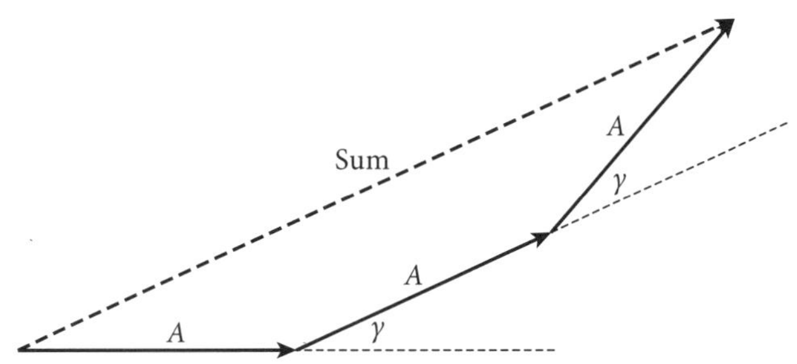

4

TWO-DIMENSIONAL KINEMATICS

In Chapter 2, we limited our discussion of motion to a single dimension. Of course, as everyone knows, we actually live in a 3-dimensional universe. What this means is that, as it takes one number to specify position on a 1-dimensional line, it takes two numbers to specify position in a 2-dimensional space and three numbers in a 3-dimensional space.

Though we live in a 3-dimensional universe, most of the new mathematics and physics that arise as we go beyond the restriction of a single dimension can already be understood and appreciated in 2 dimensions. And, as we will see, there are plenty of interesting applications that are inherently 2-dimensional anyway. So, for most of the work in this book, we will consider motion in 2 dimensions only. This will work well for us until we get to Chapter 19, where we will finally be forced to expand to 3 dimensions by the inherent 3-dimensional nature of the magnetic field.

4.1 Position Vectors and Coordinate Systems

We began Chapter 2 with a discussion of a coordinate system that we used to give an address to each point of the 1-dimensional highway along which a car was moving. How might we produce coordinate-system addresses in 2 dimensions?

In Chapter 3, we discussed how to find a pirate's buried treasure by giving its *position vector* from the tree. We saw that every point on the island could be specified by its position vector relative to the tree. We know that it will require two numbers to specify points in any 2-dimensional space, so one way of setting up a *coordinate system* for such a space would be this. Let us take some point in the 2-dimensional space as the origin and let us define two perpendicular directions to orient the system. We can then use the two numbers that define the position vectors as the coordinates of all points. If we express the vectors in curly bracket notation, with magnitude and direction as the two numbers, we are led to a plane polar coordinate system (Figure 4.1a). If we give the vector in parentheses notation, with the two components as the two numbers, we are led to a Cartesian coordinate system (Figure 4.1b).

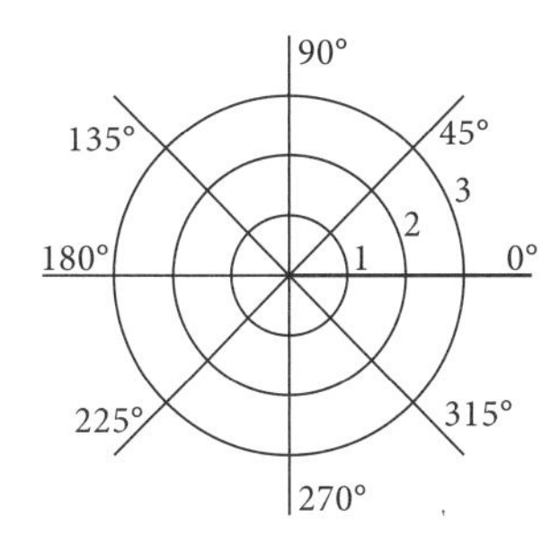

Figure 4.1a Plane polar coordinates.

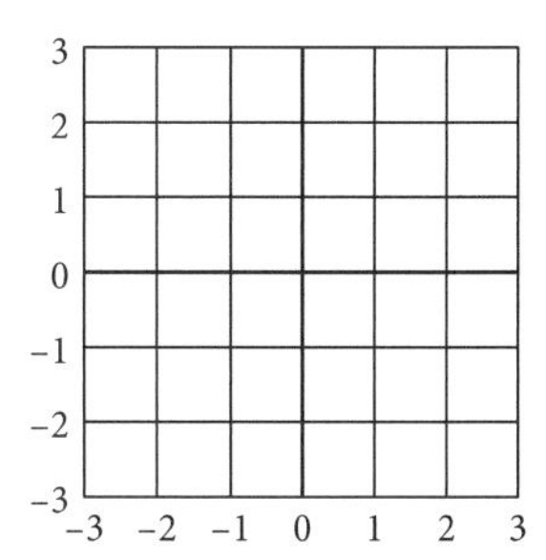

Figure 4.1b Cartesian coordinates.

4.2 Displacement and Velocity

As we discussed in Chapter 3, the concept of *displacement* in more than one dimension was the basis for developing the mathematics of vectors, so it is probably unnecessary to state that displacement is a vector, having magnitude and direction. The pirate's dilemma (Figures 3.2 and 3.7) addressed the question of how to go from the palm tree to the treasure. However, as we just saw, the position vectors from the palm tree can also be thought of as producing a *coordinate system* with origin at the tree and with axes oriented east (x-axis) and north (y-axis). The position vector $\vec{r}$ from the origin to the location of the treasure may then be expressed either as $\vec{r}$ = {50 paces, 37° east of north} or as $\vec{r}$ = (30 paces, 40 paces). And every point on the island has such an address, whether there is a treasure there or not.

If a sea turtle were to lay its eggs in the soft dirt where the treasure had just been buried, and then, after laying its eggs, it traveled to the seashore along the path shown in Figure 4.2, we could define its initial position with the vector $\vec{r}_\mathrm{i}$ = (30 paces, 40 paces) and its final position with the vector $\vec{r}_\mathrm{f}$ = (–20 paces, 50 paces). The displacement of the turtle from its initial location would then be a vector defined by

$$\vec{D} = \Delta\vec{r} = \vec{r}_\mathrm{f} - \vec{r}_\mathrm{i}. \tag{4.1}$$

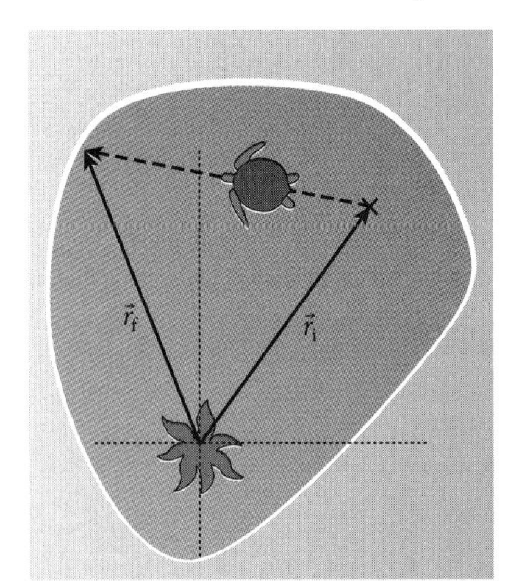

Figure 4.2 The annual migration of the sea turtle from $\vec{r}_\mathrm{i}$ to $\vec{r}_\mathrm{f}$.

This definition looks like the one in Equation 2.1, except that we have used the vector $\vec{r}$ instead of the single coordinate x. The displacement is once again the vector that must be added to the initial position to produce the final position. It is the dashed-line vector along the turtle's path in Figure 4.2.

Please think this next statement over for a moment to be sure you understand it and agree with it: If we had chosen a different point for our origin, something other than the tree, then the initial and final position vectors would have been different, but the displacement vector would have been the same.

Now let's talk about *velocity*. Back in Chapter 2, we noted that the difference between speed and velocity was that the velocity included information about the direction. Because we considered only one dimension in Chapter 2, the direction could be completely specified by giving a sign, positive or negative, depending on whether the motion was taking the thing toward more positive or more negative positions along the number line. When we go from one to two dimensions, the direction is more complex. It requires an actual number to represent the angle between the direction of the vector and one of the coordinate axes.

You may remember that, back on pages 39–40, we defined the multiplication of a vector by a scalar so as to produce a new vector whose magnitude was the value of the scalar times the magnitude of the original vector, and whose direction was the same as that of the original vector. Having defined this mathematical operation, it is now possible to define average velocity in two dimensions using a vector form of the same definition we used in one dimension. The vector form of Equation 2.3 is

$$\vec{v}_\mathrm{ave} = \frac{\vec{r}_\mathrm{f} - \vec{r}_\mathrm{i}}{t_\mathrm{f} - t_\mathrm{i}} = \frac{\Delta\vec{r}}{\Delta t}. \tag{4.2}$$

Here, $1/\Delta t$ is the scalar factor by which we multiply the vector $\Delta\vec{r}$ to get the vector $\vec{v}_\mathrm{ave}$. The result of the multiplication is, as we remember, a vector with a magnitude equal to the product of $1/\Delta t$ and the scalar magnitude of $\Delta\vec{r}$, and with a direction that is the same direction as $\Delta\vec{r}$. The displacement of the turtle is a vector, so it can be written in either curly brackets as $\Delta\vec{r}$ = {51 paces, 11.3° north of west}, or in parentheses as $\Delta\vec{r}$ = (–50 paces, 10 paces).

Then, if the turtle covers that displacement in ½ hr, we can write its average velocity as $\vec{v}_{ave}$ = {102 paces/hr, 11.3° north of west} in curly bracket notation, or we may use component notation to write $\vec{v}_{ave} = (-100, 20)$ paces/hr.

The instantaneous velocity may also be defined by analogy to Equation 2.4:

$$\vec{v} = \frac{\delta \vec{r}}{\delta t}, \tag{4.3}$$

where we have again used the calculus-style definition for a "small enough" δt, as one that goes toward a limit of a zero time interval.

To make sure we understand the idea of instantaneous velocity in two dimensions, let us consider the bird's-eye view of a marble rolling along a circular track in Figure 4.3, and let us see what happens as we take smaller and smaller intervals of time. Let's say that the speed, the rate at which the marble is covering distance along the track, is a constant 2 cm/s, and let us find the instantaneous velocity at point $t = 0$. In the figure caption, we calculate the magnitude of the average velocity between the point $t = 0$ (where the bright red marble is seen) and several choices of final point (the pale red marbles). The Δr values in the figure are the measured magnitudes of the displacements. There are two things to notice as the final point gets closer and closer to the initial point (i.e., as $\Delta t \rightarrow 0$). First, looking at the calculations in the caption, we see that the magnitude of the velocity comes closer and closer to the actual 2 cm/s uniform speed along the track. Second, we see that the direction of the displacement (and so of the velocity) comes closer and closer to being tangent to the circle at the point $t = 0$. In the limit that $\Delta t \rightarrow 0$, we conclude:

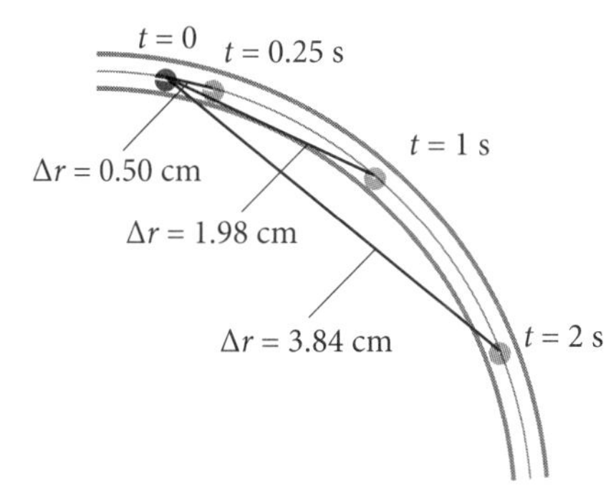

Figure 4.3 The average velocities measured between several points around a circular arc. From 0 to 2 s, v_{ave} = 1.92 cm/s. From 0 to 1 s, v_{ave} = 1.98 cm/s. From 0 to 0.25 s, v_{ave} = 2.00 cm/s. Thus, as $\Delta t \rightarrow 0$, v_{ave} approaches the uniform circular speed of the marble.

Anything moving along a curved path has an instantaneous velocity vector whose direction is tangent to the curve, at each point of the path, and whose magnitude is the instantaneous speed of the object along the path.

4.3 Acceleration in Two Dimensions

As we turn our attention to acceleration in two dimensions, we find that we must actually define this concept in a way that is contrary to most people's common automobile experience. In a car, we are used to thinking of acceleration as something that the accelerator pedal produces. We found in Chapter 2 that we had to allow both positive and negative accelerations, and this led us to recognize the brake pedal as just another "accelerator." Now we are going to have to recognize another "accelerator" in the car.

The problem is this. We are used to thinking of acceleration as the rate of change of speed. But if we were to keep that definition, it would upset the principle of superposition that allows us to break complicated 2-dimensional motion into two simpler 1-dimensional motions. This is too great a price to pay. The only way we are going to be able to solve 2-dimensional problems cleanly is to define acceleration as the rate of change of velocity, not speed. This means that the definition of average acceleration will be the vector counterpart of Equation 2.6. It is

$$\vec{a}_{ave} = \frac{\vec{v}_f - \vec{v}_i}{t_f - t_i}, \tag{4.4}$$

and the instantaneous acceleration will be defined as the vector counterpart of Equation 2.7,

$$\vec{a} = \frac{\delta \vec{v}}{\delta t}. \tag{4.5}$$

If the acceleration is uniform (constant), then Equations 4.4 and 4.5 give identical results.

The advantage of these definitions, as we said, is that they enable us use the principle of superposition. We can to take a 2-dimensional problem and write it as two independent 1-dimensional problems. To see how this works, let us write the components of a position vector as x and y, so we have $\vec{r} = (x, y)$. Then the definitions of $\vec{v}$ and $\vec{a}$ allow us to write, say, the x-components of these quantities as

$$\Delta x = x_f - x_i, \quad v_x = \frac{\delta x}{\delta t}, \quad \text{and} \quad a_x = \frac{\delta v_x}{\delta t},$$

and this is defined *without* reference to any motion that may or may not be going on at the same time in the y-direction. One component at a time; isn't that simple? However, the price we must pay for this simplicity is that our old ideas about acceleration must be revised.

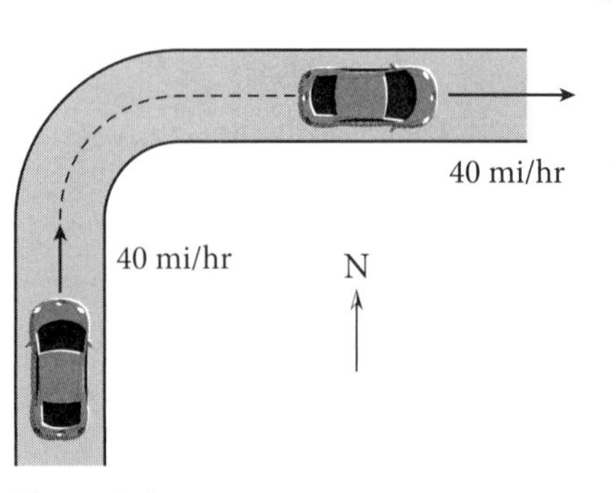

Figure 4.4 A car turning a 90° corner while always moving at constant speed is nevertheless accelerated.

Consider the car in Figure 4.4 that starts out going 40 mi/hr north and then turns a corner to head east, going a smooth 40 mi/hr all through the 5 seconds that it took to complete the turn. Did the car accelerate? Granted, the speed did not change, but did the velocity change? The car was initially moving 40 mi/hr north and it ended up moving 40 mi/hr east. The initial velocity was $\vec{v}_i = (0 \text{ mi/hr}, 40 \text{ mi/hr})$ and the final velocity ended up at $\vec{v}_f = (40 \text{ mi/hr}, 0 \text{ mi/hr})$. Those two vectors are not the same. The velocity has changed. And if the velocity changed, the car must have accelerated. It did not speed up, but it did accelerate. The average acceleration can even be calculated. Since the turn took 5 seconds to complete, the average acceleration components are

$$a_{\text{ave,east}} = \frac{\Delta v_{\text{east}}}{\Delta t} = \frac{40 \text{ mi/hr} - 0}{5 \text{ sec}} = 8 \text{ (mi/hr)/s}$$

$$a_{\text{ave,north}} = \frac{\Delta v_{\text{north}}}{\Delta t} = \frac{0 - 40 \text{ mi/hr}}{5 \text{ sec}} = -8 \text{ (mi/hr)/s}$$

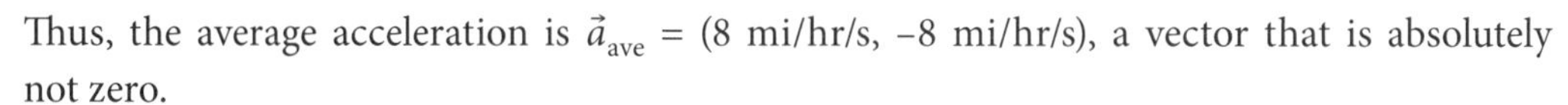

Thus, the average acceleration is $\vec{a}_{\text{ave}} = (8 \text{ mi/hr/s}, -8 \text{ mi/hr/s})$, a vector that is absolutely not zero.

It is important for you to understand the following statement and agree with it: *The car accelerated, but it did not speed up.* The thing that produced the acceleration was not the accelerator or the brake; it was the steering wheel.

For constant acceleration, the average acceleration is the same as the instantaneous acceleration, so we can write $\vec{a}_{\text{ave}} = \vec{a}$ in Equation 4.4. The equation can then be rearranged to read

$$\vec{v}_f = \vec{v}_i + \vec{a}\Delta t.$$

In this form, we see that the vector $\vec{a}\Delta t$ is added to the initial velocity vector to produce the final velocity vector. We have produced a table (Table 4.1) to help us see how the direction of $\vec{a}\Delta t$ affects the final velocity. The characteristic outcome of each case is summarized in the bottom row of the table.

TABLE 4.1 Five Cases of Relative Orientation of the Red $\vec{a}\Delta t$ Vector to the Black Initial Velocity Vector

a. Acceleration is in the same direction as the velocity.	b. Acceleration has a forward component and a perpendicular component.	c. Acceleration is perpendicular to the velocity.	d. Acceleration has a backward component and a perpendicular component.	e. Acceleration is in the opposite direction as the velocity.
$\vec{a}\Delta t$, $\vec{v}_i$, $\vec{v}_f$	$\vec{a}\Delta t$, $\vec{v}_i$, $\vec{v}_f$	$\vec{a}\Delta t$, $\vec{v}_i$, $\vec{v}_f$	$\vec{a}\Delta t$, $\vec{v}_i$, $\vec{v}_f$	$\vec{a}\Delta t$, $\vec{v}_i$, $\vec{v}_f$
Object speeds up going in the same direction.	Velocity vector gets longer and turns to the right.	Object turns right but the speed is not changed.	Velocity vector gets shorter and turns to the right.	Object slows down going in the same direction.

In each case shown in Table 4.1, it is helpful to think of the $\vec{a}\Delta t$ vector as grabbing the head of the initial velocity and dragging it to some new position (while the tail stays put). A parallel acceleration either stretches or shrinks the velocity but does not change its direction. If the acceleration is not along the velocity vector (i.e., if it has only a perpendicular component) then $\vec{a}\Delta t$ will turn the velocity vector. And, of course, the turning could also be to the left, which isn't shown in Table 4.1.

Now, if you carefully examine Table 4.1, you might find one statement you disagree with, and you'd be right. This is the claim that a perpendicular acceleration changes the direction without changing the speed. Clearly, if you add a perpendicular vector to the initial vector, the final vector must be longer than the initial vector because this resultant vector is the hypotenuse of a right triangle. So why did we write something in Table 4.1 that was clearly wrong? Well . . . it's because it's only sort of wrong.

When the acceleration is perpendicular to the velocity, the first thing it does—the thing that happens in the first small increment of time—is to change the direction of the velocity, not its magnitude. In Figure 4.5, we display the change in velocity produced by a perpendicular acceleration $\vec{a}$ that acts for shorter and shorter times Δt. We have put the calculations of the velocity changes for the three different cases into the spreadsheet next to the figure, in Table 4.2. In the table, column 1 labels the case; column 2 is the initial speed; column 3 is the product of a constant a with smaller and smaller times Δt; $\Delta\theta$ is the angle by which the initial velocity is turned; the 5th column is the final speed; the last column, labeled Δv, gives the increase in the speed. All velocity units are m/s.

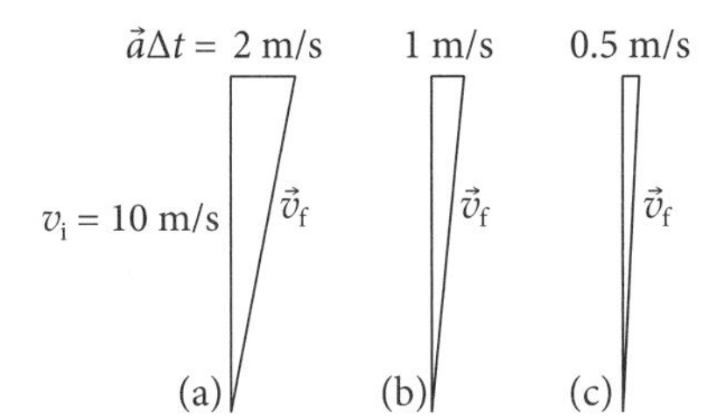

Figure 4.5 The effects of a perpendicular acceleration for smaller and smaller times.

TABLE 4.2 Values from Figure 4.5

Case	v_i	$a\Delta t$	$\Delta\theta$	v_f	Δv
(a)	10	2	11.30°	10.20	0.20
(b)	10	1	5.70°	10.05	0.05
(c)	10	0.5	2.85°	10.01	0.01

The thing to notice in Table 4.2 is that when the value of Δt is cut in half, as in going from case (a) to case (b), the angle is cut roughly in half, while the increase in speed is cut by a factor of 4. From case (b) to case (c), there is another factor of 2 decrease in $\Delta\theta$ and another factor of 4 drop in Δv. So a small perpendicular acceleration produces a small change in the *direction*, but the change in the velocity *magnitude* is tiny, essentially negligible. In that sense, what we said in Table 4.1 was true. And, in the limit that Δt goes to zero, as you can probably guess, the statement in Table 4.1 becomes exactly true.

To get a feel for 2-dimensional paths, let's follow the motion of a car along the road shown in Figure 4.6. At each labeled point, the velocity and acceleration vectors are shown. Note that the velocity vectors are always parallel to the tangents to the curve. There is no connection between the lengths of the velocity and acceleration vectors, because they have different units and cannot be compared. Each has been scaled in its own way. But the relative length of each kind of vector does indicate the relative magnitudes at the different points. This figure is worth examining carefully.

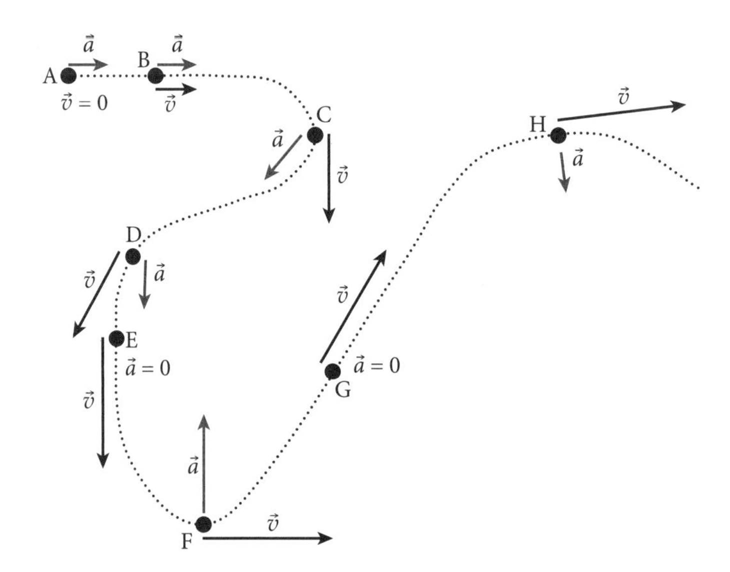

Figure 4.6 Velocity and acceleration vectors along the path of a car on a winding road:

A. The object starts at rest, so the velocity is zero. The acceleration cannot be zero; otherwise the object would simply remain at rest. The object is about to begin moving in a straight line along the track, so the acceleration points along the track.

B. The velocity, as always, points along the direction of the track (out the front window of the car). The object is speeding up but not turning, so the acceleration points in the same direction as the velocity.

C. The velocity is even longer because it has been speeding up for some time. The acceleration vector has a component along the path (speeding up) and to the right (turning), so it is angled to the right compared to the velocity.

D. The object has continued to speed up, so the velocity vector is longer still. At D, the object begins turning left, though the curve is not as sharp as at point C. The acceleration vector points slightly to the left of the velocity. Since it also has a component along the velocity, the speed continues to increase.

E. The object is moving at constant speed in a straight line. The acceleration is zero. From this point on, the acceleration vector will either be perpendicular to the path or it will be zero, meaning that the speed is constant for the rest of the trip.

F. The object's speed is constant but it is turning sharply to the left. The velocity is still tangent to the path. The acceleration is perpendicular to the velocity and relatively large.

G. The object is moving in a straight line at constant speed. The acceleration is again zero.

H. The object is traveling at constant speed and turning gently to the right. It has an acceleration that is perpendicular to the velocity and relatively small.

4.4 Projectile Motion

We now turn our attention to one of the two most common examples of kinematics in two dimensions—the motion of objects thrown through the air, or *projectile motion*. At first glance, projectile motion might appear to require a 3-dimensional analysis because objects may now move up and down, in addition to the two dimensions in which they can travel around the map. However, if you were to dip a sponge in paint and throw it through the air in a gym, you would find that the paint spots on the floor would all lie along a line. This means that the object's position can be described with only two numbers—how far it went along the floor in the direction of the paint droppings (the horizontal displacement) and its height above the floor (the vertical displacement). Only if there were some other influence, like a wind blowing the object away from the paint line, would we have to introduce a third dimension.

Speaking of wind, this seems as good a time as any to address the problem of air resistance in practical problems involving things moving near the surface of the earth. As anyone knows who has ever put a hand out the car window as they are driving down the highway, air resists anything that tries to move through it. The size of the resistance depends on the speed of the object through the air. So far, we have just ignored this. Is this a big mistake on our part? The answer is yes and no. The approximation that air resistance can be ignored turns out to be very good for many problems, but it works badly in other cases. For instance, if you toss a penny across the room, the true motion of the penny will be so close to what you would predict under the assumption of zero air resistance that you would have to carry out very precise experiments to detect any deviation from the ideal situation at all. In contrast, the trajectory of a wadded-up piece of paper lobbed toward the trash can is far from ideal, due to the low density of the wad. More relevant to common practical problems, a mortar round will not hit its target, though its density is high, without taking air resistance into account, because the velocity of the shell is so high that air resistance is important. Part of the art of physics is in understanding what simplifying assumptions can be made while still producing a viable solution in the real world.

But we don't expect you to be an artful physicist yet, so what should we do? Let's agree to this. Don't worry about air resistance or other complicating factors that may occur to you. Your trajectory calculations are not going to be used by the army to target mortar rounds. The goal of your problem-solving is to help you understand the principles. So, as we go through the book, you may assume that you can use the simplest possible model, unless we state otherwise. Later on, as you come to better understand the principles, you will be able to figure out for yourself when additional factors are required for solving some problem correctly.

Now let's get back to projectile motion. Projectile motion is the archetype for the principle of superposition, for breaking a 2-dimensional problem into two 1-dimensional problems. We learned in Chapter 2 that all objects moving freely near the earth's surface experience the same downward acceleration of 10 m/s^2. (Actually, if you remember, it was 9.8 m/s^2, but we agreed to use the pretty-good approximation of 10 m/s^2.) In the discussion accompanying Figure 3.1 of the tire being thrown off the cliff, we discussed Galileo's observation that this vertical acceleration will occur whether or not the object is also moving horizontally. So let's begin to analyze projectile motion with a generic case—a cannonball shot into the air at some initial speed v_i and direction θ above the horizontal.

Before we begin to analyze this problem, we should notice that there are actually two vectors to consider, the initial velocity $\vec{v}_\mathrm{i}$ pointing at an angle θ above the horizontal and the

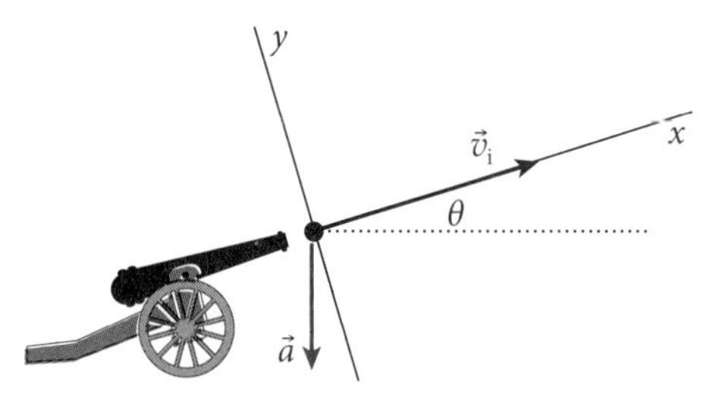

Figure 4.7 The projectile problem with an unusual choice of coordinate system.

acceleration $\vec{a}$ pointing straight down (see Figure 4.7). These two vectors do not point in the same direction as each other, nor are they perpendicular to each other. So if we are to use the standard approach to analyzing the motion, in which all vectors are resolved into components along two perpendicular directions, then at least one of the vectors is going to have to be resolved into components. Now, of course, in keeping with your inalienable mathematical rights, you can choose coordinates any way you want. So you could equally well choose the slanted coordinate directions shown in Figure 4.7 instead of the more usual horizontal and vertical directions. In the coordinate system of Figure 4.7, $\vec{v}_i$ is entirely along the x-axis and $\vec{a}$ has one component in the minus x direction and one in the minus y direction. The analysis of the motion can be done perfectly well in this coordinate system, the motion in both directions being motion in one dimension with uniform acceleration. However, most projectile problems ask for things like maximum height above the ground, horizontal range of the cannon, time in the air, etc., and these really do tie us to the usual horizontal and vertical coordinates.

In Figure 4.8, we look at the motion of a cannonball as we did in Figure 4.6, noting the directions and magnitudes of the acceleration and velocity vectors at several points along the path. The initial speed of the cannonball is v_A and the initial angle above the horizontal is θ.

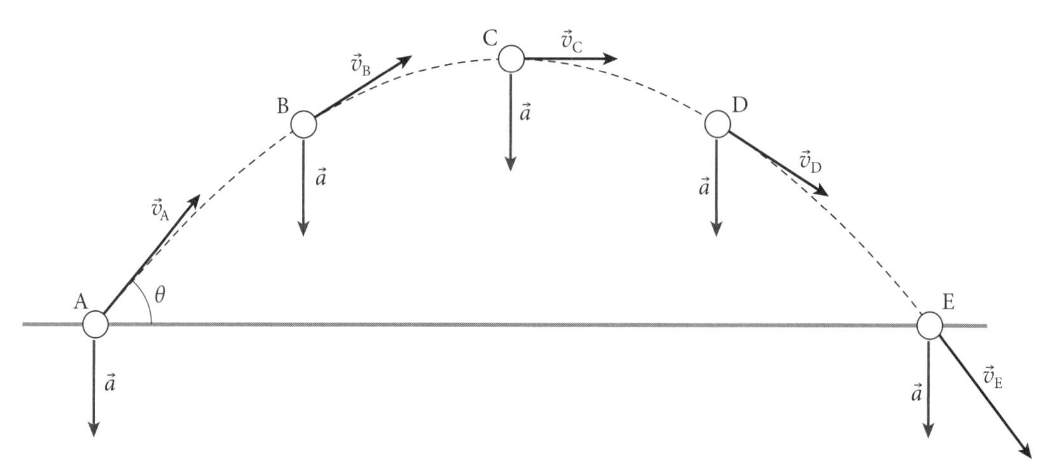

Figure 4.8 The trajectory of a cannonball with initial velocity v_A at an angle θ above the horizontal. Snapshots of the cannonball are taken at regular intervals, from $t = 0$ at point A.

For the portion of the path between the initial point A and the highest point C, the acceleration vector is always at an angle greater than 90° with respect to the velocity, meaning that the object is slowing down and turning. For the portion of the motion from point C (highest point) to point E (final), the acceleration vector is always at an angle less than 90° with respect to the velocity, meaning that the object is speeding up and changing direction. At point C, the acceleration vector is perpendicular to the velocity vector, meaning that it is neither speeding up nor slowing down. It is changing direction and keeping a speed that is instantaneously constant; it is in transition from slowing down to speeding up.

Let us also look in more detail at the velocity vectors at three of the points. These vectors are resolved into components in Figure 4.9. The principle of superposition is evident in the components. At each instant, the horizontal component is unchanged while the vertical component is exactly what you would expect for the 1-dimensional vertical velocity of a uniformly accelerated object thrown straight up into the air. During the interval between point A and C, the acceleration has slowed the vertical velocity to zero. From point C to D, it has turned the velocity around, but only to half the speed it had at point A, because it has been operating for only half as long. One helpful idea to learn from this figure, something that will be useful in solving projectile motion problems, is that when we divide the initial velocity vector into horizontal and vertical components, where only the vertical motion is accelerated by gravity, then only the vertical component of velocity will change during the motion.

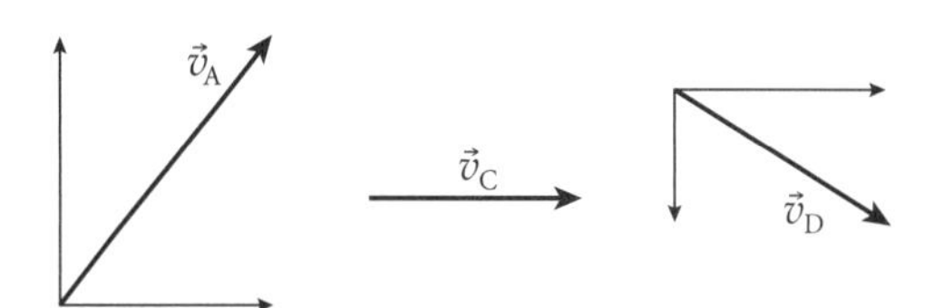

Figure 4.9 The velocity vector at three points, resolved into components.

EXAMPLE 4.1 The Horizontal Component of Projectile Motion

A ball is launched from ground level at an initial speed of 20 m/s at an angle of 37° above the horizontal. Assuming that you leave from the same place at the same time, what speed would you have to run to keep the ball directly overhead?

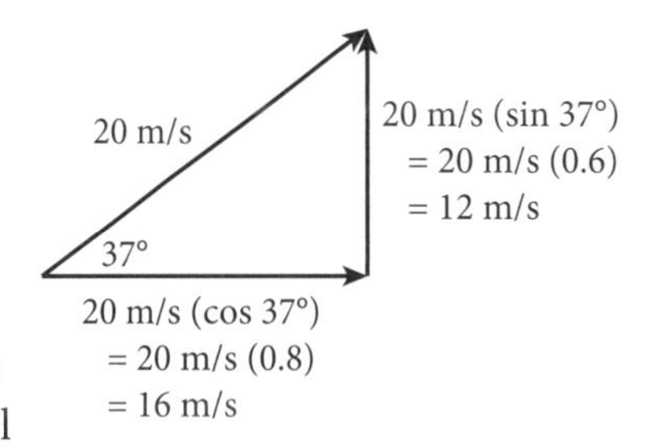

ANSWER To solve this problem, we take the object's initial velocity and divide it up into its horizontal and vertical components as shown at right. You can refer to Chapter 3 to remind yourself of the technical details of how to do this. As the figure shows, the horizontal velocity is 16 m/s, and this does not change during the motion. So, if you were to leave at the same instant running at 16 m/s (or maybe driving, as this is rather fast), you would remain directly under the ball during its entire flight. You will see the ball go straight up and down, just as it did in the 1-dimensional problems you solved in Chapter 2. Indeed, if you continued at that speed, you would get hit exactly on the head by the ball at the end of its flight.

EXAMPLE 4.2 The Vertical Component of Projectile Motion

In the previous example, suppose you wish to jump (or be launched by something) straight upward from the same location as the ball is launched, so that you are always at the same elevation as the ball during its entire flight. At what upward speed would you need to be launched?

ANSWER In the figure in Example 4.1, we calculated that the ball's initial upward velocity is 12 m/s. So, if you were launched (or could jump) upward at 12 m/s, your vertical speed would always be the same as the ball's. The vertical speed sets the maximum height, so, if you were to start with the same vertical speed, you would always be at the same elevation. When the ball hits the ground, so would you.

These last two examples have been chosen, not because they are typical of the kind of problems you will be asked to solve, but because they illustrate the essential strategy for solving projectile motion problems. That is, we divide the motion into horizontal (constant speed) and vertical (10 m/s^2 downward acceleration) components and then treat the components separately. For instance, from the above examples, we know that after 1 second, the ball will have moved 16 m horizontally, while its upward speed will have reduced from 12 m/s to 2 m/s, meaning that it is nearing the highest point in the motion (when the vertical speed is zero).

We also know that the average upward speed during the first second is 7 m/s (the average of 2 m/s and 12 m/s) so it will have risen to an elevation of 7 m after 1 second. There is more we could do with this, but this should give you the idea. When the individual kinematics problems become sufficiently complicated (especially in the vertical direction) we adopt the procedures from Chapter 2 using the equations we learned there. The following more-lengthy example is designed to demonstrate this procedure in a typical projectile motion problem.

EXAMPLE 4.3 Elements of Projectile Motion

You throw a rock with an initial speed of 15 m/s at an angle of 37° above the horizontal. Find:

a) the maximum height the rock reaches,
b) the length of time it is in the air,
c) its speed when at the maximum height,
d) its speed just before it hits the ground, and
e) the horizontal distance it will travel before it hits the ground.

ANSWER We begin by dividing the initial velocity into its horizontal and vertical components as shown below. We then proceed to analyze the two motions completely separately. The rock will always be at the same height as an imaginary rock thrown straight up at an initial speed equal to the upward component of the rock's velocity, and it will always be as far horizontally as an imaginary rock sliding along the ground with a speed equal to the rock's horizontal velocity component.

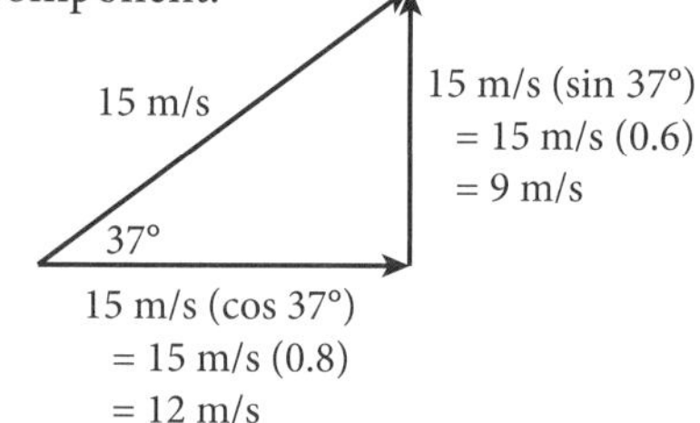

a) Clearly the question of the maximum height attained depends only on the vertical motion, so we analyze this motion first. Although there are short cuts you could use in this problem, let's analyze the uniformly accelerated 1-dimensional vertical motion using the 5-step method from Chapter 2 (page 24):

1. We take the initial point to be the point where the rock is released. We want the final point to be the top of its vertical motion, a point where it is no longer moving upward, so we pick the final point to be the point where it has zero vertical velocity.
2. We take "up" as positive, with coordinate y, and set the origin at the initial point.
3. The given information is $v_{yi} = 9$ m/s, $v_{yf} = 0$ m/s, and $a_y = -10$ m/s. (Note that the acceleration points down and so it must have a negative sign in our chosen coordinate system.) We are trying to find the displacement Δy.
4. The equation that connects these variables is $v_{yf}^2 = v_{yi}^2 + 2a_y\Delta y$.
5. We solve for Δy to find $\Delta y = \dfrac{v_{yf}^2 - v_{yi}^2}{2a_y}$, and substitute to calculate

$$\Delta y = \frac{0 - (9 \text{ m/s})^2}{2(-10 \text{ m/s}^2)} = \frac{-81 \text{ m}^2/\text{s}^2}{-20 \text{ m/s}^2} = 4.1 \text{ m}$$

Another way to answer the maximum height question would be to note that, at an acceleration of 10 m/s^2, it would take 0.9 seconds for the rock to change its upward velocity from 9 m/s to 0 m/s. During this time the average speed would be 4.5 m/s. This leads to a displacement of (0.9 s)(4.5 m/s) = 4.1 m.

b) We just worked out that the rock takes 0.9 s to reach maximum height. Because the motion is symmetric (it comes down to the same elevation from which it was launched), it would take another 0.9 s to drop from the top of its flight back down to the ground, a total time of 1.8 s. Alternatively, the 5-step program could be used to find the time at which $\Delta y = 0$. As we saw in Example 2.7, the solution will give two answers to a quadratic equation, because $\Delta y = 0$ both at $t = 0$ and at $t = 1.8$ s.

c) The speed at maximum height is now easy because we know that the vertical component of the velocity is zero while the horizontal (x-component) velocity is unchanged at $v_x = 12$ m/s. So, at the highest point, the total speed is just the horizontal 12 m/s.

d) We might (correctly) guess that the speed just before it hits the ground ($y = 0$ m) would be the same as its launch speed because of the symmetry of the motion. We can also demonstrate this rigorously by finding the x- and y-components of the velocity at time $\Delta t = 1.8$ s. The x-component is simple—it is always 12 m/s (the acceleration in the x-direction is zero). In the y-direction, we use the equation $v_{yf} = v_{yi} + a_y\Delta t$ to find the velocity after 1.8 seconds. This formula gives

$$v_{yf} = +9 \text{ m/s} + (-10 \text{ m/s}^2)(1.8 \text{ s}) = -9 \text{ m/s}.$$

The rock's velocity in the y-direction was 9 m/s (up) when first thrown, and it is –9 m/s (9 m/s down) just before hitting the ground. Combined with the constant 12 m/s velocity in the x-direction, this gives a speed of $v_f = \sqrt{12^2 + 9^2} = 15$ m/s, as expected.

e) Finally, finding the horizontal distance the rock travels represents the first time we actually have to analyze horizontal motion. Since the horizontal distance is covered at a uniform rate of 12 m/s, and since the elapsed time from the initial point to the final point is 1.8 s, the horizontal distance traveled is $\Delta x = (12 \text{ m/s})(1.8 \text{ s}) = 22$ m.

Although there are many different ways to state projectile motion problems, the solutions to all of them are more or less similar to Example 4.3. The most important things to remember are:

- Be explicit about your choices of coordinate system and variables and be careful not to make any unjustified assumptions. For instance, in Example 4.3, the final elevation was the same as the initial elevation, which meant that the rock achieved its maximum height at exactly the halfway point in time. If the final elevation is different from the initial (and it often will be) you cannot make this assumption.
- Treat the two components of motion separately, but recognize the places where the answer from the analysis of the motion in one direction may determine something about motion in the other. An example is part (e) of Example 4.3, where the time of flight was first determined from the vertical motion and, in turn, was used to find the distance traveled in the horizontal direction.
- Ultimately, the best way to become good at doing these problems is to work lots of examples. (Does it seem as if we just keep telling you that?)

4.5 Circular Motion and Centripetal Acceleration

Another common 2-dimensional kinematics application is that of uniform circular motion, or circular motion at constant speed. As we analyze this kind of motion, we must actually do the opposite kind of thing that we did in the section on projectile motion. There, we knew the acceleration and we were trying to work out the motion. Here, when we say that the motion is along a circular path at constant speed, we are saying that we already know the motion, and we will usually end up trying to work out the acceleration. Our goal in this section is to find a formula for the magnitude and direction of the acceleration vector for any object traveling at constant speed v along a circular arc of radius r. We want to find what the acceleration must be to get an object to move in a circular path like this.[1]

First, to be clear, we remember from Figure 4.4 that, even though something is moving at constant speed, it is accelerated if its direction changes. In fact, the direction of the acceleration vector is something we already know. In Figure 4.5, we learned that an acceleration perpendicular to the instantaneous velocity vector would not change the speed, but only the direction of the motion. So if an object is moving along a curved path at constant speed, the acceleration must always be perpendicular to the velocity.

In what follows in this section, we will first derive the direction of the acceleration vector that must be present when an object is moving in uniform circular motion and we will note that it satisfies the prediction from column (c) of Table 4.1. We will then derive the magnitude of the acceleration vector. The value of this derivation is as a worked example that gives insight into the meaning of instantaneous velocity and acceleration. We wouldn't take the time to show this to you if we didn't think it would be helpful on this score. However, we admit that the derivation itself is not something that you will be expected to be able to reproduce. The only thing that you will need to know is our final result for the acceleration magnitude, as it appears in Equation 4.6. This will show up many times in future chapters and will be required for solving many problems involving circular motion.

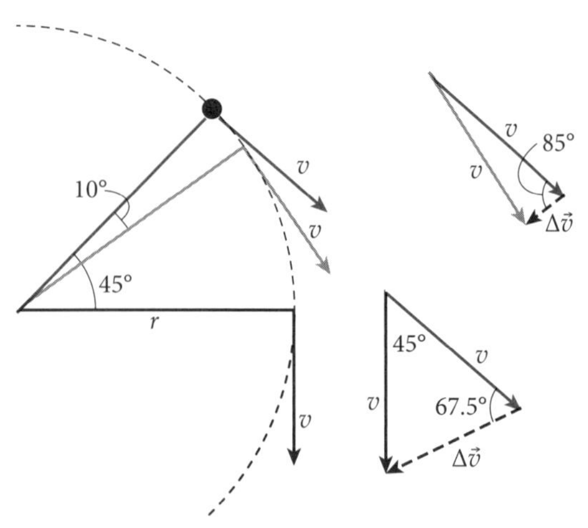

Figure 4.10 Position and velocity vectors for an object in circular motion at speed v and radius r.

Okay, let's go on. Figure 4.10 shows a little black ball moving clockwise at uniform speed v along a circular track of radius r. The object's velocity vector is shown at three points. First, there is the initial point indicated with the red radius line. Then there are two later points—a point 10° around from the initial point indicated by the gray line and a point 45° around from the initial point, indicated by the black line. These radial lines from the center of the circle are the position vectors to each point, measured from an origin at the center. Let us first look at the vector triangle at the lower right of the diagram. In circular motion, the velocity vector is always tangent to the circle, perpendicular to the position vector. So the angle between the black velocity vector and the red initial velocity vector in this triangle will be the same 45° as the angle between the red position vector and the black position vector. These two vectors form the sides of an isosceles triangle (the two sides are equal because the magnitude of the velocity vector is a constant v). The difference between the initial and final vectors is the horizontal dashed black vector labeled $\Delta\vec{v}$. We remember that the sum of the angles in a triangle is 180°, so with a 45° angle at the top, the angle between the red initial velocity and the dashed $\Delta\vec{v}$ must be $67\frac{1}{2}$°.

1. A note on notation: Remember that we use r for the magnitude of the position vector $\vec{r}$. But the position vector from the center of a circle to an object going around the circle has a constant magnitude that is equal to the radius of the circle. So r is also the radius of the circle.

Turning our attention to the triangle in the upper right of Figure 4.10, we see that the triangle with the red initial velocity vector as one side and the gray velocity vector as the other will have a 10° angle at the top, leading to an 85° angle between the $\vec{v}$ vectors and the $\Delta\vec{v}$ vector. As the final velocity vector approaches the initial vector, the angle at the vertex of the triangle will approach zero, and so the angle between the velocity vector and $\Delta\vec{v}$ will approach 90°. Since the acceleration always points in the direction of the change in the velocity, the acceleration will point toward the center of the circle, a result we might have anticipated by looking at column (c) in Table 4.1. Because this acceleration always points toward the center, the acceleration of an object undergoing uniform circular motion is called the *centripetal acceleration* (from Latin *centrum* "center" and *petere* "to seek").

Our last task is to determine the magnitude of this centripetal acceleration vector. This quantity can be worked out by referring to the two triangles in Figure 4.11. In the triangle at the left, the position vectors to two nearby points are shown, the red position vector of length r to the initial point and the gray position vector, also of length r, to the final point. The angle between these two vectors is the same θ as between the red initial velocity and the gray final

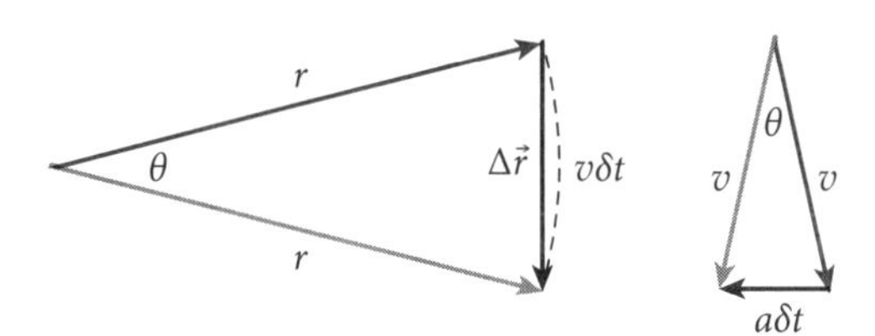

Figure 4.11 Two similar isosceles triangles, one with $v\delta t$ as the short side and the other with $a\delta t$ as the short side.

velocity in the triangle at right. Because these two triangles are both isosceles triangles with the same angle θ, they are similar, and their respective sides will all be in proportion to each other. In the triangle at left, the black downward vector is the change in the position, $\Delta\vec{r} = \vec{r}_f - \vec{r}_i$. In the limit that the time between the initial and final points is a tiny δt, the length of this displacement vector $\Delta\vec{r}$ will equal $v\delta t$.

Do you see why that last statement is true? Go back and look again at Figure 4.3. No, really . . . go back and look. Did you see that Figure 4.3 is where we found that the average speed $\delta r/\delta t$ approached the true speed v in the limit that δt went to zero? So even though the displacement is along the black straight line in the triangle in Figure 4.11, while the $v\delta t$ distance traveled is, strictly speaking, along the circular dashed path, this is not a problem. This is because, in the limit that δt goes to zero, the length of the black displacement vector becomes the same as the distance along the dashed curved path.

The hard work is now done. All that remains is to recognize that the ratio of the short side to the long side will be the same in these two triangles, so we can write

$$\frac{v\delta t}{r} = \frac{a\delta t}{v}$$

The δt cancels on each side, and a little algebra gives us

$$a = \frac{v^2}{r}. \tag{4.6}$$

Whenever you see an object moving along a circular arc of radius r, at constant speed v, you may be sure that its acceleration is the *centripetal acceleration*. It is a vector pointing exactly toward the center of the circle, and its magnitude is given by Equation 4.6.

EXAMPLE 4.4 The Moon's Acceleration

The moon takes 27.3 days (2.36×10^6 s) to go around the earth in a nearly circular orbit whose radius, centered on the earth, is 3.84×10^8 meters.

a) Assuming that the moon has constant speed in its orbit (which it very nearly does), what is the moon's orbital speed?

b) What are the magnitude and direction of the moon's acceleration?

ANSWER

a) As the moon circles the earth, it covers a total circumferential distance of $d = 2\pi r = 2(3.1416)(3.84 \times 10^8 \text{ m}) = 2.41 \times 10^9$ m. Since it covers this distance in $t = 2.36 \times 10^6$ s, its speed is $v = d/t = (2.41 \times 10^9)/(2.36 \times 10^6) = 1.02 \times 10^3$ m/s, or just over 1 km/s.

b) If any object moves in a circle at uniform speed v, it will have an acceleration toward the center of the circle of magnitude $a = \frac{v^2}{r}$. For the moon in its orbit, this gives $a = \frac{(1.02 \times 10^3 \text{ m/s})^2}{3.84 \times 10^8 \text{ m}} = 0.00266 \text{ m/s}^2$. The acceleration points toward the center of the earth, since that is the center of the circular orbit.

4.6 Relative Velocity

A Little League pitcher wants to test the speed of his fast ball on a rainy day, so he goes into a long room and measures out a distance of 44 feet from one wall to a pitcher's mound he marks on the floor (Figure 4.12). His sister stands off to the side with a stop watch and measures the time it takes from when the ball leaves his hand until it hits the wall. It comes out to be exactly 0.50 seconds (don't ask us how she could measure it that accurately). What is the speed of his fast ball?

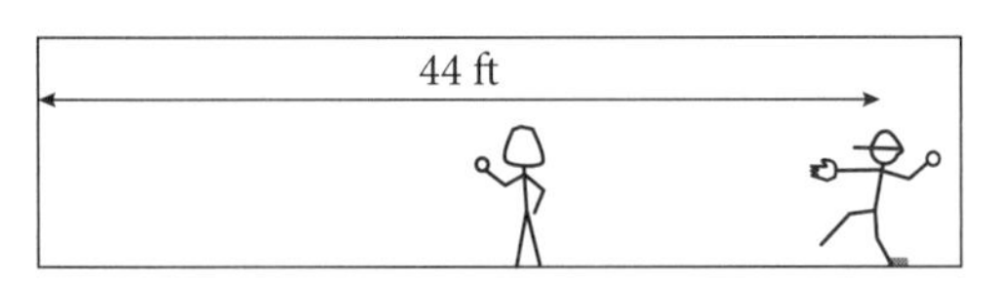

Figure 4.12 Measuring the speed of the fast ball (wheels not shown).

The speed, of course, is the distance divided by the time. This gives (44 ft)/(0.50 s) = 88 ft/s, which is almost exactly 60 mi/hr. Except for one thing. We forgot to tell you that this particular long room is actually a boxcar on a train, and that, while the boy is throwing toward the west, the train is actually traveling toward the west at 50 mi/hr. Will that change the speed of his fast ball?

The answer is yes and no. If you go back and look again at Section 4.2, you will see that the velocity is the displacement divided by the time, and that the displacement requires an origin for the measurement. There is nothing that says that the pitcher and his sister cannot choose a point on the floor of the car as their origin. In that case, the ball will go from an initial $x = 0$ to a final $x = 44$ ft, and that 44 feet will be its displacement *relative to the car.* If the boy is pitching toward the west, and if the ball travels 44 feet in a total of 0.5 seconds, then its velocity *relative to the boxcar* will indeed be 88 ft/s toward the west. The boy has a 60 mi/hr fast ball.

But is that its real velocity? Well, it depends on what you mean by "real" velocity. If you think the real velocity should be the velocity relative to the train, then its real velocity is 60 mi/hr west, just as you calculated. If you want the real velocity to be the velocity relative to the ground, then the velocity would be 60 mi/hr plus the 50 mi/hr velocity of the train, for a velocity of 110 mi/hr toward the west. Of course, we probably need to point out that this is all taking place as the earth's spin on its axis takes the tracks eastward, relative to the center of the earth, at 800 mi/hr. So the velocity relative to the center of the earth is 800 mi/hr east, minus the 110 mi/hr west, for a velocity of 690 mi/hr east. And we didn't mention that the whole thing happens at noon, at which time the earth's orbit is carrying it westward at 67,500 mi/hr relative to the sun, so maybe the "real" velocity should be . . . . We hope you get the idea.

There is no single "real" velocity of an object. All velocities are relative to the origin that has been selected. In most cases, the origin will be obvious (actually, we will almost always use an origin on the earth), but it is important to remember that velocity is defined using a displacement relative to some origin. The origin matters.

We started this discussion with the Little League baseball example, where all the displacements lay in a line, but let us also note that displacements and velocities in more dimensions than one also depend on the choice of origin. If the position of an object A relative to an origin B is $\vec{r}_{A,B}$, as in Figure 4.13, and if the position of origin B relative to another origin C is $\vec{r}_{B,C}$, then the position of object A relative to origin C is given by the vector sum

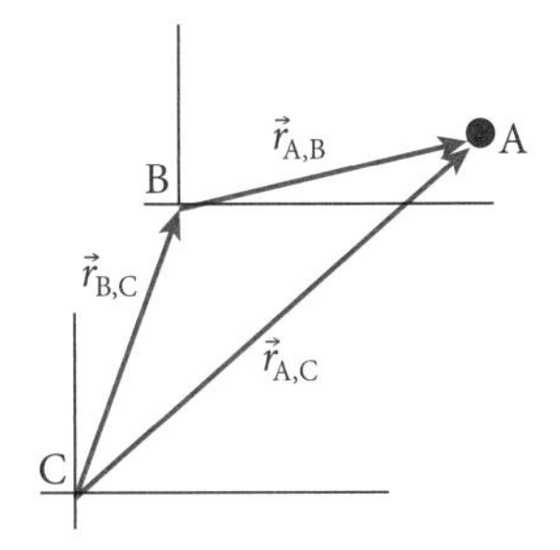

Figure 4.13 The position of object A relative to two different origins, B and C.

$$\vec{r}_{A,C} = \vec{r}_{A,B} + \vec{r}_{B,C}. \tag{4.7}$$

Note the order of the subscripts in this equation.

If the object is moving, and even if the two origins are moving relative to each other, this relationship is still valid at each point in time. If we use a superscript i for the initial time, then the relationship at that time is

$$\vec{r}^{\,i}_{A,C} = \vec{r}^{\,i}_{A,B} + \vec{r}^{\,i}_{B,C},$$

and if we use a superscript f for the final time, we get

$$\vec{r}^{\,f}_{A,C} = \vec{r}^{\,f}_{A,B} + \vec{r}^{\,f}_{B,C}.$$

The displacement of an object is defined in Equation 4.1. The displacement of object A relative to the origin C is therefore given by

$$\vec{D}_{A,C} \equiv \vec{r}^{\,f}_{A,C} - \vec{r}^{\,i}_{A,C}.$$

If we use our previous two equations to write the position vectors in each reference frame, then we see that the displacements relative to the two origins are related by

$$\vec{D}_{A,C} \equiv \vec{r}^{\,f}_{A,C} - \vec{r}^{\,i}_{A,C} = \vec{r}^{\,f}_{A,B} + \vec{r}^{\,f}_{B,C} - \vec{r}^{\,i}_{A,B} - \vec{r}^{\,i}_{B,C} = \vec{D}_{A,B} + \vec{D}_{B,C}. \tag{4.8}$$

Finally, if we divide both sides of Equation 4.8 by the elapsed time $\Delta t = t^f - t^i$, we find the relationship between the velocities measured relative to the two origins to be

$$\vec{v}_{A,C} = \vec{v}_{A,B} + \vec{v}_{B,C}, \tag{4.9}$$

an equation known as the *law of velocity addition.*

This equation relating velocities relative to two different origins is also a reflection of the law of superposition. If an object has a velocity $\vec{v}_{A,B}$ with respect to something, while that thing itself has velocity $\vec{v}_{B,C}$ relative to us, then the overall velocity of the object relative to us is also the vector sum of the two. Let's see how this works with an example.

EXAMPLE 4.5 The Airplane in the Wind

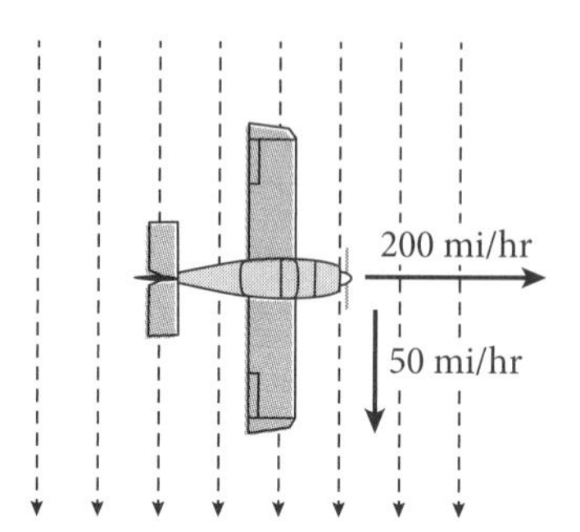

An airplane is heading directly east, flying at an air speed of 200 mi/hr. While it is flying, there is a wind from the north blowing south at a ground speed of 50 mi/hr. What is the plane's ground speed and heading?

ANSWER The bird's-eye view at right shows the direction of the airplane's velocity relative to the air and the direction of the air velocity relative to the ground. If there were no wind, the airplane would be 200 mi farther east after an hour, and if the airplane were floating in the air, it would be 50 miles farther south after an hour. The principle of superposition says that it will do both at the same time, so its velocity relative to the ground is the vector sum of the velocity of the plane relative to the air and the velocity of the air relative to the ground. Using Equation 4.9, we can write this as

$$\vec{v}_{\text{plane,ground}} = \vec{v}_{\text{plane,air}} + \vec{v}_{\text{air,ground}}.$$

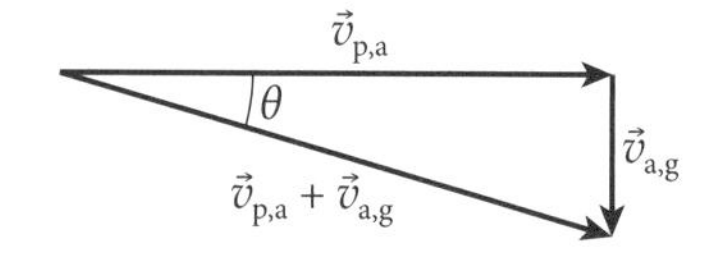

The vector addition is shown in the triangle at right. Here we see that the velocity of the plane relative to the ground will have magnitude $\sqrt{(50\text{ mi/hr})^2 + (200\text{ mi/hr})^2} = 206$ mi/hr and that it will travel along a direction $\theta = \tan^{-1}(50/200) = 14°$ south of east. The plane's velocity is therefore $\vec{v}_{\text{plane,ground}} = \{206\text{ mi/hr}, -14°\}$.

4.7 Summary

From this chapter, you should have learned the following things:

- You should know that the 2-dimensional velocity vector is always tangent to the path of an object and has a magnitude equal to the instantaneous speed of the object along the (possibly curved) path.
- You should be able to calculate average velocity between an initial and a final point and the average acceleration between an initial and a final velocity.
- You should understand how to break 2-dimensional motion into two 1-dimensional motions by choosing a coordinate system and resolving vectors into components in each of the chosen directions.
- You should understand the effect of an acceleration that lies along the instantaneous velocity vector and of one that is perpendicular to the instantaneous velocity vector.
- You should be able to solve projectile motion problems by, typically, resolving the initial velocity into horizontal and vertical components, using constant-velocity formulas to analyze the horizontal motion and using constant-acceleration formulas (see Table 2.2) to analyze the vertical motion.
- You should understand that any time an object is moving at speed v along a circular arc of radius r, you may be sure that it is being accelerated in a direction perpendicular to the instantaneous velocity vector, toward the center of the arc, and that the magnitude of the acceleration will be $a = v^2/r$.

CHAPTER FORMULAS

A 2-dimensional problem may be divided into two 1-dimensional problems. Then (for constant acceleration a_x in the x-direction and a_y in the y-direction) we have:

$$v_{xf} = v_{xi} + a_x \Delta t \qquad v_{yf} = v_{yi} + a_y \Delta t$$

$$\Delta x = \frac{1}{2}(v_{xi} + v_{xf})\Delta t \qquad \Delta y = \frac{1}{2}(v_{yi} + v_{yf})\Delta t$$

$$\Delta x = v_{xi}\Delta t + \frac{1}{2}a_x \Delta t^2 \qquad \Delta y = v_{yi}\Delta t + \frac{1}{2}a_y \Delta t^2$$

$$v_{xf}^2 = v_{xi}^2 + 2a_x \Delta x \qquad v_{yf}^2 = v_{yi}^2 + 2a_y \Delta y$$

For uniform circular motion, the acceleration toward the center of the circle is $a = \frac{v^2}{r}$.

The velocity addition formula for the velocity of A relative to two origins, B and C, is

$$\vec{v}_{A,C} = \vec{v}_{A,B} + \vec{v}_{B,C}.$$

PROBLEMS

4.2 *1. A helicopter leaves the helipad and, 25 seconds later, has traveled 150 feet upward at an angle of 50°. What is its average velocity vector in feet per second, using a curly bracket vector notation?

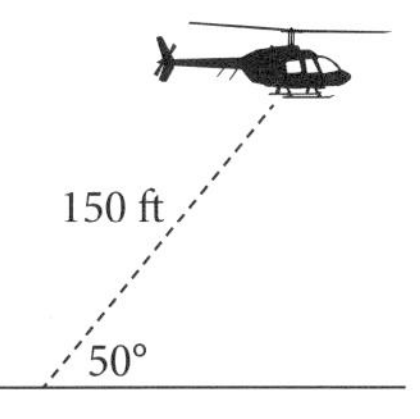

*2. A car drives 3 blocks north and then turns right and drives 4 blocks east. The trip, in traffic, takes him 15 minutes. Using a coordinate system with an x-axis pointing east and a y-axis pointing north, find the car's average velocity for this trip, in blocks per minute. Express the velocity in curly bracket notation.

*3. In Example 3.5 (Chapter 3), we found the displacement arising from a 10-mile leg at an angle 35° east of north, followed by a leg of 5 miles due east. If the entire trip took 4 hours, find the average velocity in: a) velocity magnitude and direction form, and b) eastward and northward velocity components.

**4. An ant follows the curved path across the page shown below, arriving at the points indicated at the times shown. Measuring directly from this printed page, find the magnitude of the ant's average velocity in cm/s between point A and each of the labeled points, B, C, and D. Measure the angle θ for B, C, and D, as shown, and combine magnitude and direction to give the ant's average velocity in curly bracket notation between A and each of the points B, C, and D.

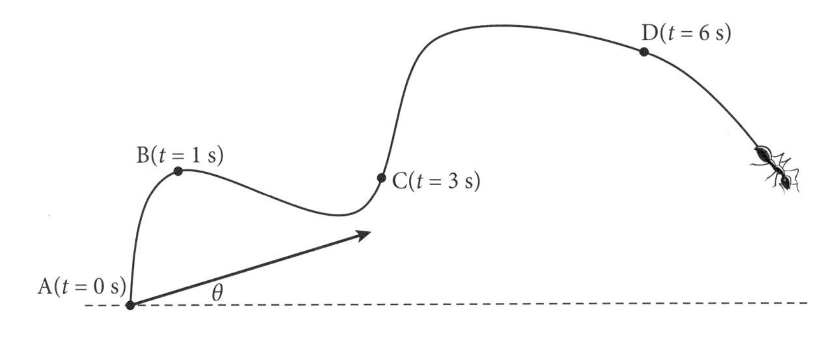

4.3 ** 5. A car traveling 30 m/s due east turns a smooth corner and heads northeast. The car maintained a constant speed throughout the 10 seconds it took to make the turn. What was the average acceleration for the 10 seconds? (Give the answer in component form for eastward and northward components.)

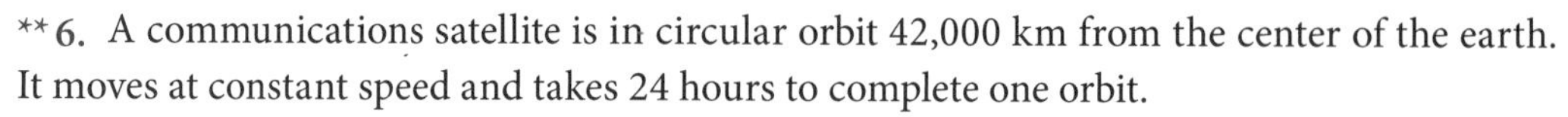

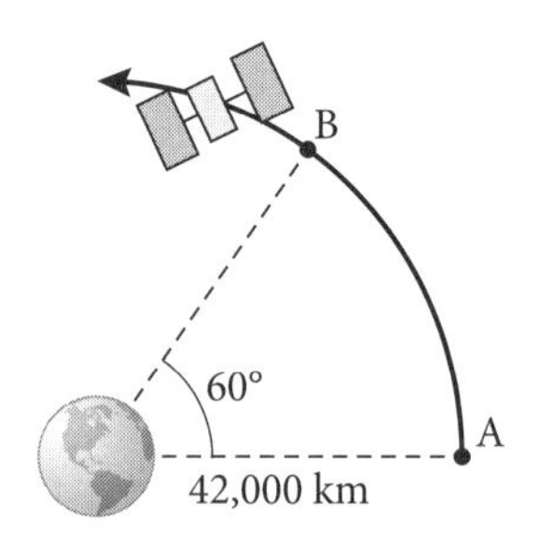

** 6. A communications satellite is in circular orbit 42,000 km from the center of the earth. It moves at constant speed and takes 24 hours to complete one orbit.

a) Find its orbital speed.

b) Find its instantaneous velocity (magnitude and direction) at the point labeled A and at the point labeled B, a point 1/6 of the way (60°) around its orbit.

c) It takes 2 hours for the satellite to travel from point A to point B. Find its average acceleration (magnitude and direction) during this interval of time.

** 7. A car climbs a 10° hill at point A at 50 mi/hr, goes over the crest, and descends the 15° downward slope at point B at 75 mi/hr. If the time between the car being at point B and at point A is 4 minutes, find the average acceleration of the car (magnitude in mi/hr/min and direction in degrees) over the 4-minute interval. [Hint: Be careful to note that the direction we want is not the direction of the car, but the direction of the average acceleration vector.]

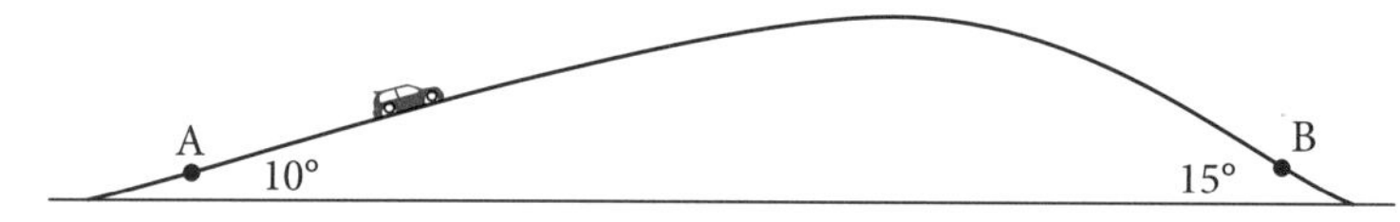

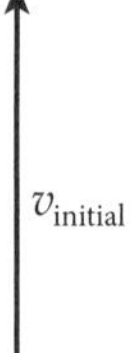

* 8. A car's velocity vector is represented by the arrow at left. Redraw the vector on your paper and then draw and label (with an A, B, or C) the directions of the acceleration vectors that would cause the car to:

A) slow down without turning,

B) turn left without changing speed, and

C) turn right and speed up.

** 9. A penguin starts from rest and accelerates across the snow at a constant rate until he hits the ice, as depicted in the figure below. He then throws himself down and begins to slide along on the frictionless ice. He then goes over the edge into an ice gully and slides up a vertical ice cliff on the other side.

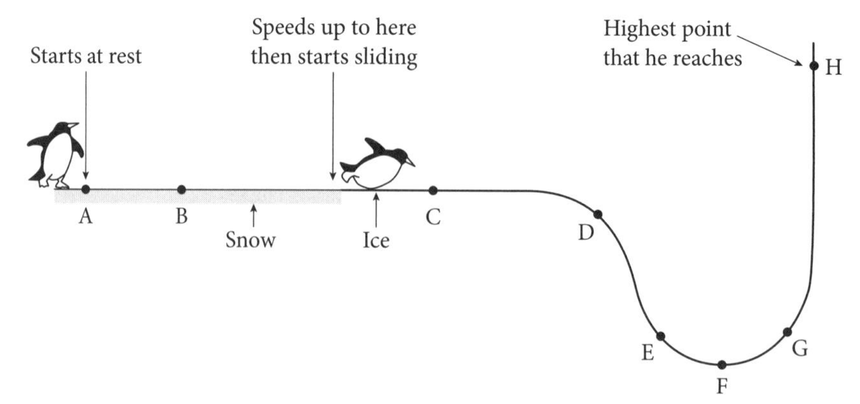

a) Draw a curved line on your paper that represents the shape of the snow and ice (you don't have to draw the penguins). Add points to your drawing that are like the labeled points above. Along your line, draw an arrow to represent the penguin's velocity at each of the labeled points, including roughly scaling the lengths of the velocity vectors to represent the penguin's speed at each point. If it is zero at any point then write "$v = 0$" by the point.

b) Draw arrows to represent the direction of the penguin's acceleration at each of the labeled points. If the acceleration is zero at any of these points, then write "$a = 0$" by the point.

** **10.** A student observes the motion of a skeeter bug gliding on the surface of a pond and carefully draws the trajectory of the bug, as shown. At certain labeled points (A–F), he has indicated the *directions* of the *velocity* and *acceleration* of the bug, knowing that the bug started from rest at point A. He shows his drawing to you.

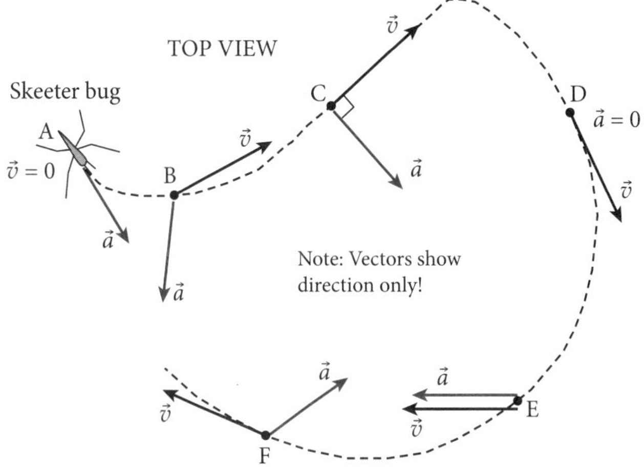

a) For points A–E only, create a table like the one below and decide whether the directions of the vectors the student drew are correct or not (given that the trajectory is correct). For each case where they are not correct, explain why the direction is incorrect.

b) Assume that the student has the correct directions of the velocity and acceleration at point F. Is the speed of the bug increasing, decreasing, or neither, at point F? Explain your reasoning.

Point	Is the velocity direction correct? (Y/N)	Is the acceleration direction correct? (Y/N)	If either is incorrect, explain.
A			
B			
C			
D			
E			

** **11.** A person sits on a tire that swings from a rope. The person is just sitting on the tire and not trying to make the swing go any higher. A strobe illustration of the motion is shown at right. The person and tire start from rest at point A, then swing to point E before reversing direction. Consider only the portion of the motion from point A to point E.

a) Trace or sketch a drawing of the locations of the tire at the five points A–E. Then, on your drawing, sketch vectors to show the velocity of the tire at each labeled point. Draw the vectors to relative scale. If the velocity equals zero at any point, state that explicitly.

b) Sketch arrows to show the approximate *direction* of the acceleration of the person at each labeled point. Do not try to draw the arrows to scale. If the acceleration equals zero at any point, state that explicitly.

4.7 ** 12. Suppose that the angle θ in Figure 4.7 is 30°.

a) Find the components of the earth's gravitational acceleration along the initial velocity vector (the x-axis in Figure 4.8) and perpendicular to the velocity vector (along the y-axis shown in the figure).

b) If the initial muzzle velocity of the cannon is 600 m/s, how far in the x-direction, along the initial velocity, will the cannonball be after 2 seconds?

** 13. A gun in a hovering helicopter fires a bullet with an initial velocity of 100 m/s at an angle of 37° below the horizontal. The bullet hits the ground 2 seconds later.

a) Find the horizontal (x) and vertical (y) components of the bullet's initial velocity.

b) How far does the bullet travel in the horizontal direction before it lands?

c) How high above the (level) ground was the plane when the bullet was fired?

d) Find the magnitude of the velocity of the bullet right before it hits the ground.

** 14. A cannonball shot from a buried cannon is fired at an angle of 37° above a level field. The initial velocity has a magnitude 50 m/s.

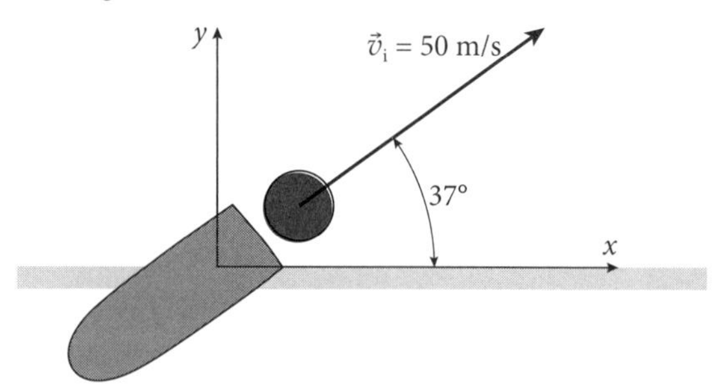

Neglect air resistance and answer the following questions:

a) Find the horizontal and vertical components of the vector v_i.

b) How many seconds does the cannonball stay in the air?

c) How far from the cannon will the ball land?

d) How high above the field is the cannonball at its highest point?

** 15. Your friend is stranded on the other side of a canyon, as shown below. He's getting hungry and so you decide to throw him a banana. You throw the banana with an initial velocity of 20 m/s at an angle 37° above the horizontal. The canyon is 8 m wide and your friend is on a plateau that is 5 m above your elevation.

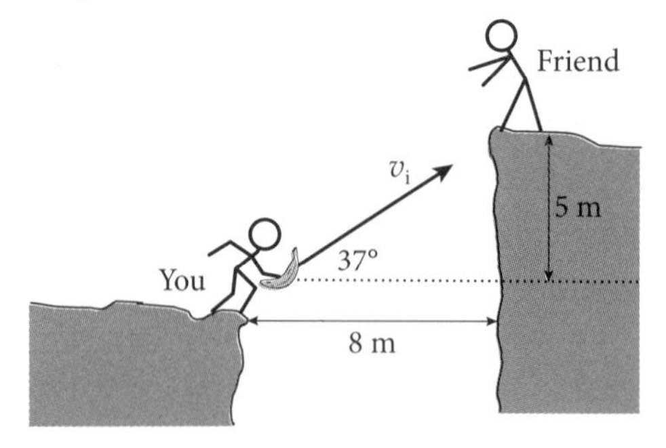

a) Find the horizontal (x) and vertical (y) components of the banana's initial velocity.

b) Determine whether or not the banana will reach your friend on his side of the canyon (i.e., will he get to eat the banana?).

** 16. In the most bizarre wildlife documentary ever made, Animal Kingdom has filmed a rainbow trout leaping from a pond with an initial velocity of 35 m/s at an angle of 37° above the horizontal. Three seconds later, the trout lands in a pickle jar filled with water that is sitting on top of a telephone pole.

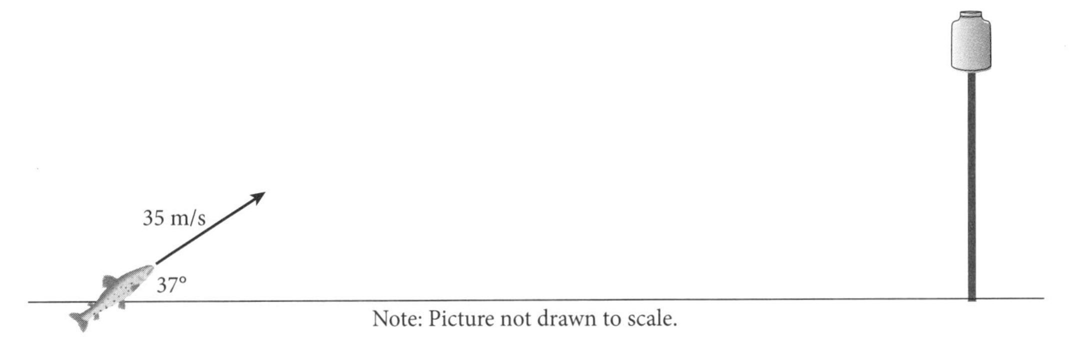

Note: Picture not drawn to scale.

a) How far is the telephone pole from the place where the trout leaves the water?

b) What is the greatest height above the water attained by the trout during the jump?

c) What is the speed of the trout when it goes through the mouth of the jar?

d) How much higher is the mouth of the pickle jar than the surface of the pond?

** 17. A careless hunter finds himself on the wrong side of a canyon with a grizzly bear. In a desperate attempt to escape from the bear, the hunter gathers his strength and leaps a mighty leap. He leaves the bear's side of the canyon at a velocity of 10 m/s at an angle of 53° with respect to the horizontal. The canyon is 18 m wide and the other side of the canyon is lower by 12 m.

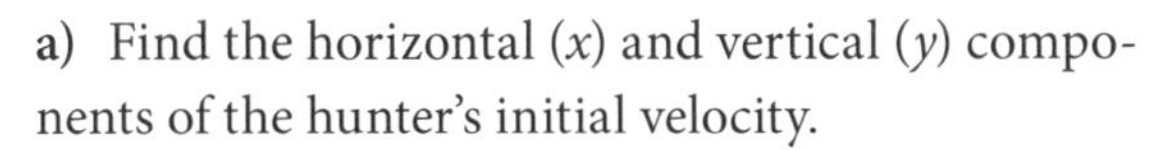

a) Find the horizontal (x) and vertical (y) components of the hunter's initial velocity.

b) Will he land on the top of the cliff on the other side or will he fall into the canyon?

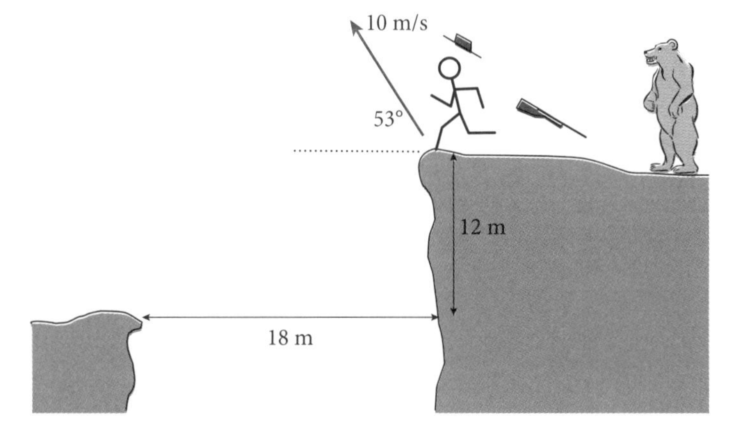

** 18. A cannon atop the wall of a fort fires a cannonball exactly horizontally at a speed 100 m/s. It hits the target 4 s later. Where was the cannon (height and horizontal distance), relative to the target? What was the speed of the cannonball as it hit the target?

** 19. Shannon is walking at a uniform speed of 2 m/s in the Amazon forest, unaware that she is headed right toward a tree containing a hungry jaguar. When she is 9 meters from the base of the tree, the jaguar jumps at her with a velocity of 6 m/s at an angle of 37° below the horizontal.

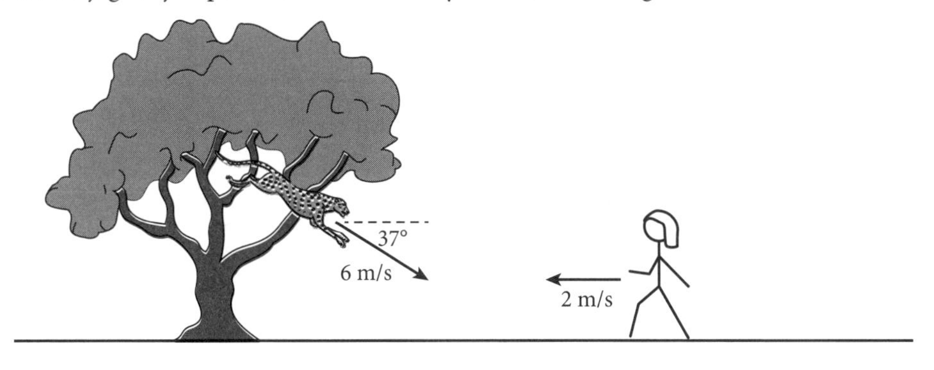

a) Find the components of the jaguar's initial velocity vector $\vec{v}_i$.

b) If the jaguar is in the air for 0.9 seconds, from how high up did it jump?

c) Find the velocity of the jaguar (magnitude and direction) right before it lands on the ground.

d) Find the distance between Shannon and the jaguar right as it lands on the ground.

** **20.** Bozeman, Montana's Rolf Wilson, the 2012 world champion in Gelande-style ski jumping, comes off the ramp at a speed of 10 m/s and at an angle of 16° above the horizontal. He hits the snow at a point 30 m below the point where he took off. If the slope is designed so that he makes a perfectly soft landing on the snow, what should be the angle of the slope where he hits? Neglect air resistance (which, of course, no successful ski jumper can afford to do). [Hint: Find the direction of Rolf's velocity vector just as he hits the snow.]

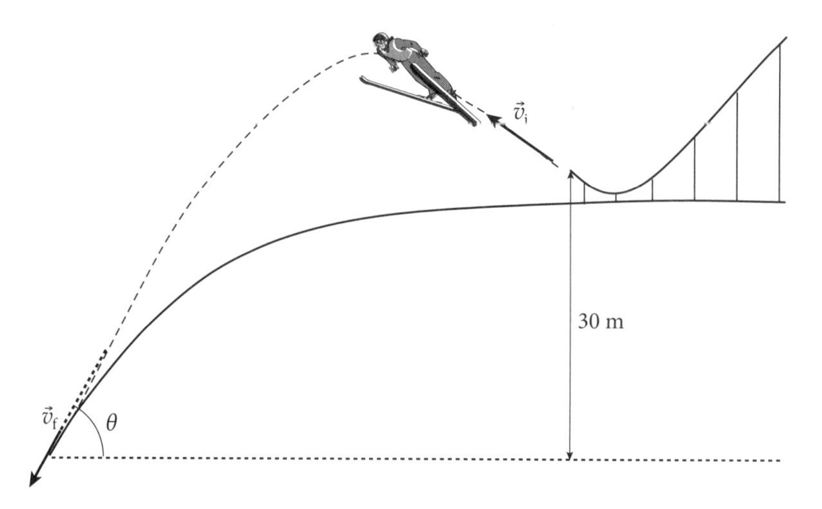

** **21.** A sailor drops his pocket knife from the top of the mast on a sailing ship heading eastward at a constant speed of 5 m/s. The mast is 20 m high. How far from the bottom of the mast does the knife hit the deck?

** **22.** Matt shoots the ball with an initial velocity of 20 m/s at an angle of 53° above the horizontal. Three seconds later, the ball swooshes through the hoop. His friend, Mo, is waiting directly under the basket. In the questions that follow, you may neglect air resistance and assume that $g = 10\ \text{m/s}^2$.

a) Find the components of the initial velocity vector $\vec{v}_0$.

b) How far is Matt from Mo when he shoots the ball?

c) What is the *y*-component of the ball's velocity (v_{yf}) when it goes through the hoop?

d) What is the speed of the ball when it goes through the hoop? [Hint: The answer is *not* 20 m/s.]

e) How much higher is the hoop than the release point of the ball?

** 23. A ping-pong ball rolls at a speed of 0.60 m/s toward the edge of a table which is 0.80 m high. The ball rolls off the table.

a) How long is the ball in the air?

b) How far out from the point directly below the edge of the tabletop does the ball hit the floor?

** 24. A small plane flying at a speed of 40 m/s is at an altitude 300 m above a flooded town where four inhabitants are stranded on the roof of their house. The plane wants to drop an inflatable life raft to the inhabitants so that they can paddle to safety. How far from the house should the pilot release the raft so that it will land on the roof of the house?

*** 25^f. [Remember that a superscript f means that the goal of the problem is to find a new formula, rather than a numerical answer. So look for given letters in the problem statement, rather than for numbers.] A cannon is fired at an angle θ above the horizontal, with an initial velocity v_0. Find the range of the cannon over level ground, i.e., the horizontal distance the cannonball will cover before it hits the ground. [Hint: If you have done the problem correctly, you will be able to use the trig identity $2\sin\theta\cos\theta = \sin(2\theta)$ to simplify your answer.]

4.5 ** 26. A car travels around a semicircular curve of radius 40 m. What would the speed of the car need to be in order to have a centripetal acceleration toward the center of the circle that is the same as the acceleration due to gravity ($g = 10$ m/s^2)?

** 27. The earth's radius is 6.4×10^6 m. If we could make the earth spin at any rate we wanted, what would be the speed (in m/s) of a point on the earth's surface, if we want the centripetal acceleration of a man standing on the equator to be 10 m/s^2? (At this speed, a man would actually be in orbit around the earth, his feet just barely touching the ground.)

** 28. The earth travels in a nearly perfect circular orbit around the sun. The radius of the earth's orbit is 1.5×10^{11} m and the length of the solar year is 365¼ days. Assume the earth's velocity around the sun is uniform. (It very nearly is.)

a) Find the velocity of the earth as it orbits the sun.

b) Find the earth's acceleration (magnitude and direction).

** 29. When a projectile is at the top of its trajectory, its velocity is horizontal and its acceleration is vertical. This condition, with the acceleration perpendicular to the velocity, is the condition for motion along a circular arc. Thus, just for an instant, the trajectory of a ball thrown into the air is approximately a small arc of a circle. Take the values from Example 4.3, in which the ball is moving horizontally at a constant 12 m/s. The acceleration due to gravity at the top of the trajectory is still 10 m/s^2 downward. What is the radius of the instantaneous circular path of the ball? Is it more or less than the actual height of 4.1 m at the top of the path?

*** 30. **Non-uniform circular motion.** Our formula for centripetal acceleration was found in the limit that δt goes to zero, so v^2/r is an instantaneous value. This will thus be the the centripetal component of acceleration for any object in circular motion, even if its speed is changing. An airplane is heading due north, when it goes into a turn along a circular path with radius 6 km. At this point, it also begins to increase its speed along the path. Just as it begins the turn, its acceleration vector has magnitude 3 m/s^2 and direction 25° north of due east.

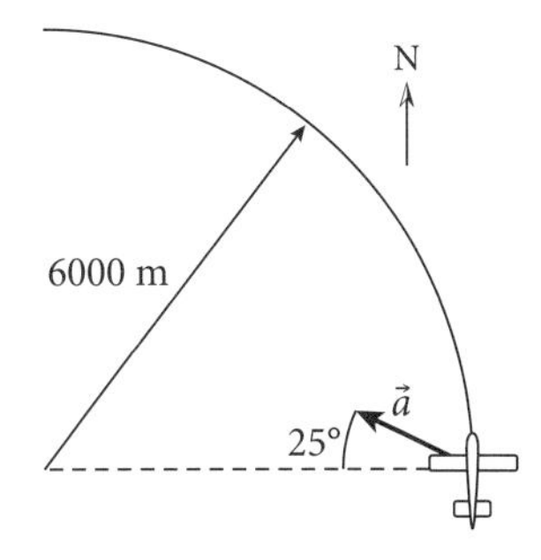

a) What is the speed of the aircraft as it begins the turn?

b) At what rate is its speed increasing at the same moment?

*** 31[f]. A car moving along a circular track of radius r starts at a speed v_0 and speeds up at a rate a. Its acceleration vector will thus have a component a along its path, and motion along the track can be analyzed using one-dimensional formulas with constant acceleration a. However, since the car is turning, it will also have a component of acceleration perpendicular to its path. Find the perpendicular component of the acceleration at time t if the car is to remain on the track.

4.6 * 32. Two boats are initially 2 miles apart and are heading directly toward each other. Boat A moves east at 15 mi/hr while Boat B travels west at 20 mi/hr. How long will it take the two boats to come together?

** 33. Boat A and Boat B start from the same dock. Boat A travels north at 15 mi/hr, while Boat B goes 20 mi/hr due east. At what rate is the distance between the two boats increasing? At what angle is Boat A moving, as seen from Boat B?

** 34. An 80-meter-long moving sidewalk travels at a speed of 1.5 m/s. If a man walks along the sidewalk at a speed of 1.7 m/s relative to the sidewalk, how long will he be on the moving sidewalk as he covers the 80-m length?

** 35. A woman puts in her rowboat directly across a 100-foot-wide river from a boat dock where her children are waiting for her. She can row at a speed of 4 mi/hr in still water, and the current in the river is flowing at a speed of 6 mi/hr. She wants to get to the other side as quickly as she can, so she points her boat directly across the river and rows as fast as she can.

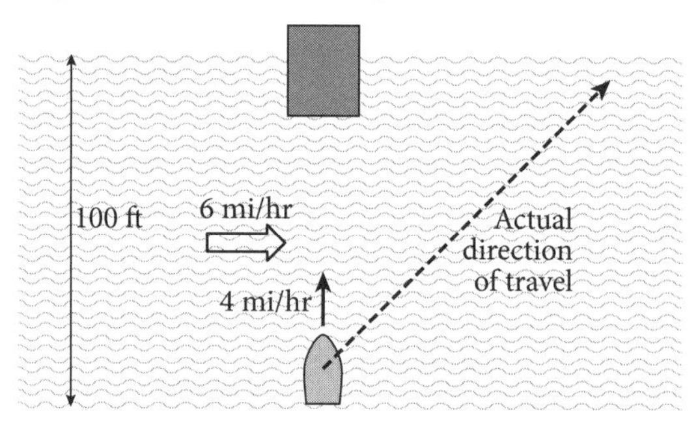

a) Explain why rowing directly across the river will bring her to the other side in the shortest time. [Hint: Consider the components of her final velocity.]

b) What will be the angle between the actual direction of the boat's velocity and the direct line across the river?

c) How far downriver will she end up when she gets to the other side?

*** 36. A ferry starts at Lakeside at noon and heads due north across Torry Lake at 6 km/hr. At the same time, a tour boat leaves Port Ellis, at a point 60 km northeast of Lakeside, and travels at a speed of 10 km/hr. What should be the heading of the tour boat if it wants to rendezvous with the ferry as quickly as it can? [Hint: It may help to note that $\sin\theta = \sqrt{1 - \cos^2\theta}$, when $\theta < 90°$. It may also help to read the warning in parentheses on page 592.]

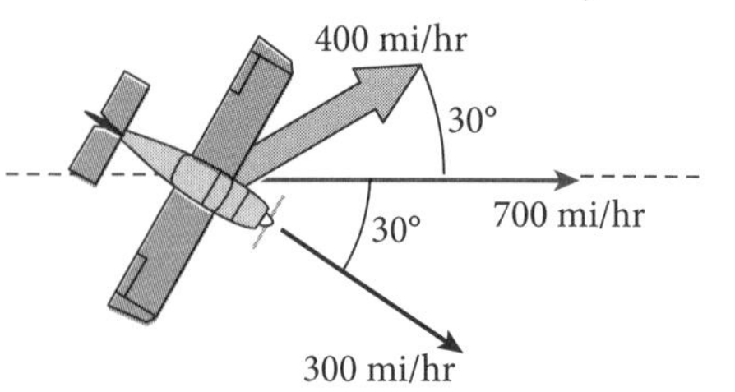

*** 37. An airplane is on a heading 30° south of east (the direction it would be flying in still air) with airspeed of 300 mi/hr. It encounters a jet stream that is moving at 400 mi/hr relative to the ground in a direction 30° north of east. Since the plane is presently at 42° north latitude, the earth's rotation carries the surface of the earth toward the east at a speed of 700 mi/hr. What is the plane's velocity (magnitude and direction) relative to the earth's axis?

5

NEWTON'S LAWS

As we stated back in Chapter 1, the essential goal of physics is to be able to take a description of the state of some system at some initial time and to predict its state at some future time. In the first half of this book, what we will mean by "the state of the system" is simply where everything is located and how it is moving. The kinematics that we have learned so far gives us the way to describe the position and motion of objects, but, with the exception of our discussion of falling bodies, we have not yet addressed what that motion should be like. In fact, if you think about it, the relationships we have found so far are just mathematics, things that cannot possibly fail to be true because they are true by definition. It is not physical law, for example, that says that the distance traveled by an object with constant velocity v in a time interval Δt is $\Delta x = v\Delta t$; it is the mathematics which *defined* velocity to *be* $v = \Delta x/\Delta t$. But now, finally, in this chapter, we are going to do some physics. We are going to explore the *causes* of motion, a subject called *mechanics.*

We begin with a friendly warning. Many of the common misconceptions of physics that people struggle with are found in this section. Often, answers that seem obvious turn out to be wrong. We will spend some time in this chapter and the next "unlearning" some of our false prejudices. Our goal is to become "Newtonian" in our thinking. Isaac Newton was the man who was able to reduce the entire intellectual content of mechanics to three elegant principles, or "laws." Learning to apply these laws in a wide variety of situations will form the foundation for most of what follows in our study of physics.

5.1 Newton's First Law

Perhaps the first question that must be answered is what we should expect mechanics to tell us. Should the laws of physics tell us where everything is? Is there a law that says that all rocks will be on the ground and all birds in the air? Or should the laws of physics specify velocities? Do all rocks of some particular mass have to fall toward the ground at some specified speed? Or should physics specify acceleration? As we shall see, the right answer is the last answer of the three. The goal of mechanics is to take what is known about a system of objects and to thereby determine what the accelerations of each of those objects will be.

It was Galileo who first stated clearly that motion at constant velocity requires no explanation. It is a natural state. He observed that a ball rolling down an inclined ramp had an acceleration related to the incline of the ramp, but, as the angle of the incline went to zero, and the cause of the acceleration went away, the subsequent velocity was constant. He summed up what he had seen in what he called the *principle of inertia*:

> A body moving on a level surface will continue in the same direction at constant speed unless disturbed.

Something certainly caused the motion of a moving object, like the balls being accelerated by the inclined ramps, but, once something has a velocity, that velocity requires no further explanation. It is what it is. Physics will not and cannot specify what the velocity of some kind of object has to be.

The first of Newton's three laws of mechanics is really just a restatement of Galileo's principle of inertia. It translates from Newton's original Latin as[1]

- **Law I** Every body perseveres in its state of rest, or of uniform motion in a straight line, unless it is compelled to change that state by forces impressed on it.

A force is the name given to a push or a pull. The phrase "uniform motion in a straight line" means "constant velocity." In fact, since a state of rest is a state of zero constant velocity, the law could also be simply restated as: *All objects move at constant velocity, unless acted upon by an outside force.*

Now we are reluctant to undercut Newton's very first law by suggesting that there are situations where it does not apply, but honesty (as well as our desire for you to really understand the first law) requires just that. So let's discuss a situation where you are seated in the driver's seat of your car, about to start a trip. Consistent with your inalienable right to define a coordinate system as you choose, you choose an origin between your seat and the passenger seat and you take the x-axis toward the front of the car. You say, "Okay, let's go," and your passenger puts her cup of coffee up on the dashboard so that she can put on her seatbelt. Just then you hit the gas. What happens? Well, in the reference frame of the car, the coffee clearly accelerates toward the back of the car even though there is no force acting on it at all. It is not the force of the dashboard on the cup that has done this, because the dashboard pushes the cup forward, while the coffee has accelerated toward the back. The problem is that, while you are free to define any coordinate system you choose and to measure position and velocity and acceleration relative to this coordinate system, you are not free to apply Newton's laws in any old reference frame. Newton's first law simply does not apply in a reference frame that is accelerating; it applies only in frames where it applies. So it is merely a definition. The reference systems where Newton's first law applies are called *inertial frames*, and inertial frames are *defined* as those frames in which Newton's first law applies. If you see that all objects with no forces on them move at constant velocity, then you are in an inertial frame, and Newton's laws will all work.

And, lest you worry about whether or not you are working in an inertial frame as you work on future problems, let's just point out that a reference frame tied to the earth may be assumed to be an inertial frame to high accuracy. Its tiny centripetal acceleration as it orbits the sun may be neglected in most of the problems we address.

5.2 Newton's Second Law

So let's assume that we have found an inertial frame by observing that objects with no force on them all move with constant velocities. What happens then when there *is* a force? This is where Newton's Second Law comes in.

Actually, Newton's second law requires a prior definition. Before he stated any of his laws, Newton defined something he called the "quantity of motion." Roughly translated from the Latin, the definition is

---

1. All quotes from Isaac Newton are the authors' rough translation of his wording in *Philosophiæ Naturalis Principia Mathematica* (Mathematical Principles of Natural Philosophy), generally just referred to as "The *Principia*." (If you want to sound like a Latin nerd, you should pronounce this as "prin-kip'-ee-ya.")

> The quantity of motion is the way to measure motion, arising as the product of mass[2] and velocity.

Newton's "quantity of motion" is what we now call "momentum." Mathematically, it is the vector $\vec{p} = m\vec{v}$. (By the way, we have never heard a good explanation of why momentum would be represented by a letter p, but it is an almost universal tradition.)

Armed with this definition, let us go on to Newton's second law. It is

- **Law II** The change in motion is proportional to the applied force, and is made in the direction of the straight line in which that force is applied.

Because of Newton's previous definition of how motion is to be measured, we see that "change in motion" means "change in momentum." And, because the mass of an object typically does not change, this means "mass times the change in velocity." In mathematical language, Newton's second law is simply

$$\vec{F} = m\vec{a}. \tag{5.1}$$

But this still leaves us a few issues to clean up.

First, it is not clear from Newton's wording of his second law if he meant to refer to the change in velocity, $\Delta\vec{v}$, or to the rate of change in velocity, $\Delta\vec{v}/\Delta t$. Later on in the *Principia*, however, Newton applies his second law to various examples; and here he uses the idea of a rate of change. So $\vec{F} = m\vec{a}$ is probably what he meant.

Second, let us note that Newton says that $m\vec{a}$ is *proportional* to the force, not *equal* to the force, as we have in Equation 5.1. The reason for this is that Newton did not specify what the units for force should be. If we were to experiment with different forces until we found one that would give an acceleration of 1 m/s^2 to an object of mass 1 kg, then we could call that force our unit of force, and Equation 5.1 would work as an equality. Actually, this is what we do. This unit of force is named the "newton," after you-know-who, and this is the unit that has been set by international agreement as the SI unit of force. So Equation 5.1 works as it is. The right-hand side of Equation 5.1 has units of kg·m/s^2, the derived unit that arises when a mass is multiplied by an acceleration. The newton must therefore have these same units. One newton (abbreviated N) is one kg·m/s^2.

Note, however, that if we had chosen to measure force in, say, pounds, then the two sides of Equation 5.1 would no longer be equal. The force required to accelerate a 1 kg object at 1 m/s^2 is not 1 pound, but 0.2248 pounds. If we wanted to measure force in pounds, instead of newtons, then the force would indeed be proportional to ma, as Newton stated, but the second law would read $\vec{F} = Km\vec{a}$, with $K = 0.2248$ lb·s^2/kg·m as the constant of proportionality.

Third, let's look at the question of the mass. The word *mass* is a familiar one. We often loosely define it as a measure of the amount of matter something contains. Two identical blocks have twice the mass of a single block. But a better definition of mass actually comes from Equation 5.1. It is a measure of an object's *inertia*, of how difficult it is to accelerate it. We know that it is more difficult to stop a supertanker moving at 2 mi/hr than to stop a small rowboat moving at the same speed. We say that the supertanker has more mass than the rowboat. But let us remember that, back in Chapter 1, we mentioned that the unit of mass, the kilogram, is a particular lump of platinum–iridium in Sèvres, France, and that the masses of other objects are found by

---

2. Remember that mass is a fundamental measurement that we defined in Chapter 1, on page 3.

balancing them against the kilogram in earth's gravity. As we said then, and as we now understand, this may not be the right process. The definition needed in Newton's laws should involve the resistance of an object to a force, not the force of gravity on the object. We will discuss this again in more detail in Chapter 6. But, while we are on the subject of mass, let's note that the resistance of a body to a force does not depend on the direction the mass is moving. Mass is a scalar; it has no direction.

Fourth, Newton's law mentions only the force $\vec{F}$. But it is often the case that several forces act simultaneously on an object. What do we do then? The answer is that, before using Equation 5.1 to predict the acceleration, we must first add up all the force vectors to produce what is called the "net force." Equation 5.1 will still apply, as long as we understand $\vec{F}$ to mean net force. If there are many forces acting, they must first be vector-summed. They may even cancel, leaving a zero net force and producing zero acceleration. In order to include cases with more than one force acting, we will, from now on, write Newton's second law as

$$\text{Law II:}\quad \vec{F}_{\text{net}} = m\vec{a}. \tag{5.2}$$

Consider a car moving along a straight highway at a speed of 60 mi/hr. If friction and air resistance were absent, the car would travel straight on forever at 60 mi/hr, without using any gasoline. Newton's first law promises us this. However, our experience with hands sticking out the car window tells us that there *is* air resistance opposing the car's motion. The only way for a car to continue at constant speed is if there is no net force acting on it, so there must be a force somewhere that cancels the air resistance. The engine produces this force, forcing the wheels to turn and the gas tank to empty.

One last thing. In applying Newton's second law, some have viewed the equals sign as meaning "is the same as," leading to the idea that mass times acceleration (ma) is somehow a force that acts on an object. This is incorrect. The first step in solving any mechanics problem is to identify all the forces acting on a body, and ma will *never* be one of them. The equals sign in Newton's second law should be read as "is numerically equal to." If we divide the net force vector by the scalar mass, we get a result that is numerically equal to the acceleration. Conversely, if we know the object's acceleration (by observing its motion), we can multiply this acceleration by the mass to get a result that is numerically equal to the net force applied.

The purpose of the next two examples is to remind us of where we need to be careful and to help us get a sense of how Newton's second law works.

EXAMPLE 5.1 Direction of the Motion

You are pushing on a wheeled cart with a forward force of 200 N to the right while your friend pushes backwards on the same cart with a force of 300 N to the left. Assuming that any frictional forces are small enough to be ignored, what can you conclude about the direction that the cart is moving?

ANSWER We can conclude that the net force on the cart is 100 N to the left and, thus, that the cart will accelerate to the left. However, this tells us nothing about the direction the cart is moving (the velocity). There is not enough information to answer the question. If the cart is moving to the left, it is moving to the left and speeding up. If it is moving to the right, it is moving to the right, but it will be slowing down. Newton's second law relates net force and acceleration, not net force and velocity.

EXAMPLE 5.2 Finding the Acceleration

If the cart in the preceding example has a mass of 40 kg, what will be the magnitude of the resulting acceleration?

ANSWER The net force of the two applied forces was found in Example 5.1 to be 100 N to the left. Newton's second law therefore predicts an acceleration of

$$\vec{a} = \frac{\vec{F}_{\text{net}}}{m} = \frac{100\text{ N (to the left)}}{40\text{ kg}} = 2.5\text{ m/s}^2\text{ (to the left)}.$$

5.3 Newton's Third Law

Two sumo wrestlers crash into each other in what can only be described as a thrilling display of athleticism. It is clear that both sumo wrestlers feel a force, and that the force felt by one wrestler will be in the opposite direction to the force felt by the other. What is not so obvious is that the magnitude of these two forces will always be the same. Even if one wrestler is clearly overmatched and losing, it is not because he is exerting less force on his opponent than his opponent is exerting on him. His problem must lie somewhere else, because Newton's third law explains that the force exerted on each wrestler is the same.

The translation of Newton's Third Law from the *Principia* reads:

- **Law III** To every action there is always opposed an equal and opposite reaction: or the mutual actions of two bodies are always equal, and directed to contrary parts.

Unfortunately, this language seems to suggest that one force (the action) is more important than the other (the reaction) when, in reality, there is no such hierarchy. It probably would have been clearer if the third law had been stated:

> Law III: If an object exerts a force on a second object, the second object exerts an equal and opposite force back on the first object.

In truth, however, the *best* way to think of this is to count interactions. There is only the one *interaction*, and that interaction will always produce two forces, one acting on each of the interacting bodies, equally and in opposite directions.

You may have noticed that this third law is different from Newton's other two laws in that it does not discuss the motion of a body, but the forces that act on a body. Actually, it does not even tell us how to figure out what the force on a body will be, but it explains, as in our example of the sumo wrestlers, how the forces on two different bodies are related to each other. Any force on some body must arise because of some other body, and the law states that, whenever there is a force on the one body, there will always be an equal and opposite force by that body back on the first. This is perhaps most clearly seen by looking at Figure 5.1. Here we show three bodies, A, B, and C. If you look at body A, you will see that it has two forces acting on it—one, F_{BA}, being the force by B on A and one, F_{CA}, being the force by C on A. Body B is similarly subject to forces F_{AB} and F_{CB}. The point of Newton's third law is that forces always occur in pairs. In Figure 5.1, there are the red pair, the pink pair, and the

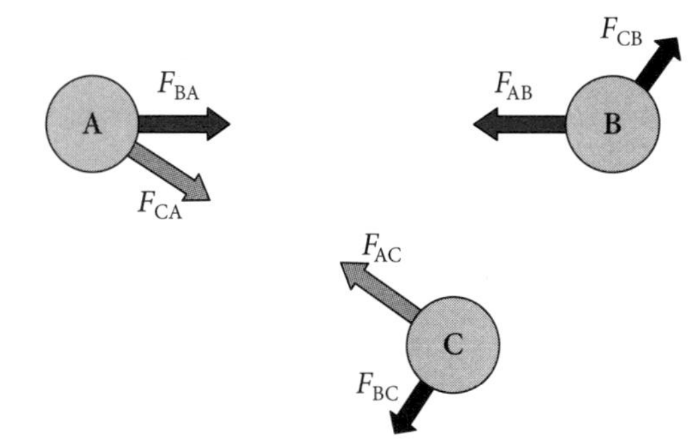

Figure 5.1 Three third-law force pairs acting on three bodies.

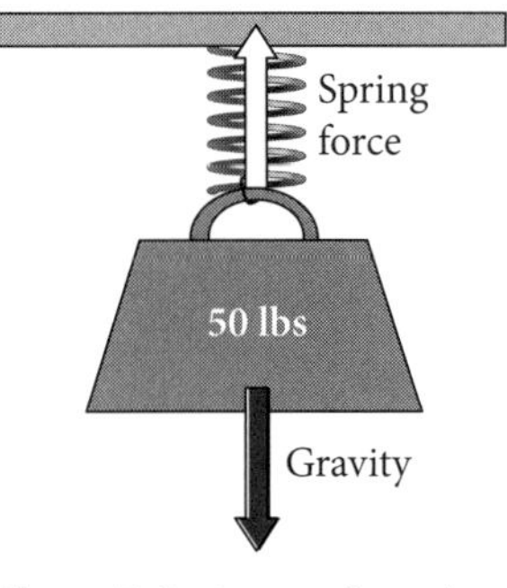

Figure 5.2 A pair of equal and opposite forces that are NOT a third-law pair.

black pair. For each pair of forces, the two forces are always equal, always oppositely directed, and always applied to opposite bodies.

In contrast, Figure 5.2 shows a 50-lb lump being held up by a stiff spring. Because the lump is at rest, the upward force exerted on it by the spring must be equal and opposite to the downward force of gravity. The two forces are equal and opposite, but *they are not a third-law pair* because they are both applied to the same body. The only reason that these two forces are equal and opposite is that the spring will always stretch until its force just balances the force of gravity, allowing the weight to remain at rest. The third-law counterpart of the red gravity force would be an upward force on the earth exerted by the 50-lb lump.

The next two examples help to clarify the conditions under which two equal and opposite forces constitute a third-law pair.

EXAMPLE 5.3 Pushing the Shop Dolly

There is a large crate sitting on a shop dolly. You are pushing to the right with a force of 150 N while your friend pushes to the left with a force of the same magnitude. (And he calls himself your friend!) The wheels on the shop dolly are very smooth so any friction can be ignored. What force is the Newton's third-law companion to the force you are applying?

ANSWER Right in the question there are two forces of equal magnitude and opposite direction, and it is awfully tempting to identify these as the third-law companions. However, these two forces both act on the same object (the crate) and therefore cannot be part of the same interaction. Third-law companion forces never act on the same object. Your pushing on the crate is part of the interaction between you and the crate. The third-law companion force is the push by the crate on you.

EXAMPLE 5.4 Daredevil Dave and the Bungee Cord

Daredevil Dave leaps off a bridge with a bungee cord tied to his left ankle. As he nears the rocks below, the cord stretches and begins to slow him down. As he slows, is the upward force by the cord on his ankle greater than, equal to, or less than the downward force by his ankle on the cord?

ANSWER Since Dave is moving downward, we might be tempted to say that any downward force must be greater than the upward force. But, of course, this would mean we have improperly connected force with velocity. So we might go a step deeper and recognize that, since Dave's downward velocity is decreasing, his net acceleration vector must point upward, meaning that the force of the cord on his ankle must be greater than the force of Dave's ankle on the cord. But, even here, we have still made the mistake of thinking of both forces as acting on Dave. In fact, only one of the two forces we asked about is actually a force on Dave. It is the force by the cord on his ankle. The other force we asked about acts on the cord. It is the other part of the cord–ankle interaction, and, as third-law companion forces, these two will always be exactly equal to each other in magnitude and will act on opposite bodies.

The forces acting on Dave, by the way, are the force of gravity pulling him down and the elastic force in the cord pulling him up. Since Dave is slowing down, the net acceleration must, in fact, be upward. This means that the force by the cord on Dave must be greater than the gravitational force on Dave, but this has nothing to do with the question that was actually asked.

In the last example, it is critical to identify, for each force, what is causing the force and what is experiencing that force. We have a notation that will help us always do this correctly. Let us label the force *by* the cord *on* Dave's ankle as $\vec{T}_{\text{cord,Dave}}$ or $\vec{T}_{\text{cD}}$. The subscript for the object applying the force comes first and the object feeling the force comes second. We thus label the other force, the one *by* Dave *on* the cord, as $\vec{T}_{\text{Dave,cord}}$ or $\vec{T}_{\text{Dc}}$. This labeling convention makes the Newton's third-law companion forces simple to identify, because they will always be the same type of force (for example, if one is a friction force the other will be a friction force), and they will always have indices reversed. It is also of great help when we apply Newton's second law to a body because it also helps us collect forces acting on a single object—they will all have the same last subscript. Although this may seem a little cumbersome at first, it greatly reduces the chance of making errors when solving force problems.

Finally, and just to make sure we understand what the third law means when it says that the forces in an interaction will be the same, consider what happens if the handle of the 50-pound lump in Figure 5.2 suddenly slips off the spring, allowing the lump to plummet toward the floor. Since Newton's third law claims that the force by the earth on the lump is the same as the force by the lump on the earth, why is it that the lump accelerates toward the earth while the earth does not accelerate toward the lump? If this troubles you at all, please remember that the acceleration depends on the force *and* on the mass. If the earth and some lump of matter are subjected to the same magnitude force, and if the lump accelerates toward the earth at 10 m/s^2, what will be the acceleration of the earth? Well, let's see. If we drop a one-kilogram lump from a height of 5 m, it will accelerate downward at 10 m/s^2. This means that there must be a force of 10 N acting on it, due to the earth. Then, by Newton's third law, it must exert a force of 10 N on the earth. The earth's mass is about 6×10^{24} kg. If you exert a force of 10 N on an object of mass 6×10^{24} kg, the acceleration will be 1.7×10^{-24} m/s^2. That's pretty small. In the time it takes the lump to fall to the ground from a height of 5 m, the earth will have moved less than 10^{-24} m. This is a distance less than a billionth the size of a nucleus. We really can't even measure a distance that small.

5.4 Forces

As we can now see, Newton's laws tell us how to calculate the acceleration of some object, given the forces acting on it, but they do not tell us what these forces will be. That is a separate problem. So Newton's laws are *not* a complete description of how the universe works; they are only the *framework* in which whatever other laws we discover can be applied. Let us briefly talk about forces.

The pushes and pulls of mechanical forces between bodies arise for a variety of reasons. In Chapter 6, we will discuss several kinds of force that typically arise in mechanics problems. In this chapter, however, we need to have a few forces available to use as we explain how to work with Newton's laws, so we will just introduce a few forces here and ask you to wait until Chapter 6 for more details about these and others.

Weight. We already know that all objects near the surface of the earth fall toward the earth at the same constant acceleration, $g = 10$ m/s^2. (Actually, it's more like 9.8 m/s^2, but we already talked about this back on page 25.) We have also already seen that, when we know the acceleration of an object of mass m, then Newton's second law tells us that there must be a force $F = ma$ acting on a free-falling body (Figure 5.3). The force of gravity on an object is called its weight W, and the acceleration of a free-falling body is $g = 10$ m/s^2. This produces a formula

$$W = mg$$

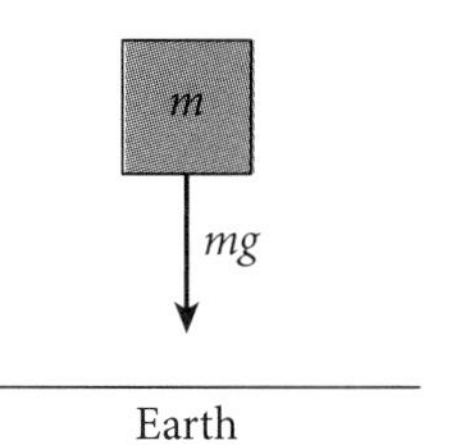

Figure 5.3 Gravity.

for the weight of any object near the surface of the earth. When g is used to determine the weight this way, it is perhaps more appropriate to think of it as having units of N/kg. (A little work with the units will convince you that N/kg are the same as m/s^2.) A 12-kg object in earth's gravity will thus have a weight $W = (12\text{ kg})(10\text{ N/kg}) = 120\text{ N}$ pulling it downward. Gravity is an example of a force that acts on an object without anything touching the object. The rest of the forces in this section require a contact.

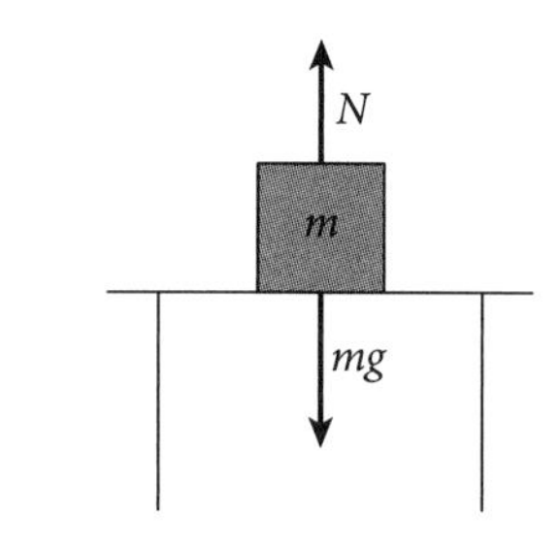

Figure 5.4 Normal force.

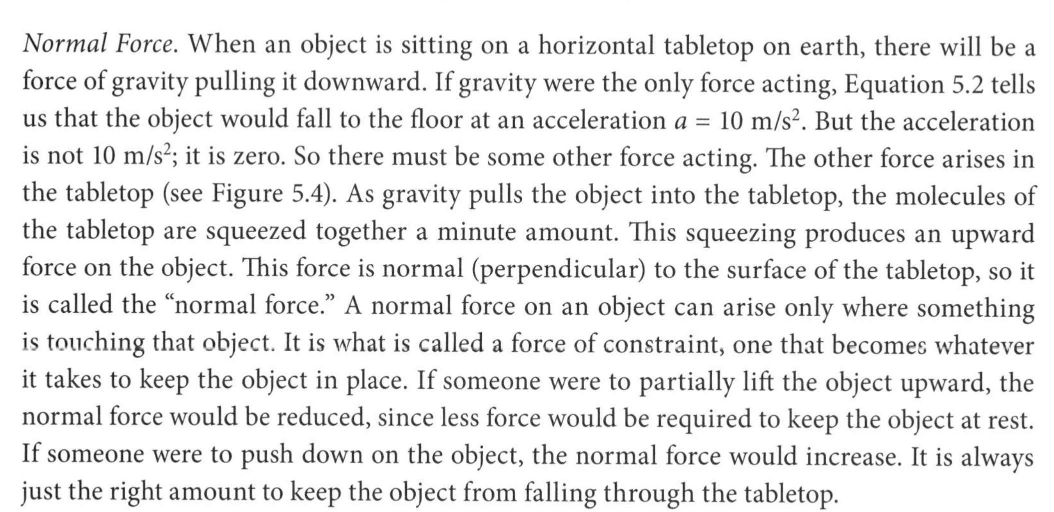

Normal Force. When an object is sitting on a horizontal tabletop on earth, there will be a force of gravity pulling it downward. If gravity were the only force acting, Equation 5.2 tells us that the object would fall to the floor at an acceleration $a = 10\text{ m/s}^2$. But the acceleration is not 10 m/s^2; it is zero. So there must be some other force acting. The other force arises in the tabletop (see Figure 5.4). As gravity pulls the object into the tabletop, the molecules of the tabletop are squeezed together a minute amount. This squeezing produces an upward force on the object. This force is normal (perpendicular) to the surface of the tabletop, so it is called the "normal force." A normal force on an object can arise only where something is touching that object. It is what is called a force of constraint, one that becomes whatever it takes to keep the object in place. If someone were to partially lift the object upward, the normal force would be reduced, since less force would be required to keep the object at rest. If someone were to push down on the object, the normal force would increase. It is always just the right amount to keep the object from falling through the tabletop.

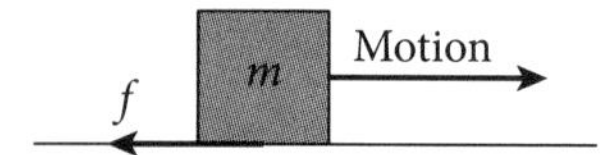

Figure 5.5 Kinetic friction.

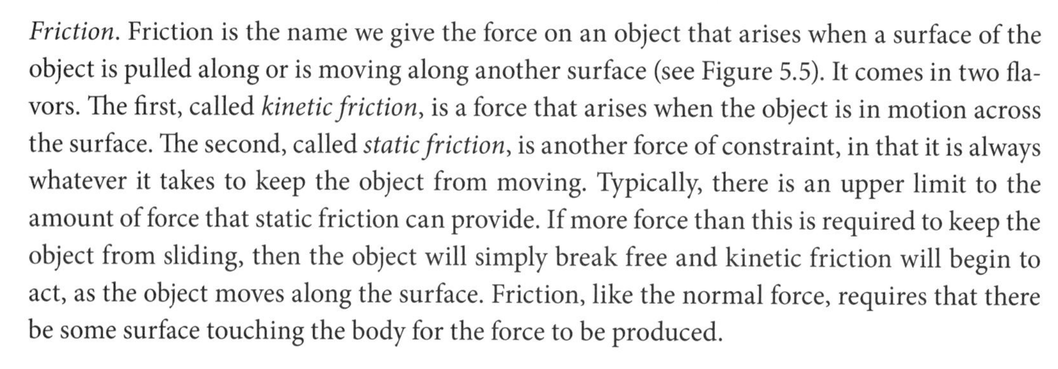

Friction. Friction is the name we give the force on an object that arises when a surface of the object is pulled along or is moving along another surface (see Figure 5.5). It comes in two flavors. The first, called *kinetic friction*, is a force that arises when the object is in motion across the surface. The second, called *static friction*, is another force of constraint, in that it is always whatever it takes to keep the object from moving. Typically, there is an upper limit to the amount of force that static friction can provide. If more force than this is required to keep the object from sliding, then the object will simply break free and kinetic friction will begin to act, as the object moves along the surface. Friction, like the normal force, requires that there be some surface touching the body for the force to be produced.

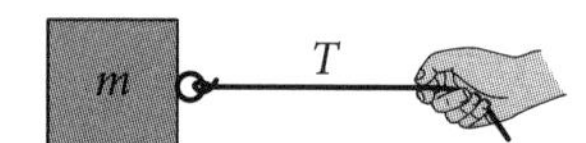

Figure 5.6 Tension force.

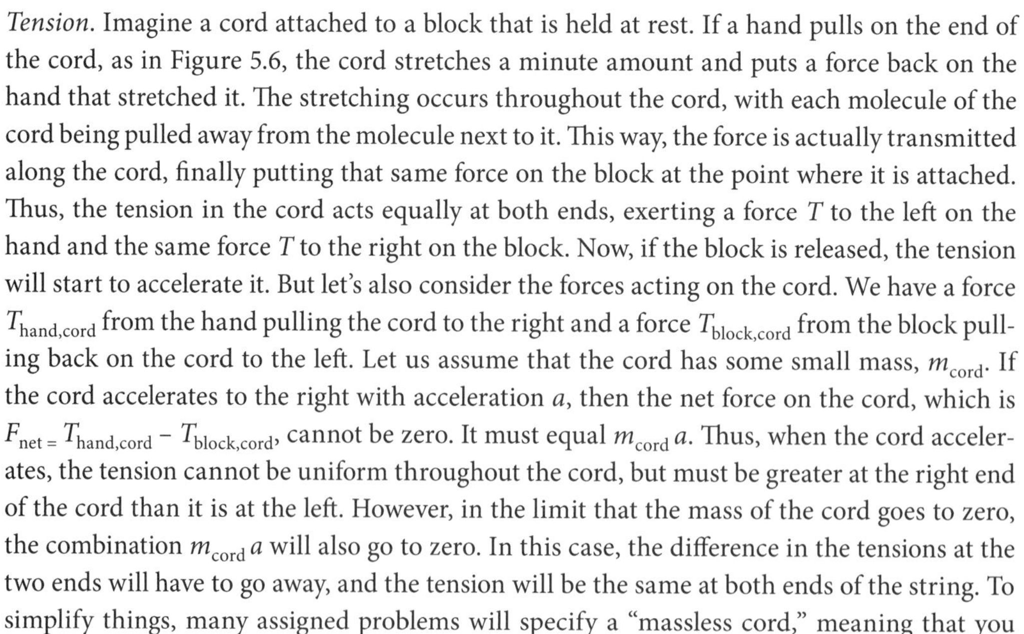

Tension. Imagine a cord attached to a block that is held at rest. If a hand pulls on the end of the cord, as in Figure 5.6, the cord stretches a minute amount and puts a force back on the hand that stretched it. The stretching occurs throughout the cord, with each molecule of the cord being pulled away from the molecule next to it. This way, the force is actually transmitted along the cord, finally putting that same force on the block at the point where it is attached. Thus, the tension in the cord acts equally at both ends, exerting a force T to the left on the hand and the same force T to the right on the block. Now, if the block is released, the tension will start to accelerate it. But let's also consider the forces acting on the cord. We have a force $T_{\text{hand,cord}}$ from the hand pulling the cord to the right and a force $T_{\text{block,cord}}$ from the block pulling back on the cord to the left. Let us assume that the cord has some small mass, m_{cord}. If the cord accelerates to the right with acceleration a, then the net force on the cord, which is $F_{\text{net}} = T_{\text{hand,cord}} - T_{\text{block,cord}}$, cannot be zero. It must equal $m_{\text{cord}}\, a$. Thus, when the cord accelerates, the tension cannot be uniform throughout the cord, but must be greater at the right end of the cord than it is at the left. However, in the limit that the mass of the cord goes to zero, the combination $m_{\text{cord}}\, a$ will also go to zero. In this case, the difference in the tensions at the two ends will have to go away, and the tension will be the same at both ends of the string. To simplify things, many assigned problems will specify a "massless cord," meaning that you may treat the tension as equal on both ends of the cord. In fact, unless we specifically mention some mass, you may assume that all ropes, cords, cables, and strings are massless.

5.5 Free-Body Diagrams

We have now introduced all three of Newton's laws of mechanics and introduced several of the forces that we will use throughout this book. It's time to put it all together to see how the motion of objects is determined. If we algebraically solve Equation 5.2 for $\vec{a}$, we find

$$\vec{a} = \frac{\vec{F}_{\text{net}}}{m}. \tag{5.3}$$

This formula is probably in the form that we will use most commonly. It gives the acceleration $\vec{a}$ of an object of mass m, when it is subjected to a net force $\vec{F}_{\text{net}}$. Note that this formula applies to a single body, the body of mass m. The formula works only when the net force in the equation is the net force acting on that body. So the vectors that are added together to give $\vec{F}_{\text{net}}$ must be only the forces that act on the body in question. When we apply Newton's second law, it is critical to begin by selecting the single body that you will apply it to. One of the most common errors that students make is to not have that clearly in their mind before they begin. Before you go to solve a problem, you *must* decide carefully which body you are considering and then consider only forces that act *on that body.*

Imagine that you are pulling your little sister on a sled and that the sled is speeding up. Think of all the forces that might be acting on the sled—on the sled, not on the sister or the snow or anything else. There would be the weight of the sled, the gravitational force of the entire earth pulling down on it, and that would be the only gravitational force that matters. All other forces would come from things touching the sled. The things touching the sled are the rope, the snow, and the sister. The rope can provide a tension force only along its length, so this force acts along the direction of the rope. The sister and the snow can both provide forces that are normal to their surfaces and parallel to their surfaces, so a normal force and a frictional force. The snow is pushing up on the sled with a normal force and resisting the forward motion of the sled with a frictional force that acts backward, parallel to the surface of the snow. Your sister is pushing down on the sled with a normal force and backward with a static frictional force. How are we to keep track of all these forces?

The best way to do this is to produce what is called a *free-body diagram*. As the name suggests, we isolate, or free, the object in question (in this case, the sled) from everything else. It is helpful to picture the body you have chosen as if it had a dashed line around it. Then the forces acting on this body must all originate from things *outside* the dashed line. If the body is near the earth, then gravity will be one of those forces. All other types of force but one (the force from a refrigerator magnet that we will discuss in Chapter 6) arise where something from outside the dashed line is touching the body. Not until we get up to Chapter 16 will we need to consider any other non-contact forces.

To draw a free-body diagram, we represent our object with a dot and draw all the forces acting on the object with the tail of each vector starting on the dot. We label each vector to indicate what type of force it represents: $\vec{W}$ for a gravitational force (or weight), $\vec{N}$ for a normal force, $\vec{f}$ for a frictional force, and $\vec{T}$ for a tension force (we discuss these types of forces again in Chapter 6). Because each force is an interaction between two objects, it is also useful, as we have previously suggested, to include two subscripts for each force label, the first to indicate which object is exerting the force and the second to indicate which object is being acted on. For example, the tension force exerted by the rope on the sled would be labeled $\vec{T}_{\text{rope,sled}}$ or $\vec{T}_{\text{rs}}$.

In Figure 5.7, we picture the sled with a red dashed line around it and, in Figure 5.8, we show the free-body diagram for the sled. Note that the second subscripts on the forces in Figure 5.8 all say "sled." Only forces acting on the sled appear on this free-body diagram. Note also that all of the first subscripts refer to things *outside* the dashed line of Figure 5.7. Figure 5.8 contains all of the relevant forces acting on the sled. If we vector-add all six of these

Figure 5.7 Isolating the sled.

Figure 5.8 The free-body diagram.

forces, and divide by the mass of the *sled*, we will find a vector that is the acceleration of the sled. This is how the process works. Drawing a free-body diagram should always be the first step in solving a problem involving Newton's second law. The time required to draw the diagram is never wasted, because most of the problems you will see will be too complicated to solve correctly without first drawing a diagram.

Once we have identified the forces, we are able to apply Newton's second law to the problem. Sometimes we know all the forces and the goal of a problem is to calculate the two components of the acceleration. Sometimes we know the acceleration, which means that we know the net force, and we need to calculate some components of some of the forces that contributed to the net force. Sometimes it is a combination of these, where we know the acceleration in one direction and want to find the acceleration in the other. An example of a situation like this would be the forces on the sled in Figure 5.8. We would know, for example, that there is no acceleration in the vertical direction, because you would never pull so hard that your sister and the sled were jerked up off the snow. So the sled does not accelerate in the vertical direction. This could tell us, say, about the tension in the rope. We could then use the information about the tension in the rope to work out the net horizontal force and thus determine the horizontal acceleration.

The following are a couple of worked examples that illustrate the use of free-body diagrams and emphasize the critical nature of the choice of the body whose motion is to be considered.

EXAMPLE 5.5 Pulling Two Crates—I

A 6-kg wooden crate is tied with a cord to a 10-kg crate. When a man pulls on the 6-kg crate, the system accelerates at 2 m/s^2. If we know that the tension in the cord connecting the crates is 50 N and that the frictional force by the floor on the 6-kg crate is 18 N, what is the force of the man on the crates?

ANSWER In this problem, we know the acceleration and we know two of the forces acting on the 6-kg crate. We need to find the remaining force on the same crate. It will obviously be simplest to choose the 6-kg crate as the body to isolate. It is pictured in the diagram below left, with a dashed line drawn around it to isolate it from the outside world. We then draw its free-body diagram below right.

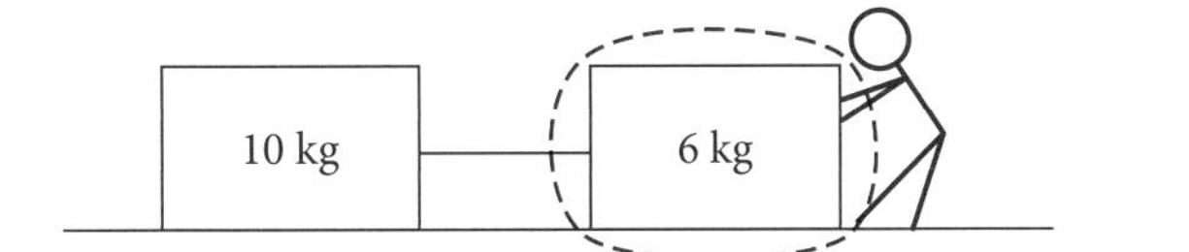

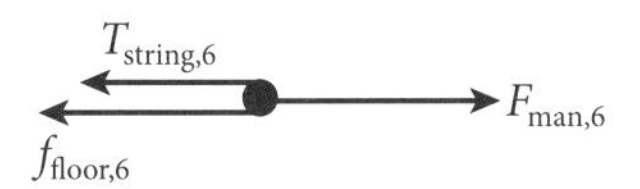

Since there is no vertical motion, and since we were not asked about any of the vertical forces, we include only forces acting in the horizontal direction. We know that the forces in the vertical direction, including normal forces, will adjust themselves to keep the crate on the floor. Note in the free-body diagram above that the final subscripts for all of the forces in the diagram are the same. We have considered only forces acting on the object inside the dashed line.

As we said, we know the acceleration and need to find one of the forces. The sum of the forces on the crate is $F_{man,6} - T_{string,6} - f_{floor,6}$, so Newton's second law gives

$$F_{man,6} - T_{string,6} - f_{floor,6} = ma.$$

Each of the variables in this equation refers only to the magnitude of the forces, since we have already taken the direction of the force explicitly into account by the use of the minus signs in the equation.

The $T_{string,6}$ and $f_{floor,6}$ forces are given. Solving for the unknown $F_{man,6}$ we find

$$F_{man,6} = T_{string,6} + f_{floor,6} + ma = 50\text{ N} + 18\text{ N} + (6\text{ kg})(2\text{ m/s}^2) = 80\text{ N}.$$

EXAMPLE 5.6 Pulling Two Crates—II

A 6-kg wooden crate is tied with a cord to a 10-kg crate. When a man pulls on the 6-kg crate, the system accelerates at 2 m/s². If we know that the frictional force acting on the 10-kg crate is 30 N and the frictional force on the 6-kg crate is 18 N, what is the force of the man on the crates?

ANSWER In this version of the problem, we know the two friction forces, one on each crate, but we do *not* know the tension in the cord joining the crates. We are again looking for the applied force of the man pulling to the right. This time it seems best to choose the combination of both crates as our body to isolate, so we picture the dashed line drawn around both crates. Note that the tension in the cord joining the two crates is still there, but it is irrelevant, since it is not a force *on* the 16-kg body formed by the two crates. It does not arise from any object outside the dashed line, but is an internal force, and internal forces do not appear in free-body diagrams. As we draw our diagram, we should, according to our subscript convention, label both frictional forces with notation like $f_{floor,crates}$, since the body in question consists of both crates. However, because there are two frictional forces given, it will be easier to differentiate between them if we write one as $f_{floor,10}$ and the other as $f_{floor,6}$, as we have done in our

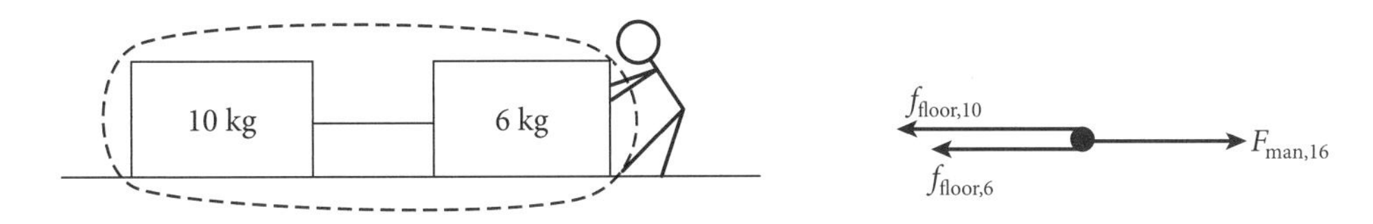

free-body diagram. We will just have to remember that both forces do act on our isolated 16-kg body. When we add up the forces, Newton's second law gives us

$$F_{\text{man},16} - f_{\text{floor},6} - f_{\text{floor},10} = ma.$$

The mass, of course, is now the 16-kg mass of the isolated body. Solving for $F_{\text{man},16}$, and remembering that the acceleration is 2 m/s², we find

$$F_{\text{man},16} = f_{\text{floor},6} + f_{\text{floor},10} + ma = 18\text{ N} + 30\text{ N} + (16\text{ kg})(2\text{ m/s}^2) = 80\text{ N}.$$

5.6 Statics

One very common use of Newton's laws is the situation where an object is *static*, not moving and not accelerating. In this case, for a two-dimensional problem, we know that two components of the acceleration will be zero. This will give us two independent equations for the static object. The sum of the forces in the x direction will be zero and the sum of the forces in the y-direction will be zero. We will always begin with a free-body diagram to be sure we have included all of the forces that might be acting on a body, and we will still have to be sure we consider only the forces acting *on that body*. But, then, the thing that separates these problems from the kinds of problems we have been looking at so far is that we will always be looking for values of the forces, not for the acceleration.

It is probably easier to just do an example of a statics problem, rather than to continue to explain things about them. So here is a generic statics problem.

EXAMPLE 5.7 The Wet Laundry Bag on the Clothesline

A 10-kg bag of wet laundry is hung from a tight clothesline. The bag pulls the clothesline down until each line segment makes a 5° angle with the horizontal, as shown. What are the tensions in the two segments of the clothesline?

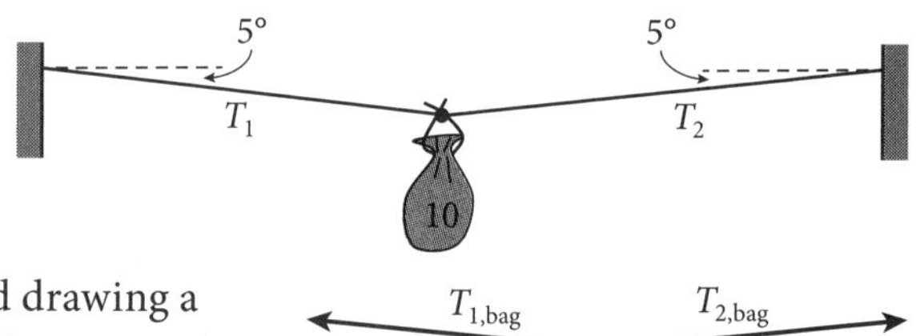

ANSWER We begin by isolating the laundry bag and drawing a free-body diagram for it. We then choose a horizontal x-axis and a vertical y-axis. The sum of the forces in the x-direction is

$$T_{2,\text{bag},x} - T_{1,\text{bag},x} = T_{2,\text{bag}} \cos 5° - T_{1,\text{bag}} \cos 5° = 0,$$

which can be satisfied only if $T_{1,\text{bag}} = T_{2,\text{bag}}$.

In the y-direction, the net force is

$$T_{2,\text{bag},y} + T_{1,\text{bag},y} - W_{\text{earth,bag}} = T_{2,\text{bag}} \sin 5° + T_{1,\text{bag}} \sin 5° - W_{\text{earth,bag}} = 0.$$

Since the x-component equation gave $T_{1,\text{bag}} = T_{2,\text{bag}}$, we can write this as

$$2T_{1,\text{bag}} \sin 5° - W_{\text{earth,bag}} = 0, \quad \text{or} \quad T_{1,\text{bag}} = \frac{W_{\text{earth,bag}}}{2 \sin 5°}$$

We know that $W_{\text{earth,bag}} = mg = (10\text{ kg})(10\text{ N/kg}) = 100\text{ N}$, so the tension $T_{1,\text{bag}}$ is

$$T_{1,\text{bag}} = \frac{W_{\text{earth,bag}}}{2\sin 5°} = \frac{100\text{ N}}{(2)(0.0872)} = 575\text{ N}.$$

The tension $T_{2,\text{bag}}$ is the same. Since only a small component of the tension acts upward, the tension has to be much larger than the weight of the bag to hold it up.

5.7 Summary

In this chapter you should have learned the following:

- You should remember to verify that the reference frame in which you choose to work out a mechanics problem is an inertial frame. In most cases, this involves verifying that the reference frame is tied to the earth, or is unaccelerated relative to the earth.
- You should be able to identify third-law pairs of forces in systems with many forces acting on the bodies of the system.
- Most importantly, you should be able to systematically solve force problems by isolating a body (or bodies, if several bodies are linked) and drawing a free-body diagram for each body, remembering that forces are 2- or 3-dimensional vectors and lead to 2- or 3-dimensional accelerations.

CHAPTER FORMULAS

Not much new in the way of formulas in this chapter. There is:

Newton's second law: $\vec{F}_{\text{net}} = m\vec{a}$ or, solved for acceleration, $\vec{a} = \dfrac{\vec{F}_{\text{net}}}{m}$

Weight: $W = mg$

And, just as a reminder:

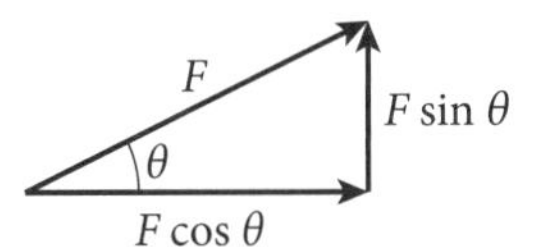

PROBLEMS

5.1 ** 1. A spaceship is coasting in empty space, somewhere between the solar system and Alpha Centauri. An astronaut wishes to test Newton's laws by making careful measurements within his spaceship. Is his spaceship an inertial frame . . .

a) when it is coasting at a constant velocity of 120 km/s? Explain.

b) when it has zero velocity relative to the solar system and to Alpha Centauri, but is momentarily accelerating at 40 m/s^2? Explain.

** 2. A car is moving along a circular path at a smooth speed of 40 mi/hr. Is a reference frame with origin in the center of the car and with the x-axis toward the front of the car, an inertial frame? Explain.

5.2 * **3.** What is your mass in kilograms? What is your weight in newtons?

* **4.** What net force is required to give a ball of mass 3 kg an acceleration of 5 m/s^2?

* **5.** What is the acceleration of a 0.5-kg book that slides across a table with a net force of 6 N on it?

** **6.** A 10-kg object on a level frictionless surface experiences a 50-N force parallel to the surface. If it starts from rest, what is the object's speed when it travels a distance of 10 m?

** **7.** An object with mass 20 kg starts from rest and experiences a constant net force of 80 N. How much time will be required for the object to travel 72 meters?

** **8.** A bullet of mass 100 grams is uniformly accelerated down an 80-cm rifle barrel, acquiring a muzzle velocity of 80 m/s.

a) What is its constant acceleration?

b) What force (assumed constant) did the gunpowder impose on the bullet?

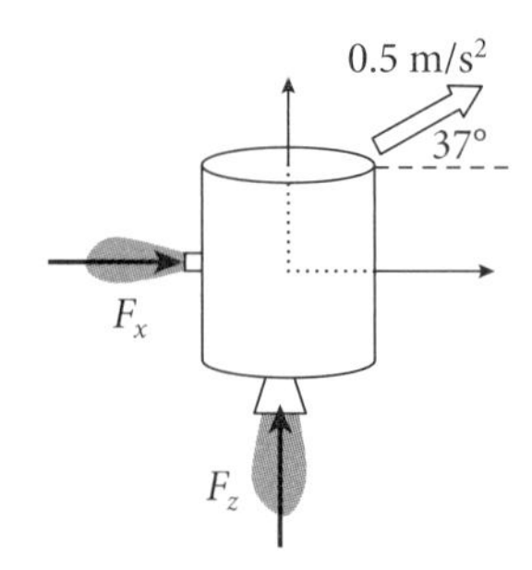

** **9.** A 200-kg spy satellite fires two thrusters to adjust its orbit to go to a new station. One thrusts along the x-axis of the satellite and one thrusts along the z-axis. The desired acceleration is 0.50 m/s^2 at 37° from the x-axis, as shown. What should be the thrust from each thruster in newtons?

5.3 ** **10.** You push on a large, heavy crate, trying to slide it across the floor, but it will not budge. Newton's third law requires that the crate push back on you just as hard as you are pushing on the crate. Some people (not us) claim that the crate cannot move because these two forces (you pushing on the crate and the crate pushing on you) cancel each other out. Explain why this reasoning is wrong.

** **11.** There is a game played on the Ile River in Strasbourg in which two boats are rowed toward each other while a man standing on a platform in each boat tries to knock the other into the water using a padded lance (see the picture below). Remembering Newton's third law, identify the things that would be important in determining who will win. Think of things like the direction of the current, speed of the boats, mass of the jouster, placement of the lance, etc., and then decide what does and does not matter in the contest. Explain your conclusions.

** **12.** An airplane is flying a level path at constant velocity. Identify all of the forces acting on the plane and identify the Newton's-third-law counterpart of each.

** **13.** A mother (50 kg) and her child (25 kg) are facing each other in the middle of a frictionless pond of ice. The mother pushes horizontally on her child with a force of 100 N.

a) What is the magnitude of the child's acceleration?

b) What is the magnitude of the mother's acceleration?

** **14.** When a baseball bat hits a 145-gram baseball, the baseball experiences an average acceleration of 1.3×10^6 m/s^2. What is the average force by the bat on the ball? What is the average force of the ball on the bat?

5.4 * **15.** When g is used to determine the weight of a body, we give it as 10 N/kg. Remember the definition of a newton and show that a N/kg is the same as a m/s^2.

** **16.** A tractor is pulling a log using a strong chain. Considering the tractor and the log as separate bodies, identify five third-law force pairs in which at least one of each pair acts either on the log or on the tractor.

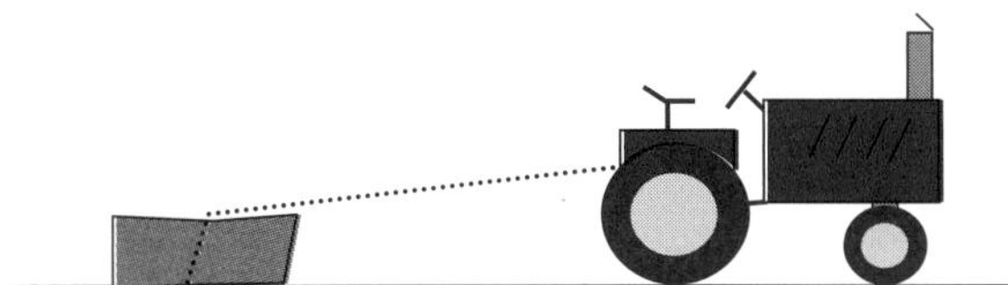

** **17.** A blackboard eraser slides across a table and falls to the floor.

a) Identify all of the forces on the eraser, while it is sliding, while it is falling, and after it comes to rest on the floor.

b) Identify the Newton's-third-law counterpart for each of the forces in part (a).

5.5 * **18.** A spider with a mass of 0.01 kg is dropping downward while playing out silk upwards from her spinnerets. If the silk strands exert an upward force on her of 0.08 N, what is the net force on the spider? [Hint: Remember to draw a free-body diagram.]

** **19.** Greg jumps off a bridge with a bungee cord tied to his ankle. At some instant when Greg is moving down and slowing, his acceleration has magnitude 4 m/s^2. If Greg has mass m_g = 80 kg, find the tension in the cord at this instant of time. [Hint: Remember to draw a free-body diagram.]

** **20.** A 24-kg block is being lowered by two massless ropes tied to the top corners of the block. When the block got to dropping too fast, the tension in the ropes was increased, forcing the block to slow down at a rate of 2 m/s^2. The ropes each make an angle of 53° with respect to the horizontal, as shown. Find the tension force provided by each rope.

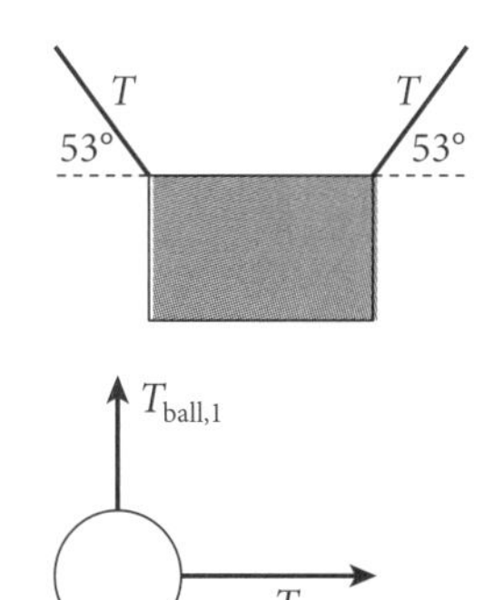

** **21.** A 2-kg ball is held in place by a vertical rope. A horizontal rope is then attached to the ball and a horizontal force is applied, causing the ball to accelerate horizontally to the right at a rate 1.5 m/s^2. What are the tensions in the two ropes?

** **22.** A sled of mass m = 30 kg (including the load sitting on the sled) is pulled by a rope across a level field of snow, as shown below. The tension in the rope has a constant magnitude of 100 newtons and makes an angle of 37° with respect to the horizontal. A friction force of 50 N acts on the sled. Follow the steps below to find the acceleration of the sled.

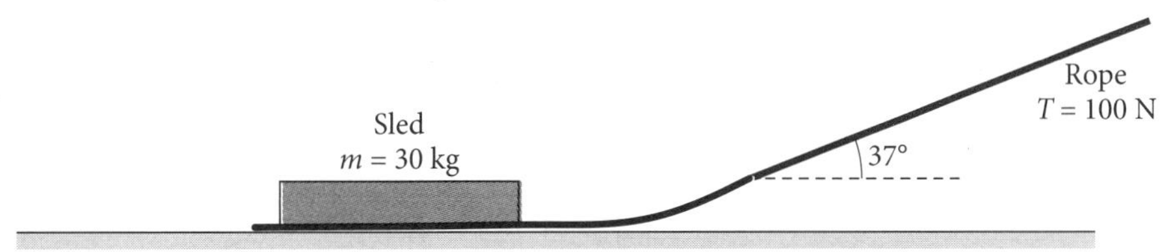

a) Draw a free-body diagram for the sled. Label each of your forces as described on pages 87–88.

b) If any of the forces on your diagram are pointing in some awkward direction (like 37° with respect to horizontal), break this force up into a horizontal component (x-part) and a vertical component (y-part).

c) Find the magnitude of the gravitational force acting on the sled. (Remember the approximation $g = 10$ N/kg.)

d) Use Newton's second law and the fact that the sled has no acceleration in the y-direction to solve for the magnitude of the normal force exerted by the snow on the sled.

e) Use Newton's second law to find the acceleration of the sled. (Remember the friction force!)

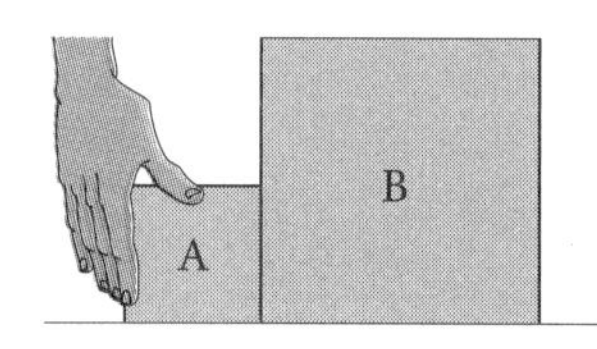

** **23.** A hand pushes two blocks across a frictionless pond of ice. Block A has mass $m_A = 1$ kg and block B has mass $m_B = 4$ kg. If the hand pushes with a constant horizontal force of 15 newtons, find the magnitude of the normal force exerted by block B on block A.

** **24.** You are riding upward in an elevator carrying a 25-kg box of rocks in your arms. As you reach your floor, the elevator begins to slow at a rate of 3 m/s^2.

a) Describe in words the direction of the acceleration.

b) Draw separate free-body diagrams for you and for the box of rocks. Label your forces as shown on page 88. Make the length of your vectors represent the relative magnitude of the forces (for example, your weight is greater than that of the box of rocks).

c) For each of the forces appearing in the free-body diagram of you (not the rocks), identify the corresponding third-law companion force.

d) Is the net force on the box of rocks greater than, equal to, or less than the net force on you? Explain your reasoning. [Hint: If you are in a college physics class, your weight is probably more than 25 kg.]

e) As the elevator is slowing down, is the force exerted by the elevator floor on your feet greater than, equal to, or less than the force exerted by your feet on the elevator floor? Explain.

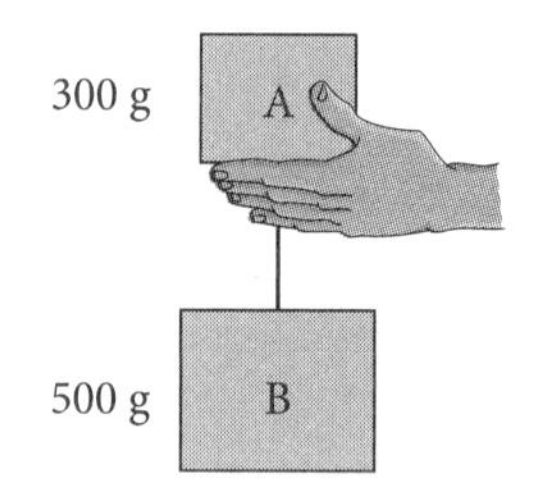

** **25.** Two blocks are tied together with a massless string. A hand is used to lift one block so that both speed up while moving upward. Block A has a mass of 300 g and block B has a mass of 500 g.

a) How do the accelerations of block A, block B, and the string between them compare? Explain your answer.

b) Draw and label separate free-body diagrams for blocks A and B.

c) Without doing any calculations, compare the magnitudes of the net force on each block. Explain the reasoning used to arrive at this comparison.

d) Calculate the magnitudes of all the forces in your diagrams for the case where the acceleration of block A has a magnitude of 4 m/s^2.

e) How does the magnitude of the force that the string exerts on block A compare to the weight of B? Explain.

** **26.** A 4-kg block, which is tied to another block, is pulled to the right by a tension force $T = 78$ N, as shown. The system accelerates to the right at 2 m/s^2. If the frictional force by the table on the 4-kg block is 20 N, what is the tension in the string connecting the two blocks?

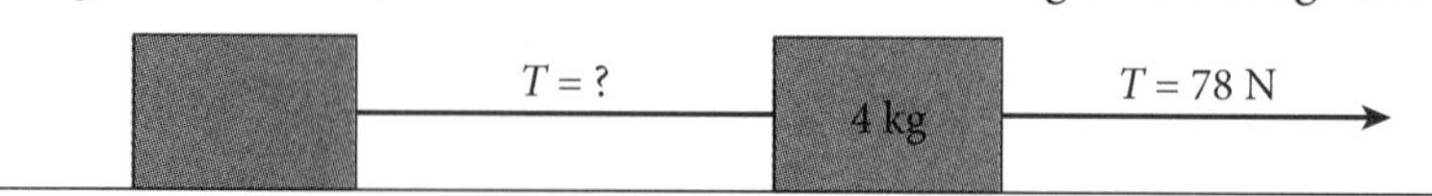

** 27. A top fuel dragster can cover the new standard distance of 305 m in 3.7 seconds, starting from rest. At such accelerations, air resistance is negligible. The dragster has a mass of 1100 kg. What is the average force supplied by the engine as the dragster covers the 305 m?

** 28. A sounding rocket is carrying its payload through the air. At the instant shown, it is thrusting with a force of 3×10^6 N at an angle of 37° above the local horizontal. The vehicle has a total mass of 200,000 kg, so its weight is 2×10^6 N. Find the acceleration of the rocket at the instant shown. [Remember: Acceleration is a vector, so both magnitude and direction must be given.]

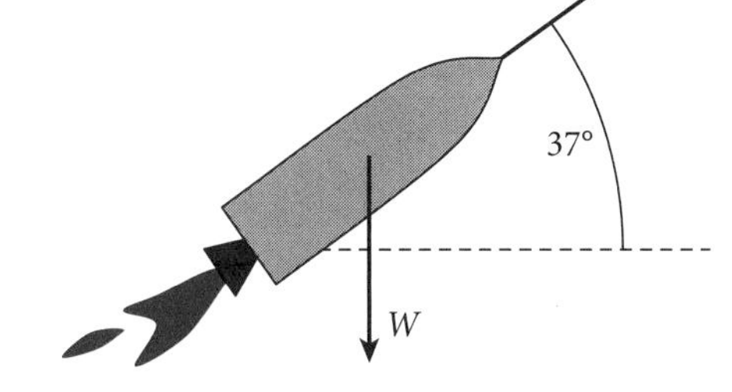

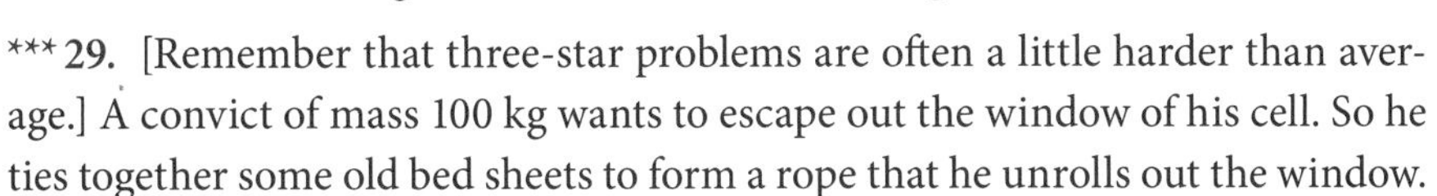

*** 29. [Remember that three-star problems are often a little harder than average.] A convict of mass 100 kg wants to escape out the window of his cell. So he ties together some old bed sheets to form a rope that he unrolls out the window. Unfortunately, the sheets can support only a tension of 800 N without tearing. The convict figures, however, that if he slides down the sheets with the right constant acceleration, he can avoid tearing the sheets.

a) What is the acceleration that would avoid tearing the sheets?

b) If his 4th-story window is 16 m above the ground, how fast will he be moving when he hits the ground?

c) For comparison, from what height could he jump *without* using the sheets (that is, with the usual acceleration of 10 m/s^2) that would give him this same final velocity as he hits the ground?

** 30. Three identical 2-kg blocks are free to move on a frictionless surface. The three blocks are connected by short massless cords and a force of 24 N is applied to the rightmost block.

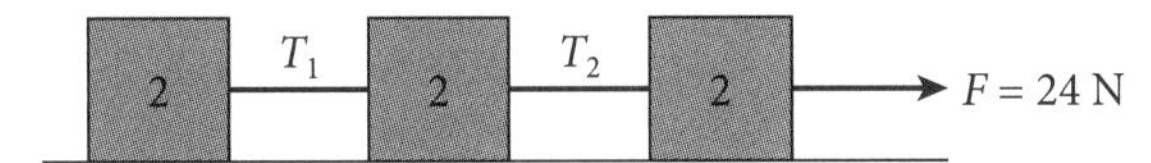

a) Find the acceleration of the three blocks.

b) Find the tensions in the two cords connecting the blocks.

5.7 * 31. Giselle pushes horizontally with a force of 400 N on a crate that is sitting on the ground. You find that it takes a force of 220 N by you, in the opposite direction, to keep the crate from moving. What is the static friction force of the floor on the crate?

* 32. A 5-kg mass is connected by a string to a 4-kg mass, and the string passes over a massless frictionless pulley so that the 5-kg mass sits on a tabletop and the 4-kg mass hangs straight down. If the system is static, what must be the frictional force acting between the tabletop and the 5-kg mass?

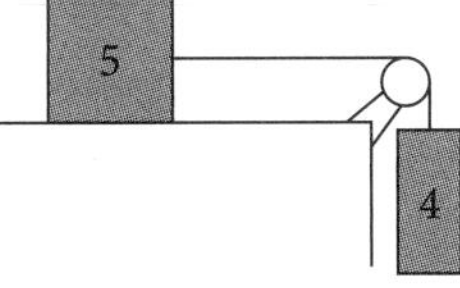

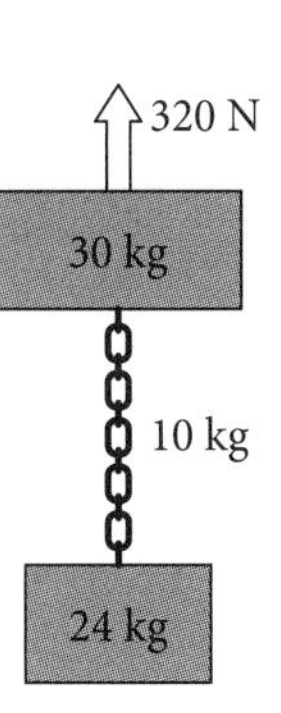

** 33. Two blocks are connected by a heavy 10-kg chain, as shown at right. The mass of the top block is 30 kg and the mass of the lower block is 24 kg. A constant upward force of 320 N is then applied to the top block.

a) What will be the acceleration of the system and will it be upward or downward?

b) What is the force of the top block on the top link of the chain?

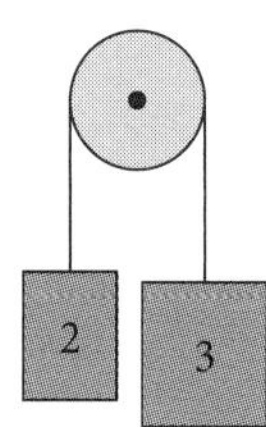

** **34. The Atwood Machine.** One of the standard physics student experiments over the last couple of *centuries* is what is referred to as the Atwood Machine. In our case, let us take a 2-kg mass attached to a 3-kg mass by a light string. The string is passed over a low-friction aluminum pulley so that one mass hangs down on each side of the pulley, as shown at left. Assume that there is no friction in the pulley and that the masses of both the string and pulley may be neglected. Find the acceleration of the two masses via the following procedure:

a) Draw a free-body diagram for each of the masses separately. Note that the tension in the massless string that pulls upward on the 3-kg mass is the same as the tension pulling upward on the 2-kg mass.

b) Find the net force on each mass and write down Newton's second law for each mass.

c) Explain why the accelerations of both masses will have the same magnitude, but opposite directions.

d) Solve the two simultaneous equations (check out Appendix A, if you want a review on how to solve simultaneous equations) and find the acceleration of the system and the tension in the string.

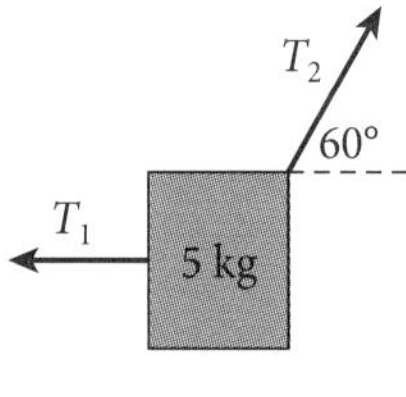

** **35.** A 5-kg block, connected to two strings, is at rest. The angles of the strings are as shown in the picture at left. Find the tensions in each of the two strings. [Hint: In this problem, you can solve for the two unknowns one at a time. One equation will give you one unknown tension by itself and that value can then be used in the other equation to give the other unknown tension.]

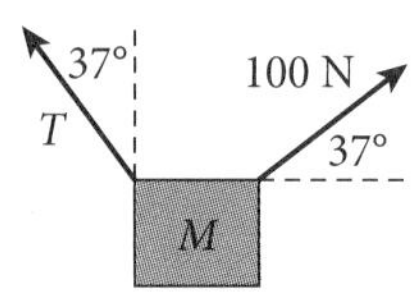

** **36.** A block of mass M is held at rest by two strings, as shown in the picture at left. The string pulling to the right has a tension of 100 N in it, as indicated. Find the tension in the other string and find the mass of the block. [Hint: In this problem, there are two unknowns, but you can solve for them one at a time. One equation will end up involving only one of the unknowns.]

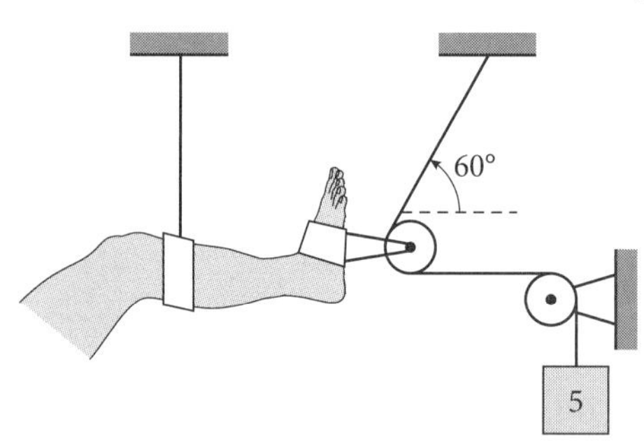

*** **37.** An unfortunate skier has had an accident and has to have his leg in traction for a few days. The leg is suspended from the ceiling, and a horizontal force is applied to the leg with the arrangement of pulleys, cables, and a weight shown below. The cable is attached to the ceiling, after which it passes through a frictionless pulley attached to his foot and then a pulley attached to the wall, before a 5-kg weight is hung on the end to provide the traction. Neglect the small mass of the cable.

a) What is the horizontal force of traction on his leg? [Hint: Think carefully about the cable tension where it is applied to the pulley that holds his foot.]

b) If the traction force were the only horizontal force acting on his leg, the leg would accelerate to the right. Where does the force come from that keeps his leg from accelerating?

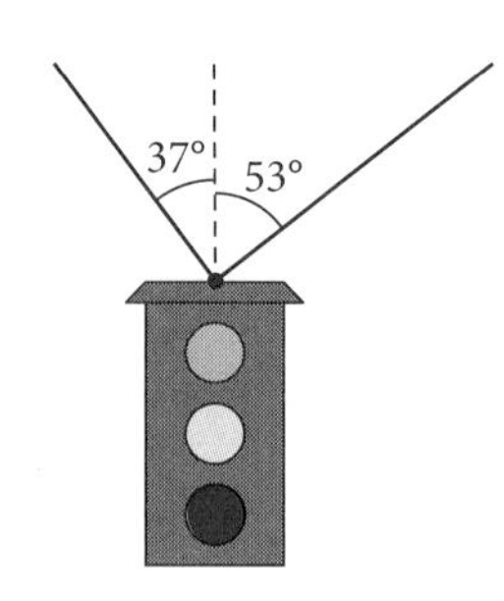

*** **38.** A 40-kg traffic light is suspended from two wires. One of the wires makes a 37° angle to the vertical and the other makes a 53° angle, as in the picture below. Find the tension in the two wires. [Hint: There are two unknowns in this problem, but the net force on the traffic light in the vertical direction and the net force in the horizontal direction are separately zero. This gives two independent equations for two unknowns. If you need a little help solving simultaneous equations, see Appendix A.6.]

6

FORCES

In the last chapter, we learned what forces are and we introduced a few of them by name. Here, we will look in more depth at each of the forces we will use in this book (until we get to Chapter 16 where we introduce the electrical force). So here it comes—everything you need to know about forces in one chapter.

6.1 Gravitational Force: Weight

In the last chapter, we introduced the idea of the gravitational force on an object, a force called *weight.* We also know from Chapter 2 that, when gravity is the only force acting, objects near the surface of the earth all fall with the same constant acceleration. In a way, this fact may now seem somewhat surprising. We remember that Newton's second law tells us that the acceleration of any object subject to a single force F is $a = F/m$, so the acceleration of an object would appear to depend on its mass—the bigger the mass the smaller the acceleration. In fact, the only way in which all falling bodies can have the *same* acceleration is if the force acting on a body gets bigger as the mass gets bigger. This is indeed the nature of gravity. The weight of an object is given by the formula $W = mg$, where g is the earth's *gravitational field strength.* As we used the approximate 10 m/s^2 for the acceleration of a falling body, instead of the more accurate 9.8 m/s^2, so will we approximate the earth's gravitational field as 10 N/kg, instead of the more accurate, and more commonly used, 9.8 N/kg. In problems, we will generally just ignore the 2% error in this assumption and pretend that the earth's gravitational field is exactly 10 N/kg. This simplifies calculations enough that it seems worthwhile. The weight of an object of mass 12 kg is thus

$$W = mg = (12 \text{ kg})(10 \text{ N/kg}) = 120 \text{ N}. \tag{6.1}$$

The acceleration of such an object, when its weight is the only force acting on it, is

$$a = \frac{W}{m} = \frac{120 \text{ N}}{12 \text{ kg}} = 10 \text{ m/s}^2,$$

as expected.

This constant force of gravity on an object near the surface of the earth is really just a manifestation of a more general *Law of Universal Gravitation.* One of the discoveries that Isaac Newton published in his *Principia* in 1687 is that any two objects with mass will attract each other with a gravitational force given by

$$F_g = G\frac{mM}{r^2}, \tag{6.2}$$

where m and M are the masses of the two objects, r is the distance between their centers, and G is a universal constant. The value of Newton's gravitational constant G was not discovered until 1798, when Henry Cavendish of England found it to be very close to the presently accepted value of $6.67 \times 10^{-11}\ \text{N}\cdot\text{m}^2/\text{kg}^2$.

Equation 6.2 says that the gravitational force between two bodies will vary with distance r between them. However, if an object of mass m is on the surface of the earth, it will always be about 6.4×10^6 m (the radius of the earth) from the center of the earth. The earth has a mass 6.0×10^{24} kg, so the force of the earth's gravity on an object of mass m on its surface will be

$$F_g = G\frac{mM}{r^2} = (6.67 \times 10^{-11}\ \text{N}\cdot\text{m}^2/\text{kg}^2)\frac{m(6.0 \times 10^{24}\ \text{kg})}{(6.4 \times 10^6\ \text{m})^2} = m(9.8\ \text{N/kg}) = mg.$$

This is Equation 6.1. If an object of mass m is on the surface of another planetary body, say the moon, then a similar calculation using the moon's radius (1.74×10^6 m) and the moon's mass (7.35×10^{22} kg) will give the gravitational field strength of the moon. The value of g on the moon turns out to be about one-sixth the gravitational field strength on earth.

This is also a good time to be sure we understand the difference between mass and weight. Mass is a measure of how difficult it is to accelerate an object, to speed it up or slow it down, or to shake it. Weight is the gravitational force exerted on that object and is proportional to the gravitational field strength g. A blacksmith's anvil, for example, would weigh a great deal on the earth, but only one-sixth as much on the moon. It would be much easier to hold it above your head on the moon. On the other hand, the mass of the anvil does not change when I transport it to the moon (or anywhere else). It would be just as difficult to shake an anvil on the moon as on earth.

While the gravitational force of a planet on a body near its surface can be large enough to produce large accelerations, the gravitational forces between laboratory-sized bodies are negligibly small due to the tiny value of G. Thus, most of the gravitational forces we will deal with will be weight forces, the gravitational force of the entire planet on some object, for which we can use the simple $W = mg$.

The gravitational force is an example of a non-contact force. The two interacting objects do not have to touch each other to exert force on each other. The other non-contact force that we will use in this chapter is the magnetic force between a refrigerator magnet and a refrigerator, discussed in Section 6.7. All other forces that we consider for the time being are contact forces, requiring that the objects touch in order to exert a force.

6.2 Normal Force

When the surfaces of two bodies are in contact with each other, there are two ways they can push on each other. One of these is called the "normal force." The normal force is the force that acts perpendicular ("normal") to the surfaces of the two objects. A book lying on a table, as in Figure 6.1, exerts a downward normal force on the table. By Newton's third law, the table exerts a normal force upward on the book. Whenever one solid object pushes on another solid object, a normal force will arise to oppose that push, keeping the two objects from passing through each other. Gravity pulls the book down into the table, so the molecules of the table respond with a normal force up on the book. There may also be a force parallel to the surface. This force is called friction and will be discussed in the next section.

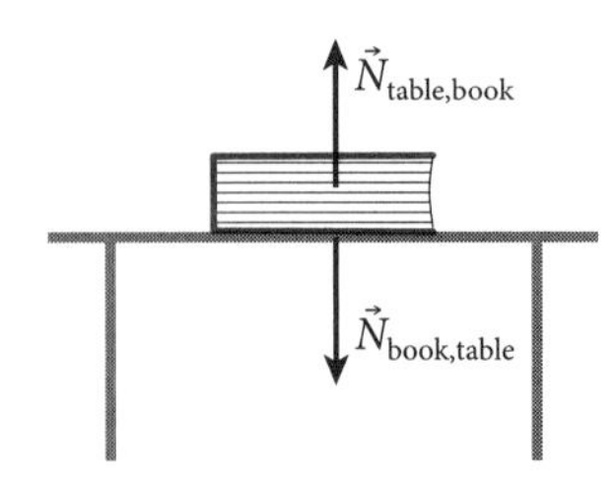

Figure 6.1 The normal forces between surfaces.

EXAMPLE 6.1 The Sister on the Sled—I

Consider the situation discussed in Section 5.5, where you are pulling your sister on a sled. If your sister has mass 21 kg, and the sled has mass 18 kg, find all of the normal forces acting on the sled, assuming that you are pulling with a force of 400 N at an angle of 37° above the horizontal.

ANSWER We start by drawing free-body diagrams for your sister and the sled. We have indicated the forces that are third-law companion forces with colors and with an equal number of tick marks on each

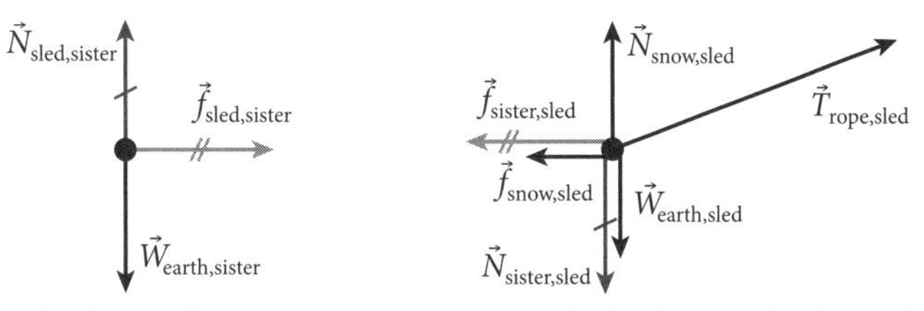

Free-body diagram for your sister | Free-body diagram for the sled

member of the pair. (Note the reversed subscripts for each pair as well.) Both the sled and your sister are accelerating to the right, so both diagrams should have more force to the right than to the left. The sled is not jumping off the snow or sinking into it, so there must be just as much upward vertical force as there is downward vertical force on each body in the two diagrams.

It is clear from the diagram on the left that the upward normal force exerted by the sled on your sister (the red vector on the left) must be equal in magnitude to her weight (since there must be zero net vertical force). This gives

$$N_{sled,sister} = W_{earth,sister} = m_{sister}g = (21\text{ kg})(10\text{ N/kg}) = 210\text{ N}.$$

If the red upward normal force on your sister is 210 N, then the normal force she exerts down on the sled must also be 210 N (by Newton's third law). This value may be carried over into the free-body diagram on the right, giving the magnitude of that red force.

We can find the normal force exerted on the sled by the snow by balancing the vertical forces acting on the sled in the sled's free-body diagram. The sled's weight is $W_{earth,sled} = m_{sled}g = (18\text{ kg})(10\text{ N/kg}) = 180\text{ N}$. If we add that to the $N_{sister,sled}$ we just calculated, we get a total downward force on the sled of $180\text{ N} + 210\text{ N} = 390\text{ N}$.

Now, if the rope were not pulling up on the sled, the normal force exerted by the snow would have to be an upward 390 N to keep everything balanced. But the rope *does* have an upward component, so it is helping the sled not to fall downward through the snow. To find the upward pull of the rope, we break this force into its vertical and horizontal components and find that the vertical component is upward at $400\text{ N}\sin(37°) = 240\text{ N}$. The normal force must therefore combine with this upward component of tension to produce a total upward force of 390 N. This means that the normal force must be $N_{snow,sled} = 390\text{ N} - 240\text{ N} = 150\text{ N}$.

It is important to note that there is no special equation that can be used to calculate a normal force. It is a force that arises when two objects are pressed together and it is always whatever it takes to keep the objects from squeezing into each other. As in the example shown above, the size of a normal force is typically found by adding up all the force components in the direction normal to the surface along which two bodies are in contact and using the fact that we know that the body remains on the surface to conclude that the net force in the direction normal to the surface is zero. We also want to emphasize that a normal force need not point vertically up or down. If the surface is not horizontal, the normal force will not be vertical, as we will see when we discuss inclined-plane problems in Section 6.4.

6.3 Frictional Forces: Kinetic and Static

The second possibility for a contact force is the friction force. This force acts parallel to the surface between the object exerting the force and the object experiencing the force. The friction force manifests itself in two very different ways. When two rough surfaces are sliding across each other (so that we hear that scraping sound), it is *kinetic friction*. If we are trying to move one rough surface across another, but the friction force between them keeps them from sliding, then it is *static friction*.

Kinetic friction is much simpler than we have any reason to expect. Laboratory measurements indicate that, in most everyday situations, this force depends on only two factors. The first factor is how rough the two surfaces are (quantified by a *coefficient of kinetic friction*, μ_K). The coefficient of kinetic friction, μ_K, simply must be measured for every combination of two surfaces. The second factor is how tightly the two surfaces are pressed together (the normal force acting between the two surfaces). If a book is sliding across a table, the kinetic friction force is given by the formula $f^K_{\text{table,book}} = \mu_K N_{\text{table,book}}$. The subscripts on the friction force will always be the same as the subscripts on the normal force.

EXAMPLE 6.2 Kinetic Friction—I

You push a 4-kg block to the right across a table with a constant force of 30 N. The coefficient of kinetic friction between the block and the table is $\mu_K = 0.25$. Find the acceleration of the block.

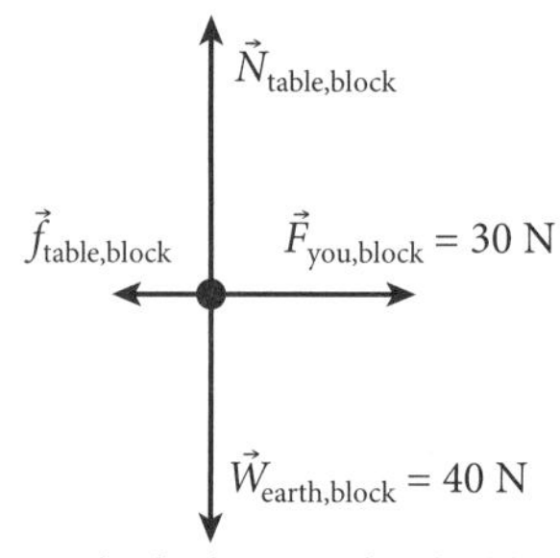

Free-body diagram for the block

ANSWER To find the acceleration of the block, we will need to know the net force on it. The horizontal forces include your 30-N force and the kinetic friction force, which we will need to figure out.

The kinetic friction force depends on the magnitude of the normal force exerted between the table and the block, so we will need to find this normal force first. The normal force must be whatever it takes to keep zero acceleration in the vertical direction. It will have to balance the downward force of gravity on the block, so the normal force will be $N_{\text{table,block}} = W_{\text{earth,block}} = 40$ N. Knowing $N_{\text{table,block}}$, we can then go on to determine the friction force from the formula. This gives

$$f^K_{\text{table,block}} = \mu_K N_{\text{table,block}} = (0.25)(40 \text{ N}) = 10 \text{ N}.$$

We now know both horizontal forces, so we can use Newton's second law to calculate the acceleration of the block. Taking positive to the right, the net horizontal force acting on the block is $F_{\text{net},x} = 30$ N − 10 N = 20 N. Newton's second law gives the acceleration as

$$a = \frac{F_{\text{net}}}{m} = \frac{20 \text{ N}}{4 \text{ kg}} = 5 \text{ m/s}^2.$$

Note especially how the value of the normal force, found by analyzing the vertical motion (or lack of motion), carried over into the analysis of the horizontal motion.

Static friction is different than kinetic friction. If you push on a crate with a force of 200 N in an effort to slide it across the floor, but it doesn't budge, then the net force on the crate must be zero. Since you are pushing to the right with 200 N of force, the floor (the only other thing touching the crate) must be pushing to the left with a force of 200 N. This is the static friction. The nooks and crannies on the bottom surface of the crate are locked into the nooks and crannies of the floor. If you increase your push to 350 N and still fail to slide the crate, then the friction force must have increased to 350 N as well. It is essentially the same friction source, but the size of the force has increased. The molecules in the nooks and crannies may be deformed a bit more, but they are still holding. Static friction adjusts itself to be whatever magnitude is required to keep the two surfaces from moving relative to each other. In this way, it is like the normal force. There is no formula for static frictional force. It is whatever it takes to keep motion from occurring.

However, at some point, you can push hard enough to deform the nooks and crannies enough that the surfaces do start to slide across each other. This breaking point, at which the static friction force reaches its maximum value, depends on the natures of the two surfaces (characterized by a coefficient of static friction μ_S) and on the normal force. It is

$$f^{S,max}_{floor,crate} = \mu_S N_{floor,crate}. \qquad (6.3)$$

This is not a formula for the static frictional force. Do not ever use it this way. This is a formula for the *maximum value* that the static frictional force can have. For a particular value of μ_S and N, the static frictional force can have any magnitude between zero and this maximum amount. The actual static friction force is determined, not from the formula, but from Newton's second law and from the observation that the two surfaces are not moving relative to each other. By the way, the coefficient of static friction for any two surfaces generally has a larger value than their coefficient of kinetic friction. We usually expect $\mu_S > \mu_K$.

EXAMPLE 6.3 Static Friction

You are trying to move a sled loaded with rocks by applying a force of 500 N to a rope that is pulling 37° above the horizontal, as shown. The mass of the sled with the rocks in it is 120 kg. The coefficient of static friction between the sled and the floor is $\mu_S = 0.9$. The sled does not move. Find the magnitude of the friction force.

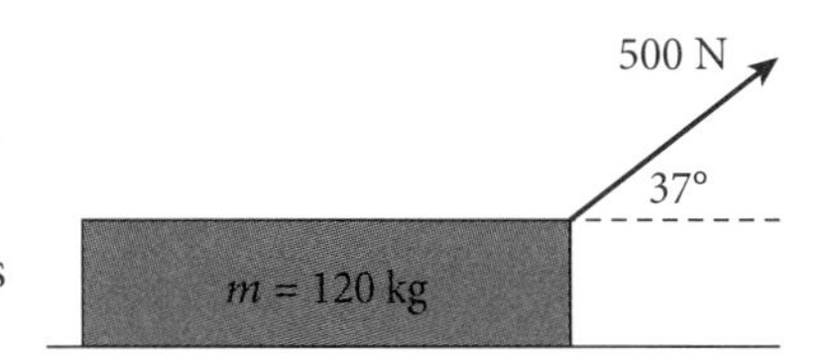

ANSWER Because the sled is not moving across the floor, we know that we are looking for a *static* friction force. The first step is always a free-body diagram.

The sled is not accelerating, so we know the net force is zero. In the horizontal direction, the static friction force to the left must balance the horizontal component of the tension force to the right. This component has magnitude (500 N)cos(37°) = 400 N, so the static friction force is 400 N to the left. We are done! *And note that we have solved the static friction problem without even looking at Equation 6.3.*

As a matter of interest, however, we can verify that static friction can provide this much force. We will first need to find the normal force.

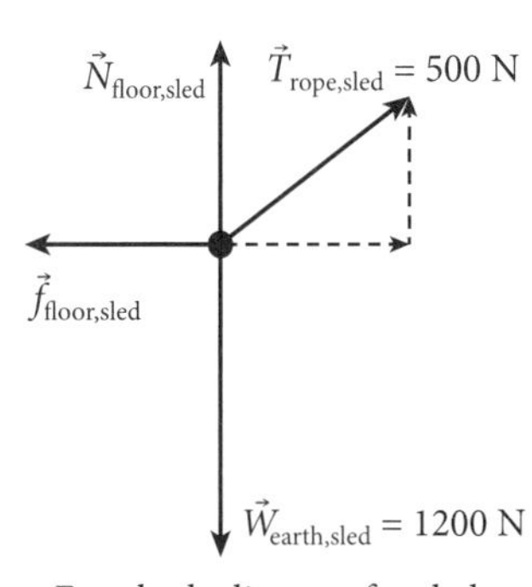

Free-body diagram for sled

$N_{\text{floor,sled}}$ may be found by noting that it, plus the vertical component of $T_{\text{rope,sled}}$, must balance the downward weight of the sled. This means that we must have $N_{\text{floor,sled}} = W_{\text{earth,sled}} - T_{\text{rope,sled}} \sin(37°) = 1200 \text{ N} - 300 \text{ N} = 900 \text{ N}$. Knowing the normal force, we can calculate the maximum force that static friction between the floor and the sled can provide. It is

$$f^{\text{S,max}}_{\text{floor,sled}} = \mu_S N_{\text{floor,sled}} = (0.9)(900 \text{ N}) = 810 \text{ N}.$$

The required 400 N of static friction is indeed less than this available amount. And, just to be clear, the static frictional force in this problem is 400 N, not 810 N.

EXAMPLE 6.4 Kinetic Friction—II

A friend comes to help you move the sled of rocks in the previous example. She pushes to the right with a horizontal force of 620 N and you continue to pull with the rope the same as before. The coefficient of kinetic friction between the sled and the floor is $\mu_K = 0.6$. Find the resulting acceleration of the sled.

ANSWER The combined horizontal push to the right supplied by you and your friend is now 1020 N. This is larger than the maximum static friction force for the sled, so the sled will break free and begin to accelerate as it scrapes across the floor. The maximum static frictional force will not matter any more, since the sled is sliding.

The friction is now kinetic friction, whose magnitude is always given by the kinetic friction formula. It is $f^{\text{K}}_{\text{floor,sled}} = \mu_K N_{\text{floor,sled}} = (0.6)(900 \text{ N}) = 540 \text{ N}$. The net force on the sled is therefore 1020 N – 540 N, or 480 N, to the right. Thus, the acceleration is to the right with magnitude

$$a = \frac{F_{\text{net}}}{m} = \frac{480 \text{ N}}{120 \text{ kg}} = 4 \text{ m/s}^2.$$

We will look at one more friction problem to show that *static* friction can be acting on an object even when the object is accelerating. Static friction must, of course, arise in a case where the acceleration of the object and that of the surface it is touching are the same, so that the two surfaces are not scraping along each other.

EXAMPLE 6.5 The Sister on the Sled—II

We return to Example 6.1, where you are pulling your sister on a sled. Recall that your sister has mass 21 kg, and the sled has mass 18 kg, and that you are pulling with a force of 400 N at an angle of 37° above the horizontal. Find the magnitude of both friction forces acting on the sled if the acceleration of the sled is 5 m/s^2.

ANSWER Because you are always kind to your sister, you will be pulling slowly enough that she will not slide backward off the sled. As the sled accelerates at 5 m/s^2 to the right, your sister will also accelerate to the right at 5 m/s^2. We begin by finding the friction force of the *sled on the sister,* since that is so easy to find. We will then use Newton's third law to note that it is also the friction force of the sister on the sled. We see from her free-body diagram (the diagram on the left in Example 6.1 on page 99) that the vertical forces will cancel, leaving the *static* friction force acting between her and the sled as the only force acting on her. So it is the net force. This force can be found simply by asking what net force is required to accelerate a 21-kg sister at an acceleration 5 m/s². Newton's second law gives

$$F_{\text{net}} = f_{\text{sled,sister}} = ma = (21 \text{ kg})(5 \text{ m/s}^2) = 105 \text{ N}.$$

Note that this is a static frictional force on an accelerating object, your sister.

Referring again to the free-body diagrams on page 99, we can now use Newton's third law to claim that the static friction force exerted by your sister on the sled must also be $f_{\text{sister,sled}} = 105$ N. This leaves one more friction force to find, $f_{\text{snow,sled}}$.

The force required to accelerate an 18-kg sled at a rate of 5 m/s² is

$$F_{\text{net}} = ma = (18 \text{ kg})(5 \text{ m/s}^2) = 90 \text{ N},$$

so the net horizontal force on the sled must be 90 N. The horizontal forces add up to

$$T_{\text{rope,sled}} \cos(37°) - f_{\text{sister,sled}} - f_{\text{snow,sled}} = 90 \text{ N}.$$

The horizontal component of the tension is 400 N cos(37°) = 320 N, and we have just seen that $f_{\text{sister,sled}} = 105$ N. This leaves

$$f_{\text{snow,sled}} = T_{\text{rope,sled}} \cos(37°) - f_{\text{sister,sled}} - 90 \text{ N} = 320 \text{ N} - 105 \text{ N} - 90 \text{ N} = 125 \text{ N}.$$

6.4 Problems on an Inclined Plane

Whenever we know that an object's motion is confined to be along a particular surface (like an inclined plane), we know that it will not accelerate into the surface or jump off of it. So the only acceleration it can have is along the surface. In situations like this, it is generally best to use a coordinate system with the *x*-axis aligned with the inclined plane. This way, a lot of the forces, like normal force and friction, end up being aligned with either the *x*-axis or the *y*-axis. If the plane is inclined, the gravitational force will be an exception. It always points toward the center of the earth, regardless of the slant of the plane, so it will typically have to be resolved into its components in the two directions.

EXAMPLE 6.6 The Rope Tow

A physics student is being pulled up the bunny slope at Bridger Bowl (the authors' favorite local ski hill) at a constant speed of 1.5 m/s, as shown. The slope of the hill is 53°, the mass of the student is 60 kg, and the coefficient of kinetic friction between her skis and the snow is 0.3. Find the magnitudes of all the forces acting on the student.

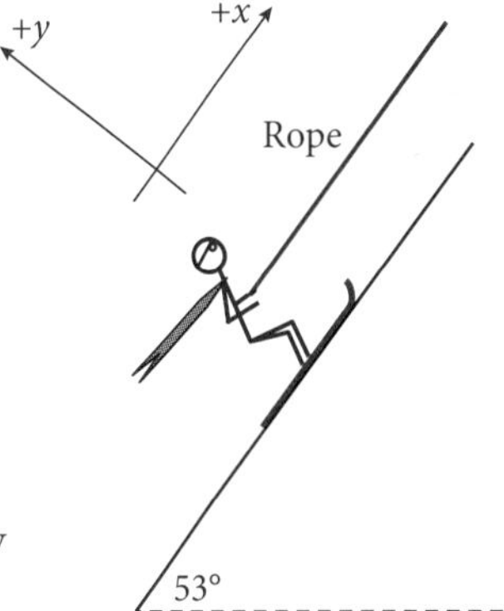

ANSWER Yes, we know very well that you don't pull a skier up a 53° slope with a rope tow. Please just humor us on this one.

Let us begin by orienting our coordinate system with the slope of the hill, with the *x*-axis going up the hill, as shown above. Then we generate a free-body diagram and go to work on the forces.

The force $W_{\text{earth,student}}$ is easy. The gravitational force on a 60-kg student will be 600 N. But our coordinate system choice means that the student's weight must be broken into components. The magnitudes of the two components in the directions we have chosen are

$$W_x = (600\ \text{N})\sin(53°) = 480\ \text{N}$$
$$W_y = (600\ \text{N})\cos(53°) = 360\ \text{N}.$$

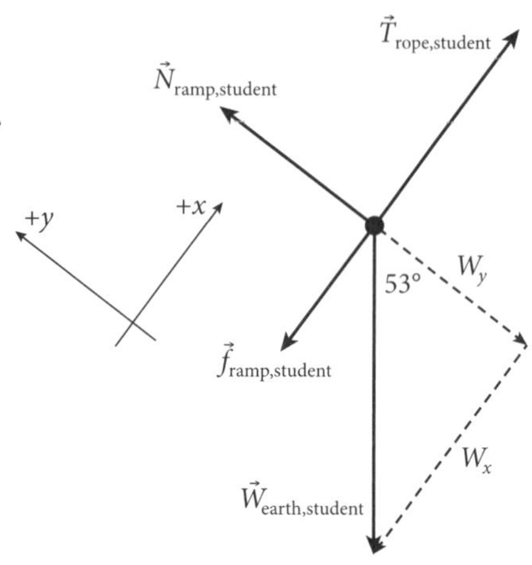

Free-body diagram for the student

We will worry about the signs of these components when we add them together. We know that the student remains on the snow, so the y-acceleration will be zero, and we know that a rope tow pulls skiers at constant speed along a straight path, so the acceleration in the x-direction will likewise be zero. The zero acceleration in the y-direction means that the normal force $N_{\text{ramp,student}}$ must balance the y-component of the weight, giving $N_{\text{ramp,student}} = W_y = 360$ N. Turning to the forces in the x-direction, we see that zero net force means that the tension force from the rope must just balance the sum of the two forces in the minus x-direction. This gives

$$T_{\text{rope,student}} = f^{\text{K}}_{\text{ramp,student}} + W_x.$$

Knowing the normal force, we can find the kinetic friction force from the kinetic friction formula. It is $f^{\text{K}}_{\text{ramp,student}} = \mu_{\text{K}} N_{\text{ramp,student}} = (0.3)(360\ \text{N}) = 108$ N. Now that we know both forces on the right hand of the equation for $T_{\text{rope,student}}$, we may add them together to find the tension to be $T_{\text{rope,student}} =$ 108 N + 480 N = 588 N.

Let us notice how we resolved the gravitational force in this last example into components. We began by drawing a vertical downward arrow to represent the weight. We then started at the dot and went in the minus-y direction, perpendicular to the slope of the hill. We had to make this component just long enough that we could take a 90° turn and draw the x-component of the weight along a line downward, parallel to the hill, and just reach the end of the vertical weight vector we drew initially. This process is well worth practicing until you feel comfortable with it.

6.5 Tension

In this section we provide some formal background to the idea of the tension force. A tension force is a contact pull exerted by a rope, chain, or string on some other object. By Newton's third law the object must also exert a tension force on the rope, chain, or string. We often talk about the tension "in" a rope. This is the force that each piece of the rope exerts on its neighbor. This is easier to visualize in a chain made up of individual links. Imagine a 5-kg block hanging from the ceiling by a chain, as shown in Figure 6.2. The tension at each link of the chain is the force with which that link pulls on its neighbor. If the chain is massive, the links closer to the ceiling will have to pull harder on their neighbors than the links near the block. The lowest link would pull upward on the block with a force equal in magnitude to the weight

Figure 6.2 The tension in the chain links will be bigger nearer the top of the chain.

of the block, 50 N. The highest link would pull downward on the ceiling with a force equal to the weight of the block plus the weight of the chain. The tension in the chain would therefore be different at the two ends. You also may remember back in Chapter 5, on page 86, that we explained how an accelerating rope with some mass must have a different tension at the two ends of the rope, or it would not be able to accelerate. We then went on to explain that only in the limit of a negligible mass for the rope could the tension be the same throughout the rope.

In most physics problems we assume that the mass of the chain (or rope) is negligible compared to the mass of the other objects in the problem. In this approximation, the pull at the two ends of the chain is the same, and we talk about a single tension in the chain that pulls at both ends. We often call this a "massless" string, but this is just a code word for saying that the mass is negligible.

EXAMPLE 6.7 An Arrangement of Blocks and Pulleys—I

The diagram at right shows a system of three blocks attached by two massless strings, one of which passes over a massless frictionless pulley. Block B has a mass that is twice the mass of block A, and the mass of block C is three times that of block A. Assume that there is no friction between the table and blocks B and C. While the hand is holding the system stationary, how does the tension in String 1 compare to the tension in String 2?

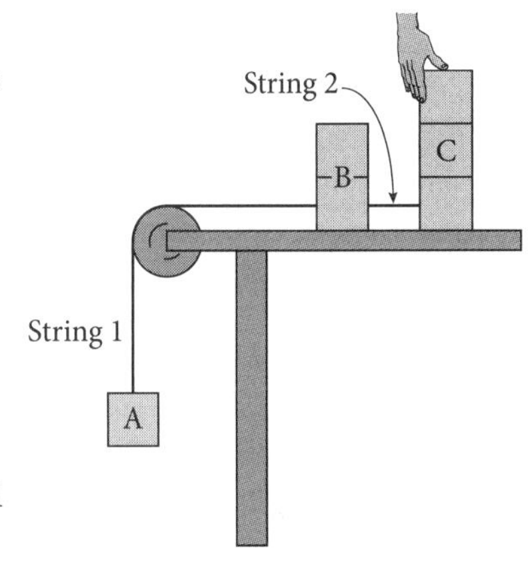

ANSWER First, let us explain that, since there is no motion in the system, the tension in String 1 cannot change as it goes over the pulley. A difference in the tensions of the vertical segment of String 1 and the horizontal segment would put a torque on the pulley and make it start to spin. Since this does not happen, the tension in String 1 must be the same throughout its length.

When comparing the magnitudes of two forces, it is often useful to find a free-body diagram in which both forces appear, and to use Newton's second law for that diagram. Both String 1 and String 2 pull on block B, so block B is a good body to choose. Because block B is not accelerating, we know the net force on it must be zero. The leftward force of String 1 and the rightward force of String 2 must cancel. The tension in the two strings must therefore be equal in magnitude.

We can also say something about the size of these two equal tensions. A free-body diagram of block A has only two forces on it: the gravitational force down and the tension force exerted by String 1. Because block A is not accelerating, these two forces must cancel. The tension in String 1 is equal to the weight of block A. But we have already seen that the tension in String 2 is equal to that in String 1, so the tension is String 2 is also equal to the weight of block A.

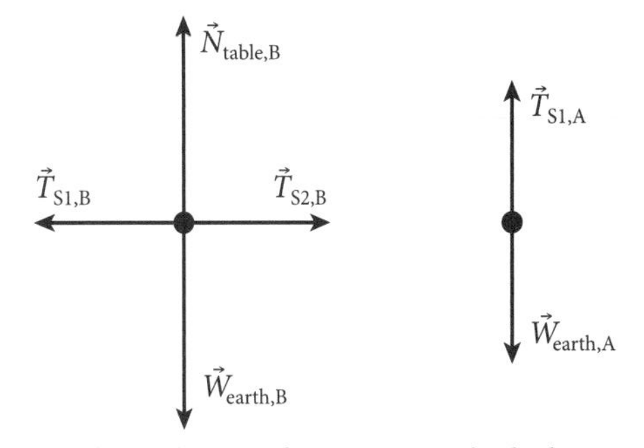

Free-body diagram for B Free-body diagram for A

Students often make the mistake of looking at surface features of the problem to compare forces. A common incorrect answer for the question in the previous example is: "String 2 is holding more mass from moving to the left, so the tension in String 2 must be bigger." This type of intuitive reasoning will hardly ever produce useful results. You really will be better off if you draw free-body diagrams for everything in the problem that has mass and then use Newton's second law. And do note that it is never useful to draw free-body diagrams for massless things like massless strings, since a zero mass in Newton's second law makes for an infinite acceleration unless the net force is also zero. The tension in a massless string will simply always be the same on both ends, without any need for a diagram to prove it.

EXAMPLE 6.8 An Arrangement of Blocks and Pulleys—II

The hand in the previous example now releases the system. Does the tension in String 1 increase, decrease, or stay the same, compared to when the system was stationary? What about the tension in String 2?

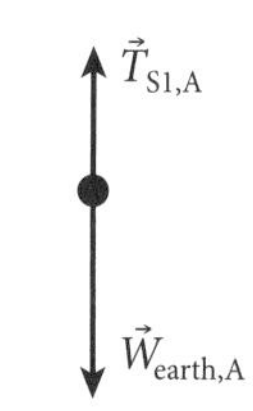

Free-body diagram for A after hand releases system

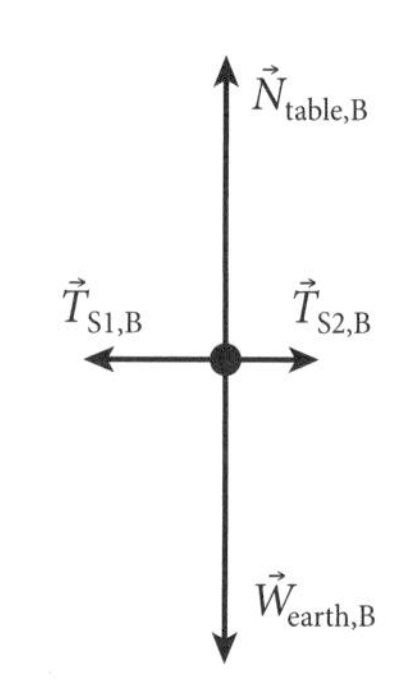

Free-body diagram for B after hand releases system

ANSWER The free-body diagram for block C would now have a single unbalanced force to the left (the hand is now gone and there is no friction between the table and the block), so block C must accelerate to the left. The blocks are connected by strings, so their accelerations must always be the same magnitude. Block A is now accelerating downward, indicating that the net force on block A must be downward. The gravitational force on block A does not change when the hand releases the system, so the tension in String 1 must decrease (as shown in the diagram at right). Since the pulley is massless, the tensions in both segments of String 1 must be the same, or the pulley would begin to spin up with infinite acceleration.

We can also explore what happens to the tension in String 2 by looking at a diagram that includes this force. The free-body diagram for block B must now allow a net force to the left (in the direction of the acceleration), so the magnitude of the tension in String 2 must be less than the magnitude of the tension in String 1 (which is already less than it used to be). Therefore, the tension in String 2 must decrease even more than did the tension in String 1.

6.6 Elastic and Spring Forces

One thing we have not mentioned about the tension in a string is that, if a string is to provide a tension force, it must actually stretch a microscopic amount. The stretch is generally so small that we ignore it in problems, but it is there. It is also true that, when a tabletop provides the normal force that holds a book up, the tabletop will actually have been compressed by a microscopic amount. Both of these effects are according to the law discovered in the 17th century by British physicist Robert Hooke. Hooke's Law states that the force by which a material body resists deformation is proportional to the size of the deformation of the body. Mathematically, this is written

$$F_S = k\Delta s \tag{6.4}$$

where F_S is the spring force, Δs is the size of the deformation, and k is a constant called the "spring constant." If we want F_S to be in newtons, when Δs is in meters, the units of k will be N/m. Note that a large value for a spring constant means that a large force is achieved with a small deformation (springs with large spring constants are referred to as being "stiff.")

The Hooke's Law force arises in all material objects that are held together by molecular forces, at least as long as the deformations are small enough that they do not exceed the "elastic limit," the point beyond which the object will not completely snap back when released. In most materials, the deformation is small. However, there are objects that are designed to obey Hooke's law for large deformations. These are called springs. The material used to construct the spring and the geometry of the spring determine the spring constant. Once the spring constant is known, the force a spring provides may be calculated from knowledge of amount of stretch or compression of the spring, as in the following example.

EXAMPLE 6.9 A Simple Spring

A spring compresses by 2 cm when a force of 10 N pushes down on it. If the spring is then released and a force of 20 N is applied to it to stretch it upward, by how much will the spring extend? (The spring constant for compression of a material is generally identical to that for extension.)

ANSWER Using Hooke's law, we first solve for the spring constant, k, by using the information about the compression of the spring,

$$k = \frac{F_S}{\Delta s} = \frac{10\text{ N}}{0.02\text{ m}} = 500\text{ N/m}.$$

We then turn to the question of how far the spring extends when it is stretched by the force. Since the stretching force is 20 N, the stretch of the spring will be

$$\Delta s = \frac{F_S}{k} = \frac{20\text{ N}}{500\text{ N/m}} = 0.04\text{ m} = 4\text{ cm}.$$

6.7 Refrigerator Magnet Force

At this stage of our discussion of Newton's laws, we think we would like to have another non-contact force, besides gravity, to give us a richer palette of forces for problem-solving purposes. We will therefore consider the force between a refrigerator and a refrigerator magnet (see Figure 6.3). But please, we don't mean to give the impression that refrigerator magnets somehow represent a separate fundamental force in physics.

If you place a magnet close to a refrigerator door, it will jump toward the door, even before there is any contact between the magnet and the refrigerator. It is an attractive force, so the direction of the magnetic force exerted by the refrigerator on the magnet must point toward the refrigerator. We may also invoke Newton's third law to say that a refrigerator that exerts a magnetic force must be able to experience a magnetic force. There will thus be a magnetic force exerted by the magnet on the refrigerator that points directly toward the magnet. It is not important at this point of our discussion to understand the details of how a refrigerator exerts a magnetic force (we will take this up in Chapter 19). However, it is important to remember that a refrigerator can exert both a non-contact magnetic pull on a magnet and a contact normal-force push on the same magnet at the same time.

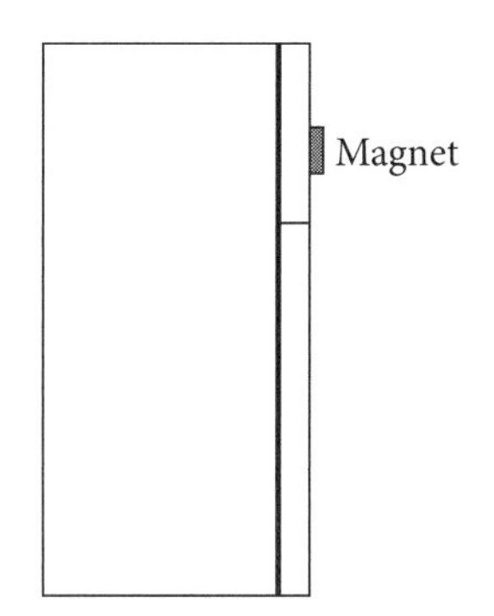

Figure 6.3 A magnet on a refrigerator.

EXAMPLE 6.10 The Refrigerator Magnet

A refrigerator magnet is firmly attached to the face of a refrigerator door, as in Figure 6.3. The magnet has a mass of 0.2 kg, and the coefficient of static friction between the magnet and the refrigerator door is $\mu_S = 0.5$. The magnet is at rest. Draw a free-body diagram for the magnet, and determine the magnitude of all the vertical forces acting on the magnet.

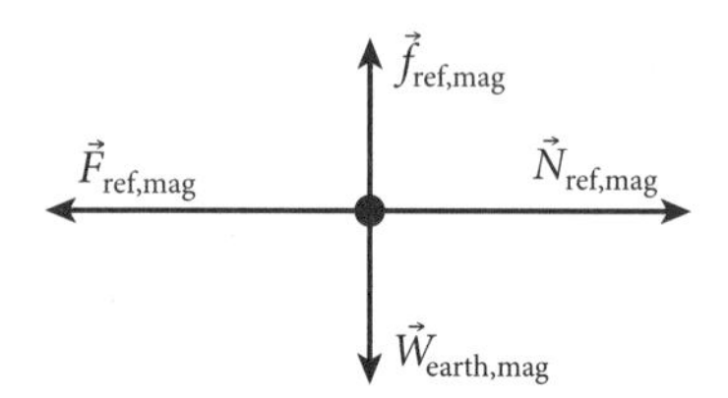

Free-body diagram for magnet

ANSWER We begin with a free-body diagram. The first force we draw is almost always the weight force exerted by the earth on the object. We then ask if there are any other non-contact forces acting on the magnet. In this case, there is a magnetic attraction between the magnet and the refrigerator. The refrigerator pulls on the magnet, and, by Newton's third law, the magnet also pulls on the refrigerator. Only one of these forces acts on the magnet, and so it goes on our free-body diagram. (The other force in this third-law pair would go on a free-body diagram of the refrigerator.) After we have drawn all relevant non-contact forces on our diagram, we look for contact forces. The only thing touching the magnet is the refrigerator. Whenever one thing touches another, there can always be a normal force that pushes the two away from each other. There may also be a friction force. In this case, we *know* there must be a static friction force, or the magnet's weight would send it sliding down the front of the refrigerator. In fact, you should note that it is the friction force, not the magnetic force, that keeps the magnet from sliding to the floor. Even the most powerful magnet cannot stay up on a frictionless refrigerator.

The weight force is easily found from $W_{\text{earth,mag}} = mg = (0.2\text{ kg})(10\text{ N/kg}) = 2\text{ N}$. Then, because the magnet is not accelerating in the vertical direction, we know from Newton's second law that the friction force must also be equal to 2 N. This is the second (and last) vertical force asked for. The question is now answered.

But let us also note that we have no way of knowing the magnitude of the horizontal forces, although we know that they must be equal in magnitude to each other (or the horizontal acceleration would not be zero). We do know that the normal force must be large enough to supply the required static frictional force on the magnet. This wasn't asked for in the question, but we can use $\mu_S = 0.5$ in the formula for the *maximum* static friction, $f^{\text{S,max}}_{\text{ref,mag}} = \mu_S N_{\text{ref,mag}}$, to say that the normal force must be *at least* 4 N for the static friction force to be 2 N. As long as the normal force of the refrigerator on the magnet is more than 4 N, the static friction force will be the 2 N required to keep the magnet from falling. This also means that the magnetic force must be more than 4 N or the normal force would not be more than 4 N. A too-weak magnet would not stay up on the refrigerator.

6.8 Apparent Weight

A person's weight is defined as the gravitational force exerted on him. It can be changed by diet or exercise or by moving to another planet. If we stand on a bathroom scale in a stationary bathroom, the reading on the scale is equal in magnitude to our true weight because the scale must push up on us as hard as the earth is pulling down on us (by Newton's second law with zero acceleration). However, if we stand on that same scale while accelerating in an elevator, we can get a reading on the scale that is different from our true weight. Newton's second law requires that a net force act in the direction of acceleration. Since there are only two forces acting on us, the gravitational force down and the scale pushing up, these two forces can no longer be equal in magnitude. Indeed, a simple application of Newton's second law allows us to predict the new reading on the scale, as the following example illustrates.

EXAMPLE 6.11 A Bathroom Scale in an Elevator

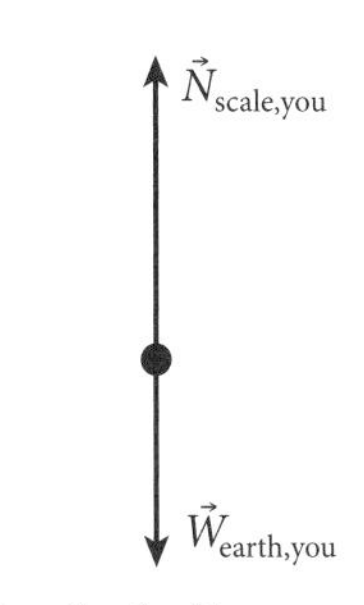

Free-body diagram for you in the elevator

When you stand on your bathroom scale in your bathroom, it reads 700 N. What would the same scale read if you stand on it while moving downward in an elevator and slowing down at a rate of 3 m/s^2, as the elevator comes to a stop?

ANSWER The elevator is moving down and slowing its speed, so your acceleration must be pointing upward. There are only two forces acting on you—the gravitational force pulling you down (your true weight) and the scale pushing up. Acceleration in the upward direction must be caused by a net force up, meaning the scale must be pushing up on your feet with a force ($N_{\text{scale,you}}$) that is greater than your weight ($W_{\text{earth,you}} = 700$ N). This upward push by the scale will be measured by the reading on the scale and is called your "apparent weight." In this case your apparent weight is larger than your true weight and you feel heavy on your feet.

We can use Newton's second law to calculate your apparent weight (the reading on the scale) in this situation. If your true weight is 700 N, then your mass must be 70 kg. Newton's second law requires that the net force needed to accelerate a mass of 70 kg at a rate of 3 m/s^2 is $F_{net} = (70\text{ kg})(3\text{ m/s}^2) = 210$ N. This means that the force pushing you up must be 210 N larger than the force pulling you down. The scale will have to push up on your feet with a force of 700 N + 210 N = 910 N. Your true weight is the 700 N force that the earth's gravity produces. Your apparent weight is the 910 N upward normal force that the scale produces.

Let us emphasize what we have said in the last example. "Apparent weight" is a contact force that keeps you from free-falling toward the center of the earth, due to the earth's gravity. It is the upward normal force by the floor on your feet or the tension force in a rope when you are hanging on to it. It is the sensation of this force that gives us the feeling of weight. If you were in a free-falling elevator (acceleration 10 m/s^2 downward) the normal force between the floor and your feet would be zero and you would have a weightless sensation. Your internal organs would not hang in the usual place; your food would not pool in the bottom of your stomach, etc. It is this strange feeling in our innards that we call "weightlessness." But you are not truly weightless, because the earth has not stopped pulling on you. Your true weight, the force of gravity by the earth on you, is still 700 N. It is only your apparent weight that goes to zero when you are free-falling.

EXAMPLE 6.12 A Fish Scale in an Elevator

A 5-kg fish is hanging from a spring scale in an elevator. If the elevator has just started downward and is increasing its speed at a rate of 2 m/s^2, find the reading on the scale.

ANSWER The elevator is moving down and speeding up, so the acceleration of the fish must be in the downward direction. In order to accelerate, there must be a net force $F_{net} = ma = (5\text{ kg})(2\text{ m/s}^2) = 10$ N acting in the downward direction. The true weight of a 5-kg fish is 50 N and the tension force pulling up on the fish must be smaller than the true weight by 10 N. The reading on the scale will be 40 N.

6.9 Centripetal Force

An object with uniform circular motion is always accelerating in a direction perpendicular to its velocity, toward the center of the circle. This acceleration is called *centripetal*, meaning "center seeking." In Section 4.5 we derived the magnitude of this acceleration and found it to be $a_{cent} = v^2/r$, where v is the speed of the object and r is the radius of the circular path. You remember that, according to Newton's second law, if there is an acceleration then there must be a net force in that same direction. So, if some object accelerates toward the center of its circular path, it can only be because it is being acted upon by a net force acting toward the center of the circular path. An object of mass m experiencing an acceleration of v^2/r must have a net center-directed centripetal force on it given by $F_{net} = F_{cent} = mv^2/r$.

Now what body is creating this force? Go back and read the last paragraph again to see if it tells you where the force comes from that acts on an object that is moving along a circular path. It doesn't, does it? That is because *centripetal force is not a force*. It is a *requirement* on the forces. Centripetal force should never appear in a free-body diagram, because it is not a force. If a rock is being swung around a circular path on the end of a string, the force that makes the rock go in a circle comes from tension in the string. Tension is a force. If a car is traveling along a circular track, the force that makes it turn instead of continue in a straight line comes from friction on the tires. (Try turning a car in a tight circle on ice.) Friction is a force. If the moon orbits the earth, the force that turns the moon's path into a circle comes from the gravity of the earth, via Newton's law of gravity (Equation 6.2). Gravity is a force. Centripetal force is not a force.

And centripetal force does not arise *because* something is moving in a circle. Circular motion does not cause centripetal force. Just the opposite. Circular motion is the result of a force satisfying the centripetal requirement. Things move in a circle because there is a real force on them (tension in a string, friction, gravity, etc.) that satisfies the centripetal requirement. Otherwise, the object will not go in a circle.

Sometimes there are many forces acting on an object that is going at uniform speed around a circular path. None of those are the centripetal force. However, if a body is moving around the circle, you may be sure that the forces on it will exactly add up to produce a net force that satisfies the centripetal requirement, $F_{net} = mv^2/r$.

By the way, if you are thinking that a section on centripetal force does not belong in a chapter about the different kinds of forces, we congratulate you. You have understood.

EXAMPLE 6.13 Test Driving a Car Around a Circle

A test driver drives a new car around a circle of radius 20 m (67 feet), laid out in a level parking lot. The fastest he is able to travel before the tires skid is 12 m/s (about 27 mi/hr). In this situation, what is the coefficient of static friction between the tires and the pavement?

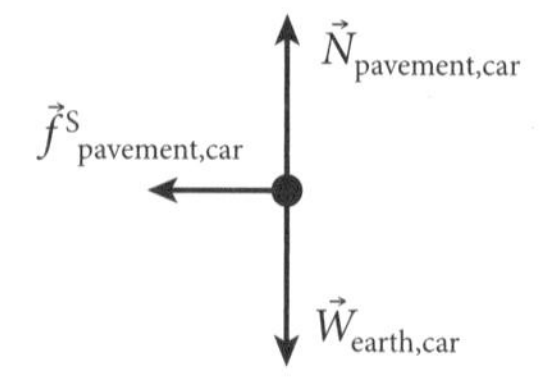

ANSWER First, let us point out why the friction in this example is static, even though the car is moving. It is because the piece of the tire that is instantaneously in contact with the road is not moving relative to the road. The tire is not skidding. As each piece of the tire comes down to contact the road, it rests on the road surface.

The free-body diagram of the car is drawn above right. The vertical forces will cancel, since the car remains on the pavement. So we can be sure that $N_{\text{pavement,car}} = W_{\text{earth,car}}$. This is one of the rare cases where Equation 6.3, the formula for the *maximum* static frictional force, gives the *actual* static frictional force. This is because we are told in the example that friction is producing the maximum force it can provide. The net force is the frictional force of the pavement on the car's tires,

$$f^{\text{S}}_{\text{pavement,car}} = \mu_{\text{S}} N_{\text{pavement,car}} = \mu_{\text{S}} W_{\text{earth,car}} = \mu_{\text{S}} mg.$$

We know the car is moving in a circle, so this net force must satisfy the centripetal requirement. Equating the frictional force and the centripetal requirement gives

$$\mu_{\text{S}} mg = m\frac{v^2}{r} \quad \Rightarrow \mu_{\text{S}} = \frac{v^2}{rg} = \frac{(12\ \text{m/s})^2}{(20\ \text{m})(10\ \text{m/s}^2)} = 0.72.$$

The static coefficient of friction between the tires and the pavement is 0.72.

EXAMPLE 6.14. The Spinning Refrigerator and the Magnet

Suppose that the refrigerator in Example 6.10 now spins about a vertical axis through its center, as shown at right. The magnet remains on the door. At the instant shown, draw a free-body diagram for the magnet. How is this free-body diagram different from the diagram in Example 6.10?

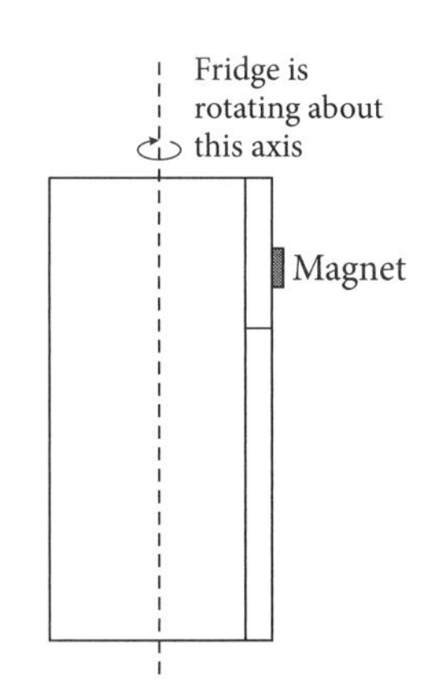

ANSWER The magnet is moving with uniform speed in a circle about the axis of rotation. This requires an acceleration toward the center of the circle, which is to the left at the instant shown in the diagram. Our free-body diagram for the magnet, shown below right, reflects this acceleration to the left by showing a longer arrow for $F_{\text{ref,mag}}$ than for $N_{\text{ref,mag}}$. Otherwise, the free-body diagram is the same as the one for Example 6.10, because there are no additional forces in the problem.

Note, in particular, that there is no centripetal force on the free-body diagram. This is because (do we need to say it again?) centripetal force is *not* a force.

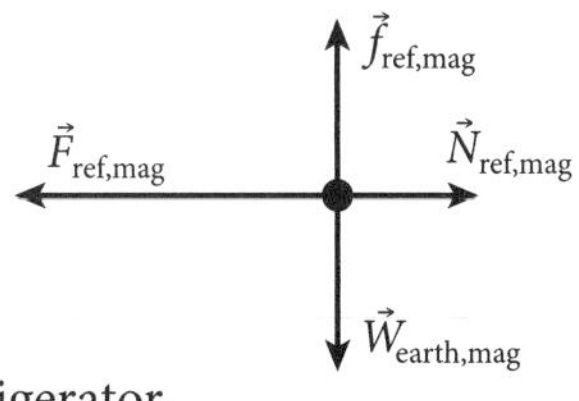

The magnetic force to the left is a function of the strength of the magnet and its proximity to the door. These are unchanged by the rotation. The only way there can be a net force to the left is if the normal force pushing to the right is reduced, as we have drawn in the diagram. This makes sense. As we increase the rate of spin, the magnet gets closer and closer to flying off the refrigerator, as the normal force goes to zero. The faster the refrigerator spins, the smaller the normal force becomes.

The vertical forces on the magnet remain unchanged. The gravitational force is still 2 N and so the frictional force must also be exactly 2 N, if the magnet is to remain stuck on the refrigerator. But you should note that the maximum force that friction can provide will now be less than it was when the refrigerator was not spinning, because the $N_{\text{ref,mag}}$ in $f^{\text{S,max}}_{\text{ref,mag}} = \mu_{\text{S}} N_{\text{ref,mag}}$ is now less than it was before. In fact, as the refrigerator spins faster and faster, the normal force will eventually drop to a value where the frictional force will not be enough to hold the magnet up, and it will slide down the door toward the floor.

6.10 Centrifugal and Other Fictitious Forces

Back in Section 5.1, we discussed Newton's first law and discovered that there exist certain reference frames in the universe that are inertial frames, frames in which objects with no forces acting on them move at constant speed in a straight line. We also discussed our common experience that there also exist frames of reference (like the inside of an accelerating car) where Newton's first law does not work. A cup of coffee sitting on the dashboard will topple over backward with no backward force acting on it from anything. A similar effect occurs when a car is going in a circular path. If you are making a hard left turn, your coffee cup may spill over to the right.

The cleanest thing to do, if you want to understand the mechanics of coffee cups sitting on dashboards, is to never analyze anything using a non-inertial reference frame. It is possible, however, to work in one of these improper reference frames if you pretend that there is another force acting on everything. We call these *fictitious forces*, because they do not arise from any physical object, but only from our choice of a reference frame. We also call them *inertial forces*, since they act to get things back to an inertial frame. You can analyze motion in the frame of an accelerating car, one accelerating at a rate a_{frame}, if you include an inertial force $\vec{F}_{\text{fict}} = -m\vec{a}_{\text{frame}}$ that acts on every body of mass m. Or you can analyze motion in a car going around a circular path of radius r at speed v if you include an inertial force of magnitude $F_{\text{fict}} = mv^2/r$, acting outward from the center of the circle on a body of mass m. This latter force is called a *centrifugal force*, because it flees (fuga) the center.

Our only reason for telling you all this is so you will know where the idea of centrifugal force comes from. But it's a bad idea, if you want to do physics problems correctly. The best practice is to work in an inertial frame and analyze real forces, the ones considered in this chapter. This way, neither centrifugal force nor centripetal force will ever appear in your force diagrams—the first because you are working in an inertial frame where fictitious forces do not appear and the second because it isn't a force, but a requirement on the forces.

6.11 Summary

In this chapter, you should have learned the following:

- The force of gravity acting on an object of mass m due to a mass M a distance r away is $F = GmM/r^2$, where G is Newton's gravitational constant, given as $G = 6.67 \times 10^{-11}$ Nm^2/kg^2.
- The force of gravity on an object of mass m near the surface of the earth is $\vec{W} = m\vec{g}$, where $\vec{g}$ is the strength of the gravitational field.
- The normal force and the static frictional force are forces of constraint. Their value is whatever it takes to constrain the motion of an object to be what we know it will be. For static friction, there is a maximum force available, given by $f^S \leq \mu_S N$.
- The force of kinetic friction may be calculated once the normal force between the surfaces is known. It is $f^K = \mu_K N$.
- Tension in a string must either be given in the problem or calculated as a consequence of the motion of the system. There is no formula for tension in a string.
- Problems on inclined planes typically require resolving vectors into components. The most useful choice of component directions is usually along the incline and perpendicular to it.

- Centripetal and centrifugal forces are not real forces and should not appear on free-body diagrams. However, whenever there is uniform circular motion taking place, you may be sure that the sum of all the forces that do appear in the diagram will satisfy the centripetal requirement, $F_{\text{cent}} = mv^2/r$.

CHAPTER FORMULAS

Gravity: $F_g = G\frac{mM}{r^2}$ and $\vec{W} = m\vec{g}$

Centripetal requirement: $F_{\text{cent}} = m\frac{v^2}{r}$

Kinetic friction: $f^{\text{K}} = \mu_{\text{K}}N$

Static friction: $f^{\text{S,max}} = \mu_{\text{S}}N$

Spring force: $F_{\text{S}} = k\Delta s$

PROBLEMS

6.1 *1. Calculate the gravitational force between your car (mass 1400 kg) and a tour bus (mass 3000 kg), when your car is 20 m from the center of the tour bus. With this magnitude of force, are you going to be able to claim that the accident wasn't your fault because the tour bus pulled you into it with its gravity?

*2. In Cavendish's laboratory measurement of the value of G, a 2-inch diameter lead sphere with mass 0.73 kg was suspended next to a much larger 160-kg stationary lead sphere so that the distance between their centers was 23.3 cm. What would be the size of the force between these two spheres?

**3. The earth has a mass $m_{\text{earth}} = 5.97 \times 10^{24}$ kg and the moon's mass is $m_{\text{moon}} = 7.35 \times 10^{22}$ kg. The earth–moon distance is 3.84×10^8 m.

a) Find the gravity force acting between the earth and the moon.

b) What is the resulting gravitational acceleration of the moon toward the earth? (This acceleration, provided by gravity, is the centripetal acceleration that curves the moon's trajectory into its circular orbit.)

c) What is the acceleration of the earth due to the moon's gravity? (This acceleration arising from the force of gravity by the moon on the earth causes the earth's center to orbit a point 1700 km below the earth's surface.)

**4. Jupiter orbits the sun once every 11.86 years (3.743×10^8 s) in a nearly circular orbit of radius 7.783×10^{11} m and at nearly constant speed.

a) Find Jupiter's speed in m/s and its acceleration toward the sun in m/s^2.

b) Use the fact that this acceleration must be provided by the gravitational pull of the sun to determine the sun's mass. Use $G = 6.674 \times 10^{-11}$ Nm^2/kg^2.

**5. The mass of the sun is 2.0×10^{30} kg and the mass of the earth is 6.0×10^{24} kg. The earth orbits the sun at a nearly constant distance of 1.5×10^{11}m.

a) Find the gravitational force of the sun on the earth and calculate the earth's resulting acceleration toward the sun.

b) Mars has a mass of 6.4×10^{23} kg. At its closest approach to the earth, it is only 7.8×10^{10} km from earth. What is the gravitational force of Mars on the earth at this point, and what acceleration of the earth toward Mars will this produce?

6.2 *6. What is the normal force of a bathroom scale on the feet of a 50-kg woman standing quietly on the scale? In what direction does the normal force act?

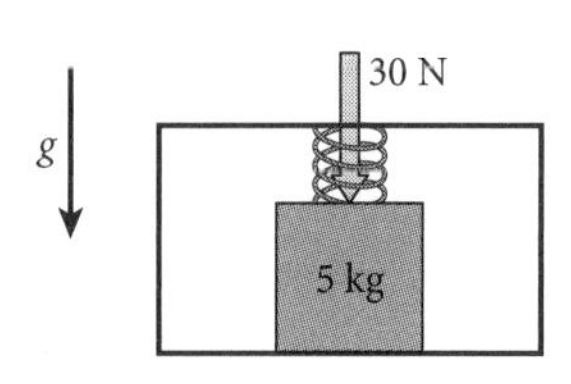

*7. A 5-kg block is placed inside a closed box and a spring above the block is pressing down on it with a force of 30 N, as shown at left. What is the normal force by the block on the bottom of the box?

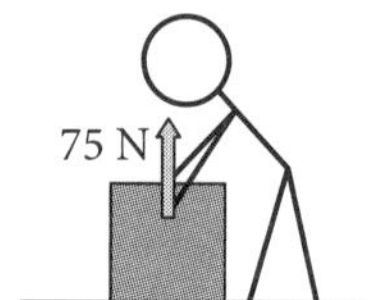

*8. A kindergartner pulls up on a 10-kg crate of alphabet blocks with a force of 75 newtons, as shown at right, but he is not able to lift the crate off the ground. What is the normal force of the floor on the crate while the kindergartner is pulling up on it?

6.3 *9. A 1.6-kg wood block, with a kinetic coefficient of friction $\mu_K = 0.75$ m between itself and a tabletop, is sliding across the level tabletop. What is the frictional force on the block by the tabletop?

*10. A 10-kg block rests on a tabletop. It is found that a horizontal force of 70 N is required in order to move the block. What is the coefficient of static friction between the block and the tabletop?

**11. A physicist is trying in vain to push a large safe filled with gold coins across his garage floor. (This isn't very realistic, is it? What would a physicist be doing with a safe full of gold coins?) The mass of the safe, with the gold, is 350 kg. The coefficient of static friction between the safe and the floor is 0.60. The physicist is pushing with a horizontal force of 900 N but the safe will not budge. What is the magnitude of the frictional force by the floor? Explain your reasoning.

**12. A girl pushes horizontally on a 100-kg crate, as shown in the two figures below. The coefficient of kinetic friction between the crate and the floor is 0.2; the coefficient of static friction is 0.4.

a) For each of the two situations described in the figures, indicate whether the frictional force is static or kinetic.

b) In each case, find the magnitude of the frictional force.

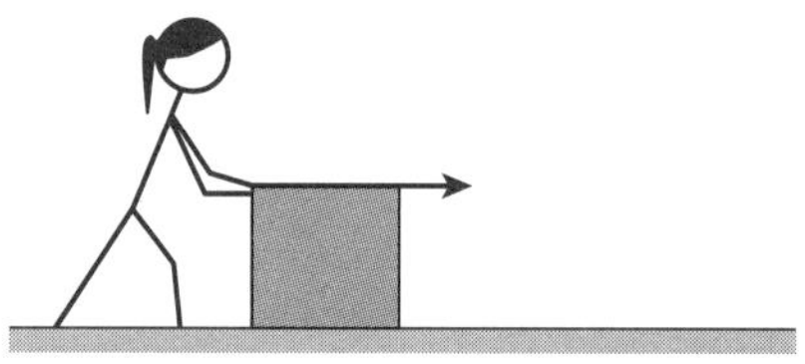
Case 1: The girl is pushing with a force of 300 N, and the crate does not move.

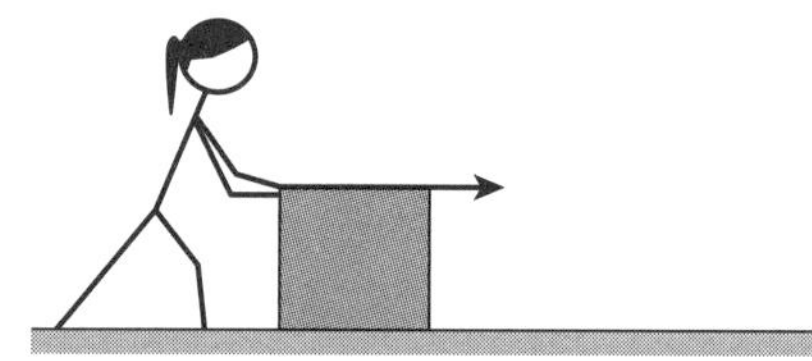
Case 2: The girl is pushing with a force of 300 N, and the crate is moving along the floor.

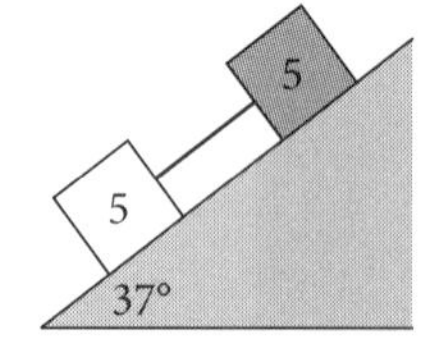

**13. Two 5-kg blocks, a white block and a red block, are made of different materials. The coefficient of kinetic friction between the white block and the slope is 0.7. When they are connected by a string on a 37° slope, as shown, they are seen to slide down the slope at constant speed. What is the coefficient of friction between the slope and the red block?

***14. A dog is standing on a frozen lake (static coefficient of friction 0.05) when he sees a squirrel 10 m away. How long will it take him to reach the squirrel if he is careful that his paws do not slip on the ice as he accelerates at a constant rate?

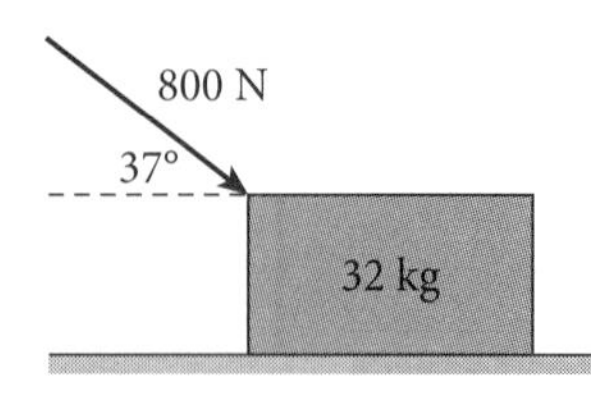

**15. Jeff is pushing a 32-kg crate across the floor, but he is making life difficult for himself by applying his 800 N force at an angle of 37° below the horizontal. The coefficient of kinetic friction between the crate and the floor is 0.50.

a) Draw a free-body diagram for the crate.

b) Find the magnitude of the normal force by the floor on the block.

c) Find the acceleration (magnitude and direction) of the crate.

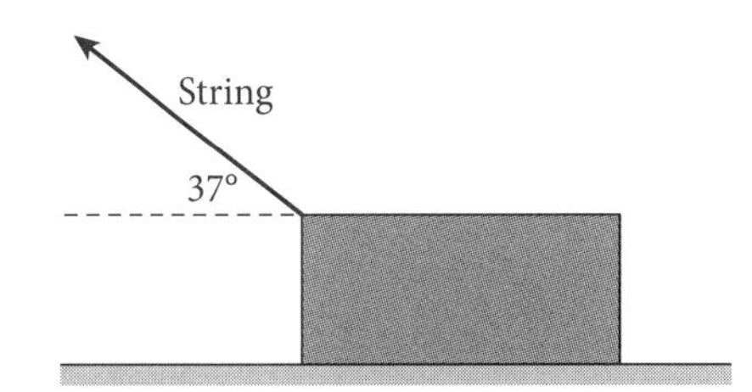

** **16.** Ron pulls with a string on a wooden block of mass $m = 80$ kg. The tension in the string is 300 N and the string makes an angle of 37° with respect to the horizontal, as shown. The floor is very rough ($\mu_K = 0.35$ and $\mu_S = 0.75$) and the block does not move.

a) Draw a free-body diagram for the block.

b) Find the magnitude of the normal force exerted by the floor on the block.

c) Find the magnitude of the frictional force exerted by the floor on the block.

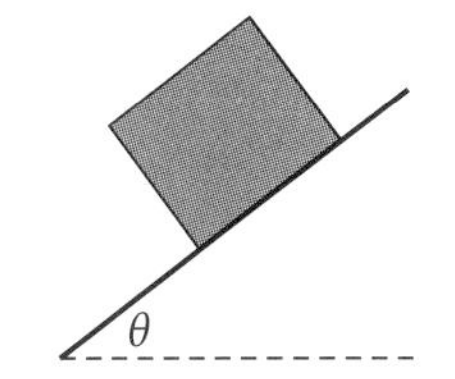

6.4 ** **17^f^.** [Remember that a problem with an f superscript is a problem where we are trying to find a formula, not a numerical answer.] An experimenter places a block on an inclined plane and adjusts the angle θ until the block is just about to slide down the ramp. Draw a free-body diagram for the block and show that the coefficient of static friction between the block and the ramp is $\mu_S = \tan\theta$.

** **18.** A block slides down an inclined ramp, speeding up as it goes. The ramp is not frictionless ($\mu_K = 0.50$). The ramp makes an angle of 53° with respect to the horizontal, as shown at right.

a) Draw a free-body diagram for the block. Label your forces using the method explained on page 85.

b) If the magnitude of the normal force is 90 N, find the mass of the block.

c) Find the acceleration (magnitude and direction) of the block.

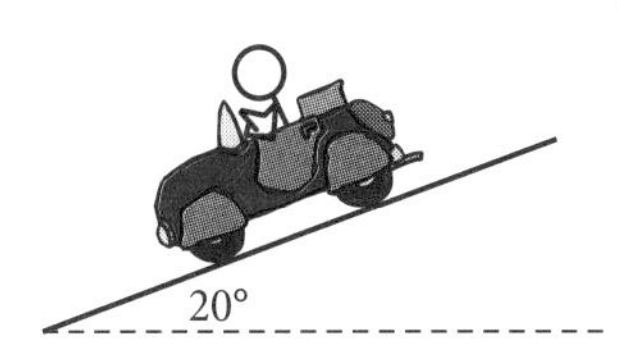

** **19.** Ron is speeding down a steep mountain pass out of Butte, going 30 m/s (about 67 mi/hr). He suddenly sees an elk on the road ahead and slams on his brakes, locking all the wheels while the car slides to a stop. The mass of the car (with Ron in it) is 1000 kg. The coefficient of kinetic friction between the tires and the road is $\mu_K = 0.80$. The road makes an angle of 20° with the horizontal.

a) What is the weight of the car (including Ron)?

b) Find the total friction force acting on the car. (You may treat the car as if it is a block sliding down a ramp, i.e., don't treat the four wheels separately.)

c) Find the acceleration (magnitude and direction) of the car after the wheels lock.

d) How far will the car slide before it comes to a stop?

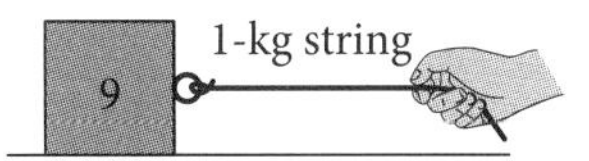
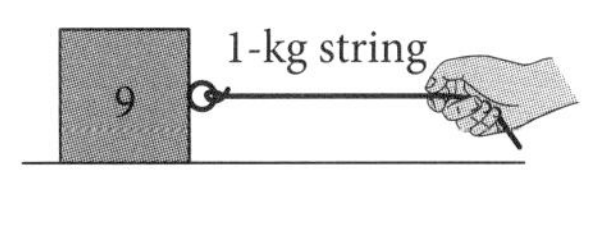

6.5 ** **20.** In Section 6.5, we explained that the tension in a massless string will be the same all along the string. In this problem we want to consider a string with mass, for which the tension will change from one end to the other. A 9-kg block has a 1-kg string attached to it, and the string is pulled, giving the block and the string an acceleration of 2 m/s^2.

a) What is the tension in the end of the string that is being pulled?

b) What is the tension in the end of the string that is attached to the block?

6.6 ** **21.** A box is placed atop a spring of spring constant $k = 250$ N/m, compressing the spring by 2 cm. What is the mass of the box?

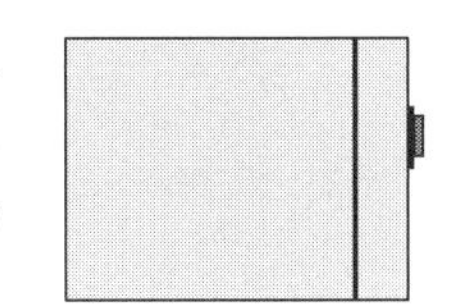

6.7 ** **22.** A paper business card of mass 25 g is held to the face of the door of an office-sized refrigerator (as shown) by a magnet of mass 50 g. The business card and magnet are at rest. Draw a free-body diagram for the business card and a separate free-body diagram for the magnet. Use the force-naming convention from page 85 to label your forces.

6.8 * **23.** In a clever attempt to lose a little weight, Edna places a bathroom scale in the elevator in her building and always weighs herself in the morning when the elevator is accelerating downward at 1 m/s^2. If her mass is 50 kg, what will the scale give as her weight in newtons during the acceleration? Will this be her true weight or her apparent weight?

** **24.** A safe filled with gold (total mass = 875 kg) is sitting on a bathroom scale in an elevator. The elevator is moving up and speeding up by 2 m/s^2. What is the reading on the scale (in newtons)?

** **25.** Zola (the wonder dog) is told to sit and stay on a bathroom scale that reads 150 newtons with her on it. But she soon gets bored with sitting on the scale and leaps straight up into the air with an acceleration of magnitude 2 m/s^2.

a) What is Zola's mass?

b) What is the reading on the scale as she launches herself upward?

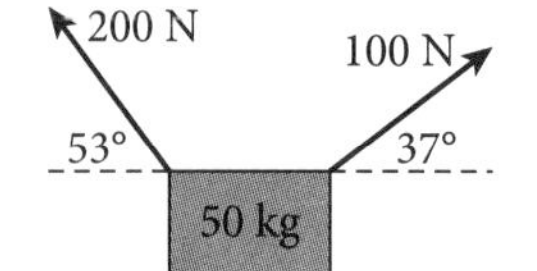

** **26.** Two strings are pulling on a 50-kg block as shown. The block does not move. The coefficient of static friction between the floor and the block is $\mu_S = 0.6$.

a) Find the magnitude of the normal force exerted by the floor on the block.

b) Find the magnitude of the static friction force by the floor on the block.

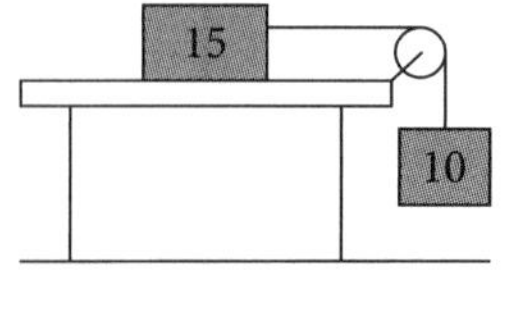

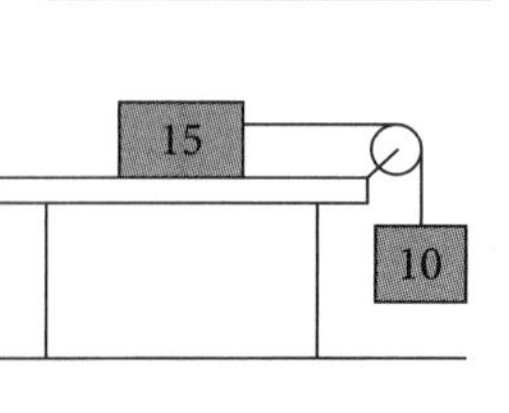

** **27.** In the diagram at left a 15-kg block slides without friction on a smooth tabletop. It is connected by a massless string that passes over a massless frictionless pulley to a 10-kg block that is hanging down. The 15-kg block is initially held in place and then released. What will be its acceleration after it is released?

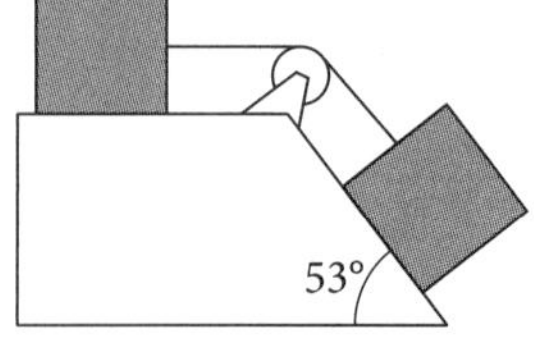

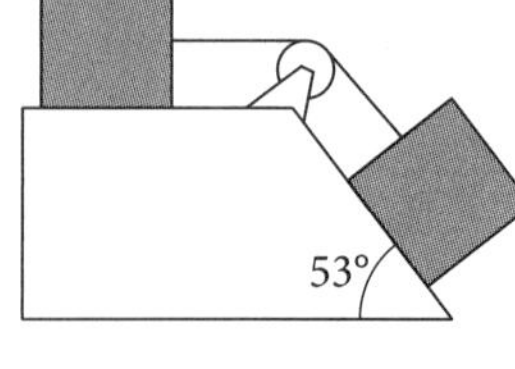

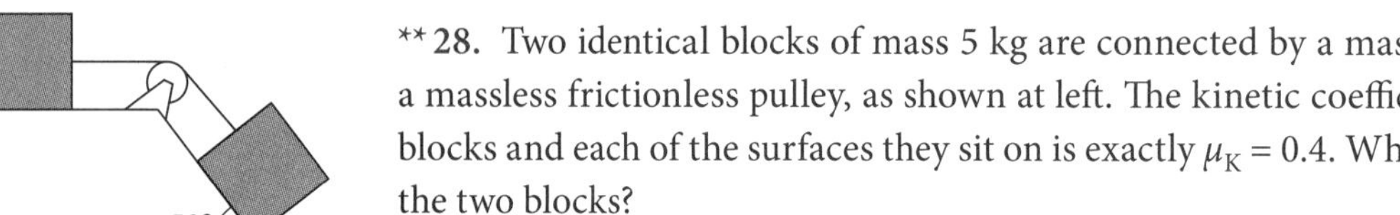

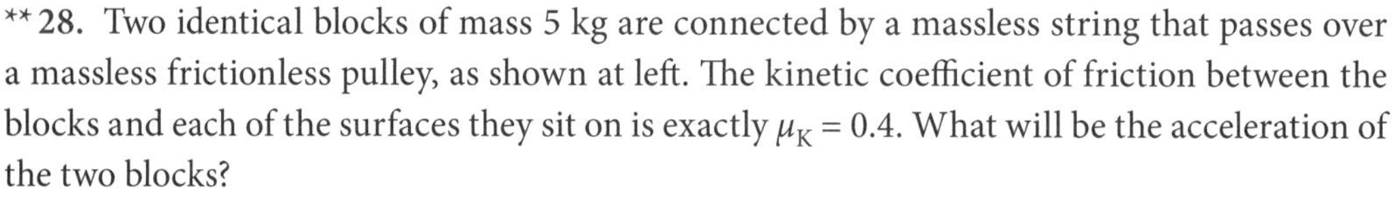

** **28.** Two identical blocks of mass 5 kg are connected by a massless string that passes over a massless frictionless pulley, as shown at left. The kinetic coefficient of friction between the blocks and each of the surfaces they sit on is exactly $\mu_K = 0.4$. What will be the acceleration of the two blocks?

6.9 ** **29.** A cylindrical room is rotating fast enough that two small blocks stacked against the wall do not drop. The mass of block A is 4 kg and that of block B is 3 kg.

The cylinder is rotating about this axis

A B

$m_A = 4$ kg
$m_B = 3$ kg

a) Draw a diagram of the wall and of blocks A and B. Indicate the direction of the acceleration of block B at the instant shown in the figure. If it is zero, state that explicitly.

b) Draw separate free-body diagrams for blocks A and B and label the forces as described on page 85.

c) Identify any third-law companion forces on your diagrams using tick marks like those used in Example 6.1.

d) Rank the magnitudes of all the horizontal forces that you identified above in order from largest to smallest. Explain your reasoning.

e) Determine the magnitude of each of the vertical forces on block A. If it is not possible to determine one of these, explain why not.

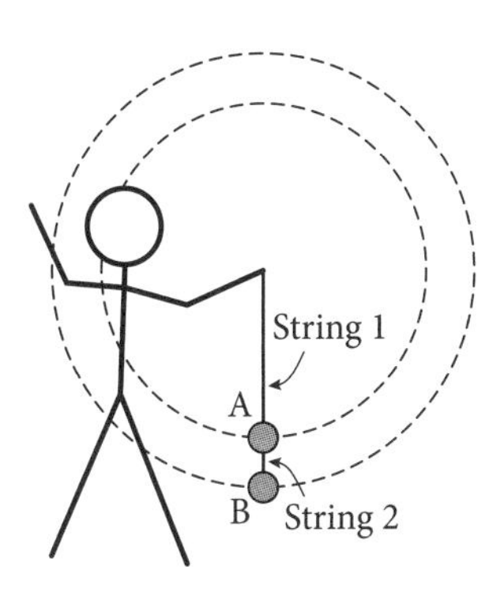

** 30. Jeff has invented a new toy that consists of two identical balls connected by two massless strings. Jeff swings the balls at constant speed in a vertical circle. In every part of the problem, consider the instant in time when the toy is in the vertical orientation shown.

a) Draw a diagram and indicate the direction of the acceleration of ball A. If the acceleration is zero, state that explicitly.

b) Is the tension in String 2 greater than, equal to, or less than the tension in String 1? Use a free-body diagram and one of Newton's laws to explain your reasoning.

c) If the mass of each ball is 1½ kg, and ball B is moving 3 m/s at the instant shown on a circular path of radius 1 m, find the tension in String 2.

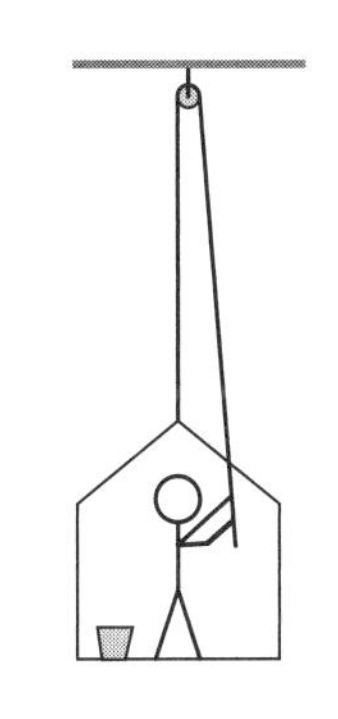

** 31. A window-washer stands on a platform that is connected at the top by a rope that goes up to a pulley and then back down to attach to the platform. The platform and the window-washer together have a mass of 120 kg (and thus a weight of 1200 newtons).

a) Draw a free-body diagram for a body that includes the platform and the window-washer.

b) What downward force should the window-washer apply to the rope if he wants to accelerate himself upward at 1 m/s^2?

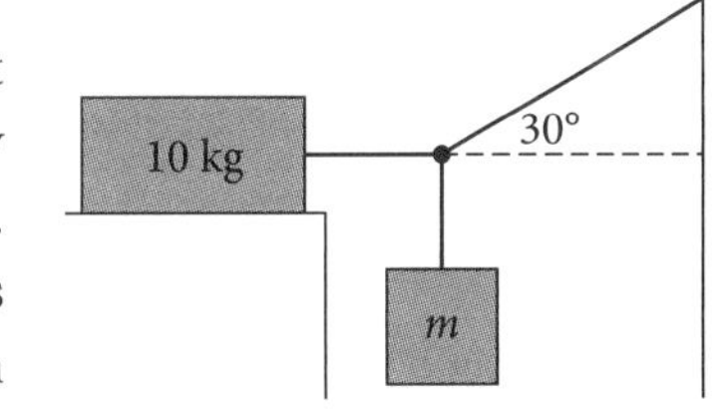

** 32. A 10-kg block rests on a table for which the coefficient of static friction between the block and the tabletop is exactly 0.4. The block is connected to a knot by a horizontal string. Another block hangs from the knot, and another string goes up at an angle 30° to a wall, as shown at right. How large can the mass m be before the 10-kg block begins to move to the right?

*** 33. A physicist has finally invented a shop dolly with "frictionless wheels," and is using the dolly to move a large safe full of gold. The mass of the safe (including gold) is 875 kg. The mass of the shop dolly (including wheels) is 5 kg. The coefficient of static friction between the safe and the dolly is large enough ($\mu_S = 1.1$) that the safe does not slip relative to the dolly. The physicist pushes on the safe with enough force to cause it to accelerate at 0.4 m/s^2. Find the friction force between the safe and the dolly.

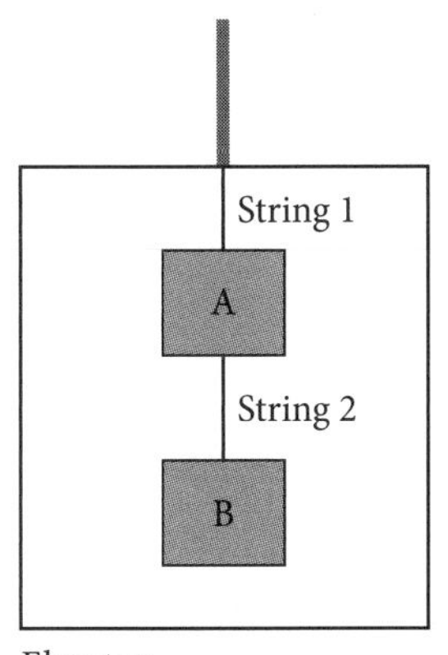

** 34. Two blocks, A and B, are suspended by massless strings inside an elevator, as shown. The mass of block B is greater than that of block A. The elevator moves downward *with increasing speed*. The strings remain taut and the blocks never swing from side to side.

a) Draw two squares, one for each block, and draw arrows showing the direction of the acceleration of each block. If the acceleration of either block is zero, state that explicitly.

b) Is the magnitude of the acceleration of block B *greater than, less than,* or *equal to* the magnitude of the acceleration of block A? Explain.

c) Is the magnitude of the net force exerted on block B *greater than, less than,* or *equal to* the net force exerted on block A? Explain your reasoning.

d) Is the magnitude of the weight of block B *greater than, less than,* or *equal to* the magnitude of the downward tension force on block A? Explain.

e) If the mass of block B is 3 kg and the magnitude of the acceleration of the elevator is 4 m/s^2, find the apparent weight of block B.

** **35.** In the figure below, an 8-kg block rests on top of a 12-kg block, and the 8-kg block is tied to the wall. The coefficient of kinetic friction is exactly 0.4 between the 8-kg block and the 12-kg block, and the coefficient of friction between the 12-kg block and the tabletop is exactly 0.6. What force must be applied to the 12-kg block in order to accelerate it to the left at 3 m/s^2?

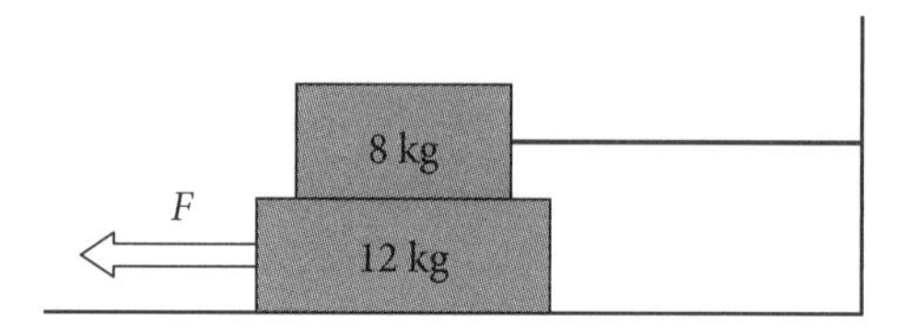

*** **36.** In the figure below, a massless string connects a 1-kg block to a 2-kg block, by passing through a massless frictionless pulley, as shown. The coefficient of kinetic friction between the two blocks is exactly 0.6 and that between the 2-kg block and the tabletop on which it rests is exactly 0.5. Find the acceleration of the two blocks when a force of 30 N is applied to the 2-kg block, as shown.

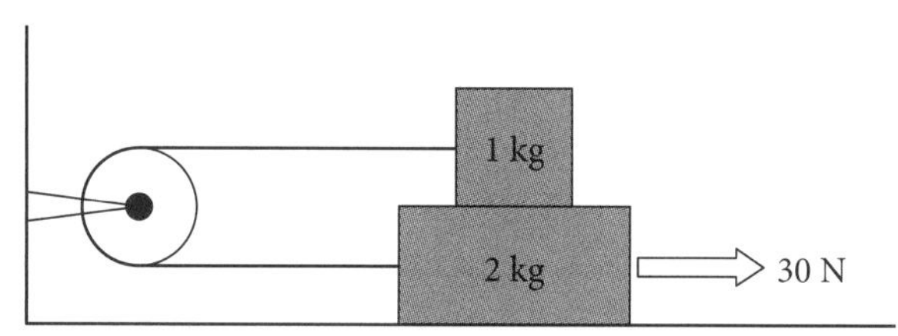

** **37.** To test the ability of a sports car to turn a corner at high speed, the car is driven around a flat circular pavement track called a skidpad. The car is driven at higher and higher speeds, until it starts to slip. At this point, the speed is noted and the centripetal acceleration calculated. On a certain skidpad of radius 103 ft (31.4 m), a GTI's top speed around the circle was found to be 38 mi/hr (17.0 m/s).

a) What was the centripetal acceleration of the GTI? What fraction is this of the gravitational acceleration, $g = 9.8$ m/s^2? (That is, how many g's is this?)

b) What is the static coefficient of friction between the pavement and the tires of the GTI? (Use $g = 9.8$ m/s^2 throughout.)

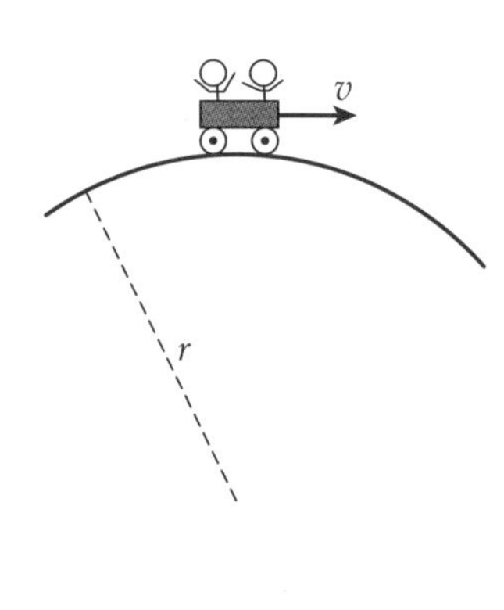

** **38**[f]. A roller coaster of mass m (when filled with passengers) goes over the crest of a circular section of track at speed v. The radius of the circular loop is r.

a) Find a formula for the normal force of the track on the roller coaster, as a function of m, g, v, and r.

b) Find a formula for the speed at which the normal force goes to zero. (By the way, at this speed, the roller coaster will start to lift off of the track and the passengers will feel completely weightless.)

c) Let's put some numbers in the formula. If the curved track at the top of a crest is 20 m (about 60 ft) in radius, find the speed in m/s and mi/hr at which the car would lift off the track.

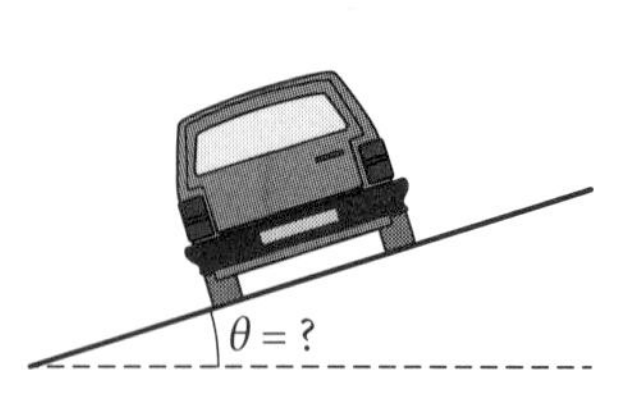

*** **39.** An engineer wants to bank the road on a tight circular turn in which the radius of the turn is 120 m. The highway is to be designed so that a car going at a speed 33 m/s (about 74 mi/hr) will round the turn safely even if the road is icy (so that there is no friction between the road and the tires). What should be the angle of the incline for the road?

*** **40.** Engineers want to create an "artificial gravity" by dividing their spacecraft into two modules, letting each out on 40-m cables, and then by spinning the entire spacecraft around the center point. Since an astronaut in the module is moving in a circle at constant speed, the normal force of the floor of the module pushing upward on him must satisfy the centripetal requirement. The rotational speed is designed so that this normal force (the astronaut's apparent weight) is equal to his usual weight on earth, using $g = 10$ N/kg. Treat each module as if its mass, with the astronaut in it, is concentrated at the floor of each module.

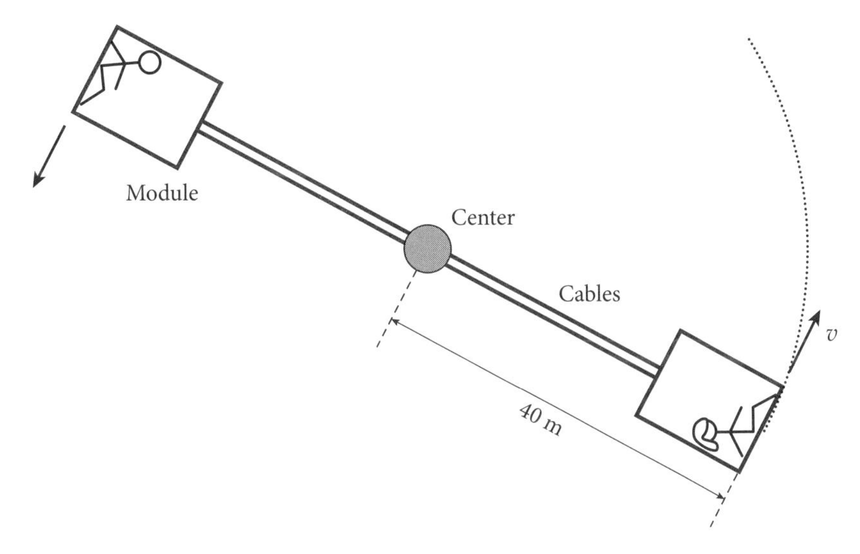

a) What must be the speed of each module around the center of the spacecraft so that the centripetal force requirement produces the right normal force?

b) One entire module (with the astronaut in it) has a mass of 800 kg. The spring constant of the cables is 4×10^6 N/m (a reasonable spring constant for steel cables). Calculate the tension in the cables and find the amount by which the cables stretch when the spacecraft is spinning.

7

ENERGY

Back when we first introduced Newton's laws of motion, on page 81, we noted that Newton defined something called the "quantity of motion" as the product of mass and velocity, $\vec{p} = m\vec{v}$, a vector we now call "momentum." In Newton's time, however, there was *another* candidate for how to measure how much motion a body had. This was a scalar quantity called the "kinetic energy," defined via the formula

$$\text{KE} = \frac{1}{2}mv^2, \tag{7.1}$$

where m is the mass of the object and v is its speed. Even though Newton's use of momentum was clearly the right choice for his laws of mechanics, the idea of energy has actually turned out to be a much more far-reaching concept, one that extends beyond mechanics to electricity, chemistry, thermodynamics, and many other fields. In this chapter we will concentrate on mechanical energy and see how it may be used to provide simple shortcuts to solving many problems.

7.1 Forms of Energy

The key to using the concept of energy to solve problems is that there are more forms of energy than the kinetic energy defined above. In particular, there are many kinds of "potential energy." In this chapter, we consider two kinds of potential energy (one additional kind, electric potential energy, will come up in Chapter 17). So, for the moment, you will need to know about only three forms of energy. They are:

- *Kinetic Energy:* This is the energy associated with motion. A bowling ball sliding down the bowling alley at 7 m/s has more energy than one at 5 m/s. To calculate an object's kinetic energy, we use Equation 7.1. Note that kinetic energy is a scalar quantity. It is just a number; it does not have a direction and it can never be negative. For a system of more than one object, the total kinetic energy is just the sum of the kinetic energies of the individual objects (just add up the numbers). It is also important to note that, if you know an object's mass, knowing its kinetic energy is equivalent to knowing its speed and vice versa.
- *Gravitational Potential Energy:* This is the energy stored in the gravitational interaction between an object and the earth. It arises when the object is at some elevation h above the surface of the earth. As long as this elevation is small compared to the radius of the earth, the value of the stored potential energy can be written as

$$\text{PE}_\text{G} = mgh, \tag{7.2}$$

where m is the object's mass, g is the gravitational field strength (normally we use 10 N/kg), and h is the object's height. Note that gravitational potential energy can be either positive or negative, depending on whether h is positive or negative. You can choose to measure h from any point you like. (You can even use the ceiling and then all values are negative.) The important thing is to pick a place and stick with it. It can be helpful to label this point on your diagram of the problem. (We normally recommend the lowest point in the problem, so that all gravitational potential energies are positive, but it's up to you.)

- *Spring Potential Energy:* When you either stretch or compress a spring by an amount Δs, you store an amount of energy in the spring equal to

$$\mathrm{PE_S} = \frac{1}{2}k\Delta s^2, \tag{7.3}$$

where k is Hooke's constant for the spring (discussed in Section 6.6 on page 106). Note that spring potential energy is always positive.

If you check, you will find that each of these prescriptions for calculating energy results in an answer that has units of newton · meters (or kg · m^2/s^2). We call this unit the joule (J). Energy is measured in joules.

The thing that makes energy such a useful concept is that the various forms of energy can be converted from one to another. Thus, potential energy is called potential energy because it has the "potential" of becoming kinetic energy, and kinetic energy can be converted back into potential energy. Under appropriate circumstances, the *mechanical energy* of a system, defined (for the needs of this chapter) as

$$E = \mathrm{KE} + \mathrm{PE_G} + \mathrm{PE_S}, \tag{7.4}$$

always stays the same. KE can become $\mathrm{PE_G}$; $\mathrm{PE_G}$ can become $\mathrm{PE_S}$; but the sum of all three will always give the same number. Physicists have a term they apply to quantities that always stay the same. They say that the quantity is *conserved*. This is an unfortunate choice of language, because "conserve" is a perfectly good English word and it doesn't mean that something stays the same. It means to save something so that you will have it later. Thus you can appropriately be reminded to conserve energy by turning out the lights when you leave a room, because of what the English word means. But a physicist might reply that energy is always conserved whether he turns out the light or not, because he means something different by the term "conserve." So remember that this is a physics book, and when we say that something is conserved we mean that its value stays constant.

You may have noticed that we were being a little cagey in the text leading up to Equation 7.4 by saying "under appropriate circumstances." Let us be more precise. In this chapter, the circumstances in which the mechanical energy defined in Equation 7.4 is conserved will be when: 1) the only forces acting on an object are gravity or springs, or 2) any other kind of force acting on the object acts only in a direction perpendicular to the object's motion. We will see why the second item is required later in this chapter.

Energy is a concept invented by scientists to aid in problem solving and, perhaps more importantly, to provide a powerful way to think about physical systems. When the amount of mechanical energy remains constant, we may think of it like a fluid whose volume always remains the same, even though it may be poured into different buckets. At any point in the motion, a system's mechanical energy can be thought of as the sum of the energies contained in each of the buckets—kinetic energy, gravitational potential energy, and spring potential energy. If one of the buckets stays at zero throughout the motion (e.g., a problem without any springs in it or one that all takes place at constant elevation) then we normally just ignore that bucket. We figure that the best explanation at this point would be to solve a few examples.

EXAMPLE 7.1 A Suction-Cup Dart Gun

A 50-gram suction-cup dart is loaded into a spring gun, compressing the spring by 2 cm. The spring has a spring constant $k = 8000$ N/m. How fast will the dart be moving as it leaves the gun? (Assume that the spring's mass is negligible compared to that of the dart, so we need not consider the kinetic energy of the front of the spring as it expels the dart.)

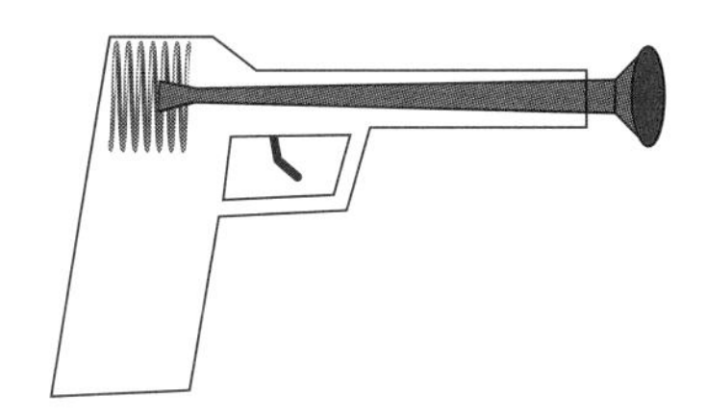

ANSWER We consider the system composed of the spring and the dart. There are two kinds of energy in this problem. The dart is free to move, so it can have kinetic energy, and there is a spring that can be compressed so as to acquire potential energy. The gun is fired horizontally, so there is no change in gravitational potential energy. We will clearly need two buckets for this problem. We take the initial point to be with the spring cocked and the dart ready to be fired. At this point, the kinetic energy bucket is empty, since there is nothing moving, and the spring potential energy bucket is full since the spring is compressed. Its initial energy is

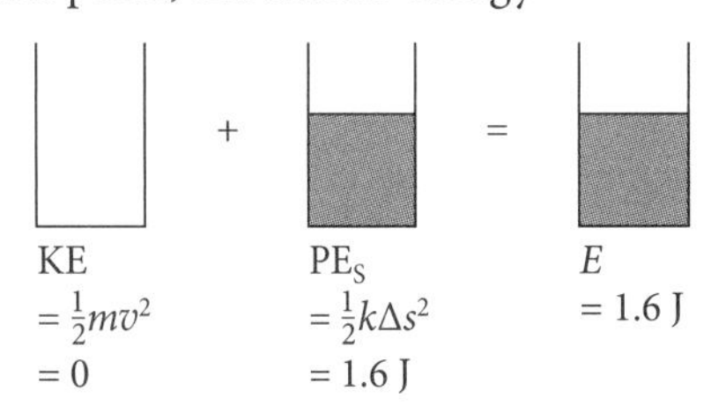

$$PE_S = \frac{1}{2}k\Delta s^2 = \frac{1}{2}(8000 \text{ N/m})(0.02 \text{ m})^2 = 1.6 \text{ J},$$

so we have a situation like the one pictured at right.

We take the final point to be the point where the dart has just left the spring, which is now uncompressed. At this point, the bucket for potential energy is empty, since the spring is uncompressed, and the bucket for the kinetic energy is full, since the dart is moving. The buckets would look as at right.

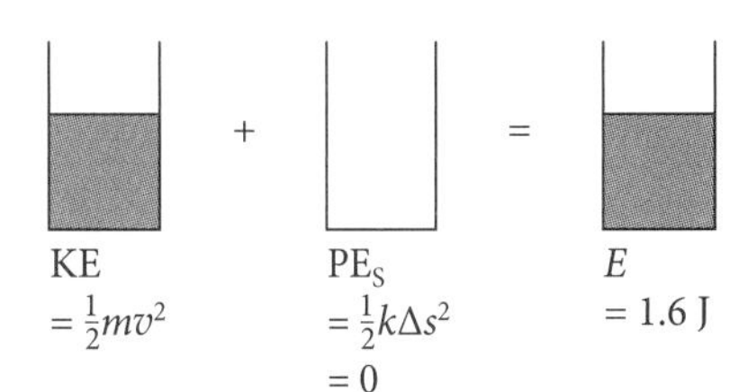

Energy is conserved, so the initial energy must equal the final energy. We write this as

$$\frac{1}{2}mv^2 = 1.6 \text{ J}.$$

Solving for the unknown v, we find

$$v^2 = \frac{2(1.6 \text{ J})}{m} = \frac{3.2 \text{ J}}{0.05 \text{ kg}} = 64 \text{ m}^2/\text{s}^2.$$

The speed of the dart as it leaves the gun is $v = \sqrt{64 \text{ m}^2/\text{s}^2} = 8$ m/s.

EXAMPLE 7.2 The Spring, the Block, and the Ramp—I

The 5.0-kg block in the figure at right is pushed down against a spring with spring constant $k = 2000$ N/m and with negligible mass. The spring compresses 0.50 m. The block is held in place briefly and then released. The spring pushes on the block so that the block slides up the 37° slope. Assume that the force of friction between the slope and the block is negligible. What is the maximum distance the block slides up the ramp from its compressed-spring position?

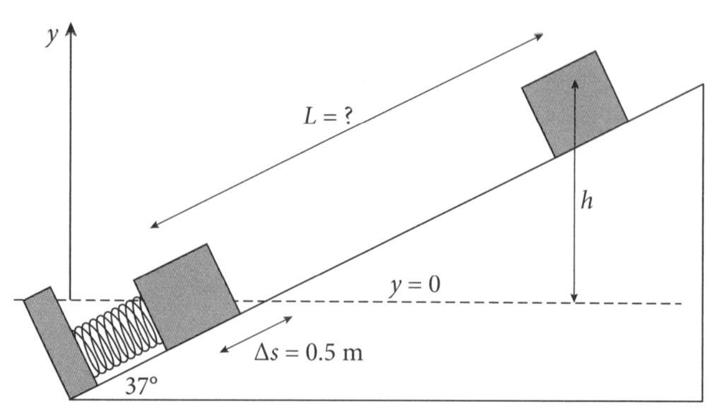

ANSWER There is no significant force of friction in this problem, and the normal force of the ramp on the block acts perpendicular to the motion of the block, so this is one of the cases where total energy is conserved. Before the block is released, it is not moving, so it initially has nothing in the kinetic energy bucket. The spring is compressed to begin with, so there is energy in the spring potential energy bucket. We choose to measure gravitational potential energy from the lowest position of the block, the dashed $y = 0$ line in the picture, so there is initially no energy in the gravitational potential energy bucket. The initial energy situation is shown in this first set of buckets. Using SI units for all quantities, we have calculated the energies in this picture and found a mechanical energy of $E = 250$ J. Since mechanical energy is conserved, this will be the energy at all points in the motion.

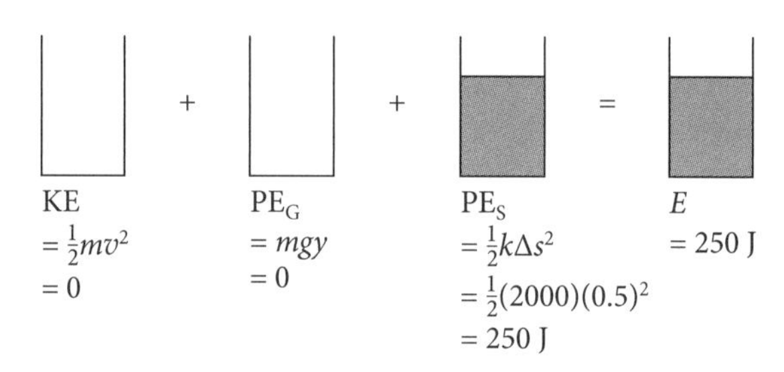

Once released, the block rises in elevation and speeds up due to the spring. As the spring potential energy bucket empties, the kinetic energy and the gravitational potential energy buckets start to fill. At the moment the block leaves the spring, the spring potential energy bucket is empty and all of the energy (still the same 250 J) is split between the kinetic and potential gravitational energy buckets, although the exact distribution between the buckets doesn't concern us here. As the block continues up the slope, the kinetic energy bucket continues to empty as more and more of the total energy shows up in the gravitational potential energy bucket. When the block reaches its maximum height up the ramp, its velocity is zero and all of the energy is in the gravitational potential energy bucket. By the way, the reason that the problem specified a spring of negligible mass is so we would not have to consider any kinetic energy the expanded spring might have.

KE $= \frac{1}{2}mv^2 = ?$ + $PE_G = mgy = ?$ + $PE_S = \frac{1}{2}k\Delta s^2 = 0$ J = $E = 250$ J

KE $= \frac{1}{2}mv^2 = 0$ + $PE_G = mgh = 50\,h$ + $PE_S = \frac{1}{2}k\Delta s^2 = 0$ = $E = 250$ J

Because the formula for gravitational potential energy is written in terms of height y above the ground, the equations will actually give us the height $y = h$ shown in the first picture. In our last step, we will use trigonometry to find L, the distance along the slope that the question asked for. We knew everything about the system at the bottom and used that to calculate the mechanical energy of 250 J. We will now use this mechanical energy value to find out about the block at the top.

At the top of the motion, all 250 J are in the potential energy bucket, which we have written $PE_G = mgh = (5\text{ kg})(10\text{ m/s}^2)h = (50\text{ kg}\cdot\text{m/s}^2)h$. This allows us to find h as

$$h = \frac{250\text{ J}}{50\text{ kg}\cdot\text{m/s}^2} = 5\text{ m}.$$

The final step is to use $\sin 37° = h/L$ to find the distance L that was asked for:

$$L = \frac{h}{\sin 37°} = \frac{5\text{ m}}{0.6} = 8.3\text{ m}.$$

EXAMPLE 7.3 The Spring, the Block, and the Ramp—II

Find the speed of the block in the previous problem when it is halfway to the top.

ANSWER When the block is halfway to the top, it is 2.5 m in elevation above the point where it started. We know that, at this point, the mechanical energy is still 250 J while the spring potential energy bucket is empty.

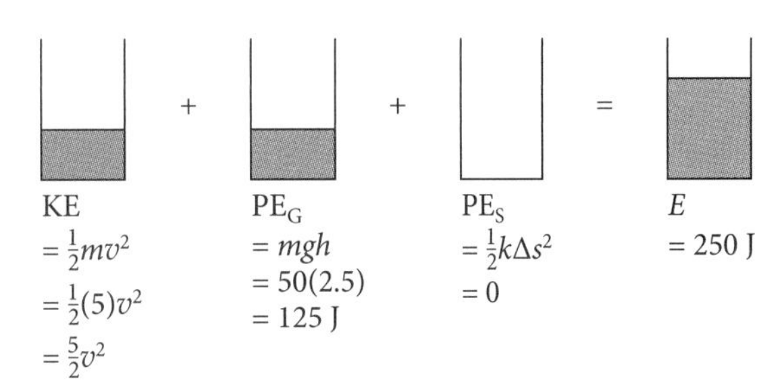

To simplify the notation, we assume SI units throughout. Consider the energy buckets when $h = 2.5$. We calculate PE_G at this point to be $PE_G = mgh = (5)(10)(2.5) = 125$ J. The rest of the 250 J must be in the KE bucket, as shown above.

The equation for the mechanical energy at this point gives

$$\frac{5}{2}v^2 + 125 = 250$$

$$\frac{5}{2}v^2 = 125$$

$$v^2 = 50$$

$$v = 7.1\text{ m/s}.$$

It is worthwhile to note that both the gravitational potential and kinetic energies are proportional to an object's mass, so that doubling an object's mass will double both of these quantities. This means that, in problems with no springs involved, the answer will turn out not to depend on the mass. In these cases, the mass does not even need to be known in order to solve the problem, as in this next example.

EXAMPLE 7.4 The Tarzan Swing

Tarzan swings on a vine, as shown in the figure. He begins standing on the riverbank, 4 m above the water. At the bottom of his swing he is 1 m above the water, and he lets go of the rope when he is 2 m above the water. Find Tarzan's speed the moment he hits the water.

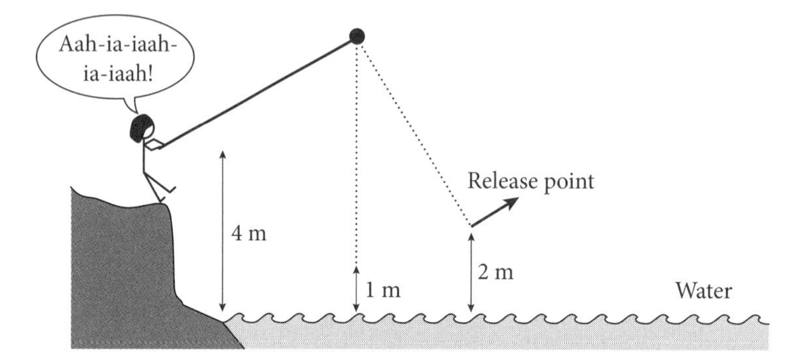

ANSWER The only forms of energy in the problem are kinetic and gravitational potential. It is natural in this problem to use the level of the lake to be $h = 0$. At the beginning, there is only gravitational potential energy because Tarzan has zero speed when he begins his swing. As he swings down, the gravitational potential energy bucket empties while his kinetic bucket fills. As we will discuss in detail later, the force from the rope of a Tarzan swing cannot change the mechanical energy, which therefore must remain constant. At the bottom of the arc, Tarzan starts slowing down as his elevation begins to increase and to fill the gravitational potential bucket. After the release point, Tarzan will continue moving energy from the kinetic bucket to the gravitational potential bucket while he rises to some maximum height before picking up speed (i.e., kinetic energy) again and finally hitting the water. At the moment he hits the water, all of his energy is in the kinetic bucket.

This problem illustrates a critical point: If the total energy is not changing, we don't need to worry about the details of how the system gets from beginning to end. In this problem, we know a lot of details about the motion between the initial position and the point of hitting the water, but these just don't matter. Let's consider the energy buckets. The initial situation is shown at right (where SI units have been assumed for all quantities). Of course, here we face the problem that we don't know Tarzan's mass, m_T. But, let us leave it as m_T and hope that we will figure out how to deal with this later. So the total mechanical energy, E, is just $40m_T$.

KE	+	PE_G	=	E
$= \frac{1}{2}m_T v^2$ $= 0$		$= m_T gh$ $= m_T(10)(4)$ $= 40m_T$		$= 40m_T$

When Tarzan hits the water, the buckets look like the picture at right.

KE	+	PE_G	=	E
$= \frac{1}{2}m_T v^2$		$= m_T gh$ $= 0$		$= 40m_T$

Now we know that the energy as Tarzan hits the water is the same as at the beginning, so we set the initial energy equal to the final energy, giving

$$\frac{1}{2}m_T v^2 = 40m_T.$$

It is at this point that we see why we don't need Tarzan's mass. Because it appears on both sides of the equation, we just divide both sides by m_T, leaving us with $\frac{1}{2}v^2 = 40$, which means $v^2 = 80$, whose solution is $v = 8.9$ m/s.

EXAMPLE 7.5 The Tarzan Swing—Maximum Height

Consider the previous example. After Tarzan lets go of the rope, does his maximum height above the water ever get as great as 4.0 meters? Explain your reasoning.

ANSWER We know that when Tarzan starts his swing he is 4.0 meters above the water, and all of his energy is in the gravitational potential bucket. But when Tarzan reaches his highest point after releasing the rope, he still has to be moving in the horizontal direction. Since some of Tarzan's mechanical energy must always be in the kinetic energy bucket, there must always be something less than that total in the gravitational energy bucket. So Tarzan must always be less than 4.0 meters above the water.

7.2 Energy Change and Work

In the previous section we saw how the concept of a conserved mechanical energy could help us visualize motion and how it can help to produce simple solutions to certain kinds of problems. But we also explained that certain conditions had to be satisfied if mechanical energy was, in fact, to be conserved. Let us look at one way in which the mechanical energy of some system might actually change.

Consider a pendulum formed by your sister swinging on a swing. She has potential energy at the top of her swing. As she descends, some of the potential energy turns into kinetic energy. If we define the system to consist of your sister, the swing, and the earth, then we have an isolated system. If there is no friction in the swing, then the mechanical energy of the sister–swing–earth system is conserved. She will go on swinging back and forth in exactly the same way forever. However, if you step up and begin to push her, then the mechanical energy of the system changes, because you are acting from outside the system. In this section, we will see how an *external force* changes the mechanical energy of a system. We begin with an example.

EXAMPLE 7.6 Creating Kinetic Energy

A 15-kg block on wheels is initially traveling 5 m/s on a level surface. A force of 2 N pushes the block forward for 10 m. How much is the kinetic energy bucket filled during this interval?

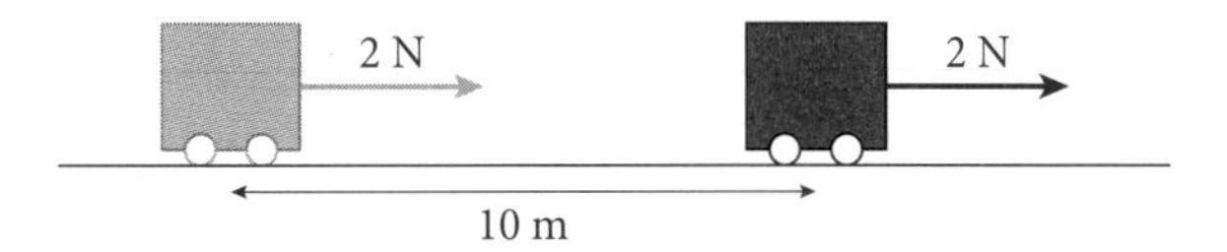

ANSWER We can easily calculate the initial kinetic energy of the block, but we need to find its final speed (after the pushing has stopped) before we can calculate the final kinetic energy. Since the net force on the block (and hence the acceleration) is constant, we can use the kinematic equation from Table 2.2 to find the final speed,

$$v_f^2 = v_i^2 + 2a\Delta x.$$

If we multiply this equation by $\frac{1}{2}m$, we can rewrite this in the form

$$\frac{1}{2}mv_f^2 - \frac{1}{2}mv_i^2 = ma\Delta x = F_{net}\Delta x,$$

where, in the last step, we have remembered that $F_{net} = ma$. The left-hand side of this equation is the change in the kinetic energy. The right-hand side must therefore be the cause of the change. If a single force of 2 N acts on the block through a distance of 10 m, the kinetic energy increases by $F\Delta x = (2\ \text{N})(10\ \text{m}) = 20\ \text{J}$.

We should note that the change in the kinetic energy in Example 7.6 is equal to the product of the applied force times the distance the object moved while the force was acting on it. This quantity is given a name. It is called *work*.[1]

The work done on a system by something external to the system can change its energy. If the force acts in the same direction as the displacement, the work is positive and the energy increases. If the force acts in the direction opposite to the displacement, the work is negative and the energy decreases. Forces that act perpendicular to the motion do not change the energy of the system, but only change the direction of the motion, as we saw in Table 4.1 (page 59). This is why the normal force exerted by the frictionless ramp and the tension force exerted by a Tarzan swing, which always act perpendicular to the motion, do no work on the system. If a force acts at some other angle relative to the direction of motion, it must first be broken into components along the direction of motion and perpendicular to the direction of motion. Only the parallel component does any work on the system. The formula for calculating the work done on an object by a force $\vec{F}$ is

$$W \equiv F_{||}\Delta x, \tag{7.5}$$

where by $F_{||}$ we mean the component of the force that lies along the direction of the motion Δx.

We have been led to the concept of work as a quantity that changes the kinetic energy of a system, but, as we shall see, the relationship between work and energy is more general than that. Indeed, there is a general connection between work and the mechanical energy of a system, including potential energy. It is called the *work–energy theorem*,

$$W = \Delta E = E_f - E_i. \tag{7.6}$$

Note that positive work—the kind that occurs when a force on one object in a system acts in the same direction the object moves—will increase the energy of the system. The final energy will be greater than the initial energy. On the other hand, negative work will decrease the energy; the final energy will be less than the initial energy. The next example demonstrates how to resolve a force into the proper components and how to determine the sign of the work that the force does.

1. It seems that the physicists are at it again. As we appropriated the term "conserved" to mean something it does not mean in English, so have we appropriated this new term "work" to mean something it does not usually mean. In English, work is effort for which you expect to be paid. A waiter holding a 5-pound tray of snacks for four hours at the entrance to the hotel ballroom would expect to be paid for his effort. However, since the tray did not move ($\Delta x = 0$) while he was holding it up with a force of 5 pounds, he has not really done any work on the tray, in the physics sense. Moral: do not take a job serving snacks at a Physics Department party.

EXAMPLE 7.7 The Force Component That Does the Work

A force of 80 N acts on a block at an angle of 60° with respect to the horizontal, as shown. Find the work done on the block by the force as the block moves through a displacement of 4 m to the left.

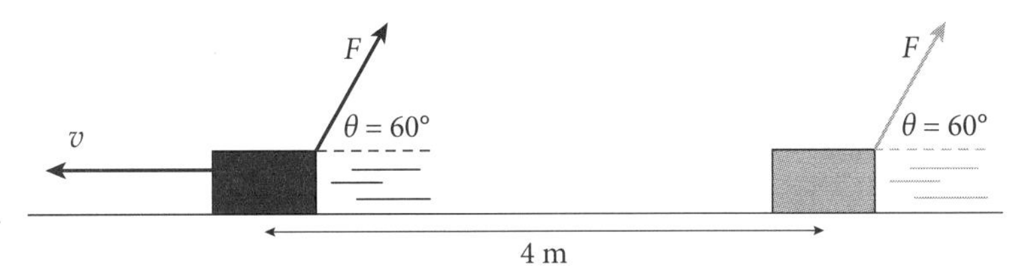

ANSWER Clearly, the block did not start from rest. The force has a component to the right, while we are told that the block moves to the left. The only way for this to occur is for the block to have already been moving to the left before the force was applied.

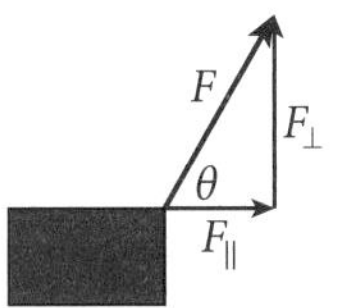

The force must be broken into components, as shown at right. The component of the force perpendicular to the motion ($F_\perp$) does no work. The component that is parallel to the motion ($F_{||}$) acts to the right in the diagram, while the motion of the block is toward the left. As we apply Equation 7.5 to find the work done, we take the direction of $F_{||}$ explicitly into account by use of a minus sign, leaving the symbols $F_{||}$ and Δx to stand for the magnitudes of these quantities. We see that the force does negative work:

$$W = -F_{||}\Delta x = -F\cos\theta(\Delta x) = -(80\text{ N})(0.500)(4\text{ m}) = -160\text{ J}.$$

As a result of the negative work done, the block's kinetic energy will *decrease* by 160 J.

You should also note the somewhat surprising fact that the amount of work done depended on the distance the object traveled, *whether or not the distance traveled was due to the force that was acting.* If the block had been initially moving at a higher speed, it would have gone farther, and so the amount of work done by the same force in the same time would have been greater.

As we said as we introduced the work–energy theorem (Equation 7.6), work can also affect the potential energy of a system, as shown in the next two examples.

EXAMPLE 7.8 Creating Gravitational Potential Energy

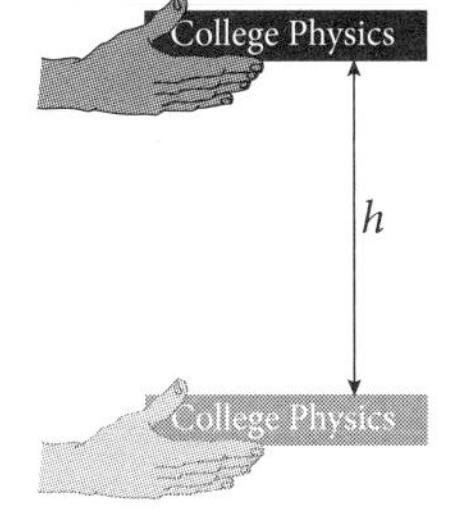

How much work is required to lift a book of mass m straight up a distance h? Assume that the book starts and ends at rest.

ANSWER We suppose that the hand moves the book upward at some slow constant velocity. The force exerted by the hand on the book must therefore be equal to the weight of the book all during the rise of the book, though the hand must push a *little* harder than mg to get the book started, and a little less than mg to bring it to a stop again at the top. The upward force of the hand acts in the same direction as the motion, so the work done by the hand is positive.

$$W = +F_{\text{hand,book}}\Delta x = +mgh.$$

This positive work increases the mechanical energy of the book–earth system by an amount mgh, and, since the kinetic energy is zero to begin with and zero at the end, it is only the gravitational potential energy of the system that increases by mgh. Indeed, this was the method used to find the equation $PE_G = mgh$ for gravitational potential energy.

EXAMPLE 7.9 Creating Spring Potential Energy

How much work is required to compress a spring of spring constant k by an amount Δs?

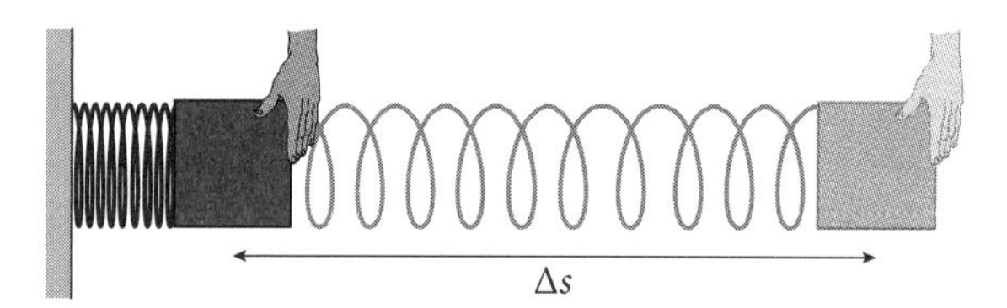

ANSWER This is more difficult than the previous example because the hand must push harder and harder the more the spring is compressed. If we suppose again that the hand pushes just hard enough to move the block at a low constant velocity, then the hand must always push just as hard as the spring to give a zero net force on the block. According to Hooke's law, the force exerted by the spring is zero before the spring is compressed and increases linearly to $k\Delta s$ when the spring is compressed by Δs. To find the work done by the hand, we find the average force over the distance Δs. This gives

$$W = +F_{\text{average}}\Delta x = +\frac{1}{2}(F_{\min} + F_{\max})\Delta s = +\frac{1}{2}(0 + k\Delta s)\Delta s = +\frac{1}{2}k\Delta s^2.$$

This positive work increases the mechanical energy of the spring–block system by an amount $\frac{1}{2}k\Delta s^2$. Since kinetic energy and gravitational potential energy do not change, the spring potential energy of the system increases by $\frac{1}{2}k\Delta s^2$, as expected. Indeed, this is the way that the formula for the spring potential energy, $PE_S = \frac{1}{2}k\Delta s^2$, was derived.

There is one last aspect of the work–energy theorem that we want you to be clear on. This is that, in this section, we considered only the change in mechanical energy produced by a force that is *external* to the system. If there is an *internal* force between elements of the system—a force whose effect on objects in the system can be represented by a potential energy—then we will want to include that potential energy as part of the mechanical energy. We will not worry about the work it does. In Example 7.8, our system consisted of the book and the earth. As the force of the hand was applied to the book, we included the potential energy stored in the gravitational interaction between the book and the earth as part of the mechanical energy. This is why we did not consider the work done by gravity on the book as we worked out the change in energy. The force of gravity on the book was an *internal* force, and its effects were completely taken into account as the potential-energy part of the mechanical energy of the earth–book system. It was for this same reason that we did not worry about the work done by the spring in Example 7.9. The spring energy was included in the mechanical energy of the system.

7.3 Power

We will need to understand the quantity known as "power" for several future applications, but it is very easy, so we can explain it simply and quickly. The rate at which the energy of some system changes, or the rate at which some agent does work on a system, is referred to as *power.* It is defined simply as

$$P = \frac{\Delta E}{t} \quad \text{or} \quad P = \frac{W}{t}, \tag{7.7}$$

where t is the time it took to make the energy change ΔE or to do the work W.

Power is measured in J/s. A joule per second is also called a watt (W). This is actually pretty easy, isn't it? Let's make sure we understand by doing an example.

EXAMPLE 7.10 Power Compressing a Spring

A spring of spring constant $k = 100$ N/m is compressed by 10 cm in 2 s. What average power was required to do the job?

ANSWER The work done will equal the change in the potential energy of the spring. We use Equation 7.3 for the change in energy, giving

$$\Delta E = \frac{1}{2}k\Delta s^2 = \frac{1}{2}(100\text{ N/m})(0.1\text{ m})^2 = 0.5\text{ J}.$$

We then use Equation 7.7 to find the power:

$$P = \frac{\Delta E}{t} = \frac{0.5\text{ J}}{2\text{ s}} = 0.25\text{ watts}.$$

You are probably most familiar with watts as units of electrical energy. An old-style incandescent Christmas tree light bulb uses about 0.25 watts.

7.4 Conservative and Non-Conservative Forces

In Section 7.2, we discussed how work done by forces external to a system could change the mechanical energy of a system. In this section, we note that there are certain kinds of *internal forces* that also change the mechanical energy of a system.

If we slide a book across a desk, it comes to a stop. We can define a system that comprises the book and the earth (with its projecting tabletop). There are no springs in the problem; the desk is level (so the gravitational potential energy does not change); and yet the kinetic energy of the book is lost as the book comes to a stop. Where did the energy go? Surprisingly, it turns out that there *is* a subtle change in the book and the desk that *can* account for the energy loss, but it is not a form of mechanical energy. It is thermal energy. After the book comes to rest, it may be observed that the temperatures of the book and the tabletop are slightly higher than they were before the book started its slide. The friction has converted kinetic energy into thermal energy. If we carefully take into account the thermal capacity of the materials that compose the book and the table, then every joule of mechanical energy lost through friction shows up in the increase of the temperatures. Even though the system is an isolated one, a force like friction can convert mechanical energy into another form of energy, in this case into thermal energy.

There are, in fact, two classes of forces in nature—conservative forces and non-conservative forces. A conservative force is one that has a potential energy associated with it. In this book, we have so far seen two conservative forces, the forces due to gravity and to springs. We have also seen one non-conservative force, the frictional force. This force does not allow a form of potential energy to be associated with it.

A ball that is thrown upward initially has kinetic energy, but that kinetic energy changes into gravitational potential energy as the ball rises (see Figure 7.1). When the ball stops at the top of its trajectory, all of the energy is in the gravitational potential energy bucket. But the next thing that will happen is that the ball will fall back again toward the earth, emptying the potential energy bucket as the kinetic energy bucket refills. Now consider a book that slides across a tabletop. As friction slows its motion, its kinetic energy is lost, but, when the book comes to rest, there is no way to get all of that energy back. As we explained above, the lost energy can actually be accounted for as thermal energy inside the book and tabletop. However, if we were to picture a thermal energy bucket, we could picture it filling, but, once energy goes into that bucket, there is no way to get it all out again. In this way, it is not like the other buckets, and so we should not think of it in the same way. The conservative forces—gravity and springs—have potential energy buckets; the frictional force does not. Work done by non-conservative forces will always change the mechanical energy of a system, even if the non-conservative force is a force that is completely internal to the system we have defined.

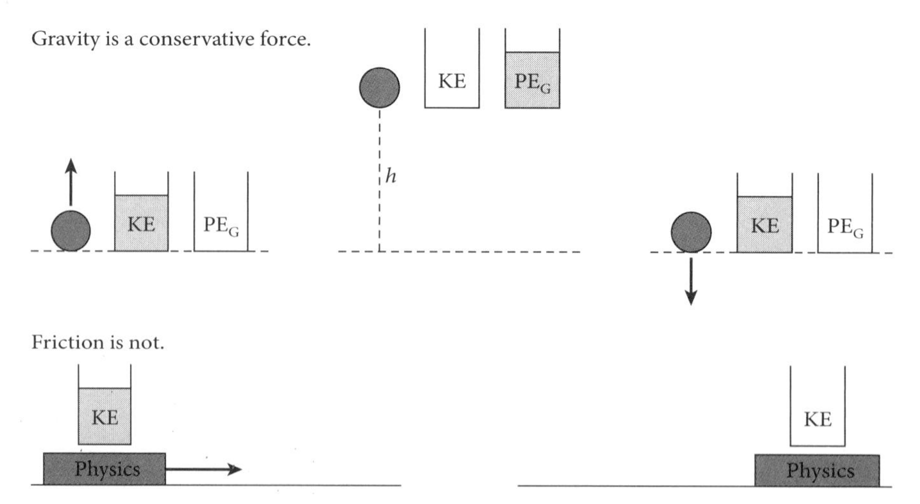

Figure 7.1 The energy buckets for a ball rising in the presence of gravity and for a book sliding across a tabletop with friction.

EXAMPLE 7.11 The Work Done by Friction

A crate with a mass of 15 kg slides 4 meters down a 37° ramp, for which the coefficient of kinetic friction between the crate and the ramp is $\mu_K = 0.40$. How much work is done by friction? By how much will the mechanical energy of the earth–block system change?

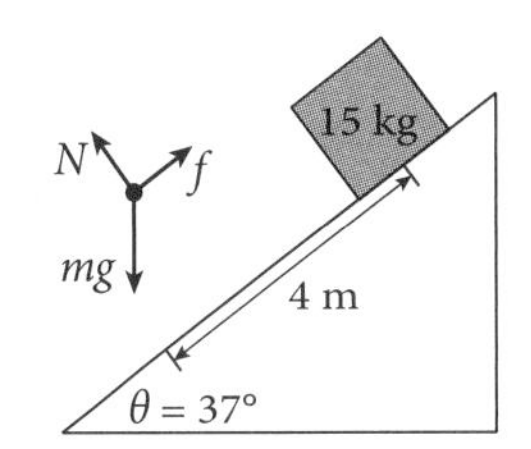

ANSWER We define the block and the earth (with its attached ramp) as the "system." Friction is an internal non-conservative force in this system. The frictional force may be found by the method familiar from Chapter 6 to be

$$f = \mu_K N = \mu_K mg\cos(37°) = (0.40)(15\text{ kg})(10\text{ N/kg})(0.80) = 48\text{ N}.$$

As the crate slides down the ramp, friction acts up the ramp, producing negative work,

$$W_{NC} = -F\Delta x = -(48\text{ N})(4\text{ m}) = -190\text{ J}.$$

The mechanical energy of the block–earth system will drop by 190 J as the block slides down the ramp. By the way, since friction always opposes the relative motion of two surfaces, the work done by friction will always be negative.

To sum up the results of this chapter so far, the mechanical energy of a system with gravitational and spring forces acting internally is given by Equation 7.4,

$$E = \text{KE} + \text{PE}_G + \text{PE}_S.$$

If there are other kinds of conservative forces acting internally to the system, other potential energy terms would be added on the right to expand the definition of the total mechanical energy. Mechanical energy can change as a result of external work done on the system, W_{Ex}, or as a result of the work done by non-conservative forces within the system, W_{NC}. We may therefore write the work–energy theorem as

$$\Delta E = E_f - E_i = W_{Ex} + W_{NC}. \tag{7.8}$$

The work on the right-hand side may be positive (when the force acts in the same direction as the motion) and add mechanical energy to the system, or it may be negative (when the force acts in the opposite direction as the motion) and subtract mechanical energy from the system.

When using energy concepts to solve problems, be sure to carefully define your "system" by deciding which elements are included as internal to the system and which elements are external. You will want to include any moving objects that have mass in your system, so that you can consider their kinetic energy. Include any springs in your system so you can account for spring potential energy. If an object changes its height above the earth, you will need to include the earth in your system, but only to properly pick up the gravitational potential energy as internal to the system. The earth is so big compared to the other objects in the system that its velocity (and thus its kinetic energy) may safely be ignored. If friction exists between parts of your system, then the work done by friction can be included in W_{NC}. If your system has been defined so that friction ends up as an external force, then the work done by friction could be included in W_{Ex}. Just make sure that you do not count the friction work twice. Other external agents that do work on your system may include things like a hand that is pushing or tension in a rope that is pulling. Frictionless ramps and Tarzan swings can also be external agents, but they do no work because they are always pushing or pulling in a direction perpendicular to the motion. Let's look at a couple of further examples.

EXAMPLE 7.12 Pulling a Block up a Ramp—I

A string attached to a winch exerts a tension force of 80 N on a 10-kg block and pulls it a distance of 5 m up a frictionless incline, as shown. The string pulls parallel to the incline. Find the change in kinetic energy for the block–earth system as the block moves 5 m up the incline.

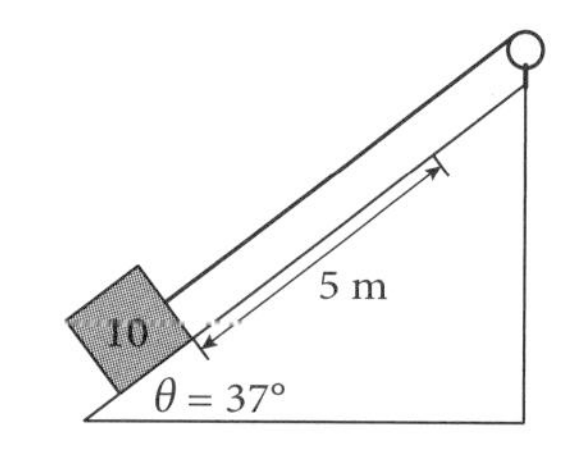

ANSWER We define a "system" composed of the block and the earth, thereby including the gravitational potential energy. Everything else is external to the system. Let us first find the total work done to change the mechanical energy of the block–earth system. A free-body diagram for the block shows three forces: the gravitational force of the earth on the block (a conservative force internal to our system), the normal force exerted by the frictionless ramp (which does no work because it is perpendicular to the motion), and the tension force by the string on the block. This last force is the only one to do external work on our system. The string pulls in the same direction as the motion, so the work is positive. The change in energy is

$$\Delta E = W_{Ex} = +T\Delta x = +(80 \text{ N})(5 \text{ m}) = +400 \text{ J}.$$

The mechanical energy of the system must therefore increase by 400 J, but how much of this increase goes into the gravitational potential energy bucket and how much goes into the kinetic energy bucket? Simple trigonometry indicates that the block rises 3 meters as it moves 5 m along the ramp. This means that the gravitational potential energy increases by $PE_G = mgh = (10 \text{ kg})(10 \text{ N/kg})(3 \text{ m}) = 300 \text{ J}$. This leaves 100 J to go into the kinetic energy bucket. The change in the block's kinetic energy is +100 J.

EXAMPLE 7.13 Pulling a Block up a Ramp—II

How would the answer to the last problem change if there were a small coefficient of friction $\mu_K = 0.25$ between the block and the ramp?

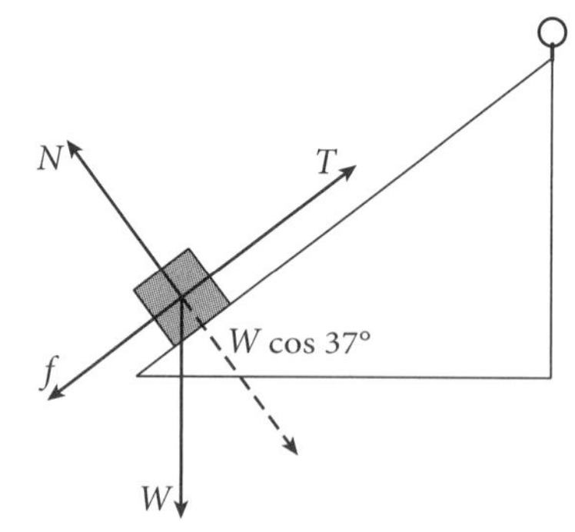

ANSWER We draw a free-body diagram of all the forces acting on the block and resolve the forces into components. The normal force on the block is $N = W \cos 37°$. The static frictional force acts down the ramp, opposing the block's upward motion, with a magnitude

$$f = \mu_K N = \mu_K mg \cos 37° = (0.25)(10 \text{ kg})(10 \text{ m/s}^2)(0.8) = 20 \text{ N}.$$

The ramp does not move, so we would not have to include it in our "system." But let us choose to do so anyway. We can consider it as part of the earth, which is itself part of the system because of the need for the gravitational potential energy to be included in the problem. Then the work done by friction is an internal non-conservative force and it is negative because it acts downward as the block is pulled upward.

$$W_{NC} = F_{\|}\Delta x = (20\text{ N})(-5\text{ m}) = -100\text{ J}.$$

The work done by the winch is the same as before, so the change in the mechanical energy is

$$\Delta E = W_{Ex} + W_{NC} = +400\text{ J} - 100\text{ J} = 300\text{ J}.$$

In Example 7.10, we have already found the increase in the gravitational potential energy as the block goes up the ramp. It was 300 J. So, in this case, we are left with *no* energy in the kinetic energy bucket. The winch force exactly balances the friction and gravity and the block reaches the top with only whatever tiny initial velocity it might have had for it to be moving up the ramp at all.

7.5 Conservation of Energy

We began this chapter by defining mechanical energy. We saw how the idea of a conserved mechanical energy provided a simple way to solve certain kinds of problems. We then noted in Section 7.2 that a non-isolated system could gain or lose mechanical energy if there were external forces that did work on the system. We later saw, in Section 7.4, that, even in an isolated system, non-conservative forces between elements of the system can upset mechanical energy conservation by transforming mechanical energy into thermal energy. We will discuss this subject in more detail in Chapter 14. However, let us point out here the important fact that, when all forms of energy are accounted for, there is a quantity that is always conserved.

Let us define the *total energy* of a system as

$$E_{TOT} = E + E_{INT} + \cdots, \tag{7.9}$$

where we continue to use E for total mechanical energy, including *all* potential energies in the system, and where we denote thermal energy as E_{INT} (thermal energy is also called internal energy because it resides *inside* the bodies of the system). The ellipsis (the "$\cdots$") in Equation 7.9 suggests that other kinds of energy can be added to this equation. We will discuss one of these, mass energy, in Chapter 24. In fact, the history of physics has been characterized by increasingly sophisticated versions of Equation 7.9, leading to the following important fact. It is one of the experimentally verified fundamental laws of physics that *the total energy of an isolated system is always exactly conserved.*

If we choose to ignore terms in Equation 7.9, then we may not expect the remaining part of the total energy to be conserved. Mechanical energy, in particular, is only the first term on the right-hand side of Equation 7.9. If we consider only the mechanical energy, then the only way we can count on mechanical energy being conserved is if there is no way for that energy to be converted to non-mechanical forms of energy.

7.6 Solving Mechanical Energy Problems

There is a simple problem-solving strategy that renders mechanical energy problems relatively straightforward:

1. Always start every energy problem by thinking about the work–energy equation, $E_{\text{initial}} + W_{\text{Ex}} + W_{\text{NC}} = E_{\text{final}}$. This states that the final energy of your system is the same as the initial energy, unless it is changed by some external agent or some non-conservative force that does work on the system.
2. Decide what you mean by the initial and final. You get to choose these events. You should generally choose one of them to be a time when you are looking for information about the system and the other to be a time when you know everything about the energy of the system. Whichever of these two events happened earlier is the initial event.
3. Ask yourself, "What flavors of energy does my system have at the initial event?" If anything with mass is moving, you have kinetic energy. If anything is off the floor, you have gravitational potential energy. If any springs are stretched or compressed, you have spring potential energy. Write an equation for your initial energy.
4. Ask yourself, "What flavors of energy does my system have at the final event?" Run through the same checklist as before. Write an equation for your final energy.
5. Calculate the work done by external agents between the initial event and the final event. It usually helps to draw a free-body diagram and ask yourself two questions: "Which of my forces on my diagram are caused by external agents?" and "Which of these external forces do work on my system?"
6. Calculate the work done by any non-conservative forces between elements you have chosen to include in your system.
7. Use the work–energy equation to relate the quantities identified in steps 3, 4, 5, and 6. This should result in a messy equation with one unknown. Solve for the unknown and you solve your problem.

One last thing should be explained. If you are given a problem without knowing that it was expressly set up to be solved using an energy approach, how do you know whether to use the energy shortcut or to work through the steps in the problem from beginning to end using Newton's laws? First, take a look at the fundamental energy equation, Equation 7.8. It is a single scalar (non-vector) equation, so it may only be used to give one number, and that number has to be one that is represented by one of the variables that appear in the various kinds of energy or in the various ways that work can be done on a system. For example, in the Tarzan swing example (Example 7.4), we could use the work–energy equation to find Tarzan's speed as he hit the water, because speed is one of the variables that enters the formula for kinetic energy. But we could not use it to determine the direction of his velocity, nor how long it took him after he swung from the riverbank until he hit, nor how far from the bank he finally hit the water. All of these things may be found by using Newton's laws, but the energy method alone will not give them to you. Second, energy methods are most useful where there are conservative forces involved (i.e., gravity and springs). Finally, look for places where you are told that a force acts for a given distance. Since the work is the product of force and distance, these tend to be work–energy problems.

EXAMPLE 7.14 Energy on a Track

A block of mass $M = 10$ kg slides from rest down a frictionless ramp such that the center of the block drops a distance of 80 cm. As the block passes from point B to point C, it encounters a 2-m section of track that is rough, having a coefficient of kinetic friction $\mu_K = 0.3$, exactly. After passing point C, the block compresses a spring with spring constant $k = 160$ N/m. How fast is the block moving when the spring has been compressed by 25 cm?

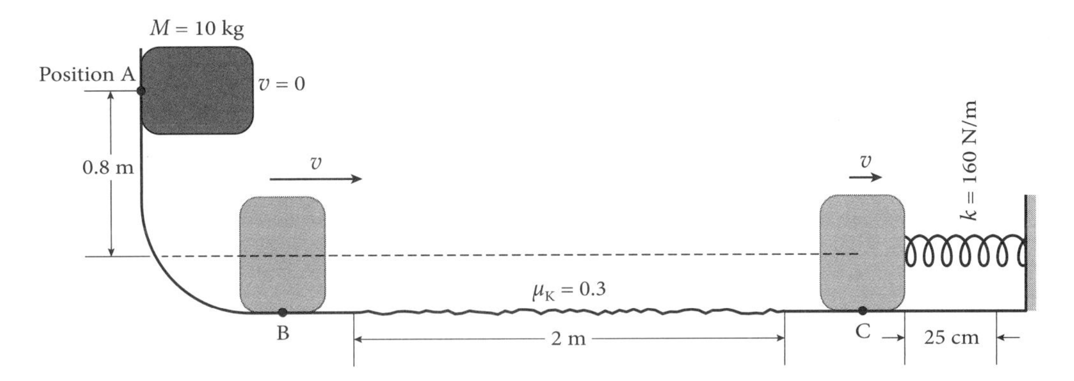

ANSWER First, we suspect that energy methods will be useful for this problem because it asks for only a single number, the speed of the block at a certain position. It also contains gravity and springs, which are conservative forces with potential energies associated with them. Lastly, we were given the distance over which the frictional force acted, giving us the ability to calculate the non-conservative or external work done. So let's use the work–energy strategy outlined above, applying it to the block–earth–spring system. We have included the earth so as to keep track of the gravitational potential energy. Since the track does not move, we can choose to exclude it from the system and count the frictional force as an external force.

- We begin with the work–energy equation: $E_{\text{initial}} + W_{\text{Ex}} + W_{\text{NC}} = E_{\text{final}}$. We see that we can calculate the energy of the system at point A. Let that be our initial event. We want to know the speed of the block when the spring is compressed 25 cm. Let that be our final event.
- At point A the block is not yet moving and the spring is not yet compressed, so the only energy at this point is gravitational potential energy, $\text{PE}_G = mgh = (10 \text{ kg})(10 \text{ N/kg})(0.8 \text{ m}) = 80 \text{ J}$.
- At the final point, we think through what kinds of energy we have. When the spring is compressed 25 cm we obviously have spring potential energy: $\text{PE}_S = \frac{1}{2}k\Delta s^2 = \frac{1}{2}(160 \text{ N/m})(0.25 \text{ m})^2 = 5 \text{ J}$. Along the horizontal floor, the gravitational potential energy has dropped to zero, but the block is still moving, so we have kinetic energy, $\text{KE}_f = \frac{1}{2}mv_f^2$. Finding the kinetic energy of an object of known mass is equivalent to finding the final speed of the block.
- There is an external force on the system from the friction between the block and the track. We will need to calculate the work done on the system by this force. A free-body diagram of the block as it crosses the rough section of track has three forces acting: a weight force, $W_{e,b} = mg = 100$ N, a normal force by the track on the block, $N_{t,b}$, and a friction force by the track on the block, $f_{t,b} = \mu_K N_{t,b}$. The block remains on the horizontal track, so the vertical forces must balance, giving $N_{t,b} = 100$ N. The resulting knowledge of the normal force lets us calculate the friction force as $f_{t,b} = (0.3)(100 \text{ N}) = 30$ N.

Only the friction force does any work, and that work is negative, since the force acts in the opposite direction as the displacement. The work is

$$W_{Ex} = -f_{t,b}\Delta x = -(30\text{ N})(2\text{ m}) = -60\text{ J}.$$

- Knowing the work done on the system, we can calculate the change in its mechanical energy and thereby determine the kinetic energy at the final point.

$$E_{initial} + W_{Ex} = E_{final}$$
$$80\text{ J} + (-60\text{ J}) = 5\text{ J} + \text{KE}_{final}$$
$$\text{KE}_{final} = 15\text{ J}$$

- Finally, once we have the kinetic energy, we can directly get the speed.

$$\frac{1}{2}(10\text{ kg})v_f^2 = 15\text{ J} \quad \Rightarrow \quad v_f = 1.7\text{ m/s}$$

7.7 Summary

In this chapter you should have learned the following:

- You should be able to recognize when the work–energy shortcut will enable a problem to be solved using work–energy methods.
- You should be able to carefully define the "system" to which your work–energy methods will be applied, deciding which elements to include in the system and where there are external forces on the system.
- You should be able to calculate the work done on a system by a force applied to an object as the object moves a particular distance.
- Most importantly, you should be able to systematically solve work–energy problems following the seven steps outlined in Section 7.6.

CHAPTER FORMULAS

A few new formulas for this chapter:

Kinetic energy: $\text{KE} = \frac{1}{2}mv^2$

Work: $W = F_{||}\Delta x$

Mechanical energy: $E = \text{KE} + \text{PE}_G + \text{PE}_S$

Gravitational potential energy: $\text{PE}_G = mgh$

Spring potential energy: $\text{PE}_S = \frac{1}{2}k\Delta s^2$

Work–energy: $\Delta E = E_{final} - E_{initial} = W_{Ex} + W_{NC}$

PROBLEMS

7.1 * **1.** The earth has a mass of 6.0×10^{24} kg and moves around the sun at an orbital speed of 30 km/s (3.0×10^4 m/s). What is the earth's kinetic energy in joules? The world uses about 7.2×10^{19} J of electrical energy in a year. If the earth's motion could all be turned into electrical energy (not a good idea), how long would it last?

*2. A $1\frac{1}{2}$-kg rock is raised from the ground and placed on a picnic table 80 cm high to hold the tablecloth down. What is the change in the rock's gravitational potential energy?

*3. A suction-cup dart gun has a spring inside whose spring constant is 2000 N/m. How much spring potential energy is stored in the spring when it is compressed 6 cm as the dart is inserted into the barrel?

*4. In an NCAA swim meet, Ruth Davis jumps from the 10-m diving board with an initial upward velocity of 2 m/s. What is her speed as she hits the water?

**5. A cannonball that is fired upward at 45 deg is seen to rise to a maximum height of 1000 m. Assume that air resistance may be ignored.

a) What is its muzzle velocity?

b) What will be its kinetic energy at the top of its trajectory?

**6. The following question appears on a physics exam: "Three bears are throwing identical rocks from a bridge to the river below. Papa Bear throws his rock upward at an angle of 30 degrees above the horizontal. Momma Bear throws hers perfectly horizontally. Baby Bear throws his at an angle of 30 degrees below the horizontal. Assuming that all three bears throw with the same speed, which rock will be traveling fastest when it hits the water?" Three students meet after the exam and argue about the correct answer.

Student 1 says, "Baby Bear's rock will be going the fastest because it starts with a downward component of velocity."

Student 2 says, "But Papa Bear's rock will stay in the air the longest, so it will have more time to speed up. I think his rock will be traveling the fastest."

Student 3 says, "Papa Bear's rock does stay in the air longer, but part of that time it is moving upward and slowing down. I think Momma Bear's rock will be traveling fastest when it hits the water because it is in the air longer than Baby Bear's and it is speeding up all of the time."

Which student (if any) do you agree with? Explain.

**7. A 38-kg child sits on a 2-kg swing held up with massless ropes. At the bottom of the swing, the child and swing are moving at 4 m/s. How high above this low point will the child rise before she stops?

**8. A pendulum is formed by a 1-kg bob on a 2-m string (a bob is the thing on the end of a pendulum). If the KE of the bob at the bottom is 5 J, what is the maximum angle of the string?

**9. A 3-kg pendulum bob hangs from a 5-m string. The bob is lifted 2 m higher than its equilibrium point and released from rest so that it swings back and forth. What is the maximum speed of the bob during each swing?

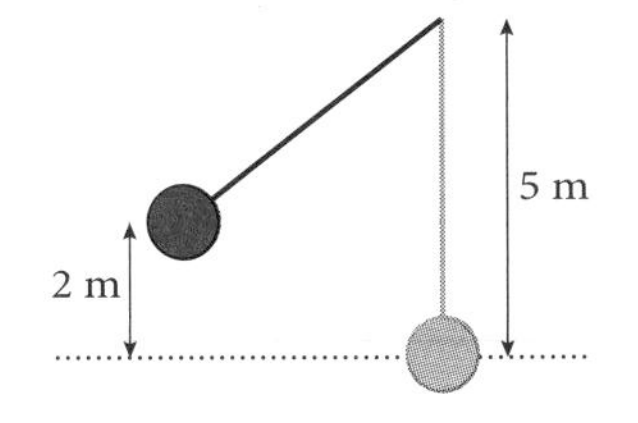

**10. A rope is suspended from a branch. You wish to raise yourself up off the ground by running along the ground, grabbing the rope, and letting the rope swing you upward. How fast would you have to run to be able to rise by 2 m?

**11. Two masses are connected by a string that is passed over a massless pulley, as in the picture at right. At an initial $t = 0$, the 6-kg mass is moving upward at 2 m/s, while the 4-kg mass is descending at the same speed. How far will the 6-kg mass rise before it stops?

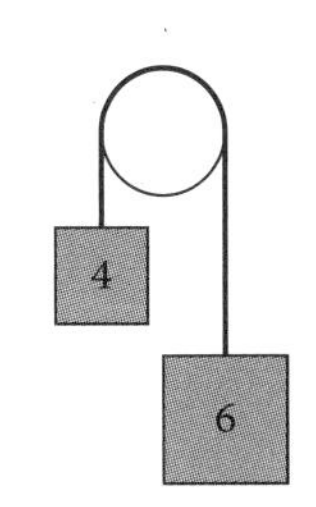

**12. A 3-kg block is held 5 meters directly over an initially uncompressed spring and is dropped straight down. The spring compresses by 20 cm as it brings the block momentarily to rest. What is the spring constant of the spring?

7.2 ** 13. A force $F = 60$ N acts on a block at an angle of $\theta = 30°$ with respect to the horizontal floor, in the direction shown below. Find the work done on the block by the force as the block moves through a displacement of $\Delta x = 2$ m to the right.

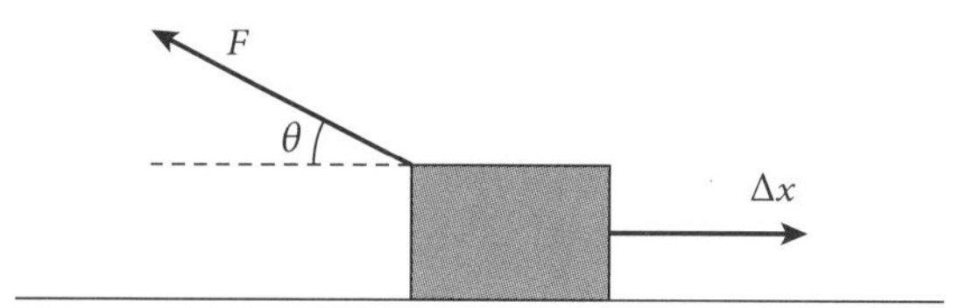

** 14. A 2.0-kg block compresses a spring 0.50 m from its equilibrium length. The mass is released and slides along a frictionless surface. When it has just lost contact with the spring, its speed is measured to be 5.0 m/s. Find the spring constant k.

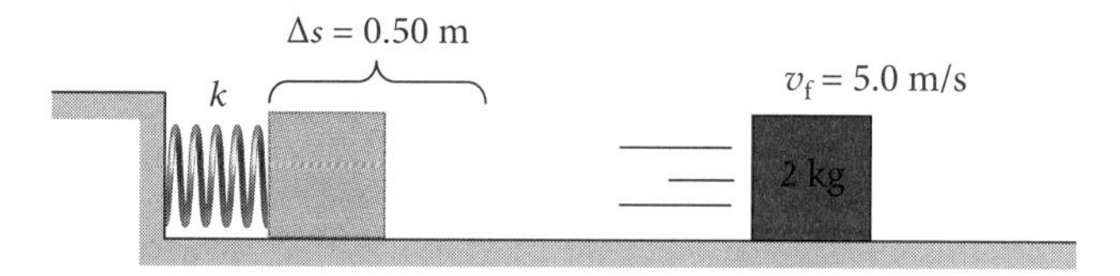

** 15. An extreme skier takes off from a jump and is going for height. If she leaves the jump traveling 15 m/s, and is traveling 13 m/s as she reaches the top of her flight, how high does she get above the top of the ramp?

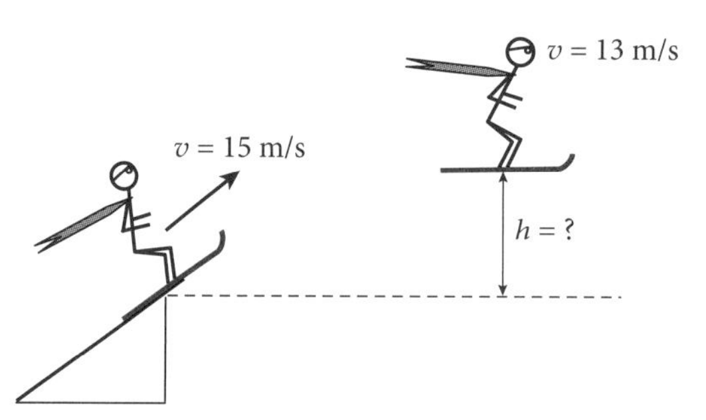

** 16. You swing on a "Tarzan Rope" from a ledge that is 5 m above the lake. Your 5-m rope is fastened to a tree limb that is 6 m above the lake's surface. When you let go, you have passed the bottom of the swing and have risen to a height of 2 m above the water. How fast are you traveling just before you finally hit the water? (Do not use the kinematic equations from Chapter 4; they do not apply.)

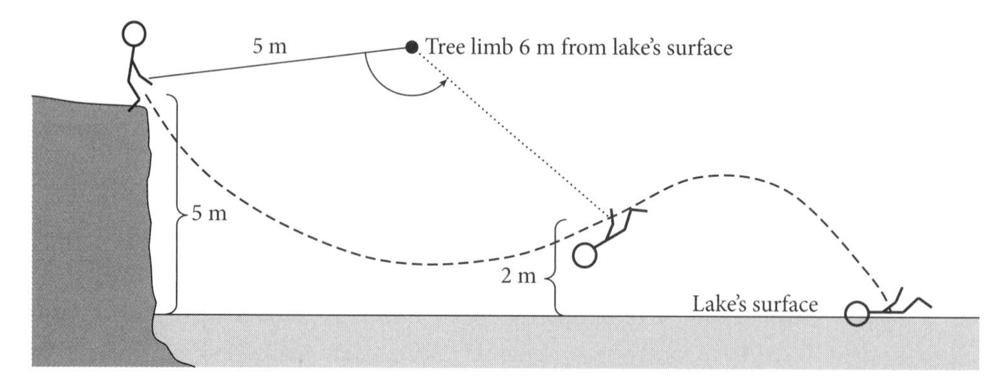

*** 17. The *Bobsled Jump* will be a new event at the next Winter Olympic Games (not really). The bobsleds are launched at point A by a large spring. The sleds then drop into a circular-shaped ramp made of frictionless ice, with a radius of 100 m, as shown below. The bobsled teams go for distance and try to display their best form as they land with a thump on the level landing field below.

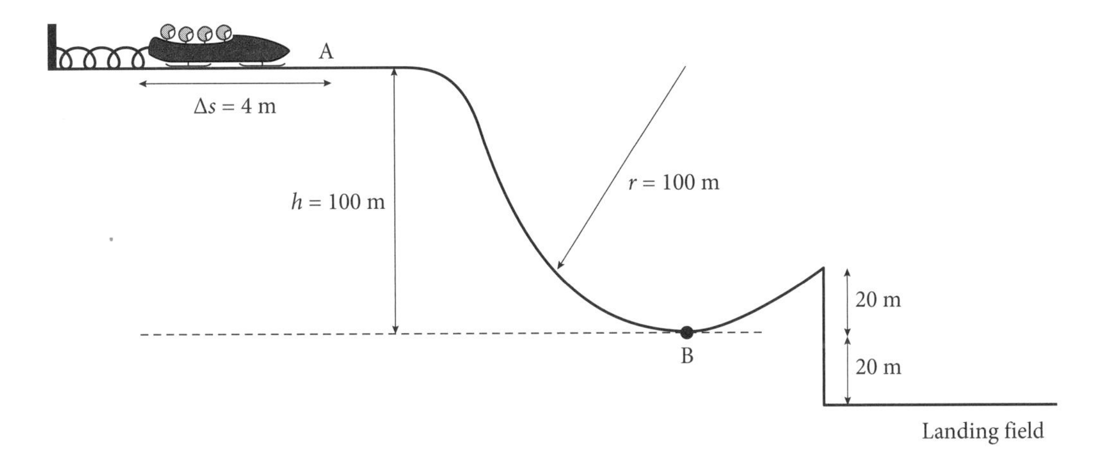

a) Find the work (if any) done by the ramp on the bobsled as it moves from point A to point B. Explain your reasoning.

b) Before each Olympics, the spring is calibrated by putting the Jamaican bobsled team (total mass 800 kg, including the bobsled) on the launcher and compressing the spring by $\Delta s = 4$ m. They then measure the speed v_B at point B and are thus able to determine the spring constant of the spring in the launcher. When the spring was calibrated, the bobsled was traveling $v_B = 50$ m/s at B. Find the spring constant for the spring.

c) Find the acceleration of the bobsled (magnitude and direction) at point B.

d) Find the magnitude of the normal force by the track on the bobsled at point B.

e) Assuming that the air resistance is negligible, how fast will the Jamaican team be traveling right before they land on the landing field?

* **18.** A 60-kg man climbs a 4-meter-high flight of stairs. How much work does he do?

* **19.** Five books, each 4 cm thick and each weighing 2 kg, are lying on the ground. Find the amount of work that is required to form them into a stack 20 cm high.

** **20.** A string exerts an upward tension force of 45 N on a 5-kg block as it lowers the block a distance of 4 meters.

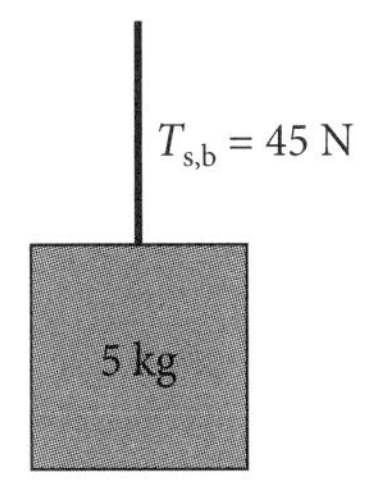

a) Find the total external work done on the block–earth system by the tension in the string as the block drops the 4 meters.

b) Find the change in gravitational potential energy for the block–earth system as the block drops 4 meters.

c) Find the change in kinetic energy for the block–earth system as the block drops 4 meters.

*** 21. A skier of mass $m = 80$ kg starts from rest at point A and skis down a hill and then up a smaller hill having radius of curvature $R = 25$ m. The skier is testing some new Frictionless Skis™ which are (strangely enough) entirely frictionless.

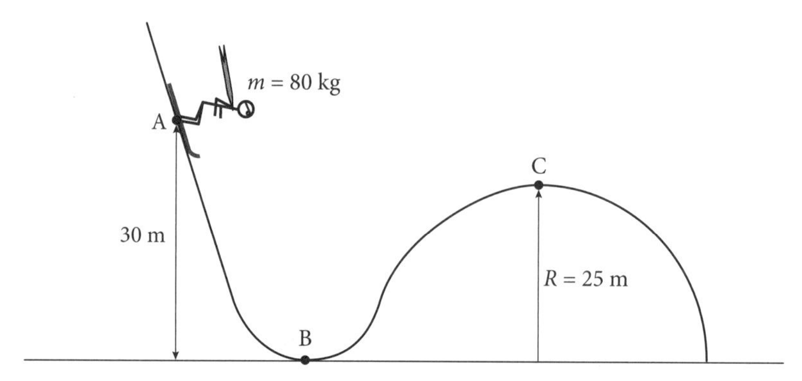

a) Find the work (if any) done by the snow on the skier with the Frictionless Skis as she moves from point A to point C. Explain your reasoning.

b) Find the kinetic energy of the skier at point B.

c) Find the speed of the skier at point C.

d) Find the acceleration of the skier (magnitude and direction) at point C.

e) Find the magnitude of the normal force by the hill on the skier at point C.

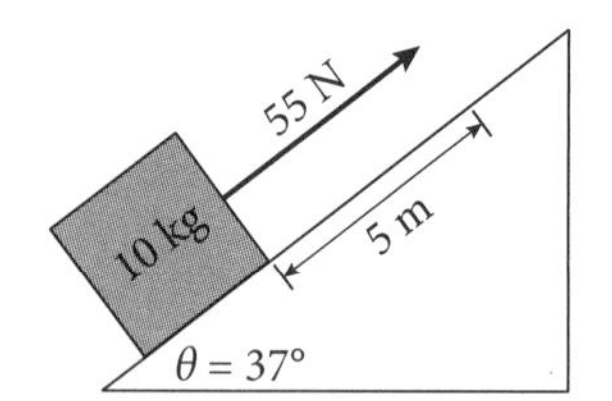

** 22. A string exerts a tension force of 55 N on a 10-kg block as it moves a distance of 5 m up a frictionless incline, as shown. The string pulls parallel to the incline.

a) Find the total external work done on the block–earth system as the block moves 5 m up the incline.

b) Find the change in gravitational potential energy for the block–earth system as the block moves 5 m up the incline.

c) Find the change in kinetic energy for the block–earth system as the block moves 5 m up the incline.

7.3 * 23. A box is pushed by a hand across a level table at constant speed. The frictional force exerted by the table on the box is 60 newtons. The box moves a distance of 20 meters in 4 seconds. What is the power being supplied by the hand?

* 24. What is the KE of a 1500-kg car moving at 80 km/hr? How long would a 1500-W hair dryer have to run to provide this much total energy?

* 25. How many joules of work can a 15,000-watt motor do in 20 minutes?

* 26. A 10-kg bucket of water is raised from the bottom of a well at a uniform speed of 0.4 m/s. What power is being expended to raise the bucket?

** 27. If a 60-kg sprinter can accelerate from a standing start to a speed of 10 m/s in 2 s, what average power will be generated in his muscles?

** 28. As the engine in a car forces its wheels to turn, static friction between the road and the wheels of a car produces a force of 1000 N acting forward on the car to keep it going at a constant speed of 25 m/s.

a) How much work does the static friction force do to the car as the car travels along the road for 10 seconds? [Note: The definition of work on page 128 means that static friction does work on the moving car, even though the point of the car where the force is applied (the bottom of the tire) is not instantaneously moving relative to the road.]

b) At what rate in watts is the power being supplied?

c) A horsepower is a U.S. unit of power, equal to 746 watts. Convert the power from part (b) into horsepower.

** 29. A golf cart with mass 800 kg (including the mass of two golfers in the cart) is powered by a 2000-watt electric motor. The cart is climbing a 10° slope.

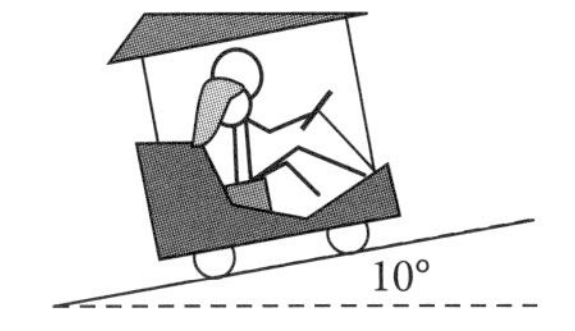

a) If the cart is moving at 1 m/s up the slope, how quickly will it be increasing its height, in m/s?

b) Neglecting friction in the wheels and in the engine, find the maximum speed at which the cart can climb the slope.

** 30. A tractor is towing a barge along a canal. The rope makes an angle of 25 degrees with the direction along the canal. It takes 2000 W of power to overcome the friction of the water on the barge when the barge is moving at 2 m/s.

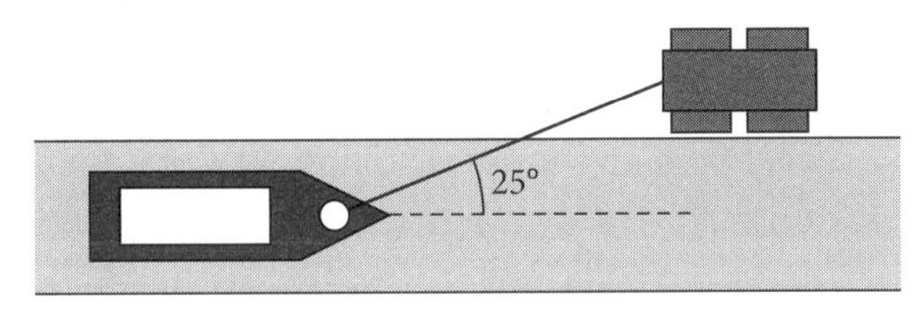

a) What is the component of the tension in the rope that lies parallel to the direction of the barge's travel?

b) What is the tension in the rope?

7.4 ** 31. A brick weighing 50 newtons slows as it slides 4 m across a horizontal floor. If the coefficient of kinetic friction between the brick and the floor is $\mu_K = 0.2$, find the change in kinetic energy.

** 32. A young girl of mass 40 kg slides down a metal slide that is 4 m long and angled 30° above the horizontal, as shown. She arrives at the bottom moving at a speed of 5 m/s. How much energy was converted by friction into thermal energy, warming the girl's jeans and the metal of the slide?

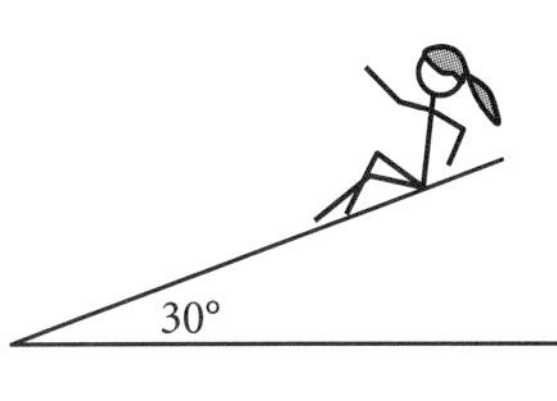

** 33. A 75-kg skier starts from rest at the top of a hill that makes an angle of 35 degrees with the horizontal. The coefficient of kinetic friction is 0.1.

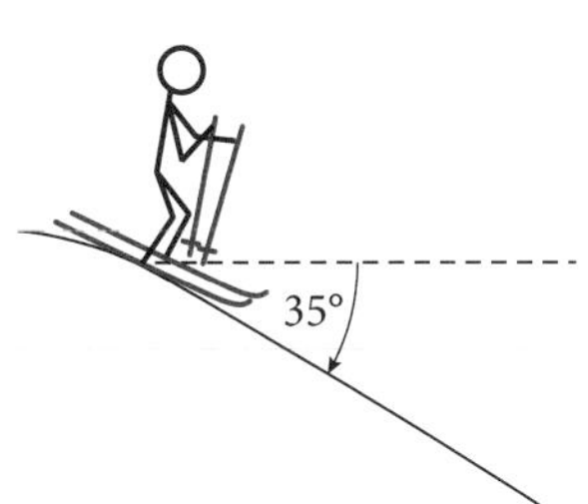

a) How much energy is lost to friction as the skier slides 25 meters downward along the slope?

b) What is the speed of the skier after he has slid those 25 meters?

** 34. Initially, a 10-kg block is sliding along a level floor at a speed of 5 m/s. The coefficient of kinetic friction between the block and the floor is exactly 0.5.

a) What are the initial kinetic, potential, and total energies of the block?

b) What is the force of friction by the floor on the block?

c) What is the work done by the frictional force as the block slides 2 m across the floor (i.e., how much energy is removed from the system)?

d) What are the final kinetic, potential, and total energies of the block?

e) What is the block's speed after sliding 2 m?

** 35. A car moving at 20 km/hr skids 10 m with locked brakes before coming to rest. How far will the car skid (on the same surface) if it is traveling at 80 km/hr?

*** 36. A block of dry ice is attached to a brick by a massless string, as shown. The system is pushed across a slate tabletop with a horizontal force, so that the system has a constant speed of 10 m/s. The dry ice floats on a cushion of gas so that friction between the dry ice and the table is negligible. The coefficient of kinetic friction between the brick and the table is $\mu_K = 0.75$.

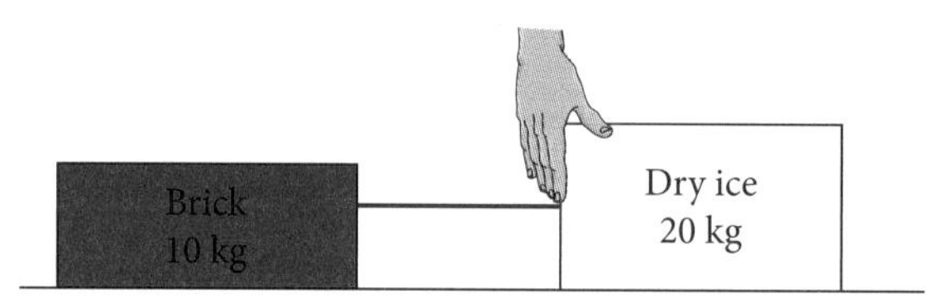

a) Find the power that must be supplied by the hand to keep the system moving at $v = 10$ m/s.

b) When the hand suddenly stops pushing on the system, the ice and brick system gradually slows and then comes to a complete stop. How far does the system slide after the hand stops pushing?

** 37. In an unfortunate hunting accident, a physics professor was shot by an arrow of mass 0.050 kg. The arrow was fired from a spring-loaded launcher in which a spring, with spring constant k = 80 N/m, was initially compressed 0.25 m. The shooter, as yet unidentified, is known to have fired from a window 10 m above where the professor was hit. The arrow penetrated the professor's body to a depth of 0.10 m. [Hint: If you are worried that we haven't told you the angle at which the arrow was shot, you may want to go back to question 6 and make sure you got that one right.]

a) What is the energy of the arrow–spring–earth system just before the spring is released?

b) How fast was the arrow traveling just before striking the professor?

c) What was the average frictional force that the professor exerted on the arrow?

** 38. A 4-kg toy jet sled is accelerated forward from rest by escaping CO_2 from the jet engine. After the sled has gone 10 m, the engine shuts off, leaving the sled going 10 m/s. The coefficient of friction between the sled and the snow is 0.2. How much force did the jet produce? How far will the sled slide after the jet shuts off?

*** 39. A boy slides down a 37° slide from 4 m above the ground. The coefficient of friction between his pants and the slide is 0.2. At the end of the slide, there is a horizontal section 1 m above the ground to slow him down. How long must the horizontal section be if the boy is to come to a full stop at the end? [Hint: The normal force will be different on the slant than on the horizontal.]

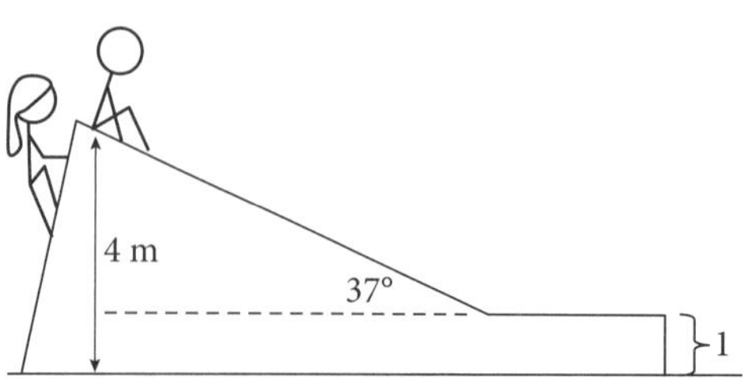

*** **40.** A student who has read *College Physics: Putting It All Together* decides to use his new physics expertise to create an amusement park ride. In this ride, a car with two riders is initially pushed up against a spring (k = 30,000 N/m), compressing it by 2 m. When the spring is released, the car goes up a ramp which is frictionless except for the 2½-meter patch shown. Next the car goes through a loop-the-loop and is eventually shot straight up into the air. He feels that everyone will pay big money to ride this ride. (Wouldn't you?)

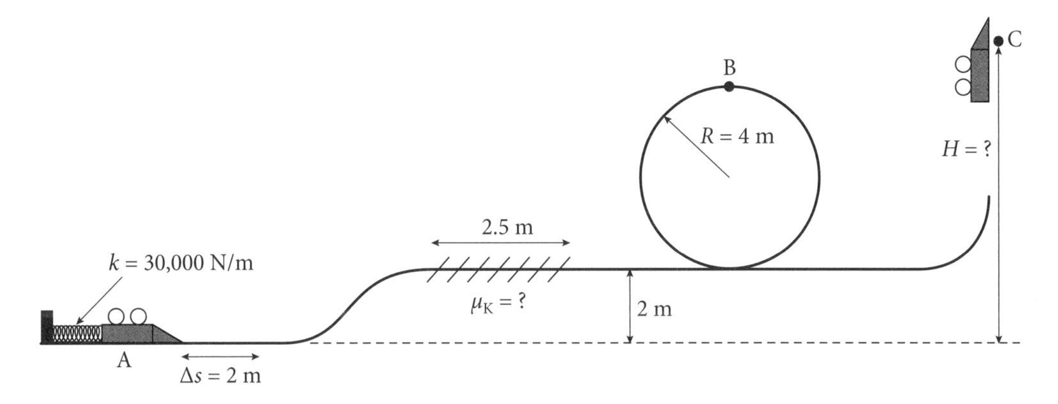

a) The mass of the car with two riders is 250 kg. If the speed of the car is 12 m/s at point B (at the top of the loop-the-loop), find the coefficient of kinetic friction for the rough patch.

b) Find the normal force (magnitude and direction) exerted by the track on the car at point B.

c) Find the maximum height h that the car (with two riders) attains at point C, before falling back onto the horizontal portion of the track.

d) If the car is launched empty to test it, will the height attained at point C be greater than, less than, or equal to the height attained with riders? Explain.

*** **41.** A block starts from rest at point A along the mostly frictionless track shown below. A hand pushes a 4-kg block with a constant force $F_{\text{hand,block}}$ for a distance 2.5 m. The block then encounters a 2-m rough section (μ_K = 0.70) starting at point B. When the block reaches point C, at the bottom of the frictionless bowl, it is traveling at a speed of 9 m/s. It then slides up the track and compresses a spring by Δs = 50 cm before stopping and turning around.

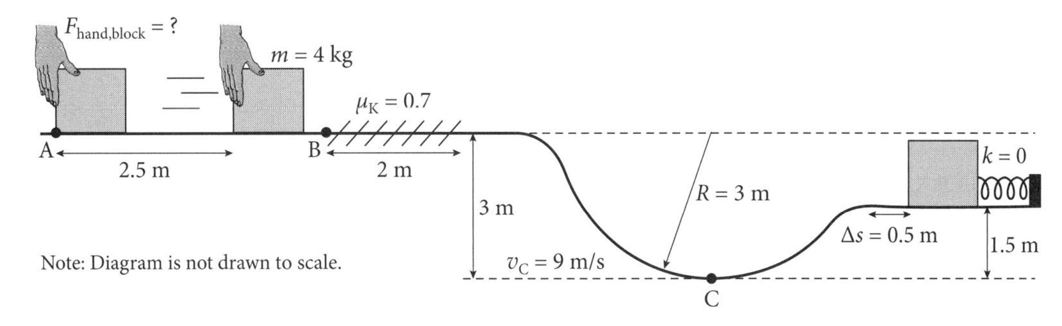

a) Find $F_{\text{hand,block}}$, the force exerted by the hand on the block.

b) Find the normal force exerted by the track on the block when it is at point C.

c) Find the spring constant of the spring.

*** 42. A 3-kg block is launched from rest at point A by a spring of spring constant $k_1 = 64$ N/m onto a mostly frictionless track, though there is a 1.5-m section with friction ($\mu_K = 0.30$) right after point B. The block slides along the track, up over a hill (point C), and then collides with another spring at point D, causing the block to turn around and go the other way.

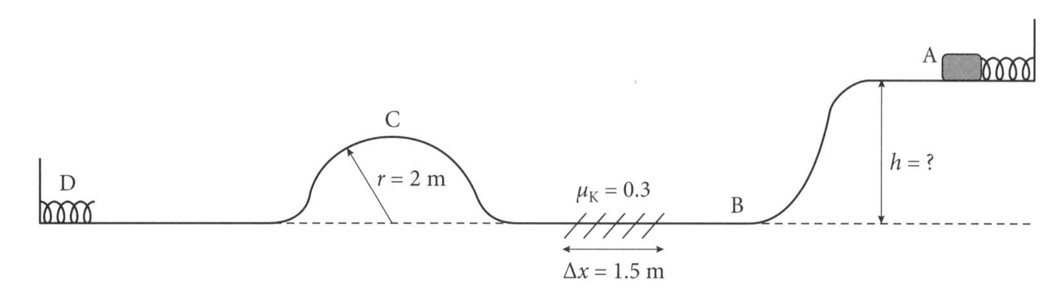

a) The block is found to be traveling at a speed of 4 m/s when it gets to point C (at the top of the hill). If the first spring is initially compressed by an amount $\Delta s = 50$ cm, find the height h of the block at point A.

b) Find the normal force (magnitude and direction) exerted by the track on the block at point C.

c) The spring at point D has spring constant $k_2 = 42$ N/m. How far will the spring be compressed before the block comes to a momentary stop?

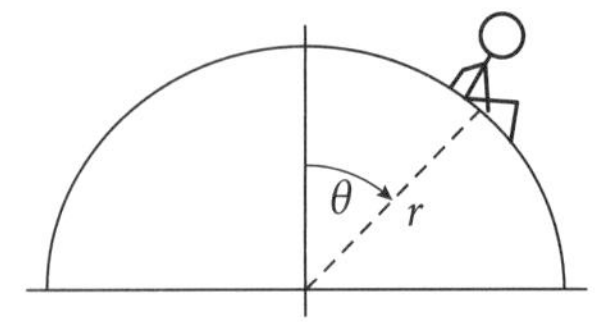

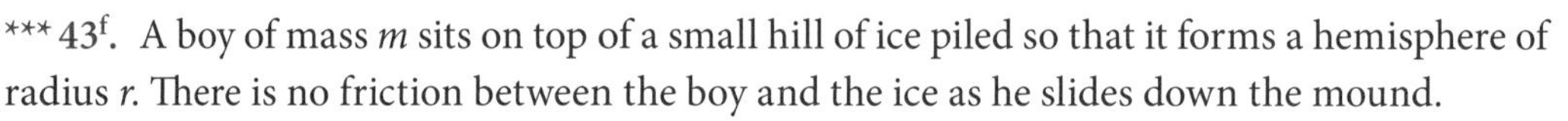

*** 43[f]. A boy of mass m sits on top of a small hill of ice piled so that it forms a hemisphere of radius r. There is no friction between the boy and the ice as he slides down the mound.

a) Find a formula for the speed of the boy as a function of g, r, m, and θ.

b) Find the centripetal component of the force on the boy, as a function of m, g, and θ.

c) Draw a free-body diagram for the boy and find the normal force of the ice on the boy as a function of m, g, and θ.

d) Find a numerical value in degrees for the angle θ at which the boy flies off the ice. [Hint: This is the angle at which the normal force of the ice on the boy drops to zero.]

8

MOMENTUM

So far, both in our discussions of Newton's laws and in our use of energy methods, we have been careful to consider interactions between objects that could be described in terms of constant or, at least, uniformly increasing forces acting over some extended time. Nothing ever banged into anything else in a kind of sharp collision. Our reasons for this were good ones. The forces involved in collisions are hard to describe, making the use of Newton's laws difficult. Also, collisions often result in forms of energy (sound, heat, deformation, etc.) that are difficult to account for, making energy calculations impossible. However, colliding objects are very common, and so we need to learn to deal with them. One of the keys to handling collisions is the concept of "impulse."

8.1 Impulse and Momentum

In our study of energy, we discovered that there is a quantity called work that changes an object's mechanical energy. The work is found by multiplying the applied force by the *distance* over which the force acts. In this section, we will find that there is another interesting and useful quantity that comes from multiplying an applied force by the *time* over which it acts.

Let us begin where we are comfortable, with a constant applied force. We then define a quantity called the *impulse* as the product of the force and the time interval over which it acts. There is no conventional symbol for impulse, so we will just write it

$$\overrightarrow{\text{Impulse}} = \vec{F}_{\text{APP}}\Delta t, \tag{8.1}$$

where $\vec{F}_{\text{APP}}$ is the constant applied force and Δt is the time over which it acts. Just as there is no conventional symbol for impulse, there is also no named unit. Impulse is measured, as can be seen from the definition, in newton · seconds, or N · s. Alternatively, since a newton is itself a $\text{kg} \cdot \text{m/s}^2$, the impulse can also be expressed as a kg · m/s. Note that impulse is a vector that points in the same direction as the applied force.

So what does this accomplish? Let us go back to Newton's second law and assume that a constant applied force acts on an object that is otherwise in equilibrium. This means that the net force is the applied force, and, if the applied force is constant, then the acceleration will be constant. For constant acceleration, the impulse of the applied force will produce a change in the motion characterized by

$$\begin{aligned}\overrightarrow{\text{Impulse}} &= \vec{F}_{\text{APP}}\Delta t = \vec{F}_{\text{NET}}\Delta t = m\vec{a}\Delta t \\ &= m\left(\frac{\vec{v}_{\text{f}} - \vec{v}_{\text{i}}}{\Delta t}\right)\Delta t \\ \overrightarrow{\text{Impulse}} &= m\vec{v}_{\text{f}} - m\vec{v}_{\text{i}}.\end{aligned}$$

Thus, an impulse (a force times the time interval over which it acts) causes a change in the quantity $m\vec{v}$. We have actually seen this quantity before. Back on page 81, we saw that this was the way Newton defined his "quantity of motion." We now call this the *momentum*, defined by the formula $\vec{p} = m\vec{v}$. This definition allows us to write the effect of an impulse on an object in a form called the impulse–momentum equation:

$$\overrightarrow{\text{Impulse}} = \vec{p}_\text{f} - \vec{p}_\text{i} \qquad \text{or} \qquad \vec{p}_\text{i} + \overrightarrow{\text{Impulse}} = \vec{p}_\text{f}. \tag{8.2}$$

The last version serves to emphasize that impulse is the thing that changes an object's momentum from its initial value to its final value. Let us also note that, since impulse and momentum are added together in Equation 8.2, they must have the same units, and indeed they do. Momentum is measured in kg·m/s (or N·s), just like impulse.

We came to Equation 8.2 by assuming a constant force, but we will insist that the impulse of a variable force be defined in such a way that Equation 8.2 always applies.

EXAMPLE 8.1 Momentum of a Bumper Car

Two children are riding in a bumper car at the fair. The mass of the car plus riders is 300 kg. The car is moving straight north at 3 m/s just before it collides head-on with the wall. After the collision, the bumper car is traveling straight south at 2 m/s. What is the impulse that acts on the car during the collision?

ANSWER Because the quantities involved are vectors, we need to be explicit about our coordinate system. Let's choose an axis along the north–south direction and take north to be positive. Then the initial velocity is northward at +3 m/s and the final velocity is –2 m/s north. Equation 8.2 then gives the impulse as

$$\begin{aligned}\overrightarrow{\text{Impulse}} &= \vec{p}_\text{f} - \vec{p}_\text{i} = m\vec{v}_\text{f} - m\vec{v}_\text{i} \\ &= 300\text{ kg }(-2\text{ m/s}) - 300\text{ kg }(+3\text{ m/s}) \\ &= -600\text{ kg}\cdot\text{m/s} - 900\text{ kg}\cdot\text{m/s} \\ &= -1500\text{ kg}\cdot\text{m/s}.\end{aligned}$$

The impulse is negative, with positive taken north, so it is toward the south. The wall stopped the northward momentum of the bumper car and then produced a momentum southward.

If you think about the force on the bumper car by the wall during the collision, it must start out small (actually zero until contact is made), grow sharply as the bumper compresses against the wall, peak at some value, and then drop quickly back to zero as the bumper car rebounds. All of this happens in perhaps a few hundredths of a second. The solid line in Figure 8.1 is a graph showing what this force might look like as a function of time.

We began this chapter by setting a goal of being able to work with forces that varied, though we derived the impulse–momentum equation by considering a constant applied force only. However, the key to working with varying forces may be seen in Figure 8.1. When the force varies, its total impulse is found by considering it as a nearly infinite number of tiny impulses of duration δt. Each small impulse starts with an initial v_i and changes it to a final v_f, which then becomes the v_i for the next small impulse, and so on. The δt widths are chosen to

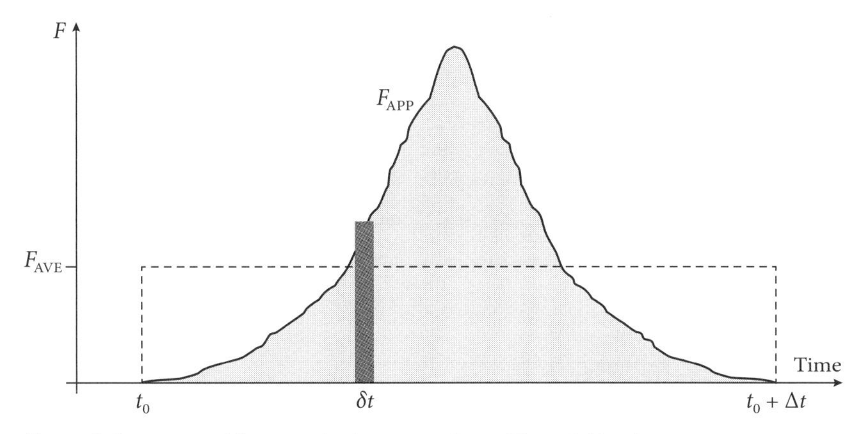

Figure 8.1 An actual force applied to some object (the solid line) over a time Δt versus the average force applied during the same time (dashed line). The red-shaded narrow rectangle is the impulse for a tiny segment of the time, δt.

be short enough that the force is nearly constant during that time, so that the tiny increment of impulse during the time δt is the instantaneous value of F_{APP} times δt. The total impulse is found by adding up all the tiny incremental impulses. When total impulse is defined in this way, it will always satisfy Equation 8.2. Note also that the impulse produced over the tiny δt shown in Figure 8.1 is the height in newtons of the narrow red-shaded rectangle in the figure times its width in seconds. So this small impulse is proportional to the area of the red-shaded rectangle in the figure, meaning that the total impulse between t_0 and $t_0 + \Delta t$ will be proportional to the gray-shaded area between the $F = 0$ axis and the F_{APP} curve.

Figure 8.1 also shows a dashed-line rectangle whose height is the average force F_{AVE} and whose width is Δt. The dashed-line rectangle has the same area as the area under the F_{APP} curve. This means that F_{AVE} is the force which, if it acted over the entire time Δt, would have the same total impulse (1500 N·s) as the actual force F_{APP}. For instance, if the bumper car collision in Example 8.1 lasted for $\Delta t = 0.05$ seconds, we would have $F_{AVE} \times 0.05\text{ s} = 1500\text{ N}\cdot\text{s}$. This would mean that the average force by the wall on the bumper car was $F_{AVE} = 1500\text{ N}\cdot\text{s}/0.05\text{ s} = 30{,}000\text{ N}$. This clearly doesn't tell us everything about the collision. For instance, if the graph is correct, the peak force acting on the bumper car was two to three times greater than this average force. However, in many cases, an estimate of the average force will be all we need.

EXAMPLE 8.2 Impulse on a Hockey Puck

A 1-kg practice hockey puck is sliding east at 5 m/s on a frozen pond. Marc deflects the puck with his stick so that the force by his stick on the puck is directed due north and hit with such a careful touch that the average force he applies is 100 N over a time of 0.05 s. What is the final velocity (speed and direction) of the puck?

ANSWER Given the average force acting on the puck and the duration of the interaction, we can find the impulse. This determines how the puck's momentum changes during the deflection. We know enough to find the initial momentum, so adding the change will give us the final momentum.

$$\vec{p}_f = \vec{p}_i + \overrightarrow{\text{Impulse}}$$

$$\vec{p}_f = m\vec{v}_i + \vec{F}_{AVE}\Delta t$$

The initial momentum is (1 kg)(5 m/s), or 5 kg·m/s, to the east. We can calculate the impulse to be (100 N)(0.05 s), or 5 N·s (or kg·m/s), to the north. If we write these vectors in component form, with the x-component toward the east, we have

$$\vec{p}_f = (5 \text{ kg}\cdot\text{m/s}, 5 \text{ kg}\cdot\text{m/s}).$$

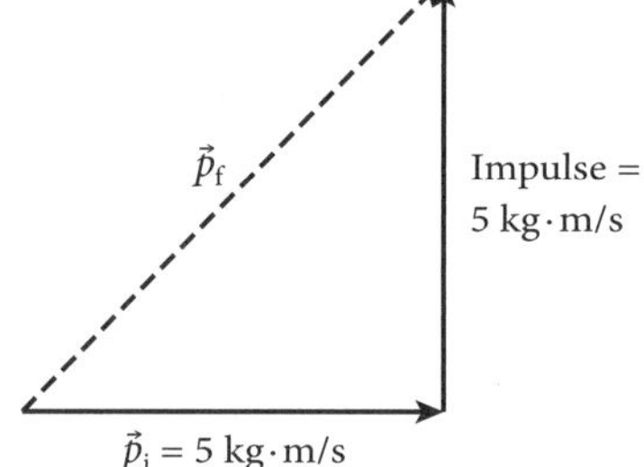

A vector diagram is shown at right. We see the direction of the final momentum is northeast. Using the Pythagorean theorem, we find the magnitude of $\vec{p}_f$ to be 7.1 kg·m/s. We have only to find the corresponding velocity. Since the puck has a mass 1 kg, its velocity will be $v = p_f/m = 7.1$ m/s to the northeast.

8.2 Collisions

In the last section we found that the idea of impulse provided a practical way to approach forces that vary in time, like those that occur during a collision between two bodies. Momentum then arose as the quantity that appeared in the impulse–momentum equation, an alternative version of Newton's second law. However, the real utility of the concept of momentum comes when we look at the total momentum of an isolated system of bodies that interact or collide with one another.

An *isolated system* is a set of bodies that have no interaction with the outside world. We recall from Chapter 7 that the *total energy* of an isolated system is strictly conserved. But we also saw that this wasn't the last word on the subject, because *mechanical energy* may or may not be conserved, depending on the types of forces acting between the bodies of the system. When there are non-conservative forces involved, mechanical energy can be converted into other forms. Mechanical energy is evident in the speed and position of the bodies in the system, but, if mechanical energy turns into thermal energy, it is no longer visible in the motion alone. This is the case for energy. However, as we shall soon see, the *total momentum* of a system has no such back-door exit strategies available. It cannot hide inside a body like thermal energy does. It is always apparent in the motion. This is what makes the idea of momentum so useful in dealing with isolated systems.

A "collision" is always divided into three phases. First, there is a phase when the bodies have no significant interaction with each other, having constant initial momenta. Second, there is a short time when the bodies interact via internal forces that produce equal and opposite impulses on each body, changing their momenta. Third, there is a final phase where the bodies are again no longer interacting with each other, and the final momenta are again constant.

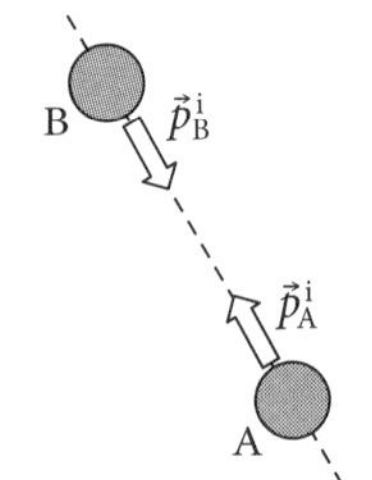

Figure 8.2a The initial momenta.

To see how this all works, let us consider a system consisting of two objects, A and B, that are about to collide with each other, as in Figure 8.2a. We assume that there are no forces acting on either of the bodies from anything outside the system, or, at least, that any external forces may be neglected because they are small compared to the collision forces. We write the initial momenta of the two bodies as $\vec{p}_A^i$ and $\vec{p}_B^i$, where the subscripts indicate the object and the superscripts indicate that this is the initial state. The objects then collide with one another for a short time Δt, after which they move off with new momenta $\vec{p}_A^f$ and $\vec{p}_B^f$ (Figure 8.2b).

Each object's change in momentum is described by the impulse–momentum theorem for that object. For object A, this gives

$$\vec{p}_A^{\,f} = \vec{p}_A^{\,i} + \vec{F}_{B,A}\Delta t, \quad (8.3)$$

where $\vec{F}_{B,A}$ is the average force by B on A during the collision. Similarly, for object B we get

$$\vec{p}_B^{\,f} = \vec{p}_B^{\,i} + \vec{F}_{A,B}\Delta t.$$

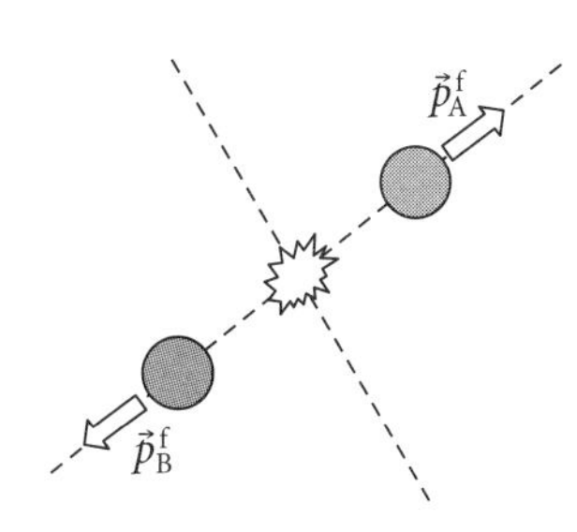

Figure 8.2b The final momenta.

Now we need to realize that $\vec{F}_{A,B}$ and $\vec{F}_{B,A}$ are Newton's third-law companion forces, so that $\vec{F}_{A,B} = -\vec{F}_{B,A}$. We can therefore rewrite the last equation as

$$\vec{p}_B^{\,f} = \vec{p}_B^{\,i} - \vec{F}_{B,A}\Delta t. \quad (8.4)$$

If we now add Equations 8.3 and 8.4 together, remembering that the time intervals are the same in both equations, the terms containing the forces will cancel, leaving

$$\vec{p}_A^{\,i} + \vec{p}_B^{\,i} = \vec{p}_A^{\,f} + \vec{p}_B^{\,f}. \quad (8.5)$$

On each side of Equation 8.5 we have a vector sum, and we certainly know how to add vectors by now. In each case, we are vector-adding the momentum vector for body A to the momentum vector for body B. The result of this operation is a new vector, which we will obviously want to call the total momentum of the system,

$$\vec{p}_{TOT} = \vec{p}_A + \vec{p}_B.$$

So, on the left-hand side of Equation 8.5, we have the initial total momentum of the system, $\vec{p}_{TOT}^{\,i}$, and, on the right-hand side, we have the final total momentum, $\vec{p}_{TOT}^{\,f}$. Then you see what Equation 8.5 means. As long as no external forces act on either body in the system, there is nothing the two bodies can do to each other—no weird non-conservative force, no ugly irregular time-varying force—that will change the total momentum of the system. *As long as a system of bodies is isolated from the outside world*, the total momentum of the system is drop-dead conserved. It is an absolutely reliable fact of the motion. Other things may change. Kinetic energy may turn into thermal energy, atomic nuclei may decay and become different nuclei with different masses, but *the total momentum of the system just never ever changes.*

EXAMPLE 8.3 The Ballistic Pendulum

A 10-g bullet is fired horizontally at 300 m/s into a 2-kg block of wood suspended on the end of a long string fastened to the ceiling. What is the speed of the block with the bullet inside immediately after the bullet embeds in the block? What is the maximum gain in height that the block+bullet achieves as it swings?

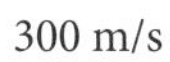

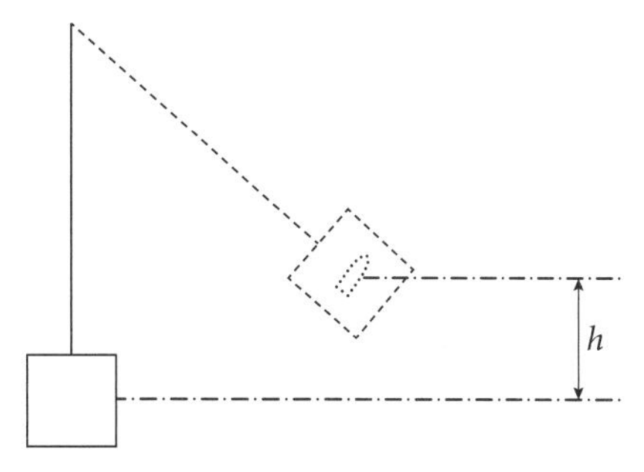

ANSWER This is really two problems in one. The purpose of the first part is to make clear that, even in a complicated setup like this one, a collision is a collision, and momentum is conserved. The second part is what happens after the collision. Here, there are external forces, gravity and rope tension, making it an energy problem like many we saw in the previous chapter. This example also demonstrates that to apply these concepts we must learn when it will be useful to apply energy concepts and when momentum concepts will be most productive. We will also give some helps for addressing this question in the next section.

First, we treat the collision using the concept of momentum conservation. We take as our system the block + bullet because these are the two things that collide. The collision itself only happens in one dimension (the two-dimensional motion doesn't start until after the collision is over), so this is a one-dimensional momentum conservation problem. We take the direction of the initial momentum of the bullet to be positive. The initial momentum of the system is found by adding up the momenta of the constituents:

$$p^{i}_{TOT} = m_{bullet}v^{i}_{bullet} + m_{block}v^{i}_{block} = (0.01\text{ kg})(300\text{ m/s}) + (2.0\text{ kg})(0\text{ m/s}) = +3\text{ kg}\cdot\text{m/s}.$$

After the collision, the total momentum must still be +3 kg · m/s although the momenta of the individual parts will have changed. If the bullet collides with a block, there are an infinite number of combinations of bullet and block speeds that would add up to +3 kg · m/s. However, in this case we have an additional constraint. We know that the bullet ends up embedded in the block, so the two move with the same final velocity, which we call v^f. Conservation of momentum then requires

$$p^{f}_{TOT} = m_{bullet}v^{f} + m_{block}v^{f} = (m_{bullet} + m_{block})v^{f} = (2.01\text{ kg})v^{f} = 3\text{ kg}\cdot\text{m/s}.$$

This gives

$$v^{f} = 1.49\text{ m/s}.$$

So, the block+bullet system moves at 1.49 m/s immediately following the collision.

In the second part, we recognize that the block+bullet system now swings on the end of a rope that does no work on the system, so energy is now conserved. As the system swings, it converts kinetic energy into gravitational potential energy. At the maximum height, all of the energy is in the potential energy bucket.

Just after the collision and before the block+bullet starts to rise, the energy is

$$E_{initial} = KE_{initial} + PE_{initial} = \frac{1}{2}(2.01\text{ kg})(1.49\text{ m/s})^2 + 0 = 2.23\text{ J}.$$

The fact that this total does not change allows us to find the maximum height of the swinging pendulum. We write

$$E_{final} = 2.23\text{ J} = KE_{final} + PE_{final} = 0 + (m_{bullet} + m_{block})gh$$

$$\text{or}\quad h = \frac{E_{final}}{(m_{bullet} + m_{block})g} = \frac{2.23\text{ J}}{(2.01\text{ kg})(10\text{ N/kg})} = 0.11\text{ m}.$$

Although it wasn't required to solve the problem in Example 8.3, we can find out how much kinetic energy was lost during this collision, if we want. Before striking the block, the bullet's kinetic energy was KE = $\frac{1}{2}mv^2 = \frac{1}{2}(0.01)(300)^2 = 450$ J, and the block was at rest. So, immediately before the collision, there were 450 J of kinetic energy in the system. Immediately after the collision there were only 2.23 J of kinetic energy, as we calculated above. Thus, about 99.5% of the initial kinetic energy was lost to other forms of energy (mostly thermal). Clearly, if we had assumed that kinetic energy would be conserved during the collision we would have badly overestimated the speed of the block after the collision (and, of course, violated the principle of the conservation of momentum). So we hope the moral of this story is clear: *You cannot count on kinetic energy being conserved in a collision, but you can always count on momentum being conserved in a collision.*

8.3 The Three Flavors of Collisions

Imagine two 1500-kg cars, each driving at 10 m/s, about to hit head-on (see Figure 8.3). The momenta of the two cars have the same magnitude but they are pointing in opposite directions. This makes the initial total momentum of the system exactly zero.

Figure 8.3 Two cars approaching each other initially.

Although the tires of the cars are in contact with the pavement when they hit, the magnitude of the force between the two cars is so much greater than friction with the pavement or other forces from the environment, that the system of the two cars is effectively an isolated system. This means that the total momentum will always be the same. In particular, it must still be zero after the collision. Figure 8.4 shows three particular cases among the infinite possibilities for the final velocities. All of these cases satisfy the conservation of a zero total momentum.

In Figure 8.4a, the cars have the same speeds they had coming into the collision, so the total kinetic energy after the collision is the same as it was before the collision. Kinetic

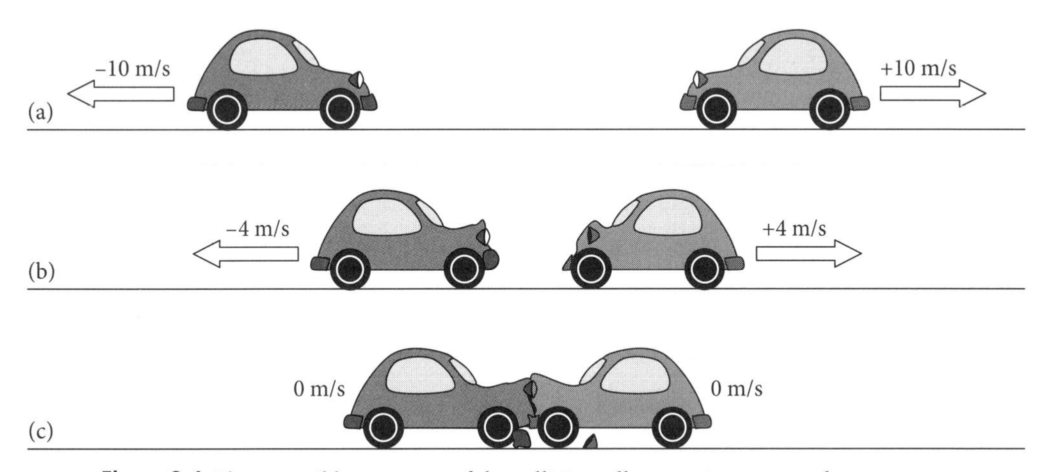

Figure 8.4 Three possible outcomes of the collision, all preserving zero total momentum.

energy has been conserved. This could happen only if the forces between the two cars were all conservative forces, like perfectly springy bumpers that could turn the cars around without any loss of energy. Collisions that conserve kinetic energy like this are called *elastic*, or *totally elastic*. Most collisions that we encounter do not meet this standard, since most collisions lose some energy (for example, sound waves carry energy, so, if you hear the collision, some of the kinetic energy has been converted to sound waves). Generally, the only way you can count on kinetic energy being conserved in a problem is if the problem tells you that the collision is elastic.

In Figure 8.4b, we see that there has been significant loss of kinetic energy during the collision. At least some of the forces that deformed the fronts of the cars were non-conservative, and a lot of the initial kinetic energy has been transformed to other forms of energy (mostly to thermal energy). We use the term *inelastic collision* to describe collisions that lose some kinetic energy.

Finally, let us consider Figure 8.4c. Here, the two cars have locked bumpers and stuck together, coming to a complete stop. The final kinetic energy is zero. The system has lost as much kinetic energy as it could (in this case, all of it). Collisions that lose the maximum possible kinetic energy are called *maximally inelastic*.[1]

Let us consider the maximally inelastic case a bit more. In the two-car collision of Figure 8.4, the total conserved momentum was always zero, so it was possible for both cars to end up at rest. All of the initial kinetic energy was lost. However, think what could have happened if the red car had initially been moving to the right at 12 m/s, instead of 10 m/s. Then the initial total momentum would have been (1500 kg)(12 m/s) – (1500 kg)(10 m/s) = +3000 kg · m/s, to the right. But, the collision must conserve momentum, since the two cars constitute an isolated system. This means that it would not be possible to end up with both cars at rest. The equal and opposite impulses acting on the two cars during the collision cannot simultaneously bring both cars to a stop. If the magnitude of the impulse is exactly the right amount to stop the gray car, it will not be the amount needed to stop the red car, and vice versa. So, if the initial momentum is non-zero, then the final momentum will be non-zero, and we must always end up with something in the system that is moving.

Thus, the way to tell if a collision is maximally inelastic is not to look for a case where everything comes to a stop, but to note that a maximally inelastic collision is one in which all the bodies involved in the collision end up stuck together, all moving in the same direction (the direction of the initial momentum vector) at the same speed. This is the situation that satisfies a non-zero momentum requirement with the least excess kinetic energy. The bullet+block collision we studied in Example 8.3 is an example of a maximally inelastic collision. We know that it is maximally inelastic because the two objects became locked together. And, as we saw, that system lost about 99.5% of its initial kinetic energy during the collision.

---

1. Many textbooks refer to these collisions as "perfectly inelastic" or "completely inelastic." We find these terms misleading, since they seem to suggest that all the initial kinetic energy is lost in the collision. Since this is not generally the case, we prefer the term "maximally inelastic."

EXAMPLE 8.4 A One-Dimensional Collision

A 10-kg block is sliding to the right across a frictionless floor at 4 m/s. A 5-kg block is traveling left at 2 m/s such that it hits the other block head-on. ("Head-on" is our way of promising that nothing in the collision will change the one-dimensional nature of this problem.) After the collision, the 10-kg block is observed moving to the right at 1 m/s. Is this collision elastic, inelastic, or maximally inelastic?

ANSWER Before we can identify the type of collision, we need to know the final velocity of the 5-kg block. For instance, if we found that it was also moving at 1 m/s to the right, we would know the blocks were stuck together (maximally inelastic). To find the 5-kg block's final velocity we use the fact that momentum is conserved. We have enough information to find the initial momentum of the system:

$$\vec{p}^{\,\mathrm{i}}_{\mathrm{TOT}} = (10\ \mathrm{kg})(+4\ \mathrm{m/s}) + (5\ \mathrm{kg})(-2\ \mathrm{m/s}) = +30\ \mathrm{kg\cdot m/s},$$

where we have chosen positive to mean to the right. After the collision, we know the momentum of the 10-kg block. It is

$$\vec{p}^{\,\mathrm{f}}_{10\ \mathrm{kg}} = (10\ \mathrm{kg})(+1\ \mathrm{m/s}) = +10\ \mathrm{kg\cdot m/s}.$$

Since the total must still equal +30 kg·m/s, the final momentum of the 5-kg block must be +20 kg·m/s. Thus, the 5-kg block is moving to the right with a speed of 4 m/s after the collision. The two final velocities are not the same, so we know that the collision was not maximally inelastic. To determine whether the collision is elastic or inelastic we must compare the kinetic energies before and after the collision. The two numbers are

$$\mathrm{KE}_{\mathrm{initial}} = \frac{1}{2}(10\ \mathrm{kg})(4\ \mathrm{m/s})^2 + \frac{1}{2}(5\ \mathrm{kg})(2\ \mathrm{m/s})^2 = 80\ \mathrm{J} + 10\ \mathrm{J} = 90\ \mathrm{J}$$

$$\mathrm{KE}_{\mathrm{final}} = \frac{1}{2}(10\ \mathrm{kg})(1\ \mathrm{m/s})^2 + \frac{1}{2}(5\ \mathrm{kg})(4\ \mathrm{m/s})^2 = 5\ \mathrm{J} + 40\ \mathrm{J} = 45\ \mathrm{J}.$$

Because the kinetic energy of the system is reduced during the collision, and because the two blocks do not become locked together, we conclude that the collision is inelastic, but not maximally inelastic. Note that knowing that the kinetic energy is reduced is enough to determine that the collision is inelastic, but it is not enough to distinguish a maximally inelastic collision. For that, we must also know if the objects end up with the same final velocity.

8.4 Collisions in Two Dimensions

Example 8.2 emphasized that the impulse–momentum equation is a vector equation. The quantities involved, forces and velocities, have directions that must be considered, in addition to the magnitudes. Yet, so far, we have dealt only with collisions that happen along a straight line—one-dimensional collisions. Although we did have to remember that momentum is a vector in solving these problems, the vector part of the problem was easily handled by picking one direction to be positive and the other negative. Now, we will expand our horizons a bit. Our real world has three dimensions, of course, but the mathematics required in a full 3-dimensional analysis of a collision is much harder than for two, and it doesn't really add very much to the physics. So let us just look at a two-dimensional problem and see how we must now be very explicit in our treatment of the vector nature of momentum.

Based on the philosophy that a long gory example is sometimes the best way to put everything together, we choose to present . . . a long gory example.

EXAMPLE 8.5 The Red Car and the Gray Truck

A red 1000-kg car is driving due east at 30 m/s. At the same time, a gray 2000-kg truck is driving due north at 20 m/s. The two collide in an intersection and lock bumpers so that they stick together. After the collision, they both have their brakes locked, so they are skidding together. Assume that the coefficient of kinetic friction between the tires and the road for both vehicles is exactly 0.9.

a) What is the speed of the pair immediately following the collision?

b) In what direction does the pair move following the collision?

c) What impulse does the car apply to the truck during the collision?

d) Is the collision elastic, inelastic, or maximally inelastic?

e) How far does the pair slide before coming to rest?

ANSWER This multi-part problem reviews most of the concepts from this chapter as well as providing a little more practice using the work–energy theorem (part e).

a) The first two parts of this problem are a direct application of the principle of momentum conservation in a collision. We begin by finding the initial momentum of the system. The *magnitudes* of the two momenta are

$$p^{\mathrm{i}}_{\mathrm{car}} = (1000 \text{ kg})(30 \text{ m/s}) = 30{,}000 \text{ kg}\cdot\text{m/s}$$
$$p^{\mathrm{i}}_{\mathrm{truck}} = (2000 \text{ kg})(20 \text{ m/s}) = 40{,}000 \text{ kg}\cdot\text{m/s}.$$

If we define a coordinate system with the x-axis to the east and the y-axis to the north, the total initial momentum can be written in component notation as

$$\vec{p}^{\,\mathrm{i}}_{\mathrm{TOT}} = (30000, 40000) \text{ kg}\cdot\text{m/s}.$$

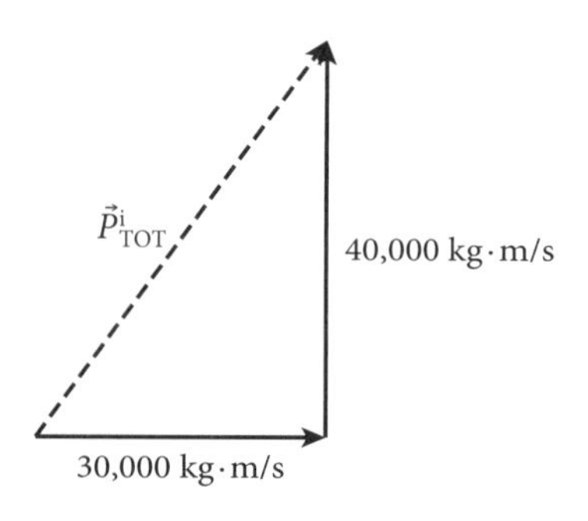

We show the components of the initial total momentum as the solid lines in the diagram at right. The vector triangle is a 3-4-5 triangle, so the magnitude of the initial momentum (and of the final momentum, for that matter) is 50,000 kg·m/s. Note also that the direction of the total momentum is 53° north of east, and that this is true even though no object is actually moving in that direction. This vector will be conserved all through the collision. As long as the system is isolated, nothing can ever happen to change the total momentum vector. After the collision, the vehicles end up locked together, creating a single object of mass 3000 kg moving at some velocity $\vec{v}^{\mathrm{f}}$. Since the mass of the two locked cars times their speed must give us $mv^{\mathrm{f}} = 50{,}000$ kg·m/s, the speed of the cars immediately after the collision must be $v^{\mathrm{f}} = (50000 \text{ kg}\cdot\text{m/s})/(3000 \text{ kg})$, or 16.7 m/s.

b) Immediately after the collision, there is only one object, so that object must be moving in the direction of the system's momentum. It will therefore move in the same direction as the initial total momentum, or 53° north of east.

c) The impulse that the car applies to the truck must, according to the impulse–momentum theorem, be equal to the change in the truck's momentum. We know that the truck's momentum before the collision is 40,000 kg·m/s north. To find the momentum of the truck after the collision, we use its mass and speed to get the magnitude of the momentum as

$$p^{\mathrm{f}}_{\mathrm{truck}} = (2000\ \mathrm{kg})(16.7\ \mathrm{m/s}) = 33{,}400\ \mathrm{kg \cdot m/s};$$

and the direction will be 53° north of east. To find the impulse, we need to find the change in momentum, the final minus the initial. It is convenient to first divide the truck's final momentum into components:

$$p^{\mathrm{f}}_{\mathrm{truck},x} = 33{,}400\cos(53°) = 33{,}400(0.6) = 20{,}000\ \mathrm{kg \cdot m/s};$$
$$p^{\mathrm{f}}_{\mathrm{truck},y} = 33{,}400\sin(53°) = 33{,}400(0.8) = 26{,}700\ \mathrm{kg \cdot m/s}.$$

We then find that

$$\mathrm{Impulse}_x = 20{,}000\ \mathrm{kg \cdot m/s} - 0 = 20{,}000\ \mathrm{kg \cdot m/s};$$
$$\mathrm{Impulse}_y = 26{,}700\ \mathrm{kg \cdot m/s} - 40{,}000\ \mathrm{kg \cdot m/s} = -13{,}300\ \mathrm{kg \cdot m/s}.$$

These impulse vector components are diagrammed at right. We use our usual vector methods to find that the magnitude of the impulse on the truck is 24,000 kg·m/s and that the angle of the impulse is $\theta = 56°$, south of east. The impulse that the truck applied to the car would have the same magnitude, but it would point in the opposite direction. You might want to verify this by redoing the above calculation for the car.

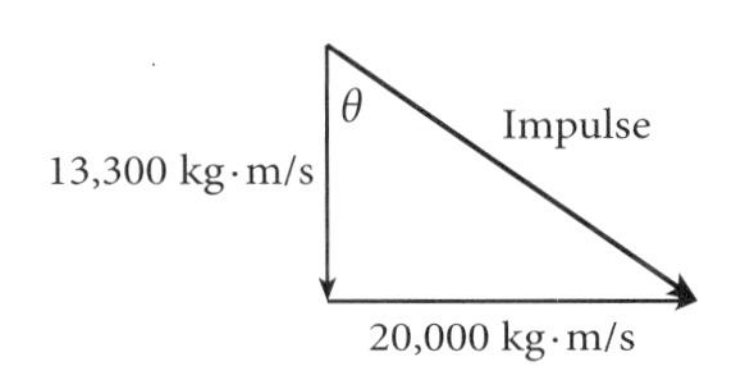

d) In this collision, the objects are locked together and have the same final velocity. Thus we know that the collision is maximally inelastic.

e) Probably the easiest way to solve this problem is to use the work–energy theorem. If we consider the motion of the car and truck from the time immediately after the collision until they have come to rest, the mechanical energy of the system is decreased through a process we understand—kinetic friction during the skid. So we can solve this part of the problem by finding the mechanical energy of the system immediately after the collision and asking how far it must slide to get rid of this energy. The only mechanical energy of interest here is the kinetic energy.

$$\mathrm{KE}_{\mathrm{after\ collision}} = \frac{1}{2}m_{\mathrm{total}}v^2 = \frac{1}{2}(3000\ \mathrm{kg})(16.7\ \mathrm{m/s})^2 = 418{,}000\ \mathrm{J}.$$

This energy is reduced to zero during the skid because of the work done by the kinetic frictional force. We can find the magnitude of this force from

$$f_{\mathrm{road,tires}} = \mu_{\mathrm{K}} N_{\mathrm{road,tires}} = \mu_{\mathrm{K}} m_{\mathrm{total}} g = (0.9)(3000\ \mathrm{kg})(10\ \mathrm{m/s^2}) = 27{,}000\ \mathrm{N}.$$

The work done by this force is the magnitude of the force multiplied by the distance that the vehicles slide. So, using d to represent the length of the skid, we can write

$$W = f_{\mathrm{road,tires}}\, d = \Delta E$$

$$d = \frac{\Delta E}{f_{\mathrm{road,tires}}} = \frac{418{,}000\ \mathrm{J}}{27{,}000\ \mathrm{N}} = 15.5\ \mathrm{m}.$$

Though complicated, this example integrates a number of important ideas from this chapter. It is worth reviewing several times to be sure that you see the "big picture" and don't get lost in the technical details. By the way, you might be interested to know that this kind of thing is exactly what police departments do in reconstructing an accident. They measure the lengths of incoming skid marks to determine how much kinetic energy was lost before the collision; they assume that momentum was conserved during the collision; and they measure the lengths of the outgoing skid marks to determine the kinetic energy lost after the collision. This way, they are able to determine each car's speed going into the accident and see if anyone was exceeding the speed limit.

So, let us sum up. Here is the procedure for handling problems involving collisions. First, make sure that you have included in the system all bodies that will bump into each other. Second, you might as well just go ahead and write down Equation 8.5, because you know you are going to have to use it eventually. Pick out the masses of whatever bodies you have, resolve all initial and final velocities into components, and then write down as many versions of Equation 8.5 as you have components in the problem. Some velocities or masses or angles may be unknown, and you will have to keep them as letters to be solved for. Note that the number of dimensions in the problem will tell you how many equations you will have and, therefore, how many unknowns will be determined. For motion that you know is confined to one dimension, Equation 8.5 will give a single equation that can determine one unknown. If the motion is going to be two-dimensional, there will be two independent equations that come out of Equation 8.5, and you will be able to determine two unknowns by using the single principle of conservation of momentum.

Finally, let us be clear on the limitation of the momentum method for solving collision problems. As we mentioned back in Section 1.1, the purpose of the laws of physics is to be able to correctly predict the future state of some system, knowing the details of an initial state. If we know the initial velocities of two objects in a collision, and the details of the forces that act during the collision, Newton's laws are sufficient to exactly predict the final velocities of both objects after the collision. In a two-dimensional collision of two bodies, that means predicting both final speeds and both final directions—four quantities. The two components of the momentum conservation equation in two dimensions provide only two independent equations, which is not enough to completely determine the final state of things. What is missing in, say, a two-car collision is the detail of the forces involved in the crunching of bumpers and grilles. While such an analysis is possible in principle, no one will do it in practice. Fortunately, the law of conservation of momentum gives a lot of information without having to go into that level of detail at all.

In our next section, however, we discuss one of the few situations where the conservation laws are sufficient to completely specify the final state after a collision. This is the case of elastic collisions in one dimension.

8.5 Elastic Collisions in One Dimension

In this section, we add no new physics, but simply consider how to use the conservation of kinetic energy in problems where we know that the collision is elastic. Just a warning—the algebra we need to do for the solution here is a little tougher than our average example. But take it slowly and you'll be OK.

EXAMPLE 8.6 An Elementary Particle Collision in One Dimension

Two elementary particles, of identical mass, travel in colliding beams from laboratory accelerators. One, call it particle A, is moving to the right at a speed 4×10^6 m/s and the other, particle B, is moving to the left at a speed of 3×10^6 m/s. The beams are exactly aligned, so that the collision is head-on and the two particles rebound back along the directions they came. What will be the final velocities of the two particles?

ANSWER First, let us point out that one of the few times that we can assume that a collision is elastic without being told is when the colliding objects are elementary particles. This is because they have no internal structure where they can hide any energy in another form. Second, although there are no details about the kind of force that acts between the two particles, the fact that the entire collision is one-dimensional already limits a lot of the details of the interaction. No non-symmetric interactions will kick these particles away from the line of their incoming directions. Since the collision is elastic, we know that kinetic energy is conserved, in addition to the usual momentum conservation we can count on in all collisions in isolated systems. This means that there will be *two* independent equations relating the initial and final velocities of the particles. And *this* means that we will have enough information from conservation laws alone to determine the two unknown final velocities.

We are not given the mass of the two identical particles, so let us designate this with a letter m and hope that it will not matter in the final solution. Conservation of momentum in one dimension lets us write

$$mv_A^i + mv_B^i = mv_A^f + mv_B^f,$$

and energy conservation requires

$$\frac{1}{2}m(v_A^i)^2 + \frac{1}{2}m(v_B^i)^2 = \frac{1}{2}m(v_A^f)^2 + \frac{1}{2}m(v_B^f)^2.$$

Sure enough, m cancels out of both of these equations. If we take velocities to the right as positive, we have $v_A^i = 4 \times 10^6$ m/s and $v_B^i = -3 \times 10^6$ m/s, so the two equations become

$$v_A^f + v_B^f = 4 \times 10^6 - 3 \times 10^6 = 1 \times 10^6$$

$$(v_A^f)^2 + (v_B^f)^2 = (-3 \times 10^6)^2 + (4 \times 10^6)^2 = 25 \times 10^{12}.$$

Please note that, after having carefully checked that all numbers are in meters and seconds, we have dropped the units in these two equations.

If you are a little rusty on how to solve two equations for two unknowns, you might want to check out the algebra review we have back in Appendix A. We can solve the first equation for v_B^f, giving $v_B^f = (1 \times 10^6) - v_A^f$, and substitute this in the second equation to get

$$(v_A^f)^2 + [1 \times 10^6 - v_A^f]^2 = 25 \times 10^{12} \quad \text{or} \quad (v_A^f)^2 = 25 \times 10^{12} - [1 \times 10^6 - v_A^f]^2.$$

Squaring the term in the square brackets turns this into

$$(v_A^f)^2 = 25 \times 10^{12} - [1 \times 10^{12} - (2 \times 10^6)v_A^f + (v_A^f)^2]$$

or

$$(v_A^f)^2 = 24 \times 10^{12} + (2 \times 10^6)v_A^f - (v_A^f)^2.$$

If we bring everything to the left side of the equation, we get

$$2(v_A^f)^2 - (2 \times 10^6)v_A^f - 24 \times 10^{12} = 0.$$

We can divide by 2, giving a simple quadratic equation:

$$(v_A^f)^2 - (10^6)v_A^f - 12 \times 10^{12} = 0.$$

There are always two solutions to a quadratic equation. We use the quadratic formula (see Appendix A.4) to get

$$v_A^f = \begin{cases} 4 \times 10^6 \\ -3 \times 10^6. \end{cases}$$

First, let's be clear why there are two solutions. The initial velocity of particle A was 4×10^6 m/s to the right and the first solution has the final velocity of particle A as 4×10^6 m/s. This would be the right answer if particle A never hit particle B. But remember that the equations only tell us what final velocities conserve momentum and kinetic energy. They have no way of knowing that a collision occurred. But *we* know that there was a collision, so we ignore the first solution and take the second. If v_A^f is -3×10^6 m/s, then the final velocity of particle B can be found from

$$v_B^f = 1 \times 10^6 - v_A^f = 1 \times 10^6 - (-3 \times 10^6) = 4 \times 10^6.$$

The two particles have exchanged velocity. Particle A started out at 4×10^6 m/s and ended up going -3×10^6 m/s, while particle B started with -3×10^6 m/s and ended up going 4×10^6 m/s.

Note that the answer did not depend on the nature of the forces between the two particles. They could have been two protons, repelling each other electrically (which we will learn about in Chapter 16), or they could have been two neutrons, interacting only at close distance via the strong nuclear force (see Chapter 26). The details of the motion during the time the forces are acting would be very different in those two cases, but, once the forces have had their effects, the conservation laws are enough to correctly predict the velocities.

8.6 Summary

In this chapter you should have learned the following:

- You should be able to calculate the impulse acting on an object by noting the change in its momentum, using Equation 8.2. You should also be able to calculate impulse by use of Equation 8.1 when the force is constant.
- You should be able to carefully define the elements of an "isolated system" in which no forces intrude from the outside world or in which outside forces may be neglected compared to the large forces involved, say, in a collision.
- You should be able to write down the equations of momentum conservation in one or two dimensions, and to use these to solve for unknowns in the initial or final state of the system.
- You should be able to recognize where, in a multi-step problem, there is a momentum-conserving collision that may be used to connect properties of the system before the collision to those after, and then to recognize where momentum conservation may no longer apply. See Examples 8.3 and 8.5.

CHAPTER FORMULAS

Impulse: $\overrightarrow{\text{Impulse}} = \vec{F}_{\text{APP}}\Delta t$

Momentum: $\vec{p} = m\vec{v}$

Impulse–momentum: $\overrightarrow{\text{Impulse}} = \vec{p}_{\text{f}} - \vec{p}_{\text{i}}$

Momentum conservation: $\vec{p}_{\text{A}}^{\text{i}} + \vec{p}_{\text{B}}^{\text{i}} = \vec{p}_{\text{A}}^{\text{f}} + \vec{p}_{\text{B}}^{\text{f}}$

PROBLEMS

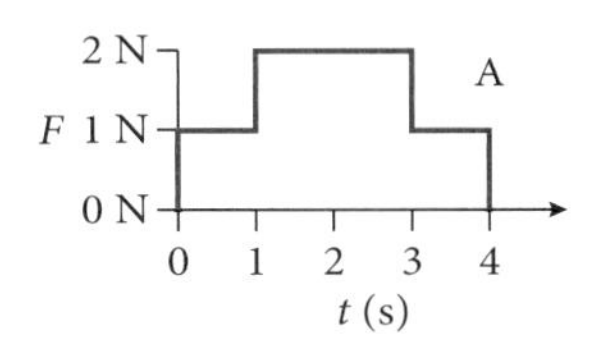

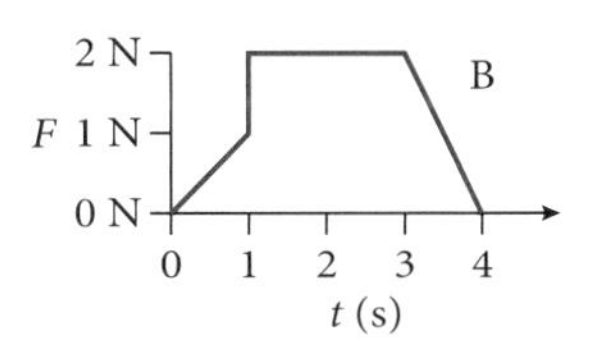

8.1 * 1. What is the impulse of a constant force of 300 N acting for 0.01 s?

* 2. What is the impulse provided by a force whose magnitude as a function of time is given by graph A at right?

** 3. What is the impulse provided by a force whose magnitude as a function of time is given by graph B at right?

* 4. In a collision of a car with a wall, the wall acts over a time 0.15 s and provides an impulse of 4000 N · s. What is the average force of the wall on the car?

* 5. What is the momentum of a 60-kg woman running north at a speed of 4 m/s?

* 6. What impulse is required to stop a 160 g pool ball moving at 2 m/s?

** 7. A softball with mass 200 g is thrown horizontally directly toward a brick wall. It is traveling at 15 m/s just before hitting the wall and rebounds from the wall at 10 m/s, still traveling horizontally. The ball is in contact with the wall for 0.02 s and the ball is compressed a maximum of 7 mm.

a) What is the magnitude of the ball's change in momentum from just before to just after striking the wall?

b) What is the magnitude of the average force of the wall on the ball?

** 8. A molecule of mass 3×10^{-26} kg is moving toward a wall at a speed of 3×10^5 m/s. It bounces back in exactly the opposite direction going at the same 3×10^5 m/s.

a) What was the magnitude of the change in momentum of the molecule?

b) What was the magnitude of the impulse applied to the molecule?

c) If the molecule interacted with the wall for 1 μs (10^{-6} s), what was the average force of the wall on the molecule?

d) What impulse is applied by the molecule to the wall?

** 9. Object A is more massive than object B, but they have the same momentum. You stop each of them with the same retarding force. Answer and explain the following:

a) Which one will stop in the shorter time, or will the times be the same?

b) Which one will stop in the shorter distance, or will the distances be the same?

** 10. A 1500-kg car and a 4000-kg truck have the same momentum. The car's kinetic energy is 6×10^5 J. What is the kinetic energy of the truck?

** 11. A 4-kg block slides without friction along a tabletop at 10 m/s. A constant force of 20 N is applied until the block is going in the opposite direction at 10 m/s. Over what period of time was the force applied?

** 12. A 100-gram beach ball is dropped from a height of 1.8 m. After it bounces off the floor, it rises to a height of only 0.8 m.

a) How fast was it going just before it hit the floor and how fast was it going just after it bounced?

b) What was the total impulse of the floor on the ball?

c) How much energy was lost in the bounce?

d) If the ball was in contact with the floor for 0.05 s, what was the average force of the floor on the ball?

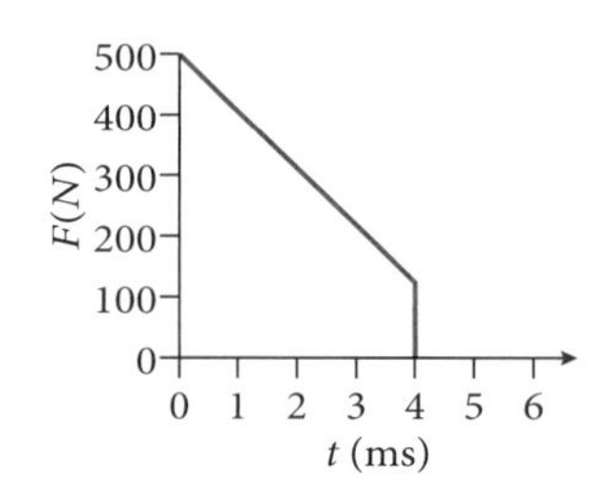

** 13. The force on a bullet in a rifle is given by the graph at left. The force starts at 500 N, falls off linearly to 100 N at the end of the barrel, and then drops quickly to zero. The mass of the bullet is 5 g. What is the muzzle velocity of the rifle?

** 14. A hockey puck of mass 0.160 kg is traveling north at a speed of 40 m/s when it is hit by a hockey stick that provides an impulse of 6.4 N·s toward the east.

a) What is the initial momentum of the hockey puck, before it hits the stick?

b) What is the final momentum (magnitude and direction) of the puck?

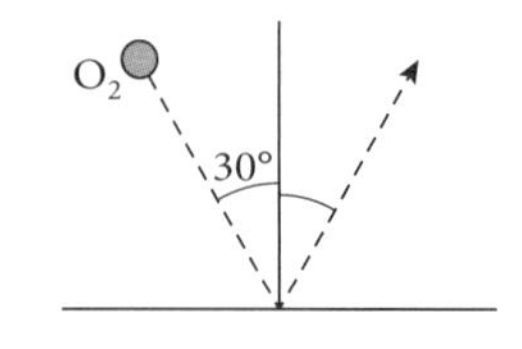

*** 15. A single oxygen molecule of mass 5.3×10^{-26} kg hits the wall of an oxygen tank moving at a speed of 480 m/s and at an angle 30° from the normal to the tank wall, as shown at left. It rebounds elastically at an equal angle and speed. What total impulse (magnitude and direction) is imparted to the molecule by the wall?

* 16. A 5-kg rifle, not held tightly against a shooter's shoulder, fires a 20-gram bullet at a speed of 450 m/s. What is the recoil speed of the rifle? (Note: This will hurt.)

8.2 * 17. An empty boxcar traveling at 10 m/s approaches a string of 4 identical boxcars sitting empty and stationary on the track. The moving boxcar collides and links with the stationary cars and the 5 move off together along the track. What is the final speed of the 5 cars immediately after the collision?

* 18. An ice fisherman slides a 0.40-kg can of bait along the ice toward a friend at a speed of 3 m/s. It hits his friend's 3.2-kg tackle box that is initially resting on the ice. Right after the collision, the bait can ends up stopped on the ice. What is the speed of the tackle box right after the collision?

* 19. A 100-kg hockey player wants to find the mass of his figure-skating girlfriend. He asks her to push him away when they are both initially at rest on the ice and he observes that he then moves backward from a mark on the ice at 2 m/s while she moves away from the same mark at 3 m/s. What is the figure skater's mass?

** 20. A 4-kg block slides to the right at 12 m/s on a frictionless table and collides head-on with a 6-kg block that is moving to the left, also at 12 m/s. Right after the collision, the small block is seen to be moving to the left at 3 m/s. Find the final velocity of the large block.

** 21. A neutron of mass 1.67×10^{-27} kg moves at a speed 4.00×10^{5} m/s and strikes a nucleus of helium-3 (mass 5.01×10^{-27} kg) that is initially at rest. The neutron is absorbed by the nucleus and forms a nucleus of helium-4 with mass 6.64×10^{-27} kg. What is the final speed of the helium-4 nucleus?

8.4 ** 22. Three children are standing together on a totally frictionless pond. At the count of three, they all push against each other in an attempt to escape to the edge of the pond. Child 1

(m_1 = 50 kg) ends up moving north at 4 m/s. Child 2 (m_2 = 30 kg) ends up moving east at 5 m/s. Child 3 has mass m_3 = 50 kg and ends up moving at an angle θ south of west, as shown.

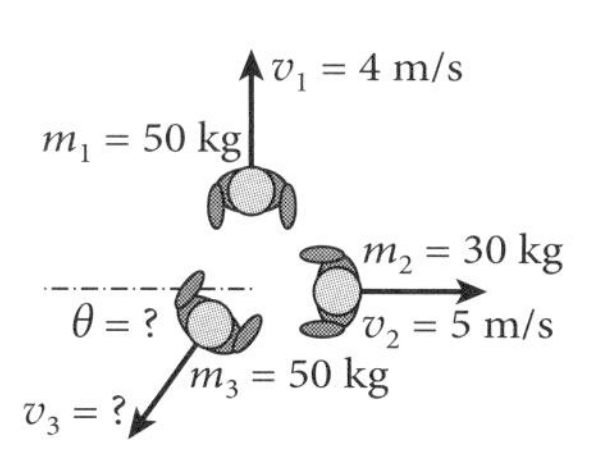

a) Is the *magnitude* of the impulse experienced by Child 1 as they push off greater than, equal to, or less than the *magnitude* of the impulse experienced by Child 2 as they push off? Explain.

b) Find the final velocity of Child 3 (magnitude and direction).

** 23. A 2-kg mortar shell in a fireworks display rises from the ground to a height at which it is moving directly upward at 10 m/s. At this instant, the shell explodes and splits into two pieces. A 0.5-kg piece heads off northward and exactly horizontally at a speed of 20 m/s. What is the velocity (magnitude and direction) of the second piece? What is the change in kinetic energy during the explosion?

** 24. A small block (m_S = 4 kg) and a large block (m_L = 12 kg) experience a head-on collision on a frictionless tabletop. Before the collision, the small block was moving 8 m/s to the right and the large block was moving at 4 m/s to the left, as shown.

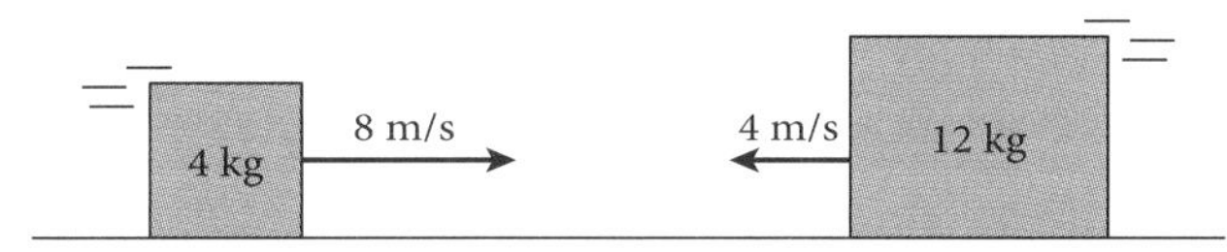

a) If the small block bounces back from the collision with a velocity 3 m/s to the left, find the final velocity (magnitude and direction) of the large block.

b) Suppose instead that the two blocks stick together when they collide. Now what will the final velocity of the two stuck-together blocks be?

c) How many joules of kinetic energy would be lost in the collision in part (b)?

** 25. A small block of mass m_1 = 2 kg is pushed by hand to compress a spring a distance Δs = 50 cm. The hand then releases the block and the spring launches it across a frictionless surface toward a large block of mass m_2 = 4 kg that initially is not moving. The small block is moving at velocity v = 6 m/s just before it strikes the large block. After the collision, the large block is observed to be moving at a velocity of 4 m/s to the right. The large block then slides over a long, rough section of the track having a coefficient of kinetic friction μ_K = 0.25 exactly, and is eventually brought to a stop after a time interval Δt.

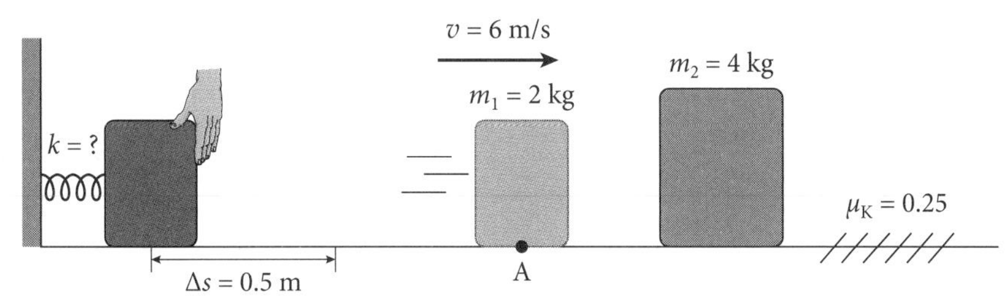

a) Find the spring constant for the spring.

b) Find the velocity (magnitude and direction) of the small block just after the collision.

c) What is the system's kinetic energy just before the collision?

d) What is the system's kinetic energy just after the collision?

e) Is this collision maximally inelastic, inelastic, or elastic? Explain your reasoning.

f) For how many seconds will the large block slide across the rough section of track before it comes to a halt? (Assume g = 10 m/s^2.)

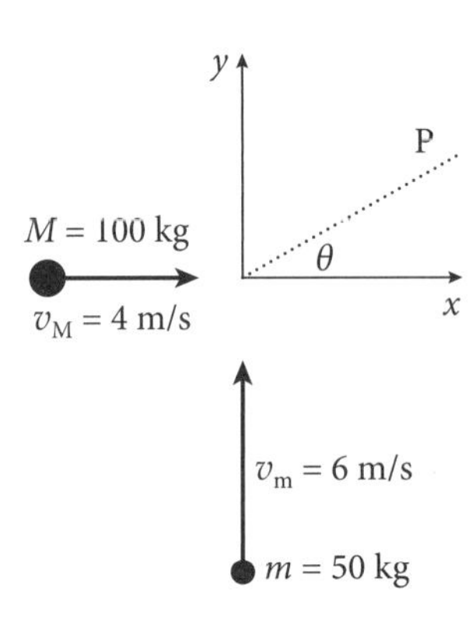

** 26. Two football players are running along perpendicular paths. The masses and velocities are given in the figure at left. The small one tackles the larger one at the point represented by the origin of the coordinate system and the two stick together, sliding in the mud along the line marked P. The coefficient of friction between the players and the mud is 0.20. You may assume that the force exerted by the ground on the players' cleats is negligible compared to the huge force they exert on each other. Find:

a) The total momentum of the system before the collision (magnitude *and* direction).

b) The final momentum of the system just after the collision (magnitude and direction).

c) The final velocity of the players (magnitude and direction) right after the collision.

d) The kinetic energy of the system just after the collision.

e) The distance the players slide before coming to rest.

** 27. A desperate Little League outfielder, seeing a fly ball going over his head, throws his mitt at it. Miraculously, the ball hits the mitt and sticks in it. If the ball is going exactly horizontally at 28 m/s at the instant it hits the mitt, and if the mitt is moving straight upward at a speed of 6 m/s when the ball hits it, find the momentum vector (magnitude and direction) of the mitt with the ball in it. The mass of a baseball is 0.143 kg and the mass of the mitt is 0.6 kg.

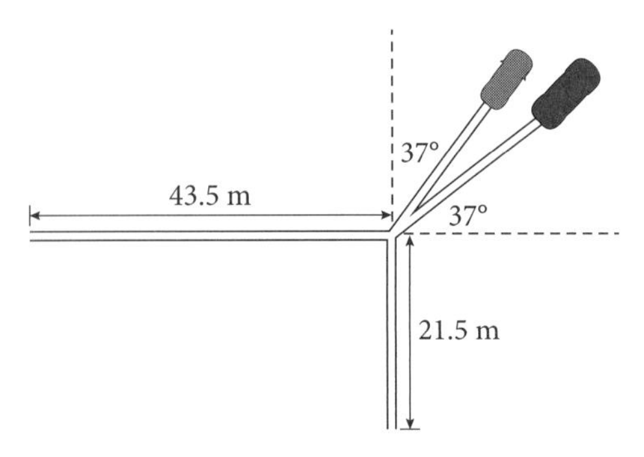

*** 28. A police investigator arrives at an accident scene where he finds that a gray 1000-kg Honda Civic has collided with a red 2000-kg SUV. From the skid marks, he is able to determine that the Honda was initially traveling due east when the driver hit the brakes, while the SUV was going due north. After the collision, the Honda skidded 37° east of north (what other angle could it be?) for a distance of 3.5 m, while the SUV skidded 37° north of east for a distance of 4.0 m. The pre-collision skid marks of the Honda are 43.5 m long and the pre-collision skid marks of the SUV are 21.5 m long, as shown. The coefficient of friction of Honda tires with the pavement is 0.7, while that of the SUV tires is 0.8. Help the investigator work out the initial speeds of the two vehicles before they hit their brakes.

a) From the lengths of the skid marks after the collision, find the speeds of each vehicle just after the collision.

b) Use conservation of momentum during the collision to find the speed of each vehicle just before the collision. [Hint: Remember that momentum is a vector.]

c) From the lengths of the skid marks before the collision, determine the initial speeds of the two vehicles before they hit their brakes.

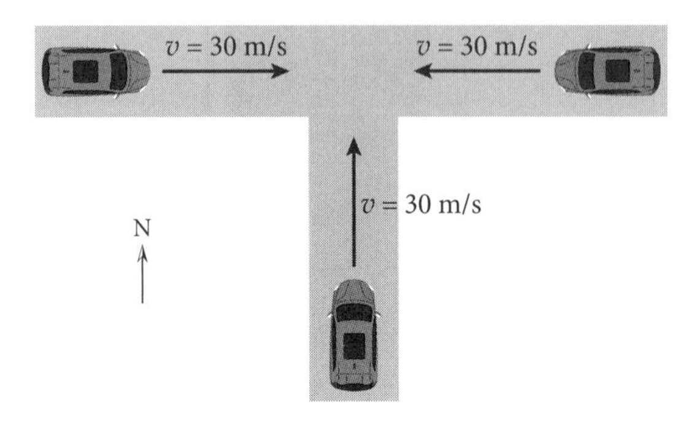

*** 29. Three identical Volvos, each of mass 2000 kg, enter an intersection, one heading west, one heading east, and one heading north. Each of them is initially traveling at 30 m/s. The three collide and lock bumpers, becoming one single mangled Volvo-thing with a dozen wheels.

a) Find the total momentum (magnitude and direction) of the three-car system just before the collision.

b) What is the total momentum (magnitude and direction) of the three-car system just after the collision? Explain your reasoning.

c) Find the velocity (magnitude and direction) of the mangled Volvo-thing just after the collision.

d) Find the kinetic energy of the Volvo-thing just after the collision.

e) If the coefficient of kinetic friction between the Volvo-thing and the road is $\mu_K = 0.2$, find the distance the Volvos will slide before coming to rest.

*** 30. A spacecraft of mass 200 kg (including the solar panels) is moving at a speed of 3000 m/s toward an initially stationary 10-kg meteoroid. The meteoroid strikes the spacecraft in the middle of the strut that holds one of the two 20-kg solar arrays, severing that solar array from the spacecraft. The meteoroid is kicked directly forward, along the initial direction of the spacecraft, at a speed of 2000 m/s. The collision knocks the spacecraft, minus one of its solar arrays, away from its initial path. It ends up moving at a speed of 2500 m/s at an angle of 37° from the initial direction. Find the velocity, magnitude, and direction of the severed solar array.

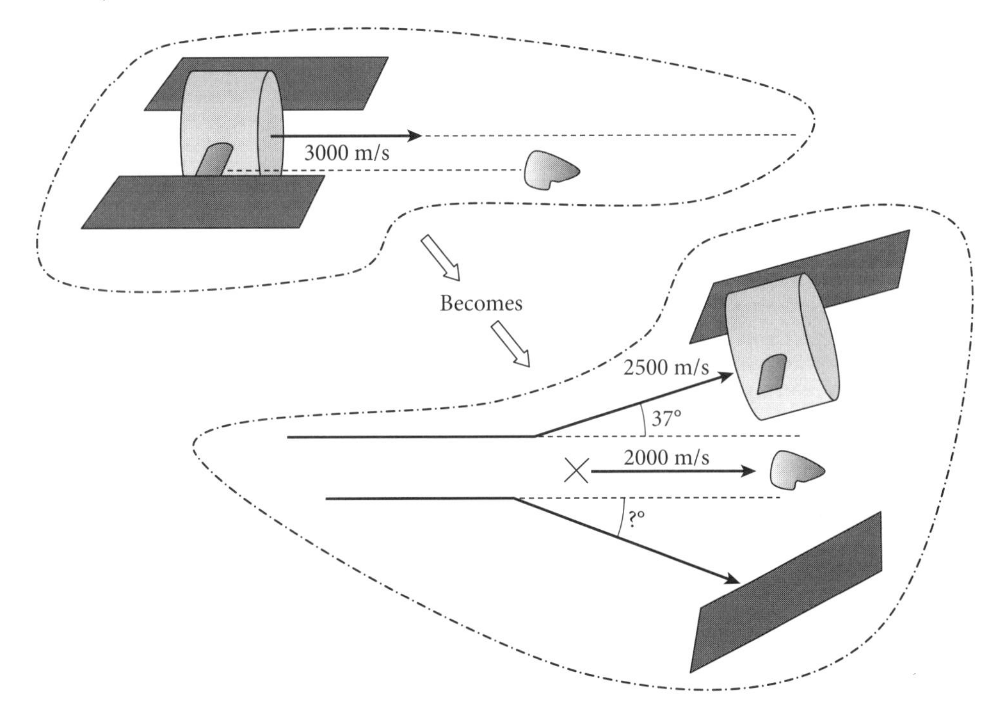

** 31. A pendulum consists of a ball of mass 3.0 kg on a string of length 4.0 m. The ball is pulled back until its center has risen to a height of 0.80 m. The ball is released such that it strikes a 1.0-kg block, as shown. The speed of the block after the collision is measured to be 6.0 m/s.

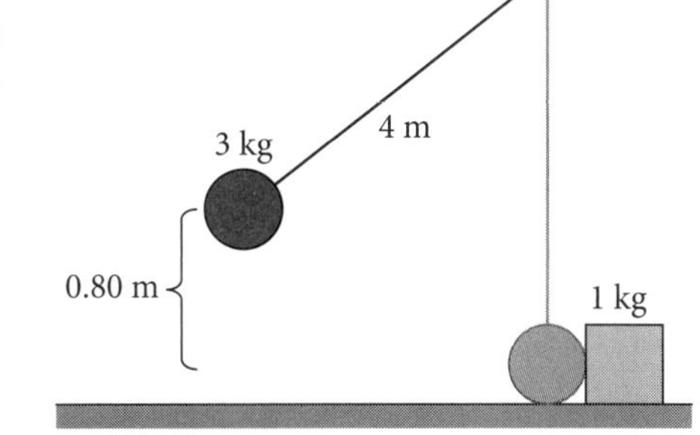

a) What is the speed of the ball just before the collision?

b) What is the speed of the ball immediately after the collision?

c) What flavor of collision is this (elastic, inelastic, or maximally inelastic)? Justify your answer.

** 32. A pendulum bob of mass 4 kg is released from height h and swings down to collide with a stationary block of mass 6 kg. Right before the collision the bob is moving to the left at speed 5 m/s. Just after the collision, the block is traveling to the left at speed 3 m/s. The block then slides over a very long, rough surface with a coefficient of kinetic friction that is exactly $\mu_K = 0.2$, bringing the block to a stop at a point Δx past the beginning of the rough section.

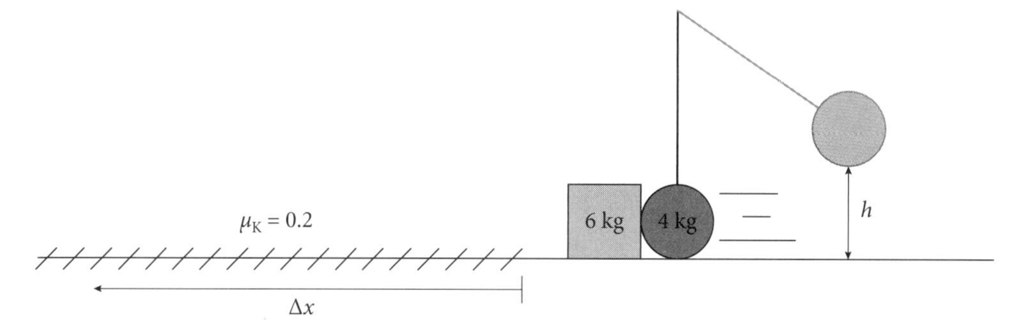

a) Find the height h from which the bob was released.

b) Find the velocity (magnitude and direction) of the bob right after the collision.

c) What is the bob–block system's kinetic energy just before the collision?

d) What is the bob–block system's kinetic energy just after the collision?

e) Is this collision maximally inelastic, inelastic, or elastic? Explain.

f) How far will the block slide across the rough section of track before it comes to a halt?

** 33. Block A, with mass $m = 2$ kg, slides down a frictionless track so that its center of mass drops 1.25 m, as shown. Block A then collides with block B, which has mass $M = 4$ kg. After the collision, block B is observed moving to the right with speed 2 m/s.

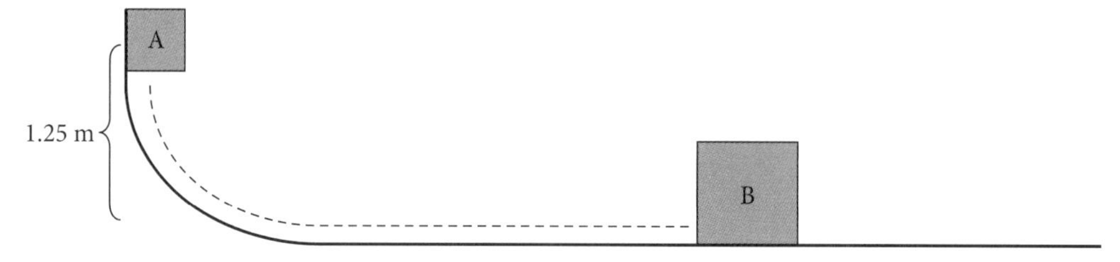

a) How fast is block A moving when it first reaches the flat portion of the ramp?

b) Find the velocity (magnitude and direction) of block A after the collision.

c) What is the system's kinetic energy just before the collision?

d) What is the system's kinetic energy just after the collision?

e) Is the collision inelastic, elastic, or maximally inelastic? Explain.

8.5 *** 34[f]. Consider the case where a hard-hit cue ball of mass m is sliding across a pool table at speed v and strikes an initially stationary 3-ball head-on in an elastic collision. The 3-ball has an identical mass m. Find a formula for the final speeds of both balls in terms of the initial speed v.

*** 35. A ball of mass 2 kg moving at a velocity 6 m/s to the right strikes a 4-kg ball that is initially stationary. Assume that the collision is head-on and elastic, and find the final velocities of the two balls.

*** 36. A 7-kg bowling ball hits a 1½-kg pin head-on. After the collision, the bowling ball is still moving down the bowling alley. Its speed is 2 m/s. Assume that the collision is *elastic*. Find the initial speed of the ball before it hit the pin and the speed of the pin just after the collision with the ball.

9

ROTATIONAL DYNAMICS

A well-thrown football follows a projectile path while spinning around its long axis. The motion of the football is easy to analyze because the spin does not affect the path (at least in the absence of air resistance) and the path does not affect the spin. The rotational and translational motions are independent of each other. In this chapter we examine rotational motion without examining any accompanying translational motion.

9.1 The Radian

The rules for rotational motion have many analogies with those for the translational motion we have been studying so far. The difference for rotation is that we measure the displacement not in ordinary distance units, like feet or meters, but in angular units such as degrees, revolutions, or radians. Just as feet and meters can be converted to each other, so may we convert between the various angular units. There are 360 degrees in a circle, so 1 degree is $\frac{1}{360} \approx 0.0028$ revolutions.

You may not have encountered the radian before. One radian is a measure defined as the angle at which the arc length around a segment of a circle equals the circle's radius (Figure 9.1). Thus, the distance around a 1-radian arc of a circle of radius r will be $d = r$. A larger angle will produce a proportionately longer arc length, so the general relationship between arc length and the internal angle in radians becomes $d = r\theta$. This also means that, since the circumference of an entire circle is 2π times the radius, there will be 2π radians in a complete 360° circle. One radian thus turns out to be $360°/2\pi \approx 57.3°$.

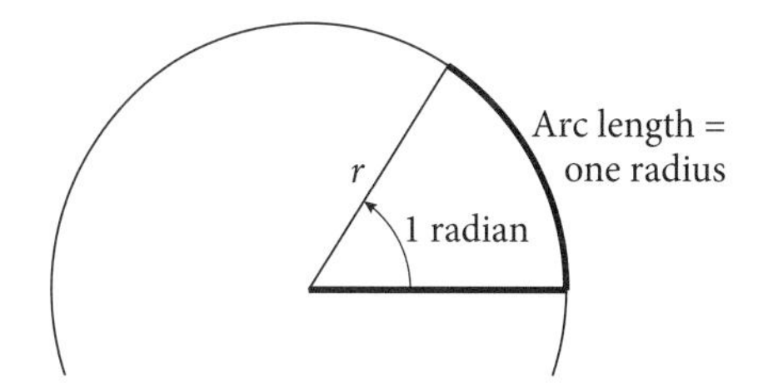

Figure 9.1 The length of a 1-radian (57.3°) arc equals the radius of the circle.

9.2 Rotational Kinematics

Back in Chapter 2, we defined translation as the motion of an object from one place to another and defined the translational speed v as the displacement Δx (change in position) divided by the elapsed time. Here, we will define an *angular speed* ω as the *angular displacement* $\Delta\theta$ (change in angular position) divided by the time. The units used to express this measurement could be radians per second, revolutions per minute (rpm), etc. We also define *angular acceleration* α to be a measure of the rate at which angular speed changes, $\alpha = \Delta\omega/\Delta t$. Here, the change in angular speed, $\Delta\omega$, is the final angular speed minus the initial angular speed, as usual. Let's see how these definitions work numerically.

A compact disc (CD) starts spinning at 500 revolutions per minute (RPM) while it is reading the data from the inside track of the disk. As time goes by, it slows its angular speed in order to keep a constant linear speed passing under the laser readout, reaching 200 RPM as it reads the last, outside track. If a CD has an hour's worth of music on it, we can calculate its average angular acceleration from the definitions:

$$\alpha = \frac{\Delta\omega}{\Delta t} = \frac{\omega_f - \omega_i}{\Delta t} = \frac{200 \text{ rpm} - 500 \text{ rpm}}{60 \text{ min}} = -5\frac{\text{rpm}}{\text{min}}.$$

The change is negative, in the opposite direction as the angular velocities, because the CD is slowing. Its angular speed decreases by 5 RPM each minute of playing time.

Back in Section 2.3, we discussed how the linear speed of an object tells how many meters it covers each second, regardless of the direction it is traveling, while the term "velocity" discriminates between directions by calling one direction positive and the opposite direction negative. For rotational motion, there is likewise a difference between something that is turning clockwise and something that is turning counterclockwise. We could define "angular velocity" by calling one of these rotation directions positive and the other negative. We live, however, in a 3-dimensional world, and a complete description of rotation should have a way to specify the axis about which an object is spinning in 3-dimensional space. So physicists and mathematicians have determined a way to assign a 3-dimensional vector to the rotation. At any given moment, a spinning body will be spinning about one particular axis. We choose the rotation axis to represent the direction of the spin, with the following additional stipulation: If you take the fingers of your *right hand* and let them curl in the direction of the rotation of the body, as shown in Figure 9.2, then your thumb will point in the direction of the vector that is associated with the rotation. If you think about it, you will see that using your left-hand fingers instead of the fingers of your right hand would point the vector in the opposite direction. So we say that the direction of the rotation is determined by a Right-Hand Rule. To summarize, the *magnitude* of the *angular velocity vector* is the angular speed ω, and its *direction* is given by the right-hand rule along the axis of rotation.

Figure 9.2 A Right-Hand-Rule gives the direction of the angular velocity vector.

This way of defining angular velocity as a vector is not just the physicist's way of making things complicated, but is required for cases where rotation axes can point in different directions in space. To help us get used to this definition, we will continue to use it throughout the book, even though we will generally only consider rotation about a single fixed axis. Let us see how it works with this example.

EXAMPLE 9.1 Angular Velocity of a Second Hand

What is the angular velocity (magnitude and direction) of the second hand of a wristwatch?

ANSWER The second hand makes one complete revolution each minute, or 2π radians every 60 seconds. The magnitude of the angular velocity is therefore

$$\omega = \frac{2\pi \text{ rad}}{60 \text{ s}} \approx \frac{6.28 \text{ rad}}{60 \text{ s}} = 0.105 \text{ rad/s}.$$

Put your right hand over your watch with your fingers curling clockwise. Your thumb points into your wrist. The direction of the angular velocity vector is into your wrist.

The method we use to determine the direction of angular velocity can also be used to give a direction to *angular acceleration*. An angular acceleration that causes a positive angular velocity in some direction to spin faster will be in the same direction as the angular velocity, while an angular acceleration that causes a positive angular velocity in some direction to spin more slowly will be a negative angular acceleration, meaning that it points in the opposite direction to the angular velocity. But you're already used to this kind of thing from Chapter 2.

In fact, given a particular axis direction for angular motion, the mathematical relationships between the angular displacement $\Delta\theta$, angular velocity ω, and angular acceleration α are exactly the same as those between linear displacement Δx, linear velocity v, and linear acceleration a. It must therefore be true that the kinematic equations we encountered in Chapter 2 will each have their rotational counterpart:

$$v_f = v_i + a\Delta t \quad \Rightarrow \quad \omega_f = \omega_i + \alpha\Delta t \tag{9.1a}$$

$$\Delta x = \frac{1}{2}(v_i + v_f)\Delta t \quad \Rightarrow \quad \Delta\theta = \frac{1}{2}(\omega_i + \omega_f)\Delta t \tag{9.1b}$$

$$\Delta x = v_i\Delta t + \frac{1}{2}a\Delta t^2 \quad \Rightarrow \quad \Delta\theta = \omega_i\Delta t + \frac{1}{2}\alpha\Delta t^2 \tag{9.1c}$$

$$v_f^2 = v_i^2 + 2a\Delta x \quad \Rightarrow \quad \omega_f^2 = \omega_i^2 + 2\alpha\Delta\theta \tag{9.1d}$$

Just as the linear kinematic equations were valid only for constant acceleration, these rotational kinematic equations can be used only when angular acceleration is constant.

EXAMPLE 9.2 Angular Kinematics of a CD

We have seen that a CD slows from 500 rpm to 200 rpm as it plays 60 minutes of music. How many revolutions does it make during this time?

ANSWER Using the second rotational kinematic equation (9.1b), we find

$$\Delta\theta = \frac{1}{2}(\omega_i + \omega_f)\Delta t = \frac{1}{2}(500 \text{ rpm} + 200 \text{ rpm})(60 \text{ min}) = 21{,}000 \text{ rev}.$$

We can also express this angular displacement in units of radians:

$$\Delta\theta = 21{,}000 \text{ rev}\left(\frac{2\pi \text{ rad}}{1 \text{ rev}}\right) = 132{,}000 \text{ rad}.$$

Let us explain one last thing about rotational motion. If a wheel of radius r turns through an angle $\Delta\theta$, then, as we noted in Section 9.1, a point on the circumference of the wheel will move a distance $\Delta d = r\Delta\theta$ around the circle. If it takes a time Δt to spin that far, then we can divide both sides of the formula by Δt to find the relationship between the angular speed of the wheel ω and the linear speed v of a point on the circumference of the wheel, a distance r away from the axis. The relationship is

$$v = r\omega. \tag{9.2}$$

Now let us imagine that the axle of this wheel is not attached to a lab bench, but is the axle of a car, as the car rolls along the highway. If the car is moving eastward with velocity v, then, mea-

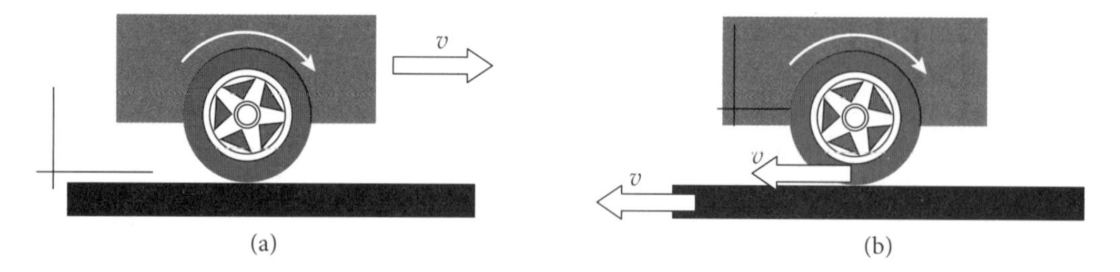

Figure 9.3 (a) A car rolls along the highway at speed v. (b) Then the highway moves relative to the car at a speed v in the opposite direction. If the wheels are not slipping on the pavement, the point of the wheel in contact with the highway will also be moving at v relative to the car.

sured relative to a coordinate system fixed to the car, the highway is actually moving to the west with velocity v (Figure 9.3). If the wheels are turning without slipping on the blacktop, then the angular speed of the wheels must be exactly what it takes to make the lowest point of the wheel, the point that is in contact with the highway, move backward with velocity v relative to the car. Therefore, we see that Equation 9.2 also serves as the relationship between the angular speed ω of a rolling wheel and the speed v of the vehicle to which the wheel is attached.

9.3 Torque

In the previous section, we discussed rotational kinematics for rotating systems with constant angular acceleration and we found that the relationships between the various angular quantities were the same as those we found for the linear quantities in Chapter 2. But we have not yet considered what *causes* an angular acceleration. We know it is force that causes linear acceleration. Is there something like force that causes angular acceleration? Let's think about this.

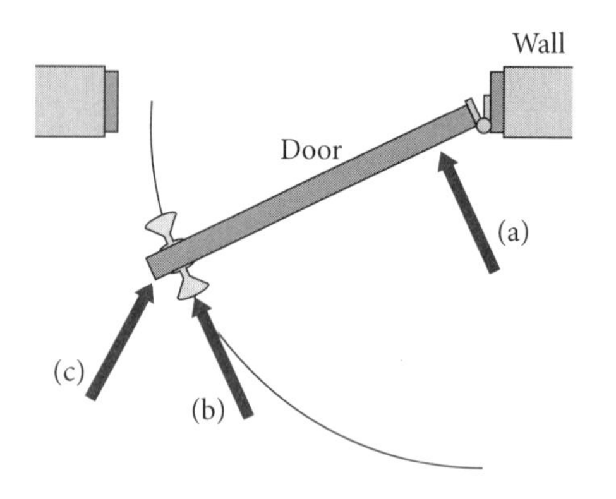

Figure 9.4 A bird's-eye view of a door with force being applied near the hinge (a), to the knob (b), or, strangely, not at right angles to the door (c).

Suppose you have a heavy door that someone left ajar and you want to close it. The door is initially not rotating and you want to change its angular velocity so that it starts to close. What does it take to create the angular acceleration that changes the angular velocity from zero to some positive value? Clearly, you must push on the door in the right direction. But, if you reach out and push on the door at a point a few inches from the hinges, like force (a) in Figure 9.4 you find that it takes a great deal of force to get it to rotate. On the other hand, if you move over and push on the doorknob, as depicted by force (b) in the figure, it closes easily. Clearly the point where the force is applied is as important as the amount of force you use. That, of course, is why they put doorknobs where they do.

The last thing we need to include in our discussion of how to create an angular acceleration is probably so obvious that it is barely worth mentioning. But, clearly, if you want to get the door moving with the least amount of effort, you should push perpendicular to the door, like forces (a) and (b) do in the figure. The direction chosen for force (c) in Figure 9.4 gives the force a component that is pushing directly toward the hinges. This component does no good at all. Only the component perpendicular to the door will get it to turn. If we put all these ideas together, we see that the thing that causes an angular acceleration of an object about some axis, the angular analog to the force in linear dynamics, is a twofold quantity called *torque*. The magnitude of the torque τ is the product of two quantities. The first is the distance from the axis of rotation out to the point where the force is applied, (r). The second quantity is the component of the applied force perpendicular to a vector from the axis to the point where the force is applied, $F_\perp$. The magnitude of the torque is thus defined to be

$$\tau = rF_\perp. \tag{9.3}$$

The SI unit for torque is the newton · meter (N · m). As we mentioned in Section 1.2, the units for torque in the various systems of units do not yet have their own names.

Torque can also be made into a vector by giving it a defined direction along the axis about which it causes things to rotate, using the right-hand rule. The torques arising from the three forces in Figure 9.4 will all cause the door to rotate clockwise in the figure. The right-hand rule (with fingers pointing clockwise around the hinge) tells us that the torque in this picture points downward, into the floor, or into the page in this bird's-eye view.

Note that the torque magnitude defined in Equation 9.3 depends on two things—the force and the distance to where the force is applied. But distance from where? If an object has a fixed point, like the hinges of the door in Figure 9.4, you will probably want to choose to measure r from that axis out to the point where the force is applied. But, strictly speaking, torque may be defined by measuring the distances from any axis you might choose. Therefore, you cannot ask, "How much torque acts on an object?" That question has no answer. You can only ask, "What is the torque on some object, measured about an axis along . . . ," and then you have to specify the axis you have chosen.

Let's learn what torque means by working a couple of short examples.

EXAMPLE 9.3 Finding the Torque—I

A 10-N force is applied 40 cm from the end of a gray metal rod and at an angle of 30° to the rod, as shown. What is the torque produced by this force, measured about an axis at the end of the rod (the dot in the figure)?

ANSWER Before using the formula for the magnitude of the torque, we must first find the component of the force that is actually perpendicular to the rod. This is the dashed red component in the figure. It has magnitude $F_\perp = (10 \text{ N})\sin 30 = 5$ N. The torque about the dot is therefore:

$$\tau = rF_\perp = (0.4 \text{ m})(5 \text{ N}) = 2 \text{ N}\cdot\text{m}.$$

Since this force would turn the rod counterclockwise around the axis (the black dot on the left end of the stick), the direction of this torque is out of the page.

EXAMPLE 9.4 Finding the Torque—II

The same force is applied to the same rod as in the previous example. Find the torque about an axis located at the point where the force is applied (the new black dot in the figure).

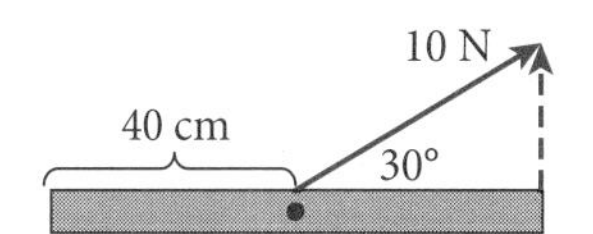

ANSWER The formula for torque is $\tau = rF_\perp$. The component of force perpendicular to the rod is still $F_\perp = 5$ N, but the distance r from the chosen axis to the point where the force is applied is zero. The value of the torque is therefore zero. Torque depends on the axis you choose.

Since it is the torque that causes an object to change its angular velocity, we may write down an analog to Newton's first law, applied to rotation:

- Law I for Rotation: The angular velocity of a rigid object remains constant unless it is acted on by a net external torque.

Remember that, for a vector to remain constant, both its magnitude and its direction must not change.

Notice that Law I for Rotation does not specify a point about which the torque must be measured, a point we will refer to as the "pivot point." If there is a net torque about *any* pivot point, then the body will experience an angular acceleration about that point. Inversely, if an object experiences no angular acceleration, then you may pick any pivot point you like and you may be sure that the net torque about that point is zero.

When there is more than one applied force, situations can arise where the net force is zero but the net torque is not zero. The two forces acting on the board in Figure 9.5 are equal in size and opposite in direction, but they do not act along the same line. With zero net force on it, the board will not accelerate. However, it will begin to spin in a clockwise direction, because the two torques about the center of the board are non-zero and act in the same rotational direction.

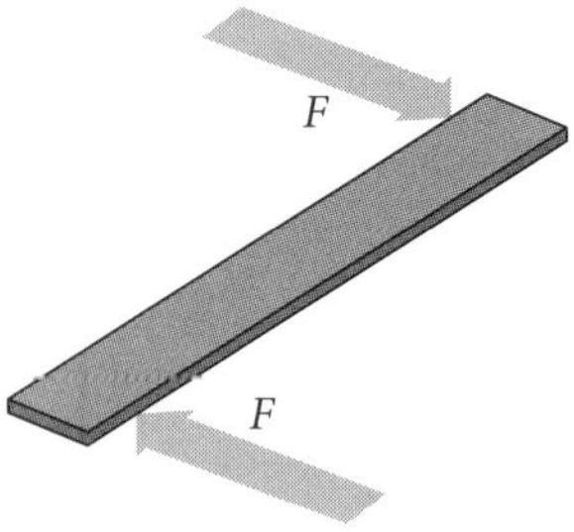

F

F

Figure 9.5 Two forces act to produce a net torque but no net force.

9.4 Center of Mass

Back in Chapter 5, where we first introduced Newton's second law, we applied the forces to blocks, carts, and sleds without worrying about *where* the force was applied. We did not point out that, if an object experiencing those same forces is able to rotate, then the motion of some parts of the body may be different than other parts (see Figure 9.6). However, if we had done

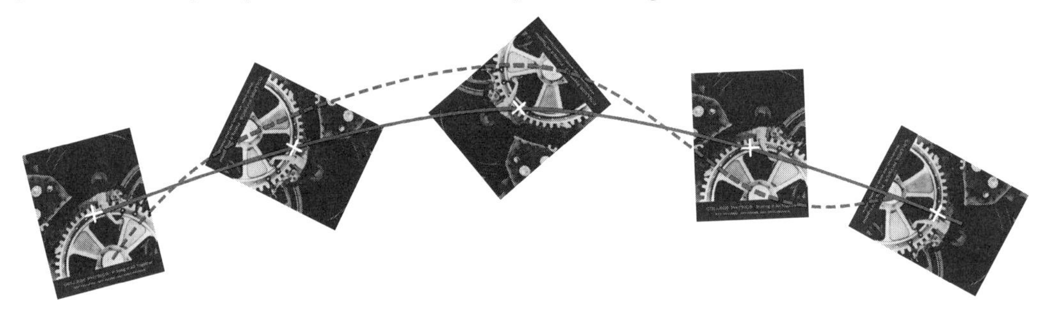

Figure 9.6 When we say that the trajectory of a book obeys Newton's second law, what we mean is that this is what the *center of mass* of the book does (the solid red line). Another point, like the hub of the big gear in the picture, follows a different path (the dashed red line).

an analysis[1] of all the forces acting on all the parts of an extended body, including internal forces, it would turn out that we did the right thing by ignoring where the forces were applied, because there will be one point in an extended spinning body that follows the simple path determined by those forces alone. That point is known as the *center of mass*. The motion of an extended body that is both translating and rotating may be described as a smooth motion of

1. A general analysis would require calculus, so we do not do it here. But it involves dividing an extended body into pieces, writing down Newton's second law for each piece, and then adding up all of the individual $F = ma$ equations. The result is that the center of mass (Equation 9.4) moves in response to the sum of the external forces, regardless of where they are applied.

the center of mass, following Newton's second law as if all the forces on the body were applied at that point, combined with rotation of the object about the center of mass.

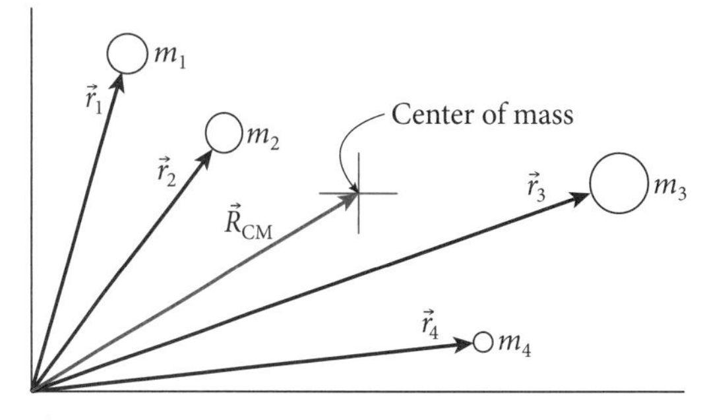

Figure 9.7 The center of mass of four particles.

The *center of mass* is defined as the average position of the mass in a system. Think of it like this. You know how to calculate your grade-point average. You multiply your numerical grade in each class by the number of units in that class. You then add these up and divide by the total number of units you've taken. Similarly, if you have a number of small masses distributed out in space, like those depicted in Figure 9.7, you first decide on an origin from which you want to define the vector to the center of mass, and you then find the average position by using the same kind of averaging procedure you would with your GPA:

$$\vec{R}_{CM} = \frac{m_1\vec{r}_1 + m_2\vec{r}_2 + m_3\vec{r}_3 + m_4\vec{r}_4}{m_1 + m_2 + m_3 + m_4}. \tag{9.4}$$

The distances are like the grades you got in each class and the masses are like the number of units each class is worth. Equation 9.4 is a vector equation, so, to find the center of mass, you will have to remember how to add vectors, resolving each vector into components and then adding the components. If all the masses lie in a single plane, the vectors will all be 2-dimensional vectors. For masses that lie along a single line, the summation in the numerator of Equation 9.4 is just an arithmetic sum.

EXAMPLE 9.5 Center of Mass in One Dimension

Three weights are distributed along a massless meter stick. There is a 100-g weight at the 20-cm mark, a 300-g weight at the 60-cm mark, and a 100-g weight at the 90-cm mark, as shown below. Find the point on the stick that is the center of mass.

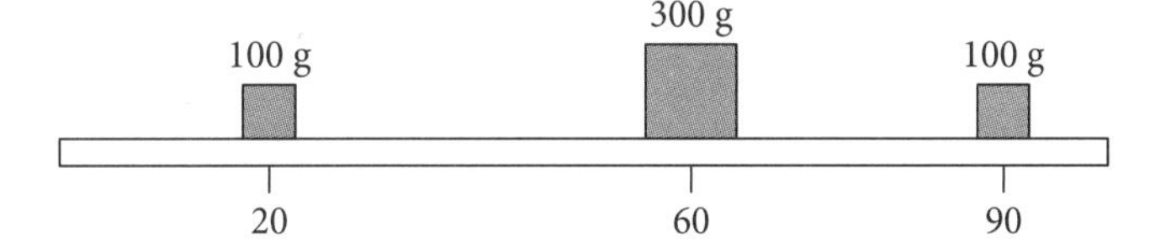

ANSWER Let us choose to express the position of the center of mass as a distance from the left side of the stick. Then the position of each mass should be measured from this same point. If we write all masses in grams and all distances in centimeters, the definition in Equation 9.4 produces

$$R_{CM} = \frac{(100)(20) + (300)(60) + (100)(90)}{100 + 300 + 100} = \frac{2000 + 18{,}000 + 9000}{500} = \frac{29{,}000}{500} = 58.$$

The center of mass will lie on the 58-cm mark, 2 cm to the left of the 300-g mass.

Equation 9.4 is valid when you have individual point masses in the system, but what if the mass is distributed over an extended body like the flying book in Figure 9.6? As long as the mass is uniformly distributed in the body, the center of mass is easily found; it will lie at the geometrical center of the object. On the other hand, if the distribution of matter in a body is *not* uniform, or if the geometry of the body is too complicated to determine where the geometrical center lies, the simple formula in Equation 9.4 will have to be replaced by some calculus, which, of course, we do not do in this book.

EXAMPLE 9.6 Center of Mass of Two Objects

A 10-cm radius wire hoop of mass 60 g has a bead of mass 20 g soldered to it. Find the position of the center of mass of the object.

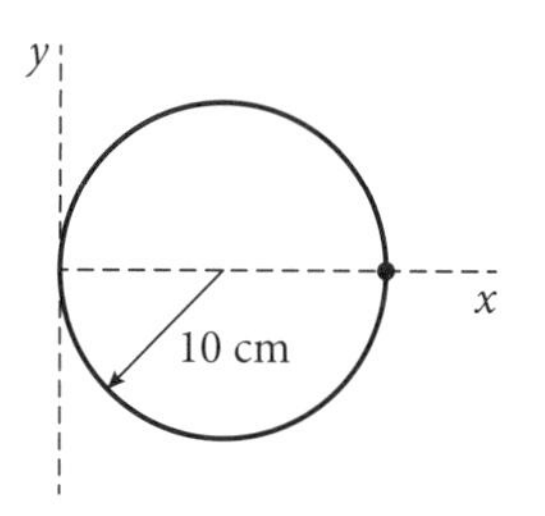

ANSWER From the symmetry of the object, we know that its center of mass must lie somewhere along the line joining the center of the hoop and the bead, because there will then be as much mass as far above that line as below it. We also know that the center of mass of the wire hoop is at its geometrical center, the center of the circle. For purposes of calculating the center of mass of the complete object, we can treat the mass of the hoop as if it were concentrated at the center. (Note that the center of mass is an abstract point and that there does not have to be any actual matter at that point.) We choose to measure $\vec{R}_{CM}$ from the origin of the coordinate system shown. Although this is a two-dimensional object, we will only need to use Equation 9.4 to find the x-component of $\vec{R}_{CM}$, since we have already explained that the symmetry requires that the y-component be zero. Again, putting masses in grams and distances in centimeters, Equation 9.4 gives

$$R_{CM,x} = \frac{(60)(10) + (20)(20)}{60 + 20} = \frac{600 + 400}{80} = \frac{1000}{80} = 12.5.$$

The center of mass will lie 2.5 cm to the right of the center of the hoop, on the x-axis. If forces of any kind are applied to the wire or to the bead, this point will accelerate according to Newton's laws, as if all of the forces applied anywhere on the body had been applied at this center of mass instead.

There is one additional property of the center of mass that we need to address. Consider the object in Figure 9.8, consisting of an 80-kg mass on one end of a rigid meter stick of negligible mass and a 20-kg mass on the other end. The center of mass of this object will be 20 cm from the 80-kg mass, as shown by the × in the figure. Let us put the stick in a horizontal orientation, as shown, and let us calculate the forces and torques that are produced by gravity acting downward on the two masses at the ends. Newton's second law says that the center of mass of this 100-kg body will move according to the net force applied, regardless of where on the body the force is applied. But how about the torque due to gravity? Will gravity cause the body to have an angular acceleration?

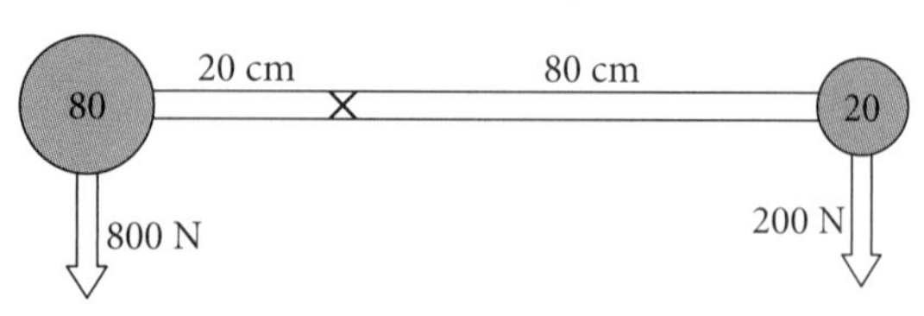

Figure 9.8 Two masses rigidly held by a massless meter stick, acted on by gravity.

If we choose the center of mass of the object in Figure 9.8 to be the point about which we want to measure the torque, we find an interesting result. With clockwise rotation positive into the page and counterclockwise rotation negative, the torque is

$$\tau = +(0.8 \text{ m})(200 \text{ N}) - (0.2 \text{ m})(800 \text{ N}) = 160 \text{ N}\cdot\text{m} - 160 \text{ N}\cdot\text{m}.$$

The net torque about the center of mass is zero, so the angular acceleration about the center of mass is zero. Whatever angular velocity the object in Figure 9.8 is released with, including if it is released with zero angular velocity, then that angular velocity will remain constant. But note that this condition of zero net torque about the center of mass is the *same* result we would get if we assumed that the force of gravity on the body was actually applied at the center of mass rather than at the two ends, since a force applied at some point produces no torque about that point. So, even though there is no actual mass at the center of mass (remember that the stick was massless), both the linear motion and the angular motion of the object in Figure 9.8 may be correctly found by assuming that *the force of gravity on a body acts at its center of mass.*

We have worked out this result for the special two-mass object of Figure 9.8, but it is, in fact, generally true.[2] It is true for a body composed of many masses, and it is even true for a continuous body, where the mass is spread out over an infinite number of small points. Because gravity always acts as if it were applied at the center of mass of an object, the center of mass is also referred to as the *center of gravity.*

9.5 Extended Free-Body Diagrams

If a painter weighing 700 N stands in the center of a 300-N platform, as in Figure 9.9a, we can easily argue that the tension in each of the support cables must be 500 N. The total downward force on the platform is its 300 N weight plus the 700-N normal push exerted by the painter. Since the platform is at rest, the downward force of 1000 N must be balanced by an upward force of 1000 N, supplied by the two cables. And it is obvious from symmetry that each cable must supply half of that force.

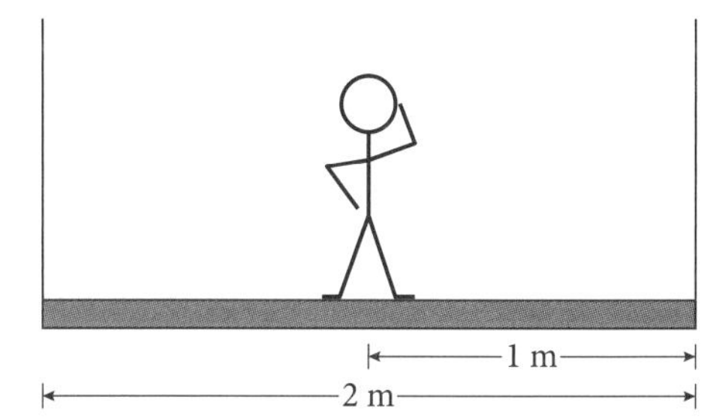

Figure 9.9a A painter stands in the exact middle of the platform.

But consider what will happen if the painter moves off to the right, as in Figure 9.9b. Here, the symmetry is broken. We expect the cable on the right to support more of the 1000 N than the cable on the left, but how much more? Newton's second law produces a single equation here, and so it does not provide enough information to solve for two tensions. But there is another condition that we know is satisfied. Since the platform is not rotating, we know from Newton's laws for rotation that the net torque acting on the platform must be zero. What's more, we are free to choose any point as our origin and then calculate the torques about that point. They must always add to zero.

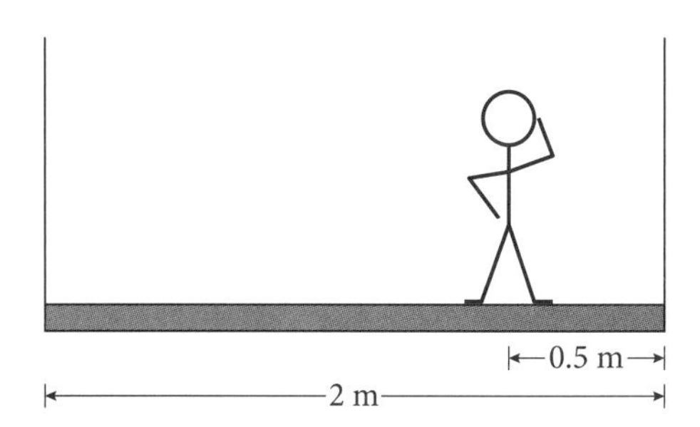

Figure 9.9b The painter has moved away from the center of the platform.

Torque depends on *where* a force is applied relative to a chosen origin, or "pivot point." In Chapter 5, our free-body diagrams did not worry about such things. All forces were drawn acting on a dot representing the center of mass of the object. But when balancing torques, we must draw what we will call an *extended free-body diagram* that shows where each force is applied. Let us see how this works.

We want to find the tensions in the two support cables for a situation like that shown in Figure 9.9b. We begin by drawing an extended diagram of the platform, indicating where each

---

2. This is probably best seen in the following way. We remember that the net torque on a body due to gravity is found by summing up the products of distances to each part of the body times the force of gravity on that part. The force always acts exactly in the same downward direction. If you factor the common gravitational field $\vec{g}$ out of the sum, the remaining factor will be a sum like $m_1\vec{r}_1 + m_2\vec{r}_2 + m_3\vec{r}_3 + \cdots$. But note that this same summation figures in the numerator of the definition of the center of mass in Equation 9.4. If we measure all the $\vec{r}$ vectors from the center of mass, and if we remember that the distance from the center of mass to the center of mass is zero, we see that the numerator will always be zero. This means that the gravitational torque on a body, measured about its center of mass, will be zero for any body of any shape.

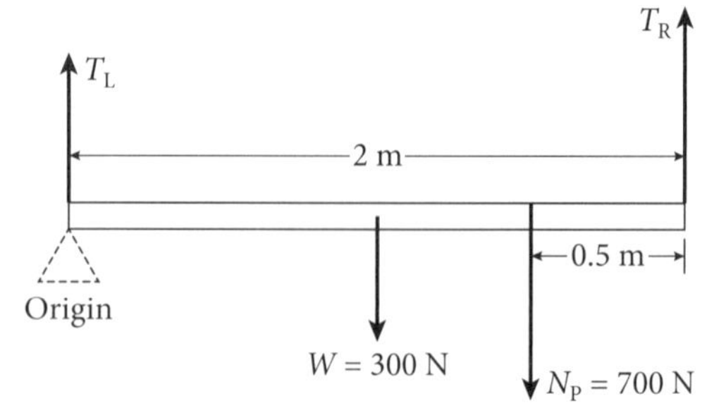

Figure 9.10 The extended free-body diagram.

of the forces is applied, as in Figure 9.10. We then arbitrarily choose a pivot point as origin, maybe the left end of the platform. This choice removes the tension in the left cable from our torque calculation, since that force acts directly through the origin ($r = 0$) and produces no torque about this point. The gravitational force at the center of the platform and the normal force of the painter pushing down on the platform both produce torques that act into the page, clockwise around our chosen pivot point. The tension in the right cable is acting in a direction that would make the platform rotate about an axis out of the page. The torques in the two opposite directions must be equal in magnitude so that they will cancel, allowing us to write

$$(300\text{ N})(1\text{ m}) + (700\text{ N})(1.5\text{ m}) - T_R(2\text{ m}) = 0 \qquad \text{or} \qquad T_R(2\text{ m}) = 1350\text{ N}\cdot\text{m}.$$

Note how the choice of the left cable attachment point as the pivot point eliminated one unknown from the torque equation, allowing us to find a single equation for a single unknown T_R. The solution is $T_R = 675$ N. The total force acting upward on the platform must still add up to 1000 N, by Newton's laws for linear motion, so the tension in the left cable must provide the remaining $T_L = 325$ N.

Now that we have the general idea of how to handle problems in which an extended body is in equilibrium (neither moving nor spinning), let us codify this into a set of steps that may be followed. These steps are useful for an area of mechanics called *statics*. In statics, we analyze the forces in situations where there is no motion—zero acceleration and zero angular acceleration. For simplicity, we will specialize to the case of motion in two dimensions only. In this case, a body can move in the x-direction or in the y-direction, or it may spin in the x–y plane about an axis in the z-direction. We thus have three conditions that will be satisfied. The steps are:

1. Isolate a body, just as you did for free-body diagrams in Chapter 5. Only this time you will draw a rough sketch of what the body actually looks like, not just a dot to represent the body.
2. Find all the forces acting *on* the body you have isolated and determine where they are applied. Draw vectors for the forces and attach them to the body at the places where the forces are applied. Remember that, if there is a physical axle in the problem, there may be forces applied by the axle on the body.
3. Decide on one point in your drawing (not necessarily in the body) to serve as the origin for angular quantities and draw vectors from the origin out to where each force is applied. A clever choice of origin (like at a point where one of the forces is applied) may simplify your analysis, but any point will work.
4. Choose directions for the x- and y-axes and resolve vectors into components along these axes. Also, for each force, find the component perpendicular to the line joining the origin to the point where the force is applied. Use these perpendicular components to find the torques produced by each of the forces.
5. For motion in two dimensions, there will be three conditions that the forces must satisfy. The sum of the forces in the x-direction must add to zero. The sum of the forces in the y-direction must add to zero. And the sum of the torques about an axis through the origin you have chosen must add to zero.

Let us do an example to see how these rules work.

EXAMPLE 9.7 Conditions of Equilibrium in Two Dimensions

A horizontal plastic meter stick, of mass 400 g, has two additional forces acting on it. There is a downward force of 2 N applied to the left end of the stick and a 6-N force to the right acting on the extreme right end of the stick. If we want to suspend the stick from a single cable, where must the force be applied? What will be the tension in the cable? What angle will the cable make with the vertical?

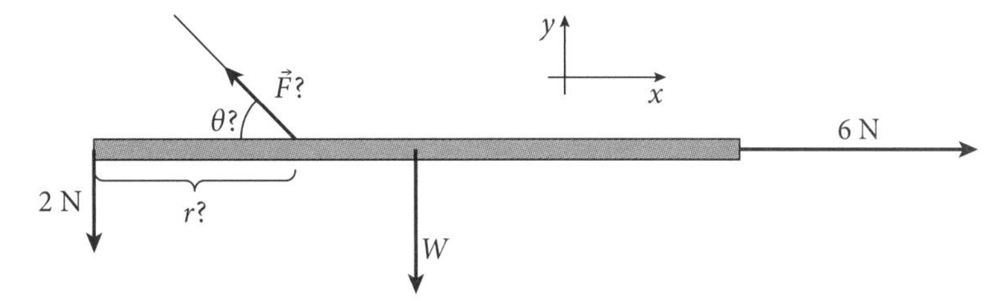

ANSWER This is one of our favorite examples, because it has three unknowns in it, all three of which may be determined from Newton's laws. If the meter stick is to be completely at rest, no movement and no rotation, then the sum of all forces in the x-direction must be zero, the sum of the forces in the y-direction must be zero, and the sum of the torques, measured about any point we please, must be zero.

Let us divide the cable force $\vec{F}$ into components F_x and F_y, along the axes shown in the picture. The net force in the x-direction will be zero, giving

$$F_x + 6\text{ N} = 0 \qquad \text{or} \qquad F_x = -6\text{ N}.$$

In the y-direction, the weight is $W = mg = (0.4\text{ kg})(10\text{ m/s}^2) = 4\text{ N}$, leading to

$$F_y - 2\text{ N} - 4\text{ N} = 0 \qquad \text{or} \qquad F_y = 6\text{ N}.$$

We now choose the left end of the stick as the origin and write down the torque equation. We take a rotation into the page as positive and remember that the only component of $\vec{F}$ that contributes to the torque is the perpendicular component, F_y. Using r for the distance from the origin out to the place where $\vec{F}$ is applied gives

$$(50\text{ cm})(4\text{ N}) - r(6\text{ N}) = 0 \qquad \text{or} \qquad r = 33\text{ cm}.$$

So the cable must be attached at the 33-cm mark of the meter stick. Looking at our two force components, $F_x = -6$ N and $F_y = 6$ N, we see that the cable must point upward and to the left at an angle of 45° and that the magnitude of the tension in the cable must be $\sqrt{(-6\text{ N})^2 + (6\text{ N})^2} = 8.5\text{ N}$.

9.6 Rotational Inertia

We have seen that the angular acceleration is like the linear acceleration and that the torque is like the linear force. For linear motion, the relation between force and acceleration is given by Newton's second law, $F = ma$. One wonders. Is there a form of Newton's second law for rotational motion and, if so, what plays the part of the mass? What is the quantity that determines the inertia of an object when we are trying to change its rotation? It is obvious that mass must play a part. It is much harder to close the door to a bank vault than it is to close the door to your closet. But a little thought will make it clear that there is more to it than that.

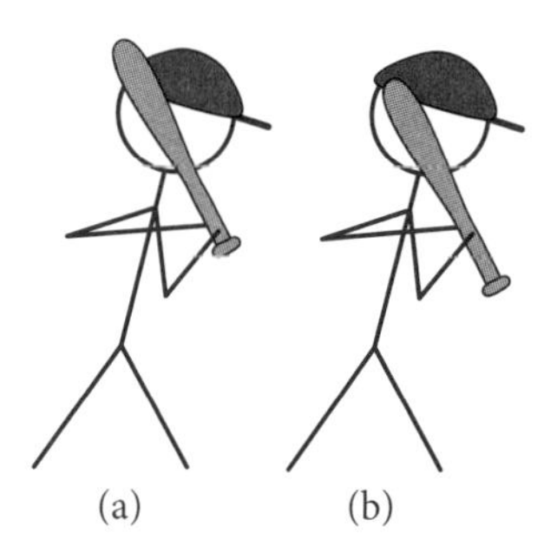

Figure 9.11 A Little-Leaguer is told to "choke up" on the bat to get his swing around faster.

Consider the case of a Little League baseball player facing a pitcher who throws a mean fast ball. By the time the poor kid gets his bat around, the ball is already in the catcher's mitt. You probably know what a good coach will tell the batter in this situation. He'll tell him to choke up on the bat—that is, he'll tell him to hold the bat closer to the middle of the bat rather than right out at the end of the handle where he was holding it before (Figure 9.11). With this change, he is able to swing more quickly. Of course, the bat has the same mass it had before. The torque produced by the batter's two hands acting where they grasp the handle is the same as it was before. But what has changed is how far most of the mass of the bat is from the place around which the torque is applied. The thing that measures the resistance of a body to torques is its *rotational inertia*,[3] a quantity that is formed by a combination of the mass of the object and how far from the chosen axis the mass is distributed.

The fundamental formula for the rotational inertia of a single mass m attached a distance r away from an axis of rotation is

$$I = mr^2.$$

The units of rotational inertia are mass-times-distance-squared, or $\mathrm{kg \cdot m^2}$ in SI units. There is no name given to this combination of units.

The combined rotational inertia of several individual masses, attached so that they will rotate about the same axis, can be found by adding up the rotational inertias contributed by each mass. Computing the rotational inertia of an extended body requires calculus, of course, so we will not be able to do this here. The calculus has been done for a variety of uniform bodies of different shapes, however, and, to give a sense of what the rotational inertias are for a few of these, we show several bodies in Figure 9.12 with formulas for their rotational inertia. For the wheel in figure (c), where the spokes are assumed to be massless, all of the mass is still a distance r from the axis, so the formula is the same as that of an individual mass. For the uniform disk in figure (b), some of the mass is less than the distance r from the axis, so the inertia is less. Figures (d) and (e) illustrate why the Little Leaguer wants to hold the bat a little closer to the middle if he wants to produce a greater angular acceleration of the bat by lowering the rotational inertia about his hands.

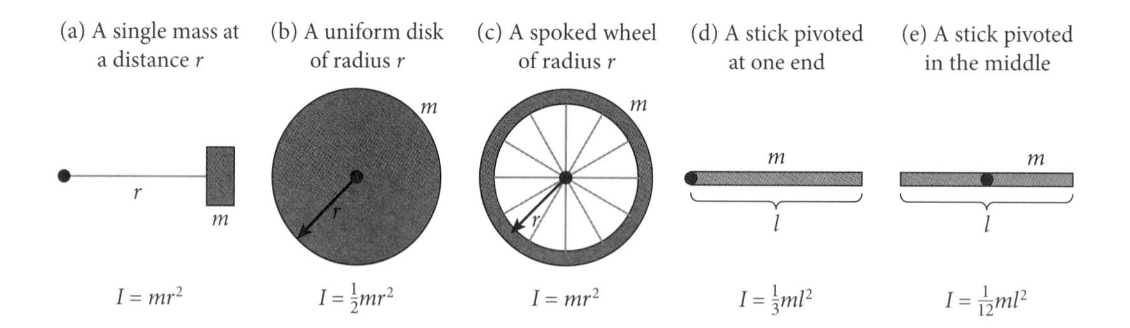

Figure 9.12 Formulas for the rotational inertia of several common types of uniform bodies. A single mass m a distance r from the axis (the black dot in each figure) gives the fundamental formula in (a). For other objects, the same mass produces a lower rotational inertia when more of the mass is closer to the axis.

3. In many (or most) textbooks, this is called the "moment of inertia." Since that name doesn't seem to us to help us remember how it is used, we prefer the term "rotational inertia."

EXAMPLE 9.8 Rotational Inertia of the Merry-Go-Round

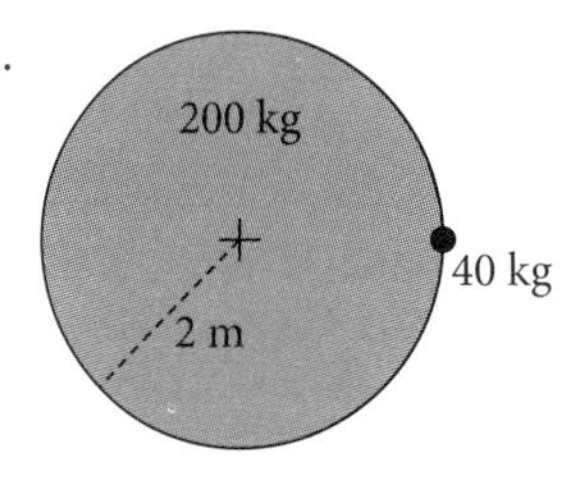

A merry-go-round is composed of a uniform disk of mass 200 kg and radius 2 m. A girl of mass 40 kg stands on the rim of the merry-go-round. What is the rotational inertia of the merry-go-round with the girl on it?

ANSWER We can add rotational inertia contributions when they are calculated relative to the same axis. The rotational inertia of the merry-go-round about an axis through its center can be found from the formula in Figure 9.12b. It is

$$I = \frac{1}{2}mr^2 = \frac{1}{2}(200\text{ kg})(2\text{ m})^2 = 400\text{ kg}\cdot\text{m}^2.$$

To this we may add the girl's rotational inertia, found in Figure 9.12a:

$$I = mr^2 = (40\text{ kg})(2\text{ m})^2 = 160\text{ kg}\cdot\text{m}^2.$$

Thus, the total rotational inertia is $I = 560\text{ kg}\cdot\text{m}^2$.

Now let's get back to the original question with which we began this section. Yes, there is a Newton's second law for rotation. For objects that spin about a single fixed axis, or for rotation referenced to the center of mass of an object, it is

$$\tau_{\text{net}} = I\alpha. \tag{9.5}$$

The torque τ is like the force F; the angular acceleration α is like the linear acceleration a; and the thing that is like the mass m is the rotational inertia, I. Just as translational acceleration must always point in the same direction as the net force, angular acceleration points in the same direction as the net torque.

You should especially note the qualification in the sentence leading up to Equation 9.5. When an object is free to move and spin in space, the rotational inertia and the form of Newton's second law for rotation depend on the point chosen as origin, the point about which rotational inertia and torque are defined. The result is a general law that is more complicated than Equation 9.5. However, the simple form in Equation 9.5 is exactly right in two cases. The first case is where there is a fixed axis pinning the rotating object in place, so that it can spin only about that axis. In this case, torque and rotational inertia must be measured relative to that axis. However, the equation is also exactly correct, even in the case of a body that can move and spin freely in space, as long as the point chosen as the origin for defining torques and rotational inertia is the center of mass of the object.

Finally, let us say something about the units for rotational motion. Torque (see Equation 9.3) has units of $\text{N}\cdot\text{m}$, or, since a newton is a $\text{kg}\cdot\text{m/s}^2$, it is in $\text{kg}\cdot\text{m}^2/\text{s}^2$. Rotational inertia has units of $\text{kg}\cdot\text{m}^2$. This leaves the angular acceleration with units of $/\text{s}^2$, or "per-seconds-squared." This may sound funny, but it is not really a problem as long as we use radians for the angular measurement. We remember that a radian is the ratio between the arc length around a circle and the radius of that circle, and ratios are dimensionless. So an angular acceleration that reads "per second squared" can always be read "radians per second squared," and vice versa. Also, we must remember that, if we *ever* use Equation 9.5, or any of the other rotational equations yet to come in this chapter, we *must* measure any angular quantities in radians.

The next example shows how Newton's second law for rotation works. It is a very practical application of this law in a real engineering environment, that of adjusting or maintaining the orientation of a spacecraft in space by using tiny thrusters in such a way as to create torques without generating any net force on the spacecraft.

EXAMPLE 9.9 Spacecraft Attitude Control with Indium Ion Thrusters

A 100-kg spacecraft in the shape of a cylinder with a 50-cm radius has a rotational inertia that is measured to be 14.1 kg·m^2 for rotation about its cylinder axis, and the spacecraft center of mass lies along this axis. In order to start to spin around this axis, it fires two indium ion thrusters at an angle 45° above the cylindrical side of the spacecraft and in opposite directions. The thrust produced by each thruster is 10 μN (remember that μ means 10^{-6}). Find the angular acceleration of the spacecraft while the thrusters are thrusting.

ANSWER Our extended free-body diagram is shown at right. As the indium ions are accelerated away from the spacecraft (the red jets), there is a Newton's-third-law companion force by the ions on the spacecraft, as shown by the heavy black arrows. The forces by the two thrusters are the same, but in opposite directions, so the net force on the spacecraft is zero. As a result, the spacecraft center of mass will not accelerate as the thrusters fire. As explained on the previous page, we may use the simple form of Newton's second law for rotation for this freely moving spacecraft, as long as we choose the center of mass of the spacecraft as the point about which we measure torques and rotational inertia. To find the torque, we need the component of each force perpendicular to the vector r from the center of mass out to the thruster,

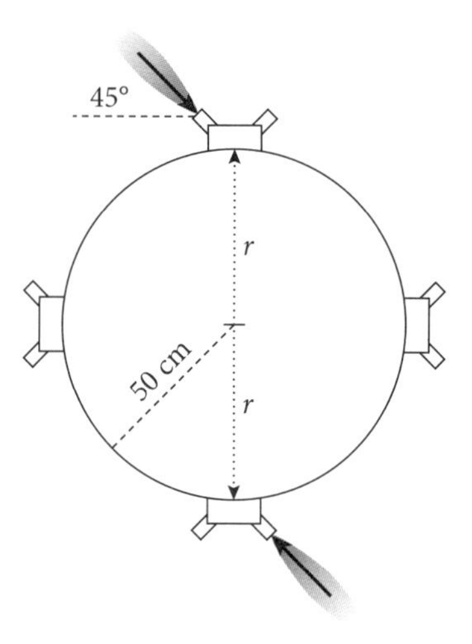

$$F_{\perp} = F\cos(45°) = (10\ \mu\text{N})(0.707) = 7.07\ \mu\text{N}.$$

If we measure torque about the center of mass of the spacecraft, the net torque will be clockwise, into the page, with magnitude

$$\tau = 2 \times (0.5\ \text{m})(7.07\ \mu\text{N}) = 7.07\ \mu\text{N}\cdot\text{m},$$

where the 2× in the equation came from the fact that there are two identical thrusters, each producing the same torque. The angular acceleration is therefore

$$\alpha = \frac{\tau}{I} = \frac{7.07\ \mu\text{N}\cdot\text{m}}{14.1\ \text{kg}\cdot\text{m}^2} = 0.50\ \text{microradians/s}^2.$$

Just in case you are interested, we can use $\Delta\theta = \frac{1}{2}\alpha\Delta t^2$ to see that this angular acceleration would require 2500 seconds (a little over 40 minutes) to turn the spacecraft by 90° ($\Delta\theta = \pi/2$ radians). This is a very tiny torque being applied, but, in the benign environment of space, this could be all that is required to turn the spacecraft to point in some desired direction.

9.7 Angular Momentum and Kinetic Energy

We saw in Chapter 8 that, when there are no external forces acting on some system of bodies, the total momentum of the system is conserved. Again one wonders. Does the analogy between linear quantities and rotational quantities go so far as to define a conserved analog of momentum? The answer is that it does. The new quantity is called *angular momentum* and it is traditionally represented by the letter L. The magnitude of the angular momentum of a single particle of mass m moving at speed v in a circle of radius r is given by

$$L = mvr. \tag{9.6}$$

Let us note that the presence of the variable r in this definition means, as we saw in the definition of torque, that defining angular momentum requires us to define the point about which the angular momentum is to be measured. You can only ask for the angular momentum of some object if you specify the point about which it is measured.

The angular momentum also has a direction associated with it. For a particular choice of origin, the velocity of the object will be instantaneously rotating it in some circular direction about that origin, as shown in Figure 9.13. If you curl the fingers of your right hand in that circular direction, your thumb will point in the direction of the angular momentum. In Figure 9.13, the angular momentum is out of the page.

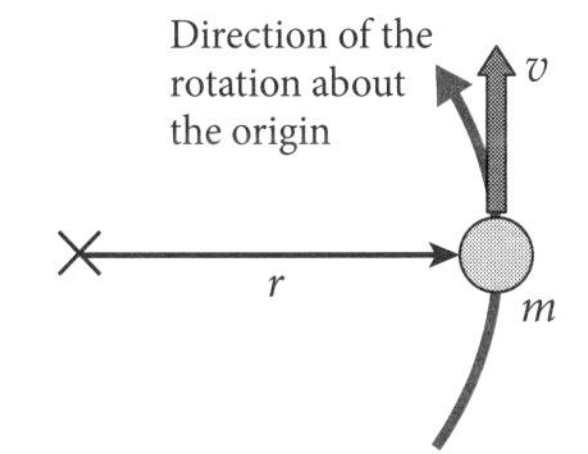

Figure 9.13 A velocity v of an object at a distance r from the origin × will rotate the object in the direcion shown in red.

An extended body that is spinning about a fixed axis has angular momentum because it is really just a collection of tiny particles, each of which is spinning about the same axis. The total angular momentum of a spinning object is the sum of the individual angular momenta of the individual particles. This sum produces an angular momentum vector in the same direction as the angular velocity ω, or[4]

$$\vec{L} = I\vec{\omega}, \tag{9.7}$$

where the rotational inertia I is measured about the same axis that we use for defining angular velocity and angular momentum. One may pleasantly note that this expression is analogous to the expression for linear momentum, $\vec{p} = m\vec{v}$, in which the angular momentum $\vec{L}$ replaces the linear momentum $\vec{p}$, the rotational inertia I replaces the mass m, and the angular velocity $\vec{\omega}$ replaces the translational velocity $\vec{v}$.

But here's a little caution. Back in Chapter 7, we introduced kinetic energy and potential energy. They were both forms of energy. They had the same units and they could be added together to form the total energy. But please note: Angular momentum is *not* a form of momentum. Momentum and angular momentum are two completely different things; they have different units; and they cannot be added together to form anything, certainly *not* the "total momentum."

As we began this section, we explained that there was a law of conservation of angular momentum. It is this:

- Conservation of Angular Momentum: If the net external torque on a system is zero, the total angular momentum of the system does not change.

---

4. This result should not be surprising. When a body is spinning about a single fixed axis, all the particles of the body have angular momentum in the same direction, so their contributions are simply added together. The contribution of each is given by Equation 9.6. We also remember that the linear speed of a particle that is a distance r from an axis, and spinning at angular speed ω about that axis, is given by $v = r\omega$ (Equation 9.2). Thus, each particle of mass m contributes $mr^2\omega$ to the angular momentum of the body. The sum of the mr^2 contributions from all of the particles in the body produces the rotational inertia I, so the magnitude of the angular momentum would be $L = I\omega$.

Note that the conservation of angular momentum does not require that there be zero net force, only zero net torque. This is the case for the football with which we began this chapter. The force of gravity acts on the center of mass of the football to change its trajectory, but, since gravity produces zero torque about the center of mass, the angular momentum of a spinning football (in the limit of no significant air resistance) remains unchanged throughout the trajectory from quarterback to receiver.

Figure 9.14 A spinning ice skater reduces her moment of inertia in order to increase her angular velocity.

There are some situations in which the angular momentum of a spinning object is conserved but where the object actually changes its angular speed. Near the end of a performance, many ice skaters go into a spin. The spin usually starts out slowly and then gets faster and faster. This might appear to be a violation of the law of conservation of angular momentum but is in fact a beautiful example of its validity. The magnitude of the angular momentum is given by the product $L = I\omega$. If the net torque on the system is zero, then L must remain constant. So, if the rotational inertia of an isolated object changes, the angular speed must also change in order to compensate. The skater in Figure 9.14 began with her arms extended. But then, as her arms were drawn in toward her body, her rotational inertia decreased because the mass of her arms was pulled closer to her axis of rotation. The result is that her angular speed increased.

Lastly, it is probably obvious to everyone that the matter in a spinning body is moving, and so it must have kinetic energy. If you have been following our analogy between linear quantities and angular quantities, it should not come as a surprise that the kinetic energy of a body with rotational inertia I, spinning at angular speed ω, is

$$\mathrm{KE} = \frac{1}{2}I\omega^2. \tag{9.8}$$

Note that this kinetic energy of a rotating body is just kinetic energy. It has units of energy and it makes perfect sense to add it to other forms of energy. If a football has mass m and rotational inertia I about its center of mass, and if it is spinning at angular speed ω as it flies through the air with velocity v, then the football's kinetic energy is

$$\mathrm{KE} = \frac{1}{2}mv^2 + \frac{1}{2}I\omega^2. \tag{9.9}$$

We should also add, although we admittedly have not explained the reason for it, that this formula only works if v is the *velocity of the center of mass* of the object and if the rotational inertia and angular velocity are measured *relative to the center of mass* of the object. But, with these conditions satisfied, the kinetic energy of an extended body is simply the translational kinetic energy *of* its center of mass plus the rotational kinetic energy *about* its center of mass.

Let us add one more point about angular momentum and rotational kinetic energy. Anyone who has ever performed something like the ice skater's trick shown in Figure 9.14 knows that it takes work to bring your arms in closer to your body, work from the force you have to apply to overcome the apparent centrifugal force that seems to pull outward on your arms. The force you apply is a real force over a real distance, and the resulting positive work on your arms (force and motion in the same direction) must increase your kinetic energy. Angular momentum is conserved when the internal force is applied by you on your arms, but kinetic energy is not. Work, even internal work like this, changes the mechanical energy of a system.

9.8 Summary

In this chapter you should have learned the following:

- You should be able to use the rotational kinematic equations in Equation 9.1 to relate angular displacement $\Delta\theta$, angular velocity ω, and angular acceleration α, for cases of constant angular acceleration.
- You should be able to use Equation 9.4 to find the location of the center of mass of a few individual bodies, as in Example 9.5, or to estimate the location of the center of mass of an extended body by its symmetry (see pages 173–174).
- You should be able to determine unknown forces or distances using extended free-body diagrams for a body in equilibrium, as discussed in Section 9.5.
- You should be able to use the formulas in Figure 9.12 to calculate the rotational inertia of bodies with the shapes shown in the figure, and to combine these as was done in Example 9.8.
- You should be able to use Equation 9.5 to determine the angular acceleration of a body with a given rotational inertia subject to a torque.
- You should be able to calculate angular momentum of a body moving in a circle using Equation 9.6, the angular momentum of a spinning body using Equation 9.7, and the kinetic energy of a spinning body using Equation 9.9.

CHAPTER FORMULAS

Newton's second law for rotation: $\tau_{\text{net}} = I\alpha$

Angular momentum for a moving body: $L = mvr$

Angular momentum for a spinning body: $L = I\omega$

Kinetic energy for a spinning body: $\text{KE} = \frac{1}{2}I\omega^2$

Rotational kinematics:

$\omega_f = \omega_i + a\Delta t$

$\Delta\theta = \frac{1}{2}(\omega_i + \omega_f)\Delta t$

$\Delta\theta = \omega_i\Delta t + \frac{1}{2}a\Delta t^2$

$\omega_f^2 = \omega_i^2 + 2a\Delta\theta$

PROBLEMS

9.2 * 1. Find the magnitude and direction of the rotational velocity (in units of rad/sec) of the minute hand on a clock on the wall.

* 2. Find the magnitude and direction of the rotational velocity (in units of rad/sec) of a person living on the equator. Then consider the angular velocity of a person living in Bozeman, Montana, halfway between the equator and the North Pole. What would be its direction? Would its magnitude be more, less, or the same as that of the person on the equator?

** 3. A wheel that is initially not rotating is turned through 30.0 revolutions during the time that it is subjected to a constant angular acceleration of 3.77 rad/s^2 (directed out of the page). Use the kinematic equations to solve the following questions.

a) How long did the acceleration last?

b) Find the final angular velocity (magnitude and direction) of the wheel.

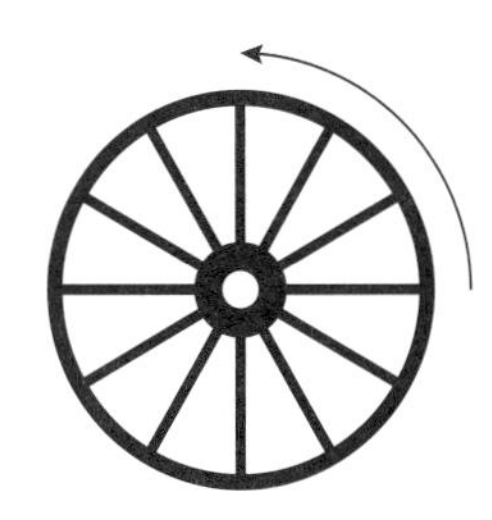

*4. A bench grinder is spinning at 3450 rpm when the power is turned off. Friction in the bearings of the wheel slows the grinder uniformly to a stop in 26 seconds. What is the magnitude of the constant angular acceleration of the grinder?

*5. What is the angular speed in revolutions per minute of a 56-cm-radius bicycle wheel when the bicycle is rolling down a street at 14 m/s? Is the direction of the angular velocity to the rider's left or right?

*6. A child grabs the edge of a merry-go-round and runs counterclockwise (looking down on the merry-go-round from above) at a speed of 1.5 m/s around the outside of the merry-go-round. If the diameter of the merry-go-round is 9 m, what are the magnitude and direction of the merry-go-round's angular velocity in radians per second?

**7. The wheel of a stationary exercise bike is initially not rotating (at $t = 0$). It is then subjected to a constant angular acceleration of 2.5 rad/s^2 as it turns through 20 revolutions. The radius of the wheel is 50 cm.

a) How long did the acceleration last?

b) What is the final angular speed of the wheel?

c) How fast would the bike be moving if it were a real bike? Assume that the wheels are rolling without slipping on the road.

**8. A truck accelerates uniformly from rest and after 5 seconds the 50-cm radius wheels on the truck have an angular velocity $\omega_f = 30$ rad/s.

a) What is the magnitude of the angular acceleration, α, of the wheels?

b) What is the magnitude of the final velocity, v_f, of the truck?

c) Through what total angle does each wheel turn in this time?

**9. A dragster with 30-inch-diameter racing slicks on the back wheels covers a 1000-foot track at constant acceleration, starting from rest, in just 4 seconds.

a) Through how many total revolutions do each of the tires rotate in this time?

b) What is the constant angular acceleration of each wheel during the acceleration?

***10. Billy is riding the Rotor ride at the carnival. This ride is a cylindrical room that spins up to a final angular speed, after which the floor drops out. The ride starts from rest, and after 15 seconds of constant angular acceleration the room has made a total of 6 revolutions. The radius of the room is 2 m.

Billy

a) Find the angular acceleration of the room in rad/s^2.

b) What is the final angular speed of the room?

c) Find the magnitude of Billy's centripetal acceleration after the 15 s. How does this compare to the 10 m/s^2 acceleration due to gravity?

d) What must be the coefficient of static friction between Billy and the wall if he is not to slide down the wall and fall out of the bottom?

**11^f. A puck of mass m slides without friction on a tabletop. There is a massless string attached to the puck that drops without friction through a hole in the tabletop and attaches to a ball of mass M. If the length of the string on the tabletop is a constant r (circular motion), find the angular speed ω in rad/s at which the puck must rotate to keep the ball from falling toward the floor or rising upward.

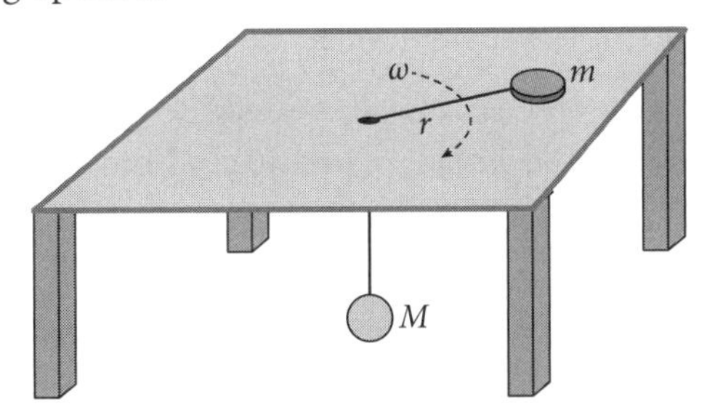

9.4 * 12. A 4-kg ball is placed at the origin of a 2-dimensional coordinate system and a 6-kg ball is placed on the x-axis at $x = 2$ m. Where is the center of mass of the system of two balls?

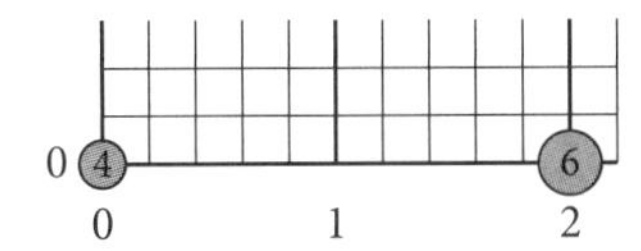

* 13. Penelope fashions her own badminton racket out of a thin steel hoop of mass 50 grams and diameter 20 cm, bolted to the end of a long piece of aluminum tubing of mass 60 grams and length 40 cm. Where is the center of mass of the racket? Neglect the mass of the strings.

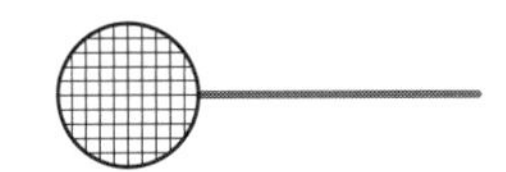

** 14. Caryn's twirling baton is composed of a 200-g stainless steel rod of length 70 cm. Rubber caps are attached on both ends so that their centers of mass are at the ends of the rod. One rubber cap has a mass of 120 g and the other has a mass of 80 g. How far from the center of the rod is the center of mass of the baton?

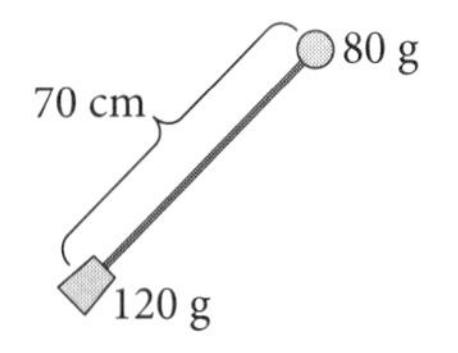

** 15. In a 2-dimensional coordinate system, a 4-kg ball is placed at the origin and a 6-kg ball is placed on the x-axis at $x = 2$ m. A third ball, with a mass of 10 kg, is placed at the point $x = 0.8$ m, $y = 1$ m. Find the location of the center of mass of the system of three balls.

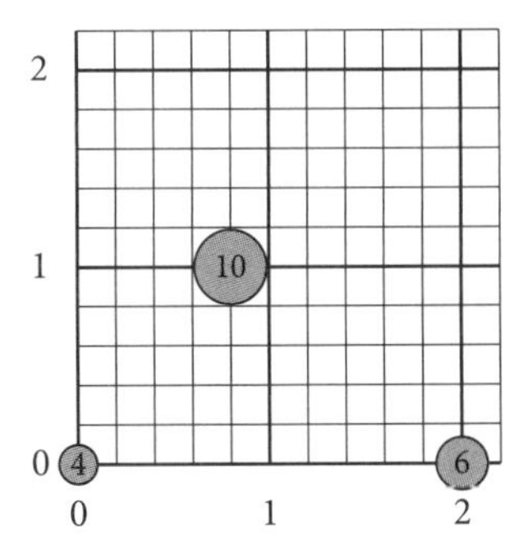

** 16. Three equal 2-kg masses sit at the vertices of a 45° right triangle, as shown. The two equal sides of the triangle are 30 cm long. Find the location of the center of mass of the three masses. Give your answer in distance from the right-angle vertex and as an angle θ from one of the equal sides, as shown.

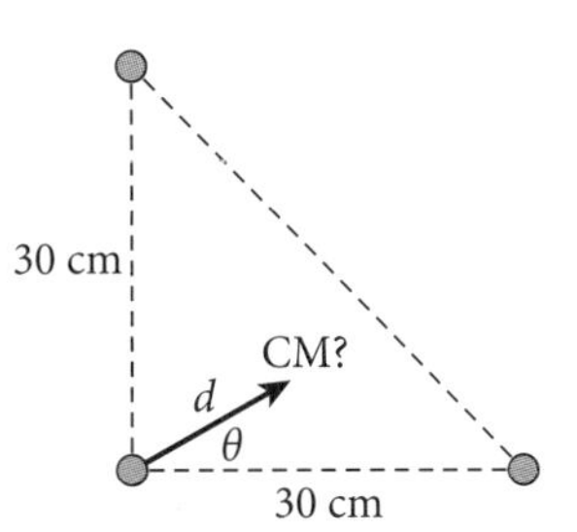

** 17. Three 10-cm wires are soldered together to form three sides of a square. Each wire has a mass of 20 g. From symmetry, we can say that the center of mass of this object must lie along the dashed x-axis shown in the picture at right. Find the distance along the dashed x-axis from the × in the middle of the leftmost wire to the center of mass of the object.

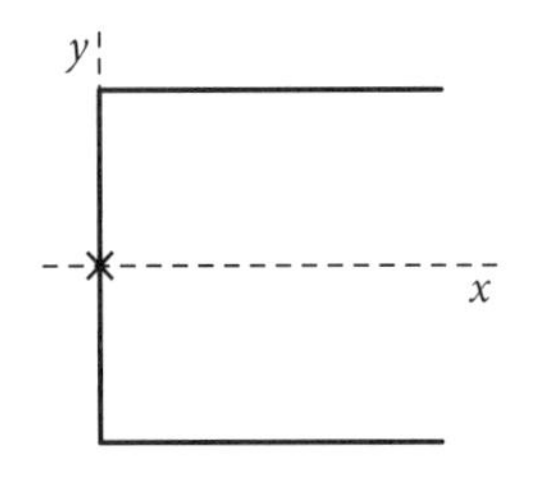

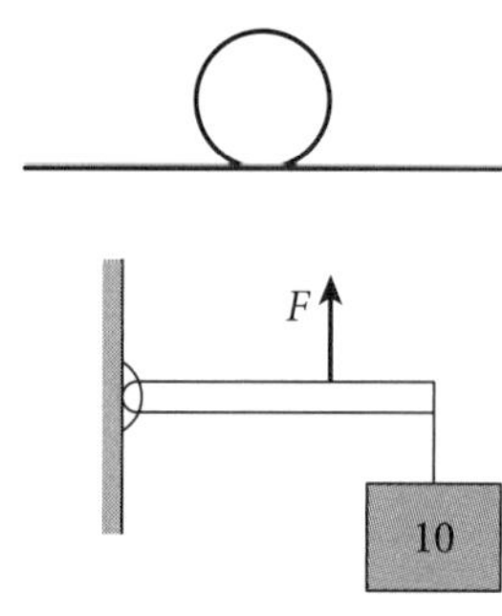

9.5 ** 18. A 1-meter length of wire of mass 1 kg is curved into a circle and another 1-kg length of wire is soldered to it, with the center of the straight wire attached tangentially to the circle, as shown at left. Where is the center of mass of the two wires?

* 19. A 10-kg mass hangs at the end of a massless 60-cm-length rod. The rod is pivoted at the other end. What upward force in newtons, applied 40 cm from the pivot, will keep the rod in equilibrium?

** 20. The 50-kg beams in the two figures below are held in equilibrium by tension in cables angled upward at 30° to the beam. The cable is attached to the end of the beam in (a) and to the middle of the beam in (b). Find the tensions in each case.

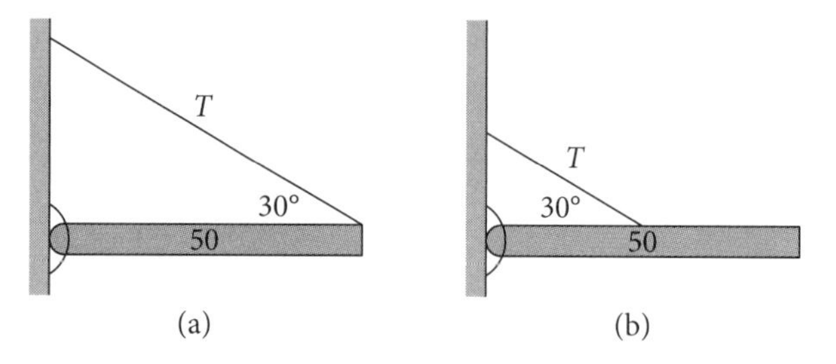

** 21. Two masses, 1 kg and 2 kg, are suspended from the ends of a 60-cm massless rod by a single vertical cable that keeps the rod and masses in equilibrium. Find where along the beam the cable must be attached and what the tension in the cable will be for the two cases shown: (a) when the rod is horizontal and (b) when the rod is at a 30° angle to the horizontal.

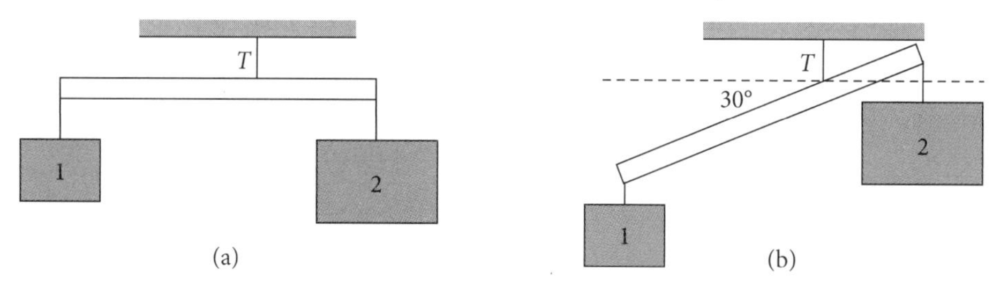

** 22. A box (mass 5.6 kg) sits 3 meters from the right-hand side of a 10-meter-long beam (mass 20 kg). The beam is supported on the left by a rope that is exactly 0.5 meters from the left end of the beam. On the right edge, Greg supports the beam on his head.

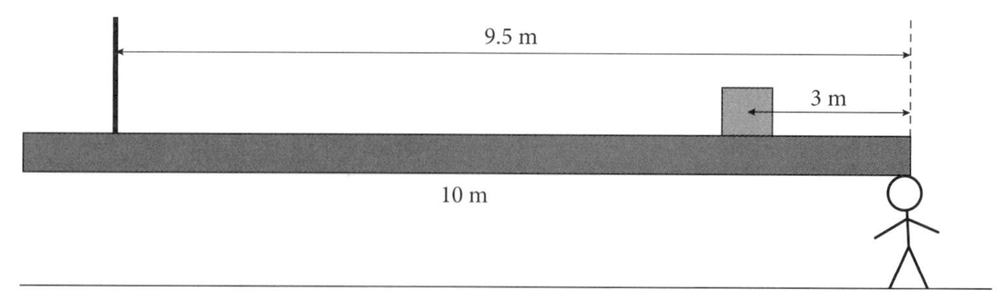

a) Draw an extended free-body diagram for the beam.

b) Determine the magnitude of the normal force by Greg's head on the beam.

c) Find the tension in the rope.

d) The rope will break if more than 150 N of force is applied to it. If we slide the box to the left, the tension in the rope increases. How far can we slide the box (from its starting position) before the rope breaks?

** 23. Bob and his lazy brother, Will, are working together on a construction site, holding a triangular pane of glass as shown. The weight of the glass is 300 N. The center of gravity is indicated by a cross on the diagram. Both of the workers are pushing directly upward on the glass and the pane of glass is not moving. Find the upward force of each worker on the glass.

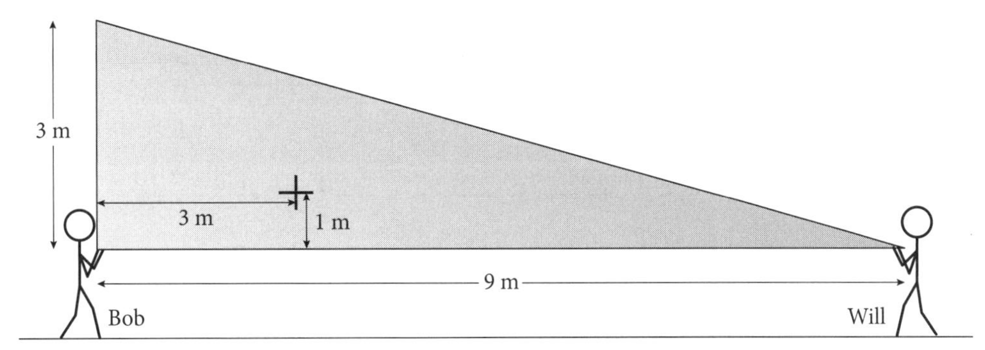

** 24. Bob and Will have been told to carry a uniform 2.0-m board of mass 71 kg. Will is supporting the board at the end while Bob is 60 cm from the other end as shown. Will has attached his lunch to Bob's end of the board, and the tension in the string supporting the lunch is 200 N. Find the upward force that each exerts on the board.

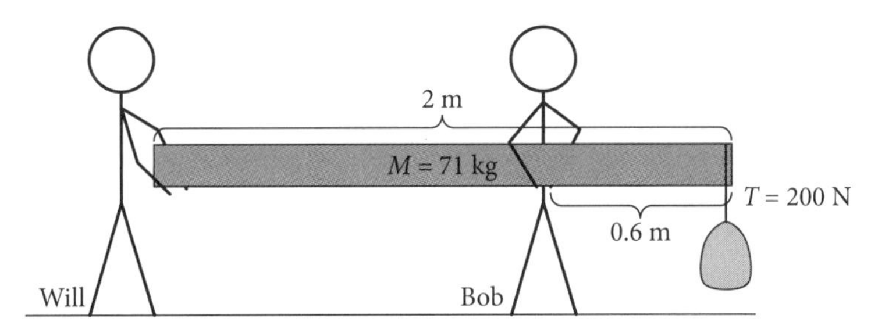

** 25. A 70-kg painter stands on a 40-kg platform that is supported by two ropes as shown. The platform is 2 m long. The left rope is tied to the platform 50 cm from the end and the painter is standing 50 cm from the right rope. Find the tension in each rope.

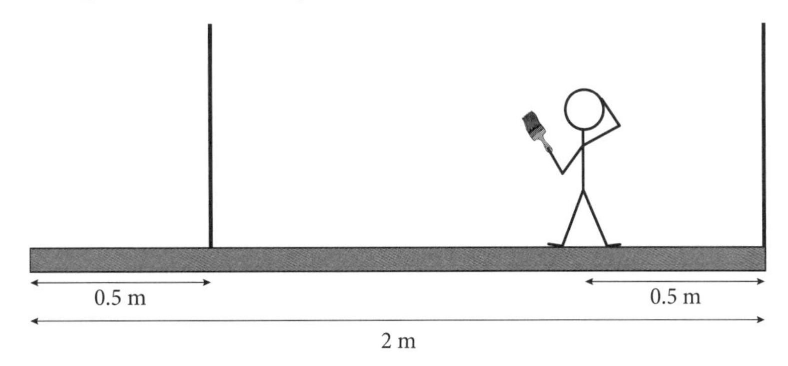

** 26. A knight is pulling straight up on Excalibur (a sword) with force of 500 N, as shown at right. The sword is tilted at an angle θ 53° above the horizontal. Find the torque (magnitude and direction) by the knight on the sword, measured about the point where the sword is attached to the stone.

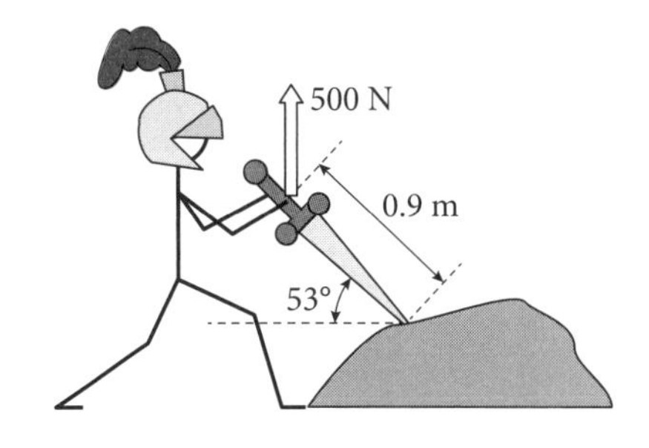

** 27. Bob and Will are at it again, working at a lumberyard holding a heavy uniform board in place so that it can be cut. The board is 2 m long and has a mass of 90 kg. It is perched on a frictionless roller 50 cm from Will's end. Will thinks he is helping by pushing horizontally on the board with a force of 400 N.

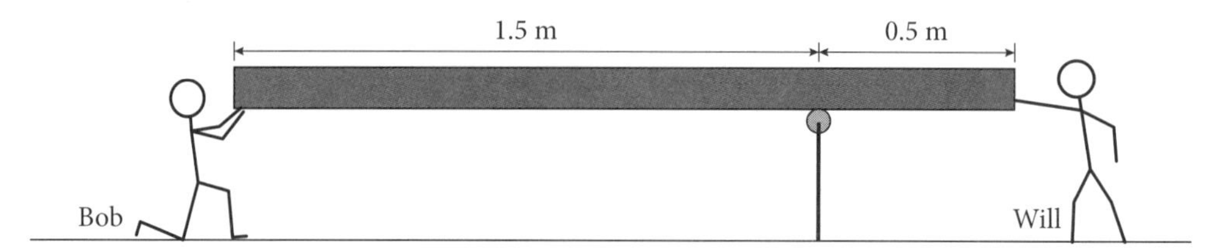

a) Draw the beam and add an arrow to it indicating the approximate direction of the force that Bob must exert on the board to keep it from moving.

b) Find the magnitude of the force that Bob must exert to keep the board from moving.

c) Suppose Bob finally gets fed up and walks away from the job. What force (magnitude and direction) would Will have to exert on his end to keep the board from moving?

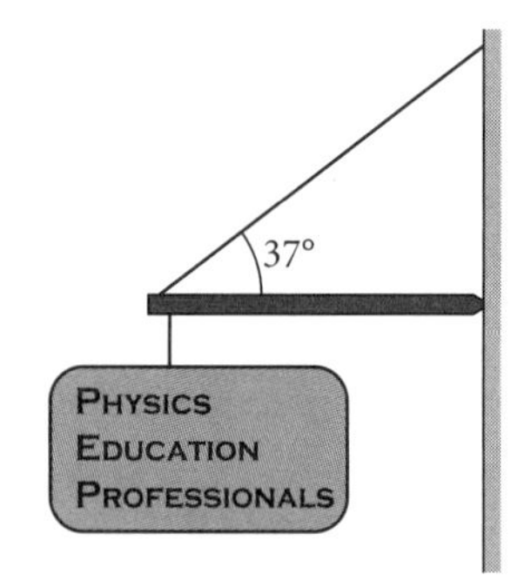

** 28. A 30-kg sign hangs at the end of a 3-meter pole of mass 12 kg (as shown). Find the tension in the support wire if it makes an angle of 37° with respect to the pole. [Hint: Don't forget that forces can be applied to the pole at the point where it is attached to the wall.]

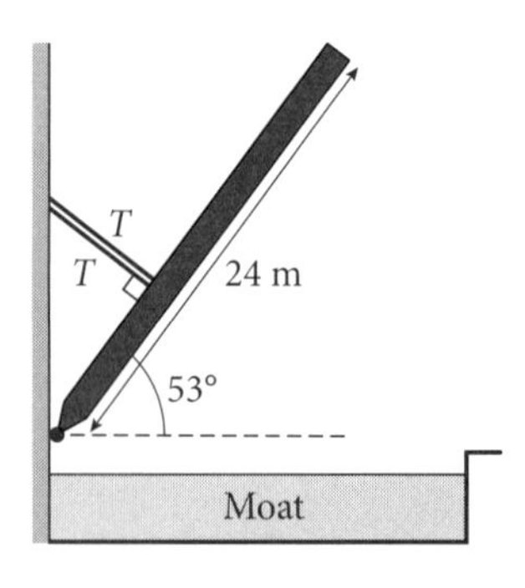

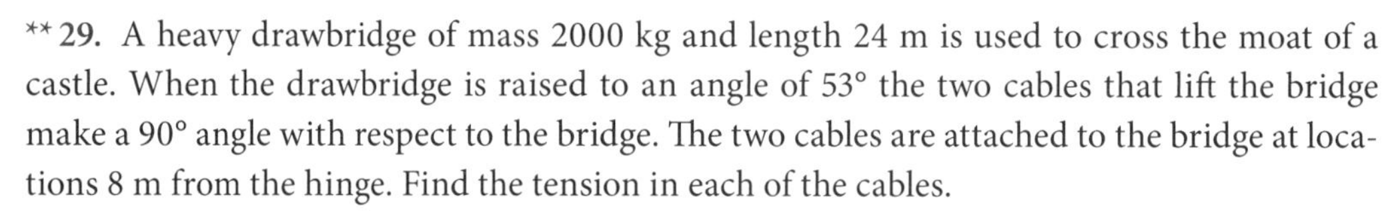

** 29. A heavy drawbridge of mass 2000 kg and length 24 m is used to cross the moat of a castle. When the drawbridge is raised to an angle of 53° the two cables that lift the bridge make a 90° angle with respect to the bridge. The two cables are attached to the bridge at locations 8 m from the hinge. Find the tension in each of the cables.

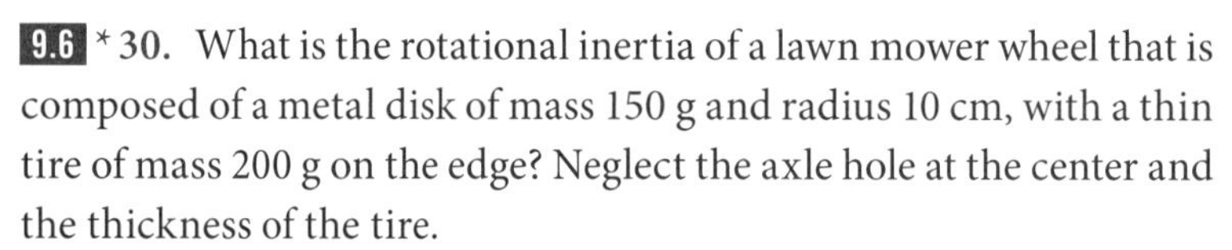

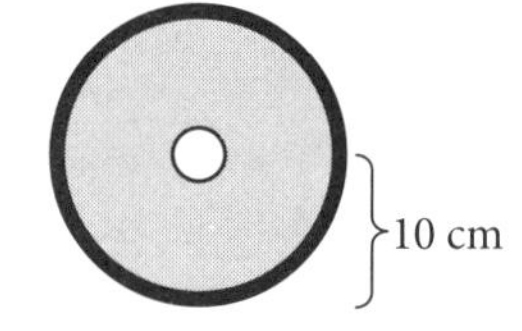

9.6 * 30. What is the rotational inertia of a lawn mower wheel that is composed of a metal disk of mass 150 g and radius 10 cm, with a thin tire of mass 200 g on the edge? Neglect the axle hole at the center and the thickness of the tire.

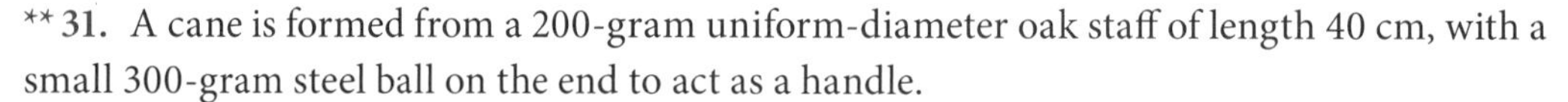

** 31. A cane is formed from a 200-gram uniform-diameter oak staff of length 40 cm, with a small 300-gram steel ball on the end to act as a handle.

a) Find the location of the center of mass of the cane.

b) A man wants to swing the cane around over his head, grasping it at the center of mass. Calculate the rotational inertia of the cane about the center of mass. [Hint: Think of the staff as two sticks, one on each side of the center of mass.]

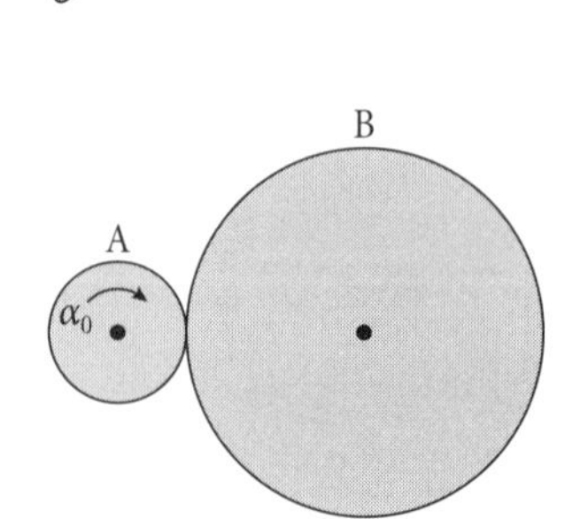

** 32. Two solid wheels, A and B, are mounted on a wall on parallel horizontal axes, as shown at left. The two wheels *roll on each other without slipping.* Both wheels are uniform disks and both wheels have the same mass, but wheel B has a greater radius than wheel A. The two wheels start out at rest, but then a motor gives wheel A a constant angular acceleration $\vec{\alpha}_0$, pointing into the page.

a) Compare the rotational inertia of wheel A to the rotational inertia of wheel B. Explain your reasoning.

b) Compare the magnitude of the frictional force by wheel A on wheel B to the magnitude of the frictional force by wheel B on A. Explain your reasoning.

c) Compare the magnitude of the torque exerted by wheel A on B to the magnitude of the torque exerted by wheel B on A. Explain your reasoning.

d) Is the magnitude of the angular acceleration of wheel B greater than, less than, or equal to $\vec{\alpha}_0$? Explain your reasoning.

e) Find the direction of the angular acceleration of wheel B. Explain your reasoning.

f) Explain why there must be a normal force on wheel B by its axle. [Hint: A free-body diagram of disk B would help.]

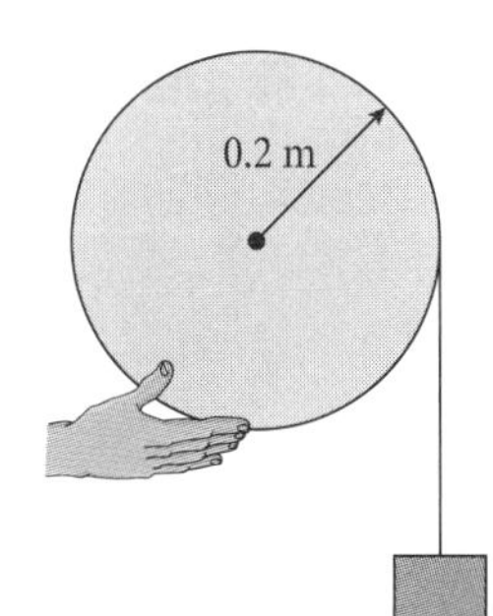

** 33. A block hangs by a massless string that is wound around a solid circular pulley of radius 20 cm. There is no friction in the pulley. The block is released from rest, and the pulley goes through three complete rotations (6π radians) in 4 seconds.

a) What is the angular acceleration of the pulley (magnitude and direction)?

b) How far does the block drop in this time?

At the end of the 4 seconds, a hand reaches out and touches the pulley, producing a constant frictional force that slows the block's descent.

c) What is the direction of the pulley's angular velocity? Explain your answer.

d) What is the direction of the pulley's angular acceleration? Explain.

e) Is the magnitude of the torque by the hand on the pulley less than, equal to, or greater than the magnitude of the torque by the string on the pulley? Explain.

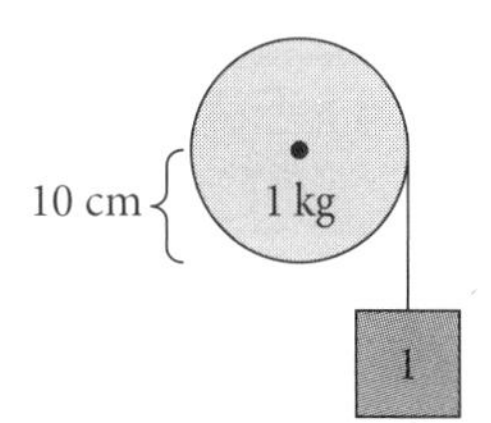

*** 34. A 1-kg block is hung from a string that wraps around a 1-kg frictionless pulley of radius 10 cm. Find the tension in the string, the angular acceleration of the pulley, and the linear acceleration of the falling block. [Hint: Remember that, if the block is accelerating, the tension in the string will not equal the weight of the block.]

*** 35. When a cue stick hits a pool ball directly in the center, the ball begins by sliding across the felt tabletop without spinning. Kinetic friction between the tabletop and the ball slows the ball down, but it also provides a torque that causes the ball to spin up (clockwise in the figure below). Eventually, the spin of the ball will be fast enough that the velocity of a point on the circumference of the ball, relative to the center of the ball, is the same as the speed of the ball. At this point, the ball will roll without slipping, and the kinetic friction will go away. The ball's mass is 0.17 kg and its radius is 2.86 cm. The rotational inertia of a sphere with these dimensions is $5.6 \times 10^{-5}\,\text{kg}\cdot\text{m}^2$. The initial velocity of the ball is 3.00 m/s and the coefficient of friction between the ball and the tabletop is $\mu_K = 0.20$.

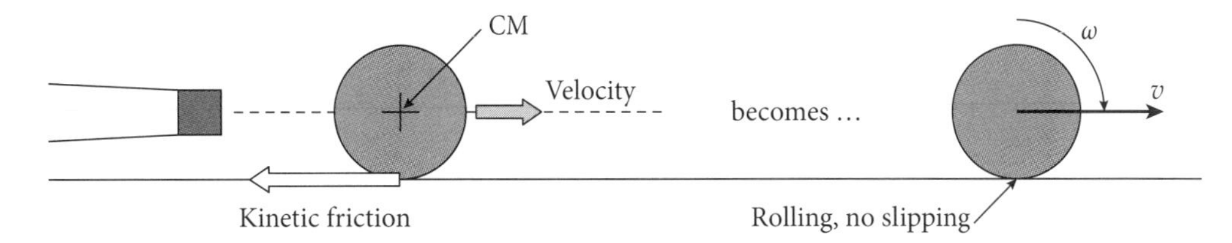

a) What is the force of friction by the table on the ball and what torque about the center of mass of the pool ball does this force create?

b) What are the linear acceleration and the angular acceleration of the ball?

c) How long will it take to get to the point where the spin of the ball stops the slipping of the ball's surface relative to the surface of the pool table?

d) What will be the ball's speed at this point?

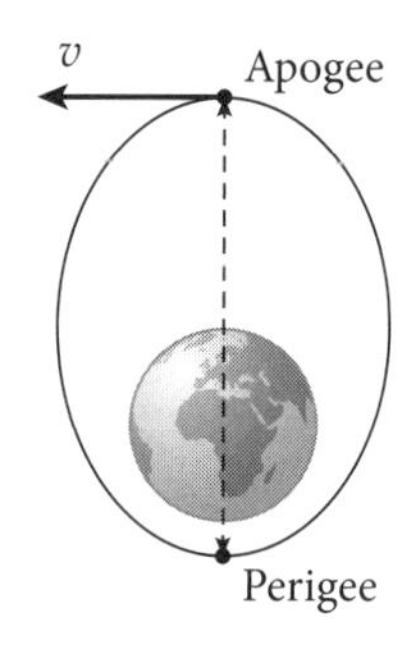

9.7 ** 36. A spacecraft orbits the earth in an elliptical orbit, going from an apogee (furthest point) 21,000 km from the center of the earth to a perigee (closest point) 7,000 km from the center of the earth. When it is at apogee, its orbital speed is 3 km/s.

a) Explain how we know that the angular momentum of the earth–spacecraft system is conserved.

b) What will be the orbital speed of the spacecraft at perigee?

** 37. Experimental tidal power stations use the rise and fall of the ocean tides to create electrical power. The source of this energy is ultimately the rotational kinetic energy of the earth. The earth's rotational inertia is $I = 8 \times 10^{37}$ kg·m^2.

a) Calculate the earth's rotational kinetic energy in joules.

b) Currently, the world's electrical energy use is 7.2×10^{22} J/yr. At this rate, how many years would the rotation of the earth last, if all the earth's electrical power were generated using tidal generators?

** 38. A quarterback fires off a pass with the football traveling at a speed of 20 m/s and with a rotational speed of 5 revolutions per second. The mass of the football is 400 g and its rotational inertia is 2×10^{-3} kg·m^2.

a) What is the total kinetic energy (translational plus rotational) of the football just as it is thrown?

b) If the ball rises to a maximum height of 10 m above the point where it left the quarterback's hand, what will be its translational and rotational kinetic energies at that point? [Neglect any effects of air resistance.]

*** 39. A 2006 Lincoln-head penny has a mass of 2.5 g (in 1982, the copper content was reduced and the mass of the penny dropped from its previous value of 3.1 g). The penny's diameter is 19 mm, giving it a rotational inertia of 1.13×10^{-7} kg·m^2. Such a penny is balanced on top of a ramp 50 cm high and it begins to descend the ramp, rolling without slipping all the way down. When it reaches the bottom of the ramp and rolls out along level ground, what will be the penny's linear velocity and what will be its angular velocity? [Hint: Friction does almost no work on a rolling penny, so total mechanical energy will be conserved.]

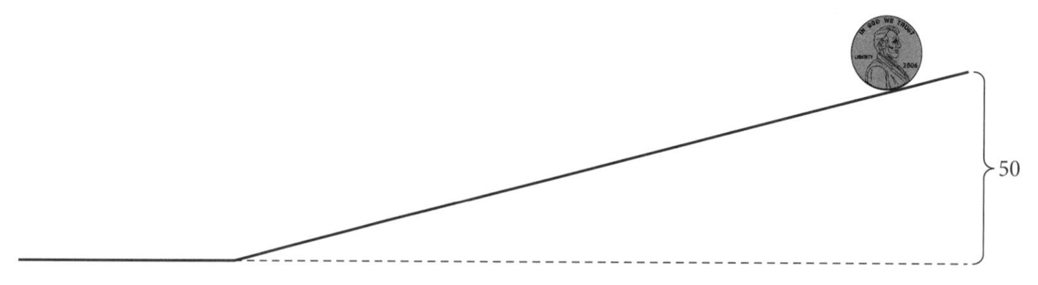

10

PRESSURE AND FLUIDS

10.1 Pressure

A fakir[1] gets up in the middle of the night from his bed of nails and, in the dark, he accidentally steps on a carpet tack that was left on the floor. Why does the carpet tack stick into his skin while the nails in his bed do not?

The answer is that the quantity that determines whether or not something can break your skin is *not* the applied force, but the applied *pressure*. Pressure is defined as the force applied to an object divided by the area over which it is applied:

$$P = \frac{F}{A}. \qquad (10.1)^2$$

In the US, we measure pressure in pounds per square inch (abbreviated PSI in most commercial uses). In SI units, the unit of pressure is a $\mathrm{N/m^2}$, also called a pascal (Pa).

EXAMPLE 10.1 The Fakir

In figure (a) below, a 60-kg (600-N) fakir is asleep on a contoured bed of nails. He is supported by a total of 500 nails, each of which has a point ¼ mm across. In figure (b), a single nail supports his entire weight where it touches his foot. Find the pressure produced by the bed of nails and by the single nail.

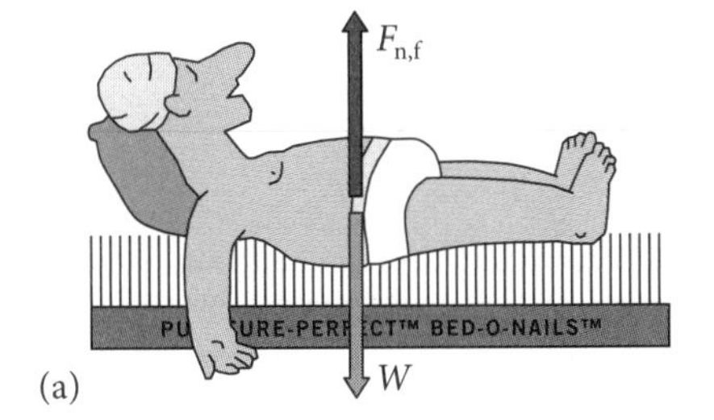

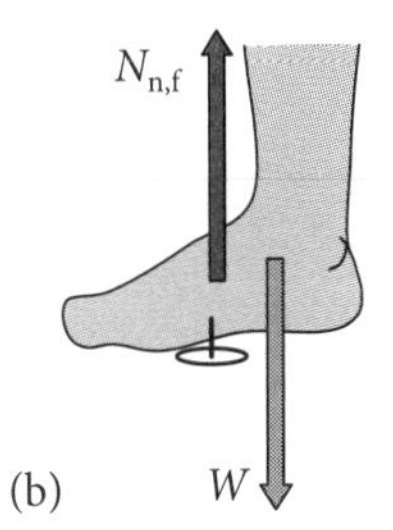

1. Pronounced feh-keer′, not fay′-ker. See, you learn all kinds of interesting things in this book.

2. Note that we have used the same symbol P for pressure that we used back on page 131 for "power." Sorry, but these are almost universally used symbols for both of these quantities.

ANSWER In each case, the normal force of the nails on the fakir $N_{n,f}$ must equal the 600-N weight of the fakir, since the fakir is at rest. But the area over which the force is applied differs by a factor of 500. Each nail has a cross-sectional area at its tip (assuming a square tip) that is approximately

$$A = (0.25 \times 10^{-3}\ \text{m})^2 = 6 \times 10^{-8}\text{m}^2.$$

For the single nail on the floor, the pressure is

$$P = \frac{N_{n,f}}{A} = \frac{600\ \text{N}}{6 \times 10^{-8}\text{m}^2} = 10^{10}\frac{\text{N}}{\text{m}^2} = 10^{10}\ \text{Pa}.$$

For the nail bed, the total area is 500 times as big, giving a pressure of

$$P = \frac{N_{n,f}}{A} = \frac{600\ \text{N}}{500 \times 6 \times 10^{-8}\text{m}^2} = 2 \times 10^{7}\frac{\text{N}}{\text{m}^2} = 2 \times 10^{7}\ \text{Pa}.$$

The pressure is 500 times less from the bed than from the tack. This is like the difference between placing a half-ounce balloon on top of your birthday cake and setting a sixteen-pound bowling ball down on it.

When you dive down to the floor of the ocean to pick up a shell, it feels like the water is trying to pop your eardrums. The deeper you go, the greater the force you feel. We will talk about why the pressure increases in the next section. Here, let us just note that our experience tells us that there is no way you can hold your head to make the force go away. If your ear is pointing downward, the force on your eardrum is up. If your ear is up, the force is down. If your ear is to the west, the force is to the east, and so on (Figure 10.1). Although the force on your eardrum is a vector, because force is always a vector, pressure can actually produce a force in any direction. It has no direction by itself. *Pressure is a scalar.*

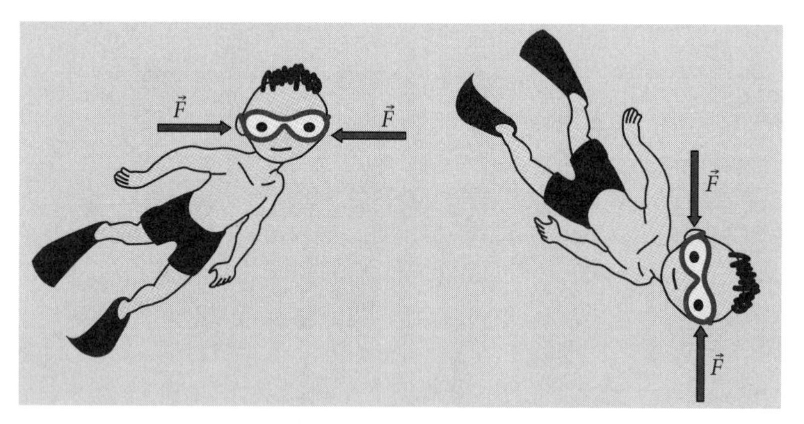

Figure 10.1 At a given depth below the surface of the water, the direction of the force of the water on your eardrum only depends on the orientation of your ear. The pressure itself does not have any direction associated with it. It is a scalar.

10.2 Pressure in Fluids

A fluid is any substance that conforms to the shape of its container. This includes liquids and gases. There is a whole branch of physics that studies the mechanics of fluids, but, in this short chapter, we limit our discussion to studying the forces exerted by fluids on other objects that are immersed in them.

When we are floating on the surface of the ocean, we are also at the bottom of a deep ocean of air that produces a pressure on us just like the water does in Figure 10.1. If we are at sea level, this pressure is called one "atmosphere" and equals 14.7 lbs/in^2 or 101,000 N/m^2 (pascals). Why does this pressure not make us uncomfortable? It is because our bodies have adjusted to it. The pressure in our lungs, blood vessels, organs, etc. adjusts itself until it equals atmospheric pressure. However, when we dive down below the surface of the water to pick up a shell, the pressure inside our bodies does not change quickly enough to compensate. We feel the pressure.

So how much does the pressure increase as you dive deeper and deeper under water? To work this out, let's consider a calm lake filled with water and let us look at a rectangular block of the water that sits just below the surface of the lake, as in Figure 10.2. Let us assume that the height of the block is h and that the area of the base is A. Since the lake is entirely placid, we know that this block of water is not spurting up out of the lake or sinking down toward the bottom or sloshing to one side or the other. This piece of water is not accelerating, so we know that there is no net force on it. The block of water is immersed in the rest of the lake, so there will be pressure from the lake water around it, creating horizontal forces on each of its four vertical sides. But, since the block remains stationary, we know that these horizontal forces must all simply cancel (so we have not shown them in Figure 10.2). We also know from the stillness of the water that the net vertical force on the block of water must likewise be zero. But here, we note that the upward force on the water from the pressure at the bottom of the block cannot simply be the same as the downward force on the block from the atmospheric pressure at the surface, because there is another force in the vertical direction that must be dealt with. The water has mass, and so the earth's gravity will pull down on it.

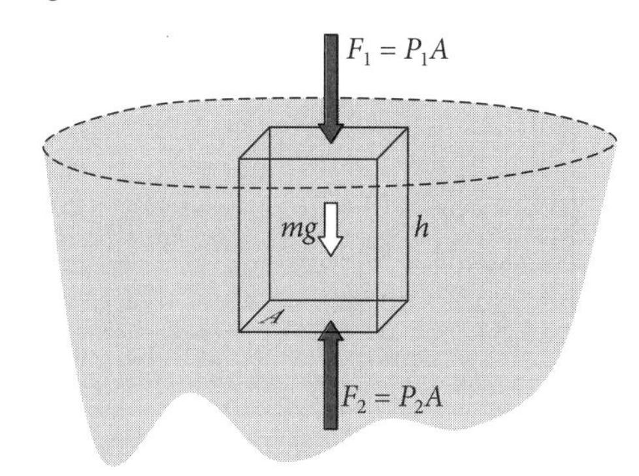

Figure 10.2 An imaginary block of water sitting at the surface of a smooth lake. The fact that the net force on the block must be zero tells us how much greater the pressure must be at the bottom of the block than the pressure at the top.

A glance at Figure 10.2 shows that the upward force caused by the pressure at the bottom must balance both the downward force from the pressure at the top plus the weight of the block of water. The weight will be mg, where m is the mass of the block of water and g is the strength of gravity. Using Equation 10.1 to relate the force to the pressure (that is, writing $F = PA$), this can be written

$$P_2A = P_1A + mg,$$

where P_1 is the pressure at the surface of the lake and P_2 is the pressure a distance h below the surface. If we divide this equation by the area A, we find

$$P_2 = P_1 + \frac{mg}{A}. \tag{10.2}$$

So the increase in pressure as we go from the top of the block to the bottom is mg/A. Let's look at this additional pressure in a different way. The mass of the block of water is equal to the density of water ρ times the volume of the block V, so we may write

$$\frac{mg}{A} = \frac{(\rho V)g}{A} = \frac{(\rho Ah)g}{A} = \rho gh,$$

where we have recognized that the volume of this rectangular block is $V = Ah$. When we make this substitution in Equation 10.2, we end up with

$$P_2 = P_1 + \rho gh. \tag{10.3}$$

You should note that, although we assumed that P_1 was the pressure at the surface of the lake, everything we have done would work just as well if it were all underwater. P_1 is the pressure at some depth and P_2 is the pressure at a distance h below that. Every time we start somewhere and go further down by an amount h through a liquid of density ρ, the pressure increases by an amount ρgh.

EXAMPLE 10.2 Finding the Pressure in a Lake

The density of pure lake water is 1000 kg/m³. How deep can you dive in a lake at sea level before the pressure in the water gets up to 3 atmospheres?

ANSWER We choose our reference pressure P_1 to be the pressure at the surface of the lake. (The problem states that the lake is at sea level, but the pressure at other places could be less than one atmosphere, if they are not as deep in the "ocean of air" as a lake surface at sea level.) We set the pressure at the deeper location to be three atmospheres and solve for h. (We also use $g = 9.81\ \text{m/s}^2$, for a change, since we want you to note some exact values for how pressure varies with depth in pure water.)

$$P_2 = P_1 + \rho gh$$

$$3(101{,}000\ \text{N/m}^2) = 101{,}000\ \text{N/m}^2 + (1000\ \text{kg/m}^3)(9.81\ \text{m/s}^2)h$$

$$h = \frac{2(101{,}000\ \text{N/m}^2)}{9.81 \times 10^3\ \text{N/m}^3} = 20.6\ \text{m}$$

The pressure at 20.6 m is 2 atmospheres higher than at the surface, so pressure increases by one atmosphere for every 10.3 m we dive in fresh water. A similar calculation will show that it only requires 10.0 m of sea water to increase the pressure by one atmosphere, since salty sea water has a slightly higher density than fresh lake water ($\rho_{\text{sea water}} = 1025\ \text{kg/m}^3$).

10.3 Pascal's Principle

Let's note again that Equation 10.1 tells us that an object of surface area A in a fluid with pressure P experiences a force $F = PA$ from the surrounding fluid. So, even if two objects are subject to the same pressure, they will experience different forces on their surfaces if the areas of the surfaces are different. The following example will help us get a feel for this difference between force and pressure.

EXAMPLE 10.3 The Leaking Water Tank

A 5-m-high water tank develops a small leak through a 1-mm-radius hole near its bottom. Could you reasonably stop the leak by pressing your thumb over the hole? Unfortunately, the 10-cm-radius outlet pipe from the bottom of the tank then breaks off and water begins to pour out of the broken pipe. Could you reasonably stuff a plastic water bucket into the hole and keep the water in the tank until help arrives?

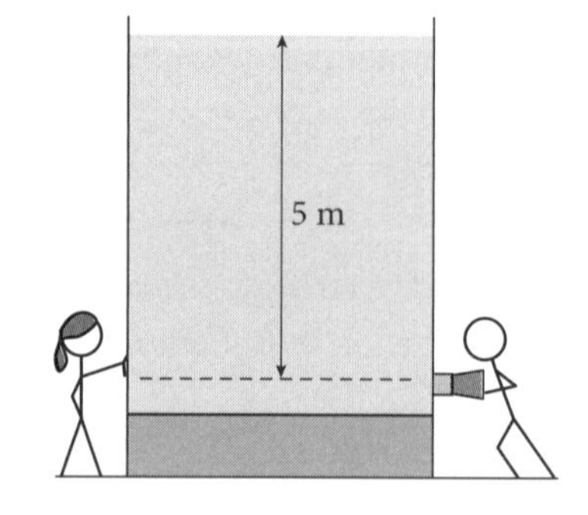

ANSWER The outside of the tank is entirely at the same one atmosphere of pressure, so we concentrate on pressure differences only. The additional pressure at the bottom of a 5-meter column of water is

$$P = \rho gh = (10^3 \text{ kg/m}^3)(10 \text{ m/s}^2)(5 \text{ m}) = 5 \times 10^4 \text{ N/m}^2 = 5 \times 10^4 \text{ Pa}.$$

(You may want to check the units in that calculation to make sure you see how they work.) Since the pressure is the same at all points at the same depth in the water (see Equation 10.3), the outward force on your thumb from the 1-mm hole in the tank would be

$$F = PA = P(\pi r^2) = (5 \times 10^4 \text{ Pa})\pi(0.001 \text{ m})^2 = 0.157 \text{ N},$$

about the same as the weight of a ½-ounce plastic bottle cap. You could easily provide that force with your thumb. On the other hand, the force required to hold the bucket against the water coming out of the big broken pipe is

$$F = PA = (5 \times 10^4 \text{ Pa})\pi(0.1 \text{ m})^2 = 1570 \text{ N},$$

or about 350 pounds. You should run before you get wet.

We have seen how the pressure differs in a fluid due to gravity acting down on the fluid. The pressure increases with increasing depth. But, once the pressure profile is established in a static fluid, the pressure in a fluid can still change if external forces are imposed on it. French philosopher Blaise Pascal (1623–1662) experimentally discovered a fact about how the pressure in a fluid changes when external forces act on the fluid. This fact is now called *Pascal's Principle*. It states

> A change in pressure at any point in an enclosed incompressible fluid at rest is transmitted undiminished to all points in the fluid.

Pascal discovered his principle experimentally, but it could easily have been worked out theoretically by applying Newton's laws to the forces on a cubic piece of the fluid, as in Figure 10.2. Once the pressures in a fluid have adjusted themselves so as to keep the fluid static, then any additional pressure applied to one side of a piece of the fluid would have to be exactly balanced by the same increase in the pressure applied to the opposite side of that piece. If not, that piece of the fluid would begin to accelerate. So, if a fluid is at rest, any increase in pressure on one part of a fluid must be transmitted uniformly throughout the fluid. Let us see one of the practical results of this principle.

EXAMPLE 10.4 The Hydraulic Lift

A 1500-kg car sits on the top of a 10-cm-radius cylinder filled with an incompressible fluid. The fluid also fills a connected reservoir and a 1-cm-radius cylinder, as shown in the figure. How much force is required to keep the fluid in the 1-cm cylinder from moving as the car pushes down on the fluid in the 10-cm cylinder? If you want to raise the car by 1 meter, how far must the fluid in the 1-cm cylinder be pushed down to do the job?

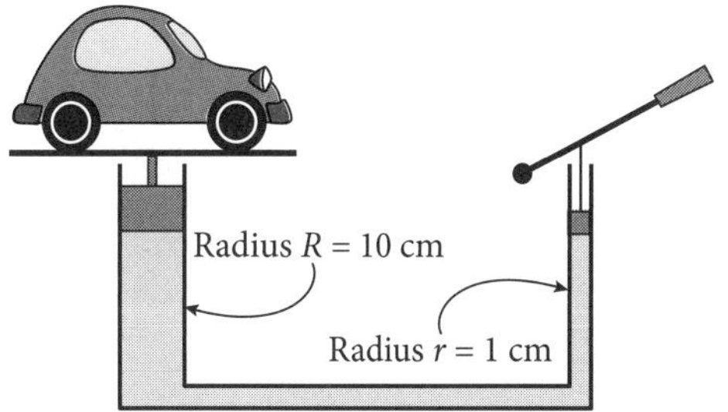

ANSWER The additional pressure the car puts on the fluid is given by its weight divided by the area of the fluid that is holding the car up. It is

$$P = \frac{F}{A} = \frac{mg}{\pi R^2} = \frac{(1500\ \text{kg})(10\ \text{m/s}^2)}{\pi(0.1\ \text{m})^2} = 4.77 \times 10^5\ \text{Pa}.$$

According to Pascal's Principle, this will also be the additional pressure at the top of the 1-cm cylinder. The radius of the small cylinder is 1 cm, so the force of the fluid on the stopper in the small cylinder will be $F = PA = (4.77 \times 10^5\ \text{Pa})\pi(0.01\ \text{m})^2 = 150$ N, or about 33 lbs.

Yes, a 33-pound force is all that is required to hold up a 3300-pound car. If we want to raise the car, a force only slightly more than 33 lb will have to be applied. But, in order to raise the car 1 meter, the volume of the fluid that must be pushed down in the 1-cm cylinder is the same volume that must be pushed up into the 10-cm cylinder. This volume is $V = \pi R^2 H = (3.14)(0.1\ \text{m})^2(1\ \text{m}) = 3.14 \times 10^{-2}\text{m}^3$. In order to provide this much fluid, the fluid in the 1-cm cylinder must be pumped a distance

$$h = \frac{V}{\pi r^2} = \frac{3.14 \times 10^{-2}\text{m}^3}{\pi(0.01\ \text{m})^2} = 100\ \text{m}.$$

In a real pump, more and more fluid is supplied to the small cylinder, rather than moving the fluid already in the cylinder such a long distance.

10.4 Archimedes' Principle

We all know that it is easier to stand on our tiptoes in a pool of water than on the sidewalk. We often say that things are "lighter" when they are submerged in water, but clearly the gravitational force has not been reduced. What happens is that the water exerts inward forces on every part of the submerged object's surface, leading to a net upward force. We call this force the *buoyant force.* It arises because the water pressure acting on the bottom of an object tends to be greater than that acting on the top (because the bottom parts are deeper in the water). The buoyant force always ends up acting straight up, toward the surface of the water.

No one would want to work out the net force on the swimmer in Figure 10.3 by adding up all the pressure forces on all the little parts of her total surface. Fortunately, we can find the magnitude of the buoyant force by using a trick that dates back to Ancient Greece and Archimedes.

Figure 10.3 A body submerged in a fluid has pressure forces that act inward along all parts of its surface.

Consider what would happen if the swimmer in Figure 10.3 had a ghost that looked exactly like her. Since ghosts are supposed to be immaterial, we could fill up the swimmer's ghost with water, as in Figure 10.4a. Now let's take the water-filled ghost and place her deep in the lake, as in Figure 10.4b. Will she float or sink? Remember that the ghost itself has no mass; it only serves to separate the water inside the ghost from the water outside the ghost. But it is all water, and so it will no more sink or float than will any other chunk of the water of the lake. The water-filled ghost just stays wherever we put it.

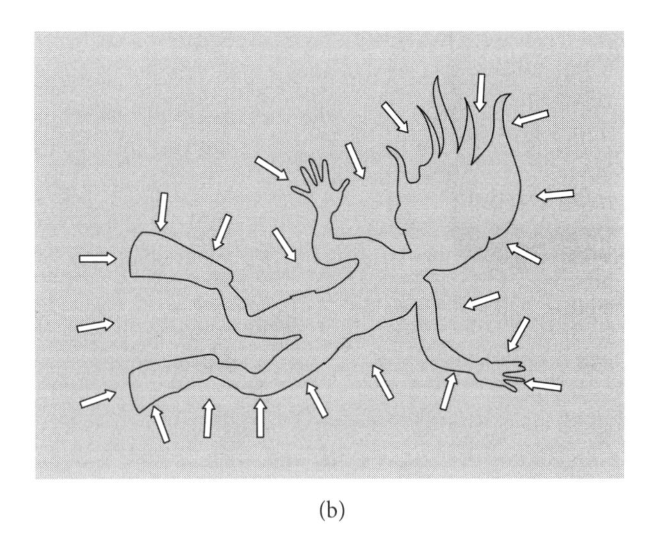

(a) (b)

Figure 10.4 A ghost who is exactly the same shape as the swimmer is filled with water and immersed in a lake. Because the only matter inside the ghost is the water, she will no more sink or float than would any other piece of water in the lake. Therefore, the buoyant force on the ghost must equal the weight of the water she contains.

Now if we were to do a free-body diagram for the water-filled ghost while it is submerged, we would see that, since the acceleration of the body is zero, the upward buoyant force must equal the downward weight of the water that is inside the ghost. But remember where the buoyant force comes from. It is the result of the pressure in the water all around the submerged body, given the particular shape of the body. It is the sum of all the little white arrows in Figure 10.4b. But it doesn't care what's inside the surface of the body, only the shape of the surface itself. So, because the swimmer has exactly the same shape as her ghost, the sum of these individual pressure forces must also be the buoyant force on the swimmer. We conclude that the buoyant force on any body submerged in a fluid is equal to the weight the body would have if it were made of the fluid instead of whatever it is actually made of.

The mass of an object is equal to its density times its volume, so the mass of the water that would fill up the ghost in Figure 10.4 is $m_{\text{fluid}} = \rho_{\text{fluid}}V$, where ρ_{fluid} is the density of the fluid and V is the volume of the ghost, or the volume of the body itself. The buoyant force is the weight of this water-filled ghost,

$$F_{\text{buoy}} = \rho_{\text{fluid}}Vg. \tag{10.4}$$

Since V is the *fluid displaced* when the body is immersed in water, we may also say that the buoyant force on any submerged body is the weight of the fluid it displaces,

$$F_{\text{buoy}} = W_{\text{fluid displaced}}. \tag{10.5}$$

Equation 10.5 is referred to as *Archimedes' Principle.*

We remember that the weight of a homogenous body with density ρ_{body} is

$$W = m_{\text{body}}g = \rho_{\text{body}}Vg. \tag{10.6}$$

Comparison of Equation 10.6 with Equation 10.4 shows that, if an object has a density greater than that of the fluid in which it is immersed, then W is greater than F_{buoy} and the object sinks. On the other hand, if $\rho_{\text{fluid}} > \rho_{\text{body}}$, then F_{buoy} is greater than W and the object will float to the top.

It is perhaps a little surprising that the various pressure forces on a submerged object, like the one in Figure 10.3, should always add up to produce a net force that is exactly upward, and that this does not depend on the shape of the object, but this is the case. As long as the volume of the swimmer does not change, there is no way that she can change her shape, no way that she might position her arms and legs or roll to the side, that will change the buoyant force on her. The buoyant force depends only on her volume and the density of the water in which she is immersed.

Let's look at an example of how to use these equations to calculate the net force on an object that is completely submerged in a fluid.

EXAMPLE 10.5 Buoyant Force on a Heavy Body

If the density of stainless steel is 7.8 g/cm^3, find the rate of acceleration of a cube of stainless steel, 10 cm on a side, that is starting to sink in a pond of fresh water.

ANSWER There are two forces acting on the cube, gravity and the buoyant force. The cube has a volume (10 cm)^3, or 1000 cm^3, so its mass is (7.8 g/cm^3)(1000 cm^3), or 7.8 kg. Its weight is mg = (7.8 kg)(10 m/s^2) = 78 N. The buoyant force is equal to the weight the cube would have if it were made of water instead of steel. Since the density of water is 1 g/cm^3, the mass of the 1000 cm^3 of water displaced is 1000 g, or 1 kg. This volume of water has a weight of mg = 10 N, so this is the buoyant force. The net force is the vector sum, 78 N down plus 10 N up, for a net force of 68 N down. The acceleration is then found from Newton's second law for the cube,

$$a = \frac{F_{\text{net}}}{m} = \frac{68 \text{ N}}{7.8 \text{ kg}} = 8.7 \text{ m/s}^2.$$

This result is correct only when the frictional force of the water on the cube (also called water resistance) is unimportant. Water resistance is typically proportional to the speed of the object moving through it, meaning that the force will be nearly zero when the cube begins to drop but will increase as the speed of the sinking cube increases. Eventually the cube will reach a speed called the *terminal velocity*, at which the sum of gravity, buoyant force, and frictional force all add to zero. The velocity-dependent water resistance force is a little hard for us to work with here, so we are not prepared at this point to work out what this final velocity would be.

EXAMPLE 10.6 Buoyant Force on a Light Body

A party balloon has a volume of exactly 0.1 m^3 when it is inflated with helium. The mass of the balloon by itself (not counting what's inside it) is 20 g. Helium at one atmosphere of pressure has a density of 0.16 kg/m^3, and air has a density of 1.38 kg/m^3 at the same pressure. What is the upward acceleration of the balloon?

ANSWER The downward force on the balloon is due to its weight. The mass of the helium inside the balloon is $m = \rho_{He}V = (0.16\ kg/m^3)(0.1\ m^3) = 0.016$ kg. The total mass is therefore 0.036 kg, for a weight of $mg = 0.36$ N. There is also an upward buoyant force equal to the weight of 0.1 m^3 of the fluid (the air) in which the balloon is immersed. This weight is $B = \rho_{air}Vg = (1.38\ kg/m^3)(0.1\ m^3)(10\ N/m) = 1.38$ N. This gives a net upward force of $F_{net} = 1.38\ N - 0.36\ N = 1.02$ N and an acceleration of $a = F_{net}/m = (1.02\ N)/(0.036\ kg) = 28\ m/s^2$. Note that it was the entire mass of the balloon that was accelerated, not just the helium, and not just the mass of the skin of the balloon.

Our next example asks us to analyze the buoyant force on two objects in the same fluid, leading to a result that may be somewhat surprising.

EXAMPLE 10.7 The Soda Cans in the Aquarium

A can of Soda and a can of Diet Soda are both put into an aquarium filled with water. The can of Soda sinks to the bottom of the aquarium, but the can of Diet Soda floats to the top. Which can experiences the greater buoyant force?

ANSWER It is actually the can that sinks which has the greater buoyant force acting on it. To see why this is so, let us calculate the pressure forces directly. The can is nearly a perfect cylinder, so the only pressure forces that matter are the upward force on the bottom of the can and the downward force on the top. If the Soda can is completely submerged, the difference in the pressures between these two surfaces is given by Equation 10.3:

$$P_{bottom} - P_{top} = \rho_{water}gh,$$

with h being the height of the Soda can.

The top of the Diet Soda can sits in an ocean of air, and the difference between the pressure at the top of the can and that at the surface of the water is negligible for a fluid as light as air. The depth of the bottom of the Diet Soda can below the surface of the water is less than h, so the pressure difference between the top and the bottom of the Diet Soda can will be less than for the Soda can. Thus, the buoyant force on the Diet Soda can will be less than the buoyant force on the submerged Soda can.

Since the Soda can sinks, its density must be greater than the density of water, while the density of the floating Diet Soda can must be less than the density of water. (It's actually the mass of all that sugar that makes the difference.)

10.5 And What If It Floats?

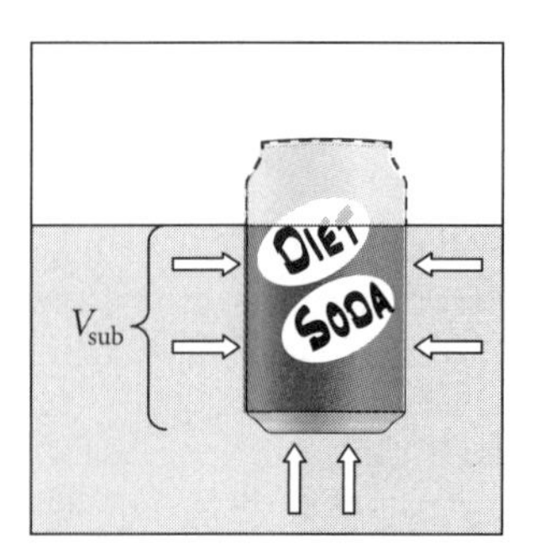

Figure 10.5 The buoyant force only depends on the submerged volume V_{sub}.

The case of the floating Diet Soda can in Example 10.7 raises a problem. We have found that the buoyant force on an object that is completely submerged in a fluid is equal to the weight of the fluid that would fill up the interior of the object. But how do we handle an object that is only partially submerged in the fluid? What is the buoyant force in that case?

Well, let us remember that the buoyant force arises from the pressure acting all around the surface of an object. If an object is floating, then only that portion of the surface that is actually in the fluid is subject to pressure forces. The portion not immersed in the fluid plays no part at all. So the true buoyant force could be found if one did the same analysis we did for a submerged object (Figure 10.5), but using an imaginary object consisting of only the submerged part of the original object. If a can of Diet Soda has a total volume V, but if the part that is submerged when it is floating at the surface only has a volume V_{sub}, then we consider only the portion V_{sub} and conclude, using the same method we did in the last section, that the formula for the buoyant force in Equation 10.4 will have to be modified. The *buoyant force on a floating body* is given by

$$F_{buoy} = \rho_{fluid} V_{sub} g. \tag{10.7}$$

Of course, if an object is floating at rest, then it is not sinking or jumping out of the water. And so we know that the net force on it must be zero. This means that the buoyant force must be equal to its weight. Thus, Equation 10.7 is not useful so much for finding the buoyant force on a floating body, since it is always equal to the body's weight, as it is useful for finding V_{sub}, the volume of the body that will actually be submerged. In fact, this is what determines how a floating object sits in the water. It always sinks down until it displaces an amount of fluid equal to its own mass. A ten-ton boat floats by displacing exactly ten tons of water.

Our next two examples are conceptual examples whose goal is to clarify the character of the buoyant force on floating objects.

EXAMPLE 10.8 Ice in a Bucket of Water

A large bucket is filled to the brim with water. A five-pound block of ice is then lowered carefully into the bucket. Anyone who has ever seen a glass of ice water knows that ice cubes float in water. Of course, water will spill out of the bucket as the ice is lowered to its equilibrium position, but the bucket will still be full to the brim. If we let the ice melt, will the water level in the bucket go down, stay the same, or will more water spill from the bucket?

ANSWER Surprisingly to some of us, the water level stays the same. The five-pound block of ice requires a five-pound buoyant force to float it (by Newton's first law). Archimedes' Principle dictates that a five-pound buoyant force must be caused by the ice displacing five pounds of water. Since the bucket starts out full to the brim, the five pounds of displaced water must spill out of the bucket and onto the floor. If we now lift the five-pound block of ice out of the bucket, there is room in the bucket for five more pounds of water (the amount that spilled out). If we hold the five-pound block of ice above the bucket and allow it to melt, turning into five pounds of water, the melted water will fill up the bucket. If the ice actually melts while floating in the water, the result is the same.

By the way, the reason that ice floats in water is that water expands as it freezes, giving a lower density for ice than for water. So another way to look at the problem is to recognize that, even though some of the ice stuck up above the water when it was frozen, it will, as it melts, contract back to exactly fill the volume it originally displaced.

EXAMPLE 10.9 The Archimedes' Principle Debate

Three students have just come from an interesting lecture on Archimedes' Principle, where they were shown two wooden blocks of the same size and shape floating in a bucket of water, as shown at right. They could see that block A floated low in the water and block B floated high. They are discussing the buoyant forces on the blocks.

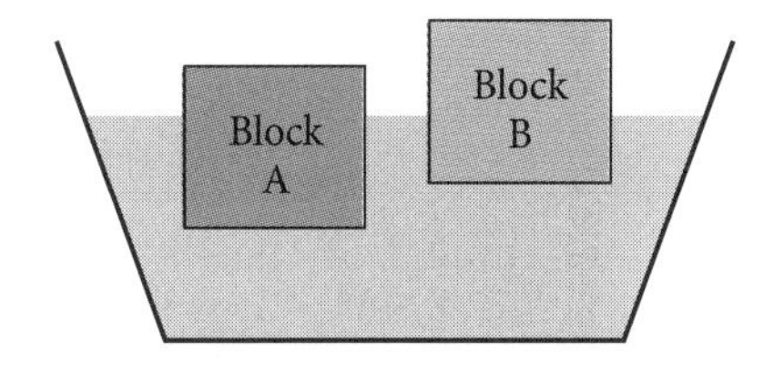

Student 1: Block B is floating higher in the water. It must have the greater buoyant force acting on it.

Student 2: You are forgetting Newton's first law. Neither block is moving, so the buoyant force must balance the gravitational force in both cases. The buoyant forces must be equal to each other.

Student 3: Archimedes taught us that the buoyant force is always equal to the weight of the fluid displaced. Block A is displacing a lot more water than block B, so block A has the larger buoyant force.

Do you agree with any of these students?

ANSWER Student 3 is correct. Archimedes' Principle *always* applies, regardless of whether an object sinks or floats. Block A displaces the most water, so it experiences the larger buoyant force. Student 2 starts out with correct ideas but then draws a faulty conclusion. The buoyant force on either block must equal the gravitational force on that block (by Newton's first law), so block A must be heavier. Since both blocks have the same volume, block A must be made of a denser wood. Block A might be made of oak, and block B made of pine, or something like that.

10.6 Summary

In this chapter you should have learned the following:

- You should be able to use Equation 10.1 to calculate the pressure on an object, given the applied force and the area over which the force is applied, or to calculate the force on an object, knowing the pressure and the total area over which the pressure acts.
- You should be able to use Equation 10.3 to determine the pressure at some depth h in a fluid of density ρ or to calculate the depth at which a particular pressure will be found.
- You should be able to calculate the buoyant force on a submerged object, using Equation 10.4 or 10.5, to calculate the buoyant force on a floating object, using Equation 10.7, or, more commonly, to use Equation 10.7 to find the portion of a volume that will be submerged when we know that a body will float in a liquid.

CHAPTER FORMULAS

Pressure definition: $P = \frac{F}{A}$

Force due to fluid pressure: $F = PA$

Pressure in a fluid: $P_2 = P_1 + \rho gh$

Buoyant force: $F_{buoy} = m_{fluid}g = \rho_{fluid}Vg$ or $F_{buoy} = \rho_{fluid}gV_{sub}$

PROBLEMS

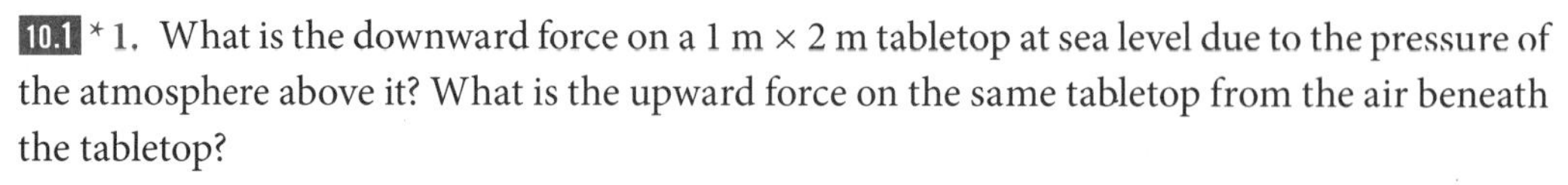

10.1 * 1. What is the downward force on a 1 m × 2 m tabletop at sea level due to the pressure of the atmosphere above it? What is the upward force on the same tabletop from the air beneath the tabletop?

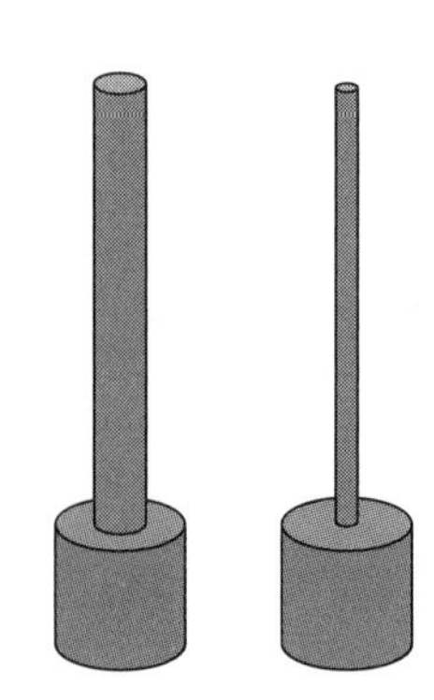

** 2. Two identical wooden barrels are fitted with long pipes extending out their tops, as shown. The pipe on the first barrel is 6 inches in diameter, and the pipe on the second barrel is only 1 inch in diameter. When the large pipe is filled with water to a height of 15 feet, the barrel bursts. To burst the second barrel, will water have to be added to a height less than, equal to, or greater than 15 feet? Explain.

10.2 * 3. How deep must you dive in a lake of alcohol ($\rho = 790\ \text{kg/m}^3$) to increase the pressure by one atmosphere? (Use $g = 9.8$ N/kg here to get the correct value for alcohol.)

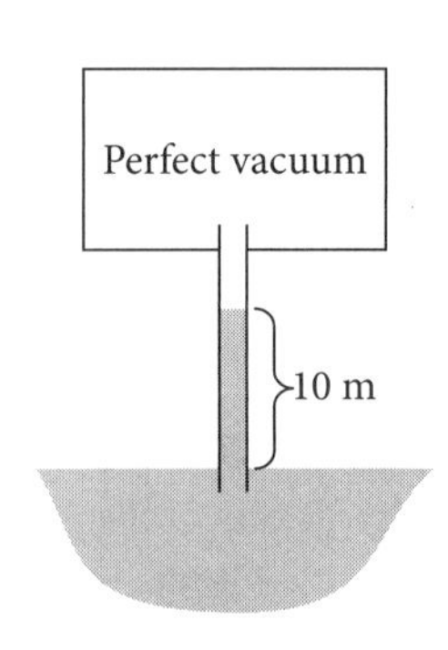

** 4. At sea level even a "perfect" vacuum at the top of a pipe can only suck water up the pipe to a height of 10 meters. At an elevation of 4800 feet (the elevation in Bozeman, Montana), will a complete vacuum raise water to more than 10 meters, exactly 10 meters, or less than 10 meters? Explain your reasoning.

** 5. When you blow air out of your lungs, you are changing both your mass and your volume. Your mass decreases by a little bit (because air has left the "system") and your volume decreases because your chest is not puffed out. Which of these effects explains why this causes you to sink to the bottom of the swimming pool?

** 6. A diver experiences a doubling of the pressure by descending about 10.3 meters in fresh water (P_{atm} = 101,000 Pa). How deep must you dive in a lake of alcohol (ρ = 790 kg/m^3) to achieve a doubling of the pressure? (Use $g = 9.8\ \text{m/s}^2$ here.)

** 7. When a deep sea robotic explorer is tested in a lab, it is submerged in a liquid with a density of 1.5 g/cm^3 and is seen to descend to a depth of 120 m before the structure of the explorer fails. How deep can it go in water ($\rho = 1\ \text{g/cm}^3$) before its structure fails?

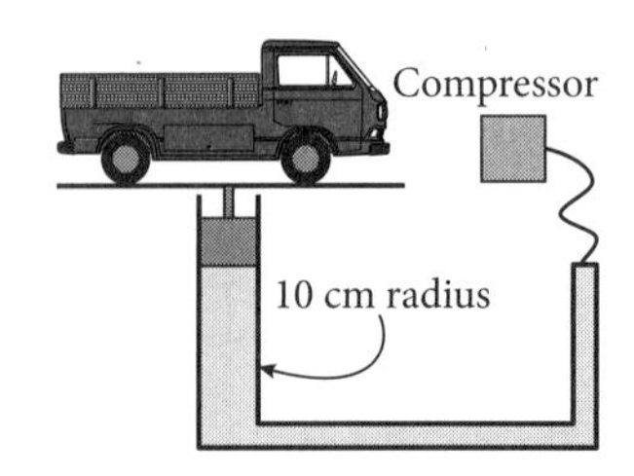

10.3 ** 8. An auto mechanic wants to raise a 10,000 kg truck on a hydraulic lift in which the large main cylinder has a radius of 10 cm. He uses compressed air to drive the small cylinder, as shown. What must the pressure of the compressed air be? How many atmospheres of pressure is this?

** 9. An underwater diver lifts a 0.25 m^3 pirate chest by applying an upward force of 500 N. But when he tries to lift the same chest out of the water, he finds he is unable to do so. To see why, find the force he would have to apply to lift it in air.

** 10. A ball of mass 400 g and a volume 100 cm^3 is released from rest just under the surface of a lake of fresh water.

a) Find the buoyant force acting on the ball. [Hint: Be careful with units. To find a weight in newtons, you must use SI units for g (10 m/s^2) and for the mass.]

b) Find the acceleration of the ball right after it has been released.

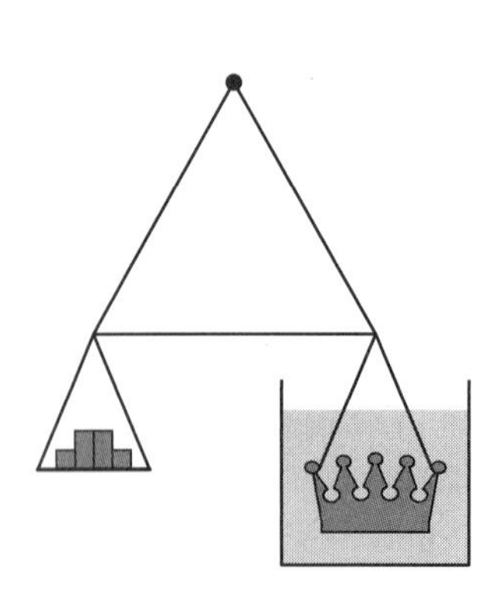

10.4 ** 11. Hiero II, King of Syracuse, gave a certain quantity of gold to his goldsmith to make a solid gold votive crown that he planned to donate to the temple. Thinking that his goldsmith may have been cheating him by stealing some of the gold and substituting silver, he asked Archimedes to devise a way to determine the density of the crown, without changing its shape. Although Archimedes' solution was arrived at using a different method (he placed the crown in a container filled to the brim with water and measured the volume of water that was displaced by the gold), he could have solved the problem by using Archimedes' Principle.

a) Suppose Archimedes had weighed the crown and found that its mass was 800 grams in air, and that it weighed 750 grams when submerged in water. What was the buoyant force on the crown when it was submerged?

b) What was the volume of the crown, and what was its density?

c) The density of gold is 19.3 g/cm^3 and that of silver is 10.5 g/cm^3. Was King Hiero II being cheated by his goldsmith?

10.5 ** 12. You have a block of unknown material of mass 18 grams and volume 20 cm^3. You have a bucket of water ($\rho = 1$ g/cm^3) and a bucket of alcohol ($\rho = 0.79$ g/cm^3). Both buckets are filled right to the brim.

a) What is the density of the block?

b) Which of the buckets, if any, does it float in? Explain your reasoning.

c) When the block is placed slowly and carefully in the bucket of water, how many cm^3 of water spill over the edge?

d) When the block is placed slowly and carefully in the bucket of alcohol, how many cm^3 of alcohol spill over the edge?

e) If you had a 37-gram chunk of this material (the same material as the block), what would its volume be?

** 13. The two objects described in the table at right are each dropped into a deep lake of fresh water. The density of the water is 1 g/cm^3 throughout the lake (i.e., ignore temperature effects).

	Mass	Volume
Green object	600 g	800 cm^3
Yellow object	600 g	400 cm^3

a) Explain whether each object will sink or float in the lake.

b) Find the buoyant force exerted on each of the objects if each is completely submerged.

c) Find the acceleration of each of the objects if they are set down on the surface of the lake and then released. If the acceleration is zero, state that explicitly.

d) A crown is made out of the same material as the yellow object. When the crown is placed gently into a full bucket of water, 360 cm^3 of water spills out. What is the mass of the crown?

** 14. You are presented with three beakers, each filled to the brim, one with Liquid X, one with Liquid Y, and one with Liquid Z. You do not know the densities of the three liquids. You have a ball of clay with volume 5.0 cm^3 that you carefully lower into each of the beakers in turn and carefully measure the mass and volume of liquid that spills over the edge of each beaker. The results of your experiment are listed in the table below.

	Mass of fluid displaced	Volume of fluid displaced	Sinks or floats?
Liquid X	4.0 g	5.0 cm^3	
Liquid Y	3.0 g	5.0 cm^3	
Liquid Z	6.0 g	3.0 cm^3	

a) Finish the table above by indicating whether the clay ball sinks or floats in each of the liquids. Explain your reasoning.

b) Use the information in the table above to answer the following questions.

i. What is the mass of the ball of clay? Explain your reasoning.

ii. What is the density of the clay used to make the ball?

c) Which of the three fluids exerts the smallest buoyant force on the ball of clay? Explain your reasoning.

d) How much extra clay (in cm^3) would you have to add to increase the mass of the ball by 4.0 g?

** 15. A cylindrical piece of wood of length 10.0 cm and density 0.75 g/cm^3 is placed in a pond with water density 1 g/cm^3. How many centimeters of wood will poke up above the surface of the pond?

** 16. A cube of ice floats in a glass of fresh water with 90% of its volume below the surface (ρ_{water} = 1 g/cm^3). How much of the cube would be below the surface if it was put into a glass of alcohol instead ($\rho_{alcohol}$ = 0.8 g/cm^3)?

** 17. You are presented with two beakers filled to the brim, one with water and one with Liquid Z. The density of water is 1 g/cm^3, but you don't know the density of Liquid Z. You are then presented with two objects of identical volume and different masses and find the following sinking and floating behavior:

	Water	Liquid Z
Blue object	floats	sinks
Red object	sinks	sinks

a) What, if anything, can you say about the density of Liquid Z? Explain.

b) You carefully place the blue object into the beaker of water (where it floats) and find that 25 cm^3 of water spills over the edge of the beaker. When you place the blue object into Liquid Z, 30 cm^3 spills out.

i. What is the mass of the blue object? Explain your reasoning.

ii. What is the volume of the blue object? Explain your reasoning.

iii. What is the density of the blue object?

c) Compare the buoyant force on the red object in the water to the buoyant force on the blue object in the water. Explain your reasoning.

d) Compare the buoyant force on the blue object in the water to the buoyant force on the blue object in Liquid Z. Explain your reasoning.

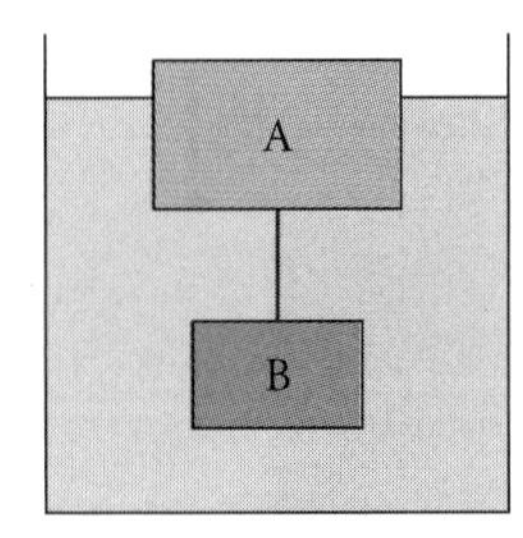

** 18. Two blocks with the same mass of 500 g are connected to each other with a massless string and slowly lowered into a container of water (ρ = 1 g/cm^3). The system remains at rest as shown in the picture with B fully submerged and A with ¾ of its volume under water. The volume of block B is 200 cm^3. [Hint: A free-body diagram for each of the blocks will be very useful.]

a) Determine the buoyant force on block B.

b) Determine the tension in the string.

c) Determine the buoyant force on block A.

d) Determine the volume of block A.

e) Determine the volume of a block that is made from the same material as A but has a mass of 800 g.

f) If the string between the two blocks is cut, what will be the acceleration (magnitude and direction) of block B?

** 19. A sphere is machined from pure unobtainium, which is known to float in water. The sphere has a mass of 1.2 kg. When the ball is completely submerged in fresh water, it experiences a buoyant force of 48 N.

a) Find the magnitude of the acceleration of the ball just after it has been released from rest, deep below the surface of the water.

b) Find the volume (in cm^3) of the ball.

c) The ball reaches the surface of the pond and floats. Find the magnitude of the buoyant force on the ball while it floats.

d) What fraction of the ball's volume is now above the surface of the pond?

** 20. A big letter G with a volume of 400 cm^3 is fully submerged in water. Two strings hold the G in place, as shown, and they each have a tension of 1.5 N. The density of water is 1 g/cm^3.

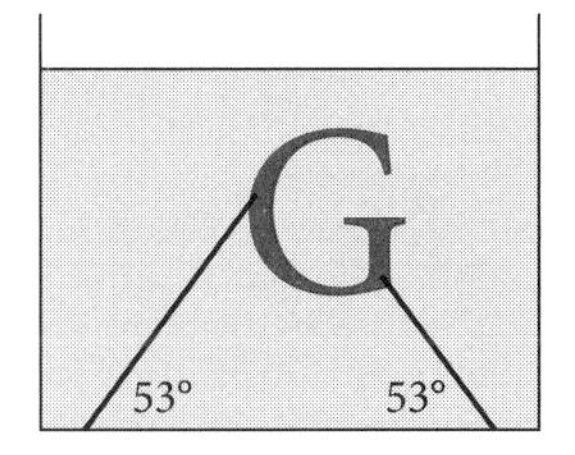

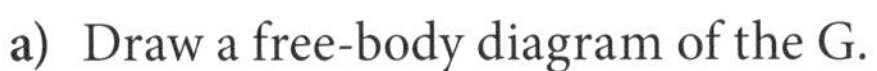

a) Draw a free-body diagram of the G.

b) Find the magnitude of the buoyant force exerted on the G by the water.

c) What is the mass of the G?

d) What is the density of the G?

e) The two strings are cut simultaneously. Find the acceleration of the G (direction and magnitude) right after the strings are cut.

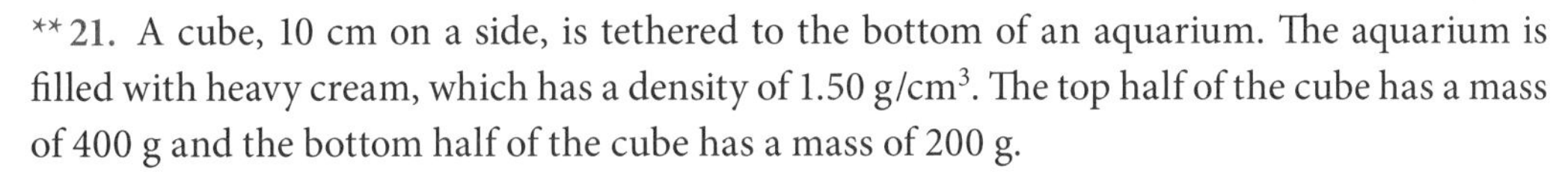

** 21. A cube, 10 cm on a side, is tethered to the bottom of an aquarium. The aquarium is filled with heavy cream, which has a density of 1.50 g/cm^3. The top half of the cube has a mass of 400 g and the bottom half of the cube has a mass of 200 g.

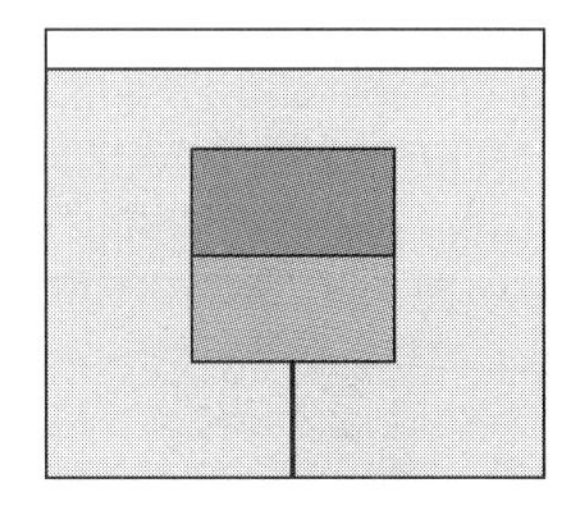

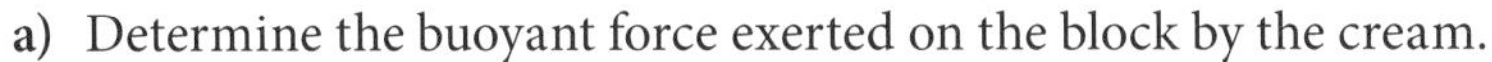

a) Determine the buoyant force exerted on the block by the cream.

b) Determine the tension in the string.

c) The top half of the block suddenly breaks free from the bottom half and accelerates toward the surface. Find the magnitude of its acceleration right after it breaks free.

d) When the top half of the block reaches the surface, it floats. What volume of the top half is now above the surface of the cream? Explain your reasoning.

e) Find the new tension in the string, now that the top half of the block is gone.

*** 22. A horizontal *underwater* plastic meter stick, of mass 400 g and volume 3×10^{-4} m^3, has two additional forces acting on it. There is a downward force of 2 N applied to the left end of the stick and a 6 N force to the right acting on the extreme right end of the stick. If we want to suspend the stick from a single cable, how far from the left end of the stick must the force be applied? What will be the tension in the cable? What angle will the cable make with the vertical? [Hint: The buoyant force on a symmetric body of uniform density acts at the geometric center of the body.]

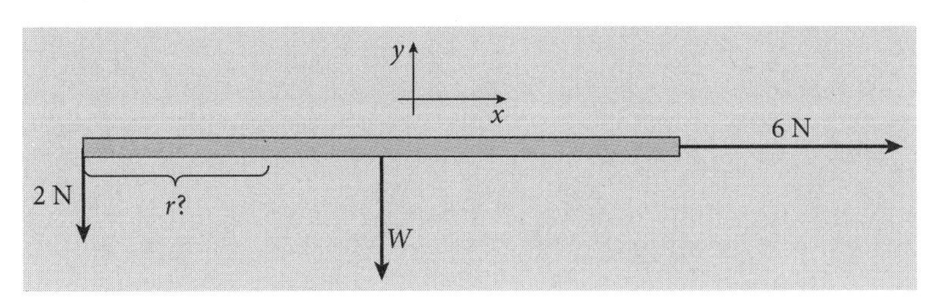

*** 23. A 70-kg painter wearing a 20-kg weighted belt stands *under water* on a 40-kg platform that is supported by two ropes as shown. The platform is 2 m long and has a volume of 0.025 m^3. The painter's volume, including his belt, is 0.080 m^3. The left rope is tied to the platform 50 cm from the end and the painter is standing 50 cm from the right rope. Find the tension in each rope. Assume the ropes have zero volume and mass. [Hint: See the hint for Problem 22.]

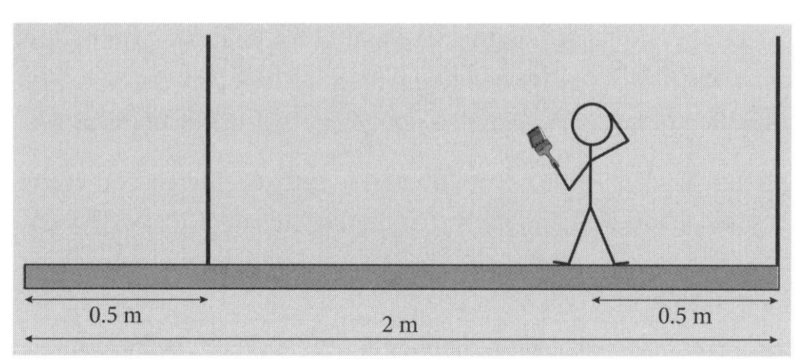

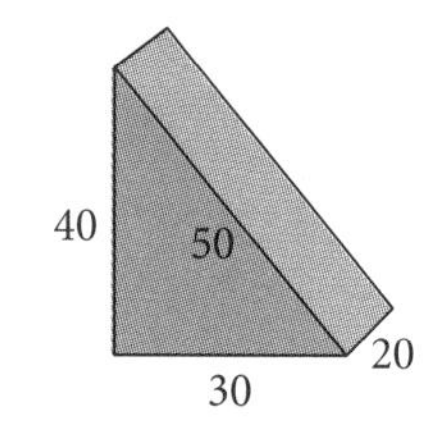

*** 24. A plastic triangular wedge has dimensions 30 cm across, 40 cm high, and 20 cm deep, as shown at right. When it is placed in water, it sinks to the bottom.

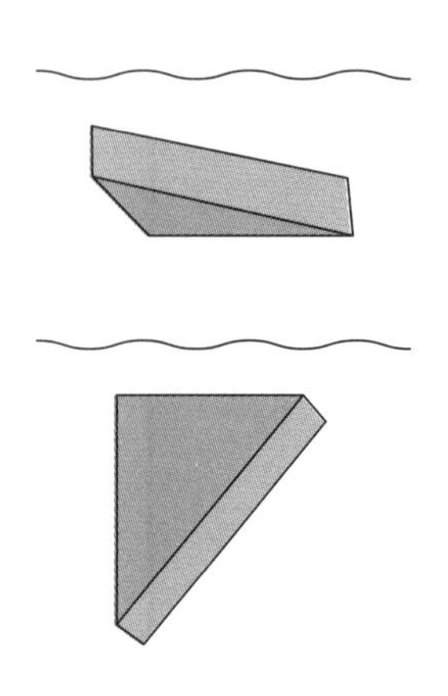

a) The wedge is first placed in the water, oriented as shown at left, with its triangular sides on the top and bottom. Find the net buoyant force on the wedge by calculating the actual pressure forces on the top and bottom surfaces. [Hint: The area of a triangle is $A = \frac{1}{2}bh$.]

b) The wedge is then placed with its 30-cm rectangular surface on the top and its 40-cm side vertical, as shown at left. The pressure on the 50-cm rectangular surface will vary across that surface, since it is not at a single depth in the water, but the average pressure on the surface will be the pressure halfway down from the top. Find the buoyant force on the wedge by calculating the net upward and downward pressure forces on the wedge surfaces that experience upward or downward components of force. Remember that pressure forces are always normal to the surfaces on which they press.

11

SIMPLE HARMONIC MOTION

In Chapter 7, we pointed out that we had seen two conservative forces so far in this book, the gravitational force and the spring force. For the gravitational force, we know that, near the surface of the earth, the force is approximately constant, so the motion of an object subject to this constant gravitational force could be analyzed using the simple constant-acceleration formulas we developed in Chapter 2. But, as we saw in Chapter 6 (Equation 6.4), the spring force is a force that obeys Hooke's law:

$$F = k\Delta s. \tag{11.1}$$

Because this force is not constant, but varies with the position of the object that is stretching or compressing the spring, the motion of an object subject to a Hooke's law force is much more difficult to analyze. Indeed, we will not go so far as to provide the kind of complete solution that we were able to provide for motion under a constant force motion (see Table 2.2), but we still will be able to learn a great deal about how an object moves when there is a Hooke's law force acting on it.

One thing that is important to know about Hooke's law is how often it arises in different kinds of physical systems. If an object is at rest and if there are no forces acting on it, we say it is in *equilibrium*. If, when the object is moved away from the point of equilibrium, a force arises that tends to push it back toward the point of equilibrium, we say that the object is in *stable equilibrium*. Most objects around us that are at rest are at rest because they are in stable equilibrium, and most objects in stable equilibrium end up being subject to forces that are proportional to the amount of the displacement, at least in situations where the displacement from equilibrium is small. A force proportional to the displacement is a Hooke's law force.

So Hooke's law seems to be everywhere. When the atoms in a complex molecule vibrate, the restoring forces that create the vibrations are proportional to the displacement of the atoms from their equilibrium distances in the molecule. When the surface of a star contracts inward, a restoring pressure arises that is proportional to the size of the contraction. If your sister sits on a swing and you give her a little push, gravity produces a restoring force that is proportional to how far she is from the vertical. If you clamp a popsicle stick in your teeth and pluck the free end of the stick, the molecular forces in the stick are proportional to how far the stick was bent.

11.1 Period and Frequency

Whenever the motion of some object repeats itself in a regular fashion, the motion is called *periodic*, with the length of time between repeats of the motion being called the *period* of the motion. Examples of periodic motion include the motion of a clock pendulum or the orbit of a planet around the sun. When periodic motion is strictly oscillatory, like the up and down bounce of a basketball or the back-and-forth motion of a low-hanging branch caught in the current of a creek, the motion is called *harmonic motion*. This name arises because of its connection with the kind of vibrations that create sounds in musical instruments. Finally, if the only force acting on an oscillating object is a Hooke's law force, like the motion of a mass on the end of a spring, the motion is called *simple harmonic motion*.

Figure 11.1 shows the motion of a mass hanging from a spring. The dashed line in the figure shows the equilibrium point, the point where the spring is stretched enough that it exactly balances the gravitational force to provide no net force. If the mass were placed at rest at that point, it would remain at rest. In figure (a), however, the mass has been raised to a height A and dropped. The mass gains speed as it falls, passing its equilibrium position at some nonzero speed and continuing downward until the spring is stretched enough to finally check the downward motion and bring the mass to rest at its lowest point. At this point, the extension of the spring still provides a net upward force. The mass will therefore rise again, eventually attaining the height it had at the start, and the motion will repeat itself.

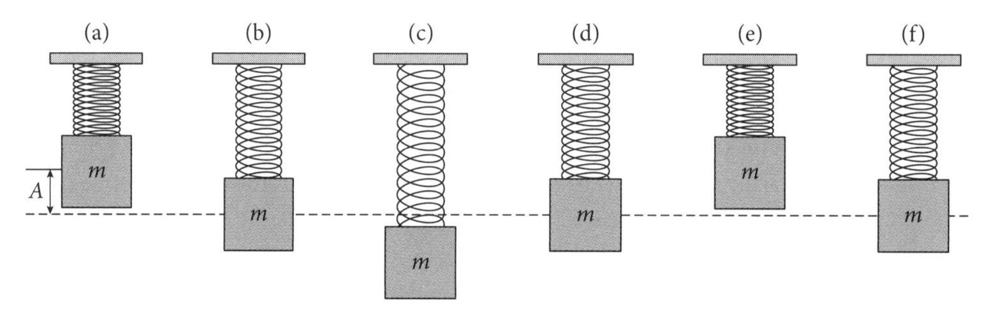

Figure 11.1 A mass hanging from a spring begins at its maximum height (a), descends to the equilibrium point (b), continues on to its lowest point (c), moves upward back to equilibrium (d), rises again to the top of its motion (e), and continues on to repeat the sequence (f).

In Figure 11.2, we graph the height y of the mass above the equilibrium point for a case where the initial height is $A = 10$ cm and where the period, the time required for the motion to repeat itself, is 0.8 s. Note that the period identified in the figure is the time from the top of the motion to the next top of the motion, and that this is the same as the time between when the mass drops through the equilibrium point at $t = 0.2$ s till the time when it again drops through

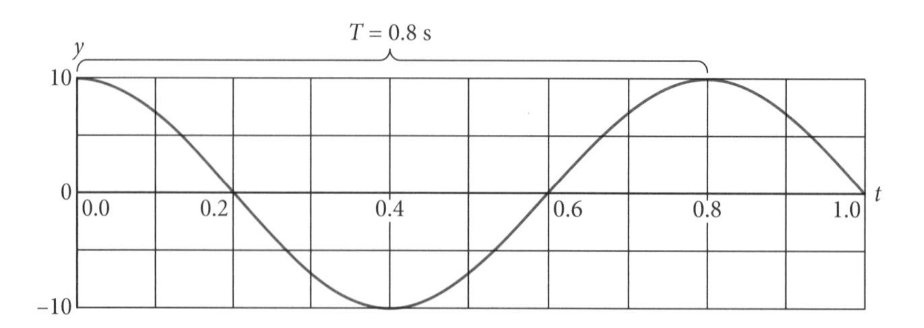

Figure 11.2 A graph of the vertical displacement y of the mass above its equilibrium point as a function of time t. The amplitude of the motion is $A = 10$ cm. The period is identified at the time for one complete cycle of the motion. It is $T = 0.8$ s.

the equilibrium point at $t = 1$ s. But note that 0.8 s is *not* the time between the point where the mass is at the equilibrium point (at $t = 0.2$ s) to the point when it is again at the equilibrium point (at $t = 0.6$ s). This is because the period is not the time to get back to some point, but the time for the motion to exactly repeat itself. Upward motion through the equilibrium point is not the same as downward motion through the equilibrium point.

Finally, let us introduce another quantity that characterizes harmonic motion, one that carries the same information as the period. Systems that oscillate rapidly have very short periods—much less than a second. For this reason, it is often convenient to describe how rapidly a system is oscillating by stating its *frequency*. As the period of an oscillating system is the number of seconds per cycle of the oscillatory motion, the frequency is the number of cycles per second. A system that has period $T = 0.1$ s will be able to complete 10 oscillations in one second, so, instead of giving the system's period, we could just as well state its frequency, 10 cycles per second. Because of the importance of electrical oscillations in radio science, the unit "cycles per second" was named after Heinrich Hertz (1867–1894), who did important early experiments with electromagnetic waves. A cycle per second is a *hertz* (abbreviated Hz). Formally, we can get from period to frequency or back by recognizing that they are reciprocals:

$$f = \frac{1}{T} \quad \text{or} \quad T = \frac{1}{f}. \tag{11.2}$$

Cycles per second are the reciprocal of seconds per cycle.

EXAMPLE 11.1 Computer Clock Speed

There is an oscillator inside each computer that times the operations the computer performs. The so-called "clock speed" of a computer is the number of computational cycles that the oscillator drives in one second. A certain computer's clock speed is advertised to be 3.3 gigahertz (3.3×10^9 Hz). How much time does each clocked computational cycle of the computer take?

ANSWER The clock speed of a computer is a frequency, a number of cycles per second of an electrical oscillator. The time it takes for each cycle is the period of the oscillator. These two quantities will be reciprocals, giving

$$T = \frac{1}{f} = \frac{1}{3.3 \times 10^9 \text{ Hz}} = 3 \times 10^{-10} \text{ s} = 300 \text{ picoseconds}.$$

At present, clock speeds are very close to the maximum allowed by the speed of electrical currents in the processor, given the physical dimensions of the processor. Therefore, attempts to make a faster computer now usually concentrate on parallel computing, in which calculations run on multiple processors at once.

11.2 Simple Harmonic Motion

Consider a block of mass m, resting on a frictionless horizontal tabletop and connected to a spring of spring constant k, as shown in Figure 11.3. The only horizontal force acting on the block will be the force of the spring, so the equilibrium position of the block on the tabletop will be the point where the spring is both uncompressed and unstretched. Since the only force acting on the

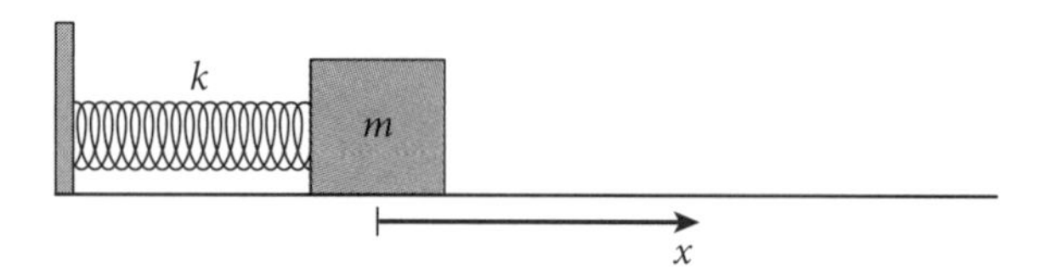

Figure 11.3 A block of mass m resting on a frictionless tabletop, connected to a wall by a spring of spring constant k.

block obeys Hooke's law, the motion will be *simple harmonic motion*. Let us take the equilibrium point as the origin of our coordinates, and let us use a variable x to define the position of the block, with positive values of x to the right. Then, if the block moves to a point x, the spring will be stretched by an amount $\Delta s = x$ and there will be a force on the block by the spring equal to $F = -kx$, the minus sign indicating that the force will be to the left when the block is to the right. If the block moves to the left of the origin, to a negative value of x, then the spring will be compressed, producing a force to the right. So for any position, positive or negative, it will be true that the force on the block is given by $F = -kx$. The acceleration of the block at a position x will therefore be

$$a = \frac{F}{m} = -\frac{k}{m}x. \tag{11.3}$$

Let's think for a minute about the kind of motion this equation predicts. What is simple harmonic motion like? If we take the block out to a point to the right, to a point we will call $x = A$, and then release it, the acceleration will move it back toward the left. Therefore, that initial point will be the largest positive position the block will ever have. Also, Equation 11.3 indicates that the acceleration at that point will be the greatest negative acceleration the block will ever experience. As the block starts from rest there and begins to move to the left, the leftward acceleration will continually increase the speed of the block, until the block reaches the equilibrium point at $x = 0$. Here the acceleration, according to Equation 11.3, will drop to zero. The block is still moving to the left at this point, so it will continue to do so. But, as it moves to negative values of x, the force, and thus the acceleration, will begin to be positive, slowing the leftward speed. This means that the speed the block had at $x = 0$ is the greatest speed the block will have at any point in its motion. As the block continues to the left, the acceleration will become more and more positive and will eventually bring the block to rest at the point $x = -A$. From this leftmost position, the second half of the motion will be the mirror image of the first half. When the block is back again, momentarily at rest at the point $x = A$, it will have completed one complete oscillation.

The exact position of the block at each point in time cannot be solved without calculus, so we will not consider that problem yet. But let us share the results of the calculus by drawing a graph (Figure 11.4) of the horizontal position of the block as a function of time, for a system with $A = 10$ cm and a period of 0.8 seconds.

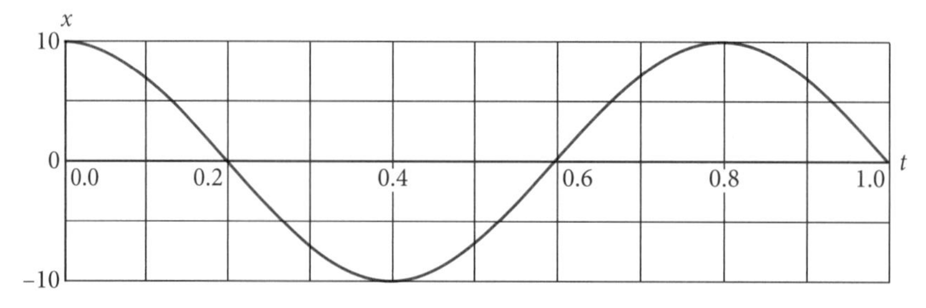

Figure 11.4 The horizontal position of the block x as a function of time t.

Honesty requires us to admit that we copied this graph from Figure 11.2. But, surprisingly, it's still correct. In the next section, we will explain why that system with two forces, the spring and gravity, exhibits the same simple harmonic motion as this system with only a single horizontal spring force. But, for now, let's just take a minute to think about the simple harmonic motion exhibited in the graph in Figure 11.4.

First, let us note that the graph (if continued, of course) assigns a value of x to every value of t. So, once again, we see that Newton's laws are sufficient to exactly specify the future motion of an object, given its initial position and velocity.

Second, a characteristic of simple harmonic motion apparent in Figure 11.4 is that the magnitude of the maximum displacement of the block to the left of the origin is the same as that to the right of the origin. This maximum displacement is called the *amplitude* of the motion, A. In Figure 11.4, the amplitude is $A = 10$ cm. Note that the distance between the leftmost position and the rightmost position is equal to $2A$, but that distance is not what we mean by amplitude. The amplitude is defined to be the distance from the equilibrium point to either extreme of the motion.

Finally, let us notice the symmetry in the shape of the curve in Figure 11.4. The complete period naturally divides itself into four equal sections. The time for the block to get from $x = 10$ to $x = 0$ is the same as the time for the block to get from $x = 0$ to $x = -10$, which is the same as the time for the block to get from there back to $x = 0$, etc. Within each section, the shapes of the segments of the curves are the same. Some may be upside-down or mirror images of each other, but the shapes are the same. As you can see, the time for the block to get from $x = 10$ to $x = 5$ is longer than half the 0.2-second segment, longer than 0.1 s. But, whatever that time turns out to be, it will be the same as the time it takes the block to go from $x = -10$ at $t = 0.4$ s to $x = -5$, because of the identical shapes of the curves in each section of the motion.

11.3 The Hanging Block on a Spring

In Figure 11.5, a spring of spring constant k is connected to a block of mass m. When the spring is unstretched, as in figure (a), the center of mass of the block is level with the dashed line. However, if we slowly let gravity pull the block down, it will drop until its center of mass reaches a new lower level at the solid line in figure (b), a distance d below the original dashed line. Here the block will be in equilibrium, with the downward force of gravity mg being balanced by the upward spring force kd. Let us now choose this equilibrium point as the origin for locating the vertical position of the block during any subsequent motion, using a variable z to indicate the position of the block, measured positive downward from the solid line in Figure 11.5.

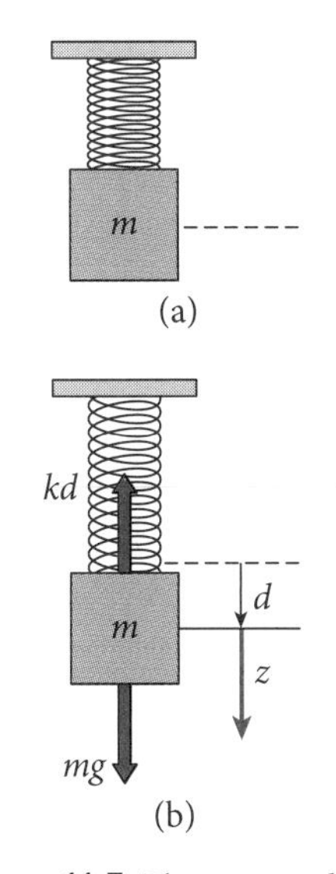

Figure 11.5 The unstretched spring in figure (a) stretches in figure (b) by an amount d, until $kd = mg$.

If the block is now pulled downward by an amount z, then the forces acting on the block will be the positive downward force of gravity and the upward negative force of the stretched spring, giving a net force (taking positive downward) of

$$F = mg - k(d + z),$$

where we have recognized that the spring's total stretch is $\Delta s = d + z$. If we expand the term inside the parentheses, this becomes

$$F = mg - kd - kz. \tag{11.4}$$

The distance d is just the distance between the dashed line and the solid line, and it does not change if the block moves, so kd is constant, as is mg. In fact, d was the distance that was required in order to have $kd = mg$. Therefore, the two constant terms in Equation 11.4 will always exactly cancel, leaving $F = -kz$. Using Newton's second law, we may then find the formula for the acceleration as

$$a = -\frac{k}{m}z. \tag{11.5}$$

Since Equation 11.5 is identical to Equation 11.3, the motion of a block hanging from a spring in the presence of gravity will be the same simple harmonic motion as for a horizontal block, as long as the position of the hanging block is measured from the point of equilibrium where the initial stretch of the spring balances the force of gravity. Except for having to choose the origin at this particular point, the effect of gravity may be ignored in describing the block's subsequent motion.

EXAMPLE 11.2 Analyzing a Block Hanging from a Spring

A block hanging from the ceiling by a spring is lifted 20 cm from its equilibrium position and released. The block takes 2 seconds to drop down and then finally return to the height from which it was released. What is the frequency of the motion? What is the total distance that the block moves during the first 10 seconds after it is released? What is the block's displacement at the end of that 10 seconds?

ANSWER The block completes one cycle in 2 seconds, so that is its period. The frequency is the reciprocal of the period,

$$f = \frac{1}{T} = \frac{1}{2\text{ s}} = 0.5\text{ oscillations/second} = 0.5\text{ Hz}.$$

During the first 2 seconds, the block moves down 20 cm to equilibrium, another 20 cm to the bottom of the motion, 20 back up to equilibrium and, finally, 20 back up to the release point. This is a total distance of 80 cm (4 times the amplitude) in a single cycle. In 10 seconds, the system completes 5 cycles, so the block moves a total distance of 5×80 cm = 400 cm = 4 m. However, because the block returns to the release point after each complete cycle, the block will be back where it started after 10 seconds. The distance traveled is 4 m but the final displacement is zero.

11.4 The Period Formula

As we mentioned in Section 11.2, the only way to work out the motion of an object with a Hooke's law force acting is through the use of calculus, due to the fact that the force varies with position. But the calculus does always give a solution. Let's see what this means.

Let us begin by remembering that velocity tells us how the position is changing, and that acceleration tells us how the velocity is changing. Newton's second law for a block on a spring, Equation 11.3, requires that the motion of the block be such that, as the position and acceleration both change, the acceleration is always proportional to minus the position, with proportionality constant k/m. Calculus allows us to find the mathematical functions that satisfy this requirement, and it tells us two things.

First, calculus tells us the period (or the frequency) of the motion. For a spring of spring constant k attached to a block of mass m, the period will be

$$T = 2\pi\sqrt{\frac{m}{k}}. \tag{11.6}$$

The way the period depends on m and k will probably make sense, if you think about it. For a spring of a given stiffness k, it makes sense that the larger the mass on the end of the spring, the more sluggishly the system will perform. It should take longer to go through its cycle, and the period will be longer. Then, for a block of a given mass m, a greater stiffness in the spring (a bigger spring constant) will provide more acceleration and run the system through its cycle more quickly, thereby reducing the period.

But let us also point out one thing that is missing in Equation 11.6. There is no mention of the amplitude. It is perhaps a little surprising that pulling the spring further away from the equilibrium point does not increase the period by increasing the distance the block has to travel in each oscillation, nor does it decrease the period by increasing the acceleration of the block when it is pulled back further. The period of a simple harmonic oscillator simply does not depend on the amplitude at all.

Before we go on to the second thing we learn from doing the calculus, let's check out an example of how to work with the period formula.

EXAMPLE 11.3 The Period of a Simple Harmonic Oscillator

A 2-kg block is suspended by a spring of unknown spring constant. At equilibrium, the spring is stretched by 10 cm. The block is pulled down an additional 5 cm and released. What is the period of the subsequent oscillations?

ANSWER We were not told anything like the number of oscillations completed in a fixed time, so we cannot determine the period from its definition. However, Equation 11.6 offers a way to determine the period if we know the mass and the spring constant. We were given the mass in the problem statement, but we do not yet know the value of the spring constant. However, we were told something about the spring's stiffness: We know that the spring stretches 10 cm when a 2-kg block hangs from it. This is like the Hooke's law examples we solved back in Chapter 6. At equilibrium, the force of the spring on the block must equal the force of the earth on the block. The weight is $W_{\text{e,b}} = mg = (2\text{ kg})(10\text{ N/kg}) = 20\text{ N}$, so the magnitude of the force by the spring on the block will be

$$F_{\text{s,b}} = k\Delta s = 20\text{ N}.$$

We know that the stretch of the spring is 0.1 m, so we have

$$k = \frac{20\text{ N}}{0.1\text{ m}} = 200\text{ N/m}.$$

With values of both m and k in hand, we can find the period from Equation 11.6:

$$T = 2\pi\sqrt{\frac{m}{k}} = 2\pi\sqrt{\frac{2}{200}} = 2\pi\sqrt{0.01} = 2\pi(0.1) = 0.63\text{ s}.$$

It is worth noting again that the additional distance that the object was pulled down (the 5 cm) did not figure in the solution. Period does not depend on amplitude.

The second thing we learn from applying calculus to the problem of simple harmonic motion is that, if a simple harmonic oscillator starts out at rest at positive displacement A, as we assumed for the block in Figure 11.3, then the motion that satisfies the relationship between acceleration and position required by Equation 11.3 is expressed mathematically by a cosine function. That was the function we graphed in Figure 11.4. The formula that gives the position x at any time t may be written

$$x = A\cos\left(2\pi\frac{t}{T}\right). \tag{11.7}$$

If you know the amplitude A and the period T of the simple harmonic motion, you can plug in any value of t you desire and find out exactly where the block will be found at that time. But we do need to be clear that, if the block does *not* start out at rest at its extreme positive position A, then the solution for its motion will *not* be Equation 11.7. The solution can still be found exactly, but it will generally be a bit more complicated.

Equation 11.7 may actually be a little harder to work with than other formulas we have presented in this book, so let's make sure we understand how to use it. If we want to find the value of x at time t, the first thing we would have to do is to figure out the quantity inside the parentheses, the angle whose cosine we will need to get. And let us be clear that the quantity $2\pi t/T$ really *is* an angle. The time t is in seconds and the period T is in seconds, so t/T has no units. When we multiply by 2π, the result still has no units. But we might remember from page 167 that, technically, an angle in radians has no units. So as long as we have set our calculator to find the cosine of an angle that is given in radians, we will get the right answer for the cosine.

Now let's see what Equation 11.7 tells us about simple harmonic motion. First, at the initial $t = 0$, the quantity $2\pi t/T$ will be zero, and the cosine of zero is one. So, at $t = 0$, we will have $x = A$, as we expected. When we have completed one cycle, that is, when the time is $t = T$, the quantity $2\pi t/T = 2\pi$. The cosine of 2π radians is again one, so we will again have $x = A$. But this is just where we would expect to be after one complete cycle. In fact, t/T is really just a way of expressing how many cycles we have been through, including fractions of a cycle. If we then multiply this quotient by 2π, we convert the number of cycles into the number of radians in that many cycles.

Time for an example.

EXAMPLE 11.4 Using the Simple Harmonic Oscillator Solution

On page 211, we discussed the time it would take for our block on a spring to get from $x = A$ to $x = \frac{1}{2}A$ and noted that it is obvious from the graph that the time is not simply one-eighth of the total period (or 0.1 s). Let us find the actual time required. A block of mass 1.6 kg is attached to a spring of spring constant 100 N/m, and the block is pulled 10 cm in the positive x-direction and released. Find the period of the subsequent simple harmonic motion and determine how long after it is released it will first reach the point $x = +5$ cm.

ANSWER The period of the motion is easily determined from the parameters of the problem. A 1.6-kg mass attached to a 100-N/m spring will produce simple harmonic motion with a period

$$T = 2\pi\sqrt{\frac{m}{k}} = 2\pi\sqrt{\frac{1.6}{100}} = 0.8 \text{ s}.$$

A 10-cm initial displacement from rest gives an amplitude of the motion $A = 10$ cm. Note that these parameters are the same ones used in Figure 11.4. Knowing the period and amplitude, the position of the block at any time t may be found from

$$x = (10\text{ cm})\cos\left(\frac{2\pi t}{0.8\text{ s}}\right).$$

In this problem, we need to find the time t at which the position will be $x = 5$ cm. We set x to 5 cm and then solve for the unknown t.

$$5\text{ cm} = (10\text{ cm})\cos\left(2\pi\frac{t}{0.8\text{ s}}\right) \quad \text{or} \quad \cos\left(2\pi\frac{t}{0.8\text{ s}}\right) = 0.5$$

The requirement that determines t is that $(2\pi t)/(0.8\text{ s})$ must be the angle in radians whose cosine is 0.5. To find what that angle is, I put 0.5 into my calculator, make sure that my calculator is set to work with angles in radians, and then hit the $\cos^{-1}$ button (or however your calculator finds the inverse cosine). If this is done correctly, the number displayed will be 1.0472. So I am able to find the time t from

$$2\pi\frac{t}{0.8\text{ s}} = 1.0472 \quad \Rightarrow \quad t = \frac{(1.0472)(0.8\text{ s})}{2\pi} = 0.133\text{ s}.$$

You may want to look back at Figure 11.4 to make sure that this answer is about what you would expect from looking at the graph.

Before leaving the subject of the period of a simple harmonic oscillator, let us note one more thing about the fact that the period formula does not depend on the amplitude. If a simple harmonic oscillator begins at rest and starts to vibrate at its natural period, as given by Equation 11.6, and if there is a repeated force on the object with this same period, then the amplitude of the motion will increase. But, as the amplitude increases, the natural period will not change, because amplitude does not affect the frequency. So this regular periodic force will simply continue to add more and more amplitude, ultimately without limit. This phenomenon is called "resonance" and is responsible for the operation of many systems in which a large amplitude of motion is desired. But it is also responsible for dangerous effects on systems where a large amplitude of motion could be disastrous, such as oscillations of a power line by low-frequency variation in high winds.

11.5 Energy in Simple Harmonic Motion

When there is negligible friction in a harmonic oscillator—no friction between a block and the tabletop, no creation of thermal energy in a spring from the constant flexing, no air resistance acting on the oscillating mass—then the total mechanical energy of an object subject only to a Hooke's law spring force will be conserved, as we saw in Chapter 7. For our simple block on a horizontal frictionless tabletop, the total energy will consist of the kinetic energy of the block (assuming that the spring's mass may be neglected) and the potential energy of the spring. If we use a variable x to locate the block, measured from the block's location when the spring is unstretched, then we can write the total conserved mechanical energy as

$$E = \frac{1}{2}mv^2 + \frac{1}{2}kx^2. \tag{11.8}$$

Note that the Δs^2 that gives the energy stored in a spring is the same as x^2, and that this is true regardless of whether x is positive or negative, stretching or compressing.

What Equation 11.8 accomplishes is that it allows us to take what we know of the position and velocity of the system at any initial point, determine the total energy from that, and then relate the position to the velocity at any other point. Going back to the bucket model of Chapter 7, once we are able to find the total energy available to both the kinetic energy and the spring potential energy buckets, we may then always know how much is in one bucket as long as we know how much is in the other.

Let's see the kind of problems this now allows us to solve.

EXAMPLE 11.5 Using Energy Conservation in Simple Harmonic Motion

A 4-kg block attached to a spring is oscillating horizontally on a frictionless table. The period of the motion is ½ s and the amplitude is 10 cm. Find the maximum acceleration magnitude and the maximum speed of the block.

ANSWER The only horizontal force acting on the block is exerted by the spring, so the maximum acceleration will occur when the spring force is a maximum. This occurs when the spring is stretched 10 cm from equilibrium. To find the force by the spring on the block at this point, we need to know the spring constant, k, but it isn't given in the problem. However, we know the period, so we know that the spring must have the correct stiffness to produce a ½-s period for a 4-kg block. We write

$$T = 2\pi\sqrt{\frac{m}{k}},$$

and solve for k, giving

$$k = \frac{4\pi^2 m}{T^2} = \frac{4\pi^2(4\text{ kg})}{(0.5\text{ s})^2} = 632\text{ N/m}.$$

The maximum acceleration of the block will occur at the maximum extension of the spring. Since we now know k, we can use Equation 11.3 to calculate this acceleration:

$$a = \frac{k}{m}x = \frac{632\text{ N/m}}{4\text{ kg}}(0.1\text{ m}) = 16\text{ m/s}^2.$$

Now how do we find the maximum speed? Let's begin by thinking about the potential energy stored in the spring. At the end points of the motion (all the way left or right), the speed of the block is instantaneously zero, so all of the mechanical energy is in the form of spring potential energy. On the other hand, when the block is in the middle, at $x = 0$, the spring is unstretched, and the potential energy is zero. All of the conserved energy is in the kinetic energy bucket at this point. The place where the speed will be maximum is where the kinetic energy is a maximum, and this will clearly be where the potential energy is zero.

To find the total energy, we consider the instant when the block is at rest at the extreme ends of its motion. In fact, it is generally true that, once one knows the amplitude of the motion of a simple harmonic oscillator, one knows the total energy, because the point where $x = A$ will always be the point where the kinetic energy is zero. Thus, we will always be able to relate the energy to the amplitude via

$$E = \text{KE} + \text{PE} = 0 + \frac{1}{2}kA^2 = \frac{1}{2}kA^2.$$

The total energy in this problem is thus

$$E = \frac{1}{2}kA^2 = \frac{1}{2}(632\ \text{N/m})(0.1\ \text{m/s})^2 = 3.16\ \text{J}.$$

At $x = 0$, all this energy becomes kinetic energy, allowing us to find the speed from

$$\frac{1}{2}mv^2 = E \quad \Rightarrow \quad v^2 = \frac{2E}{m},$$

giving

$$v^2 = \frac{2(3.16\ \text{J})}{4\ \text{kg}} = 1.58\frac{\text{m}^2}{\text{s}^2} \quad \Rightarrow \quad v = 1.26\ \text{m/s}.$$

Before we leave this example, let us just mention that, once we had calculated the maximum acceleration as 16 m/s^2, we might have been tempted to use the equation $v = at$ to get the speed of the block after it had completed ¼ of its ½-second period, that is, to find the speed at $t = 0.125$ s. However, we must remember that 16 m/s^2 is only the acceleration when the spring is stretched by 10 cm. As the block moves toward the center, the stretch in the spring decreases, the force decreases, and, thus, the acceleration decreases. In fact, as the block passes through the center (equilibrium) position, the acceleration has dropped to zero. We cannot use the constant-acceleration kinematic equations from Chapter 2 to solve for motion when the acceleration is not constant. The speed question we had here was clearly a job for an energy approach, not for Newton's second law and kinematics.

EXAMPLE 11.6 More Energy Considerations in Simple Harmonic Motion

For the system in Example 11.5, find the speed of the block when the block is halfway between the center point and the maximum amplitude.

ANSWER We already know that the total energy is 3.16 J throughout the motion, which includes the point $x = 5$ cm. At $x = 5$ cm, the energy is divided between the kinetic and potential buckets, but it is important not to jump to any conclusions about how it is divided. In particular, the energy is not divided 50/50. The spring potential energy at $x = 5$ cm is $\frac{1}{2}kx^2 = \frac{1}{2}(632\ \text{N/m})(0.05\ \text{m})^2 = 0.79\ \text{J}$. The total energy is 3.16 J, so the kinetic energy must be 3.16 J – 0.79 J = 2.37 J. We are therefore able to find the speed at the 5-cm location from

$$\frac{1}{2}mv^2 = \frac{1}{2}(4\ \text{kg})v^2 = 2.37\ \text{J} \quad \Rightarrow \quad v = 1.09\ \text{m/s}.$$

Note that this 1.09 m/s is close to the 1.26 m/s maximum speed from Example 11.5. The block acquires kinetic energy quickly (and loses potential energy quickly) when the spring is close to its maximum extension and has its maximum acceleration.

11.6 Other Harmonic Oscillators

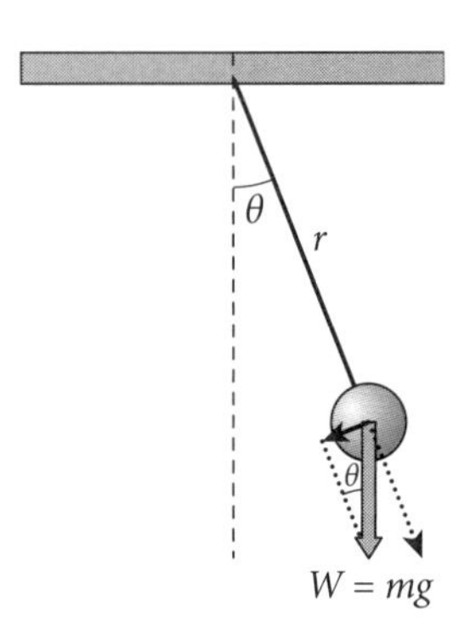

Figure 11.6 The restoring force on a pendulum when it is displaced from the vertical by an angle θ.

We have said that simple harmonic oscillators seem to be found everywhere, leading you to wonder, perhaps, what kind of a thing would *not* be a simple harmonic oscillator. So let's discuss something that is not a simple harmonic oscillator.

In Figure 11.6, we show a laboratory pendulum composed of a bob of mass m at the end of a massless rod of length r. This configuration is known as a *simple pendulum*. We specify the orientation of the pendulum with an angular variable θ and note that, when the pendulum is lifted away from the vertical, the force of gravity on the bob (the gray arrow pointing downward) will have a component along the line of the rod (the dotted arrow in the force diagram) and a component perpendicular to it (the little black arrow). Tension in the rod will balance the radial component of the weight *and* it will provide enough additional force to satisfy the centripetal requirement for this circular path. But we know all about this part of the motion and can ignore it. This leaves us with the motion tangent to the circle to consider. The force tangent to the circular motion of the pendulum bob is the perpendicular component of the weight, the little black arrow in Figure 11.6.

Of course, since we are considering angular motion, we will not be concerned with forces as much as with torques. Let us therefore apply Newton's second law for rotation (Equation 9.5) to this pendulum.

The perpendicular component of the weight may be seen in the tiny triangle of the force diagram. It is $F_\perp = mg \sin\theta$ and produces a torque $\tau = rF_\perp = rmg \sin\theta$, measured about the pendulum pivot. This torque will act on the pendulum to bring it back toward its vertical equilibrium configuration. For a positive angular displacement out of the page, as shown in Figure 11.6, the torque will be into the page, or negative. The angular acceleration out of the page will thus be given by Equation 9.5:

$$\alpha = \frac{\tau}{I} = -\frac{rmg}{I}\sin\theta. \tag{11.9}$$

Back in Figure 9.12a, we gave the formula for the rotational inertia of a single mass m a distance r from the center of rotation. It is $I = mr^2$. Making this substitution yields

$$\alpha = -\frac{g}{r}\sin\theta. \tag{11.10}$$

Now, if Equation 11.10 had been

$$\alpha = -\frac{g}{r}\theta, \tag{11.11}$$

then it would have looked just like Equation 11.3, with θ in the place of x, α in the place of a, and g/r in the place of k/m. The mathematics would have been exactly the same as Equation 11.3, and so the angular motion would have been simple harmonic motion. But $\sin\theta$ is *not* the same as θ, and so Equation 11.10 is *not* the same as Equation 11.11. The motion of a pendulum is *not* simple harmonic motion. It *is* periodic, but the motion does not follow the cosine shape of Figure 11.2, the period is complicated to find, and the period actually does depend on the amplitude.

But here's an interesting observation. Sometime, when you have a little time to kill, try setting your calculator to work in radians and find the sine of 0.5 radians (a little less than 30°). Your calculator will tell you the sine is 0.48. Then try 0.05 radians. The answer will be 0.04998.

For 0.005 radians, about a third of a degree, the sine will be 0.0049999. Do you see where this is going? As the angle in radians gets smaller and smaller, the sine of the angle gets closer and closer to the angle itself. For small enough angles, we have $\sin\theta \to \theta$, and Equation 11.10 *will become* Equation 11.11. When the amplitude is small, a pendulum *is* a simple harmonic oscillator. The position is θ instead of x; the acceleration is α in place of a; and g/r replaces k/m. Therefore, the period of a pendulum, in the limit of small amplitude of vibration, is

$$T = 2\pi\sqrt{\frac{r}{g}}. \tag{11.12}$$

Note that this formula for the period does not depend on the amplitude of vibration or even on the mass. In the limit of small-amplitude oscillations, a pendulum's period on earth (where the gravitational field is g) is completely determined by its length.

We hope that you can see how to generalize what we did for a pendulum. Let's summarize. If the acceleration of some object located with a variable z can be written

$$a_z = -qz,$$

where q is some constant, then the motion of the object will be simple harmonic motion, with period $T = 2\pi\sqrt{1/q}$. Let's see one more example of how this works.

EXAMPLE 11.7 The Bobbing Buoy

A cylindrical buoy of mass $m = 40$ kg floats in a lake. Its cross-sectional area is exactly $A = 0.03\ \text{m}^2$. If the buoy is pushed down 15 cm and released, what will be the period of oscillation of the buoy? Neglect friction between the water and the buoy.

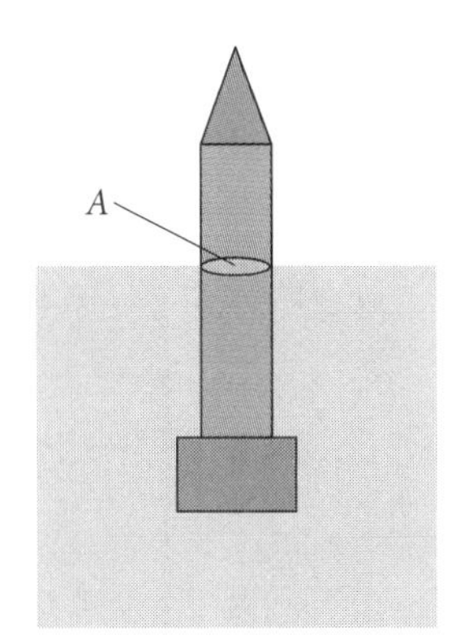

ANSWER The equilibrium position for the buoy will be the point where its submerged volume displaces enough water to just balance its weight. If the buoy is then pushed down a small additional amount x, the additional buoyant force will equal the weight of the additional water displaced. This additional volume displaced will be $V = Ax$. The buoyant force will therefore be an upward force of magnitude $B = \rho_{\text{water}}gAx$. Newton's second law then gives the equation for the buoy's acceleration as

$$a_x = -\frac{\rho_{\text{water}}gA}{m}x.$$

Because the x-acceleration is proportional to minus x, this is the equation of a simple harmonic oscillator. The period will be determined from the inverse of the constant of proportionality, giving

$$T = 2\pi\sqrt{\frac{m}{\rho_{\text{water}}gA}} = 2\pi\sqrt{\frac{40\ \text{kg}}{(1000\ \text{kg/m}^3)(10\ \text{m/s}^2)(0.03\ \text{m}^2)}} = 2.3\ \text{s},$$

Perhaps you have seen a buoy move up and down like this after a speed boat has passed by and can picture the slow motion of the resulting oscillations.

11.7 Summary

From this chapter, you should have learned the following things:

- You should know how to identify the period of harmonic motion and to switch between characterizing the motion by its period or by its frequency.
- The fact that gravity can be ignored in describing the motion of a mass hanging from a spring is contrary to many people's intuition. You should understand how this arises and, especially, how the restoring force arises when the block is above the equilibrium position. In the presence of gravity, a spring that is stretched *less* than it is at its equilibrium position produces the same force as compressing the spring in a situation with no gravity.
- You should be familiar with the formula for the period of a mass attached to a spring, $T = 2\pi\sqrt{m/k}$, and be able to use it to find the period, or the mass, or the spring constant, given the other two quantities in the formula.
- You should recognize that the formula $x = A\cos(2\pi t/T)$ is valid only when a system starts at rest at the point $x = A$. But, when those *are* the initial conditions, you should be able to use the formula to relate the position to the time.
- You should be able to use conservation of energy in simple harmonic motion to relate position, amplitude, and velocity at any point in the motion.
- You should recognize that any equation for the acceleration of a quantity z that is in the form $a_z = -qz$ produces simple harmonic motion for z, and that the period of the motion will be $T = 2\pi\sqrt{1/q}$.

CHAPTER FORMULAS

The relationship between period and frequency: $f = \frac{1}{T}$ or $T = \frac{1}{f}$

Equations for Simple Harmonic Motion (SHM):

Equations producing SHM: $a = -\frac{k}{m}z$, $\alpha = -\frac{g}{r}\theta$, or any $a_z = -qz$

Equations for the period in SHM: $T = 2\pi\sqrt{\frac{m}{k}}$, $T = 2\pi\sqrt{\frac{r}{g}}$, or $T = 2\pi\sqrt{\frac{1}{q}}$

Conservation of energy for a mass/spring system: $E = \frac{1}{2}mv^2 + \frac{1}{2}kx^2 = \frac{1}{2}kA^2$

Solution for SHM, when $x = A$ and $v = 0$ at $t = 0$: $x = A\cos\left(2\pi\frac{t}{T}\right)$

PROBLEMS

*1. The four pictures below represent a ball resting on ramps of different shapes. Which of these represent a situation of stable equilibrium? Explain your answers. [Hint: You may want to check the definitions of equilibrium on page 207.]

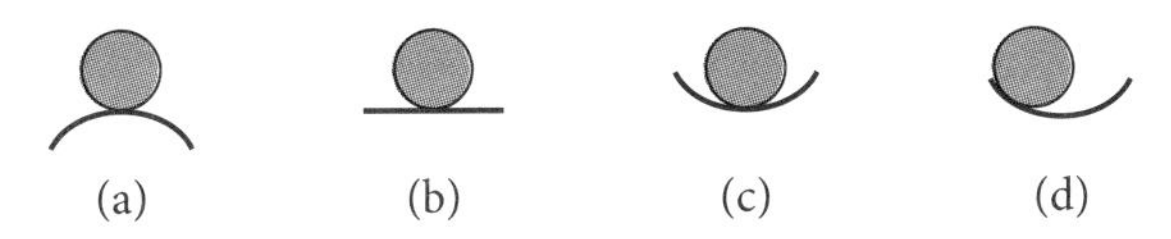

11.2 *2. Only one of the following statements about an object executing simple harmonic motion is true. Tell whether each is true or false and explain why for each.

a) Its velocity is never zero.
b) Its acceleration is never zero.
c) Its velocity and acceleration are zero at the same time.
d) Its velocity is zero when its acceleration is a maximum.
e) Its maximum acceleration is equal to its maximum velocity.

11.3 *3. You make a careful study of the motion of a mass suspended from a spring that oscillates up and down. The same apparatus is then transported to the moon where the gravitational field strength is considerably smaller ($g_{\text{moon}} \cong \frac{1}{6}g_{\text{earth}}$). Is the time for the system to complete one cycle of its motion on the moon (a) less than on earth, (b) the same as on earth, or (c) greater than on earth? Explain your answer.

*4. A 2.0-kg mass is suspended from the end of a spring whose spring constant is 350 N/m. The mass is pulled down an additional 10 cm and released. What is the period of the subsequent oscillations?

**5. A 2-kg block is suspended by a spring of unknown spring constant. At equilibrium (i.e., just hanging there but not moving), the spring is stretched 10 cm. The block is pulled down an additional 5 cm and released.

a) What is the spring constant of the spring?
b) What value of spring constant would be required to double the period compared to the original situation?

**6. A 4-kg mass is suspended from a spring. At its equilibrium position, it is found that the spring is stretched 17 cm from its unstretched length. If the mass is then pulled down an additional 10 cm and released, what will be the period of the subsequent oscillations?

**7. A block of mass $m = 6$ kg is suspended vertically from a spring. The time it takes for the block to complete 10 oscillations is 30 s. The amplitude of oscillation is $A = 15$ cm.

a) What is the frequency of the simple harmonic motion?
b) What is the spring constant of the spring?

**8. A 500-g block is suspended from a spring. With the block at rest, the spring stretches 15 cm. The block is pulled down an additional 10 cm and released.

a) What is the spring constant of the spring?
b) How many complete oscillations will the system complete in 20 s?
c) Find one other combination of mass and spring constant that would produce the same period.
d) By how much would the spring stretch at rest for this new case?

11.4 ** 9. Having successfully survived his first bungee jump, Ron is hanging at rest from the end of the bungee cord. "I wonder what the spring constant of this bungee cord is?" he asks himself. So he decides to perform an experiment. He pulls himself up about 75 cm and lets go. As he begins to bob up and down, he times his bounces and finds that he completes 4 oscillations in 20 seconds. He knows his mass is 75 kg.

a) What is the period of his motion?

b) What is the frequency of his motion?

c) What total distance does Ron travel in 20 seconds? Explain your reasoning.

d) What is the spring constant of the bungee cord?

** 10. A 100-kg man gets inside an old clunker car that he is thinking of buying, and he determines that the car drops 1.0 cm closer to the ground when he is in it.

a) What is the effective spring constant of the springs in the car?

b) As he test drives the car, he hits a bump and discovers that the shock absorbers are shot, so there is no damping of the vibrations. The car with him inside it vibrates with a period of 0.75 s. What is the mass of the car alone?

*** 11. A boy on a carnival bungee ride is released from rest at the bottom of the ride and oscillates up and down with an amplitude of 8 meters and a period of 4 s. How long after he is released does he first get to a height 10 m above the place where he started? [Hint: Choose the top of the poles as $y = 0$, and measure positive downward.]

** 12. A block of mass 2 kg hangs from a spring of spring constant 18 N/m. It is lifted up to a height of $y = 20$ cm above its equilibrium point and dropped from rest.

a) What will be the period and amplitude of the motion?

b) After 1.2 seconds where will the block be located?

** 13. A 25-kg block on the end of a spring is lifted upward to a point 2 m away from its equilibrium position and released from rest. As the block heads downward, back toward its equilibrium position, it first arrives at a point 0.6 m above the equilibrium position after a time 0.3 seconds has elapsed.

a) What is the period of the simple harmonic motion of the block?

b) What is the spring constant of the spring?

*** 14. **The General Simple Harmonic Oscillator Solution.** If a block on the end of a horizontal spring is not released from rest, but is given an initial velocity at some intermediate initial position, its motion will not be given by Equation 11.7, but by the more complicated formula: $x = A\cos(2\pi t/T + \phi)$, in which ϕ is the so-called "phase constant" whose value is set so that the formula will correctly give x as a function of t. A certain spring/block combination has amplitude 10 cm and period 2 s, and it begins its motion at the point $x = 5$ cm. Find an equation that correctly gives x as a function of t. [Hint: Find the value of ϕ that makes $x = 5$ at $t = 0$.]

11.5 ** 15. A 4.0-kg block is fastened to a spring with spring constant 100 N/m and is oscillating horizontally on a frictionless tabletop with an amplitude of 0.16 m.

a) Draw the free-body diagram for the block when its displacement from equilibrium (distance from the center) is 0.12 m. Label all of the forces.

b) Use Newton's second law to find the acceleration of the block at this point.

c) What is the total mechanical energy of the oscillating block (potential plus kinetic)? Explain your answer.

** 16. A 500 g block is fastened to a spring with spring constant 100 N/m and is oscillating horizontally with an amplitude of 0.16 m. What is the maximum potential energy that is stored in the spring during the motion?

** 17. The Simple Harmonic Oscillator is a new ride at Wally World. A cart with a single rider is released from rest at point A, 50 m above the ground. The cart then slides without friction and eventually collides with a large massless spring that is initially uncompressed. The cart then grabs the spring and oscillates back and forth, causing the rider to squeal with delight. The spring constant is 1000 N/m. The cart and rider have a combined mass of 400 kg.

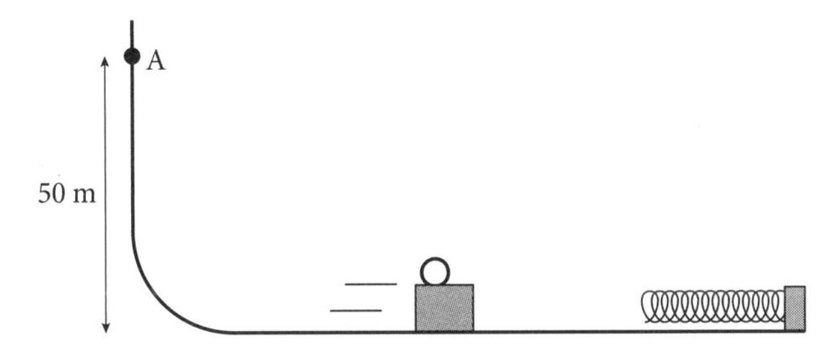

a) Find the kinetic energy of the cart and rider right before they hit the spring.

b) Find the amplitude of oscillation for the cart and rider.

c) When the spring is compressed by $\Delta s = 6$ m, find the speed of the cart and rider.

d) Find the maximum acceleration experienced by the rider.

e) The rider gets tired of the oscillation and hits the panic button, which ejects the rider and the seat straight up into the air (there is a parachute). If the rider ejects at one of the turn-around points, when the cart is momentarily stopped, indicate what will happen to the motion of the cart? Will the amplitude increase, decrease, or stay the same? Will the period increase, decrease, or stay the same? Will the speed of the cart at equilibrium increase, decrease, or stay the same? Explain all of your answers.

** 18. A 5.0-kg mass is attached to a spring and executes simple harmonic motion with a period of 2.0 s. If the total energy of the system is 10 J, what is the amplitude of the motion? [Hint: Start by finding k.]

** 19. A 10-kg mass is at rest at the end of an unstretched spring with a spring constant of 4000 N/m. The mass is struck with a hammer, giving it a velocity of 6 m/s to the right. What is the amplitude of the resulting oscillations?

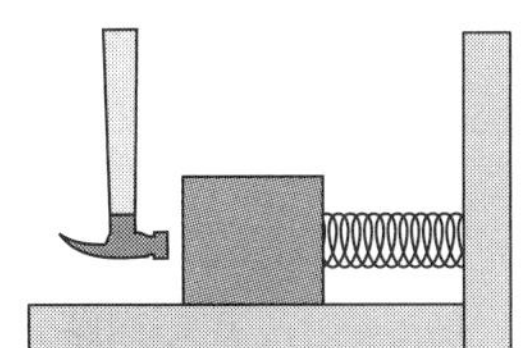

** 20. A 4-kg block is attached to a spring, and oscillates horizontally on a frictionless tabletop. The spring has constant $k = 1875$ N/m. The maximum speed measured for the block is $v = 10$ m/s.

a) What is the total mechanical energy of the block–spring system? Explain.

b) When the block is moving at a speed of $v = 5$ m/s, how far is the spring stretched (or compressed)?

** 21. A 6-kg block is attached to a spring, and oscillates horizontally on a frictionless tabletop. The spring has constant $k = 1000$ N/m. The block has speed $v = 5$ m/s when the spring is stretched a distance $\Delta s = 10$ cm from equilibrium.

a) What is the total mechanical energy of the block–spring system?

b) What is the amplitude of the motion?

** 22. A 4-kg block is fastened to a spring and oscillates horizontally on a frictionless surface. The block has speed 6 m/s when it is at a point 20 cm from equilibrium. If the greatest speed achieved by the block during its oscillation is 8 m/s, find the spring constant of the spring.

11.6 ** 23. A pendulum clock that ticks once per second (that being the period of one complete cycle of the pendulum) is taken to the moon to use as a timekeeper for a lunar colony. What will be the period of the pendulum when it is on the moon? [Hint: Remember that the gravitational field on the moon is 1/6 the strength of the gravitational field on the earth.]

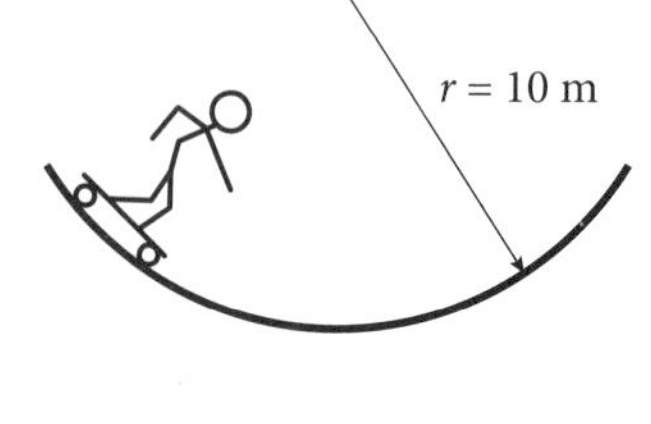

** 24. A skateboarder is rolling back and forth inside a "half-pipe" that is a part of a circle of radius 10 m. His motion is driven by gravity exactly like the motion of a pendulum bob at the end of a pendulum. If he starts from rest, 2 meters from the bottom of the pipe, how long will it take him to get back to where he started? (Assume small-amplitude angular motion of this "pendulum.")

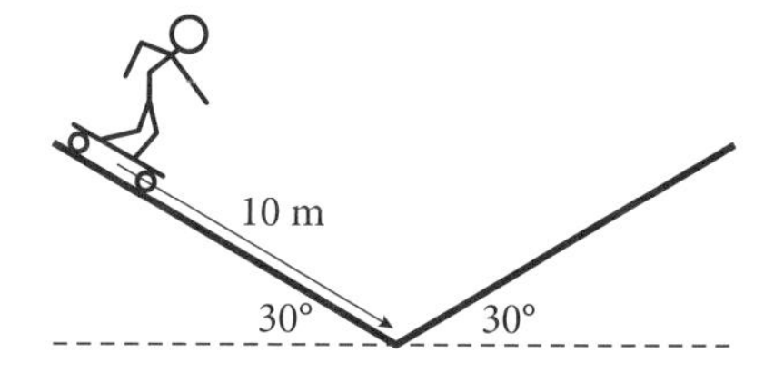

*** 25. The motion of the skateboarder on the track shown at right is an example of motion that is harmonic, but not simple harmonic. The skateboarder starts from rest 10 m from the bottom of the track (and 5 m above the ground) and oscillates back and forth. Find the period of his motion. (Ignore the little jolt he will get as the track changes slope at the bottom.) [Hint: Each side of this track produces motion with constant acceleration, so the kinematic formulas from Chapter 2 may be used.]

** 26[f]. We saw in Section 11.6 that a pendulum formed by a bob of mass m on the end of a massless string of length L is a harmonic oscillator, in the limit of small oscillations. Let us use an angle θ to locate the position of the pendulum bob, as shown in Figure 11.6, and let us use ω for the angular speed of the pendulum. Consider the variables m, g, L, θ, and ω, and find an expression for the total energy of a pendulum in the limit of small oscillations. [Hint: This problem applies Chapter 7 and Chapter 9 principles to the simple harmonic motion of a pendulum.]

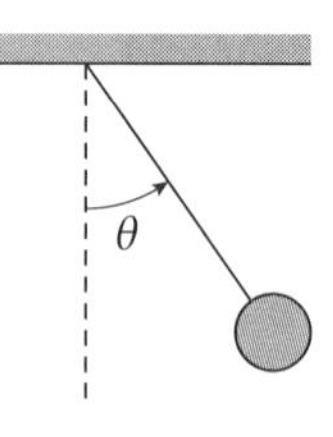

** 27. Even though the motion of a simple pendulum is not simple harmonic motion when the amplitude of oscillation is large, gravity is still a conservative force, so the total energy will be conserved. A pendulum of mass 2 kg and length 0.6 m is raised to an initial angular displacement $\theta = 45°$ and released.

a) What is its total energy?

b) What will be the angular velocity of the pendulum when it is at the angle $\theta = 30°$?

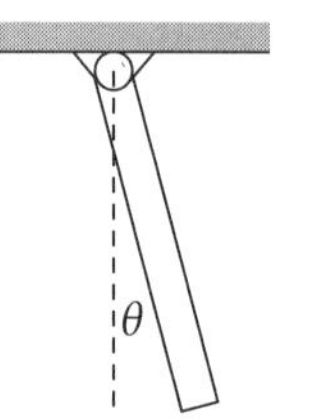

*** 28. **The Physical Pendulum.** A pendulum is formed by hanging a 60-cm-long stick of mass m from a pivot, as shown at left. Find the angular acceleration of the pendulum when the pendulum is displaced by a small angle θ, as shown, and show that the motion is simple harmonic motion in the limit of small oscillations. Find the period of the pendulum in this limit. (An extended body that is free to rotate like this is called a "physical pendulum." The solution may require rereading a little from Chapter 9.)

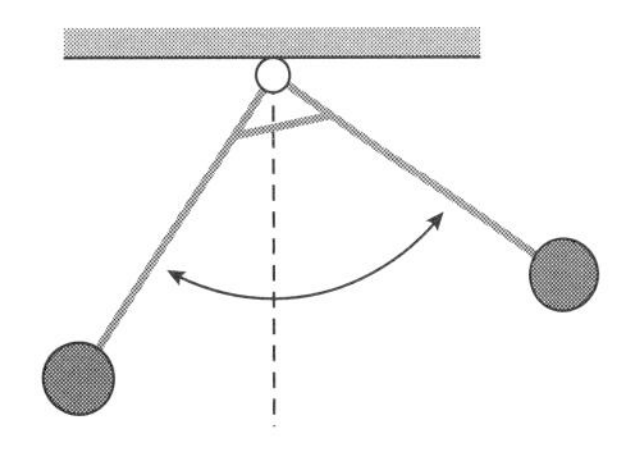

*** 29. Two thin aluminum rods are braced and welded rigidly together at 90°, and then pivoted at the weld so that they can rotate in the plane of the two rods, as shown in the figure at right. The rods are each 80 cm long and have negligible mass. At the end of each rod is a small iron ball of mass 500 g. The equilibrium position for this combination of rods and balls is when each rod is 45° from the vertical. Find the period of small oscillations of this object about its equilibrium position.

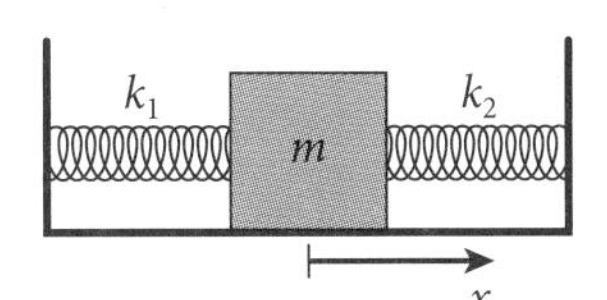

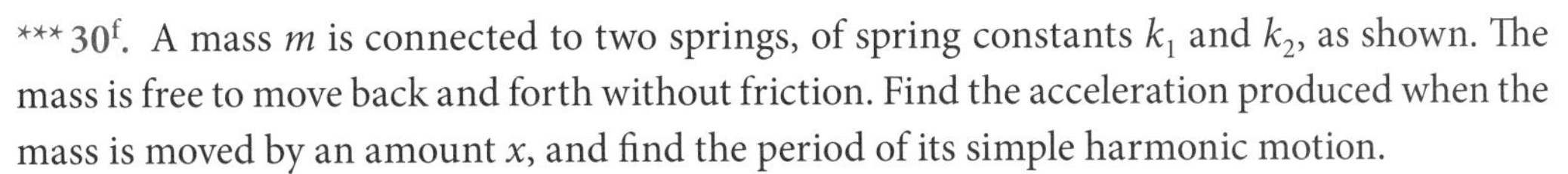

*** 30[f]. A mass m is connected to two springs, of spring constants k_1 and k_2, as shown. The mass is free to move back and forth without friction. Find the acceleration produced when the mass is moved by an amount x, and find the period of its simple harmonic motion.

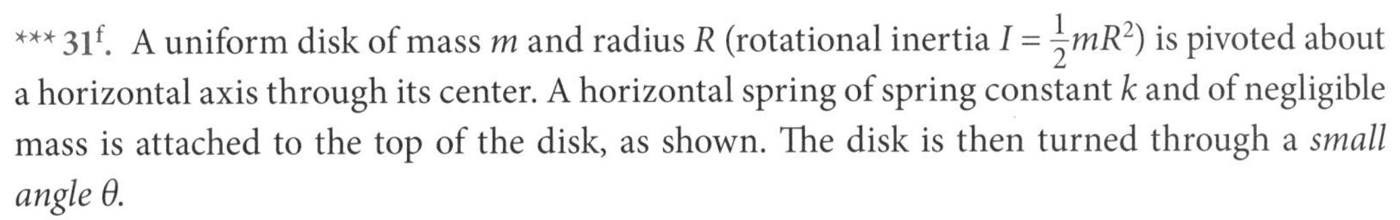

*** 31[f]. A uniform disk of mass m and radius R (rotational inertia $I = \frac{1}{2}mR^2$) is pivoted about a horizontal axis through its center. A horizontal spring of spring constant k and of negligible mass is attached to the top of the disk, as shown. The disk is then turned through a *small angle* θ.

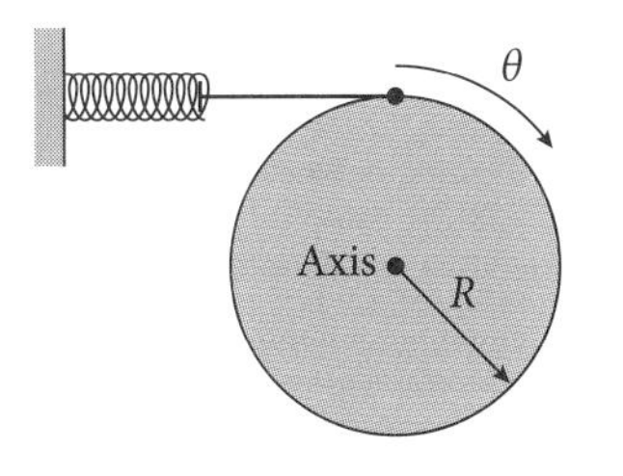

a) Find the restoring torque exerted by the spring on the rim of the disk and the resulting angular acceleration of the disk.

b) The disk is now released from rest. Find the period of its small rotational oscillations, considering the variables k, m, and R.

*** 32[f]. The fluid in an open U-shaped tube is in equilibrium when the water level is the same in both sides of the tube. The fluid in the left side of the tube is pushed downward with a piston by an amount x and the piston is then taken away, releasing the fluid to oscillate up and down in the tube. Assume that the tube has cross-sectional area A and that the fluid has a density ρ, and assume that the column of fluid in the tube is of total length L. Consider a case where the fluid in the right tube is raised by an amount x, as shown at right. In this case, there will be a restoring force on the fluid due to the unbalanced force of gravity acting downward on the plug of the fluid of height $2x$ shown in the figure.

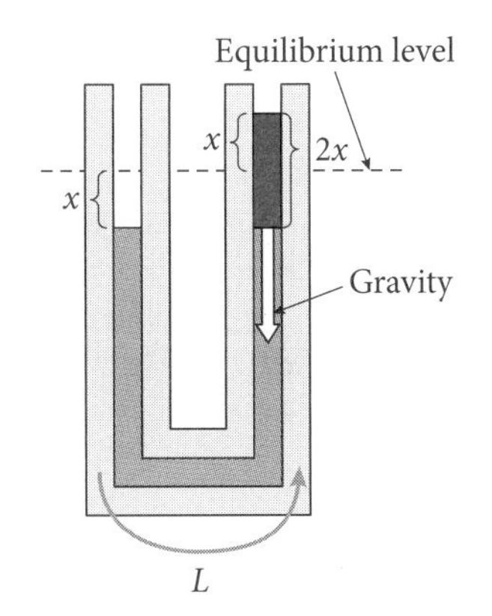

a) Use Newton's second law to find the acceleration of the fluid when the right surface is raised by an amount x, as shown.

b) Explain why the motion of the fluid back and forth in the tube will be simple harmonic motion and find the period of the motion, considering the variables g, L, ρ, and A.

12

PULSES AND WAVES

Mary and Paul are holding the two ends of a long coil spring stretched between them (Figure 12.1). Mary wants to get Paul's attention, but he is looking the other way. Now, if Mary is holding a baseball, in addition to holding the end of the spring, she can get Paul's attention by throwing the ball at him, in which case she transmits energy to Paul in the form of a moving mass. Alternatively, Mary could start shaking the end of the spring, back and forth or up and down, and Paul will soon feel the traveling disturbance in his hand. In this case, Mary has again transmitted energy to Paul, but the physical object that is transmitting the energy, the spring, hasn't moved from one place to the other. Whenever something transmits energy without transmitting mass, we say that the energy has been transmitted by a *wave*.

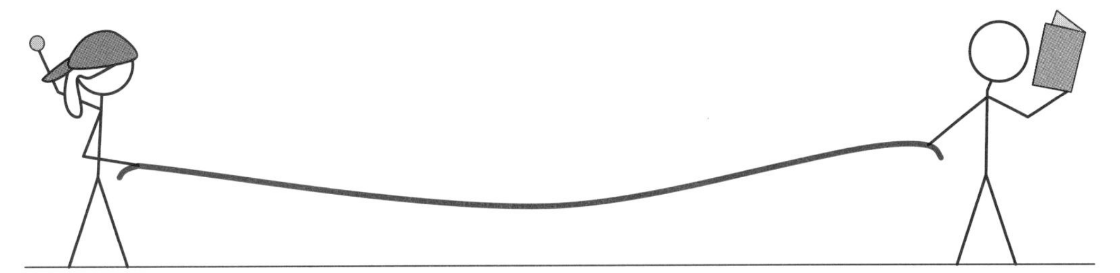

Figure 12.1 Mary is trying to get Paul's attention. She can either throw a baseball at him of she can shake the spring they are both holding. The second method involves a wave.

In our previous chapter, we examined the simple harmonic motion of an object that vibrates back and forth with some period. In this chapter on waves, we extend this idea to systems where there are many objects in a line, and where they are connected in such a way that, if one of them is moved away from its equilibrium position, forces will arise that act on nearby objects, causing them to move as well. Thus, these connected objects will each vibrate back and forth, but the motions will be offset from each other in such a way that the peak of the vibrational motion moves in a particular direction. This is a wave. It is a traveling disturbance in a *medium*—sometimes solid, as in the case of Paul and Mary's stretched spring, and sometimes fluid, as in the case of sound waves traveling in air. Each element of the medium, either an individual coil in the spring or a particular molecule in the air, moves back and forth or up and down, and each one induces its neighbor to move, just a little bit later. This way, energy is transmitted through the medium, even though the individual elements of the medium experience no net displacement in the direction the wave propagates.

12.1 Definitions

The technical definition of a *wave* is as we have given it above—it is something that transmits energy without transmitting mass. In this book, we will continue to use it in this general sense. But we will also discriminate between a *continuous wave* and a *pulse*. In the previous example, if Mary had just moved the spring sharply up and down once, a single crest would have moved along the spring, as shown in Figure 12.2a. This is what we will call a "pulse." In contrast, if she had moved her end of the spring up and down in a regular fashion, she would have created a series of crests and troughs on the spring, with the whole pattern appearing to move toward Paul, as in Figure 12.2b. This is what we will call a "continuous wave." Figure 12.2b represents a "snapshot" of what the spring might look like some time after Mary begins shaking the spring, but before the front of the continuous wave reaches Paul. For comparison, Figure 12.2a shows the pulse at the same elapsed time. Of course, a continuous wave may also be thought of as a series of pulses, first in one direction and then in the other.

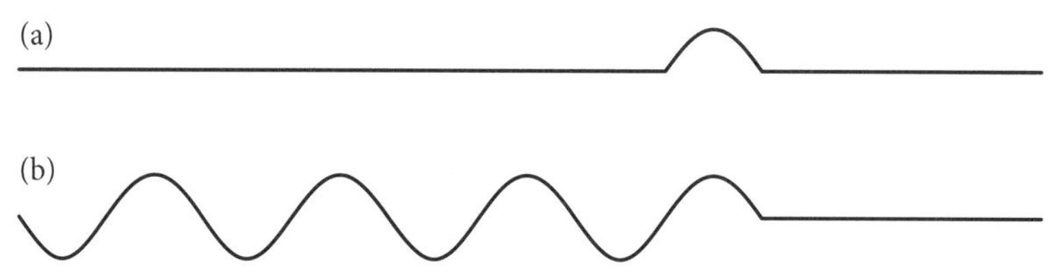

Figure 12.2 Our definitions for a *pulse* (a) and a *continuous wave* (b).

The main thing that distinguishes the continuous wave from the pulse is that the motion is periodic for a continuous wave. If Mary had sent a continuous wave down the spring, and if she were to watch only one coil in the spring, she would see that coil continue to move up and down in a regular fashion with a fixed period. In contrast, were she to send only a single pulse down the spring, she would only see a single up-and-down motion, and it would not be possible to define a period because the motion would not be periodic.

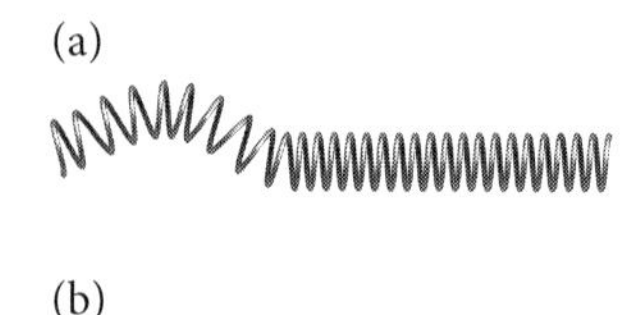

Figure 12.3 A transverse wave on a coil spring (a) and a longitudinal wave in a coil spring (b).

In the beginning of the chapter, we said that Mary could get Paul's attention by moving her end of the spring either up and down or back and forth. If she initially moves her hand up, then the leading edge of the wave (continuous or pulse) will move each spring coil upward as the wave moves along the spring. The motion of the coils (the up and down) is perpendicular to the direction that the wave is moving (from Mary to Paul). We call this a *transverse* wave. If, instead, Mary begins by moving her hand directly toward or away from Paul, parallel to the spring, then each coil of the spring will move back and forth along the spring's length as the wave passes. We call this a *longitudinal* wave. Some media, like springs (Figure 12.3), can support either type, whereas some media, like air, can only support longitudinal waves. The maximum displacement of the elements of the medium—either in a pulse or in a continuous wave, in a transverse or a longitudinal wave—will be referred to as the *amplitude* of the wave.

And let us just mention here that the sound waves we hear are longitudinal waves in air. The molecules of the air move back and forth in the direction the sound is moving, but there is no net movement of the air as the sound wave travels (i.e., there is no wind). Each molecule of the air just does its little forward-and-back dance and then ends up right back where it began.

Finally, let us point out that we use the idea of a stretched coil spring, like the one shown in Figure 12.4, to develop a lot of the wave concepts discussed in this chapter. The reason for this is that wave speeds along these springs are typically very slow and easy to picture, if you've ever seen a real one. However, our conclusions for stretched springs apply equally well to stretched strings, wires, ropes, or garden hoses, anything that can support a tension along its length. We will often refer to these other string-like things as well, in which case, instead of referring to a particular coil in a coil spring, we may often refer to a particular segment in a string or a rope.

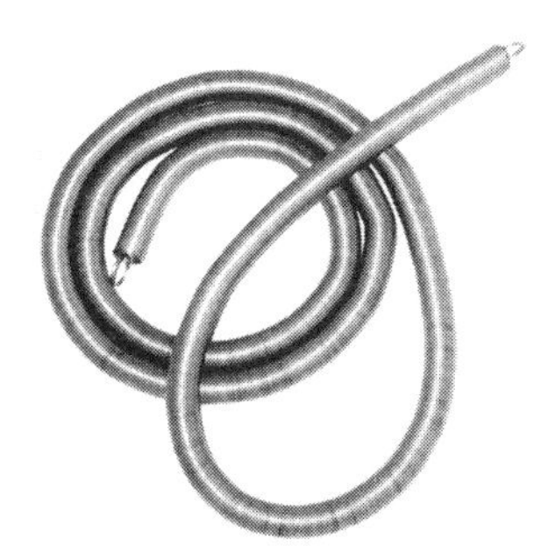

Figure 12.4 A coil spring often used for classroom wave motion demos.

12.2 Wave Speed

The meaning of the speed of a pulse along a medium is fairly obvious. It is the speed you would have to run alongside to stay even with the leading edge, and this speed is a property of the medium. It does not vary with the size of the pulse. Big pulses and small pulses will travel along the spring at the same speed. We should point out, however, that this speed is unrelated to the speed of the individual particles in the medium as they vibrate up and down or back and forth. Imagine the stretched spring connecting Mary and Paul in Figure 12.1, and imagine a piece of ribbon tied to one coil to identify it. If Mary were to start a small pulse with a narrow width going down the spring, then the ribbon, when the pulse got to it, would jump up and back down during the time the pulse passed. It would not go very far up and down, because it was a small pulse, and so it would not have to move up and down at a very high speed. If, on the other hand, Mary had put in a *large* amplitude pulse *of the same width*, the ribbon would have to go up and down a large amount. Since the speed of the pulse along the spring is the same in both cases, the time it takes that pulse to pass would be the same in both cases. So the medium would have to move at a higher up-and-down speed. The speeds of the two waves are the same, because they are moving along the same spring, but the transverse speeds of the individual coils are different in the two cases.

In the simple harmonic motion we discussed in Chapter 11, we saw that the period of the motion depended on only two properties, the inertia (mass) and the restoring force (spring constant). The speed of a transverse wave on a string is similar in its simplicity. It depends on only two properties of the medium—the inertia (the mass per unit length of the string) and the restoring force (the tension in the string). The formal derivation of the relationship takes more time than it is worth, and it is only exactly determined using concepts from calculus anyway, so let us simply state the result. We assure you that the formula *can* be derived and that it is easily verified experimentally, if you want. The formula for the speed of a wave along a stretched string (or spring, or rope, or cable, etc.), is

$$v = \sqrt{\frac{F}{\mu}}, \tag{12.1}$$

in which F is the tension in the string, measured in newtons, and μ is the mass per unit length,[1] in kg/m. The concept of tension in a string is a familiar one from Chapter 6, though in

1. We try pretty hard to use different letters for different things in this book, but, if we stuck to this, we could only teach you 50 different things (26 Latin letters plus 24 Greek letters). Also, there *are* conventional letters for certain quantities, and these often double up the usage. All of this is to be sure you understand that the μ in Equation 12.1 is not the same as the coefficient of friction introduced in Chapter 6. Actually, there should never really be any confusion.

Chapter 6, we used a T for tension. Here we have chosen F so as to avoid confusion with the period, T. The easiest way to understand μ (and a straightforward way to measure it) would be to cut off a 1-m length of the string and find its mass. Alternatively, you could take a string of length L, find its mass m, and calculate a density from $\mu = m/L$. The quantity μ is known as the *linear density* of a string. Finally, let us be clear that this formula is valid only for a transverse wave on a string, not for other kinds of waves. It does not work for ocean waves or sound waves through the air, and it does not even work for longitudinal waves along the same stretched string.

EXAMPLE 12.1 The Blip on the Wire

A steel wire of mass 0.30 kg and length 0.60 m supports a 100-kg block as shown. The wire is struck exactly at its midpoint, causing a small blip in the shape of the wire. As a result, pulses will move upward toward the top of the wire and downward toward the hanging mass. How long does it take the pulse to reach the top of the wire?

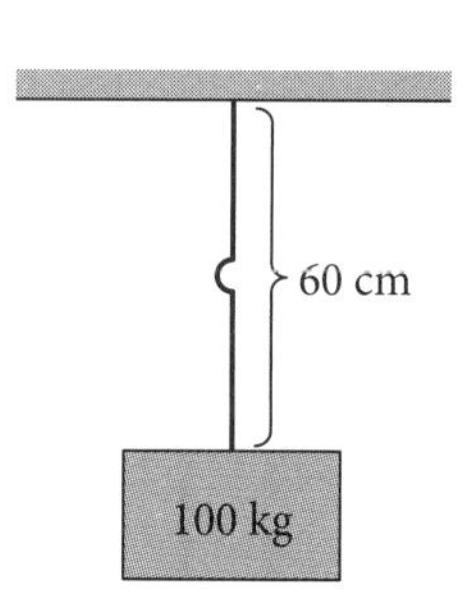

ANSWER The pulse has to move 30 cm to reach the top of the wire. We can find the time this will take if we know the speed at which the pulse travels. This speed depends on only two things: the mass per unit length of the wire (a property of the wire) and the tension under which the wire is stretched. The mass per unit length, μ, is found from the fact that we are given the mass and length of a piece of the wire. It is

$$\mu = \frac{m}{L} = \frac{0.30 \text{ kg}}{0.60 \text{ m}} = 0.50 \text{ kg/m}.$$

We also need the tension in the wire. The wire is supporting a 100-kg block. A free-body diagram of the block quickly reveals that, since the block has zero acceleration, the tension in the wire must equal the weight of the block, which is 1000 N. It is true that the cable itself has mass and, therefore, the tension changes from bottom to top. It is 1000 N where the cable attaches to the block and 1003 N (the wire weighs 0.3 kg) where it attaches to the ceiling. But, since this difference appears only out in the 4th decimal place, we will ignore it here.

We can find the speed of the pulse using Equation 12.1.

$$v = \sqrt{\frac{F}{\mu}} = \sqrt{\frac{1000 \text{ N}}{0.50 \text{ kg/m}}} \approx 45 \text{ m/s}.$$

So the pulse is moving pretty fast along the wire (45 m/s ≈ 100 mi/hr). The distance divided by the speed gives the time for the pulse to reach the ceiling:

$$t = \frac{d}{v} = \frac{0.30 \text{ m}}{45 \text{ m/s}} = 0.0067 \text{ s}.$$

With a time less than a hundredth of a second, you probably would not be able to follow the motion with your eye.

12.3 Superposition of Pulses

If you take a snapshot of a pulse moving along a stretched spring, you can describe the shape of the pulse by noting the displacement from equilibrium for each point along the spring. For a single narrow pulse, most points will actually have zero displacement. However, a complete description of the shape of an arbitrarily shaped pulse would require a fine grid and lots of values to be given at lots of different places. We don't want to worry about this complication, so, in this section, we will consider only pulses that can be described by a small number of straight line segments. Figure 12.5 shows such a pulse, with the corresponding displacements in centimeters listed next to the key points.

If the speed of this pulse along the spring is 10 cm/s, and if the pulse is moving to the right, then, 0.1 second later, the entire pulse would be translated 1 cm to the right. The coils of the spring themselves do not move to the right, but the future shape of the spring is found by picturing the displacements shifting to the right. Figure 12.5b shows the spring 0.1 seconds after Figure 12.5a and Figure 12.5c shows the spring after 0.2 s. The pulse has moved 2 cm to the right.

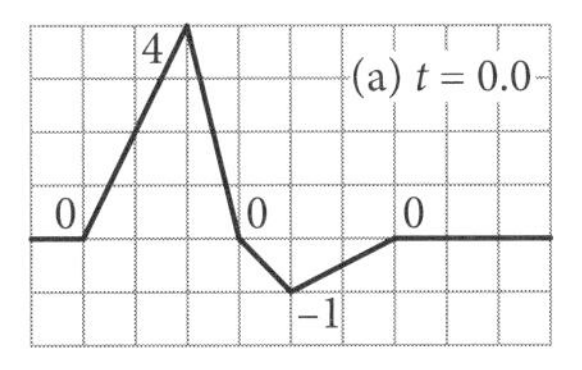

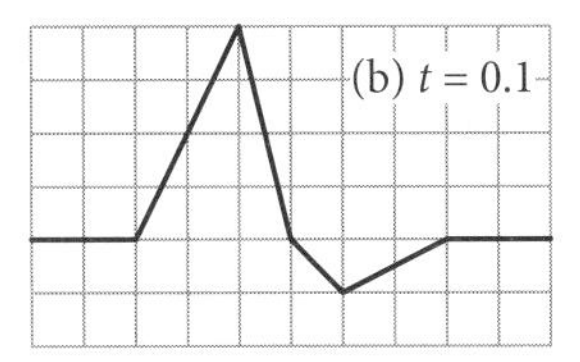

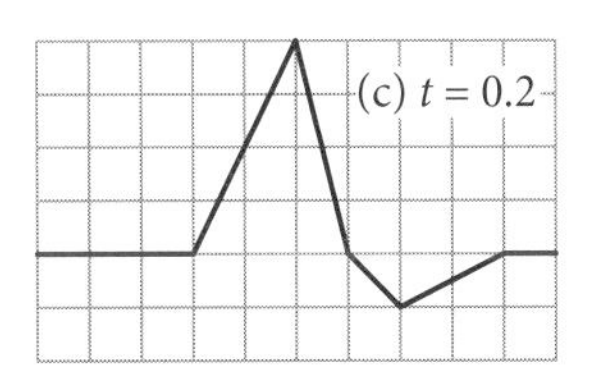

Figure 12.5 A pulse moving to the right at 10 cm/s. In each successive picture, 0.1 additional second elapses. (1 square = 1 cm.)

We next consider what happens when two pulses, traveling toward one another on the same spring, come together at some point and begin to overlap. It is tempting to call this event a "collision," and yet the results are exactly the opposite of what usually happens during a collision. When two cars hit one another head on, they can either bounce back from each other or stick together. But, when two pulses "collide" on a spring, they just keep right on going. During the overlapping of the pulses, the shape of the spring is modified, but the pulses emerge from the collision with exactly the same shapes they had before.

What's more, predicting the shape of the pulse during the interaction turns out to be quite simple, because it relies on a simple but powerful principle that we have already encountered in this text—superposition. A spring with two waves passing along it responds to each wave without modifying its response to the other. If one wave tells it to move upward 1 mm and the other wave tells it to move upward 2 mm, the spring will move upward 3 mm. The individual pulses still exist, but, where they overlap, we only observe their combined effects. Figure 12.6a shows two pulses approaching each other along a spring on which waves move at 10 cm/s. Figure 12.6b shows what the individual pulses would look like as they "collide," 0.2 s later, but without adding the pulses together. Finally, Figure 12.6c shows the net displacement of the spring when the two pulses are added together in the collision.

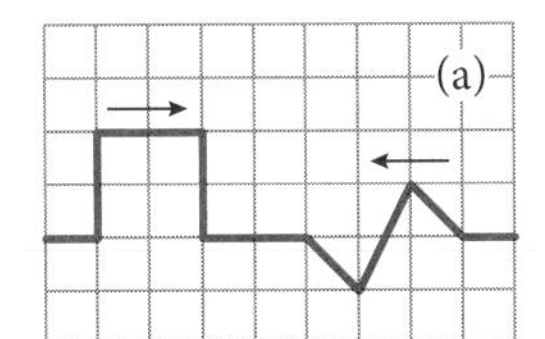

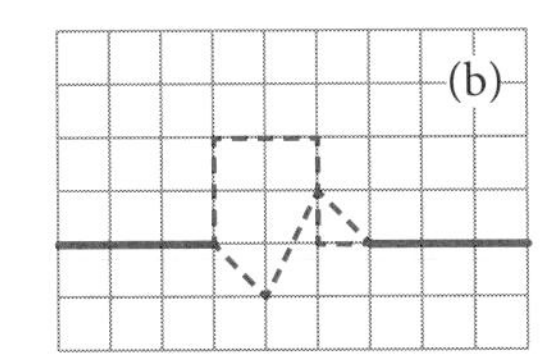

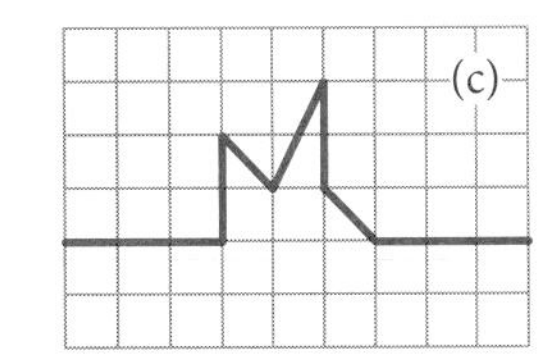

Figure 12.6 (a) The two pulses coming toward each other. (b) The two pulses shown where they would overlap. (c) The two pulses added together where they overlap and interfere. (1 square = 1 cm.)

The superposition of two waves when they combine together is called *interference*. Notice that there are places in the combined shape in Figure 12.6c where the spring has a greater displacement than either of the individual pulses would create. We call these regions of *constructive interference*. Other points have a displacement that is less than the individual pulses would create on their own. We call these regions of *destructive interference*.

12.4 Reflection of Pulses at a Boundary

So far, we have considered pulses traveling along a spring, but we have not considered what happens when the pulse comes to the end of the spring, or to the boundary of whatever medium it is traveling through. A boundary may be fixed, say where you fasten the end of the spring firmly to a wall. Or a boundary may be free. A perfectly free boundary is a little harder to realize in practice than a fixed boundary. It is a case where the spring's end is free to move up and down but continues to be held tight under tension. Such a condition can be approximated by attaching the end of the spring to a light pulley that is free to move on a rod held perpendicular to the spring as shown in Figure 12.7.

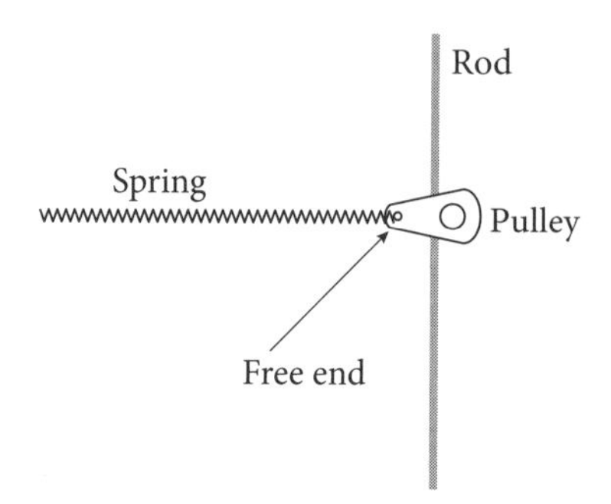

Figure 12.7 A free boundary for a stretched spring.

When a pulse hits a boundary, fixed or free, it is reflected. It turns around and travels back the way it came. The first part of the pulse to reach the boundary is the first part to bounce and head the other way, so the pulse retains its same leading edge. The shape can look pretty complicated during the brief time that the leading part of the pulse is moving backward while the trailing part of the pulse is still moving forward, but eventually a reflected pulse emerges that looks just like the original (or incident) pulse. Except, if the incident pulse bounces off a fixed boundary, the reflected pulse will be inverted and travel on the opposite side of the spring. On the other hand, if the boundary is free, the reflected pulse travels on the same side of the spring as the incident pulse, as shown in Figure 12.8.

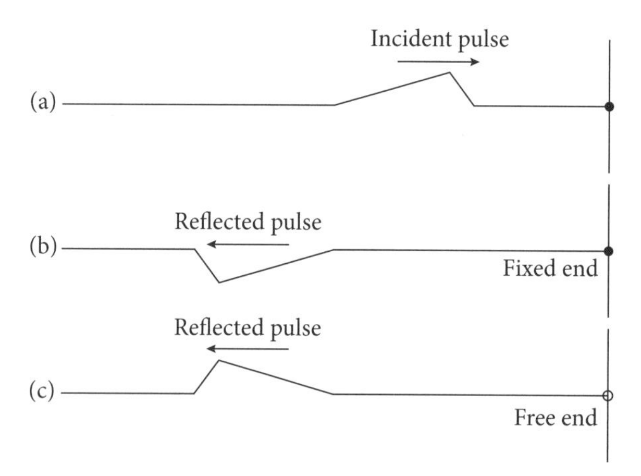

Figure 12.8 In (a) the incident pulse is moving to the right. In (b) we see the inverted shape of the reflected pulse from the fixed end. In (c) the reflected pulse is erect after it has reflected from the free end.

In the last paragraph, we simply told you what happens when a pulse reaches a boundary. You could just trust us on this, or, if you would like, you can test the fixed-end prediction yourself. Next time you are trying to straighten the kinks out of your 100-foot outdoor extension cord, try tying one end to a fencepost and pulling it tight to put a tension in the cord. Then, if you put a little up-blip in the cord, you can watch it travel along to the fencepost and then reflect back. You will see for yourself that the up-blip turns into a down-blip after reflection.

Finally, you may believe us and you may believe your own eyes, but if you still want to understand *why* a pulse is inverted when it reflects from a fixed boundary, here is an explanation that may help.

Consider a tight string carrying a pulse moving to the right, as shown in Figure 12.9. In figure (a), the black particle is a typical particle in the middle of the string, shown just as a pulse reaches it. The tension in the string, acting on the black particle along the direction of the gray

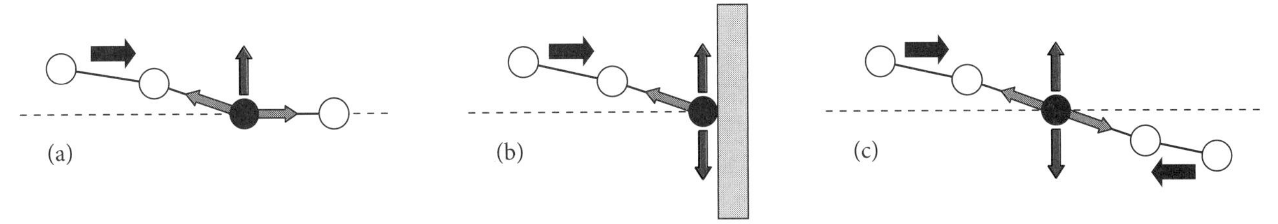

Figure 12.9 In (a), a pulse, moving to the right, arrives at the black particle and accelerates it upward in the direction of the red arrow. In (b), the black particle is rigidly tied to a wall and the wall produces a force that keeps the black particle stationary. In (c), the same force on the black particle is produced by imagining an extension to the string that carries an identical but opposite pulse moving to the left.

leftward arrow, has an upward vertical component (the red arrow). There is no downward component of tension coming from the white particle to the right of the black particle, so the black particle accelerates upward due to the net unbalanced upward force on it, and the wave continues to propagate to the right. In figure (b), the black particle is fixed to a wall. When the tension in the string to the left of the black particle creates an upward force, the wall produces a downward force that keeps the black particle at rest (the downward red arrow). However, as shown in figure (c), the same downward force that the wall produces could be provided if there were an imaginary extension of the string to the right and if there were an inverted pulse moving to the left along this imaginary part of the string. At each moment in time, the upward tension component due to the shape of the rightward-moving pulse would be exactly compensated by a downward tension component produced by the shape of the leftward-moving pulse, leaving the black particle at rest. Because the boundary no longer exists in this string with an imaginary extension, the rightward-moving pulse does not reflect. It continues moving to the right as the leftward-moving pulse travels to the left. The two pulses pass through each other, the inverted leftward-moving pulse continuing along the real string while the rightward-moving incident pulse exists only on the imaginary string, after it passes the end of the real string. This situation is shown in Figure 12.10a.

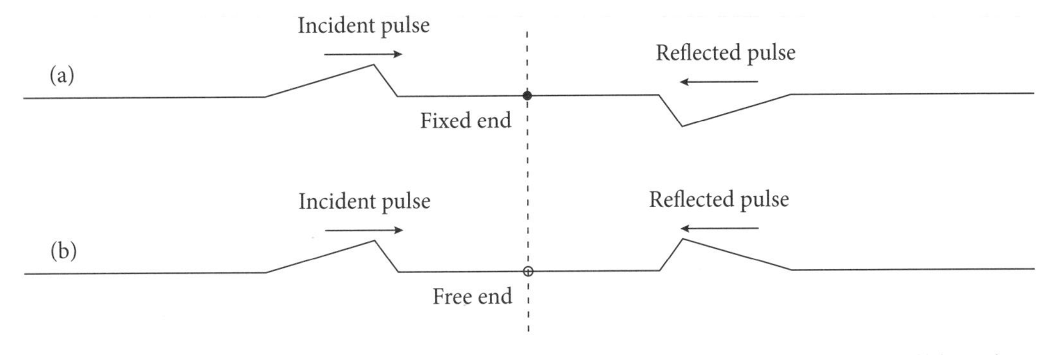

Figure 12.10 The imaginary-extension model for reflection of a pulse from a fixed (a) and free (b) boundary.

Also shown, in Figure 12.10b, is the reflection from a free end, where, as we have said, the reflected wave comes back upright relative to the incoming wave. As you can see, the imaginary-extension model with a leftward-moving upright pulse on the imaginary portion of the string predicts the correct behavior for the reflected pulse. This model also predicts the movement of the free end when the two pulses overlap. It will move twice the maximum height of the incident pulse. For an explanation of this behavior, look back at Figure 12.9a. As the black particle in the middle of the string starts to move upward in response to the upward component of the tension from the molecule on its left, the white particle on its right will begin to produce a tension force component downward. But if the black particle is on the free end of a string, there will be no restoring force from particles to its right, and the resulting displacement of the black particle will be greater than that of the rest of the particles on the string. When worked out in detail, the displacement of the end of the string turns out to be exactly twice the amplitude of the traveling pulse, as predicted by Figure 12.10b.

With the model of reflection shown in Figure 12.10, it is easy to calculate the shape of a string during the period of time when the two pulses overlap. We just add the contributions of each pulse at every point along the real string, remembering that we need not even think about the part of the string that extends off to the right of the boundary point, since it is not part of the real string.

EXAMPLE 12.2 Combining Incident and Reflected Pulses at a Boundary

A stretched spring is fastened to a vertical rod with a frictionless pulley, so that the boundary is a free end. A pulse is incident on the boundary, as shown (note that the pulse is not symmetric). The pulse travels at a speed of 10 cm/s, and each square in the figure represents 10 cm. Draw the spring's shape 4 seconds later.

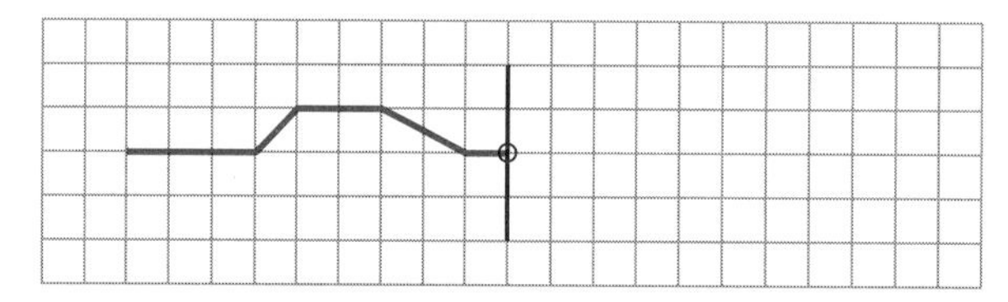

ANSWER We model the free-boundary situation by replacing the boundary with an imaginary extension that carries a second pulse, traveling in the opposite direction. It has the same shape as the initial pulse and is on the same side (up) of the spring. In the figure below, the solid lines show the initial situation at $t = 0$, the solid red line representing the incident pulse on the real spring and the solid black line representing the imaginary reflected pulse. Then dashed lines depict the red incident pulse and the black reflected pulse 4 seconds later, but the two pulses have not been added together in the region where they now overlap.

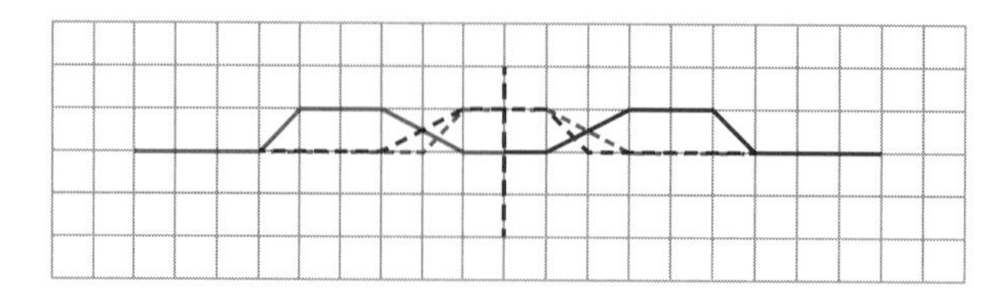

We then use the principle of superposition to add the two pulses at $t = 4$ s and find the actual shape of the spring. At this point, we put the boundary back in and only include the real spring in our diagram.

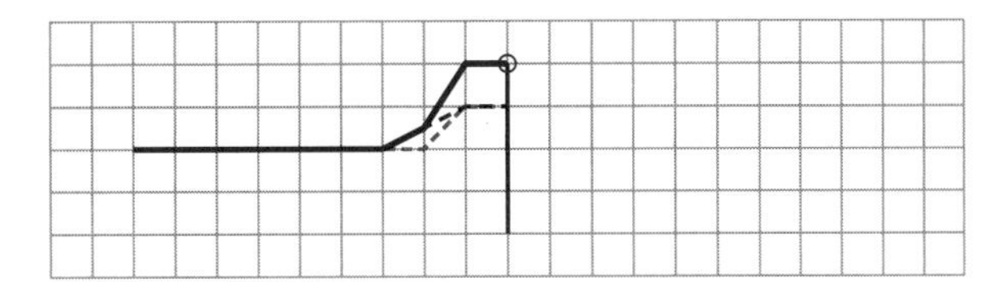

If we were to repeat this process at 1-second intervals, over a period from $t = 0$ to $t = 6$ s, drawing a red line moving to the right and the black line to the left, we could generate a series of snapshots showing the spring's shape during the entire reflection process.

12.5 Reflection of Pulses at a Boundary Between Two Media

We saw in Equation 12.1 that the speed of a wave along a string depended on two things, the tension in the string and the linear density μ. The string itself does not move left or right, so the tension must be the same throughout the string. Then let us consider what would happen if we connected two strings of different linear density and sent a pulse down the composite string, as shown in Figure 12.11. Since the speed of the pulse is $v = \sqrt{F/\mu}$, the pulse speed in the heavier section must be less than in the lighter section. So we wonder what will happen when the pulse strikes the boundary between the two media?

Figure 12.11 A composite string formed by joining two segments of different μ. The speed of the pulse in the heavier right side of the string is less than that in the left.

It probably contradicts no one's intuition to suggest that the pulse continues from the lighter faster section into the heavier slower section. There is still tension in the string, after all. But it may not be intuitively obvious that there will also be a pulse reflected from the boundary, a pulse that moves back in the opposite direction as the incident pulse. When a pulse meets a boundary between two media that have different pulse speeds, part of the pulse energy will continue forward into the new medium (the transmitted pulse) and the rest of the pulse energy will be reflected back into the original medium (the reflected pulse). Although we have introduced this idea by considering transverse waves and pulses on a string, the principle is generally true and applies equally well to ocean waves, sound waves, light waves, etc.

When an incident pulse hits a boundary between segments with different wave speeds, the transmitted pulse always maintains the same orientation as the incident pulse. If it is on a section of string with a faster pulse speed, the transmitted pulse will be wider than the incident pulse. If the transmitted pulse is on a string segment with a lower pulse speed, it is narrower than the incident pulse (though it still has the same basic shape). But the reflected pulse is more complicated. When a pulse hits the boundary from a fast medium to a slow medium, the reflected pulse is inverted. It looks as if it had reflected from a fixed end (Figure 12.8b).

However, when an incident pulse goes from a slow medium to a fast medium, the reflected pulse is not inverted. It looks as if there had been a reflection off a free end (Figure 12.8c). In fact, we can think of our fixed-end and free-end conditions from the last section as special cases of reflection off of an interface between two media. A fixed boundary is like going into a medium with infinite μ and a free boundary is like interfacing with a medium of zero μ.

Please note that we have not explained how to figure out the relative amplitudes of the transmitted and reflected waves, since that formula is fairly complicated to derive. But we can still discuss the shapes of the pulses, as in the next example.

EXAMPLE 12.3 Analyzing Pulses from a Boundary in a Wave Medium

The diagram below shows pulses on two coil springs joined together, but we don't know which way the pulses are moving or which side has the faster pulse speed. We know that a single pulse was created on one side traveling toward the junction between the springs, but we don't know on which side of the spring the original pulse was created. Because we see two pulses, we know that the picture shows the springs *after* the pulse has hit the boundary between the two. Determine whether the pulse came in from the right side or left, and draw the shape of the springs (one side will be flat) before the incident pulse hit the boundary.

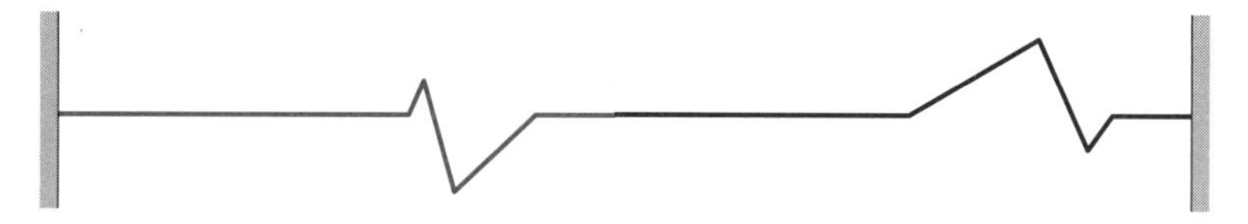

ANSWER When a pulse travels from a slow (heavy) spring to a fast (light) spring, the boundary acts like a free end, and the reflected pulse comes back on the same side of the spring as the incident pulse. The transmitted pulse, of course, is always on the same side as the incident pulse. So, when a pulse goes from a slow medium to a fast medium, the reflected and transmitted pulses end up on the same side of the spring. But this is not what we observe here. When a pulse travels from a fast (light) spring to a slow (heavy) spring, the boundary acts like a fixed end. This causes the reflected pulse to be inverted, so the reflected and transmitted pulses end up on opposite sides of the spring, as we observe in this example. We therefore know that the incident pulse had to start on the side with the greater wave speed, but we don't yet know which side this is. However, we do know that the two pulses we observe in the "after" picture started at the boundary at the same time, so the one on the right must be moving on the faster segment of the spring. This is confirmed by the fact that the pulse on the right is wider, and the wider pulse will be on the faster spring. We therefore conclude that the original pulse was on the right-hand spring, moving to the left. To draw the incident pulse, we need to make sure the leading part of the pulse is the smaller of the two humps, since that is the case for both pulses in the figure above. The initial pulse also needs to be inverted compared to the reflected pulse. This gives a picture that looks like this.

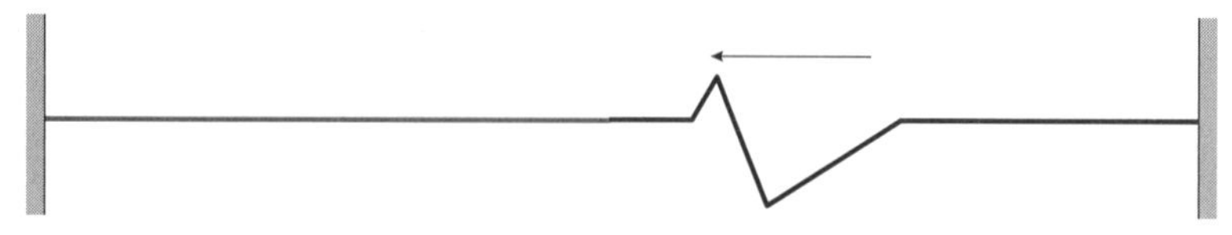

Note that the width of the pulse is the same for the incident pulse as it was for the reflected pulse, because both pulses are in the same medium. It is only in the slower medium on the left that the pulse width is affected; in this case it is narrower.

12.6 Properties of Continuous Waves

In Section 12.1 we defined the term "continuous wave" to apply to a case where each point of the medium moves up and down or back and forth in periodic motion, repeating itself multiple times. As we saw in Chapter 11, we can specify the period of such motion in seconds, or we can give its inverse, the frequency, in hertz. The period of oscillation of each point of the wave medium will be the time it takes one full cycle of the wave to pass a given point. Let's see why this is.

Because a wave moves at uniform speed along a medium, there is a similarity between the shape of the wave—its physical appearance in a snapshot taken at a single instant in time—and the shape of a graph of the periodic motion of any given point in the medium as a function of time. In Figure 12.12, we show snapshots of two waves propagating along a spring and compare the shapes with the graphs of the motion of the flagged point on the spring. If you picture what the flag will do in each case, as the wave passes along, you should see your expectations verified in the two graphs of displacement versus time on the right side of the figure.

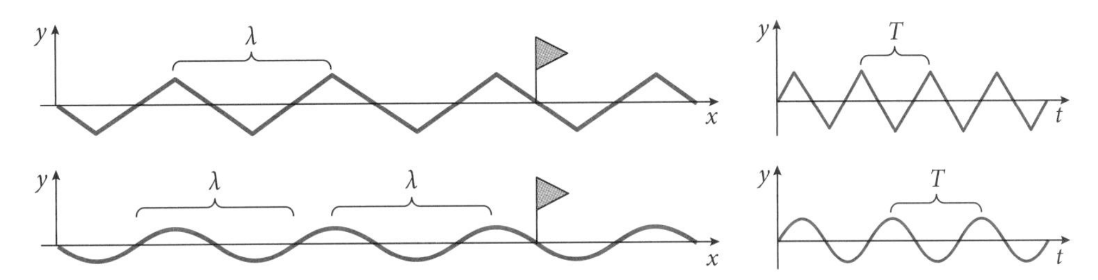

Figure 12.12 The instantaneous shape of a triangular wave (top left). As the wave moves to the right, the flagged point will rise and fall, as shown in the graph of the displacement vs. time, shown on the top right. On the lower left is a sinusoidal (sine or cosine) wave, with its graph shown on the lower right. The wavelengths are shown in the snapshots and the periods are shown in the graphs.

In each of the snapshots on the left side of Figure 12.12, we identify the distance between repetitions of the wave shape (from one peak to the next, from one rising point of zero displacement to the next, etc.). This distance is called the *wavelength* and is traditionally designated with a Greek letter λ (lambda). We also show the familiar period T in the graphs of the periodic motions of the flagged point in the spring. Note that the general shapes of the snapshots and the graphs are the same, but that there was no attempt made to make the widths of the oscillations the same. The wavelengths and the periods are in different units (distance for the wavelength, time for the periods), and we can use whatever scale we want for the time axes in the graphs.

There is a simple relationship between the wavelength of a wave, the period of the associated periodic motion, and the speed of the wave. If you consider the motion of the flag in each case shown in Figure 12.12, it should be obvious that the period of the flag's motion will be the time it takes the wave to move one complete wavelength. If the wave is moving at speed v, then the time it takes to cover the distance λ is

$$T = \frac{\lambda}{v}. \tag{12.2}$$

If we combine this with the familiar $f = 1/T$, we get a very useful equation that is often referred to as "the wave equation",[2]

2. There is another equation in physics that is commonly called "the wave equation." It is a partial differential equation that must be satisfied by the displacement ψ as a wave travels through a medium with wave speed v. It is written $v^2\nabla^2\psi - (\partial^2\psi/\partial t^2) = 0$, but we thought we wouldn't bother you with this one.

$$v = \lambda f. \quad (12.3)$$

Finally, let us point out that, while Figure 12.12 showed transverse waves on a spring, the ideas could equally well apply to longitudinal waves on a spring, or even to longitudinal waves in a fluid like air. To highlight the similarity, we show two waves in Figure 12.13, a transverse wave on a string and a longitudinal sound wave in air. In each case, we graph the displacement of a segment of the medium (of the spring or of the air) from its equilibrium position. Note that the graphs shown along the bottom of Figure 12.13 are not graphs of displacement versus time, like the graphs on the right side of Figure 12.12, but are graphs of displacement versus distance in the medium, all taken at a single point in time. In the transverse wave, the displacement is in the y-direction, while, in the longitudinal wave, the displacement is in the same x-direction in which the wave is moving. But the graphs for the two cases look alike.

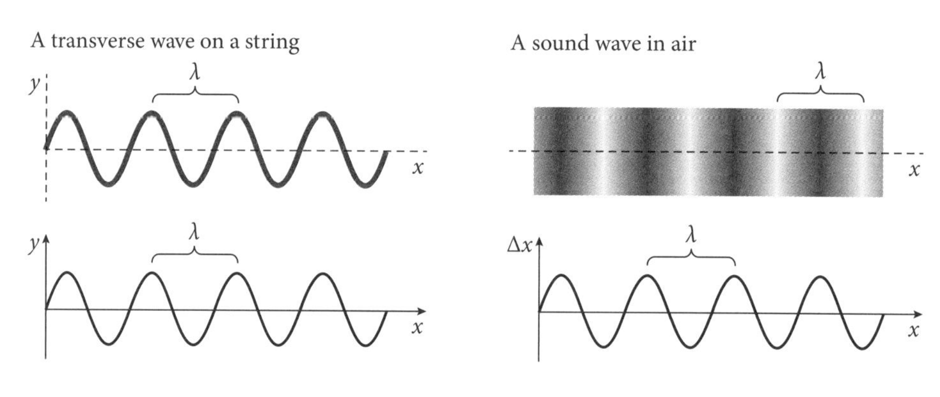

Figure 12.13 Displacement graphs for transverse and longitudinal waves look alike. The dark areas in the sound wave picture represent high density and the light areas represent low density.

EXAMPLE 12.4 Travels of the Yarn on the Spring

A wave with wavelength 2½ m is traveling along a stretched spring at a speed of 50 m/s. A piece of yarn tied to the spring moves over a total distance of 10 meters in a 5-second interval. What is the amplitude A of the wave?

ANSWER We begin by making sure we understand the statement, "... moves over a total distance of 10 meters." First, the yarn moves only in the direction perpendicular to the direction in which the wave is moving, not along the wave. Second, the displacement of the piece of yarn is exactly zero after each complete period. However, each time a wavelength goes by, the yarn first moves a distance A away from equilibrium, then moves a distance A back to equilibrium, for a total distance of $2A$, so far. But there is still the second half of the wave to go. When the rest of the wave passes the yarn, it will move it an additional $2A$ for a total distance traveled of $4A$ as one wavelength went by.

If we had been told that one wavelength passed in the 5-second interval, then we would say that the total distance traveled was 10 m = $4A$, giving A = 2½ m. But the problem does not say that. In fact, the question of how many wavelengths pass the yarn in the 5 seconds is one we first need to answer. We remember that the wave equation, $v = f/\lambda$, relates speed, wavelength, and frequency. We were given the wavelength and the speed, so we can calculate $f = v/\lambda$ = (50 ms)/(2½ m) = 20 Hz. In one second, 20 wavelengths will pass the piece of yarn, and, in 5 seconds, 100 wavelengths will pass. If 100 complete oscillations move the yarn a total of 10 m, 1 oscillation must move it 10 cm. We know that one wavelength causes a point on the spring to move 4 times the amplitude of the wave, so the amplitude must be 2½ cm.

12.7 Sound

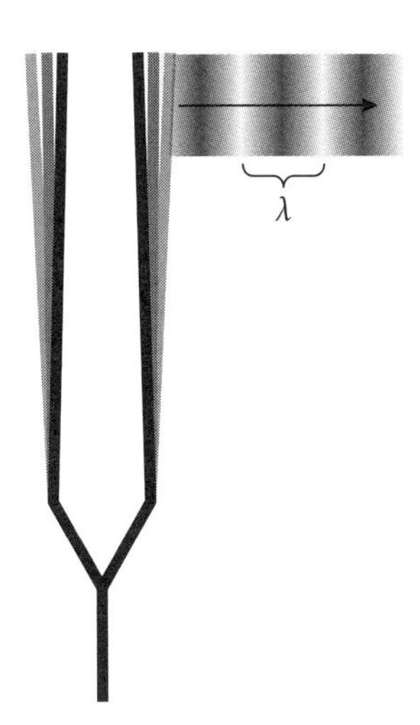

Figure 12.14 The longitudinal sound wave emitted by a tuning fork. Another similar wave also travels to the left.

Imagine how the single right-side tine of a tuning fork creates a sound wave (Figure 12.14). When the tine moves to the right, it pushes the air molecules together, causing the air to compress. The increased density of the air molecules in that region causes an increased pressure in the gas. This presses on the molecules to the right of that region, causing them to compress in turn. As the compression moves to the right, the tine that caused it moves back to the left, leaving a region where the molecules are now less densely packed than in undisturbed air. This is called a rarefaction. As the compression moves to the right, the rarefaction trails right behind, creating a sound wave. On average, the displacement of individual molecules is zero, but energy is carried from the tuning fork out into the room. The wavelength of this longitudinal wave is the distance between adjacent compressions (or rarefactions).

Eventually, that train of compressions and rarefactions may reach your ear, and your eardrum will move in response to these varying pressures. The frequency of the tuning fork determines the frequency at which your eardrum will vibrate, and this frequency is converted (in a complicated process in the inner ear) into the sensation of pitch. The high notes you hear are sound waves whose frequencies are higher than the low notes you hear.

Sound is a pressure wave, but it is important to remember that the changes in pressure required for the sound to be heard are tiny. For sound waves you normally hear around you, the change in pressure is measured in hundredths of a pascal, whereas the ambient atmospheric pressure is 100,000 Pa. So the pressure change in typical sound waves is only a few parts in ten million. Nevertheless, it is the size of the pressure change, the amplitude of pressure variation, that determines how loud we perceive a sound to be.

The frequency of the vibrating source creating the sound wave determines the pitch of the sound, and the amplitude of the vibration of the source determines the amplitude of the pressure changes, and thus the loudness of the wave. And, as with all waves, Equation 12.3 relates the frequency and the wavelength to the speed of the waves. Wave speed is a property of the medium and is not influenced by the wave's amplitude. In air (at 20° C, to be precise), this speed is about 343 m/s. The speed does depend on temperature, however, rising by about 0.6 m/s for every Celsius degree of temperature increase. But, once we know the speed of a wave, the wavelength may be determined from the frequency, and vice versa. For example, a 343-Hz note sung by an alto, when the air is at 20° C, creates a sound wave of wavelength 1 m.

One last aspect of our perception of sound that we should mention is what is called the "quality" of the sound. The difference between a clarinet playing a B-flat and a violin playing the same note is a difference of quality. Our ear responds most simply to a sinusoidal wave of the shape graphed in Figure 12.13; we perceive this as a "pure" tone at a particular frequency or pitch. If the sound wave has a different shape, even at the same frequency, or if there are other frequencies being carried through the medium from the same source, we perceive the sound as having a different quality. This is one of the ways we tell one musical instrument from another.

12.8 Beats

Let's consider another case of what happens when different tones are heard at the same time. Suppose that an oboe is playing one note and a violin right next to it is playing a slightly higher note. It turns out that the combination of the two sounds produces an interesting phenomenon. First, a listener will hear a note with a pitch that is an average of the two frequencies. Second, and perhaps surprisingly, the note will no longer seem to have a constant amplitude, even though the two sources are actually completely steady in amplitude. Instead, the sound we hear goes through rapid loud-soft-loud cycles called *beats*. In fact, if you wanted to adjust your violin to match the oboe's pitch, you would listen for the beats and then loosen the violin string until the beats went away.

Figure 12.15 shows how the beats arise. Here, a 10-Hz sound wave and a 12-Hz sound wave (black lines) are superimposed at a detector (the red line). At the time of the first dashed vertical line, the amplitudes are adding destructively, creating a signal of very small amplitude.

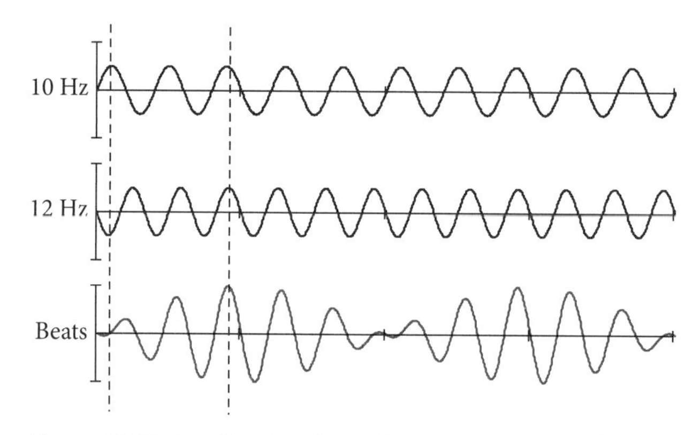

Figure 12.15 Amplitudes of sound waves versus time, showing how beats arise. A 10-Hz sound is superimposed on a 12-Hz sound. The result is a sound at 11 Hz with an amplitude that varies twice per second.

However, by the time of the second dashed line, the two waves are interfering constructively, producing a wave with twice the amplitude of either individual wave. The result is a wave whose amplitude rises and falls.

The "beat frequency" is the number of times per second that a combined sound goes through its loud-soft-loud cycles. The formula for the beat frequency between two tones of frequency f_1 and f_2 is

$$f_{\text{beat}} = |f_1 - f_2|, \tag{12.4}$$

the absolute value signs || being needed because it doesn't matter which frequency is larger, the number of beats is the same (and a negative frequency would be nonsense anyway).

To make sure we understand this, let's suppose that our oboe is playing 440 Hz and that the violin is playing 442 Hz. Then the loudness of the sound we hear will fluctuate twice per second. However, if the violin were to play 438 Hz, instead of 442 Hz, the beat note would be the same. So you can't tell from the beat note whether your violin is playing too low or too high. However, if you slowly increase the pitch of the violin's note from 438 Hz, the beat frequency will get smaller and smaller, making the note sound steadier and steadier, until the two frequencies match. At this point, the beat frequency will go to zero and the loud-soft-loud cycles will disappear. It turns out, by the way, that if a beat frequency gets much greater than 8 or 10 Hz, your ear can no longer hear it as loud-soft-loud cycles of a single average frequency, but you hear the sound as two separate tones. Therefore, the beat phenomenon applied to our ears is limited to frequencies that are close but not identical.

EXAMPLE 12.5 Beats Between Tuning Forks

You play an unknown tuning fork alongside a 440-Hz tuning fork. The resulting sound has a beat frequency of 3 Hz. You then play the unknown alongside a 445-Hz tuning fork and hear a beat frequency of 8 Hz. What is the frequency of the unknown tuning fork?

ANSWER It is tempting to try writing down the beat frequency equation and solving the problem algebraically. But the absolute value signs in the equation actually make the algebra quite difficult. Instead, we just reason through a solution based on the idea that the beat frequency is the difference in the two source frequencies. From the first experiment we know that the unknown tuning fork must be different from 440 Hz by 3 Hz. The choices are 437 Hz and 443 Hz, but we do not yet know which choice is the correct one. The second experiment tells us that the unknown tuning fork must be different from 445 Hz by 8 Hz. The choices are 437 Hz and 453 Hz. The choice that is consistent with both experiments is 437 Hz.

12.9 The Doppler Effect

When you are standing at the side of the road and hear an ambulance go by, you hear an unmistakable change in the pitch of the siren as the ambulance passes. If you were to listen to the same siren with the ambulance stopped at a light, you would find that the approaching siren had been sounding a slightly higher pitch than it did when at rest and that the receding siren had sounded a lower pitch than the siren at rest. When sources are moving, the listener will hear a frequency that is different from the frequency of the source, the magnitude and sign of the difference depending on the velocity of the source. This is called the *Doppler effect*.

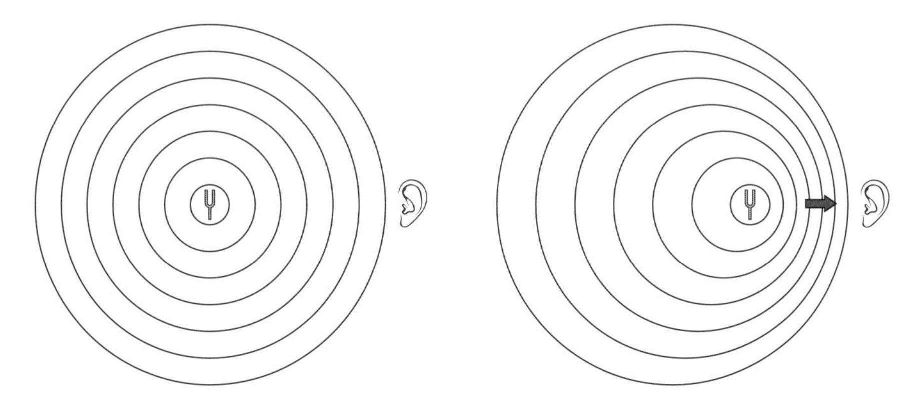

Figure 12.16 Compression sound wave peaks from two tuning forks, the one at the left being stationary relative to the observer and the one on the right moving toward the observer. For the moving tuning fork, the wavelength is shorter in the direction of its velocity and longer in the direction opposite to the tuning fork's velocity.

Figure 12.16 shows two sources, two tuning forks, each emitting sound waves at the same frequency. The tuning fork on the left is stationary while the one on the right is moving to the right. The circles that you see represent the expanding pressure maxima. Once a source creates a pressure front, that front expands at the speed of sound in all directions, forming a sphere around the origin. If the source is stationary, then all of the pressure fronts will be centered on the same point, as shown in the picture on the left. But, if the source is moving, each subsequent sphere is centered on a different point. When the source is stationary (the picture on the left), listeners on all sides of the source intercept the wavefronts at the same rate, equal to the frequency of the source. But for the moving source (the picture on the right), the situation is different. The wavefronts in front of the source bunch up, creating waves of a shorter wavelength. The speed of the waves in the air has not changed, so the shorter wavelength means that these wavefronts must reach the observer at a higher rate than the rate at which they were emitted. We say that the frequency is Doppler-shifted upward. Observers behind the moving source hear a sound that is Doppler-shifted down. The size of the effect depends on the speed of the source—the faster the source is moving, the greater the shift in frequency.

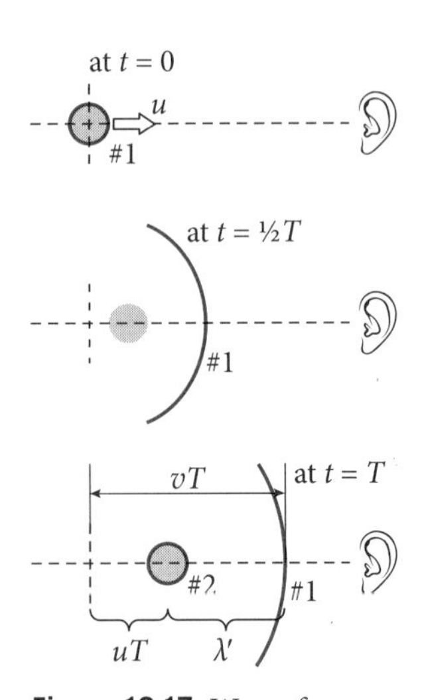

Figure 12.17 Waves from a moving source. The $t = 0$ picture shows a wave (the red circle around the gray source) just being emitted. The $t = ½T$ picture shows that first wave heading toward the observer. The $t = T$ picture shows the next wave being emitted by the source.

The formula for the Doppler shift of a moving source is actually easy to work out using simple kinematics. Let us consider a situation like that shown in Figure 12.17, in which a source of sound waves with period T (the gray ball) is moving to the right at velocity u, producing sound waves that move through the air at a speed v. We orient our coordinates so that velocities to the right are positive. There are three pictures shown in the figure. The first, at $t = 0$, shows a red wave crest, a pressure maximum, just as it leaves the source to expand away in all directions. The second picture, at $t = ½\,T$, shows the wave crest traveling toward the observer. The third picture, at $t = T$, represents the situation after one complete period has elapsed and the next wave crest is just leaving the source. In this picture, the source has moved a distance uT from the origin where it started. We remember that the wavelength is *defined* to be the distance between one pressure maximum and the next. Therefore, because of the motion of the source, this distance is not the usual $\lambda = vT$, but it has been reduced to

$$\lambda' = vT - uT, \tag{12.5}$$

as shown in the figure. If the source had not been moving, the wavelength observed would have been $\lambda = vT$. If we use this relation to replace T in Equation 12.5, we get

$$\lambda' = v\frac{\lambda}{v} - u\frac{\lambda}{v} = \lambda - \frac{u}{v}\lambda.$$

Thus, the fractional shift in the wavelength of a wave from a moving source is

$$\frac{\lambda' - \lambda}{\lambda} = -\frac{u}{v}. \tag{12.6}$$

If u is positive, the source moving toward the observer, the change will be negative (the observed wavelength is shorter), and if u is negative (the source moving away from the observer), the wavelength is longer.

Let's take a look at a couple of examples.

EXAMPLE 12.6 Doppler Shift in a Slowing Ambulance Siren

An ambulance, with siren blaring at a single frequency, is driving toward the scene of a minor accident. The ambulance slows and stops at the scene. While the ambulance is slowing down, is the siren pitch that the accident victim hears increasing or decreasing?

ANSWER There are two important points we must remember to solve this problem: 1) when a source is approaching, the sound you hear is Doppler-shifted up from the rest frequency; and 2) the amount that the frequency is shifted depends on the speed. If the speed of the approaching source drops, then the frequency becomes less Doppler-shifted upward, so it actually decreases. It is important to distinguish between the property itself (the frequency is increased) from how that property is changing (the increase is less, so the frequency is actually decreasing).

EXAMPLE 12.7 The Doppler Shift of Frequency

A bullet train, moving at 43 m/s (about 96 mi/hr), sounds its 300-Hz whistle. If you are sitting in a car at the crossing, find the whistle frequency you will hear when the train is coming toward you and the frequency you will hear as it is going away.

ANSWER Equation 12.6 gives the change in the wavelength of a wave from a moving source, but we are being asked about the frequency. We could easily use the information we have to find wavelengths and then find the associated frequency. But, instead, let's begin with Equation 12.5 and find a new formula for the frequency that will be observed from a moving source. This way, we will have a new formula we can use if we ever need it. First, we write Equation 12.5 in terms of the frequency emitted by the source (remembering that $T = 1/f$),

$$\lambda' = \frac{v}{f} - \frac{u}{f}.$$

We then recognize that the frequency that will actually be observed will be the rate at which these bunched-up waves hit the ear of the observer. Since the waves still move at speed v, this relationship is $f' = v/\lambda'$. Making this substitution for λ', we get

$$\frac{v}{f'} = \frac{v}{f} - \frac{u}{f}.$$

A little algebra (well . . . a moderate amount of algebra) turns this into

$$\frac{f' - f}{f} = \frac{u}{v - u}.$$

We take u to be positive when the train is coming toward us, and we consider only the case where the speed of the train is less than the speed of sound. Then the frequency change is positive, indicating a higher frequency (as we expected).

With the numbers we have in this question, the shift in the frequency is thus

$$f' - f = \frac{uf}{v - u} = \frac{(43 \text{ m/s})(300 \text{ Hz})}{343 \text{ m/s} - 43 \text{ m/s}} = \frac{(43 \text{ m/s})(300 \text{ Hz})}{300 \text{ m/s}} = 43 \text{ Hz}.$$

The frequency will be 300 + 43 = 343 Hz when the train is approaching.

If the train is going away, the velocity will change to −43 m/s. This will not only change the sign of $f' - f$ in the formula we have derived, but it will also change its magnitude, since there will be a new value for the denominator in the formula. In this case, the formula gives

$$f' - f = \frac{uf}{v - u} = \frac{(-43 \text{ m/s})(300 \text{ Hz})}{343 \text{ m/s} + 43 \text{ m/s}} = \frac{(-43 \text{ m/s})(300 \text{ Hz})}{386 \text{ m/s}} = -33 \text{ Hz}.$$

The frequency will be 300 − 33 = 267 Hz when the train is going away.

12.10 Summary

From this chapter, you should remember the following things:

- We use the term "pulse" for a single non-repetitive disturbance in a medium, and we use the term "continuous wave" for a periodic disturbance moving through a medium. We use the term "wave" for anything that transfers energy without transferring mass, including pulses and continuous waves.
- The speed of a transverse wave on a stretched string is given by $v = \sqrt{F/\mu}$, where F is the tension in the string and μ is the linear mass density. For a segment of the string of mass m and length L, we have $\mu = m/L$.
- You should be comfortable with the principle of superposition for a wave or a pulse, remembering that the total disturbance when two waves are at the same point in a medium is the sum of the disturbances of the two individual waves.
- You need to be familiar with what happens when a wave or a pulse hits a boundary, including a boundary between two wave carrying media, and be sure to know how to figure out the sense of the wave reflected from the boundary.
- You should be able to use Equations 12.2 and 12.3 to relate period T, frequency f, wavelength λ, and wave speed v.
- You should understand where the phenomenon of beats comes from and be able to predict the beat frequency of two close frequencies using Equation 12.4.
- The Doppler effect is an important concept, and is derived from kinematics and the wave equation. You should understand and be able to use Equation 12.6 to find the observed wavelength of a wave emitted by a moving source. You should also feel comfortable with the kind of procedure we went through in Example 12.7 to find the frequency.

CHAPTER FORMULAS

The speed of a transverse wave or pulse on a string: $v = \sqrt{\dfrac{F}{\mu}}$

Relationships between wave quantities: $f = \dfrac{1}{T}$ and $v = \lambda f$

Beats: $f_{beat} = |f_1 - f_2|$

Doppler shift: $\dfrac{\lambda' - \lambda}{\lambda} = -\dfrac{u}{v}$ or $\dfrac{f' - f}{f} = \dfrac{u}{v - u}$

PROBLEMS

12.2 * 1. A long coil spring is under a tension of 50 N. The spring has a mass per unit length $\mu = 0.075$ kg/m. What is the speed of a pulse on the spring?

* 2. The C string on a piano has a mass per unit length of 8.7×10^{-3} kg/m. Under what tension must a 3.0-m length of this string be placed so that transverse waves will travel at a speed of 330 m/s?

* 3. A steel wire of mass 0.40 kg and length 0.64 m is stretched between two pegs and the wire is tightened until the tension in the wire is 1000 N. The wire is struck exactly at its midpoint, causing a small displacement. How long does it take the displacement to reach each end of the string?

** 4. A 23-kg crate is suspended from the ceiling by a 3-m piece of piano wire with a linear mass density of 3.31×10^{-4} kg/m. A pulse is sent up the wire toward the ceiling. The time required for the pulse to reach the ceiling is carefully measured at 3.64×10^{-3} s, which gives a velocity for the pulse of 825 m/s. If this apparatus could be taken to the moon, would you expect the time required for the pulse to travel from the crate to the ceiling to increase, decrease, or stay the same? Justify your answer.

** 5. A weight is hung over a pulley and attached to a string composed of two parts, each made of the same material but one having 4 times the mass per unit length of the other. The string is plucked so that a pulse moves along it at speed v_1 in the heavier part and at speed v_2 in the lighter part. What is the ratio v_1/v_2?

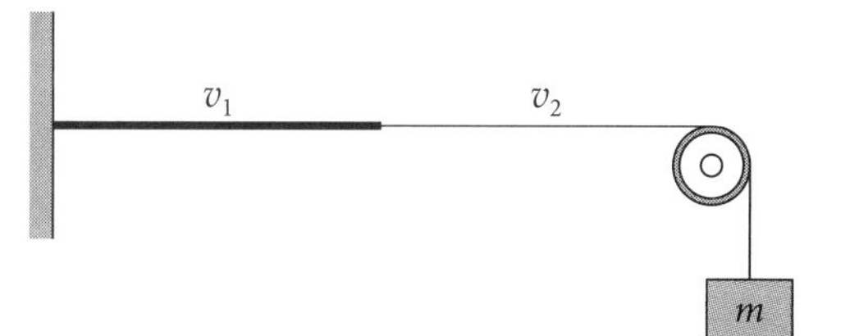

12.3 * 6. Two triangular pulses are traveling toward each other along a string, as shown in the picture below. Each pulse is 1 cm wide and 1 cm high.

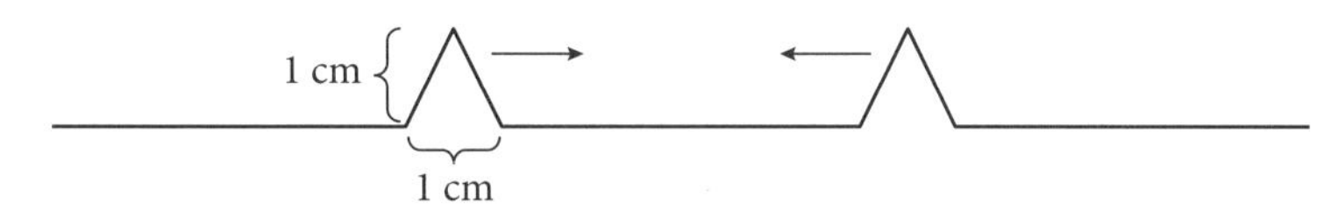

a) What is the maximum height of the string when the two pulses merge?
b) What will be the width of the pulse at this time?

** 7. Two pulses are traveling in opposite directions along a stretched spring, as shown in the diagram below. The speed of the transverse waves in the coil spring is 100 cm/s. The rightward-moving pulse is a square pulse 4 cm high and 20 cm wide. The leftward moving pulse is a triangular pulse 8 cm high and 10 cm wide. At $t = 0$, the leading edges of the two pulses are 20 cm apart, as shown. Draw four sketches, showing the shapes of the spring at times:

a) $t = 0.1$ s b) $t = 0.2$ s c) $t = 0.3$ s d) $t = 0.4$ s

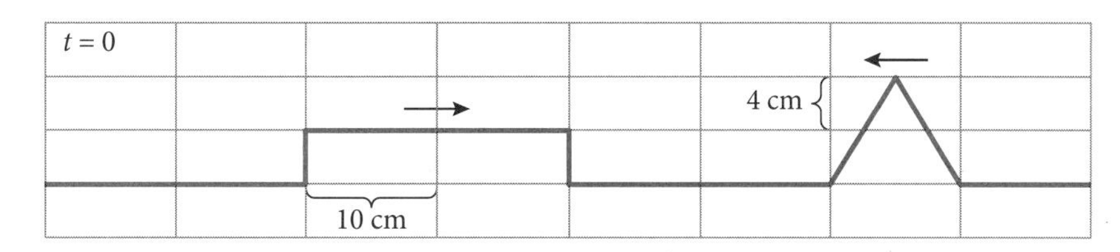

** **8.** Two pulses are traveling in opposite directions along a stretched spring, as shown in the diagram below. The speed of the transverse waves in the coil spring is 100 cm/s. The rightward-moving pulse is a semicircular pulse 5 cm high and 10 cm wide. The leftward-moving pulse is a triangular pulse 5 cm high and 10 cm wide. At $t = 0$, the leading edges of the two pulses are 20 cm apart, as shown. Draw five sketches, showing the shapes of the spring at times:

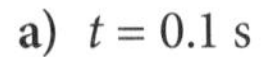

a) $t = 0.1$ s **b)** $t = 0.15$ s **c)** $t = 0.2$ s **d)** $t = 0.25$ s **e)** $t = 0.3$ s

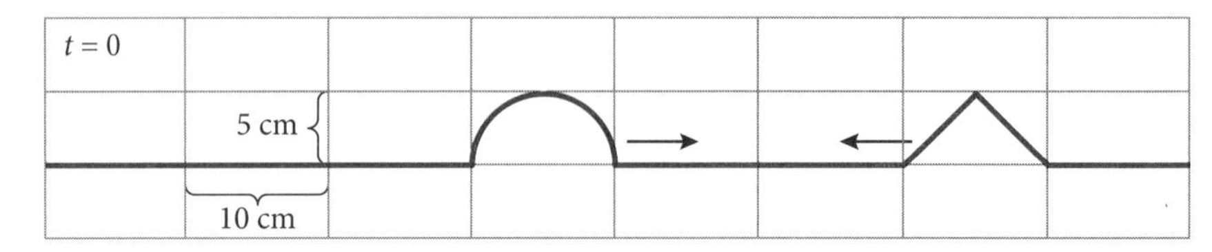

12.5 ** **9.** The diagram below is a snapshot of two connected springs held at their ends by two students. One of the students created a pulse. The snapshot is taken some time *after* the pulse has reached the junction between the springs. One spring is steel; the other is brass, and the linear density is greater for steel than for brass. [Note: The shades of gray of the springs in the diagram are not meant to convey any information.]

a) Determine which is the brass spring and which is the steel spring. Explain your reasoning.

b) Determine which pulse is the reflected pulse and which is the transmitted pulse. Explain your reasoning

c) Sketch a diagram that shows the shape and direction of motion of the incident pulse.

** **10.** Two springs of different mass densities are fastened end-to-end. The spring on the left has the greater mass density. Two identical pulses are initiated 10 cm apart on the far left and are traveling toward the boundary as shown. Consider the situation *after* both pulses have passed the boundary and entered the new section.

a) In which spring, left or right, do the pulses travel faster?

b) Will the separation between the transmitted pulses on the right spring be less than, equal to, or greater than 10 cm? Explain your reasoning.

c) Will the reflected pulses (left spring) be on the top or the bottom of the spring? Explain.

** **11.** A long string is pulled into the shape shown below, a triangle of width 10 cm and height 0.8 cm. The string is maintained quiescent for a moment, with nothing moving, and the string is suddenly released.

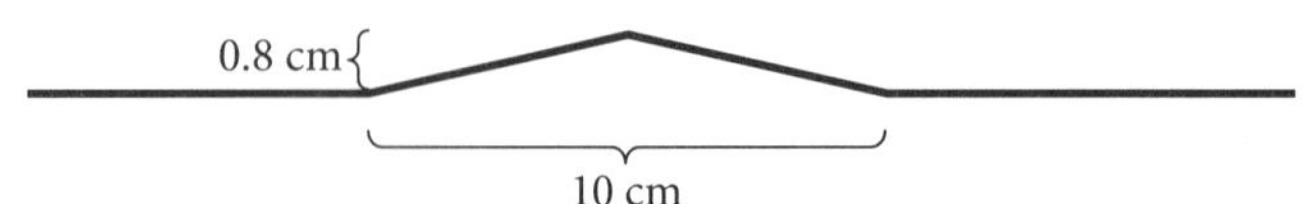

a) Explain why the string will end up with two pulses on it, one moving to the right and one to the left.

b) What will be the width and height of each pulse?

12.7 * 12. What are the wavelengths of sound waves of frequency 262 Hz (middle C), 1000 Hz, and 15,000 Hz?

* 13. The human ear can hear sound waves whose frequencies are between 20 Hz and 20,000 Hz. What is the shortest *wavelength* of sound that a human ear can hear? What is the longest?

* 14. An FM radio signal has a frequency of 90.9 MHz (90.9×10^6 Hz) and a wavelength of 3.30 meters. What is the speed of the radio wave?

* 15. You stand on a cliff and note that waves are coming directly into the cliff and are hitting the bottom of the cliff every 8 s. Looking a little ways out, you compare the distance between wave crests with a standard 12-meter racing-class sailboat and determine that the wavelength of the incoming waves is 18 m. What is the speed of the water waves on the ocean off the cliffs?

** 16. You are standing at an ocean beach knee-deep in the water watching the waves rolling in. From a snapshot taken by your friend, you can estimate the distance between wave crests to be 3 m and the wave amplitude to be 10 cm. As you watch the water move up and down your legs, you see that the point where the water touches your legs has traveled a total distance of 1.2 m in a time of 10 s.

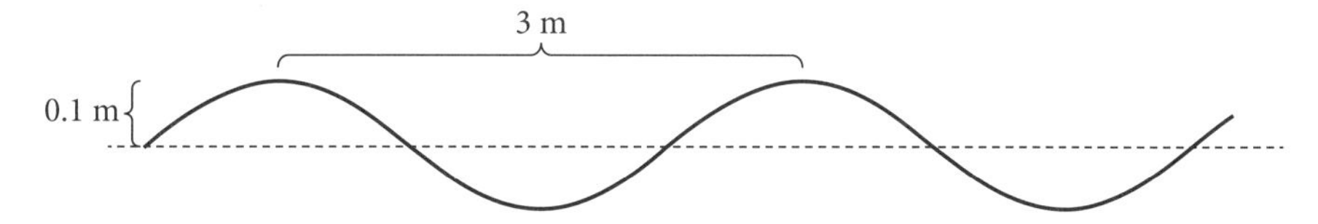

a) What is the difference between the lowest point on your legs where the water hits and the highest as one wavelength rolls by? Explain.

b) How many wave crests arrive and pass you during those 10 s?

c) What is the speed of the waves?

* 17. A periodic wave is sent down a stretched spring in the direction shown by the arrow. The sketch shows a snapshot of the pulse at a certain instant in time. Points A and B are points on the string. Which of the following statements correctly describes how points A and B are moving at the moment shown? Explain your choice.

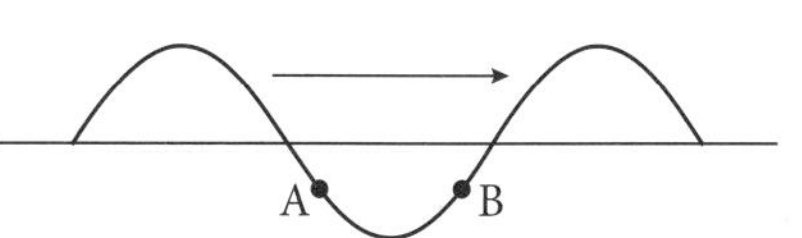

a) A is moving up and B is moving down.

b) A is moving down and B is moving up.

c) A and B are both moving down.

d) A and B are both moving up.

e) A is moving left and B is moving right.

** 18. There is a limit to how short a musical note an orchestra can play because of the difference in the time it takes the sound from the front of the orchestra to get to the listener, compared to the time it takes sound from the back of the orchestra to arrive. If an orchestra is seated on a stage that is 12 meters from front to back, and if the entire orchestra plays a very short note on a visible signal from the conductor, how long must that note be extended in time due to the physical size of the orchestra? (The speed of sound is given on page 239.)

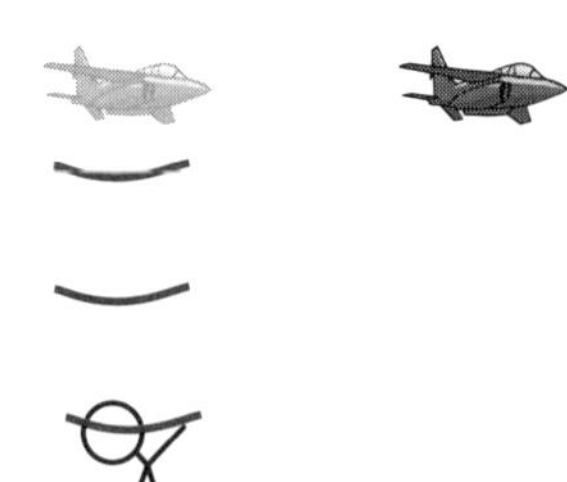

** 19. You look directly overhead and see a plane exactly 1 km above the ground flying faster than the speed of sound. By the time you hear the sound from the plane, the plane has traveled a horizontal distance of 1.4 km. Find the speed of the plane in units of meters per second. (The speed of sound is $v_s = 340$ m/s.)

***20. **Energy in Waves.** As a sine-shaped wave moves along a stretched spring, each coil of the spring will execute simple harmonic motion, as depicted in the bottom two graphs in Figure 12.12. Suppose that a spring with linear density $\mu = 0.40$ kg/m has a tension $T = 1.6$ N in it, and that a sine-shaped wave of amplitude 10 cm and wavelength 1.5 m is moving along the spring. Consider a 1-cm segment of the spring, like the one carrying the flag in Figure 12.12.

a) What is the mass of the 1-cm segment of the spring?

b) What is the speed of the wave along the spring?

c) What is the period of the simple harmonic motion executed by this segment of the spring?

d) Knowing the mass of the segment and the period of its simple harmonic motion, find the effective spring constant of the coil spring for transverse displacements away from its equilibrium. [Hint: See Equation 11.6.]

e) What is the total energy of the simple harmonic oscillator of the segment?

f) There are 150 1-cm segments within the 1.5-meter wavelength. What will be the total harmonic oscillation energy in one wavelength of the wave?

g) At what rate in joules/second (or with what power in watts) will this energy be transferred past a given point on the spring?

12.8 * 21. Aaron, the piano tuner, hits a key on the piano and strikes a 220-Hz tuning fork at the same time. He hears a 2-Hz beat note. As he tightens the piano string, the beat frequency increases. What was the original frequency played by the piano string?

** 22. Two identical tuning forks vibrate at 587 Hz. Small pieces of clay are placed on the tines of one of the forks, after which 8 beats per second are heard. What is the period of the tuning fork that holds the clay? Explain how you determined your answer.

** 23. A guitar string produces 4 beats/second when played along with a 250-Hz tuning fork and 9 beats/second when played along with a 255-Hz tuning fork. What is the vibrational frequency of the guitar string?

** 24. Yuki plays the A string on her violin at 440 Hz. When Nihel simultaneously plays his A string, he hears 2.2 beats per second. He can tell that his note is lower. By what percent should he increase the tension in his string so that the two notes sound the same?

12.9 * 25. If an electronic oscillator produces a sound of wavelength 0.60 meters, what will be the wavelength produced when that oscillator is placed on a lab cart moving away from you at 17 m/s? Assume that the speed of sound in the lab is 340 m/s.

* 26. You and your friend are standing at opposite ends of a city block. A police car with its siren blaring is driving down the street toward you and away from your friend.

a) Is the frequency that you hear higher than, lower than, or the same as the frequency your friend hears? Explain.

b) The car now passes you and continues down the street. Is the frequency that you hear higher than, lower than, or the same as the one your friend hears? Explain.

* 27. Bill and Elaine are holding portable beepers that sound at slightly different frequencies. When they are both standing still, the two beepers produce a beat frequency of 4 hertz. When Elaine begins to run away from Bill, the beat frequency increases. Whose beeper was at the higher initial frequency? Explain.

** 28. An echocardiogram can measure the Doppler shift of the sound reflected from flowing blood. A sound wave of 5 MHz is used to image the heart and a shift of 100 Hz is detected from a region of flowing blood. What is the blood's speed? [Note: The speed of sound in heart tissue is about 1500 m/s.]

** 29. In order to measure the speed of her turtle, Nancy places an oscillator whose frequency is 1000.00 Hz on the turtle's back and an identical oscillator on the ground. When the turtle is coming directly toward her, she hears the loudness of the combined sound go through a loud-soft-loud variation that takes 5 seconds. The speed of sound is known to be 340 m/s. What is the speed of the turtle?

** 30. A siren is fastened atop a mass oscillating in simple harmonic motion between points A and C, as shown. The amplitude of the oscillations is 50 cm and the maximum velocity is 10 m/s. An observer stands to the right of the system and listens to the siren. At which point (A, B, or C) will the mass be located when the observer hears the highest frequency? Explain.

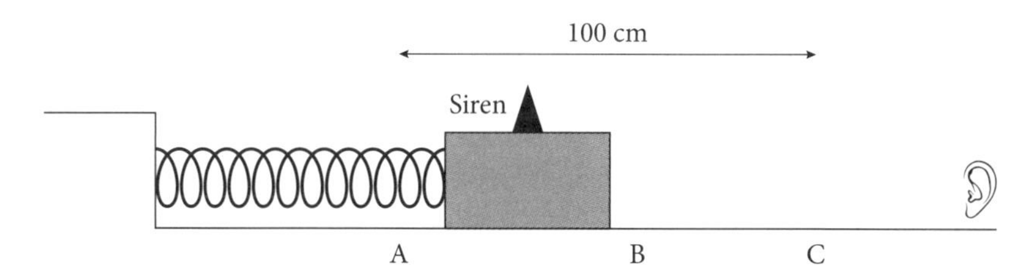

** 31. On a cold day when the speed of sound is 330 m/s, a police car is traveling directly toward you in a 35 mi/hr speed zone. You think that he is traveling faster than he needs to, so you want to estimate his speed and see if you can give *him* a ticket, for a change. You record the fact that the siren on his police car, whose frequency is 500 Hz when the car is not moving, actually produced a frequency of 550 Hz. How fast was the police car going? What is this in mi/hr?

*** 32[f]. **The Doppler Shift for a Moving Observer.** The formula for the Doppler shift shown in Equation 12.6 is valid when a source of sound is moving through the air and the observer is at rest in still air. Suppose, instead, that a source of sound producing wavelength λ is at rest in the air and that you, the observer, are moving toward the source at a speed u, while the speed of sound in the still air is 2. Find a formula for the relative shift in frequency of the sound you observe by answering the following questions:

a) What frequency f would be measured by an observer at rest relative to the source, written in terms of v and λ? [Yes, this part is easy.]

b) As you travel toward the source, and as the waves travel toward you, how long will it take you to move from one wave crest to the next? [This will be the period of the wave you observe.]

c) Find the frequency f' of the waves that you observe, written in terms of u, v, and f.

d) Find the ratio $(f' - f)/f$ in terms of u and v.

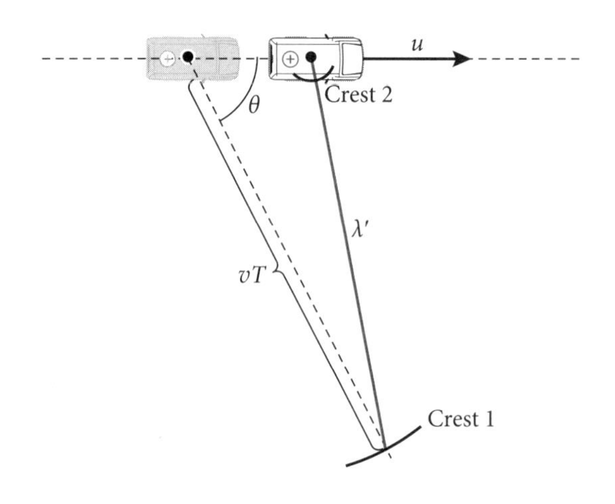

*** 33[f]. **The Doppler Shift at an Oblique Angle.** In Section 12.7, we considered a source of sound whose wavelength is λ when the source is stationary and worked out the Doppler shift in the wavelength of the sound when the source is moving directly toward the observer or directly away. There will always be a time, however (unless the source runs right over you), when the source is not coming directly at you or going directly away. Consider a siren on an ambulance that produces a sound with wavelength λ when it is stationary. Then suppose that the ambulance is moving at a speed u, but that the sound wave you hear is one that is traveling at an angle θ to the direction the ambulance is moving, as shown above. The picture shows the situation after one period T has elapsed between crest 1 and crest 2. Crest 1 has traveled a distance vT at an angle θ, and crest 2 is just leaving the siren. The observed wavelength is still the distance between successive wavefronts of the sound, the distance labeled λ' in the drawing at left. Find a formula for the Doppler shift in wavelength when the sound from the source is traveling to the observer at an angle θ from the direction of the source's velocity. [Hint: This will require a little (?) trigonometry.]

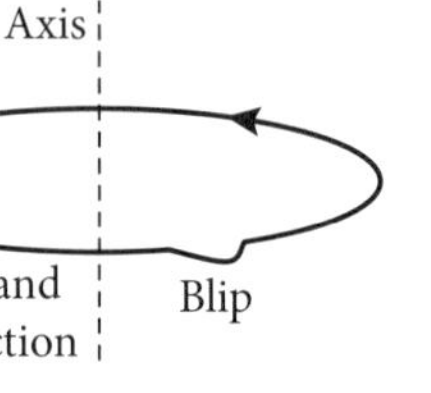

*** 34[f]. This one probably deserves four stars, but we'll talk you through it. Consider a circular rubber band floating in space inside the space shuttle and suppose that the rubber band spins about an axis through its center, as in the picture at left. The rubber band has a linear mass density μ and, because the spinning will stretch the rubber band by some amount, there will be a tension F in the rubber band. Show that a small blip in the rubber band, like the one shown, will remain fixed in space as the rubber band spins. To do this, go through the following steps:

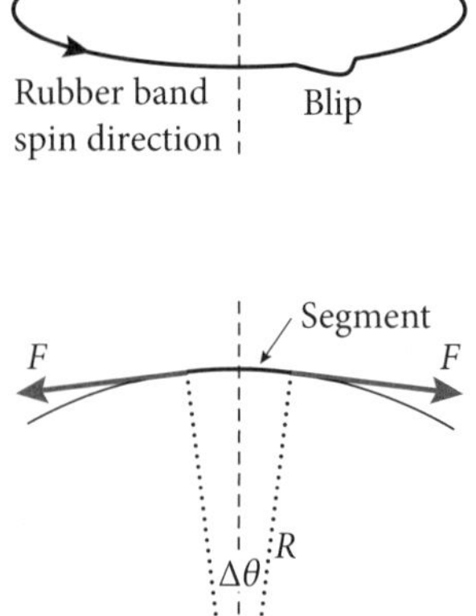

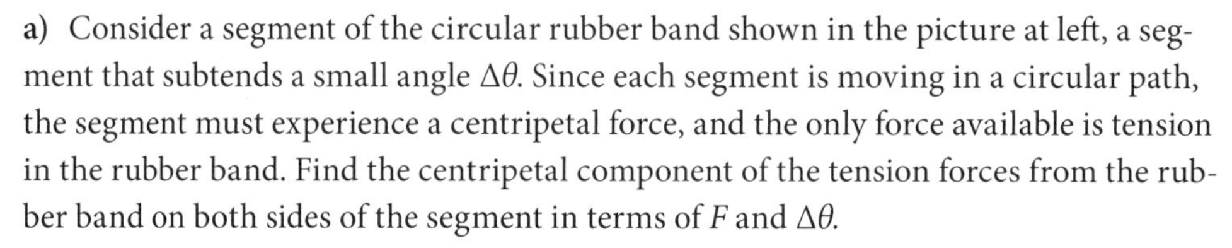

a) Consider a segment of the circular rubber band shown in the picture at left, a segment that subtends a small angle $\Delta\theta$. Since each segment is moving in a circular path, the segment must experience a centripetal force, and the only force available is tension in the rubber band. Find the centripetal component of the tension forces from the rubber band on both sides of the segment in terms of F and $\Delta\theta$.

b) If the rubber band has linear density μ, the segment will have mass $\mu(R\Delta\theta)$. Use Newton's Second Law to determine the acceleration of this segment.

c) Equate the acceleration found in part (b) to the formula for centripetal acceleration, and thus find the speed of the rubber band segment for which this centripetal acceleration provides the circular motion.

d) Compare this speed with the pulse speed around the rubber band, found from the tension in the rubber band and its linear mass density.

13

STANDING WAVES

Figure 13.1 shows a photograph of a vibrating guitar string. The string is made of brass and it has some linear mass density. The guitarist tunes his instrument by putting tension in the string, by twisting the tuning keys on the guitar head that stretch the string. The two quantities, mass per unit length and tension, determine the speed of transverse waves along the string, and any disturbance in any segment of the string will propagate along the string at that speed. And yet, when we look at the vibrating string in the figure, all of the segments appear to just be going back and forth with a constant amplitude. There is no sense of anything traveling along the string at all. What we are looking at, in Figure 13.1, is what is called a *standing wave*.

Figure 13.1 The low E-string of a guitar vibrating in what is called a "standing wave."

Our discussion of waves began with the notion that a wave transfers energy without transmitting mass. The direction of a wave is identified with the direction of energy transfer. Now we introduce the idea of a standing wave, which at first glance appears to be an oxymoron. Can a wave really exist if it is not transferring energy from one place to another? The basic answer is no, but let's look more carefully at waves or pulses that might be traveling along the guitar string in Figure 13.1. If we displaced a small segment near one end of the string, and let it go, that pulse would indeed travel down along the string, carrying energy toward the other end. But what happens when the pulse hits the far end of the string? As we saw in the last chapter, it will reflect off the fixed end, keeping its shape but inverted relative to the initial pulse. The energy the pulse carries would now be going the opposite direction. If we were to quickly add a second pulse at the initial end, identical to the original pulse we created, we would now have energy going in both directions on the string. In fact, as much energy would be going up the string as there is going down the string. Each pulse is doing what pulses do (transferring energy), but the net energy transferred is zero. And consider what happens when the returning inverted pulse collides with the new pulse we created. When the two meet, they will interfere, and the motion of the string will be whatever is required by superposition.

It is the combination of incident and reflected waves, traveling in the same medium, that produces the phenomenon of standing waves. The standing waves may exist on strings, like the one shown in Figure 13.1, and they may also exist in the columns of air enclosed inside pipes, with waves of density propagating through them. We will discuss sound in air columns in Section 13.2, but we begin in the next section by looking at the standing transverse waves on a stretched string.

13.1 Standing Wave on a String

Let us use a stretched coil spring like the one we introduced in the last chapter as our "string" and let us see how we might create a standing wave on it. Imagine holding a spring of length L that extends from your hand to a fixed endpoint, and suppose you move your hand up and down at a set frequency f. If you move your hand only slightly, in the presence of a standing wave of large amplitude, your hand will actually behave as a fixed endpoint to a very good approximation. The wave you have created will travel down the spring at a speed v (determined by the tension and the mass density of the spring). When it reaches the far end, the wave reflects back on the opposite side of the spring (fixed-end reflection), and, when the reflected wave returns to your hand (which is approximately a fixed end), it will reflect back on the original side of the spring. The time it takes the leading edge of the initial wave to travel down the spring and back is found by dividing the distance ($2L$) by the wave speed v,

$$T_{2L} = \frac{2L}{v}. \tag{13.1}$$

If the frequency with which you are shaking the end is just right, the wave you are just then generating will be on the same side as the twice-reflected wave and the amplitude will grow. We say that the wave we just generated is *in phase* with this reflected wave, as the two travel together along the spring. The frequencies that allow this kind of synchronization of the reflected waves moving up and down the spring are called the *natural frequencies* of the spring. When we drive the waves with one of these frequencies, we say that the driving force is in *resonance* with one of the natural frequencies of the system.

With a little thought, you will see that the frequency we will have to choose, in order to keep these waves adding to each other every time, is one in which the time it takes the wave to make the round trip on the spring is an integer multiple of the period of our harmonic-oscillating hand. That way, every time a wave reflects and is ready to start again, our hand will be giving it a tiny boost in the right direction. We must choose a frequency for which the period T satisfies the requirement that T_{2L}/T = some integer. The longest period to satisfy this condition is obviously where the integer is one, corresponding to $T = T_{2L} = 2L/v$. This largest value for T will correspond to the lowest frequency, so the lowest resonant frequency our standing wave can have is

$$f_1 = \frac{v}{2L}. \tag{13.2}$$

Figure 13.2 shows the shape of the standing wave with the choice of this lowest frequency of vibration. Note that in this figure the spring is represented by the dark solid line. At the time of the figure, all points on the spring are at their maximum displacements. The dotted line represents where the spring will be one half-cycle later. This picture resembles the shape of a football (at least it would if it were a little fatter). It is wide in the middle and pointed at the ends. When a spring is vibrating quickly enough, the whole pattern even blurs enough to look almost solid, enhancing the resemblance to a thin football. Figure 13.1 showed a guitar string vibrating this way in its lowest-frequency mode.

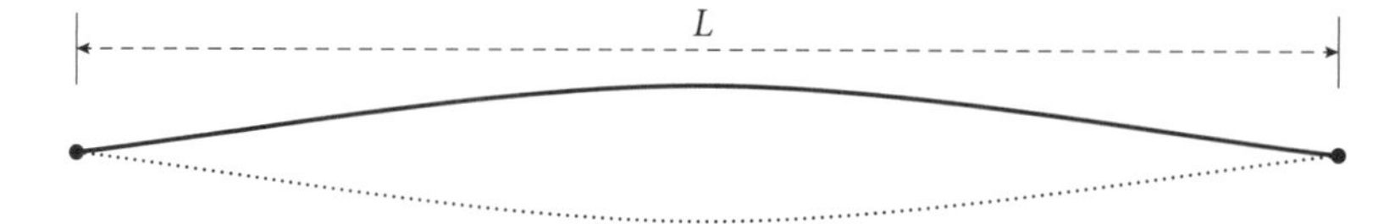

Figure 13.2 A diagram of a transverse standing wave in its lowest mode of vibration.

EXAMPLE 13.1 Changing the Pitch on a Guitar

Use Equation 13.2 to explain what happens to the pitch of a guitar string vibrating in its lowest frequency mode when you (a) tighten it, and (b) press your finger on the fingerboard to effectively shorten the string.

ANSWER This is a question where it's likely that you already know the answer from your own experience, but we need to provide a physical basis for what we know.

a) Tightening the string doesn't change the length, but it does increase the speed by increasing the tension. (It also reduces the mass density a small amount, but this is a minor effect.) From Equation 13.2, we see that increasing the speed increases the frequency. The pitch gets higher.

b) Pushing the string against the fingerboard shortens the portion of the string that vibrates, thus reducing L in the equation. Therefore, shortening the vibrating portion of the string increases the frequency and the pitch rises.

As we derived Equations 13.1 and 13.2, we chose to seek the lowest frequency of vibration that would satisfy the condition of our driving hand being synchronized, so that it is in phase with the twice-reflected wave heading back up the spring. However, as we have already noted, any situation where T_{2L}/T has an integer value would satisfy the requirement. This corresponds to driving frequencies given by

$$f_n = n\frac{v}{2L}, \tag{13.3}$$

where n is some integer (1, 2, 3, etc.). Equation 13.2 is the $n = 1$ example, called the *fundamental frequency*, f_1. Modes of vibration corresponding to higher values of n are called *harmonics* of the fundamental frequency (the fundamental frequency is also called the first harmonic). If we were to drive our end of the spring at $f_2 = 2f_1$, we would get a wave shape that looks like Figure 13.3. The spring vibrates up and down in two segments with a stationary point in the middle. It looks like two footballs end-to-end. In this second-harmonic case, the frequency is twice the fundamental, or twice the first harmonic. In musical terms, this corresponds to a one-octave increase in the pitch. Vibrating the end at 3 times the fundamental, $3f_1$, produces a resonance pattern with three lobes (or a 3-football pattern), as shown in Figure 13.4.

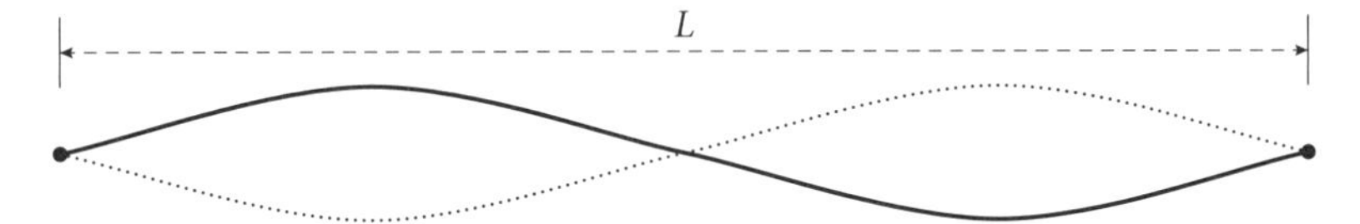

Figure 13.3 The second harmonic or "2-football" mode of vibration.

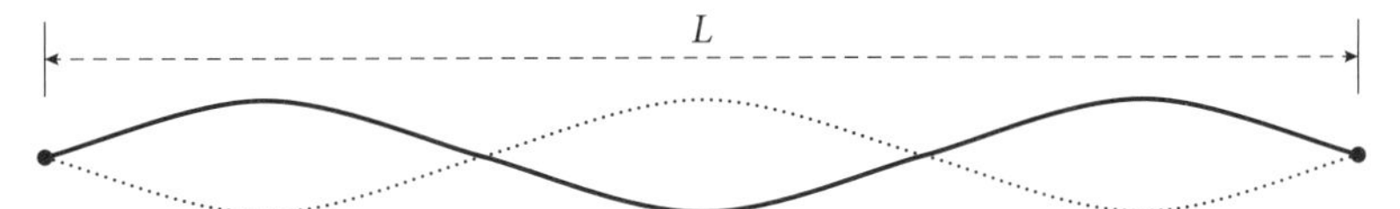

Figure 13.4 The "3-football" mode at the third harmonic of the fundamental.

We derived the natural frequencies of the standing waves by considering the transit times for the waves, being sure that our driving frequency is properly timed to be in phase with the twice-reflected waves. However, a look back at Figure 13.2 shows another method we could have used. The solid line in Figure 13.2 represents one half of a complete sine wave; it is the peak and not the trough. Figure 13.3 shows an entire wavelength fitting into the length of the spring and Figure 13.4 shows 1½ waves fitting on the string. The condition for a standing wave is thus that an integer number of half-wavelengths fit into the length of the string. The wavelengths that satisfy this condition are given by

$$n\frac{\lambda_n}{2} = L \qquad \text{or} \qquad \lambda_n = \frac{2L}{n}, \tag{13.4}$$

where n is a positive integer. Another way to think about this is to recognize that each "football" represents one half of a wavelength, and so the one-half-wavelength length of each football, times the number of footballs in the resonance pattern, must equal the length of the string. When n is the integer number of footballs on the string, this again gives us Equation 13.4.

EXAMPLE 13.2 Harmonics of a Vibrating String

A string of unknown mass per unit length is fastened between two fixed ends a distance 1½ meters apart. The string is plucked in such a way that it vibrates in its 3rd harmonic. The frequency is measured at 450 Hz when stretched to a tension of 2025 N. Find the string's mass per unit length.

ANSWER We are given the tension. If we also knew the wave speed, we could use Equation 12.1 to get the linear mass density. So how do we find the wave speed?

The standing wave pattern in the third harmonic is as shown back in Figure 13.4. From this picture, we see that one wavelength (2 footballs) is ⅔ the 1½-m length of the string. We thus arrive at $\lambda = 1$ m. We could also have gone back to Equation 13.4, setting $n = 3$ for the three footballs in the picture, to find λ from

$$1\tfrac{1}{2}\text{ m} = 3 \times \frac{\lambda}{2} \qquad \Rightarrow \qquad \lambda = 1\text{ m}.$$

With the wavelength in hand, we can use the wave equation to find the wave speed:

$$v = \lambda f = (1\text{ m})(450\text{ Hz}) = 450\text{ m/s}.$$

So, finally, the mass per unit length may be found from

$$v = \sqrt{\frac{F}{\mu}}$$

$$450\text{ m/s} = \sqrt{2025\text{ N}/\mu}.$$

If we square each side and solve for μ, we get $\mu = 0.01$ kg/m.

It is important to note that, for the first part of our solution, it was critical to know which harmonic was being excited. Our answer would have been different if the string had produced its 450-Hz tone in a different mode. Once the wavelength was known, however, the wave speed could be found and the linear density could be determined. It no longer mattered exactly how the string was vibrating.

Before we go on to the next section, let's summarize what we have learned so far about standing waves. In a standing wave, the elements of the medium just move back and forth at fixed amplitude, and there is no sense of anything moving along the medium or of energy being transferred. A standing wave cannot occur, however, unless there are boundaries in the medium that reflect waves back and forth, because a standing wave is actually the result of waves traveling in both directions in a medium. It is because the waves actually do travel that we can use the wave equation to relate the frequency and wavelength of a standing wave to the speed of traveling waves in the medium. As the waves interfere, they produce constructive interference at certain places in the medium and destructive interference in others.

We developed the equations for standing waves in this section by considering a stretched spring held in our hand. If the motion of our hand was phased correctly, a large standing wave would be created, but, if driven at the wrong frequency, the result would only be tiny waves running up and down the spring with no coherent pattern to them. We began this chapter, however, with a picture of a standing wave on a guitar string. Here there is no continual driving force, just an initial pluck. But the equations still apply. What happens after the string is plucked is that the parts of the initial displacement with wavelengths given by Equation 13.4 survive to interfere constructively and produce large amplitudes. The rest just remain tiny wiggles moving up and down the string, too small to be noticed. A similar situation occurs when a violin string is scraped by a bow. The scraping excites all frequencies, but only the frequencies satisfying Equation 13.4 attain large enough amplitudes to be heard.

13.2 Reflection of Sound in an Air Column

Standing waves are not only responsible for the musical notes produced by stringed instruments, but they are also responsible for the tones produced by pipe organs and wind instruments. The source of sound in a wind instrument is a standing longitudinal sound wave in the air column contained inside the instrument. The tube of the organ pipe or other musical instrument constrains the air molecules against motion transverse to the tube and allows motion more freely along the tube. However, if we are going to produce standing waves in an air column, we will need some kind of boundaries at the ends of the tube to reflect waves back and forth. So let's begin by understanding how longitudinal sound waves can reflect off of a boundary in an air column.

The reflection of a pulse from a closed fixed end in an air column is easy to understand. The actual molecules in a gas do not sit still, of course. They have large random motions as they collide with each other and with the walls of their enclosure, but, in Figure 13.5, we ignore this motion and use an orderly set of red dots to represent the *average* positions of molecules at different locations inside the tube. If there were no wave in the column, the average positions of the molecules at some location would not move.

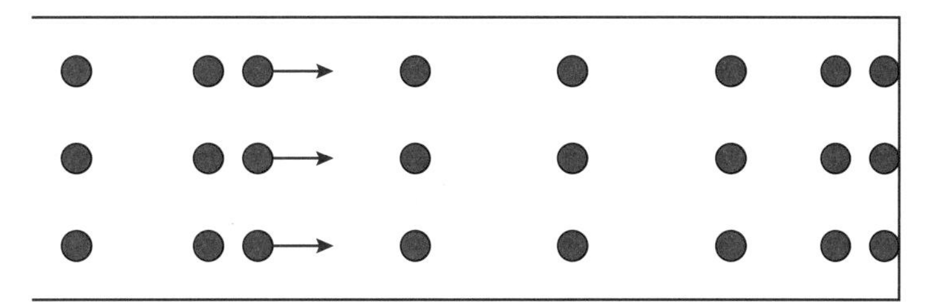

Figure 13.5 Reflection of a longitudinal wave from a fixed end. As molecules within the air column (the leftmost set of dots) move to the right, the molecules ahead of them are pressed forward, continuing the wave. But those at the end of the column (the rightmost dots) cannot move to the right. The increased pressure required to stop the motion produces a reflected inverted pulse traveling back to the left.

Figure 13.5 shows two density pulses moving to the right down the tube. The increased density produces an increased pressure on molecules to the right of the pulse. These molecules will therefore move to the right themselves, as indicated by the arrows, and the pulse propagates down the tube. The density pulse at the end of the tube, however, sees a different situation. Here, the molecules are not able to move any further to the right because of the closed end of the tube, and the same logic we used for a transverse pulse reflected at a fixed end in Figure 12.10a applies. The result is that there will be an increased pressure at the closed end of the tube, a pressure large enough to force the nearby molecules to move back to the left. This creates a reflected pulse that will head in the opposite direction to the incident pulse. And note that the motion of the red dots representing the actual air molecules is also to the left in the reflected pulse, in contrast to the incident pulse where the red dots all moved to the right. The disturbance in the medium due to the reflected pulse is inverted from that of the incident pulse. Molecules are pushed toward the left as the reflected pulse moves to the left.

An effect that may be harder to see intuitively is what happens as a pulse traveling down an enclosed tube reaches an open end (see Figure 13.6). There, the air column is suddenly able to expand in three dimensions, instead of the single dimension that was available inside the tube. The result is that the back pressure from the air to the right of the pulse is significantly

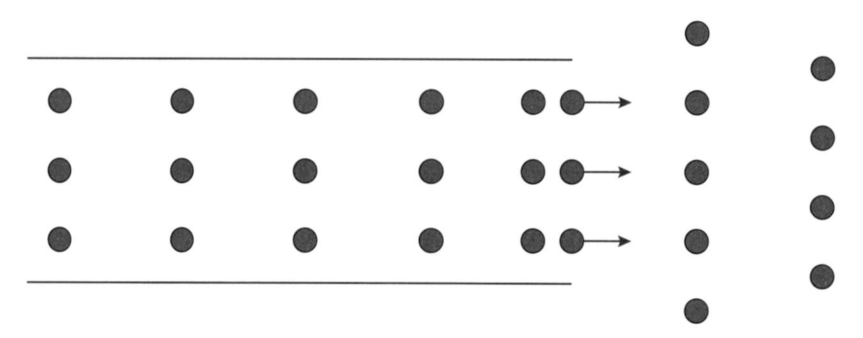

Figure 13.6 Reflection of a longitudinal wave from an open end. Representative molecules at the end of the column (rightmost arrows) are freer to move than they were inside the tube, producing a larger average displacement from their original positions.

lower, producing a greater displacement for the molecules at the rightmost boundary. This large motion produces a reflected pulse that travels leftward, back along the pipe. The molecules of the reflected pulse actually move in the same direction as they did in the incident pulse, for reasons similar to what we showed in Figure 12.10b. If the incident pulse pushed molecules to the right, the reflected pulse will pull molecules to the right as it itself propagates toward the left.

13.3 Longitudinal Standing Waves: Open–Open Tube

Now that we understand how pulses are reflected from boundaries in an enclosed tube, let us go on to see how standing waves are produced. In Figure 13.7 we represent the molecules in a tube that is open at both ends, in which there is a standing wave. Waves are reflected from both open ends in the manner of the pulse reflections shown in Figure 13.6, and we assume that the frequency of the waves is just what it takes to synchronize the waves traveling in both directions so as to create a large-amplitude standing wave.

One way to produce a standing wave in an open–open tube like the one in Figure 13.7 would be to blow across one end of the tube (not into it; that would just create a wind through the tube). This way, all frequencies are excited, just as a violin bow excites all frequencies on a string, but only the frequencies that produce standing waves have enough amplitude to be heard. By the way, the model of standing waves in a tube open at both ends is a good model of the frequencies that are heard from a flute. Yes, we know that a flute is closed at one end, but there is a hole near the closed end of a flute that the flutist blows across without impeding the free motion of the molecules near the hole. That hole acts as an open end of the tube.

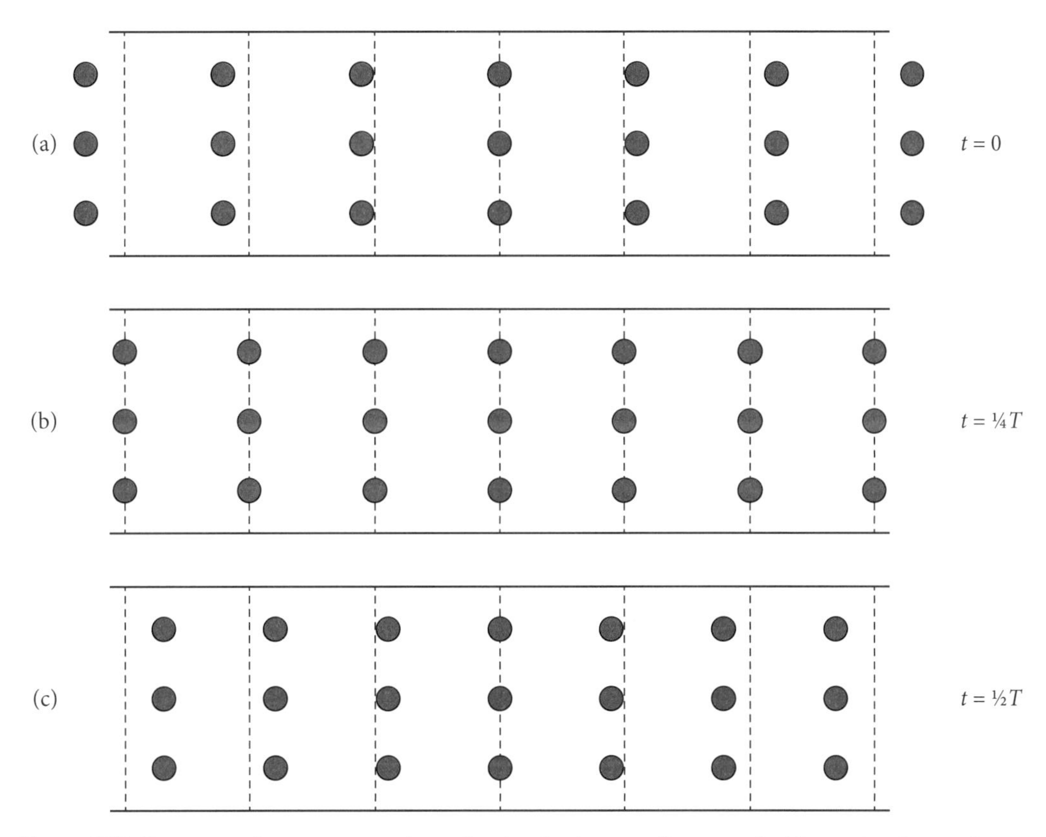

Figure 13.7 Representative average motions of molecules in a standing wave inside an air column where the tube is open at both ends. The molecules near the ends of the tube oscillate back and forth with large amplitude. The molecules in the center do not move. Figure (a) is at $t = 0$; figure (b) is at $t = \frac{1}{4}$ a period; and figure (c) is at $t = \frac{1}{2}$ a period.

Just like any standing wave, the wave depicted in Figure 13.7 has no remaining sense of motion of waves back and forth along the air column. Each molecule, on average, simply vibrates left and right with some fixed amplitude. The amplitude of the oscillation of molecules at the center of the tube is zero, and it remains zero at all times. Here, at the center, the leftward-traveling sound wave and the rightward-traveling sound wave will always be phased so as to completely cancel in destructive interference. This point, at which there is no motion, is called a *node* of the standing wave pattern. In contrast, the molecules at the two ends of the tube have the largest amplitudes of oscillation, and that amplitude remains constant. Places where the amplitude is a maximum are called *antinodes* of the standing wave. The standing wave shown in Figure 13.7 has a node at the center and antinodes at the two ends.

Figure 13.7 represents three snapshots of the positions of the air molecules in a standing wave in an open–open pipe. The three shots cover one-half of the full period. These drawings, however, are a pretty cumbersome way to show the motion of the molecules in a longitudinal standing wave. If you will look back to Figure 12.13, on page 238, you will see how the motion of the molecules in a longitudinal wave can be graphed, and this graph is easier to analyze than the picture of the longitudinal wave. In using the graph, however, one has to be careful to remember that, unlike the case for transverse waves on a string, the wave does not actually look like its graph. Still, as long as we are careful not to confuse ourselves, it is often easier to work with the graphs. To see how to do this, let us graph the average displacements in Figures 13.7a, b, and c, on the same graph (Figure 13.8).

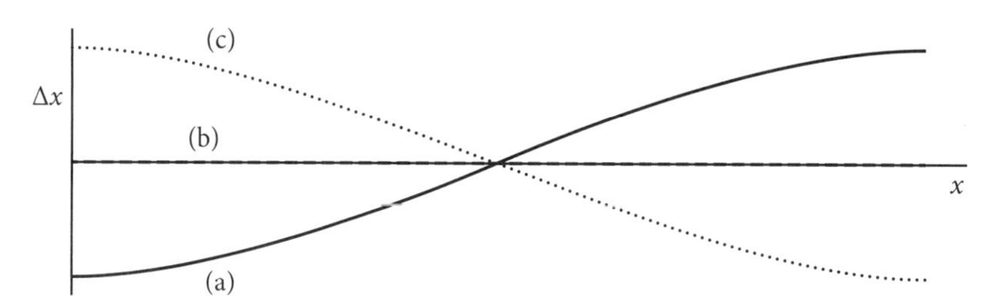

Figure 13.8 A graph of the displacements of the air molecules in the open–open pipe. The solid line is at $t = 0$, the dashed line at $t = ¼$ period, the dotted line at $t = ½$ period.

Figure 13.8 is easy to use to see the displacements of the molecules in the air column at the chosen points in time, as long as we remember that the motion will not be up and down, as it appears in the graph, but back and forth along the x-direction, as in Figure 13.7. If you are clear on this point, we are going to do one additional thing to our graph. Instead of graphing the displacements along an x-axis, as in Figure 13.8, let us superimpose the graph on a picture of the tube. This way, you can *see* where you are in the tube, instead of having to connect a value of x with position in the tube. If we do this with Figure 13.8, we produce Figure 13.9.

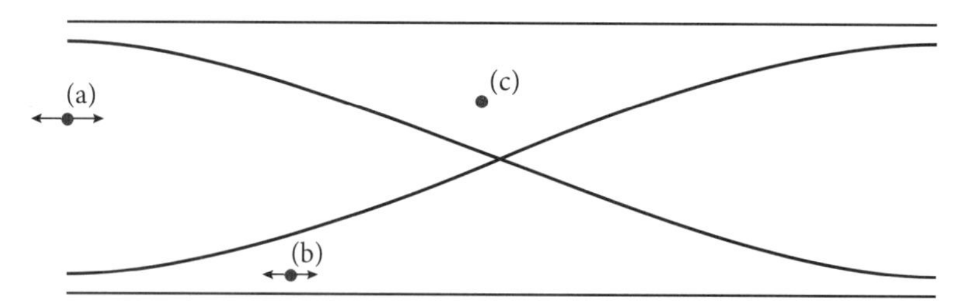

Figure 13.9 A graph of the average displacements of the air molecules in the pipe, the two solid lines showing the extreme displacements at $t = 0$ and at $t = ½ T$. Also shown are three red dots representing average molecules and the amplitudes of their horizontal motion.

In Figure 13.9, there is an almost overwhelming urge to see molecules in the air column as moving up and down between the two solid lines representing the extremes of the motion. But the solid lines in Figure 13.9 are still just curves in a *graph* of *horizontal* displacement versus position along the tube. To help us understand, we have depicted three typical red dots in the air column, with arrows showing the amplitudes and directions of their periodic motion. Note that we show dots outside the region between the two curves.

The tube is actually completely filled with air, both between and outside the curves we drew. The graphs are *not* pictures of anything. They are only graphs, on some vertical scale, of the maximum average displacements of the molecules as a function of their horizontal location along the pipe.

Having established a way to depict longitudinal standing waves, we turn our attention to the frequencies of these modes. Although the motion of the longitudinal sound waves in an open-ended pipe is quite different from that of the transverse waves in strings we considered in Section 13.1, the good news is that we can borrow heavily from what we learned there. The major difference, of course, is that we now have antinodes at the two boundaries of the air column, instead of the nodes at the two ends of the string. And there is a node of the standing sound wave in the middle of the tube, replacing the antinode in the middle of the standing wave on a string.

Under what conditions can there be a standing wave in a tube open at both ends? First, there must be antinodes at the two ends. Second, since you can't just have a single continuous antinode all along the tube, there must be a node between any two antinodes. The longest wavelength for a resonant standing wave in an air column must therefore be one that looks like Figure 13.9 above, with one node in the middle and antinodes at the two ends. This pattern looks like two half footballs, so it is a case where one complete football would fit in the tube. The wavelength of this resonance is therefore $\lambda_1 = 2L$, and the wave equation will give a frequency $f_1 = v/\lambda_1 = v/2L$. This formula is the same as the formula for the longest-wavelength standing wave on a string, in which there is a node at each end and an antinode in the middle.

Just as we did with standing waves on a string, we can also create standing waves of higher frequency in the tube, those where an integer number of footballs of the pattern in the graph exactly fit into the tube as we introduce more nodes into the standing wave. The two end points must always be antinodes, so the next possible higher-frequency standing wave pattern is shown in Figure 13.10. To generalize, we may say that the standing wave frequencies

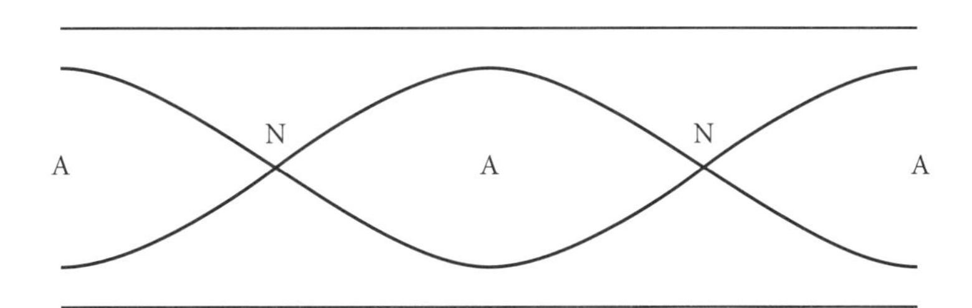

Figure 13.10 The second harmonic of an open–open air column. The nodes in the pattern are indicated with an N and the antinodes with an A.

produced in an open–open air column are those where an integer number of half-wavelength footballs fit into the air column. This gives the following formulas for the resonant wavelengths and frequencies:

$$\lambda_n = \frac{2L}{n} \quad \text{and} \quad f_n = \frac{v}{\lambda_n} = n\frac{v}{2L}. \tag{13.5}$$

One last difference between longitudinal waves in air and transverse waves on a string should be mentioned. This is that, whereas the speed of waves on a string can be adjusted by tightening the string to increase the tension, the speed of waves in air is fixed by the speed of sound in air. So, while a violin can be tuned by tightening the strings and the notes changed by shortening the length of the vibrating string, the only way to tune a wind instrument is by adjusting the length of the column (there is no practical way to manipulate the speed of sound in air, except for the minor effect of "warming up" the instrument by blowing through it). Once the overall length of a wind instrument is adjusted, the different notes are produced by opening holes in the column, thereby changing the effective length of the air column.

EXAMPLE 13.3 Harmonics in an Open–Open Tube

The effective length of a flute is 66.0 cm. Find the frequency of the lowest note that a flute can play when the speed of sound in the warm air in the flute is 344.5 m/s. By blowing slightly into the mouthpiece of a flute (instead of directly across) the fundamental frequency is suppressed, and only the second harmonic may be heard. This is referred to as "overblowing." What is the frequency of the overblown second harmonic of the lowest note?

ANSWER The flute is an open–open air column. Its fundamental wavelength, the lowest note it can play, is given by Equation 13.5, with $n = 1$. The result is $\lambda_1 = 2L = 2 \times 66$ cm $= 1.32$ m. The wave equation gives the frequency of a sound with this wavelength. It is

$$f_1 = \frac{v}{\lambda_1} = \frac{344.5 \text{ m/s}}{1.32 \text{ m}} = 261 \text{ Hz}.$$

This note is a middle C, the lowest note a flute plays.

The wavelength of the second harmonic is found using $n = 2$ in Equation 13.5. It gives $\lambda_2 = L = 0.66$ m. The frequency is

$$f_2 = \frac{v}{\lambda_2} = \frac{344.5 \text{ m/s}}{0.66 \text{ m}} = 522 \text{ Hz}.$$

This is twice the fundamental frequency. It is a high C, one octave above middle C.

13.4 Longitudinal Standing Waves: Open–Closed Tube

Most wind instruments create their sounds in an air column that is open at one end and closed at the other. An oboe, for instance, has a bell at the bottom and a small double reed at the top (Figure 13.11). The bell is obviously an open end, and it may seem that the reed should be treated like the open end as well, but it is not. When being played, the reed is enclosed tightly between the oboist's lips. It provides a vibration that drives the waves in the air column, but there is very little actual motion of the air at that point. In fact, when an oboist stops for a breath after playing a long passage of music, it is not to get air into his lungs, but to let it out.

So let us analyze oboes and trumpets and the like by considering a tube with one end open and the other end closed. The open end must be an antinode, just like the case we considered in the last section, but the air molecules at the closed end are no longer able to move. The closed end is forced to be a node. This change in boundary conditions from the case we considered in the previous section changes the possible modes of oscillation. The simplest mode, with one end a node and the other an antinode, is shown in Figure 13.12.

Reed

Bell

Figure 13.11
An oboe has an open–closed air column inside it.

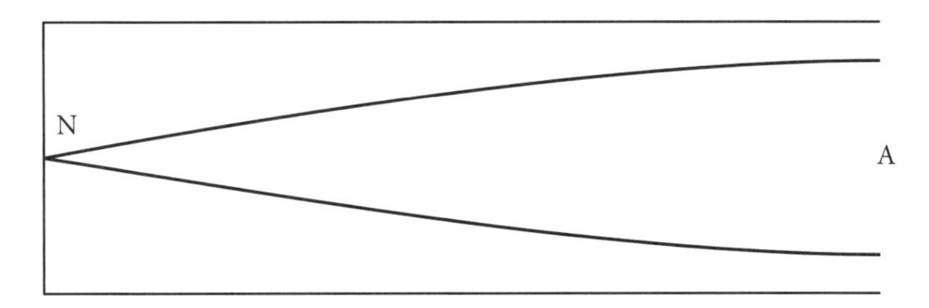

Figure 13.12 The longest wavelength for a standing wave in an open–closed tube. The closed end is a node and the open end will be an antinode. Looking at either curve makes it clear that we have one-fourth of a wave contained inside the tube.

As is clear from the diagram, this longest-wavelength, lowest-frequency mode has half a football fitting into the length of the tube, and each football again represents half a wavelength. We are thus led to conclude that this wavelength and the length of the oboe are related by $L = \frac{1}{4}\lambda_1$, giving a fundamental wavelength of $\lambda_1 = 4L$. The fundamental frequency comes from the wave equation as $f_1 = v/\lambda_1 = v/4L$.

The higher harmonics of this fundamental frequency are a little more difficult to work out. Since there must be a node at the closed end of the tube and an antinode at the open end, the next longest wavelength to fit into the tube would have the graph shown in Figure 13.13.

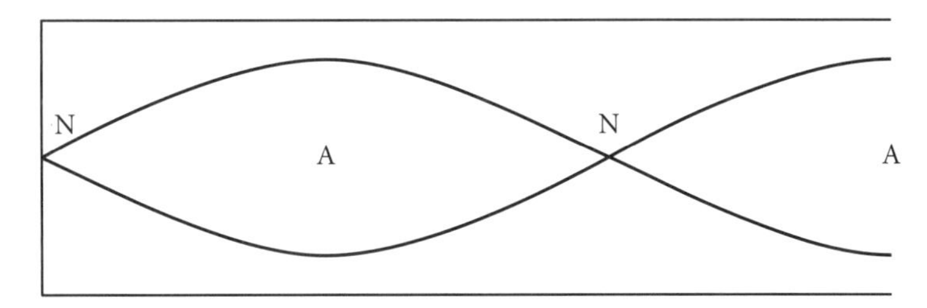

Figure 13.13 The next harmonic in an open–closed pipe. A node and an antinode have been added to the pattern in Figure 13.12. Consideration of either displacement curve makes clear that three-quarters of a complete wavelength fit inside the tube.

We see from Figure 13.13 that we have added one complete football to the pattern in Figure 13.12, for a total of 1½ footballs. Since each football represents half a wavelength, we have

$$L = \text{number of footballs} \times \frac{\lambda}{2} = \frac{3}{2}\frac{\lambda}{2}.$$

The wavelength of this standing wave is therefore $\lambda = (4/3)L$ and the frequency of this harmonic is

$$f = \frac{v}{\lambda} = \frac{v}{4/3\,\lambda} = \frac{3v}{4\lambda} = 3f_1.$$

What is interesting about this is that it tells us that there is no mode that produces a $2f_1$ harmonic. This is different from the case of the open–open tube. Because the next mode above the fundamental is $3f_1$, we will call this mode the third harmonic and say that there is no second harmonic for an open–closed tube. Going on from there, we see that each higher harmonic in an open–closed tube adds one more football to the pattern and creates an odd multiple of the fundamental. The harmonics are therefore

$$f_n = n\frac{v}{4L} = nf_1, \tag{13.6}$$

but the values of n can only be the odd integers. The harmonics of a closed-open tube are therefore $f_1, f_3, f_5, \dots$ only.

EXAMPLE 13.4 Characterizing a Tube from Its Harmonics

A tube is found to produce resonances at 150 Hz and 250 Hz but at no frequency in-between. Is this an open–open tube or an open–closed tube? What is the tube's length?

ANSWER There is really no formula to answer the first part of this question. What we have to ask is whether this appears to be part of a series of all multiples of a single frequency or just the odd multiples. The highest frequency that goes evenly into both values is 50 Hz. An open–open tube with a 50-Hz fundamental would produce 100 Hz, 150 Hz, 200 Hz, 250 Hz, etc. Even though it would produce the two required frequencies, it would also produce a resonance at 200 Hz, which the question statement tells us this tube does not do. An open–closed tube would produce 50 Hz, 150 Hz, 250 Hz, 350 Hz, etc., as required. This is therefore an open–closed tube with a fundamental frequency of 50 Hz.

To find the tube's length, we first find the wavelength of the fundamental 50-Hz wave. The formula $\lambda = v/f$ gives $\lambda = 343$ m/s/50 Hz = 6.86 m. In the fundamental mode, the length of the tube is one quarter of a wavelength, so $L = \lambda/4 = 1.72$ m.

EXAMPLE 13.5 Harmonics of an Open–Closed Tube

A 0.75 m tube with one end open and one end closed resonates at 343 Hz, but we do not know if this is its fundamental harmonic. Determine the harmonic of this oscillation and draw a figure representing the standing wave. If the tube could be extended, what would be the next greater length of tube that could support a standing wave at this same frequency?

ANSWER A 343-Hz standing wave will have a wavelength given by

$$\lambda = \frac{v}{f} = \frac{343 \text{ m/s}}{343 \text{ Hz}} = 1.0 \text{ m}.$$

If a wave has a wavelength of 1 m, the half-wavelength footballs will be 0.5 m long. So the question is whether or not we can draw a 0.75-m tube with 0.5-m footballs in it in a way that represents a valid standing wave pattern. A little thought tells us that there is such a picture and that it would look like this:

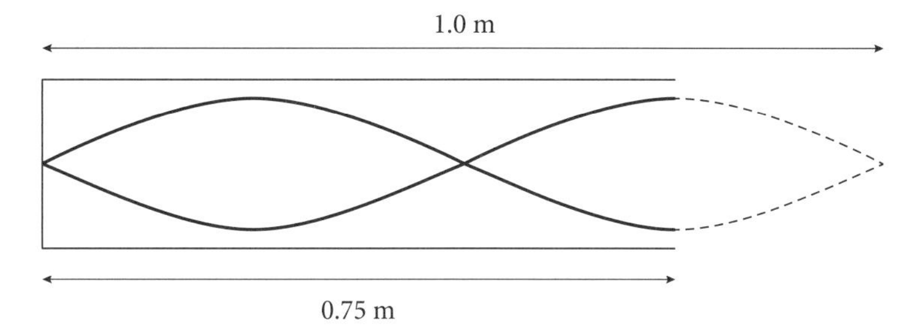

This is a perfectly reasonable picture for a third harmonic, with ¾ of a wavelength inside the tube. (The dashed line outside the tube is just drawn for reference.) But note that if the wavelength had been different, say 1.2 m, it would have been impossible to find a valid pattern that would fit in a 0.75-m tube. A 0.75-meter tube open at one end will simply not produce resonant sound of wavelength 1.2 m.

For the second part of this question, it is critical for us to see that, while the wavelength does not change, any number of longer tubes can create standing waves of that same wavelength (and frequency). All you must do is to add more and more half-wavelength footballs, so that sound of the same frequency emerges as a higher harmonic of a longer tube. For example, the next greater length that would produce the same 343-Hz frequency sound would be the one in which we extend the length of the tube by one football (half a wavelength) to get the following pattern:

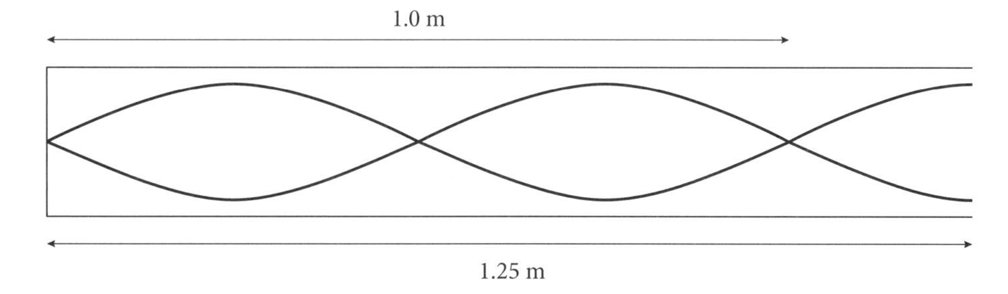

This 1.25-m tube plays a 343-Hz tone as its 5th harmonic. We could also have used tubes of length 1.75 m, 2.25 m, 2.75 m, etc. If we had been asked, we could have also noted that a 0.25-m tube would also play a 343-Hz note, this time as its fundamental.

13.5 Summary

From this chapter, you should take away the following:

- You should understand how a standing wave is produced by equal-amplitude waves of the same frequency traveling in opposite directions in a medium, how the distance between the boundaries of the medium gives rise to resonance, and what is meant by the terms *nodes* and *antinodes*.
- You should be able to use Equation 13.4 to determine the wavelengths of standing waves on a string.

- You should be comfortable with the concepts of longitudinal waves in a tube of gas, including the behavior of a pulse that reflects off closed or open ends of the tube. This is analogous to the inversion of pulses on a string.
- You should be comfortable with the process we used to find the wavelengths and frequencies of standing waves in air columns, open–open and open–closed. If you were asked to determine the modes of vibration of standing waves in a tube closed at both ends (which we did not address in this chapter), you should be confident that you could figure it out.
- You should be able to use Equations 13.5 and 13.6 to find the wavelengths and frequencies of the harmonics of standing waves in air columns, understanding why Equation 13.5 is the same as Equation 13.4, and understanding why Equation 13.6 is different from Equation 13.5.

CHAPTER FORMULAS

Harmonics of standing waves on a string: $\lambda_n = \frac{2L}{n}$ and $f_n = n\frac{v}{2L}$ in which $n = 1, 2, 3, \ldots$

Harmonics of standing waves in an open–open tube: $\lambda_n = \frac{2L}{n}$ and $f_n = n\frac{v}{2L}$

in which $n = 1, 2, 3, \ldots$

Harmonics of standing waves in an open–closed tube: $\lambda_n = \frac{4L}{n}$ and $f_n = n\frac{v}{4L} = nf_1$

in which $n = 1, 3, 5, \ldots$

PROBLEMS

13.1 * 1. Find the wavelength of the fifth harmonic of a string 40 cm long. Sketch the shape of a string that is resonating at this wavelength.

* 2. If the third harmonic for a vibrating string is 660 Hz, what will be the frequency of the fourth harmonic?

* 3. If the lowest frequency produced by a vibrating ukulele string of length 0.40 m is 400 Hz, what is the speed of a wave along the string?

** 4. The A string on a cello plays a frequency of 220 Hz. The length of the vibrating string is 70 cm, and the mass of this vibrating section is 1.2 grams. What is the tension in the A string in newtons?

** 5. The note played on a violin is determined by pressing a finger firmly on the fingerboard to set the length of the string that vibrates. Often, a violinist will play "vibrato" by rocking this finger back and forth while the note is being played. When the vibrating portion of a particular string is 26.7 cm long, the violin will play 392 Hz (G). If the finger is moved back and forth by 0.5 cm to produce the vibrato, what will be the range of frequencies that are played?

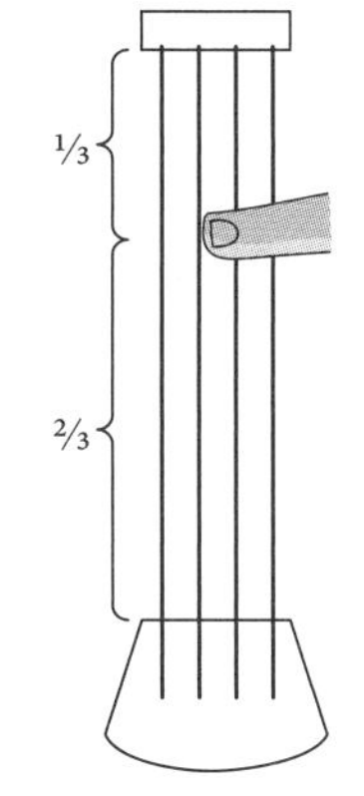

** 6. On occasion, a violin player will *lightly* press her finger ⅓ of the way up the A string (which would otherwise vibrate in its fundamental mode at 440 Hz). This light pressure suppresses any motion of the string at this point and forces this point to be a node of the string's vibration, while still allowing the short ⅓ of the string and the longer ⅔ of the string to vibrate at the same frequency as each other. Instead of 440 Hz, what resonant frequency will the string now produce? Explain why this is the frequency produced.

** 7. A guitar is plucked near the end of the string, so that the lower harmonics of the string are not excited. As a result, the lowest two frequencies heard from the string are 612 Hz and 816 Hz. What is the fundamental frequency of the string?

*** 8. A man wants to get the kinks out of his 30-meter outdoor extension cord, so he wedges one end between the post and one rail of his fence and pulls on the other end to straighten out the cord. Just for fun, he starts to wiggle his hand up and down until he sees that he has set up a standing wave that looks like the picture below. He notes that his hand moves at 4 oscillations per second and he knows that the total mass of his cord is 1.5 kg. What is the tension in the cord?

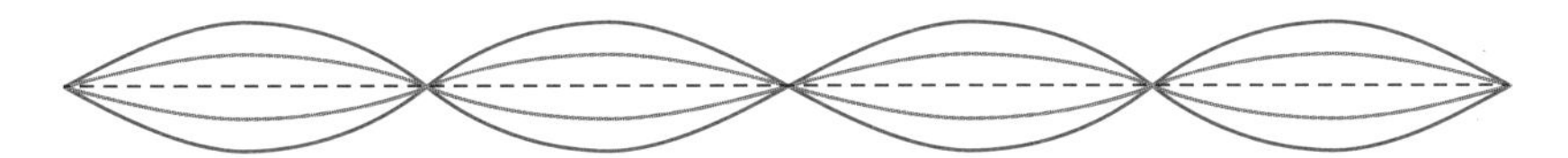

** 9. A small bird is jumping up and down on an electrical transmission wire at the location indicated below, setting up a transverse standing wave that is the fourth harmonic for the stretched wire. (The bird is obviously not drawn to scale. Assume that his mass is small compared to the mass of the wire.)

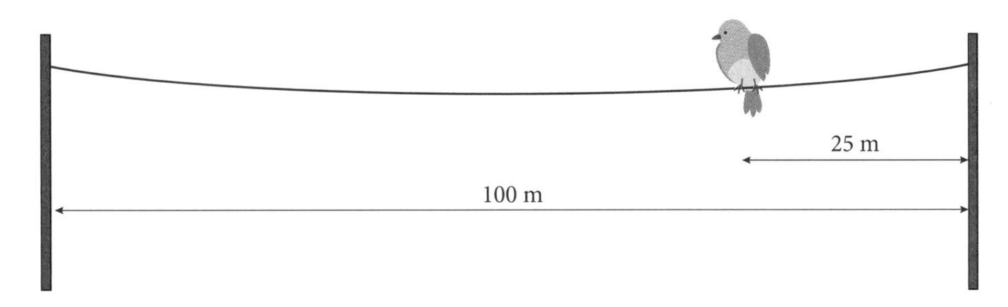

a) Is the bird hopping at a node or an antinode? Draw a picture of this standing wave to justify your answer.

b) Find the wavelength for this standing wave.

c) If the bird is hopping with a period of $T = 0.5$ s, find the speed of traveling waves on this wire.

d) If 100 m of wire has a mass of 60 kg, what is the tension F in the wire?

** 10. A string is stretched between two supports that are 0.80 m apart. The string has a linear density of 0.40 g/m. What tension in the string will give the string a fundamental frequency of vibration of 200 Hz?

** 11. If you know the wavelength of a wave on a string that produces a certain musical note, is it possible, with that information alone, to calculate the wavelength in air of the sound wave that carries that sound to your ears? Explain your answer.

13.4 * 12. What is the lowest resonant frequency of sound that may be produced in a tube 70 cm long that is closed at one end?

* 13. An air column, closed at one end and open at the other, has a fundamental frequency for standing waves of 343 Hz. What is the length of the column?

** 14. An organ pipe has an open end at the lower end of the air column (the slot shown in the picture at left) and the top end can be open (like the picture) or closed. If an open pipe 8 feet long plays a 65-Hz frequency, how long will a closed organ pipe have to be to play the same note?

** 15. Two organ pipes resonate at the *same* fundamental frequency. Pipe A is 4 m long and is *open at both ends.* Pipe B has one end open and the other closed.

a) Find the fundamental frequency of pipe A.

b) Find the length of pipe B.

** 16. A boy plays a long piece of pipe by buzzing his lips at one end and adjusting the frequency of the buzz until he hears a resonant loud note out of the pipe. He finds that the lowest frequency he can play this way (though it is not the fundamental frequency of the pipe) is 63 Hz. He finds that he can also play a 105-Hz note. What is the fundamental frequency of the pipe? [Hint: His lips close the pipe at the end where he is buzzing.]

** 17. The lowest 4 frequencies at which the air column contained inside a tube will support standing waves are found to be 75, 225, 375, and 525 Hz. What is the length of this tube and what combination of closed and open ends must it have?

** 18. Two pipes of different lengths are placed side by side. One of the pipes is open at both ends; the other is open at one end. The lowest resonant frequency produced in each pipe is 200 Hz.

a) Which pipe is the longer of the two?

b) Give all the resonant frequencies of the two pipes, up to 1000 Hz.

c) On a cold day, the wavelength of the 200-Hz frequency is 1.5 m. What are the wavelengths of all the other frequencies found in part (b)?

** 19. An alto saxophone has a key that opens a hole very close to the reed end of the air column inside the sax. Opening this hole changes the saxophone from a closed–open tube to an open–open tube. If a saxophone plays D above middle C (294 Hz) with the hole closed, what frequency will it play with the hole open?

** 20. A member of the marching band adjusts the length of her flute so that it plays exactly the correct fundamental frequency indoors. Outside, she finds that the air is much colder and therefore the speed of sound is reduced well below what it was indoors. Is the fundamental frequency of the flute now too high or too low? Explain your reasoning. (Thermal contraction of the flute itself is negligible.)

*** 21. Members of the high school orchestra tune their instruments in the warm rehearsal room behind the open-air band shell in the city park. When they get out to the band shell, the air is 5.7 degrees colder than it was in the rehearsal room. As a result, the speed of sound out on the field is 340.0 m/s instead of the 343.4 m/s speed that sound had in the air in the locker room.

a) The oboe was tuned to play concert A (440 Hz) in the rehearsal room. What will its frequency be when it is taken out into the cold air?

b) The violins' resonant frequencies are unaffected by the speed of sound in air. If the violins and the oboe now play the same concert A, how many beats per second will be heard? (By the way, this does sound terrible.)

** 22. Two air columns resonate at the same fundamental frequency. Column A is open at both ends and column B has one end open and the other one closed. The length of column B is 0.40 m.

a) What is the fundamental frequency of column B?

b) What is the next highest frequency above the fundamental that column B can produce?

c) What is the length of column A? Justify your answer.

** 23. It is possible to establish standing waves inside a cylindrical tube with both ends closed by placing a small speaker inside the tube to excite the resonances. The boundary condition is that both ends must be nodes of the motion (the air molecules at the ends cannot move back and forth at all). Draw the standing wave patterns and determine the lowest two resonance frequencies for a double closed-end tube of length 0.60 m. Take the speed of sound to be 340 m/s.

** 24. A standing wave is set up in a long air column that is open at both ends. The resonance that is excited is shown schematically below.

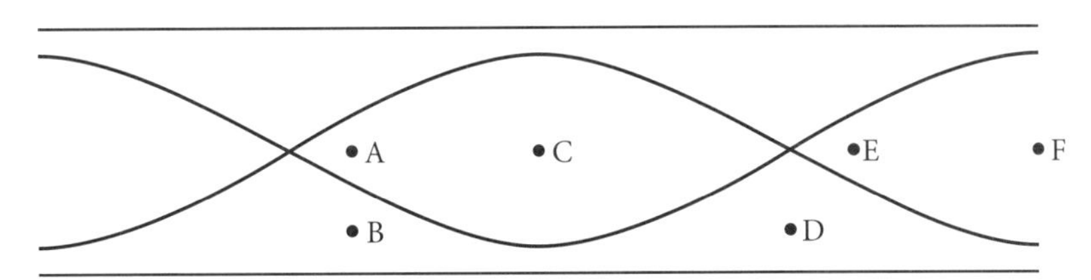

At the instant represented, let us say that the displacement at point F is large and to the right. For each of the remaining points in the figure (A through E), indicate the size (large, small, or none) and direction (right, left, or none) of the displacement at this instant.

** 25. An air column with one end open and one end closed supports a standing wave at its *fundamental frequency*. Describe the motion of the air right at the middle of the air column (equal distance from both ends) by answering each question below and justifying your answer for each.

a) Does the air move side to side, perpendicular to the length of the pipe, or back and forth, parallel to the length of the pipe?

b) Is the amplitude a maximum, a zero, or something in-between?

** 26. We have estimated the wavelengths of standing waves in tubes by assuming that there is an antinode at the open end of a tube. This is a good approximation, but it is not exact. In truth, the antinode at an open end sits a small amount beyond the end of the tube. The distance from the end of the tube to the true antinode is called the *end correction*. If a tube of length 60 cm produces a fundamental frequency of 137 Hz when the speed of sound in air is 340 m/s, what is the end correction for this tube? (By the way, please ignore this end correction as you solve any other problems in this chapter.)

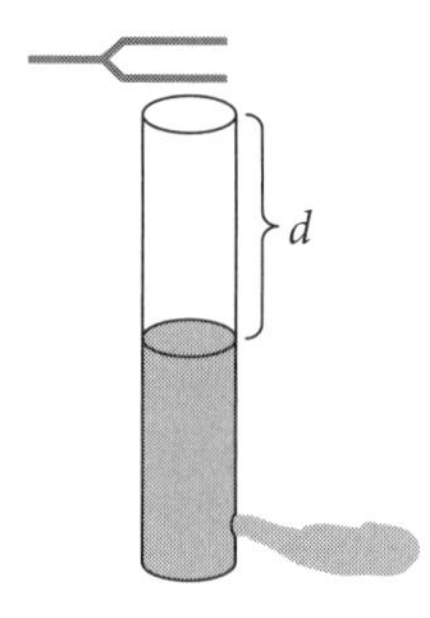

** 27. A tuning fork that vibrates at a frequency 680 Hz is placed over an 80-centimeter-long cylinder that is completely filled with water. The water is then allowed to drain out, creating an air column of varying length above the surface of the water and inside the cylinder. Find the distances d between the top of the cylinder and the surface of the water for which the air column will resonate at 680 Hz. Assume that the speed of sound is 340 m/s.

** 28. A open–open pipe on a truck driving down the street has a total length of 21 cm. It produces a sound wave that is the 6th harmonic of the pipe. An identical pipe on the ground produces a sound wave that is its 7th harmonic. If the speed of sound is 343 m/s, how fast would the truck have to be moving toward the stationary pipe for the two notes to sound the same? Would the truck be breaking the speed limit?

** 29. Two wires are stretched between two pegs so that the lengths of the wires are the same and the tensions in the strings are the same. However, one wire's linear mass density is 5% greater than the other. When the two strings are plucked together, their average frequency is 540 Hz. What beat frequency will be heard between the sounds from the two wires?

*** 30. The strings of a violin are tuned to G, D, A, and E in their lowest harmonics. G is the lowest frequency string. The frequency of the D string is $1\frac{1}{2}$ times the frequency of the G string; that of the A string is $1\frac{1}{2}$ times the frequency of the D string; and the frequency of the E string is $1\frac{1}{2}$ times that of the D string. The linear density of the G string is 3 g/m. If, to avoid putting a torque on the bridge that holds the strings, the tension is to be the same on each string, find the preferred linear densities of the remaining three strings.

14

THERMAL PHYSICS

Back in Section 7.4, we first mentioned that mechanical energy is lost as a book slides across a tabletop. But we also explained that the mechanical energy lost through the action of the non-conservative frictional force could actually be accounted for if we realized that both the book and the tabletop warm up as the mechanical energy disappears. We will begin this chapter by seeing in more detail how this happens.

The friction between a book and a tabletop turns out to be a very complicated subject, but let us take a view that attributes friction to the irregularities in the two surfaces. Then, as the surfaces interact with each other, they will actually do so only at the few points that are in close contact, as shown in Figure 14.1a. As the book moves across the tabletop, contact forces between the surfaces at those points may elastically deform the surfaces, as with the point of the tabletop that protrudes up in the middle of the dashed red circle in Figure 14.1b. As the book continues to move to the right, that point, like others in the book and tabletop, will eventually spring back and be left vibrating back and forth, as shown in Figure 14.1c. Because

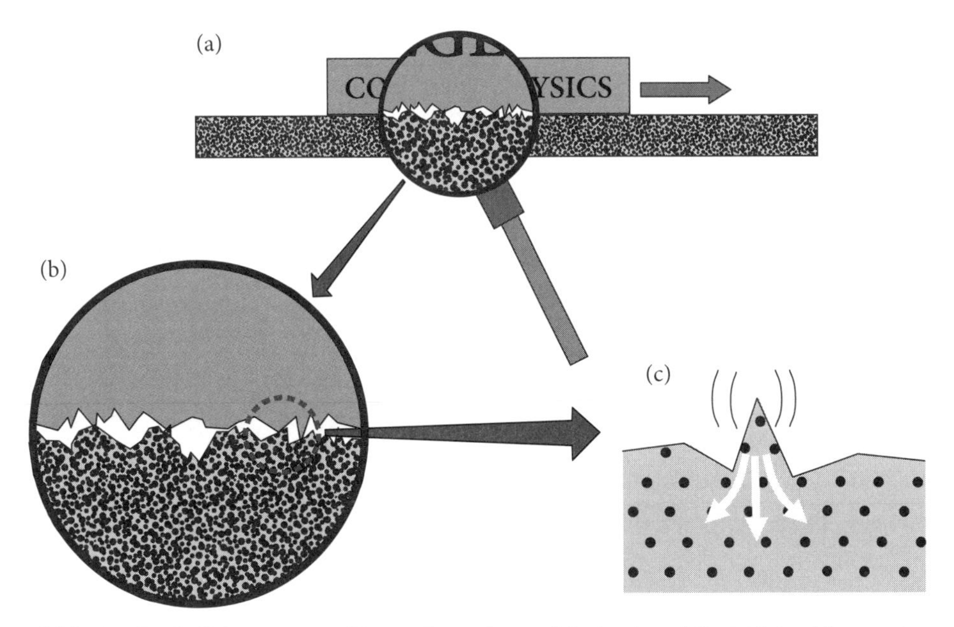

Figure 14.1 As a book slides across a tabletop, the surfaces of the book and the tabletop (a) are not perfectly smooth. Frictional forces arise as contact deforms the surfaces (b). After the book passes by, the deformation snaps back and continues to vibrate (c). Since the molecules of the tabletop (the black dots in c) are held in place by molecular forces, the kinetic energy of the vibrating molecules near the surface is quickly transferred to the rest of the tabletop (as indicated by the white arrows in c).

the molecules of the tabletop are held in place in the solid metal tabletop by molecular forces, and because motion of one of the molecules changes the forces on nearby molecules and accelerates them, the vibration is ultimately shared by all the molecules in the tabletop. The same kind of process occurs within the molecules of the book.

The kinetic energy of all of the molecules of a body is called the body's *internal energy*. Note that a body has this internal energy even though the body as a whole (its center of mass) is not moving. Because of the connection between internal energy and temperature, as we will see in the next section, internal energy is also called *thermal energy*. What has happened in Figure 14.1 is that the non-conservative frictional force has converted kinetic energy of the book into thermal energy shared by the molecules of the book and the tabletop.

14.1 Temperature

As energy is transferred from molecule to molecule in the tabletop, it may show up in several different forms. The molecule itself may vibrate by changing its size or shape; it may rotate; or it may move back and forth, changing its position relative to nearby molecules of the metal. But if you think about it, only one of these kinds of motion will actually transfer any energy to a nearby molecule. If a molecule is only vibrating within itself or rotating about its center of mass—if it isn't actually moving to a different place within the structure of the material—then it will not collide with any nearby molecules or change the forces on them, and so its energy cannot really be transferred. As we mentioned back in Chapter 2, the change in the position of an object is referred to as *translation*, so we may say that internal energy in a material is transferred by the *translational kinetic energy* of the molecules of the material.

As we saw in Chapter 8, a fast-moving particle that collides with a particle that is originally stationary will transfer momentum to the stationary particle. In the process, the fast-moving particle will lose some of its kinetic energy while the particle that was originally stationary will acquire kinetic energy. If the target particle had been moving slowly instead of being completely stationary, the result would be pretty much the same. The slow-moving particle would acquire kinetic energy from the collision while the originally fast-moving particle would lose some of its energy. In fact, on average, particles with high kinetic energy tend to lose energy in collisions while particles with lower kinetic energy tend to gain energy. If there are many collisions in a system of particles, then energy will be exchanged between them until the average translational kinetic energy of the particles is the same throughout the system.

If cube A, whose molecules are vibrating with high average translational kinetic energy, is placed in contact with cube B, in which the molecules are vibrating with low average translational kinetic energy, then the collisions of the molecules in the surfaces of the two cubes will, on average, raise the translational kinetic energy of the molecules of cube B and lower the translational kinetic energy of the molecules of cube A. This process will continue until the average translational kinetic energies of the molecules in cube A and that of the molecules in cube B are the same.

If we *define* the *average translational kinetic energy* of the molecules of a body to be what we will call the *temperature* of the body, then we may conclude that internal energy will always be transferred from an object with higher temperature to an object with lower temperature, and that the process will continue until the two temperatures are the same.

Historically, the idea of temperature, including the observation that energy flows from a hot body to a cold body, was known long before the molecular view of matter was understood.

We are therefore left with several historical artifacts in the definitions and units of thermal physics. As you can see from the definition of temperature in the previous paragraph, the proper standard unit of temperature should be the standard unit of energy—the joule. And the temperature of a body should only go to zero in the limit that the body has no more internal translational kinetic energy to transfer away. However, you are probably aware of several other commonly used temperature scales, such as the Fahrenheit scale, the Celsius scale, and the Absolute, or Kelvin, scale. Each of these uses energy units that are different from the energy in joules, and two of them, the Fahrenheit and Celsius scales, go to zero in the wrong places.

Figure 14.2 displays the three temperature scales. The Celsius and Kelvin scales use the same unit of temperature, but set their zeroes at different places. The Fahrenheit scale uses a different unit *and* sets its zero at a different place from the other two. The relation between the units may be seen by noting that there are 100 Celsius (or Kelvin) units between the point where ice melts and the point where water boils, while the Fahrenheit scale has 180 units (212 minus 32) between these same two points. So it must be that 9 Fahrenheit degrees correspond to 5 Celsius degrees. If you examine Figure 14.2 for a moment, you could probably determine for yourself that a formula for converting Celsius temperature to Fahrenheit temperature would be

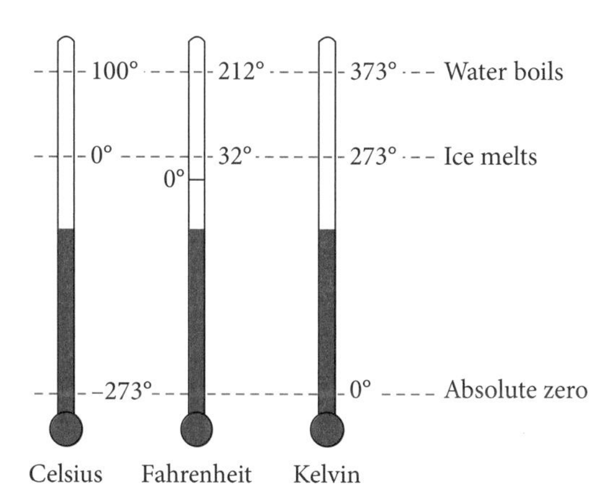

Figure 14.2 The three different temperature scales. We suggest that you study this picture for a few minutes.

$$T_F = \left(\frac{9}{5}\right)T_C + 32. \tag{14.1}$$

Finally, conversion between the Celsius and Kelvin scales consists simply in adding or subtracting the 273 K offset, as appropriate.[1]

We emphasize that both the Fahrenheit degree and the Celsius degree (also called the "kelvin" and abbreviated K) are units of average translational kinetic energy, though only the Kelvin scale goes to zero as this available energy of a molecule goes to zero. If we use angle brackets like $\langle\ \rangle$ to mean that some quantity is averaged over all the molecules of the system, we can write the relationship between the temperature properly measured in joules and the temperature measured in kelvins as

$$\langle \text{ke} \rangle = \frac{3}{2}kT, \tag{14.2}$$

where $\langle \text{ke} \rangle$ is the average translational kinetic energy per molecule and k is one of the fundamental conversion factors of physics, known as Boltzmann's constant. Its vlue is

$$k = 1.38 \times 10^{-23}\ \text{J/K}.$$

Joules, kelvins, and degrees Celsius are all considered SI units of temperature.

Since temperature is really the same thing as average kinetic energy per molecule, the factor $\frac{3}{2}k$ in Equation 14.2 may be seen simply as a conversion factor between kelvins and joules. The reason it is so small is because joules were defined so they would be useful for describing the energies of kilogram-sized objects moving at speeds of meters per second, while a molecule is much much smaller than that. Finally, the reason for not absorbing the $\frac{3}{2}$ in Equation 14.2 into the definition of Boltzmann's constant will be explained in the next chapter.

1. The actual offset between Celsius and Kelvin is 273.15 K, though we will always approximate this as 273 K.

14.2 Thermal Expansion

Gases expand to fill the space available to them, but if liquids and solids are maintained at a constant temperature, they will have a fixed volume, and there will thus be a fixed distance between the atoms or molecules of the substance. On the other hand, if the temperature of a liquid or a solid goes up, then some of the internal energy of the substance will go into increasing the potential energy between the molecules by increasing the average distance between them. The substance will expand.

Let us picture a column of atoms in a solid, separated from each other by a distance ℓ, as in Figure 14.3. Let us label each atom with a number—0, 1, 2, 3, etc. As the temperature of the solid increases, the distance between each atom and its neighbor increases from ℓ to $\ell + \Delta\ell$. If we hold atom 0 in place, then atom 1 will move away until it is $\ell + \Delta\ell$ from atom 0. But atom 2 will have to be $\ell + \Delta\ell$ away from atom 1, so it will have to move $2\Delta\ell$ away from where it started. The sequence will continue until the nth atom in the column is $n\Delta\ell$ from its original position. The distance moved by the last atom in the chain is also the change in the overall length of the column ΔL.

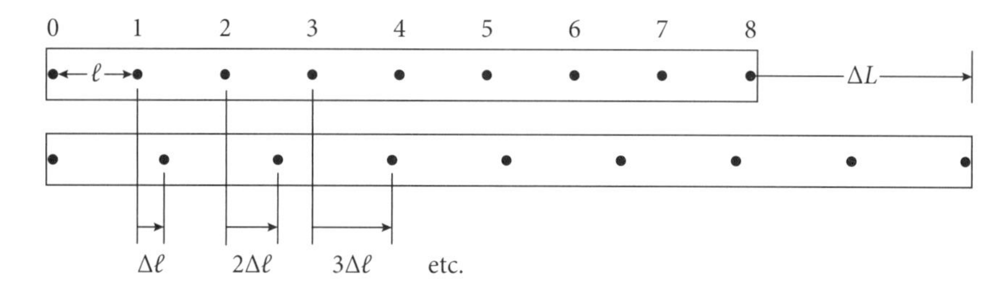

Figure 14.3 The expansion of a solid rod ΔL depends on the amount by which the increase in temperature increases the distance between atoms ($\Delta\ell$) and on how many atoms there are in the chain—i.e., on the length of the rod L.

Now, in practice, physicists do not calculate the theoretical increase in inter-atomic distance for a particular change in temperature (this is really very complicated) and they do not count the number of atoms in a column (because this is very difficult to do). Rather, they observe experimentally that the expansion of a solid of length L when there is a change ΔT in its temperature is given by a simple formula:

$$\Delta L = \alpha L \Delta T, \tag{14.3}$$

where α is an experimentally measured *coefficient of thermal expansion*. The fact that ΔL is proportional to the original length L is not surprising, given Figure 14.3 (a longer column means more atoms in the chain). Less obvious, perhaps, is that ΔL is proportional to the change in temperature. To be honest, experiments show that Equation 14.3 is only true if the temperature does not change *too* much, but it does turn out to be approximately valid over a wide range of temperatures for most materials. Table 14.1 contains values of α for a few useful materials.

TABLE 14.1 Coefficients of Thermal Expansion (parts per million per K, at 0° C)

Material	Aluminum	Iron or steel	Concrete	Glass	Fused quartz	Zerodur
α ($\times 10^{-6}$/K)	26	11	12	8.5	0.4	0.02

EXAMPLE 14.1 A Glass Lens in an Aluminum Lens Holder

A 10-cm-diameter glass lens is held in place with a 10-cm-diameter aluminum band around its circumference. If the temperature of the lens and its holder increases by 20° C, how much of a gap will be created between the outer edge of the lens and its holder?

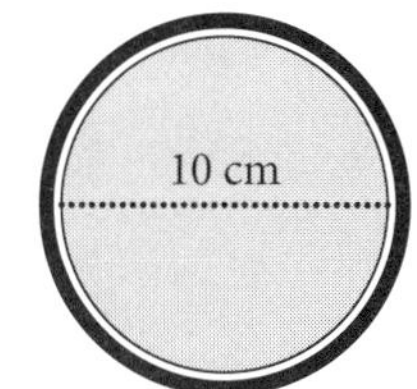

ANSWER Although the lens is a circular disk, there is a line of atoms across the diameter that obey the expansion relation given in Equation 14.3. We may thus calculate the magnitude of the expansion of the glass to be

$$\Delta L = \alpha L \Delta T = (8.5 \times 10^{-6}/\text{K})(0.10 \text{ m})(20 \text{ K}) = 17 \ \mu\text{m}.$$

A μm, called a micrometer or a micron, is 10^{-6} m. The aluminum band does not extend across the diameter, but the circumference of a circle is always π times its diameter. So, as the chain of molecules around the circumference grows due to the temperature increase, the diameter will increase proportionately. If we use Equation 14.3 for the increase, picking up the coefficient of expansion for aluminum, we find

$$\Delta L = \alpha L \Delta T = (26 \times 10^{-6}/\text{K})(0.10 \text{ m})(20 \text{ K}) = 52 \ \mu\text{m}.$$

This difference between the two diameters will be spread over gaps on both sides, giving a gap between the glass and the aluminum of $\Delta r = \frac{1}{2}(52 - 17) = \frac{1}{2}(35) = 17.5$ microns. At the warmer temperature, the lens will not be as securely held in place. This kind of effect on lenses and mirrors that are mounted on a spacecraft and then flown into the cold temperatures of space, creates a design problem for engineers who work on mountings for spacecraft optics.

EXAMPLE 14.2 The Volume Coefficient of Expansion

In addition to the linear coefficient of expansion, scientists also define a volume coefficient of thermal expansion β via the equation $\Delta V = \beta V \Delta T$. Find the volume thermal coefficient of expansion for steel.

ANSWER The volume coefficient may be found from the linear coefficient. For steel, Table 14.1 gives $\alpha = 11 \times 10^{-6}$. Now consider a steel block, 1 meter on a side, with volume 1 m^3. If the temperature is raised by 1 K, each dimension will increase by

$$\Delta L = \alpha L \Delta T = (11 \times 10^{-6}/\text{K})(1 \text{ m})(1 \text{ K}) = 11 \ \mu\text{m},$$

meaning that each side will go from 1 meter to 1.000011 meters. The volume of the cube will then be $(1.000011)^3 = 1.000033$ m^3, so the change is $\Delta V = 33 \times 10^{-6}$ m^3. We then use $\Delta V = \beta V \Delta T$ to find that $\beta = 33 \times 10^{-6}/\text{K}$. In fact, it is a general rule that the volume coefficient of expansion is always just 3 times the linear coefficient of expansion.

14.3 Specific Heat

In Section 14.1, we explained that the internal energy that is transferred to the molecules of a substance may show up as translational kinetic energy, vibrational kinetic energy, or rotational kinetic energy, as depicted in Figure 14.4, but that only translational kinetic energy transfers thermal energy from molecule to molecule. An individual molecule may, at any given time, have some of its energy in any or all of these modes. However, there are so many molecules in a laboratory-sized piece of the material that the proportion of internal energy in each mode, on average, is very precisely determined.

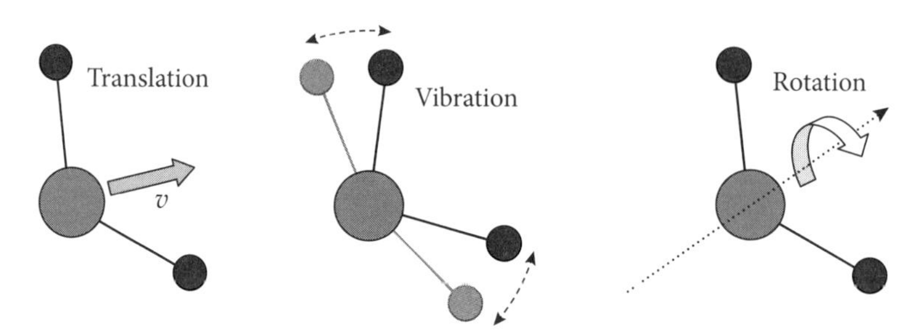

Figure 14.4 Motions of a water molecule (gray spheres for the oxygen atoms and red spheres for the hydrogen atoms).

The available modes of thermal energy are fairly simple for a gas, but they are more complicated for liquids and more complicated still for solids. Nevertheless, it is found that, over a fairly wide range of temperatures for most substances, the relation between the thermal energy added to a substance and the proportion of that energy that shows up as translational kinetic energy of its molecules (thereby increasing the temperature of the substance) may be expressed by the following formula:

$$Q = mc\Delta T, \tag{14.4}$$

where Q is the thermal energy added (also referred to as the *heat* added), m is the mass of the substance absorbing the heat, ΔT is the increase in temperature when the heat is absorbed, and c is an experimentally determined quantity called the *specific heat* of the substance. Examination of the units of the known quantities in Equation 14.4 shows that the SI units for specific heat must be J/(kg·K).

The specific heats of a few common solids and liquids are given in Table 14.2. Note that a high specific heat means that a lot of heat may be absorbed with very little change in temperature. The differences in the values of c are mainly due to the fact that Equation 14.4 depends on the mass, so a substance with more massive atoms generally has a lower specific heat per kilogram. But the value is also affected by the extent to which the substance is able to absorb energy in ways that do not show up as translational kinetic energy, but represent energy internal to the molecule itself.

TABLE 14.2 Specific Heats at 20° C (except as specified)

Substance	Silver	Copper	Glass	Wood	Ice (–10°)	Alcohol	Water
c (J/[kg·K])	233	386	840	1640	2050	2400	4186

EXAMPLE 14.3 Calorimetry

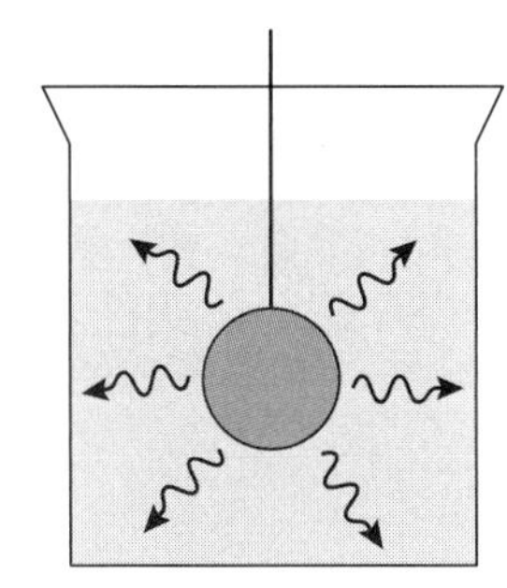

A 400-g copper sphere is heated to 100° C and then suspended in 1200 g of water that is initially at 10° C. The water is contained inside an insulated can of negligible mass so that no heat is lost to the outside environment and no significant amount of heat is absorbed by the can. Find the final temperature of the water and the copper sphere when thermal equilibrium is obtained.

ANSWER If we define our system to consist of the water and the sphere (the can may be neglected because of its small mass), then the total energy contained in the water and the sphere will be conserved. The internal energy lost by the sphere will be absorbed by the water. We may therefore use Equation 14.4 to relate the heat that goes out of the sphere to the heat that goes into the water:

$$Q_{out} = m_{copper}\, c_{copper}\, \Delta T_{copper} = Q_{in} = m_{water}\, c_{water}\, \Delta T_{water}.$$

We know both masses, both specific heats (found in Table 14.2), and the initial temperatures of each substance. We also know that the final temperatures of the copper and the water will be the same temperature T when the heat transfer is complete. We can substitute the quantities we know (in SI units) to produce

$$m_{copper} c_{copper}(T_{copper} - T) = m_{water} c_{water}(T - T_{water})$$
$$(0.4)(386)(100 - T) = (1.2)(4186)(T - 10).$$

We then solve for the unknown T, with algebra that goes like this:

$$(154.4)(100 - T) = (5023.2)(T - 10)$$
$$15{,}440 - 154.4T = 5023.2T - 50{,}232$$
$$65{,}672 = 5177.6T$$
$$T = \frac{65{,}672}{5177.6} = 12.7.$$

The final temperature of both water and sphere is 12.7° C. The water has warmed 2.7°, from 10° C to 12.7° C. The sphere has cooled by 87.3° from its initial 100° C down to 12.7° C. The water's smaller change in temperature is due to there being more of it to absorb the heat and to its much higher specific heat. This kind of analysis of heat transferred between materials is referred to as *calorimetry.*

14.4 Latent Heats

Consider an experiment where an electric coffee heater is embedded in a 1-kg block of ice, as shown in Figure 14.5. The initial temperature of the ice is –30° C (a block just brought inside on a typical very cold day in Bozeman, Montana). The ice is insulated so that no heat is exchanged with the environment, and we assume that the heat is added through the coffee heater slowly enough that the internal energy has time to spread out, keeping the temperature the same everywhere in the block. By measuring the voltage and the electrical current in the heater, we can figure out how many joules of heat are added to the ice by the heater. (We will see how this can be done once we get to Chapter 18.) If we were to plot the temperature of the ice as a function of the total heat added, we would get the graph shown in Figure 14.6.

Figure 14.5 A heater inside an ice block.

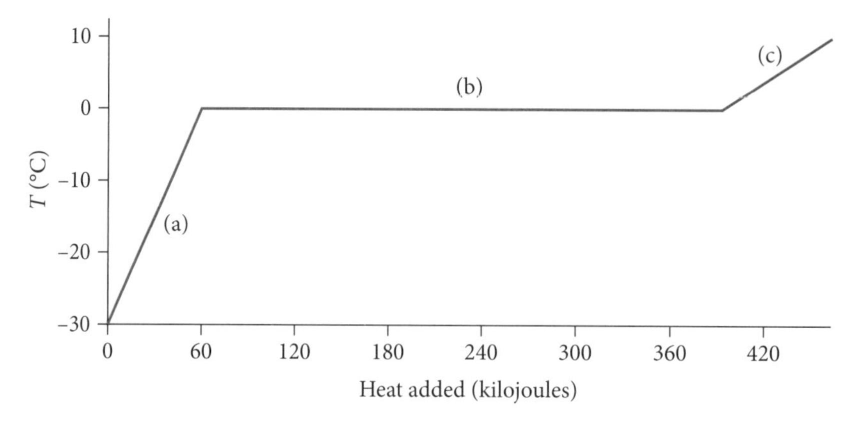

Figure 14.6 A plot of the temperature of a 1-kg block of ice as a function of the heat added by an electric heater embedded in the ice. The temperature rises (a) until the melting point of ice is reached, after which the temperature remains constant while the ice melts (b). Once all the ice is melted, the temperature of the now-liquid water again rises (c). Note that the slope of segment (c) is less than that of segment (a), since the specific heat of water is higher than ice and the heat is therefore absorbed with a smaller change in the water's temperature.

The specific heat of ice is 2050 J/(kg·K), so the 1-kg block of ice requires 61.5 kJ of heat to go from a temperature of –30° C up to 0° C. But then what happens? Once the ice reaches 0° C, which is the temperature at which ice melts, more and more heat is provided by the coffee heater, but the temperature doesn't budge. Thermal energy is going into the ice, but none of it is showing up as increased translational kinetic energy of the molecules of the ice. What is happening instead is that the energy is all going into potential energy stored in the intermolecular forces in the ice. The rigid structure of the ice is being broken apart and each water molecule in the ice becomes free to move around as liquid water. Let us be clear. While the ice is melting, the temperature of the ice and melted water does not change; both the ice and the water remain exactly at 0° C. You do not melt ice by raising its temperature above the melting point. You melt ice by adding heat while the temperature remains constant.

If you could measure carefully from the graph in Figure 14.6, you would see that it took precisely 334 kJ of heat to melt the entire 1-kg block of ice. This quantity is known as the *latent heat of melting* for ice.[2] The melting points and latent heats of melting for a few common elements are shown in Table 14.3.

TABLE 14.3 Melting Points and Latent Heats of Melting in kJ/kg

Substance	Nitrogen	Alcohol	Water	Lead
Melting point (°C)	–210.0	–114.0	0	327.5
Latent heat (kJ/kg)	25.7	108	334	28.5

2. This quantity is also (and most commonly) known as the *latent heat of fusion*, the term fusion referring to freezing. The same amount of heat must be removed from a substance to freeze it as must be added to the substance to melt it, so either term is acceptable.

Using the latent heats from Table 14.3, a simple formula for the heat required to melt a mass m of some substance is

$$Q = mh, \tag{14.5}$$

where h is a common symbol for latent heat.

A similar thing occurs when the temperature of a liquid reaches its vaporization (or boiling) point. For water, this is at 100° C. The *latent heat of vaporization* of water is 2.26 MJ/kg. Once water reaches 100° C, additional heat added to the water will not change its temperature. Vaporizing water will remain exactly at 100° C while all of the liquid water turns into water vapor at 100° C. By the way, the huge value for the latent heat of vaporization for water, compared to ice's latent heat of melting, reflects the fact that much more energy is required to enable water molecules to escape the liquid and expand freely to fill their container than is required to allow the molecules in ice to simply move around within what is about the same volume. We're telling you about this latent heat so that you will know, but we won't really worry about vaporization in this book, since, if you understand the latent heat of melting, you understand the latent heat of vaporization well enough.

These transformations from solid to liquid and back, or liquid to gas and back, are referred to as *changes of phase.*

The following example addresses the idea of mechanical energy turning into thermal energy and shows how to calculate the resulting change in temperature or phase.

EXAMPLE 14.4 The Silver Bullet Problem

The Lone Ranger fires a 20-g silver bullet at a speed of 200 m/s. The initial temperature of the bullet as it flies through the air is 30° C. The bullet hits a 10-kg block of ice that is at a temperature of 0° C and the bullet comes to rest, embedded inside the block. How much of the ice is melted by the bullet?

ANSWER The initial kinetic energy of the bullet is entirely lost when it stops in the ice. But total energy is always conserved, so this kinetic energy must be accounted for as thermal energy inside the bullet and the ice. The temperature of the bullet, even before it encounters friction inside the ice, is higher than the temperature of the ice, so thermal energy will tend to flow from the bullet into the ice anyway.

Let us begin our solution by assessing the sizes of things. The kinetic energy that will be lost by the stopped bullet is $\text{KE} = \frac{1}{2}mv^2 = \frac{1}{2}(0.02\text{ kg})(200\text{ m/s})^2 = 400\text{ J}$. The thermal energy that the bullet can give up if it were to drop down to the initial temperature of the ice is $Q = mc\Delta T = (0.02\text{ kg})(233\text{ J/kg}\cdot\text{K})(30\text{ K}) = 140\text{ J}$. The latent heat of fusion for ice is 334,000 J/kg, so the 10 kg of ice would require 3.34 million joules for it to melt completely. We conclude that only a fraction of the ice will be melted by absorbing the energy of the bullet and that the final temperature of the bullet and the ice and melted water will thus be 0° C. The total energy available from stopping the bullet is 400 J and the energy available from cooling it down to 0° C is 140 J, so we have a total 540 J available to melt ice. The mass of ice that may be melted by this amount of thermal energy is found from Equation 14.5 to be

$$m = \frac{Q}{h} = \frac{540\text{ J}}{334{,}000\text{ J/kg}} = 1.6\text{ grams.}$$

This water will be found along the path the bullet takes through the ice. It will not refreeze again, no matter how long it sits there, since there is no transfer of heat from the 0° C ice block into the 0° C water.

14.5 Summary

In this chapter you should have learned the following:

- You should understand temperature as the average translational kinetic energy per molecule of a substance and be able to convert between this average kinetic energy in joules and the absolute Kelvin temperature using Equation 14.2.
- You should be familiar enough with Figure 14.2 to convert between the three temperature scales shown.
- You should be able to use Equation 14.3 and Table 14.1 to calculate the thermal expansion of a substance.
- You need to understand specific heats, as defined in Equation 14.4 and Table 14.2, and be able to use them to solve problems like Example 14.3.
- You need to understand changes of phase and latent heats, being able to use Equation 14.5 to calculate the heat required in each case.
- Finally, you should be clear on the usage of the new terms from this chapter. *Thermal energy* is the total internal energy of a body (p. 270). *Temperature* is the average translational kinetic per molecule (p. 270). *Heat* is the thermal energy transferred from one body to another by contact between the molecules (p. 274).

CHAPTER FORMULAS

Conversion from Celsius to Fahrenheit temperature: $T_F = \left(\frac{9}{5}\right)T_C + 32$

Conversion between temperatures in kelvins and in joules: $\langle \text{ke} \rangle = \frac{3}{2}kT_K$

Thermal expansion: $\Delta L = \alpha L \Delta T$

Specific heat equation: $Q = mc\Delta T$

Latent heat equation: $Q = mh$

PROBLEMS

14.1 * 1. A medical assistant takes your temperature using a hospital thermometer and writes down 40.0° C on your chart. What is your Fahrenheit temperature? Should you stay in the hospital?

** 2. In Figure 14.2, we did not label absolute zero for the Fahrenheit scale. What is the temperature of absolute zero, measured in the Fahrenheit scale?

** 3. A blacksmith heats a piece of iron until it glows bright yellow at a temperature of 2200° F. What is its temperature in the Kelvin scale?

** 4. The helium gas in the outer layer of the sun has an average kinetic energy per molecule (temperature) of about 6000 K.

a) What is this energy in joules?

b) The mass of an atom of helium is 6.6×10^{-27} kg. What is the average speed of a helium atom at this temperature?

** 5. A helium ion (an ion is an electrically charged atom) of mass 6.65×10^{-27} kg is accelerated to a speed of 5×10^{7} m/s. Find its kinetic energy and calculate the Kelvin temperature that this kinetic energy would correspond to.

** 6. A bricklayer drops a lump of mortar from a point 30 meters up the side of a brick building he is working on and it hits the ground with a splat. Mortar has a specific heat of 1200 J/(kg·K). By how many degrees will the mortar be warmer, assuming that all the energy it acquires during the drop is converted to internal energy inside the mortar and not the ground?

** 7. If a lead bullet is at a temperature of 27° C (300 K), then each atom has a random translational velocity due to its temperature. If the bullet is at rest, the average *speed* of the atoms is not zero, because of the non-zero temperature, but the average *velocity* of the atoms *is* zero, because the center of mass of the bullet does not move. However, if a bullet is moving at a speed of 300 m/s, the average of the velocities of the atoms is 300 m/s in the direction of the motion. The mass of each lead atom is 3.44×10^{-25} kg. Find the average speed of the lead atoms when the bullet is at rest and compare with the 300 m/s common velocity that the atoms acquire as a result of the bullet's motion.

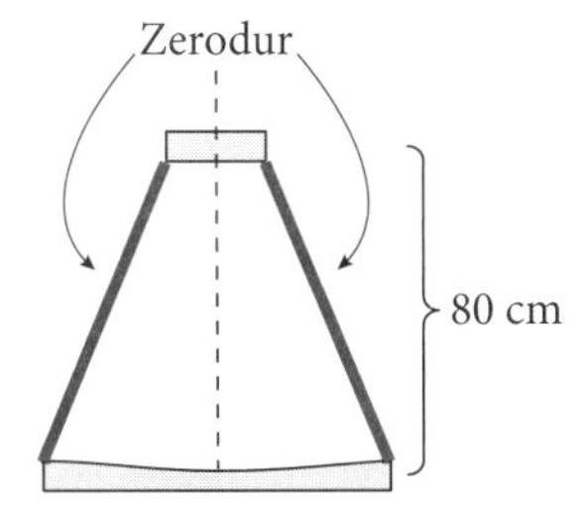

14.2 ** 8. The primary and secondary mirrors of a space telescope are assembled using zerodur spacers (see Table 14.2) in an engineering laboratory in which the room temperature is 20° C. In space the telescope will drop to –30° C. If the distance between the mirrors is exactly 80 cm in the lab, what will it be in space?

** 9. Steel rails 20 m long are laid on a cold winter day at –10° C. How much space must be left between the rails to allow for expansion at a summer temperature of 40° C?

** 10. If the aluminum cap is too hard to unscrew on a new jelly jar, explain why running the cap under hot water will help to loosen it.

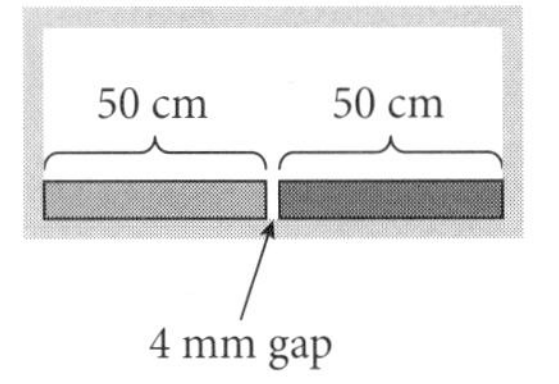

** 11. Two bars, each of length 50.0 cm, are placed inside a concrete box whose inside dimension is 1.004 m. One of the bars is made of steel and the other is made of aluminum. The initial temperature of the box and the base is 20° C. The box and its contents are then heated up. At what temperature will the bars just touch together? [Hint: Do not forget the expansion of the concrete.]

** 12. When the pioneers were crossing the plains in covered wagons, a metal rim could be attached to a wooden wagon wheel by heating the rim to expand it and then placing it over the wheel before it cooled. The shrink-attachment would firmly hold the rim in place and strengthen the wooden wheel. Suppose a certain wooden wagon wheel has an outer circumference of 300.0 cm and that its iron rim has an inside circumference of 299.5 cm when it is at 15° C. To what temperature must the rim be heated so that it will just fit over the wheel?

*** **13.** A steel cup at 20° C contains exactly 200 cm^3 of water at 20° C when it is full to the brim. If the cup is put in an oven that raises the temperature to 80° C, what volume of water will overflow the cup? The volume coefficient of thermal expansion for water is $\beta = 207 \times 10^{-6}$/K. [Hint: See Example 14.2 for the meaning of β.]

** **14.** It is possible to define a coefficient of surface area expansion γ via the equation

$$\Delta A = \gamma A \Delta T.$$

Find the coefficient of surface expansion for aluminum by considering the linear expansion of an aluminum plate 1 meter on a side.

14.3 ** **15.** One historical unit of thermal energy that most people have heard of is the calorie, defined as the amount of heat required to raise 1 g of water by 1 K. Find the number of joules in a calorie. [By the way, the food calorie that dieters know so well is actually a kilocalorie—the amount of heat required to raise 1 *kg* of water by 1 K.]

* **16.** How much heat is required to raise 0.50 kg of alcohol from room temperature of 20° C to the boiling point of alcohol at 78° C?

* **17.** If it takes 1800 joules of heat to raise 180 grams of some substance by 20 K, what is the specific heat of the substance?

** **18.** An ornamental pond is 80 cm deep. If the sunlight falling on the pond transfers energy to the surface of the pond at an average rate of 200 watts/m^2, find the time it will take to raise the temperature of the pond from 5° C to 20° C. Assume that all of the heat absorbed by the water remains in the water. (But this is a very bad assumption!)

** **19.** A woman driving a 1600-kilogram minivan at a speed of 20 m/s (about 45 mi/hr) applies the brakes and brings the car to a stop without skidding. The frictional force that slows the car is the friction between the two small ceramic pads and steel disk that comprise each disk brake. Assume that the car's energy goes into heating the disks, and that none goes into the pads. Also assume that none of the energy is conducted away by the air around the brakes (though that is exactly what disk brakes are expected to do). If each disk of the 4 disks on the car has a mass of 5 kg, what will be the increase in temperature of the disks as the car stops? (The specific heat of steel is 490 J/kg·K.)

** **20.** When a 100-gram cylinder of copper, initially at 100° C, is placed in 100 grams of household ammonia, originally at 20° C, the final temperature of the ammonia is measured to be 26° C. What is the specific heat of the household ammonia?

** **21.** A metal cylinder is made of 50 g of silver and 120 g of copper. It is heated to 100° C and then placed in 400 grams of water at 10° C, which is contained in a well-insulated can of negligible mass. What will be the final temperature of the cylinder in the can of water?

14.4 ** **22.** How much heat is required to melt 0.040 kg of ice at 0° C and then raise the temperature of the melted water to 30° C?

** **23.** An ice cube of mass 0.500 kg is initially at a temperature of –10° C. If the ice cube absorbs 15,000 joules of heat, what will be its final temperature?

** **24.** A 10-gram ice cube at 0° C is dropped into 100 grams of water at 10° C. The water and ice cube are well insulated from the outside environment. Will the ice cube melt completely or not? Justify your answer.

** 25. A 2-kg block of copper is heated to 400° C and dropped into an insulated bucket that contains a mixture of ice and water at 0° C. The final temperature of the copper is 0° C. How much ice melted in the process?

*** 26. A man in a sidewalk café complains to the waiter that his coffee is too hot. The waiter informs him that the coffee they serve is always served at 96° C. The man explains that his name is "Bond, James Bond," and that he always takes his coffee at exactly 72° C. The waiter knows that the café's cups hold exactly 180 cm^3 of coffee (about 6 ounces), and he knows that the density and specific heat of coffee are the same as that of water. How many 10-gram ice cubes initially at 0° C must the waiter place in the coffee to bring the temperature of the mixture down to 72° C? Please be accurate to the nearest tenth of an ice cube.

*** 27. A section of iron railroad track 20 m long resists expansion or compression with a spring constant $k = 6.00 \times 10^6$ N/m. The track was laid on a winter day at 0° C and nailed tightly to wooden ties that were themselves anchored into the ground. On a hot 40° C summer day, what sideways force will the expanding track put against the nails in the wood? Assume that the ground and the wood do not significantly expand with the change in temperature.

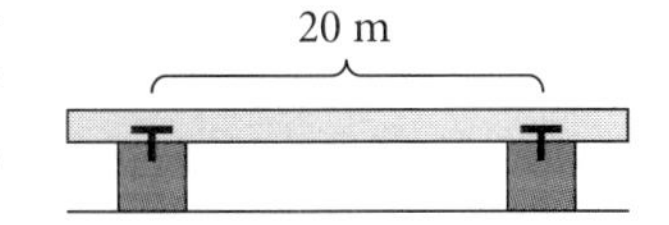

*** 28. A 200-gram chunk of aluminum alloy, with specific heat $c = 1000$ J/(kg·K), is heated to a temperature of 467° C and dropped into a Styrofoam can containing 0.100 kg of ice mixed with 0.90 kg of water, initially at 0° C. Assume that the Styrofoam completely insulates the ice water and aluminum combination and that the Styrofoam has negligible mass of its own. Find the final temperature of the combination.

15

THERMODYNAMICS

In our introduction to thermal physics in the last chapter, we consistently approached the topic from the point of view of our modern understanding that matter is made up of atoms and molecules and that thermal energy and temperature are just statistical ways of accounting for the internal motions of those atoms and molecules. But this was not always the case. In the past, theories that explained heat as a fluid contained in warm bodies, a fluid called *caloric*, were able to successfully explain many laboratory observations. However, one of the great successes of the statistical view of thermal physics was the derivation of the ideal gas law from the laws of mechanics alone. We begin this chapter by showing how this is done.

15.1 The Kinetic Theory of Gases

We should probably point out that this derivation is not something you will need to be able to reproduce and that the problems for this chapter will actually only use our final result, Equation 15.7. But we do hope that you have become enough of a physicist by now that this use of mechanical laws may appeal to you. And, anyway, a general understanding of the statistical nature of thermal physics is one of the things expected of students in physics classes at our level.

Let us begin by considering N molecules of a gas that are free to move inside a rectangular box of cross-sectional area A and length L (see Figure 15.1). Let all the molecules have the

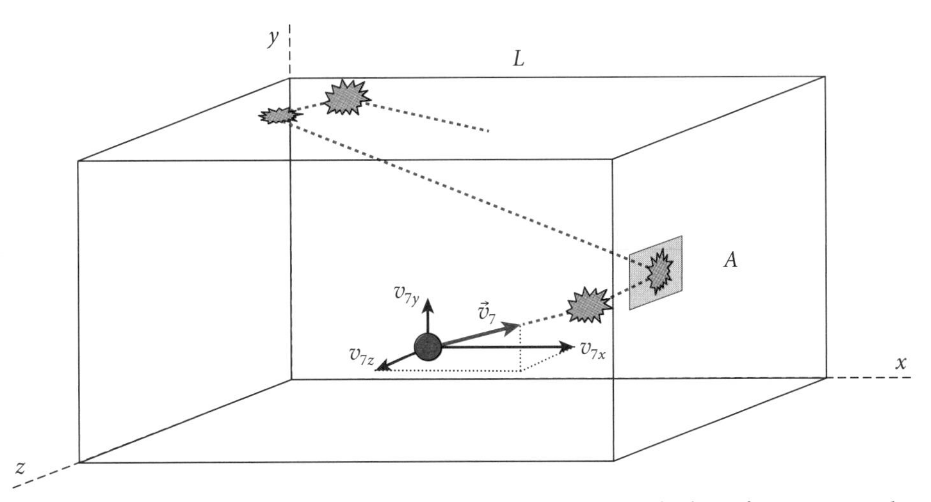

Figure 15.1 One of the gas molecules (molecule number 7) in a rectangular box of cross-sectional area A and length L. The molecule's velocity is shown as a red arrow. The velocity components are shown in black. The path of the molecule is the dotted red line, bouncing off the walls of the box at the little red splats. Its bounce off the rightmost wall (in the small gray rectangle) puts an impulse of pressure on that wall.

same mass m, and let us number the molecules from 1 to N. Molecule number 7 is shown as a typical molecule in Figure 15.1. It has velocity $\vec{v}_7$, with components (v_{7x}, v_{7y}, v_{7z}) in the three directions. The path of this molecule over time is indicated by the red dotted line. As it bounces off the right wall of the box (the bounce in the little gray rectangle), the x-component of its velocity will change from v_{7x} to $-v_{7x}$, the same magnitude but the opposite direction. The other velocity components remain what they were before hitting that wall.

Every time molecule number 7 bounces off the rightmost wall, it experiences a sharp change in its momentum of $\Delta p_x = -2mv_{7x}$. The change is due to an impulse numerically equal to the change in its momentum. Newton's Third Law requires that this molecule must impart an equal and opposite impulse to the wall, $2mv_{7x}$ in the positive x-direction. If the force acts during a short time δt, then the average force of the molecule on the wall during the time it acts on the wall is $\hat{F}_{7x} = 2mv_{7x}/\delta t$. (See Equations 8.1 and 8.2 for the relationship between force and the change in the momentum.)

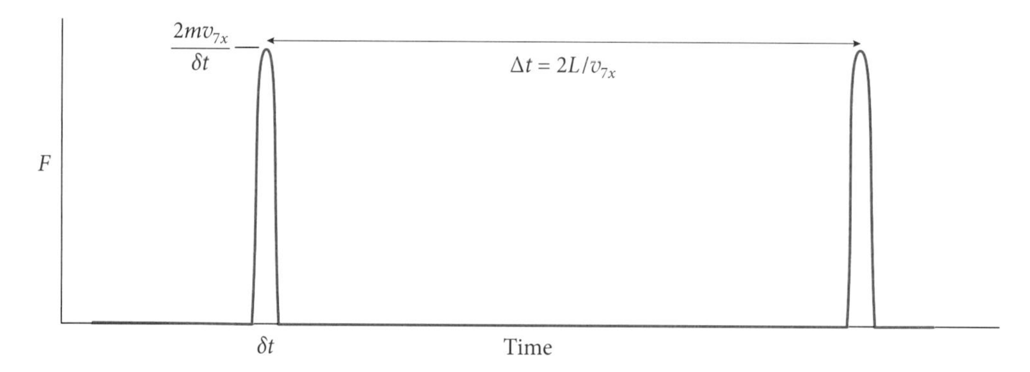

Figure 15.2 The force produced by molecule number 7 on the rightmost wall of the box as a function of time.

The magnitude of the force that the wall imposes on molecule number 7, and therefore the force of the molecule on the wall, is graphed in Figure 15.2. There is a sharp force on the wall as the molecule strikes it. Then there is a long gap, as that molecule travels back to the left, bounces off the left wall, and travels back toward the right wall to strike it again. We assume that the wall is at the same temperature as the gas, so the molecule does not lose energy to the walls. Therefore, the collisions will be elastic, and the x-component of the velocity will not change magnitude, only direction. Since the total x-distance the molecule has to travel between bounces off the right wall is $2L$, the time between bounces is $\Delta t = 2L/v_{7x}$. Therefore, the average force on the right wall, averaged over time in the long term, is

$$F_{7x} = \frac{2mv_{7x}}{\Delta t} = \frac{2mv_{7x}}{2L/v_{7x}} = \frac{m}{L}v_{7x}^2. \tag{15.1}$$

This is the time-averaged force produced by the one molecule, molecule number 7. The *total* force on the rightmost wall will be due to all of the N molecules in the box. One mathematical way of saying that we add everything up is to write

$$F_{\text{tot},x} = F_{1x} + F_{2x} + F_{3x} + \cdots + F_{Nx} = \frac{m}{L}\left(v_{1x}^2 + v_{2x}^2 + v_{3x}^2 + \cdots + v_{Nx}^2\right). \tag{15.2}$$

The three dots in this equation mean "and so on, up to the last item."

There are two ways we want to look at the summation on the right-hand side of Equation 15.2. First, let us note that the definition of the average of some quantity that is a property of the particles consists in adding the quantities up and dividing by the total number of particles. Thus the average value of v_x^2 is

$$\langle v_x^2 \rangle = (v_{1x}^2 + v_{2x}^2 + v_{3x}^2 + \cdots + v_{Nx}^2)/N, \tag{15.3}$$

where we write $\langle v_x^2 \rangle$ to mean the average of the quantity v_x^2, averaged over all the molecules of the gas. Equation 15.3 allows us to also write

$$(v_{1x}^2 + v_{2x}^2 + v_{3x}^2 + \cdots + v_{Nx}^2) = N\langle v_x^2 \rangle. \tag{15.4}$$

Second, let us note that the kinetic energy of any one of the molecules, say molecule number 7, is $\mathrm{ke}_7 = \frac{1}{2}mv_7^2 = \frac{1}{2}m(v_{7x}^2 + v_{7y}^2 + v_{7z}^2)$. Then let us point out that there is nothing special about the x-direction, compared to any of the other two directions. So, on average, it must be true that $\langle v_x^2 \rangle = \langle v_y^2 \rangle = \langle v_z^2 \rangle$. This means that we can write the average kinetic energy over all of the molecules of the gas as

$$\langle \mathrm{ke} \rangle = \frac{1}{2}m\langle v^2 \rangle = \frac{1}{2}m\left(\langle v_x^2 \rangle + \langle v_y^2 \rangle + \langle v_z^2 \rangle\right) = \frac{1}{2}m\left(3\langle v_x^2 \rangle\right)$$

or

$$\langle \mathrm{ke} \rangle = \frac{3}{2}m\langle v_x^2 \rangle. \tag{15.5}$$

If we put Equations 15.2, 15.4, and 15.5 together, we find

$$F_{\mathrm{tot},x} = \frac{m}{L}\left(v_{1x}^2 + v_{2x}^2 + v_{3x}^2 + \cdots + v_{Nx}^2\right) = \frac{m}{L}N\langle v_x^2 \rangle = \frac{2N}{3L}\langle \mathrm{ke} \rangle.$$

The pressure of the gas on the rightmost face of the box is equal to the force on that side divided by the area of the side. This gives

$$P = \frac{F_{\mathrm{tot},x}}{A} = \frac{2N}{3LA}\langle \mathrm{ke} \rangle = \frac{2N}{3V}\langle \mathrm{ke} \rangle,$$

where we noted in the last step that $V = LA$ is the volume of the box. If we multiply both sides of this equation by V, we get

$$PV = N\left(\frac{2}{3}\langle \mathrm{ke} \rangle\right). \tag{15.6}$$

We recall from Equation 14.2 that the conversion from average translational kinetic energy in joules to temperature in kelvins is $\langle \mathrm{ke} \rangle = \frac{3}{2}kT$, where k is Boltzmann's constant (1.38×10^{-23} J/K). Making this substitution turns Equation 15.6 into

$$PV = NkT. \tag{15.7}$$

This relation between the pressure, volume, and temperature in a gas containing N molecules is known as the *ideal gas law*. It was, in fact, the desire to not carry along the $\frac{2}{3}$ factor that appears in Equation 15.6 that led to keeping the explicit $\frac{3}{2}$ factor back in Equation 14.2.

That was a lot more algebra than we usually take you through, wasn't it? Let's see what we have accomplished with all of this.

15.2 The Ideal Gas Law

First, let's notice what assumptions we made as we derived Equation 15.7 in the last section. We first worked out the average force that a single molecule put on a single wall of a box of volume V. When we applied this result to the case with N molecules in the box, we did not adjust the volume. So, in effect, we have assumed that the gas is rare[1] enough that the additional molecules do not significantly reduce the volume available for any single molecule. We also talked only about kinetic energy, ignoring any possible potential energy that might exist between the molecules. So the result we got assumes that forces between the molecules of the gas are negligible until the molecules collide with the box walls or with each other. These are the assumptions that define what is called an *ideal gas*, and this is the only kind of gas to which Equation 15.7 applies.

Now some of you may be used to seeing the ideal gas law written like this:

$$PV = nRT, \tag{15.8}$$

where n is the number of "moles" of the gas. Equation 15.8 is really the same equation as Equation 15.7. If, in Equation 15.7, we had chosen to count molecules by dozens, then the ideal gas law would read $PV = dQT$, where d is how many dozen molecules you have ($d = N/12$) and Q would be a new constant, equal to $12k$. But it would still be the same equation. Like a dozen, a "mole" is a particular number, in this case Avogadro's number, 6.02×10^{23}. If we want to count molecules by moles instead of by dozens, then the gas law can be written $PV = nRT$, where n is the number of moles of the gas ($n = N/6.02 \times 10^{23}$), and where R is a new constant, called the "gas constant," given by $R = (6.02 \times 10^{23})\ k = 8.31$ J/mole. But it is still the same equation.

While we're at it, we might as well point out where Avogadro's number came from. The mass of an atom is a mass, of course, and so it should be measured in kilograms (or grams, if you like). But when you are working with atomic or molecular masses, it is annoying to keep having to carry around the 10^{-23} on everything. So a new unit for mass was defined, called the *atomic mass unit* (clever name), abbreviated u. The atomic mass of an atom is close to the number of protons and neutrons in its nucleus. However, even though the number of protons and neutrons in a nucleus is always an integer, the masses of most atoms and molecules are not exact integers, for reasons we will discuss in Chapters 24 and 26. The only exception to this non-integer value for atomic mass is the case of carbon. This is because the atomic mass unit is *defined* so that the mass of a carbon-12 atom is exactly 12 u.

Avogadro's number was defined so that the mass in grams of Avogadro's number of the molecules was numerically equal to the mass of one molecule of the substance in atomic mass units. Carbon has an atomic mass of 12 u, so the mass of Avogadro's number of molecules is 12 g. If I have 24 g of carbon, I have exactly 2 moles of carbon. A mole is 6.02×10^{23}, so I have 12.04×10^{23} atoms of carbon. The atomic mass of hydrogen is 1.008 u, so one mole of hydrogen will have a mass of 1.008 g. Thus, if I have 24 g of hydrogen, I have (24 g)/(1.008 g/mole) = 23.8 moles of hydrogen, or 1.43×10^{25} atoms of hydrogen. In our problems and examples, we will sometimes measure the number of atoms or molecules in plain numbers or in moles, so you will need to be able to use both Equations 15.7 and 15.8.

We finish this section with a couple of examples that show how to use the ideal gas law. The first example is just to help us see the meaning of the quantities in the equation.

---

1. The term "rare" does not just mean "unusual," it also means the opposite of "dense."

EXAMPLE 15.1 Using the Ideal Gas Law—I

Twenty-two grams of a particular gas consist of 3×10^{23} molecules of the gas. The gas fills a cubical metal box 10 cm on a side, maintained at room temperature of 20° C. What pressure will the gas exert on the walls of the container?

ANSWER The problem gives N, V, and T, and asks for the pressure P. We can solve Equation 15.7 for pressure, to get

$$P = \frac{NkT}{V}.$$

Everything on the right-hand side is known, but we must still be careful with the quantities we put in the equation. First, the question gives the temperature in Celsius degrees, but we must remember that the T in the gas law needs to be in the Kelvin scale. (Remember that we used $\langle \text{ke} \rangle = \frac{3}{2}kT$ in deriving the ideal gas law.) We convert the Celsius scale to the Kelvin scale by adding 273, giving a temperature $T = 293$ K. The volume of the box needs to be in m^3, so we calculate it as $V = (0.1 \text{ m})(0.1 \text{ m})(0.1 \text{ m}) = 10^{-3} \text{ m}^3$. Using these numbers in the equation for P, we get

$$P = \frac{(3 \times 10^{23})(1.38 \times 10^{-23} \text{ J/K})(293 \text{ K})}{10^{-3}\, m^3} = 1.2 \times 10^6 \text{ N/m}^2.$$

The pressure is 1.2×10^6 N/m^2, or 1.2 million pascals. If you remember that one atmosphere of pressure is about 10^5 Pa (see page 193), you will note that the pressure in the box is about 12 atmospheres.

Our next example shows how to use the gas law when not all of the quantities in the equation are given in the problem. In this case, as you will see, it is possible to take two versions of the gas law and divide one by the other. This gives us an equation in which we can cancel quantities that are the same on each side of the equation.

EXAMPLE 15.2 Using the Ideal Gas Law—II

On a day when the air temperature and the water temperature are the same, a man descends in a 3-meter-high cylindrical diving bell to a depth of 20 m. At this depth, how high is the remaining air space in the diving bell?

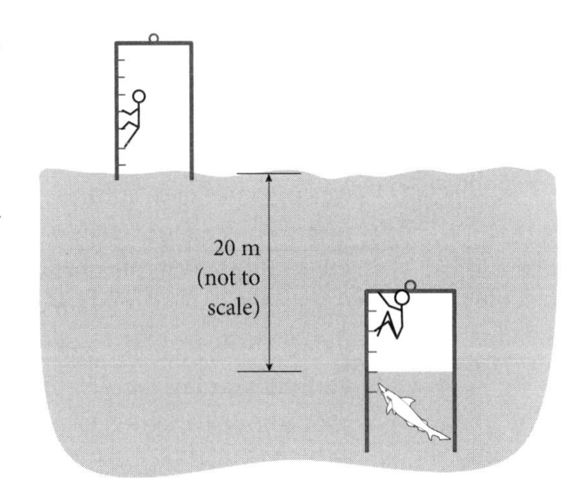

ANSWER The number of air molecules and the temperature are not given in this problem, but we know that they both stay the same as the diving bell submerges. If we use a subscript 1 for the picture at sea level and a subscript 2 for when the diving bell is submerged, we can write the gas law for each case:

$$P_1V_1 = N_1kT_1 \qquad \text{and} \qquad P_2V_2 = N_2kT_2.$$

If we divide one of these equations by the other, we get

$$\frac{P_1V_1}{P_2V_2} = \frac{N_1T_1}{N_2T_2}.$$

In our case, $N_1 = N_2$, since there is no way for the air to get out of the diving bell. We also know that the temperatures are the same before and after the diving bell submerges, so $T_1 = T_2$. This means that the right-hand side is just 1. If we then multiply both sides of this equation by P_2V_2, we get

$$P_1V_1 = P_2V_2.$$

The pressure at sea level is $P_1 = 10^5$ Pa. The pressure inside the air pocket in the diving bell is the same as water pressure at the depth of 20 m, which we can calculate from Equation 10.3:

$$P_2 = P_1 + \rho gh.$$

This yields a value $P_2 = 10^5$ PA + (1000 kg/m^3)(10 m/s^2)(20 m) = 3×10^5 Pa. Solving for V_2, we find

$$V_2 = V_1\frac{P_1}{P_2} = V_1\frac{1 \times 10^5 \text{ Pa}}{3 \times 10^5 \text{ Pa}} = \frac{1}{3}V_1.$$

The cylinder is uniform in cross-sectional area, so if the air space fills the entire 3 m at sea level, there will only be 1 meter of air available at a depth of 20 m.

15.3 Thermodynamics

You have now read 288 pages of physics, and we figure that you are now wise enough to know the truth. You remember back in Chapter 7, where we said that the energy lost to friction should not be thought of as going into a thermal energy bucket? You probably got the impression that the reason was that you could never get any energy back out of that thermal bucket. Well, we thought that it would just confuse things to bring this up back in Chapter 7, but it turns out that this isn't completely true. You will never be able to get *all* the energy out of the bucket, as we said, but, with the right setup, it is possible to convert *some* thermal energy back into mechanical energy. Let's see how.

Let us look again at the sliding book with which we began Chapter 14. There, we discussed how the frictional force transforms the kinetic energy of the book into thermal energy in the book and the tabletop. Although our discussion on page 269 emphasized the internal energy in the tabletop, the same analysis would apply to the molecules of the book, and we would conclude that some of the initial kinetic energy will end up as thermal energy in the book, thereby raising the temperature of the book.

Now let's imagine the apparatus shown in Figure 15.3. A closed rectangular box is separated into two halves by a movable piston, and there is an adjustable spring in the right half, as shown. The left half is open to the air in the room, but the right half of the box is pumped out until no air remains. The spring is then adjusted until the restoring force of the compressed spring just balances the force on the piston from the atmospheric gas pressure in the left half of the box. After the sliding book has come to a stop, it will be warmer than the air in the room. So, if the book is placed inside the left half of the box, and the box closed, the book will lose thermal energy as its vibrating molecules collide with the air molecules in the left half of the box and raise their temperature. The increased air temperature will raise the pressure, as predicted by Equation 15.7, and the higher pressure will make the gas in the left half of the box expand, doing work as it compresses the spring a little more. Thus, we have taken some of the thermal energy contained in the book and used it to do work on the spring. In the process, the temperature of the air in the left half of the box will have increased overall while the temperature of the book decreased after it was put in the box.

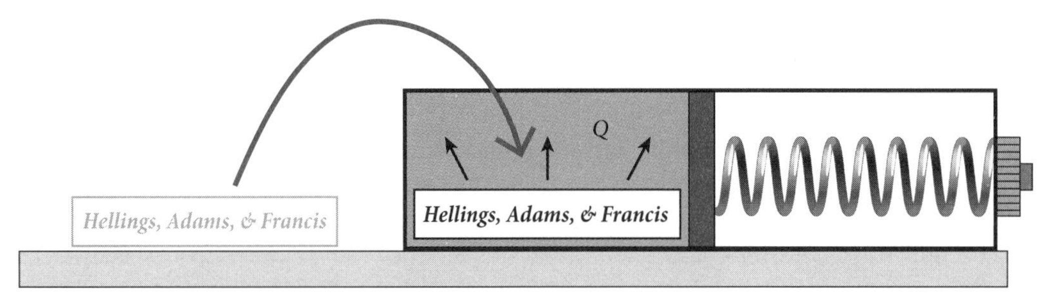

Figure 15.3 The warm book is placed in the left half of the box. As heat is transferred from the book to the surrounding air, the pressure in the air increases and does mechanical work on the spring.

Of the total thermal energy transferred into the gas, some has gone into the work done by the gas as it expanded to compress the spring, and some will show up as increased internal energy of the gas. But, because total energy is always conserved, it will all end up in one place or the other. We may express this energy conservation as

$$Q_{in} = \Delta E_{int} + W_{by}. \tag{15.9}$$

The heat transferred *into* the gas Q_{in} shows up as internal energy of the gas ΔE_{int} plus the work done *by* the gas on outside elements W_{by}. Traditionally, this important equation of thermodynamics is written in terms of the work done *on* the gas, rather than the work done *by* the gas, so this equation representing the conservation of energy in a thermal system is generally written

$$\Delta E_{int} = Q_{in} + W_{on}, \tag{15.10}$$

where $W_{on} = -W_{by}$ is work done *on* the gas by pressing on it to decrease its volume. Equation 15.10 is known as the *First Law of Thermodynamics*. The First Law of Thermodynamics, in the form of Equation 15.10, says that the two ways to increase the internal energy of a gas are to do work *on* the gas or to transfer heat *into* the gas.

The book-in-the-box setup of Figure 15.3 is an example of a *heat engine*. A heat engine is an apparatus in which heat is removed from some object at a high temperature and transferred to some other object that is at low temperature, and where, in the process, mechanical work is done by the apparatus.

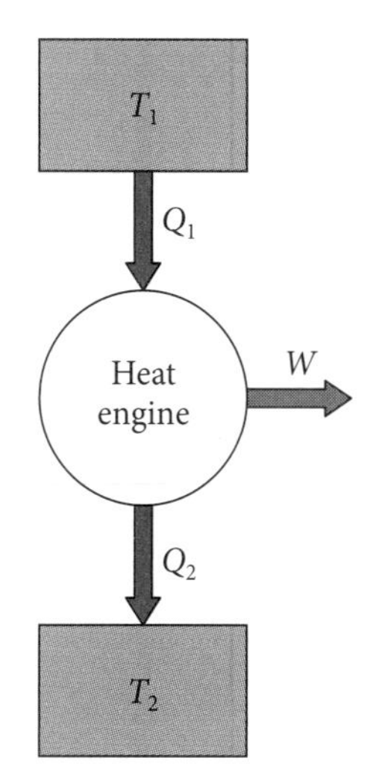

Figure 15.4 A schematic diagram of a heat engine.

Figure 15.4 is a schematic diagram for a heat engine. It shows a reservoir[2] of thermal energy at a high temperature T_1, another reservoir of thermal energy at a lower temperature T_2, and some kind of engine in-between that can convert thermal energy into mechanical

2. By "reservoir," we mean some system that is large enough that the addition or subtraction of heat from it does not significantly change its properties, in this case its temperature.

work. As heat flows from high temperature to low temperature, a quantity of heat Q_1 will leave the high-temperature reservoir, a quantity Q_2 will be dumped into the low-temperature reservoir, and some mechanical work W will be done by the engine. The *efficiency* of a heat engine is a measure of how much work can be extracted from a given amount of heat taken from the high-temperature reservoir:

$$\varepsilon = \frac{W}{Q_1}. \tag{15.11}$$

We assume that no internal energy builds up inside the heat engine, so the First Law of Thermodynamics applied to the heat engine requires $Q_1 = W + Q_2$. If a heat engine were to be 100% efficient, Equation 15.11 would require $W = Q_1$, meaning that the exhaust heat Q_2 would be zero. In this case, all of the heat energy could be converted into mechanical energy. Unfortunately, there is a *Second Law of Thermodynamics*, which, when applied to heat engines, may be stated as saying that "*in any thermodynamic process between a high-temperature reservoir and low-temperature reservoir, some heat will always be lost to the low-temperature reservoir.*"[3]

Because of the Second Law, all heat engines will be less than 100% efficient, but how good can they get? Well, if one is careful to add heat at just the right times and allow the working gas to expand in just the right conditions, it is possible to reach the efficiency of what is called a Carnot engine, named after French physicist Sadi Carnot.[4] It has been proven that the efficiency of a Carnot engine is the maximum efficiency that can be obtained by any cyclical heat engine operating between two heat reservoirs, one at T_1 and one at T_2. The theoretical Carnot efficiency is

$$\varepsilon_{max} = \frac{T_1 - T_2}{T_1}, \tag{15.12}$$

with the temperatures being measured in the Kelvin scale.

Now there is probably someone out there reading Equation 15.12 who suddenly realizes,

> Aha! I see a loophole in the Second Law! If I can create a reservoir with $T_2 = 0$, I can build a Carnot engine with 100% efficiency. What's more, since a 100% efficient engine would have $Q_2 = 0$, my low-temperature reservoir would never heat up. If I can just find a tiny BB at zero kelvins, I can use it to drain all the heat in all the oceans of the world to provide virtually unlimited power. I will be rich beyond my wildest dreams. Bwaa-ha-ha-ha!

Unfortunately, there is also a *Third Law of Thermodynamics*, which may be stated as "*no physical system may ever be brought to a temperature of absolute zero in less than an infinite number of steps.*"

The following example shows how the Carnot efficiency formula works.

---

3. Actually, the Second Law of Thermodynamics is generally stated in terms of a quantity called *entropy*. Entropy measures the number of ways that the molecules of a system can share the energy of the system, subject to the constraint of maintaining well-defined temperatures. But this concept is not necessary for understanding the limited thermodynamics we do here.

4. The last name is pronounced car-no′. It is not nice to pronounce it "care not."

EXAMPLE 15.3 The Efficiency of a Carnot Heat Engine

A Carnot engine operates between 400 K and 300 K.

a) What is its theoretical efficiency?
b) How much waste heat will be dumped into the low-temperature reservoir for each 1000 J of useful work?

ANSWER The Carnot efficiency is

$$\varepsilon_{max} = \frac{T_1 - T_2}{T_1} = \frac{400\text{ K} - 300\text{ K}}{400\text{ K}} = 0.25.$$

The heat that must be taken from the high-temperature reservoir to provide 1000 J of work is found from Equation 15.11. It is

$$Q_1 = \frac{W}{\varepsilon} = \frac{1000\text{ J}}{0.25} = 4000\text{ J}.$$

So the waste heat must be $Q_2 = Q_1 - W = 4000\text{ J} - 1000\text{ J} = 3000\text{ J}$. Waste heat is one of the sad physics realities of the use of energy. The more energy we produce, the more thermal pollution we must create as a by-product.

15.4 Specific Heats of a Gas

When we first introduced specific heats, back on page 274, we explained that the relationship between the temperature of a substance and its total internal energy was complicated for liquids and solids, but not so difficult to understand for gases. Now that we have studied gases a bit, let's examine how this relationship arises.

Let us first consider a *monatomic* gas like helium, where each molecule consists of only a single atom. The mass of a single atom is concentrated in its tiny nucleus, producing a rotational inertia that is very small. We should pause for a moment to explain what this implies. When we finally get to the subject of quantum mechanics in Chapter 25, we will be able to see why it is that, at normal laboratory temperatures, thermal energy cannot show up as rotational kinetic energy if a particle has a small rotational inertia. (In fact, you will explain this result yourself if you do problem 14 in Chapter 25.) For now, you must simply trust us that rotational kinetic energy is not available as a place to find thermal energy in a gas of monatomic molecules, because of their small rotational inertia. In addition, the single atoms in a monatomic molecule have no neighboring atom in the molecule to provide a restoring force, so there can be no molecular vibrations. As a result, the only thing single atoms can do with their internal energy is translate from one place to another. Their total internal energy will always be the number of molecules in the gas times the average translational kinetic energy per molecule. For N molecules of a *monatomic gas*, the internal energy is

$$E_{\text{int}} = N\langle\text{ke}\rangle = \frac{3}{2}NkT, \tag{15.13}$$

where we have remembered the Equation 14.2 relation between average translational kinetic energy per molecule and the Kelvin temperature.

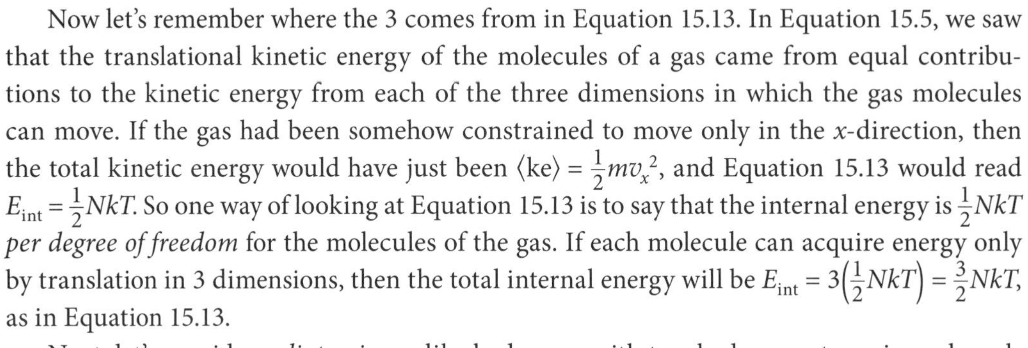

Now let's remember where the 3 comes from in Equation 15.13. In Equation 15.5, we saw that the translational kinetic energy of the molecules of a gas came from equal contributions to the kinetic energy from each of the three dimensions in which the gas molecules can move. If the gas had been somehow constrained to move only in the x-direction, then the total kinetic energy would have just been $\langle \text{ke} \rangle = \frac{1}{2}mv_x^2$, and Equation 15.13 would read $E_{\text{int}} = \frac{1}{2}NkT$. So one way of looking at Equation 15.13 is to say that the internal energy is $\frac{1}{2}NkT$ *per degree of freedom* for the molecules of the gas. If each molecule can acquire energy only by translation in 3 dimensions, then the total internal energy will be $E_{\text{int}} = 3\left(\frac{1}{2}NkT\right) = \frac{3}{2}NkT$, as in Equation 15.13.

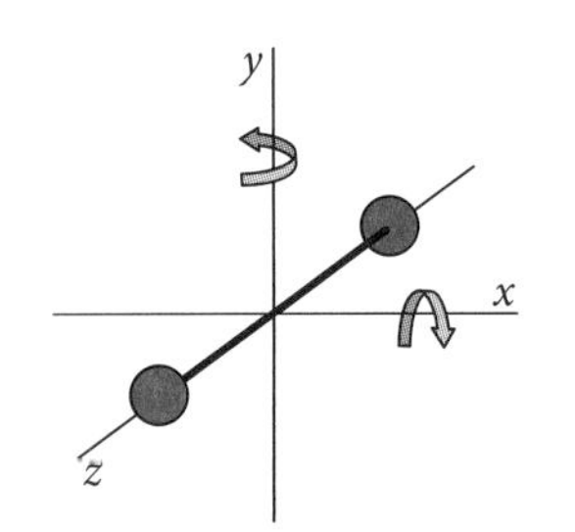

Figure 15.5 The two rotational degrees of freedom for diatomic molecules like H_2.

Next, let's consider a *diatomic* gas like hydrogen, with two hydrogen atoms in each molecule as depicted in Figure 15.5. It will, of course, have the same three translational degrees of freedom that all molecules have. But a diatomic molecule will have significant rotational inertia for rotation about the x-axis or the y-axis, as shown by the curved arrows in the figure. About the z-axis, however, the small rotational inertia will not allow for rotation to be excited, due to the same quantum mechanical reason we mentioned above. In principle, the molecule could vibrate along the z-axis, but, because the inter-atomic forces along the line joining the atoms are very great, the energy required to excite vibration is typically very high compared to that required for rotation or translation, meaning that a vibrational degree of freedom typically comes into play only at very high temperatures. Therefore, at room temperatures, only the three translational degrees of freedom and the two rotational degrees of freedom will carry significant internal energy, and the relation between internal energy and temperature for a *diatomic gas* will be approximately

$$E_{\text{int}} \approx \frac{5}{2}NkT. \tag{15.14}$$

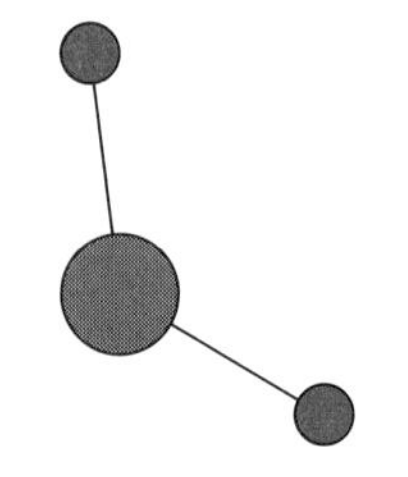

Figure 15.6 Water (H_2O) molecules are triatomic molecules.

Finally, a *polyatomic* molecule with three or more atoms (like Figure 15.6) will be able to absorb significant kinetic energy by rotating about any of the three axes. In addition, vibrations are easier to excite in these more complicated molecules, so the internal energy in these cases will satisfy

$$E_{\text{int}} \geq \frac{6}{2}NkT, \tag{15.15}$$

where the $>$ possibilities come into play as the vibrational modes are excited.

Let's make sure we are clear on the meaning of Equations 15.13, 15.14, and 15.15. If a monatomic gas, a diatomic gas, and a polyatomic gas are all at the same temperature, the internal energy in the polyatomic gas will be greatest. This is because temperature measures only the translational kinetic energy of the molecules of the gas, while there are many ways to store energy in the more complicated polyatomic molecules.

We are now ready to tackle the specific heats of an ideal gas. In Section 14.3, we introduced the specific heat as part of the relationship between the heat added to some system and the resulting change in its temperature. Equations 15.13 to 15.15 give the connection between temperature and *internal energy* in a gas. However, the relationship between the *heat added* to a gas and its resulting change in temperature is complicated by the fact that some of the heat may go into work done as the gas expands. So, to begin with, let's eliminate that possibility by restricting the gas to a closed container with *constant volume*. Then the gas can do no work, and all of the heat added will raise the internal energy of the gas.

When we wrote $Q = mc\Delta T$ back in Section 14.3, the specific heat c was the specific heat per kilogram of the material. For a gas, however, we usually use the specific heat per mole, so let us define the *molar specific heat at constant volume* via

$$Q = nC_V\Delta T, \tag{15.16}$$

where C_V is the molar specific heat at constant volume, the subscript telling us which quantity is held constant and a capital C being used for *molar specific heat* in place of the small c we used for mass specific heat back in Equation 14.4.

We can find specific heats at constant volume for the different kinds of gas molecules by use of Equations 15.13 to 15.15. When no work is done, Equation 15.10 becomes $Q = \Delta E_{\text{int}}$. For a monatomic gas, Equation 15.13 allows us to write $Q = \Delta E_{\text{int}} = \frac{3}{2}Nk\Delta T$. If we then remember that Equation 15.8 showed that $Nk = nR$, we see that we can also write this relationship in terms of the gas constant as $Q = \frac{3}{2}nR\Delta T$. Comparison of this result with Equation 15.16 allows us to read off the value of C_V for monatomic molecules. It is $C_V = \frac{3}{2}R$. The same process for other kinds of gas molecules gives us the molar specific heats at constant volume for the three cases as

$$\text{Monatomic: } C_V = \frac{3}{2}R \qquad \text{Diatomic: } C_V \approx \frac{5}{2}R \qquad \text{Polyatomic: } C_V \geq 3R \tag{15.17}$$

Although the values given in Equation 15.17 for diatomic and polyatomic molecules are only approximate, there is always an exact value of C_V for any gas. Now let us remember that the C_V in Equation 15.16 relates the heat added to a gas to the resulting temperature change in that gas, in the special case where all of the heat goes into the internal energy. This means that it is C_V that determines how the change in internal energy of a gas relates to the change in the temperature of the gas, in all cases. It is simply *always* true that

$$\Delta E_{\text{int}} = nC_V\Delta T, \tag{15.18}$$

whether the volume is actually constant or not. If you know the value of C_V for a gas, you know the relationship between its temperature and its internal energy in all conditions.

In addition to finding specific heats at constant volume, we can also calculate specific heats at constant pressure. For example, we can consider a system like the one shown in Figure 15.7, where a mass sitting on a movable piston maintains a force mg on the gas. If the area of the piston is A, then the pressure on the gas will be a constant $P = mg/A$. Thus, as the gas is heated, it will expand in a state of constant pressure. If the piston rises by Δh, then the work done *by* the gas is $W_{\text{by}} = mg\Delta h$. Written in terms of the pressure, this is $W_{\text{by}} = PA\Delta h$. But $A\Delta h$ is also the change in the volume of the gas, so we can write the work as $W_{\text{by}} = P\Delta V$. The First Law of Thermodynamics in the form of Equation 15.9 allows us to write the equation,

Figure 15.7 A mass on the piston keeps a constant pressure on the gas inside the cylinder.

$$Q = \Delta E_{\text{int}} + W_{\text{by}} = nC_V\Delta T + P\Delta V, \tag{15.19}$$

in which, in the second step, we also used Equation 15.18 to write ΔE_{int} in terms of C_V. Now, since P and n remain constant as the gas expands, the ideal gas law requires $P\Delta V = nR\Delta T$, so Equation 15.19 can also be written

$$Q = nC_V\Delta T + nR\Delta T = n(C_V + R)\Delta T. \tag{15.20}$$

If we define the *molar specific heat at constant pressure* via the equation $Q = nC_P\Delta T$, we may read off the value for the molar specific heat at constant pressure, giving

$$C_P = C_V + R. \tag{15.21}$$

It's probably time for an example.

EXAMPLE 15.4 Thermodynamics with a Gas at Constant Pressure

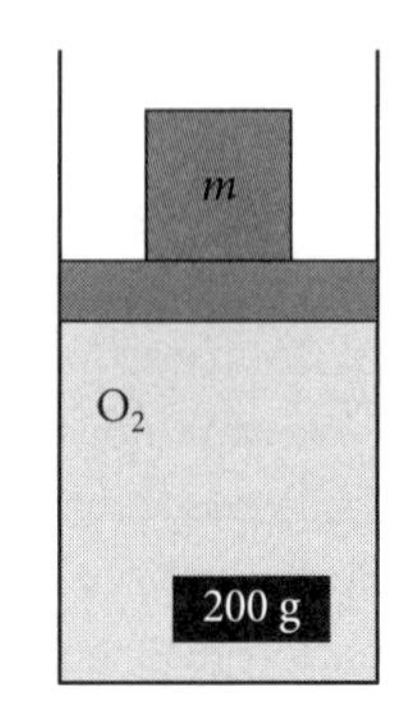

A piston holds 2 moles of oxygen (with $C_V = \frac{5}{2}R$) in an insulated cylinder, as shown in the figure at right. The oxygen is initially at a temperature of 300 K. A 200-g slug of an aluminum alloy with specific heat $c = 900$ J/kg·K is heated to 400 K and placed in contact with the oxygen.

a) What will be the final temperature of the oxygen and the aluminum?

b) How much work will the gas do as it expands at constant pressure?

c) What is the efficiency of this apparatus in generating work?

ANSWER a) The heat that leaves the aluminum will be absorbed by the oxygen gas, where it will both do work on the piston and increase the internal energy of the gas. Let us use T_{Al} for the initial temperature of the aluminum, T_{O_2} for the initial temperature of the oxygen, and T for the final temperature of both. Then we have

$$Q_{out} = mc(T_{Al} - T) = Q_{in} = nC_P(T - T_{O_2}).$$

We used C_P in this equation because the piston maintains constant pressure as the gas expands. The value of C_P will be $C_P = \frac{7}{2}R$ because O_2 is a diatomic molecule with $C_V = \frac{5}{2}R$, so that $C_P = C_V + R = \frac{7}{2}R$. If we put in all the values we know, our previous equation becomes

$$(0.2 \text{ kg})(900 \text{ J/kg}\cdot\text{K})(400 \text{ K} - T) = (2 \text{ moles})\left(\frac{7}{2}8.31 \text{ J/mole}\cdot\text{K}\right)(T - 300 \text{ K}).$$

We do the same kind of algebra we used in Example 14.3, and we find $T = 375$ K.

b) The work done at constant pressure is $W = P\Delta V$, and the ideal gas law informs us that $P\Delta V = nR\Delta T$. We may thus calculate

$$W = nR\Delta T = (2 \text{ moles})(8.31 \text{ J/mole}\cdot\text{K})(375 \text{ K} - 300 \text{ K}) = 1250 \text{ J}.$$

c) To find the efficiency, we treat the slug of aluminum as if it had acquired its temperature by being put in contact with a heat reservoir at 400 K, and find how much of that energy was lost when it was placed in the oxygen. This will be our Q_1.

$$Q_1 = Q_{out} = mc(T_{Al} - T) = (0.2 \text{ moles})(870 \text{ J/mole}\cdot\text{K})(400 \text{ K} - 375 \text{ K}) = 4350 \text{ J}.$$

The efficiency is therefore

$$\varepsilon = \frac{W}{Q_1} = \frac{1250 \text{ J}}{4350 \text{ J}} = 0.29.$$

You may remember that Example 15.3 worked out the theoretical Carnot efficiency for a heat engine that operated between these same two temperatures, and the answer came out 25%. So why does this engine do better? The discrepancy is resolved by pointing out that the Carnot efficiency is for a complete cycle of an engine, one that produces work but leaves the system in the same condition it had to begin with. A complete cycle for our engine in this example would have to involve cooling the oxygen back to 300 K by allowing it to exchange energy with a low-temperature reservoir, like the room it is sitting in, and then reheating the aluminum back to 400 K by putting it back in contact again with the high-temperature reservoir.

15.5 Summary

In this chapter you should have learned the following:

- You should be able to use the gas law, Equation 15.7 or 15.8, to determine one of the quantities, P, V, N (or n), and T, given the other three quantities. You should also be able to connect initial and final states of a gas, using the gas law in the form found by dividing $P_1V_1 = n_1RT_1$ by $P_2V_2 = n_2RT_2$:

$$\frac{P_1V_1}{P_2V_2} = \frac{n_1T_1}{n_2T_2}.$$

- You should understand the First Law of Thermodynamics as the law of conservation of energy, where the total energy consists of both mechanical and thermal energy. In this, be sure you understand the meaning of Q_{in} and the difference between W_{on} and W_{by}, as discussed on page 289.
- You should understand heat engine efficiencies, especially how to use Equations 15.11, 15.12, and the relation $Q_1 = W + Q_2$.
- You need to understand that changes in the internal energy of a gas are related to changes in its temperature by Equation 15.18, whether the volume of the gas remains constant or not.
- You need to be able to use the concept of molar specific heats and see how they differ from mass specific heat in their usage. You should also understand why the work done by a gas as it absorbs heat affects the value of the specific heat, all of this as given in Equations 15.16–15.21.

CHAPTER FORMULAS

Ideal gas law using the number of molecules: $PV = NkT$ or $\frac{P_1V_1}{P_2V_2} = \frac{N_1T_1}{N_2T_2}$

Ideal gas law using the number of moles: $PV = nRT$ or $\frac{P_1V_1}{P_2V_2} = \frac{n_1T_1}{n_2T_2}$

First Law of Thermodynamics: $Q_{in} = \Delta E_{int} + W_{by}$ or $\Delta E_{int} = Q_{in} + W_{on}$

Efficiency: $\varepsilon = \frac{W}{Q_1}$ or $\varepsilon = \frac{Q_1 - Q_2}{Q_1}$ Carnot efficiency: $\varepsilon_{max} = \frac{T_1 - T_2}{T_1}$

Molar specific heat equation: $Q = nC_V\Delta T$ and $Q = nC_P\Delta T$

Internal energy: $E_{int} = nC_VT$

Molar specific heats at constant pressure: $C_P = C_V + R$

Work done by the expansion of a gas at constant volume: $W_{by} = P\Delta V$

PROBLEMS

15.2 * 1. A small cylinder of nearly pure silver (atomic number 108 u) weighs 27 grams. How many silver atoms are there in the cylinder?

* 2. Show that the volume of one mole of any ideal gas at one atmosphere of pressure and at 0° C is 22.4 liters. These temperature and pressure conditions are called "standard temperature and pressure," abbreviated STP. [Hint: See Appendix C for the volume in m^3 of one liter.]

* 3. Suppose that someone discovered that everyone had been making a big mistake all these years, and that Avogadro's number should have been 9.0×10^{23}/mole all along. What would be the new, improved value of the gas constant R?

** 4. Water has an atomic mass of 18.015 u. How many molecules of water are contained in a teaspoon of water? (One teaspoon is 5 cm^3 in volume.) [Hint: First find the mass of the water contained in one teaspoon.]

** 5. The total internal energy of 5 gallons of water is 2×10^8 J. The total internal energy of the water in a 100,000-gallon swimming pool is 5×10^{11} J. Which body of water is at the higher temperature? Explain your answer.

** 6. By measuring the pressure and temperature of a gas in a large container, it is found that there are 1.5 moles of the gas present. If the gas in the container is found to weigh 108 grams, what is the mass of one molecule of the gas in atomic mass units?

** 7. A 0.050-m^3 tank holds hydrogen (H_2) at a pressure of 4×10^6 Pa at a temperature of 300 K. How many molecules (that's molecules, not moles) are contained in the tank?

* 8. Gas in a propellant tank on board a spacecraft registers a pressure of 8,000 Pa when it is at 20° C. When the spacecraft turns so that the tank is in direct sunlight, its temperature rises to 40° C. What is the new pressure in the tank? [Hint: If you got 16,000 Pa you had better think about the question again.]

* 9. A can filled with air is heated in boiling water until its temperature is 100° C. The can is then sealed by closing it with a cap, and the can is removed from the water. When the temperature of the gas drops to room temperature of 20° C, what will be the pressure in the can?

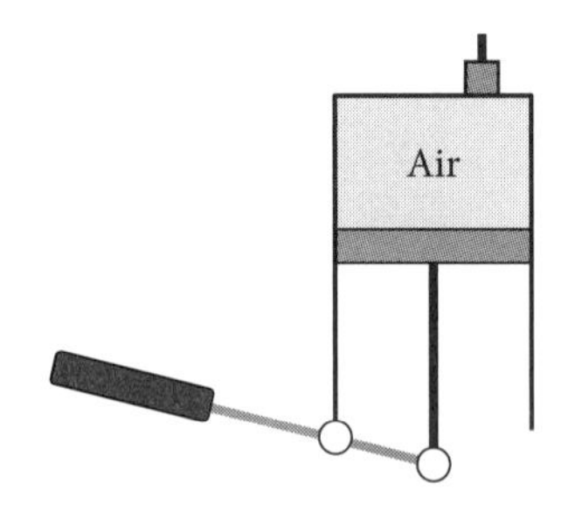

** 10. The air reservoir in a BB gun has an escape valve that limits the pressure difference across it to three atmospheres (see diagram at left). If the maximum pressure is to be achieved with a single pump, by what factor must the volume of the air in the reservoir be reduced?

** 11. The tires on a race car are filled with air at 10° C, to a gauge pressure of 35 lb/in^2 (about 2.4×10^5 Pa). The gauge pressure is the difference between the pressure in the tire and the atmospheric pressure outside the tire. During the race, the tires heat up to 120° C. What will be the gauge reading for the tire pressure at this temperature?

** 12. A beach ball has a volume of 0.100 m^3. The pressure of the air in the ball is one atmosphere and the temperature is 27° C. When a diver pulls it 20 meters under water, its volume is 0.032 m^3. What is the temperature of the water at this depth?

** 13. Hydrogen cyanide (HCN) is a gas at temperatures above 26° C (79° F). Its atomic mass is 27 u. A chemist stores 100 grams of HCN in a tank at a pressure of 2×10^5 Pa and keeps the tank at a constant temperature well above 26° C. During one 24-hour period, the pressure was observed to drop to 1.8×10^5 Pa. What mass of gas remained in the tank and what mass of the gas had leaked out into the room?

** 14. A canister containing 8 grams of helium gas (atomic mass 4.0 u) was used to fill up party balloons. When the canister was full, the pressure in the gas was 3 atmospheres and its temperature was 27° C. After filling the balloons, the pressure was seen to have dropped to 2 atmospheres and the temperature dropped to 17° C. How many grams of helium gas were used to fill the balloons?

** 15. A good laboratory vacuum system can typically reduce the air pressure to 100 Pa at 20° C. The average atomic weight of air molecules is 28.8 u. What is the typical number density (molecules per cubic meter) of air molecules inside such a vacuum system?

** 16. If nitrogen gas and oxygen gas are combined in a tank, they will be at the same temperature and fill the same volume.

a) Basing your explanation on the kinetic theory of gases, as derived in Section 15.1, explain why you would find the total pressure in the tank by adding together the individual pressures of the nitrogen and oxygen acting alone. This conclusion is known as Dalton's Law of Partial Pressures.

b) A 1-m^3 tank contains 0.8 moles of nitrogen and 0.2 moles of oxygen at 300 K. What is the total pressure in the tank?

*** 17. A frontier woman cans raspberry jam by filling a Mason jar with jam at room temperature 20° C. She then seals the jar with a rubber gasket and a metal lid that allows the air in the jar to escape, but not to re-enter. When the jar, with about 10 cm^3 of air in it, is placed in 100° C boiling water, the jam cooks and air escapes from the jar, leaving a lower pressure in the jar after it cools. This lower pressure holds the lid on tight. The radius of a normal Mason jar opening is 3 cm. How much force would it take to lift the lid off the jar, after it has cooled back to 20° C? [Hint: The pressure will be a constant one atmosphere while the jam is being heated, and the number of moles of the air in the jar will be constant while the jam is cooling.]

*** 18. A Mylar balloon is filled with 400 grams of helium gas (atomic weight 4 u) at an atmospheric pressure of 1×10^5 Pa. The temperature of the helium is 0° C.

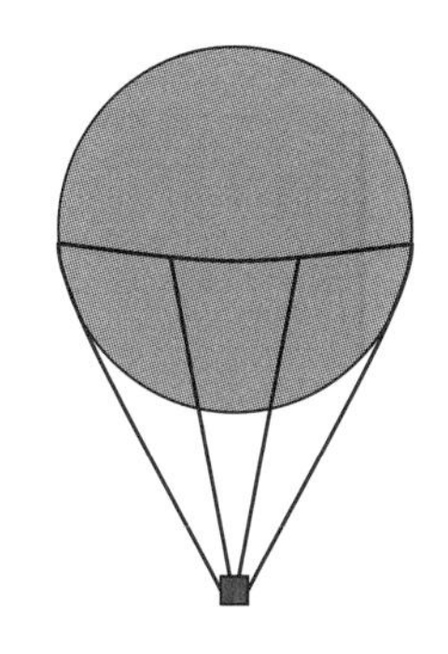

a) How many moles of helium are contained in the balloon?

b) What is the volume of the balloon?

c) Air has an average density of 1.2 kg/m^3. Neglecting the weight of the Mylar and neglecting the volume of the little payload hanging down from the balloon, find the mass of the payload that the balloon can lift into the air. [Hint: Remember how to calculate the buoyant force on an object immersed in a fluid, in this case a fluid composed of air.]

*** 19. A scuba diver swimming at a depth of 40 meters releases an air bubble whose volume is 8 cm^3 and whose temperature is 35° C. As it rises to the surface of the water, its temperature drops to 20° C. Assume that the atmospheric pressure at the water surface is 100,000 Pa and find the volume of the bubble just before it arrives at the surface.

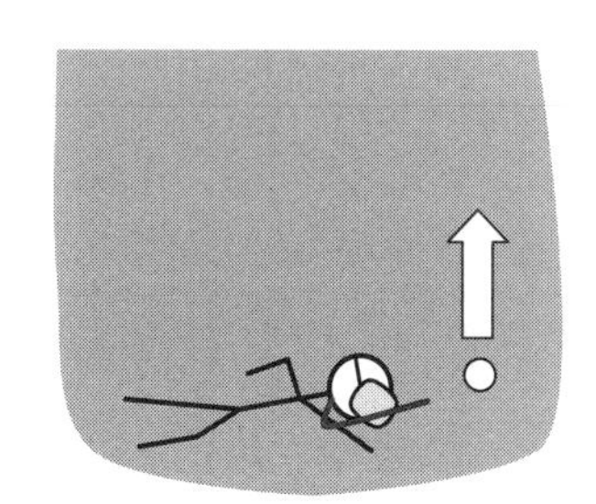

15.3 * 20. If a particular heat engine produces 300 J of work as it dumps 2700 J of waste heat to a cold reservoir, what is the efficiency of the engine?

* 21. A Carnot engine, working at an efficiency of 30%, uses energy from a high-temperature reservoir at 400 K. What is the temperature of its low-temperature reservoir?

* 22. In a steam turbine, superheated steam enters the turbine at 1000 K and is exhausted at 500 K. What is the theoretical maximum efficiency of such an engine?

** 23. Most real engines lose heat due to friction, in addition to their theoretical heat losses coming from the laws of thermodynamics. The frictional loss is always eventually dumped to the low-temperature reservoir, along with the waste heat required by the theoretical efficiency of the engine. If a Carnot engine operating between temperatures of 800 K and 300 K produces 20 kW of useful output power, but also loses energy to internal friction at a rate of 4 kW, find the overall efficiency of the engine. [Hint: Remember that power was defined in Section 7.3.]

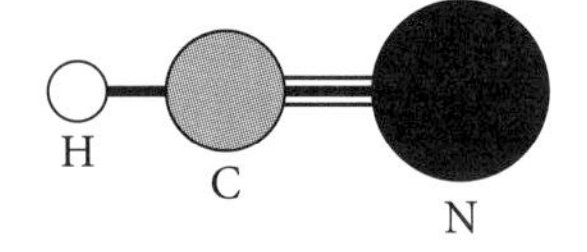

15.4 * 24. Hydrogen cyanide is a triatomic gas in which the molecules are aligned along a single line, as shown at left. Estimate the molar specific heat at constant volume for HCN by considering the number of degrees of freedom for the molecule. Assume that no molecular vibrational modes are excited.

** 25. At normal room temperatures, a water molecule, pictured below, can vibrate in such a way that both hydrogen atoms (white) approach the oxygen atom (red) together, as in picture (a), or in such a way that one hydrogen atom approaches while the other recedes, as in picture (b). A third mode of vibration, in which the two hydrogen atoms come toward each other, as shown in picture (c), actually takes significantly more energy to excite, and so water molecules do not vibrate in this mode at room temperature. Find the number of degrees of freedom for a gas of water vapor at normal room temperatures and estimate its molar specific heat.

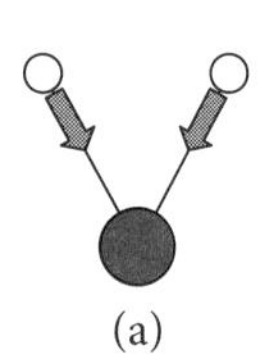
(a)

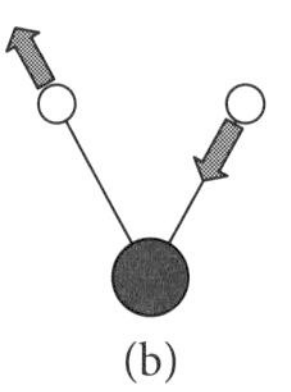
(b)

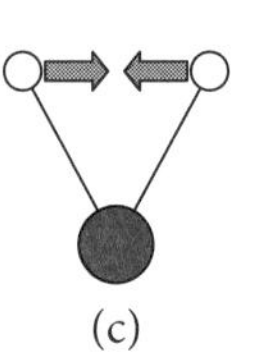
(c)

** 26. One gas canister holds 10^8 molecules of neon (a monatomic gas) and another holds the same number of molecules of oxygen (a diatomic gas). Both canisters start off at 20° C and each absorbs 1000 J of heat. Which gas will end up at the higher temperature? Explain your reasoning.

*** 27. A physics professor has a small cabin in his back yard to use as an office. Its floor area is 4 m × 5 m and the ceiling is 3 m high. It is very well insulated, so no heat is lost through walls or windows. During the winter, it is heated by a 1500-watt electric heater. The professor goes in the cabin on a cold morning (10° C) and turns on the heater. The initial atmospheric pressure in the cabin is 10^5 Pa. Assume that the windows and doors are tight, so that no air enters or leaves the cabin. Air is composed almost entirely of the diatomic molecules N_2 and O_2. Estimate the molar specific heat at constant volume for the air, and calculate how long it will take the heater to raise the temperature to a more comfortable 20° C.

** 28. An *adiabatic process* for some system is one in which no heat flows into or out of the system (i.e., $Q = 0$). Typically, this occurs when a gas is compressed or expanded so rapidly that the heat does not have time to transfer from molecule to molecule, or when a system is thermally insulated from the outside world. As a shock wave from a supersonic jet passes through the atmosphere, the air at a point along the jet's path of the wave is compressed adiabatically to 90% of its previous volume. Will the temperature of the compressed air increase, decrease, or stay the same? Justify your conclusion.

** 29. Five moles of neon (a monatomic gas) at 400 K are compressed adiabatically, reaching a final temperature of 450 K. [Note: Problem 28 explains the meaning of the term "adiabatic."]

a) What is the value of C_V for this gas?

b) By what amount did the internal energy of the neon increase?

c) How much heat was added during this process? [Hint: Reread the note.]

d) How much work was done in compressing the gas?

*** 30. An *isothermal process* is one in which a system remains at constant temperature while its other parameters change. Isothermal processes typically occur when the system is in thermal contact with a heat reservoir at constant temperature (like the air in a large room) that can absorb or supply heat to the system in question.

a) If a gas is compressed isothermally from an initial 5×10^{-3} m^3 volume and 3×10^5 Pa pressure to a final volume of 2.5×10^{-3} m^3, what will be its final pressure?

b) Will the internal energy of the gas change during an isothermal process? Explain.

c) If it required 10^4 J of work to compress the gas in part (a), will heat have been added to the gas or lost from it, and by what amount?

*** 31. Sixteen grams of oxygen (O_2) gas are sealed inside a cylinder in which a constant pressure is provided by a mass m, as shown. The oxygen begins at a temperature of 400 K and loses heat through the walls of the cylinder, ending up at the room temperature of 300 K. The atomic weight of O_2 is 32.00 u.

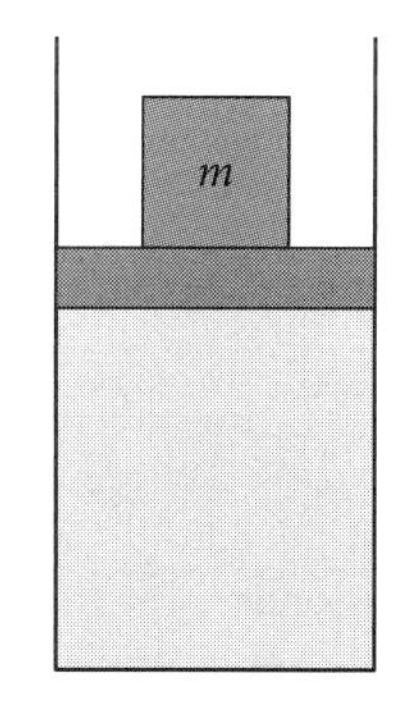

a) What are C_V and C_P for this diatomic gas, estimated from the number of degrees of freedom available to the molecules of the gas?

b) How much heat is lost by the oxygen as it cools?

c) How much internal energy does the oxygen lose as it cools?

d) How much work is done on the oxygen as the piston compresses it?

e) If the pressure provided by the mass on the piston is 1000 Pa, by what amount in m^3 will the volume of the oxygen decrease?

*** 32. A double-chambered container contains one mole of helium in one of its 1000 cm^3 volume chambers. The container is well-insulated, and of low specific heat, so that no appreciable heat is added to the gas during the process we describe. The gas is initially at a temperature of 300 K and a pressure of 1 atmosphere. The partition between the two chambers is then quickly raised, and the gas expands freely to fill the entire container. Whenever a monatomic gas like helium doubles its volume adiabatically like this, the pressure in the gas will drop to 0.315 of what it was before (for reasons that we did not explain in this chapter), so the final pressure of the expanded gas will be 0.315 atmospheres.

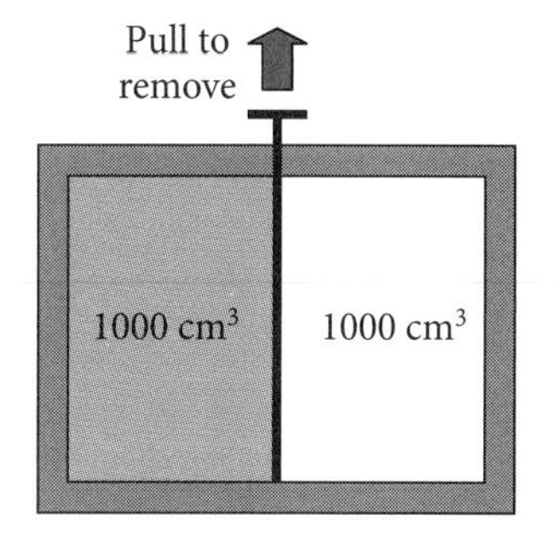

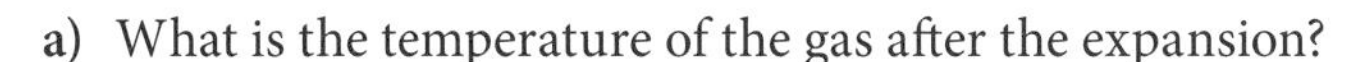

a) What is the temperature of the gas after the expansion?

b) What is the change in the internal energy of the gas?

c) How much heat is added to the gas? [Hint: Maybe read the problem again.]

d) How much work is done by the gas as it expands? [Hint: The pressure is not constant, so you cannot use the formula $W = p\Delta V$.]

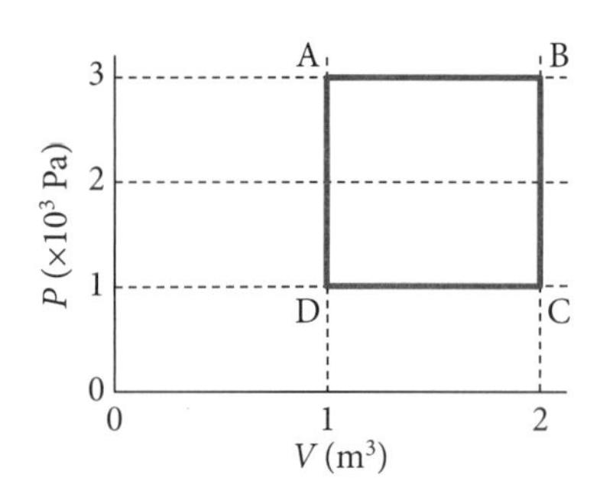

*** 33. A heat engine uses *one mole* of a monatomic gas to convert heat energy into work. In the process, the gas follows the cycle shown in the graph at left. It begins at point A with volume 1 m^3 and pressure 3×10^3 Pa. It then expands at constant pressure to a volume of 2 m^3, as at point B. The pressure then drops to 1×10^3 Pa, with no change in volume, thus arriving at point C. The pressure is then kept constant at 1×10^3 Pa while the volume drops back to 1 m^3, at point D. Finally, the volume is again kept constant while the pressure rises back to 3×10^3 Pa at point A, completing one full cycle.

a) Use the ideal gas law (Equation 15.8) to find the temperatures of the gas at each of the four points, A, B, C, and D.

b) Find the work done *by the gas* along each of the processes – from A to B, B to C, C to D, and D to A. [Hint: Check the Chapter Formulas box.]

c) Find the heat added to the gas along each of the same 4 processes. [Hint: Remember that this is a monatomic gas, whose specific heats are known.]

d) Find the total work done by the gas during the complete 4-process cycle.

e) Find the sum of the positive heats added to the gas. (Whenever the heat added is positive, it is coming from some high temperature reservoir, and is part of Q_1. The negative heats added represent heat lost to a low-temperature reservoir, contributing to Q_2.)

f) Use your results to calculate the efficiency of the engine.

16

ELECTRIC FORCES AND FIELDS

16.1 Charge

This is a good time to take stock of where we are in our study of physics. We have now been through fifteen chapters, all basically analyzing quantities measured with units of time, distance, and mass. In this chapter we introduce a new fundamental unit, one that measures a property of matter known as *electric charge*. The building blocks of matter are atoms, and atoms are composed of protons and neutrons, clumped together in a nucleus and surrounded by a cloud of electrons. Protons and electrons have electric charge, and so all matter contains electric charge.

In our first few chapters, we used the concepts of distance and time to describe the motion of objects. In Chapter 5, we introduced the idea of mass and explained its place in the framework of Newtonian mechanics. In Chapter 6, we presented several of the forces we would use. Since Chapter 6, you may have noticed that we were able to use Newtonian mechanics to explain a lot of things. We *proved* that momentum is conserved in isolated systems; we *explained* how a wave propagates in a medium; we *derived* the ideal gas law. But that's all over for the moment. Here we are again, with something that we cannot explain based on other things. Why does a proton have an electric charge? Because it's a proton, and that's what protons have.

The word electricity comes from *electron*, the Greek word for "amber," a petrified wood resin. It was known anciently that if you rubbed a piece of amber with a piece of fur, it would attract small pieces of feather due to its *static electricity*. Over the years it was discovered that there were, in fact, two kinds of static electricity. One was created by rubbing a rod of amber (or rubber) with a piece of fur. The second was created by rubbing a glass rod with a piece of silk. It was found that rubbed glass rods would attract rubbed rubber rods, that rubbed glass rods would repel other rubbed glass rods, and that rubbed rubber rods repel rubbed rubber rods (try saying that three times fast).

In the early 18th century, it was proposed that there were simply two kinds of electricity—resinous (amber) electricity and vitreous (glass) electricity—and that the two kinds attracted each other. Then, in the late 18th century, Benjamin Franklin (yes, the same Benjamin Franklin) proposed that there was only one kind of electricity. He called vitreous electricity "positive" and suggested that resinous electricity was a lack of positive electricity, so it was "negative." What Franklin thought was that rubbing glass with silk would transfer a little positive electricity onto the glass, and that rubbing amber with fur would steal a little positive electricity fluid from the amber, leaving it negative. Ben was very close to being right, at least for normal materials.[1]

Everything is composed of atoms with nuclei, which are held more-or-less rigidly in the material, and electrons, which can move more freely and can thus be added or subtracted by rubbing. Unfortunately for all of us, Franklin guessed the wrong sign. When you rub a glass rod with silk, electrons are removed, not added. So when he called the charge on the glass rod "positive," he made it so that the charge on an electron, the thing that carries charge from place to place, was negative. As a result, generations of physics and engineering students have had to deal with the fact that a wire, in which there are electrons moving to the right, is, in fact, carrying a positive current to the left. Be warned. We deal with currents in Chapter 18.

Electric charge is a property of matter, and the existence of charge needs no explanation; it's just the way the world is. But how do we specify *how much* charge something has? What should we use as units for electric charge? Based on the history of electricity, we might think to define the fundamental unit of charge as the charge acquired by a standard piece of glass after one rub with a standard piece of silk. But such a unit would be a nightmare to reproduce, since the actual charge imparted depends on the humidity of the air, on how hard you hold the silk, on how fast you move the silk—in short, on so many uncontrollable things as to make it impractical.

Alternatively, we could lean on the fundamental nature of charge. The electric charge on an electron is equal and opposite to that on the proton, so we could define the unit of charge to be the elementary charge (e) on a proton.[2] The charge on an electron would then be –1 e, and we could measure charge by counting electrons or protons. This unit is actually used for elementary particles and atoms, where it has a very practical size. But it is not well-suited to normal-sized objects, with many atoms, where counting fundamental particles is completely impractical.

Or a charge unit could be based on the force between charges. There *does* exist a unit of charge that is based on this kind of definition. One *statcoulomb* of charge is defined as the amount of charge that, if placed on two objects 1 cm apart, will produce a repulsive force between them of 1 dyne (the cgs unit of force, equal to 10^{-5} N). This would be a *great* unit of charge for us to use. It is the right size to use in laboratory situations and homework problems and it produces very simple formulas for force and field calculations. Unfortunately, this isn't the unit we use.

Probably because the most common practical use of electricity is in electric circuits, not in introductory physics classes, the SI unit of charge is based on the standard unit of electric current, the ampere (which we will see in Chapter 18). If a current of one ampere is flowing along a wire, then the amount of charge that passes a given point on the wire in one second is defined to be one *coulomb* (C), named after the physicist whose work we will introduce in the next section. Unfortunately (have you noticed that we keep using that word?), the coulomb is a *very* impractical unit for discussing static charges. It is huge (see

1. As we will see in Chapter 26, elementary particles carry positive or negative charge, so, in this sense, Franklin was wrong. There *are* two kinds of charge. However, in ordinary matter, it is the electrons that carry the charge from place to place, so he was also correct in a way.

2. There is a tendency to think that the unit e stands for "electron." This is a mistake. Think of it as "e for elementary." The charge on the proton is +1 e; the charge on an electron is –1 e.

Figure 16.1). If we placed one coulomb of charge in the middle of the room you're sitting in, everything that isn't nailed down (including you) would be dragged toward it. So, we will not talk about coulombs of charge, except in the abstract. Our most useful units of charge will be microcoulombs (1 μC = 10^{-6} C), nanocoulombs (1 nC = 10^{-9} C), or picocoulombs (1 pC = 10^{-12} C).

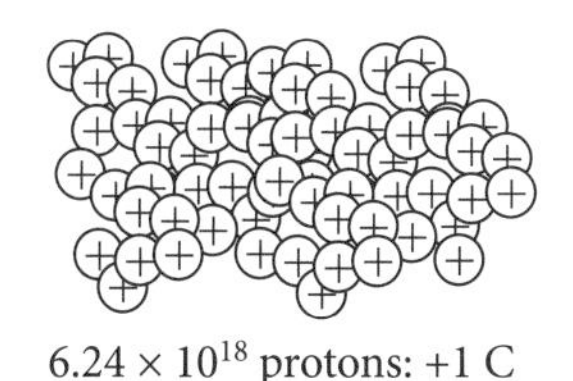

Figure 16.1 Relative sizes of the two main units of charge.

Having defined two basic units of charge, the microscopic elementary charge e and the inappropriately macroscopic coulomb C, we can convert between them based on the fact that the elementary charge has been measured to be 1 e = 1.602×10^{-19} C.

Let's also be sure we understand how the charge on a normal-sized object arises. Most objects, like bananas and baseball bats, are electrically neutral. This is because they contain equal numbers of protons and electrons in them. The way an object becomes electrically charged, then, is either to gain a few electrons, in which case it will end up negatively charged, or to lose a few electrons, in which case it will be positively charged. The mass of an individual atom is tiny, so there are something like Avogadro's number of atoms in each object. But the charge on an individual electron is also tiny. So the only way to get even a picocoulomb of charge on a normal-sized object is for millions of electrons to be added to it or subtracted from it.

Finally, let us be clear that the electric charge on an object is a property of the object, like its shape or its temperature or its mass. It is not a force on the object. In fact, there is not necessarily *any* force on a charged object. The relationship between the electric charge on an object and whatever force might act on it because of its charge is discussed in the next section.

Our first example shows how charge arises from the elementary charges an object contains.

EXAMPLE 16.1 A Static Charge Due to Excess Electrons

An electrically neutral piece of hard rubber contains 1×10^{27} protons in the nuclei of all of its atoms (and even more neutrons, which don't concern us here). You then add 4×10^{15} extra electrons to the object by rubbing it with fur. How many electrons were initially on the object and what is the final charge of the object, expressed in coulombs?

ANSWER Because the piece of rubber is initially neutral, it must contain equal numbers of protons and electrons. So the number of electrons initially contained in the piece of rubber must have been the same as the number of protons, or 1×10^{27}.

An object's charge, called the net charge, depends on the difference in the number of protons and electrons in it. Because, after rubbing it with fur, the piece of rubber contains 4×10^{15} more electrons than protons, its charge is -4×10^{15} e. Since each elementary charge represents 1.6×10^{-19} C, this net charge can also be written as

$$(-4 \times 10^{15}\ \text{e}) \times \frac{1.6 \times 10^{-19}\ \text{C}}{\text{e}} = -6.4 \times 10^{-4}\ \text{C}.$$

Note how we converted the units from elementary charges (e) to coulombs (C). It is customary to use a lowercase q (or a capital Q) to represent charge, so we report the result as $q = -6.4 \times 10^{-4}$ C, or -640 μC.

16.2 Coulomb's Law

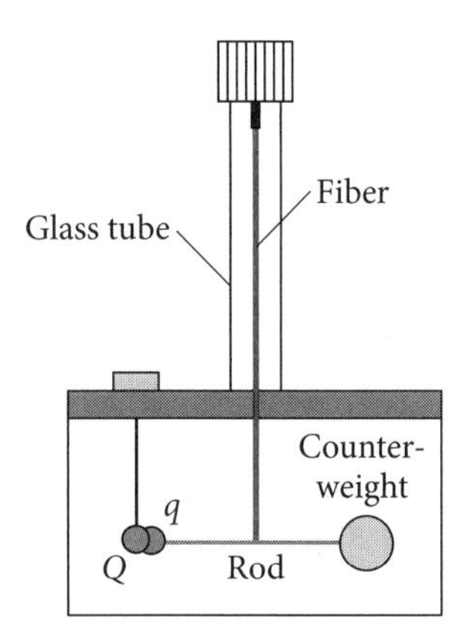

Figure 16.2 A diagram of Coulomb's apparatus. The small red spheres were electrically charged, and the distance between them adjusted by turning a knob holding the fiber.

In 1784, Charles-Augustin de Coulomb published the results of experiments he had done with a small electrically charged sphere attached to the end of an insulating rod, suspended by a thin fiber (see Figure 16.2). A similar electrically charged sphere was held in place, and the distance between the two charges could be controlled by turning the fiber support on top of the apparatus. The torque produced when the fiber twisted through a small angle was proportional to the size of the angle (good old Hooke's law again), so, by measuring the angle through which the fiber twisted, Coulomb could determine the small electric force that arose between the charged spheres. He found that, if the two charges were Q and q, and if the distance between them was r, then the electric force between the charged spheres was given by what is now known as *Coulomb's Law*:

$$F_E = k\frac{Qq}{r^2}, \tag{16.1}$$

where k is a universal constant called "Coulomb's constant."[3] For charges measured in coulombs and distance in meters, Coulomb's constant was eventually found to have the value $k = 9.0 \times 10^9\ \text{N}\cdot\text{m}^2/\text{C}^2$. Let's be clear. Equation 16.1 represents a new piece of fundamental physics, the electric force between two electrically charged objects. It is discovered experimentally, not derived from other principles, and it is what it is.

Equation 16.1 gives the magnitude of the Coulomb force, but not the direction. To find the direction, it is probably easiest to simply remember that *like charges repel* and *unlike charges attract*, as we explained in the last section. In Figure 16.3, if Q and q are either both positive or both negative, then the directions of the forces on each will be given by the red arrows. If either one is positive and the other negative, the directions of the forces will be given by the gray arrows.

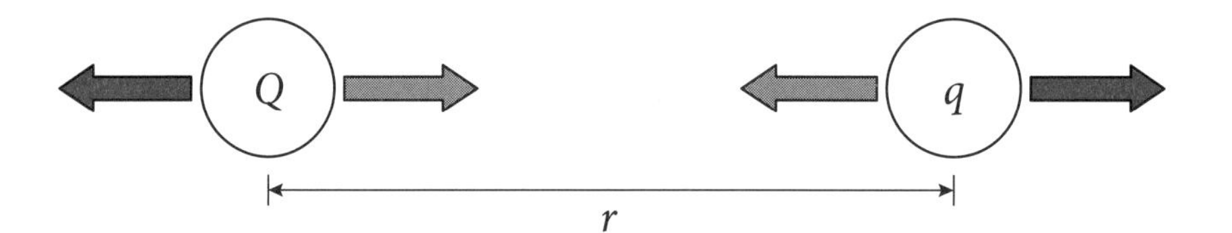

Figure 16.3 The geometry for using Coulomb's law and the force directions for like charges (red arrows) and unlike charges (gray arrows).

We're sure everyone sees the similarity between Coulomb's law for electric force (Equation 16.1) and Newton's law of gravitation from Chapter 6:

$$F_g = G\frac{Mm}{r^2}. \tag{16.2}$$

Both F_g and F_E are fundamental forces, each arising from a property of matter—mass for gravity and charge for electricity. Both are proportional to the inverse square of the distance between the centers of the objects, so that the force quickly gets weaker as the objects separate.

3. We apologize for using the k that has already stood for the spring constant (Chapter 6) and for Boltzmann's constant (Chapter 14) as the symbol for Coulomb's constant. The problem is that these are all traditional, and we want you to feel comfortable if you read other books.

And both incorporate a universal constant (k or G) whose value gives the size of the force that nature produces when the amounts of mass or charge are measured in the units that physicists have chosen. But there is one huge difference between gravitational force and the Coulomb force. This is that the force of gravity can only be attractive; it can only point in the directions of the gray arrows in Figure 16.3. Electric charges can be either positive or negative, and so the directions of the forces can be either in the directions of the red arrows or the gray arrows, depending on whether the charges have the same sign or opposite signs.

EXAMPLE 16.2 The Forces Between Two Charges—I

A gray object with charge $Q_{gray} = 5 \times 10^{-8}$ C and a red object with charge $Q_{red} = -2 \times 10^{-8}$ C are separated by a distance $r = 0.50$ m. The red object is due east of the gray object. Find the electric force on each object.

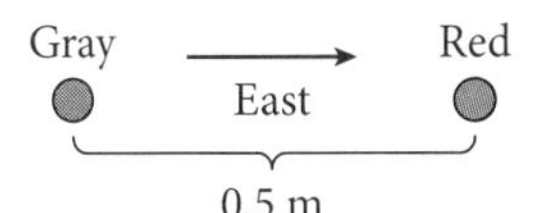

ANSWER The magnitude of either force can be found by a straightforward use of Equation 16.1. It gives

$$F_E = k\frac{Q_{gray}Q_{red}}{r^2} = \left(9.0 \times 10^9\ \text{N}\frac{\text{m}^2}{\text{C}^2}\right)\frac{(5 \times 10^{-8}\ \text{C})(2 \times 10^{-8}\ \text{C})}{(0.50\ \text{m})^2} = 3.6 \times 10^{-5}\ \text{N}.$$

Notice that we ignored the fact that the red object's charge was not 2×10^{-8} C, but -2×10^{-8} C. This is because we are quite capable of figuring out the directions of the forces by ourselves, knowing whether the charges are like or unlike, and we would rather not let the algebra confuse us. If we had put -2×10^{-8} C in the equation, our answer would have been -3.6×10^{-5} N, and what would that mean? We know that a minus force means that the force is in the opposite direction of something, but of what? We find it simpler and consistently less confusing to ignore the sign of the charges in Coulomb's formula and work out the directions for ourselves. Because we know the forces are attractive, the electric force on the red object due to the gray object's charge is

$$\vec{F}_{gray,red} = \{F_E, \text{west}\} = \{3.6 \times 10^{-5}\ \text{N, west}\},$$

and the force on the gray object by the red object is

$$\vec{F}_{red,gray} = \{F_E, \text{east}\} = \{3.6 \times 10^{-5}\ \text{N, east}\}.$$

There is one caution we should add, however, about applying Coulomb's law. This is that it works fine in the limit that the charged particles are small relative to the distance between them, but it is not always true if you try to apply it to large bodies close together. For example, if we were to ask you to determine the electric force between a 4-cm-diameter metal sphere carrying a charge $Q = 3\ \mu$C and another 4-cm-diameter metal sphere, 10 cm away, with zero net charge ($q = 0$), you might look at Equation 16.1 and conclude that the force is zero. And you would be wrong. Let's see why.

16.3 Induction and Polarization

All matter is composed of molecules, and molecules are composed of atoms, with positively charged protons in the nucleus and clouds of electrons surrounding the nucleus. Thus, everything has electric charge inside it. When we say that an object has zero charge, we mean that this is its net charge—that it is neutral. We do not mean that there are no electric charges inside it.

In some kinds of materials (typically metals), there are electrons that are free to move throughout the object. These materials are called *conductors*. In other kinds of materials (wood, glass, rubber), called *insulators*, the electrons are more tightly bound to their individual molecule.

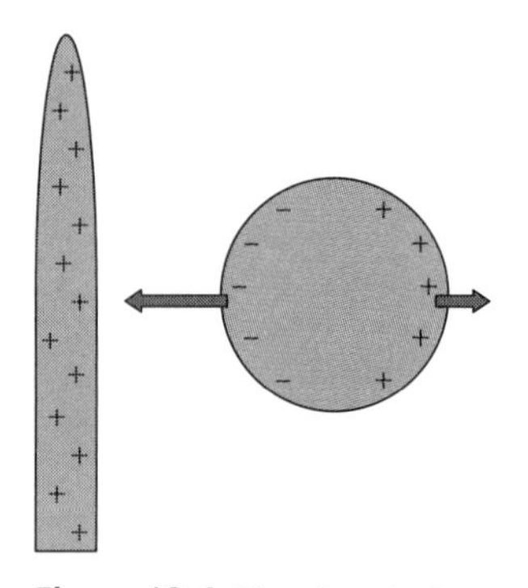

Figure 16.4 The electrical polarization of a conducting sphere that is brought close to a positive charge, and the resulting forces on each charge component.

Now consider what will happen if a glass rod with a static positive charge, created by rubbing it with silk, is placed near a conducting sphere, as shown in Figure 16.4. The electrons inside the conducting sphere feel an attractive force toward the glass rod. They are free to move, and so they move toward the rod. This places a net negative charge on the near side of the sphere, and leaves behind a net positive charge on the far side. We say that the sphere has become *polarized* and that the charge in the rod has *induced* the polarization. The sphere is still neutral—the net charge on it is still zero—and there will be both an attractive force on the sphere because of its electrons and a repulsive force on the sphere, due to its isolated protons. But, because of the distance dependence in Coulomb's law, the attractive force on the nearby negative charge is greater than the repulsive force on the farther-away positive charge, producing a net attractive force on the neutral conducting sphere.

So a neutral conductor can experience an electric force from a nearby charge, because it will become polarized. How about an insulator? In an insulator, the electrons do not move, but remain in their molecule. But even an insulator can become polarized. In each atom, the electrons, although tied to their nucleus, still see the force of a nearby charge and respond to that force by a distortion of the electron cloud. This polarizes each atom or molecule, as shown in Figure 16.5. The electrons spend more time closer to the rod, leaving a separation between the positively charged nucleus and the average position of the negative electrons in the atom.

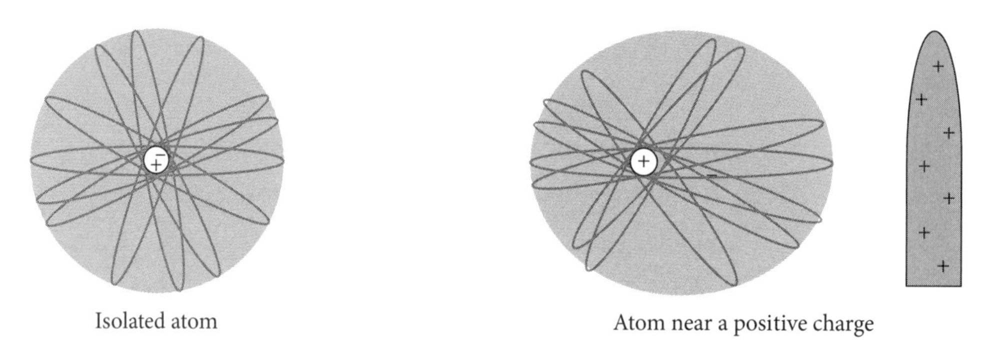

Figure 16.5 Polarization of an unpolarized atom by the presence of a nearby electric charge.

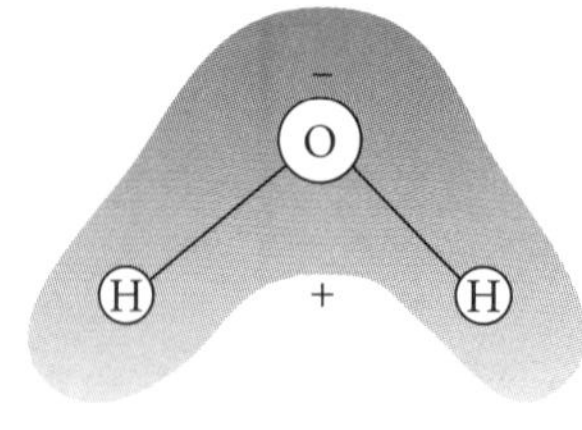

Figure 16.6 The natural polarization of a water molecule, H_2O.

In addition, some molecules are naturally polarized, simply because of how the atoms are arranged in the molecule. In a water molecule, for example, the positions of the oxygen and hydrogen atoms, coupled with the fact that the electrons spend more time associated with the oxygen atom, produce a net negative charge on one end of the molecule and a net positive charge on the other (see Figure 16.6).

Whether its molecules are polarized by induction or are naturally polarized, an insulator will experience a force in the presence of a nearby charge. The force is not due to the polarization of the individual molecules (the plus and negative points in each molecule are so close together that the difference in attractive and repulsive force on each molecule is negligible). Rather, the force arises as the average orientation of each molecule is influenced by the nearby charge. In Figure 16.7a, we show the random orientation of polarized molecules in an insulator with no nearby charge. In Figure 16.7b, we see the effect of the forced re-orientation, with more of the negative ends of each molecule attracted toward the positive rod. The interior of the insulator remains electrically neutral, on average, but there is a net negative charge on the *surface* closest to the rod and a net positive charge on the *surface* furthest away. Again, the attractive electric force on the nearby surface charge is greater than the repulsive electric force on the far surface charge, because of the difference in distance between the two. There is thus a small attractive force between a charged glass rod and a neutral insulator.

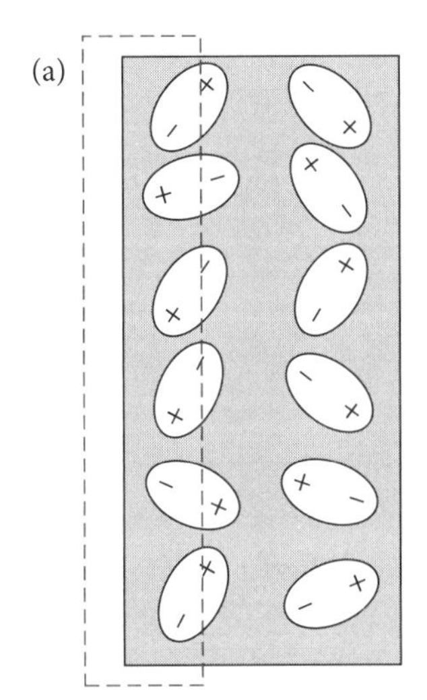

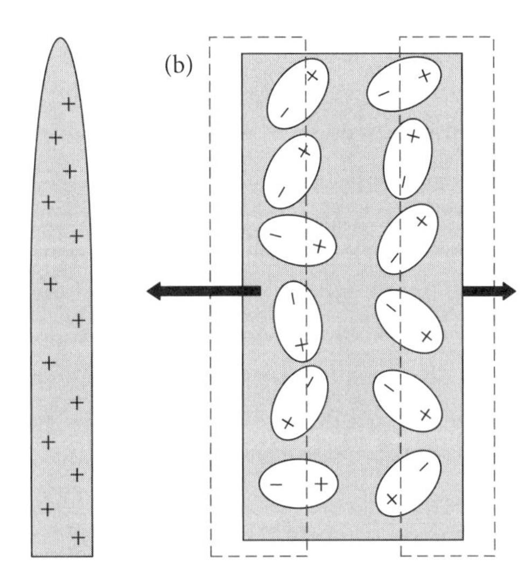

Figure 16.7 The polarization of an insulator by a nearby positive charge on a glass rod, creating a negative surface charge nearest the rod and a positive surface charge away from the rod. The result is a net attractive force on the insulator.

However, as the charge creating the force moves further away, or as the size of the neutral object becomes smaller, the difference in the distance to the two sides becomes proportionally less, and the forces on the two sides become more the same. The net force thus tends toward zero as the two objects separate, in keeping with Coulomb's law when one of the charges is zero.

16.4 Electric Fields

In Chapter 6, we introduced the fundamental formula for the gravitational force between two masses. It was Newton's law of gravity, reproduced in this chapter as Equation 16.2, and again below:

$$F_g = G\frac{Mm}{r^2}.$$

In practice, however, we used this formula very little. For example, suppose we were asked to find the weight of a 2-kg block resting on a tabletop on the surface of the earth. We could have gone off to look up the mass and radius of the earth and then calculated the answer using Newton's formula,

$$F_g = G\frac{Mm}{r^2} = \left(6.67 \times 10^{-11}\ \mathrm{N}\frac{\mathrm{m}^2}{\mathrm{kg}^2}\right)\frac{(6.0 \times 10^{24}\ \mathrm{kg})}{(6.4 \times 10^6\ \mathrm{m})^2}(2\ \mathrm{kg}) = \left(9.8\frac{\mathrm{N}}{\mathrm{kg}}\right)(2\ \mathrm{kg}) = 19.6\ \mathrm{N}.$$

But notice the parts of the calculation inside the shaded area. For a point on the surface of the earth, this part of the calculation will always give the same 9.8 N/kg. And so the force on an object on the tabletop will *always* be 9.8 N/kg times the mass of the object in kg. Thus, we wouldn't bother with that whole calculation every time we needed to calculate a weight. We would just remember that the calculation inside the shaded area always gives the same quantity, one we called the *gravitational field*,

$$g = \frac{F_g}{m} = G\frac{M}{r^2} = \left(6.67 \times 10^{-11}\ \mathrm{N}\frac{\mathrm{m}^2}{\mathrm{kg}^2}\right)\frac{(6.0 \times 10^{24}\ \mathrm{kg})}{(6.4 \times 10^6\ \mathrm{m})^2} = 9.8\frac{\mathrm{N}}{\mathrm{kg}}, \quad (16.3)$$

though, as you remember, we decided to generally use the approximation g = 10 N/kg. We should also remember that the gravitational field is always downward, and so we would find the weight force from $\vec{W} = m\vec{g}$.

In this chapter, in Equation 16.1, we introduced the fundamental formula for the electric force between two charges. It was Coulomb's law,

$$F_E = k\frac{Qq}{r^2}.$$

And, just as we introduced the idea of a gravitational field at a point a distance r from an object of mass M using Equation 16.3, so we can introduce the idea of an electrical field at a point a distance r from an object of charge Q with the formula

$$E = \frac{F_E}{q} = k\frac{Q}{r^2}. \quad (16.4)$$

Note that the units of the electric field are the units of F_E/q, or newtons per coulomb. Equation 16.4 gives the magnitude of the electric field, but we also want the field to carry information about direction. If we decide how to do this, then we could write the electric force that a charge Q exerts on a charge q as

$$\vec{F}_{Q,q} = q\vec{E}, \quad (16.5)$$

where the subscripts on the force indicate the force by Q on q, as usual, and where the electric field due to the charge Q is evaluated at the point, called the *field point*, where q is located. For Equation 16.5 to work as a vector equation, we must define the correct direction for the electric field. Here is the definition we require. The direction of an electric field due to a positive charge is defined to be directly away from the charge and the direction of the field from a negative charge points toward the charge (Figure 16.8).

Figure 16.8 Electric field directions for a positive charge (a) and a negative charge (b).

We may also say that the electric field at some field point is the force that an object carrying 1 C of positive charge *would* feel *if* it were at that point. Note the conditional words in that last sentence—would and if. If there is no object at that point, there is no force, but the field still exists. It is a property of the point in space. The electric field at a point is a vector whose magnitude is the magnitude of the force on a 1 C charge, if it were at that point, whose direction is the direction of the force on a +1 C charge, if it were at that point.

The next example shows how to determine the electric field at several field points near a single charge.

EXAMPLE 16.3 The Electric Field at Different Field Points

Find the electric field due to a charge $Q = 16$ pC (picocoulombs) at field points 2 cm due east of the charge, 3 cm exactly northeast of the charge, and 4 cm due north of the charge, as shown in the figure at right.

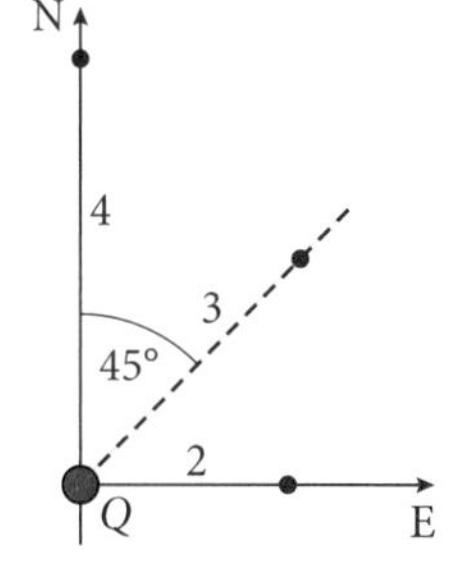

ANSWER The magnitudes of the electric field are found from Equation 16.4. At the point 2 cm to the east, the magnitude is

$$E = k\frac{Q}{r^2} = \left(9 \times 10^9\ \mathrm{N}\frac{\mathrm{m}^2}{\mathrm{C}^2}\right)\frac{(16 \times 10^{-12}\ \mathrm{C})}{(0.02\ \mathrm{m})^2} = 360\frac{\mathrm{N}}{\mathrm{C}}.$$

Because the electric field from a positive charge is everywhere outward from the charge, the field at the field point 2 cm to the east is $\vec{E} = \{360\ \mathrm{N/C},\ \text{east}\}$.

At the point 3 cm to the northeast, the field is

$$E = k\frac{Q}{r^2} = \left(9 \times 10^9\ \mathrm{N}\frac{\mathrm{m}^2}{\mathrm{C}^2}\right)\frac{(16 \times 10^{-12}\ \mathrm{C})}{(0.03\ \mathrm{m})^2} = 160\frac{\mathrm{N}}{\mathrm{C}} \quad \Rightarrow \quad \vec{E} = \left\{160\,\frac{\mathrm{N}}{\mathrm{C}},\ \text{northeast}\right\}.$$

And the field 4 cm north of the charge is

$$E = k\frac{Q}{r^2} = \left(9 \times 10^9\ \mathrm{N}\frac{\mathrm{m}^2}{\mathrm{C}^2}\right)\frac{(16 \times 10^{-12}\ \mathrm{C})}{(0.04\ \mathrm{m})^2} = 90\frac{\mathrm{N}}{\mathrm{C}} \quad \Rightarrow \quad \vec{E} = \left\{90\,\frac{\mathrm{N}}{\mathrm{C}},\ \text{north}\right\}.$$

Note that the field is defined at each of these points, even though there is nothing at any of the points specified. The field is a property of space. It is the set of vectors defined at every point in the space surrounding the charge Q.

Now we need to admit that the surface of the earth provided a place where we would often want to know the gravitational force, so that working out the gravitational field there first was very efficient. There is no such obvious distance from any particular charge where knowing the electric field at that point will commonly come up in problems. Nevertheless, the way we solve electric force problems in this book is to first determine the electric field at the appropriate point and to then find the electric force afterwards. It may seem that this unduly complicates the process, but it actually simplifies the calculation in many cases and, most importantly, it is the only way to solve some of the kinds of problems we deal with later in this chapter. It will also be required for developing the concept of voltage in the next chapter. So let's do an example to see how we calculate electric forces by first finding the electric field.

EXAMPLE 16.4 The Forces Between Two Charges—II

A gray object with charge $Q_{\mathrm{gray}} = 5 \times 10^{-8}$ C and a red object with charge $Q_{\mathrm{red}} = -2 \times 10^{-8}$ C are separated by 50 cm. The red object is due east of the gray object. Find the electric force on each object by first finding the electric field at the location of each charged object due to the presence of the other charged object.

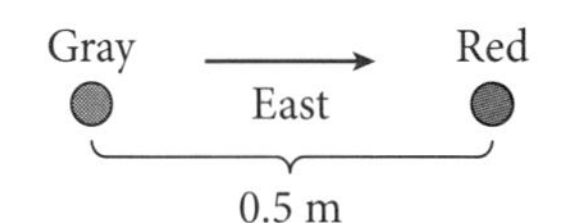

ANSWER Yes, this is the same example we worked before, but this time we are asked to start by finding the electric field at the location of each object. Once we know the field at each object's location, we can find the force on each object due to the field where it is located. To find the field due to the gray object at the location of the red object, we ignore the red object and use Equation 16.4 to get

$$E_{\text{gray}} = k\frac{Q_{\text{gray}}}{r^2} = \left(9 \times 10^9 \frac{\text{N}\cdot\text{m}^2}{\text{C}^2}\right)\frac{5 \times 10^{-8}\ \text{C}}{(0.5\ \text{m})^2} = 1800\frac{\text{N}}{\text{C}}.$$

We now remember that the charge on the gray object is positive and that the electric field from a positive charge always points away from the charge. At the position of the red object, this would make the field point east, so we would have

$$\vec{E}_{\text{gray}} = \left\{1800\ \frac{\text{N}}{\text{C}},\ \text{east}\right\}.$$

To find the force on the red object, we apply Equation 16.5 to get

$$\vec{F}_{\text{gray,red}} = q_{\text{red}}\vec{E}_{\text{gray}} = (-2 \times 10^{-8}\ \text{C})\left\{1800\ \frac{\text{N}}{\text{C}},\ \text{east}\right\} = \{-3.6 \times 10^{-5}\ \text{N, east}\}.$$

Since a force in the minus-east direction is the same as a force in the west direction, we can also write this force as $\vec{F}_{\text{gray,red}} = \{3.6 \times 10^{-5}\ \text{N, west}\}$. For the gray object, we could simply use Newton's Third Law to get $\vec{F}_{\text{red,gray}} = \{3.6 \times 10^{-5}\ \text{N, east}\}$, but, since we were supposed to find the electric field at the location of the gray charge anyway, let's just start from scratch.

To find the electric field at the location of the gray charge, due to the red charge, we again apply Equation 16.4 to the point where the red charge is located, to find

$$E_{\text{red}} = k\frac{Q_{\text{red}}}{r^2} = \left(9 \times 10^9 \frac{\text{N}\cdot\text{m}^2}{\text{C}^2}\right)\frac{2 \times 10^{-8}\ \text{C}}{(0.5\ \text{m})^2} = 720\frac{\text{N}}{\text{C}}.$$

Note that we have ignored the sign of the -2×10^{-8} C charge since we only care about the magnitude of the field in this calculation. We will figure out the direction for ourselves. The direction for a negative charge is toward the negative charge, so, at the location of the gray charge, it is to the east,

$$\vec{E}_{\text{red}} = \left\{720\ \frac{\text{N}}{\text{C}},\ \text{east}\right\}.$$

To find the force, we again use Equation 16.5 to give

$$\vec{F}_{\text{red,gray}} = q_{\text{gray}}\vec{E}_{\text{red}} = (5 \times 10^{-8}\ \text{C})\left\{720\ \frac{\text{N}}{\text{C}},\ \text{east}\right\} = \{3.6 \times 10^{-5}\ \text{N, east}\}.$$

It is worth emphasizing that, as in this last example, we do not calculate the direction of the electric field by worrying about the sign of the Q in Equation 16.4. Our experience tells us that trying to rely on the equation to give the direction can confuse our answer. Maybe this is because we often interpret a positive result as meaning "right" and a negative result as meaning "left." Or maybe it is because we know the direction should be inward and we forget if the minus sign means inward or if it means the opposite of the inward direction we expected. On the other hand, we *do* generally use the sign of the charge on some object in an electric field to tell us whether the force is along the field or in the opposite direction as the field. And, of course, we can always check. The force on a positive charge is always along the local electric field direction and the force on a negative charge is always opposite to the local field direction. If your answer disagrees with this, you have done it wrong.

16.5 Superposition

In the last section, we learned how to use the electric field to find the force on one charged object due to another charged object. Here, we extend this concept to find the force on one charged object due to several other charged objects. As we will see, this is a place where the concept of an electric field begins to simplify our calculation.

Let us consider the black marble carrying charge q, as shown in Figure 16.9. It sits in the presence of three other objects carrying charges, Q_1, Q_2, and Q_3. We know that each charge will put a force on the charged marble and that the net force will be the sum of the three forces, each force being the result of the charge q interacting with the local electric fields due to the three charges. Let us label the three fields $\vec{E}_1$, $\vec{E}_2$, and $\vec{E}_3$, as shown in the figure. (From the direction of $\vec{E}_1$, we know that charge Q_1 must be negative.) Since forces are vectors, the net force on the marble is

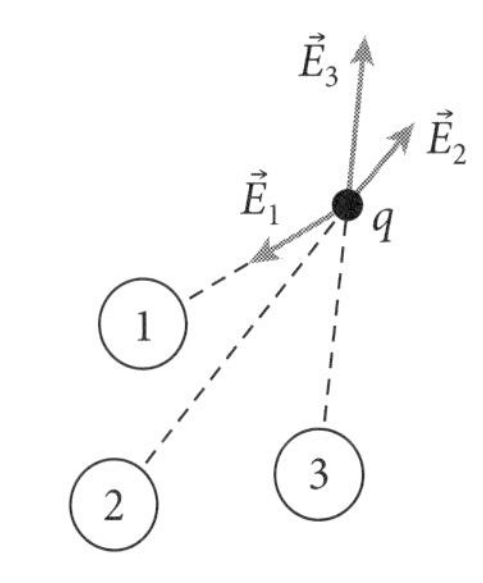

Figure 16.9 Three charges each create electric fields at the point where a charge q is located. (Q_1 is obviously negative.)

$$\vec{F}_{\text{net,marble}} = \vec{F}_{1,\text{marble}} + \vec{F}_{2,\text{marble}} + \vec{F}_{3,\text{marble}} = q\vec{E}_1 + q\vec{E}_2 + q\vec{E}_3.$$

However, instead of multiplying each field by q and then adding the three force vectors, we could more easily factor out the q, giving

$$\vec{F}_{\text{net,marble}} = q(\vec{E}_1 + \vec{E}_2 + \vec{E}_3),$$

meaning that we may vector-add the electric field vectors first and then multiply the total electric field by q to get the force. This simplifies the calculation a little bit.

We define the total electric field at some point in space, due to multiple charges, by the vector sum:

$$\vec{E}_{\text{net}} = \vec{E}_1 + \vec{E}_2 + \vec{E}_3 + \ldots \quad (16.6)$$

Note that here, as in Equations 16.4 and 16.5, the electric field is well defined *only* after one gives the point in space where it is to be evaluated. The question, "What is the electric field due to a +2 C charge at $x = 0$ and a –1 C charge at $x = 1$ m?" has no answer, because we have not specified the point at which we want to know the field. If we were to ask for the electric field at the point $x = 2$ m, produced by those same two charges, then this is a question that has an answer. Which leads us to . . .

EXAMPLE 16.5 The Electric Field Due to Co-Linear Charges

What is the electric field at the point $x = 2$ m, due to a +2 C charge located at $x = 0$ and a –1 C charge at $x = 1$ m?

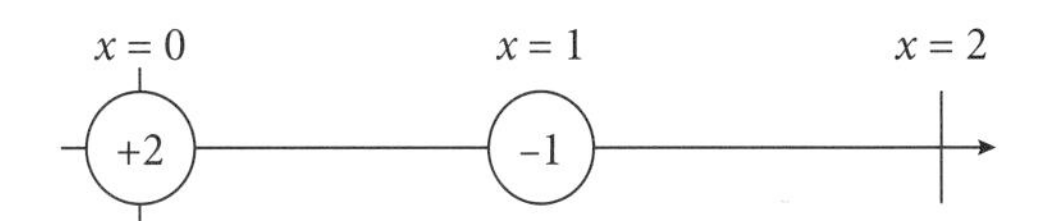

ANSWER We first find the individual fields due to the two individual charges. The charge at the origin is $r = 2$ m from the point where we need to find the field, so

$$E_{+2} = k\frac{Q}{r^2} = \left(9 \times 10^9 \frac{\text{N}\cdot\text{m}^2}{\text{C}^2}\right)\frac{2\text{ C}}{(2\text{ m})^2} = 4.5 \times 10^9 \frac{\text{N}}{\text{C}}.$$

The direction of this field is toward the right, away from the positive charge, in the plus-x direction. Thus, the field is $E_{+2} = +4.5 \times 10^9$ N/C. The charge at $x = 1$ m is only $r = 1$ m from $x = 2$ m, so the field magnitude due to this charge is

$$E_{-1} = k\frac{Q}{r^2} = \left(9 \times 10^9 \frac{\text{N}\cdot\text{m}^2}{\text{C}^2}\right)\frac{1\ \text{C}}{(1\ \text{m})^2} = 9 \times 10^9 \frac{\text{N}}{\text{C}},$$

where, once again, we ignore the sign on the –1 C charge as we get the magnitude of the field. The direction of this field at the point $x = 2$ m is toward the –1 C charge, in the minus-x direction, giving $E_{-1} = -9 \times 10^9$ N/C.

The total electric field in the $+x$-direction at the point $x = 2$ m is therefore

$$\vec{E}_{\text{net}} = \vec{E}_{+2} + \vec{E}_{-1} = +4.5 \times 10^9\ \text{N/C} - 9 \times 10^9\ \text{N/C} = -4.5 \times 10^9\ \text{N/C}.$$

Whether or not there is any real object at the $x = 2$ m, there is a field there, and it points to the left with magnitude 4.5×10^9 N/C.

Our next example determines the electric field vector due to charges that do not all lie in a line. Here, we will have to remember how to do vector addition.

EXAMPLE 16.6 The Electric Field Due to Non-Co-Linear Charges

Three small charged insulating spheres with the following charges are located at the following coordinates:

Sphere 1: $Q_1 = 5 \times 10^{-9}$ C at (–1 m, 0)
Sphere 2: $Q_2 = 2 \times 10^{-9}$ C at (1 m, 0)
Sphere 3: $Q_3 = -6 \times 10^{-9}$ C at (1 m, 1 m)

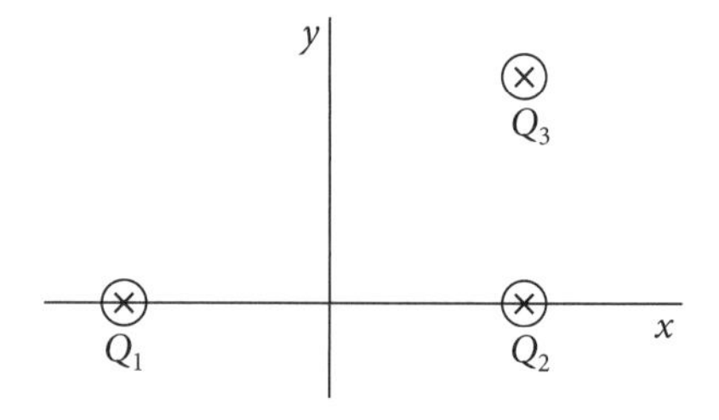

Find the magnitude and direction of the electric field at the origin.

ANSWER For each charged sphere, we first find the magnitude of the electric field it produces at the origin and then use what we know about the directions of electric fields to find the direction. We then resolve the electric field vector into components in the horizontal and vertical directions, preparatory to adding the vectors together.

Sphere 1: The charge is $Q_1 = 5 \times 10^{-9}$ C and the distance from the charge to the origin is 1 m. The magnitude of the electric field is thus

$$E_1 = k\frac{Q_1}{r_1^2} = \left(9 \times 10^9 \frac{\text{N}\cdot\text{m}^2}{\text{C}^2}\right)\frac{5 \times 10^{-9}\ \text{C}}{(1\ \text{m})^2} = 45\frac{\text{N}}{\text{C}}.$$

The field due to a positive charge always points directly away from the charge. At the origin, this would be to the right. This $\vec{E}_1$ vector is drawn in the figure below.

Sphere 2: The distance from sphere 2 to the origin is likewise 1 m, so the field is

$$E_2 = k\frac{Q_2}{r_2^2} = \left(9 \times 10^9 \frac{\text{N}\cdot\text{m}^2}{\text{C}^2}\right)\frac{2 \times 10^{-9}\ \text{C}}{(1\ \text{m})^2} = 18\frac{\text{N}}{\text{C}}.$$

Here, the direction that is away from the positive charge on sphere 2 is to the left at the origin, leading to the vector labeled $\vec{E}_2$ in the diagram below.

Sphere 3: The Pythagorean theorem gives the distance from sphere 3 to the origin as 1.41 m, so the magnitude of the electric field at the origin due to sphere 3 is

$$E_3 = k\frac{Q_3}{r_3^2} = \left(9 \times 10^9 \frac{\text{N}\cdot\text{m}^2}{\text{C}^2}\right)\frac{6 \times 10^{-9}\text{ C}}{(1.41\text{ m})^2} = 27\frac{\text{N}}{\text{C}}.$$

You may be tired of our pointing out that we didn't worry here about the fact that the charge on sphere 3 is negative, so we'll assume you get this now and we will quit repeating ourselves. To find the direction, we remember that the electric field due to a negative charge always points toward the negative charge, so, at the origin, the electric field due to sphere 3 is directed toward sphere 3, at 45° up from the x-axis, as in the figure.

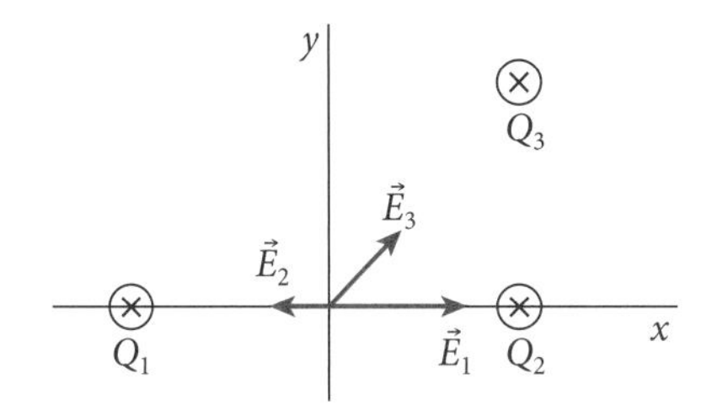

To find the net field, we add these three vectors by breaking them into components. The electric field vectors due to spheres 1 and 2 are already in the x-direction. The field due to sphere 3 will have to be resolved into components by the usual method, $E_{3,x} = 27$ N/C cos(45°) and $E_{3,y} = 27$ N/C sin(45°), giving

Sphere	E_x (N/C)	E_y (N/C)
1	45	0
2	−18	0
3	19.1	19.1
SUM	46.1	19.1

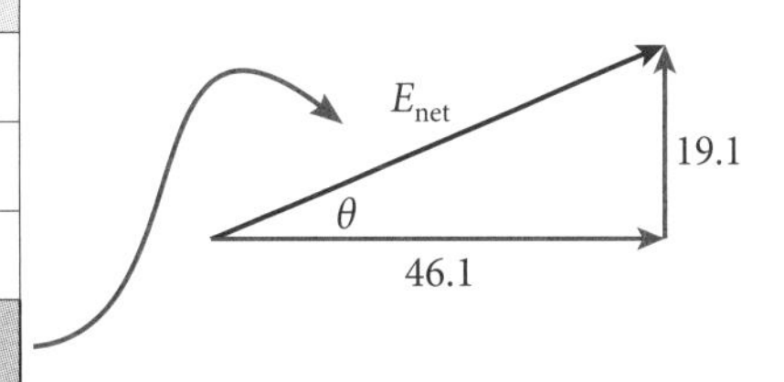

The final vector is the black arrow in the drawing to the right of the table. Its magnitude is 49.9 N/C at an angle $\theta = \tan^{-1}(19.1/46.1) = 22.5°$ above the x-axis.

Before leaving this example, let's be clear that this answer only gives the field at the origin. If we wanted to know the electric field at any other point, we would have to begin again. If more charged spheres were added to the problem, we would just find the electric field for each additional sphere, break that vector into components, and add it to the ones that we already have. As with all of the vector addition we have done before, this is not a job for a brilliant physicist, but for a careful bookkeeper. Be careful and turn the crank slowly.

16.6 Electric Field Lines

If we are only interested in the electric field at a single point, we can draw the electric field vector at that point, as we have done in the last two sections. However, if we are interested in the field in a region of space, drawing the vectors becomes very cumbersome. Well, actually, it becomes impossible, doesn't it? To do this, you would have to show a different vector at each point and, even if you decided to show only representative points, the vectors would surely tend to overlap and become confused.

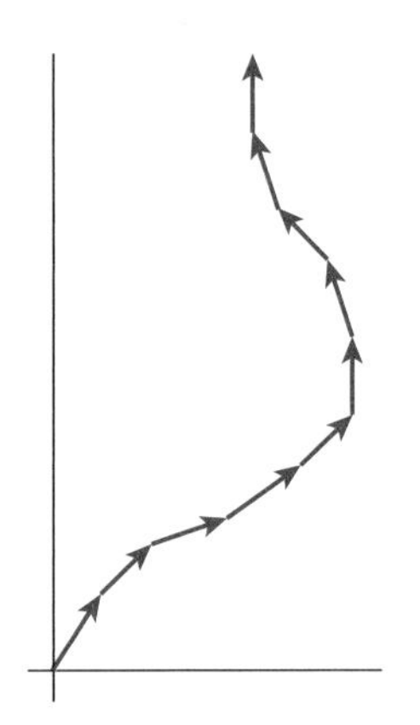

Figure 16.10 The evolving directions of an electric field as one travels a path that follows these directions from point to point.

To deal with this, we now introduce an alternative representation of the electric field. This is one that represents the electric field in some region of space by drawing *electric field lines.* We draw an electric field line by choosing a point in a region containing charged particles that are fixed in place. We find the direction of the electric field at that starting point and we then take a small step in the direction of that vector. At the new point, we again find the direction of the electric field and take a small step in that new direction. Continuing this process creates a path like that shown in Figure 16.10. Note that the arrows in the diagram are not electric field vectors, but are the small displacements that are made parallel to the field lines at each point. If we connect the little steps smoothly, we call this resulting curve an electric field line. The final step is to put a small arrow on the line to indicate the direction of travel (and of the electric field at that point). We can then do this again, starting at a new point in the region, and generate another electric field line. If we do this from enough different starting points, and in enough different directions, we can get a sample of lines throughout the region, as shown in Figure 16.11.

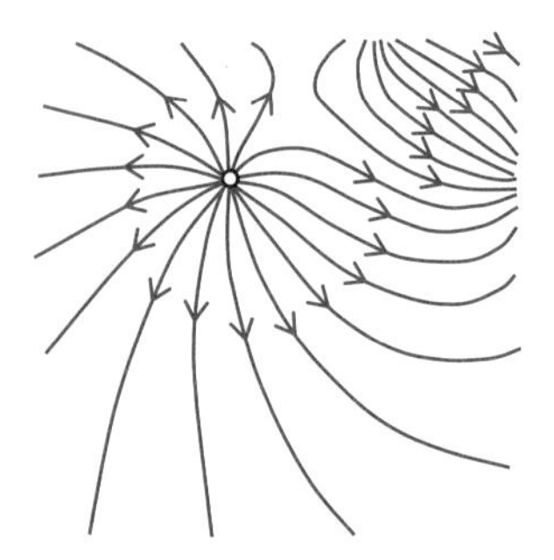

Figure 16.11 A sample of electric field lines in a region of space.

This sounds like a complicated process—and it is. The good news is that we can often learn a good deal through qualitative (non-numerical) sketches of electric field lines, without having to do any calculations. Your choice of the number of lines to draw is arbitrary, but, once a choice is made, the following rules should be obeyed:

1. Electric field lines never cross.
2. Electric field lines begin on positive charges and end on negative charges. If the region you are considering contains more positive than negative charges, some lines will leave the region, and never return. If the region contains more negative charge, some lines will come in from outside the region.
3. The lines entering or leaving a point charge are distributed evenly around the charge.

One thing that is perhaps a little surprising about the sketches of electric field lines that are produced following these rules and the procedure of Figure 16.10 is that the relative strength of the electric field at each point in space turns out to be strictly proportional to how densely packed the lines are in the region around that point. To see what this means, consider the 3-dimensional picture shown in Figure 16.12. Here, we follow four diverging field lines as they first intersect the small circular surface A_1 and then the larger circular surface A_2. The size of each surface is chosen to just catch these four lines and no more. It then turns out that the electric field in the region near each of the surfaces is exactly proportional to the density of field lines intersecting the surface. Once we have decided how many field lines to draw to represent the field due to a particular charge, the proportionality may be written:

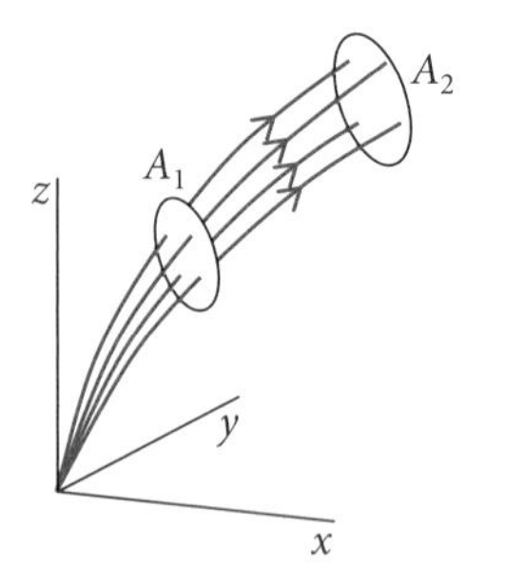

Figure 16.12 Electric field lines (in red) intersecting two successive surfaces.

$$E \propto \frac{\text{number of field lines crossing a surface}}{\text{area of the surface}}. \tag{16.7}$$

Equation 16.7 says only that the electric field is proportional to this density of lines. We will not be able to get the numerical value of the electric field directly from the field lines. In fact, the number of field lines we drew was our own arbitrary choice. Nevertheless, these sketches are useful for understanding a lot of qualitative things about the electric field. In the future, we just need to remember that: first, the electric field at a point is tangent to the field line passing through that point; second, the strength of the electric field is proportional to the density of electric field lines, as in Equation 16.7; and third, for a point that does not lie exactly on one of the electric field lines we have drawn, we can interpolate from the surrounding lines.

Using the rules for generating field lines, we may learn a lot about the field due to some distribution of charge. Let us take, as an example, the field due to two charges, $+q$ and $-q$, separated by some distance, as shown in Figure 16.13. We choose to draw eight lines in the plane of the page for each charge. (There would also be additional lines going into and out of the page at different angles, but we will get to these in a moment.) This will give us Figure 16.13a for the field lines for points very close to each charge. We have already used rule 3 in drawing the ends of these eight lines in Figure 16.13a. We now apply the remaining rules, at the same time thinking about symmetries of the picture. The line that leaves the $+q$ charge going directly to the right has no reason to turn upward or downward, since nothing prefers either direction, so it must go straight to the right and become the line that goes directly into the $-q$ charge. The line that leaves the $+q$ charge just above or below it must never cross another line, so these two must stay fairly close to that first horizontal line. Applying these same considerations to the rest of the lines gives us the drawing shown in Figure 16.13b. Finally, to see the 3-dimensional view of the electric field lines, we picture Figure 16.13b spun about the axis joining the two charges to provide the different orientations.

Figure 16.13a The electric field lines equally distributed around each point charge.

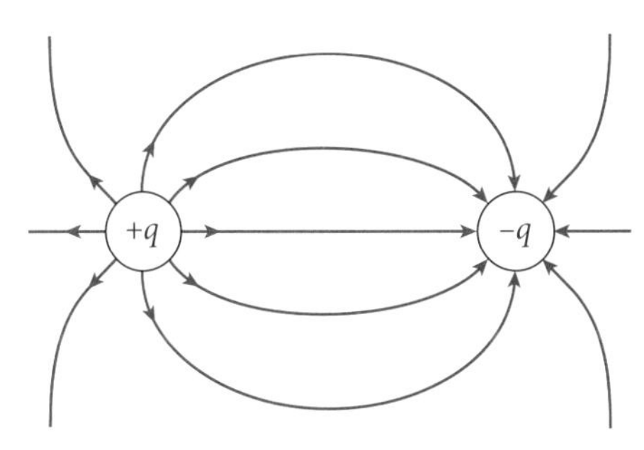

Figure 16.13b Connected electric field lines, generated using the rules on p. 314.

16.7 Uniformly Charged Sheets

Another example of how a combination of thinking about the symmetry of the charge configuration and the rules for drawing electric field lines can give us a very good qualitative understanding of the electric field is the case of a *uniformly charged sheet*. As we will see, this configuration has several practical applications.

We consider the case (and this is the only case we will consider) in which the region where we want to find the field is close to the charged sheet, that is to say that the distance from the field point to the sheet is small compared to the sheet's size. This is often called the "infinite sheet" approximation. This does not mean that the sheet has to actually be infinitely large, only that we are interested only in regions that are relatively very close to the sheet and very far from the edges.

Figure 16.14 shows an edge-on view of a section of a gray charged sheet that extends off to the left and right beyond the page boundary. The charge is uniformly distributed on the sheet, which means that each square centimeter contains the same net charge, and so the electric field lines must be uniformly distributed. We can determine the direction of the field lines

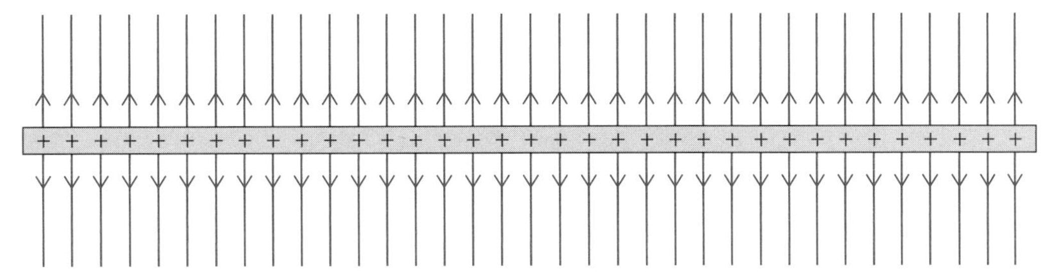

Figure 16.14 A cross-sectional view of an "infinite sheet" of charge and its electric field lines.

from symmetry. If we consider an electric field line starting on the sheet, we see that the line must be perpendicular to the sheet because, from that point in space, all directions look the same. There is no reason for the line to favor left over right. This is similar to being on the ocean on a cloudy night, out of sight of land. Your only direction reference is to keep your boat's wake directly behind you, to continue along in the direction you were going. Similarly, as the electric field line moves away from the sheet, its direction will not change, and so it remains perpendicular to the sheet. If we consider the next line over, the situation is the same, as long as it is also far from the edges of the sheet. The result is a series of equally spaced field lines, parallel to each other and perpendicular to the sheet, as in Figure 16.14. However, if we look at the points further from the sheet, the lines will start to diverge because the sheet no longer appears to extend off to infinity in all directions; edge effects become important. An example of the edge effect is shown in Figure 16.15. Note how the field lines near the center of Figure 16.15, only a couple of millimeters above or below the sheet, look the same as those in Figure 16.13. But they begin to diverge as you go near the edge or away from the sheet.

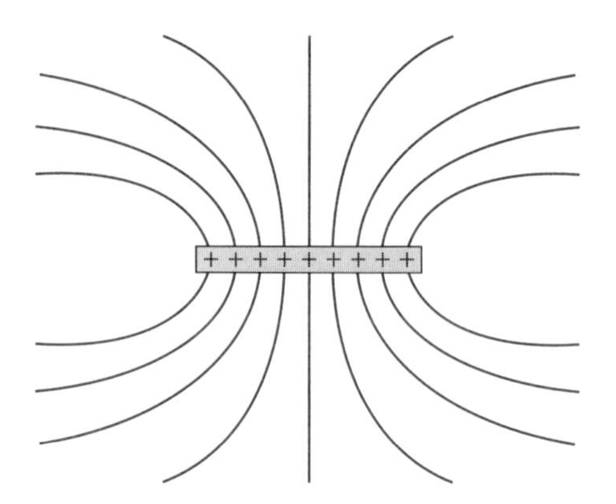

Figure 16.15 Electric field lines far from a finite-length charged sheet.

Thus, if we restrict ourselves to a region close to the sheet, as in Figure 16.14, we discover what may be a surprising result. Because the electric field lines remain parallel, their density does not change as we move away from the sheet, and so the strength of the electric field remains constant. The electric field near the middle of a square meter of charged sheet is the same 1 mm above the sheet as it is 2 mm above the sheet. Only when we got sufficiently far from the sheet—perhaps 10 or 20 cm above or below it—would we notice that the strength of the field is decreasing. Uniformly charged sheets are used in several practical applications ranging from television sets to capacitors, and using the "infinite sheet" approximation that predicts a uniform electric field proves to be remarkably useful (because it is easy and it really is very accurate for many cases).

But, while all of this tells us that the electric field close to a charged sheet is perpendicular to the sheet and constant in magnitude, it does not tell us the value of the electric field. This will obviously depend on how much charge is on the sheet—more charge, more field. But the field seems like it would be difficult to calculate. If we had a square sheet, 1 m on a side, that carried a total charge Q, and if we went to a point 1 cm above the sheet, the electric field would be due to the field from the tiny charge 1 cm directly below it, plus the fields from the charges a little bit off to the side on all sides, plus the fields from the charges even further off to the side, etc. While this calculation may seem impossible, it isn't really. In fact, we will eventually derive the field for a charged sheet in Section 16.9. But, for now, let us just state the result.

The strength of the electric field close to a charged sheet depends only on the density of the charge on the sheet. For a total charge Q residing on a sheet of area A, the surface charge density is defined as $\sigma = Q/A$. As we will see in Section 16.9, the field near the sheet, is

$$E = 2\pi k\sigma, \tag{16.8}$$

as long as we are not too near the edge of the sheet or too far above or below the sheet. The k in Equation 16.8 is Coulomb's constant from Equation 16.1.

Before going on, it is probably time we told you something that we have been avoiding for several pages now. But we know we ought to mention this eventually, so why not now? Many treatments of Coulomb's law do not write Coulomb's constant as $k = 9.0 \times 10^9\ \mathrm{N \cdot m^2/C^2}$. Instead, they replace it with $k = 1/(4\pi\varepsilon_0)$, where ε_0 (pronounced "epsilon-nought") is a fundamental constant called the *permittivity of free space*. It has the value $\varepsilon_0 = 8.85 \times 10^{-12}$ C²/N·m². This really changes nothing, but we didn't want you to be embarrassed some day if you saw this somewhere and didn't know what it was. To help you become familiar with this form of the constant, we will use it from time to time. If we use ε_0 instead of k, the Coulomb field due to a point charge (Equation 16.4) becomes

$$E = \frac{1}{4\pi\varepsilon_0}\frac{Q}{r^2}, \tag{16.9}$$

and Equation 16.8 would be changed to read $E = \sigma/2\varepsilon_0$.

EXAMPLE 16.7 A Charged Dust Particle Near a Charged Sheet

A dust particle with mass $m = 3 \times 10^{-4}$ kg and charge $q = -6 \times 10^{-8}$ C sits directly above the center of a uniformly charged sheet with total area 2 m². The total charge on the sheet, Q, is adjusted until the dust particle remains stationary 1 cm above the sheet's surface. Find the total charge on the sheet and describe what would happen if the dust particle were then lifted to a height of 2 cm and released.

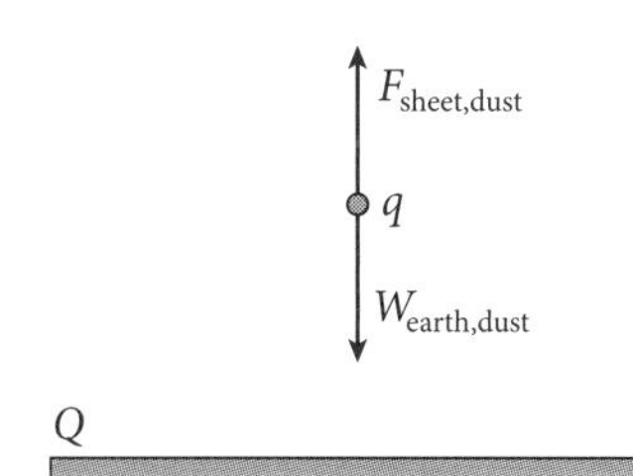

ANSWER The dust particle will remain stationary when the gravitational force (down) is exactly balanced by an electrical force (up). Since the dust particle has a negative charge, an electric field that would lift it upward would have to point downward. (You may want to read that last sentence again to be sure you agree with it.) This tells us that the sheet will have to be negatively charged.

The gravitational force is $W_{\text{earth,dust}} = mg = (3 \times 10^{-4}\text{ kg})(10\text{ N/kg})$, producing a force vector $\{3 \times 10^{-3}\text{ N, down}\}$. So the electrical force must balance the gravitational force with a force vector $\vec{F}_{\text{sheet,dust}} = \{3 \times 10^{-3}\text{ N, up}\}$.

If this is the upward force on a $q = -6 \times 10^{-8}$ C charge, the downward electric field must be of strength

$$E_{\text{sheet}} = \frac{F_{\text{sheet,dust}}}{q} = \frac{3 \times 10^{-3}\text{ N}}{6 \times 10^{-8}\text{ C}} = 5 \times 10^4\,\frac{\text{N}}{\text{C}}.$$

All that remains is to figure out what charge on the sheet will produce an electric field of this magnitude at a point 1 cm above the sheet.

The strength of this electric field depends only on the surface charge density on the sheet, so we can use our just-found value for E_{sheet} to get the charge density, and we can then find the total charge on the sheet from this (because we know the area of the sheet). We already know that the charge is negative (to produce a downward field) so we only need to find its magnitude. The equation for the electric field near a charged sheet is $E = \sigma/2\varepsilon_0$. (Notice that we are giving you a little practice using ε_0 here.) The charge density must therefore be

$$\sigma = 2\varepsilon_0 E = 2\left(8.85 \times 10^{-12}\,\frac{\text{C}^2}{\text{N}\cdot\text{m}^2}\right)\left(5 \times 10^4\,\frac{\text{N}}{\text{C}}\right) = 8.85 \times 10^{-7}\,\frac{\text{C}}{\text{m}^2}.$$

To be completely accurate, we should probably include the sign of the charge and write $\sigma = -8.85 \times 10^{-7}$ C/m². Then, as there are -8.85×10^{-7} C per m², and as the sheet has 2 m² of surface area, the total charge on the sheet must be -1.77×10^{-6} C.

The question also asks what would happen if we were to lift the dust particle to 2 cm from the sheet. It is tempting to think that the particle would fall back down to its previous location or perhaps oscillate back and forth. However, the electric field does not change significantly as we move upward (because we are still close to the sheet) and so the forces remain balanced at any position. If you move the dust particle to some other place, it just stays where you put it.

16.8 Charged Conducting Plates

Although we didn't state it explicitly, we have been using the word "sheet" to describe a very thin layer of charge, perhaps spread on the surface of a piece of insulating material, like a sheet of plastic. By "plate" we will mean a flat piece of material with small but significant thickness. In this section, we will consider plates made of a conducting material like iron or silver or copper.

The first thing we need to know about any conductor is that the electric field inside a conductor must be zero whenever there are no charges moving through the conductor. Think why this is. Conductors contain free electrons. If there is any net electric field inside a conductor, the electrons will move in response to it. Only when the electrons have distributed themselves so that the field inside the conductor is dead zero will the charge distribution remain static.

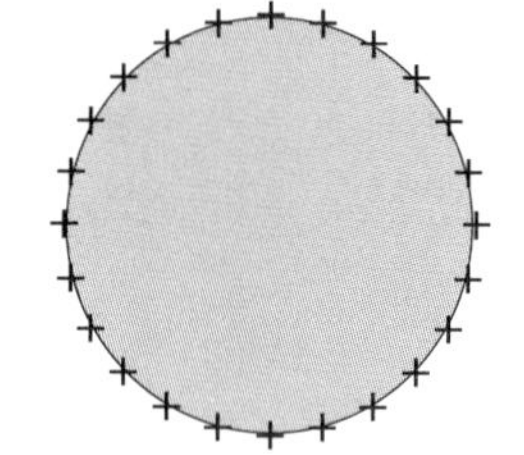

Figure 16.16 The way excess charge is distributed on the surface of a simple conductor.

One consequence of this fact is what happens when there is a net electric charge on a conductor. All of the excess charge will end up on the surface. The reason for this is that the individual charges will repel each other, and sitting on the surface is the way that they can get as far from each other as possible without actually leaving the conductor (atomic forces will keep them tied to the material). For a symmetric object, like the sphere in Figure 16.16, the charge distribution will be uniform around the outside surface. However, for an irregularly shaped conductor, the distribution will have to be non-uniform, with the charges adjusting themselves so as to keep the electric field zero everywhere inside the conductor.

Another consequence of the electric field having to be zero inside a conductor is what happens when an external field is applied to the conductor. If there is an external field, the charges in the otherwise neutral conductor will redistribute themselves in the material until those charges themselves create an electric field that cancels the external electric field. In Figure 16.17 we have drawn a neutral conducting plate placed perpendicular to a uniform external electric field. Initially, the right-pointing electric field creates a force to the left on the free electrons within the conductor, pulling electrons to this side. This process will continue until this induced charge distribution is sufficient to counteract the external electric field and produce a zero net field within the conductor. At this point, the charges will cease moving around. We can think of the positive induced charge on the right surface of the conductor as a positive induced sheet of charge with surface charge density σ_+ and the negative charge on the left as a negative sheet of charge, with density $\sigma_- = -\sigma_+$. For a uniform external electric field E_{ext}, the field induced inside the conductor by the two surface charge densities will have to exactly compensate the applied field inside the material, so that $E_{induced} = E_+ + E_- = -E_{ext.}$ This requirement will tell us what charge density we may expect on each surface. Let's see how this works by presenting an example.

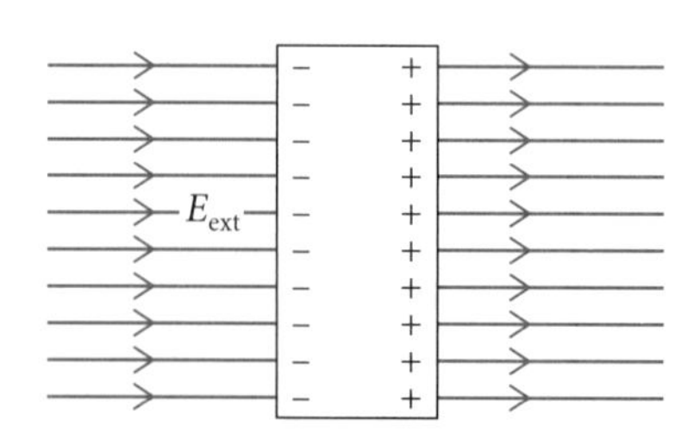

Figure 16.17 The induced charge distribution in a conducting plate placed in a uniform external field.

EXAMPLE 16.8 A Conducting Plate in an External Electric Field

A conducting plate is placed perpendicular to a uniform external electric field of magnitude $E_{ext} = 2 \times 10^5$ N/C to the right. Find the charge density on the surfaces of the plate. How does this charge distribution affect the electric field outside the plate?

ANSWER The net electric field inside the plate must be zero, so the induced charge on each side must create an electric field inside the conductor that exactly cancels the external field, as in Figure 16.17. Because the net charge in the electrically neutral conductor is zero, a surface charge density σ induced on the right surface of the conducting plate will be accompanied by a charge density $-\sigma$ induced on the left surface of the conducting plate.

The total field inside the conductor will be composed of E_{ext}, taken positive pointing to the right, and two fields E_+ and E_- created by the positive and negative surface charges. Both surface charge densities will be uniform sheets of charge and will each produce a field toward the left at points inside the conductor. So the net field to the right is

$$E_{net} = E_{ext} + E_+ + E_- = E_{ext} - \frac{\sigma}{2\varepsilon_0} - \frac{\sigma}{2\varepsilon_0} = 0,$$

where we have again used Equation 16.8 in its ε_0 form for the fields due to the two surface sheets of charge. This equation may be solved for σ, giving

$$E_{ext} - \frac{\sigma}{\varepsilon_0} = 0 \quad \Rightarrow \quad \sigma = \varepsilon_0 E_{ext}.$$

The surface density is thus $\sigma = (8.85 \times 10^{-12}\ \text{C}^2/\text{N}\cdot\text{m}^2)(2 \times 10^5\ \text{N/C}) = 1.77 \times 10^{-6}\ \text{C/m}^2$. The right side of the plate will have a charge density of $+1.77 \times 10^{-6}\ \text{C/m}^2$, while the left will have a charge density $-1.77 \times 10^{-6}\ \text{C/m}^2$.

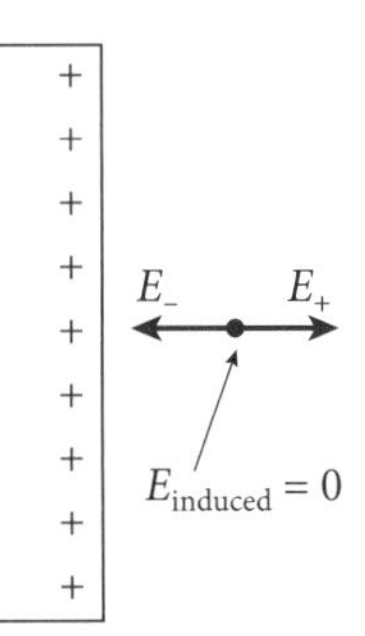

If we look at a point just outside the conductor's positively charged side, we can find the effect of the rearranged charges from the conductor. To the right of the plate, E_+ will point away from the plate while E_- will point toward the plate. The charge densities are the same magnitude and, in the infinite sheet approximation, the strength of the field does not depend on the distance from the surface, so the two induced electric fields exactly cancel. The field outside the sheet is unaffected by the redistribution of charges within the plate.

16.9 Electric Flux and Gauss's Law

If you remember back to the first part of this book, you will recall that, after introducing Newton's laws of motion and using them in free-body diagrams, we then went on to define several new quantities. We defined work as the product of force and distance and we defined momentum as the mass times the velocity. These definitions did not provide any new fundamental physics, but both of these concepts provided us convenient methods for solving problems in mechanics. In this section, we introduce a new quantity that will help us to solve problems involving electric fields. It is called the *electric flux*.

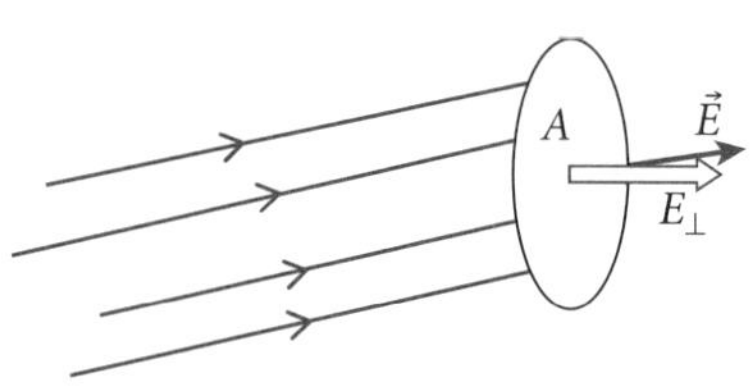
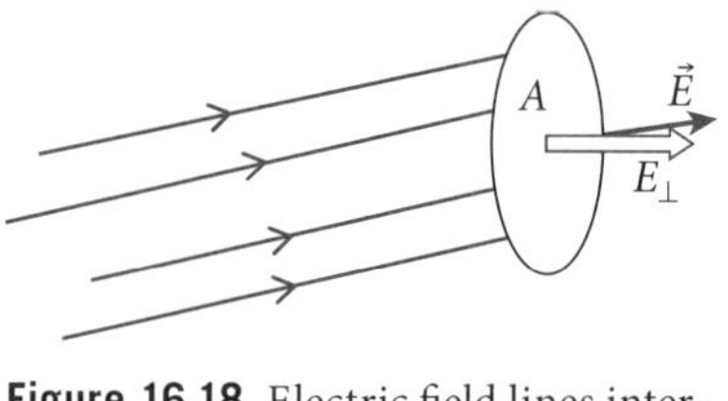

Figure 16.18 Electric field lines intersect area A. The component of $\vec{E}$ perpendicular to A is $E_\perp$.

The electric flux is defined where there is an electric field in a region of space. It is found by selecting some imaginary surface in the region and then defining the *electric flux* as the product of the component of the electric field perpendicular to that surface times the area of the surface. By taking only the perpendicular component, we are choosing only that component that penetrates the surface, ignoring components that lie entirely in the surface. The electric flux is

$$\Phi_E = E_\perp A. \tag{16.10}$$

The symbol Φ is a Greek capital "phi" and is the traditional symbol for flux. A graphic representation of the quantities in this definition is shown in Figure 16.18.

If the electric field is not constant everywhere on the surface we chose, we would have to do something like divide the surface into a bunch of smaller surfaces, each small enough that the field is roughly constant on each one, and we would then add up the flux from all of the little surfaces. That would be complicated, of course. So let us calculate the electric flux for one configuration that we know we can work with.

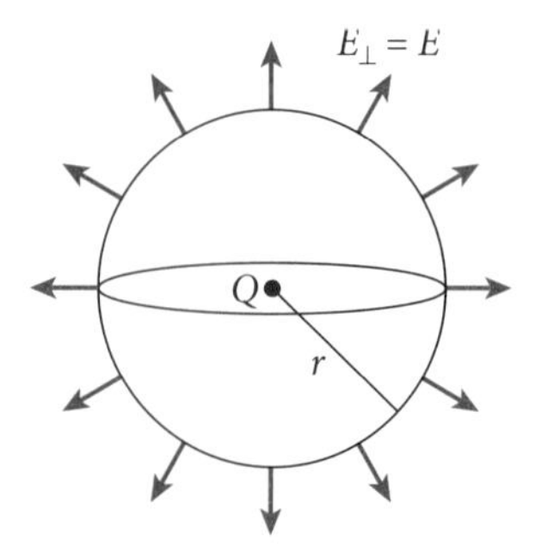

Figure 16.19a The electric field due to a charge Q inside a sphere of radius r.

Let us imagine the electric field produced by a single positive charge $+Q$, and let us find its flux through a spherical surface of radius r surrounding that charge. The magnitude of the electric field is the same at each point of the surface, and the direction is always perpendicular to the surface (see Figure 16.19a), so the flux is

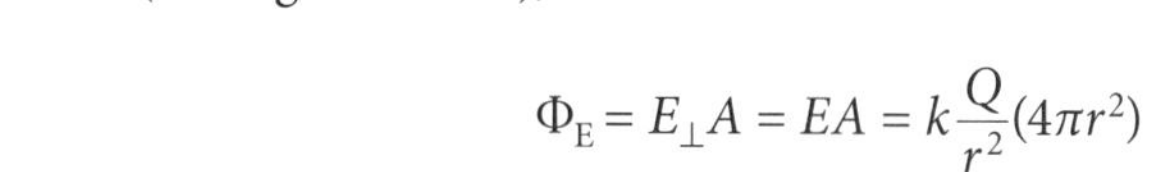

$$\Phi_E = E_\perp A = EA = k\frac{Q}{r^2}(4\pi r^2)$$

in which we have remembered that the area of the surface of a sphere is $A = 4\pi r^2$. If we cancel the r^2 terms, we get the following formula for the electric flux through a sphere containing a charge Q at its center:

$$\Phi_E = 4\pi kQ = \frac{Q}{\varepsilon_0}. \tag{16.11}$$

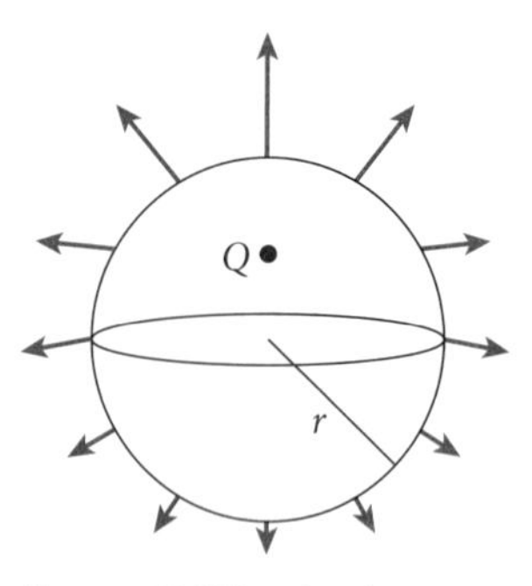

Figure 16.19b The electric field due to an offset charge Q inside a sphere of radius r.

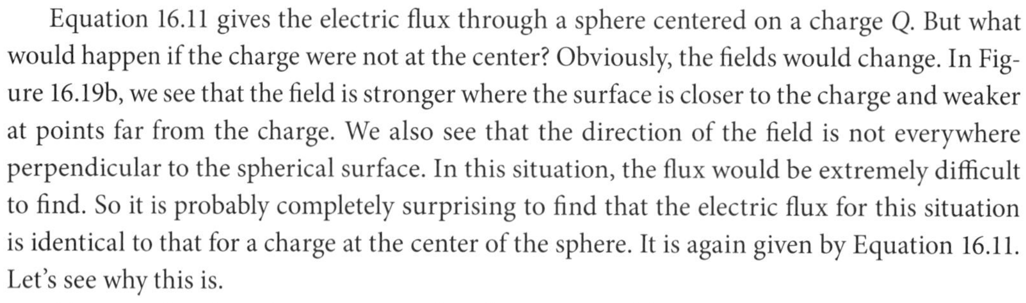

Equation 16.11 gives the electric flux through a sphere centered on a charge Q. But what would happen if the charge were not at the center? Obviously, the fields would change. In Figure 16.19b, we see that the field is stronger where the surface is closer to the charge and weaker at points far from the charge. We also see that the direction of the field is not everywhere perpendicular to the spherical surface. In this situation, the flux would be extremely difficult to find. So it is probably completely surprising to find that the electric flux for this situation is identical to that for a charge at the center of the sphere. It is again given by Equation 16.11. Let's see why this is.

We remember from Equation 16.7 that the electric field is proportional to the number of field lines crossing a surface divided by the area of that surface. If we begin with Equation 16.7 and multiply the electric field by the area of a surface through which the field lines are passing, we find

$$EA \propto \text{number of field lines crossing a surface.}$$

But the product EA is the electric flux through that surface. So we may also write

$$\Phi_E \propto \text{number of field lines crossing a surface.}$$

If we choose to draw 144 field lines emerging from a charge Q, and if we consider the closed surface of Figure 16.19a, then all 144 lines will cut through that surface. But the number of field lines cutting through the surface in Figure 16.19b is also all 144 lines, since they all still have to get out of the sphere. Because the number of field lines cutting through the surface is the same for a charge at the center of a sphere as it is when the charge is not at the center, and because the flux through any surface is proportional to the number of lines cutting the surface, the flux through the sphere will be the same for a charge anywhere inside the sphere as it is for the charge at the center of the sphere. If we cut our charge into pieces, ½Q + ½Q, and put the two pieces at different locations inside the sphere, there would still be as many field lines drawn, and there would thus be the same total flux. If we deformed our sphere into a square box, then the same number of field lines would still have to get out of the box, and so the flux through the box would be the same as through the sphere.

This then is a general law that all electric fields obey. It is called *Gauss's Law* (named after Carl Friedrich Gauss who discovered it in 1835). It states that the electric flux through any closed surface is equal to $1/\varepsilon_0$ times the charge contained inside that closed surface,

$$\Phi_E = \frac{Q_{\text{inside}}}{\varepsilon_0}, \tag{16.12}$$

and it works however the charge is distributed inside any surface you might choose.

16.10 Using Gauss's Law

If you have stuck with us through all of these steps deriving Gauss's law, you now may be wondering where the payoff is. Gauss's law gives a simple relationship between the electric flux through a closed surface and the total charge inside that surface, but what good is that? Why would I want to know the electric flux? Here is your answer.

Gauss's law is a single scalar equation relating one number, the total charge, to another number, the flux. All of the direction information contained in the electric field vectors has been lost, because the electric flux counts only the perpendicular component of the electric field. And any variation of the electric field around the surface has been folded away because the flux is only the total flux over the whole surface. But, often, when we use the symmetry of some system of charges to help us find the electric field lines, the one thing we are missing is the actual strength of the field—a single number. If the symmetry of the problem lets me find a surface on which the electric field is constant, even though I do not know its strength, and if I can determine the direction of the field all over that surface, then I can write down an equation for the flux through that closed surface that is only missing the strength of the electric field. And, in this case, Gauss's law will provide the missing number.

Here is the example that we have been waiting to do for about five pages now.

EXAMPLE 16.9 The Electric Field Near a Sheet of Charge

A thin sheet of area $A_{tot} = 2\ m^2$ carries $4\ \mu C$ of charge, spread evenly over the sheet. If we remain far from the edges of the sheet, what is the electric field a distance 1 cm away from the sheet?

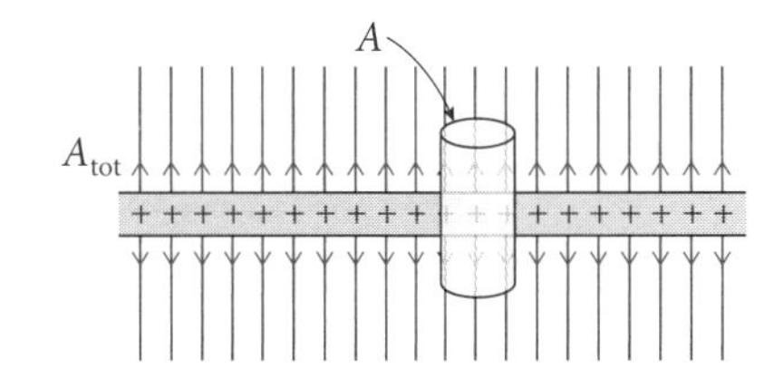

ANSWER We know from our discussion of the symmetry back on page 315 that electric field lines near the center of the sheet must be uniformly distributed and perpendicular to the sheet, as long as we are close to the sheet and far from the edges of the sheet. At a point 1 cm above or below the sheet, the electric field must point directly away from the sheet and must not change as we move horizontally to a nearby point. Let us therefore choose a closed surface that exploits what we know about the field. If we pick a cylinder of height 2 cm and a cross-sectional area A, then we may picture this cylinder sitting centered on the sheet and protruding 1 cm above and 1 cm below the sheet.

Let us apply Gauss's law to this surface. There are three parts to the surface of the cylinder—the curved side of the cylinder and the two caps of area A. Along the side, the electric field lies entirely in the surface, and so it has no component that pierces it. The perpendicular component of the electric field is zero, so the flux through this part of the closed surface is zero. Over the two caps, the field is perpendicular to the surfaces, penetrating outward, and it is constant over the surfaces. From symmetry, we know that the magnitude of the electric field must be the same below the sheet as it is above the sheet. If we call the magnitude of the field on the caps E, then the flux through each cap is EA, and the total flux through the cylinder is

$$\Phi_E = \Phi_{E,\text{side}} + \Phi_{E,\text{top}} + \Phi_{E,\text{bottom}} = 0 + EA + EA = 2EA$$

To use Gauss's law, we will have to determine the total charge contained inside this cylinder. We know that the total charge, $Q = 2\ \mu C$, is spread over the entire surface, so the portion inside the cylinder is only a small part of that. The surface charge density is $\sigma = Q/A_{\text{tot}}$, and so the amount inside the cylinder of area A is $Q_{\text{inside}} = \sigma A$. Gauss's law then gives us

$$\Phi_E = \frac{Q_{\text{inside}}}{\varepsilon_0}$$

$$2EA = \frac{\sigma A}{\varepsilon_0}.$$

Solving the last equation for E gives us Equation 16.8, $E = \sigma/2\varepsilon_0$, now derived from Gauss's law and not simply stated without proof, as we did back on page 316.

In our case, $\sigma = (2 \times 10^{-6}\ C)/(2\ m^2) = 10^{-6}\ C/m^2$, and so the numerical value of E is

$$E = (10^{-6}\ C/m^2)/(2 \times 8.85 \times 10^{-12}\ C^2/N \cdot m^2) = 5.65 \times 10^4\ N/C.$$

This will be the electric field magnitude above and below the plate, as long as we do not go far enough away from the sheet that the assumptions for the direction and the constancy of the field are violated.

Let us reiterate that Gauss's law is always true, but it can be used to determine the electric field at some point only if we can find some surface over which the variation of the electric field in magnitude and direction is known, and where only the single number corresponding to the magnitude of the field is needed.

Finally, let us point out that we have introduced only one new piece of physics in this chapter. It is contained in Coulomb's law (Equation 16.1). The electric field was a defined quantity, based on Coulomb's law, and Gauss's law was derived in this section from the formula for the electric field.

16.11 Summary

There were quite a few new things to take away from this chapter, since it involved the introduction of an entirely new law of physics.

- You should understand the two units of electric charge and the conversion between them: 1 e = 1.6×10^{-19} C.
- Coulomb's law (Equation 16.1) is a new fundamental law of physics. You should make sure you understand the quantities in the equation and how to use them to find the electric force between two charges.
- The electric field is defined as $\vec{E} \equiv \vec{F}_E / q$. It subdivides the process of finding a force into two steps—first find the field and then find the force via $\vec{F}_E = q\vec{E}$. But finding the electric field is an important task on its own. You may often be asked not to "find the force," but to just "find the field."
- The electric field due to a point charge is given by Equation 16.4.
- The electric field due to many charges is found by vector-adding the fields due to the individual charges. Make sure you remember your vector addition.
- You should be able to sketch electric field lines in systems with a high degree of symmetry.
- You should understand how electric fields and surface charges are induced in conductors and insulators.
- Finally, you should understand Gauss's law (Equation 16.12) and be able to follow the way it was used to determine the electric field strength in Example 16.9.

CHAPTER FORMULAS

Constants: $1\text{ e} = 1.6 \times 10^{-19}\text{ C}$ $\quad k = 9.0 \times 10^9\text{ N}\cdot\text{m}^2/\text{C}^2$ $\quad \varepsilon_0 = 8.85 \times 10^{-12}\text{ C}^2/(\text{N}\cdot\text{m}^2)$

Coulomb's Law: $F_E = k\dfrac{Qq}{r^2}$

Electric field due to a point charge Q: $E = k\dfrac{Q}{r^2}$

Electric field close to a uniform sheet of charge density σ: $E = \dfrac{\sigma}{2\varepsilon_0}$

Flux: $\Phi_E \equiv E_\perp A$ $\qquad$ Gauss's Law: $\Phi_E = \dfrac{Q_{\text{inside}}}{\varepsilon_0}$

PROBLEMS

16.1 *1. How many excess electrons are required for an object to carry a charge of 1 pC (10^{-12} C)?

*2. A penny initially contains 1.76×10^{23} protons and an equal number of electrons. The penny is then charged by removing 1.50×10^{15} electrons. What is the final net charge on the penny?

*3. You remove 2×10^{10} electrons from a quarter and the same number from a penny. Do the coins end up with a net positive charge or a net negative charge? Which of the coins ends up with the greater net charge? Explain.

**4. If all of the atoms in 12 grams of carbon (see page 286 for a reminder of how many atoms this is) lost just one electron each, how many coulombs of charge would the carbon carry? Is this a lot?

16.2 *5. Two uniformly charged spheres made of an insulating material are held atop fixed posts. The charge on the sphere on the left is three times the charge on the sphere on the right. Which diagram correctly represents the magnitude and direction of the electrostatic forces on the two spheres? Explain your answer.

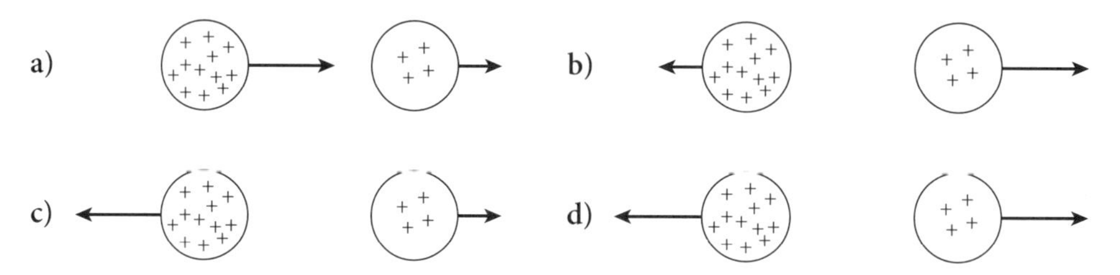

*6. What force is exerted by a charge of 5×10^{-11} C on a 3×10^{-10} C charge 2 m away?

*7. Two tiny charged spheres, separated by 2 m, exert a 36 mN force on each other. What will be the force between them if the distance is increased to 4 m?

*8. Two dust particles carry identical charges of 2 nC. How far apart must they be placed for the force between them to be 10 μN?

**9. A statcoulomb of charge is defined as the amount of charge that you would put on each of two conducting spheres, separated by 1 cm, to produce a dyne (10^{-5} N) of electrostatic force. How many coulombs in a statcoulomb?

**10. Assume that the electron in a hydrogen atom is 53.0 pm from the nucleus of the atom, which consists of a single proton.

a) Calculate the electrical force between the electron and the nucleus.

b) Calculate the gravitational force between the electron and the nucleus.

c) What is the ratio of the gravitational force to the electrical force?

Use values found in Appendix B, as needed.

**11. It is 15 km from downtown Bozeman, Montana, to The Ridge, atop the Bridger Bowl Ski area. If a 1 C positive charge is placed on the top of The Ridge and a 1 C positive charge is placed atop the Baxter Hotel in downtown Bozeman, what will be the force of repulsion between the two charges?

**12. How far apart must two electrons be placed if the force of repulsion between them is equal in magnitude to the gravitational force on an electron on the surface of the earth?

**13. Two alpha particles (helium nuclei, each with a charge of +2 e) approach each other to within a distance of 100 pm (three times the radius of the helium atom). What will be the force of repulsion between the two particles?

** 14. Two small charged spheres are mounted on insulating bases as shown below. The spheres are small enough that they can be treated as point charges. Both spheres have a negative charge, but the magnitude of the charge on sphere 1 is greater than that on sphere 2.

a) Draw a free-body diagram for sphere 2. For each of the forces on the diagram indicate:
- whether the force is contact or non-contact,
- the object exerting the force, and
- the object on which the force acts.

b) Compare the magnitude of the electric force by sphere 1 on sphere 2 to the magnitude of the electric force by sphere 2 on sphere 1. Explain your reasoning.

16.3 ** 15. A negatively charged rubber rod is brought near a neutral empty soda can. Explain how the rod can attract the can toward it.

** 16. Three charges are located at different positions on a 2-dimensional grid. There is a $+5\ \mu C$ charge at the origin, a $+2\ \mu C$ charge on the x-axis at $x = 2$ m, and a $-4\ \mu C$ charge on the y-axis at the point $y = 3$ m. Find the forces on the $5\ \mu C$ charge at the origin.

** 17. In the diagram below, what is the force (magnitude and direction) on the 30 nC charge?

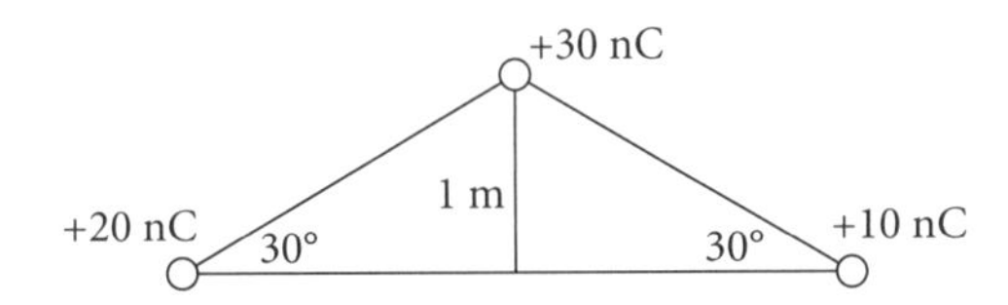

** 18. A charge $Q = +30$ nC is placed at the origin (remember a nC is 10^{-9} C). A charge -60 nC is placed at coordinate (10 cm, 0). For this problem, express forces to the right as positive and to the left as negative. Find:

a) the net force on the $+30$ nC charge, and

b) the net force on the -60 nC.

c) Explain how you determined the directions of the forces.

*** 19. Two small copper balls, each of mass 100 g, are suspended from a single point on threads 30 cm long. The two balls are then given identical electric charges, and it is found that they come to rest when they are 30 cm apart, as shown. What is the charge on each ball?

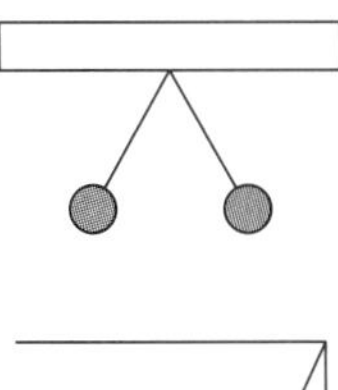

*** 20. A small ball of mass 50 g carries a $+2\ \mu C$ positive charge. It is suspended from the ceiling by a massless thread 50 cm long attached where the ceiling joins the wall, as shown. What uniform static charge density, placed on the surface of the wall, will cause the charged ball to come to equilibrium at a distance 30 cm from the wall?

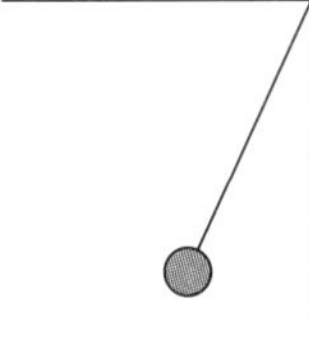

16.4 * 21. An electron (charge $-e$) experiences a force of 1.5×10^{-15} N directed due north. What is the electric field (magnitude and direction) at the point where the electron is located?

* 22. The force on a charge of -5 pC is measured to be 2.5 N upward. What are the magnitude and direction of the electric field at this point?

* 23. Find the force on a 5 pC charge in a place where the electric field is 400 N/C.

* 24. At what distance from a charge $Q = -3.0 \times 10^{-6}$ C would a charge $q = +6.0 \times 10^{-6}$ C experience an electric field whose strength is 5.0×10^{4} N/C?

** 25. Two tiny spheres, A and B, are charged to $Q_A = -4.0 \times 10^{-9}$ C and $Q_B = +7.0 \times 10^{-9}$ C. With the spheres held 0.23 m apart, what electric field (strength and direction) is experienced by sphere A (express the direction of the field as either toward or away from sphere B)?

16.5 * 26. Charge $Q_1 = 2\,\mu$C is located at (–5 cm, 0) while charge $Q_2 = -4\,\mu$C is located at (+5 cm, 0), as shown. In which of the three regions below (a, b, or c) is it possible for all other charges to experience a net force of zero due to Q_1 and Q_2?

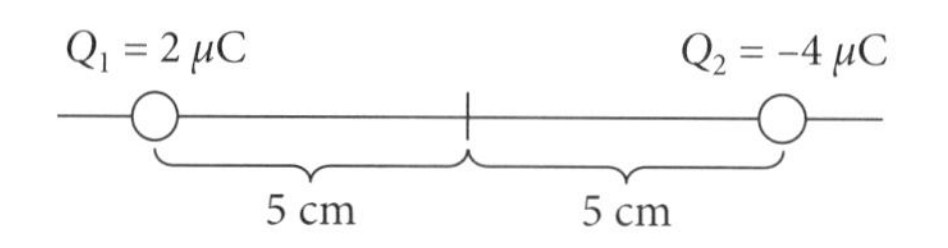

a) To the left of Q_1.

b) Between the two charges.

c) To the right of Q_2.

Explain and justify your answer.

** 27. Two charges, $Q_1 = 6 \times 10^{-8}$ C and $Q_2 = -6 \times 10^{-8}$ C, are held in place at a separation of 0.16 m. What is the strength of the electric field due to Q_1 and Q_2 at the point halfway between the two charges?

** 28. A charge $Q_A = 4.0 \times 10^{-9}$ C experiences a net electrostatic force of 1.2×10^{-6} N toward the north.

a) Find the electric field (magnitude and direction) at the location of Q_A.

b) If the charge Q_A is replaced by a charge $Q_B = -1.5 \times 10^{-9}$ C, what force (magnitude and direction) will Q_B experience?

** 29. A charge $Q_1 = 5.0\,\mu$C is in a location where there is an electric field of 1500 N/C directed away from another charge $Q_2 = 2.5\,\mu$C. What is the separation between the two charges?

** 30. Two fixed charges, Q_A and Q_B, are located along the x-axis, as shown.

$Q_A = +4 \times 10^{-8}$ C $Q_B = -2 \times 10^{-8}$ C

× 5 cm   10 cm

a) Find the net electric field (magnitude and direction) at the point labeled ×.

b) What force (magnitude and direction) would a charge $q = -5 \times 10^{-9}$ C experience if placed at ×?

** 31. Two charges, Q_A and Q_B, are located 20 cm apart as shown. When a third charge $Q_C = 4 \times 10^{-9}$ C is placed at the ×, 10 cm from Q_B, the electric force on Q_C is zero.

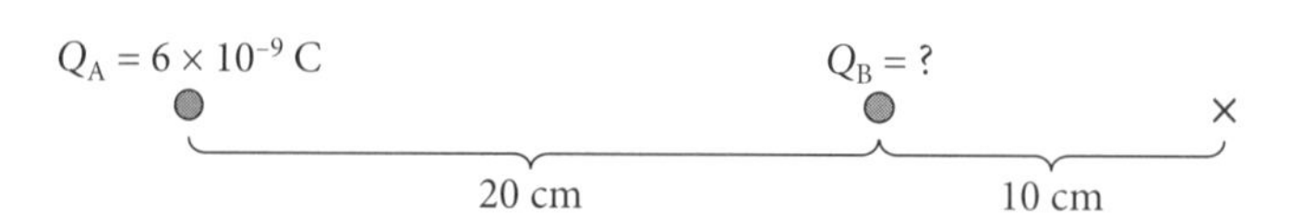

a) What is the net electric field at ×?

b) What is the electric field at × due to Q_A alone (magnitude and direction)?

c) Find the value (including sign) of Q_B.

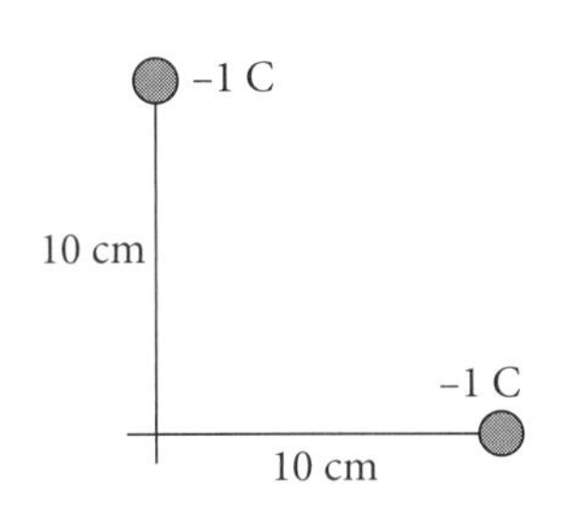

* 32. A −1 C charge is placed 10 cm above the origin and another −1 C charge is placed 10 cm to the right of the origin. What is the electric field (magnitude and direction) at the origin?

** 33. Three charges are placed on a grid in which each little square is 1 cm on a side. All charges are the same size, q, but are negative or positive, as indicated. Draw a vector diagram that shows the individual electric fields at the point × due to each of the three charges and then determine the direction of the E-field at that point. Be clear in labeling all vectors that you draw and be sure to indicate the vector that represents your final answer.

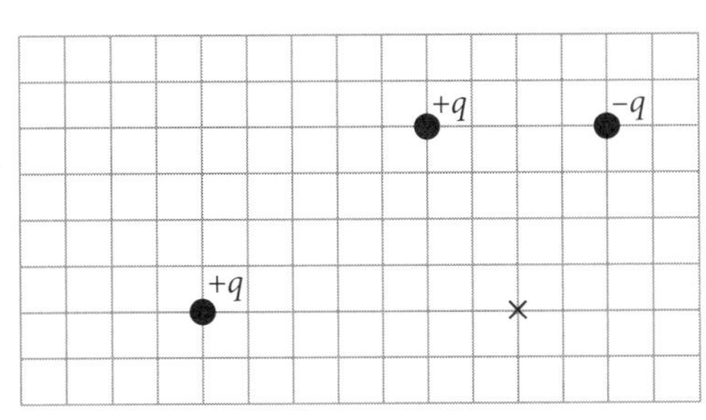

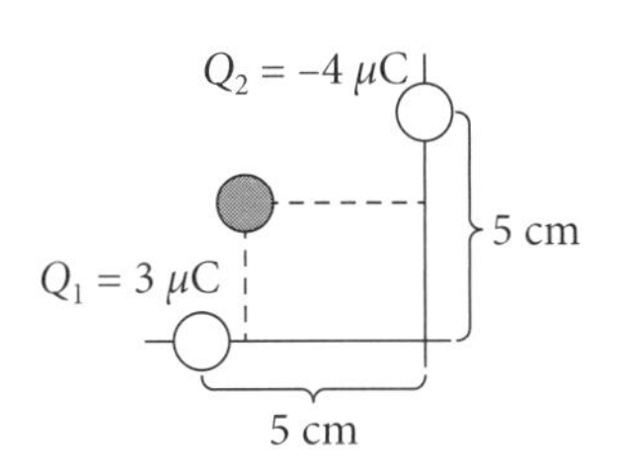

** 34. A charge $Q_1 = +3\ \mu C$ is located at (−5 cm, 0) while charge $Q_2 = -4\ \mu C$ is located at (0, +5 cm). These are the two white charges in the diagram at right.

a) What is the electric field at the origin (0, 0) due to these two charges?

b) If a third charge of 5 μC (the gray charge) is placed at the point (−4 cm, 3 cm), what will now be the electric field at the origin?

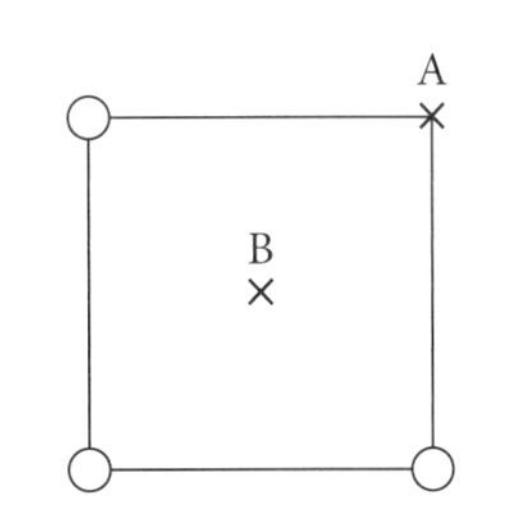

** 35. Three +3 μC charges are placed at three corners of a square 2 m on a side, as shown at right. Find the electric field (magnitude and direction) at the two field points designated with the ×, Point A at the fourth corner of the square and Point B at the center of the square.

***36[f]. An *electric dipole* is formed by two equal and opposite charges, +Q and −Q, separated by a vector $\vec{d}$, as shown in the figure below. The negative charge is on left. We define a as half the distance between the two charges, as shown.

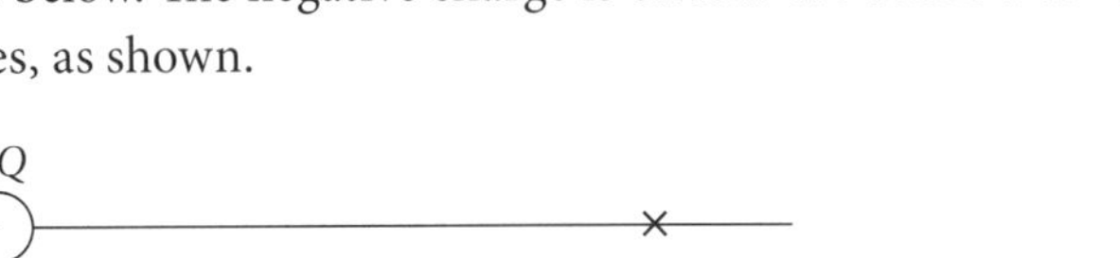

a) Find the electric field along a line parallel to $\vec{d}$ and at a distance r from the center of the dipole, as shown above.

b) Note that $(r \pm a)^2 = r^2(1 \pm a/r)^2$ and apply the binomial approximation (see Appendix A.5) for the small quantity, a/r when $a \ll r$, to show that the electric field due to the dipole is $\vec{E} \approx 2kQ(\vec{d}/r^3)$.

** 37. A large uncharged conducting ball is placed to the right of a small negative point charge −Q, as shown below.

a) Draw a vector diagram that shows the direction of the electric field at × due to $-Q$ alone.

b) Draw a circle representing the uncharged conducting sphere and add little "+" and "–" signs to your drawing to indicate roughly how the induced charge distribution on the sphere will be distributed.

c) Is the magnitude of the net electric field at × greater than, equal to, or less than the electric field due to $-Q$ alone? Carefully explain your reasoning.

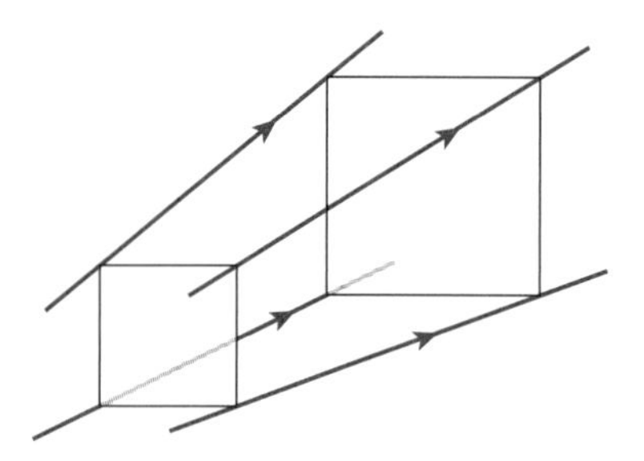

16.6 ** **38.** Electric field lines are diverging from each other in a region of space, as shown in the diagram at left. Four adjacent field lines intersect the corners of a square with area 2 cm^2, the lines being perpendicular to the surface of the square. The magnitude of the electric field on this square is 10 N/C. The same four lines intersect the four corners of a larger square, again at right angles to the surface of the square. The area of the second square is 5 cm^2. What is the electric field strength on the 5 cm^2 surface?

** **39.** A portion of an electric field is shown below. Four locations within the field are marked with black dots and labeled a, b, c, and d. Draw a diagram showing only the four dots, and then draw four vectors, one for each location, that estimate the electric field vector at each of these locations. (Think of the force that would be exerted on a positive test charge $+q_0$ if it were placed at each of these locations.) Your vectors should include both direction and relative magnitude, and you should explain your reasoning for each of the vectors you have drawn.

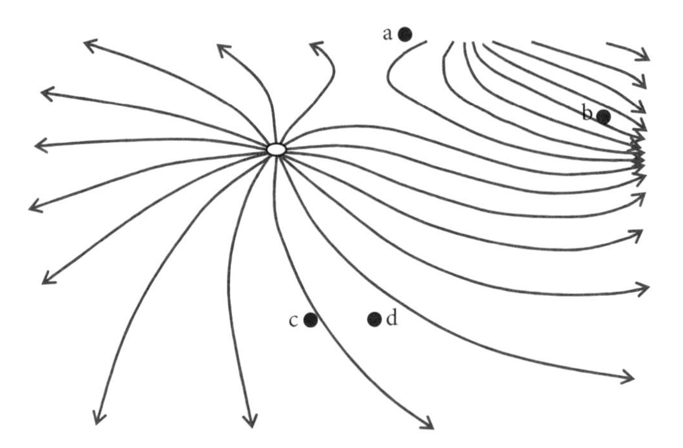

** **40.** The electric field lines in a region of space are shown in the diagram below. The dashed line indicates the path of a positively charged particle that is free to move in this field, subject only to the electric field represented by the field lines.

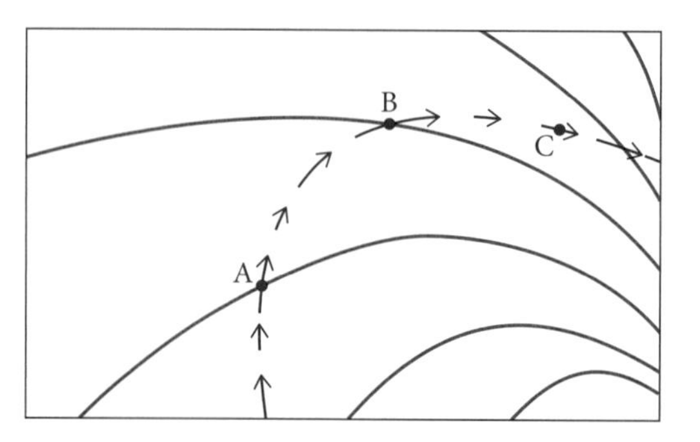

a) Consider the path of the charged particle and decide if the field along each line should be roughly left to right or right to left. Explain how you arrived at your answer.

b) Place three points in a drawing representing points A, B, and C, and draw arrows at each point representing the electric field there. Make the arrow point in the direction of the field and make its magnitude roughly represent the magnitude of the field (i.e., small, intermediate, and large).

** 41. The field lines (solid lines) shown below represent the electric field in a certain region of space. A charged particle is released with an initial velocity and follows the trajectory from points 1 to 4 as shown by the black line.

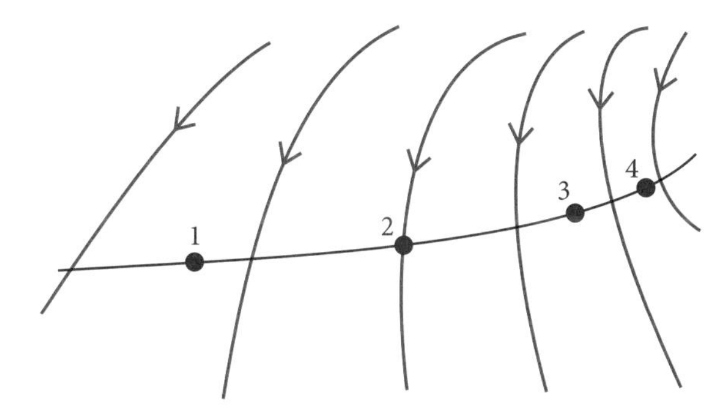

a) Place four points in a drawing representing points 1, 2, 3, and 4. Draw a vector to represent the force on the charged particle at each of the four positions. Explain your reasoning.

b) What is the sign of the charged particle? Explain how you could tell.

** 42. Use the rules described on page 314 to sketch the electric field lines between a uniform sheet carrying charge Q and a small uniformly charged sphere carrying charge –Q. Be sure to indicate the direction of the electric field lines.

*** 43. Use the rules described on page 314 to sketch the electric field lines in the region surrounding a –1 C charge and a +2 C charge, separated by a short distance, as shown. Be sure to indicate the direction of the electric field lines. [Hint: Rule number 2 on page 314 means that you will have more field lines leaving the +2 C charge than you have entering the –1 C charge.]

16.7 ** 44. A square insulating plate, 2 meters by 2 meters, lies flat on the floor. A total charge $Q = 8 \times 10^{-8}$ C is distributed uniformly on the plate. A pith ball (a small non-conducting ball made of a light spongy plant material called "pith"), also with charge $Q = 8 \times 10^{-8}$ C, is fixed 50 cm directly above the center of the plate. Point A is located 5 cm from the plate, directly below the pith ball, as shown.

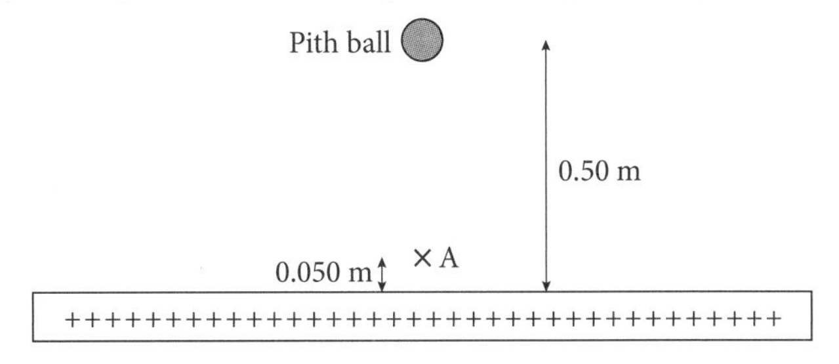

a) What is the electric field at A due to the plate (magnitude and direction)?

b) What is the electric field at A due to the pith ball (magnitude and direction)?

c) What is the net electric field at A (magnitude and direction)?

d) What force (magnitude and direction) would an electron experience if it were located at point A?

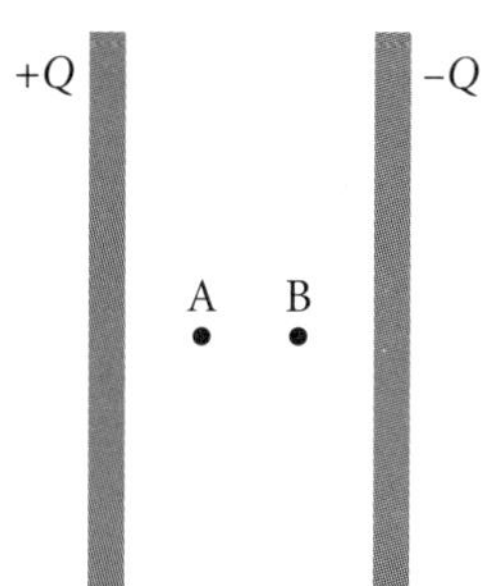

16.8 ** 45. Points A and B are located as shown between two large uniformly charged parallel plates that are close enough together that all points between the two may be considered as being near the plates. Which of the following correctly describes what happens to the electric fields at A and B if the plate on the right (−Q) is removed? Explain your answer.

a) The field at A drops more than the field at B.

b) The field at B drops more than the field at A.

c) Both fields remain that same.

d) Both fields drop by the same amount.

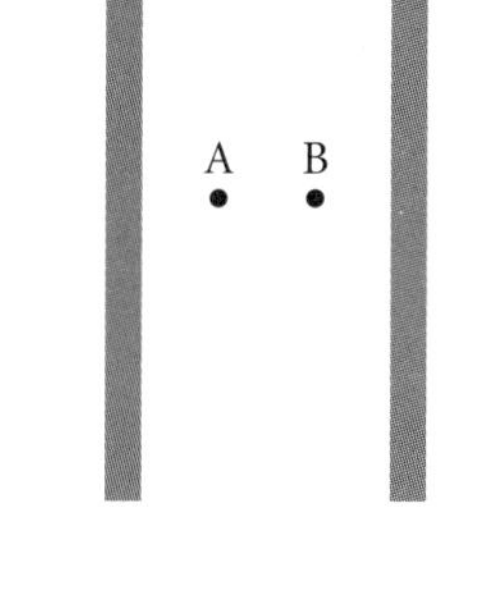

** 46. Large parallel conducting plates, each with area 1.0 m^2, are separated by 5 cm. The plates carry opposite charges of $\pm 7.5 \times 10^{-6}$ C. What is the magnitude of the force on a proton placed between the plates at 1.5 cm from the positive plate?

** 47. Two large charged sheets are placed a distance D apart. The top sheet has uniform surface charge density $+\sigma$ while the bottom sheet has uniform charge density -2σ. The sheets are large compared to their separation.

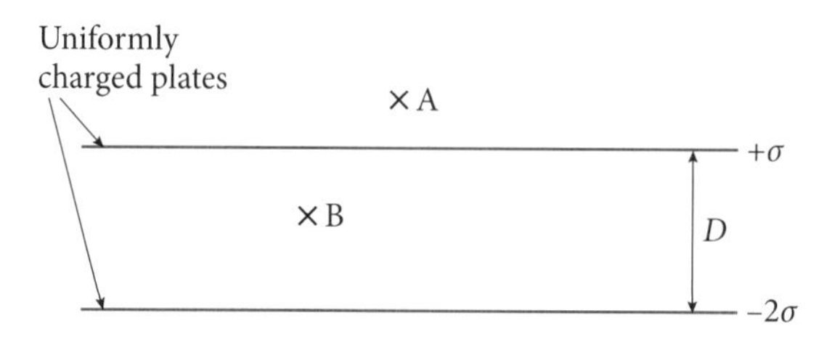

a) What direction is the electric field at point A? Justify your answer.

b) At which point, A or B, is the strength of the electric field greater? Explain your reasoning.

c) If the bottom plate were suddenly removed, would the *direction* and the *magnitude* of the electric field at point A change? Explain your reasoning.

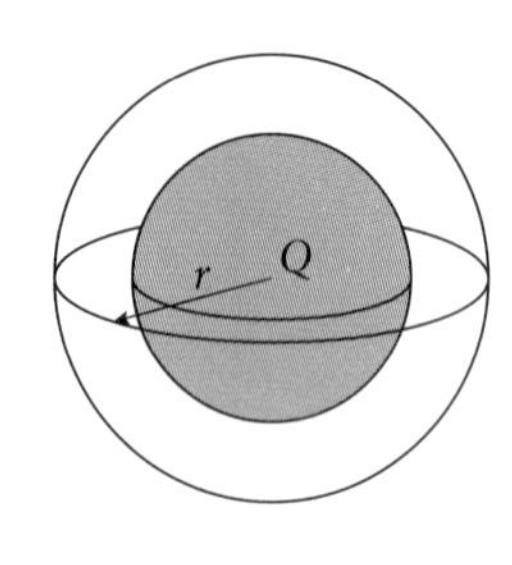

16.9 *** 48[f]. In this problem, we will derive a formula for the electric field outside a spherical conductor with a net charge Q placed on it.

a) First, choose a Gaussian surface that is a sphere of radius r, centered on the center of the spherical conductor. Then explain how the symmetry of the source can be used to determine if the electric field can vary over the sphere you have chosen and in what direction the electric field must point on your chosen surface.

b) Show that the electric flux through the closed surface of the Gaussian surface you have chosen will be $4\pi r^2 E$, where E is the magnitude of the electric field a distance r from the center of the conducting sphere.

c) Use Gauss's law to find a formula for the electric field a distance r from the center of a conducting sphere carrying a net charge Q.

*** 49[f]. A conductor of arbitrary shape carries a net charge, distributed in such a way that the surface charge density in a small region of the surface is σ. For points close to the surface of the conductor, the surface looks approximately flat.

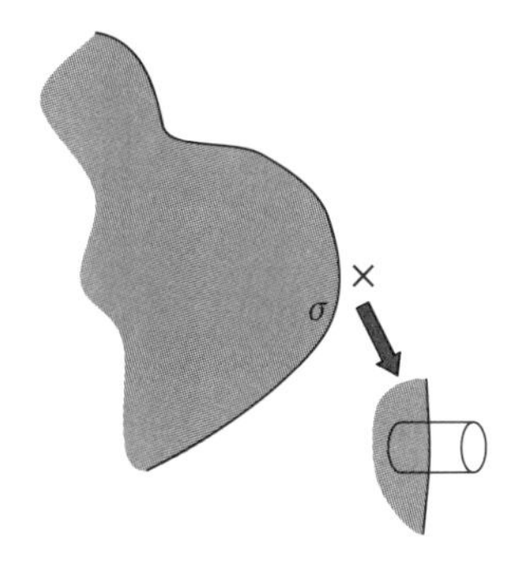

a) Consider the small cylindrical Gaussian surface in the figure at right, which has one cap inside the conductor and one just outside the conductor, and explain why the electric field will be constant over the outside cap, zero on the inside cap, and why the field will be parallel to the sides of the cylinder.

b) Use Gauss's law to find the electric field just outside the surface of a charge conductor carrying surface charge density σ.

c) Explain why this result is twice the field outside a sheet of charge with the same surface charge density. [Hint: Consider how the electric field inside the conductor can be zero, even though there is a surface charge σ nearby.]

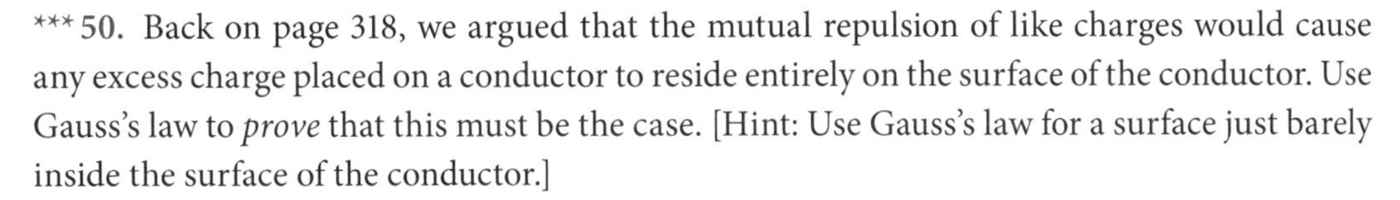

*** 50. Back on page 318, we argued that the mutual repulsion of like charges would cause any excess charge placed on a conductor to reside entirely on the surface of the conductor. Use Gauss's law to *prove* that this must be the case. [Hint: Use Gauss's law for a surface just barely inside the surface of the conductor.]

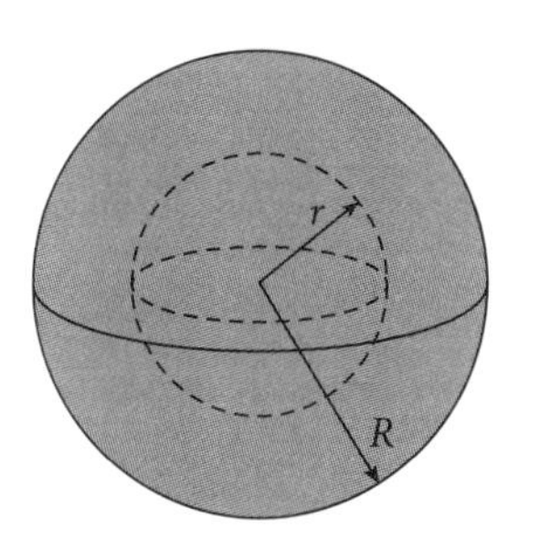

*** 51[f]. An insulating material in the shape of a sphere of radius R has a total charge Q spread uniformly throughout its volume. The charge density, in coulombs per cubic meter, is thus a uniform $\rho = (3Q)/(4\pi R^3)$ everywhere inside the sphere. (We have used $V = (4/3)\pi R^3$ for the volume of the sphere.) Find a formula for the electric field strength inside the sphere, at a radial distance r from the center of the sphere. In order to find this, consider an imaginary sphere of radius r inside the insulating sphere, as shown at right, and use Gauss's law applied to this imaginary sphere. Explain what symmetry arguments you used to allow you to use Gauss's law.

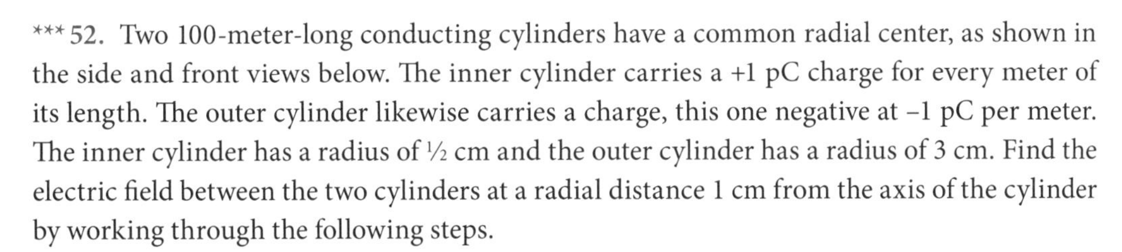

*** 52. Two 100-meter-long conducting cylinders have a common radial center, as shown in the side and front views below. The inner cylinder carries a +1 pC charge for every meter of its length. The outer cylinder likewise carries a charge, this one negative at −1 pC per meter. The inner cylinder has a radius of ½ cm and the outer cylinder has a radius of 3 cm. Find the electric field between the two cylinders at a radial distance 1 cm from the axis of the cylinder by working through the following steps.

(a)

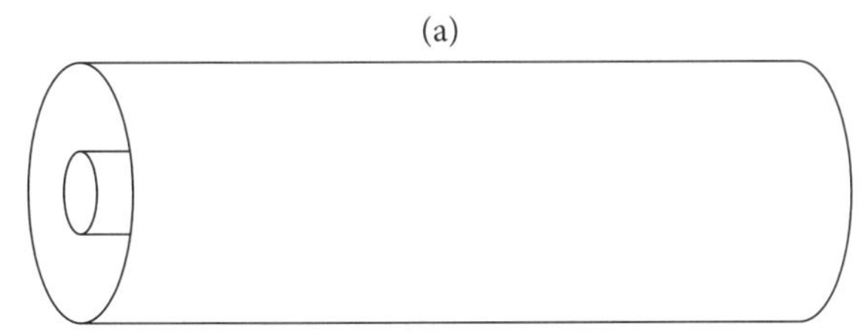

Side view of a section of the cylinders

(b)

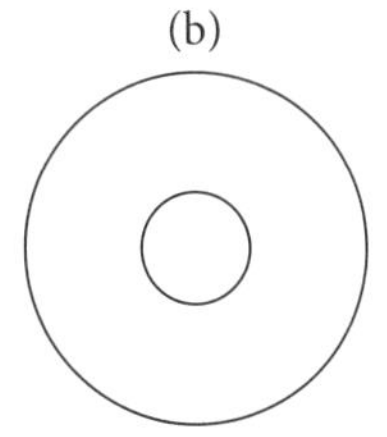

Front view

(c)

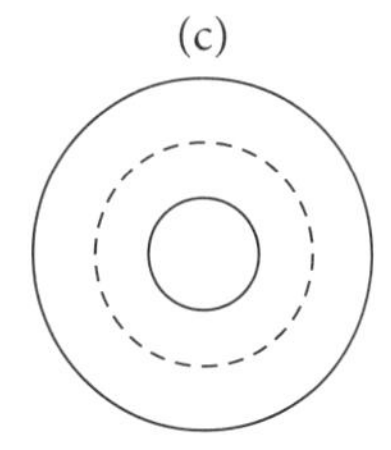

Front view with a Gaussian cylinder

a) Draw the two concentric cylinders as shown in the front view (picture b) and sketch the electric field lines in the region between the two cylinders.

b) Use the fact that the cylinders are very long to explain how the electric field behaves as one moves along the cylinder axis. Will the field direction or strength vary with position along the axis or will it be the same?

c) Consider a Gaussian surface that is a cylinder with a radius of 1 cm and a length of 1 m. A front view of the Gaussian surface is shown as the dashed circle in picture c. Use what you know about the directions of the fields around the side of the cylinder and on the two end caps to work out the electric flux through the Gaussian surface. Leave the strength of the field, E, as an unknown in the expression you work out.

d) Calculate the total charge contained inside the 1-meter length of the cylinders and use Gauss's law to find the strength of the electric field at a radial distance 1 cm from the center axis.

e) How would the field change if you went out to a distance 2 cm from the center axis? Comment on how the field varies with radial distance from the axis and how this result compares with what you would expect from considering the density of field lines, as discussed on page 314.

17

VOLTAGE

17.1 The Gravitational Potential

We begin this chapter by taking a step backward to reconsider the gravitational potential energy. We think it will be easier to understand the new idea we will present in this chapter, one involving our new electrical force, if we start first with a similar concept based on the more familiar gravitational force.

Back in Chapter 7, we defined the potential energy stored in the gravitational interaction between the earth and an object of mass m to be $PE_G = mgh$, where g is the earth's constant gravitational field and h is the height of the object above some level that we defined to be the "ground" at $h = 0$. This potential energy is measured in joules (J) and is equivalent to the work that an external agent would have to do to move the object at constant speed from the ground to the final height. If an object is moved from point A at height h_A to point B at height h_B, then there will be a change in the gravitational potential energy given by

$$\Delta PE_G^{A\to B} = mgh_B - mgh_A = mg\Delta h_{A\to B}.$$

Now, imagine that we place a number of different objects on a shelf that is a height h above the ground and that we want to know the gravitational potential energy stored in the object-earth system for each object. Clearly, we would have to multiply the mass of each object by gh to find each potential energy. But, since gh is the same for each case, we could save some effort by first calculating gh and then multiplying each mass by the resulting number. This new number is given a name—it is called the *gravitational potential*. Note that this new quantity is a property of a point in space. A point h meters above the ground will have a gravitational potential $U = gh$, even if there is no object with mass at that point to produce any gravitational potential energy. Also, you should be careful that you do not confuse *gravitational potential* ($U = gh$) with *gravitational potential energy* ($PE_G = mU = mgh$). They are related to each other, but they are not the same. They even have different units. Gravitational potential energy is measured in joules, while gravitational potential is measured in J/kg.

This gravitational potential—the gravitational potential energy per kilogram—is closely analogous to a similar quantity called the electrical potential that is the focus of this chapter. Once you know more about this, it would be valuable to review the previous paragraphs to see the connections. There is, however, one major difference between electrical potential and gravitational potential. This is the fact that gravitational potential energy is proportional to mass, whereas electrical potential energy is proportional to an object's charge, which can be either positive or negative. Without a certain amount of care, this difference can be the source of some confusion.

17.2 Electric Potential Energy and Voltage

Back in Chapter 7, we identified two conservative forces—forces that led to a potential energy that could be added to the kinetic energy to give a conserved total mechanical energy. They were the gravitational force and the spring force. We are now ready to add a third force to the list. This is the electric force we presented in Chapter 16. The Coulomb force is conservative, so there will be a potential energy associated with it, the *electric potential energy.* It is a true energy, measured in joules, and it can be added to other forms of energy in the work–energy theorem.

Coulomb's law shows that the electric force on an object is proportional to the charge on the object, and we have said that there is a potential energy associated with this force. This means that we can create an electrical analogy to the gravitational potential discussed in the last section. We can define an *electric potential* in the following way. If the electric potential energy of a charge q changes by an amount $\Delta PE_E^{A\to B}$ as it moves from point A to point B, then the electric potential difference between the two points is defined to be

$$\Delta V_{A\to B} = \frac{\Delta PE_E^{A\to B}}{q} = \frac{PE_E^{B} - PE_E^{A}}{q}. \tag{17.1}$$

The units of the electric potential are obvious from Equation 17.1; they are joules per coulomb. Because the electric potential is used so often, this unit is given a name. It is called a volt (V), after Allesandro Volta, the Italian physicist who invented the chemical battery in 1800. The electric potential is also referred to by the name of its units; it is called the *voltage.* To avoid confusion with the electric potential energy, and because it slips a little easier off the tongue, we will refer to the electric potential as the "voltage" throughout the rest of this book.

But why do we care about the voltage anyway? First, as we will see in the next chapter, it is the critical quantity used in the analysis of electric circuits. Second, it provides a simple way to address questions of electrical energies, as we can see in the following two examples.

EXAMPLE 17.1 A Proton Passing Through a Voltage Difference

A proton in an electric field begins at rest and moves through a voltage difference of –1000 V. Find the final speed of the proton.

ANSWER As a proton moves from a point with a higher voltage to a point with lower voltage, its electrical potential energy must drop. But electric force is conservative, so the total energy remains constant. Any electric potential energy lost must show up as an increase in the kinetic energy of the proton.

The problem statement gives us the potential difference through which the proton moves, giving the change in electric potential energy from Equation 17.1 as

$$\Delta PE_E^{A\to B} = q\,\Delta V_{A\to B} = (1.6 \times 10^{-19}\ \text{C})(-1000\ \text{J/C}) = -1.6 \times 10^{-16}\ \text{J}.$$

Since the proton has lost 1.6×10^{-16} J of electric potential energy, it must have gained the same amount of kinetic energy. The proton started from rest, so the change in its kinetic energy is its final kinetic energy. This is

$$\frac{1}{2}mv^2 = 1.6 \times 10^{-16}\text{ J}.$$

Solving for v (and knowing that the mass of a proton is 1.67×10^{-27} kg), we find

$$v^2 = \frac{2(1.6 \times 10^{-16}\text{ J})}{1.67 \times 10^{-27}\text{ kg}} \quad \Rightarrow \quad v = 4.4 \times 10^5\text{ m/s}.$$

Having done this calculation, let us think for a minute about how we worked with the units. A joule, we remember, is a $\text{kg} \cdot \text{m}^2/\text{s}^2$. When this is divided by kilograms, we get units of m^2/s^2 for v^2. The square root produces the appropriate m/s units for v.

The previous example represents a common usage of the concept of voltage. When an electric field is used to accelerate particles, it is the voltage difference that is generally known, and the final speed of the particle can then be calculated from that knowledge. The change in voltage in the previous example was negative, but a common way to state this would be to say, "The proton was accelerated through 1000 volts." If a positive charge, like a proton, is accelerated, it means that it must have gone from higher voltage to a lower one, a voltage difference that is negative.

EXAMPLE 17.2 An Electron Passing Through a Voltage Difference

An electron has a velocity of 5.00×10^6 m/s when observed at point A. When observed at point B, its speed is found to be 1.00×10^7 m/s. Find the change in voltage from point A to point B.

ANSWER The particle is speeding up, which means that it is gaining kinetic energy. But it is an electron, so care must be taken with the signs of things. Because the electric force is conservative, the electric potential energy must decrease in order to compensate for the increase in kinetic energy. The kinetic energies are

$$\text{KE}_\text{A} = \frac{1}{2}mv_\text{A}^2 = \frac{1}{2}(9.11 \times 10^{-31}\text{ kg})\,(5.0 \times 10^6\text{ m/s})^2 = 1.14 \times 10^{-17}\text{ J}$$

$$\text{KE}_\text{B} = \frac{1}{2}mv_\text{B}^2 = \frac{1}{2}(9.11 \times 10^{-31}\text{ kg})\,(1.0 \times 10^7\text{ m/s})^2 = 4.56 \times 10^{-17}\text{ J}$$

$$\Delta\text{KE}_{\text{A}\to\text{B}} = \text{KE}_\text{B} - \text{KE}_\text{A} = 4.56 \times 10^{-17}\text{ J} - 1.14 \times 10^{-17}\text{ J} = 3.42 \times 10^{-17}\text{ J}.$$

We know that the change in total energy must be zero ($\Delta\text{KE}_{\text{A}\to\text{B}} + \Delta\text{PE}_\text{E}^{\text{A}\to\text{B}} = 0$), from the work–energy theorem. We must therefore have $\Delta\text{PE}_\text{E}^{\text{A}\to\text{B}} = -3.42 \times 10^{-17}$ J. To find the change in voltage, we return to Equation 17.1 and write

$$\Delta V_{\text{A}\to\text{B}} = \frac{\Delta\text{PE}_\text{E}^{\text{A}\to\text{B}}}{q} = \frac{-3.42 \times 10^{-17}\text{ J}}{-1.60 \times 10^{-19}\text{ C}} = 214\text{ J/C} = 214\text{ V}.$$

This is a positive number. The voltage at point B must be 214 volts higher than the voltage at point A.

Example 17.2 is worth reviewing several times because it deals with a number of critical elements. The first important point is the meaning of the negative signs that show up. In the case of the charge of the electron, the negative sign tells us that the property itself is negative. In the case of electric potential energy, the negative sign is telling us that a property is decreasing (i.e., the change is negative). We also see that, although the change in electric potential energy was negative, the change in voltage (the property of space along the path of the electron) was positive. Our negatively charged electron lost potential energy (and therefore sped up) in moving from point A to point B, but a positive 1-coulomb charge making the same journey would *gain* 214 J of potential energy and would therefore slow down. This was to be expected, of course, because the electric force acting on a positive charge in that region of space would be in the opposite direction as the force on a negative particle in the same region. So we again emphasize that the voltage is a property of space and does not depend on what is moving between the points. An elephant moving from A to B would move through a voltage difference of +214 volts, but it would experience no change in electric potential energy, because it is electrically neutral.

Our two previous examples used the concept of the change in voltage between two points. Sometimes, however, it is convenient to define the voltage at a specific point to have some value and then find particular voltages at other points, based on this choice. This is equivalent to defining the gravitational potential energy of a block on the floor to be zero. In Example 17.2, if we had said that point A is at 1000 volts, then point B would be at 1214 volts. If we said that point B was at 0 volts, then point A would be at –214 volts. We can assign voltage values, but only voltage differences have any physical significance. This is why we often find it easier to talk about voltage differences, rather than absolute voltages.

17.3 Uniformly Charged Parallel Plates

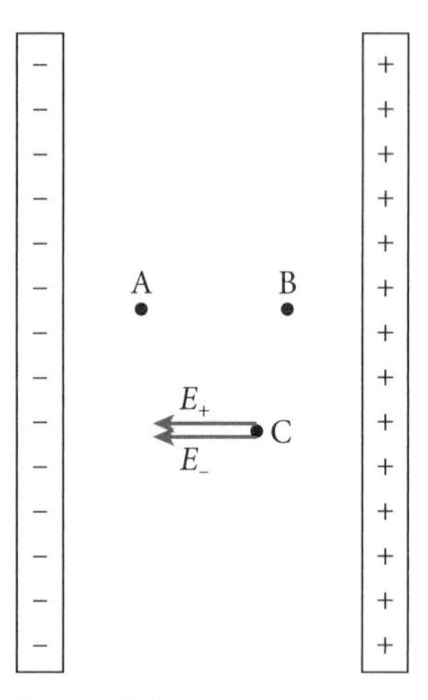

Figure 17.1 At point C, the electric fields (in red) from the positive and negative charge distributions add in the region between the plates.

A common configuration for many electrical applications consists of two large conducting plates, one carrying a charge $+Q$ and one carrying a charge $-Q$, where the distance between them is small compared to the size of the plates. In this region, the electric field from each plate will be uniform and perpendicular to the surface of the plate. We recall from Equation 16.8 (with $1/4\pi\varepsilon_0$ replacing k) that the field due to a single plate is given by $E = \sigma/2\varepsilon_0$, where $\varepsilon_0 = 8.85 \times 10^{-12}\ \text{C}^2/\text{N}\cdot\text{m}^2$, and $\sigma = Q/A$ is the surface charge density. Let us consider a point between the two plates, like the point labeled C in Figure 17.1. The field at this point, due to the positive charge on the right plate, points to the left. The field due to the negatively charged plate will point in that same direction. The *electric field between two conducting plates* is therefore the sum of the two fields, $E = \sigma/2\varepsilon_0 + \sigma/2\varepsilon_0$, or

$$E = \sigma/\varepsilon_0. \tag{17.2}$$

Now consider again the charged parallel plates of Figure 17.1. Since we said that the plates are close to each other and very large in area, the electric field will have essentially the same value and will point in the same direction everywhere between the plates. This parallel-plate configuration thus produces a nearly constant electric field, and we have a close analogy to the case of the gravitational field g on the surface of the earth, since that field is also constant in both magnitude and direction. Therefore, in this parallel-plate case, we are able to easily find the potential energy by calculating the work that would be required to move a charge from one point to another, just like what we were able to do for a mass near the surface of the earth.

Looking again at Figure 17.1, let us imagine pushing a positive charge q from point A to point B at constant speed. By Newton's Second Law (for zero acceleration), the external force we would need to apply would have to have the same size as the electric force on the object. The magnitude of the force we would have to apply would be $F_{ext} = qE$, where E is the electric field, and its direction would have to be opposite to the direction of the field. So the work done by this constant external force would be

$$\Delta PE_E^{A \to B} = W_{A \to B} = F_{ext}\, D_{A \to B}, \tag{17.3}$$

where $D_{A \to B}$ is the displacement from A to B.

In Equation 17.3, we have assumed that F_{ext} points along the path from A to B, so that the work is positive, indicating a higher potential energy at point B than at point A. If we go on, however, to replace F_{ext} by qE in this equation, there arise many places for negative signs to become confused—opposite directions for the field and negative charges instead of positive. We therefore prefer to produce an equation that only gives the magnitudes, regardless of whether it is A → B or B → A. We write the magnitude of the potential energy difference between points A and B, separated by a distance $D_{A \leftrightarrow B}$ in an electric field E, as

$$\Delta PE_E^{A \leftrightarrow B} = qED_{A \leftrightarrow B}.$$

Then, when we use Equation 17.1 to find the voltage difference between points A and B, we will choose to write a formula only for the *magnitude* of the difference. It is

$$\Delta V_{A \leftrightarrow B} = \frac{\Delta PE_E^{A \leftrightarrow B}}{q} = \frac{qED_{A \leftrightarrow B}}{q} = ED_{A \leftrightarrow B}. \tag{17.4}$$

This formula will give us the magnitude of the voltage difference between A and B. To find which point, A or B, is at the higher voltage, we simply need to remember two things from Figure 17.1. *The point closer to the more positive charge is at the higher voltage and the electric field always points from higher voltage to lower voltage.*

We often refer to Equation 17.4 as the "ED" equation for the voltage. There are two important things to remember about this formula. The first is that the ED equation only applies in places where the electric field is uniform (like the region between parallel charged plates). It does not, for example, give voltage differences in the region around a point charge. The second thing to remember is that it is critical to always identify the beginning and end points between which the voltage difference is being measured, as the following example illustrates.

EXAMPLE 17.3 Finding the Voltage Difference from Test Particle Motion

Two large, uniformly charged sheets are placed $D = 6$ cm apart. The plates are connected to a power supply that keeps the voltage difference between the plates at a constant voltage. It is found that if electrons at point A (1 cm from the positive plate) are moving straight downward with a speed of 6×10^6 m/s, they will slow down and momentarily stop at point B (1 cm from the negative plate). What is the voltage difference between the plates?

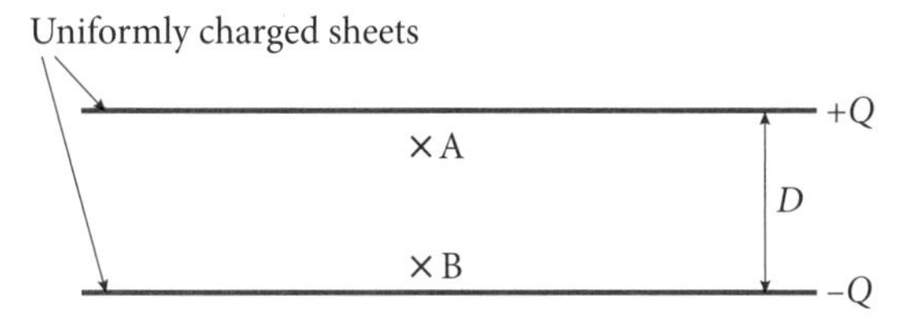

ANSWER If we knew the electric field between the plates, we could find the voltage difference between them by applying the ED equation, with the initial and final points being at the two plates (a distance 0.06 m apart). But we haven't been given the electric field. However, we can use what we know about electrons moving between the plates to find it. Let us do this first.

The electron has a mass 9.1×10^{-31} kg. As electrons move from A to B (a distance of 0.04 m), they come to a stop at B, losing all of the kinetic energy they once had. The kinetic energy lost is $\Delta KE_{A\rightarrow B} = ½(9.1 \times 10^{-31}\ \text{kg})(6 \times 10^{6}\ \text{m/s})^2 = 1.6 \times 10^{-17}$ J. Since total energy is conserved, the electric potential energy must have increased by this amount. We use the first equality in Equation 17.4 to find the magnitude of the voltage difference between points A and B. It is

$$\Delta V_{A\leftrightarrow B} = \frac{\Delta PE_E^{A\leftrightarrow B}}{q} = \frac{1.6 \times 10^{-17}\ \text{J}}{1.6 \times 10^{-19}\ \text{C}} = 100\ \text{volts.}$$

Note that we have dropped the minus sign on q, since we are only looking for the *magnitude* of the voltage difference, $\Delta V_{A\leftrightarrow B}$, independent of whether we have A→B or B→A. Knowing the voltage difference between these two points, the ED equation will be able to give the magnitude of the electric field between the plates as

$$\Rightarrow \quad E = \frac{\Delta V_{A\leftrightarrow B}}{D_{A\leftrightarrow B}} = \frac{100\ \text{V}}{0.04\ \text{m}} = 2500\ \text{V/m.}$$

You might want to check that V/m is the same as N/C, our Chapter 16 unit for electric field. Now that we know the electric field, we can find the voltage difference between the plates from

$$\Delta V = ED$$

$$\Rightarrow \quad \Delta V = 2500\ \text{V/m} \times 0.06\ \text{m} = 150\ \text{V.}$$

In this example, we have used the notation D, without any subscripts, to mean the distance between the plates and we write ΔV, without any subscripts, to mean the voltage difference between the negative plate and the positive plate. This ΔV is also the voltage of the power supply that is connected to the plates.

This last example demonstrates the need to be careful about where the voltage difference is being measured. We first found a value of 100 V, but that was the voltage difference between points A and B. When we needed the voltage difference between the plates, we found a different value, 150 V. We also mentioned that the voltage difference between plates connected to a power supply or a battery is always just the voltage of the power supply or battery. Finally, if we know the voltage difference ΔV between the plates and the distance D between the plates, it is easy to find the electric field in the region between the plates from $E = \Delta V/D$.

A couple of additional examples should help make these relationships clearer. In both examples, the distance between two charged parallel plates is changed. The goal of these examples is to help us see how the voltage between the plates, the electric field between the plates, the charge on the plates, and the voltage differences between points in the region between the plates are related to each other.

EXAMPLE 17.4 Parallel Plates Connected to a Constant Voltage Supply

The terminals of a twelve-volt battery are connected to two parallel conducting plates as shown in Figure 1. [Note: The symbol with two vertical lines, one short and one slightly longer, is the common circuit symbol for a battery.] A proton is now released from rest at point A and is measured to be moving with speed 4×10^4 m/s at point B. The right plate is then moved closer to the left plate as shown in Figure 2 and the experiment is repeated.

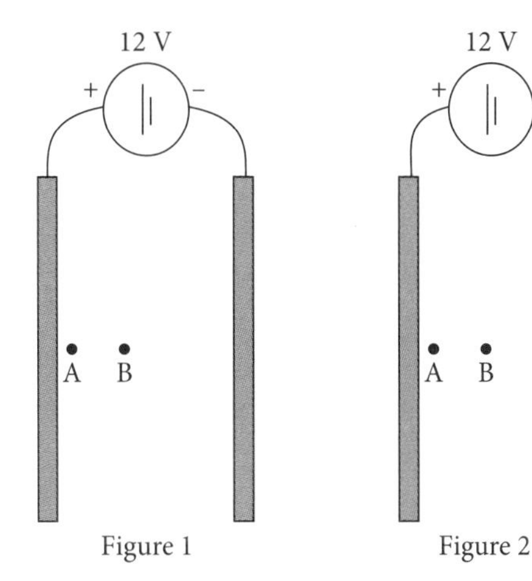

Figure 1 Figure 2

a) Will the magnitude of the electric field at point B increase, remain the same, or decrease because the right plate is moved to the left?

b) Will the proton's speed at point B in Figure 2 be less than, equal to, or greater than 4×10^4 m/s?

c) An electron released from rest at the negative plate in Figure 1 is found to be moving at $v_f = 5 \times 10^6$ m/s just before it hits the positive plate. If an electron is released from rest near the negative plate in Figure 2, will its speed be less than, equal to, or greater than 5×10^6 m/s just before it hits the positive plate?

ANSWER

a) Since the plates remain connected to the 12-volt battery, the voltage difference between the plates stays the same, no matter how far apart the plates are. Then, since the left side of $\Delta V = ED$ has to remain constant, we know that E must increase as D decreases. We also know, though the question did not ask about it, that the only way for the field to increase is for the charge density on the plates to increase (see Equation 17.2), and, since the area of the plates does not change, the charge on the plates will have to increase. The battery will have to move charge from the negative plate to the positive plate. This means that changing the separation of the plates will actually cause a small electric current to flow through the battery. That current will drop to zero when the plates stop moving. (We will discuss currents in much more detail in Chapter 18.)

b) We just saw that the field between the plates increased in Figure 2, so the force on the proton will be greater. Since it moves the same distance as it did in Figure 1, the proton will acquire more kinetic energy and end up moving at a speed *greater* than 4×10^4 m/s.

c) We cannot use the same reasoning in part c as we did in part b, because the distance the particle travels is different in the two situations. However, we know that ΔV is the same in both cases. In Figure 1, the energy gained by the electron will be $e\Delta V$. In Figure 2, the electron will gain the same amount of kinetic energy and so must strike the positive plate at the same speed as it did in Figure 1.

In the next example, the plates are the same and the change in distance is the same as in the previous example. The two chosen intermediate points between the plates are also the same. But what is different is that there is now no battery attached to the plates while the distance between them is changed. You should note carefully what difference that makes in how the voltages and fields behave as the plates are moved.

EXAMPLE 17.5 Isolated Charged Parallel Plates

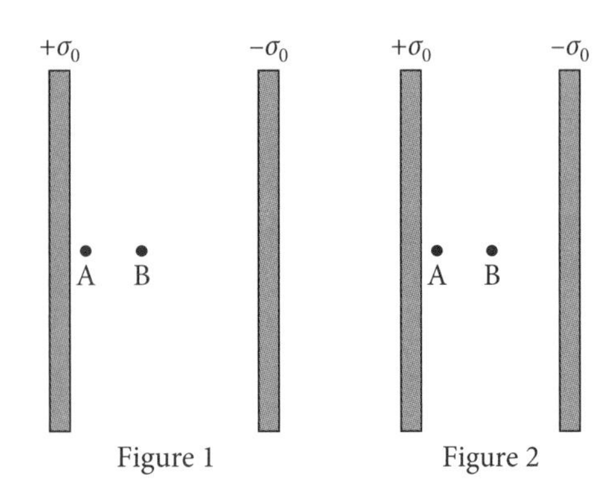

Figure 1 Figure 2

Two very large closely spaced parallel conducting plates carry charge densities $+\sigma_0$ on the inner surface of one plate and $-\sigma_0$ on the inner surface of the other. A proton is released from rest at point A and ends up moving with speed $v_f = 4 \times 10^4$ m/s at point B. The right plate is then moved closer to the left plate, as shown in Figure 2, and the experiment is repeated.

a) Will the magnitude of the electric field at point B increase, remain the same, or decrease when the right plate is moved to the left?

b) Will the proton's speed at point B in Figure 2 be less than, equal to, or greater than 4×10^4 m/s?

c) An electron is now released from rest near the negative plate in Figure 1 and is found to be moving at $v_f = 5 \times 10^6$ m/s just before it hits the positive plate. If an electron is released from rest near the negative plate in Figure 2, will its speed be less than, equal to, or greater than 5×10^6 m/s just before it hits the positive plate?

ANSWER

a) This example is different from the previous one because, in this case, the charge on the plates remains fixed, since there are no wires attached to the plates and therefore no way to add or subtract electrons. The strength of the electric field depends only on the charge density, and not on the plate separation, and, therefore, the electric field cannot change. It is the same in both figures.

b) Neither the electric field nor the distance traveled from A to B changes between the two situations. Therefore the final speed of the proton in Figure 2 will be *equal* to the final speed in Figure 1.

c) In this case, unlike what we saw in Example 17.4, the voltage difference between the plates is not held constant by a battery. It will change, and we can apply the ED equation to see how. We know that E remains constant in the equation $\Delta V = ED$. Therefore, as the plate separation decreases, ΔV must decrease as well. The kinetic energy gained by the electron is $e\Delta V$, so decreasing ΔV decreases the final kinetic energy of the electron. Therefore, the speed of the electron just before hitting the positive plate in Figure 2 is *less* than the 5×10^6 m/s it had in Figure 1.

Examples 17.4 and 17.5 were presented because they illustrate a number of important points about electric voltage and the use of the ED equation. Perhaps the most important point is the need to always specify from where and to where you are measuring the change in voltage. In Example 17.5, for instance, you cannot claim that the voltage is constant (or not constant) without specifying from where to where. In that example, the voltage between the plates changed when the plate separation was decreased while the voltage between points A and B did not. These two examples are probably worth your looking over a second time.

17.4 Equipotential Surfaces

There is another important aspect of the ED equation, which you may have figured out yourself, but which we have not yet emphasized. This is that the distance D in the ED equation must be measured parallel to the direction of the electric field (i.e., perpendicular to the plates). To see why, let us look at Figure 17.2 and ask this question: How much work would be required to move a +1 C charge from point A to point B at constant speed? To overcome the electric force on the object, which points to the right, an external agent would have to apply an equal force to the left. But this force would be perpendicular to the direction the charge was moving. And, as we're sure you remember, the work done by a component perpendicular to the motion is zero. (And, if you didn't remember, you can always go back and look at Equation 7.5.) As a result, an object's electric potential energy is unchanged as it moves from point A to point B. The voltage difference between two points is the potential energy difference for a charge q moving between the two points, divided by q itself. This means that the voltage at the two points would also be the same. In fact, all points on the dashed line in the figure (which would actually be a flat plane in 3-dimensional space) are points that are at the same voltage as each other. We call this an *equipotential* (or equal-potential) surface.

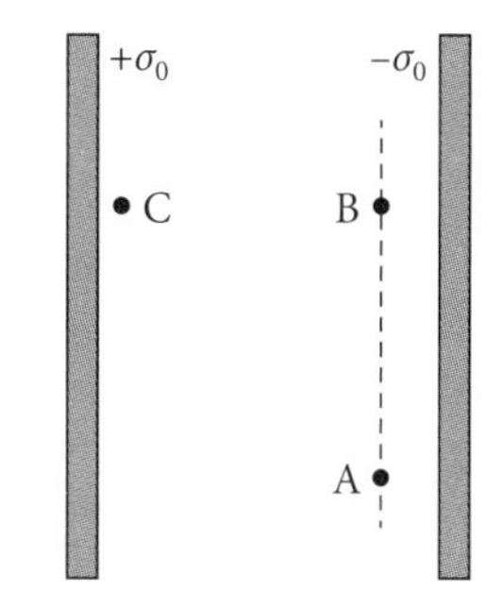

Figure 17.2 Several points in the electric field between two charged conducting plates. The dashed line is an equipotential.

Now suppose we needed to find the difference in voltage going from point A to point C. First, we would remember that the voltages at the two points are properties of those two points only, being the difference in the potential energy of a charge q at the two locations, divided by the charge itself. The difference in the potential energies of a charge q at point A and the same charge at point C is equal to the work done by an external force that moves q slowly from one point to the other. But note that there is nothing that specifies the path that the charge q must follow in getting from A to C. Indeed, if the voltage is to be a well-defined property of points in space, the voltage differences must be independent of whatever path one might take to define them. So, in order to find the potential difference between points A and C, we do not have to follow a straight line between A and C, but we can take a two-step path that first follows the equipotential line from A to B and then goes along the electric field line from B to C. So we write

$$\Delta V_{\mathrm{A}\to\mathrm{C}} = \Delta V_{\mathrm{A}\to\mathrm{B}} + \Delta V_{\mathrm{B}\to\mathrm{C}}.$$

But, since the path from A to B lies in an equipotential surface, we must have $\Delta V_{\mathrm{A}\to\mathrm{B}} = 0$. The path from B to C is parallel to the electric field, so the ED equation applies. Thus the total voltage change from A to C is just that from B to C, giving

$$\Delta V_{\mathrm{A}\to\mathrm{C}} = \Delta V_{\mathrm{B}\to\mathrm{C}} = ED_{\mathrm{B}\to\mathrm{C}}.$$

Notice that the distance from A to C doesn't matter. It is only the distance parallel to the field that is important here.

In the case of gravitational potential, an equipotential would be a collection of all points at the same elevation. If we were considering regions near earth's surface, these gravitational equipotential surfaces would be flat (e.g., the floor on the 20th story of a high-rise). However, if we were considering regions far from earth's surface, these equipotential surfaces would be spheres centered on earth. An important point to note is that in both the electrical and gravitational cases the equipotential surfaces always intersect the field lines at 90°, because it is only for paths that reside entirely in such a surface that the work done on a charge moving along the path will be zero.

EXAMPLE 17.6 Potential Differences in Electric Fields

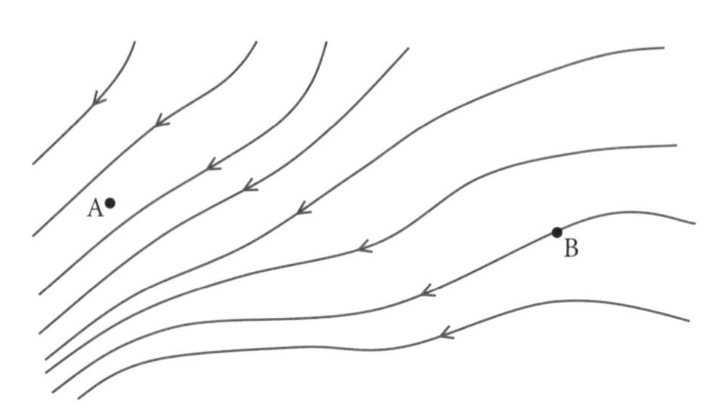

The electric field lines in a region of space are shown in the figure at right. Is $\Delta V_{A\to B}$ positive, negative, or zero?

ANSWER The potential difference between points along a single electric field line is easy to determine. We know that electric field lines always point from high voltage to low voltage, so, as we move against the direction of an electric field line, ΔV will be positive, or, if we move in the direction of the field, ΔV will be negative. The difficulty in this problem is that the direct path from A to B cuts obliquely across many field lines. However, we can get around this by dividing our path into two segments. The first segment will be chosen to be an equipotential line. We begin at point A, taking care to sketch a line that is always perpendicular to the local field lines, and we continue until we reach a point on the same field line as point B. We call this point C, and we know $\Delta V_{A\to C} = 0$. From point C, we move along a single field line to point B. Here, we are moving in the opposite direction as the direction of the field, so $\Delta V_{C\to B}$ will be positive. We conclude that $\Delta V_{A\to B} = \Delta V_{A\to C} + \Delta V_{C\to B} = \Delta V_{C\to B}$ must be positive.

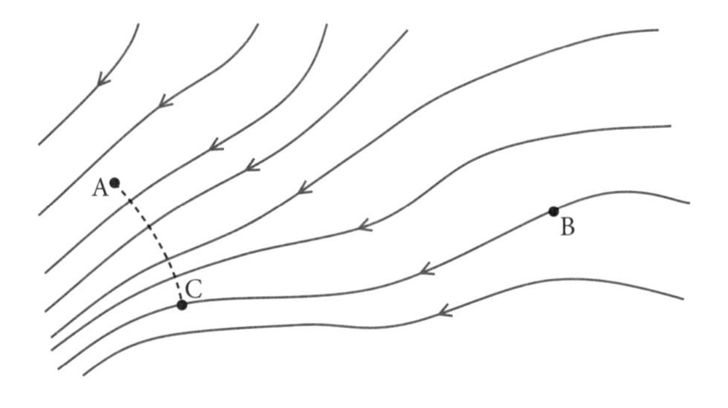

So far, we have been using the electric field lines to tell us something about the equipotential lines. We can also use the equipotential lines to learn about the electric field. In a region of uniform electric field, we know that the magnitude of the electric field between two points is given by $E = \Delta V_{A\leftrightarrow B}/D_{A\leftrightarrow B}$, which is just the ED equation rearranged. Written this way, we see that the electric field can be thought of as a measure of how rapidly the voltage is changing as you move through the region. If the voltage is constant as you go in some direction ($\Delta V = 0$), then the component of the electric field in that direction is zero. In regions with high electric fields, the voltage changes a great deal over small distances. (In this case, the unit for the electric field that expresses it as volts/meter seems quite natural.)

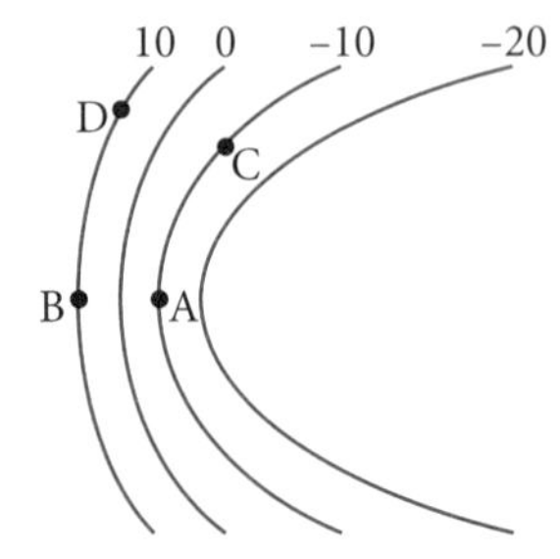

Figure 17.3 Equipotential lines in a region of space, each labeled by the voltage that characterizes the line.

Even in regions where the electric field is not uniform, we can still use this interpretation as a guide, remembering that it is the average electric field that is given by $E = \Delta V_{A\leftrightarrow B}/D_{A\leftrightarrow B}$. As an example, Figure 17.3 shows the equipotential lines in a region of space. The lines are labeled by the voltages along those lines. Points A and C are at –10 V and points B and D are at +10 V, so both $\Delta V_{A\to B}$ and $\Delta V_{C\to D}$ are equal to +20 volts. But the distance between points A and B is less than that between points C and D. From this we may conclude that the average electric field in the region between A and B is greater than that in the region between C and D. If A and B are separated by 1 cm, we know that the electric field near A or B is approximately 20 V/0.01 m = 2000 V/m (or 2000 N/C). Note that the line that passes between points A and

B is labeled 0 volts, and yet we just said that the electric field there is roughly 2000 V/m. This serves to remind us that it is not the value of the voltage that matters in determining the electric field; what matters is how the voltage differs from point to point.

17.5 Capacitance

Without actually calling them that, we have already introduced and done some analysis of standard electric circuit components called *capacitors*. In its simplest form, a capacitor is two parallel conducting plates separated by a gap. In practice, the real capacitors that commonly serve as elements in electric circuits are comprised of two thin conducting sheets separated by a thin layer of material, all of which is then rolled up into a compact package. Wires connected to the two sheets come out of the unit to be connected into circuits. Taking off the back of electronic devices will often reveal capacitors, which normally look like small cylinders. For our purposes, we will limit our capacitors to the parallel plates we have already been considering, but most of what we learn applies very well to the more common capacitor forms.

When you think about the term capacitance, it suggests the ability to contain something—in this case, electric charge. However, it is not really the ability to contain a total amount of charge that we will want to measure, but the ability to separate the electric charge into two "containers." In normal applications, capacitor plates carry charges of the same magnitudes but opposite signs. For instance, when capacitor plates are hooked to a battery, the battery pulls electrons off of one plate, passes them through the battery, and puts them on the other plate. If one plate of a capacitor carries $+Q$ of charge and the other carries $-Q$, we say that the charge on the capacitor is Q, even though the net charge on both plates is actually zero. Capacitors separate electric charge, and capacitance must be a measure of how much charge they can separate until. . . . Until what? For your gas tank, the capacity is how much gas it will hold until it overflows. But for capacitance, we need a somewhat different analogy.

Picture, then, a water tank standing by the shores of a lake. How many gallons of water the tank can hold before it overflows is one question. But suppose that water must be pumped out of the lake to fill the tank. In this case, another important question is how much water you can put in the tank for each foot that the pump has to lift the water to add it to the water already in the tank. A wide, short tank would hold more water for each foot of height than a narrow, tall tank.

For a capacitor, the thing we measure is not how much charge you can put on each plate before it sparks across the gap, but how much charge you can store on the plates for each volt of voltage difference between them. If we have a capacitor with a voltage difference ΔV between its two plates, when charges of $\pm Q$ reside on the plates, then we define the capacitance of that capacitor to be[1]

$$C = \frac{Q}{\Delta V}. \tag{17.5}$$

To get a little intuition for what this definition implies, consider parallel plates of area A charged to $\pm Q$, where the voltage between the plates is 1 volt. Then imagine a second capacitor with the same separation between the plates and carrying the same $\pm Q$ charge, but having twice the area. If the charge is the same and the area is doubled, the charge density on this second set of plates is half of that on the first. From Equation 17.2, we know that this will

1. Please be careful to differentiate between C (italic) for capacitance and C for coulomb.

reduce the electric field between the plates by a factor of 2, and, from the ED equation, we see that this will reduce the voltage difference by a factor of 2. So the voltage between the plates would be 0.5 volts. Only if we doubled the charge on the second set of plates to $\pm 2\,Q$ would we get back a voltage difference of 1 volt across the plates. The second set of plates, the one with the greater area, has a greater capacitance because it can store more charge (actually separate more charge) before reaching a voltage of 1 volt.

Although we exclusively consider parallel conducting plates in this book, any two conductors actually form a capacitor. And the capacitance of that capacitor is determined exclusively by the geometry of the conductors, not by any charge on them. Nevertheless, to calculate the capacitance, we must begin by imagining a charge $+Q$ on one conductor and $-Q$ on the other. If we can find the resulting voltage difference between the conductors, we can then use Equation 17.5 to calculate the capacitance. Typically, finding the voltage difference involves finding the electric field due to the $\pm Q$ charges on the conductors and then finding the work required to move a test charge q from one conductor to the other in this field. As indicated by Equation 17.1, we then divide that amount of work by the size of the test charge q to find the ΔV between the conductors. With ΔV in hand, Equation 17.5 then gives the calculated capacitance. We can see how some of this works as part of the following example.

EXAMPLE 17.7 The Capacitance of a Parallel-Plate Capacitor

What is the capacitance of a capacitor formed from two parallel square conducting plates, each with an area of exactly 0.1 m^2, separated by 1 mm? What charge would be necessary to create a voltage difference of 2 volts between the plates?

ANSWER To find the capacitance, we imagine charges $\pm Q$ on the two plates, giving a surface charge density on each plate of $\sigma = Q/A$, where $A = 0.1$ m^2 is the area of each plate. The resulting electric field between the plates is given by Equation 17.2. It is $E = \sigma/\varepsilon_0 = Q/A\varepsilon_0$. To get the voltage difference across the plates, we do not need to consider a test charge q, since we already know the relation between the electric field and the voltage difference for parallel plates. It is the ED equation,

$$\Delta V = ED = \frac{Q}{\varepsilon_0 A} D.$$

Equation 17.5 then allows us to state the capacitance of a parallel-plate capacitor of area A and plate separation D as

$$C = \frac{Q}{\Delta V} = Q\frac{1}{\Delta V} = Q\frac{\varepsilon_0 A}{QD} = \frac{\varepsilon_0 A}{D}.$$

For the parameters of the particular capacitor in the problem, we have

$$C = \frac{\varepsilon_0 A}{D} = \frac{(8.85 \times 10^{-12}\ \mathrm{C^2/N \cdot m^2})(0.1\ \mathrm{m^2})}{0.001\ \mathrm{m}} = 8.85 \times 10^{-10}\ \mathrm{C/V}.$$

Finally, knowing the capacitance, we can find the charge required to produce the requested 2-volt difference across the plates. We solve Equation 17.5 for Q to get

$$Q = C\Delta V = (8.85 \times 10^{-10}\ \mathrm{C/V})(2\ \mathrm{V}) = 1.77 \times 10^{-9}\ \mathrm{C}.$$

As we found in this last example, Equation 17.5 sets the units for capacitance as coulombs per volt, and, as usual, this unit has been given a name. A coulomb per volt is called a farad (F), named after English physicist Michael *Faraday*. ("Why didn't they call it a faraday?" you ask. We don't know.) Anyway, 1 C/V is 1 farad, so we should express the capacitance in Example 17.7 as 8.85×10^{-10} F, or 885 pF.

Finally, let us point out that we just derived a very useful formula in Example 17.7. There, in the middle of the calculation, we found a formula for the capacitance of a parallel-plate capacitor of plate area A and plate separation D. It was

$$C = \frac{\varepsilon_0 A}{D}. \tag{17.6}$$

You should also note, in Example 17.7, how the charge Q that we used to determine the capacitance canceled out of the final calculation of C, as we predicted it would, and as it always does. Capacitance is determined by the geometry of the conductors—in this case, by the area and the separation. The capacitance of a parallel plate-capacitor is increased by either increasing the area of the plates or by decreasing the separation between them, and by nothing else.

17.6 Polarization and Dielectrics

Well, almost nothing else.

Up till now, our capacitors have had nothing at all in-between the plates—a perfect vacuum. But what would happen if we filled the region between the plates with some material? If the material were a conductor, we already know what would happen. Each charge on the plates of the capacitor would attract an opposite charge to its side, so that the electric field inside the conductor would be dead zero, as shown in Figure 17.4. Since the only work that would be done, as a test charge q is dragged from one plate to another, is the work done in the gaps between the plates and the conducting slab, we would have created a capacitor with an effective separation of $2d$, where d is the gap between each side of the slab and the plate nearest to it. But, while that narrow gap would produce a capacitor with high capacitance, it would also be susceptible to sparking, as charge jumps a distance d from the plate to the nearby slab. On this score, we would have been better off with a simple capacitor with a single gap of $D = 2d$.

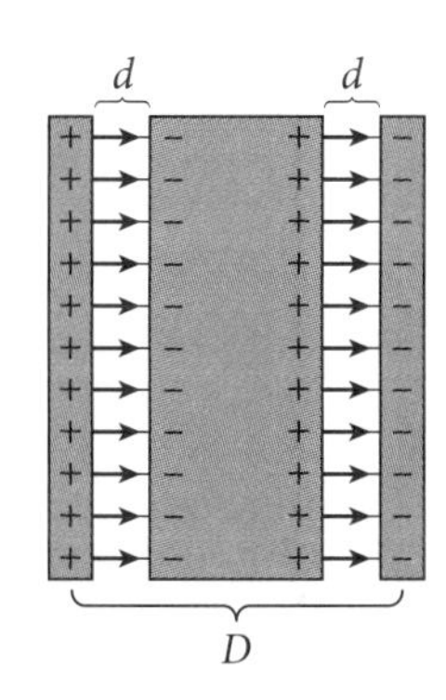

Figure 17.4 Capacitor with a conducting slab between the plates.

But what if the material between the plates is an insulator? First, there would be much less tendency for charge to jump from a plate to the slab of insulating material. Second, if you remember our discussion from Section 16.3, the electric field between the plates of the capacitor will polarize the insulating material in the slab. A material that becomes highly polarized in an external electric field is called a *dielectric*. When a dielectric is placed between the charged plates of a capacitor, the dielectric will become polarized. The surface of the dielectric facing the plate with a positive charge on it will acquire a net negative charge, while the surface of the dielectric facing the negative side of the capacitor will acquire a net positive charge, as shown in Figure 17.5. The surface charge will be large, but it will not be the 100% of the facing plate charge that appears when the slab between the plates is a conductor (Figure 17.4).

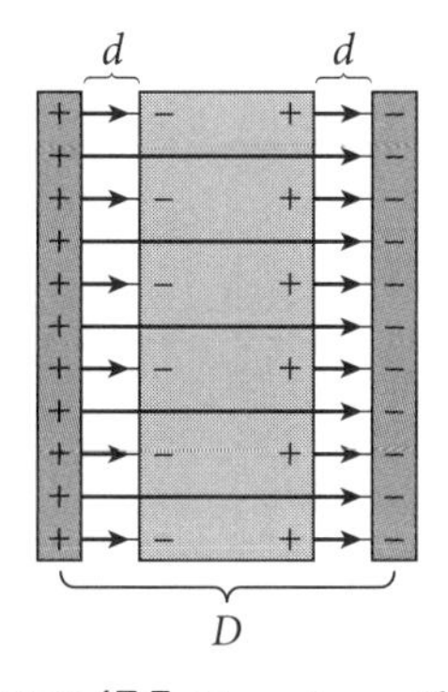

Figure 17.5 Capacitor with an insulating slab between the plates.

As Figure 17.5 illustrates, some of the field lines originating on a plus charge on the left plate terminate on a negative charge on the surface of the dielectric, while other field lines pass through the dielectric to terminate on a negative charge on the right plate. This leaves some, but not all, of the field lines inside the dielectric. Thus there is a lower density of field lines inside the dielectric, and, therefore, a lower, but non-zero, electric field inside the dielectric. With charges $\pm Q$ on the two plates of the capacitor, there will be a non-uniform electric field between the plates. The fact that there is a lower electric field inside the dielectric means that the work required to take a positive test charge q from the negative plate to the positive plate of the capacitor will be less than it would be if the dielectric material were not there. The resulting lower potential energy difference experienced by a test charge moving between the plates means a lower voltage difference between the plates, which means a higher capacitance for the capacitor. And the presence of the insulator inhibits the sparking of charge from plate to plate that causes a capacitor to break down. So the use of a dielectric material between the plates of a capacitor is a winning situation all around.

In order to calculate the capacitance of a capacitor with a dielectric between the plates, we must know how much the electric field is reduced inside the dielectric. An actual calculation of the degree of polarization of any given dielectric material is very difficult to do, but experiment has shown that the electric field inside the dielectric is proportional to the external field E_0 and inversely proportional to an experimentally determined constant called the dielectric constant κ (Greek letter kappa),

$$E = \frac{E_0}{\kappa}. \tag{17.7}$$

Since the fields in the material E are always less than the imposed field E_0, the dielectric constants will always be greater than one. A table of the dielectric constants for a few materials is given in Table 17.1.

TABLE 17.1 Dielectric Constants of a Few Common Materials at 20° C

Material	κ
Vacuum	1
Dry air at 1 atmosphere pressure	1.0006
Glycerin	42.0
Methyl alcohol	33.0
Glass	7.0
Paper	3.0
Paraffin (wax)	2.1

According to Equations 17.2 and 17.7, the electric field between the plates of a parallel-plate capacitor, filled with a dielectric of dielectric constant κ and carrying charge Q, is given by

$$E = \frac{\sigma}{\kappa\varepsilon_0}.$$

When working with a dielectric, the combination $\kappa\varepsilon_0$ consistently shows up in formulas in place of ε_0, leading physicists to define a new quantity

$$\varepsilon \equiv \kappa\varepsilon_0. \tag{17.8}$$

This ε is called the *permittivity*. In fact, the reason for calling ε_0 the *permittivity of free space*, as we defined the term back on page 317, is that it is the permittivity in vacuum, where the dielectric constant is $\kappa = 1$. Formulas that contain ε_0 when they are describing vacuum situations will instead have an ε in them when they apply to conditions inside a dielectric. In particular, the capacitance of a parallel-plate capacitor filled with a dielectric of dielectric constant κ may be written

$$C = \frac{\varepsilon A}{D}, \tag{17.9}$$

which should be compared with Equation 17.6 to see how ε replaces ε_0.

Since the value of $\varepsilon = \kappa\varepsilon_0$ for, say, paper is three times the value of ε_0, the capacitance of a capacitor with paper as a dielectric between the plates will have three times the capacitance of a parallel-plate capacitor filled with vacuum or air. One common way to create a capacitor, then, is to separate two long strips of foil with a strip of paper, to cover the three strips with another strip of paper, for insulation, and to roll all the strips up into a cylinder. The paper increases the capacitance and (being an insulator) keeps charge from sparking from one of the foil sheets to the other.

17.7 Energy in a Capacitor

When a capacitor is charged with a charge $+Q$ on one plate and $-Q$ on the other, there is a strong electric field in the region between the plates. If some of the positive charge could leave the positive plate, it would be accelerated toward the negative plate and would acquire kinetic energy in the process. This means that there is electric potential energy stored in a charged capacitor. Let's see how much.

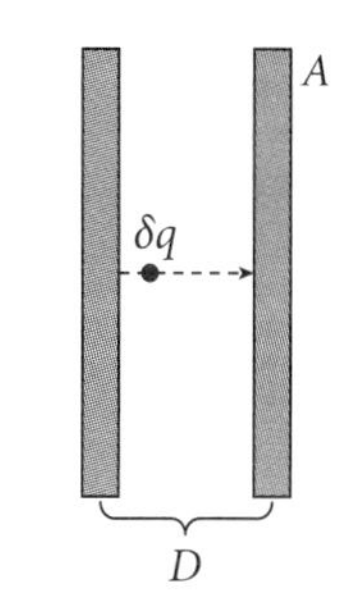

Figure 17.6 Charging a parallel-plate capacitor.

Let us consider an initially uncharged parallel-plate capacitor, like the one in Figure 17.6, and let us imagine a process that slowly charges the capacitor up to a final charge Q by taking tiny increments of positive charge δq from the plate on the left to the plate on the right. To begin with there is no charge on the plates, so no electric field between the plates, and so no work is required to move that first tiny charge from what will eventually be the negative plate to what will be the positive plate. However, at the end of the process, the last little δq we have to move to get the total charge up to Q will have to be moved against the field that has been created when almost all of the total charge is already on the plates. The final field is nearly

$$E = \frac{\sigma}{\varepsilon_0} = \frac{Q}{\varepsilon_0 A},$$

where we have written $\sigma = Q/A$. At any intermediate point in the charging process, when the charges on the plates have risen to $\pm q$, the field between the plates will be given by $E = q/(\varepsilon_0 A)$. As more and more charge moves from the left plate to the right plate, q will increase in proportion. In Figure 17.7, we graph the electric field as a function of the charge q on the plates at any given time. The graph follows a straight line. Some of the charge will move against a small field and some a larger field. The average field that the charge will see as it moves from the negative plate to the positive plate is shown by the dashed line in Figure 17.7. This is the average "over the charge," where as much charge sees a field as far below the average as charge sees a field above the average. Clearly, this average field will be half the final electric field, $E_{\text{AVE}} = Q/(2\varepsilon_0 A)$. The average force that acts on the total charge Q, as it moves from one plate to the other, will be $F_{\text{AVE}} = QE_{\text{AVE}} = Q^2/(2\varepsilon_0 A)$. So the work done in taking a total charge Q across the distance D between the plates will be

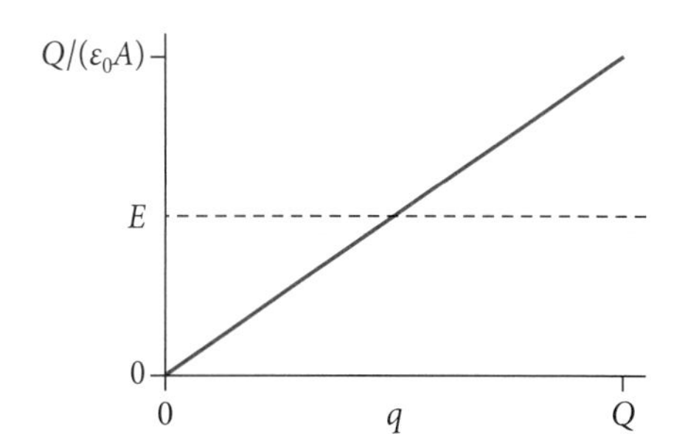

Figure 17.7 The electric field between the plates as the charge on the plates increases. The dashed line gives the average electric field.

$$W = F_{\text{AVE}} D = \frac{Q^2 D}{2\varepsilon_0 A}.$$

With a little staring at this equation, we might recognize that the combination $D/(\varepsilon_0 A)$ looks a little familiar. It is, in fact, the inverse of the capacitance of the parallel-plate capacitor. So we conclude that the electric potential energy stored in a capacitor, being equal to the work done in charging the capacitor from zero to Q, will be

$$\text{PE}_{\text{C}} = \frac{1}{2}\frac{Q^2}{C}. \tag{17.10}$$

While we are at it, let us just present this energy equation in one more form, one that will be useful in a later chapter. If we relate the charge on the plates to the *electric field* in the capacitor by writing $Q = \varepsilon_0 AE$, and if we write the capacitance in the usual form $C = (\varepsilon_0 A)/D$, we can substitute these formulas into Equation 17.10 to get

$$\text{PE}_{\text{C}} = \left(\frac{1}{2}\varepsilon_0 E^2\right)(AD). \tag{17.11}$$

In this form, we see that the energy in the capacitor is the product of two terms. The second term, the simple product AD, is actually the volume inside the capacitor. The quantity that it multiplies, which we write $\rho_{\text{E}} = ½\varepsilon_0 E^2$, must therefore be the *energy density* generated by the electric field inside the capacitor. It will have units of J/m^3.

Let us end this chapter with one of our favorite examples. We like this example because our concluding explanation ends up being a bit surprising.

EXAMPLE 17.8 A Dielectric Slab Inserted into a Capacitor

A parallel-plate capacitor has plates that are 10 cm × 10 cm square. The separation between its plates is 1 cm. It is then hooked to a battery and charged up until there is a charge of $Q = \pm 8\ \mu\text{C}$ on the plates.

a) How much energy is stored in the capacitor?

The battery is then disconnected, leaving a fixed charge on the plates. A glass slab that just fits into the gap is now inserted between the plates, while carefully not allowing any charge to leave the plates.

b) What is the new potential energy stored in the capacitor?

c) Explain.

ANSWER

a) The first part is pretty easy. We have to first find the capacitance of the capacitor, and we can then use Equation 17.10 to get the potential energy stored. We use Equation 17.6 to find the capacitance without the dielectric. It is

$$C_0 = \frac{\varepsilon_0 A}{D} = \frac{(8.85 \times 10^{-12}\ \text{m}^2/\text{N}\cdot\text{C}^2)(0.01\ \text{m}^2)}{0.01\ \text{m}} = 8.85 \times 10^{-12}\ \text{F}.$$

With a charge $Q = \pm 8\ \mu\text{C}$ on the plates, the energy stored is found to be

$$\text{PE}_{\text{C}_0} = \frac{1}{2}\frac{Q^2}{C} = \frac{1}{2}\frac{(8 \times 10^{-6}\ \text{C})^2}{8.85 \times 10^{-12}\ \text{F}} = 3.62\ \text{J}.$$

b) Now what happens when the dielectric is inserted? Equations 17.8 and 17.9 tell us that the presence of the dielectric will increase the capacitance by a factor equal to the dielectric constant for glass, which is $\kappa = 7.0$. The new capacitance is

$$C = \kappa C_0 = (7.0)(8.85 \times 10^{-12}\ \mathrm{F}) = 6.2 \times 10^{-11}\ \mathrm{F}.$$

With a new capacitance and with the same old charge on the plates, the potential energy stored in the capacitor after the glass slab is inserted between the plates becomes

$$\mathrm{PE_C} = \frac{1}{2}\frac{Q^2}{C} = \frac{1}{2}\frac{(8 \times 10^{-6}\ \mathrm{C})^2}{6.2 \times 10^{-11}\ \mathrm{F}} = 0.52\ \mathrm{J}.$$

c) The last part of the problem asks us to explain. Explain what? Well, we know that the electric force is a conservative force, and we don't see any obvious friction in the problem. So, since we started with 3.62 J, we should end up with 3.62 J, not 0.52 J. Where did the 3.10 J of energy go?

There is an answer to this.

As the dielectric slab is placed next to the capacitor, ready to be inserted between the plates, there will be a charge induced on the surfaces of the glass. This induced charge will create an attractive force by the charged plates on the glass, as shown in the picture below.

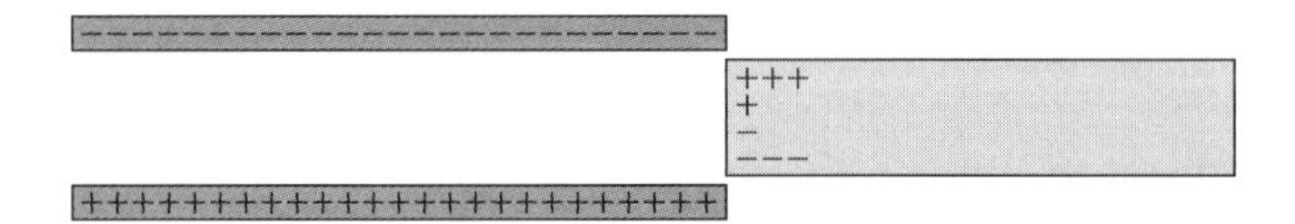

If the glass is released at this point, it will accelerate into the gap. When it is all the way between the plates, the plate-slab configuration will have 3.1 joules less potential energy, as we calculated above. So, if there is no significant friction between the glass slab and the plates, this energy will show up as kinetic energy of the slab. When it reaches the center of the capacitor, it will be moving to the left with 3.1 J of kinetic energy. As time goes on, its motion will cause it to emerge from the capacitor to the left. When it is completely outside the plates, it will again be at rest, but it will then be pulled once again back into the capacitor. Except for eventual losses to friction of one kind or another, the oscillation of the slab back and forth in the capacitor will continue forever.

17.8 Summary

There is no new fundamental physics in this chapter, but we have introduced new concepts of electric potential energy and voltage which are very valuable in practical problem solving. From this chapter, the following ideas are important:

- The change in the electric potential energy of a charge q as it moves between points A and B is the work required to move the charge at constant velocity between those two points.
- The voltage difference between two points is related to the change in the potential energy of a charge q as it moves between the two points. The relationship is given by $\Delta \mathrm{PE_E}^{\mathrm{A}\to\mathrm{B}} = q\Delta V_{\mathrm{A}\to\mathrm{B}}$.
- The electric field between two charged plates is constant, with magnitude $E = \sigma/\varepsilon_0$.

- Equipotential surfaces consist of all points with the same electric potential, and they are always perpendicular to electric field lines that intersect them.
- A capacitor is a device composed of two conductors. Its capacitance is found by imagining a charge $+Q$ on one conductor and $-Q$ on the other, by determining the voltage difference between the two conductors, and by then calculating $C = Q/\Delta V$. The result will always be independent of the assumed charge Q.
- When a dielectric material fills the space between the plates of a capacitor, the polarization of the dielectric will reduce the voltage difference between the plates and will thereby increase the capacitance. For a dielectric of constant κ, the capacitance will increase by a factor κ.
- The energy stored in a capacitor is $PE_C = \frac{1}{2}CQ^2$.

CHAPTER FORMULAS

Voltage definition: $\Delta V_{A\to B} = \dfrac{\Delta PE_E^{A\to B}}{q}$

Capacitance definition: $C = \dfrac{Q}{\Delta V}$

Equations for a parallel-plate capacitor: $E = \sigma/\varepsilon_0$ $\quad \Delta V_{A\leftrightarrow B} = ED_{A\leftrightarrow B}$ $\quad C = \dfrac{\varepsilon_0 A}{D}$

In dielectrics: $E = \dfrac{E_0}{\kappa}$ $\quad \varepsilon = \kappa\varepsilon_0$ $\quad C = \dfrac{\varepsilon A}{D}$ (for a parallel-plate capacitor)

Energy in a capacitor: $PE_C = \dfrac{1}{2}\dfrac{Q^2}{C}$ $\quad$ Energy density in a capacitor: $\rho = \dfrac{1}{2}\varepsilon_0 E^2$

PROBLEMS

17.1 * 1. There is a unit of energy that is a more appropriate size for elementary particles like electrons and protons. It is defined as the energy acquired by an electron that is accelerated through a voltage difference of 1 volt. The unit is called an electron-volt (abbreviated eV). How many joules in 1 eV?

* 2. A small sphere carrying a charge of –3 C at point A has a potential energy of 200 J. When it is moved to point B, its potential energy is 500 J. What is the voltage difference between points A and B? Which point is at a higher voltage?

* 3. How much work is required to move 5 μC of charge from a point where the voltage is 200 V to a point where it is 204 V?

* 4. If a +2-coulomb charge is moved from a place where the voltage is –100 V to a place where the voltage is +100 V, what has been the change in its potential energy? Has the potential energy increased or decreased?

* 5. In a region containing an electric field, point A is at 750 volts and point B is at 950 volts. An electron is observed moving at some speed at point A. When it reaches point B, will the electron's speed have increased, decreased, or stayed the same? Explain your answer.

*6. If a proton gains 6×10^{-12} J of electrical potential energy as it moves from point A to point B, what is the voltage difference between points A and B, and which point is at the higher potential?

**7. A proton is accelerated from rest through a voltage difference of 1500 V. What is its final speed?

**8. A proton has a velocity of 1×10^5 m/s when observed at point A and a velocity of 3×10^5 m/s when observed at point B. If the voltage at point A is 1000 volts, what is the voltage at point B?

**9. An electron is released from rest in a uniform electric field whose magnitude is 2.5×10^5 N/C. What is the kinetic energy of the electron after it has moved a distance of 1.0 cm?

17.3 **10. Two parallel conducting plates are placed close together. A charge of +1 μC is placed on one plate and a charge of −1 μC is placed on the other. The area of each plate is exactly 4 cm^2. What is the electric field between the plates?

**11. Two parallel plates are placed 1 mm apart from each other. Equal and opposite charges are placed on the two plates until there is a voltage difference of 50 volts between them.

a) What is the electric field between the plates?

b) What is the charge density (in C/m^2) on each plate?

**12. Two parallel plates, each of area 0.10 m^2, are held 0.010 m apart. The parallel plates are uniformly charged to $Q = \pm 7.0 \times 10^{-7}$ C. What is the voltage difference ΔV between the plates?

**13. Two large parallel conducting plates, each with an area of 500 cm^2, are separated by 0.0060 m. If the plates are connected to the opposite terminals of a 12-V battery, what will be the strength of the electric field in the region between the plates?

**14. The terminals of a 12-V battery are connected to two parallel conducting plates separated by some distance D. How far apart do the plates need to be (i.e., what is the value of D) to produce an electric field between the plates of 1200 N/C?

**15. Two large conducting plates on insulating stands are placed a distance $D = 50$ cm apart, as shown at right. The inner surface of one has a charge density of $+\sigma_0$; the other, $-\sigma_0$. The charge density on the outer surface of each plate is zero.

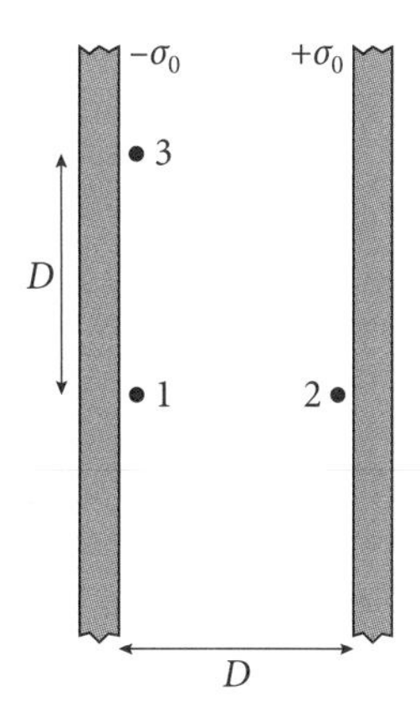

a) Sketch the configuration of the plates as shown and draw field lines on your diagram to represent the electric field everywhere that a field exists.

b) An electron is released at point 1 and accelerates to the right. When the electron reaches point 2, its horizontal component of velocity is equal to 1.45×10^6 m/s. Find the voltage difference between the two plates.

c) Find the magnitude of the electric field at a point halfway between points 1 and 2.

d) How much work must be done by an external agent to move an electron from rest at point 2 to rest at point 3?

**16. Two large parallel plates are separated by a 0.015-m gap. The plates are connected to the terminals of a 12-V battery, which remains connected.

a) What is the strength of the electric field in the region between the plates?

b) What happens to the strength of the electric field if the gap between the plates is reduced to 0.010 m? Justify your answer.

c) Does charge flow through the battery as the separation between the plates is being reduced? Explain your answer.

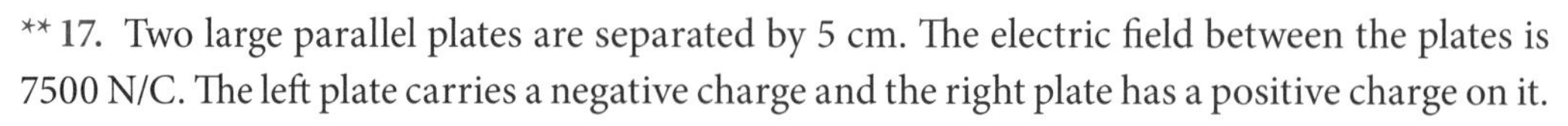

** 17. Two large parallel plates are separated by 5 cm. The electric field between the plates is 7500 N/C. The left plate carries a negative charge and the right plate has a positive charge on it.

a) What is the voltage difference between the plates?

b) Which plate (left or right) is at the higher voltage?

c) If a proton is fired through a small hole in the negative plate straight toward the positive plate, how much would the electric potential energy of the proton increase in traveling between the plates?

d) If the proton is initially moving at a speed of 2×10^5 m/s, will it be able to reach the positive plate? Explain your answer.

** 18. The electric field between two charged parallel plates is 3500 N/C. A pith ball with mass $m = 1.2 \times 10^{-5}$ kg carries a charge $Q = 6 \times 10^{-11}$ C. The pith ball is held at rest at the midpoint between the plates. If the pith ball is released, how much kinetic energy does it gain in traveling 2 cm toward the negative plate?

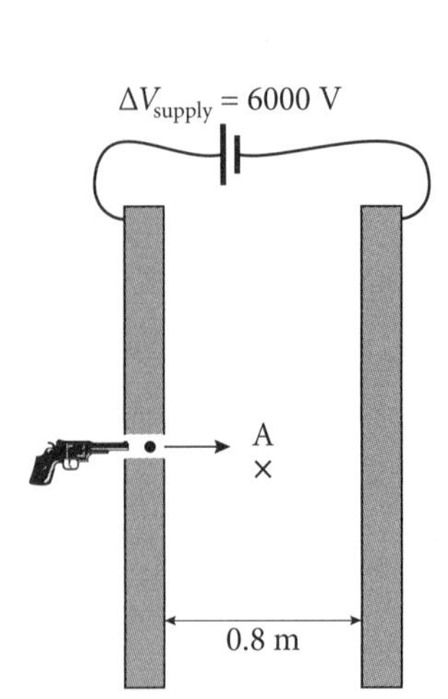

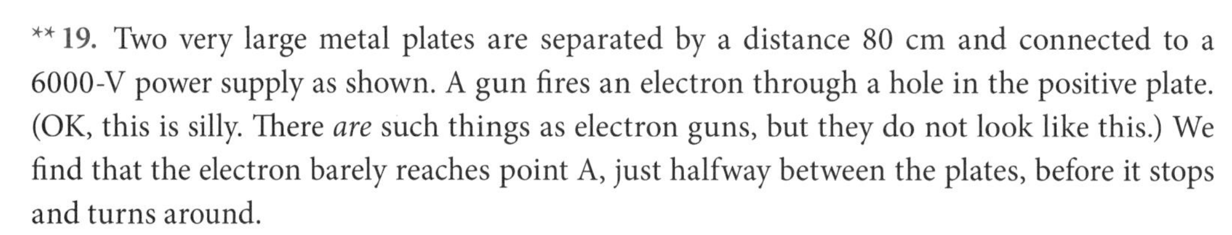

** 19. Two very large metal plates are separated by a distance 80 cm and connected to a 6000-V power supply as shown. A gun fires an electron through a hole in the positive plate. (OK, this is silly. There *are* such things as electron guns, but they do not look like this.) We find that the electron barely reaches point A, just halfway between the plates, before it stops and turns around.

a) Sketch field lines to represent the electric field everywhere that a field exists.

b) Find the electric field at point A.

c) Find the initial velocity of the electron as it enters the region between the plates.

d) The right plate is now moved closer so that the distance between the plates is only 40 cm. If the gun fires the electron the same as before, will the electron reach the negative plate? Explain why or why not.

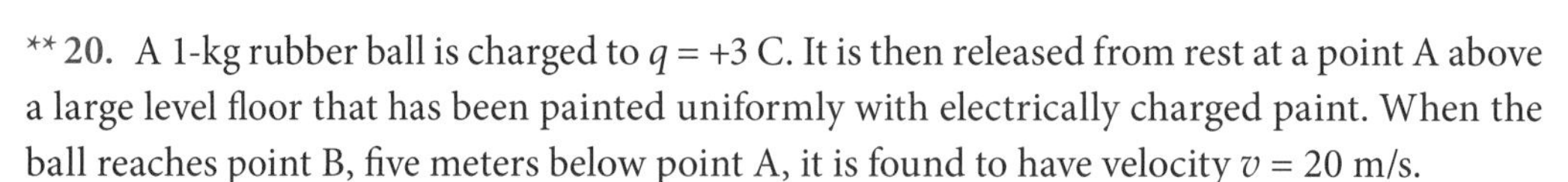

** 20. A 1-kg rubber ball is charged to $q = +3$ C. It is then released from rest at a point A above a large level floor that has been painted uniformly with electrically charged paint. When the ball reaches point B, five meters below point A, it is found to have velocity $v = 20$ m/s.

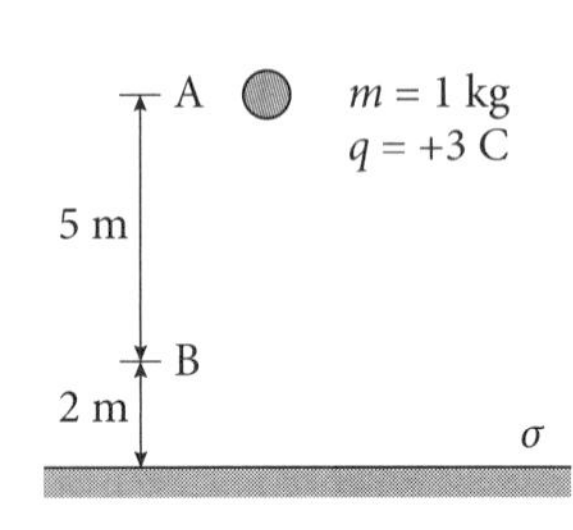

a) Find the change in kinetic energy of the ball as it travels from point A to point B.

b) Find the change in gravitational potential energy of the ball as it travels from point A to point B.

c) Find the change in voltage, $\Delta V = V_B - V_A$.

d) Find the electric field (magnitude and direction) at point B due to the sheet of charge (the floor).

*** 21. A storm cloud 200 m off the ground sits over a fish pond that is initially at 20° C. The pond is defined to be at voltage zero and the cloud found to be at 250 MV. A lightning strike carrying 50 C of charge leaves the cloud and hits the pond. Assuming that all of the energy of the transferred charge is absorbed by the water, how many kg of water can be raised to 100° and then boiled away?

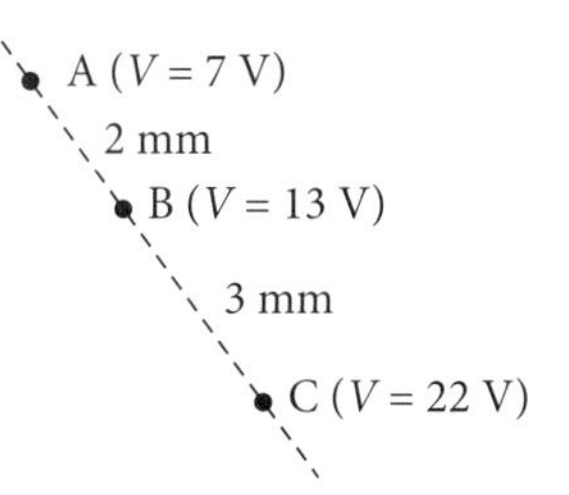

17.4 ** 22. The voltage at point A is 7 V. The voltage 2 mm away at point B is 13 V. If we continue the straight line between points A and B until we get to point C 3 mm beyond point B, we find a voltage at point C of 22 V. What is the electric field (assumed constant) in the region between points A and C?

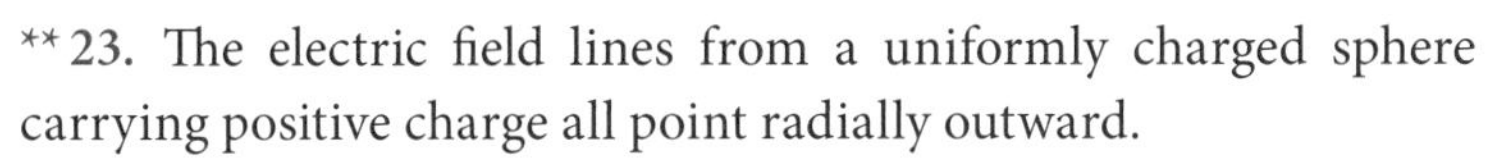

** 23. The electric field lines from a uniformly charged sphere carrying positive charge all point radially outward.

a) Draw the picture at right on your own paper and then sketch in equipotential lines in the region around the charge.

b) Consider points A and B in the diagram and decide which is at the higher voltage. Explain your answer.

** 24. In Figure 16.13b from the last chapter, reproduced below, we estimated the shapes of the electric field lines for an electric dipole.

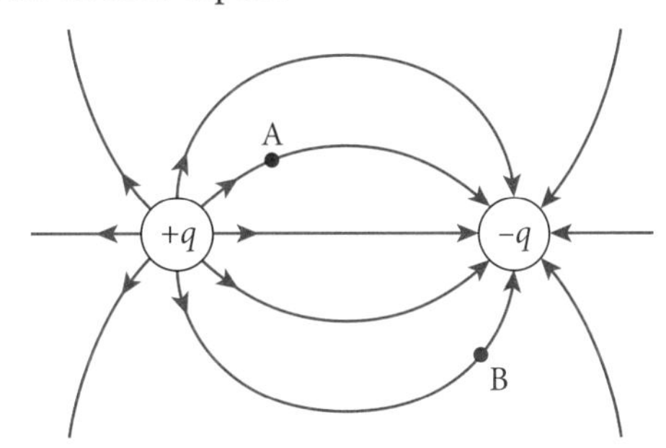

a) Which point, A or B, is at the higher voltage? Explain how you came to this conclusion.

b) Redraw the figure above and sketch equipotential lines for the electrical dipole.

17.5 * 25. When charges of ± 3.60 mC are placed on the two plates of a parallel-plate capacitor, a voltage difference of 60.0 V is measured across the plates. What is the capacitance of this capacitor?

* 26. A 6-μF capacitor is charged by a 12-V battery. What charge is on the plates of the capacitor?

* 27. What is the capacitance of a parallel-plate capacitor made up of two plates of area 4 cm^2 separated by a 0.20-mm gap?

** 28. A parallel-plate capacitor is charged to $Q = \pm 8 \times 10^{-8}$ C. The resulting electric field between the plates is 9.0×10^5 V/m. If the plates are separated by 0.015 m, what is the voltage difference between the plates?

17.6 * 29. A parallel-plate capacitor is formed of two square metal plates 10 cm × 10 cm, separated by a 2-mm-thick slab of glass. What is its capacitance?

** 30. There is an electric field 350 V/m between two parallel plates in the lab. When the plates are submerged in glycerin, what does the magnitude of the electric field between the plates become?

** 31. An experimenter wants to make a 150-pF capacitor out of two thin pieces of copper foil, separated by a piece of paper. Each piece of foil has an area that is exactly 3 cm^2. What thickness of paper should the experimenter use?

** 32. Two parallel plates are separated by a gap of 3 mm. Equal-magnitude positive and negative charges are then placed on the plates so that each has a surface charge density of magnitude $\sigma = 0.50\ \text{C/m}^2$.

a) What is the electric field between the plates?

b) A 1-mm-thick slab of glass is then inserted between the plates. What is the electric field inside the glass?

c) What surface charge density is induced on the two surfaces of the glass? [Hint: Find the surface charge densities that would reduce the electric field inside the glass to the level you calculated for part (b).]

** 33. In very high-accuracy electric circuits, the performance of the electronics can be marred by what are called "stray capacitances" between the wires connecting the elements of the circuit. If the voltage difference between two parallel wires is 3 mV when the wires each have a net ±1 nC of charge on them, what is the stray capacitance of these two wires in the circuit?

** 34. As mentioned on page 343, a capacitor will break down by sparking across its plates if the electric field between the plates becomes too high. Capacitors carry a rating for the maximum voltage that may be created across the capacitor before breakdown. A charged capacitor is found on the shelves of the electronics lab and found to have a voltage of 20 volts across its terminals. The label on the capacitor reads: "$C = 0.5$ F, $V_{max} = 100$ volts."

a) How much charge is stored in the capacitor?

b) How much additional charge could be placed in the capacitor before it breaks down?

** 35. A parallel-plate capacitor is formed of two plates, each of area 5 cm^2, separated by a wax-filled gap of 0.50 mm.

a) What is the capacitance of the capacitor?

b) What is the electric field between the plates when there is a 10-V voltage difference across the plates?

** 36. A parallel-plate capacitor with nothing between the plates acquires 1 μC of charge when it is connected to a 2-kV battery. With the plates still connected to the battery, the capacitor is then immersed in a container of methyl alcohol. How much charge flows from the battery to the plates when the capacitor is put into the alcohol?

17.7 * 37. What energy is stored inside a 12-μF capacitor that carries a charge of 4 mC?

** 38. Two charged parallel plates have an electric field of 400 V/C in the region between the plates. What is the energy density in J/m^3 stored in the electric field between the plates?

** 39. A 35-μF parallel-plate capacitor has a slab of glass separating the plates. The capacitor is first connected to a 6-V battery to put a 6-V voltage difference across the plates. The battery is then disconnected, and the slab of glass is pulled out.

a) What voltage will now be seen across the plates of the capacitor?

b) How much work was required to pull the slab of glass out of the capacitor?

** 40[f]. Equation 17.10 gives the energy stored inside a parallel-plate capacitor in terms of the charge Q on the capacitor and the capacitance of the capacitor C. The voltage difference across a capacitor ΔV is also determined by C and Q. Find a formula that gives the energy stored in a capacitor in terms of Q and ΔV, and find one that gives the energy in terms of C and ΔV.

18

CIRCUITS

In the last two chapters, we have concentrated on electrical charge that is fixed in space. In this chapter, we will look at moving electrical charge, constrained to move in a single dimension because it is contained inside a thin wire. If a wire is made of a conducting material, many of the electrons will be free to move within the wire, but they will be held inside the material of the wire by the positively charged nuclei of the atoms that make up the wire. The change in the energy of the charges as they move through a circuit is determined by the voltage at each point. A typical electrical circuit will have some source of energy (a battery or a power supply), elements that absorb electrical energy (like a light bulb or a resistor), and wires that connect them to each other. The purpose of this chapter is to enable you to analyze an electrical circuit to determine at what rate electrical charge will flow through the various elements, what the voltages will be at various places around the circuit, and what energy will be taken from a battery or dissipated in some other circuit element.

18.1 Current and Wires

Electric *current* is a measure of the rate of flow of electric charge. We measure the quantity of charge in coulombs (C) and the interval of time in seconds (s), so the unit for current is a C/s. If the current in a thin wire is 4 C/s, then, if you could see the charged particles flowing inside the wire, you could observe that 4 C of charge pass any point in the wire every second. As we mentioned back where we first defined[1] the units of electrical charge (Section 16.1), the C/s is also called an ampere (A), or an "amp." But remember this about the current—it is not a measure of how much you have, but a measure of the rate at which it is flowing. There is no way to create or store charge along the length of the wire, so if 4 C flows into one end of a wire each second, then there must be 4 C flowing through the midpoint each second and 4 C flowing out of the other end each second. We normally just describe this by saying that the current "in the wire" is 4 A.

In the last paragraph, we have defined the magnitude of the current passing along a wire, but we have not defined the direction of the current. If a wire is laid out on the table in

1. You may remember from Chapter 16 that we said that the definition of the coulomb was based on the definition of the ampere, and we now seem to be defining the ampere in terms of the coulomb. You may be wondering if our definitions are circular. Don't worry. The ampere is indeed one coulomb per second, and the coulomb is an ampere·second. But the *definition* of the ampere comes from a different effect altogether, one we will not see until Chapter 19.

front of you, there may be current flowing to the right or to the left. Defining the direction may actually seem a little complicated at first, for the reasons we warned you about back in Section 16.1, but it will turn out in practice to present no real problems. The point is that it is very useful, in order to avoid sign problems in our analysis of electrical currents in wires, to define the direction of the current to be the direction in which positive charge flows. The conceptual difficulty is that the charged particles that actually flow in a piece of wire are electrons, and these carry negative charge. The current is created by free negatively charged electrons flowing through a matrix of fixed positively charged particles. But, when a current is flowing, the wire does not acquire a net electrical charge, so the number of positive and negative charges in any piece of the wire must be equal. Therefore, having −4 C of electrons flowing to the left each second is equivalent, for all of the applications we will care about in this book, to having +4 C of positive charge moving to the right each second. So, as you think about current, it is perfectly all right to imagine positive charge moving along the wires. This imaginary flow of positive charge is called the *conventional current*, because it is used almost universally in analyzing electrical circuits. For purposes of understanding circuits, it is therefore best to think of wires as tubes that direct the flow of positive charge. The direction of the current is the direction in which this positive conventional current is flowing.

18.2 Batteries

In 1800, the Italian physicist, Alessandro Volta, discovered that many dissimilar metals could be used to create a voltage difference. He placed strips of the metals (called *electrodes*) in contact with a liquid or other material (called an *electrolyte*), in which electrons and charged atoms (ions) were free to move. The voltages arose from the production of electrons and ions in chemical reactions with the metals. As the charges migrated from one metal to the other, an excess of positive charge would appear at one electrode and an excess negative charge would appear at the other.

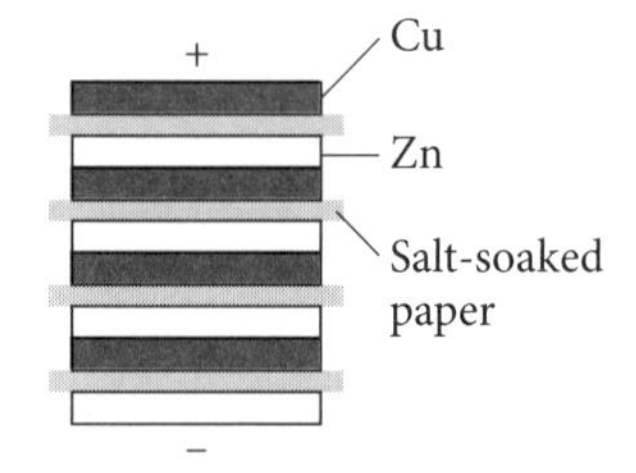

Figure 18.1 A *pile* made from four zinc (Zn) and copper (Cu) plate combinations, *in series*.

Volta made a single cell by putting a piece of paper that had been soaked in a salt solution between pieces of copper and zinc. He then stacked these cells in a pile like the one shown in Figure 18.1. Such a stack was called a "pile" or a "battery." When the zinc electrode of each cell is put in direct contact with the copper electrode of the cell before it, the voltages across the combinations add together. When the cells are arranged like this, one after the other, it is called placing the cells *in series*. The series arrangement produces a higher voltage, but the total amount of charge available cannot be greater than the charge that can be provided by a single cell.

The voltage produced by a single cell depends only on the materials used and not on the size of the cell. However, the size of the electrodes does determine the amount of charge a cell can provide. One way to produce a greater total charge is to place cells side by side, or *in parallel*, as shown in Figure 18.2. This increases the effective size of the battery. The parallel arrangement provides more total charge, but the voltage produced is the same as the voltage from a single cell.

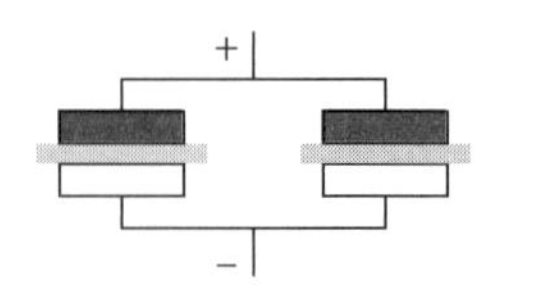

Figure 18.2 Two silver-zinc cells, wired *in parallel*.

To be clear, batteries *in series* sit one after the other on the same path, with no wire branching out between the batteries that can carry current away. Batteries *in parallel* are batteries on separate wires, and the wires that come off each side of the batteries come together at single junction points.

EXAMPLE 18.1 Batteries in Series and in Parallel

A typical "C" battery produces 1½ volts and can supply 8 amp·hours of total charge. What voltage and what charge is provided by 4 "C" batteries wired in series? What would be the voltage and charge provided if the 4 batteries are wired in parallel?

ANSWER First, let us say a word about the charge units in this problem. We know that 1 C is the amount of charge that passes a point in a wire in 1 s when a current of 1 A is flowing. So a coulomb is an amp·second. An amp·hour is the charge that flows past a point in the wire in one hour, or in 3600 seconds. So 1 A·hr = 3600 C.

When the batteries are wired in series, as in the leftmost picture in the diagram below, the 4 voltages add, giving a total voltage of $4 \times 1.5 = 6$ V. However, the 8 A·hr (28,800 C) of charge produced by the bottom battery continues on, flowing through the batteries above it. The energy produced by the other batteries does not increase the current through the stack, but only increases the voltage. So this series configuration will only produce 8 A·hr of total charge.

When the batteries are connected as in the rightmost picture in the diagram, each coulomb of charge that arrives at the plus end of the battery (the end with the little bump on it) will be 1.5 V higher in voltage than at the smooth end of the battery. So the entire set of battery tops will be the same 1.5 V higher than the battery bottoms. On the other hand, the current produced by each battery will be combined by the wires connecting the top ends (and the bottom ends), giving a total current four times the current produced by a single battery. The total charge available is thus 4×8 A·hr = 32 A·hr.

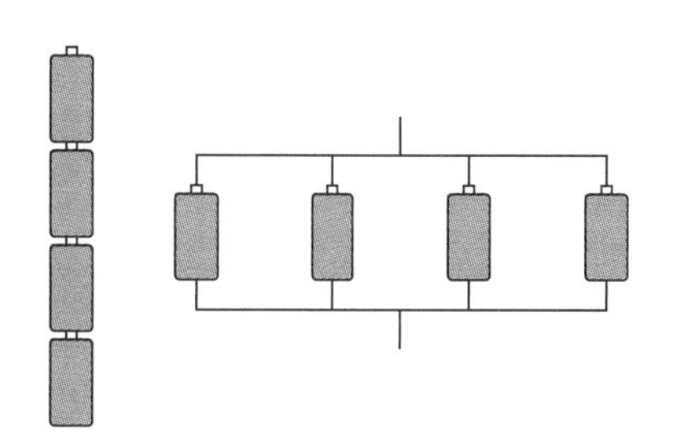

18.3 Resistors and Light Bulbs

As electrons move along a wire, they occasionally collide with the atoms in the matrix of the conducting material. As they do so, they typically lose some of their kinetic energy while the atoms gain thermal energy.[2] The average amount of energy lost by a single charge carrier depends on its average speed in the material. If a current of one ampere is moving through a wire with a large diameter, then the one ampere of current may be achieved with its many electrons and with a low average speed. There will thus be less energy lost in collisions. Wires used in a circuit are designed to have large enough diameters that the energy lost inside them is negligible, compared to other elements in the circuit, though the energy lost will always rise in proportion to the length of the wire (as this provides proportionately more collisions). A wire with a large cross-sectional area and a short length will provide only a small resistance to a current flowing through it.

In contrast, some elements of an electrical circuit are designed to have high resistance to the current flowing through them. For example, a long thin wire, coiled up and encased in some insulating material, is called a *resistor*. Its goal is to dissipate some of the energy carried

2. Actually, the atoms of the material are so tightly bound to each other that the collisions of the electrons in a current are really with collective motions of the atoms called "phonons."

by a current. Another type of resistor, whose purpose is obvious, is formed by the very thin wire in the filament of a light bulb. Here, the goal is for the material in the thin wire to absorb so much energy from the passing current that it heats up enough to glow and provide light where we need it.

We have used the term "resistance" in a general sense in the two paragraphs above. Here, we want to give the term a more precise mathematical definition. Let us suppose that there is a constant current flowing through some resistor. As the charges move through the resistor, they will lose energy by colliding with the resistor material. But, since the current is constant, the energy loss can only be in the form of potential energy, not kinetic energy, because a loss of kinetic energy would entail a loss of speed by the charge carriers and a resulting lowering of the current. So the charges moving through the resistor will move at constant average speed and the energy they lose must make it so that they have lower potential energy when they leave the resistor than when they entered it. There must thus be a voltage difference between the two ends—a higher voltage where the charges begin their path through the resistor and a lower voltage at the point where the current leaves the resistor. We define the *resistance* of a resistor to be the ratio of the voltage difference across the resistor (V) to the current flowing in the resistor (I):[3]

$$R = \frac{V}{I}. \tag{18.1}$$

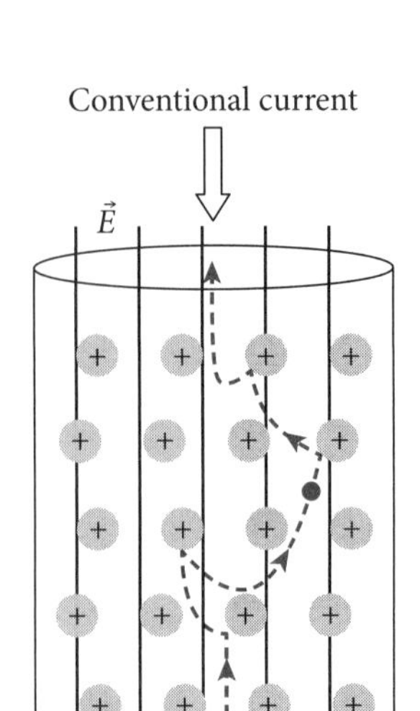

Figure 18.3 An electron (red dot) is accelerated upward between collisions with the atoms in a wire containing a downward electric field.

The units of resistance are obviously volts per ampere, also called ohms (abbreviated with a Greek capital omega, Ω).

We can create a model of the current through the wire in a resistor by reference to Figure 18.3. If the conventional current is flowing downward through this piece of wire, then the voltage will be higher at the top than it is at the bottom. Equation 17.3 then requires that there must be an electric field inside the conducting wire, pointing downward from the high voltage end to the low voltage end of the resistor.[4] This downward electric field will accelerate the free electrons in the wire (the actual charge carriers) upward. A particular electron will start at the bottom of the wire and be accelerated upward until it hits an atom of the wire material, as shown in Figure 18.3. The collision will remove some of the electron's kinetic energy and will heat up the material of the wire. After each collision, the slowed electron will again be accelerated upward, when it will again collide with one of the atoms, losing kinetic energy again and heating the material more. In this way, the electrons in the resistor maintain a constant average velocity through the resistor and provide a constant current.

If we rearrange the variables in Equation 18.1, we get a formula that gives the voltage difference across a resistor of resistance R when it carries a current I:

$$V = IR. \tag{18.2}$$

When the resistance is constant as the current changes, this equation is known as *Ohm's Law*. A device that obeys this law, with this constant-resistance restriction, is called an *ohmic device*. Ohm's law is experimentally verified in that most resistors are approximately ohmic within some specified current range. But experiment has also shown that a material typically increases its resistance as it heats up, meaning that the passage of current through such a

3. I stands for current by long physics convention. Sorry that it was also rotational inertia.

4. You may remember that we said in Section 16.8 that the electrical field inside a conductor must be zero, but that is only true if there is no current flowing in the conductor.

material will actually increase the material's resistance a little bit. So Ohm's law is not an exact law of physics. Nevertheless, the approximation of constant resistance is a good one for most resistors, and you may assume that, if we give you the resistance of something in a problem, we mean that the resistance will remain constant at this value, even if the current changes.

However, there is one class of objects we discuss in this chapter that are significantly non-ohmic. These are light bulbs.

Consider a flashlight bulb that is hooked up to a 3-V battery and suppose that, at this voltage, it carries a current of 3 A. Equation 18.1 then tells us that the resistance of the light bulb must have been 1 Ω. Now suppose that, at this current, the light bulb will not light up. So, in an effort to raise the temperature of the filament of the bulb and light it up, we decide to increase the battery voltage to 6 V. Suppose that the bulb then begins to glow brightly because its temperature has increased significantly. Since the temperature is now much higher, the resistance of the filament will also be higher. If R had not changed, Equation 18.2 would have said that our 6-volt battery would push 6 amps of current through the bulb. But the resistance did change when the temperature changed; it got higher. Therefore, the current flowing through the bulb will be less than 6 amps—maybe 5 amps, corresponding to a hot resistance of 1.2 Ω. When the resistance rises as the current changes like this, the simple proportionality given in Equation 18.2 for ohmic devices may not be relied on.

18.4 Energy and Power in Circuit Elements

The purpose of a battery is to take quantities of conventional (positive) charge from the negative electrode of the battery and to move it through the battery to the positive electrode. The source of the force that does this is in the complicated chemical reactions that take place at each electrode. The effect is that a positive charge q will be moved from a lower voltage to a higher one. For a voltage difference V_0 across the terminals of a battery, Chapter 17's Equation 17.1 tells us that the change in potential energy when a charge q moves through the battery will be $\Delta PE_E = qV_0$.

We also remember that we defined a quantity called *power* back in Section 7.3. It was the *rate* at which work is done or the rate at which energy changes. So, if it takes t seconds to move a charge q through a battery producing V_0 volts, then the power produced by the battery—the rate at which it produces electrical potential energy—will be given by

$$P = \frac{\Delta PE_E}{t} = \frac{qV_0}{t} = IV_0, \tag{18.3}$$

where we have recognized in the last step that the rate at which charge moves through the battery is the current through the battery, $I = q/t$.

In a similar calculation, we consider the loss of electrical potential energy by the charges moving through a resistor. If a current I is flowing through a resistor, and if the voltage difference between the two ends of the resistor is V, then the electrical potential energy lost in time t is $\Delta PE_E = qV = ItV$, and the power lost will be

$$P = IV. \tag{18.4}$$

If resistance is defined by Equation 18.1, we may write $V = IR$ (even when the resistance may significantly depend on temperature). Equation 18.4 then becomes

$$P = I^2R. \tag{18.5}$$

Let's look at an example that demonstrates how R, I, V, and P relate to each other.

EXAMPLE 18.2 Hot and Cold Resistances of a Light Bulb

A student hooks a 100-watt light bulb up to a 3-volt battery and measures a current of 300 mA passing through the bulb. The bulb is not lit. When the same bulb is hooked up to a 120-volt power supply, it produces its full 100 watts of power.

a) What is the resistance of the light bulb when it is hooked up to the battery?
b) What power is being dissipated when the bulb is hooked up to the battery?
c) What is the resistance of the light bulb when it is producing 100 watts?
d) What current does the bulb draw when it is completely lit up?

ANSWER We begin by noting that, even though it is the same bulb hooked up to the battery and the power supply, its properties are different when it is hot enough to emit light from its filament. This means that we may not use any results from parts (a) and (b) to help in the solution of parts (c) and (d).

a) We were given the voltage across a resistive element and the current passing through it. It is straightforward to find the resistance of the device from Equation 18.1. It is

$$R = \frac{V}{I} = \frac{3\text{ V}}{0.3\text{ A}} = 10\text{ V/A} = 10\ \Omega.$$

This is what is called the *cold resistance* of a light bulb.

b) The power dissipated by the light bulb, even though "cold," is not zero. It is found most easily from Equation 18.4:

$$P = IV = (0.3\text{ A})(3\text{ V}) = 0.9\text{ V}\cdot\text{A} = 900\text{ mW}.$$

c) As we said, we must now start fresh, since the resistance has changed. With the bulb lit up, we know the power dissipated in the filament of the light bulb and the voltage across it, but not the current flowing through it. Since both of our formulas that involve power require knowing the current, we will need to do a little algebra. Even though the resistance has changed, its new value is still defined by Equation 18.1. Solving this equation for I, we write $I = V/R$ and use this result in Equation 18.4 to find a formula for the power dissipated in a device of resistance R when there is a voltage V across it. The formula for the power dissipated is

$$P = IV = \frac{V^2}{R}.$$

Solving for R and substituting the known values of V and P, we find

$$R = \frac{V^2}{P} = \frac{(120\text{ V})^2}{100\text{ W}} = 144\ \Omega.$$

Notice that this resistance is 14.4 times higher than the cold resistance of the bulb.

d. The current may now be found from several formulas. We choose to use Equation 18.4, solved for I:

$$I = \frac{P}{V} = \frac{100\text{ W}}{120\text{ V}} = 0.833\text{ A}.$$

Compared with the situation with the bulb hooked up to the battery, we see that the current has now increased by less than a factor of 3, while the power dissipated in the bulb has increased by a factor of over 100.

18.5 Circuits

One analogy which is often used in explaining electrical circuits is what is called the "water analogy" of Figure 18.4. This helps us to get a feel for some things, but it cannot be pushed too far, or it begins to produce nonsense. Nevertheless, it is useful to see wires as pipes that carry water, batteries as pumps that lift water up to a higher gravitational potential energy, and resistors as pipes filled with sand or gravel that resist the flow of water falling through them. The water pump analogy is helpful in that it makes it clear that batteries do not create charge, like pumps that do not create water. Batteries provide energy to the charge only by moving it from the bottom of the "pump" to the top.

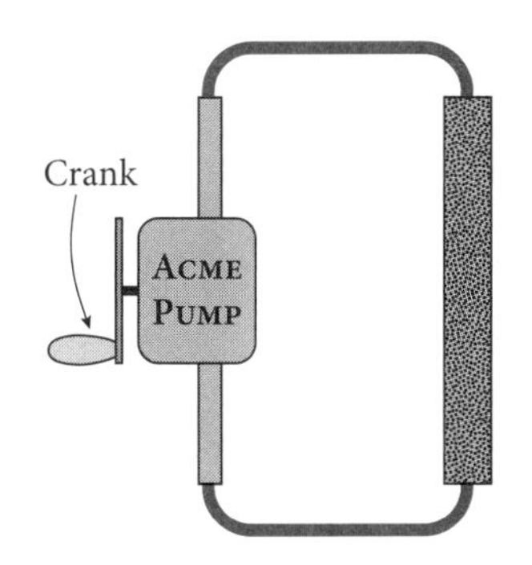

Figure 18.4 The "water analogy" for electrical circuits.

The analogy breaks down, however, in many respects. One of the most obvious is that, if a water pipe breaks, the pump goes on pumping water as the pipe leaks water all over the room. If a wire is severed, the battery maintains a static voltage difference between its two electrodes, but no current flows and the charges do not leak out of the wires. For any current to flow in an electrical circuit, there must exist a closed path for the current to follow. Any unconnected end to any wire or other element means that no current will flow through that element.

Consider, for example, the circuit diagrams shown in Figures 18.5a and 18.5b. In Figure 18.5a, the battery in the center pumps charge to higher voltage and the current flows through the battery, dividing itself at the point labeled A in the diagram and flowing through both resistors, R_1 and R_2, then back to the battery. In Figure 18.5b, there is a break in the wire connecting the resistor R_2 back to the battery. As a result, no current flows through the right branch of the circuit at all. The wire on the right and the resistor R_2 are just funny-shaped lumps sticking out to the right of the wire that is carrying current around the loop through R_1 on the left side of the diagram.

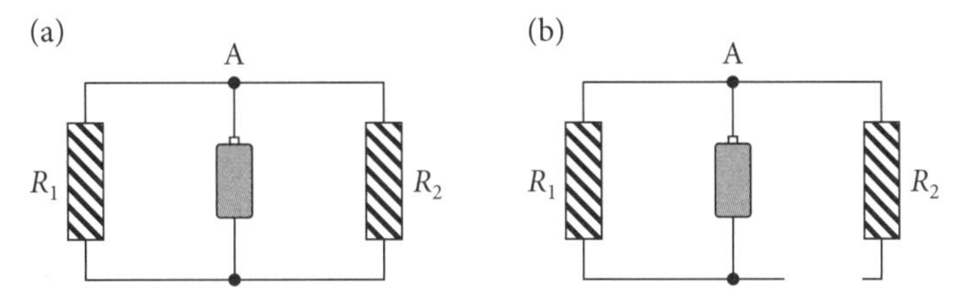

Figure 18.5 Two double-loop circuits. In (a), current flows in both loops. In (b), the break in the circuit keeps current from flowing through the right loop.

In Figure 18.5, we have used pictures to represent the battery and the resistors. While the battery may look like a battery to you, it is far from clear that the rectangles filled with the diagonal lines should represent resistors. In order to make circuit diagrams like this understandable, scientists and engineers use standard symbols to represent different kinds of circuit elements. We have already seen the standard symbol for a battery or power supply back on page 339. Figure 18.6 (on the next page) shows this symbol along with several others of the standard symbols we will use.

In the last line of Figure 18.6 we have shown the symbols for two meters—the voltmeter that measures voltage differences between two points and the ammeter that measures current along a segment of a wire. Let us end this section by saying a word about how these meters are to be placed in a circuit so that they can do their jobs.

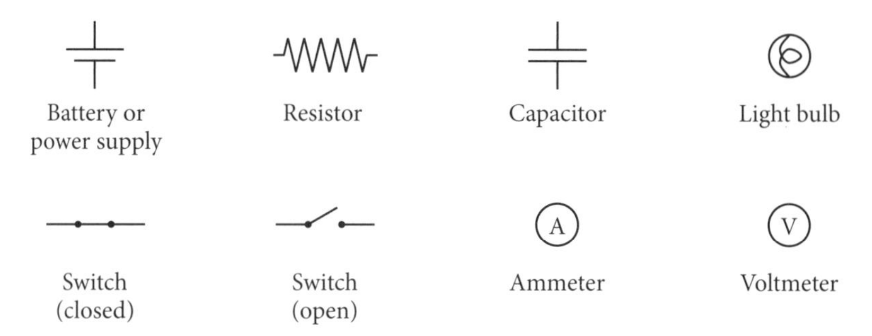

Figure 18.6 Standard symbols for electrical circuit elements.

We can use an ammeter to measure the current passing through some circuit element, but the placing of the meter has to be done in the right way. For example, if we want to measure the current flowing through the bulb shown in Figure 18.7a, we need to make sure that all of the current that flows through the bulb also flows through the ammeter. A little thought will make it clear that the only way to do this is to break the wire leading into or out of the bulb and to place the ammeter in series with the bulb, as shown in Figure 18.7b, so there is no intervening branch of the circuit that might steal current away from the meter, after it passes the bulb. We also want to make sure that the ammeter's resistance does not significantly lower the amount of current that was initially flowing through this segment of the wire. This is done by making an ammeter's resistance very small.

Figure 18.7 The correct placement of an ammeter in series with the circuit element whose current we wish to measure. Note that inserting the ammeter in an existing circuit requires breaking the wire to the element so as to allow the meter to be placed in series with it.

A voltmeter measures the voltage difference between two separated points in a circuit. Often the two points are on the two sides of a circuit element. The voltmeter, which has a known resistance inside it, works by siphoning off a tiny amount of the current passing through the element and measuring that small current. In order for the meter to correctly measure the voltage difference, it must not significantly perturb the circuit. As we will see in the next section, this means that a good voltmeter will have a very high internal resistance so that, given the choice of flowing through the element or through the voltmeter, almost all of the current will continue to flow through the original circuit element. Figure 18.8 shows the correct placement of a voltmeter when we want to measure the voltage drop across a resistor in which a current is flowing.

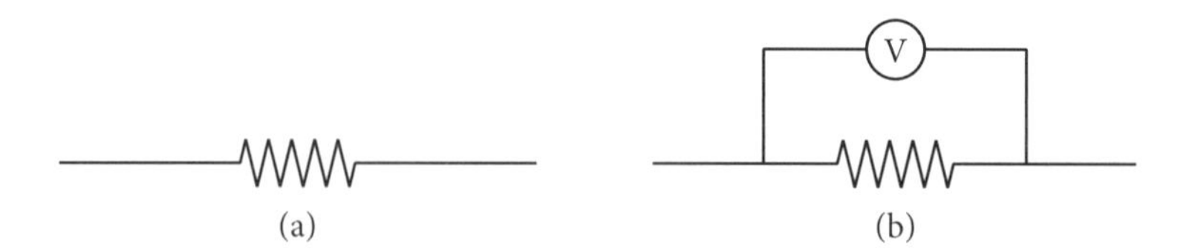

Figure 18.8 The correct placement of a voltmeter to measure the voltage drop across a resistor. The high resistance of the voltmeter will make it so that only a tiny fraction of the original current through the resistor will be picked off for the voltage measurement.

18.6 Experiments with Batteries and Light Bulbs

We often find that students, even when they are able to solve circuit problems numerically, have very little intuition for what they expect to happen in a complicated circuit. But intuition helps a lot in the solution of problems. It helps guide our thinking as we solve a problem and can help us catch a simple mistake if we have calculated something incorrectly. Our goal in this section is to help you develop a bit more of this intuition. To this end, we will describe a set of exploratory experiments and observable outcomes. You could actually do these on your own, if you wished, and you would see the same results that we will describe.

We use flashlight bulbs in our experiments because the brightness of each bulb gives an indication of how much current is passing through it. We will not try to make a quantitative connection between the brightness and the current because brightness is not easily measured and because the exact relationship between current and brightness is complicated for these non-ohmic devices. Nevertheless, we may rely on the fact that if one bulb is glowing more brightly than an identical bulb, it has more current.[5]

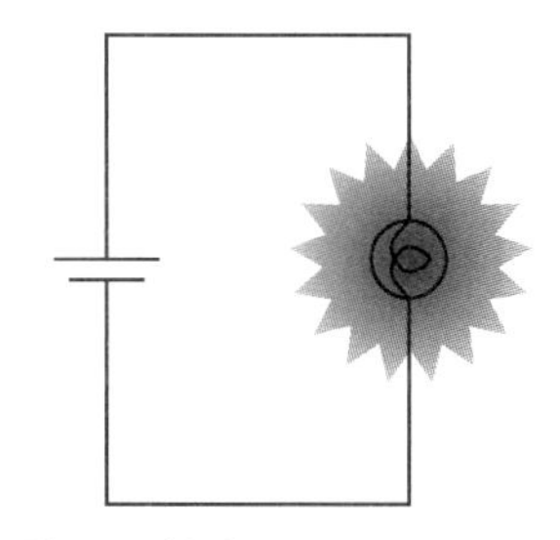

Figure 18.9 The standard circuit—one battery, one bright bulb.

We begin by creating a "standard" to which we can refer. The brightness of a single bulb connected to a single standard battery, as in Figure 18.9, will represent a standard current. We will also assume that all batteries and bulbs are identical.

Two bulbs can be connected to a battery so that there is a single path from the battery through one bulb, through the second bulb, and back to the other end of the battery. In this arrangement (Figure 18.10), the two bulbs are said to be in *series* with each other. We observe that the two bulbs have the same brightness as each other but that they are dimmer than the single bulb in the standard circuit. This decrease in brightness indicates that there is less current flowing around this series circuit than there was in our standard circuit. Since the voltage of the battery is the same, we may conclude that the resistance of two bulbs in series is greater than that of a single bulb. (It is tempting to say that there is twice as much resistance in the series circuit because there are the two bulbs. However, since the resistance of the light bulbs changes with temperature and brightness, the resistance does not exactly double.)

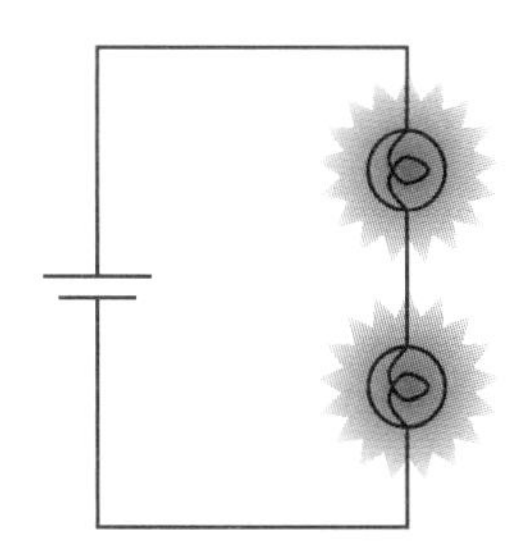

Figure 18.10 Two bulbs in series. Both are dimmer than in the standard circuit.

The fact that the same current flows through each bulb in Figure 18.10 is not unexpected. Electric charge is not being used up or lost along the pathway, and the same amount of charge must flow back to the battery as left it. If the same amount of charge flows through each bulb each second, the two identical filaments must carry the same current, must reach the same temperature, and must glow with the same brightness.

Two bulbs can also be connected so that each bulb has its own path from one end of the battery to the other, as in Figure 18.11. In this arrangement, the two bulbs are said to be wired in *parallel*. In contrast to a series connection, the current in one bulb does not pass through the other. In fact, if we were to disconnect either bulb, we would observe that the remaining bulb burns just as brightly as before. The two bulbs in parallel are equally bright, and each is as bright as the bulb in our standard circuit of Figure 18.9. Because each bulb has its own path to the battery, the battery will now supply twice as much total current as the battery did in the circuit of Figure 18.10.

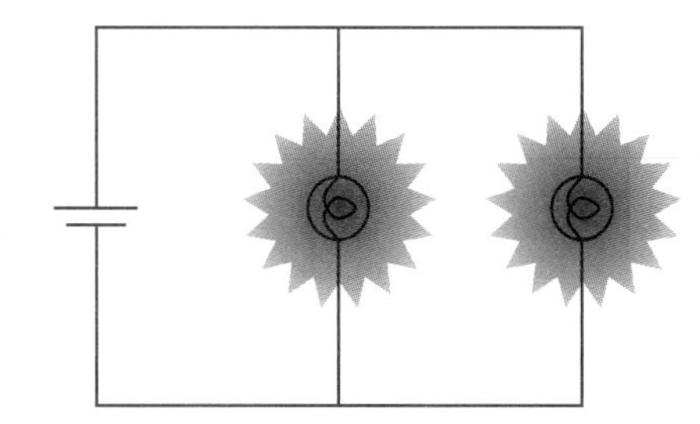

Figure 18.11 Two bulbs in parallel. Each bulb has the same brightness as in the standard circuit.

By the way, in Figure 18.11 we have assumed that the battery is an *ideal battery*, one that provides a set voltage regardless of the current through the battery. In real batteries, the voltage is provided by a chemical reaction that cannot keep up when

5. You might say, "more flow makes more glow." . . . Sorry.

the demand is too great, and so the voltage drops. This is why your headlights dim as you try to start your car. The demand for current by the starter motor reduces the battery's effective voltage, and so the lights dim. However, unless explicitly stated otherwise, we will assume ideal batteries. This is actually a good approximation in many cases.

Adding the extra bulb in parallel has increased the total current provided by the battery. But, since the voltage of the battery is the same, this means that the resistance of the circuit must have decreased. Although a bulb can be correctly thought of as providing resistance to the flow of charge, the addition of an additional bulb in a circuit can either increase or decrease the total resistance of the circuit, depending on how we add the bulb. When the new bulb is added in series (adding a new resistance on an existing line), the resistance of the circuit does indeed increase and the current through the battery decreases. But when the new bulb is added in parallel (creating a new path that did not exist before), the resistance of the circuit actually goes down and the current from the battery increases. Even though the new path contains new resistance (the additional bulb), it presents a new opportunity for flow that did not exist before, so the overall resistance drops.

When charge can go one of two ways, more charge will flow through the easier branch. We say that "current favors the path of least resistance." This does not mean that all of the current takes the easier path. It is like what happens at the end of a movie. People jostle to get out of the theater through the main door, the finite width of the door representing resistance to the flow of people. If the theater manager also opens the back door, then the flow of people out of the building increases. Even if the back door resists the flow of people because it is narrow, it still presents a new opportunity for flow, and people will take it in proportion to its width compared to the width of the main door—whatever keeps the crowds by both doors as small as possible. Similarly, a resistive path added in parallel will always lower the total resistance and, if parallel path 1 has twice the resistance as path 2, then path 1 will have one-half the current of path 2.

In the rest of this section, we will apply these analysis techniques we have learned to circuits of more and more complexity. As we said at the beginning of this section, the goal is to develop our intuition for resistors in circuits with batteries. As you read through these, you will find that you are acquiring a better and better sense of how the brightnesses of the light bulbs should behave.

EXAMPLE 18.3 Comparing Light Bulbs in Circuits—I

Two circuits with identical batteries and identical bulbs are shown at right. Which circuit presents the greatest equivalent resistance to the current coming out of the battery?

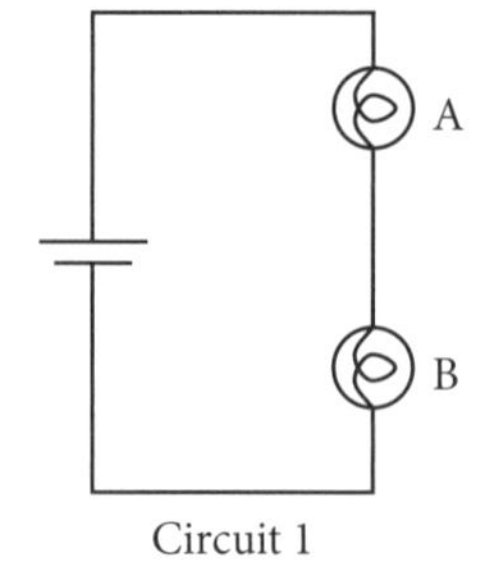

Circuit 1

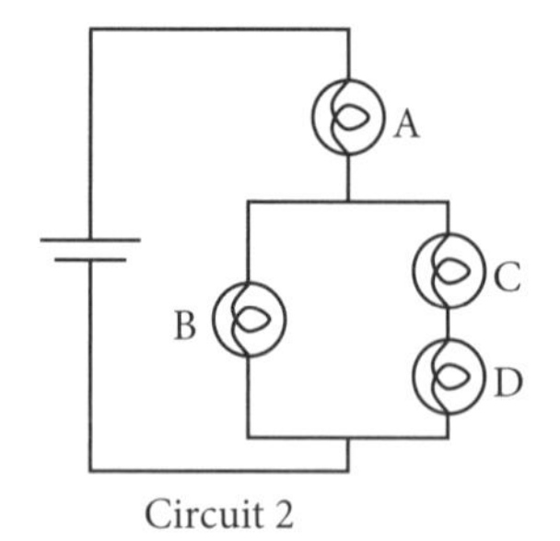

Circuit 2

ANSWER Circuit 2 certainly has more bulbs, so we are tempted to say that it presents the greater equivalent resistance to its battery. But we must remember that it is not just what is added that matters, but how it is added.

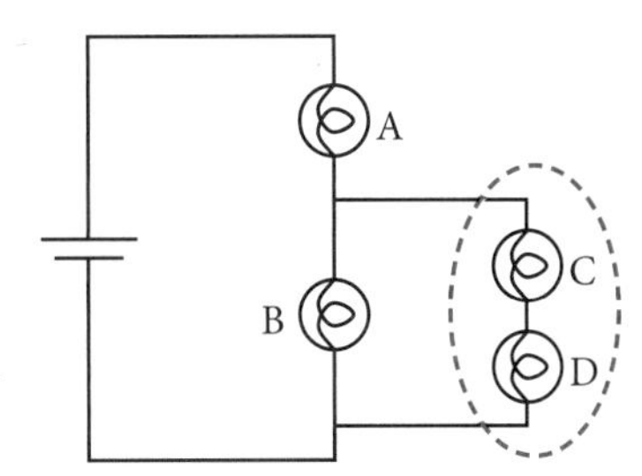

It is helpful to redraw circuit 2, to make it clearer how it differs from circuit 1, and to then draw a circle around the new part that was added, as shown. We then ask, "Was the new part inserted into an existing path to clog it up, or was it attached as a new flow opportunity that did not exist before?" As we can now plainly see, the two bulbs were added on a new path, so more charge will now flow in the circuit. The combination of B, C, and D has less resistance than B alone.

Since all of the current from the battery must pass through bulb A, in either circuit, we can think of bulb A as the indicator for the circuit, telling us about the total flow through the battery. If we set up these two circuits, we will find that the indicator bulb in circuit 2 is slightly brighter than the indicator bulb in circuit 1.

EXAMPLE 18.4 Comparing Light Bulbs in Circuits—II

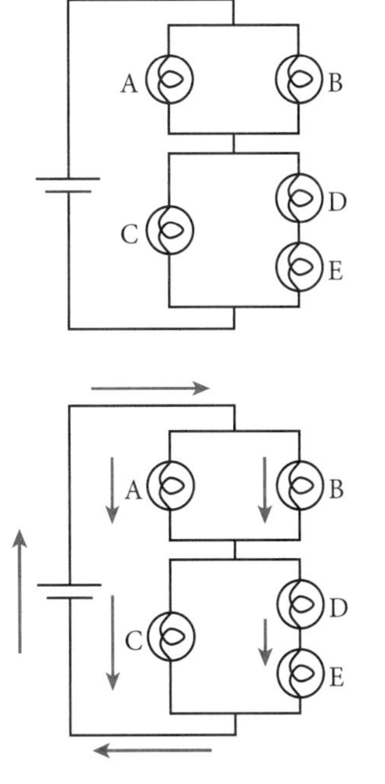

Rank the brightnesses of the five identical bulbs in the circuit shown.

ANSWER The key idea is that charge is neither created nor destroyed as it flows around a circuit. It must continue flowing, around and around, and at each junction it must choose which path to follow. If two possible paths (like the path through bulb A and the path through bulb B) are identical, then half the current will choose each path. Bulb A and bulb B are therefore equally bright and each has half the current through the battery.

When the two possible paths are not identical, then current favors the path of lesser resistance. The path containing bulb C is less cluttered than the path containing both bulbs D and E, so the current through C will be greater than the current through D (which is the same as the current through E). All of the current from the battery must go either through C or through D and E, so the current through bulb C must be more than half as big as the current through the battery, and therefore greater than the current through A or B. Our ranking is: C > A = B > D = E.

EXAMPLE 18.5 Comparing Light Bulbs in Circuits—III

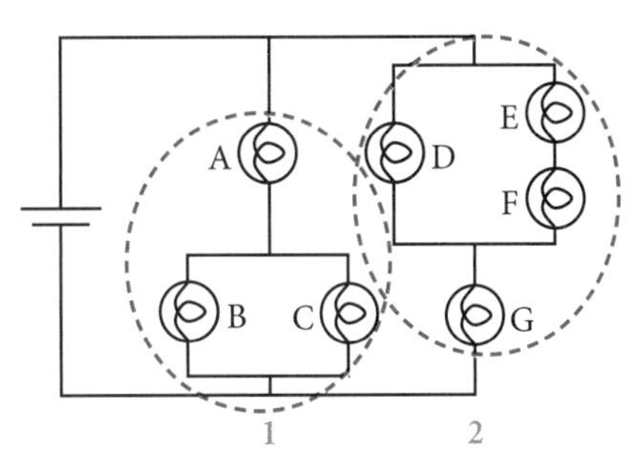

Rank the brightnesses of the identical bulbs in the circuit shown.

ANSWER This complicated circuit is simplified by recognizing that it actually contains two independent branches, labeled 1 and 2. The two branches each have their own connections to both terminals of the battery, so each will act as if it had its own private battery. It is easy to rank the bulbs within each branch against the others in that branch. First, we must have A > B = C, because B and C present equal resistance, and hence split the current through A evenly. Similar reasoning leads to G > D > E = F. We can even rank some of the bulbs in branch 1 against some of the bulbs in branch 2. We recognize that branch 2 is the same as branch 1 except for one extra bulb (bulb F), which has been added in series with another (bulb E), thus clogging an existing path. This allows us to conclude that the indicator bulb for branch 1 (bulb A) must be brighter than the indicator bulb for branch 2 (bulb G).

We can also compare the currents through bulb B and bulb E. To create an analogy, the larger current through A is like a large pizza, and B gets exactly half of the large pizza. The current through bulb G is like a small pizza and E gets less than half of the small pizza. So, we conclude that the current in bulb B will be greater than the current through bulb E (or F, since its current is the same as E).

It becomes trickier, however, when we try to compare bulbs B and D. Bulb B gets half of the large pizza, while bulb D gets more than half of a small pizza. Which is bigger? Our approach does not allow us to predict which of these bulbs will be brighter. In fact, we cannot really even rank bulbs B and G. Our analogy for these two bulbs leaves us comparing half of a large pizza with all of a small pizza. If we don't know the size of the pizzas, we cannot compare.

Of course, nature knows how much current will be going through each bulb in the previous example, and so we must also be able to figure it out. But we are going to have to look at what determines the brightness of the bulbs in a different way. We know that the brightness of a bulb depends on the current through it, and the current is given by Ohm's law as $I = V/R$. We also remember that the resistance of a light bulb can only grow if the current increases. So we can safely say that any bulb 1 will have more current through it, and will therefore be brighter, than an identical bulb 2, if the voltage difference across bulb 1 is greater than across bulb 2. We also know that a battery produces a particular voltage difference across its electrodes. So any charge arriving at the bottom of the battery must arrive at the same voltage, regardless of whether it went through branch 1 or branch 2 in the previous example. Let's keep these things in mind and take another look at Example 18.5.

EXAMPLE 18.6 Comparing Light Bulbs in Circuits—IV

Compare the brightness of bulbs B and D in the circuit at right.

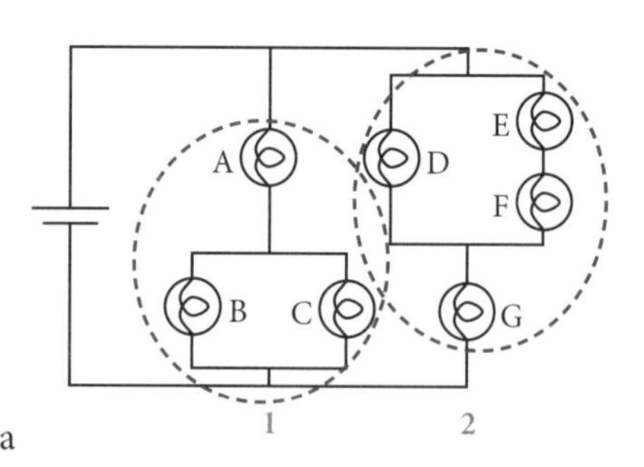

ANSWER Using current branching alone, we were unable to predict which of the two was brighter. We knew that bulb B received exactly half the current through bulb A, while bulb D received more than half of the current through bulb G. If bulbs A and G had been receiving the same amount of current, we could easily have ranked the bulbs. However, Example 18.5 told us that bulb A was brighter than bulb G, so bulb B got half of a large pizza and bulb D got more than half of a small pizza, and we were stuck.

But let us change our approach. We know that the voltage differences along *any* path from the top of the battery to the bottom must add up to the same number, the battery voltage V_0. Since we want to compare the brightnesses of bulbs B and D, we choose paths that contain each of these bulbs:

$$V_A + V_B = V_0 \quad \text{and} \quad V_D + V_G = V_0.$$

This means that

$$V_A + V_B = V_D + V_G.$$

Now we make use of the previously determined fact that bulb A is brighter than bulb G to tell us that $V_A > V_G$. Therefore, the equality above can only be satisfied if $V_B < V_D$. This means that bulb D will be brighter than bulb B. (If the math here is confusing, try this analogy. If my brother and I together make exactly the same amount of money as you and your sister do together, and if my brother makes more money than your sister, then you must be making more money than I am. Right?)

While we're at it, we can also compare the brightness of bulbs B and G. We know that bulb A is brighter than bulb B, so $V_A > V_B$. Since the sum of these voltage drops must add to V_0, V_A must be greater than $\frac{1}{2}V_0$ and V_B must be less than $\frac{1}{2}V_0$. Similar reasoning requires that V_G be greater than $\frac{1}{2}V_0$ and V_D be less than $\frac{1}{2}V_0$. This means that $V_G > V_B$, so bulb G must be brighter than bulb B.

With that example under our belt, let's tackle another tough one.

EXAMPLE 18.7 Comparing Light Bulbs in Circuits—V

Returning to the circuit in the last example, predict what will happen to bulbs G, A, and B (do they get brighter, dimmer, or stay the same) if we smash bulb C with a hammer (or just remove it and leave the empty socket behind).

ANSWER We begin by using current branching rules alone. First, we can see that bulb G stays the same. It is in branch 2, which is completely independent of branch 1 (assuming an ideal battery). Likewise, bulbs D, E, and F will be unaffected.

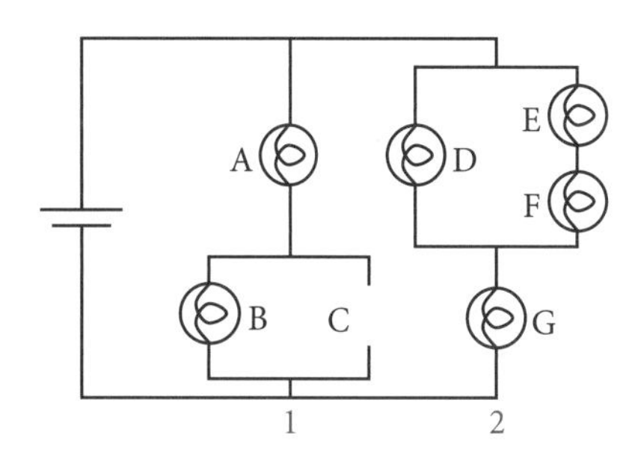

We can also see that bulb A will get dimmer. When bulb C is smashed, we remove a path from branch 1 and the total resistance of branch 1 increases. (If adding a path decreases resistance, then removing a path must increase it.) Increasing the total resistance of the whole circuit decreases the total current through the battery. Branch 2 is unaffected by the loss of the bulb, so branch 1 must be the place where the current decreases. Bulb A is still an indicator bulb for branch 1, since all of the current through branch 1 must pass through bulb A. So bulb A must get dimmer.

We run into problems when we try to predict what will happen to bulb B, using the current dividing alone. Two competing changes occur at the same time. With bulb C removed, B now gets 100% of the current through branch 1, instead of its previous 50%. The total current through branch 1, however, just got smaller. Before the change, B was getting half of a large pizza. After the change, it gets all of a small pizza, but we still don't know how big the small pizza is.

However, we just saw that bulb A gets dimmer when bulb C is removed, meaning that V_A must decrease. But, since we must always have $V_A + V_B = V_0$, V_B must increase as V_A decreases due to bulb C being smashed. We conclude that bulb B gets brighter.

18.7 Equivalent Resistance

In the previous section, we compared the resistances of paths with elements connected in series and in parallel, and used this to determine where the most current would flow. Here, we want to be more quantitative with this. When we connect two resistors in series, exactly how much resistance do they produce in that segment of the circuit? And how much resistance is produced when two resistors are connected in parallel?

We begin by considering a circuit with a 12-volt battery connected in series to a 4-Ω resistor and a 2-Ω resistor, as shown in Figure 18.12. If asked to guess what single resistor, attached to the same battery, would produce the same current in the battery as this series combination, you would probably guess 6 Ω, and you would be right. Your guess is easily checked. A 12-V battery connected to a single 6-Ω resistor would produce a current of 2 A. If the series resistors in Figure 18.12 had a current of 2 A running through them, the voltage drop across the 4-Ω resistor would be 8 V and the voltage drop across the 2-Ω resistor would be 4 V, for a total of 12 V, the voltage of the battery. If you replaced the battery with a 24-V model, the current would be doubled, but 6 Ω of resistance would still be the value that gave the right current. The generalization to any number of resistances is obvious. It is

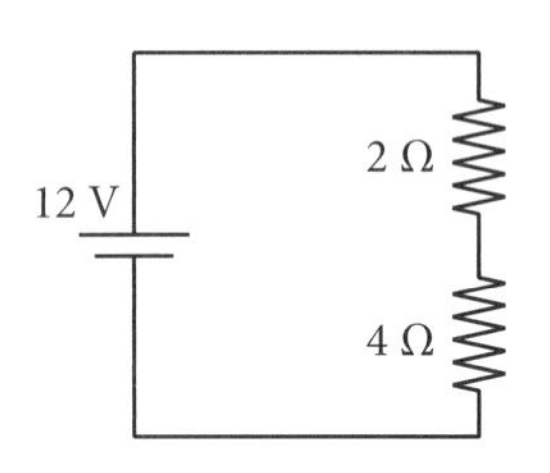

Figure 18.12 Two resistors in series.

- The equivalent resistance of any series combination of resistances is the sum of the individual resistances in the series.

The equivalent resistance of resistors in parallel is harder to see intuitively and harder to work out mathematically. We will, in fact, devote the rest of the section to this problem.

We begin with a configuration of resistors in parallel for which we have an easy intuitive solution. This is the case of two identical resistors connected in parallel, as depicted in Figure 18.13.

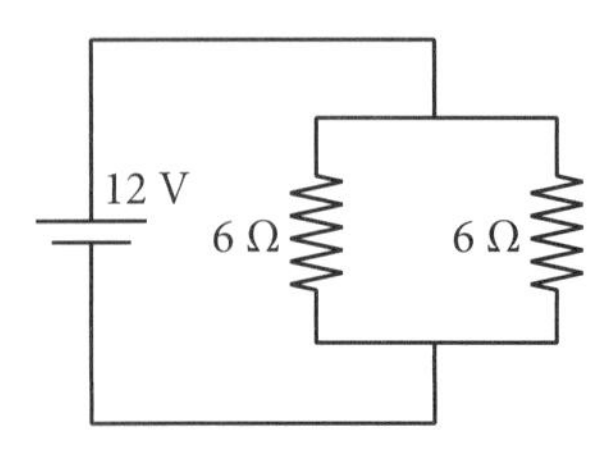

Figure 18.13 Two resistors in parallel.

We have already seen, in a case like this, that each 6-Ω resistor will have its own path to the battery and, thus, that each will carry the same 2 A of current it would have even if the other branch did not exist. The total current from the battery is therefore 4 A. A single resistor that would draw the same 4 A of current from the battery would be one with 3 Ω of resistance. So the equivalent resistance of two 6-Ω resistors in parallel is 3 Ω.

If, instead of Figure 18.13's two identical resistors in parallel, we had three identical resistors in parallel, as in Figure 18.14, the same argument would predict a total current of 6 A and a corresponding equivalent resistance of 2 Ω. By this time, we should see a pattern here. Let us state it as

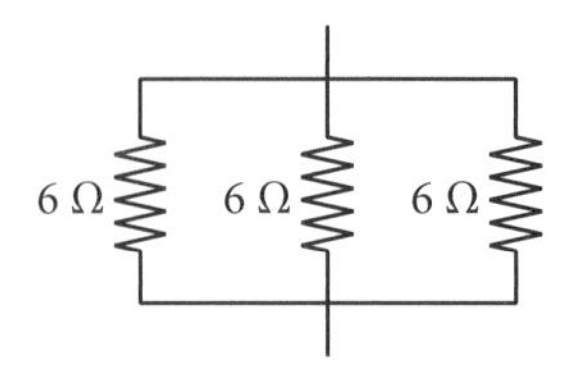

Figure 18.14 Three resistors in parallel.

- The equivalent resistance of N identical resistors of resistance R, connected in parallel, is $R_{\text{equivalent}} = R/N$.

"Very well," you may say, "but what if I don't have identical resistors? What if I have a 6-Ω resistor in parallel with a 12-Ω resistor, as in Figure 18.15a? What then?" No worries. We can simply replace the 6-Ω resistor with two parallel 12-Ω resistors. This gives us three identical resistors in parallel to which we can apply the R/N rule. The answer is 12 Ω/3 = 4 Ω.

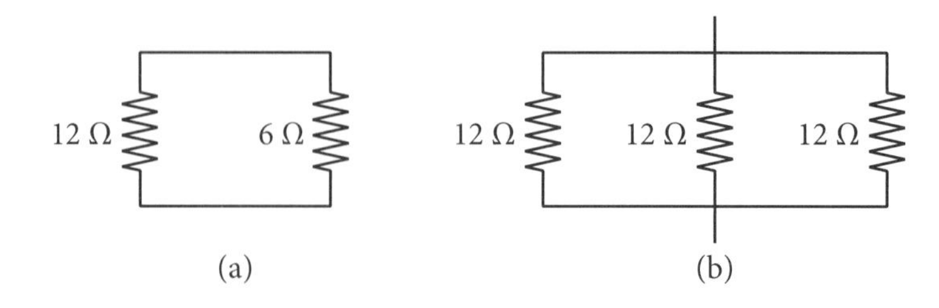

Figure 18.15 The 6-Ω resistor is equivalent to two 12-Ω resistors.

"Wait a minute," you might object, with a hint of skepticism in your voice. "That only works because the 12-Ω resistance is an even multiple of the 6-Ω resistance. What if it isn't that simple? What if I had a 3-Ω resistor in parallel with a 4-Ω resistor, as in Figure 18.16?" No worries. We just need to find a number that both 3 and 4 will go into evenly, an answer we can always find as the product of 3 and 4, giving 12. Then we could replace the 3-Ω resistor with four 12-Ω resistors, and the 4-Ω resistor with three 12-Ω resistors, and we would have an equivalent combination of seven identical 12-Ω resistors in parallel. By the *R*/*N* rule, this would give an equivalent resistance of 12/7 Ω, or 1.7 Ω.

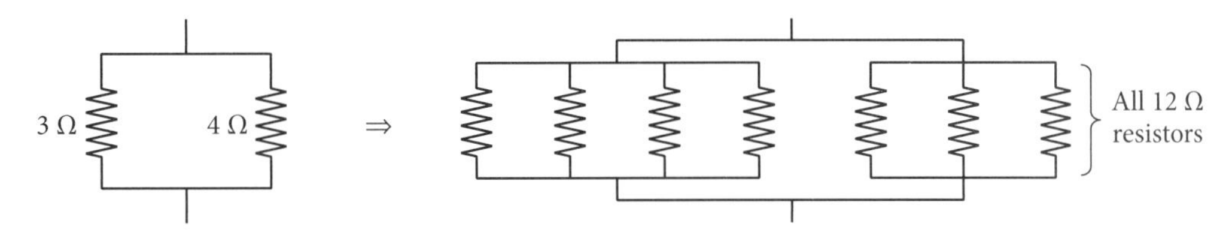

Figure 18.16 A 3-Ω resistor can be replaced by four 12-Ω resistors and a 4-Ω resistor can be replaced by three 12-Ω resistors. The equivalent resistance of the seven is 12/7 = 1.7 Ω.

What we have done, of course, is show that parallel resistors of resistance R_1 and R_2 can always be replaced by a single equivalent resistor of resistance

$$R_{\text{equivalent}} = \frac{R_1 R_2}{R_1 + R_2}. \tag{18.6}$$

The numerator is the size of the individual resistors needed and the denominator is the number of them required. If you think this through, you will see that it always works.

Finally, let us point out that, if we take the inverse of Equation 18.6, we get

$$\frac{1}{R_{\text{equivalent}}} = \frac{R_1 + R_2}{R_1 R_2} = \frac{1}{R_2} + \frac{1}{R_1}. \tag{18.7}$$

This form of the equation for resistors in parallel is perhaps more commonly found in textbooks than is Equation 18.6. And it does extend more simply to multiple resistors in parallel. If we had three resistors in parallel, with resistances R_1, R_2, and R_3, we could take the first two and find the resistance R_{12} that is equivalent to R_1 and R_2 by requiring that it satisfy $1/R_{12} = 1/R_1 + 1/R_2$ (Equation 18.7). Then this R_{12} resistance can be considered as an actual resistor that is in parallel with R_3, and Equation 18.7 can be used again to give the easy-to-remember formula

$$\frac{1}{R_{\text{equivalent}}} = \frac{1}{R_1} + \frac{1}{R_2} + \frac{1}{R_3}. \tag{18.8}$$

This form is also easy to work with on a calculator, if you are *very* careful. Suppose you wanted to find the equivalent resistance of a 2-Ω resistor, a 3-Ω resistor, and a 4-Ω resistor in parallel. On a normal scientific calculator with a 1/*x* button on it, you would input the following keystrokes: 2, 1/*x*, +, 3, 1/*x*, =, +, 4, 1/*x*, =. By this point, you would have added up all the inverses on the right-hand side of Equation 18.8. But you must then be careful to remember that this is not the equivalent resistance; it is its inverse. You must hit the 1/*x* button again to find $R_{\text{equivalent}}$.

Let's see how to use equivalent resistances to solve circuits.

EXAMPLE 18.8 Finding a Complicated Equivalent Resistance

For the circuit at right, find the current through the battery.

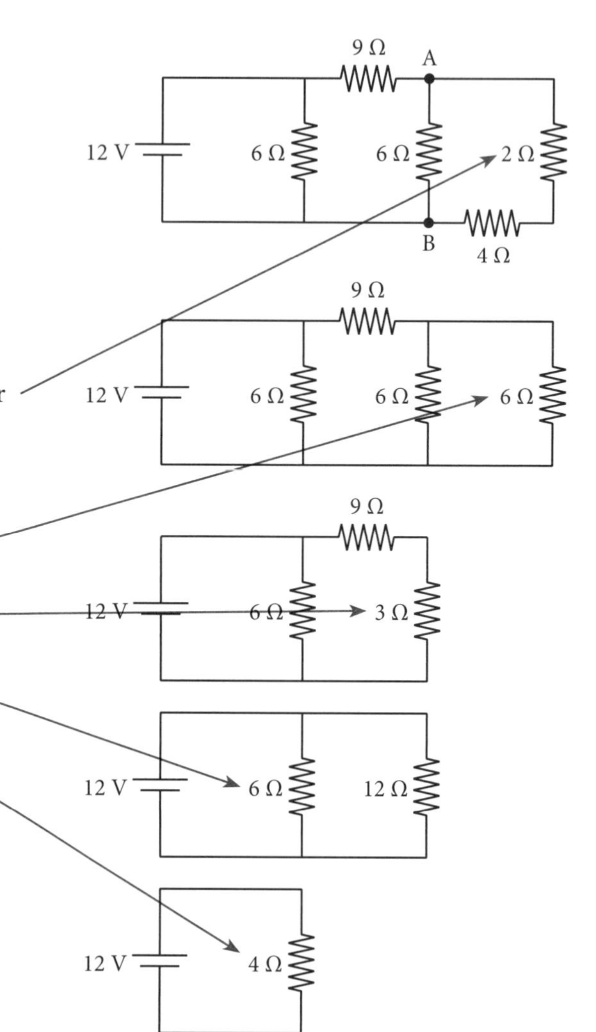

ANSWER We solve this problem by finding the equivalent resistance of the network of resistors to which the battery is connected. The idea is to comb through the circuit and replace any series or parallel sub-networks with equivalent resistors, redrawing the circuit each time. We continue the process until we are down to a single resistance.

Our first step is to recognize that the 2-Ω resistor and the 4-Ω resistor are in a simple series arrangement. We can replace them with a single equivalent 6-Ω resistor. Next, we recognize a parallel combination in the two rightmost 6-Ω resistors. These can thus be replaced by a single 3-Ω resistor. The 3-Ω resistor and the 9-Ω resistor are in series, giving an equivalent resistance of 12 Ω. The parallel 6-Ω and 12-Ω resistors look just like Figure 18.15, whose equivalent resistance was found on page 368 to be 4 Ω.

The resistance of this complicated network of resistors is therefore 4 Ω, and the current pulled from a 12-V battery by a 4-Ω resistor is

$$I = \frac{V}{R} = \frac{12\ \text{V}}{4\ \Omega} = 3\ \text{A}.$$

18.8 Kirchhoff's Laws

In this section, we present the physical laws that govern the distribution of currents and voltages in an electrical circuit when steady currents flow through all elements of the circuit. These two laws were first stated clearly by German physicist Gustav Kirchhoff in 1845, and are known as Kirchhoff's Laws.

The first law is basically a restatement of the principle that electrical charge can neither be created or destroyed. It describes what happens at a junction where two or more wires come together. *Kirchhoff's Current Law* reads

- At any junction in an electrical circuit, the sum of currents flowing into that junction is equal to the sum of currents flowing out of that junction.

The second law is a restatement of the law of conservation of energy. Energy is dissipated in resistors and provided by batteries. But any charge that travels any closed path in a circuit, upon returning to the point where it left, comes back to the same energy it started with. Otherwise, the speed of the charges and the currents they represent would not be steady, constant currents. *Kirchhoff's Voltage Law* says

- The sum of all the voltage differences around any closed loop is equal to zero.

We have essentially been using Kirchhoff's laws in Section 18.6 as we went through our experiments with batteries and light bulbs. But here, we want to set up a more systematic way to use these to completely solve general electrical circuits.

Let's begin by considering a circuit of moderate complexity, the one shown on the left side of Figure 18.17. You should always begin by drawing your own diagram of a circuit, including how you will solve it, as we do on the right of Figure 18.17.

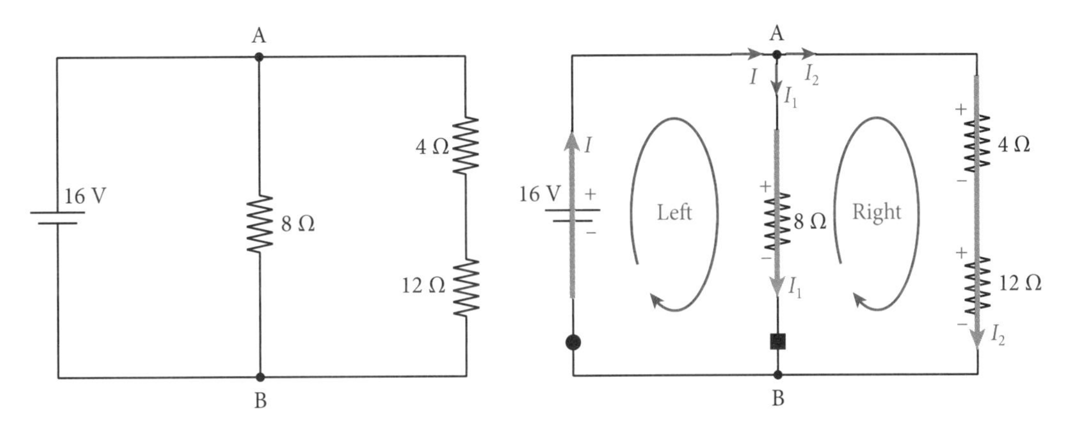

Figure 18.17 A demonstration circuit for the use of Kirchhoff's laws. The diagram on the left is the circuit diagram given in the problem. The diagram on the right is marked up preparatory to the use of Kirchhoff's laws. The currents needed for Kirchhoff's current law are shown at junction A. The two loops used for Kirchhoff's voltage law are the two clockwise loops labeled "left" and "right."

We need to identify the number of different currents there will be in the circuit and draw an arrow for the assumed direction of each one. Because there is a branch at point A, we know that the current coming out of the battery will divide at A. Let us call the current through the battery I, the current through the 8-Ω resistor I_1, and the current through the 4-Ω and 12-Ω resistors I_2. Although the direction of the current through the resistors is obvious in this problem, we can actually assume any direction we want for any current. The mathematics arising out of Kirchhoff's laws will just give us a negative value for the current if our initial guess was wrong. We also note that we do not need to define separate current values for elements of the circuit that are in series, since the currents through them will be the same. This is why we did not need to specify a different current for the 4-Ω resistor than for the 12-Ω resistor.

Once currents have been identified for all segments of the circuit, we need to determine the voltage differences across each element. The difference across a battery is the voltage rating of the battery and the difference across a resistor is given by Ohm's law, $V = IR$. We also need to note which side of a resistor or a battery is at the higher voltage, given the directions assumed for the currents through them. The high side of the voltage difference across a battery will always be at the positive electrode, regardless of the direction of the current. (Sometimes another battery somewhere can force current in the opposite direction to the battery's polarity, but that doesn't change the polarity of the battery.) For a resistor, the high-voltage side is always the side at which the assumed current enters. Based on your assumptions for the directions of the currents, you should write down a plus at the high side and a minus sign at the low side of each resistor and battery. We have done this in Figure 18.17.

Looking at the problem diagram on the left of Figure 18.17, we see that the characteristics of all of the elements are given—the voltage for the battery and the resistances for the resistors—but that we have not been given any of the currents. This is actually the most common circuit analysis you will have to perform. Let us note that there are formally three unknowns in this circuit, I, I_1, and I_2. Since we formally have three unknowns, we may expect Kirchhoff's laws to give us three independent equations relating them (though I could be seen almost intuitively to be equal to the sum of I_1 and I_2).

Anyway, one equation will come from Kirchhoff's current law at junction A. The current coming into A must equal the current going out of A:

$$I = I_1 + I_2.$$

There is another junction at point B, but a moment's thought will make it clear that Kirchhoff's current law applied to that junction just produces the same equation. We still need two more independent equations.

The other equations are provided by Kirchhoff's voltage law. One available voltage loop is the one labeled "left" in Figure 18.17. We must begin the loop at some point, so we arbitrarily choose the black circle just below the battery and go around the loop clockwise, as shown. As we pass upward through the battery, the voltage changes by +16 V. We then pass point A and drop through the 8-Ω resistor from + to –, for a change of $-I_1R = -(8\ \Omega)\ I_1$. The total voltage around the loop is thus

$$+16 - 8I_1 = 0.$$

In this equation, we should have kept a volt unit on the 16 and an ohm unit on the 8, but, since we will typically need to do a little algebra in these kinds of problems, let us just agree to simplify our equations by putting all voltages in volts, all resistances in ohms, and all currents in amps. Thus, the previous equation tells us that $I_1 = 2$ amps.

To find I_2, we will need to find a loop that includes the rightmost branch of the circuit, the segment with the 4-Ω and 12-Ω resistors. The loop labeled "right" should suffice. Starting at the little black square and going clockwise around this loop, we first encounter the 8-Ω resistor. Because we assumed that I_1 was going downward through this resistor, the high-voltage side is at the top. So, as we go from the bottom of this resistor to the top, we gain a voltage of $+8I_1$. Then, as we drop through the other two resistors, the changes are $-4I_2$ and $-12I_2$. The loop equation is thus

$$+8I_1 - 4I_2 - 12I_2 = 0,$$

or $+8I_1 - 16I_2 = 0$. Since we already know that $I_1 = 2$ A, this gives $(8)(2) = 16I_2$, for a result $I_2 = 1$ A. The currents through the resistors are thus $I_1 = 2$ A and $I_2 = 1$ A, and the current provided by the battery is $I = I_1 + I_2 = 3$ A.

Now we are well aware that we could have solved this example as an equivalent resistor problem, as we did in the last section. However, there are circuits that are not susceptible to solution that way. In all cases, the application of Kirchhoff's laws can solve the problem. Let's try an example where this approach is *required*.

EXAMPLE 18.9 Using Kirchhoff's Laws

Find the currents in each element of the circuit below.

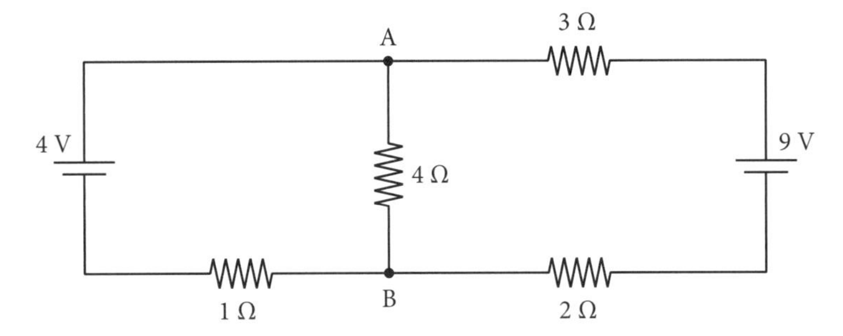

ANSWER Let's run the program. We begin by assigning a current to each element in the diagram and then write in the pluses and minuses on each element.

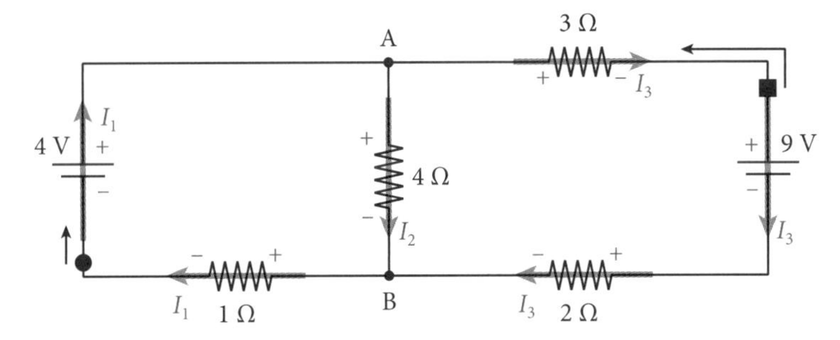

We have assigned our currents in a manner similar to Figure 18.18, but there was no special reason why we had to do that. Kirchhoff's current law applied to point A gives $I_1 = I_2 + I_3$. For loops, we take the leftmost loop, starting at the black dot and going in the direction of the little black arrow, clockwise around the loop. The voltage changes around the loop are

$$+4 - 4I_2 - 1I_1 = 0.$$

Now, just to show that we can do it, we choose, for our second loop, not the right loop, but the perimeter loop. This one goes all around the outside of the circuit and skips the 4-Ω resistor in the middle of the circuit. Let's go counterclockwise, for a change, starting at the black square at the upper right and going in the direction of the little black arrow next to it. As we work our way around the perimeter, we find

$$+3I_3 - 4 + 1I_1 + 2I_3 + 9 = 0.$$

Note that we have gone backward through all the resistors (in the opposite direction we had assumed for the currents), and so they were all positive voltage gains in the loop equation. Note also that our taking the direction we chose made for a voltage drop across the 4-V battery instead of a gain.

If we use $I_1 = I_2 + I_3$ in each of the loop equations, we end up with two equations for the two unknowns, I_2 and I_3:

$$+4 - 4I_2 - 1(I_2 + I_3) = 0$$
$$+3I_3 - 4 + (I_2 + I_3) + 2I_3 + 9 = 0.$$

Or, expanding and rearranging, we find

$$5I_2 + I_3 = 4$$
$$I_2 + 6I_3 = -5.$$

There are many ways to solve simultaneous equations like this, but the easiest to remember (see Appendix A) is probably to solve one equation for one variable, say, the first equation for I_3. This gives $I_3 = 4 - 5I_2$. This result can then be substituted into the second equation to give

$$I_2 + 6(4 - 5I_2) = -5 \quad \Rightarrow \quad I_2 + 24 - 30I_2 = -5$$

or

$$-29I_2 = -29 \quad \Rightarrow \quad I_2 = 1 \text{ A}.$$

Since the relation we used to substitute for I_3 was $I_3 = 4 - 5I_2$, we also find

$$I_3 = 4 - 5(1) = -1 \text{ A}.$$

The negative answer for I_3 tells us that our assumption of current going clockwise through the 3-Ω resistor, the 9-V battery, and the 2-Ω resistor was wrong. We had the current I_3 going clockwise through those three elements, and so the current is actually going counterclockwise in the right-hand loop. Note also that the direction we chose to go around the perimeter loop had nothing to do with the directions chosen for the currents.

The current going through the 4-V battery is $I_1 = I_2 + I_3 = 1 - 1 = 0$. This means that the 4-V battery produces no current, but just matches the 4-volt difference it sees between points A and B. It is as if it were hooked up to another 4-V battery, in which case no current would flow through it. There is also no current flowing through the 1-Ω resistor, and so there will be no voltage difference across that element either.

18.9 RC Circuits

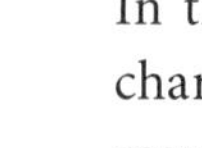

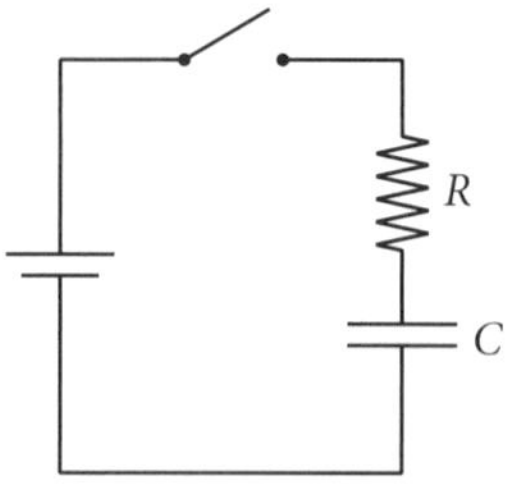

Figure 18.18 An RC circuit. The switch at the top has just been closed and the current is starting to charge the capacitor.

In the previous chapter, in Section 17.5, we introduced capacitors as elements for storing charge, a plus charge on one plate and an equal negative charge on the other. The important result from that section was that capacitors carrying charges $\pm Q$ on their two plates have a voltage difference across them. We found that larger plates, or plates separated by a smaller distance, can hold a greater charge for a given voltage difference—a property we quantified as capacitance, C. Based on the definition of C, we found that the charge stored on a given capacitor can be written as $Q = C\Delta V$.

In this section we will be analyzing circuits that contain capacitors as well as resistors. These are known as "RC circuits." To see how a capacitor is integrated into circuits like those we have been working with, let us rewrite the capacitor equation as

$$V = \frac{Q}{C}.$$[6]

Let us imagine hooking a capacitor and a resistor and a switch in series to the terminals of a battery, as in Figure 18.18. As the switch closes, the capacitor will initially

6. Note that we have replaced Chapter 17's ΔV with a simple V. We have been using V for voltage differences throughout this chapter, and want to continue to do so for capacitors.

have zero charge and, therefore, zero voltage difference across its terminals. However, as charge from the battery is pushed onto the positive plate of the capacitor (the top capacitor plate in Figure 18.18), an equal charge will flow off the negative plate and out the negative terminal of the capacitor. As far as the circuit is concerned, this uncharged capacitor is behaving exactly like a segment of wire. It has zero voltage difference across it and, for every coulomb that flows in one side, one coulomb flows out the other. Of course, this is only true at the start, when the capacitor is uncharged. As more and more charge flows onto the positive terminal, and an equal amount flows off of the negative terminal, the capacitor begins to collect charge on its plates and develop an electric potential difference across its terminals.

Kirchhoff's voltage law tells us that, as the voltage difference across the capacitor begins to grow, the potential difference across the series resistor must drop. This is because the voltage differences around the closed loop must still sum to zero, meaning that the voltages across the capacitor and across the resistor must still add up to the battery voltage. As the capacitor charges, its voltage grows and the voltage across the resistor must decrease. But since, by Ohm's law, the voltage difference across the resistor is proportional to the current through it, the current through the resistor will decrease. This means that the current drawn from the battery will decrease as well.

This process of charging the capacitor with a diminishing current continues until the voltage across the capacitor is finally equal to the battery voltage. At this point, the voltage across the resistor has dropped to zero, and the current in the circuit goes to zero. We then say that the capacitor is "fully charged." The capacitor, of course, is not "full." If the battery voltage were increased, more charge could flow onto the positive plate and off the negative plate. But "fully charged" means that current is no longer flowing and that the capacitor has reached sufficient voltage to make this happen.

In many ways, a "fully charged" capacitor is like an open circuit. There is a voltage difference across the battery terminals, but charge is no longer moving. What makes the capacitor different from a broken wire is that it has charge stored on it. Let us picture opening the switch, disconnecting the battery (Figure 18.19), and replacing the battery with a wire (Figure 18.20). With the switch still open, the capacitor is still fully charged at the same voltage as the original battery. Once the switch is closed (Figure 18.20), the resistor prevents the charge from instantaneously moving off the plates, so the initial voltage across the capacitor is still there for an instant. There can be no voltage difference across the newly inserted wire, so the voltage difference across the capacitor must always equal the voltage difference across the resistor, by Kirchhoff's voltage law. The only way for there to be a voltage difference across a resistor is for a current to flow through it, but the source of the current in this case is not a battery, but the capacitor, with charge coming off the positive plate, passing through the resistor, and flowing onto the negative plate. As this current bleeds charge off the plates, the voltage across the capacitor drops, producing a steadily decreasing current through the resistor, as current and voltage eventually drop to zero together.

Although no charge actually passes through a capacitor (i.e., jumps the gap), a charging or discharging capacitor does appear to have a current through it, in that charge enters one side and an equal amount emerges from the other side. It is common to talk about the "current" through a capacitor, though we should recognize the loose way in which this needs to be understood.

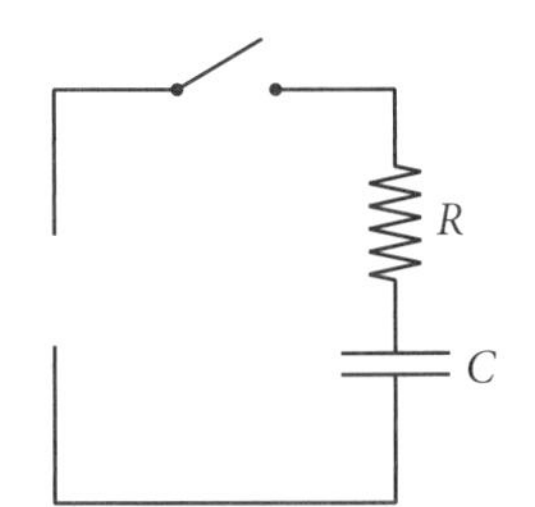

Figure 18.19 The switch is opened, leaving the charge on the capacitor, and the battery is then disconnected.

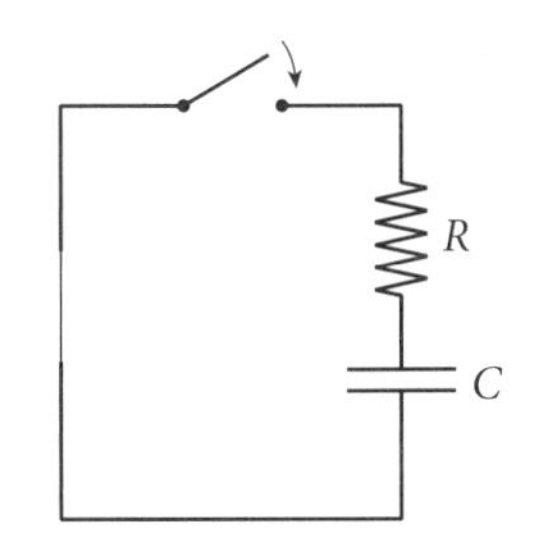

Figure 18.20 The switch is closed and the capacitor discharges through the resistor.

EXAMPLE 18.10 Capacitors in a Circuit

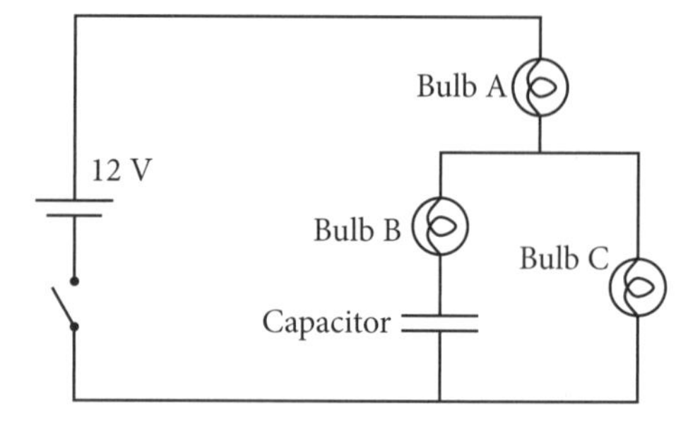

The circuit at right comprises 3 identical bulbs, a capacitor, an ideal 12-volt battery, and a switch, which is initially open.
a) The switch is now closed. Rank the voltage drops across each of the three bulbs and the capacitor immediately after the closing of the switch.
b) After the switch has been closed a long time, determine the voltage drops (numerical values) across each of the three bulbs and the capacitor.
c) The switch is now reopened. Determine the voltage drops (numerical values) across each of the bulbs and the capacitor immediately after opening the switch.

ANSWER

a) With the switch open, the capacitor has a direct path between its terminals, passing through bulbs B and C. This guarantees that it will be uncharged when the switch is first closed. At the instant the switch is closed, the uncharged capacitor has zero volts across it and acts just like a wire. We think of it this way and analyze the bulbs the way we did back in Section 18.6. From the analysis of the currents alone, we know that bulb A is the brightest (the same current as through the battery) and that bulbs B and C are equally bright but dimmer than A (each has half the battery current). The voltage drop across the capacitor, of course, is initially zero. This tells us that the voltage ranking is

$$V_A > V_B = V_C > V_{CAP} = 0.$$

b) As the capacitor charges, the voltage across its terminals grows and the current through it decreases. At some point (less than a couple of seconds in real circuits, and often much faster), the capacitor becomes fully charged and acts like an open switch, with no more current through it. At this point, bulb B goes out, which means it has zero voltage drop across it. With no current in the branch containing bulb B, bulbs A and C must now have the same current as each other (as no current now goes through B), and the voltage drop across each will therefore be the same as across the other. The voltage drops across bulbs A and C must add to the battery voltage (consider the loop containing the battery, bulb A, and bulb C). Combining these results, we see that the voltages across bulbs A and C must be 6 volts each. If we consider the loop containing the capacitor and bulbs B and C, we know that the voltages across bulbs B and C must add to the capacitor voltage. We know that the voltage across bulb B is zero, since the current has stopped flowing in that branch of the circuit. Therefore the capacitor voltage must be the same as the voltage across bulb C, or 6 volts. Our conclusions are therefore:

$$V_{CAP} = 6\text{ V};\ V_B = 0\text{ V};\ V_A = V_C = 6\text{ V}.$$

c) Opening the switch removes the battery from the circuit and causes the battery current to drop to zero immediately. This means that bulb A immediately goes out. But the capacitor cannot change its charge (and thus its voltage) instantaneously. So, right after the switch is opened, the capacitor voltage is still 6 volts. If we apply Kirchhoff's loop theorem to the loop containing the capacitor and bulbs B and C, we know that the voltages across bulbs B and C must still add up to equal the capacitor voltage. As bulbs B and C must be the same brightness (same current), they must each have a drop of 3 volts. This gives us

$$V_{CAP} = 6\text{ V};\ V_A = 0\text{ V};\ V_B = V_C = 3\text{ V}.$$

Although not asked in the problem, we can use these results to describe the behavior of the bulbs throughout the process. When the switch is first closed, bulb A comes on the brightest, with bulbs B and C equal to each other and dimmer than A. As the capacitor charges, bulb A dims (it drops from more than 6 volts to exactly 6 volts), bulb B gradually goes out, and bulb C gradually gets brighter (it increases from less than 6 volts to exactly 6 volts). When the switch is opened, bulb A immediately goes out. Bulb B gets immediately brighter and then gradually goes out as the capacitor discharges. Bulb C immediately gets dimmer as it drops from 6 volts to 3 volts. It then gradually dims and goes out.

Before leaving this section, let us point out that, except for this last section, all of the circuits we have worked with in this chapter have produced constant currents. The currents in RC circuits will vary with time, for the reasons we have explained. However, even when the currents are not constant, Kirchhoff's laws still apply to the instantaneous values of currents and voltages in the circuit. Because the voltage across a capacitor depends on the charge on the capacitor, while the voltage difference across a resistor depends on the rate at which charge is moving (the current), it is not possible to work out the time-varying currents and voltages in the circuit without recourse to calculus and the techniques of differential equations. We are absolutely not going to do that here, but we thought you might like to know that it is doable, and that the fundamental physics you need is still the good old Kirchhoff's laws.

18.10 Summary

From this chapter, you should have learned the following things:

- Current is the rate of charge flow, $I = q/t$, and it is measured in amperes.
- Batteries produce a constant voltage across their terminals and an ideal battery is one that can produce unlimited current without its voltage dropping.
- Ohm's law gives the relationship between the voltage difference across the ends of a resistor and the current flowing through the resistor: $V = IR$. An "ohmic" device is one in which the resistance remains constant as the current changes.
- The power produced by a battery is $P = IV$ and the power dissipated in a resistor is $P = IV = I^2R$.
- When two elements are connected in series, the current through one element also passes through the other element. When two elements are connected in parallel, the current divides itself between the two elements so that the currents are inversely proportional to the resistances of the two elements.
- The equivalent resistance of two resistors in series is $R_{eq} = R_1 + R_2$. The equivalent resistance of two resistors in parallel is given by $1/R_{eq} = 1/R_1 + 1/R_2$.
- The straightforward procedure for solving circuit problems via Kirchhoff's laws is as outlined in Section 18.8. You should make sure you can define the currents and determine the voltage drops across elements using this procedure.
- The voltage difference across a capacitor depends on the instantaneous value of the charge stored on the capacitor: $V = Q/C$.

CHAPTER FORMULAS

There are not very many new ones this time.

Ohm's law: $V = IR$

Power: $P = IV = I^2R$

Series resistors: $R_{eq} = R_1 + R_2 + R_3 + \cdots$

Parallel resistors: $\frac{1}{R_{eq}} = \frac{1}{R_1} + \frac{1}{R_2} + \frac{1}{R_3} + \cdots$

Capacitors (again): $V = \frac{Q}{C}$

PROBLEMS

18.1 * 1. If, during a period of 5 seconds, 10^{17} electrons travel to the right past a given point in a copper wire, what is the current in the wire in amperes and in which direction is the current flowing?

** 2. A lightning strike consists of charge moving from a cloud to the ground. In one particular case, the electrons in a lightning strike are confined to travel within a radius of 1.5 m and to move at a speed of 5.0×10^4 m/s. If the density of electrons in the strike is 10^{15} electrons/m^3, what current is carried by the lightning?

** 3. A positive "ion" is an atom that has had one or more electrons removed from it, leaving it with a net positive charge. Suppose that positive ions are created in the lab by removing a single electron from a number of atoms, and that the ions are then accelerated by a high voltage to a speed of 1⁄10 the speed of light (so they are moving at 3×10^7 m/s). As they move along a long vacuum tube, this short pulse of ions is seen to uniformly fill a cylinder 2 cm in diameter and 10 cm long. The density of the ions is 10^8 ions/cm^3. What current is flowing along the tube as the cylindrical collection of ions moves past?

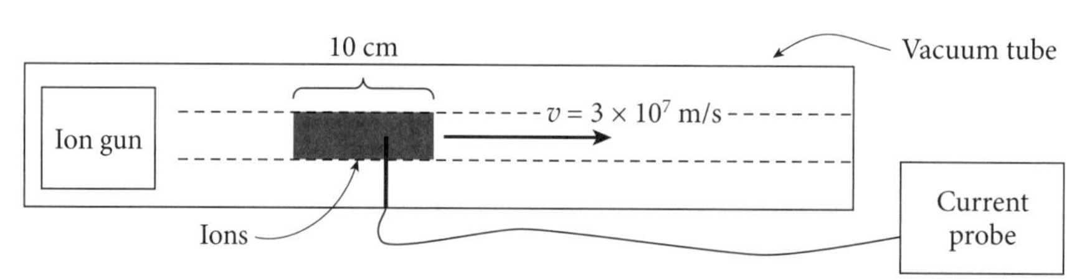

18.2 * 4. A kitchen timer requires a 9-volt battery, but the battery is dead. You need to run the timer, but all you have is a box full of 1.5-volt flashlight batteries. How many would you need and how would you connect them to power the kitchen timer?

* 5. Diesel trucks use two 12-volt automobile batteries instead of one so that they can provide more current to the starter motor. How should the batteries be wired (series or parallel)? What will be the voltage applied to the starter motor as a result?

** 6. A wildlife biologist has invented a motion detector that requires 6 volts to operate and draws 200 mA of current. He has a number of alkaline AA batteries that can each provide 50 mA of current without dropping below their nominal 1.5-V voltage. How many AA batteries must he use to power his motion detector and how should they be wired?

** 7. A watchman's flashlight requires 5 alkaline D batteries in series to operate. Each alkaline battery provides 1.5 V of voltage. If he switches to a flashlight that uses rechargeable nickel–metal–hydride batteries (that each provide 1.25 V of voltage), how many NiMH D cells will be required to power the same flashlight bulb?

18.3 * 8. What voltage difference is required to cause a current of 2 mA to flow in a wire with resistance 4 MΩ? (Remember that a megaohm, MΩ, is 10^6 Ω.)

* 9. When an emergency radio is connected to a 9-V battery, a current of 25 mA flows through the radio. What is the internal resistance of the radio?

** 10. What is the current through a cold light bulb with a resistance of 2 Ω when connected to a 3-volt battery? The light bulb requires a current of 6 amps to produce light, but at the temperature it must have to light up, its resistance increases to 5 Ω. What voltage is required to produce a 6-amp current in the bulb?

** 11. A 6-Ω resistor is connected to a 3-volt battery. How many electrons flow through the light bulb in 30 minutes?

18.4 ** 12. A 60-watt light bulb shines at full power when it is connected to a constant power supply producing 100 V. What current is flowing in the bulb?

** 13. A 4.4-Ω flashlight bulb is connected to a 3.0-volt battery and left on for 20 minutes. How much charge flows through the bulb in this time? What total battery energy has been used?

** 14. A 1500-W electric heater is connected to the same 110-V line as a computer that draws a peak power of 300 W. How much current is drawn on the line when the computer is operating at its peak power usage?

** 15. What is the efficiency of an electric motor that produces ⅓ horsepower (about 250 W) of useful power while drawing 3 A of current from a 110-V line?

** 16. The tungsten filament of a light bulb uses 60 W when connected to a 120-V power line. When the bulb is cold, its resistance is 24 Ω.

a) What is the hot resistance of the light bulb?

b) What current flows through the bulb when it is cold?

c) What current flows through the bulb when it is hot?

*** 17. A coil of wire is placed in a cup of water and, when it is plugged into the wall, it draws 5 A of current. Once the water starts boiling, it is found that 100 g of water is boiled off in 10 minutes. What is the resistance of the coil?

18.5 * 18. The batteries in the three circuits shown below are all 5-volt batteries and the resistors are 50-ohm resistors. How much current will flow through the resistor in each of the cases shown?

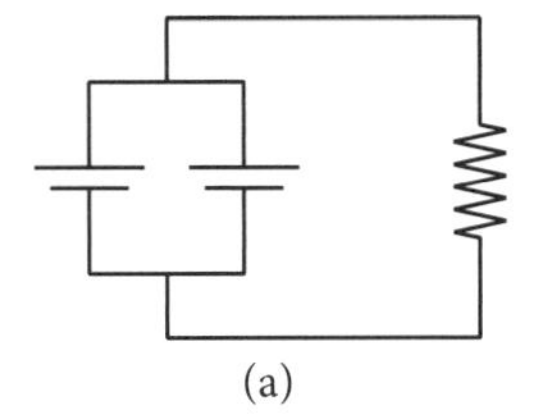

(a)

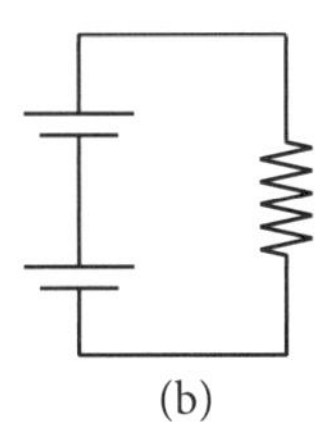

(b)

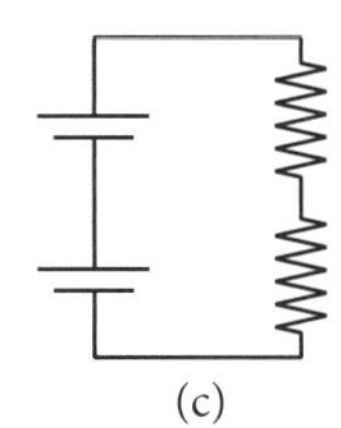

(c)

** 19. A 100-Ω resistor has a current of 2 A passing through it. We want to measure the voltage drop across the resistor by attaching a voltmeter in parallel with the resistor, as shown. We want the voltmeter to give the voltage across the resistor to within 1% of the actual voltage difference across it. To do this, the voltmeter must have an internal resistance *r*, as shown, that will restrict the current that the voltmeter branch takes away from the resistor.

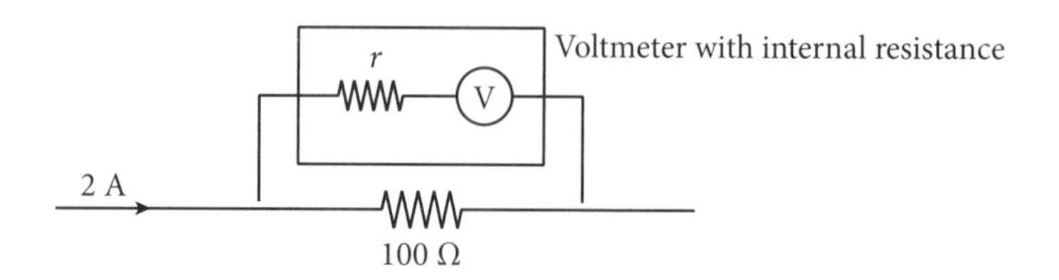

a) Calculate the voltage drop across the resistor before the voltmeter is added to the circuit.

b) Find the voltmeter resistance *r* for which the current through the voltmeter is less than 1% of the original current through the resistor.

** 20. Ellen's voltmeter has just broken, but she still has a working ammeter.

a) Explain why the ammeter in a normal usage must have a low internal resistance.

b) Devise a way that Ellen can use her ammeter, along with a 10,000-Ω resistor, to create a voltmeter that can determine the voltage drop across a resistor by measuring the current through the ammeter.

** 21. What is the voltage difference across each resistor in the diagram at right?

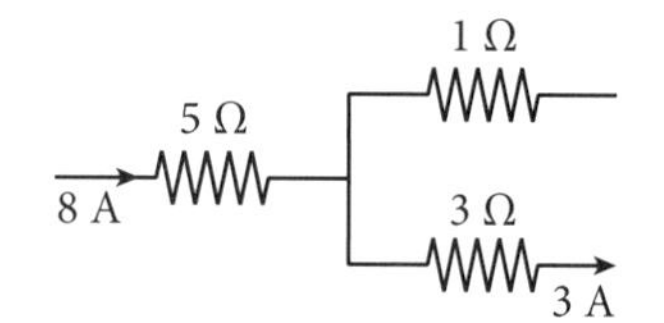

18.6 ** 22. The circuit at right contains 4 identical bulbs. Rank the bulbs in order of increasing brightness. Justify your answer.

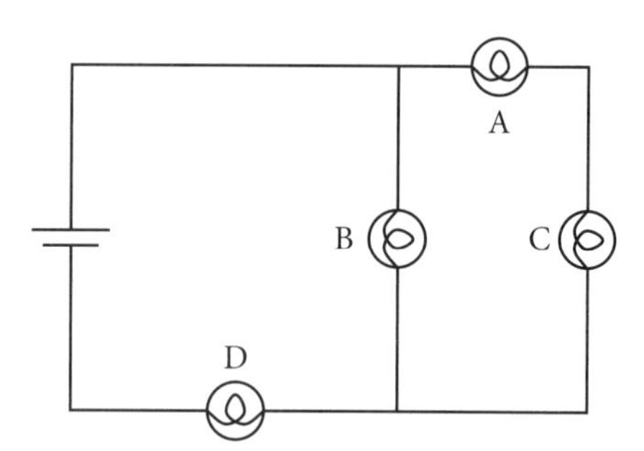

** 23. In the circuit at right, two identical bulbs are connected in series to a single 12-volt battery. Both bulbs are glowing. When the switch is closed, describe what happens to the brightness of each of the two bulbs. Justify your answers.

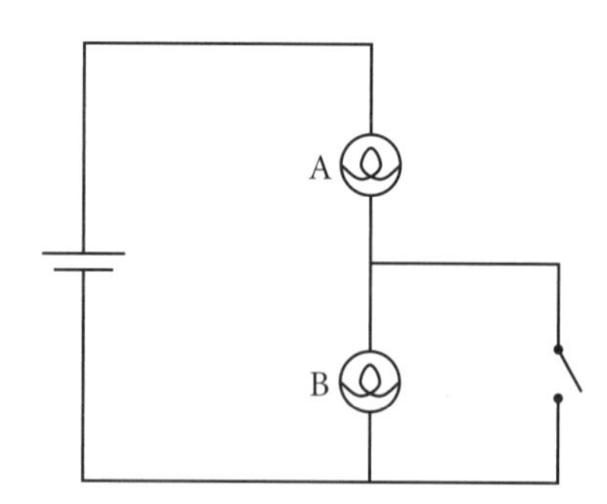

** 24. What happens to the brightness of bulbs A and B in the circuit at right when a wire is connected between points 1 and 2? Explain your reasoning.

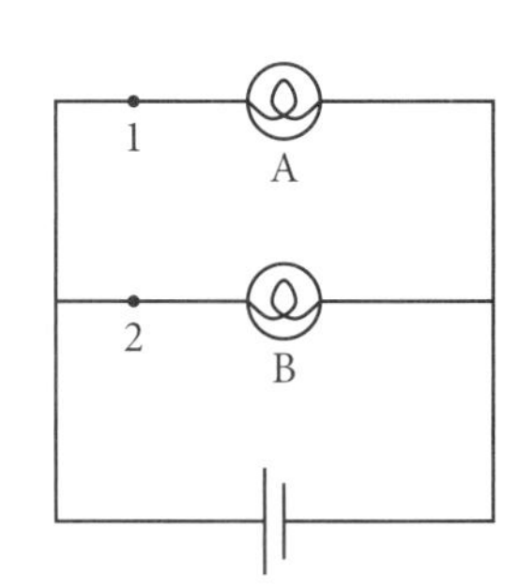

** 25. The boxes in the circuit in Arrangement 1 contain unknown resistors. Bulb A is brighter than bulb B, which in turn is brighter than bulb C. The bulbs are identical. Use this information to explain how the boxes are ranked according to their resistance. Now, these same boxes are connected as shown in Arrangement 2 and Arrangement 3. For each arrangement, rank the brightness of the bulbs. Explain your reasoning.

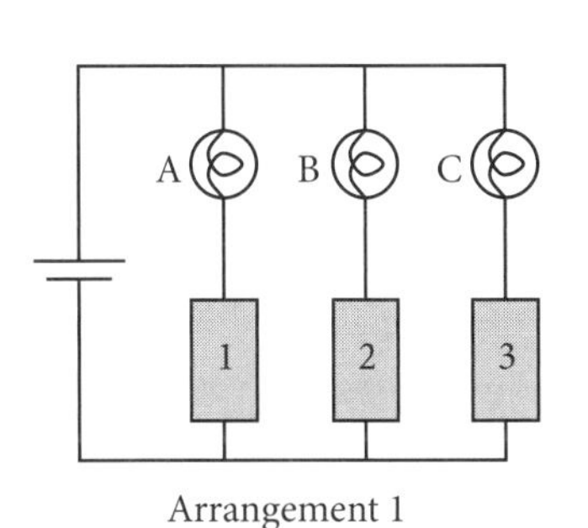

Arrangement 1

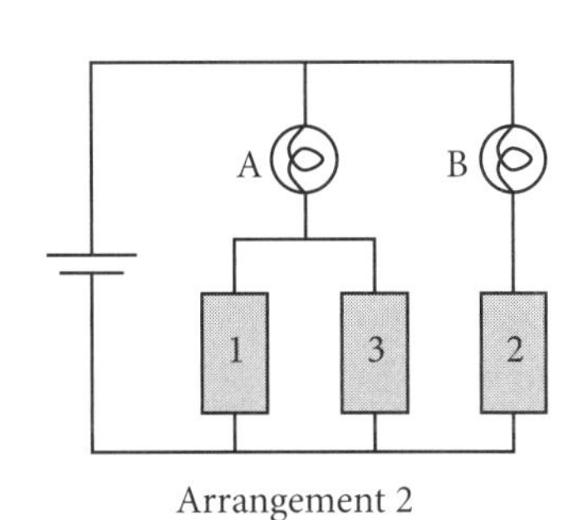

Arrangement 2

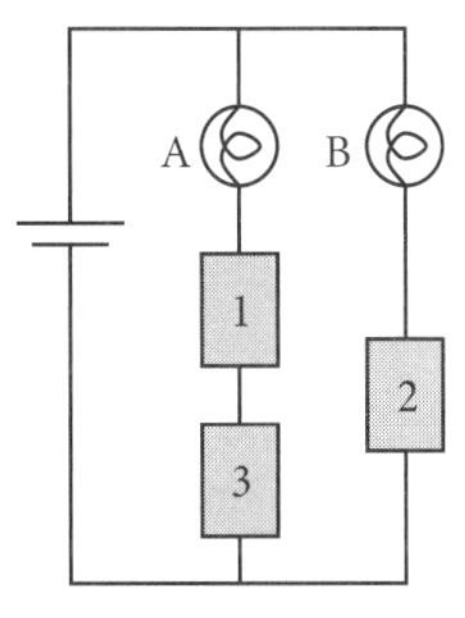

Arrangement 3

** 26. In the circuit shown at right, the switch is initially open. When the switch is closed, does the voltage drop across bulb A increase, decrease, or stay the same? Explain your answer.

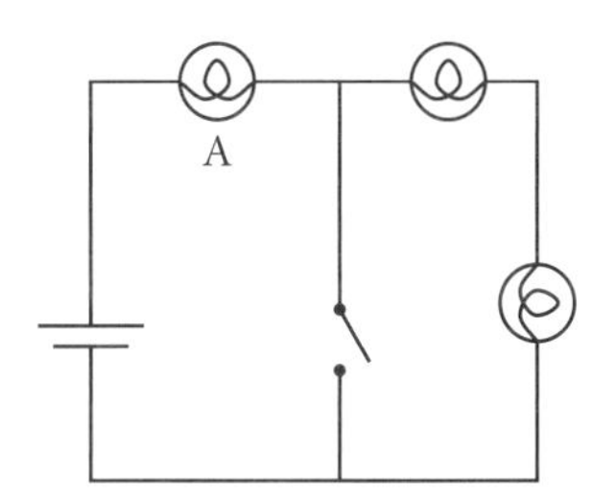

** 27. Compare the relative readings on the ideal ammeters in each of the four circuits (a–d). The light bulbs and the batteries are identical. Explain.

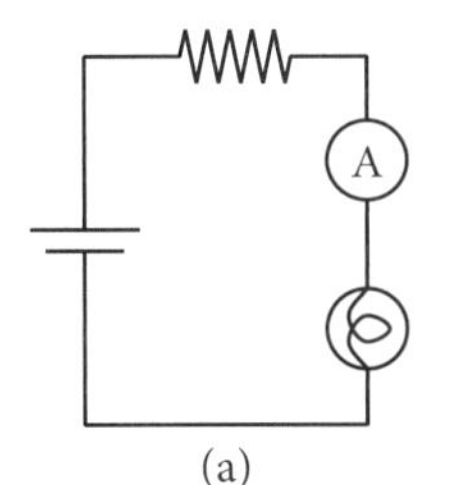

(a)

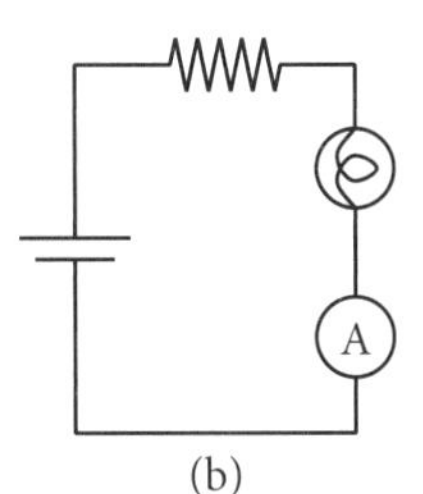

(b)

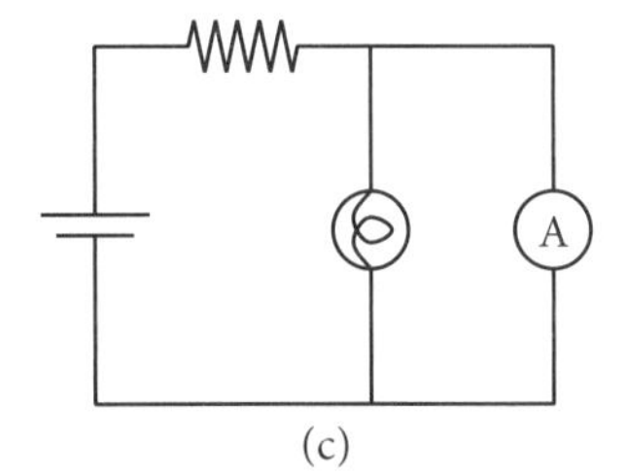

(c)

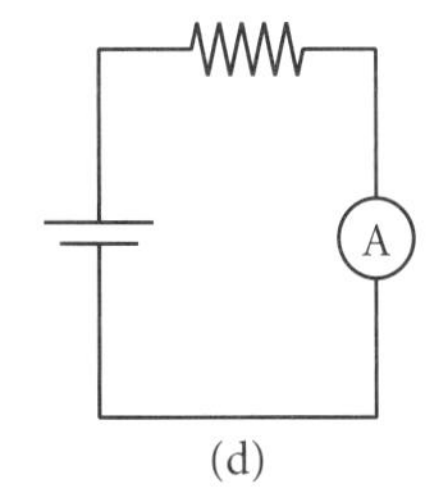

(d)

** 28. The bulbs in the circuit shown at right are identical. Treat the battery as ideal.

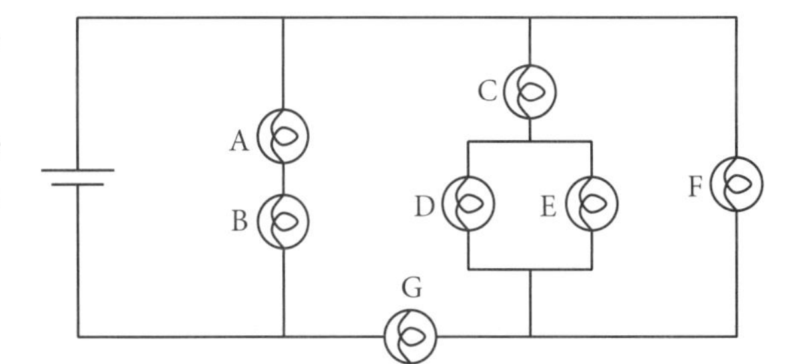

a) For each pair of bulbs listed below, indicate which bulb will be brighter. If the two have the same brightness, state that explicitly. In each case, explain your reasoning.

- A and G. (Which is brighter and why?)
- A and B.
- C and F.
- G and F.
- B and F.

b) How will the brightness of the following bulbs change if we unscrew bulb C (and leave the empty socket in the circuit)?

- Will bulb B become brighter, dimmer, or stay the same? Explain.
- Will bulb G become brighter, dimmer, or stay the same? Explain.
- Will bulb F become brighter, dimmer, or stay the same? Explain.

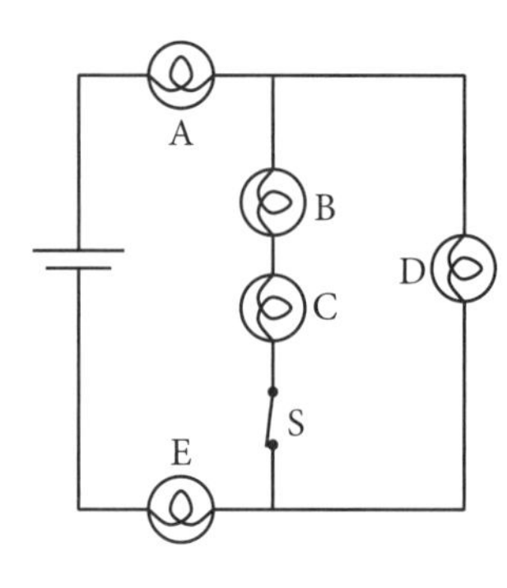

** 29. The bulbs in the circuit shown are identical. Switch S is initially closed.

a) Rank the bulbs in order of brightness from dimmest to brightest. Indicate which bulbs are the same brightness.

b) Give a specific reason for your choice of brightest bulb. The switch S is then opened.

c) Does the brightness of bulb A increase, decrease, or stay the same? Explain your answer.

d) Does the voltage drop across bulb A increase, decrease, or stay the same? Explain your answer.

e) Does the brightness of bulb D increase, decrease, or stay the same? Explain your answer.

18.7 ** 30. A resistor draws 2 A when connected across the terminals of a 12-V battery. What additional single resistor could you add to the circuit to increase the current through the battery to 8 A? Should you add this resistor in series or in parallel?

** 31. What different values of resistance can you get by combining three 6-Ω resistors? For each value, draw a diagram showing how the resistors are connected.

** 32. Four 10-Ω resistors are connected in series and in parallel. What is the equivalent resistance of each combination?

** 33. For each arrangement of resistors, find the equivalent resistance. In other words, what single resistor would offer the same obstacle to the flow of current?

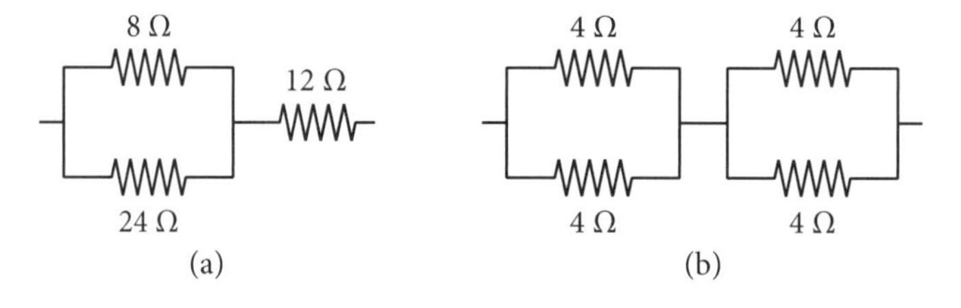

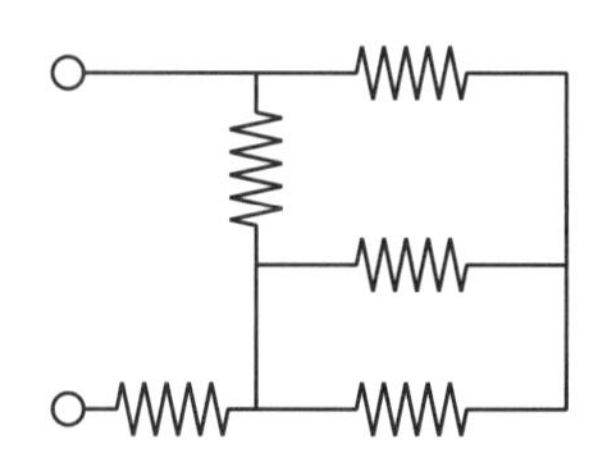

** 34. Each of the resistors shown in the circuit diagram at left is a 6-Ω resistor. Find the equivalent resistance of the network. [Hint: You might want to redraw the circuit in a way that makes the parallel and series combinations clearer.]

** 35. In the circuit shown below,

a) find the equivalent resistance of the network of resistors;

b) find the current through the battery;

c) find the current through the 2-Ω resistor. Explain your reasoning.

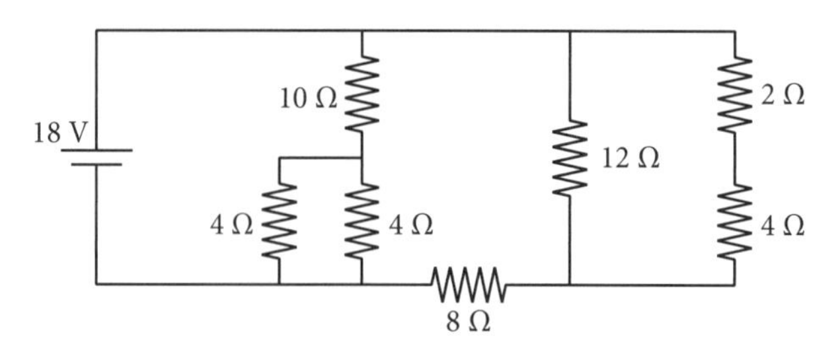

The next three problems involve non-simple circuits that may not be solved without the full machinery of Kirchhoff's laws. They may or may not require the algebra of solving simultaneous equations.

18.8 *** 36. As we mentioned on page 364, real batteries experience a voltage drop across their terminals when a current is required of them that they cannot supply. This effect may be modeled by assigning an "internal resistance" r to each battery, so that the battery voltage between the battery terminals of a non-ideal battery is the voltage difference across the ideal battery minus the voltage loss across an internal resistance that acts like a resistor connected in series. In the circuit diagrammed below, two non-ideal batteries, with internal resistances explicitly shown in the circuit diagram, are connected to three other resistors:

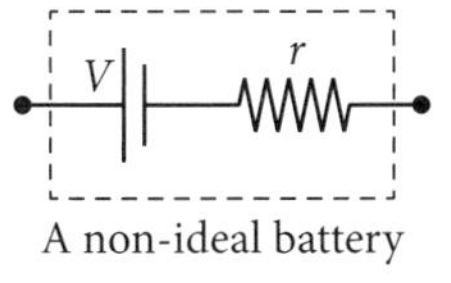

A non-ideal battery

a) Find the current supplied by each battery.

b) Find the voltage difference across the 8-Ω resistor in the middle branch.

c) What is the direction of the current through the 4-V battery? Explain the consequences of this answer.

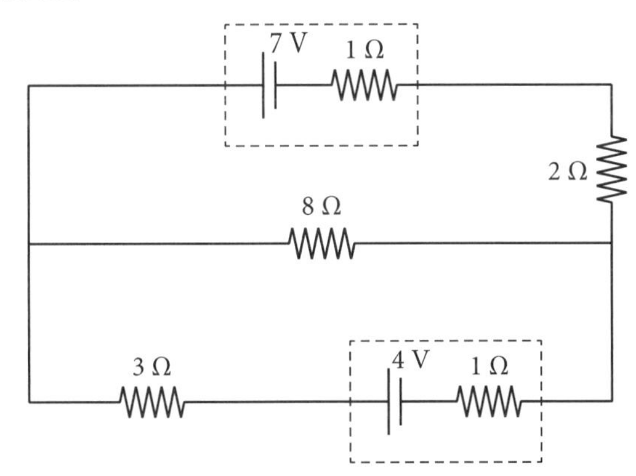

*** 37. Even though the circuit diagrammed below is composed of a single battery and five resistors, the resistors are connected in a non-simple configuration, in which it is not possible to find an equivalent resistance by using the rules of parallel and series resistors.

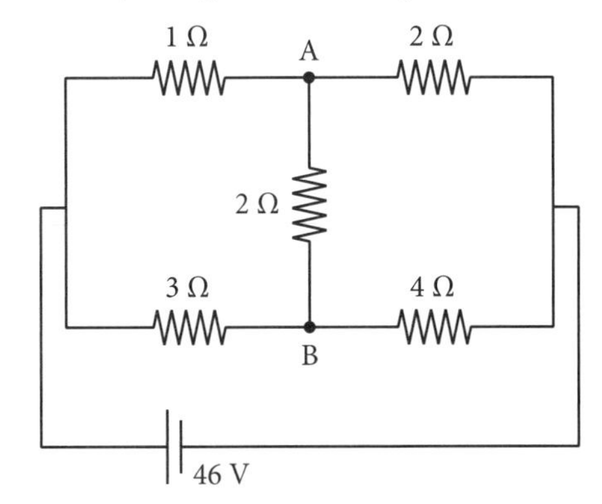

a) Explain why, in applying Kirchhoff's laws to this circuit, each resistor will need to have its own current defined. In other words, explain why no two of the resistors in this circuit can be assumed to have the same current flowing through them.

b) For the five unknown currents in this circuit (one through each resistor), apply Kirchhoff's laws to find five simultaneous independent equations.

c) Solve the equations found in part (b) to find the current through the 2-Ω resistor that sits in the branch connecting points A and B.

*** 38. The circuit diagrammed below has a break in the middle branch. [Carefully note the orientation of the batteries.] Find the voltage difference between points A and B and tell which point is at the higher voltage.

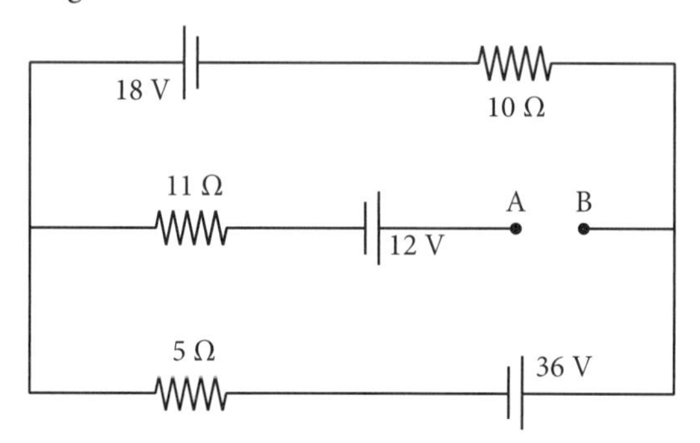

18.9 ** 39. Four bulbs and a capacitor are connected to an ideal battery as shown. The capacitor is initially uncharged.

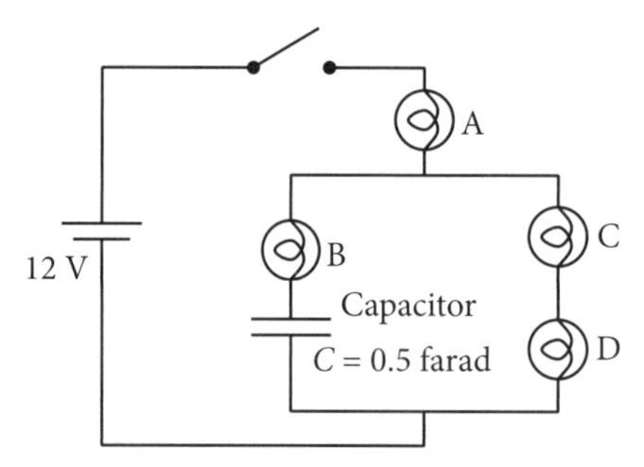

Immediately after the switch is closed:

a) What is the potential difference across the capacitor? Explain.

b) Is bulb B brighter than, equal to, or dimmer than bulb C? Explain your reasoning.

A very long time after the switch is closed (when the capacitor is essentially fully charged):

c) Is bulb B brighter than, equal to, or dimmer than bulb C? Explain your reasoning.

d) Find and explain the potential difference across: bulb A, bulb B, and the capacitor.

e) Find the charge stored on the capacitor.

19

MAGNETIC FORCES AND FIELDS

19.1 Magnetic Poles

Most of us have grown up knowing that magnets come with two ends, one labeled "north" and one labeled "south." We have also probably fooled around enough with magnets to know that ends labeled differently attract one another and that ends labeled the same repel. Further, we probably know that if we pivot a magnet about its center (and keep it away from other magnets), the magnet will eventually line up with its "north" end (sometimes painted red or sometimes labeled "N") pointing roughly toward geographic north. We call this the "north" pole of the magnet because it points north, though, to be precise, we should probably call it the "north-seeking" pole of the magnet.

Let us call on one more thing you may have already figured out about magnets. If you think back, you will realize that you have never held a magnet that had just one kind of pole in it. Every magnet, no matter how short, has a north and a south end. In fact, some of us (all right, it was Ron) have even tried breaking a magnet in two, only to discover that we were left with two small magnets, each with its own north end and south end.

19.2 Magnetic Fields

Just as the concept of the electric field was useful for understanding electric interactions, the concept of the magnetic field proves to be invaluable in making sense of magnetic phenomena. As you remember, we defined the electric field as the ratio of the force on a charge q to the charge itself: $\vec{E} = \vec{F}_E/q$. Unfortunately, the definition of the magnetic field is more complex. Let us begin by just defining the direction of the field. *The direction of the magnetic field at any point in space is given by the direction that a compass needle would point at that location.* For each point in space we associate a vector, called the magnetic field, whose direction is the direction a compass needle points. At this stage, we won't worry about the magnitude of the vector, although you might correctly assume that it will be greater when created by a stronger magnet or when the point in question is closer to the magnet. The symbol for the magnetic field vector is $\vec{B}$, so we often call the magnetic field the "B-field."

In a manner similar to the way that electric field lines are used to represent electric fields in a region of space, we define a method of determining magnetic field lines created by a magnet. A magnetic field line is the line that you would trace out if you were to hold a compass needle and just keep walking in the direction that the needle was pointing. At any point on the line, the B-field vector is tangent to the line. In addition, though we have given you no reason that you should expect this behavior, the density of magnetic field lines created this way turns out to be proportional to the magnitude of the magnetic field, just as we saw for electric fields.

Figure 19.1 shows how three compasses would be oriented around a bar magnet. It also shows B-field vectors at a few additional points, as well as several magnetic field lines. Note that the north ends of the compasses (the arrowheads) are attracted to the south end of the bar magnet and repelled by the north end of the magnet. This tells us that *magnetic field lines come out of north poles and go into south poles.*

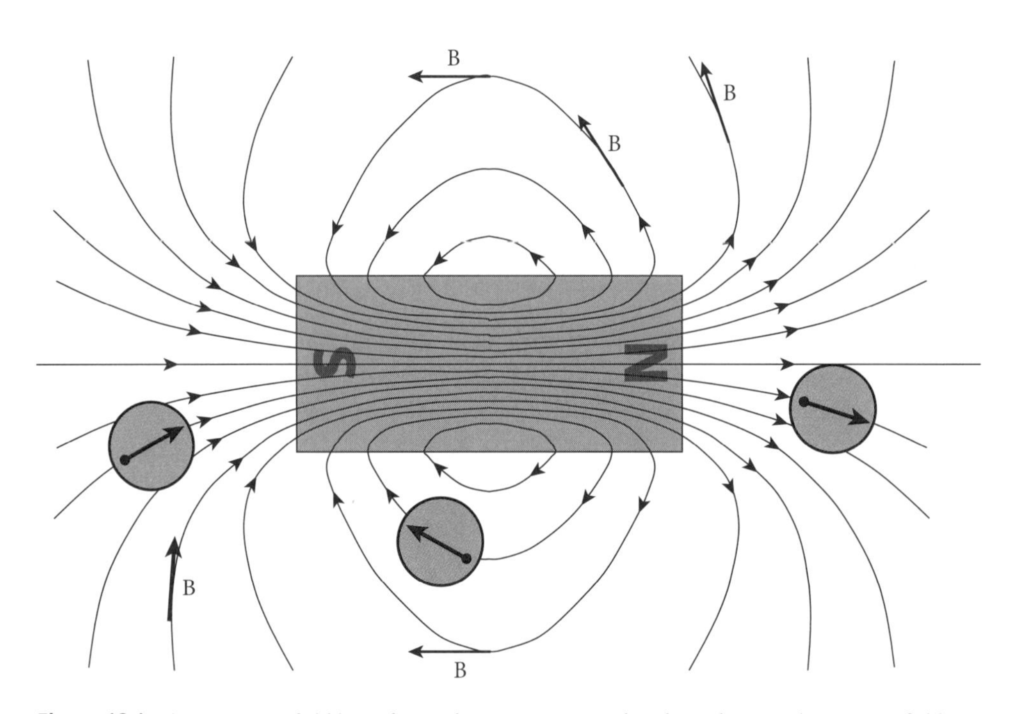

Figure 19.1 The magnetic field lines from a bar magnet. We also show the actual magnetic field vectors at five points and compass needle orientations at three additional points.

As shown in Figure 19.1, the north poles of compasses point toward the south pole of a bar magnet. But, as we all know, the north ends of compasses also point toward the earth's geographic North Pole. What this tells us is that, if the earth is thought of as a giant magnet, it is the *south* pole of that magnet that is near the earth's north geographic pole. Figure 19.2 shows the earth's magnetic field lines, along with an imaginary bar magnet that mimics the earth's magnetic field. The magnetic poles do not align with the geographic poles (the earth's spin axis). The earth's magnetic field lines emerge from Antarctica, go around the earth from south to north, and cross the earth's surface again in northern Greenland. What's more, because the hand-held compasses we use to find north are always held level, parallel to earth's surface, these compasses are really only detecting the horizontal component of earth's magnetic field. In fact, if a compass needle is well-balanced on its pivot, it can be held vertically instead of horizontally, and we will find that the compass needle will point downward by some angle in the Northern Hemisphere and upward in the Southern Hemisphere. In Montana, the earth's magnetic field points about 70° below the horizontal.

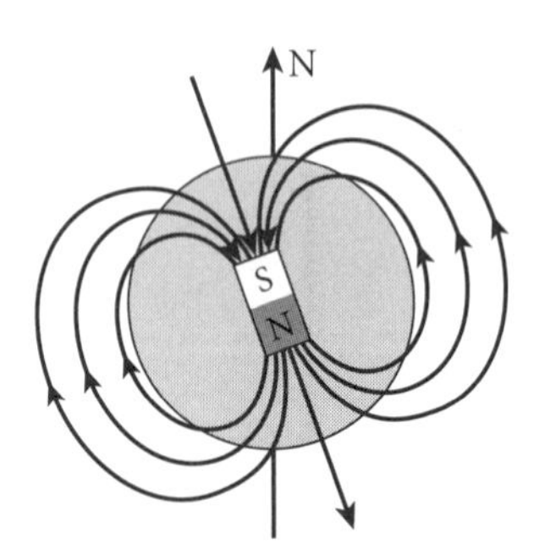

Figure 19.2 Sketch of Earth's magnetic field lines. The pole near the geographic North Pole is a south magnetic pole.

19.3 Magnetic Force on a Charged Particle

If we were to rub a glass rod with a piece of silk and move it near the north end of a magnet, there would be no force.[1] Historically, magnetic and electric forces were long known to be completely independent of each other. So the nearly simultaneous experiments of Hans Christian Œrsted and Andre-Marie Ampère in 1820 signaled a revolution in physics. Œrsted showed that a moving electric charge actually created a magnetic field and Ampère showed that a magnetic field could impose a force on an electric charge after all. But here's the catch for Ampère's result; there is no force on the charge if it is stationary, only if it is moving. In this section, we discuss the force produced by a magnetic field on a moving electrical charge.

When we introduced electric fields, back in Section 16.4, there seemed to be little distinction between electrical fields and electric forces. They either pointed in the same direction or, in the case of a negative charge, in opposite directions. However, in the case of magnetic fields, there is no missing the distinction between the magnetic field and the magnetic force it produces on a moving electrical charge. This is because *magnetic forces on charged particles and the magnetic fields that cause the forces never point in the same direction.* Through a series of experiments it is possible to establish the following properties of the magnetic force that a charged particle experiences in the presence of a magnetic field:

1. To experience a force, the particle must be moving. A stationary charged particle feels no force no matter how large the B-field is.
2. The charged particle feels no force if it is moving in the same direction or in the opposite direction as the direction of the magnetic field. In fact . . .
3. The size of the force on the charged particle is proportional to the component of velocity that is perpendicular to the magnetic field. This can be thought of as a measure of how quickly the particle is cutting *across* B-field lines.
4. The size of the force increases in proportion to the charge of the particle and a particle with a negative charge experiences a force in the opposite direction as a particle with a positive charge.
5. The direction of the force is always perpendicular to the direction of the magnetic field vector and perpendicular to the direction of the velocity vector.

We will begin with how to find the *direction* of the magnetic force. Point number 5 tells us that the directions of the B-field vector and the velocity vector narrow the choice for the direction of the force to only two possibilities. This is because two non-coincident vectors define a plane in space (that is, there is only one plane that contains both vectors), and if the force vector is perpendicular to both vectors, it must be perpendicular to this plane. There are only two vectors that satisfy this requirement, and they point in opposite directions. Of the two possibilities, the direction of the force is found from the $\vec{v}$ and $\vec{B}$ vectors by using a right-hand rule. (We will call this Right-Hand Rule #1 to discriminate it from two other right-hand rules we will introduce a little later in this chapter.) By the way, although your right and left hands look identical, they are actually different in one way. If a right hand were placed on the

1. If the magnet were made of iron, or some other conducting material, there would, of course, be a weak induced electrical force, as we discussed in Section 16.3, but this would be electrical force between electrical charges in both objects, not magnetic force on the charge in the rod.

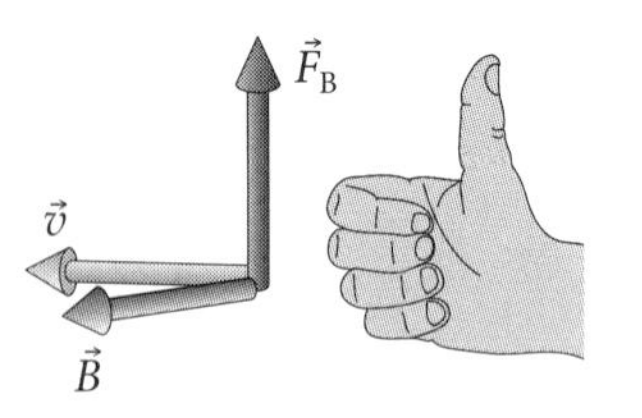

Figure 19.3 Right-Hand Rule #1. Start with your fingers pointing along $\vec{v}$. Then curl them toward $\vec{B}$. Your thumb points in the direction of $\vec{F}_B$.

left arm of a department store mannequin by mistake, you could easily spot the error. It is this breaking of the symmetry between right and left that allows us to discriminate between the two possible directions of the force on a charged particle moving in a magnetic field.

To apply Right-Hand Rule #1 (see Figure 19.3), you begin by clearly identifying the directions of the $\vec{v}$ and $\vec{B}$ vectors. Then point the fingers of your right hand in the direction of the $\vec{v}$ vector and rotate your arm until your palm is facing roughly in the direction of the $\vec{B}$ vector (so $\vec{B}$ seems to be coming more or less out of your palm). Now, curl your fingers toward the $\vec{B}$ vector and extend your thumb. Your thumb is pointing in the direction of the force on a positively charged particle. For a negatively charged particle, you start by finding the direction of the force as if the charge were positive and then just reverse the direction. (Actually, this is equivalent to using a left-hand rule for negative particles, so you can also do that if you want to.)

EXAMPLE 19.1 Directions of the Magnetic Field and the Magnetic Force

A proton is moving horizontally, traveling due north in a region of the laboratory containing a uniform electric field and a uniform magnetic field. The electric field is directed east. If the proton is seen to have a constant velocity, what must be the direction of the magnetic field? Assume that the magnetic field has no northward component (which would have no effect anyway, since it would be along the velocity) and ignore the effects of gravity and the earth's magnetic field.

ANSWER To travel at constant velocity, the proton must experience zero net force. The electric force on a positive charge is directed east, the same direction as the electric field. This means that the magnetic *force* must be directed west. But note that the question asks for the direction of the *field*, not the force. The proton is heading north, but the problem says the B-field has no north component, so it will be perpendicular to the velocity vector. We also know that the B-field is always perpendicular to the force (which must be west). So the only choices are straight up and straight down. Let's guess that it is directed up and see what happens. If you start with your fingers pointing north with your palm up, curling your fingers up shows that the force on a positive particle is east (oops). So the B-field must point down, as you can then confirm.

One of the things that this example demonstrates is that we must use a three-dimensional coordinate system when considering magnetic forces and fields. The two systems we will generally use are a cardinal earth-referenced system and a book page-referenced system. In the cardinal system, we use the four standard compass directions (north, south, east, and west, all parallel to the ground), plus up (toward the sky) and down (into the ground). In the page-referenced system, directions are given relative to the paper page. The directions we then have are right, left, top of the page, bottom of the page, into the page, and out of the page. It is important, when starting any problem, to begin by making sure you are clear on your choice of a coordinate system.

Now let us address the question of the magnitude of the magnetic force on a charged particle moving in a B-field. Figure 19.4 shows a positively charged particle moving at speed v, at an angle θ relative to the direction of the magnetic field. Only the component of the velocity perpendicular to $\vec{B}$ matters, so we divide the velocity vector into two components. The compo-

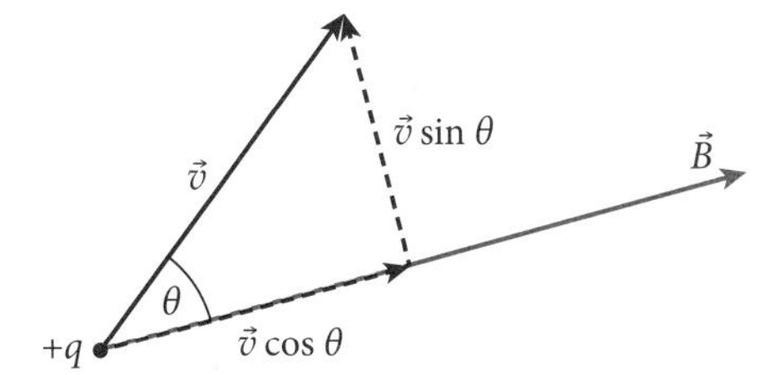

Figure 19.4 The magnitude of the magnetic force is the product of the component of the velocity perpendicular to the magnetic field, as shown, and the magnitude of the B-field.

nent parallel to $\vec{B}$ is $v \cos\theta$ and the one perpendicular to $\vec{B}$ is $v \sin\theta$. Note that the θ you choose must always be the smallest angle from $\vec{v}$ to $\vec{B}$. The absolute value of $\sin\theta$ will not change if you use the bigger angle, but the sign will.

The magnitude of the magnetic force, F_B, is proportional to the charge on the object (q), the perpendicular component of the velocity ($v \sin\theta$), and the strength of the field (B). Because we haven't actually defined the units for field strength yet, we are free to choose them so that the formula for the force has no additional numerical constant in it. The formula is therefore

$$F_B = qvB \sin\theta. \tag{19.1}$$

Note that by choosing this formula, we have effectively defined the basic unit of the magnetic field. One SI unit of B-field is the field required to create a 1-N force on a 1-C charge moving at a speed of 1 m/s in a direction perpendicular ($\sin 90° = 1$) to the field. We call this unit the tesla and give it the symbol T. One tesla is actually a very strong magnetic field. The earth's field is about 10^{-4} T, and the high magnetic field inside an MRI medical imager is still only about 2 to 3 T.

EXAMPLE 19.2 The Magnetic Force on a Moving Electron

An electron is moving at 5×10^6 m/s straight north and parallel to the ground. The earth's magnetic field at that location is 5×10^{-5} T. The field's horizontal component is directed due north, but the total vector dips sharply down so that it is 75° below the horizontal (i.e., it is pointing nearly straight down). What is the force (magnitude and direction) on the electron?

ANSWER We know the particle's speed and charge, as well as the size of the magnetic field. We thus have all we need to use Equation 19.1 but the angle between the $\vec{v}$ and $\vec{B}$ vectors. The B-field points north, but it is 75° below the horizontal. The velocity is also north, and horizontal, so the angle between them is 75°. Putting this all into the equation we find

$$F_B = qvB \sin\theta = (1.6 \times 10^{-19}\ \text{C})(5 \times 10^6\ \text{m/s})(5 \times 10^{-5}\ \text{T})\sin 75° = 3.9 \times 10^{-17}\ \text{N}.$$

Note that we have not included a minus sign for the charge on the electron. As usual, we prefer to get the magnitudes first and work out the directions later.

By the way, the force we have found may seem small, but the thing that makes it so small is the tiny charge on the electron. Recall, however, that the electron's mass is 9.11×10^{-31} kg, so that the acceleration is actually a huge 4.3×10^{14} m/s^2.

Since we're doing "by-the-ways" here, anyone worried that an acceleration this large might quickly produce a speed that exceeds the speed of light should note that the magnetic force is always perpendicular to the velocity. As a result, the magnetic field cannot speed a particle up or slow it down; it can only change its direction. It is always a steering-wheel acceleration, not an accelerator-pedal acceleration.

Finally, to finish the problem, we remember that force is a vector, and so we still need to specify the direction of $\vec{F}_B$. We place the fingers of our right hand pointing north along the velocity vector and then rotate our arm until our palm points down (roughly along the B-field vector). When we bend our fingers down, our thumb points west. This would be the direction of the force on a positive particle. But because we are dealing with a negatively charged electron, the magnetic force must be directed east.

19.4 Magnetic Force on a Current-Carrying Wire

We have already seen that a single charged particle can experience a force when moving in a magnetic field. If we think of a current-carrying wire as nothing more than a long line of charged particles, all moving in the same direction, then it seems reasonable that the wire itself should feel a force, and indeed it does. The direction of the force on a current-carrying wire is found in exactly the same way as for a positively charged particle, except that now you begin by pointing your fingers in the direction of the current rather than the velocity.

Now, at this point, you may be thinking, "Wait a minute. I know that positive charge doesn't really move in a wire. It's really electrons flowing the other direction." And you would be correct. The good news is that if you find the force on those electrons (remembering to reverse the result from Right-Hand Rule #1 because they are negative particles), you get exactly the same result for the force on the wire that you get by assuming motion of positive charges. In fact, finding the force on a current-carrying wire was just one more in a list of experiments that failed to tell physicists whether it was positive or negative charge that flowed when there was a current in a wire (but see Problem 20).

So let us go on to find the force on a current-carrying wire in the presence of a magnetic field B. Equation 19.1 gave the magnetic force on a single moving charge. Since a wire contains many moving charges, the total force on a wire will depend on how much total charge is moving and at what speed. These quantities would be hard to figure out individually, but, fortunately, the formula for the force (Equation 19.1) depends only on the *product* of q and v, and this is easier to find. Let us consider a segment of wire of length L that contains a total charge q carried by its charge carriers, as in Figure 19.5. If we watch the charge carriers moving along the wire and see that it takes a time t for the charge that started at the far left of the segment to cross the entire segment and exit to the right, then we know that the average speed of the charge carriers along the wire is $v = L/t$. For a total charge q moving at average speed v along the wire, the force on the wire will be $F_B = qvB \sin \theta$, where θ is the angle between the direction of the current and the B-field. However, since $v = L/t$, this may also be written $F_B = q(L/t)B \sin \theta$. If we now recognize that the quantity q/t is the current flowing along the wire, we find the formula for the force on a segment of wire of length L, carrying a current I, in the presence of a magnetic field B that points at an angle θ to the direction of the current. The formula is

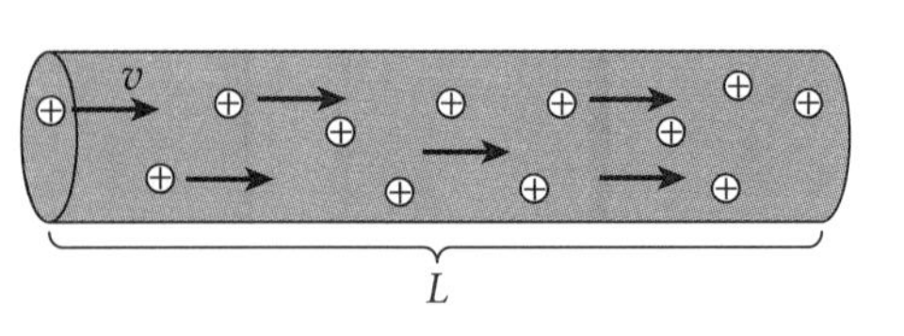

Figure 19.5 Positive charge-carriers in a segment of current-carrying wire of length L.

$$F_B = ILB \sin \theta. \tag{19.2}$$

Let us do a quick example that shows how this formula is used.

EXAMPLE 19.3 Magnetic Force on a Current-Carrying Wire

A straight wire is held in the 2 T magnetic field between the poles of a strong permanent magnet, with the wire held at an angle of 20 degrees away from being perpendicular to the magnetic field lines, as shown at right. If a 1.5-A current flows along the wire in the direction indicated by the arrow, what will be the magnetic force on a 10-cm segment of the wire?

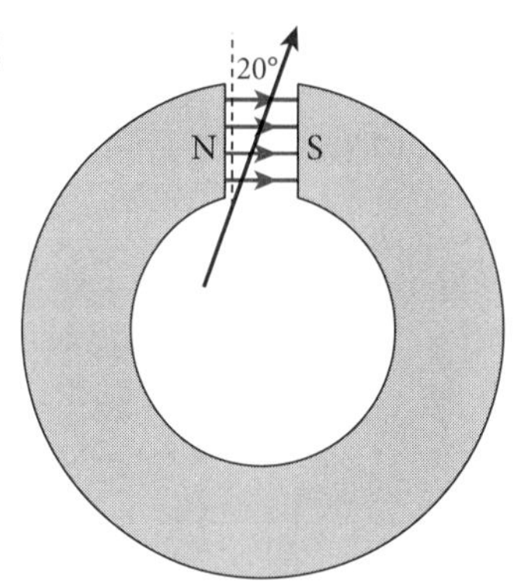

ANSWER The magnitude may be found by a straightforward application of Equation 19.2. The angle between the direction of the current in the wire and the magnetic field direction is not the 20° shown in the picture, but the 70° angle between the solid black arrow representing the current in the wire and the lines representing the magnetic field. The magnitude of the force is thus

$$F_B = ILB \sin \theta = (1.5 \text{ A})(0.1 \text{ m})(2 \text{ T}) \sin(70°) = 0.28 \text{ N}.$$

To find the direction of the force, we curl the fingers of our right hand along the 70° angle from the wire to the B-field. The thumb of the right hand points down, into the page, so this is the direction of the force on a conventional positive current.

19.5 Magnetic Field Due to a Wire

In the beginning of Section 19.3, we mentioned Hans Christian Oersted's contribution to the development of the theory of magnetism. What Oersted actually discovered was that a compass needle is deflected when it is held near a current-carrying wire. This is evidence that a current-carrying wire not only feels a force when it is in a magnetic field, but that it also generates a magnetic field.

The magnetic field a perpendicular distance r from a current-carrying wire is depicted in Figure 19.6a. The gray arrows are the magnetic field vectors. Note that, at points that are the same distance from the wire, the magnitudes of the field are the same. Note also that the direction of the field is always tangent to circles that lie in planes perpendicular to the wire and centered on the wire. Figure 19.6b shows how the magnitude of the field changes with distance from the wire. If we go out to a point a distance $2r$ from the wire, then the magnitude of the field drops to half what it was before, while the directions are still tangent to a circle centered on the wire.

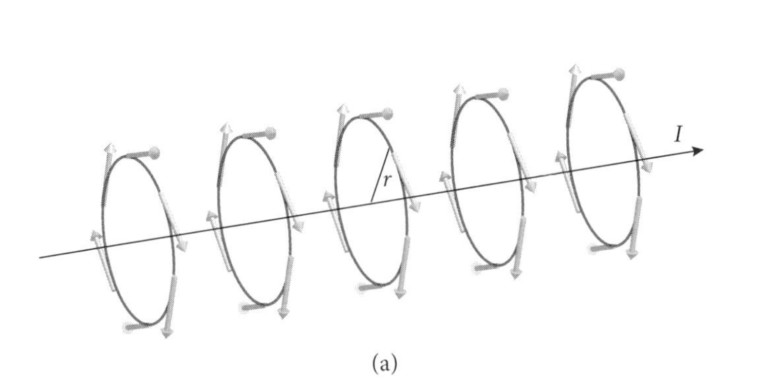

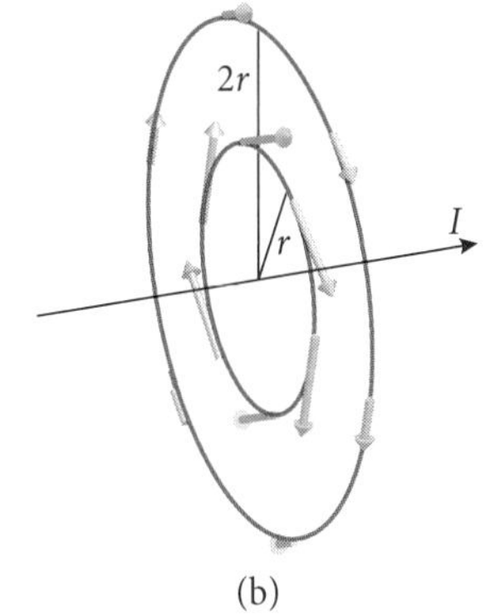

Figure 19.6 The magnetic field (gray vectors) around wire carrying a current I. Part (a) shows that the field has the same magnitude at equal distances r from the wire and that it points everywhere tangent to circles centered on the wire. Part (b) shows that the field at a distance $2r$ is half that at a distance r.

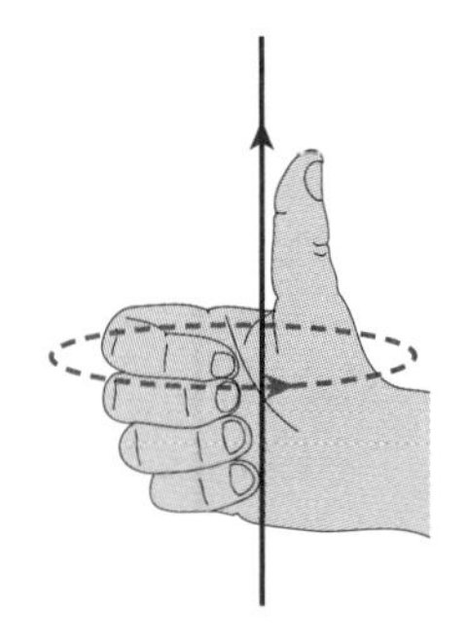

Figure 19.7 Right-Hand Rule #2.

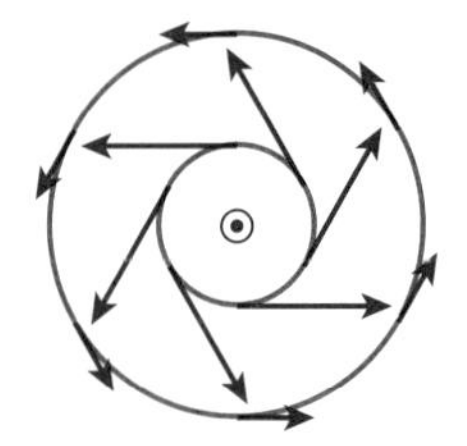

Figure 19.8 The B-field near a wire carrying current out of the page.

Next, let us think about the *magnetic field lines* around a wire. As explained on page 385, magnetic field lines are generated by starting with the magnetic field at one point and then going forward, following the direction of the field, to the next point. A moment's thought will make it clear that the magnetic field lines that we generate by this procedure are the same red circles that we drew in Figure 19.6. *The magnetic field lines around a wire form closed circles, with the wire at the center.*

Knowing that the magnetic field lines around a current-carrying wire are circles is not quite enough, of course. We also need to know which direction the field lines are going. This we find using Right-Hand Rule #2 (Figure 19.7). We imagine holding the wire with our right hand, with our thumb pointing in the direction of the current. The direction that our fingers are curled around the wire indicates the direction of the magnetic field lines. You should check that Figure 19.6 is consistent with this rule.

Before we go on to do an example, let us just say that we know it is difficult to draw two-dimensional diagrams for problems involving magnetic fields, because they are inherently three-dimensional. So we will often use a convention in which vectors pointing out of a page are circles with dots in their centers (the pointy end of an arrow, or ⊙) and vectors pointing into the page are circles with crosses in their centers (a view of the feathers on an arrow as it moves away, or ⊗). If we diagram the magnetic field of Figure 19.6b from a perspective looking down the wire as the current comes out of the page, we will get Figure 19.8. Note how the direction for a current out of the page is represented as the point of an arrow—a circle with a dot in the center.

EXAMPLE 19.4 The B-Field Direction from a Current-Carrying Wire

An experimenter fires a proton horizontally due east in a laboratory in which there is a horizontal rod that carries a current due north, as shown in the bird's-eye view at right. In what direction (north, east, south, west, up, or down) is the force on the proton?

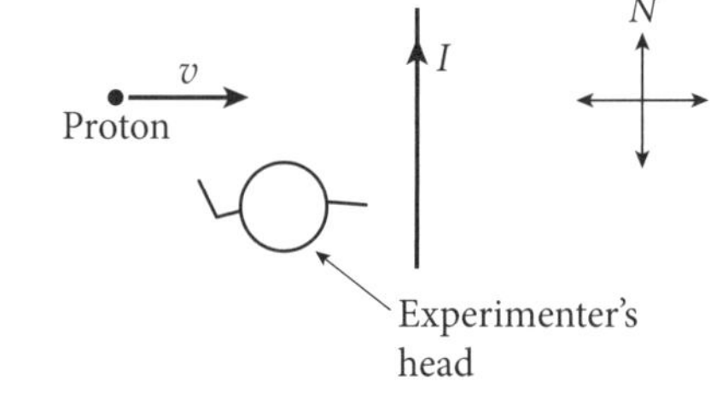

ANSWER The moving proton experiences a force because there is a magnetic field at its location, the field being due to the current-carrying rod. The first step is to find the direction of the B-field that the current-carrying rod creates at the proton's location. This is easier to see if we redraw the figure, imagining that we are standing near the south wall of the lab, looking horizontally straight north, as shown. Note how the current direction is represented by the feathers of an arrow going away from us, the × inside the little circle. We then draw in the magnetic field line that passes through the proton's position; this is a circle centered on the rod. If we imagine holding the rod, with our thumb pointing in the direction of the current, our fingers indicate that the direction of the field line is clockwise around the rod. At the proton's location, the B-field vector is tangent to the circle, pointing up (i.e., toward the sky). From here on, we

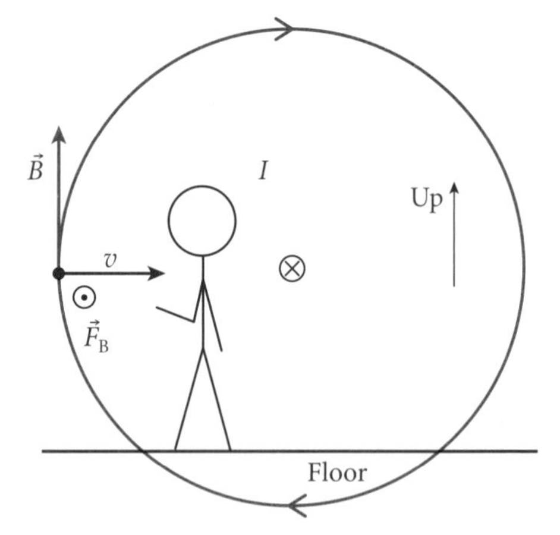

can ignore the rod, and our problem is reduced to finding the force on a proton in a known magnetic field. For this, we use Right-Hand Rule #1. A right hand with its fingers pointing east, the palm upward in the direction of the B-field, has its thumb pointing horizontally toward the south. In the picture above, where we were looking horizontally north, the direction of the force is therefore out of the page (the ⊙).

In the original bird's-eye drawing, which we reproduce at right, we see the B-field coming upward toward us, out of the page, and the force is to the south.

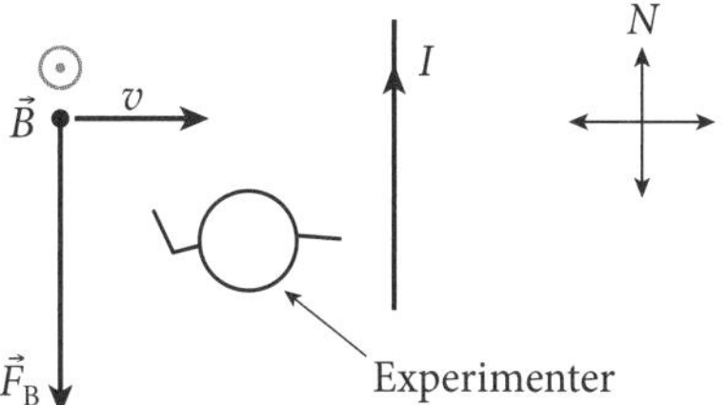

You may want to review the last example again to make sure that all of the directions are clear to you, and that everything makes sense when viewed horizontally from the south and in the bird's-eye view, seen from above, as in the last picture.

Note that, in this last example, we have worked out the force created by a charge in motion (the current in the rod) on another charge in motion (the moving proton). This is what the magnetic force represents. We remember that an *electric charge* creates an *electric field* that produces a force on another nearby *electric charge*. But, when an *electric charge is moving*, there is an additional force on *any other moving charge* nearby, due to the *magnetic field*.

Finally, as we finish this section, we need to be clear that we realize we have not yet told you how to calculate the *magnitude* of the B-field around a current-carrying wire; we have only worried about the direction. We will fill in this gap in Section 19.7. Before we do that, however, we want to investigate the magnetic field produced by a different kind of current-carrying wire.

19.6 Loops and Solenoids

In Figure 19.9, we show a wire loop carrying a current counterclockwise. We divide the loop into tiny segments and see that the magnetic field produced by each segment will point in a direction determined by the right-hand rule for that segment. Inside the loop, the contributions to the field from all segments around the loop point out of the page, as shown in the left-hand picture of Figure 19.9. They will add together to produce a strong field out of the page. Outside the loop, the contributions of some segments will be into the page while contributions from other segments will be out of the page, producing a weaker net field, as shown in the right-hand picture in Figure 19.9. (You might want to use your right hand and verify how this works for the segments shown in each picture.)

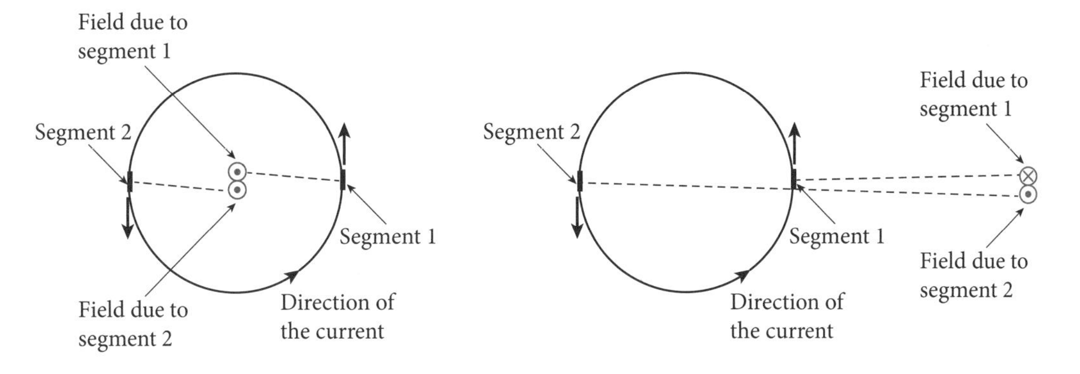

Figure 19.9 Inside a current loop, the contribution from each segment to the total magnetic field lies in the same direction as that from all other segments. Outside the loop, some segments create fields out of the page while others create fields into the page, essentially canceling each other out.

If we look at this wire loop from the side, we get Figure 19.10. In this picture, the current is going into the page at the bottom of the loop and out of the page at the top. Also shown are the magnetic field lines (in red). The addition of all the contributions from all the segments of the current loop produces a strong field pointing from left to right in the center of the loop, the strength being shown by the high density of field lines. The field is stronger at the center of the loop than it is near the wire, and it is *much* weaker outside the loop, for the reason explained in Figure 19.9.

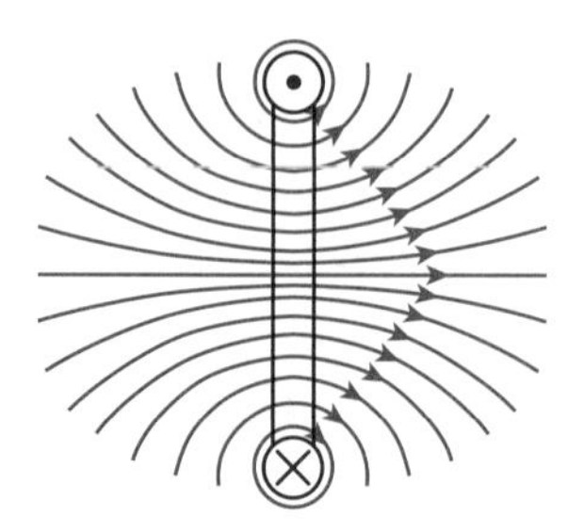

Figure 19.10 The field due to a single current-carrying loop.

Figure 19.10 may remind you of the field lines produced by the bar magnet in Figure 19.1. The directions of the field lines do look as if there were a magnet piercing the wire loop, left to right, with the north pole of the magnet on the right. There is, in fact, a quick way to find the effective north pole of a current-carrying loop. It is Right-Hand Rule #3. Imagine wrapping the fingers of your right hand around the loop, with your fingers pointing in the direction of the current flow. Your extended thumb indicates the direction of the north pole, as shown in Figure 19.11.

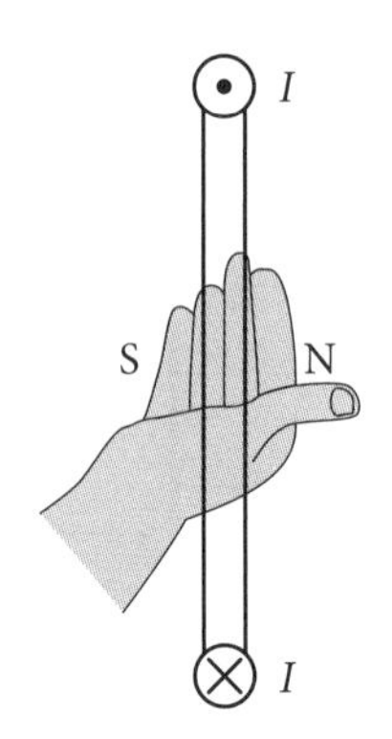

Figure 19.11 Right-Hand Rule #3.

EXAMPLE 19.5 Interacting Current-Carrying Loops

Shown at right is a side view of two current-carrying loops. The loop on the left is fixed in place while the one on the right is free to rotate about a horizontal axle. The current directions are shown using the usual convention of arrowheads and arrow tails. What direction, clockwise or counterclockwise, will the right loop pivot?

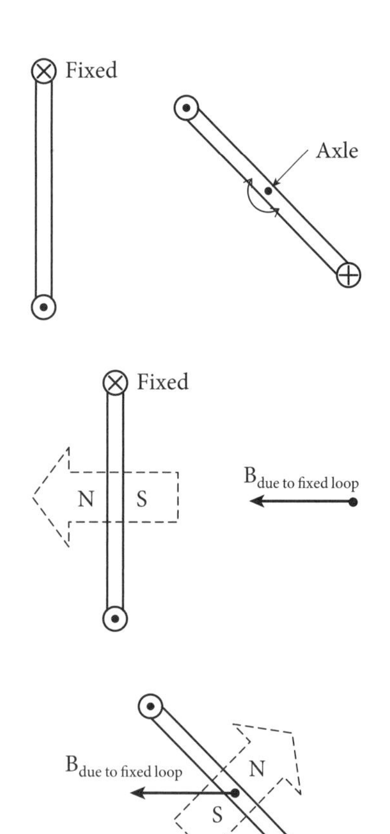

ANSWER Current-carrying loops act just like bar magnets in many ways, once one is able to establish the orientation of the magnetic poles. So let us think of the problem in this way. The left loop generates a magnetic field similar to that of a bar magnet. The right loop then acts like a bar magnet that is free to rotate in this magnetic field. If it is free to rotate, it will turn like a compass needle, aligning itself with the local magnetic field. The first step is to find the direction of the magnetic field at the location of the right loop, as a result of the left loop. Using Right-Hand Rule #3, we find the effective north pole of the fixed loop to be toward the left, as shown. This means that the B-field at the location of the rotating loop points to the left, as shown.

At this point, we can ignore the fixed loop because we know the magnetic field it creates at the location of the rotating loop. We now find the effective north pole of the rotating loop, again using Right-Hand Rule #3. The result is shown in the figure at right. The rotating loop will act just like a compass needle. It will rotate until its north pole points along the local external magnetic field direction. This means that the loop will rotate counterclockwise in the picture at right.

The magnetic field due to a single loop of current-carrying wire was shown in Figure 19.10. Now consider what the field would be like if we put several loops of wire side-by-side. Two things would happen. First, the field in the region in the middle of the loops would become stronger, as long as we arrange for the current to all go around in the same direction, so that the fields from each loop add together. Second, field lines going through the region inside the loops will have to go farther before they can spread out into the space outside the loops and eventually reconnect to themselves. A picture of such a configuration is shown in Figure 19.12a. A set of stacked loops whose currents enhance the field in the region between them in this way is called a *solenoid*. A solenoid is typically formed by wrapping a wire around a cylinder several times, so that the loops lie close together and the currents all circulate in the same direction.

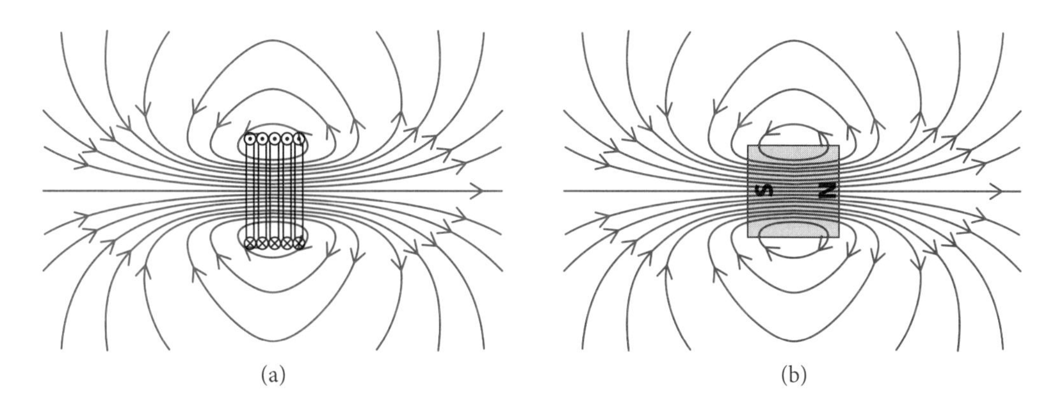

Figure 19.12 The figure on the left (a) shows the magnetic field lines in the vicinity of a current-carrying solenoid. Not shown is the connection to a battery or power supply that maintains the current in the loops. The density of field lines indicates the strength of the field. In the figure on the right (b), we show, for comparison, the magnetic field lines for a bar magnet (also called a *dipole magnet*).

In Figure 19.12b, we show the magnetic field lines from a bar magnet, allowing us to compare this field with the field from the solenoid. Because of the similarity between the field structure due to a current loop and that due to a bar magnet, with a north and south magnetic pole separated by some distance, this pattern of field lines is called a *dipole field*, regardless of its source.

If we compare the dipole field in Figure 19.12 with that from a single current loop in Figure 19.10, we see that the field from a single loop is like that of a bar magnet in which the length of the bar is nearly zero. We are also able to see a couple of important differences between the solenoid field and that from a single loop. First, because the solenoid's field lines have to go so far horizontally before they can spread out, the field lines in the region outside the solenoid will be much further apart, meaning that the field outside the solenoid will be much weaker. Second, the confinement of the field lines in the region inside a long solenoid means that the lines will be almost straight, parallel to each other and pointing in the direction given by Right-Hand Rule #3. This is in contrast to the field from a single loop, where the field quickly diverges as we go beyond the plane of the loop.

19.7 Calculating the Magnetic Field

We haven't mentioned it yet in this chapter, but we hope that you have noticed that the knowledge of how magnetic fields behave represents a piece of new physics, not derivable from anything we have presented before, and only discovered by experimental investigation of the way things actually are. Thus, when we say that the magnetic field lines around a current-carrying wire form closed circles around the wire, you should not expect to see *why* this is true, only to remember that it *is* true.

As we have explained, magnetic fields are created by electric charge in motion, and, as you may expect, there are laws of nature that predict how strong the magnetic field will be as the result of a particular configuration of moving charge. The most fundamental law is probably the one that gives the magnetic field at a field point a distance r away from a charge q that is moving at velocity v. This formula is a form of a law known as the *Biot–Savart Law* (pronounced bee-oh sav-arr′). It is

$$B = \frac{\mu_0}{4\pi}\frac{qv}{r^2}\sin\theta. \tag{19.3}$$

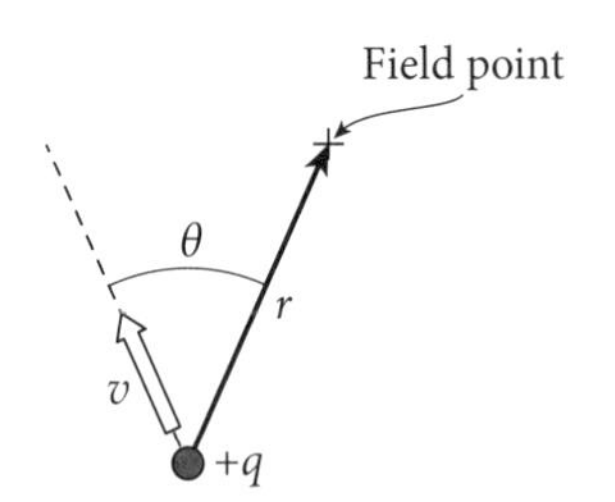

Figure 19.13 The Biot–Savart law gives a B-field into the page at the field point.

Here, θ is the angle between the charge's velocity vector and the vector that extends from the charge out to the point where we want to evaluate the field. The geometry of the situation is shown in Figure 19.13. The direction of the field is still given by Right-Hand Rule #2 with the thumb pointing along the current (the velocity vector). In Equation 19.3, we will want to measure all quantities in the units we have already defined (coulombs for charge, teslas for magnetic field, etc.). Therefore, the fact that nature, not some physicist, specifies how strong the field due to a particular moving charge will be means that we may expect a new fundamental constant to appear. This is what the μ_0 is doing in the equation. This constant is called the *permeability of free space* and its value is an even $\mu_0 = 4\pi \times 10^{-7}\ \text{T}\cdot\text{m/A}$.

We will not use the Biot–Savart law very often, since most of its applications require the use of calculus to add up tiny contributions to the field from an infinite number of segments of charge moving along a wire. But it is useful for one particular application, which does not require calculus, as we will now see.

EXAMPLE 19.6 The Magnitude of the B-Field at the Center of a Loop

What is the magnetic field at the center of a circular loop of wire of radius r, carrying a current I?

ANSWER As we showed in Figure 19.9, the field at the center of a current-carrying loop is produced by each segment of wire adding its contribution to that from all other segments. We now also know that we can use the Biot–Savart law to find the field due to each tiny segment. As we use this law, let us remember that we have already noted, in the discussion of Figure 19.5, that the charge flowing in a short wire of length L satisfies the relation $qv = IL$.

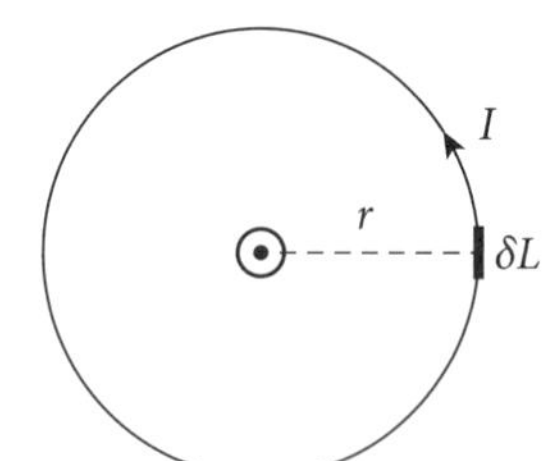

With this identification, we can rewrite the Biot–Savart law so that it gives the field produced by a tiny straight segment of wire of length δL carrying a current I:

$$B_{\text{segment}} = \frac{\mu_0}{4\pi}\frac{I\delta L}{r^2}\sin\theta.$$

Since the field due to each segment of length δL is in the same direction at the center of the loop (it is always out of the page), the contributions of the segments can simply be added up. The distance to the field point is the same r for all segments, and $\theta = 90°$ is also the same for all segments, so we can easily add up the contributions around the loop. We thus have

$$B_{\text{loop}} = \frac{\mu_0}{4\pi}\frac{I(2\pi r)}{r^2}\sin 90° = \frac{\mu_0}{2}\frac{I}{r},$$

where we noted that the sum of all the δL segments is $2\pi r$ and that $\sin 90° = 1$. This important result gives the magnetic field strength at the center of a single loop of radius r carrying current I.

In Section 19.5, we explained that the magnetic field lines around a single long straight wire form closed circles, with the magnetic field constant around the circle and tangent to the circle at each point. But we have not yet presented the formula for the magnitude of the magnetic field produced a distance r from the wire. It is possible, using calculus and the Biot–Savart law, to find the *field due to an infinite straight wire* carrying current I. It is

$$B = \frac{\mu_0 I}{2\pi r}, \tag{19.4}$$

the direction of the field being given by Right-Hand Rule #2, as before. But there is another way we could have found this result, without resorting to calculus.

19.8 Ampère's Law

You probably remember that, back near the end of Chapter 16, we introduced a new quantity called the electric flux, formed by multiplying the electric field on some surface by the area of that surface. With this quantity defined, we were able to derive a formula known as Gauss's Law that enabled us, when we knew that the electric field was constant over the surface, to find that constant magnitude. Here, we want to define another new quantity, one that will enable us to find the magnetic field along a line where the field is known to be constant. The *magnetic circulation* along a closed path P is defined by multiplying the constant component of the magnetic field parallel to the path by the length of the path. There is no standard symbol for circulation, so let us choose a C and write $C_B = B_{||}L_P$ for the magnetic circulation around a closed path of length L_P. The key to this procedure working, of course, is to find a path on which $B_{||}$ is constant.

One situation where we know that there exists such a path is the case of the circular magnetic field surrounding a current-carrying wire. If we use Equation 19.4 for the magnetic field of a wire carrying current I, we calculate the magnetic circulation around a circle of radius r surrounding the wire to be

$$C_B = B_{||}(2\pi r) = \frac{\mu_0 I}{2\pi r}(2\pi r) = \mu_0 I.$$

This equation was derived for the case of a single wire passing through the center of the circular path (see Figure 19.14), but it is generally true for any path that bounds a surface penetrated anywhere by a total current I_{CUTTING}. It is known as *Ampère's Law*, and it is written

$$C_B = \mu_0 I_{\text{CUTTING}}. \tag{19.5}$$

If we know from symmetry that the field is constant along some path P, then we can use Ampère's law to find the magnetic field strength.

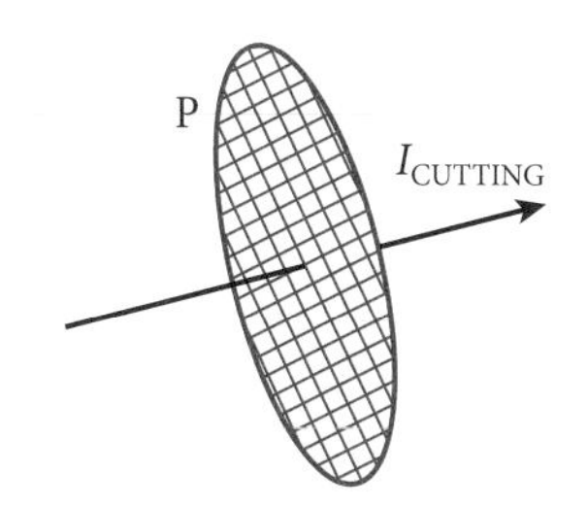

Figure 19.14 A (red) Ampère's law path and the (cross-hatched) surface that it bounds. I_{CUTTING} is the total current cutting the surface.

As a trivial application of Ampère's law, we can reverse the process of Equation 19.5 and find the magnetic field a distance r from a wire carrying current I. We know from Figure 19.6 that the magnetic field on a circle of radius r is constant at all points on the circle and is tangent to the circle. So we choose the path P to be this circle and apply Ampère's law, $C_B = \mu_0 I$, to write

$$C_B = \mu_0 I \quad \Rightarrow \quad B_{||} L_P = B(2\pi r) = \mu_0 I \quad \Rightarrow \quad B = \frac{\mu_0 I}{2\pi r},$$

which is the same as the result from the Biot–Savart law, in Equation 19.4.

Ampère's law also allows us to calculate a result that would otherwise be very difficult to derive.

EXAMPLE 19.7 The Magnetic Field inside a Solenoid

Find the magnetic field in the region inside the coils of a long solenoid having N total loops wrapped on a cylinder of radius r and length L, and carrying current I.

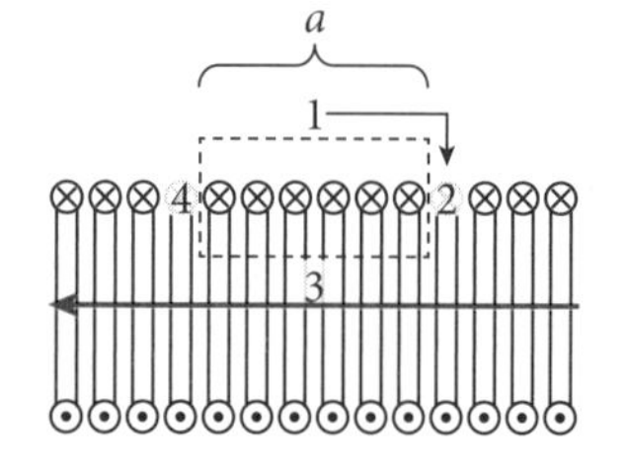

ANSWER Here is another case where we know a great deal about the magnetic field direction from previous analysis; only the magnitude of the field is missing. We have already seen in Section 19.6 that the magnetic field lines inside a solenoid will be nearly parallel and tightly packed together. As the field lines emerge from the interior of the solenoid, they will expand and fill the space outside. For the solenoid pictured at right, with current going into the page at the top of the picture and out of the page at the bottom, this allows us to approximate the field of the solenoid as being straight and horizontal to the left inside the coils. Although the field is not zero outside the solenoid, it is much weaker than inside.

Ampère's law must apply to the rectangular path of length a, shown as a dashed line in the figure above. We divide the path up into its four straight segments. Along the top segment (segment 1) the field is approximately zero, so there is no contribution to the magnetic circulation from this part of the path. Along the up-and-down portions of the path (segments 2 and 4) the field is either zero, for the portion outside the solenoid, or perpendicular to the path, for the portion inside the solenoid. So there is again no contribution to the circulation from either of these two segments. Only along segment 3, where the field is parallel to the path, will there be a contribution to the magnetic circulation. If we call the magnetic field B along this segment, then the total circulation around the square loop is $C_B = C_1 + C_2 + C_3 + C_4 = 0 + 0 + Ba + 0 = Ba$.

Ampère's law relates the magnetic circulation to the total current passing through the surface bounded by the dashed line. Since there are N total coils in the solenoid of length L, the number of coils per meter of length will be N/L, and the number of times that the wire will pierce the surface inside the dashed path is $(N/L)\,a$. The total current cutting through the surface will therefore be $(N/L)\,aI$. If we then use Ampère's law, $C_B = \mu_0 I_{\text{CUTTING}}$, we find

$$Ba = \mu_0 \frac{N}{L} aI \quad \Rightarrow \quad B = \frac{\mu_0 NI}{L}.$$

This is the magnetic field inside a solenoid. As may be seen, it depends on how tightly packed the wires are (N/L) and on the current in the solenoid.

Note that this result does not depend on the radius of the coils or on how far we are from the centerline of the solenoid. This is a little surprising and could not have been predicted without the aid of Ampère's law. As long as we stay away from the ends of the solenoid (so that our approximations are valid), the magnetic field will be uniform everywhere inside the solenoid. This is in contrast with the case for the single loop of current-carrying wire (Figure 19.10), where the field was greatest near the center of the circle and dropped off at points further from the center and closer to the wire loop.

Now that we have several formulas for the magnetic fields produced by various configurations of wires carrying electrical current, let us do a couple of examples of the kinds of problems we are equipped to solve.

EXAMPLE 19.8 Adding Magnetic Fields from Current-Carrying Wires

A single coil of wire of radius 10 cm lies in the x–y plane and carries a current of 1.00 amp clockwise, as shown in the figure at right. At the same time, a long straight wire that just touches the rim at the back of the wire loop carries a 0.318-amp current in the $+z$ direction, as shown. Find the magnetic field at the center of the wire loop.

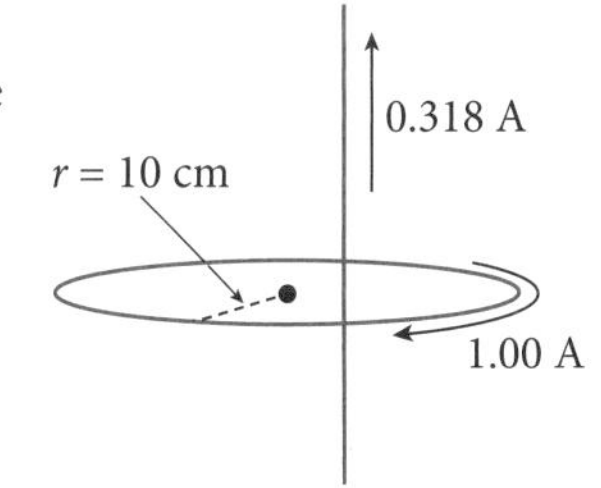

ANSWER Our first step is to find the magnitudes and directions of the two magnetic fields produced by the two wires. For the straight wire, the field at a point 10 cm from the wire is found from Equation 19.4:

$$B = \frac{\mu_0 I}{2\pi r} = \frac{(4\pi \times 10^{-7}\ \text{T}\cdot\text{m/A})(1.00\ \text{A})}{2\pi(0.1\ \text{m})} = 2.00 \times 10^{-6}\ \text{T} = 2.00\ \mu\text{T}.$$

Right-Hand Rule #2 gives the direction of the field at the dot at the center of the loop as roughly left-to-right across the page. A formula for the magnetic field at the center of a circular loop was found in Example 19.6. It gives

$$B = \frac{\mu_0 I}{2r} = \frac{(4\pi \times 10^{-7}\ \text{T}\cdot\text{m/A})(0.318\ \text{A})}{(2)(0.1\ \text{m})} = 2.00 \times 10^{-6}\ \text{T} = 2.00\ \mu\text{T}.$$

Right-Hand Rule #3 gives the direction as downward, in the minus z-direction.

The total magnetic field at the center of the loop is the sum of the two fields. We cannot just add the two numbers together, of course, because magnetic fields are vectors and must be added like vectors. The two contributions to the total magnetic field at the dot in the center of the loop are depicted in the figure at right. The vector sum of the two is a vector of magnitude $B = \sqrt{(2.00)^2 + (2.00)^2} = 2.83\ \mu\text{T}$ and direction 45° below the horizontal plane.

$B_{wire} = 2\ \mu\text{T}$

$B_{loop} = 2\ \mu\text{T}$

$\vec{B}$

EXAMPLE 19.9 The Magnetic Force between Two Current-Carrying Wires

A straight wire, 1 meter long and carrying a current of 1 amp, is placed 1 meter above a very long straight wire, also carrying a current of 1 amp in the same direction as the first wire. What is the magnetic force (magnitude and direction) acting on the top wire by the wire on the bottom?

ANSWER Each current-carrying wire will produce a magnetic field that will then put a magnetic force on the other current-carrying wire. However, it is easiest to find the field produced by the long wire (because this is one of the configurations where we know a formula for the field), evaluated at the location of the short wire. We can then find the force on the short segment. Equation 19.4 gives the magnitude of the field 1 m from the lower wire as

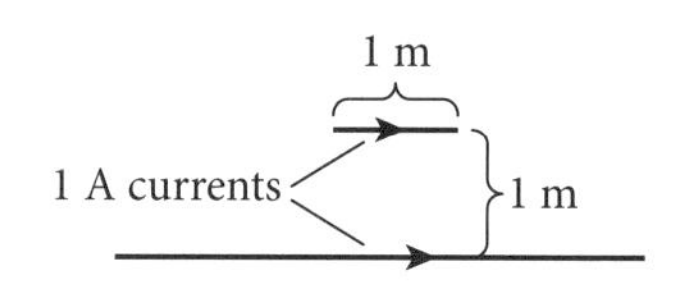

$$B = \frac{\mu_0 I}{2\pi r} = \frac{(4\pi \times 10^{-7}\ \text{T}\cdot\text{m/A})(1\ \text{A})}{2\pi(1\ \text{m})} = 2 \times 10^{-7}\ \text{T}.$$

The direction of the field produced by the lower wire at the location of the upper wire is found by the right-hand rule to be out of the page.

A constant current cannot just flow along a 1-meter segment of wire, of course, since the current has to come from somewhere. It might, for example, come from a far-away battery along long wires perpendicular to the segment. But let us just concentrate on the 1-m segment, as asked. If a 1-meter length of wire carries a current of 1 amp, there will be a force on it whose magnitude is given by Equation 19.2. The angle between the magnetic field out of the page and the current that is going from left to right is $\theta = 90°$, so we have

$$F_B = ILB \sin\theta = (1\ \text{A})(1\ \text{m})(2 \times 10^{-7}\ \text{T})(1) = 2 \times 10^{-7}\ \text{N}.$$

The direction of the force is found from Right-Hand Rule #1. We point our fingers in the direction of the current flow and hold our right hand so that the magnetic field out of the page is coming out of our palm. Then, as we curl our fingers toward the B-field, our thumb ends up pointing down. This is the direction of the force on the upper wire carrying positive current to the right.

Note that we have found that the force between two wires carrying current in the same direction is attractive.

This last example is actually a very important one. Back in the footnote on page 355, we promised that we would eventually explain how the unit of the ampere is defined. Well, here it is: *One ampere is defined to be the current that, when it is flowing in two wires 1 meter apart, produces a* 2×10^{-7} *N magnetic force on each 1-meter segment of the wire.* This may leave you wondering why 2×10^{-7} N, and not some other number. The answer is that, while this definition is the present international definition of the ampere, the history of the unit is long and complicated. When scientists decided on the present definition in 1948, they wanted to choose the size of the unit so that it would agree with previous definitions as closely as possible.

19.9 Magnetic Polarization and Refrigerator Magnets

We began this chapter with a discussion of magnets, and we would like to end there as well. There are several models that explain how magnetic fields develop inside a permanent magnet. None of the models that are at a level we can discuss in this textbook are completely satisfactory, but one model that is very helpful involves the realization that matter consists of atoms with negatively charged electrons moving around a positively charged nucleus. The currents produced by the electrons provide a kind of explanation of the observed fact that individual atoms typically have small magnetic fields around them, as if they were tiny atom-sized coils acting like atom-sized magnets. In most materials, these fields point in random directions, and so the material does not have an overall magnetic field. In some materials, however, small domains can be created in which the dipoles are aligned, and the sum of the fields in these domains creates a permanent magnet of the sort we discussed in Section 19.1.

Even in materials with no permanent magnetization, an external magnetic field applied to these materials will cause the little magnetic dipoles to rotate and align themselves with that external field. For example, if a north pole of a permanent magnet is brought near a magnetizable material, like iron, then the atomic dipoles in the iron will be rotated so that the south poles of their magnetic fields are close to the north pole of the external magnet. This magnetic polarization is similar to the electric polarization we discussed in Section 16.3, and it causes a permanent magnet to be attracted to an otherwise unmagnetized polarizable material. This is the mechanism, of course, that is responsible for refrigerator magnets sticking to metal refrigerator doors. It also causes an external magnetic field to be enhanced in a region inside and near some polarizable material.

This explanation of a magnet is also the explanation for why breaking a magnet in two will never isolate a pure magnetic pole. It is because a magnetic pole is not a thing. In a magnet, it is the result of the motion of charges in the atoms of the material, and it always appears as a vector, an arrow that always has both a head and a tail.

19.10 Summary

As a result of reading this chapter, you should be able to do the following:

- Using Equation 19.1 and Right-Hand Rule #1, you should be able to relate the force on a moving charge—magnitude and direction—to the magnetic field.
- With Equation 19.2 and the same Right-Hand Rule #1, you should be able to determine the force on a current-carrying wire.
- You should be able to use Right-Hand Rule #2 to determine the direction of the circular magnetic field around a current-carrying wire.
- Using Right-Hand Rule #3, you should be able to determine the direction of the magnetic field at the center of a current-carrying loop or inside a solenoid.
- You should be able to use Equation 19.4 to find the magnitude of the circular magnetic field around a current-carrying wire.
- You should understand how Ampère's law is defined and how it may be used to find the magnetic field in situations with high symmetry, as in Example 19.7.

CHAPTER FORMULAS

Force: $F_B = qvB \sin\theta \quad F_B = ILB \sin\theta$ Biot–Savart law: $B = \dfrac{\mu_0}{4\pi}\dfrac{qv}{r^2}\sin\theta$

Fields: Center of loop: $B = \dfrac{\mu_0 I}{2r}$ Wire: $B = \dfrac{\mu_0 I}{2\pi r}$ Solenoid: $B = \dfrac{\mu_0 NI}{L}$

Ampère's law: $C_B = B_{||}L_P \quad C_B = \mu_0 I_{\text{CUTTING}}$

PROBLEMS

19.3 *1. A magnetic field points horizontally and due north, while a positively charged particle in a cosmic ray moves vertically downward at high speed. In what direction will the path of the cosmic ray deflect?

*2. An electron is moving through a region with a uniform magnetic field, as shown below. Which arrow represents the direction of the force on the electron at the instant shown? Justify your answer.

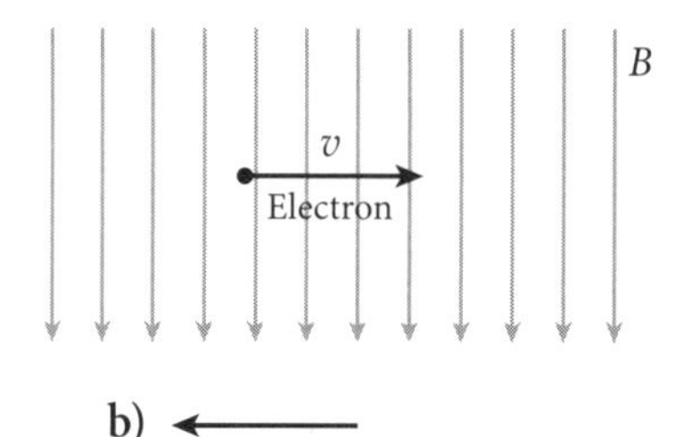

a) $\longrightarrow$ b) $\longleftarrow$ c) 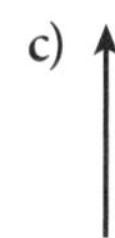

d) (into the page) $\otimes$ e) (out of the page) $\odot$

*3. A magnetic field is created in a laboratory with a field strength of 0.25 T, pointing horizontal and due north. A proton beam is directed horizontally northwest, with the protons all moving at a speed of 6.0×10^4 m/s. What are the magnitude and direction of the force on the protons in the beam?

*4. What is the force (magnitude and direction) on an electron moving horizontally west at a speed of 3×10^6 m/s (1% the speed of light) in an upward vertical magnetic field of strength 0.4 T?

*5. A region of space contains a uniform electric field to the right as shown below. A proton is observed to move at constant velocity through this region, in the direction shown. The constant velocity is made possible by there being an external magnetic field that cancels the force caused by the E-field. Find the direction in which the B-field is pointing and explain your reasoning.

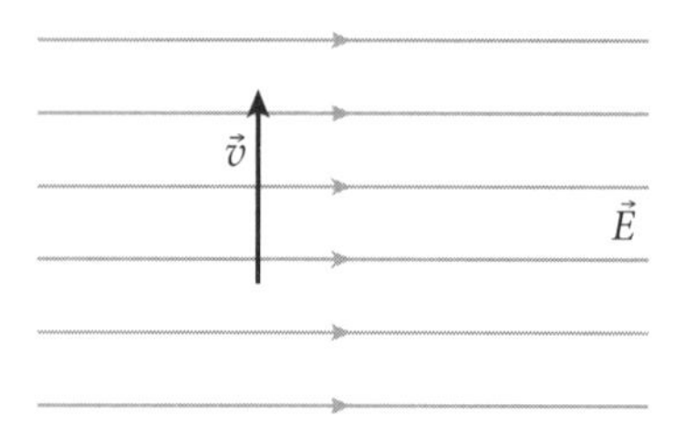

*6. A proton traveling horizontally due east enters a region that contains both a uniform magnetic field and a uniform electric field. (We are ignoring the earth's magnetic field and assuming that the uniform magnetic field is provided by a strong magnet.) The electric field lines point horizontally due north. It is observed that the proton continues to travel in a straight line toward the east.

a) What direction is the electric force on the proton? Explain.

b) What direction is the magnetic force on the proton? Explain.

c) In what direction must the magnetic field lines point? Explain.

**7. In a place where the earth's magnetic field is 6.0×10^{-5} T and points exactly horizontally due north, a hunter fires a bullet horizontally toward the northeast at a speed of 300 m/s, trying to hit an aardvark. Unfortunately, as the bullet moved down the barrel, it acquired a net electrical charge of +2.0 nC. What is the magnitude of the magnetic force on the bullet due to the earth's magnetic field?

**8. If you live in Bozeman, Montana, you will find that the magnetic field has a strength of about 5×10^{-5} T. The vector is directed northward, but it is not parallel to the ground. It actually dips 70° below the horizontal (i.e., it is only 20° away from pointing straight into the ground). In a laboratory in Bozeman, Montana, a proton is moving horizontally, parallel to the floor, and toward magnetic east at a speed of 5×10^{6} m/s. What are the magnitude and direction of the magnetic force on the proton? [Hint: Start by drawing a picture.]

**9. Before liquid crystal display (LCD) computer and television screens, the screens were cathode-ray tubes (CRTs), in which a beam of electrons was shot toward a phosphorescent screen at high speed and deflected by electric fields in order to direct the beam to the right place on the screen so that it would form the picture. In a typical television, the speed of the electrons is 1×10^{8} m/s.

a) If the television set is placed so that the CRT beam is horizontal, heading east, in a place where the earth's magnetic field is 5×10^{-5} T and points horizontally northward, what will be the magnetic force on the electron?

b) The mass of the electron can be found in Appendix B. What is the constant acceleration (magnitude and direction) of an electron in a television due to the earth's magnetic field?

c) How long will it take an electron to cover the 30 cm from the accelerator to the screen?

d) By how much will the electron deflect as it makes its way to the screen?

**10. A spacecraft in a circular geosynchronous orbit at a distance from the earth of 42,000 km encounters a magnetic field of 100 nT. At this distance from the earth, the spacecraft is also subject to charging up by absorbing cosmic rays. If a spacecraft has acquired a charge of 2.5 μC and is moving at a speed of 3000 m/s, cutting across the magnetic field lines at 90° to the field, what force will there be on the spacecraft due to its charge and the earth's magnetic field?

In a few of the following problems, we should note that the force a magnetic field exerts on a charged particle is always perpendicular to the motion of the particle, thus doing no work so that the particle's speed remains constant. We also use the fact that motion at constant speed, experiencing a constant acceleration that is perpendicular to the motion, will be circular motion.

** 11. An alpha particle (symbol α), having a mass 6.64×10^{-27} kg and a charge $+3.2 \times 10^{-19}$ C, is accelerated in the laboratory to a speed of 2.0×10^{7} m/s. It passes into a region of a 0.50-T external magnetic field that is oriented perpendicular to the α's initial velocity.

a) What is the magnetic force on the α?

b) What is the acceleration of the α?

c) The force on the α will produce circular motion at constant speed (see the note above). Find the radius of the circular motion of the α.

**12. In 1932, Ernest Lawrence invented the cyclotron, in which a strong vertical magnetic field acted on high-speed protons. As the protons gained energy, they spiraled out into horizontal circular paths of greater and greater radii (see the note above). In the original cyclotron the radius of the outer circle was 35 cm. Find the magnetic field strength required to turn the path of a proton moving at 2×10^{7} m/s into a circle of radius 35 cm. [Hint: The charge and mass of the proton can be found in Appendix B.]

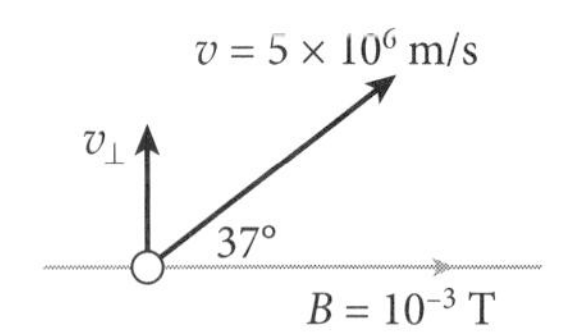

*** 13. In the magnetic field of a star, a proton moves at a speed of 5×10^{6} m/s at an angle of nearly 37° to the star's magnetic field lines, as shown, giving it velocity components 4×10^{6} m/s along the magnetic field and 3×10^{6} m/s perpendicular to the magnetic field. The magnetic field strength of the star is exactly 10^{-3} T. In a plane perpendicular to the field, the magnetic field will curve the motion into a circle (see the note above). In the direction parallel to the field, the velocity will be constant.

a) Find the radius of the circular motion in the plane perpendicular to the field.

b) Find the time it will take the proton to go around one complete circle in the plane perpendicular to the B-field.

c) In the time it takes for a complete circle in the perpendicular plane, how far in the parallel direction will the proton travel?

d) The 3-D geometric shape in which an object moves in a circle where the center of the circle moves perpendicular to the plane of the circle is called a helix. The pitch of a helix is defined as the ratio of the distance the circle center moves to the radius of the circle. What is the pitch of this helix?

*** 14[f]. **Mass Spectrometer.** One way in which experimenters can measure the mass of a charged particle is to accelerate it through a known voltage V, beaming it into a region where a perpendicular magnetic field B bends the beam into a circle of radius R. Find a formula for the mass of an ion via the following two steps:

a) Find a formula for the velocity of the charged particle in terms of its charge q, mass m, and the voltage V through which it is accelerated.

b) Knowing that the centripetal acceleration required is provided by the magnetic force on the particle, find a formula for the mass of the particle in terms of q, B, V, and R. [Hint: This will take a little algebra.]

** 15[f]. Charged particles emitted from a radioactive material will have a wide range of velocities. In order to do experiments with particles of a particular velocity, the particles are first passed through a "velocity selector" in which particles with the wrong velocities are deflected out of the beam of particles that hit the selector. In the selector, a strong electric field points perpendicular to the direction of the beam. A strong magnetic field is also created at right angles to the beam and to the electric field. If the electric field is of strength E (in volts/meter) and the magnetic field is of strength B (in teslas), find the velocity v of the particles that will pass undeflected through the selector. [Hint: Assume a charge q on the particle, but q will cancel out of the final formula for v.]

19.4 * 16. The diagram at right shows a top view of a current-carrying wire lying on a table next to a permanent magnet. The current flows in the direction indicated. In what direction is the force on the wire? Explain your answer.

a) to the right b) to the left
c) into the page d) out of the page

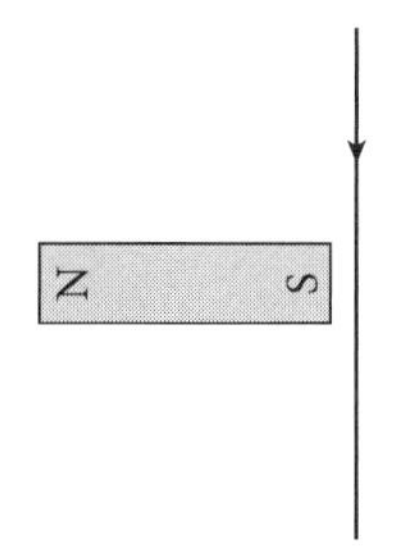

* 17. What is the magnitude of the force on 2 meters of wire carrying 0.5 A of current at an angle of 37° to a 0.05-T magnetic field?

* 18. A vertical wire is connected to a battery in such a way that a current flows from the floor upward to the ceiling, as shown in the bird's-eye view at right. A standard bar magnet is held parallel to the floor, with the north pole being closest to the wire as shown in the figure. In what direction is the force on the wire?

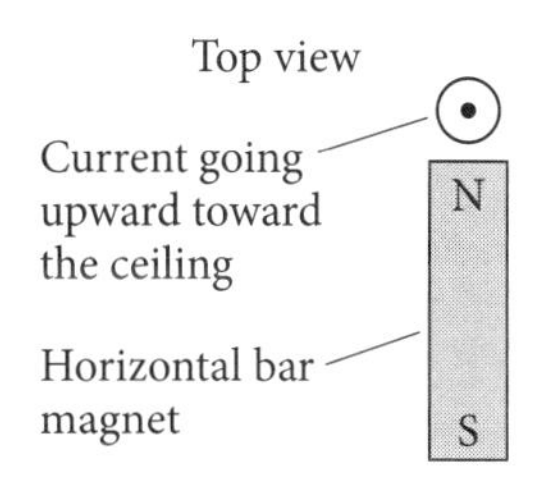

** 19. A straight wire 0.2 m long carries a current 20 mA at an angle 30° to a magnetic field of 0.1 T, as shown below. The wire and the field both lie in the plane of the page. What will be the magnitude and direction of the force on the wire?

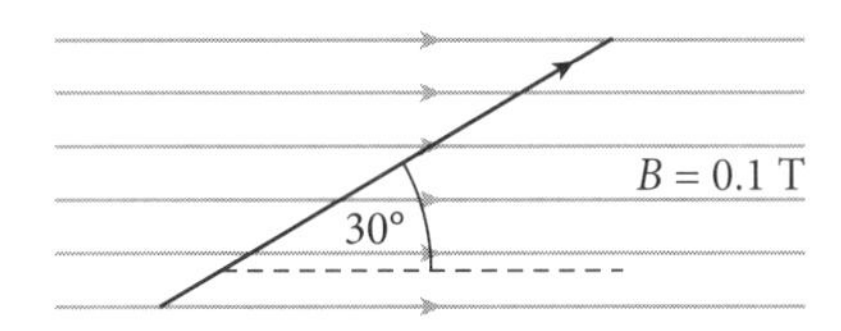

*** 20. **The Hall Effect.** On page 390, we noted that the force on a current-carrying wire in a magnetic field would be in the same direction whether the charge carriers were positive or negative. However, an 1879 experiment by E. H. Hall showed that charge carriers in a metal wire were, in fact, negative. In Hall's experiment, a current was carried along a wide flat wire in a magnetic field that was normal to the flat surface, as in the figure. The force on the charge carriers was to the right, regardless of their charge (verify this with a right-hand rule). If negative charges were the charge carriers, then they would pile up on the right edge of the strip, leaving a net positive charge on the left side. Charge would continue to build up on the two sides of the strip until a nearly constant electric field was created across the wire, whose force on a charge carrier q canceled the magnetic force on the same charge carrier. The electric field would create a detectable voltage difference, left to right across the strip. Find the magnitude of the voltage that would be produced when this equilibrium is attained in a strip of width 0.02 m in a 3-T field. (Assume the charge carriers have average velocity $v = 10^{-3}$ m/s, a typical drift velocity of charge carriers in a large current.)

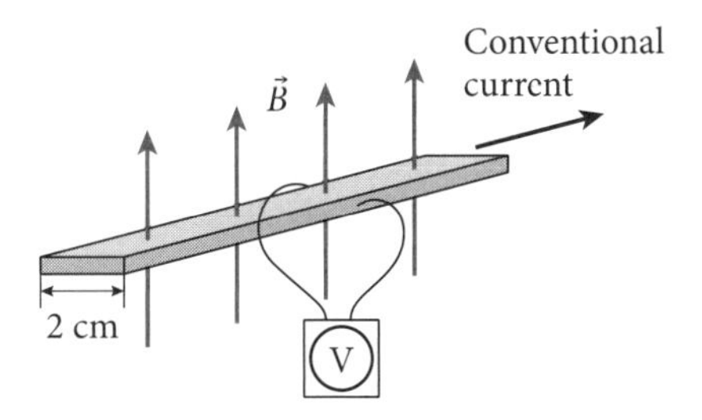

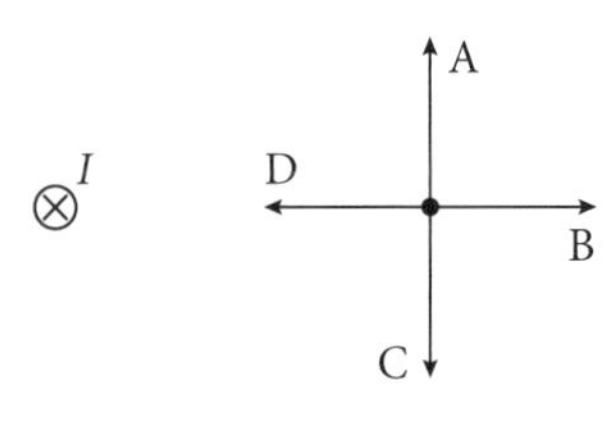

19.5 * 21. The current in a wire is directed into the page as shown at left. The magnetic field at a point to the right of the wire is represented by which arrow? Justify your answer.

** 22. The diagram at right shows a magnet that has been placed near a wire. When the switch is closed, the magnet is deflected as shown, with the south pole dropping into the page and the north pole coming out of it.

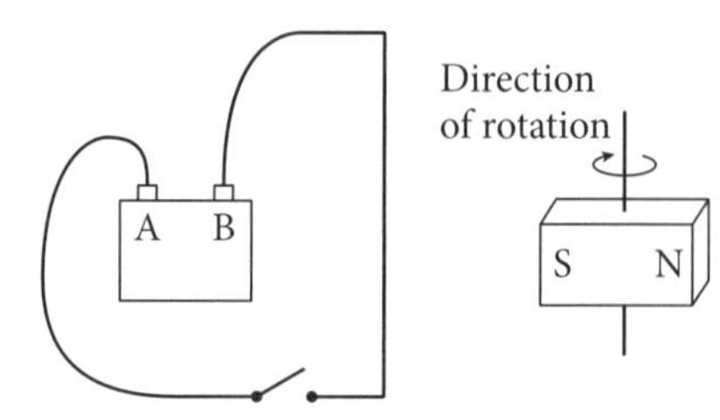

a) Find the direction that the current must be moving along the wire. Explain your reasoning.

b) Which terminal of the battery, A or B, is the positive terminal of the battery?

c) Find the direction of the net force exerted by the magnet on the current at the instant shown in the diagram. Explain.

** 23. A horseshoe magnet is sitting on a table with its north pole on the left, as shown. A current-carrying wire also lies on the table, with the current passing between the poles of the magnet at 45° to the magnetic field, as shown. In what direction will there be a force acting on the wire?

** 24. The figure at right shows the top view of a compass next to a wire. The north pole of the compass is indicated by an arrow (i.e., north is to the top of the page in the diagram). When a switch is closed, a current is carried through the wire, downward into the floor as indicated. In which direction, clockwise or counterclockwise, does the compass rotate? Explain.

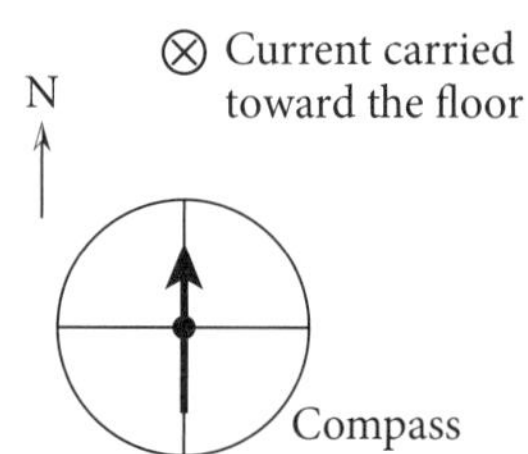

** 25. Two wires carry current into the page as shown below left. One wire carries a current I and the other carries a current $2I$. Which of the arrows below right best represents the magnetic field at point A, a point that is the same distance from both wires? Justify your answer.

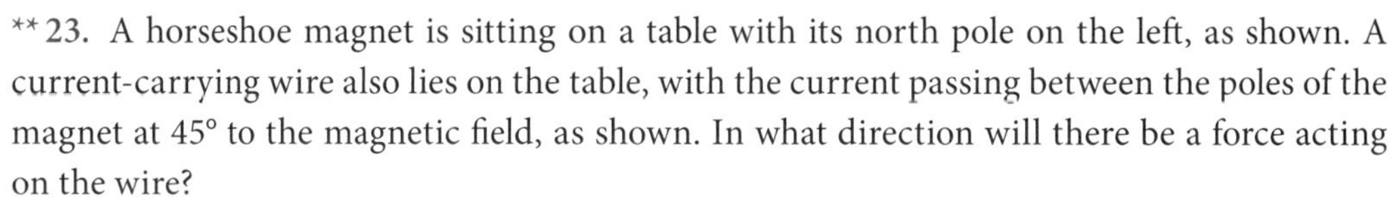

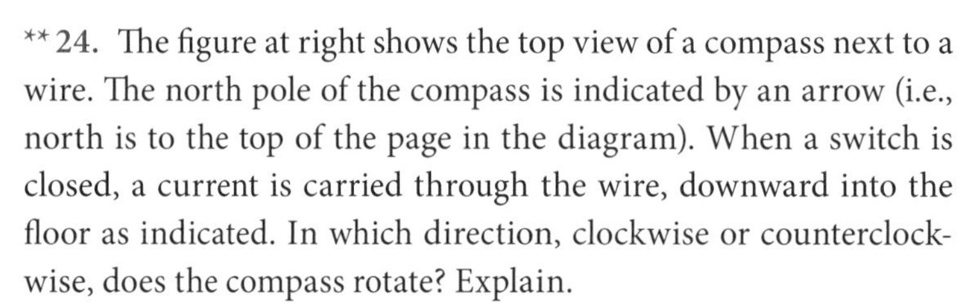

** 26. The figure below shows a wire lying flat on the surface of a table. The wire is connected to a battery so that the current flows left to right. A bar magnet is to be placed flat on the surface of the table in such a way that the wire experiences a force that lifts it up off of the table. Draw a picture showing the wire and the location and orientation of the magnet (including poles) and explain carefully why this would result in an upward force on the wire.

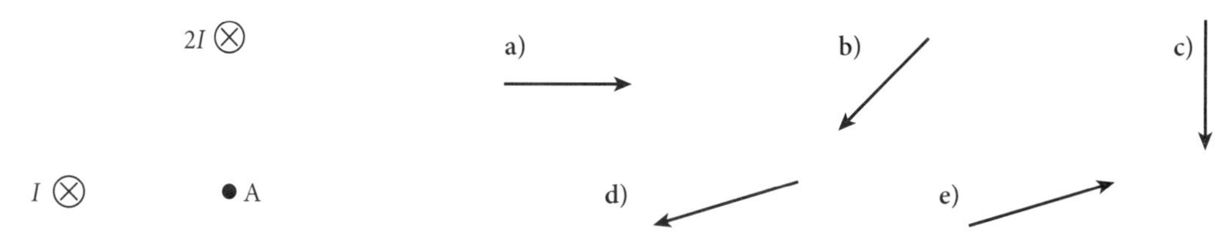

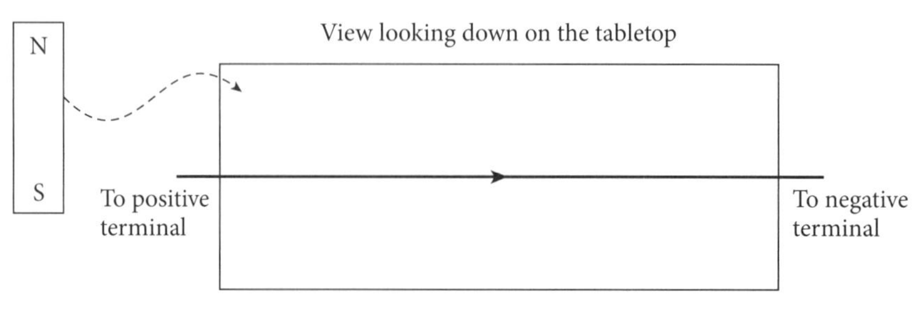

** 27. The figure below represents wires carrying equal currents into or out of the page as indicated. Your job is to determine the direction of the magnetic force on wire 3 due to wires 1 and 2. Redraw the figure and add vectors showing the fields and forces acting on wire 3, labeling each one to show what they represent (a force or a field) and what is causing them. Explain your reasoning.

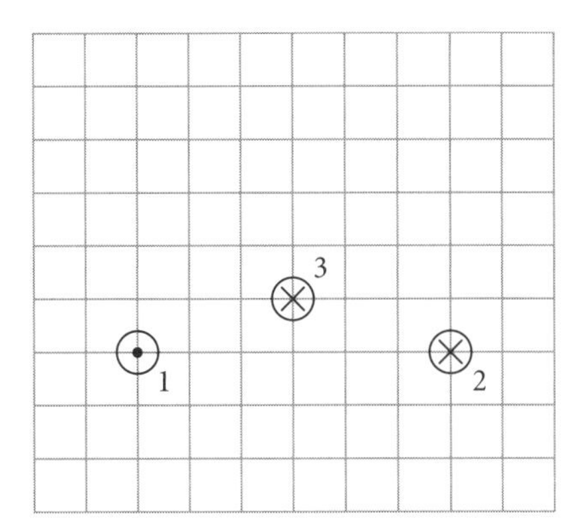

** 28. A black wire attached to the tabletop carries a current in the direction indicated in the picture at right. It is crossed at a 30° angle by a gray wire carrying a current in the direction shown.

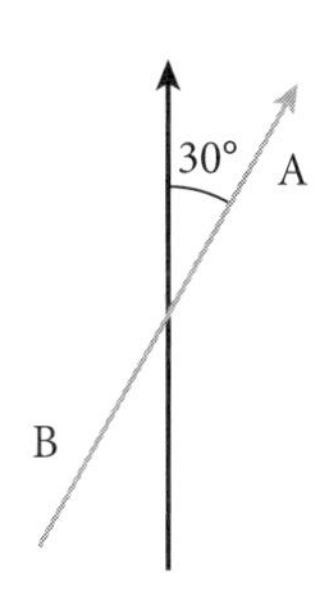

a) In what general direction will there be a force on the segment of the gray wire labeled A in the diagram? In what general direction will there be a force on the segment labeled B in the diagram?

b) As a result of these forces, what will happen to the gray wire in the picture?

** 29. Two wires are oriented perpendicular to each other as shown, carrying current in the directions shown by the arrows. One lies on top of the other, so that they are both essentially in the plane of the page. Wire B is fixed and wire A is free to move. Is there a net force on wire A? If so, in what direction? Is there a net torque on wire A? If so, will wire A begin to rotate clockwise or counterclockwise? Explain your reason for each answer.

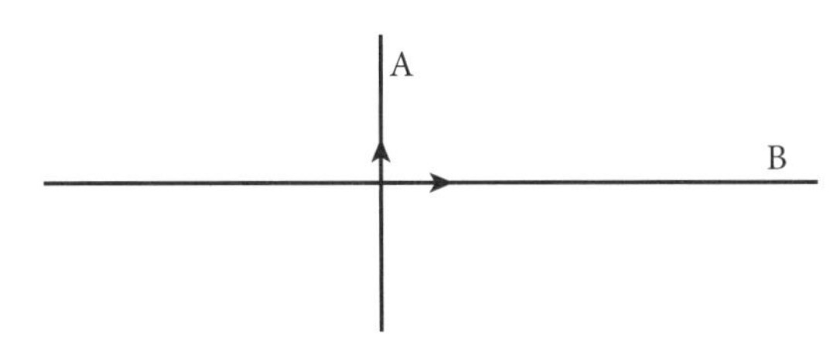

19.6 ** 30. A loop is connected to a battery so that a current will flow in the direction shown. Initially, the switch is open and there is no current flowing in the loop. A small compass (represented by an arrow) is placed below the loop, pointing to the right, as shown. When the current in the loop is turned on, in what direction will the compass turn (clockwise or counterclockwise) and what direction will it finally point? (Neglect any magnetic field from the earth.) Draw a sketch and explain your answers.

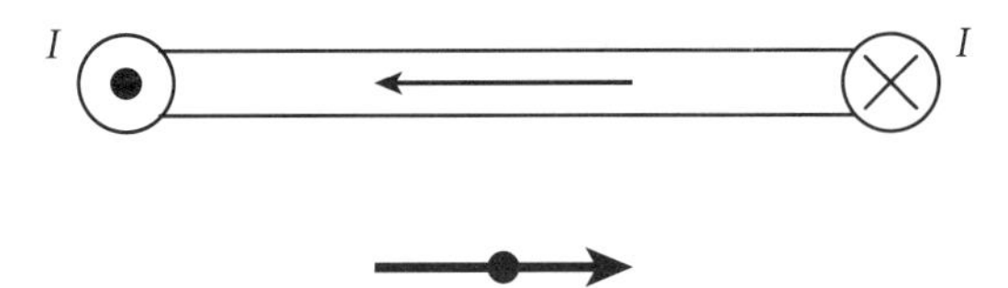

** 31. Below, we show a top view of two current-carrying loops. The loop on the left is fixed in place while the one on the right is free to rotate about a vertical axle. The current directions are shown using the usual convention.

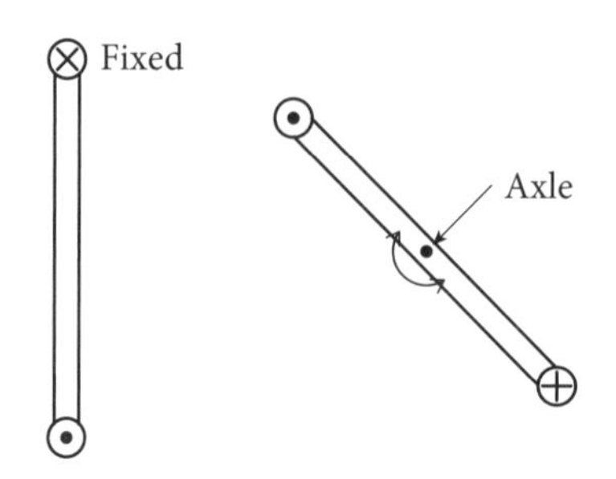

a) Redraw the picture and indicate with an arrow the direction of the magnetic field (B-field) at the center of the right loop due to the left loop.

b) What direction, clockwise or counterclockwise, will the right loop pivot in response to this field? Explain how you determined your answer.

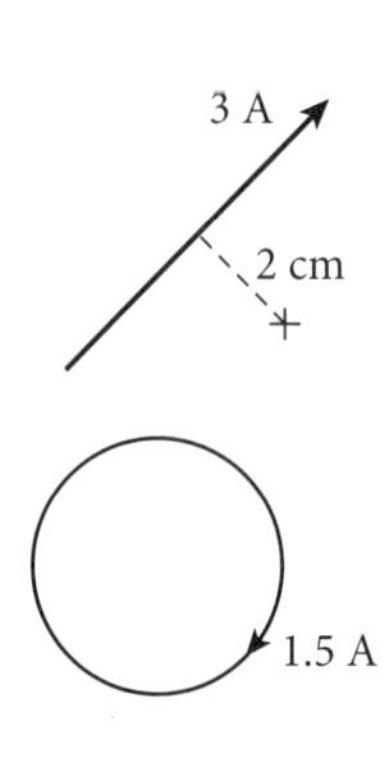

19.7 * 32. A long horizontal straight wire carries a 3-ampere current toward the northeast, as shown at left. What will be the magnitude and direction of the magnetic field at a point 2 cm southeast of the wire?

* 33. A single circular wire loop of radius 10 cm lies on a tabletop, carrying a current of 1.5 A clockwise around the loop. What will be the magnitude and direction of the magnetic field at the center of the loop due to the current in the loop?

19.8 * 34. A solenoid has 100 loops of wire wrapped around a cylinder of radius 1 mm and length 10 cm. If the wire carries a current of 5 mA, what is the magnitude of the magnetic field inside the solenoid?

** 35. In the classical picture of a hydrogen atom, a single electron moves in a circle of radius 53 pm around the nucleus of the atom. The speed of the electron in this classical picture corresponds to an average current of 10^{-3} A. (These values are approximately correct for a hydrogen atom in its lowest-energy state.) What is the magnitude of the magnetic field created for a single hydrogen atom by its circulating electron?

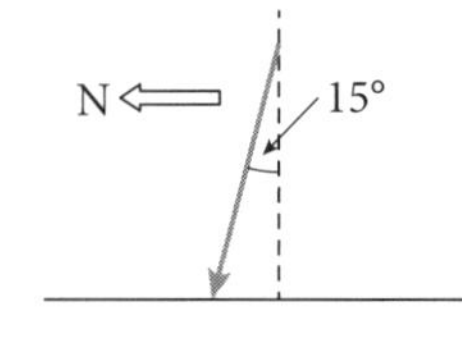

** 36. In Bozeman, Montana, the earth's magnetic field is 5×10^{-5} T, pointing north and downward at an angle 70° below the horizontal.

a) In what direction should a current-carrying wire be oriented if we want to produce the greatest possible magnetic force on the wire?

b) What will be the force on a 1-cm segment of such a wire if the wire carries a current of 20 A?

** 37. Two long parallel wires 10 cm apart each carry a current of 8 mA in the same direction. What is the magnetic field at the point shown, 2 cm from one wire and 8 cm from the other?

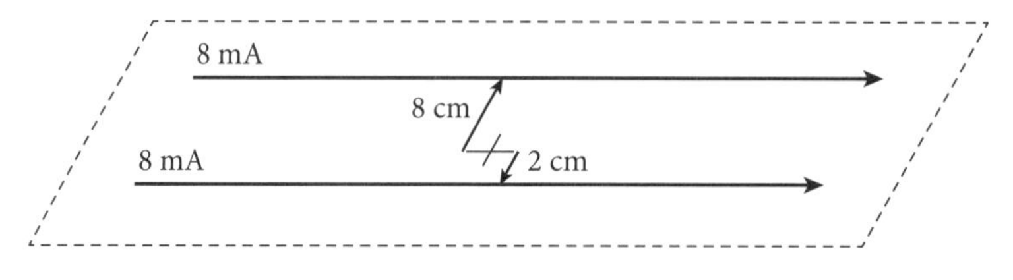

*** 38. A long straight wire carrying a current of 10 amps is bent so that the current passes around a loop of 10-cm diameter before continuing on in the original direction. What is the magnetic field at the center of the loop?

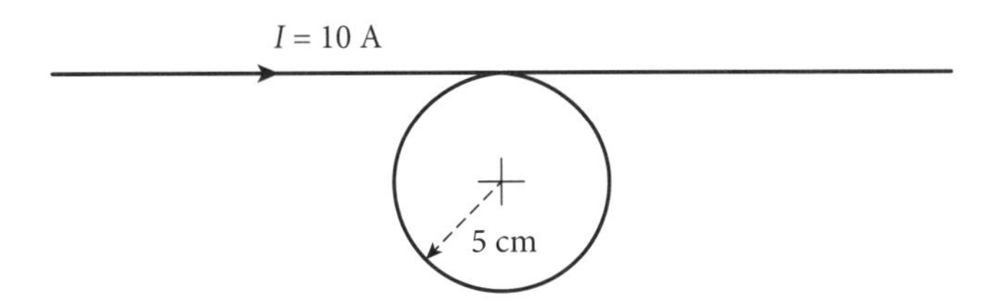

*** 39. A current of 15 A flows along a straight wire that goes down the cylinder axis of a solenoid, as shown in the top picture at right. The solenoid has 100 coils in it, uniformly wound along a total length of 0.314 m, and carries a current of 1 A. The radius of the solenoid is 2 cm. The current in the wire flows from left to right and the current in the solenoid goes around clockwise, as seen from the left. What is the total magnetic field (magnitude and direction) at a point 1 cm above the center of the straight wire, the point labeled with a dot in the bottom picture?

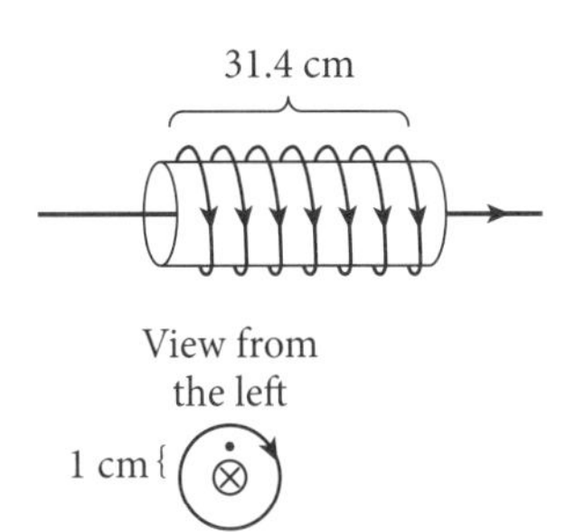

*** 40. Two long wires are 10 cm apart at their closest point. One of the wires is vertical and carries current 1 amp upward from the floor, while the other is horizontal and carries current 3 amps left to right. What is the magnetic field halfway between the two wires at their closest point? [Hint: Do not forget that magnetic fields are vectors.]

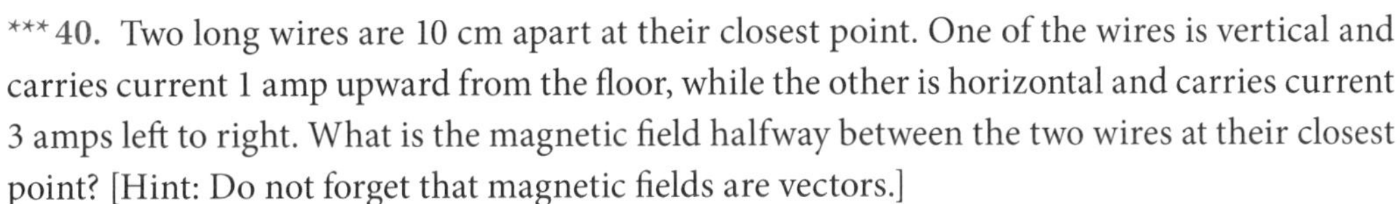

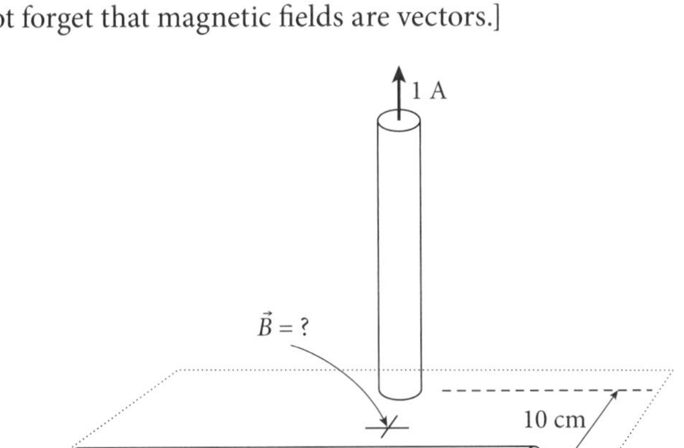

*** 41[f]. A toroidal solenoid is formed by a wire wrapped around a toroid (a shape like a donut or a bagel), as seen top right. If you picture slicing the bagel in half (including all the wires), you get the bottom picture, with current coming out of the page on the inside of the bagel and going into the page on the outside. There are N total loops of the wire around the toroid and the wire carries a current I. Symmetry indicates that the magnetic field must be tangent to circles like the dashed circle shown and of uniform magnitude around the circle. By considering a single loop of the wire, with its upward current on the inside and its downward current on the outside, we know that the direction of the magnetic field is everywhere counterclockwise, as indicated by the red arrows in the bottom picture. So everything is known about the magnetic field inside the toroid except its magnitude. Your assignment is to use Ampère's law to find the magnitude of the magnetic field at any point a distance R from the center of the toroid, as shown in the bottom picture. To do this, you must:

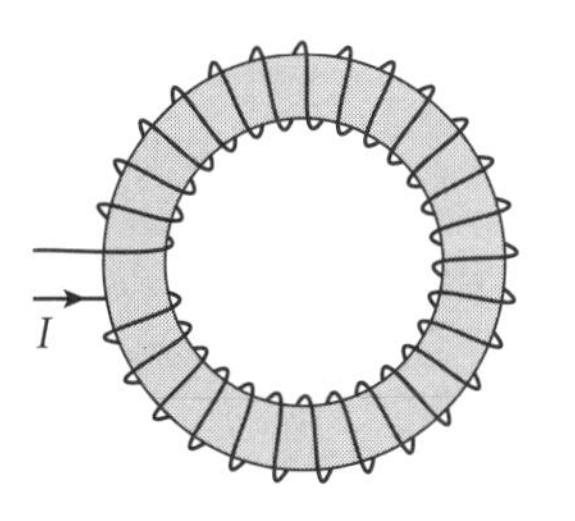

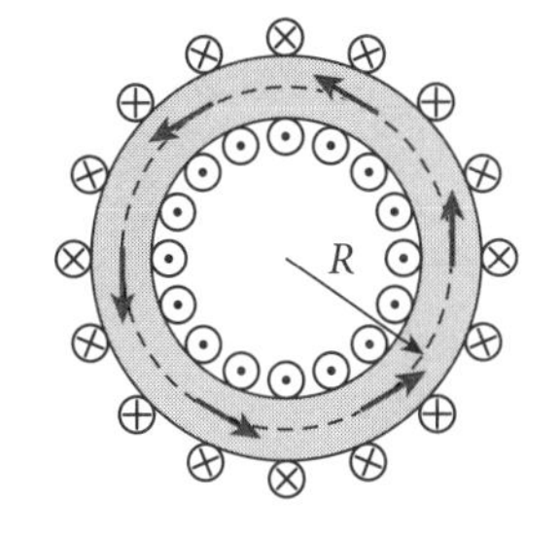

a) At a radius R, choose your path P to be the dashed circle and, calling the magnetic field strength B, calculate the magnetic circulation around the path.

b) Find the total current cutting through the interior of your chosen path.

c) Use Ampère's law to relate the magnetic circulation around the path P to the current cutting through the interior of the path, and solve for the magnetic field at a radius R within the solenoid. [As a check, your answer should be $B = (\mu_0 NI)/(2\pi R)$.]

d) Use Ampère's law to prove that the field outside the toroid is zero.

20

INDUCTION

We have seen that a battery can create a flow of charge (a current) in a wire. Nearly everything in our city runs as a result of this kind of electric flow, and yet none of us remembers a school field trip to the city battery depot. Many of us, however, have taken the "dam field trip" to see the huge hydroelectric generators at the base of a tall dam. These generators, turned by water pressure, are what creates electrical current in our cities using *magnetic induction*. Induction is the subject of this chapter.

20.1 Faraday's Law

We already have learned enough physics to let us understand how magnetic induction arises, at least in one scenario.

Let's consider an electrical circuit composed of a resistor attached to one of three arms of a wire rectangle, as shown in Figure 20.1. Also shown in the figure is a movable metal bar that rides on the side arms of the rectangle and makes electrical contact with them. We also impose a uniform magnetic field going down into the page, as shown. Now consider what happens if we pull the bar to the right with a constant velocity v. Each positive charge inside the bar is an electrical charge moving in a magnetic field. A little working with Right-Hand Rule #1 should convince you that there will thus be a force on each positive charge in the direction of the red arrow. So think what will happen. Positive charges will be forced to the top of the bar, while negative charges flow to the bottom. The bar thus acts exactly like a battery. We can even figure out the voltage of this "battery."

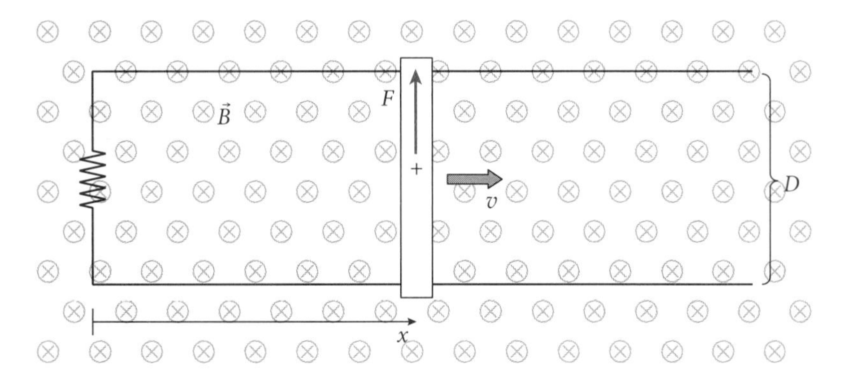

Figure 20.1 A complete circuit is formed by three wires and a movable bar. As the bar moves to the right, positive charges in the bar experience a force in the direction of the arrow. This produces a counterclockwise current in the circuit.

Because of the motion of the bar in the magnetic field, each charge in the bar experiences a force, exactly as if it were sitting in an electric field. In fact, we may say that the motion of the bar creates an *induced electrical field* in the bar. Let us calculate this field.

The force on a charge q in the moving bar is given by Equation 19.1 as

$$F = qvB,$$

and so the induced electric field associated with that force would be

$$E = \frac{F}{q} = vB.$$

When there is constant electric field in some direction, the voltage difference between two points a distance D apart is given by the "ED equation" (Equation 17.4),

$$V = ED = vBD. \tag{20.1}$$

If D is the distance between the two parallel wires, then Equation 20.1 is the "battery" voltage induced by the moving bar. But this may also be thought of in another way.

If the bar's velocity is constant, and if we use a variable x to locate the bar (as in Figure 20.1), then we may write $v = \Delta x/\Delta t$, where Δt is the time it takes the bar to move a distance Δx. If we make this substitution in Equation 20.1, we get

$$V = \frac{BD\Delta x}{\Delta t}.$$

But we could also notice that the three wires and the bar form a rectangle whose area is $A = Dx$, so that $D\Delta x$ is ΔA, the change in the area inside the rectangular current loop formed by the three wires and the bar. If we make this identification, we can write

$$V = \frac{B\Delta A}{\Delta t}. \tag{20.2}$$

Now, at this point, we would like to remind you of a concept from Chapter 16. In Equation 16.10, we defined the electric flux through a surface as the component of the electric field vector perpendicular to the surface, times the area of the surface, $\Phi_E = E_\perp A$. Well, in the same way, we could define a *magnetic flux* as the component of the magnetic field vector perpendicular to a surface times the area of the surface, giving $\Phi_B = B_\perp A$. You can probably see where we're going with this. In Equation 20.2, we ended up with the combination $B\Delta A$. Since the magnetic field in Figure 20.1 is exactly perpendicular to the rectangular surface enclosed by the current loop, $B\Delta A$ is actually the change in the magnetic flux $\Delta\Phi_B$ through that surface. We can therefore write Equation 20.2 as

$$V = \frac{\Delta\Phi_B}{\Delta t}. \tag{20.3}$$

You remember that the electric flux through some surface was proportional to the number of electric field lines cutting the surface. Well, the magnetic flux is likewise proportional to the number of magnetic field lines cutting the surface, for the same reasons. Therefore, Equation 20.3 means that the voltage induced in the moving rod in Figure 20.1 can be thought of as being due to the change in the number of magnetic field lines passing inside the rectangular loop, as the size of that loop increases.

Even though we derived Equation 20.3 for the particular case of a moving rod that changes the area of a loop, it actually works in all cases. Any time the number of magnetic field lines passing through a loop (any loop) changes, there will be an induced battery set up around the loop that will try to push charge around the loop. We call the effective voltage of the induced battery an *electromotive force* (or *emf*).[1] If that loop happens to be made of a conducting wire,

1. This is the traditional name, but it's a bad name. The *emf* is not really a force. It's not even in newtons, but in volts.

then a current will flow in the conductor. Figure 20.2 shows two ways in which the magnetic flux through a wire loop can change. All such ways induce a current in the loop due to an induced *emf* whose magnitude in volts is given by the same Equation 20.3.

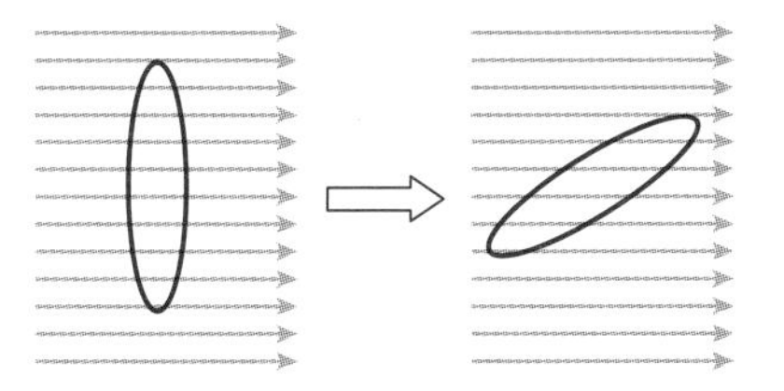

(a) Change the orientation of the loop relative to the magnetic field direction.

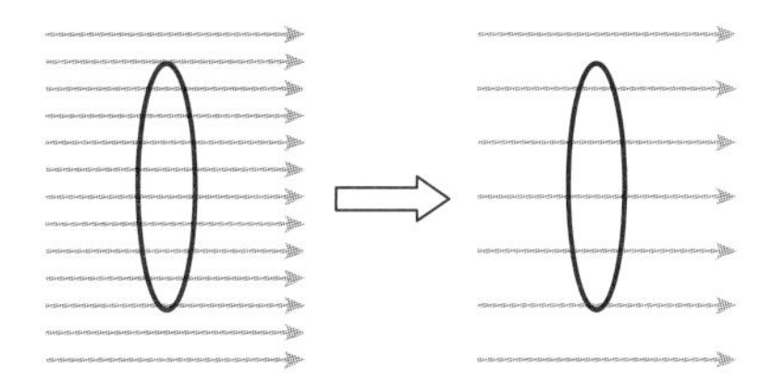

(b) Change the strength of the magnetic field through the loop.

Figure 20.2 Two ways to *reduce* the magnetic flux through a circular current loop. In each case, there will be an *emf* induced in the loop equal to the rate at which the flux is changing.

Of particular interest is the method of changing the flux by changing the orientation of the loop, as shown in Figure 20.2a. A nice analogy is to imagine that you are trying to catch bees in a butterfly net and that there are 20 bees coming toward you in a swarm. If you hold your net so that the plane of the opening is perpendicular to the direction the bees are flying, you will catch more bees than if you hold the plane of the opening parallel to the path of the bees. With the plane parallel, the bees don't see an opening at all. The change in flux due to a change in the orientation of the loop is the basic principle used in electric generators. Large coils of wire are turned by the moving water in the presence of a constant magnetic field. When the coil turns, the number of field lines that pass through the coil is constantly changing, alternately increasing and then decreasing. This alternating *emf* drives an induced alternating current that is carried by power lines to the places where it will be used.

EXAMPLE 20.1 An *emf* from a Rotating Coil

A circular loop of wire of radius 5 cm sits horizontally in a place where there is a magnetic field of strength $B = 2 \times 10^{-6}$ T, pointing directly upward. The loop is then turned slowly so that, after 2 seconds, it lies in a vertical plane, as shown at right. What is the average *emf* induced in the wire loop?

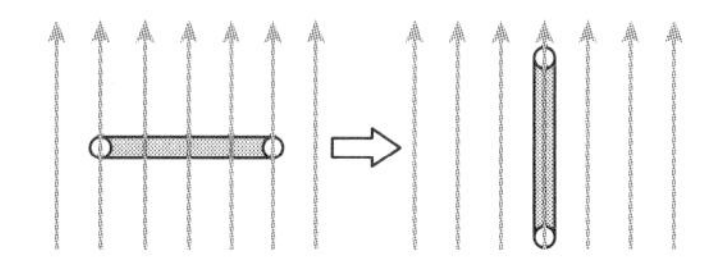

ANSWER In the first picture, there is a magnetic flux through the wire loop, given by

$$\Phi_B = BA = B(\pi r^2) = (4 \times 10^{-6}\ \text{T})(3.14)(0.05\ \text{m})^2 = 3.14 \times 10^{-8}\ \text{T}\cdot\text{m}^2.$$

A $\text{T}\cdot\text{m}^2$ is also called a weber (Wb), though our use of this unit will be limited.

In the picture above right, there are no magnetic field lines cutting through the interior of the loop, so the flux is zero. The magnitude of the change in total flux (not worrying about the sign) is therefore $\Delta\Phi_B = 3.14 \times 10^{-8}$ Wb. The time it took to make the change was $\Delta t = 2$ seconds. So, the magnitude of the average *emf* is

$$V = \frac{\Delta\Phi_B}{\Delta t} = \frac{3.14 \times 10^{-8}\ \text{Wb}}{2\ \text{s}} = 1.57 \times 10^{-8}\ \text{V}.$$

In the last step, we have noted that a weber/s is also a volt.

The purpose of this next example is to make clear that there does not have to be any wire loop or even any charged particles in the place where an electric field is induced by a change in the magnetic flux through an imaginary loop.

EXAMPLE 20.2 An *emf* from a Changing Magnetic Field

The current through a solenoid is slowly being increased. As a result, the magnetic field inside the solenoid is growing at a regular rate. Find a formula for the magnitude of the electric field induced by the changing magnetic flux inside the solenoid, at a distance r from its center.

ANSWER Let us consider the imaginary loop that we show as a dashed circle of radius r in the picture above. Because there is a magnetic field passing through the interior of the dashed loop, there will be a magnetic flux through the loop. As the current in the solenoid grows, the magnetic field strength will increase, and the flux through the loop will increase. We can even calculate the rate at which this occurs. The magnetic field inside a solenoid was found back in Example 19.6. It is

$$B = \frac{\mu_0 NI}{L}.$$

The rate of change in the magnetic field is just due to the changing current,

$$\frac{\Delta B}{\Delta t} = \frac{\mu_0 N(\Delta I/\Delta t)}{L},$$

and the rate of change of the flux through the interior of the imaginary loop is

$$\frac{\Delta \Phi_B}{\Delta t} = \frac{\Delta B}{\Delta t}A = \frac{\mu_0 N(\Delta I/\Delta t)}{L}(\pi r^2).$$

The changing magnetic flux will produce an *emf* in . . . in what? There is no wire in this problem, just empty space. If there *were* a wire, the *emf* would act throughout the circumference of the loop. Any charge making a complete loop around the wire would pass through a voltage difference $V = \Delta\Phi_B/\Delta t$. But we should remember that an increase in voltage is the result of the action of an electric field, in this case, the induced electric field. The relation between the voltage difference and the constant electric field that causes it is the ED equation, where, for this circular path, D would be the distance around the loop. Thus we can calculate the induced electric field that circles around the dashed loop from the induced voltage of Equation 20.3:

$$E = \frac{V}{D} = \frac{\Delta\Phi_B/\Delta t}{2\pi r} = \frac{\mu_0 N(\Delta I/\Delta t)}{2\pi r\, L}(\pi r^2) = \frac{\mu_0 N(\Delta I/\Delta t)}{2L} r.$$

Let us think a bit about what this answer means. We examined an imaginary circle inside a solenoid in which the magnetic field was increasing. If that circle had been a real wire, instead of just an imaginary circle, a current would have begun to flow around the wire, as if it had been accelerated by a circular electric field in the wire. This circular electric field was induced by the changing magnetic flux, not by the wire, so the circular electric field must exist whether the wire is there or not. A changing magnetic flux induces electric field lines that form closed circles in the region where the magnetic flux is changing.

In the previous chapter, we found that the magnetic field inside a solenoid is constant at all points inside the solenoid. However, from the formula we have just found, we see that the induced electric field is not constant. It grows in magnitude as one goes out from the center of the solenoid. The induced electric field produced by a changing magnetic field is proportional to the distance *r* from the center.

There are just a few final items about induced *emf*s we should mention.

First, Equation 20.3 gives the induced *emf* produced by a changing magnetic flux inside a single closed loop. Consider instead a coil of many superimposed loops, as seen in Figure 20.3, with loops of approximately the same radius in nearly the same plane. As a charge goes from one end of the wire to the other, it travels around the loop multiple times. The total *emf* induced in the coil of loops would thus be

Figure 20.3 A coil of many connected loops with the same radius and approximate location.

$$emf = N\frac{\Delta\Phi}{\Delta t} = N\frac{\Delta B_{\perp}A}{\Delta t}, \tag{20.4}$$

where *N* is the number of loops in the coil. The result is to magnify the *emf* induced in the coil by the factor *N*.

Second, let us be clear that *any* change that produces a change in the magnetic flux through a loop induces an *emf* in the loop. If, for example, we move a bar magnet toward a loop of wire, an induced current suddenly appears in the wire. This current flows only while the magnet is moving toward the loop (changing the strength of the B-field in the loop). As soon as the magnet stops, the current stops. It doesn't matter how strong the magnetic field from the magnet is, no current will be induced until the flux changes. If we now pull the magnet away, the current resumes in the opposite direction. Let us also note that Equations 20.3 and 20.4 involve the time Δt in which the flux changes. If we move a bar magnet quickly toward a wire loop, we find that the induced current is much greater than if we move the magnet slowly. In both cases we change the magnetic flux through the loop by the same amount, but in the first case we are changing the flux through the loop by a greater amount *per second*.

Finally, let us give credit to the discoverer of this principle. British physicist Michael Faraday discovered magnetic induction in an almost perfect ignorance of the mathematics. He simply stated the connection between the change in the number of field lines through a loop and the induced *emf* around in the loop. His words can be translated into a compact equation that is known as Faraday's Law:

$$emf = -\frac{\Delta\Phi_{\mathrm{B}}}{\Delta t}, \tag{20.5}$$

where Φ_{B} is the magnetic flux through a single loop of the wire. If there are *N* loops in a coil, as in Figure 20.3, the *emf* will be multiplied by *N*.

20.2 Lenz's Law

There is one new feature as we have written Faraday's law that needs a little discussion. Equation 20.5 looks just like 20.3, except that in Equation 20.5 we have introduced a minus sign. What does this mean? In a way, this is one of the more important minus signs in physics. It even has its own name. It is called Lenz's Law.

It is possible to use Equation 20.5, complete with its minus sign, to determine the direction in which an induced electric field will point or in which an induced current will flow. The geometry needed, however, involves carefully defining the orientation of the surface bounded by the circuit and connecting it to the sense in which one goes around the circuit, using a (you guessed it!) right-hand rule. We have included the minus sign in Equation 20.5 because this is the standard form of Faraday's law and because its being there serves as a reminder that the induced electric field has a direction that may be determined. But there is a simpler way of expressing and using Lenz's law. Let us share this with you.

Before we can do this, we need to distinguish between two different kinds of magnetic flux. There is, of course, the magnetic flux created by an external magnetic field passing through the interior of a circuit loop. We call this the "external flux." However, as the changing external flux through the interior of the loop induces an *emf* and an associated current in the loop, that induced current is a real current and will itself create magnetic field lines that will pass through the interior of the loop. The flux produced by this induced magnetic field is what we will call the "induced flux."

With this distinction in mind, we may express Lenz's law in this way:

- Lenz's law: Any change in the external magnetic flux through a circuit loop will induce a current in the loop that will flow in such a direction as to produce an induced flux through the loop that opposes the change in the external flux.

We suspect that a few examples will be the best way to see how this works.

EXAMPLE 20.3 Lenz's Law for a Changing Field—I

A copper loop is lying horizontally on a table. There is a uniform magnetic field in the region of the loop that points downward, perpendicular to the table as shown. If the magnetic field is getting weaker each second, find the direction of the induced current in the loop.

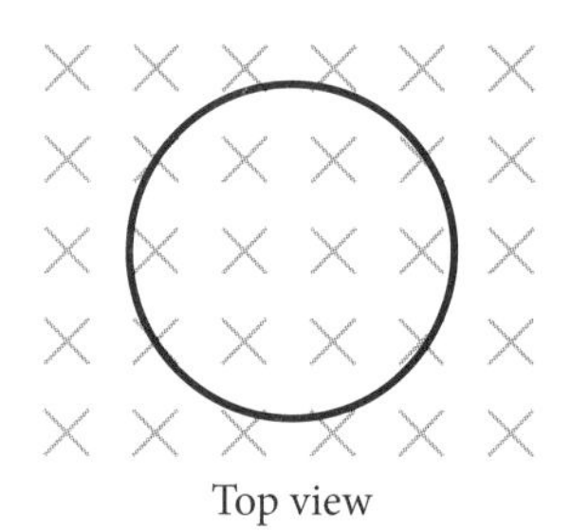
Top view

ANSWER The change that must be opposed is the loss of B-field lines going down into the table. The induced *emf* will therefore oppose this loss by providing an increased induced flux into the table. To produce a new magnetic field into the tabletop, the induced current must be clockwise around the loop, as given by Right-Hand Rule #3 (see page 394).

EXAMPLE 20.4 Lenz's Law for a Changing Field—II

What if the external magnetic field in the previous example were getting stronger each second? Would the direction of the induced current still flow clockwise?

ANSWER The change that must be opposed in this case is the sudden appearance of new B-field lines going down into the table. The induced *emf* can oppose this increase in the external flux if its own induced field is in the opposite direction as the external field. It must therefore drive an induced current counterclockwise around the loop, not clockwise. The direction is again found from Right-Hand Rule #3.

EXAMPLE 20.5 Lenz's Law for a Coil Moving in an External B-Field

The figure at right shows a loop of wire lying flat on a tabletop next to a section of current-carrying wire. An external agent is pushing the loop along the tabletop toward the wire as shown. The resulting induced current in the loop is clockwise, as indicated. Use this information to determine the direction of the current in the straight wire segment.

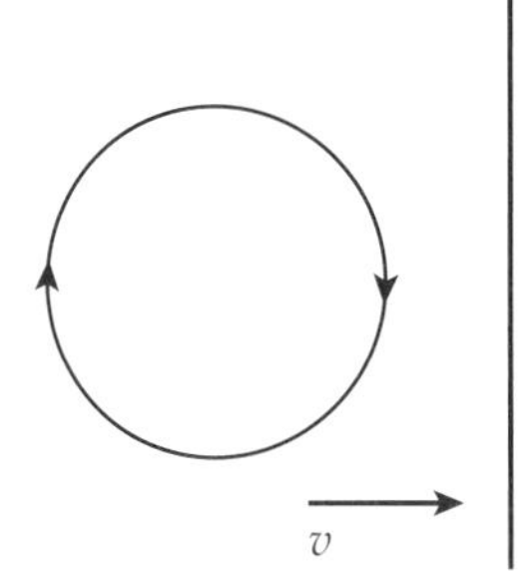

ANSWER This example is a little tricky. If there is an induced clockwise current in the loop, then Right-Hand Rule #3 tells us that the *induced field* in the loop is directed into the page. According to Lenz's law, this means that the *external field* in the loop must be changing in a way that reduces the flux into the page through the loop. A reduction of the external flux into the page could be produced in one of two ways:

1. The external magnetic field is directed *into* the page and is decreasing in strength as the loop moves toward the wire.
2. The external magnetic field is directed *out of* the page and increasing in strength as the loop moves toward the wire.

However, we know from Equation 19.4 that a loop moving toward the wire will see a magnetic field due to the wire that is getting stronger and stronger. This eliminates option 1 and tells us that the external magnetic field from the wire must be directed out of the page. Right-Hand Rule #2 (see page 392) tells us that the current that produces a magnetic field out of the page to the left of the wire must be moving upward, directed toward the top of the page in the figure above.

Once we are familiar with using Lenz's law in this way, we can easily drop the minus sign in Faraday's law (Equation 20.5), except as a reminder that we need to apply Lenz's law to get the direction in which charges would move around the loop. So, from now on, we will use Faraday's law *without* the minus sign to calculate the *magnitude* of an *emf* produced by a changing magnetic flux and then use our version of Lenz's law to figure out the *direction*.

Finally, let us notice that Lenz's law also represents an application of the law of conservation of energy. Consider a situation where the north pole of a bar magnet is brought toward a copper loop (Figure 20.4). As the magnet is pushed toward the loop, an induced current suddenly appears in the loop. This current represents new kinetic energy that didn't exist in the loop before we started moving the magnet, and this energy must have come from somewhere. If energy is being added to the system, it must be that work is being done on the system. But how can that be? All we have done is to move a bar magnet closer to an inert loop of wire. Well, we need to remember that an induced current flowing in a wire loop will create a magnetic field that looks like the magnetic field of a bar magnet. Lenz's law tells us that the induced magnetic field will oppose the increasing external magnetic field, so the north pole of the induced magnetic field will point toward the north pole of the bar magnet and repel the bar magnet. The work it takes to move the bar magnet against this real force from the induced magnetic field is the source of the energy that shows up as current in the wire. Similarly, it

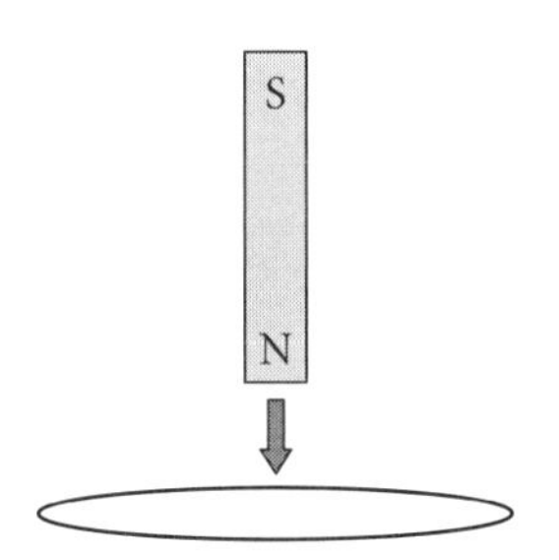

Figure 20.4 A magnet moving toward a loop feels a resisting force from the induced current in the loop.

takes work to spin a loop of wire in the static magnetic field found in a hydroelectric generating plant, since the magnetic field from the induced current in the loop experiences a torque that opposes its spin. Without the force of the water flowing from behind the dam, the wire coil would quickly be brought to rest by this torque.

20.3 Inductors

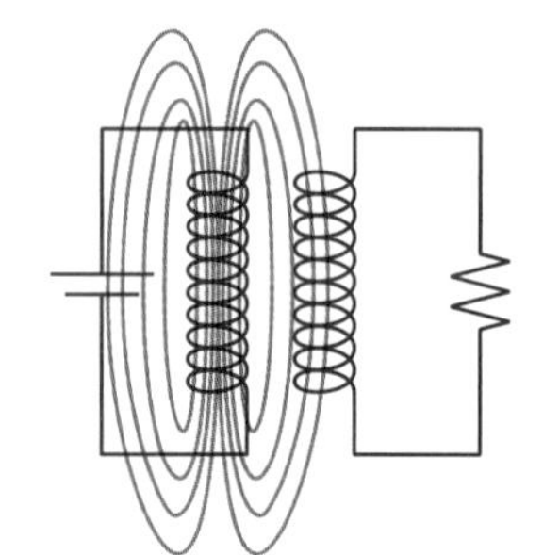

Figure 20.5 A changing field in one coil can induce a current in a nearby coil.

In Example 20.5 in the previous section, we saw how a current in one circuit (the straight wire) can induce a current in another circuit (the loop). In Figure 20.5 we show a similar situation in which there are two coils, one in which there is a current flowing and one in which there is no initial current. If the two coils are near each other, a changing current in the first coil (the primary coil) will produce a changing magnetic flux (number of field lines) in the nearby coil (the secondary coil). By Faraday's law, this changing flux creates an induced *emf* (voltage) that drives an induced current. In accord with Lenz's law, the direction of the current through the secondary coil will be the one which produces an induced magnetic field that opposes the change in the external magnetic flux through itself. We know that the field outside a solenoid is very weak, so the connection depicted in Figure 20.5 will not be very great. But if the secondary coil is closer, or is actually wrapped right over the primary coil, then more of the field lines from the primary will pass through the secondary. In such a case, we say that the two coils are "more strongly coupled."

Let us quantify the magnetic coupling between two coils by defining a quantity called the *mutual inductance*, *M*, between the coils. It will be defined as the constant of proportionality between the current flowing in the primary coil and the magnetic flux that appears in the secondary coil:

$$\Phi_2 = MI_1. \tag{20.6}$$

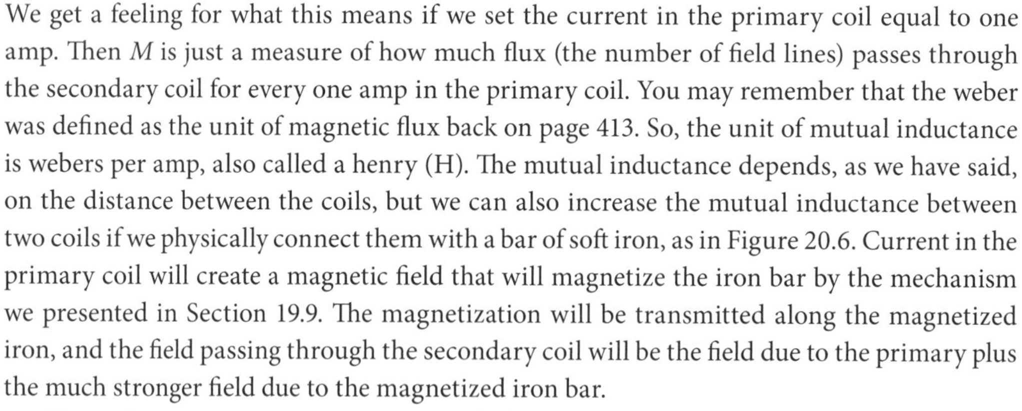

We get a feeling for what this means if we set the current in the primary coil equal to one amp. Then M is just a measure of how much flux (the number of field lines) passes through the secondary coil for every one amp in the primary coil. You may remember that the weber was defined as the unit of magnetic flux back on page 413. So, the unit of mutual inductance is webers per amp, also called a henry (H). The mutual inductance depends, as we have said, on the distance between the coils, but we can also increase the mutual inductance between two coils if we physically connect them with a bar of soft iron, as in Figure 20.6. Current in the primary coil will create a magnetic field that will magnetize the iron bar by the mechanism we presented in Section 19.9. The magnetization will be transmitted along the magnetized iron, and the field passing through the secondary coil will be the field due to the primary plus the much stronger field due to the magnetized iron bar.

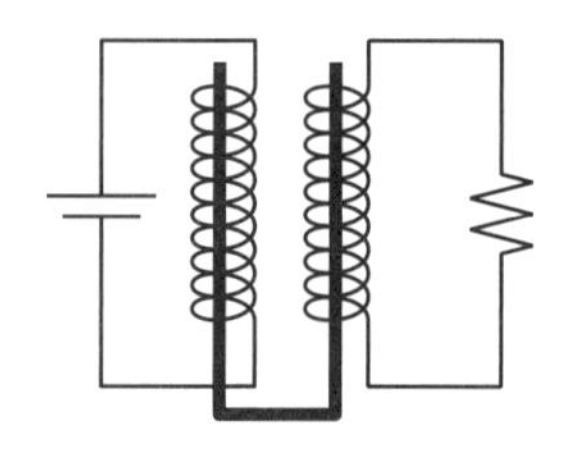

Figure 20.6 A U-shaped iron bar increases the mutual induction of the two coils.

If we substitute Equation 20.6 into Faraday's law (Equation 20.5, but leaving out the minus sign), we find:

$$emf_2 = \frac{\Delta\Phi_2}{\Delta t} = M\frac{\Delta I_1}{\Delta t}.$$

The strength of the induced *emf* in the secondary coil is proportional to how quickly we change the current in the primary coil, with the coefficient of proportionality being the mutual inductance of the coils.

EXAMPLE 20.6 Mutual Inductance of a Solenoid and a Single Wire Loop

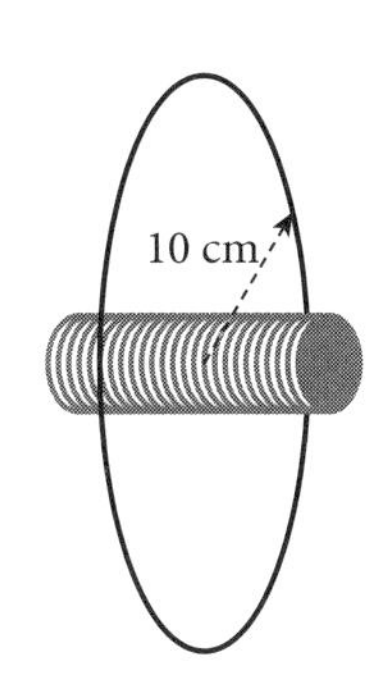

A 100-turn solenoid of length 10 cm and radius 1 cm pierces through the center of a single loop wire of radius 10 cm, as shown. Find the mutual inductance of the solenoid and the loop.

ANSWER We found the magnetic field inside a solenoid in the last chapter, in Example 19.7, and found that it was constant, directed along the solenoid axis inside the solenoid, and nearly zero outside. Since the magnetic field inside the solenoid is so easy, it would be best to take the solenoid as the primary coil and the single loop as the secondary coil. In order to find the mutual inductance, using Equation 20.6, we assume an unspecified current I_1 flowing in the primary coil (the solenoid) and calculate the subsequent magnetic flux Φ_2 through the secondary coil.

The magnetic field inside a solenoid is a constant

$$B = \frac{\mu_0 N I_1}{L}.$$

For the particular values of this solenoid, this gives (using SI units throughout)

$$B = \frac{(4\pi \times 10^{-7})(100)}{0.1} I_1 = 1.26 \times 10^{-3}\, I_1.$$

To find the flux through the single loop, the secondary coil, we note that the field is zero except inside the solenoid. So the only area with any magnetic flux through it is the area inside a circle of radius 1 cm, the area being $\pi r^2 = (3.14)(0.01\text{ m})^2 = 3.14 \times 10^{-4}\text{ m}^2$. The total flux through the interior of the loop is therefore

$$\Phi_2 = B(\pi r^2) = (1.26 \times 10^{-3}\, I_1)(3.14 \times 10^{-4}) = 3.9 \times 10^{-7}\, I_1.$$

Comparison with Equation 20.6 shows that the mutual inductance is

$$M = \frac{\Phi_2}{I_1} = 3.9 \times 10^{-7}\text{ H}.$$

Notice how the assumed I_1 current canceled out of the final calculation. This will always happen when we do such a calculation, because the mutual inductance is completely determined by the geometry of the inductors alone.

So far in our discussion of induction, we have dealt with the changing flux through some circuit due to some external agent, like the changing current in another nearby circuit. The coil, however, cannot distinguish between field lines that are generated by an external agent and field lines generated by its own current. Field lines are field lines (and flux is flux). If the current through a coil is constant, the magnetic field inside the coil will be whatever it is going to be. But, if the current starts to change, the coil will fight the resulting change in the number of field lines passing through it, regardless of the origin of those field lines. This fact leads us to the investigation of an effect that is very important for circuits in which the current varies, the case of a coil resisting a change in its own magnetic flux.

We begin by defining the self-inductance, L, of a single coil in the same way we defined mutual inductance:

$$\Phi_1 = LI_1. \tag{20.7}$$

The magnetic flux through the single coil is proportional to the current in that coil, with the coefficient of proportionality being the self-inductance of the coil. We again get a feeling for what this means if we set the current in the coil equal to one amp. In this case, L is a measure of how many field lines (flux) pass through the interior of the coil for every amp that passes through the coil. The unit of self-inductance is also the henry (H). We can increase the coil's self-inductance by increasing the number of turns in the coil. We may also increase the inductance of a in the manner of Figure 20.6 by wrapping the coil around a bar of iron.

If we substitute Equation 20.7 into Faraday's law, we find

$$emf_1 = \frac{\Delta\Phi_1}{\Delta t} = L\frac{\Delta I_1}{\Delta t}. \tag{20.8}$$

The strength of the induced *emf* in a single coil is proportional to how quickly we change the current in that coil, with the coefficient of proportionality being the self-inductance of the coil.

EXAMPLE 20.7 The Self-Inductance of a Solenoid

What is the self-inductance of a solenoid of radius R having N total turns in a total length ℓ?

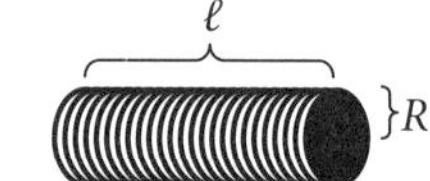

ANSWER To find the inductance of any circuit element, we assume a changing current $\Delta I/\Delta t$ and find the *emf* induced in the element by the changing flux. Then, using Equation 20.8, we can read off the self-inductance as whatever the coefficient of $\Delta I/\Delta t$ comes out to be. The field inside a solenoid composed of N/ℓ turns per meter was found in Example 19.7 to be $B = \mu_0 NI/\ell$, where I is the current flowing in the solenoid. The flux through a *single loop* of the solenoid is $\Phi = BA$, the area being $A = \pi R^2$. The flux can thus be written $\Phi = (\pi R^2 \mu_0 NI)/\ell$. It is only the current that changes, so, for a solenoid of N total turns, the *emf* induced will be $emf = N(\Delta\Phi/\Delta t) = (\mu_0 \pi R^2 N^2/\ell)(\Delta I/\Delta t)$. Comparison with Equation 20.8 allows us to read off the self-inductance of a solenoid as

$$L = \mu_0 \frac{\pi R^2 N^2}{\ell}.$$

We remember that the minus sign in Faraday's law tells us that the induced *emf* in the coil will be in the direction that fights any change in flux through the coil (Lenz's law), including changes in its own current. If the current through a coil were to change *instantly,* then $\Delta I_1/\Delta t$ would be infinite and Faraday's law would predict that an infinite *emf* would be generated in the coil to fight the change. So, in real life, we can never instantly change the current through a coil. The current through a coil must gradually increase or gradually decrease. You have probably seen evidence of this before. If you pull the electric plug on a waffle iron, you get a spark from the plug to the socket. The waffle iron contains heating coils, and they have self-inductance, so, when you pull the plug, you are trying to instantly change the current through those coils to zero. Faraday's law will not allow this, so an induced *emf* is generated by the coils to force the current to gradually decrease, leading to the spark.

Let's look at an example of a simple circuit that contains a coil (also called an inductor). A real inductor is actually two circuit elements in one—the self-inductance, represented by the coil symbol, and an internal resistance in series with it, arising from the fact that the coil is formed from a long thin wire which has resistance.

EXAMPLE 20.8 An Electric Circuit Containing an Inductor

The following questions refer to the circuit represented by the diagram below. Think of R_3 as the internal resistance of the inductor L.

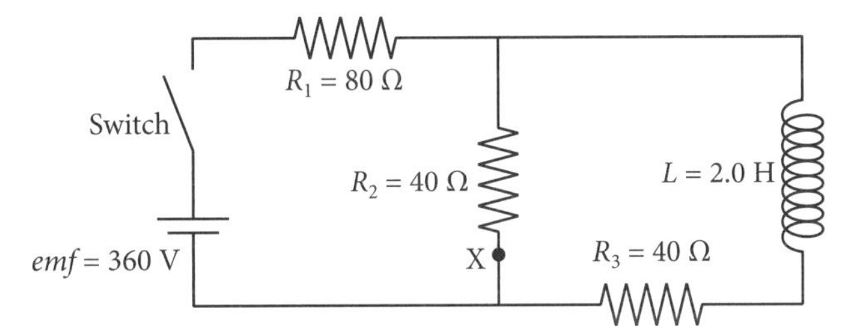

A. Immediately after the switch is closed, find:
   i. the magnitude and direction of the current (if any) through the battery;
   ii. the voltage difference across the inductor (coil).

ANSWER A With the switch open there is no current in the circuit. When the switch is closed, the current through the resistors can change instantly, but the current through the inductor must gradually increase from zero to its final value. Since we are looking "immediately after the switch is closed," the inductor will initially provide whatever backward *emf* is required for the current through it to remain zero. The inductor thus acts like an open switch, cutting off the loop on the right. The total resistance of the circuit is then just that in the other branch, $R_1 + R_2 = 120\ \Omega$. By Ohm's law, the conventional current will be 360 V/120 Ω = 3 A, passing upward through the battery. We can find the voltage difference across the inductor by using Kirchhoff's loop theorem on the right loop, going clockwise from point X: This gives $+V_2 - V_L - V_3 = 0$. Since no current initially flows through R_3, V_3 will be zero, and so the voltage across the inductor will be $V_L = V_2 = (3\text{ A})(40\text{ W}) = 120$ volts.

B. A long time after the switch is closed, find:
   i. the magnitude and direction of the current (if any) through the battery;
   ii. the voltage difference across the ideal inductor, that is, ignoring the internal resistance R_3.

ANSWER B After the switch has been closed for a long time, the current through the inductor has ramped up to its final value and the inductor has stopped opposing a changing current through itself. The inductor now just acts like any other wire and the voltage difference across the ideal coil will be zero. The two 40-Ω resistors (R_3 being the internal resistance of the inductor, which is always there as long as current flows through it) are now in parallel, and so they are equivalent to a single 20-Ω resistor. The total resistance of the circuit is then 100 Ω, and the conventional current through the battery will be 360 V/100 Ω = 3.6 A.

C. After the switch has been closed for a long time, it is suddenly opened. Immediately after the switch is opened, find:

i. the magnitude and direction of the current (if any) through the resistor R_2;
ii. the voltage difference across the inductor.

ANSWER C With the switch closed for a long time, we found the current through the battery to be 3.6 A, with half of it, 1.8 A, passing down through the inductor. But the current through an inductor cannot change instantly. So, right after the switch is opened, the current through the inductor must remain at 1.8 A flowing downward. With the switch open, current no longer flows through the battery (because of the open switch). However, the inductor will act like a battery to keep the current through it from changing instantly. But it is only the current through the inductor that cannot change instantaneously. The current through the 80-ohm resistor is cut off by the open switch, so the 1.8 A that continues to flow down through the inductor and through R_3 will now suddenly all flow upward through R_2 on its way back to the inductor. R_3 and R_2 are now in series, carrying a 1.8-A current through them. The inductor must initially look like a battery strong enough to drive 1.8 A through two 40-Ω resistances in series, so we must have $V_L = (1.8\text{ A})(80\ \Omega) = 144$ volts.

20.4 Induced Magnetic Fields

We saw in Section 20.1 that a changing magnetic field induces a circular electric field in its vicinity. So what about the inverse? Will a changing electric field induce a circular magnetic field in the area around it? Let's look into this. We begin with one method we know for creating a circular magnetic field, a long current-carrying wire.

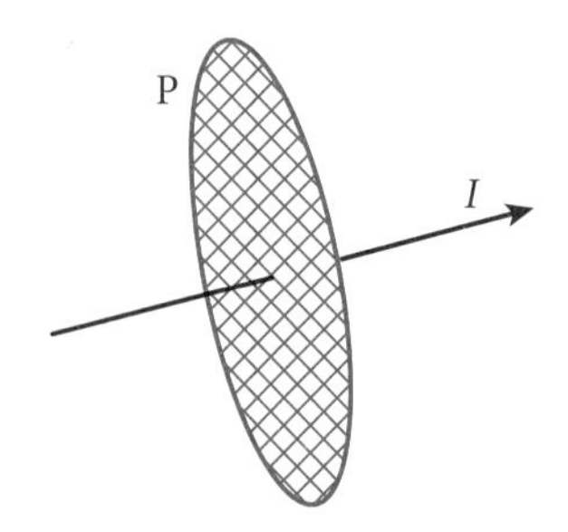

Figure 20.7 An Ampère's law path and surface.

Back in Chapter 19, after we first derived Ampère's law (Equation 19.5), we discussed how it could have been used to find the magnetic field a distance r from a long current-carrying wire. We chose a circle of radius r, centered on the wire, and used the fact that the current cutting through the surface that is bounded by the chosen circle is related to the circulation of the magnetic field around the circle. Figure 20.7 is identical to Figure 19.14, with the surface bounded by the red Ampère's law path P being the plane cross-hatched disk inside the circle and with the current cutting the surface being the current I carried by the wire. The current cutting the surface is related to the magnetic circulation around the path P by Ampère's law:

$$C_B = 2\pi r B = \mu_0 I_{\text{CUTTING}}. \tag{20.9}$$

But the simple disk that filled the inside of the circle in Figure 20.7 is not the only surface for which this law works. Figure 20.8 shows a net in the shape of a topless cylinder—curved sides and a flat bottom—where the rim of the cylinder is the path P that bounds the surface. All the current that cut through the disk in Figure 20.7 will also pass into this open cylindrical net in Figure 20.8, and must eventually cut through the cross-hatched surface of the net somewhere in order to pass out of it. So you don't have to use a disk as your surface in Ampère's law. It will work just as well with any surface that has the path P as its boundary, because any current that passes into the mouth of the net will have to cut through it somewhere.

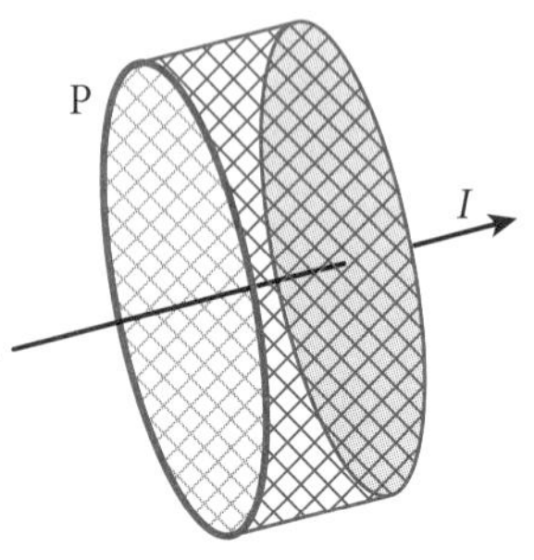

Figure 20.8 The same path P bounding a different surface.

Or will it? Suppose we cut the current-carrying wire in Figure 20.8 and insert two capacitor plates of area A, as seen in Figure 20.9. Now there will be no current cutting through the interior surface of the net, because the charge will all pile up on the leftmost plate of the capacitor and drive positive charge off of the rightmost plate. The current will continue into the opening of the net as it flows along the wire, but, as positive and negative charge build up on the capacitor plates, no actual charge will pass through the cylindrical surface bounded by P. We could simply amend Ampère's law to state that you can pick any surface you want that has the chosen path as its boundary, as long as there are no places, like a capacitor, where the charge can stop. But this would be an ugly revision. And we don't have to do it.

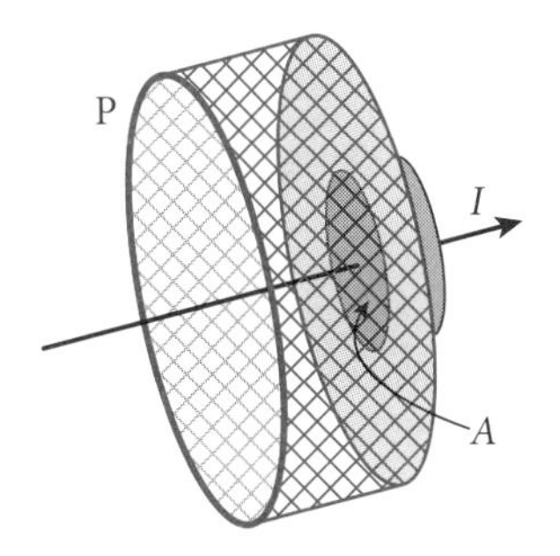

Figure 20.9 The same wire with two capacitor plates inserted.

So let us notice that, while there is no current cutting through the surface anywhere in Figure 20.9, there is *something* going on in the region where the surface passes between the two capacitor plates. Between the two charged plates, there is an electric field, and, as the charge builds up on the plates, the electric field between the plates will increase. We might wonder. Is it possible to use the changing electric field through the surface of the net to give the same magnetic field around the circular rim of the net that Ampère's law gave using the current? It turns out that it is.

Let's start with the field. The electric field in a parallel-plate capacitor of area A carrying charge Q was derived back in Chapter 17 (Equation 17.2). It was

$$E = \frac{\sigma}{\varepsilon_0} = \frac{Q}{A\varepsilon_0}.$$

So, if the charge on the plates changes, the rate at which the electric field changes is

$$\frac{\Delta E}{\Delta t} = \frac{\Delta Q/\Delta t}{A\varepsilon_0} = \frac{I}{A\varepsilon_0}, \qquad (20.10)$$

where, in the last step, we noted that the rate of change of the charge on the plates is equal to the current flowing along the wire. If we solve this equation for I, we find

$$I = \varepsilon_0 A\frac{\Delta E}{\Delta t} = \varepsilon_0\frac{\Delta\Phi_{\text{E,CAP}}}{\Delta t}, \qquad (20.11)$$

where we have recognized AE as the flux $\Phi_{\text{E,CAP}}$ through the portion of the surface that lies between the capacitor plates.

Since the electric field *outside* the region between the plates is nearly zero, this $\Phi_{\text{E,CAP}}$ flux calculated for the area between the plates is also the total flux Φ_{E} for the entire cross-hatched surface in Figure 20.9. So Equation 20.11 means that ε_0 times the rate of change of the total electric flux through the entire surface bounded by the dark red path in Figure 20.9 is equal to the current in the wire that cuts through the simple surface of Figure 20.7. The quantity on the right side of Equation 20.11 is called the *displacement current.* It has units of current (amps), though it is not a real current.

The point of all this is that, if we add the displacement current to the true current cutting through whatever surface we might want to pick for applying Ampère's law, we get a version of the law that works in all cases. We get

$$C_{\text{B}} = \mu_0 I_{\text{CUTTING}} + \mu_0\varepsilon_0\frac{\Delta\Phi_{\text{E}}}{\Delta t}, \qquad (20.12)$$

where I_{CUTTING} is the true current cutting through the surface and Φ_{E} is the total electric flux through the surface. In Figures 20.7 and 20.8, there is a current cutting the surface bounded by the circular path, but there is no charge built up anywhere, and so there is no electric field or electric flux. In Figure 20.9, no current cuts through the surface, but there is a changing electric flux through the surface. Equation 20.12 gives the correct magnetic circulation and the correct magnetic field for all cases.

Equation 20.12 is known as Maxwell's extension of Ampère's law. It tells us that there are two ways to create a magnetic field. A magnetic field may be created either by moving charge in an electric current, or by a changing electric flux, or both. And, like Gauss's law, it also lets us solve problems where there is a lot of symmetry.

EXAMPLE 20.9 The Magnetic Field from the End of a Current-Carrying Wire

A uniform current I flows in a wire that terminates in a small bead. Find the magnetic field at a point a perpendicular distance r from the wire in the plane that includes the bead.

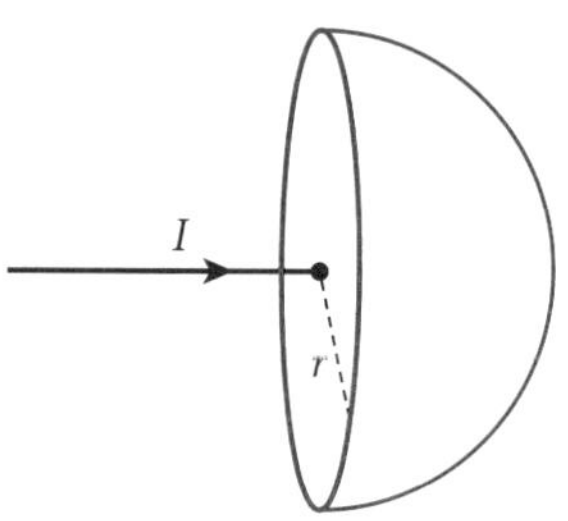

ANSWER We draw the wire in the diagram at right and choose an imaginary closed circular path that is centered on the bead and whose circumference passes through the point where we want to evaluate the magnetic field. The current in the wire does not actually penetrate the disk inside the circular path, so the I in Ampère's law is zero. However, as charge builds up on the bead, there will be an increasing electric field all around the bead. We know the electric field due to a point source (like the bead) from Coulomb's law. The magnitude of the electric field will be uniform everywhere on the surface of the half-sphere drawn in the diagram, and that electric field is everywhere outward. So, if we choose a hemispherical surface bounded by the circular path of radius r, we can use Ampère's law for this path and this surface. This will relate the magnetic circulation around the path to the changing electric flux through the surface.

When the charge on the bead is q, the electric field on the hemispherical surface bounded by the circular path is of constant magnitude

$$E = \frac{1}{4\pi\varepsilon_0}\frac{q}{r^2}.$$

Since the area of a hemisphere is $2\pi r^2$, the electric flux through the hemispherical surface is $\Phi_E = (2\pi r^2)E$. The flux through the surface can therefore be written as

$$\Phi_E = (2\pi r^2)E = \frac{2\pi r^2}{4\pi\varepsilon_0}\frac{q}{r^2} = \frac{q}{2\varepsilon_0}.$$

Since the charge on the bead changes at a rate $\Delta q/\Delta t = I$, the rate of change in the flux is $\Delta\Phi_E/\Delta t = I/2\varepsilon_0$. The circulation of the magnetic field around the red circular path we chose is $C_B = 2\pi rB$, and Ampère's law gives us

$$2\pi rB = \mu_0 I_{\text{CUTTING}} + \mu_0\varepsilon_0\frac{\Delta\Phi_E}{\Delta t} = 0 + \mu_0\varepsilon_0\frac{I}{2\varepsilon_0} = \frac{\mu_0 I}{2} \quad \Rightarrow \quad B = \frac{\mu_0 I}{4\pi r}.$$

Note that this is just half the magnetic field that Equation 19.4 gave for a complete long wire carrying a current I. This is obviously because the present problem only has one half of a long wire as the source of the magnetic field.

20.5 Maxwell's Equations

In the last five chapters, we have presented the fundamental new physics related to electric and magnetic fields. We began with Coulomb's law in Chapter 16 and have now concluded with Maxwell's extension to Ampère's law. This is a good time for us to mention the definitive work of the English physicist, James Clerk Maxwell, who put all of this together in the 1860s in a complete structure for finding and relating electric and magnetic fields.

Maxwell's work was a work of synthesis in which the fundamental content of electromagnetism was shown to be the result of a succinct set of four laws, known as Maxwell's Equations. They are a little too abstract to be of much use in solving the kinds of problems you have been and will be solving. Nevertheless, they provide a way of looking at electric and magnetic fields that will help us understand how electromagnetic waves arise, as we discuss them in our next chapter.

So let us summarize Maxwell's equations here, bringing together the relevant equations from previous chapters, as needed.

- A charge Q_{inside} contained inside an imaginary closed surface will produce an outward electric flux through the surface given by Gauss's law:

$$\Phi_{\text{E}} = \frac{Q_{\text{inside}}}{\varepsilon_0}. \tag{16.12}$$

  Because the flux (the net number of field lines crossing the closed surface) is not zero, we know that electric field lines can originate on positive charges and diverge outward or terminate on negative charges by converging inward.

- There are no isolated magnetic poles, so the magnetic flux through any closed imaginary surface will be zero. We express this with a new equation:

$$\Phi_{\text{B}} = 0. \tag{20.13}$$

  This means that there can be no radially diverging magnetic fields. Magnetic field lines must be closed curves, without origin or termination. Any magnetic field line that crosses a Gaussian surface going inward will have to cross it again somewhere going outward.

- Closed electric field lines can be created by changing magnetic fields, as given by Faraday's law, Equation 20.5. If we recognize that the *emf* induced around a closed electric field line of spatial length L_{P} follows the "ED equation," so that $emf = EL_{\text{P}} = C_{\text{E}}$, then Equation 20.5 can be written in a new form:

$$C_{\text{E}} = -\frac{\Delta\Phi_{\text{B}}}{\Delta t}, \tag{20.14}$$

  where we have included the minus sign from Lenz's law.

- Closed magnetic field lines can be created by electrical currents or by changing electric fields, according to Maxwell's extended version of Ampère's law:

$$C_{\text{B}} = \mu_0 I_{\text{CUTTING}} + \mu_0\varepsilon_0\frac{\Delta\Phi_{\text{E}}}{\Delta t}. \tag{20.12}$$

  Note that the changing electric flux enters the equation with a plus sign, instead of the minus sign of Equation 20.14. The direction of the induced magnetic field will not oppose the change in flux; it will enhance it. This means it follows a right-hand rule. If the electric field is increasing in the direction of your right thumb, the induced circular magnetic field will be in the direction your fingers curl.

The power of Maxwell's equations is that, given a particular distribution of electrical charges and currents, these equations are sufficient to determine the electric and magnetic fields throughout all space. The mathematics required for this effort is the area of advanced calculus known as partial differential equations, so you can feel quite safe that we are not going to take you any further into this. But we felt you would like to know this important fact.

Before we close out this section and this chapter, we would like to mention an idea which is presently only a speculation, but which arises in many modern theories of fundamental particles. This is the idea that there might exist isolated magnetic poles in nature—the so-called "magnetic monopoles." These would be particles that carry the fundamental magnetic pole, just like electrons and protons carry the fundamental electrical charge. The existence of isolated magnetic poles would, among other things, create a kind of symmetry in Maxwell's equations. Instead of Equation 20.13, stating that the magnetic flux through a closed surface must be zero, there would be a Gauss-like law relating a radially diverging magnetic field to the pole strength inside the surface, more like Equation 16.12. There would also be magnetic pole currents that would act as additional sources of electric field circulation by adding a magnetic-current term to the right-hand side of Equation 20.14, so it would look more like Equation 20.12. Therefore, the resulting set of Maxwell's equations would end up being symmetric under the interchange of electric and magnetic charge, electric and magnetic fields. It is important to be clear, however, that, although many theoretical physicists expect magnetic monopoles to exist because they are predicted in so many modern unified theories of fundamental particles, there have been many experimental searches for the monopole, and they have all turned out negative. At the time of this writing, there is no experimental evidence for the existence of free magnetic poles.

20.6 Summary

From this chapter, you should have learned the following things:

- You should be able to use Faraday's law (Equation 20.3 for a single loop or N times Equation 20.3 for multiple loops) to find the *emf* induced in a current loop by a changing magnetic flux through the surface bounded by the loop.
- You should be able to use Lenz's law to find the direction in which an induced current will flow.
- You should be able to use the definitions of mutual inductance and self-inductance to relate the induced *emf* to the changing magnetic flux and to the current that creates it.
- You should be able to use Maxwell's extension of Ampère's law to find the magnetic field around a closed loop, when there is sufficient symmetry to relate the magnetic field to the magnetic circulation around the loop, as was done in Example 20.8.

CHAPTER FORMULAS

Faraday's law: $C_E = -\dfrac{\Delta\Phi_B}{\Delta t}$ or $emf = -N\dfrac{\Delta\Phi_B}{\Delta t}$

Induction: $\Phi_2 = MI_1$ $\Phi_1 = LI_1$ Displacement current: $I_d = \varepsilon_0\dfrac{\Delta\Phi_E}{\Delta t}$

Ampère's law (with Maxwell's extension): $C_B = \mu_0 I_{CUTTING} + \mu_0\varepsilon_0\dfrac{\Delta\Phi_E}{\Delta t}$

PROBLEMS

20.1 * 1. A constant magnetic field of strength $B = 10^{-4}$ T points at an angle of 37° into a plane surface 10 cm on a side, as depicted at right. What is the total magnetic flux through the surface?

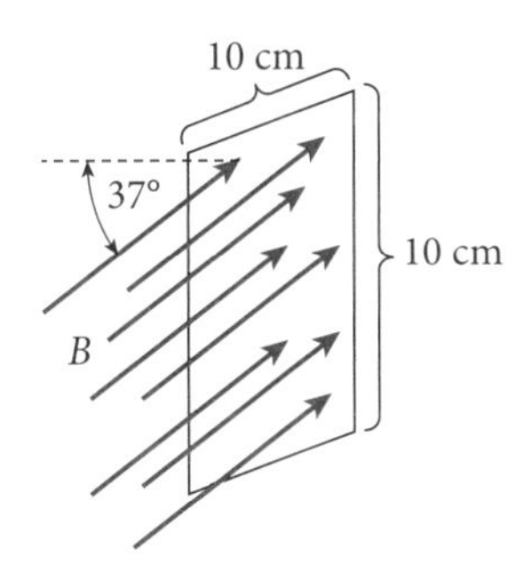

** 2. It is desired to use a magnetic field of 2.00×10^{-4} teslas to produce a total magnetic flux of 6.28×10^{-8} webers through a circular loop of wire. How should the loop be oriented relative to the field if we want to achieve this with the smallest loop possible? (Draw a picture to express your answer.) What will be the radius of the wire loop required?

** 3. We introduced a couple of new units in the last two chapters, teslas and webers, and we have occasionally just stated how these units relate to other units. Let us now ask you to work these out.

a) Work out what the tesla is (see Equation 19.1) in units of meters, seconds, kilograms, and coulombs. Work out the units of the weber (Example 20.1) in the same basic units.

b) Show that a weber/second is indeed the same as a volt.

* 4. A wire is looped ten times around a square insulator 2 cm on a side. If we start up a magnetic field into the page in the diagram at right, so that it increases at a rate 10^{-5} T/s, what *emf* will be induced between the ends of the wire loop?

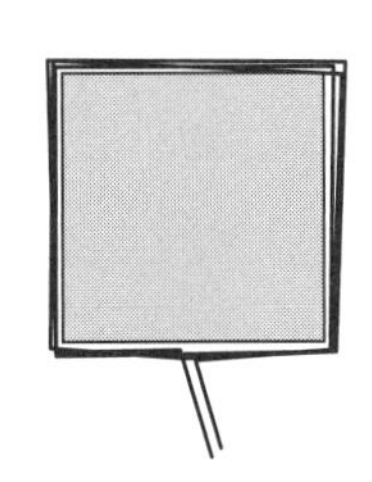

** 5. Suppose that the magnetic field in Figure 20.1 is $B = 0.01$ T, that the distance between the two side wires is $D = 0.5$ m, that the rod is moving at 3 m/s, and that the resistor in the circuit is a 10-Ω resistor.

a) Find the current in the circuit.

b) Find the force required to pull the rod along in the magnetic field.

** 6. A square loop of wire connects the ends of a 10-ohm resistor, as shown. The loop sits perpendicular to a magnetic field of strength 0.6 T that is directed into the page in the picture at right. The field then drops to zero over a time of 3 seconds. If the loop is 10 cm high and 15 cm wide, what is the current induced in the loop?

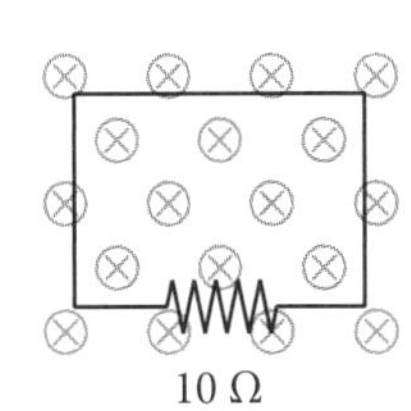

*** 7. An airplane flies directly northward over the northwest tip of Greenland at a speed of 150 m/s. The magnetic field at this north geomagnetic pole points exactly vertically downward with magnitude 5×10^{-5} T. What voltage difference is developed between the tips of the plane's 20-meter-long wingspan?

*** 8. In 1996, a NASA space shuttle experiment played out a long electrically conducting cable attached to a spacecraft built by the Italian Space Agency. Its goal was to test the possibility of generating electrical energy by induction in the earth's magnetic field. The cable was built to withstand the current generated by a voltage difference of 15,000 V between the shuttle and the spacecraft. However, as the 20-km-long cable was extended and current started flowing, it suddenly burned through and snapped, and the satellite was pulled away from the space shuttle. Experimenters decided that the cause of the excess heat was that the cable encountered an unexpectedly high magnetic field. If the space shuttle and the cable were moving at 8000 m/s in their orbit around the earth, what magnetic field would have produced a voltage difference of 15,000 V?

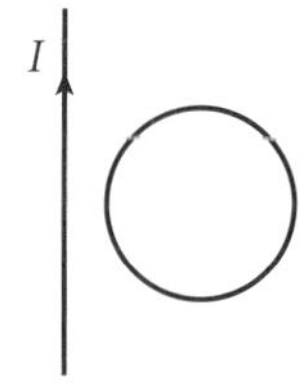

20.2 * 9. A circular conducting loop is lying on a table next to a wire that is originally carrying no current. When the current in the wire is turned on in the direction shown, it induces a current in the conducting loop. Determine the direction that the current flows in the conducting loop (clockwise or counterclockwise) and explain your answer.

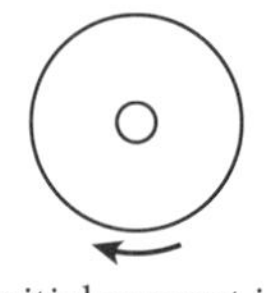

Initial current in the outer loop

* 10. Two current-carrying coils lie in the same plane as shown at left. The outer coil is connected to a battery and carries a large clockwise current. The inner coil initially carries no current. If the current in the outer coil is suddenly shut off, in what direction would current flow in the inner coil? Explain.

** 11. A small bar magnet is held stationary as shown. A wire loop is held stationary just below the end of the magnet, the loop parallel to the floor. The loop is then released and it falls to the floor. As the loop is falling, in what direction is the induced current in the loop (clockwise or counterclockwise when viewed from above)? Explain your reasoning carefully.

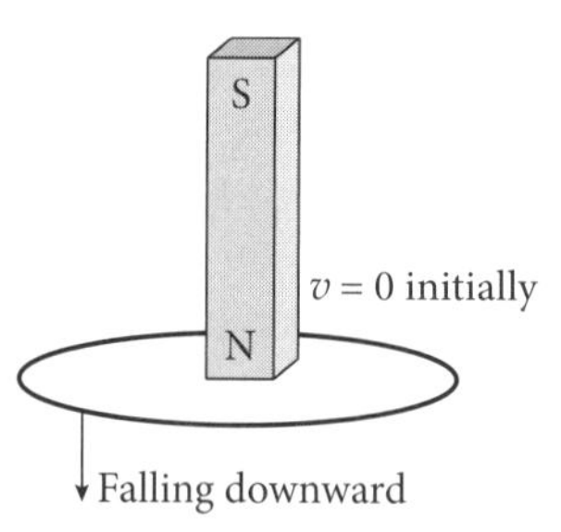

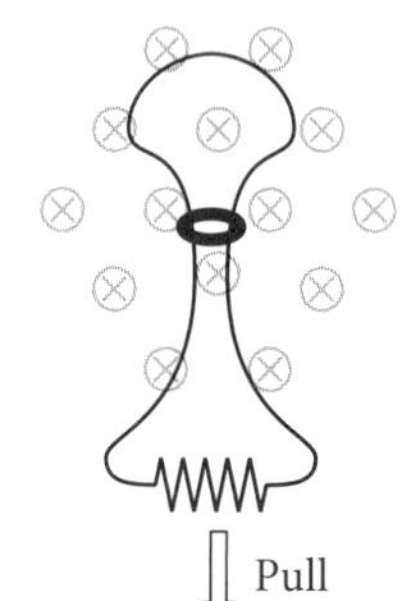

** 12. A flexible copper cable forms a conducting loop in the presence of a magnetic field of strength 5.0×10^{-3} T into the page, as shown at left. The ends of the wire pass through a narrow rubber ring and are finally attached to a 10-Ω resistor. The resistor is then pulled downward, pulling the wire through the ring and reducing the size of the loop. Assume that the resistor is carefully pulled at a speed so that the area of the loop diminishes at a rate of 0.10 m²/s.

a) What *emf* is induced in the wire?

b) In which direction, clockwise or counterclockwise, will the induced current start to flow in the wire loop?

c) Assuming the resistance of the copper cable is negligible, what is the current through the resistor?

** 13. A circular loop of wire sits face on to a 2-tesla magnetic field that points into the page, as shown in the upper diagram at right. The wire loop is turned a quarter turn by a crank, so that the right edge of the loop drops into the page and the left side of the loop comes out of the page. The radius of the loop is 20 cm and the quarter turn takes 0.5 seconds.

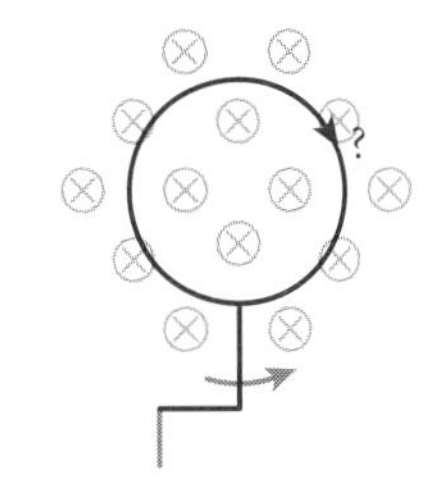

a) What is the average *emf* induced in the loop, and does the current in the wire flow in the direction of the arrow on the loop or in the opposite direction?

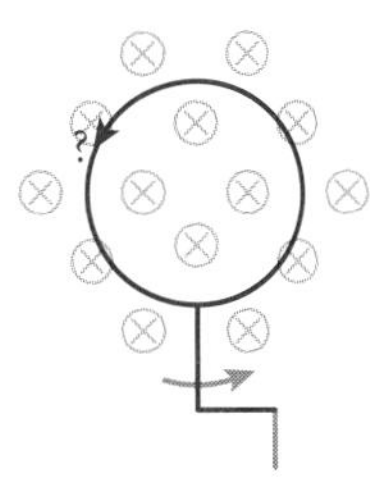

b) If the crank continues to turn the loop another quarter turn in 0.5 seconds in the same direction, until it looks like the bottom picture, what will be the average *emf* induced in the wire, and does the current in the wire flow in the direction of the arrow or in the opposite direction?

** 14. A conducting wire loop is being moved by a hand, but you don't know if it is to the left or to the right. A current carrying wire to the right of the loop carries a current as shown. At the instant shown, there is an induced current in the loop flowing clockwise. Is the hand moving the loop to the right (toward the wire) or to the left (away from the wire)? Explain your reasoning.

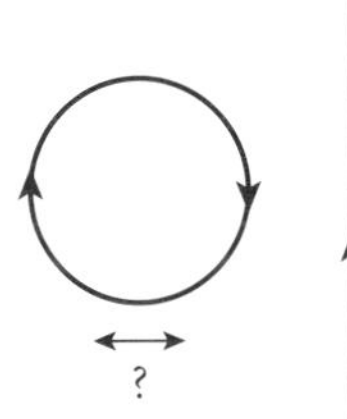

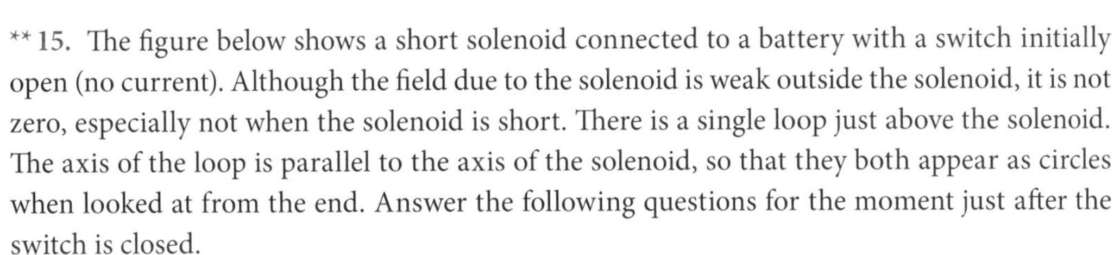

** 15. The figure below shows a short solenoid connected to a battery with a switch initially open (no current). Although the field due to the solenoid is weak outside the solenoid, it is not zero, especially not when the solenoid is short. There is a single loop just above the solenoid. The axis of the loop is parallel to the axis of the solenoid, so that they both appear as circles when looked at from the end. Answer the following questions for the moment just after the switch is closed.

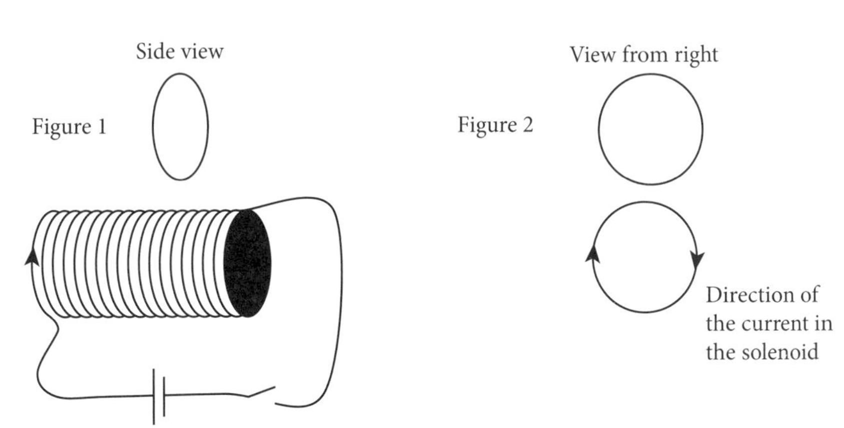

a) Draw a diagram of the solenoid as in Figure 1 and indicate on it the direction of the effective north and south poles of the solenoid.

b) Also draw a copy of Figure 2 and indicate on both of your Figure 1 and Figure 2 the direction of the magnetic field inside the loop due to the increasing solenoid current.

c) Indicate on your Figure 2 the direction of the induced current in the loop as the switch is closed. Explain your reasoning.

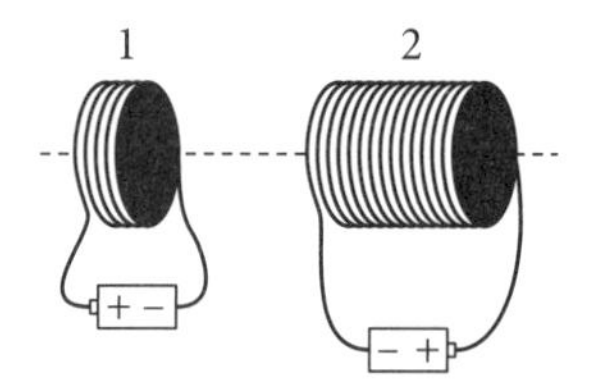

** **16.** Shown at left is a small solenoid with five loops (1) and a larger solenoid (2), each connected to a battery. A constant current has been generated in each one. Note the orientation of the battery in each solenoid and the direction of the current around the coils of each solenoid.

a) If both coils are stationary:
- Is there an induced current through solenoid 1? If yes, indicate the direction of the induced current. Explain your reasoning.
- Is solenoid 1 attracted to or repelled by 2, or is there no force at all between them? Explain your reasoning.

b) If solenoid 1 is moving toward solenoid 2:
- Is there an induced current through solenoid 1? If yes, indicate the direction of the induced current. Explain your reasoning.
- Is any force by solenoid 2 on solenoid 1 larger, smaller, or the same as in part (a)? Explain your reasoning.

** 17. A galvanometer is like an ammeter, in that it measures current, except it typically measures very small currents and its indicator needle can deflect either right or left, depending on the direction in which the current flows. A large solenoid is connected to a galvanometer as in the figure below left. A clockwise current (as viewed from the top) will deflect the needle to the left and a counterclockwise current will deflect the needle to the right.

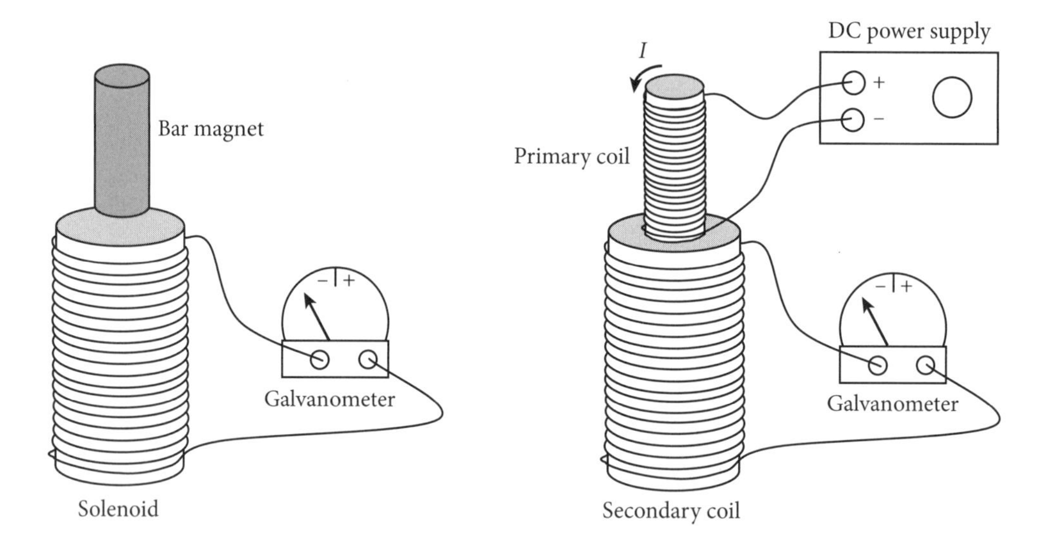

a) If the needle deflects to the right while a bar magnet is lowered into the solenoid, as at left, determine whether it is the north pole or the south pole that goes into the solenoid first. Explain how you arrived at your answer.

b) A second ("primary coil") is connected to a DC power supply as shown above right, causing a current to flow in the direction indicated. As the primary coil is inserted into the secondary coil, which direction will the needle on the galvanometer deflect? Explain your reasoning.

c) With the primary coil resting inside the secondary coil, suppose the power supply is turned off. Will the galvanometer needle deflect? If so, which direction? Explain your reasoning.

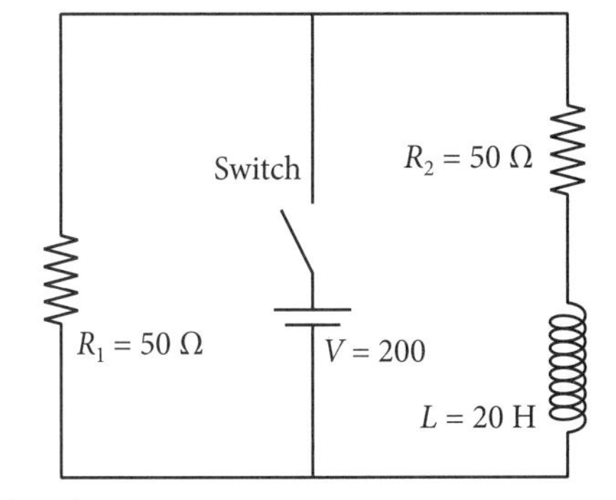

20.3 ** 18. The following questions refer to the circuit represented by the diagram at right. Think of R_2 as the internal resistance of the inductor.

a) Immediately after the switch is closed, find the magnitude and direction of the conventional current (if any) through the battery, and find the potential difference across the inductor.

b) A long time after the switch is closed, find the magnitude and direction of the conventional current (if any) through the battery, and find the potential difference across the inductor.

c) After the switch has been closed for a long time it is suddenly opened. Immediately after the switch is opened, find the magnitude and direction of the conventional current (if any) through the resistor R_1 and the potential difference across the inductor.

*** 19. A 40-kV high-voltage power line carries 10^4 A of current along wires that are 5 m off the ground. The average rate at which the current changes in a 60 Hz, 10^4 amp, AC line is 3.77×10^6 A/s. A farmer wants to run a 110-volt electric pump just under the power lines, but it is way too far from his house for him to run his own power line out to the pump. So he has an idea. He extends a long narrow square loop of wire along the ground so that the magnetic field produced by the power line cuts through the inside of the loop. The loop is 10 cm high, as shown below. How long must the loop be if he wants to generate 110 volts for his water pump? (By the way, the farmer was not the first person to think of this, and such power-stealing is illegal.)

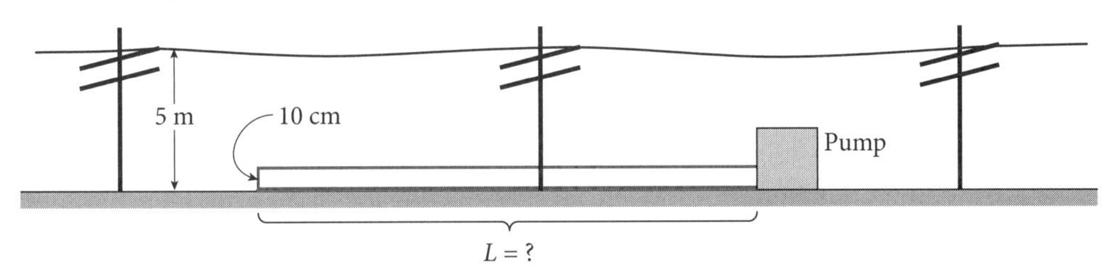

* 20. Find the self-inductance of a 1000-turn solenoid of length 5 cm and radius 0.3 cm. [Hint: See Example 20.7.]

** 21. A large inductor is to be constructed by wrapping wire around an empty paper towel roll of length 30 cm and radius 2 cm. An inductance of $L = 30$ mH is desired. How many turns of wire will have to be wound on the cylinder? [Hint: See Example 20.7.]

** 22. **Magnetic Permeability.** In the last chapter, on page 401, we explained how tiny magnetic domains in certain materials would align with an external magnetic field. On page 418, we mentioned that this could be used to enhance the field inside a solenoid. Here we will see how this effect can be used to increase the inductance of a solenoid. Iron, for example, possesses a strong magnetizability. In response to an external magnetic field B_0, the magnetic field inside the iron will be $B = kB_0$, where k is called the *relative permeability*. If the field inside an inductor is stronger, the flux will be greater. Find the inductance of a coil with 200 turns of radius 2 mm in a length of 1 cm, wound around an iron core (relative permeability $k = 200$).

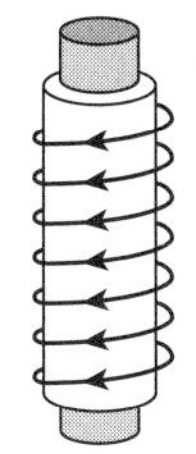

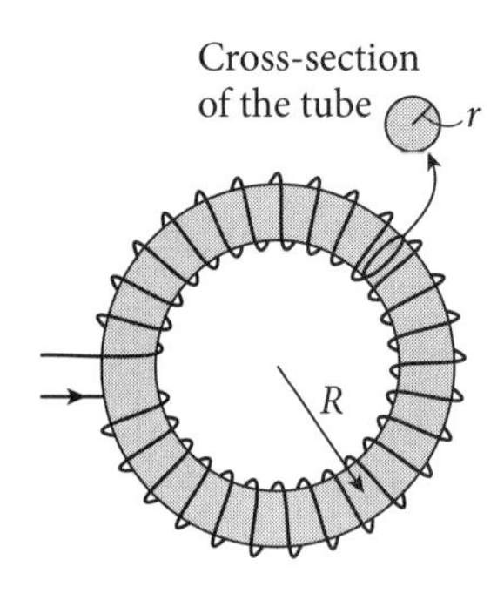

*** 23[f]. In Problem 41 of the last chapter, we asked you to find the magnetic field inside a toroidal solenoid. The answer to the problem was given as a check in part (c). If the solenoid tube that is bent into a toroid is of small enough radius, compared to the radius R of the toroid, then R will be approximately constant over the entire interior of the tube. Find the inductance of a toroidal solenoid of large toroidal radius R and small tube radius r, with N total turns around the solenoid.

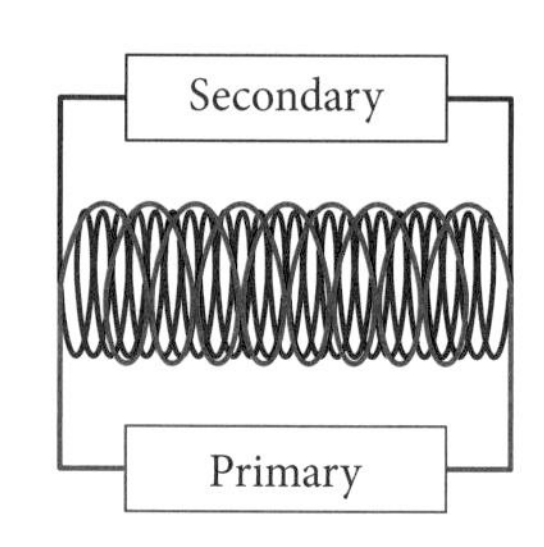

** 24. A transformer is created by wrapping two solenoids around the same cylinder, as shown in the figure at right. In one transformer, the primary (red) coil of the transformer has 20 loops wound around a 0.10-meter-long cylinder of cross-sectional area 3.0×10^{-4} m^2. The secondary (black) coil has 50 loops wound around the same cylinder. Find the mutual inductance of the two solenoids by considering a current I_1 in the primary coil and finding the total flux that it creates through the loops of the secondary coil.

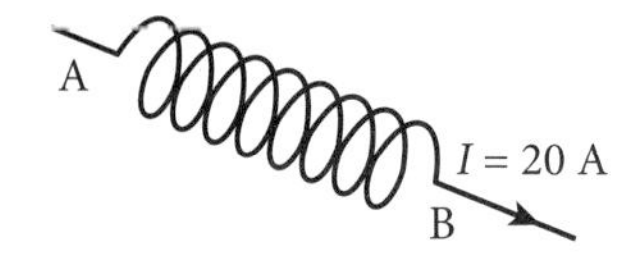

*** 25. An inductor with a self-inductance of 4 H and an internal resistance of 1.2 Ω is connected to a circuit in which there is a 20-amp current flowing in the direction shown, and the current is decreasing at an instantaneous rate of 80 A/s. What is the voltage difference across the inductor and which side, A or B, is at the higher potential? [Hint: Be careful with the directions of voltage differences.]

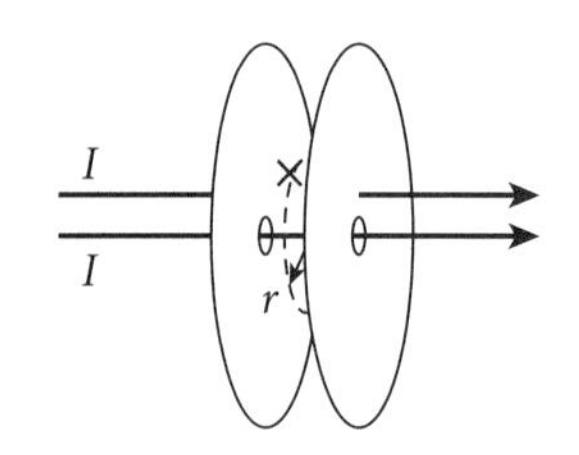

20.4 *** 26[f]. Two holes are cut in the center of the circular plates of a parallel plate capacitor, as shown in the figure at left, and a wire carrying current I passes through the holes. A second wire, also carrying current I, is attached to the left plate of the capacitor, and a third wire is attached to the right plate of the capacitor, as shown. Find the magnetic field at a point between the plates, a radial distance r from the centerline of the plates, the point labeled × in the figure.

*** 27. Two circular discs form the plates of a capacitor that is placed in a circuit in which a current $I = 3$ amps is going to the right. Each plate has an area of 10^{-2} m^2. The distance between the plates is 1 cm.

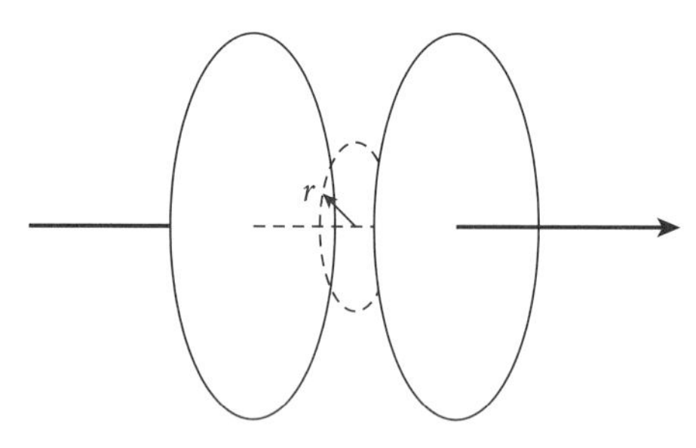

a) At a time when the charge on the plates is 6 C, find the electric field in the region between the plates.

b) If a positive current is flowing, the charge on the capacitor will be increasing. Use the appropriate right-hand rule to find the direction of the magnetic field in the region between the plates of the capacitor.

c) Find the magnitude of the magnetic field between the two plates at a distance $r = 3$ cm from the centerline of the circular plates, as shown. [Hint: You may assume that the magnetic field is constant along the dashed circular path of radius r shown in the figure and that it is everywhere tangent to the circle.]

*** 28. A bar is moved along conducting rails at a constant 2.0 m/s in the presence of a 1-tesla magnetic field into the page, as shown. A 4-ohm resistor is connected into the circuit.

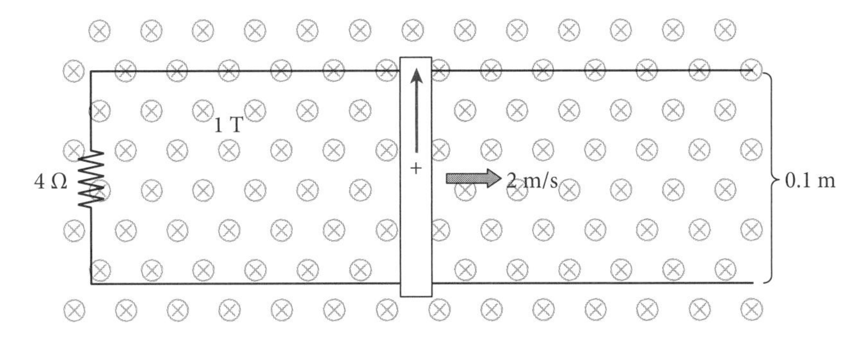

a) What current flows in the loop?

b) What force is required to maintain the constant 2 m/s velocity?

c) How much work must be done by this force in 1 second?

d) Calculate the power dissipated by the resistor and compare the energy dissipated by the resistor with the work done in this same time.

*** 29[f]. A conducting bar rides on two of the rails of a 3-rail open circuit, closing the circuit into a rectangular loop. The plane of the rails is inclined by an angle θ to the horizontal. The distance between the rails is D, and there is a vertical upward magnetic field of magnitude B, as shown at right.

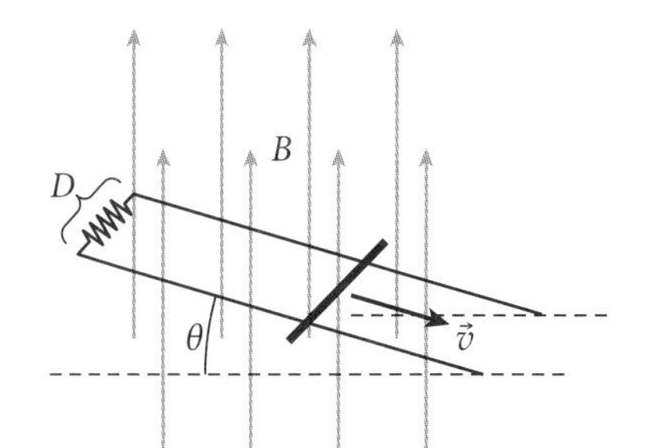

a) Assume that the bar is sliding down the rails at speed v and find a formula for the *emf* in the wire as a function of B, D, θ, and v.

b) If the resistor in the circuit is R, find the current that will flow through the circuit, in terms of B, D, θ, v, and R.

c) Find a formula for the magnetic force on the current-carrying bar (again in terms of B, D, θ, v, and R) and give the direction in which this force will act.

d) There is a component of gravity that acts down the sloping rails and there is a component of the magnetic force on the current-carrying wire (calculated in part c) that acts up the sloping rails. Find the terminal speed at which the bar will slide when these two forces are balanced.

** 30. There is a uniform electric field perpendicular to a disk of radius 20 m. Over an elapsed time of 2 s, the field changes from a value of 32 V/m to 16 V/m.

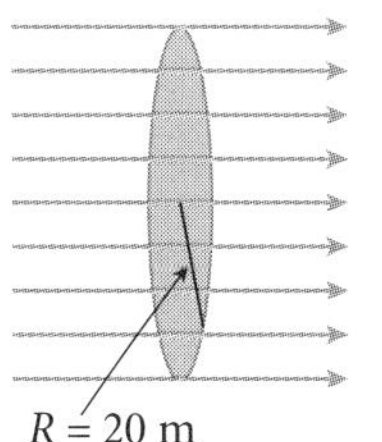

a) Find the displacement current through the surface of the disk.

b) Carefully analyze the units to show that the displacement current is in amps.

*** 31. A cardboard disk of radius 20 cm is oriented perpendicular to a radial line from a small conducting sphere that carries 3 mC of positive charge, as shown at right. The disk is moving toward the charge at 10 m/s. At $t = 0$, the loop is 3 m away from the charge and at $t = 1/10$ s, the loop is 2 m away from the charge. What is the strength of the average induced magnetic field around the circumference of the disk? [Hint: Remember that the electric field due to a point charge is given by Equation 16.4.]

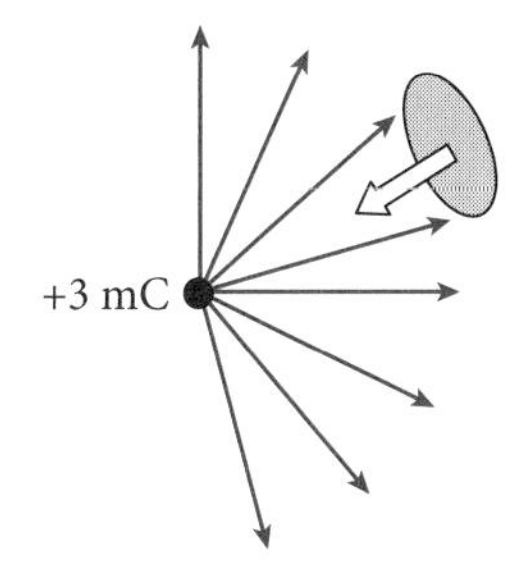

20.5 * 32. Use Maxwell's second equation (Equation 20.13) to prove that there must exist magnetic field lines inside a bar magnet. This is to say, prove that the magnetic field lines can never really look like figure (a) below, but only like figure (b).

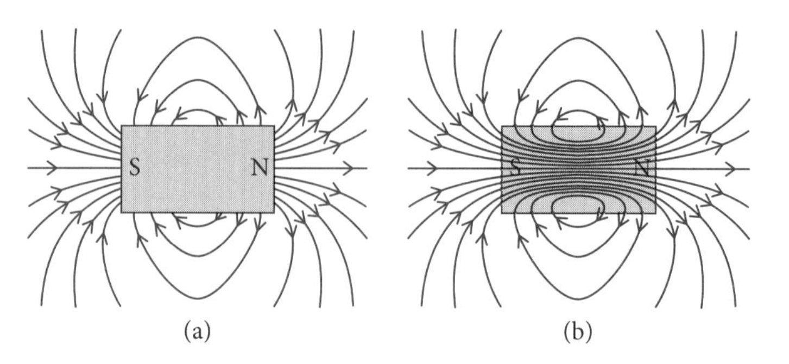

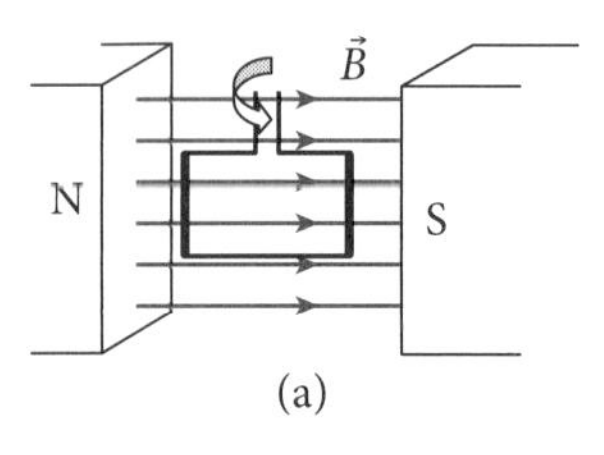

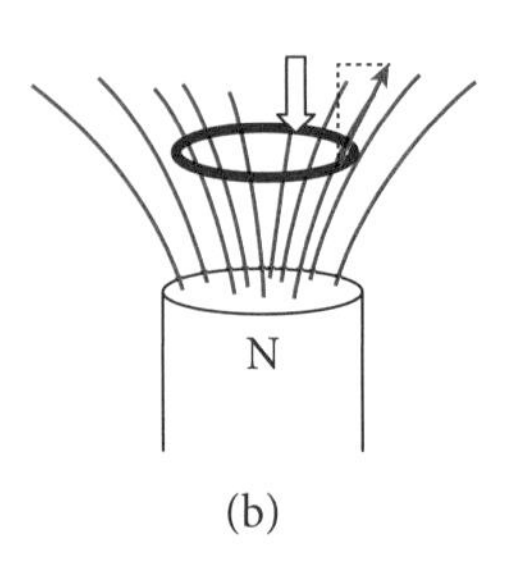

*** 33. Back on page 412, after deriving Faraday's law for the flux through a loop containing a moving rod, we simply stated without proof that the law works for any situation where there is a change in magnetic flux through a closed loop. Here, we want you to support this statement by considering two other situations.

a) Consider a fixed square loop that turns in a fixed magnetic field. Explain how the magnetic force on positive charge in the moving side-wires (the two heavy line segments) creates an *emf* around the loop, in agreement with the Lenz's law direction.

b) Support Faraday's law being valid in the case of an increasing magnetic field through a fixed loop by thinking of a loop moving closer to a magnetic pole from which magnetic field lines diverge. Show that the motion of each piece of the loop moving downward, where the magnetic field has a horizontal component (the horizontal dotted black line at left), will induce an *emf* that agrees with the Lenz's law direction.

21

LIGHT AND ELECTROMAGNETIC WAVES

One of the more interesting and useful aspects of Maxwell's equations is that they predict the existence of electromagnetic waves, waves of electric and magnetic fields that travel through empty space. This prediction was experimentally verified in 1887 by the German physicist, Heinrich Hertz. But Hertz was also responsible for one of history's most delightful errors in foresight when he famously answered, having been asked about possible practical applications of his discovery,

> "It's of no use whatsoever ... this is just an experiment that proves Maestro Maxwell was right."[1]

21.1 Electromagnetic Waves

Let us try to give a simple explanation of what an electromagnetic wave is. Back on page 308, where we introduced the electric field, we pointed out that an electric field is a property of space. The same is true of a magnetic field. An *electromagnetic wave*, then, is a set of electric and magnetic fields at various places in space, and it is a wave because the values of the fields are required by Maxwell's equations to move through space at a particular speed, the speed of light. Consider, for example, a wave consisting of large values for $\vec{E}$ and $\vec{B}$ in the gray blob in Figure 21.1. After a short time, the region that was initially to the right of the blob will acquire those large electric and magnetic field values, while the region where the blob was originally sitting will acquire the zero field values that were initially to the left of the blob. The wave *is* the moving values of electric and magnetic fields in space. In Figure 21.1, the values are moving to the right.

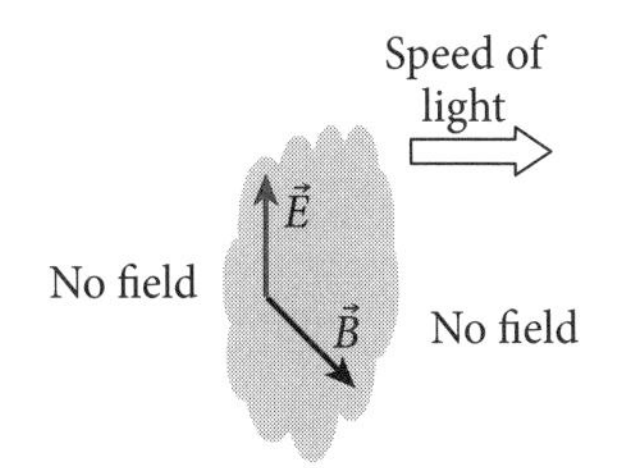

Figure 21.1 Concept of an electromagnetic wave as a moving spatial region with electric and magnetic fields inside it.

Let us be clear that electromagnetic waves are waves; they transfer energy without transferring mass, as was required by the definition in Chapter 12. But they are not like the mechanical waves discussed in that chapter. Electromagnetic waves require no medium; there is nothing that has to vibrate as the wave goes by. The energy in an electromagnetic wave is, as we will see in Section 21.5, the energy in the electric and magnetic fields themselves, not the kinetic energy of vibrating atoms. An electromagnetic wave can exist in the total vacuum of empty space.

But why do the waves exist at all? According to Maxwell's equations, electric fields are produced by electrical charge and magnetic fields are produced by electrical current. But the

1. If you missed the irony of this statement, it might help to know that radio waves are electromagnetic waves. Think radio, TV, cell phones, spacecraft communications, etc.

equations also predict another way to produce these fields. As we saw in Chapter 20, electric fields are created by changing magnetic fields and magnetic fields are produced by changing electric fields. If these fields are made to change in the right way, they end up supporting each other, acting as sources for each other. As a result, while electric fields from static charges (Coulomb's law, Equation 16.4) and magnetic fields from moving charges (the Biot–Savart law, Equation 19.3) fall off like $1/r^2$ as one moves away from the electrical charges, the mutual support of the changing electric and magnetic fields means that the field strengths in electromagnetic waves fall off less sharply, dropping instead like $1/r$. The waves do need to be created by some charged particle somewhere, however, and that particle must move in such a way that it will create fields that vary in the right ways for the magnetic and electric fields to act as sources for each other. It turns out that a charge moving at constant velocity will not do the job. All that is required, however, is an acceleration. *Any accelerated charged particle produces electromagnetic waves.*

We have said that the speed of the wave will be the speed of light. This is not something that was imposed on Maxwell's equations, but something that arose out of them. When Maxwell's equations are differentiated (yes, that would mean calculus) and combined together, they can produce an equation that electric and magnetic fields must satisfy that requires the values of the field to move in some direction as time goes on. The speed at which the values of the fields propagate through empty space turns out to be determined by the two fundamental constants in Maxwell's Equations. That speed is $c = 1/\sqrt{\mu_0\varepsilon_0}$. We know the values of these two constants from experiments with static electric and magnetic fields, so we can calculate the speed of the waves from what is already known. The result of the calculation gives

$$\frac{1}{\sqrt{\mu_0\varepsilon_0}} = \frac{1}{\sqrt{\left(4\pi \times 10^{-7}\,\frac{\mathrm{N \cdot s^2}}{\mathrm{C^2}}\right)\left(8.854 \times 10^{-12}\,\frac{\mathrm{C^2}}{\mathrm{N \cdot m^2}}\right)}} = \frac{1}{3.336 \times 10^{-10}\,\frac{\mathrm{s}}{\mathrm{m}}} = 2.998 \times 10^8\ \mathrm{m/s}.$$

Let us be clear what this calculation has accomplished. The two constants, μ_0 and ε_0, arise in describing the strengths of *static* electric and magnetic fields, but their product determines the speed of *propagating* electromagnetic waves. Historically, this calculation first made clear that light is an electromagnetic wave. When the predicted speed for electromagnetic waves turned out to be exactly the measured speed at which light was known to propagate, the evidence for the connection could not be denied.

21.2 The Electromagnetic Spectrum

The single isolated region of electric and magnetic fields shown in Figure 21.1 is an example of a wave, but it would produce only a single pulse as it moved past some point in space. Back in Chapter 12, where we first introduced waves, we agreed to use the term "continuous wave" for a periodic string of disturbances and to call a single disturbance a "pulse." We will continue that convention here.

Remember that all electromagnetic waves or pulses are created by accelerated charged particles. So, one way to produce a continuous electromagnetic wave would be to accelerate an electron up and down inside a metal rod. We could do this by arranging some electronics to produce voltages in the rod that would first accelerate the electron upward, then, by changing the polarity, it would accelerate it back downward, bringing it to a stop and starting it moving in the opposite direction. We might even arrange, by carefully choosing how we vary the imposed voltage, to make the electron move in the simple harmonic motion of Chapter 11. Figure 21.2 depicts the resulting electric field as it propagates off to the right.

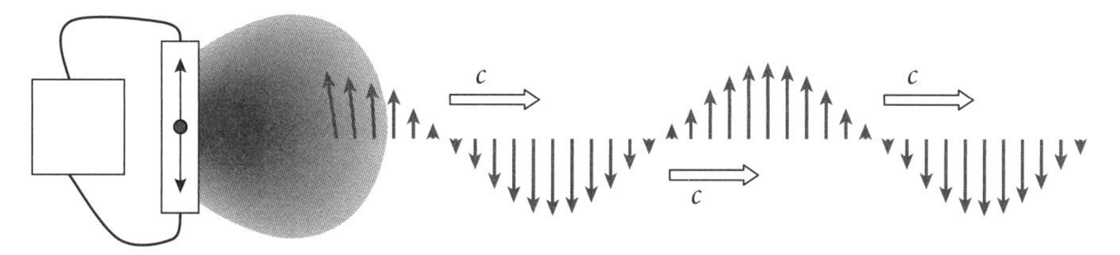

Figure 21.2 An electron in simple harmonic motion and the varying electric field it creates.

One must be careful in interpreting the diagram of Figure 21.2. Note that the propagating electric field vectors to the far right of the diagram do not point toward the electron, as the static field from a point charge would, but they point parallel to the rod in which the electron is moving. The *radial* fields produced by charged particles do point toward the electrons, but they fall off quickly ($1/r^2$) compared to the falloff with distance of the electromagnetic wave ($1/r$), so the field vectors in the wave that we see to the far right only represent the electric field in what is called the "far-field" limit, after the radial fields have dropped off enough to be neglected. The shaded region next to the rod is meant to represent the near-field region where the fields are in fact very complicated.

There is also a problem with Figure 21.2 in that it shows a varying electric field, but no magnetic field. We know, from Maxwell's equations, that those two kinds of fields accompany each other in a wave. If we remember that Equation 20.12 connects the changing electric flux through a surface to the magnetic field around the edge of the surface, we get a hint that *the electric and magnetic fields induced in this way will be perpendicular to each other.* A picture of the perpendicular electric and magnetic fields for an electromagnetic wave propagating toward the right is shown in Figure 21.3. If you start with the $\vec{E}$ vector and curl the fingers of your right hand toward the $\vec{B}$ vector, as you did with $\vec{v}$ and $\vec{B}$ when using Right-Hand Rule #1 (page 388), then your thumb will point in the direction in which the electromagnetic wave is traveling.

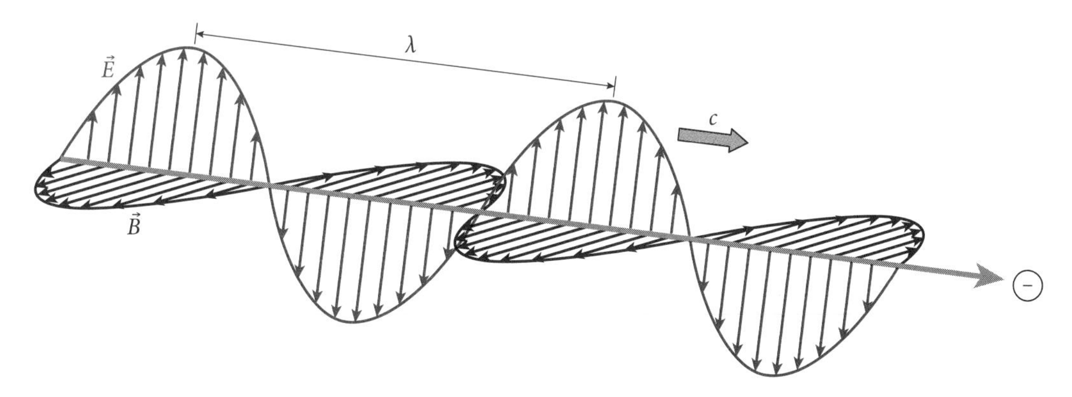

Figure 21.3 A snapshot of the electric (red) and magnetic (black) fields in an electromagnetic wave propagating toward the right. The distance between the points with identical fields is the wavelength, as with the Chapter 12 waves.

Figure 21.3 also shows an electron sitting off to the right of the propagating wave. This was meant as a reminder that these are real electric and magnetic fields. When the leading edge of the wave, with its downward-pointing electric field vectors, reaches the electron, there will be an upward force on the electron (remember the electron has a negative charge). The electron will thus begin to accelerate upward. When the upward-pointing electric field vectors reach the electron, however, the resulting downward force will slow the electron, stop it, and then move it downward.

Figure 21.3 also identifies the wavelength of the electromagnetic wave. From what we learned about waves in Chapter 12, we are able to connect the wavelength of the wave with the simple-harmonic-motion frequency f or period T that the electron will experience as the wave passes it. It is the good old "wave equation,"

$$c = \frac{\lambda}{T} = \lambda f, \tag{21.1}$$

with c as the speed of the waves. The actual value for c is 2.99792458×10^8 m/s, by SI definition, but we will invariably approximate this as $c = 3 \times 10^8$ m/s.

We have depicted electromagnetic waves as being produced by accelerated electrons inside a conducting rod, and so they are. But there are many other ways to produce electromagnetic waves, and the wavelengths of the waves routinely produced by various means in the laboratory and elsewhere vary over a wide range of values. In fact, since electromagnetic waves are typically produced by such different methods, they have been given different names, depending on their wavelengths. Table 21.1 gives the standard list of the wavelength ranges of electromagnetic waves.

TABLE 21.1 The Electromagnetic Spectrum, from Long Wavelength to Short

Name	Wavelength Range	Typical Production Methods
Radio	1 m – 100,000 km	Antennas driven by electronics
Microwave	1 mm – 1 m	Antennas driven by electronics
Infrared light	750 nm – 1 mm	Warm bodies
Visible light	380 nm – 750 nm	Hot bodies, electron transitions in atoms
Ultraviolet light	10 nm – 380 nm	Very hot bodies, higher energy electron transitions in atoms
X-rays	10 pm – 10 nm	Slamming a fast electron into a metal target
Gamma rays	0 – 10 pm	Nuclear reactions and transitions

In each case, the last column gives some prominent means by which this range of electromagnetic wave is produced, and, for each, you should be able to identify an accelerated charged particle somewhere. The references to warm, hot, and very hot bodies as sources of infrared, visible, and ultraviolet light refer to electromagnetic waves that are created by the random accelerations of molecules whose motion is due to the body's temperature. This process is called "blackbody radiation," since its ideal wavelength spectrum is achieved when the hot body producing it is also a perfect absorber of any light falling on it (so it is black in that it reflects no light). The "transitions in atoms" that are mentioned as another source for infrared, visible, and ultraviolet light refer to changes in the energy of an electron in an atom, a subject we will discuss in Chapter 25.

We also need to mention that, within the visible light spectrum from 380 nm to 750 nm, the differences between the waves of different wavelength are seen by the human eye as colors, with the longest wavelength being seen as red and the shortest being seen as violet. The correlation between color and wavelength is shown in Table 21.2. When light contains all wavelengths, our eyes perceive the color of the light as white. Black is what we perceive when there is no light.

Finally, we want to emphasize that everything mentioned in Table 21.1 and Table 21.2 is basically the same phenomenon. Radio waves, light waves, X-rays, and gamma rays are all the same thing—they are electric and magnetic fields moving at the speed of light. The only difference between them is their wavelength, the spatial distance between one peak value of the electric and magnetic fields and the next peak value of the electric and magnetic fields.

TABLE 21.2 Visible Light Spectrum

Color	Wavelength Range
Red	620 nm – 750 nm
Orange	590 nm – 620 nm
Yellow	570 nm – 590 nm
Green	490 nm – 570 nm
Blue	450 nm – 490 nm
Violet	380 nm – 450 nm

EXAMPLE 21.1 The Frequency Range of Visible Light

What is the range of frequencies of electromagnetic waves that fall in the visible range of the spectrum?

ANSWER We see from the tables that the shortest visible wavelength is that of violet light at 380 nm. We can calculate the corresponding frequency using the wave equation (Equation 21.1). The wave equation gives

$$f = \frac{c}{\lambda} = \frac{3 \times 10^{8}\ \text{m/s}}{380 \times 10^{-9}\ \text{m}} = 7.9 \times 10^{14}\ \text{Hz} = 790\ \text{THz},$$

where the "T" in THz stands for "tera," a prefix meaning 10^{12}.

Similarly, the longest wavelength is the red color at 750 nm, corresponding to a frequency

$$f = \frac{c}{\lambda} = \frac{3 \times 10^{8}\ \text{m/s}}{750 \times 10^{-9}\ \text{m}} = 400\ \text{THz}.$$

Visible light has frequencies ranging from 400 THz to 790 THz.

21.3 Huygens' Principle

Back in Chapter 12, especially in the discussion on page 233, we saw how a wave traveling along a string was propagated forward due to the tension in the string and due to the shape of the wave. Thus it is the distortion in the medium that propagates the wave forward. For an electromagnetic wave, there is no medium to be deformed, no string or tube full of air. The wave is just a wave of values of electric and magnetic field vectors, and the existence of an electric field at one point does not itself create an electric field at a point next to it. However, when an electric field at some point is changing, then the induction that arises in accordance with Maxwell's equations *does* give a mechanism for creating changing electric and magnetic fields at nearby points. Thus, in an electromagnetic wave, the relationship between the field at each point and the rate at which it is changing at that point causes the electromagnetic wave to propagate forward. This general idea—that each point of a wave serves as a source of a propagating wave—is known as *Huygens' Principle* (discovered by the Dutch physicist Christiaan Huygens in 1678). It is valid for sound waves or ripples in a pond or for electromagnetic waves. It says that you can ignore the past history of a wave, look only at a particular slice of the wave at one particular time, and treat that slice as the source of the wave going forward.[2]

In Figure 21.4, we demonstrate how Huygens' principle works. The crest of a wave (where the fields have their largest values) is shown in (a) in heavy red, with two trailing wave crests behind it. In (b), we treat each point of the chosen wavefront as a source of a spherical wave propagating forward in all directions. We call each of these pieces of the wave a *wavelet*. One complete period has elapsed between (a) and (b), so that these wavelets will each have propagated outward a distance λ. In (c), we simply connect all the wavelets to get the wave shape after one period has elapsed. As may be seen, Huygens' principle explains two things. First, it shows how a plane wave moves forward without changing its basic shape, even though it is generated by many separate points. Second, it shows what happens if the wave is really of finite extent, as it is in these pictures. We see that there is, in fact, a slight change in the shape of the wave, because it will spread outward a bit at its edges. The amplitude of the spreading edge will be much less than the amplitude at the center of the wave, because fewer wavelets are contributing to it, but it *will* divert some of the energy in the wave away from the main direction. This last effect is a small one and we will not worry about it again until Chapter 23.

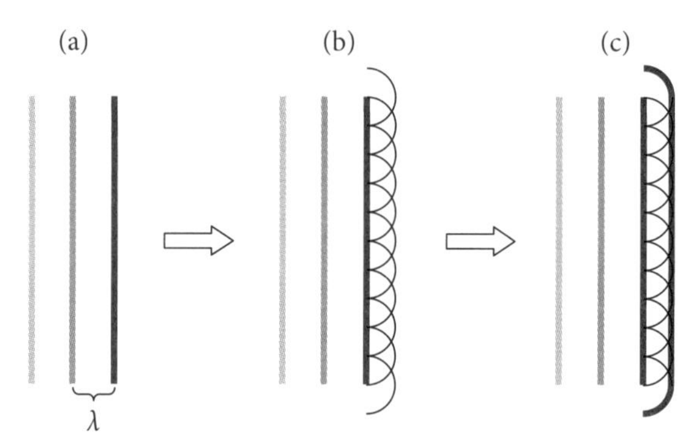

Figure 21.4 A demonstration of Huygens' principle. In (a), we choose one of the wavefronts of the wave. In (b), each point of the chosen wavefront serves as a wave source that propagates out in all directions. In (c), the superposition of all propagated wavelets forms the new wavefront.

21.4 Reflection

In this section and the next, we apply Huygens' principle to a couple of useful examples.

First, let us explain that, for the moment, we will concentrate on light waves and we will work only with plane waves. Plane waves are waves in regions so far from the source of the waves that the $1/r$ decrease in the amplitudes of the fields is negligible, and the amplitudes of the electric and magnetic fields will have approximately the same values on successive

2. It is the combination of the instantaneous wave amplitude *and* the rate of amplitude change at neighboring points that determines that only forward-going waves propagate out from each point.

flat two-dimensional surfaces drawn perpendicular to the propagation direction of the wave. Figure 21.5 depicts a plane wave. We will relax the restriction to plane waves when we get to Chapter 22.

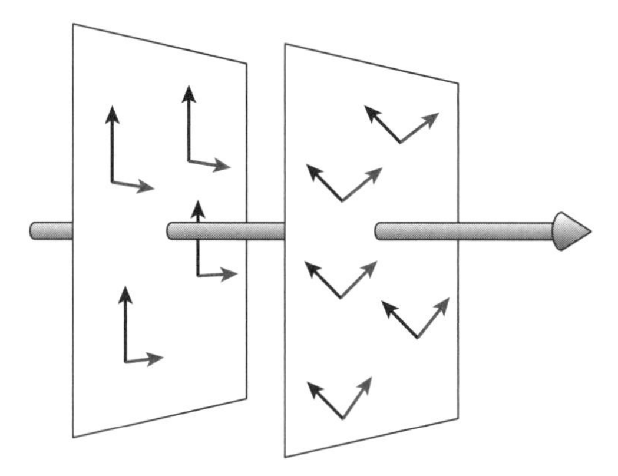

Figure 21.5 The electric (red) and magnetic (black) vectors at a few points along a propagating plane wave. The magnetic fields are always perpendicular to the electric fields, and all fields lie entirely in planes perpendicular to the gray propagation vector (also called a "ray"). Note that the fields may vary from plane to plane, but are constant over each plane.

Now, let us consider a beam of plane waves approaching a mirror at an angle θ from the line perpendicular to the mirror, as shown in Figure 21.6. The picture depicts the direction of the waves with a black arrow (called a *ray*) and the wave crests as the red planes perpendicular to the rays. In Figure 21.6a, the leftmost side of the beam has just arrived at the mirror, while the right side of the beam still has a distance d to travel. We will call the time of this picture $t = 0$. Figure 21.6b depicts the situation where the right edge of the beam has finally arrived at the mirror, a time $t = d/c$ later. At this time, the wavelet from the reflected left edge of the beam will have moved upward by a distance $ct = c(d/c) = d$. The two dotted lines of length d are each perpendicular to the wavefront. The width w of the section of the mirror illuminated by the beam is the same in each picture. This means that the two shaded triangles in the two pictures are right triangles with two sides equal. They are therefore identical triangles, with all of their angles the same. The rays are the solid black arrows that are perpendicular to the incoming and outgoing wavefronts. They make the same angle θ relative to the dashed line that was drawn perpendicular to the mirror. We conclude that a beam of light hitting a mirror at some angle will be reflected from the mirror at an equal angle.

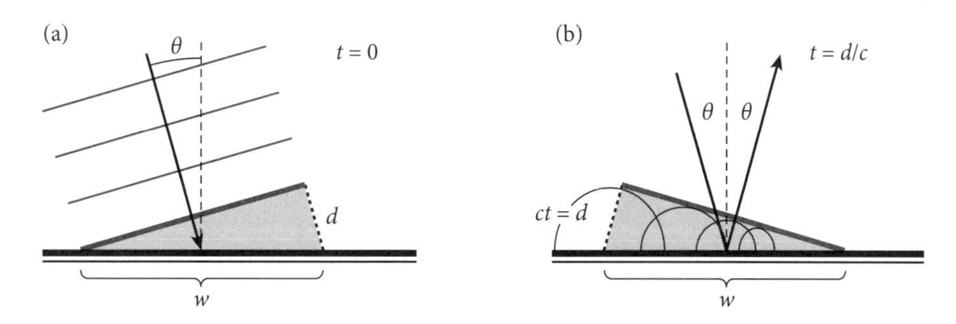

Figure 21.6 A demonstration, using Huygens' principle, of how a beam of light is reflected from a mirror at an angle equal to its initial angle.

21.5 Refraction

Before going on to our next example of the use of Huygens' principle, we need to discuss the propagation of light in a material that is not just empty space. When a light wave moves in a material, filled with atoms, its electric and magnetic fields interact with the atoms, causing them to emit their own light waves. The original wave and the re-emitted wave interfere to produce a combined wave whose crest lags a little behind where the original wave would have been. The further the wave passes into the material, the more the wave crests are delayed. The overall effect is that the light travels more slowly inside the medium than it would in vacuum. The speed may be found experimentally for different kinds of material and characterized by an index of refraction, n, defined so that the speed of a light wave passing through a material is given by the formula $c_n = c/n$. Since all speeds in a material are less than the speed of light in vacuum, all indices of refraction are greater than $n = 1$. In Table 21.3, we list indices of refraction for a few common transparent materials.

TABLE 21.3 Index of Refraction

Material	n
Vacuum	1
Air	1.0003
Ice	1.31
Water	1.33
Crown Glass	1.5
Flint Glass	1.6
Diamond	2.42

EXAMPLE 21.2 The Wavelength of Light inside an Eyeball

We say that a particular violet color of light has a wavelength 400 nm, but what we mean is that this is its wavelength *in vacuum*. Before that light hits our retina, it has to pass into the vitreous humor that fills our eyeball. If this fluid has an index of refraction $n = 1.34$, what is the wavelength of this light inside the vitreous humor?

ANSWER Of the three quantities in the wave equation, c, λ, and f, the one that cannot change as the light wave enters the eyeball is the frequency. If 750 wave crests hit the eyeball in a picosecond, all 750 must enter the eyeball in that time. They cannot pile up waiting for their turn to go in. So a change in speed inside the vitreous humor must be accompanied by a change in wavelength, not frequency. The speed of the wave in the vitreous humor is $c_n = c/n = (3 \times 10^8\ \text{m/s})/1.34 = 2.24 \times 10^8\ \text{m/s}$. The wavelength inside the eyeball can then be found from

$$f = \frac{c}{\lambda} = \frac{c_n}{\lambda_n} \quad \Rightarrow \quad \lambda_n = \lambda\frac{c_n}{c} = (400 \times 10^{-9}\ \text{nm})\frac{2.24 \times 10^8}{3 \times 10^8} = 298\ \text{nm}.$$

This should make it clear that there is nothing about a wavelength of 400 nm that is inherently "violet," but only that this is the color we perceive when 400-nm light passes into the eyeball, changes wavelength to 298 nm, and then strikes the retina.

Now we are prepared to look at the next situation where we want to apply Huygens' principle. This is when light goes from one medium into another, bending the beam in a process known as *refraction*. You have probably experienced this in a swimming pool when you reached down to get something on the bottom of the pool and realized that it wasn't where it appeared to be, or as you looked at a straw in a glass of water and noticed that it seemed to bend where it went into the water. These things occur because light beams bend as they enter or exit the water. Let us see why.

Figure 21.7 shows a beam of light passing from empty space, where the initial index of refraction is $n_1 = 1$, into a medium with index of refraction n_2. Figure 21.7a is identical to Figure 21.6a, except for the mirror being replaced with the surface of a transparent medium. Figure 21.7b is drawn at time $t = d/c$, as before, just as the right side of the wavefront reaches the new medium. The differences between Figure 21.7b and Figure 21.6b are, obviously, that the wave travels into the new medium instead of bouncing off of it, but also that the distance traveled by the wavelet from the left side of the wave in the time d/c is less than d, because it is traveling at a slower speed in the new medium. It is traveling at $c_n = c/n_2$. As a result, the transmitted wavefront is more horizontal than the incoming wavefront, meaning that the ray has been bent *toward* the normal to the surface as it entered this higher-index medium.

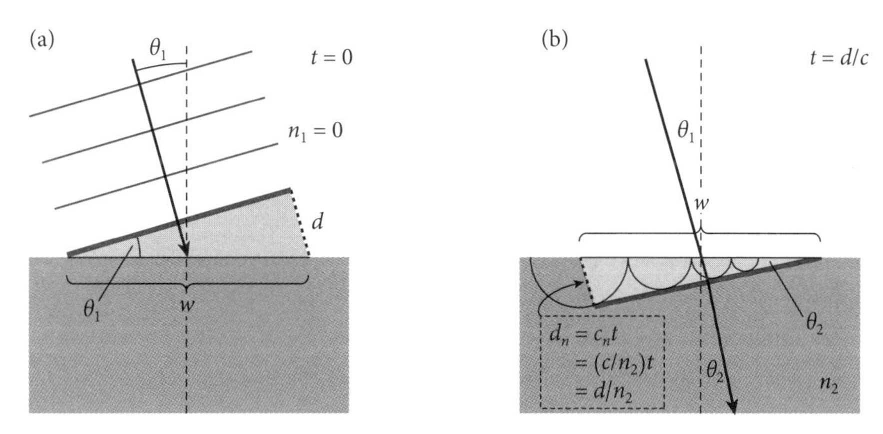

Figure 21.7 A demonstration, using Huygens' principle, of how a beam of light is refracted (bent) as it enters a material with a different index of refraction.

A little trigonometry will give us the angle θ_2 at which the beam will enter the new medium. Since $d_n < d$, the sides of the two shaded triangles in the two pictures are not all the same, so the triangles this time are not identical. The angle the wavefront makes with the surface is θ_1 in Figure 21.7a and θ_2 in Figure 21.7b. Both triangles are right triangles, with w as the hypotenuse, so we can write the sine of each angle as

$$\sin\theta_1 = \frac{d}{w} \quad \text{and} \quad \sin\theta_2 = \frac{d_n}{w}.$$

If we divide the second equation by the first, and remember that $d_n = d/n_2$, we find

$$\sin\theta_2 = \sin\theta_1\left(\frac{d_n}{d}\right) = \frac{\sin\theta_1}{n_2}. \tag{21.2}$$

Simple geometry shows that the θ in each triangle is the same as the θ between the ray and the dashed line normal to the surface in each picture, as shown in Figure 21.7.

Equation 21.2 can be brought into a form that is easier to remember if we first multiply both sides by n_2, giving

$$n_2 \sin\theta_2 = \sin\theta_1.$$

We then note that the initial vacuum had an index of refraction $n_1 = 1$, so we could always insert an n_1 in front of the $\sin\theta_1$, giving

$$n_1 \sin\theta_1 = n_2 \sin\theta_2. \tag{21.3}$$

This easy-to-remember equation, with all the ones on one side and all the twos on the other, is known as *Snell's Law*. It gives the relation between the angles of light beams traveling between two different media with two different indices of refraction. By the way, if you want to make sure that Snell's law works even if the first medium does not have $n_1 = 1$, you can do a quick review of the derivation leading up to Equation 21.2 and verify that you would still get Equation 21.3.

EXAMPLE 21.3 Suzy and the Trout

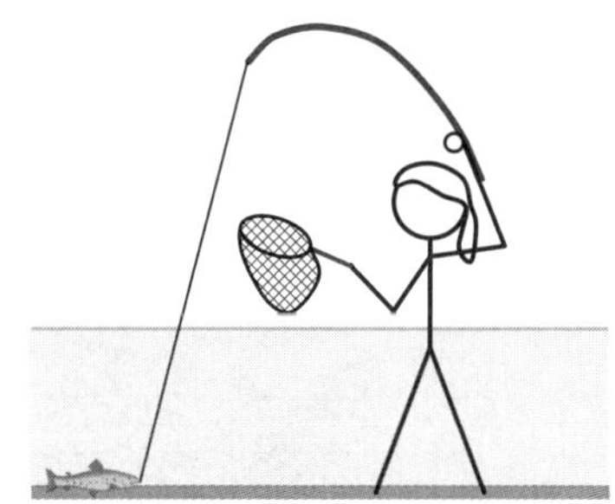

Suzy is standing in 3 feet of water, fishing a clear pool on the Gallatin River. She has managed to hook a 14-inch cutthroat on an elk hair caddis and, after a long fight, sees that the fish is resting on the bottom. Her eyes are 2 feet above the surface and the fish appears to be just 5 feet away from where she is standing. Where is the fish really?

ANSWER Since Suzy's eyes are 5 feet from the bottom and the fish *seems* to be 5 feet horizontally away from her, she must be looking down 45° away from the local vertical. This geometry is shown in the figure at right. But this figure cannot be correct, because it shows an unbent light beam emerging from the water. The true picture must be the one below, in which the light first travels upward at some unknown angle θ_1 until it reaches the surface of the water, where it is bent away from the local vertical. (If we didn't know it was bent away like this, we would find out as soon as we applied Snell's law.)

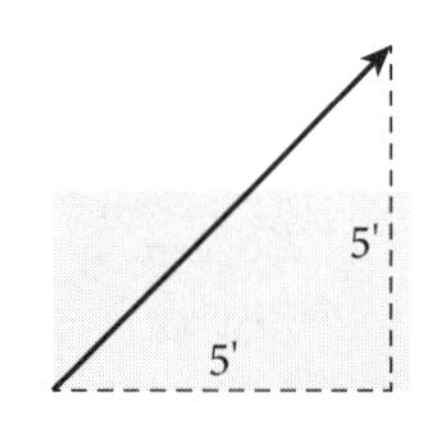

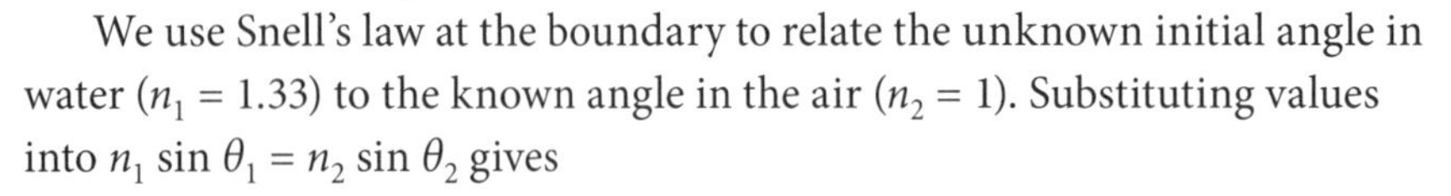

We use Snell's law at the boundary to relate the unknown initial angle in water ($n_1 = 1.33$) to the known angle in the air ($n_2 = 1$). Substituting values into $n_1 \sin\theta_1 = n_2 \sin\theta_2$ gives

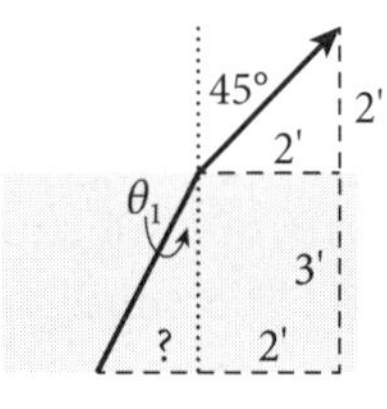

$$(1.33)\sin\theta_1 = (1)\sin 45^\circ.$$

Solving for θ_1, we find

$$\sin\theta_1 = \frac{\sin 45^\circ}{1.33} = 0.53 \quad \Rightarrow \quad \theta_1 = 32^\circ.$$

Note that this angle is indeed less than the 45° angle at which the beam emerges into the air.

From the geometry of the correct diagram we drew, we see that the distance from Suzy to the fish is 2 ft plus the distance labeled "?" in the diagram. We know the depth of the water is 3 ft and that the ratio of the side opposite the angle θ_1 to the side adjacent to it is $\tan\theta = \tan(32^\circ) = 0.63$. This gives the unknown distance as (3 ft)(0.63) = 1.9 ft. So the total distance from Suzy to the fish is not 5 ft, but 2 ft + 1.9 ft = 3.9 ft. Suzy, however, is an experienced fisherwoman, so she takes this into account. She reaches down at an angle $\tan^{-1}(3.9/5) = 38^\circ$ from the vertical, picks up the fish in a soft, fine-mesh, catch-and-release net, holds the fish gently in the current for a minute, and then responsibly releases the trout without harm.

Notice how we have cleverly worked another Montana reference into our textbook *and* added a pitch for wildlife-friendly catch-and-release fishing into the bargain.

EXAMPLE 21.4 Tracing a Ray Through a Prism

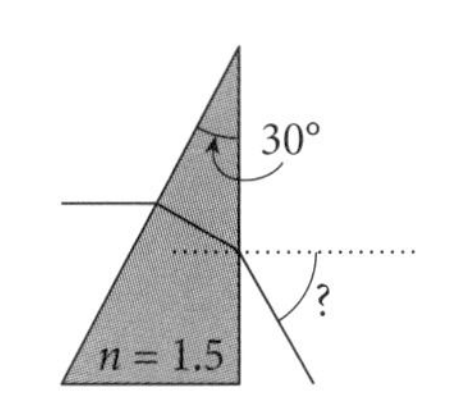

A horizontal beam of light strikes the slanted face of a 30° right-angle prism, as shown. The prism is made of glass of index of refraction $n = 1.50$. At what angle will it emerge from the prism (the question mark in the diagram)?

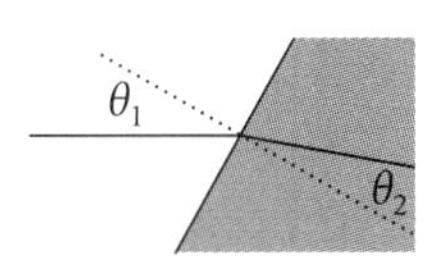

ANSWER As the incoming beam hits the face of the glass prism, it does so at an angle of $\theta_1 = 30°$ with the local normal to the face of the prism, as shown at right. Snell's law $n_1 \sin \theta_1 = n_2 \sin \theta_2$ then produces a value for θ_2,

$$(1)\sin 30° = (1.50) \sin \theta_2 \quad \Rightarrow \quad \sin \theta_2 = \frac{0.500}{1.50} = 0.333$$

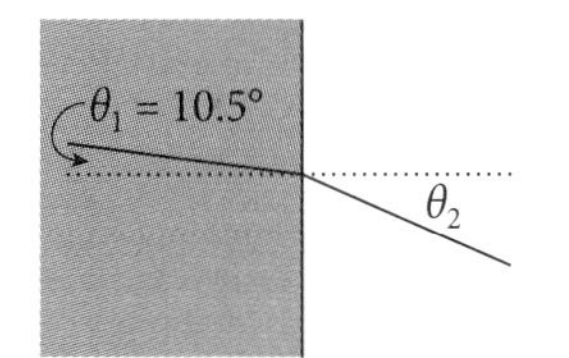

or $\theta_2 = \sin^{-1}(0.333) = 19.5°$. If the beam is 19.5° above the dotted line in the figure above, a line that itself points downward at 30°, then it must be traveling through the prism at an angle 10.5° below the horizontal. When the beam reaches the right side of the prism, it will arrive 10.5° away from the normal to the final surface, as shown at right. Snell's law (with θ_1 and θ_2 redefined), applied to this interface, gives

$$(1.50) \sin 10.5° = (1)\sin \theta_2 \quad \Rightarrow \quad \sin \theta_2 = 0.273$$

for an angle $\theta_2 = 16°$. The net effect of the light passing through the prism is to deflect the light from a horizontal beam to a beam 16° below the horizontal.

There is one last aspect of refraction that we should mention before we finish. If we have a beam of light inside a medium of high index of refraction, incident on a medium with a lower index of refraction, the beam will be bent away from the normal, as with the light from the fish in Example 21.3. Therefore, when the initial angle θ_1 gets big enough, one can reach the point where the transmitted angle θ_2 is 90°, a situation where it doesn't really enter the second medium at all, but is bent so that it travels along in the boundary between the two media. If we set the final angle to 90° and solve Snell's law for the initial angle, we find that this will happen at an initial angle given by

$$\sin \theta = \left(\frac{n_2}{n_1}\right) \sin \theta_2 = \left(\frac{n_2}{n_1}\right) \sin(90°) = \frac{n_2}{n_1}.$$

This angle is called the *critical angle*. It is the angle that satisfies

$$\sin \theta_C = \frac{n_2}{n_1}. \tag{21.4}$$

If the initial angle in the initial higher-index medium is larger than this, then the light transmitted into the lower-index medium would have to be at an angle whose sine is larger than 1, and there is no such angle. So what actually happens?

Well, there are two important things about light at the boundary between two transparent media that we haven't told you yet. First, even though the light is entering another transparent medium, some of the light is reflected. You may remember how, in Chapter 12, we discussed the fact that, when a wave on a string hits a section of the string where the wave speed is different, there is a reflected wave. The same is true for light waves. At a boundary between two transparent media with different indices of refraction, there is always a reflected wave as well as the refracted wave. The second thing you need to know is that, as the incident angle of a light wave at a boundary increases, the amplitudes of the electric and magnetic fields in the transmitted beam decrease. When the initial angle is the critical angle, the amount of light transmitted drops to zero. This means that the refracted beam that we said would skim along the boundary between the two media actually has no electric field amplitude to it at all. At the critical angle, and at any incident angle greater than the critical angle, all of the light is reflected back into the initial medium, with no loss of amplitude.

This phenomenon of *total internal reflection* at angles greater than the critical angle is what allows for fiber optics. A light wave is injected in one end of the fiber and, as long as the fiber is not crimped too sharply, the light may be directed along the fiber with no light loss as the light is reflected over and over again inside the fiber at angles greater than the critical angle for the material.

EXAMPLE 21.5 Design Considerations for Fiber Optics

An optical fiber is composed of a glass core with index of refraction $n = 1.6$, surrounded by a material called the "cladding," with index $n = 1.5$. What is the greatest angle at which the fiber may be bent without losing light into the cladding?

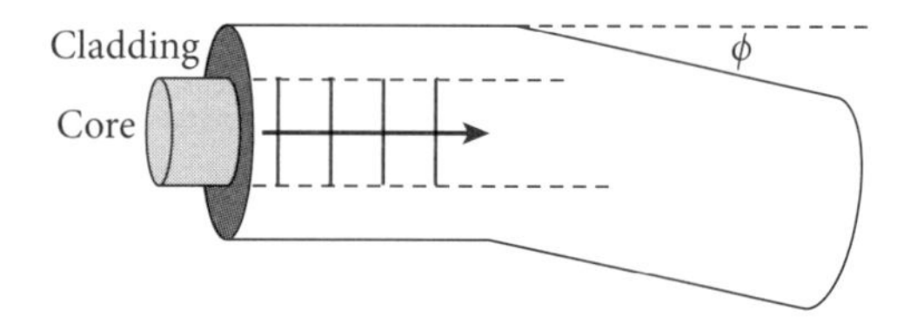

ANSWER A fiber is typically very long compared to its diameter, so light moving down the fiber is typically propagating parallel to the axis of the fiber. The geometry showing the angle between the wave direction and the normal to the core-cladding interface is therefore as shown in the picture at right.

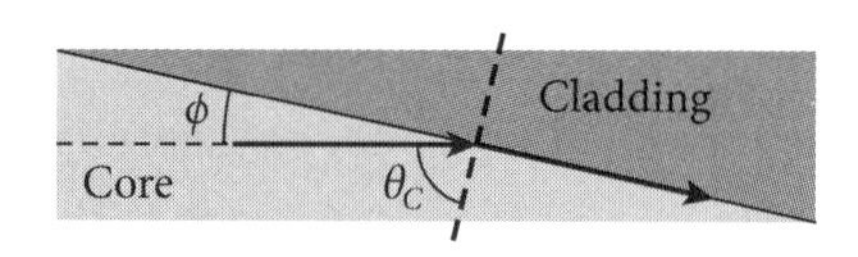

The core has $n_1 = 1.6$ and the cladding has $n_2 = 1.5$, so the critical angle is

$$\sin\theta_C = \frac{n_2}{n_1} = \frac{1.5}{1.6} = 0.9375 \quad \Rightarrow \quad \theta_C = 70°.$$

This would be the angle labeled θ_C in the figure. The rest is geometry. Since the heavy dashed line is perpendicular to the interface between the two materials, the angle labeled ϕ in the diagram we drew above is

$$\phi = 90° - \theta_C \quad \Rightarrow \quad \phi = 20°.$$

But the ϕ in the drawing above is the same as the ϕ in the drawing in the question statement, since both sides of the two angles are parallel. So the solution is that the fiber optic cable should not be bent by more than 20° if the fiber is to work properly.

21.6 Energy in Electromagnetic Waves

Back in Chapter 17, we saw that the energy stored in a capacitor in which there is an electric field E could be expressed as the product of the energy density $\frac{1}{2}\varepsilon_0 E^2$ and the volume of the capacitor (this was Equation 17.11). In fact, an electric field carries this density of energy wherever the field exists, whether inside a capacitor or in an electromagnetic wave. We have also seen in this chapter that Maxwell's equations require the electric field in an electromagnetic wave to be accompanied by a magnetic field as well.

We don't want to take the time to derive a formula for the energy carried by a magnetic field, or to work out the relative magnitudes of the electric and magnetic fields in an electromagnetic wave. But let us just point out that the magnetic field in an electromagnetic wave carries its own energy density, just like the electric field, and that the magnetic field energy density is numerically equal to the energy density in the electric field. We therefore conclude that an electromagnetic wave transports energy as it propagates, and that the total energy per cubic meter is given by

$$\rho_{\text{EM}} = \rho_{\text{E}} + \rho_{\text{M}} = 2\rho_{\text{E}} = \varepsilon_0 E^2. \tag{21.5}$$

If we want to find the total energy carried by the wave, we only have to know the electric field for the wave. Maybe an example would help us see how to use this.

EXAMPLE 21.6 The Energy and the Electric Field in a Beam

A laser beam 2 mm in diameter carries 10 watts of power. What is the magnitude of the electric field in the beam?

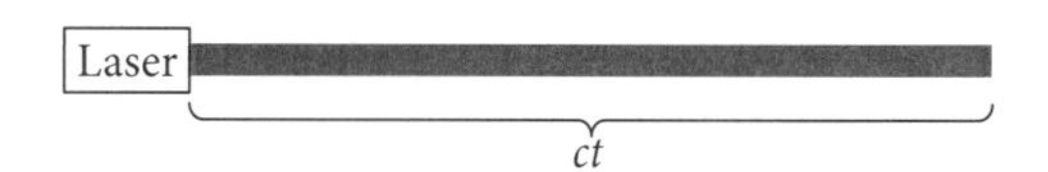

ANSWER Power (P) is measured in watts, which are joules per second. In time t, the distance covered by the wave will be ct, as shown in the figure above. So the volume of a beam of radius r that moves past a given point in time t will be $V = \pi r^2 ct$. The total electromagnetic energy carried in this volume of the beam is Pt, so the density of energy in a wave carrying this energy past a given point in an elapsed time t is

$$\rho_{\text{EM}} = \frac{Pt}{V} = \frac{Pt}{\pi r^2 ct} = \frac{P}{\pi r^2 c}.$$

Putting in values, this equation becomes

$$\rho_{\text{EM}} = \frac{10 \text{ J/s}}{(3.14)(1 \times 10^{-3} \text{ m})^2 (3 \times 10^8 \text{ m/s})} = 1.06 \times 10^{-2} \text{ J/m}^3.$$

The energy density is related to the strength of the electric field in the wave by Equation 21.5. Solving for E^2 gives

$$E^2 = \frac{\rho_{\text{EM}}}{\varepsilon_0} = \frac{1.06 \times 10^{-2} \text{ J/m}^3}{8.854 \times 10^{-12} \text{ C}^2/\text{N}\cdot\text{m}^2} = 1.20 \times 10^9 \text{ (N/C)}^2.$$

The electric field is the square root of this number, or

$$E = 3.46 \times 10^4 \text{ N/C}.$$

These values of the laser power and the radius of a laser beam are realistic, so this electric field magnitude of many thousands of N/C is a realistic value for the field strength in a laser beam.

21.7 Summary

In this chapter, you should have learned the following:

- You should feel that you understand what an electromagnetic wave is and how it propagates according to Huygens' principle.
- You should be able to connect wavelength, frequency, and speed of an electromagnetic wave with the wave equation, $c = \lambda f$.
- You should remember that light reflects at an angle equal to the incident angle.
- You should know that light slows in a transparent medium by interacting with the atoms of the medium, and that its speed in the medium is $c_n = c/n$, where n is the index of refraction.
- You should be able to use Snell's law to relate the angles of incident and transmitted light beams at an interface between two materials.
- You should be able to use the critical angle formula to find the incident angle at which light will be totally reflected internally in a high-index medium.
- You should be able to connect electric field magnitude in an electromagnetic wave with the energy density carried by the wave.

CHAPTER FORMULAS

Wave equation for light: $c = \frac{\lambda}{T} = \lambda f$ $c = 3 \times 10^8$ m/s

Speed of light in a medium with index of refraction n: $c_n = \frac{c}{n}$

Snell's law: $n_1 \sin\theta_1 = n_2 \sin\theta_2$ Critical angle: $\sin\theta_C = \frac{n_2}{n_1}$

PROBLEMS

21.2 * 1. What is the wavelength of a radio wave of frequency 101.1 MHz (one of the frequencies in the FM band)?

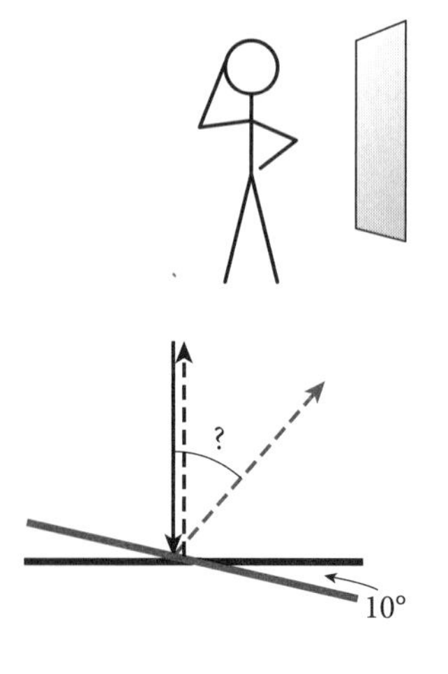

* 2. An electromagnetic wave has a frequency of 2×10^{15} Hz. What is its wavelength and what is its category in the electromagnetic spectrum?

21.4 ** 3. A man wants to hang a mirror on the wall in which he can see from the top of his head down to his feet. If the man is 6 feet tall, what is the smallest height of the mirror that will do the job? Explain exactly how high the mirror should be hung.

** 4. A beam of light travels vertically downward and strikes a horizontal mirror, reflecting directly back vertically upward, as indicated by the black dashed line in the diagram at left. The mirror is now rotated, so that it is 10° away from horizontal, as is the red mirror in the diagram. The incident solid black ray is the same in both cases.

a) At what angle from the vertical will the reflected beam (the red dashed arrow) now be seen?

b) If the mirror is further rotated until it is 20° from the horizontal, what will be the new angle between the reflected beam and the vertical?

** 5. Two beams from the left hit a polished sphere of radius 10 cm. One beam hits the sphere directly at its central point. The other beam is 3 cm above the first. In what direction will each beam reflect off of the sphere?

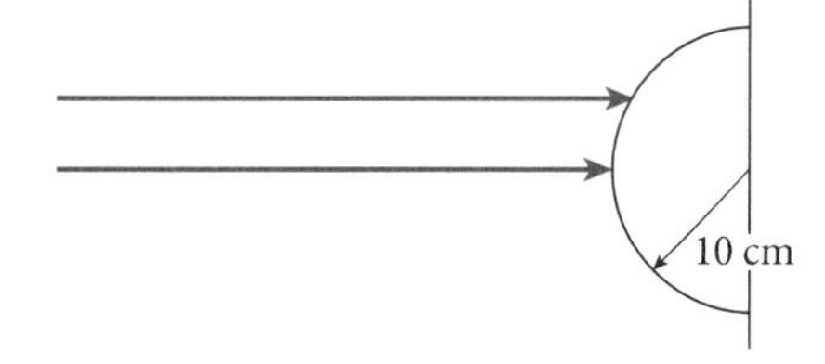

*** 6. Two mirrors are set at 90° to each other, as shown in the diagram below. Prove that, in the plane of the paper, any ray coming at the mirrors at an angle θ less than 45° from the centerline will be reflected back in exactly the direction it came from. [Hint: There is a lot of geometry of angles and triangles in this.]

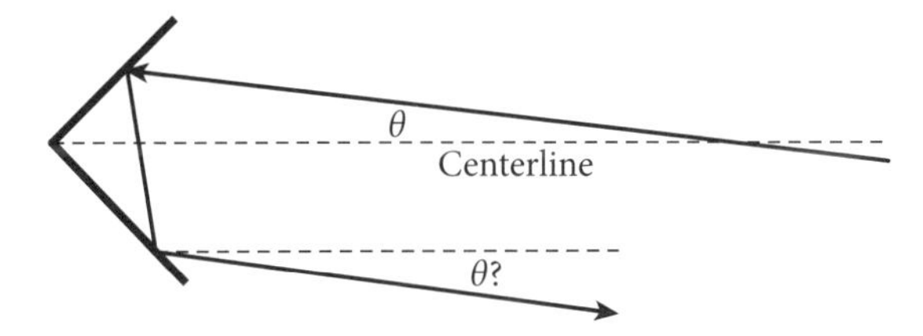

21.5 * 7. What is the speed of light in a piece of glass with index of refraction $n = 1.38$?

* 8. A beam of light in air enters the water in a glass of water, and then passes into the glass itself. What is the speed of the light in the water? What is the speed in the glass?

* 9. Light of wavelength 555 nm passes through the facet of a diamond. What are its wavelength and frequency in the diamond?

* 10. Light of frequency 5×10^{14} Hz (500 terahertz) enters a plastic whose index of refraction is 2.00. Find the wavelength and frequency of the light in the plastic.

* 11. Two beams hit a point on a slab of glass ($n = 1.5$) from different angles and pass into the glass. One of the beams (A) comes in along the normal to the slab and the other (B) comes in at an angle of 32° from the normal. At what angles to the normal will each of the beams travel through the glass?

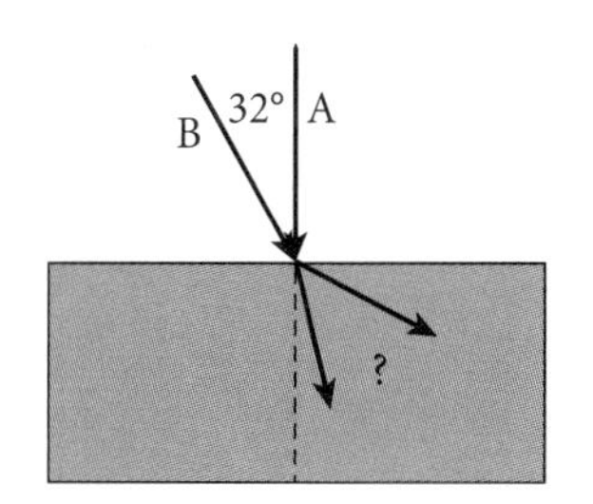

** 12. Wilmington, North Carolina, is 5024 km due east of Santa Barbara, California. High-speed communication between the two cities can be accomplished by use of satellite communication to a satellite that is a distance of 3456 km away from each city or by direct fiber optic cable between the two cities in a fiber whose index of refraction is $n = 1.62$. How long would it take each signal to go from Santa Barbara to Wilmington?

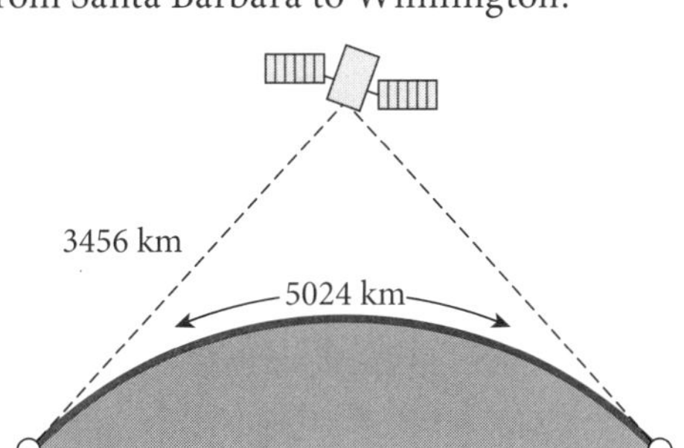

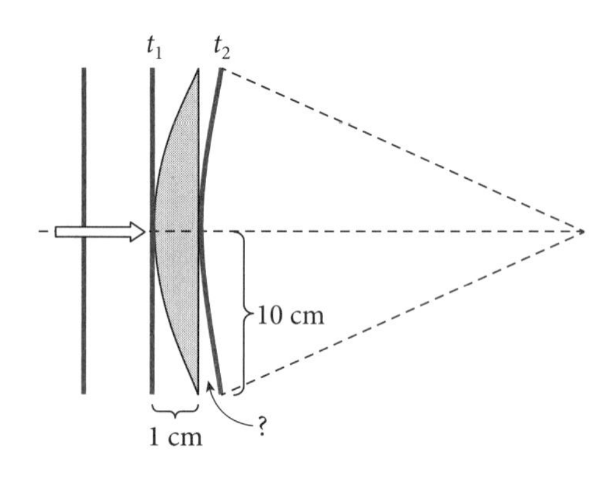

** 13. A plane wave hits a piece of glass (index of refraction 1.5) whose front surface is spherical and whose back surface is plane. The radius of the lens is 10 cm and the thickness of the glass is 1 cm at the center, as shown in the diagram at right. At time t_1, the center of the plane wavefront has just reached the lens. A short time later, at time t_2, the center of the wavefront will have passed completely through the glass, as shown.

a) Find the time that elapses between t_1 and t_2, the time it takes the center of the wavefront to pass through the middle 1 cm of the glass.

b) Find the amount by which the edges of the wavefront at t_2 will be ahead of the center of the wavefront, due to the fact that these edges passed through empty space, with no glass in their paths.

The actual shape of the emerging wavefront, if we worked through all of the locations of all the Huygens wavelets, would turn out to be a section of a sphere. We will see this idea again in Chapter 22.

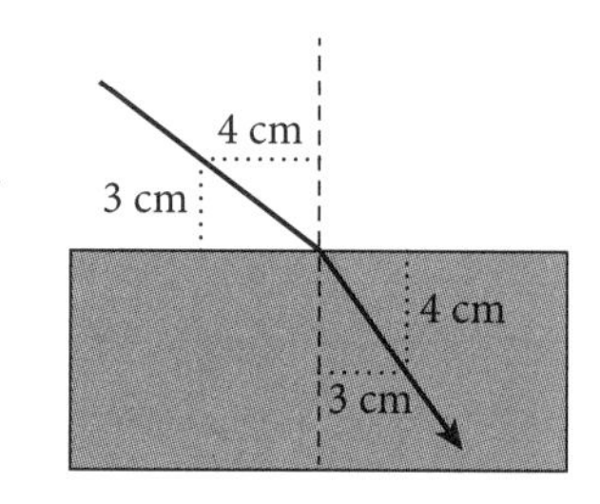

** 14. A ray of light goes from vacuum into a transparent medium along the path shown in the figure at right, with measured distances along the path geometry as shown. What is the index of refraction of the transparent medium?

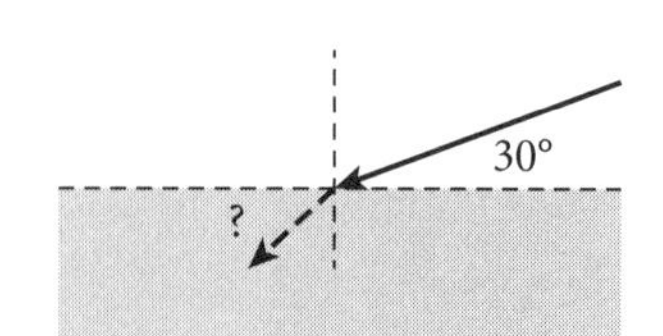

** 15. The rays of the afternoon sun strike the surface of a lake at an angle 30° above the surface of the lake. What angle will the ray that enters the water make with the surface of the lake?

** 16. A fish sits at the bottom of a pond and looks upward. He is able, because of refraction, to see everything in the entire hemisphere above the water condensed down to within a narrower cone. What is the full angle of this cone?

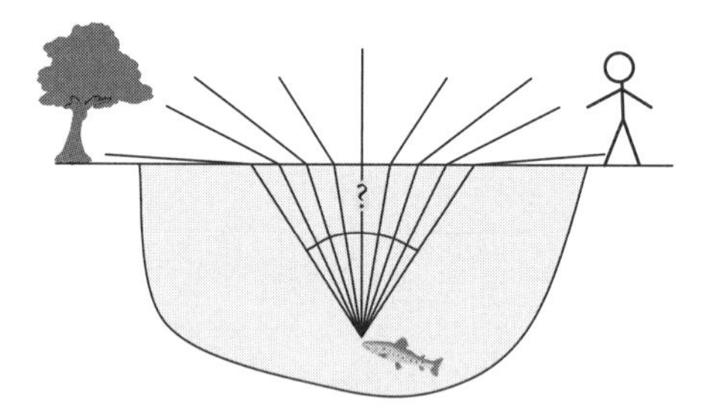

** 17. A 3-meter pole pokes up from the bottom of a pond that is 2 m deep. The sun is 45° above the horizon. What is the length of the shadow of the pole on the bottom of the pond?

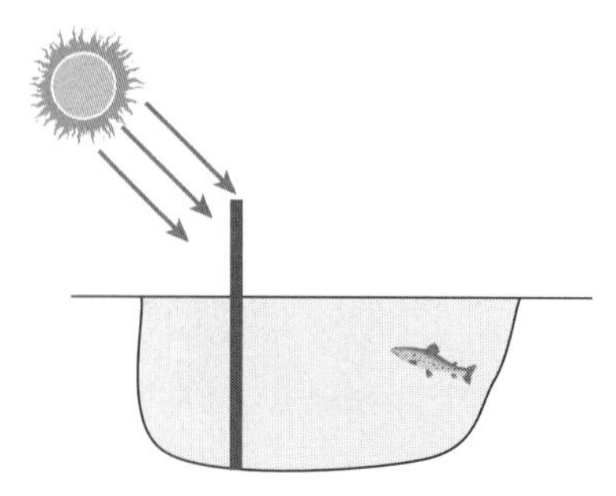

** 18. A glass prism with index of refraction $n = 1.6$ has a 60° angle at the top of the prism. If a beam of light shines onto the prism from below the horizontal, it will pass through the prism and emerge out the other side, as shown. At what angle θ below the horizontal should the light come in, if we want it to emerge going downward at the same angle?

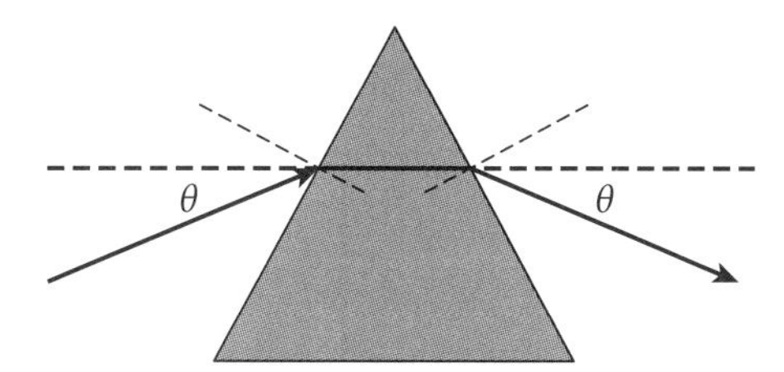

** 19. A beam of light from a Coast Guard searchlight is seen by a scuba diver to be coming downward at an angle 30° from vertical. At what angle from the vertical does the searchlight hit the water?

** 20. Light strikes the vertical side of a right-angle prism containing a 60° angle, as shown. The index of refraction of the glass is $n = 1.5$. At what angle will the ray emerge?

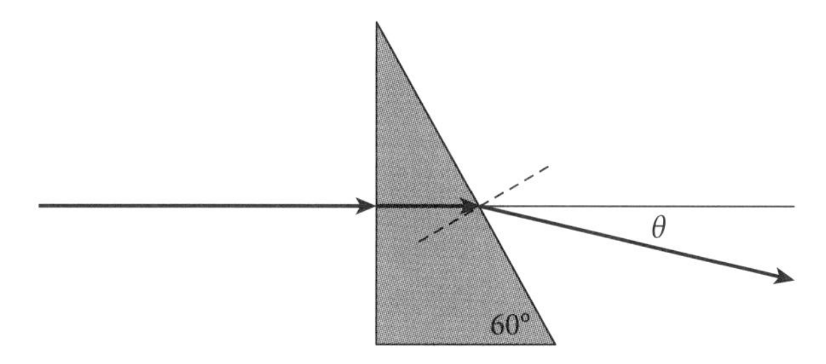

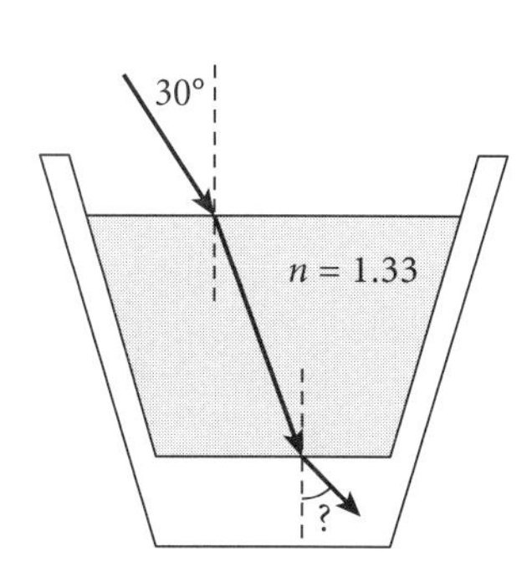

** 21. Water ($n = 1.33$) fills a shot glass ($n = 1.5$), as shown. A beam of light strikes the water at an angle of 30° to the normal to the water's surface, as shown. The beam passes through the water and into the glass base of the shot glass. At what angle will it travel through the base?

* 22. What is the critical angle for diamond in air?

** 23. What is the critical angle for a piece of glass ($n = 1.5$) immersed in water? In which material must the initial ray be located for total internal reflection to occur?

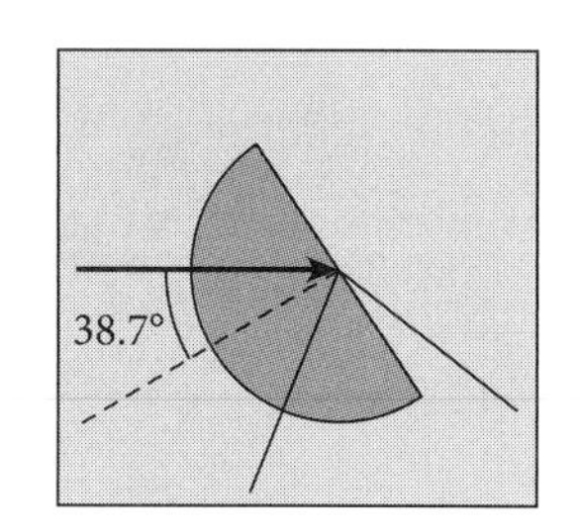

** 24. To measure the critical angle in a sample of glass, a slab of the glass is cut into a semicircle, and a beam of light is aimed at the center of the semicircle, as shown. The slab is then rotated about an axis through the center. When the transmitted ray disappears, the critical angle is the measured angle between the normal to the flat surface (the dashed line) of the semicircle and the incident ray. If this angle is measured to be 38.7°, what is the index of refraction of the glass?

** 25. Consider a long cylinder of glass ($n = 1.5$) with polished flat ends. Prove that any light ray from the air outside the glass that enters one end of the cylinder will travel along the cylinder with nothing but complete internal reflections, until it reaches the other end.

** 26. As we saw on page 436, the speed of an electromagnetic wave in vacuum can be calculated from the constants in Maxwell's equations as $c = 1/\sqrt{\mu_0 \varepsilon_0}$. In non-magnetic materials, the magnetic permeability is very close to the permeability of free space, μ_0. However, the permittivity in a medium may differ significantly from the ε_0 of free space (you might want to take another look at Equation 17.8 to remind yourself about the meaning of ε). In principle, the speed of light in a medium could be calculated from μ_0 and the ε of the medium. However, the dielectric constants (Table 17.1) that determine the values of ε are appropriate only for fields that vary slowly. At optical frequencies (the frequencies of visible light waves), the polarized molecules are not able to respond quickly enough to completely polarize the material. The effect is that the value of ε becomes frequency dependent. Use the relationship between the permittivity of a medium and the speed of light in the medium to find the permittivity of glass when an imposed electric field is changing at optical frequencies. [Hint: The index of refraction for glass is given in Table 21.3.]

** 27. **Dispersion.** Because a medium in which there is an electromagnetic wave cannot instantaneously polarize in response to rapid changes in the direction of the electric field, as explained for Problem 26, the index of refraction in a material like glass is not constant at all frequencies, but has a slight variation, known as *dispersion.* As a result of dispersion, white light passing through a prism will be refracted into different directions for light of differing frequencies (and different wavelengths). The following graph gives the index of refraction for wavelengths in the range 400 to 800 nm. Find the angles (shown in the diagram below) at which a violet ray (400 nm) and a red ray (700 nm) will emerge from a 60° prism. Which color is bent the most?

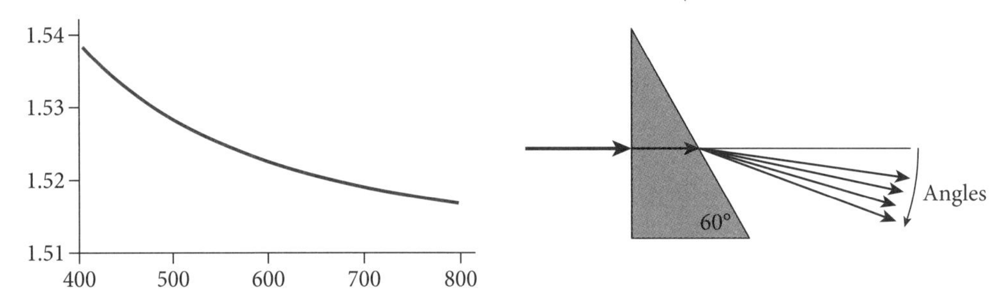

21.6 ** 28. **Poynting Vector.** There is a vector defined by John Henry Poynting in 1884 whose direction gives the direction that an electromagnetic wave transmits its field energy and whose magnitude is the rate at which energy is being transmitted per square meter of the wavefront. For mutually perpendicular E- and B-fields in the wave, the magnitude of the Poynting vector is $S = EB/\mu_0$ and the direction is given by a right-hand rule, curling fingers from $\vec{E}$ to $\vec{B}$, as we explained on page 437.

a) Carefully analyze the units of the Poynting vector and show that they are indeed watts per square meter. [Hint: Remember that electric fields are measured in N/C and magnetic fields in T.]

b) Find the intensity of the light in a beam of light whose electric field strength is 360 N/C and whose magnetic field is 1.2 μT.

** 29. **Solar Intensity.** The intensity of a light wave is the way we characterize how bright light appears. The intensity gives the number of watts of power the light wave carries in each square meter of the surface of the wave. The total intensity of sunlight striking the earth is 1340 watts/m^2.

a) How many watts of sunlight are received by a square array of solar cells, with dimensions 2 m × 2 m?

b) If the efficiency (the ratio of electric power generated to the total power hitting the cells) is 38%, how many watts of power will this solar array generate?

c) What is the electric field strength in the sunlight hitting the earth? [Hint: See Example 21.6.]

*** **30.** In the discussion of Figure 21.2 on page 437, we stated that the electric field in an electromagnetic wave decreased more slowly than the $1/r^2$ drop-off of the static field given by Coulomb's law. In this problem we will derive how the magnitude of the electric field in a spherical electromagnetic wave depends on the distance from the source of the waves. As a spherical electromagnetic wave carries energy away from a source, the conservation of energy requires that all of the energy leaving the source must pass through spheres of larger and larger radius. Consider an omni-directional radio antenna (one that emits equal power in all directions) emitting 10 kW of power:

a) How many watts will hit a one-square-meter antenna located 10 km away?

b) What is the energy density in a wave that produces electromagnetic radiation that transfers this much power (watts) to the antenna?

c) What will be the magnitude of the electric field in the wave at this distance?

d) Compare the formula for the area of the sphere of radius r over which the energy is spread with Equation 21.5 for the energy density in a wave. Then conclude how the electric field in an electromagnetic wave must depend on the distance from the source.

*** **31[f].** A beam of light hits a curved surface of glass of index of refraction n. The shape of the surface is that of a portion of the surface of a sphere of radius R. The back side of the glass is flat. A ray from far off to the left comes in horizontally, a distance d above the centerline of the glass. Find the distance f from the point where the ray first hits the glass to the point where this ray finally crosses the centerline. Assume that all angles are small, so that you can make the approximation $\sin\theta \approx \theta$ for all angles in the problem.

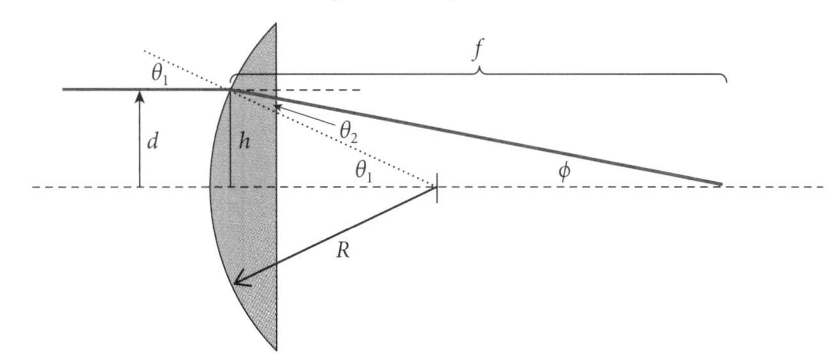

[Hint: There is a lot of geometry in this problem. It may help to consider the thin line labeled h in the diagram.] [Hint #2: The details of the light passing the two surfaces of the glass are particularly important. You might want to consider this:]

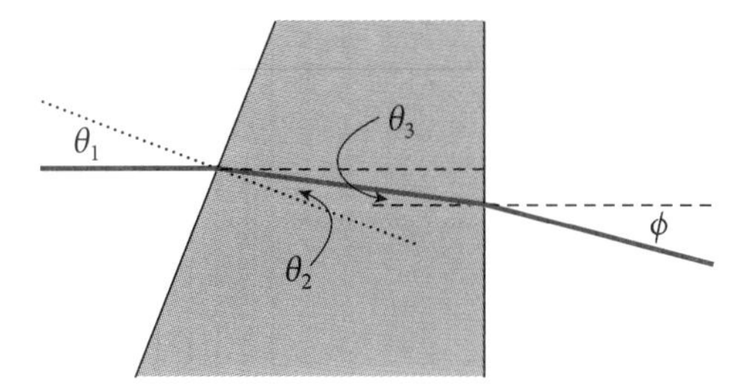

22

GEOMETRICAL OPTICS

In Figure 21.4 in the last chapter, we saw how Huygens' principle is used to determine the future shape of a plane wave. Most of us are probably more familiar with another kind of wave—the ripples on the surface of a pond when a pebble is dropped into the pond. In this case, the ripples are not straight lines; they are circles. But Huygens' principle can again be used to describe these widening circles, as we show in Figure 22.1.

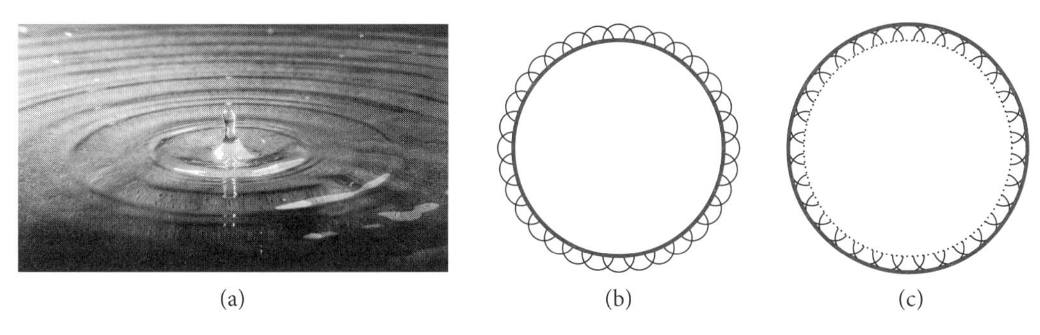

(a) (b) (c)

Figure 22.1 Part (a) shows surface ripples on a pond. Part (b) shows one ripple crest with Huygens wavelets radiating out from each point on the ripple. Part (c) shows the propagated circular wave crest as the combination of all of the wavelets.

The waves in Figure 22.1 are circular waves on a two-dimensional surface. There are analogous waves in three dimensions called *spherical waves.* In this case, Huygens' principle involves hemispherical wavelets that radiate outward from each point on an initial spherical surface to propagate the wave forward, forming a new wavefront that is a larger spherical surface.

When electromagnetic waves are created by electrons accelerated in a metal rod, as discussed in the previous chapter (see Figure 21.2), the pattern of the waves created are not truly spherical, but produce what is called a dipole pattern, depicted in Figure 22.2. Along the equator of the pattern, the shape of the surface is indeed a portion of a sphere centered on the antenna. However, as one goes away from the equator, the deviations from sphericity become large, with no electromagnetic field at all being radiated directly over the rod. *However, in the rest of these chapters, we will assume that we are working in a region of the radiation pattern where we can treat the light waves or other electromagnetic waves as if they were completely spherical.*

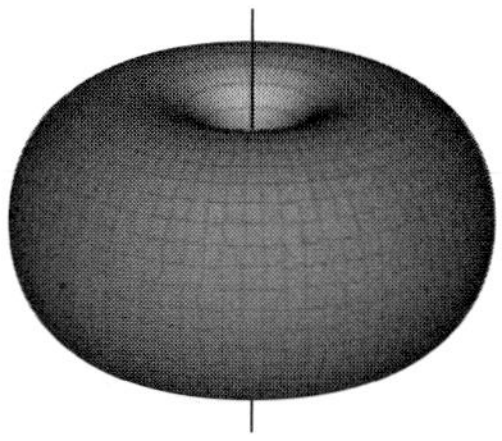

Figure 22.2 Radiation pattern from a dipole antenna.

22.1 Spherical Waves and Curvature

In the last chapter, we drew Figure 21.5 for plane waves, showing electric and magnetic fields that were constant on each plane. For a spherical wave, the wavefront is not a plane, but a sphere. The electric and magnetic fields are still perpendicular to each other, and they are still constant in magnitude everywhere on the sphere. However, their directions now change from point to point. The electric fields lie in the surface of the sphere, perpendicular to the outward radial direction in which the wave is propagating. Figure 22.3 shows the fields on one spherical surface at one instant of time. At the same time, the fields on a concentric sphere of greater radius will both have zero magnitude, and on a sphere of even larger radius, the fields will point in the opposite direction from what is shown. In this way, the oscillating nature of the traveling electric and magnetic fields will be the same as for a plane wave (Figure 21.3). The magnetic field is always oriented so that Right-Hand Rule #1, with fingers curling from $\vec{E}$ to $\vec{B}$, gives the outward radial direction (see page 437).

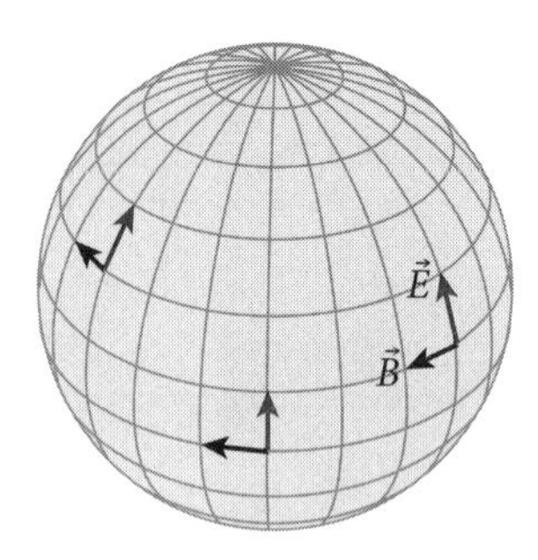

Figure 22.3 The fields lie in the spherical wavefront and are uniform magnitude on the sphere.

As we said, the wavefronts of spherical waves are surfaces of spheres in three dimensions, but, for simplicity, let us only look at a two-dimensional slice through the spheres. Figure 22.4 shows such a slice through the successive spherical wavefronts propagating away from a point source of light. The red lines represent "crests," points where the electric and magnetic fields have their maximum values. The spheres in between the red lines (and thus not shown) are where the fields are weaker. Huygens' principle generates spherical wavefronts of larger radius as the wave propagates radially away from the source. The outgoing rays (the black arrows) are everywhere perpendicular to the wavefronts, just as they were in the plane waves of Figure 21.5.

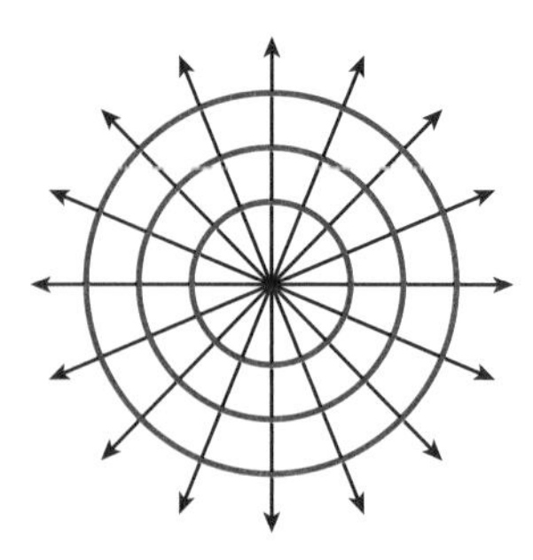

Figure 22.4 Spherical waves with wavefronts perpendicular to the propagation vectors.

As the wave moves away from its source, what is there about it that changes? In Figure 22.5, we have drawn two patches of a spherical wavefront having equal area (equal length in this cross-section drawing). The difference between the two pictures is that the patch in picture (a) is drawn at a point when the wavefront was close to the source, while the second patch (b) is at a point when the wavefront had moved some distance from the source. The question we would like to ask is this: In which picture does the wave have a greater curvature? We expect it is obvious that the wavefront in picture (a) is more curved than the wavefront in picture (b). But how can we characterize the curvature mathematically?

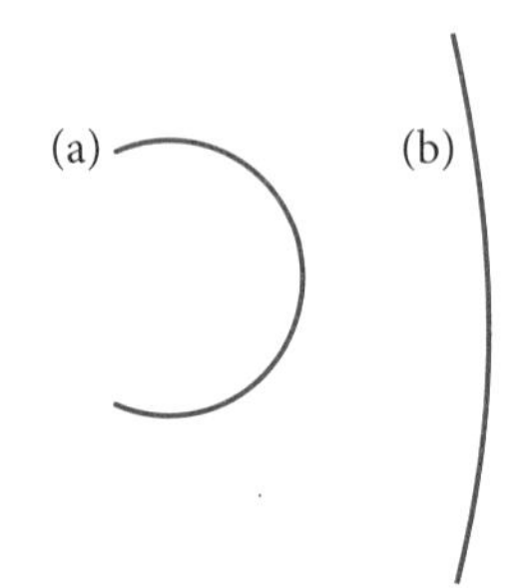

Figure 22.5 Two waves of different curvature.

Since each wavefront in Figure 22.5 is a piece of a sphere of a particular radius, there is an obvious way to define the *curvature of the surface*. If we choose a letter K for this curvature, we can define the curvature of the wavefront as $K = 1/R$, where R is the radius of the surface (also called the "radius of curvature"). This way, a high curvature will be associated with a small radius, as we want. If the radius of curvature R is measured in meters, then the units of the curvature K will be m^{-1}.

There is one last aspect of the curvature of a spherical wavefront that we would like to address. In Figure 22.6a, we show a portion of a spherical wavefront that propagates away from a point source via Huygens' principle, producing wavefronts of larger and larger radius, as we would expect. But in Figure 22.6b, we show a portion of a wavefront where the wave is still moving to the right, but where the wavefront is concave instead of convex. Huygens' principle

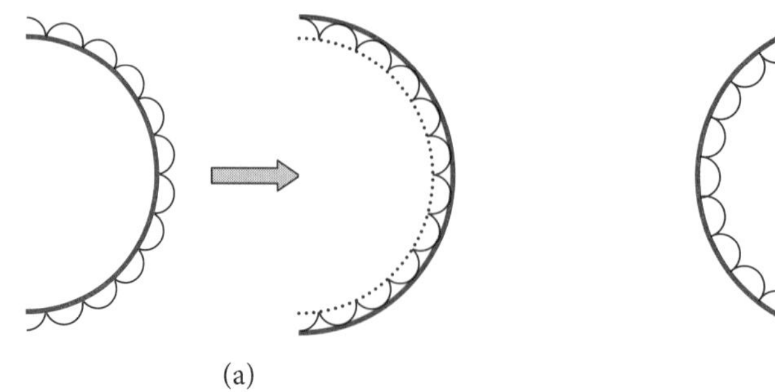

Figure 22.6 A diverging spherical wave (a) and a converging spherical wave (b).

predicts that these waves will crunch together as time goes by, with the radius of curvature getting smaller and smaller and the curvature becoming larger and larger. (By the way, you might think that a converging wavefront like this would never actually occur, but you would be wrong. We explain how this might arise in our next section.)

In order to discriminate between the two cases shown in Figure 22.6, we will choose to add to the definition of the curvature of a wavefront by giving the curvature a sign. We define the curvature of a *diverging* wave, like the wave in Figure 22.6a, to be *negative* and the curvature of a *converging* wave, like the one in Figure 22.6b, to be *positive*. The formula for the *curvature of a wave* will therefore be written

$$K = \pm\frac{1}{R}, \quad (22.1)$$

where R, being a radius, is always positive. The sign is attached to the curvature based on what we know about the wave—plus for converging and minus for diverging.

EXAMPLE 22.1 The Evolution of Curvature in a Diverging Wave

A tiny light source produces spherical light waves that radiate out to fill a room. What is the curvature of the waves 2 cm from the source and 1 m from the source?

ANSWER After the spherical wavefronts from the light source have reached a point 2 cm away from the source, the wavefront will be a sphere of radius 0.02 m. Therefore, the curvature of the wave at this point is

$$K = -\frac{1}{R} = -\frac{1}{0.02\text{ m}} = -50\text{ m}^{-1}$$

in which we have added a minus sign by hand because we know that the wave is a *diverging* wave. By the time the light has reached a point 1 m away from the source, the curvature of the wave will be $K = -(1/1\text{ m}) = -1\text{ m}^{-1}$.

Note that the curvature starts off at a large negative value (-50 m^{-1}) and increases, becoming less negative (-1 m^{-1}) as the wave travels. As the process continues for a diverging wave, the limit of the curvature will be $K \to 0$ as $R \to \infty$.

If we had been asked about a *converging* wave, like the one shown in Figure 22.6b, the wave would have started off at a small positive value, $K = +(1/R)$. Then, as the radius of the wave continued to contract, the curvature would have increased to larger and larger positive values, with the limit $K \to +\infty$ as $R \to 0$.

22.2 Thin Lenses

In this section, we want to take two concepts from previous discussions and put them together to understand how lenses work. A lens is a piece of glass or other transparent material that has one or more curved surfaces on it, as in Figure 22.7. A thin lens is one where we ignore the distance between the middle of the lens and the point where the wave enters or emerges from the lens. We approximate everything as if it happened in a single plane at the middle of the lens. Consider what happens when a diverging wave from some source encounters a lens like that in the figure. We know from Section 21.5 that light will travel more slowly inside the lens than it does in the air outside the lens. So, because of the shape of the lens, the center portion of the wavefront will hit the lens first and that portion will have to travel through a greater thickness of glass than will the outer portions of the wavefront. If we apply Huygens' principle to the incident wavefront in Figure 22.7, we will predict a transmitted wavefront whose center has been held back relative to its edges. If the difference in the thickness of the glass between the center and the edges is great enough, the wavefront can actually change from convex to concave, from negative to positive curvature, as we show in Figure 22.7. The result is that the

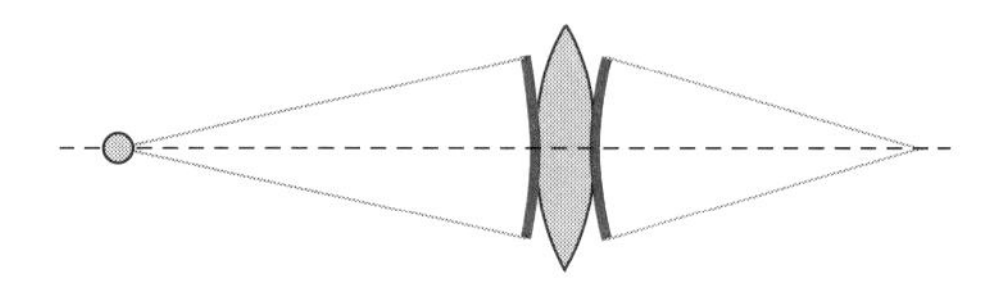

Figure 22.7 A diagram of how a lens acts to change the curvature of a wave.

spherical wave that was initially diverging from a point on the left will end up converging to a point on the right. The lens has taken a beam of diverging light from one point and focused it all down to another point. The place where the wave converges is called the *image* of the source point. The light converging to the point carries real electromagnetic energy. If you placed a 3 × 5 card at the point where the wave converges, the tiny spot would start to glow. The spot is referred to as a *real image*, because there is a real focusing of the light to this point.

The question that now arises is how to characterize a lens's ability to do this changing of the curvature. If we knew the exact shapes of the two surfaces of the lens, we could work through the geometry of each wavelet to determine what the final wave curvature would be after it leaves the lens, given the initial wave curvature before it hit the lens. But that is a lot of geometry. Let us instead describe a method to characterize a lens experimentally. If the source of a diverging wave is very far away, then the curvature of the wave when it hits our lens will be nearly zero; it will be a plane wave. In Figure 22.8, we show such a plane wave hitting a lens, and we see the lens delaying the center of the wavefront relative to the edges. The result is a wave that focuses to a point a distance f from the middle of the lens. We use the plane wave as a standard and use the resulting distance f, called the *focal length* of the lens, as a way to characterize the lens itself, in all situations.

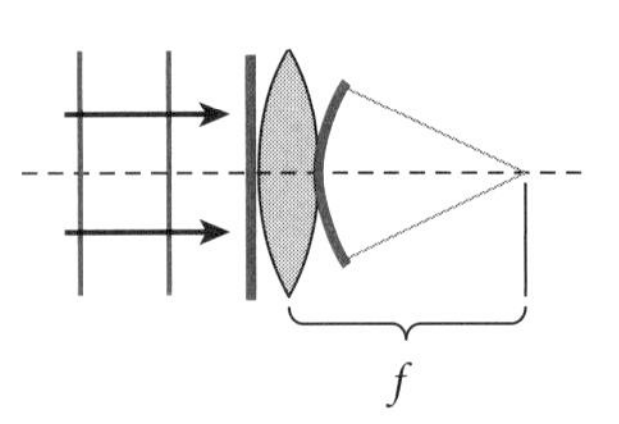

Figure 22.8 A lens of focal length f focuses a plane wave to a distance f from the lens.

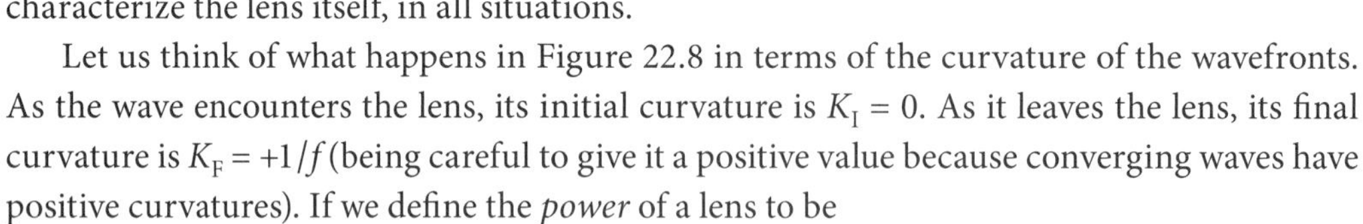

Let us think of what happens in Figure 22.8 in terms of the curvature of the wavefronts. As the wave encounters the lens, its initial curvature is $K_I = 0$. As it leaves the lens, its final curvature is $K_F = +1/f$ (being careful to give it a positive value because converging waves have positive curvatures). If we define the *power* of a lens to be

$$P = \frac{1}{f}, \tag{22.2}$$

then we can think of the effect of a lens as adding its power to an initial curvature K_I to produce a final curvature K_F, in what we will call "the lens equation:"

$$K_F = K_I + P. \tag{22.3}$$

As Equation 22.2 shows, the units for lens power are the same as those for curvature; they are m^{-1}. For lens power, however, this unit is also known as a diopter (D). If you wear glasses, your prescription will be given in diopters, which means it is in m^{-1}.

We defined Equation 22.3 so that it would work for the plane wave shown in Figure 22.8. But if we were to work through the geometry of a non-plane wave, either a converging or a diverging wave, we would discover that this same formula works just as well in all those cases. So let us summarize the process we have described.

- Light from a source a distance R_I away from a lens moves toward the lens. As it travels, its curvature increases from $-\infty$ (a diverging wave at zero distance from the source) to $K_I = -1/R_I$ (a diverging wave with curvature radius R_I).
- A lens that is shaped so that it is fatter in the center than at the edges will slow the center of the wave relative to the edges, thereby increasing the curvature of the transmitted wave. For this reason, we say that such a lens has *positive power* and is known as a *converging lens*. The lens power is $P = +1/f$.
- The effect of the lens on the incident wavefront is to act on the initial curvature to produce a final curvature $K_F = K_I + P$. If the positive power P is more positive than initial curvature K_I is negative, then K_F will end up positive and the wave will converge, as shown in Figure 22.7. The radius of curvature of the wavefront that emerges from the lens is $R_F = 1/K_F$, and so the light will focus to a point a distance R_F from the lens. Note that this distance is NOT the focal length f. The focal length is a property of the lens, and it only tells us where a *plane wave* will focus. If a wave hitting a lens is *not* a plane wave, it will *not* focus at the focal length.
- As the light continues to the right of the lens, always converging to the point R_F from the lens, its curvature will increase toward the limit $+\infty$ at the focus.

Note that the curvature of a traveling wave is always changing toward the positive. A negative diverging wave becomes less negative, while a positive converging wave becomes more positive. Note also that the only calculation we actually perform is a local one, right at the lens. The initial K_I is related to where the wave came from and the final K_F is related to where the wave will focus, but the effect of the lens only relates the initial curvature *at the lens* to the final curvature *at the lens*.

As we went through the four stages above, we considered the case where P is more positive than K_I is negative. However, if the initial point is too close to the lens, so that $K_I = -1/R_I$ is too negative, then the power of the lens added to the initial curvature will not change the sign of the final curvature. The wave leaving the lens will still be negative and it will still be a diverging wave. The only difference in the wave is that its curvature will be less negative than it was. Its radius of curvature will consequently be larger after the lens than before it, and so the wave will look as if it came from further to the left of the diagram. This geometry is seen in Figure 22.9.

In Figure 22.9, light originally diverged from the red dot. After it passed through the lens, it looked like it came instead from the white dot to the left. When the light coming from a lens looks like it came from somewhere it did not, we say that the lens has formed a *virtual image* of the point. There is no concentration of light at the white dot that is the virtual image of the red dot. In fact, we can imagine a cardboard shield to the left of the red dot, making it so that there is no light at all at the virtual image. So, in contrast to the real image discussed on page 458, a virtual image is not a place where light *converges*, but a place from which light *seems to diverge*.

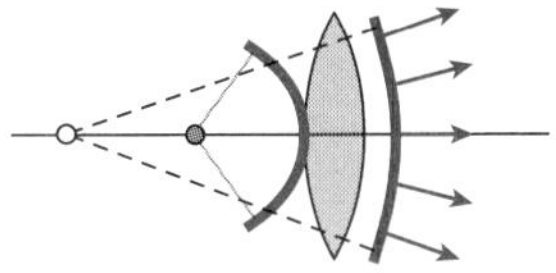

Figure 22.9 A wave can diverge too strongly for the power of the lens to focus.

EXAMPLE 22.2 Looking at a Bug

A small bug sits 25 cm from a converging lens of power $P = +4$ diopters. How far to the right of the lens will the light from the bug focus?

25 cm

ANSWER As the light from the bug reaches the lens, its radius of curvature will be 25 cm, and so its curvature will be $K_I = -1/(0.25 \text{ m}) = -4 \text{ m}^{-1}$, where we have put in the minus sign because the wave is diverging away from the bug as it hits the lens. As the wavefront travels through the lens, its center is delayed relative to its edges, and Equation 22.3 will give the final curvature of the wavefront as it leaves the lens. The result is

$$K_F = K_I + P \quad \Rightarrow \quad K_F = -4 \text{ m}^{-1} + 4 \text{ m}^{-1} = 0.$$

The final curvature is zero, meaning that the lens has produced waves going away to the right of the lens that are plane waves. Note that this is the intermediate case between that of Figure 22.7 and Figure 22.9. The lens has delayed all portions of the wavefront just enough to flatten it out completely. After the light leaves the lens, it does not converge to a point, but looks like it has come from infinitely far to the left. We say that the lens has formed a virtual image at infinity.

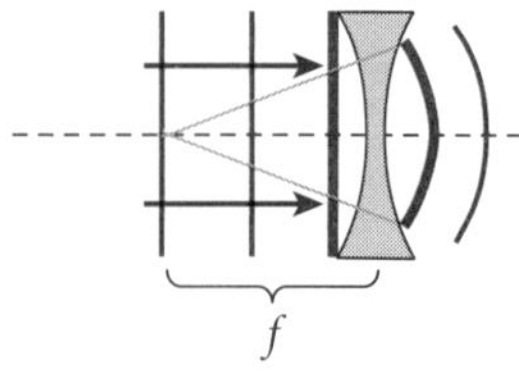

Figure 22.10 A negative lens makes a plane wave diverge with radius f.

We have defined a lens shaped so that it is fatter in the center than at the edges, one that adds to the curvature of any wave hitting it, to be a positive lens. As you may know, it is also possible to make a lens that is fatter at the edges than it is at the center (Figure 22.10). In this case, a plane wave hitting the lens will have its edges delayed relative to its center, so that it will end up diverging. It will have a negative curvature as it leaves the lens and will look as if it came from a point a little to the left of the lens instead of coming from infinitely far away, as it actually did. We call this a *diverging lens* and we define its power to be negative, equal to

$$P = -\frac{1}{f}, \tag{22.4}$$

where f is the radius of curvature of the diverging wave formed after a plane wave has passed through the diverging lens.

EXAMPLE 22.3 The Effect of a Negative Lens

A light wave from a previous positive lens is converging down to a point. Just 10 cm away from the point where it would have focused, we insert a diverging lens of power $P = -6$ diopters. Where does the wave leaving the lens focus, if at all?

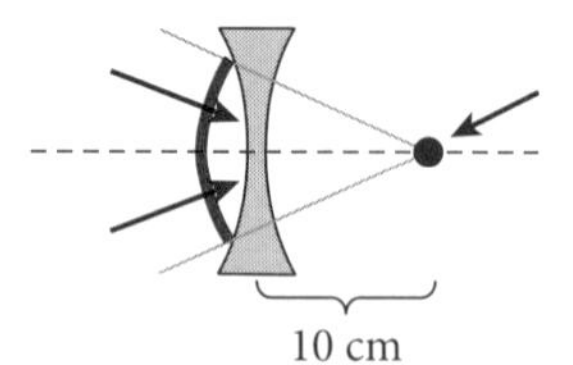

ANSWER At the lens, the curvature of the incoming wavefront is $K_I = +1/(0.1 \text{ m}) = +10 \text{ m}^{-1}$. A diverging lens of power $P = -6 \text{ m}^{-1}$ will produce a final curvature

$$K_F = K_I + P \quad \Rightarrow \quad K_F = 10 \text{ m}^{-1} + (-6 \text{ m}^{-1}) = 4 \text{ m}^{-1}.$$

Because the curvature is still positive, the wave is still converging. Its radius of curvature as it leaves the lens, and therefore the distance at which it will now come to a focus, is $R_F = 1/K_F = 1/(4\ \text{m}^{-1}) = 25$ cm. The situation will appear as in the picture at right, the solid gray lines depicting the bending of the rays normal to the wavefront.

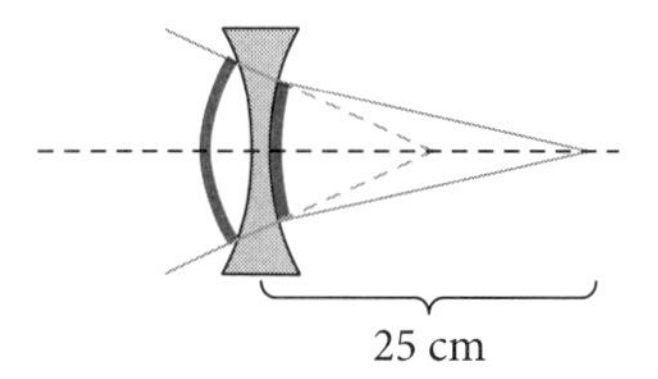

There is one thing you should notice that makes the solution method for lens problems pretty easy to remember. This is that everything that is diverging is negative—negative curvature for a diverging wave and negative power for a diverging lens—while everything that is converging is positive—positive curvature for a converging wave and positive power for a converging lens. With this in mind, let us summarize the steps we use for solving lens problems.

1. Determine the curvature K_I that a wave has as it arrives at a lens.
2. Apply the lens formula $K_F = K_I + P$ to find the final curvature K_F.
3. Use the final curvature to determine where the final wave will focus (for a positive K_F) or the point from which it will now look like it is diverging (for a negative K_F).

Finally, for contrast, let us write Equation 22.3 in a slightly different way—one using distances and focal lengths instead of curvatures and powers. If we assume a diverging initial wave, a positive lens, and a converging final wave, then Equation 22.3 becomes

$$\frac{1}{R_F} = -\frac{1}{R_I} + \frac{1}{f}.$$

This can be rearranged to give

$$\frac{1}{f} = \frac{1}{R_I} + \frac{1}{R_F}.$$

This equation is what is called the lens equation in most textbooks.[1] But this equation works only if the initial wave is diverging, the lens is a converging lens, and the final wave is converging. For other situations, one must memorize a complicated list of how the signs of R_I, R_F, and f must change to make the equation work. We have always found these rules harder to remember than the simple 1–2–3 rules we have presented for how to use Equation 22.3, so we have chosen to follow the approach we take in this book. It is also generally easier to work with in situations with multiple optical elements, as we will see in Section 22.5.

22.3 Ray Tracing and Image Size

In the last section, we saw how the waves from a point source of light lying on the axis of a converging lens will be brought to a focus at a single point on the same axis. As we have drawn in Figure 22.11a, rays diverge in all directions from a source and all rays hitting the lens will

1. Most texts actually use something like o (for object distance), instead of R_I, and i (for image distance), instead of R_F. Thus, their form of the lens equation is $(1/f) = (1/o) + (1/i)$. If the initial wavefront is converging, instead of diverging, the source is called a "virtual object" and o must be defined to be negative. The focal length is said to be negative for a diverging lens.

converge to a single focal point. But what of a point source that does not lie on the axis of the lens, like the point shown in Figure 22.11b?

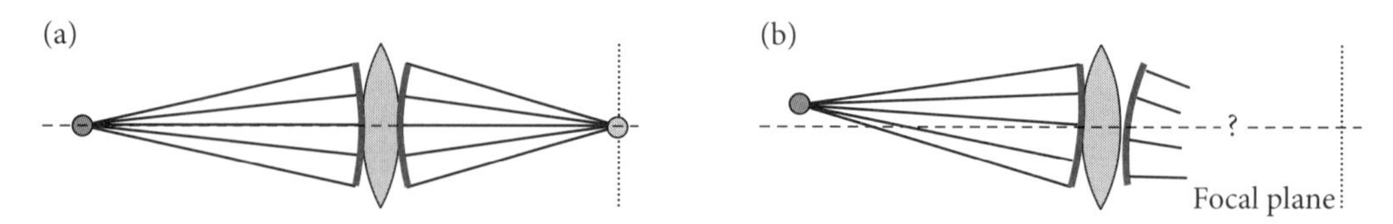

Figure 22.11 Divergent rays from an on-axis point converge to a point on the axis (a). Where do divergent rays from an off-axis point (b) converge to?

First, let us note that if light diverges from a point that is not far from the axis of the lens, then the curvature of the wavefront as the light hits the lens will be the same as that for a point on the axis. The lens will add the same amount of power to the initial wavefront to produce the final wavefront, so the wave from an off-axis point will have the same final curvature as does the wave from an on-axis point. It will therefore converge to a point the same distance away from the lens. Thus we see that the light from some extended object possesses not just a focal point, but a *focal plane* into which all near-axis points the same initial distance from the lens will be focused. We know that the light from a point on the axis will focus to the point in the focal plane that is on the axis. But where will the light from an off-axis point be focused?

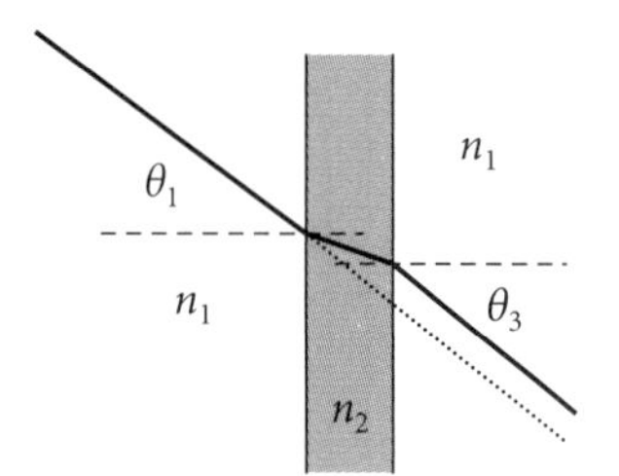

Figure 22.12 Passage of a light beam through a flat slab of glass.

To answer this question, we first step back to consider what happens when a beam of light encounters a flat piece of glass. Such a situation is depicted in Figure 22.12, where we follow a ray as it enters the glass and is bent toward the normal and then leaves the glass and is bent away from the normal. If the first and second surfaces are parallel, as they would be for a flat piece of glass, then Snell's law may be applied successively at the two surfaces to give

$$n_1 \sin\theta_1 = n_2 \sin\theta_2 \quad \text{and} \quad n_2 \sin\theta_2 = n_3 \sin\theta_3 \quad \Rightarrow \quad n_1 \sin\theta_1 = n_3 \sin\theta_3.$$

Since $n_3 = n_1$ (both beams are in air), we must have $\theta_3 = \theta_1$. The beam will emerge from the glass at the same angle it entered. Thus, as we see from Figure 22.12, the only effect of passing through the glass will be that the beam has been offset a tiny amount above its initial line. And, if the glass is thin, this offset is negligible.

The reason for our taking this little side trip becomes clear when we look back at Figure 22.11. In Figure 22.11a, all of the rays hitting the lens are bent by the lens to focus to a single point on the axis. Likewise, all of the rays diverging from the off-axis point in Figure 22.11b are also bent by the lens, and they will all converge to another single point somewhere in the focal plane. But here's the important thing. There is one ray that leaves the off-axis point in Figure 22.11b that is *not* bent as it passes through the lens. To a ray that hits the lens right at its center, the lens looks exactly like the flat piece of glass in Figure 22.12. We can use this fact to discover where the light from an off-axis point will focus. To say that the light all focuses to a point is to say that, when any ray arrives at a point in the focal plane, all of the other rays will be there too. This means that we can find where that point is located by following any one of the rays. The easiest one to follow is surely the straight, unbent ray from the source point, through the lens, to the focal plane, as shown in Figure 22.13. So where in the focal plane will we find the image of an off-axis source? It will be at the point where the *unbent* ray from the source through the center of the lens intersects the focal plane.

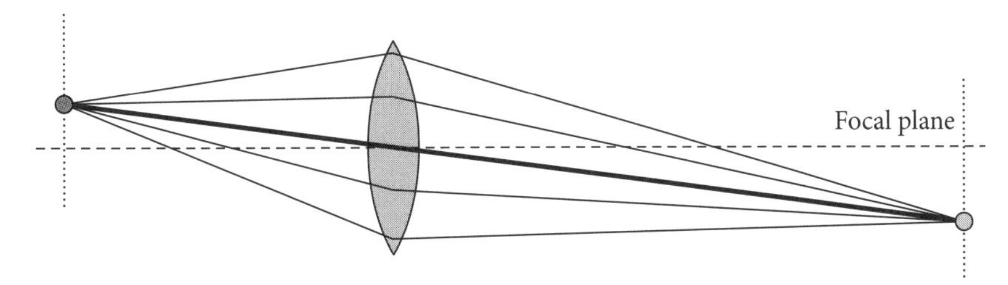

Figure 22.13 Rays from an off-axis point in the original plane are all focused to a point in the focal plane, but only the ray through the center of the lens is unbent. This can be used to find the corresponding image point in the focal plane.

So far we have discussed light that diverges from two points, one on-axis and one off-axis. Let us also consider what will happen if we put an extended object at some distance from a converging lens, like the little red and black arrow depicted on the left of Figure 22.14. Light will be emitted from each point of the arrow. Rays from the base of the arrow will focus at a point on the axis of the lens (the red rays). Rays from the tip of the arrow will focus to a point below the axis in the focal plane (the black rays). Points between the base and the tip of the arrow will focus to points in the focal plane between the images of the base and the tip. In this way, a complete image of the arrow is formed. This is real light, passing through the lens and focusing to points in the focal plane. If you put a 3 × 5 card in the focal plane, you would see a real image of the arrow glowing on the card.

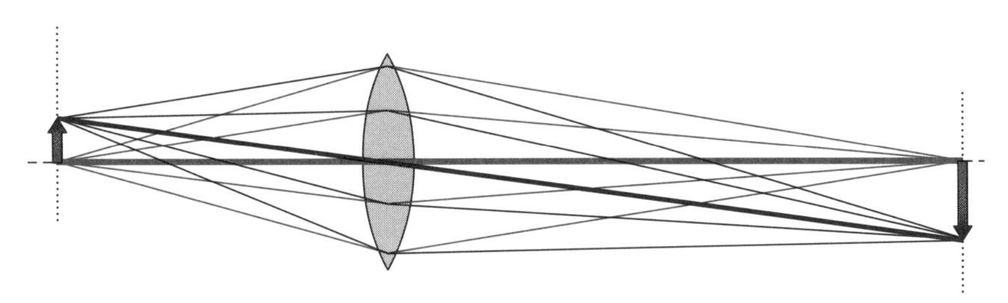

Figure 22.14 Rays from each point of the object at left diverge toward the lens and are bent to converge at corresponding points in the focal plane. The red rays diverge from the base of the arrow and the black rays diverge from the tip.

It may be hard to see, with all the rays passing through the lens in Figure 22.14, but if you concentrate on the two heavy rays, the red one from the base of the arrow and the black one from the tip, you will see that there are two right triangles in the figure. These are the triangles formed by these two heavy rays and the source arrow and the image arrow, as depicted in Figure 22.15. The two heavy rays intersect to form "opposite angles," and opposite angles are always equal. So the two right triangles will be similar. As a result, we know that

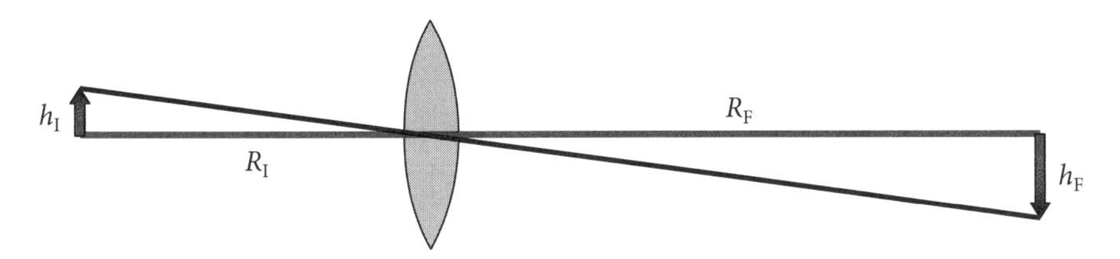

Figure 22.15 The similar right triangles formed from the rays from the top and bottom of the object passing through the center of the lens.

corresponding sides of the triangles will be in equal ratios. We denote the heights of the two arrows with h_I and h_F, and the similarity of the triangles allows us to write the ratios of the sides as

$$\frac{h_I}{R_I} = \frac{h_F}{R_F}.$$

This enables us to write down a formula for the size of the image in terms of the size of the object and the distances to the object and the image. The formula is

$$h_F = \frac{R_F}{R_I} h_I. \tag{22.5}$$

If $R_F > R_I$, the image will be bigger than the object and if $R_I > R_F$, it will be smaller.

So far, in our discussion of ray tracing, we have concentrated on the properties of real images formed by converging lenses. Let us apply the same principles to the virtual images from a diverging lens.

In Figure 22.16a, we show a big arrow to the left of a diverging lens. Three red rays diverge from the base of the arrow, the heavy one passing straight through the center of the lens, and the other two rays are dispersed by their passage through the lens. There are also three black rays diverging out of the tip of the arrow. The heavy one passes through the center

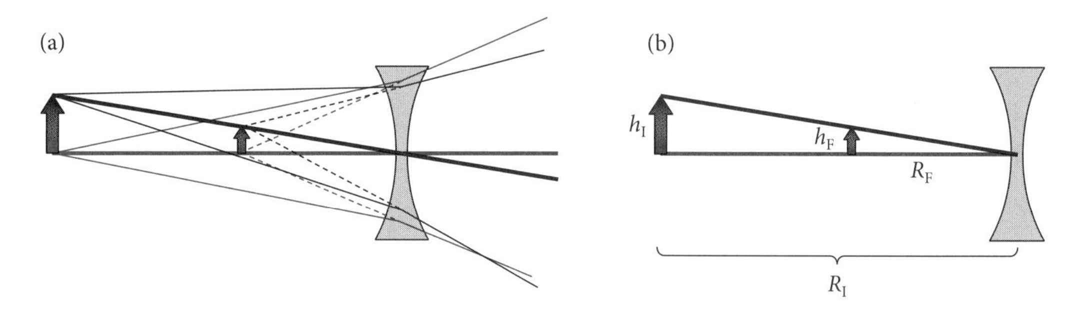

Figure 22.16 (a) A negative lens disperses rays from the base and the tip of the smaller arrow to form a virtual image. (b) The triangle with height h_I and the triangle with height h_F are similar.

of the lens, and the other two are dispersed by their passage through the lens. After passing through the lens, the three red rays look like they came from the base of the smaller arrow and the three black rays look like they came from the tip of the same arrow. In Figure 22.16b, the heavy rays are isolated to form the sides of two triangles, a big one with h_I as its height and a small one with h_F as its height. The triangles are again similar, allowing us to relate the final height to the initial height of the arrow, giving

$$h_F = \frac{R_F}{R_I} h_I,$$

the identical formula we derived as Equation 22.5.

Let us be clear about the importance of the ray that goes from the tip of an object through the center of the lens to the focal plane. *This is the ray we use to characterize an image.* It is this ray that produces the two triangles, and it is these similar triangles that give us the size of the image. By showing where the image of the tip is located, they also tell us whether the image is upside-down, which we call *inverted*, or right side up, which we call *erect*.

Finally, let us summarize our procedure for solving optics problems with lenses. We assume that the waves travel from left to right and that the goal is to describe the final image. This procedure adds a step to the 3-step procedure from page 461.

1. From the geometry of the problem, determine the curvature of the wavefront just before the wave strikes the lens of power P.
2. Use the lens equation, $K_F = K_I + P$, to determine the curvature of the wave leaving the lens.
3. If the final curvature is positive, the wave will form a real image in the focal plane, a distance $R_F = 1/K_F$ to the right of the lens. If the final curvature is negative, there will be a virtual image in a focal plane a distance $R_F = 1/|K_F|$ to the left of the lens.
4. Draw rays from the top and the bottom of the object, through the center of the lens, to where they intersect the focal plane. Use the top and bottom points of the image to determine if the image is erect or inverted, and use Equation 22.5 for these similar triangles to determine the size of the image.

We probably need an example just now to see how this all fits together.

EXAMPLE 22.4 An Arrowhead in Front of a Converging Lens

An arrowhead of height 1 cm is placed 33 cm to the left of a converging lens of power +2 diopters. Describe the final image of the arrowhead.

ANSWER We begin by sketching the geometry of the problem in the picture below.

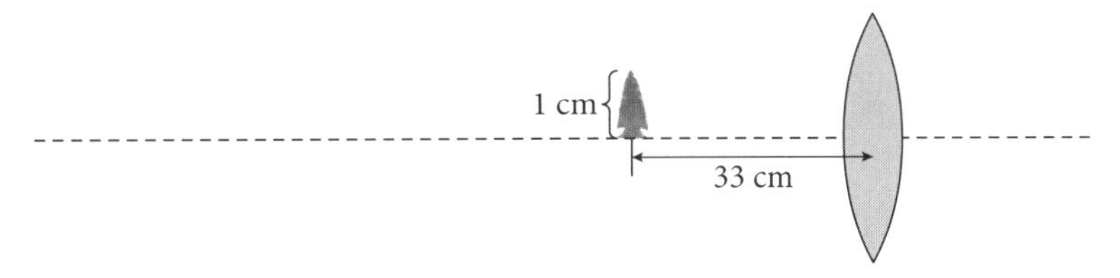

The initial curvature of the wavefront as the diverging wave hits the lens is

$$K_I = -\frac{1}{0.33 \text{ m}} = -3 \text{ m}^{-1}.$$

We use the lens equation, Equation 22.3, to find the final curvature:

$$K_F = K_I + P = -3 + 2 = -1 \text{ m}^{-1}.$$

The negative final curvature tells us that this is one of the cases where the wave is diverging too strongly for the converging lens to be able to turn it around and make it converge. It will therefore continue to diverge, forming a virtual image in a plane a distance $R_F = 1/(1 \text{ m}^{-1}) = 1$ m to the left of the lens. We add the virtual focal plane to the sketch, and we draw rays connecting the top of the arrowhead and the base of the arrowhead to the center of the lens, continuing them back until they intersect the focal plane, which is 1 m to the left of the lens. The image of the arrowhead will fill the focal plane region between the points where the red ray and the black ray intersect the plane.

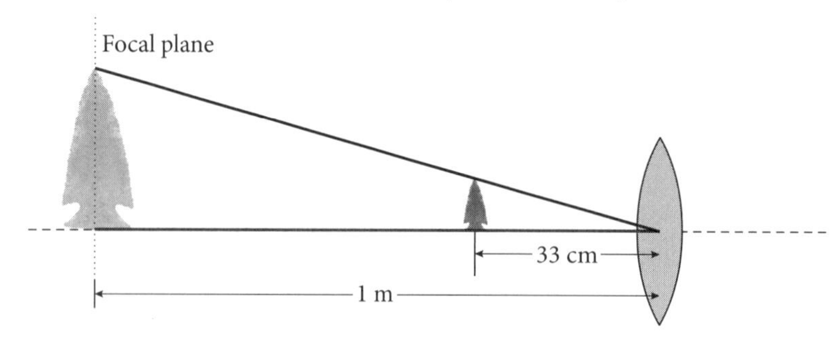

The height of the image is $h_F = (R_F/R_I)h_I = (1 \text{ m}/0.33 \text{ m})1 \text{ cm} = 3 \text{ cm}$, and the image is obviously erect.

22.4 Spherical Mirrors

Back on page 441 we discussed how a beam of light reflects from a plane mirror—the outgoing beam and the incoming beam making equal angles to a line normal to the surface of the mirror. In this section, we want to discuss how light reflects off of a spherical mirror. A spherical mirror is a curved mirror shaped like a small section of the surface of a sphere. The right-side rear-view mirror on your car (the "objects are closer than they appear" mirror) is generally a spherical mirror.

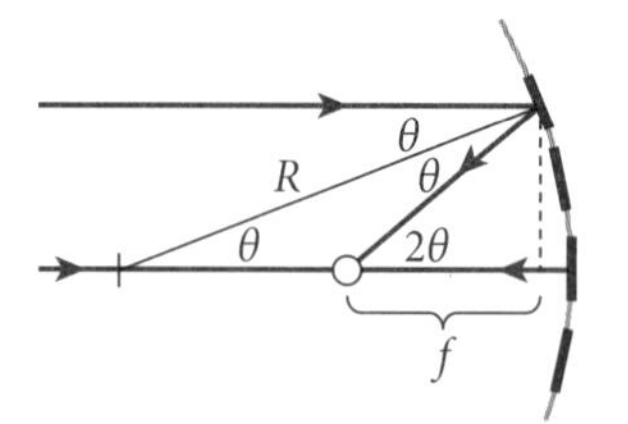

Figure 22.17 Ray-tracing demonstration that the focal length of a spherical mirror is $f = ½R$.

In Figure 22.17, plane waves are incident from the left onto a spherical mirror of radius R. We show four segments of the mirror as heavy black lines. Even though each segment is part of a spherical mirror, it will reflect like a tiny plane mirror. We concentrate on two of the rays, one that hits the segment in the center of the mirror and reflects straight back and one that hits the segment an angle θ above the axis, as shown. The two incoming red rays are parallel, and the black radius line intersects them, so the angle between the top ray and the radius line is the same θ as the angle between the central ray and the radius line. When the upper ray is reflected, it makes an equal angle θ to the radius and ends up at the white dot. Because this reflected ray likewise intersects two parallel lines, the angle between the reflected top ray and the reflected central ray is 2θ, as shown. We now call your attention to the two right triangles in the figure, both with the dashed line as their heights. From the triangle with the leftmost angle θ in it, we see that the height of the dashed line is $R \sin \theta$. If we call the distance from the mirror to the white dot f, then, looking at the triangle with 2θ as its angle, we can express the same height as $f \tan 2\theta$. If we equate the two, we find

$$R \sin \theta = f \tan 2\theta. \tag{22.6}$$

Now way back on page 219 we explained how, when an angle θ is very small, we can approximate $\sin \theta \approx \theta$. It is also true, in the same limit, that $\tan \theta \approx \theta$ (or that $\tan 2\theta \approx 2\theta$, of course). This means that Equation 22.6 can be approximated for small angles as $R\theta \approx f(2\theta)$. Therefore, for rays that hit the spherical mirror close enough to the central axis of the mirror for θ to be a small angle, we find that the ray will cross the axis of the mirror a distance

$$f = \frac{1}{2}R \tag{22.7}$$

from the mirror. Note that this answer does not depend on the angle θ, meaning that all parallel rays will hit the mirror at the same place. The white dot in Figure 22.17 is thus the place where plane waves focus, and the f in Equation 22.7 is the *focal length* of a spherical mirror of radius R. The mirror's power is defined as we defined power for lenses:

$$P = +\frac{1}{f} = +\frac{2}{R}.$$

A concave mirror like the one shown in Figure 22.17 is a *converging* mirror that brings wavefronts to a real focus, so its power is positive. If we were to instead bounce a wave off a convex mirror, its power would be negative, $P = -1/f = -2/R$. The geometry for a *diverging* mirror is shown in Figure 22.18.

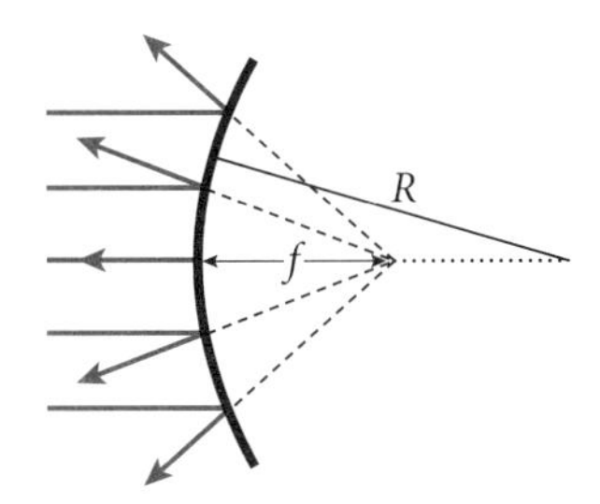

Figure 22.18 Ray-tracing for a diverging mirror.

It should be apparent that a spherical mirror is a lot like a lens. It changes the curvature of a wavefront that hits it, just like a lens does. And like lenses, there can be positive converging mirrors or negative diverging mirrors. The main difference is that a mirror reflects instead of transmitting. Essentially, it takes whatever happens on the right side of a lens (assuming incoming light from the left) and folds it back to the left. At this point, it is probably more confusing for you to have us go on explaining things; an example would probably be clearer.

EXAMPLE 22.5 An Arrowhead in Front of a Converging Mirror

An arrowhead of height 1 cm is placed 33 cm to the left of a converging mirror of radius 100 cm. Describe the final image of the arrowhead.

ANSWER A converging mirror of radius 100 cm has a focal length $f = 50$ cm, giving a power $P = +1/(0.5\text{ m}) = +2\text{ m}^{-1}$. So, yes, this is the same problem we solved in Example 22.4. Diverging waves from all points of the arrowhead will have curvature

$$K_I = -\frac{1}{0.33\text{ m}} = -3\text{ m}^{-1}$$

as they hit the mirror. The arrowhead is positioned inside the focal length of the mirror, just like it was inside the focal length of the lens in Example 22.4. So, once again, the effect of the converging mirror will be to produce a final curvature that is negative,

$$K_F = K_I + P = -3 + 2 = -1\text{ m}^{-1}.$$

As the wave reflects back to the left, it will still be a diverging wave, with a red ray reflected from the bottom of the arrowhead and a black ray from the top. With this as the wave's curvature, it will look as if it had come from a point $R_F = 1/|K_F| = 1$ m behind the mirror, as diagrammed.

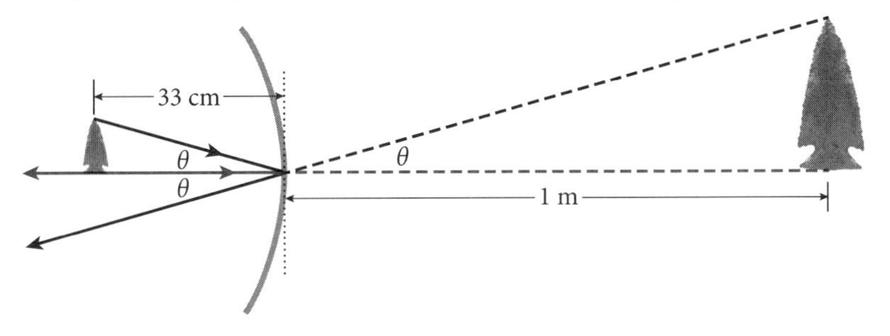

As with a lens, we have used two rays, the red ray that hits the mirror on its axis and bounces straight back and the black ray that also hits the center of the mirror at an angle. This ray bounces back at an angle equal to its incoming angle. As the outgoing red ray and outgoing black ray move back to the left, they will appear to come from a focal plane 1 m behind the lens. No light ever passed behind the mirror, of course, it's just that the reflected light will look as if it did. Tracing these rays backward to form a similar triangle behind the mirror makes it clear that the image will be erect and magnified from its actual 1 cm to h_F = (1 m/0.33 m)1 cm = 3 cm.

22.5 Combining Optical Elements

Many optical instruments and many optics problems involve multiple optical elements like lenses and mirrors. In this section we will think through the procedures needed to analyze how combinations of optical elements produce their final image products.

The advantage of the curvature approach to optics problems we have taken is that it forces us to think locally, to look at each optical element in its turn. In each case, we first do a little geometry to figure the curvature of the wavefront as it hits the element, then we apply the lens equation to get the final curvature, and then we do a little more geometry to see where the final focus will be or what the curvature will be when the wave reaches the next element. Let us see how this works with a couple of examples.

EXAMPLE 22.6 A Diverging Lens and a Converging Mirror in Sequence

A −4-diopter diverging lens is placed 25 cm in front of a converging mirror of power +5 diopters. If plane waves from a point source are incident from the left on the two-lens combination, where will a final image be formed?

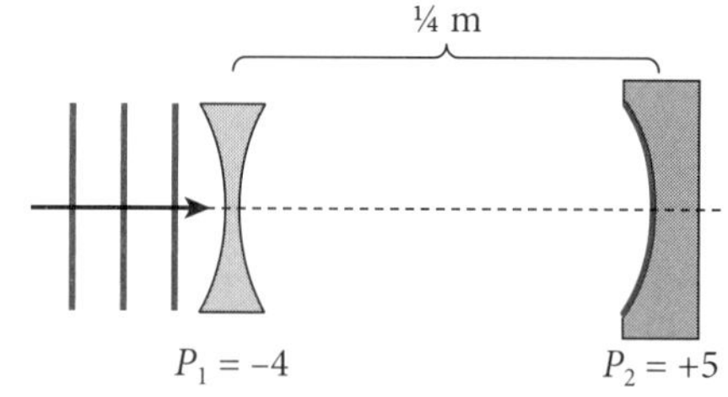

ANSWER We work one step at a time. As the plane wave encounters P_1, it will have zero initial curvature and a final curvature $K_F = 0 + (-4\text{ m}^{-1}) = -4\text{ m}^{-1}$. The negative curvature means that the wave is diverging after leaving the lens, so it looks as if it came from $R_F = 1/(4\text{ m}^{-1}) =$ ¼ m to the left of the first lens. Now one important thing to remember about using our curvature method is that the curvature of a wave evolves as it travels. The curvature of the wavefront leaving the first lens is -4 m^{-1}, but, when this same wave arrives at the mirror, having gone another ¼ m, its radius of curvature will be ½ m and so its curvature will be $K_I = -1/(½\text{ m}) = -2\text{ m}^{-1}$.

Now we are ready to consider the effect of the second optical element, the mirror. When a wave of curvature $K_I = -2\text{ m}^{-1}$ hits the mirror, the mirror will reflect the wave backward and change its curvature to $K_F = K_I + P_2 = -2\text{ m}^{-1} + 5\text{ m}^{-1} = 3\text{ m}^{-1}$. So, the reflected wavefront will be moving to the left, and it will be converging toward a point $R_F = 1/K_F = 1/(3\text{ m}^{-1}) =$ ⅓ m to the left of the mirror.

As the converging wave encounters the P_1 lens for the second time, the point toward which it is still converging will be only ⅓ m − ¼ m = 1⁄12 m away. Thus, this is the radius of curvature of the wave, corresponding to an initial curvature $K_I = +1/(1⁄12\text{ m}) = +12\text{ m}^{-1}$. The plus sign, of course, is because it is still converging when it hits the lens. As it passes through the lens, its curvature will change to $K_F = K_I + P_1 = 12\ m^{-1} - 4\ m^{-1} = 8\ m^{-1}$, meaning that it will converge to a point $R_F = 1/K_F = 1/(8\ m^{-1}) = \frac{1}{8}$ m, or 12.5 cm, to the left of the lens. It is a real image. A pinhead at this point (you wouldn't use a 3 × 5 card, since it would block too much of the incoming light) would glow due to the light focused to that point.

EXAMPLE 22.7 A Converging Lens and Diverging Lens in Sequence

A converging lens with power 10 diopters sits 15 cm in front of a diverging lens of power −30 diopters. A 2-cm-high bug is placed 20 cm to the left of the converging lens. Describe the location, the nature (real or virtual, erect or inverted), and the size of the final image of the bug.

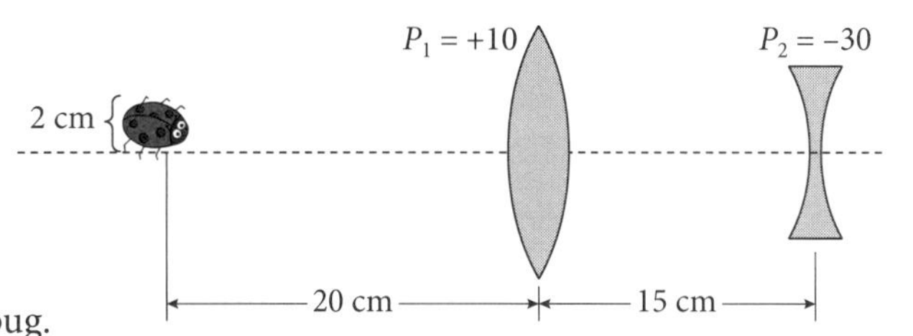

ANSWER By now, we should be pretty good at finding *where* the final image will be, so let's get this out of the way quickly. At the first lens, the lens equation gives $K_F = K_I + P_1 = -(1/0.2) + 10 = +5$, so the wave from the first lens will be converging toward a point $R_F = 1/5 = 0.2$ m = 20 cm to the right of the first lens. However, before the wave can actually come to a focus, the second lens intervenes, 5 cm in front of the original focal point. At this point, the wave is still converging, so its curvature is $K_I = +1/(0.05\ m) = +20\ m^{-1}$. After passing through the lens, the final curvature will be $K_F = K_I + P_2 = +20 - 30 = -10\ m^{-1}$. Thus, the final wave leaving the 2-lens combination will be a wave diverging from a point $R_F = 1/(10\ m^{-1}) = 0.1$ m = 10 cm to the left of the second lens (and therefore between the two lenses). Since the final wave is diverging, the image will be *virtual*.

The size and orientation of the image is a little harder to discover. Since we know that the final image will be in a plane 10 cm to the left of the diverging lens, the first thing we do is draw in the focal plane as the dotted line in the figure below. Then we locate the unrealized image from the first lens, 20 cm to the right of the first lens. For simplicity, we use an arrow to represent the bug, and we draw in the ray from the top of the white arrow that is unbent by the first lens, to locate the top of the gray arrow. Until they hit the second lens, all rays from the top of the bug were converging to this point. In particular, one of these (assuming that the first lens extended as high as we need) is the ray that passes through the center of the second lens. We have drawn this ray in dark red in the diagram. Since that ray is unbent by the second lens, we can follow it backward to find the image of the top of the bug in the focal plane. The final image is erect. The size of the first (gray) image is 2 cm. The size of the final image is (10 cm/5 cm)(2 cm) = 4 cm.

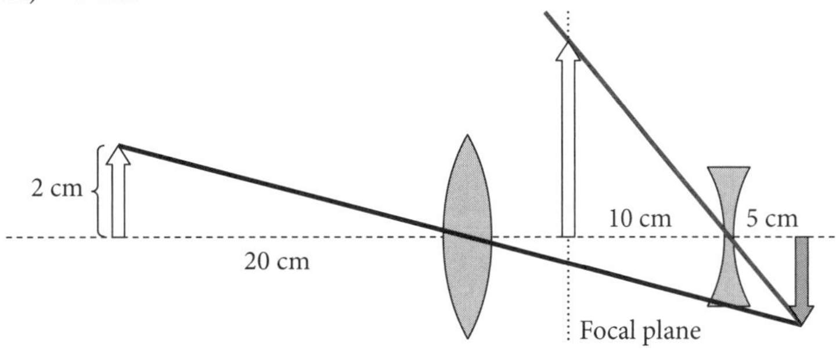

Our use of the red ray in the last example probably needs a further word of explanation. Let us explain by concentrating on the final image in an additional two-lens example. In Figure 22.19, we have two +10-diopter lenses that are 10 cm apart. A 4-cm-high object sits 20 cm to the left of the first lens and would form a 4-cm-high image 20 cm from the first lens—the gray arrow—except for the presence of the second lens. The second lens instead brings the waves to a focus 5 cm to the right of the second lens. In the figure, we show the complete paths of three of the rays from the tip of the object to the tip of the final image. On their way toward

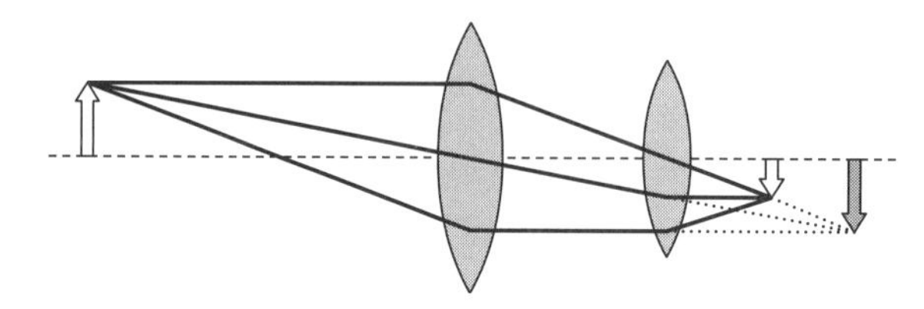

Figure 22.19 Rays from an object pass through a two-lens combination. The ray that is not bent by the first lens sets the tip of the gray image. The ray that is not bent by the second lens sets the tip on the final image.

the unrealized gray image, two were bent by the first lens and one was not. The unbent ray is the one that passed through the center of the first lens. In fact, this is how we determined what the gray arrow would look like. However, before that ray could reach the gray image, it was bent by the second lens, because it did not pass through the center of the *second* lens. But an upper ray from the object *did* pass through the center of the second lens, so we choose it as the one to concentrate on, knowing that the final image of the tip of the arrow will lie along it. There are, of course, an infinite number of rays that left the tip of the object, but we were careful to pick the one to display that ended up passing through the center of the second lens, because this is the one that allows us to see the final image properties. There will always be such a ray, and this will always be the one we will use to see the size of the final image and to see if it is erect or inverted.

22.6 Optical Instruments

The human eye (Figure 22.20) is a remarkable optical instrument. In a normal eye, plane waves from objects far away are focused to a point on the retina by a two-lens combination. The first lens is composed of the cornea and the fluid in the aqueous chamber, for a fixed power of +40 diopters. The second is a variable lens with power from +20 to +24 diopters. When the eye is relaxed, the total power is +60 diopters, focusing a plane wave to a point $f = 1/P = 1.66$ cm from the lens. When an object is closer than this, muscles around the lens contract to fatten the lens, increasing its power to +24 diopters, for a total power of 64 diopters. At this power, the eye can focus on objects as close as R_I away, where R_I is determined by the lens equation, $+1/R_F = -1/R_I + P$. The distance from the lens to the normal retina remains $R_F = 1.66$ cm, so the equation is $1/R_I = -1/0.0166\text{ m} + 64\text{ m}^{-1} = -4\text{ m}^{-1}$, corresponding to a wave coming from $R_I = 0.25$ m. The normal human adult eye, then, can see

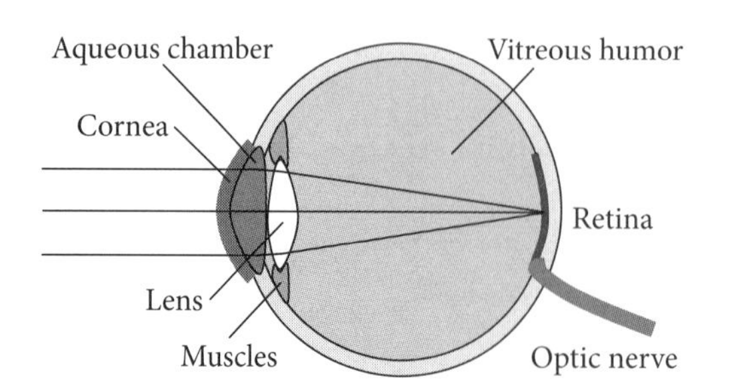

Figure 22.20 The human eye is an optical instrument.

clearly everything from a distance of infinity to a distance of 25 cm. Light waves from an object closer than 25 cm will diverge too strongly for the lens system of the eye to focus them onto the retina.

In many people, the connection between the relaxed power of the lenses and the distance to the retina is not the normal one shown in Figure 22.20. When the relaxed eye focuses the light from infinity to a point in front of the retina, as shown in Figure 22.21a, a person is said to be *nearsighted*. Light from infinity will be spread over a range of neighboring points on the retina and a person who is nearsighted cannot relax the lens enough to focus each point of a far-away object onto a single point of the retina. The object will appear fuzzy. A nearsighted person, on the other hand, will typically have very acute vision for nearby objects; hence the name "near-sighted." When the relaxed eye focuses light from infinity to a point behind the retina, as shown in Figure 22.21b, the lens will have to be fattened by the eye muscles to focus points of a far-away object onto the retina. While this situation allows the person to clearly see things far away, hence the term *farsighted* for this condition, it does so at the expense of constant eyestrain on the muscles around the lens. And a farsighted person cannot increase the power of the lens enough to focus on things that are very close.

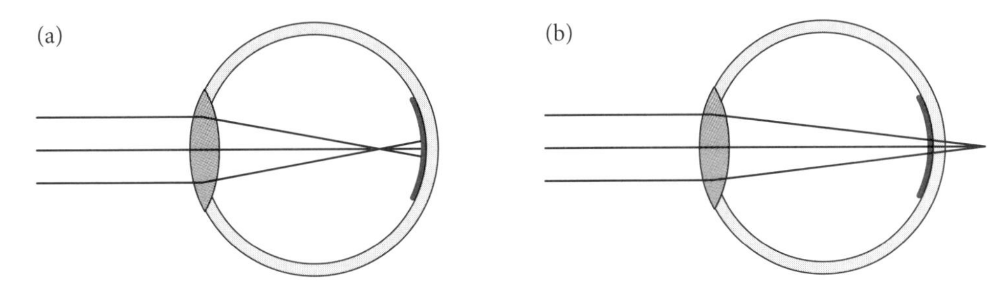

Figure 22.21 Picture (a) shows the optical geometry of a nearsighted eye, while picture (b) shows the situation for a farsighted eye.

The solution for nearsightedness and farsightedness is to prescribe glasses or contact lenses. A lens just in front of the eye can prepare the incoming beam so that the natural eye focuses it exactly on the retina. The problem with the nearsighted eye is that its lens is too high-power for the position of the retina, so a negative lens can defocus the beam before it hits the eye, as shown in Figure 22.22a. A farsighted eye needs help to focus a plane wave onto the retina, so a positive lens can take part of the task of converging the light, as shown in Figure 22.22b.

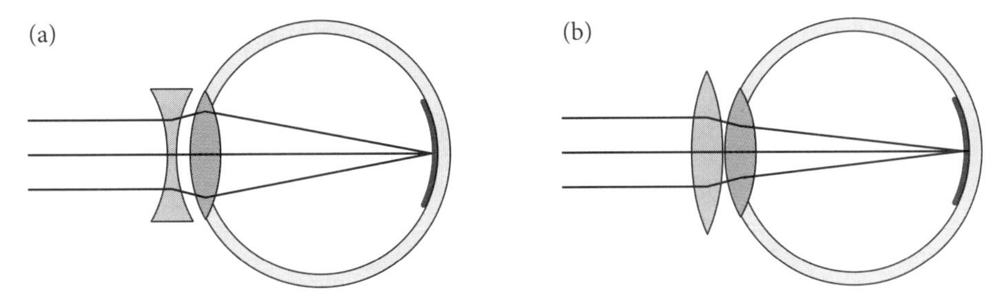

Figure 22.22 Picture (a) shows the diverging lens required for nearsightedness, while picture (b) shows the converging lens required for a farsighted eye.

EXAMPLE 22.8 The Farsighted Piano Player

A farsighted piano player cannot focus on the music if it is closer than 1 meter. This is very bad because his hands don't reach that far. So he has to sit back, read some music, scoot forward, play it, and then scoot back again to read some more. He plays very slowly. Fortunately, his optician recommends glasses so he can see the music at an ordinary distance of 25 cm. What prescription (power of the lenses) will allow him to focus on music at 25 cm?

ANSWER The piano player is able to focus on the music if it comes from a distance of 1 m, so he is able to focus a wavefront of curvature $K = -1/R = -1\ \text{m}^{-1}$ onto his retina. A wavefront from 25 cm would have a curvature $K' = -1/0.25 = -4\ \text{m}^{-1}$. The goal of his glasses is to take this wave of curvature $K_I = -4\ \text{m}^{-1}$ and turn it into the wave of curvature $K_F = -1\ \text{m}^{-1}$ that his eyes can handle. The final curvature is related to the initial curvature by the lens equation $K_F = K_I + P$, or $-1 = -4 + P$. Obviously, it will require a converging lens of power $P = +3$ diopters to do the job.

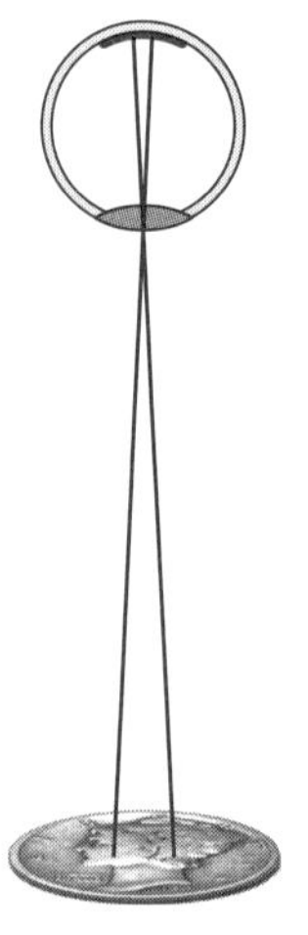

Figure 22.23 The further an object is from the eye, the closer adjacent points will appear on the retina.

The key to being able to read the fine print in a cell phone company contract or to determine the mint mark on a 1937 Mercury dime is to be able to produce an image on your retina in which different points of the object focus to different receptor cells (the rods and cones) of the retina. This way the detail can be detected. But when the object is far away, as shown in Figure 22.23, points that are well-separated on the object may show up very close or at the same receptor on the retina. If the object is brought closer to the eye, the angular separation of the images of the two points on the object becomes greater and the images become well separated on the retina. However, there is a limit to how close an object can be brought to the eye, while still allowing the eye to focus the diverging wave onto the retina. It must be roughly 25 cm or more away. If this distance is not sufficient to separate points of the image on the retina, we will need to have recourse to another optical instrument, the magnifying glass.

The goal of a *magnifying glass* is to pre-converge the light from a very close object, so the light can be focused onto the retina, without decreasing the angular size of the image. If we take a converging lens of focal length $f = 5$ cm ($P = +20$ diopters), and place a tiny object at a distance $R_I = f$ from the lens, as shown in Figure 22.24, the final curvature of the wave leaving the magnifying glass will be $K_F = K_I + P = -(1/f) + (1/f) = 0$. Therefore, this wave will look like it

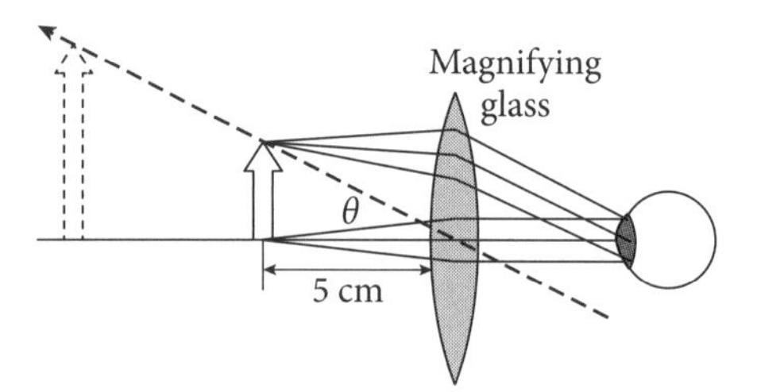

Figure 22.24 All diverging rays from a point on the object end up parallel.

came from infinity, so it can be focused by a relaxed eye. The size of the virtual image will be infinite, proportional to the infinite distance to the image off to the left of the lens. But the key is the fact that the image preserves the *angular* size of the object at 5 cm, as seen by the dashed line in the figure. The magnifying glass thus provides a way to hold an object very close to the eye, so as to be able to see fine detail in the object, and still focus the light onto the retina.

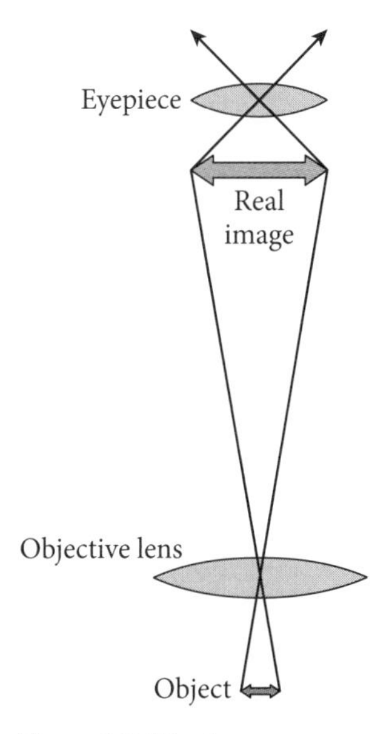

Figure 22.25 The microscope consists of two lenses.

But there is a limit to how much magnifying you can do with a magnifying glass. The object being examined in a magnifying glass must be placed at the focal length of the lens, and, in order to get the lens close to the object to produce a large angular size for the image, the focal length should be very short. But a short focal length means a fat lens and, if the lens is not thin compared to the focal length, spherical lenses no longer focus all the rays from points of the object to points of the image and no longer preserve the distances between image points. The image will be fuzzy and distorted.

If you need to see finer detail than a magnifying glass can produce, you will have to go to a *microscope*. A microscope divides the task of magnifying the angular size of the image of an object into two steps, as depicted in Figure 22.25. Because the problem is that the object is too small, the first step is to produce a bigger object, or, at least, to produce a bigger real image of the small object. When a lens creates a real image of something, the light is really there and it does not go away after the image is formed. Each point of the image becomes a source of a new diverging wave and the light leaves the image just like it would from a real object.

If you were to put a small object at the focal length of a lens, the wavefront passing beyond the lens would have zero curvature and would never focus. But, if you were to put an object just beyond the focal length of the lens, a real image will form a long way from the lens. The size of the image, as we explained in Figure 22.15, is proportional to the ratio of the image distance to the object distance, so the real image formed can be much larger than the object. A second lens, called an eyepiece, can then be used like an ordinary magnifying glass to examine the faithful magnified image, just as if it were a much larger object.

The purpose of an *astronomical telescope* is fundamentally different from the purpose of a microscope. For a telescope, the problem is not that the object we want to see is too small—astronomical objects are huge—it's that the objects are too far away. When we look at the moon with the unaided eye, two features that are a mile apart on the moon will be perceived at the same point on our retina. If we want to see more detail on the moon, we will need to bring it closer. Surprisingly, we can do that.

The purpose of the primary lens or mirror in an astronomical telescope is to produce a tiny real image of the huge moon, but to produce it there in the observatory, where it can be examined up close with a magnifying glass. The secondary lens of the telescope, also called the eyepiece, is the magnifying glass. For minimum eyestrain, it is placed so that it produces a virtual image at infinity (i.e., plane waves emerge from the eyepiece). For this, the real image we want to examine must sit at its f_2 focal length. Figure 22.26 shows rays coming from close points on the moon's surface, passing through the center of the primary lens, and focusing to separated points in the focal plane. Note that the two intersecting rays make the same angle on the moon side and on the image side of the primary lens, so the angular separation between the points is the same in the image as on the moon. The advantage of the telescope is that this image is close to us and may be examined in great detail with the eyepiece.

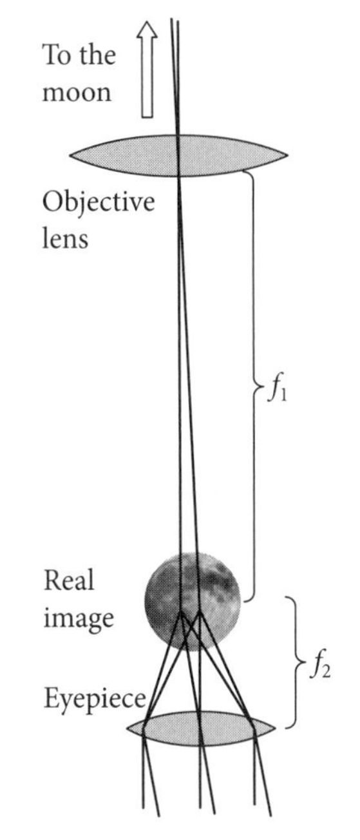

Figure 22.26 The astronomical telescope.

EXAMPLE 22.9 Telescope Observations of the Moon

Two mountain peaks on the moon, 3.85 km apart, are observed in a telescope formed from a 1-m focal length primary lens and a 5-mm eyepiece. The human eye can see detail separated by 1 arcminute (1/60 of a degree). Can the human eye see details in these mountains through the telescope?

ANSWER The moon is 385,000 km from earth. At this distance, light entering the telescope will consist of essentially plane waves. After passing through the primary lens of focal length $f = 1$ m, they will therefore focus to a point 1 m from the primary lens. The geometry of the object and the image is shown below, not to scale.

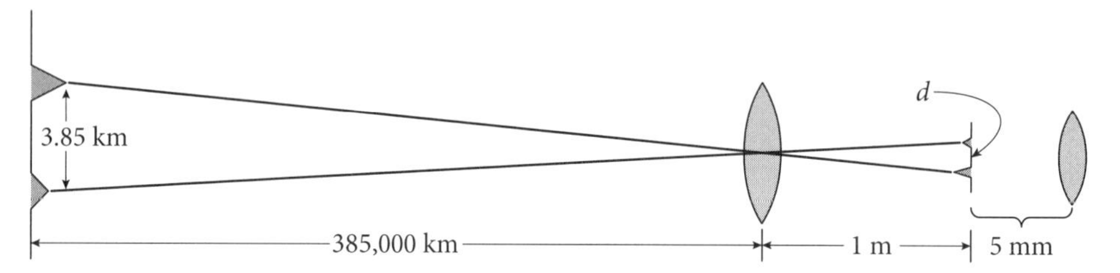

The two triangles formed by the intersecting lines are similar and so their sides are in proportion. This gives the distance between the peaks in the image as

$$\frac{d}{1\text{ m}} = \frac{3.85\text{ km}}{385{,}000\text{ km}} \quad \Rightarrow \quad d = 10^{-5}\text{ m} = 10\ \mu\text{m}.$$

That's 10 microns, or 1/100 of a millimeter. You could not resolve those mountains, even in the image, with the naked eye. However, if we mount the eyepiece so the primary image is at its focal length of 5 mm, the virtual image at infinity formed by this magnifying glass will have angular size $\theta = 10^{-5}/(5 \times 10^{-3}\text{ m}) = 2 \times 10^{-3}$ radians, where we have remembered that a small angle in radians satisfies $\theta \approx \sin\theta$. We can convert this angle to degrees using $180° = \pi$ radians, giving 0.115°. Or, finally, we convert to arcminutes using $1° = 60$ arcmins to get $\theta = 16.9$ arcmins. Those peaks will be easily visible to the astronomer in the telescope eyepiece.

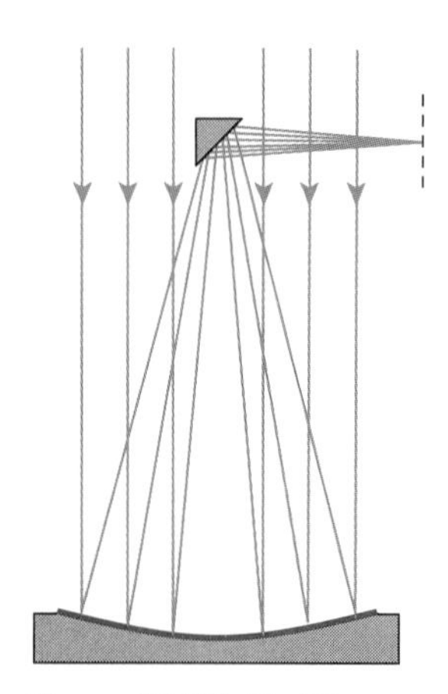

Figure 22.27 The Newtonian telescope.

As one can see in the last example, or by comparing Figures 22.25 and 22.26, the first lens in a telescope serves a very different function than the first lens in a microscope; and the two have very different properties. In a microscope, the goal is to be able to set the object very close to the objective lens, a goal that is accomplished by a first lens with a very short focal length, and thus very high power. In a telescope, the goal is to create a real image that is very large, meaning that the focal length should be as long as possible, or that the power should be very low. The flattening in the middle of a primary lens is so small that the lens looks almost like a flat plate of glass. Most astronomical telescopes actually use a primary mirror instead of a primary lens, and the mirror too looks barely curved. A particular configuration for a reflector, known as a Newtonian telescope, is shown in Figure 22.27.

Examination of Figure 22.27 often leads to the question, or misconception, with which we would like to end this chapter. The flat-angled mirror mounted in the middle of the incoming beam of a Newtonian reflector takes the upward-reflected rays that are converging toward a point and deflects them out the side of the telescope, toward an eyepiece or perhaps a photographic film in the focal plane (the dashed line in Figure 22.27). But what is the effect of the obscuration that this object produces in the beam? It will clearly block any light that comes directly down the center of the telescope tube. Won't this leave a hole in the middle of the telescope image?

The answer is shown in Figure 22.28. In diagram (a) all of the rays hitting the primary mirror are the nearly parallel rays that are coming from a point, a single point directly on the axis of the telescope and far away to the left. They might be coming from a single star in the direction the telescope is pointed. In diagram (a), with no obscuration in the beam, all of the rays from the star focus to a point, as shown. In diagram (b), the ray in the very center of the telescope is indeed blocked by the angle mirror, but, as is easily seen, there remain plenty of rays from the star that focus to the center point of the image. The same will be true for any other point in the object, say a neighboring star in the same field of view. The rays will come in parallel to each other, but at a slight angle from the rays from the first star. Some of these also will hit the angle mirror and be taken out of the beam, but there will be many other rays from the second star that pass beside the angle mirror and are focused to another point in the image plane. There is no hole in the image.

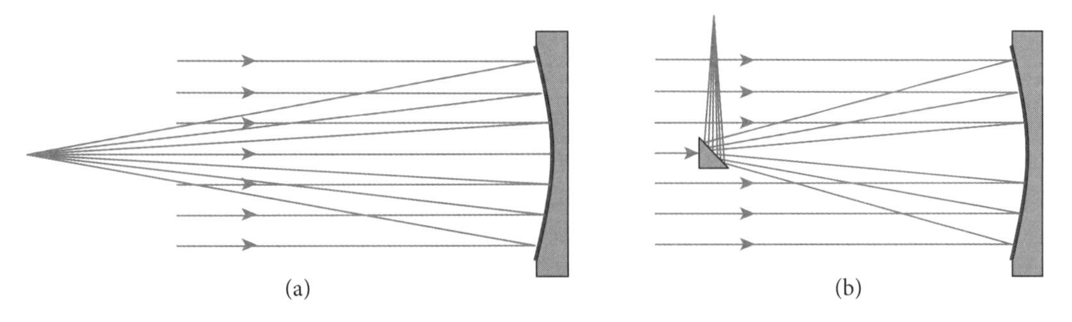

Figure 22.28 Although the central ray in diagram (a) is absorbed by the obscuration in diagram (b), many rays from the point at the center of the astronomical object off to the left still reach the focus and form an image of the point of the object that lies at the center of the telescope field of view.

Let us just add that, if this no-hole-in-the-image result seems a little unbelievable to you, you should try it yourself. Get a positive lens (a pair of reading glasses should work) and focus a scene from outside your window onto a white 3 × 5 card. Then get a friend (you shouldn't have any hands left) to take another card and cover half of the lens. You will be able to confirm that the entire picture is still there. It is not missing its left half or its right half. The only thing you might notice, if you are really discerning, is that your picture may look a little dimmer, because there are fewer light rays contributing to each point in the picture.

22.7 Summary

The things you should retain from this chapter are really all elements of a single procedure.

- You should understand how the curvature of a wavefront is determined from the distance to its source or to its focus and how it changes as it propagates.
- You should remember the convention that a wavefront converging to a point has a positive curvature and that a wavefront diverging from a point has a negative curvature.

- You should be able to use the lens equation $K_F = K_I + P$ to determine how lenses and mirrors change the curvature of a light wave.
- You should understand the effect of a spherical mirror on a wavefront and be able to determine the power of a mirror from its radius of curvature.
- You should be able to use the lens equation through multiple elements by concentrating on the curvatures as the wavefront encounters each one.
- You should be able to locate the ray through the center of the lens to determine the nature of the image formed by a lens, with special emphasis on how the process works for lenses in combination, as explained in Example 22.7 and the example shown in Figure 22.19. You should be able to use the single ray that bounces off the center of a spherical mirror for the same purpose.
- You should understand the purposes of the optical elements of the classic optical instruments.

CHAPTER FORMULAS

Curvature: $K = \pm\frac{1}{R} \begin{cases} + \text{ converging} \\ - \text{ diverging} \end{cases}$

Power: $P = \pm\frac{1}{f} \begin{cases} + \text{ converging} \\ - \text{ diverging} \end{cases}$

Mirror: $f = \frac{1}{2}R$

Lens equation: $K_F = K_I + P$

Images: $\frac{h_I}{R_I} = \frac{h_F}{R_F}$

PROBLEMS

22.1 *1. Which has the greater curvature, a baseball or a basketball? Explain your answer.

*2. The earth is roughly a sphere of radius 6400 km. What is the curvature of the surface of the earth in m^{-1}?

**3. A plane light wave hits a piece of cardboard with a tiny pinhole in it. Use Huygens' principle to explain what you would expect the shape of the light wave to look like as it emerges beyond the pinhole.

**4. A spherical wave diverges from a bug. Find the curvature of the wave (including the correct sign) when the wavefront is

a) 1 cm from the bug, b) 10 cm from the bug,
c) 20 cm from the bug, d) 1 m from the bug.

**5. A wavefront comes away from a converging lens with a curvature of $+5\ m^{-1}$. Find its curvature after it has traveled

a) 10 cm b) 20 cm c) 60 cm from the lens.

22.6 *6. A lens focuses light from very far away to a plane 10 cm from the lens. What is the power of the lens (include the sign of the power)?

*7. Light from a coin that is 5 cm from a lens is focused to a point 50 cm on the other side of the lens. What is the power of the lens (with its sign)?

** 8. A lens of power +4 diopters is held 20 cm above the print in a book. At what distance from the lens is the image formed? Is the image real or virtual?

** 9. A diverging lens of power −3 diopters is placed 50 cm to the right of an object. Where will the image be formed? Is it real or virtual?

** 10. A camera lens has a focal length of 5.0 cm (a power of +20 diopters). If the manufacturer wants it to be able to focus on objects from infinity down to 1 meter, over what range of distance must the camera lens be able to move?

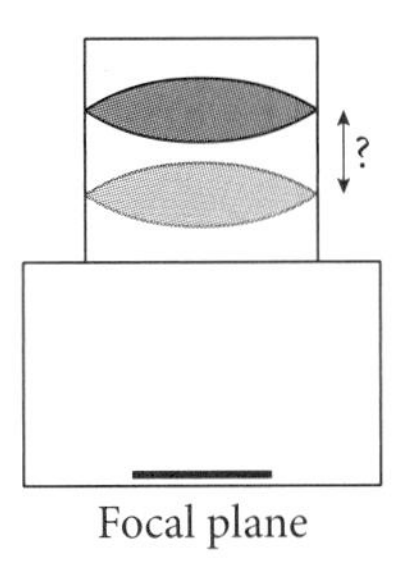

** 11. A lens is placed 20 cm to the right of a small pebble and forms a virtual image of the pebble 30 cm to the left of the lens. What is the power of the lens? Is it a converging or a diverging lens?

22.3 * 12. The tail of a tiny shrimp sits with its tail on the centerline of a converging lens, 10 cm to the left of the lens. The head of the shrimp is 1 mm above the centerline. A real image of the shrimp is formed 30 cm to the right of the lens. Draw a diagram showing the rays from head and tail of the shrimp that pass through the center of the lens, and use it to find the size and orientation of the image.

** 13. A +30-diopter converging lens forms a real image of a variegated leaf on a sheet of photographic film that is 5 cm behind the lens.

a) What is the curvature of the wavefront as it leaves the lens?

b) What is the distance from the leaf to the lens?

c) If the leaf is 4 cm high, what is the height of the image?

** 14. An entomologist observes a 1-mm-long bug with a lens of focal length 5 cm.

a) The bug is placed 10 cm from the lens. What will be the location, size, and orientation (erect or inverted) of the image?

b) The bug is placed 5 cm from the lens. What will be the location, size, and orientation of the image?

c) The bug is placed 4 cm from the lens. What will be the location, size, and orientation of the image?

** 15. A camera lens focuses light from a flower 8 cm in diameter and 30 cm from the lens on to a focal plane 6 cm behind the lens. There, a charge-coupled device (CCD) builds up charge in individual squares (pixels) of the CCD in proportion to the light received. The camera electronics read the charges to form an image.

a) What is the focal length of the lens?

b) Is the image on the CCD erect or inverted?

c) What will be the diameter of the image on the CCD array?

** 16. A hobbyist has just painted a toy soldier and wants to use a +25-diopter lens to form a virtual image that is 3 times the size of the actual soldier.

a) How far from the soldier must the lens be placed?

b) Will the image be erect or inverted?

** 17. Grandma looks at the fine print in the sales contract for her Harley-Davidson Iron 883.

a) She uses a magnifying glass with a focal length of 10 cm and holds the glass 9 cm from the paper. Will the image be erect or inverted?

b) She moves the magnifying glass back a bit, till the glass is 11 cm from the paper. Will the image be erect or inverted?

c) At which of these positions should she hold the glass if she wants to read the contract?

*** 18. Using a converging lens of focal length 30 cm, an experimenter wants to create a real image of a printed circuit board that is 1/3 the size of the actual circuit board.

a) How far from the circuit board must the lens be placed? [Hint: This requires a little algebra.]

b) Will the image be erect or inverted?

22.4 * 19. What is the power of a converging mirror whose radius of curvature is 40 cm?

* 20. A telescope mirror is to be built so that it will focus light from astronomical objects onto a focal plane that is 2 m from the mirror. What should be the radius of curvature of the mirror?

** 21. A man wants to check his face for unsightly hairs by looking in a converging shaving mirror (a concave mirror) of focal length 10 cm.

a) If he stands 30 cm away from the mirror, what will be the location and orientation (erect or inverted) of the image of his face?

b) If he stands 10 cm from the mirror, what will be the location and orientation of the image of his face?

c) If he stands 9 cm from the mirror, what will be the location and orientation of the image of his face?

** 22. A butterfly hovers 40 cm from a polished glass garden sphere of radius 10 cm.

a) How far away will its image appear to the butterfly to be located?

b) If the butterfly is 4 cm tall, how big will its refection be?

** 23. A concave (converging) dentist's mirror, with a radius of curvature of 20 cm, is held inside a patient's mouth at a distance 5 cm from a cavity in the number 18 molar. The cavity is 0.8 mm across.

a) Where will the image of the cavity appear, relative to the position of the mirror, and will it be real or virtual?

b) How big will the cavity appear in the image, and will it be erect or inverted?

** 24. An object 3 mm high is placed 20 cm from a converging mirror whose radius of curvature is 20 cm.

a) Where is the final image, and is it real or virtual?

b) How big is the image, and is it erect or inverted?

** 25. What would be the radius of curvature of a make-up mirror that produces a virtual image of your face that is twice the size of your actual face, when you are standing 20 cm in front of the mirror?

22.5 ** 26. Two +5-diopter lenses are placed 10 cm apart, and an object is placed 10 cm to the left of the first lens.

a) Where will the final image be formed?

b) The same two lenses are now placed 5 cm apart, with the object placed as before. Where will the new final image be formed?

[Note: The purpose of this problem is to emphasize the evolution of a wave between two optical elements and the effect it has on the outcome.]

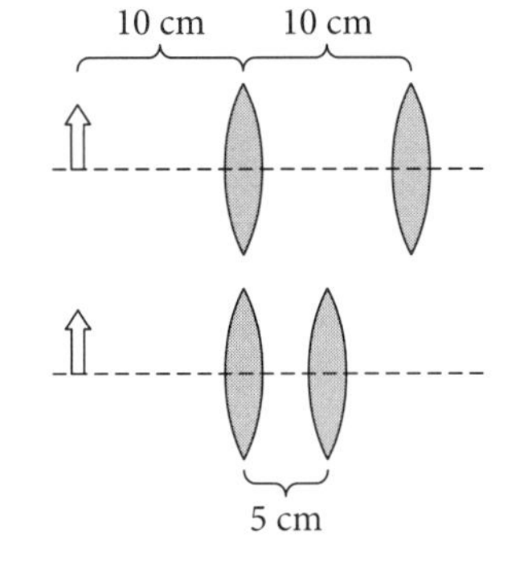

** 27. Three lenses are placed in a row, separated by 12 cm. The first lens is a diverging lens with focal length 12 cm. The second is a converging lens with focal length 10 cm. The last is another diverging lens with focal length 8 cm. If an object is placed 24 cm to the left of the first lens, where will the final image occur? Will it be real or virtual?

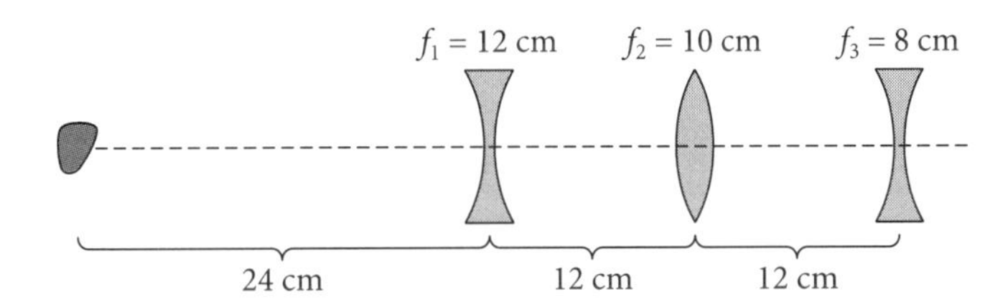

** 28. **Beam Expander.** Plane waves (zero curvature) from a laser are made to first hit a diverging lens of power $P_1 = -5$ D, causing the beam to diverge. The beam then encounters a converging lens of power $P_2 = +2$ D. If the final laser beam is to again be a plane wave, what should be the distance between the two lenses?

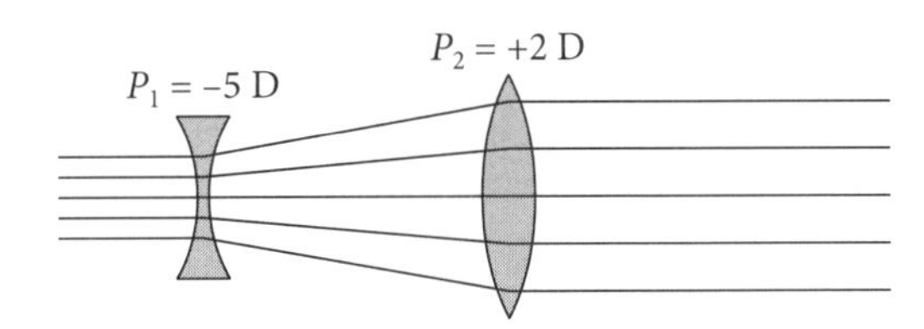

** 29. An object 3 mm high is placed 50 cm to the left of a converging lens of focal length 40 cm. A second converging lens with a focal length 10 cm is placed 220 cm to the right of the first lens. Where is the final image? Is it real or virtual? What will be the image size? Is it erect or inverted?

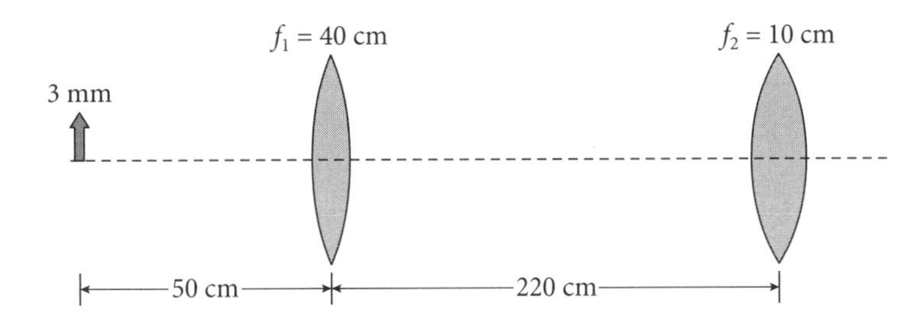

** 30. A pebble 1 cm tall is placed 40 cm to the left of a +5-diopter lens. A –5-diopter lens is placed 30 cm to the right of the first lens. Where will the final image be located, and will it be real or virtual? What will be the size of the final image, and will it be erect or inverted?

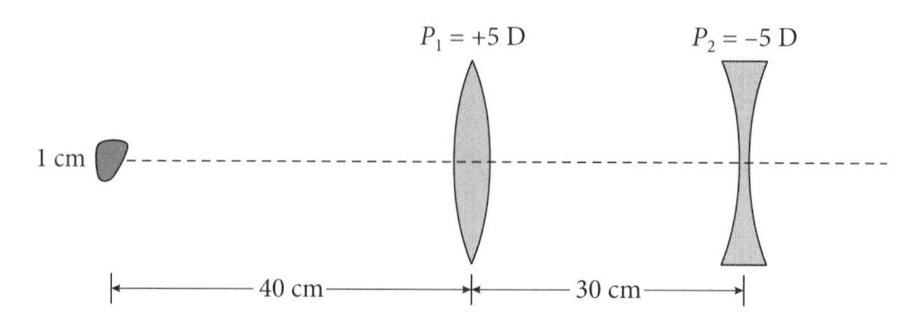

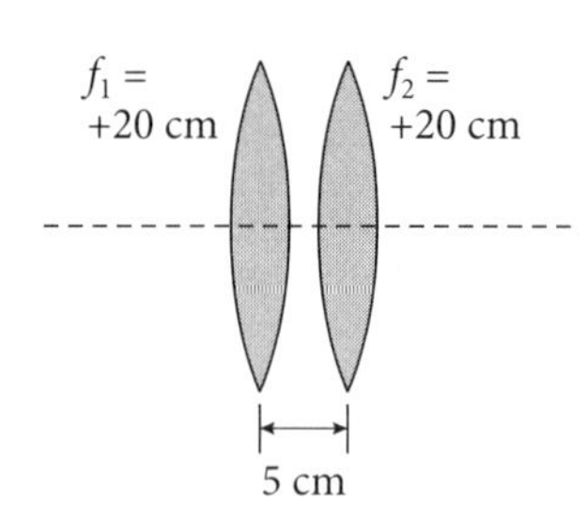

** 31. Two converging lenses, each of focal length 20 cm, are placed 5 cm apart. What is the effective focal length of this lens combination? (That is, what is the distance from the second lens to the image, if an object is infinitely far from the lenses?) Will the final image from an object at infinity be real or virtual, erect or inverted?

*** 32. A diverging lens of focal length 30 cm stands 105 cm in front of a converging mirror of focal length 15 cm, as shown. Light comes in from an object 30 cm to the left of the first mirror, passes through the lens, hits the mirror, passes back through the lens, and continues back leftward. Where will the final image appear, and will it be real or virtual?

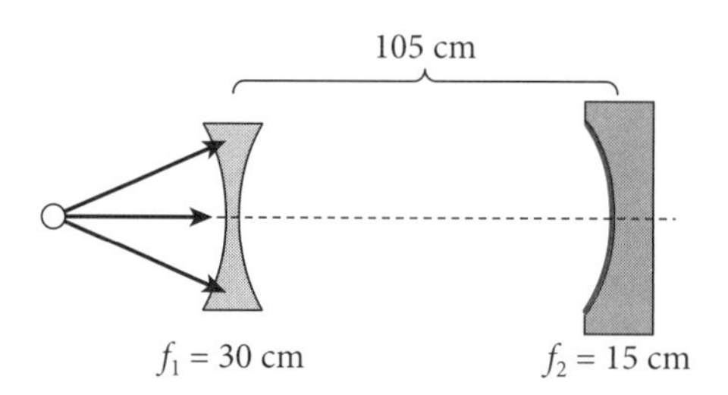

** 33. Find the angular size of a 6-foot-tall football player seen from your seat 80 yards away. Then find the angular size of the 3-inch-tall image of the same player seen on a television set that is 10 feet away.

22.6 ** 34. In children's eyes, the lens is more flexible than in adults and may be fattened enough to focus on objects that are very close. This near point for one young girl is 8 cm, while her far point (the furthest point she can focus with her lens completely relaxed) is 2 m. The distance from her lens to her retina is 2.00 cm.

a) Between what limits, measured in diopters, does the power of her lens–cornea combination vary?

b) Find the power of the lenses that an optician would prescribe so that she is able to focus on all distant objects.

c) When she is wearing her glasses, what will be her new near point?

** 35. The magnifying power of a lens is defined to be the angular size of an object seen through the lens divided by the angular size of the same object when it is sitting 25 cm from the eye. In 1675, Dutch biologist Antony van Leeuwenhoek used a lens of focal length 1.25 mm to discover bacteria. What was the magnifying power of this lens?

** 36. Binoculars are often rated by their magnifying power, which is defined as the ratio of the angular size of an object seen through the binoculars compared with the angular size seen by the naked eye. The magnifying power is followed by an X in the binocular specs. A 6-inch tall bird is perched on a branch 20 feet away.

a) What is the angular size in radians of the bird as seen by the naked eye?

b) What will be the angular size seen through 20X binoculars?

c) How tall would the bird have to be to have this angular size as seen by the naked eye from 20 feet away?

** 37. The Canada-France-Hawaii Telescope on top of Mauna Kea has a primary mirror that is 3.6 m in diameter with a power of 0.074 diopters.

a) What is the focal length of the primary lens?

b) The moon covers an angle of ½ degree in the sky, as seen from the surface of the earth. How big will the primary real image of the moon be in centimeters, inside this telescope?

** 38. A certain optical instrument is formed of two lenses, each with focal length 2 cm. When assembled, the distance between the lenses is 22 cm. What kind of instrument is this? Explain.

** 39. A sailor shipwrecked on a desert island needs to be able to see far out to sea. Fortunately, he has managed to save his reading glasses whose two lenses are of power +2 D and +20 D, and he decides to make them into a telescope.

a) Which lens should be the primary lens?

b) When the two lenses are assembled, what is the distance between the two lenses?

c) What is the magnifying power of the telescope (see Problems 35 and 36 for the definition of magnifying power)?

*** 40. Your backyard astronomical telescope has a focal length $f = 1.2$ m. The telescope eyepiece is exactly in focus, meaning that light from a far-away star passes out of the eyepiece as a plane wave. You now want to use your telescope to spy on your neighbor who is in his backyard 50 m away. By what distance will the eyepiece need to be moved to bring the neighbor exactly into focus?

*** 41. A Galilean telescope is formed by intercepting the converging light from the primary lens before it forms a real image and diverging it to produce plane waves (so that the image looks like it came from infinitely far away). Light from far away is focused by a +4 D primary lens and a diverging lens of power –20 D is placed 20 cm from the primary, as shown in the diagram below.

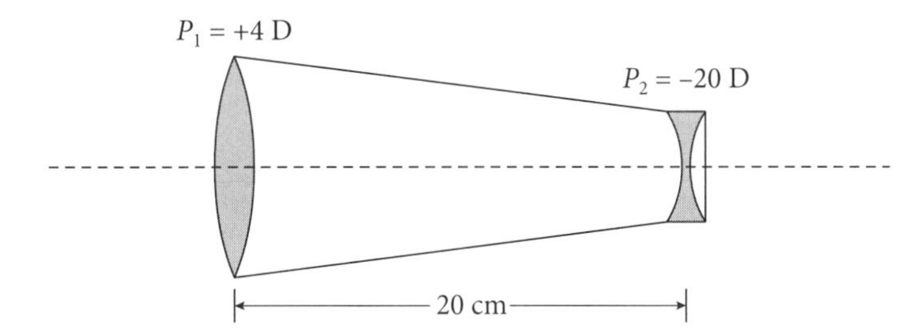

a) Work through the two-lens combination to find the location of the final image.

b) Is the final image erect or inverted?

c) The magnifying power of a telescope is defined as the angular size of the image of some object divided by the angular size of the same object seen with the naked eye. Find the magnifying power of the Galilean telescope above by choosing an object with a small angular size when seen with the naked eye and calculating its angular size in the final image of the telescope.

*** 42. The primary mirror of a telescope has a focal length of 80 cm. The secondary mirror, with power –20 D, is placed 75 cm to the left of the primary. As the beam passes through a hole in the primary, it encounters an eyepiece of power +20 D. How far from the last lens will starlight be focused?

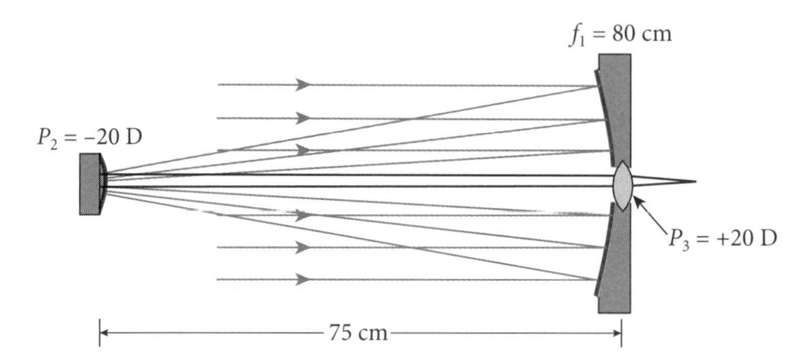

*** 43. A Cassegrain telescope has a primary mirror of focal length f_1 = 1.0 m. However, 15 cm before the image of a star can come to focus in the primary focal plane, a secondary lens of power $P_2 = -1/f_2$ intervenes and diverges the beam through a hole in the primary and to a final focus 25 cm behind the primary mirror. What must the focal length of the secondary mirror be?

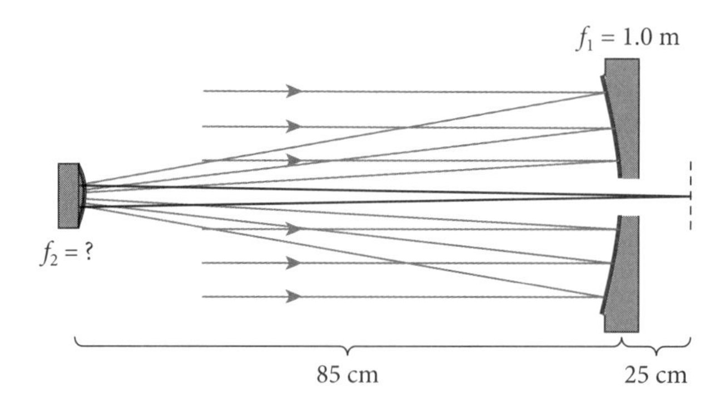

23

WAVE OPTICS

In our introduction to light in Chapter 21, we began with Maxwell's Equations and recognized that light was a type of electromagnetic wave. But this was not always the way physicists thought. By the mid 1600s, Isaac Newton had performed several important optics experiments that had led him to his own theory of light. In fact, in the late 1600s, there was a classic debate between Isaac Newton and our friend from Chapter 21, Christiaan Huygens, about the nature of light. As we remember, Huygens had proposed that light was a wave, with each point on the wavefront acting as a source for propagating the wave forward. Isaac Newton, on the other hand, was completely convinced that light consisted of a beam of particles, with the color of the light being determined by the mass of the particle.

There were two observations that Huygens felt were strongly in favor of his wave theory. First, there was the observed fact that light that passed through a narrow slit in an otherwise opaque material would spread, as described back in Figure 21.4. But Newton could also explain this observation by suggesting that particles passing near the edges of the slit would be attracted to the nearby material, giving each particle a small sideways velocity and spreading the beam. The second observation involved Huygens' explanation for the bending of light. As we learned in Chapter 21, Huygens predicted that a beam entering a higher-index material like glass would slow its speed, causing the beam to bend toward the normal to the glass. This bending, of course, is observed. But Newton had an explanation for the bending as well. He suggested that particles of light would be attracted toward the glass, and that the resulting increased component of velocity toward the glass would bend the rays toward the normal. The question could have been resolved by measuring the speed of light in glass (it would be higher in glass than in a vacuum if Newton was right and lower if Huygens was right), but the technology to make the speed measurement did not yet exist.[1]

The experiment that ultimately discriminated between the two theories once and for all was not performed for over a hundred years after the famous debate. The key to the experiment lay in the fact that particles can either be present or not present, but they cannot be "unpresent." That is, there is no such thing as a negative particle number. A wave, on the other hand, can have either positive or negative values. As a result, if light is coming from two different sources, and if light is composed of particles, then the light can only increase in brightness due to the particles from the second source. If light is a wave, however, there can be places where the superposition allows positive and negative values to cancel, producing a zero

1. The technology to measure the speed of light in a material did become available in the late 19th century, and Huygens, not Newton, turned out to be right.

amplitude for the light. We saw this kind of wave behavior as we studied interference back in Section 12.3, and we saw in Chapter 13 how a combination of waves with just the right phase relationship would produce zero-amplitude nodes at certain places.

In 1803, British physicist Thomas Young performed a double-slit experiment like the one we describe in the next section. The experiment produced an interference pattern, with nodes and antinodes, that could only be explained by a wave theory of light. At first, physicists were reluctant to believe his results since they contradicted Isaac Newton, and Isaac Newton had never been wrong before. But, eventually, it became clear that the evidence was inescapable. Huygens had been right all along.

23.1 Double-Slit Interference

A schematic diagram of Thomas Young's double-slit experiment[2] is shown in Figure 23.1. In order to have a consistent phase between the light from the two slits, so that the interference pattern will not fluctuate over time, light of a single color (wavelength) is first passed through a narrow slit. This guarantees that the light in the wavefronts hitting the two slits of the double-slit element will always have the same phase.[3] We call such phase-coordinated light *coherent light*. As the light from the two slits emerges into the region beyond, there are lines, more or less apparent in Figure 23.1, where the light from the two slits always interferes destructively. These are places where the amplitude of the light is zero. At other places, where the interference is not constructive, there will be large-amplitude oscillations as the wave moves along its way. When the light finally makes its way to a screen far off to the right of the figure, this interference pattern creates patches of bright light (the red patches) and patches of zero light on the screen, as shown in the strip at the right of Figure 23.1.

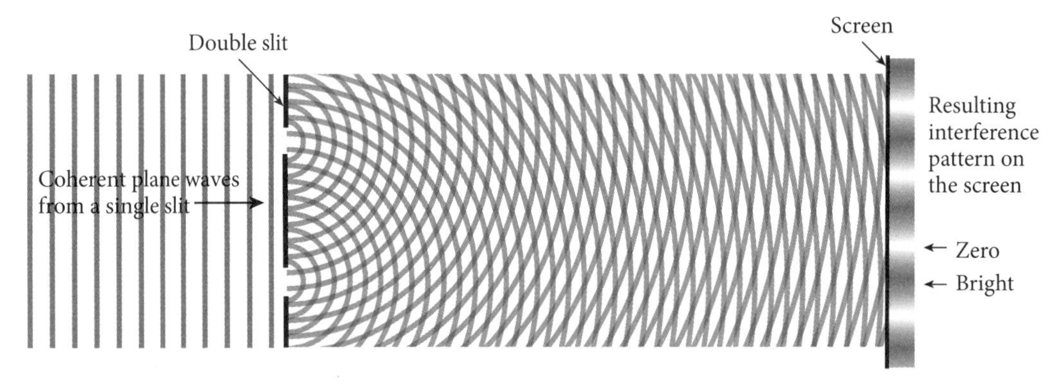

Figure 23.1 Light of a single wavelength emerges from a single slit far to the left of the diagram, restricting the light to a small wavefront whose phase is essentially uniform. The nearly plane wave then encounters the double slits and expands into the region where interference between the beams from the two slits produces a pattern on the screen on the right.

Figure 23.1 shows what we would expect from Thomas Young's experiment if Huygens' wave theory of light were true. On the contrary, if Newton's particle theory were correct, the prediction for the pattern of light that hits the screen in a double-slit experiment would be

2. Young's experiment actually used double pinholes rather than double slits, but the general outcome of the experiment is the same. Light travels in waves.

3. Another way to guarantee that the light striking two slits is of a single wavelength and in phase is to generate the light in a laser. A short description of lasers may be found in Section 23.6.

very different. Figure 23.2 shows a beam of light particles from a source far to the left of the picture, all moving directly to the right. As they reach the two slits in the figure, the particles that pass through the slits acquire a component of velocity transverse to their initial direction, and so they spread into a widened beam. The beams can overlap, but the pattern we would expect is still that of two broad patches of light, centered on each slit. This prediction is what is shown in the strip at the right of Figure 23.2. Young's experiment, as you may have guessed, showed a pattern like the one in Figure 23.1, not like Figure 23.2.

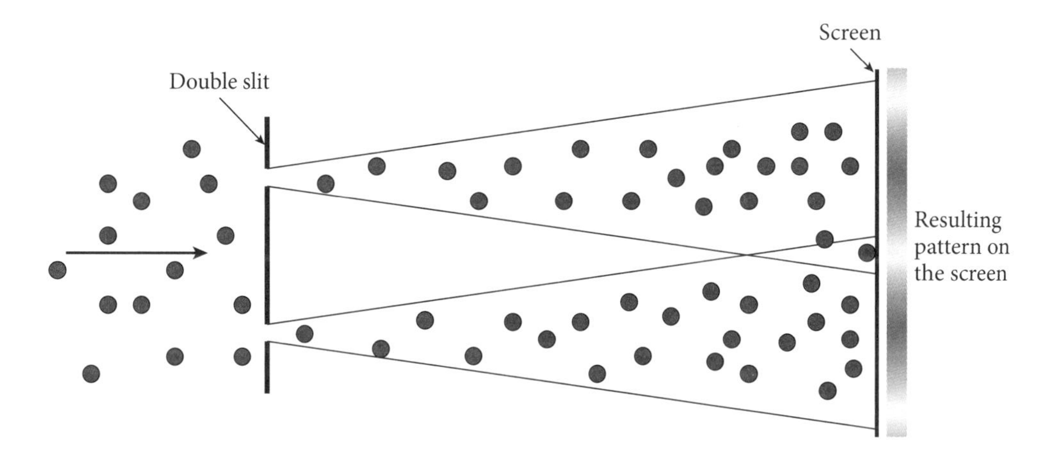

Figure 23.2 In a particle model of light, particles hitting two slits will be expanded by the passage through the narrow slits and will form two broad patches centered on the two slits.

We can actually analyze the relative phase of the waves from the two slits in Young's experiment to see exactly where the pattern on the screen should produce bright light and where it should produce no light at all. Let us look at the situation shown in Figure 23.3, where coherent plane waves of wavelength λ, coming either from a laser or from a narrow slit far to the left of the diagram, encounter two slits that are separated by a distance d. We assume that the slits are narrow enough that the light spreads evenly into the region to the right of the slits (more on this in Section 23.3). We concentrate, however, only on the light propagating along the two rays shown at an angle θ in Figure 23.3. If the two beams from the two slits are to eventually interfere, they will have to converge slightly, but we assume that the screen where the interference takes place is far enough to the right that the two beams are almost parallel.

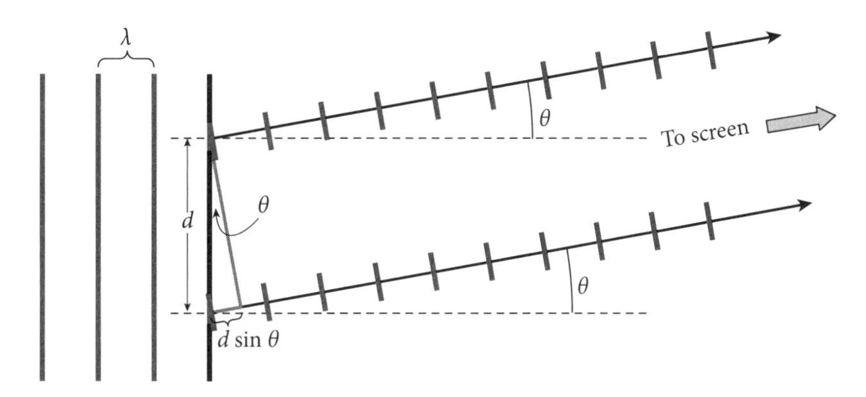

Figure 23.3 The geometry of double-slit interference. Plane waves from the left pass through the slits and emerge at all angles into the region beyond, on their way to combine at a screen.

As the beams emerge from the two slits and travel upward at an angle θ to the incoming direction, the beam from the lower slit will have farther to travel before arriving at the screen. To see this, note the gray line that connects the two outgoing rays at right angles to the two beams. From the gray line onward, the distance traveled by both beams will be the same, and the same number of wavelengths will fit between the gray line and the screen in each beam. Since the wave has the same phase in both slits, the lower beam will be delayed in phase by its additional path length, and that amount of delay will be carried all the way to the screen where the beams combine. The additional distance the lower beam has to travel is $d \sin \theta$. Now, if exactly one wavelength were to fit into that distance, so that $d \sin \theta = \lambda$, then both beams would be in phase at the points joined by the gray line, and so they will remain in phase until they arrive at the screen. When they add, they will add in phase—constructively—and will produce bright light at an angle θ from the initial forward direction. Obviously, the same in-phase behavior will occur if exactly two wavelengths fit into the $d \sin \theta$ distance, or any integer number of wavelengths. The condition for bright light is thus $d \sin \theta = n\lambda$, where n is any integer. We may summarize by saying that, when light of wavelength λ falls on two narrow slits separated by a distance d, there will be constructive interference and bright light at angles θ_n, where these angles satisfy

$$\sin \theta_n = n\frac{\lambda}{d} \qquad n = 0, 1, 2, 3, \ldots \tag{23.1}$$

Figure 23.4 The interference pattern resulting from red light of wavelength 480 nm passing through two narrow slits.

In common physics usage, the alternating light and dark patches in an interference pattern are referred to as "fringes" (probably because the alternating bright and dark bands look like fringes hanging down from a piece of cloth). So we would say that θ_1, corresponding to $n = 1$, gives the location of the "first-order bright fringe" of the interference pattern. When $n = 0$, Equation 23.1 predicts a bright fringe at the corresponding $\theta_0 = 0$, due to the zero path difference between the two beams and the resulting zero phase difference between them. A picture of a two-slit interference pattern is shown in Figure 23.4.

It is equally simple to determine the angles at which no light will appear. This would be a case where exactly one-half of a wavelength fits into the additional distance that the bottom beam in Figure 23.3 has to travel. This way the phase of the bottom beam at the level of the gray line will be at a minimum when the top beam is at a maximum and vice versa. They will always add up to exactly zero. A little thought should convince you that the same situation will also occur at an angle where 1½ wavelengths fit into the extra distance, or 2½ wavelengths or 3½ wavelengths, etc. A general formula for the zeros of the interference pattern can therefore be written

$$\sin \theta_n = \frac{n\lambda}{2d} \qquad n = 1, 3, 5, \ldots \tag{23.2}$$

It is important to look carefully at Equations 23.1 and 23.2 so that you do not get them confused. For light of wavelength λ falling on two slits separated by a distance d, the angles at which there will be no light will be those that satisfy Equation 23.2, and the angles at which there will be maximum light will be those that satisfy Equation 23.1.

We want to add just a couple of comments on these results. First, in order to have complete cancellation at the angles given by Equation 23.2, the amplitude of the light must be the same from both slits. Generally, it is sufficient to make the widths of the two slits the same.

If this is not the case, the light will be minimal, but not zero, at the Equation 23.2 angles. The maximum-light angles given by Equation 23.1 will always produce a relative maximum of the pattern. Our second observation is for those who may wonder why it is that two light bulbs illuminating the same room do not produce patterns of light and dark on the walls of the room. The answer is that they do, but only for a split second. This is for two reasons. First, light bulbs do not have a single wavelength, so each color of light they produce would have a different interference pattern and different colors would appear in different places. Second, and most importantly, the phases of the light from two light bulbs are not coherent, not constant in time. So whatever interference pattern might be set up at one instant will change in the next instant, faster than the eyeball or any other detector can follow.

EXAMPLE 23.1 A Two-Slit Interference Pattern

Coherent plane waves of light of wavelength 500 nm fall on a pair of slits that are 0.01 mm apart. The interference pattern falls on a screen that is 2 meters from the slits. Give the locations of the central bright fringe and of a few more bright fringes.

ANSWER First, though the question did not ask it, let us verify the validity of using the approximation assumed in Equations 23.1 and 23.2 in which the two rays from the two slits that converge to the screen are taken to be parallel lines. A picture of the rays converging to the central fringe is shown below.

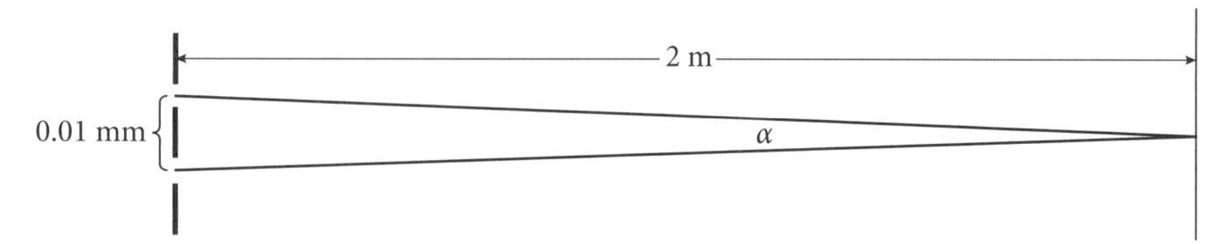

The angle in the picture is $\alpha \approx (0.01 \text{ mm})/(2 \text{ m}) = 5 \times 10^{-6}$ radians. Thus, the deviation from perfect parallelism is only about one arcsecond (1/3600 degree) and may safely be ignored in the problems we need to solve.

At the point directly in line with the slits there will, of course, be a bright central fringe at $\theta = 0$, corresponding to $n = 0$ in Equation 23.1. The next bright fringe will be the $n = 1$ fringe on either side of the central fringe, at an angle satisfying

$$\sin \theta_1 = \frac{\lambda}{d} = \frac{500 \text{ nm}}{0.01 \text{ mm}} = 0.005,$$

for a θ of 3°. At a distance of 2 m, this bright fringe will be found at a vertical distance $\Delta y = 2 \tan \theta = (2 \text{ m})(0.005) = 1$ cm from the central fringe. There will thus be two bright fringes, 1 cm on either side of the central fringe. The next closest fringes will be the $n = 2$ fringes that correspond to $\sin \theta_1 = 2\lambda/d = 2 \cdot 0.005 = 0.01$, producing bright light 2 cm to each side of the central fringe.

Our next example is not directly an example of light interference from two slits, but an example of an analogous situation for radio wave interference.

EXAMPLE 23.2 Interference in Radio Beams

Radio stations beam their signals toward a town by using two antennas in a line and by delaying the output from the back antenna so that the signals from the two antennas are *in phase* in the direction of the town and *out of phase* in the opposite direction. If an AM radio station broadcasts at 1000 kHz, what should be the delay of Antenna A relative to Antenna B and what should be the distance between the two antennas?

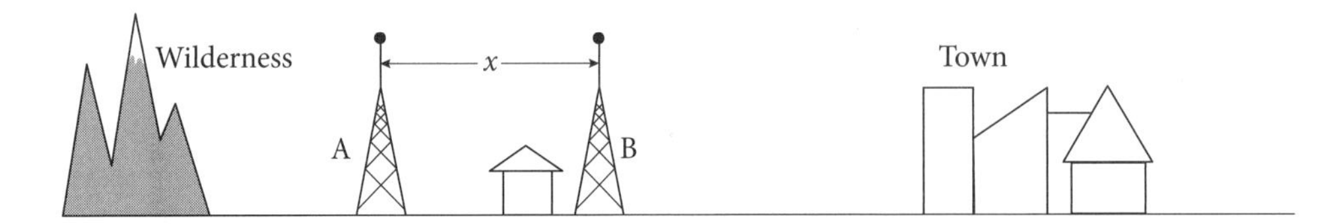

ANSWER Let us use a variable x for the distance between the antennas and let us assume that the signal from antenna A is delayed by the station's electronics by a fraction $\Delta\phi$ of a cycle. Then, by the time that signal reaches antenna B, moving toward the town, it will be delayed by $\Delta\phi + x/\lambda$ cycles. If the two signals are to add constructively, this phase difference must equal one complete cycle. On the other hand, if we don't want to waste power by sending signals away from the town, then the two signals should be made to add destructively when moving away from town. When the leftward-going signal from Antenna B reaches Antenna A, traveling to the left, its phase will be delayed by x/λ cycles. However, the signal from Antenna A is already delayed in phase by $\Delta\phi$ cycles. Thus, the relative phase between the leftward-moving signals will be $\Delta\phi - x/\lambda$. For destructive interference, this phase difference should be ½ cycle. We conclude that $\Delta\phi$ and x should be set so that both conditions are realized:

$$\Delta\phi + x/\lambda = 1$$
$$\Delta\phi - x/\lambda = \frac{1}{2}.$$

If we add these equations together, we get $2\Delta\phi = \frac{3}{2}$, or $\Delta\phi = \frac{3}{4}$ as the phase offset. Finding x requires us to first solve for x/λ. The first equation gives $x/\lambda = 1 - \Delta\phi = 1 - \frac{3}{4}$, or $x/\lambda = \frac{1}{4}$. The wavelength of the signal produced by the radio station is

$$\lambda = \frac{c}{f} = \frac{3 \times 10^8 \text{ m/s}}{1 \times 10^6 \text{ Hz}} = 300 \text{ m},$$

meaning that the separation between the antennas should be $x = \frac{\lambda}{4} = 75$ m. As strange as it might seem, though both antennas are putting out maximum power in all directions, none of that power will end up being broadcast directly into the mountains on the left.

23.2 Gratings

In the last section, we discussed the interference pattern that is formed when coherent light of wavelength λ encounters two narrow slits separated by a distance d. There is, however, a very useful optical instrument that is composed of many slits. It is called a *grating* or a *diffraction grating*. The way that such an instrument works is depicted in Figure 23.5. Here, coherent plane waves of wavelength λ are incident on many slits (six are shown in the figure), each separated from its neighbor by a distance d. Clearly, there will be a very bright patch created at $\theta = 0$ where all of the signals from all of the slits are in phase and the interference is maximally constructive. But what about the interference at other angles? Well, the geometry of Figure 23.5 is just a generalization of that of Figure 23.3 in which each slit's signal is delayed compared to the slit above it by $d \sin \theta$. If an integer number of wavelengths fits into this distance, then the signal from slit 6 will be in phase with the light from slit 5, which will

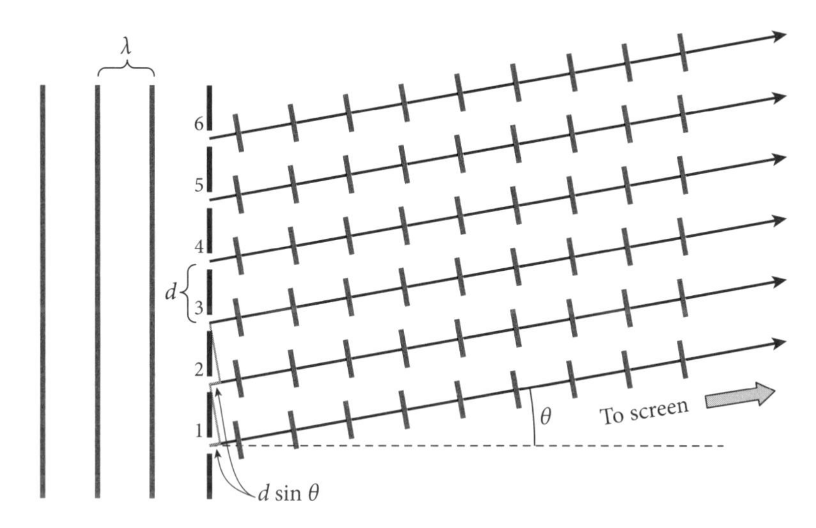

Figure 23.5 The geometry of the interference from a 6-slit grating. The brightest fringe will be found at angles where light from all six slits is in phase.

be in phase with the light from slit 4, etc. So the formula for locating the bright fringes from a multi-slit grating is the same as Equation 23.1. When light of wavelength λ encounters a grating, with multiple slits a distance d apart, the angles at which bright fringes are seen will be those that satisfy

$$\sin \theta_n = n\frac{\lambda}{d} \qquad n = 0, 1, 2, 3, \ldots \tag{23.3}$$

Equation 23.3 gives the locations of the *maxima* of the interference pattern, but how about the *zeros*? It is obvious that Equation 23.2, which gave the zeros of the two-slit interference, will also give zeros for a 6-slit grating, since a situation where half a wavelength fits into the distance $d \sin \theta$ in Figure 23.5 means that the light from slit 1 cancels the light from slit 2, the light from slit 3 cancels that of slit 4, etc. But a little thought will let us see that there is an angle smaller than this, an angle closer to $\theta = 0$, at which there will also be complete destructive interference. If we have six slits, then we can see from Figure 23.6 that, at an angle that satisfies

$$3d \sin \theta = \frac{\lambda}{2} \quad \text{or} \quad \sin \theta = \frac{\lambda}{6d}, \tag{23.4}$$

the light from slit 4 will be half a cycle out of phase with that from slit 1 and will cancel it completely. But it will also be true, at this angle, that light from slit 5 cancels that from slit 2 and light from slit 6 cancels that from slit 3. If we start at $\theta = 0$, in the middle of the central bright maximum, and work our way outward, then the first zero we will come to will be the one at the angle given by Equation 23.4. We contrast this with the case of two slits, for which Equation 23.2 gave the first zero of the pattern at an angle where $\sin \theta = \lambda/(2d)$. The angle

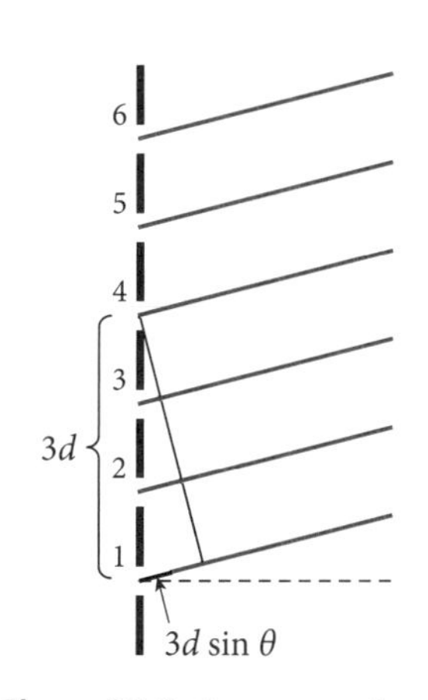

Figure 23.6 Geometry of the smallest angle θ at which there is complete destructive interference.

where $\sin\theta = \lambda/(6d)$ will be smaller than the angle where $\sin\theta = \lambda/(2d)$, and so the 6-slit central bright spot will not be as wide.

Let's think about this. Our 6-slit grating will have a central bright fringe at $\theta = 0$ and bright fringes on either side, where $\sin\theta = \lambda/d$. The first zero on each side of the central bright fringe will be at the small angle where $\sin\theta = \lambda/6d$. But we also know, from Equation 23.2, that there is a zero of the pattern halfway between the central maximum and the first maximum, at $\sin\theta = \lambda/(2d)$, or, in other words, at the angle $\sin\theta = 3\lambda/(6d)$. So we see that we have zeros at $\sin\theta = \lambda/(6d)$ and $3\lambda/(6d)$. This leads us to wonder. What will we see at $\sin\theta = 2\lambda/(6d)$, or $4\lambda/(6d)$, or $5\lambda/(6d)$? The answer is that they are all zeros. At each of these angles, the light from all six slits will add up to zero amplitude. But if these five locations are zeros, what happens at angles between them? Certainly, the light from all six slits will not add completely in phase again until we get out to $\sin\theta = \lambda/d$, but neither will the light add to zero. At intermediate points, there will be a little light, just not as much as at $\sin\theta_n = n\lambda/d$.

In Figure 23.7, we have graphed the amplitude of light on a screen illuminated by the interference pattern from a 6-slit grating. The central $\theta = 0$ peak is shown in the center of the picture and tall bright fringes corresponding to $\sin\theta_n = n\lambda/d$ appear to the left and right

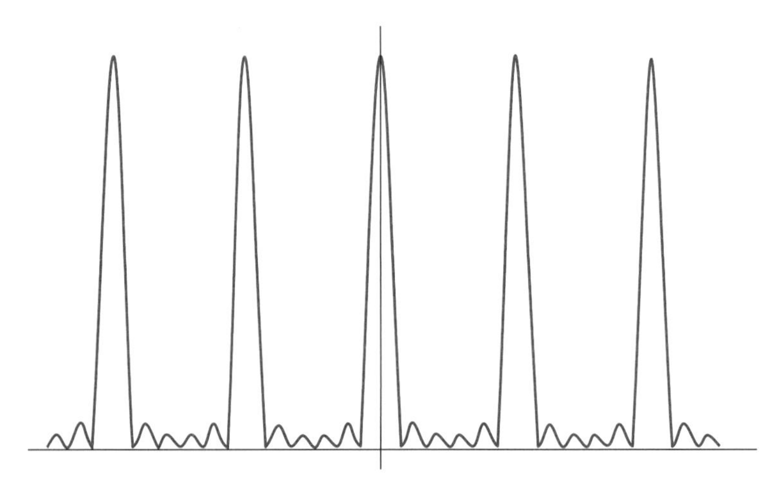

Figure 23.7 The amplitude of light on a screen from a 6-slit diffraction grating.

of the center. Between the bright fringes, there are five zeros of the pattern, with tiny local maxima between them. Because the first zero on each side of a bright fringe occurs so close to the peak, the bright fringes will be narrow, well-separated, and distinct. The actual pattern one would see is shown in Figure 23.8 (though the intermediate maxima are too dim to show up in the picture). This picture should be compared with Figure 23.4, whose double-slit arrangement had the same distance between the slits. As you can see, the maxima occur at the same angles as in the 2-slit pattern, but the bright fringes of the 6-slit pattern are narrower. This means that their locations can be more accurately measured. This last property of the light from a grating leads to one of its most useful applications, as we now discuss.

Figure 23.8 The interference pattern from a 6-slit grating.

Equation 23.3 indicates that there will always be a central bright spot at $\theta = 0$, corresponding to $n = 0$, regardless of the wavelength of the light. However, for the fringe corresponding to $n = 1$, Equation 23.3 gives

$$\sin\theta_1 = \frac{\lambda}{d},$$

so that the angle at which the first-order bright fringe appears *does* depend on the wavelength of the light—the greater the wavelength, the greater the angle. If a beam of light containing many different wavelengths of light hits a grating, the first ($n = 1$) bright fringes will occur at different angles for different wavelengths. This will spread the light from the source out into a spectrum, with the different colors appearing in different places. Knowing d and the angle at which the bright fringe appears for a particular color, the wavelength of that light may be very accurately determined.

Before going on to do some examples showing how a grating is used to measure wavelength, let us point out what the width of each fringe will be. In Equation 23.4, we found the first zero for a 6-slit grating at the angle where $\sin\theta = \lambda/6d$. Similarly, the first zero for an M-slit grating would be where $\sin\theta = \lambda/Md$. Figure 23.7 shows that the full width of the central maximum is twice the angle to the first zero and is the same as the widths of the other maxima, so all bright fringes for an M-slit grating will have widths given in radians by the formula

$$\Delta\theta \approx \sin(\Delta\theta) = \frac{2\lambda}{Md}. \tag{23.5}$$

EXAMPLE 23.3 Measuring Wavelength with a Diffraction Grating

When an electrical current passes through a tube filled with hydrogen gas, visible light is emitted by the gas at four different wavelengths. If the light is passed through a slit to make the light coherent and then through a grating having exactly 200 slits per cm, a bright red light is seen 3.938 cm from the central bright fringe on a screen exactly 3 m from the grating. What is the wavelength of this red light?

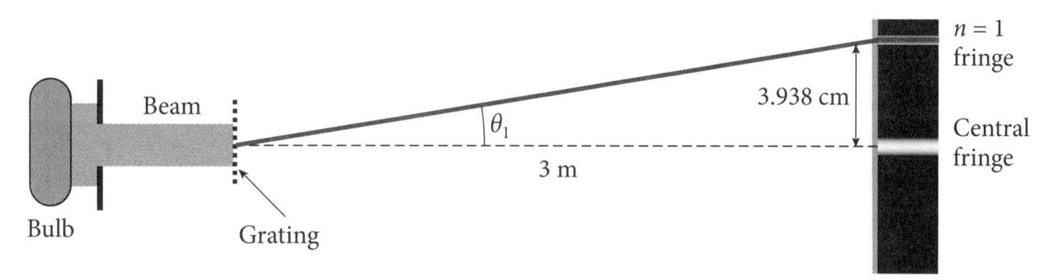

ANSWER To use our equations for the locations of bright fringes for a grating, we must know the spacing between the slits, which we can find from the number of slits per cm. In a grating with 200 slits per cm, this spacing is $d = 1/200$ cm, or $d = 5 \times 10^{-3}$ cm, or 5×10^{-5} m. Then, from the location of the red fringe, we find that the angle θ_1 must be the one whose tangent is

$$\tan\theta_1 = \frac{3.938 \times 10^{-2}\text{ m}}{3\text{ m}} = 1.3127 \times 10^{-2}.$$

The first bright fringe will therefore be seen at an angle

$$\theta_1 = \tan^{-1}(1.3127 \times 10^{-2}) = 0.07521°.$$

Knowing θ_1 and d, the wavelength is found from

$$\lambda = d\sin\theta_1 = (5 \times 10^{-5}\text{ m})(\sin 0.07521°) = (5 \times 10^{-5}\text{ m})(1.3127 \times 10^{-2}) = 656.3\text{ nm}.$$

This exact wavelength is part of hydrogen's fingerprint. If we see light of wavelength $\lambda = 656.3$ nm coming from some sample of gas, we have a telltale indicator for the presence of hydrogen in the sample.

EXAMPLE 23.4 Designing an Astronomical Grating Spectrometer

Alice the astronomer needs to accurately determine the wavelength of red light coming from a cloud of gas near an energetic star. She wants to design a grating that will give her the highest precision possible. What design characteristics should she consider that would increase the precision of the measurement?

ANSWER If the entire spectrum is bunched up within, say, one millimeter on the screen, it will be very difficult to measure where the bright fringe is located. So two things can be done to expand the first-order spectrum over a wider range. First, the formula for the angle of the first-order fringe

$$\sin\theta_1 = \frac{\lambda}{d},$$

shows that, for a given wavelength, a smaller spacing between the slits produces a larger angle for the first-order fringe. In Example 23.3, a spacing of 200 slits per cm produced a bright fringe at 0.07521°. If a grating with 6000 slits per cm had been used instead, the bright fringe would have been where

$$\sin\theta_1 = \frac{656.3 \text{ nm}}{1.67 \times 10^{-8} \text{ m}} = 0.3930,$$

or at $\theta_1 = 23.2°$. Once the angle to the fringes is determined, a second design question is the distance to the screen. The linear position of the fringes on the screen is proportional to the distance to the screen, as shown at right. Moving the screen back gives a more measurable distance to the first bright fringe.

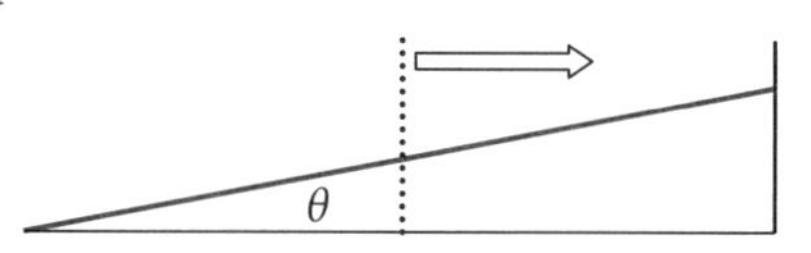

Finally, there is the question of the number of slits. If you want to measure the location of a bright fringe, it is best if the fringe itself is narrow. Figure 23.7 shows that the width of the central maximum is the angle between the first zeros on each side of the center, and that the widths of all maxima are equal to the width of the central one. If M slits are illuminated, the width of each bright fringe would be given by Equation 23.5 as $\Delta\theta = (2\lambda)/(Md)$. Thus, many slits narrow the width of each fringe so that its location may be more easily measured. In fact, light is often spread by a lens before it reaches the grating in an astronomical spectrometer in order to illuminate more lines of the grating and narrow the fringes, improving the accuracy of the wavelength measurements.

23.3 Diffraction

Back in Chapter 21, as we first described Huygens' principle, we explained that a wavefront passing through a barrier will spread out on the far side of the barrier. This is called *diffraction*. We have already used this fact to explain why there is light at angles $\theta > 0$, and it is high time we explained how this spreading takes place.

So let us look at a wavefront that passes through a slit of width w, as shown in Figure 23.9. Each point of this wavefront acts as a source of light that propagates forward with uniform amplitude into the region to the right of the slit. If the wavefront is coherent—all points going through crests and troughs at the same time—then the wavelets from each point will add constructively in the forward ($\theta = 0$) direction. This point will therefore be the brightest point on the screen, as the light from all points in the slit opening comes together in phase. At all other angles, light from some pieces of the wavefront will be slightly out of phase with the light from other pieces, and the less-than-perfect constructive interference will produce light that is not as bright as the light in the center of the pattern. In fact, there will even be places where the interference is completely destructive. Let's see how this occurs.

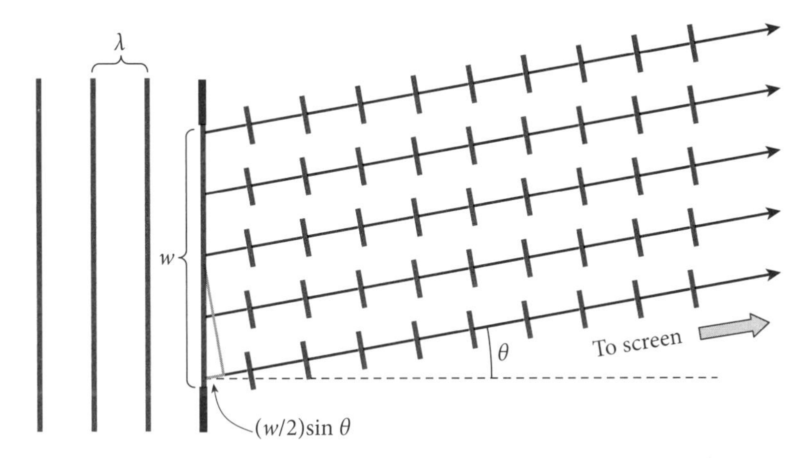

Figure 23.9 The geometry of the interference from the points in a single slit. The brightest light will be found at $\theta = 0$, where wavelets from all points on the wave are in phase. There will be no light when each point of the wave cancels another point of the wave.

We have drawn the single slit diagram in Figure 23.9 to look a little like the diagram of Figure 23.6. At an angle θ, light from the bottom of the slit will have to travel a distance $(w/2)\sin\theta$ farther than the light from the center of the slit. When exactly $\frac{1}{2}\lambda$ fits into this distance, the wavelets from these two points will cancel. But the same thing will occur for the point just above the bottom of the slit and the point just above the center of the slit, and so on and so on, until the light from the bottom half of the slit has, point-by-point, completely canceled the light from the top half of the slit. A wavefront of width w will therefore produce zero light at an angle θ_1 where

$$\frac{w}{2}\sin\theta_1 = \frac{\lambda}{2} \quad \text{or} \quad \sin\theta_1 = \frac{\lambda}{w}. \tag{23.6}$$

Another way to think of this, suggested by the second version of Equation 23.6, is that whenever the light from the top edge of a patch of a wavefront of width w is *in phase* with that from the bottom edge of the patch, then the light from the bottom half of that patch of wave will completely cancel the light from the top half of the patch.

But what happens if we look at an angle greater than the θ_1 of Equation 23.6? When there is no longer complete cancellation of the interfering wavelets, there will again be some light, though never as much as there is at $\theta = 0$, the place where all points of the wavefront showed up at the screen in phase with each other. But there will also be other angles, greater than θ_1, where all of the wavelets again cancel. Let us consider the relative phase of light from the top of the slit and from the bottom of the slit as it is depicted in Figure 23.10. When there are exactly two wavelengths in the extra distance that the light from the bottom of the slit has to travel, then the light from the bottom half of the wavefront (the red double-line wavefront) will produce zero light, since its two halves cancel each other point by point, as in Figure 23.9, and the light from the top half of the wavefront (the solid red wavefront) will be also zero, for the same reason. In fact, whenever an integer number n of wavelengths fits into this extra $w \sin\theta$ of distance, the

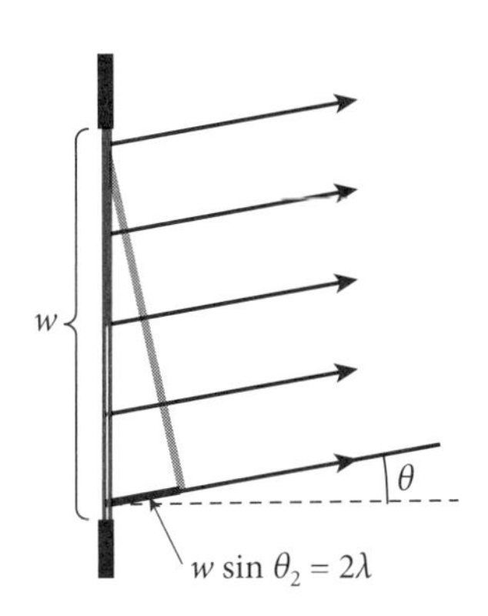

Figure 23.10 The path difference between the top of the slit and the bottom of the slit.

wavefront will divide itself into n patches, and each patch will interfere within itself to produce zero total light. The general condition for complete destructive interference is therefore

$$\sin\theta_n = n\frac{\lambda}{w} \qquad n = 1, 2, 3, \ldots \tag{23.7}$$

It might be a good idea to stop at this point to compare Equations 23.7 and 23.1. Yes, they are almost the same equation. There is a w in the single slit formula and the d in the double-slit formula, and there is no $n = 0$ in Equation 23.7. But the surprising thing is that Equation 23.1 gives *maxima* of the 2-slit interference pattern while Equation 23.7 gives the *zeros* for the single slit. You might want to think about this for a minute to be sure it doesn't trouble you. In Figure 23.11, we imagine a mask plugging up the center of the single slit, so we end up with two very narrow slits separated by a distance w. With the mask in place, there will be bright fringes at the angles given by Equation 23.1. If we remove the mask, then we open up the center of the slit of width w and all the points in the slit are now allowed to create wavelets and interfere. The result is that the maxima are changed to zeros at those same angles.

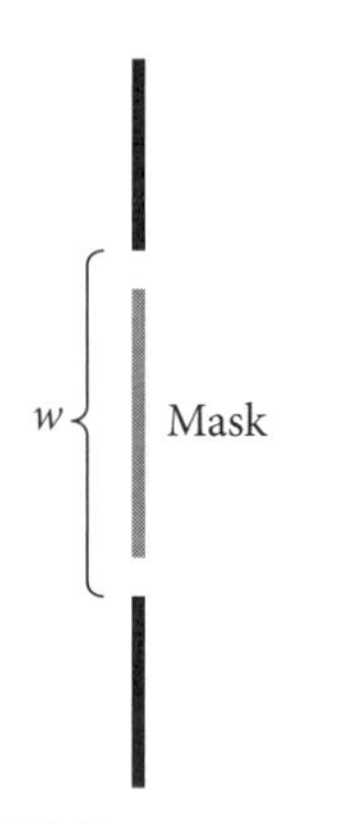

Figure 23.11 Turning a slit into a double slit.

A graph of the brightness of the light in a single-slit diffraction pattern is shown in Figure 23.12. As shown, almost all of the light arrives inside the central maximum between $-\lambda/w$ and λ/w. Some light does shine outside this bright patch and other zeros of the light pattern occur at the points characterized by Equation 23.6. It is tempting to guess that the small maxima between higher-order zeros occur halfway between the zeros, but the situation is actually more complicated than that (and we don't really want to do that calculation here). However, when we talk about the spreading of light as it passes through a single slit, we generally refer to the width of the central bright maximum. It is important to note that Equation 23.6 indicates that the narrower the slit, the wider the central maximum. The more we squeeze light as it passes through a small opening, the *wider* the beam of light we will see on the far side of the opening.

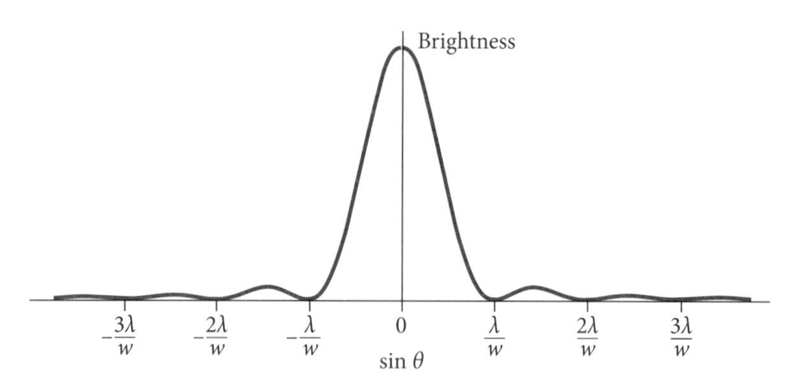

Figure 23.12 The brightness of the light from a single-slit pattern as a function of $\sin\theta$.

But let us be clear that this principle is important only for slits whose widths are on the scale of the wavelength of the light. We don't have much experience with slits as small as this. We know that if you shine a floodlight toward a wall through a 1-inch circular hole, you will find a 1-inch bright spot on the wall beyond. If you reduce the size of the hole to ½ inch, the size of the spot on the wall will reduce to ½ inch. Our intuition, based on these results, is that a smaller hole produces a smaller beam. However, if you were to look at the edges of both spots with a magnifying glass, you would see a few tiny concentric rings going all the way around the spots. These are the fringes that interference creates as the light spreads a tiny amount into the region of shadow. And, if the size of the hole were to be reduced until its diameter is only a few times the wavelength of the light, the pattern that is observed would be dominated by this spreading and the interference.

Let's take a look at an example that shows how much a beam will spread as a coherent wave passes through a narrow slit.

EXAMPLE 23.5 A Single-Slit Diffraction Pattern

Light of wavelength 600 nm falls perpendicularly on a single narrow slit of width 800 nm. How wide will the central maximum be on a screen 2 m from the slit?

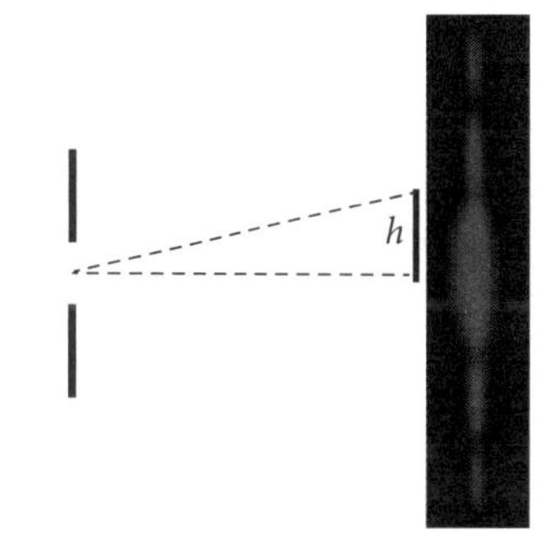

ANSWER Equation 23.6 will give the angle at which the first zero of the diffraction pattern will occur. For the parameters of this slit, the equation gives

$$\sin\theta = \frac{\lambda}{w} = \frac{600\text{ nm}}{800\text{ nm}} = 0.750,$$

which corresponds to an angle $\theta = 48.6°$. At a distance of 2 meters, the distance from the central $\theta = 0$ to the light's edge at $\theta = 48.6°$ is $h = (2\text{ m})\tan(48.6°) \approx 2.3$ m. We must also remember that the width of the central maximum is twice this distance, giving a width of 4.6 m. That's a 4.6-m beam of light coming from an 800-nm slit.

When the slit width is roughly the same size as the wavelength like this, the emerging wave fills most of the space beyond the slit. In fact, if the slit width is less than the wavelength, the equation would require $\sin\theta = \lambda/w > 1$. There is no angle whose sine is greater than 1, so this is telling us that there will be no zero of the diffraction pattern. The light will completely fill the entire 180 degrees beyond the slit. It will be dimmer at larger angles, but it will not drop to zero anywhere.

23.4 Resolution

One of the questions that is related to the size of the central maximum created when light passes through an opening is how well we can resolve two objects that are separated by a small angle θ. Light that is focused by a telescope, for example, must first pass within the diameter of the objective lens. Although a telescope opening hardly seems like a "narrow slit," the spreading of the light as it passes through the opening does prevent the light from a pinpoint source, like a star, from being focused down to a point in the focal plane. The pattern produced by a star in the telescope will, instead, be a central-maximum disk, filling out to the radius of a first interference zero, surrounded by concentric light rings, corresponding to the small maxima shown in Figure 23.12. Figure 23.13 shows a picture of the image of a star in a telescope.

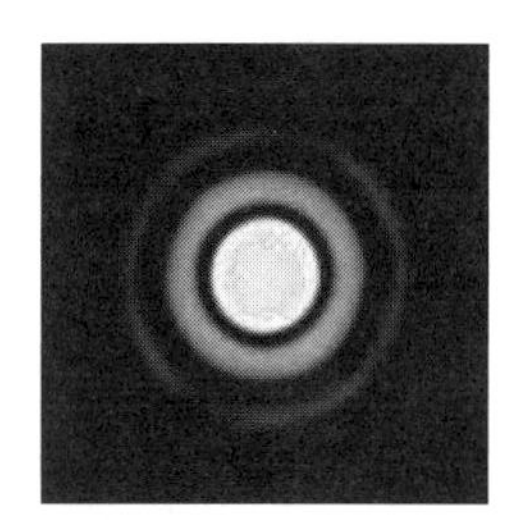

Figure 23.13 The image of a point star source in a telescope.

Given that two nearby points in space will produce images in a telescope that are actually two disks, each like the pattern shown in Figure 23.13, it is often difficult to tell whether there are two sources present or only a single source, at least if the two sources are close together in space. The ability to determine the existence of two separate sources in a telescope image is called the *resolution* of the telescope. While the ability to resolve two nearby points depends on many things, like the brightness of the star or the sensitivity of the detector, a standard criterion has been agreed on to characterize the resolution of an optical system like a telescope. This is the so-called *Rayleigh Criterion*, which states that two sources are considered resolved if the center of the maximum of the pattern from one source falls on the first zero of the pattern from the other, the situation depicted in Figure 23.14. According to this criterion, the smallest angular separation that one can observe is one where θ is the angle from the center of the pattern for one source out to its first minimum at $\sin \theta_1 = \lambda/w$ (Equation 23.6, in which w is the opening of the instrument).

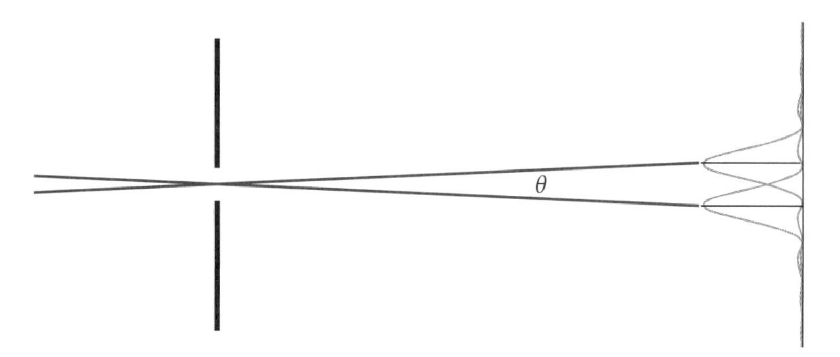

Figure 23.14 The Rayleigh Criterion for the angular resolution of an instrument.

However, we need to remember that Equation 23.6 was derived for a slit, an opening which is much smaller in one direction than in the other. For a circular opening (whose diameter is called the *aperture*), the task of determining the interference pattern involves summing up the phases from all the tiny points on the surface of a two-dimensional wavefront, a calculation that would require us to use the calculus that we do not want to do here. So, let us just tell you that the result of that calculation is this: the Rayleigh Criterion for the resolution of an instrument with a circular aperture of diameter D is

$$\theta \approx 1.22\frac{\lambda}{D}, \tag{23.8}$$

where θ is measured in radians. The following is an example of how one calculates the resolution using the Rayleigh Criterion.

EXAMPLE 23.6 The Resolution of the Hubble Space Telescope

The Hubble Space Telescope has a main telescope aperture of 2.4 m. What is its angular resolution for visible light ($\lambda \sim 550$ nm)?

ANSWER The aperture of the Hubble Space Telescope is circular, so Equation 23.8 applies. It gives

$$\theta \approx 1.22\frac{\lambda}{D} = \frac{(1.22)(550 \text{ nm})}{2.4 \text{ m}} = 2.8 \times 10^{-7} \text{ radians.}$$

If we convert from radians, we get 0.06 arcseconds. If two stars are closer than this to each other, Hubble will not be able to tell that there are two stars present, but they will look like the light from a single star, at least according to the Rayleigh Criterion.

23.5 Thin Films

Back in Chapter 21, as we discussed the index of refraction, we noted that a beam of light entering a transparent material would have a reduced speed inside the material. We also noted, on page 446, that if a light wave hits a boundary between two media, the light would be partly transmitted and partly reflected. But we had actually seen something like that before, back in Chapter 12. There, as you remember, we discussed the passage of a wave along a string and saw that a wave that hit a boundary between two media would be partly transmitted and partly reflected at the boundary.

This behavior makes for a very interesting way in which interference can occur. Consider a beam of light of wavelength λ that goes from vacuum into a thin piece of glass or a thin film of any transparent material. Some of the light will bounce off the first surface of the glass, the beam labeled beam 1 in Figure 23.15, and some will be transmitted into the glass, as shown. Of the light that goes into the glass, some will be transmitted out of the glass at the far surface, and some will be reflected back into the glass. The part reflected back into the glass will then strike the front surface of the glass. There, again, part will be reflected and part, the part represented by beam 2, will be transmitted out of the glass, heading in the same direction as beam 1. The difference between beams 1 and 2 is that beam 2 has passed through the thickness of the glass twice, once going down and once coming back up, and it will have been delayed by the additional path length it has been through. Since beams 1 and 2 are actually parts of the same initial beam, there will be a constant phase relationship between them, and so they will interfere, either constructively or destructively, as they finally combine high above the glass.

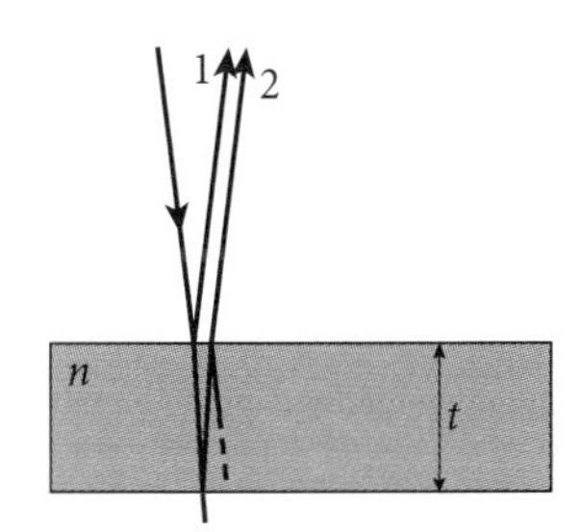

Figure 23.15 Two beams are reflected from a thin slab of glass, one delayed in phase relative to the other.

We should be able to easily work out the phase difference caused by the path delay through the material. If the angle of incidence is very small, as it is in Figure 23.15, then the additional path through the material that beam number 2 travels will simply be $2t$, where t is the thickness of the material. There will be constructive interference between beam 1 and beam 2 if both of them are going through crests and troughs of the wave at the same time. For this to occur, the phase difference between beam 1 and beam 2 must be an integer number of wavelengths of the light. But we need to be careful. As we discussed near the end of Example 12.3, the wavelength of a wave will be smaller in a medium in which the speed is lower. Since the speed of light in a medium with index of refraction n is $c_n = c/n$ (see page 442), the wavelength of light inside the glass will be $\lambda_n = \lambda/n$. So the condition in which an integer number of wavelengths will just fit into the path difference inside a medium of index n will be $2t = m\lambda/n$, where m is an integer and λ is the wavelength of the light in vacuum.

However, we have not yet considered what effect the reflections might have on the phase of the waves. Here again, we refer back to Sections 12.4 and 12.5. When a wave bounces off a medium in which the speed of the waves is lower, the reflected wave comes back upside-down relative to the incoming wave. For a sine wave, like the light waves we have been working with, turning the wave upside down is the same thing as delaying it by $\frac{1}{2}$ of a wavelength, as shown in Figure 23.16. And, as we also learned in Chapter 12, a wave that bounces off a medium in which the speed of the wave is greater will be reflected without inversion, meaning that it will remain in phase with the incoming wave.

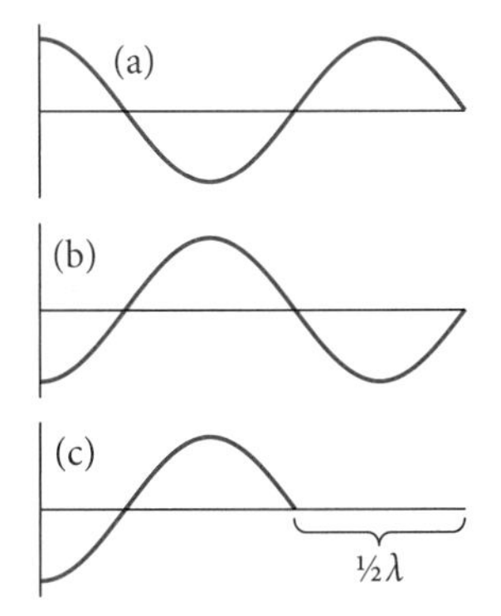

Figure 23.16 An inverted sine wave (b) is identical to a wave delayed by $\frac{1}{2}\lambda$ (c).

The result of these complications is that we cannot simply write down a single formula for the wavelengths of light that will interfere constructively or destructively in a thin slab of some material. What we can do is write down a conditional formula for the effective path difference between two rays reflected off of a thin film of the material, a formula in which we must take possible inversions of the reflected waves into account by hand. Such a formula gives the phase difference *in cycles* between the two reflected beams as

$$\phi = \frac{2nt}{\lambda} + \begin{Bmatrix} 0 \\ 1/2 \end{Bmatrix}, \tag{23.9}$$

where we have to choose either the 0 or the ½ in the curly brackets by thinking through the possible phase changes at the first and second surfaces as the beam encounters the material. Also we need to remember that the λ that appears in Equation 23.9 is the wavelength of the wave *in vacuum*, since the shortening of the wavelength in the material is taken into account by our use of the n in the formula.

The two beams reflected off the two surfaces of the thin film will be in phase, and will produce bright light when they combine, if the phase given by Equation 23.9 is an integer ($\phi = 0, 1, 2, 3, \ldots$). They will produce minimum light if the phase is a half-integer ($\phi = 1/2, 1\,1/2, 2\,1/2, 3\,1/2, \ldots$). Here are a couple of examples to show how to analyze these thin films.

EXAMPLE 23.7 Transmissive Coatings on a Lens

A lens maker wishes to design a coating for a lens so that it will provide nearly perfect transmission at a wavelength 532 nm. He chooses magnesium fluoride (MgF_2) with an index of refraction 1.38, applied in a thin layer over glass of index $n = 1.50$. What is the thinnest layer that can be applied to do the job?

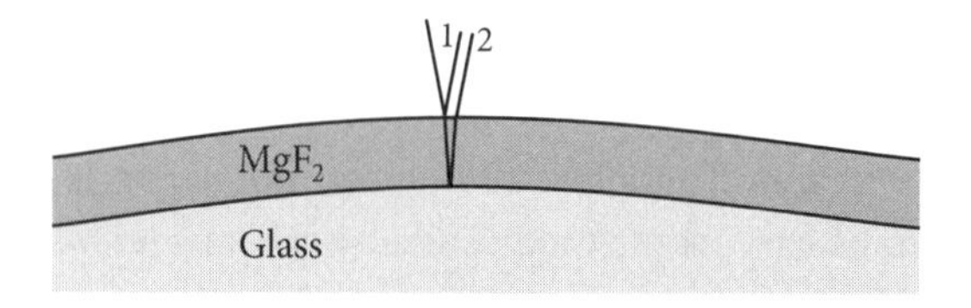

ANSWER To get the best transmission of a particular wavelength, we want to arrange a thin film of MgF_2 over glass so that no light of that wavelength will be reflected. This means that the overall phase difference in the two reflected beams should be a half-integer number. We thus want to have

$$\phi = \frac{2nt}{\lambda} + \begin{Bmatrix} 0 \\ 1/2 \end{Bmatrix} = \frac{1}{2}, \frac{3}{2}, \frac{5}{2}, \ldots$$

Let us begin by seeing if we need the 0 or the ½ option in this formula. The initial reflected beam (beam 1) bounces off of a higher index of refraction (going from $n = 1$ to $n = 1.38$), so it will experience a phase change of ½ a cycle. But the second reflected beam (beam 2) is also going from a lower-index medium ($n = 1.38$) to a higher-index medium ($n = 1.50$), so it will also experience a phase change on reflection. Since both beams are inverted on reflection, the net phase difference is zero, so we pick the 0 option in the formula and write

$$\frac{2nt}{\lambda} = \frac{1}{2}, \frac{3}{2}, \frac{5}{2}, \ldots$$

The thinnest layer that satisfies this equation comes from the lowest half-integer,

$$\frac{2nt}{\lambda} = \frac{1}{2} \quad \text{or} \quad t = \frac{\lambda}{4n}.$$

If we put in the numbers, we get a thickness of

$$t = \frac{532 \text{ nm}}{4(1.38)} = 96 \text{ nm}.$$

A 96-nm layer of magnesium fluoride deposited on the front surface of a lens will see to it that all the 532-nm light hitting the lens passes into the lens.

EXAMPLE 23.8 Reflected Wavelengths from a Thin Film of Oil on Water

White light from directly overhead hits a thin film of oil ($n = 1.48$) that has spread out over a puddle of water ($n = 1.33$). If the thickness of the oil film is 0.3 mm, what visible wavelengths of light will be brightly reflected by the oil?

ANSWER Once again, we should begin by determining what phase changes there are at the two boundaries of the thin film that is producing the interference. When the initial beam bounces off the front surface of the oil, it experiences a ½-cycle phase shift. When the beam transmitted through the oil reaches the boundary with the water, it is going from a medium with higher index of refraction ($n = 1.48$) into one with lower index ($n = 1.33$). Therefore, the beam reflected from this second surface will *not* experience the phase change on reflection. As it continues back up to the first surface and is transmitted out into the vacuum again (there is never a phase change for a transmitted wave), the relative phase of the two beams will end up with a ½-cycle difference. Equation 23.9 then becomes

$$\phi = \frac{2nt}{\lambda} + \frac{1}{2}.$$

For bright reflection, we will need wavelengths that correspond to this phase being an integer value, $\phi = 1, 2, 3, \ldots$. The phase ϕ starts with that half-integer difference, so the path difference, $2nt/\lambda$, need contribute only an additional half-integer amount. We must therefore have

$$\frac{2nt}{\lambda} = \frac{m}{2} \quad m = 1, 3, 5, \ldots \quad \text{or} \quad \lambda = \frac{4nt}{m} \quad m = 1, 3, 5, \ldots$$

If we put in the numbers, we get

$$\lambda = \frac{4(1.48)(0.3\ \mu\text{m})}{m} = \frac{1776 \text{ nm}}{m} \quad m = 1, 3, 5, \ldots$$

Every possible odd-integer value of m will produce a wavelength of light that will be brightly reflected, but visible light ranges only from 380 nm to 750 nm. If we try $m = 1$, we get a wavelength too long to be visible. For $m = 3$, the resulting wavelength, $\lambda = 1776 \text{ nm}/3 = 592$ nm, lies in the visible. But $\lambda = 1776 \text{ nm}/5 = 355$ nm does not. So only the yellow–orange (592 nm) wavelength can be seen reflected brightly from the oil film.

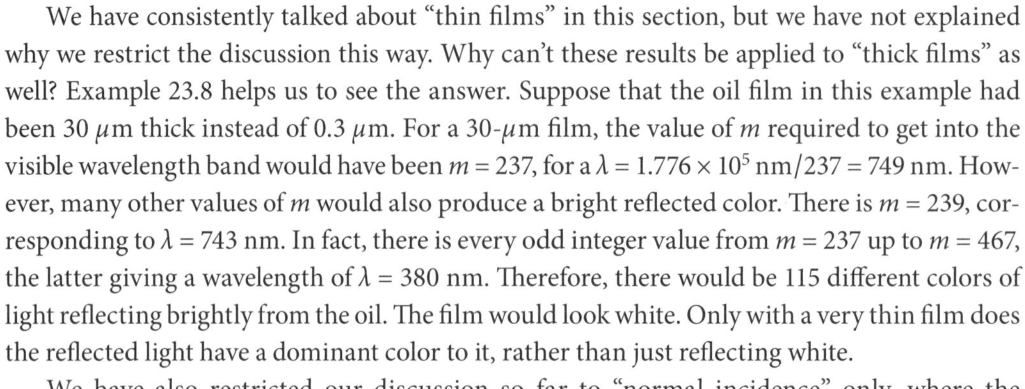

We have consistently talked about "thin films" in this section, but we have not explained why we restrict the discussion this way. Why can't these results be applied to "thick films" as well? Example 23.8 helps us to see the answer. Suppose that the oil film in this example had been 30 μm thick instead of 0.3 μm. For a 30-μm film, the value of m required to get into the visible wavelength band would have been $m = 237$, for a $\lambda = 1.776 \times 10^5$ nm/237 = 749 nm. However, many other values of m would also produce a bright reflected color. There is $m = 239$, corresponding to $\lambda = 743$ nm. In fact, there is every odd integer value from $m = 237$ up to $m = 467$, the latter giving a wavelength of $\lambda = 380$ nm. Therefore, there would be 115 different colors of light reflecting brightly from the oil. The film would look white. Only with a very thin film does the reflected light have a dominant color to it, rather than just reflecting white.

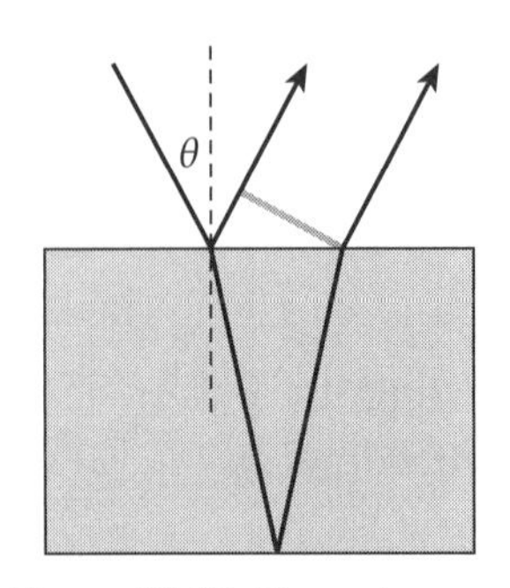

Figure 23.17 The path lengths for beams in a thin film at angular incidence.

We have also restricted our discussion so far to "normal incidence" only, where the incoming light comes in perpendicular to the surface of the glass or other material. The main reason was that the normal incidence simplified the analysis. Figure 23.17 shows light hitting a piece of glass at a slanted angle. The angle inside the glass will have to satisfy Snell's law, and some geometry will be required to determine the distance traveled by the beam reflected off the back surface. A different geometry problem will have to be solved to determine the path length for the beam that reflects off of the front surface before it gets to the gray perpendicular line between the beams, after which the path lengths will be the same. Suffice it to say that the color of the light that interferes constructively when seen at some angle away from normal incidence will not be the same as that at normal incidence. This difference in brightly reflected wavelengths at different angles is responsible for the "rainbow" of colors seen in an oil slick. The light brightly reflected from different places in the oil will come to our eye from different angles and each will carry the colors representing constructive interference within the film in that direction.

23.6 Polarization

Since light waves are waves of electromagnetic field, and since electric and magnetic fields are vectors, we can associate a field direction with an electromagnetic wave by choosing the instantaneous direction of its electric field. We call this the wave's *polarization*. If a beam of light from some source maintains its polarization over time, we say that the light in the beam is *polarized*. A source like a dipole radio antenna, for example, produces an electromagnetic wave in which the electric field maintains a consistent orientation. The radio waves are polarized. Visible light, however, is typically produced by random collisions of atoms or by individual atoms that radiate for a short time only, as in a fluorescent light bulb. We would not expect such a light source to produce polarized waves. An important exception is the light from a *laser* (laser stands for "light amplification by stimulated emission of radiation"). In a laser, an initial flash of light excites the atoms, kicking them up to a state of higher energy. Then, as a wave of light sweeps through the material of the laser, each atom is stimulated to radiate in phase with, and with the same polarization as, the light that stimulated it. Lasers produce polarized light, but most other sources of visible light do not.

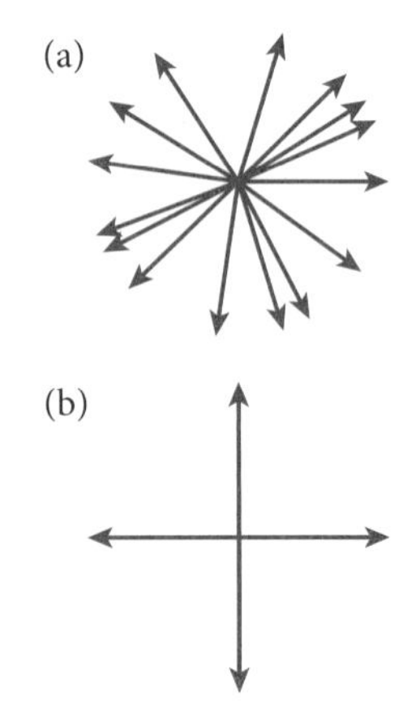

Figure 23.18 Two ways of depicting unpolarized light.

If we were to look into a beam of unpolarized light, the electric field vector would always be found in a plane perpendicular to the direction of propagation of the waves, but we would see the vector jump around the plane from one instant to the next. In a plane perpendicular to the direction of propagation of the light beam, a set of polarization vectors would be built up that might look like Figure 23.18a. Because of the randomness of the instantaneous polarizations, the average polarization will be equally likely to point left as right, or up, or down.

Therefore, we can also characterize the average polarization of the light wave with a diagram like that in Figure 23.18b. The average horizontal component of the polarization vectors is as likely to be to the left as to the right, and the average vertical component is as likely to be up as down. Although the natural origin of most visible light makes it unpolarized, it is possible to polarize light in a number of ways.

If we stretch a sheet of plastic in one direction, long chains of molecules are formed within the material of the plastic. A light wave whose electric field lies along the direction of the chains is absorbed in the plastic, while waves with polarizations perpendicular to the chains of molecules are transmitted through it. The direction in which polarized waves are passed is called the *polarization axis* or the *optic axis* of the polarizer (the black double arrows in Figure 23.19). Optic axes are generally perpendicular to the directions of the molecular chains in plastic polarizers.

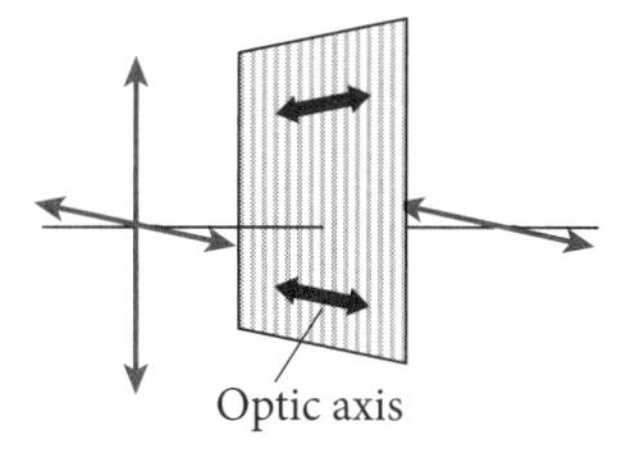

Figure 23.19 The action of a polarizer.

Two concepts are needed to be able to predict the amount of light that passes through a polarizer. First, if polarized light hits a polarizer at an angle θ from the optic axis, as shown in Figure 23.20, then the electric field $\vec{E}$ in the wave can be resolved into components along the optic axis and perpendicular to the optic axis. The component along the optic axis will be transmitted and the perpendicular component will be absorbed. The second thing we need to know is that the intensity of light is determined by the energy density in the wave, and we remember that the energy density in an electromagnetic wave is given by Equation 21.5 and is proportional to the square of the magnitude of the electric field. These facts enable us to figure the brightness of the light that passes through a polarizer.

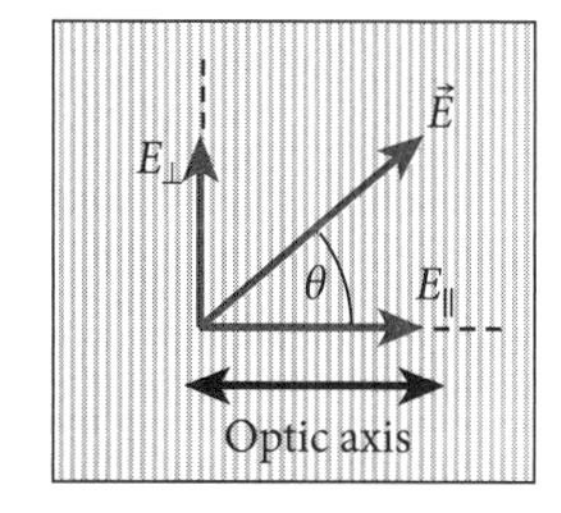

Figure 23.20 Resolving the E-field vector into components.

EXAMPLE 23.9 The Fraction of Light Passing Through a Polarizer

Light from a 60-mW laser hits a polarizer in such a way that the polarization of the light is at 30° to the optic axis. How much power makes it through the polarizer?

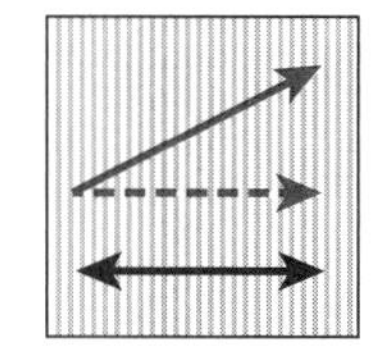

ANSWER If the electric field of magnitude E is oriented 30° to the optic axis, then the component along the optic axis will be $E\cos(30°) = (0.866)E$. Only this component, the dashed component in the figure at right, will be transmitted through the polarizer. The energy density in the initial beam is $\rho_{EM} = \varepsilon_0 E^2$ and the energy density after the light passes through the polarizer is $\rho'_{EM} = \varepsilon_0[E\cos(30°)]^2 = \varepsilon_0(0.866)^2E^2 = \varepsilon_0(0.75)E^2$. Thus, the ratio between the energy carried in the initial beam and that in the beam passed through the polarizer is 0.75. If the initial power was $\rho'_{EM} = 60$ mW, the transmitted power will be $\rho'_{EM} = (0.75)(60\text{ mW})$, or $\rho'_{EM} = 45$ mW.

One of the classic applications of polarizers is to use two polarizers whose optic axes are at an angle to each other to control the intensity of the light that passes through the combination. The first polarizer determines the polarization of the light through it and the second selects the amount of that light that passes through the combination. When the second polarizer axis is in the same direction as the first, all of the light that comes through the first polarizer passes essentially undiminished through the second. But when the axis of the second polarizer is oriented so that it is at 90° to the optic axis of the first, then none of the light passed through the first polarizer will make it through the second.

EXAMPLE 23.10 The Amazing Intermediate Polarizer

Two polarizers are aligned with their optic axes at 90° to each other, so that no light passes through the combination. A third polarizer with its optic axis at 45° to the first polarizer is placed between the other two. How much light is now passed through the 3-polarizer combination?

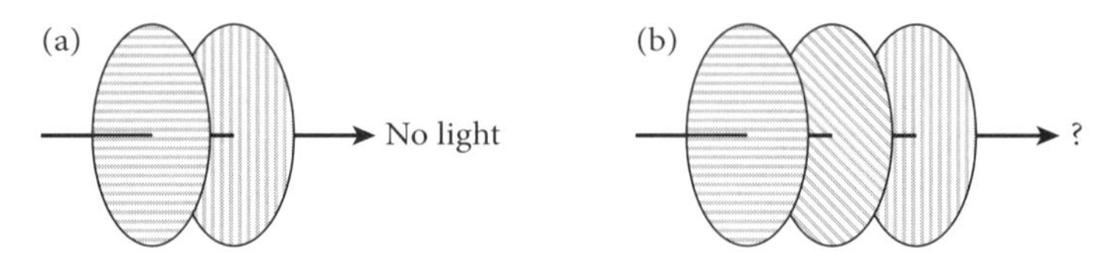

ANSWER As discussed in the last paragraph, the situation in picture (a) produces no light. The first polarizer produces light that is polarized up and down (remember that the optic axis is perpendicular to the directions of the chains of molecules), and the second polarizer will not pass any light that has this polarization.

In picture (b), the first polarizer again produces light that is polarized up and down. But, as this light strikes the intermediate polarizer, it will have a component along the optic axis and a component perpendicular to the axis, as shown at right. The component along the optic axis will be $E_1 \cos 45°$. This component, the one labeled in the top figure at right, will be completely transmitted through the intermediate polarizer and will, of course, point up at 45° to the vertical direction.

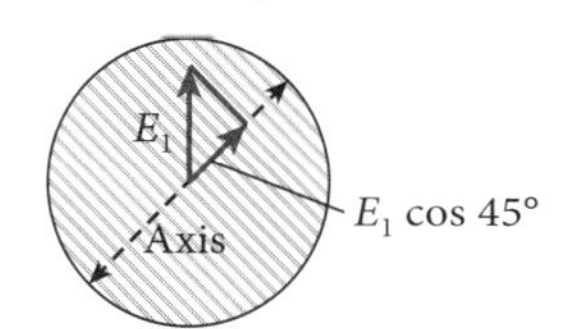

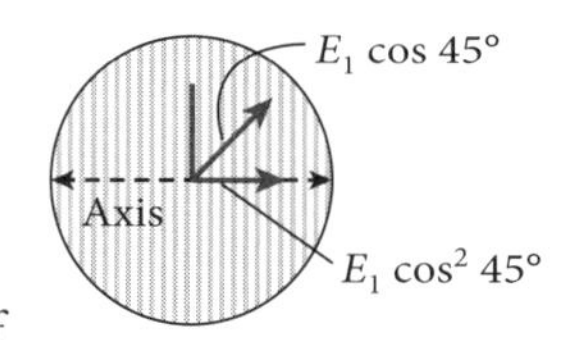

Now as this transmitted component hits the last polarizer, whose polarization axis is horizontal, the polarization vector will again have a component along the optic axis and this horizontal vector of magnitude $E_1 \cos^2 45°$ will be passed. The intensity of the light passed will be $E_1^{\,2} \cos^4 45°$, or ¼ the intensity of the light that passed through the first polarizer.

While this all should be clear enough when you go through the steps, there is something very surprising about the result. When you have just the two polarizers, no light gets through the crossed-axis pair. Then, if you *add* an additional element in between the two, some light now gets through. What's more, the 45° axis polarizer succeeds in taking the entirely up-and-down polarized light that passed through the first polarizer and rotates it into light that is polarized left-and-right. Isn't that amazing?

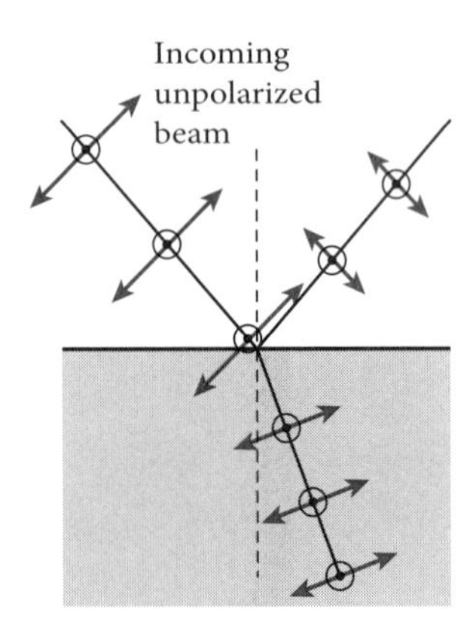

Figure 23.21 A beam is polarized by reflection.

Our last topic we want to mention in this section is what happens to the polarization of a beam of light when the beam is reflected from the surface of a medium. Figure 23.21 shows an unpolarized beam of light striking such a boundary. The two orthogonal polarizations are depicted in Figure 23.21 as double-ended red arrows in the plane of the page and as black tips of double arrows going into and coming out of the page. The key to understanding polarization by reflection is to realize that an electric field that lies in the surface of the boundary of the new medium (the black vectors pointing in and out of the page) is more efficiently scattered than an electric field with a component pointing partially into or out of the material (like the red arrows). As a result, light reflected from a surface tends to be polarized, with its electric field pointing horizontally, parallel to the surface. As this particular polarization is reflected out of the incoming beam, the transmitted beam will be preferentially polarized in the direction of the red arrows, while the reflected beam will be strongly polarized in the direction of the black arrow tips in and out of the page.

One of the practical results of this polarization is that the reflected glare from a surface like water or the shiny paint of the car in front of you will be more or less polarized and will have its electric field vector oriented preferentially back and forth in a horizontal plane. This is how Polaroid sunglasses work. They have a polarization axis oriented up and down, to block reflected glare, while allowing non-reflected light to pass more easily through the lens.

Finally, let us point out that there is an angle, known as Brewster's angle, that is defined as the incident angle which produces a completely polarized reflected beam (see Figure 23.22). We discuss additional aspects of this angle and ask you to derive a formula for it in Problem 48.

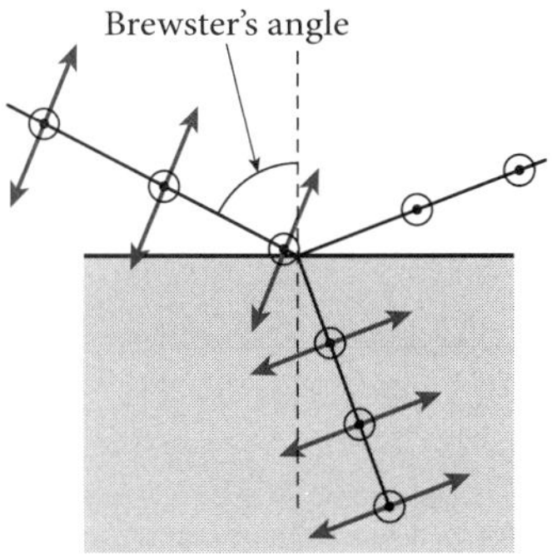

Brewster's angle

Figure 23.22 Conditions at Brewster's angle.

23.7 Summary

Many of the formulas in this chapter's material look alike. We strongly suggest that you review the chapter to see how each equation came about and how it is used. To help you, we have put together a somewhat more detailed formula box for this chapter. Anyway, here is the list of things you want to make sure you are clear on.

- Maxima of a 2-slit interference pattern occur at angles where there are an integer number of wavelengths that fit into the path delay of the light from one slit relative to the other. Zeros of the pattern occur when an odd integer number of half wavelengths fit into this same distance.
- Maxima of a grating with slit separation d occur at exactly the same angles as in the 2-slit pattern. Between these maxima there are many zeros, separated by tiny local maxima. The first zero on each side of a maximum is found at an angle offset of $\Delta\theta \approx \lambda/(Md)$, as discussed at the end of Example 23.4. This angle is half the width of each narrow bright fringe.
- Light from a single slit is brightest at $\theta = 0$ and then declines in brightness as you move away toward either side. The first zero on each side of the central bright region is at an angle where $\sin\theta_1 = \lambda/w$ for a slit of width w. This means that the spread of the central maximum is twice as wide as this. Outside the central bright region there are a few smaller fringes of light and dark.
- The image resolution of a telescope or other optical instrument is defined by the Rayleigh Criterion in which the center of the diffraction pattern from one point lies on the first zero of the pattern from the other point. The defined angular resolution is a small angle $\theta \approx \lambda/w$ for rectangular apertures with a narrowest dimension w and $\theta \approx 1.22(\lambda/D)$ for circular apertures of diameter D.
- There is no single formula explaining what wavelengths produce bright or dark interference from a thin film, due to the possibility of phase changes of reflected beams from a surface. The best thing is to determine the phase change due to the time delay (the first term in Equation 23.9) and then think through the possible phase changes due to reflections to determine what relative phase will finally be produced.
- A wave approaching a polarizer should have its polarization vector resolved into components along, and perpendicular to, the polarization axis. Only the component *along* the axis will be transmitted through the polarizer.

CHAPTER FORMULAS

Double slit: Maxima: $\sin\theta_n = n\dfrac{\lambda}{d}$ $n = 0, 1, 2, 3, \ldots$

Zeros: $\sin\theta_n = \dfrac{n\lambda}{2d}$ $n = 1, 3, 5, \ldots$

Grating: Maxima: $\sin\theta_n = n\dfrac{\lambda}{d}$ $n = 0, 1, 2, 3, \ldots$

First zero on either side of the maximum:

$\Delta\theta \approx \pm\dfrac{\lambda}{Md}$ for M slits illuminated

Single slit: Zeros on either side of the central maximum: $\sin\theta_1 = \dfrac{\lambda}{w}$

Other zeros: $\sin\theta_n = n\dfrac{\lambda}{w}$

Resolution: Circular: $\theta \approx 1.22\dfrac{\lambda}{D}$ Rectangular: $\theta \approx \dfrac{\lambda}{w}$

Phase delay between a beam that bounces off the front surface of a thin film and the beam that bounces off the second surface of the film: $\phi = \dfrac{2nt}{\lambda} + \begin{Bmatrix} 0 \\ 1/2 \end{Bmatrix}$

PROBLEMS

23.1 * 1. Two narrow slits, 6 μm apart, are illuminated by green light of unknown frequency. It is found that the 1st-order fringes are located at exactly 6° on either side of the central maximum. What is the wavelength of the light?

* 2. A two-slit interference pattern with maxima spaced every 0.064° is seen when the wavelength of the light is 450 nm. What is the separation between the slits?

* 3. When two narrow slits separated by 40 μm are illuminated by light of a single wavelength, there are bright fringes seen every 1.5 cm near the central area of a screen that is 2 m from the double slit. What is the wavelength of the light?

** 4. Coherent light of wavelength 600 nm illuminates two slits that are 3 μm apart. If we observe at an angle of 30° to the incoming direction, will we see a maximum, a zero, or something in between?

** 5. Light of wavelength 640 nm falls on two slits that are separated by exactly 0.32 mm.

a) What is the difference in angle between the locations of the first and the second fringes?

b) What is the difference in angle between locations of the seventh and eighth fringes?

c) What is the percent difference between these two numbers?

[Hint: You may have to carry your calculations out to 4 or 5 decimal places to see the difference between the two differences.]

** 6. Coherent light of wavelength 500 nm strikes two slits that are separated by 0.25 mm. If we look at the light arriving at an angle 0.5°, what will be the phase difference in cycles between the light from the two slits?

** 7. Coherent light of wavelength 600 nm falls on two slits separated by 0.30 mm. At what angle will the phase difference between the two beams be ¼ of a cycle?

** 8. One side of a block of glass with index of refraction $n = 1.5$ is painted black and two thin slits are scratched in the paint a distance 50 μm apart. When 500-nm light illuminates the two slits, at what angle in the glass will the first-order bright fringe be found? How did the glass change this result from the result in air?

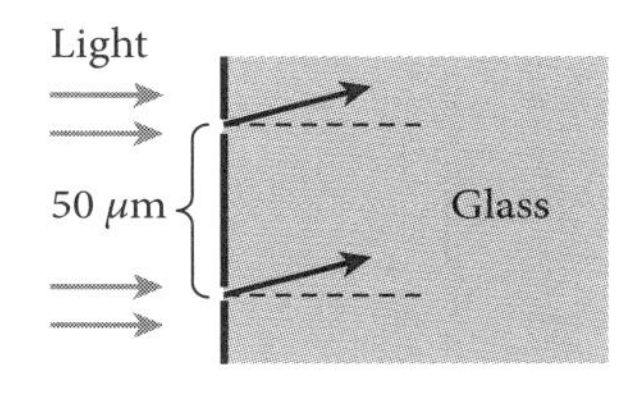

** 9. The top slit of a double-slit apparatus is filled with a 1.00-μm-thick wafer of glass of index of refraction $n = 1.50$. The distance between the slits is 10.0 μm. At what angle θ will the central maximum for all wavelengths of light be found?

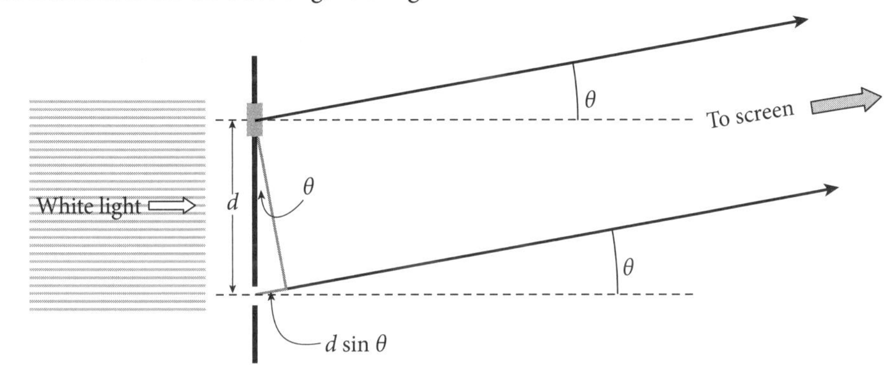

*** 10. In very-long-baseline interferometry (VLBI) tracking of interplanetary spacecraft, a 8.4-GHz radio signal is sent from a spacecraft toward the earth. In order to determine the angular position of the spacecraft, the signal is received simultaneously at two Deep Space Network (DSN) stations and the phase of the received signal at each station is measured relative to a stable clock. The distance between the antenna in Australia and the one in California is 10,588,966.336 m.

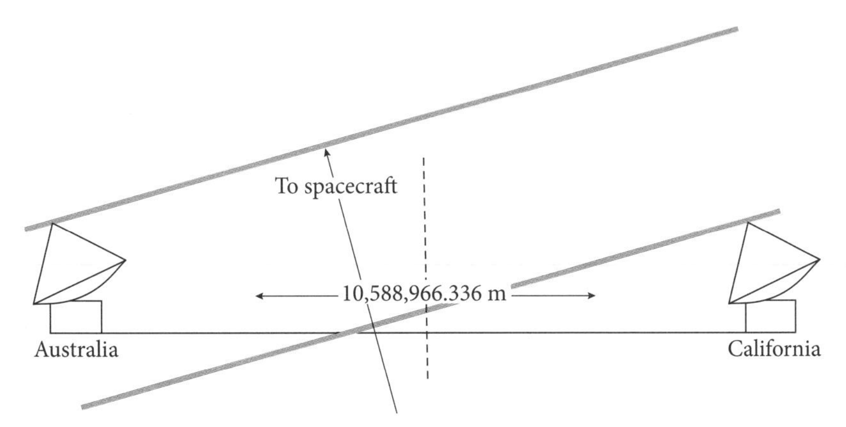

a) What is the wavelength of the X-band radio signal?

b) If the phase of the signal in the Australia antenna is found to be 10,458.6 cycles ahead of the signal in the California antenna, what must be the angle between the direction to the spacecraft and the normal to the baseline (the dashed line in the figure above)? [Check: The answer here is 0.002°.]

c) If the relative phase of the two signals in the antennas can be measured to 0.1 cycles, what is the angular accuracy of this measurement technique? [Check: The angular position accuracy of this technique is 20 nanodegrees.]

23.2 * 11. Light of wavelength 600 nm strikes a grating with 1000 lines per cm. At what angle will the first bright fringe be observed?

** 12. Light of wavelength 500 nm illuminates 10 lines of a grating with a line spacing of 10 μm. What is the width of the central maximum?

** 13. A beam of light of wavelength 400 nm is wide enough to illuminate 10 lines of a grating spectrometer. If the beam is spread so that it illuminates 20 lines, what will happen to the interference pattern after the light passes through the grating?

** 14. Light from a star is spread to cover 1000 slits of a grating whose distance from slit to slit is 2 μm. Over what range of angles will the visible first-order (n = 1) spectrum from 400 nm to 750 nm be spread?

** 15. When a grating with 4000 lines per cm is illuminated by white light, it is found that red light of wavelength 680 nm appears at exactly the same place in the interference pattern as green light of wavelength 510 nm. What are the orders of the 680-nm pattern and of the 510-nm pattern that produce this phenomenon?

** 16[f]. In Section 23.3, we found that the width of the central maximum for a 6-slit pattern was limited by the first zero of the pattern where $\sin\theta = \lambda/6d$, and we stated without proof that the higher-order maxima would have the same width. Let us prove that this is true. (Assume that all angles are small, so that $\sin\theta \approx \theta$.)

a) The location of the first bright fringe for grating is given by Equation 23.3, with n = 1. At this angle, what will be the difference between the phase of the wave from slit 1 and that from slit number 4, as shown in Figure 23.6?

b) If, at a slightly larger angle than this, there is total cancellation from all slits in the grating, what will be the difference in phase between the light from slit 1 and that from slit 4 of the 6-slit grating?

c) Find the angle at which this cancellation occurs and show that the resulting angular width of the first bright fringe is the same as that of the central fringe.

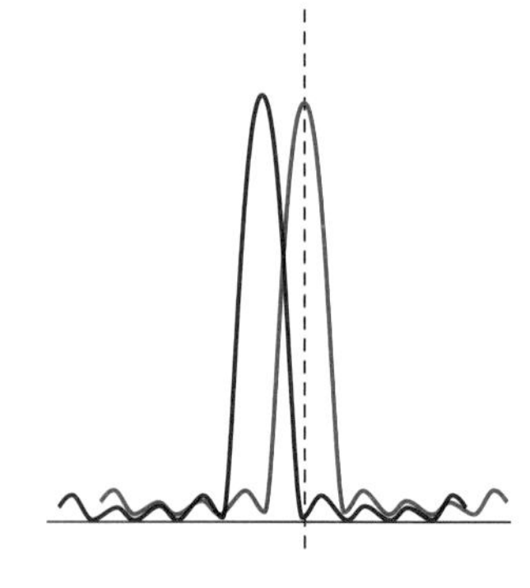

*** 17. **Wavelength Resolution of a Grating.** The wavelength resolution of a grating is defined to be the wavelength difference in two wavelengths of light where the maximum of the one wavelength lies on the first minimum on either side of the other wavelength, as shown. This is the Rayleigh Criterion applied to wavelength resolution in a spectrograph instead of images in a telescope. A particular grating whose distance between slits is 2 μm has 1000 slits illuminated. Find the wavelength resolution $\Delta\lambda$ of the grating for light near 500-nm wavelength. [Hint: Find the angle where the first maximum of 500-nm light will appear and then find the wavelength whose maximum will be at the first zero of the 500-nm spectrum beyond this. The wavelength difference is the resolution. You will have to carry your angle calculations out to several decimals to see this wavelength difference.]

*** 18. There are two wavelengths of sodium, called the "sodium doublet," that are very close together and very pronounced in M-type stars. Their wavelengths are 589.0 nm and 589.8 nm. The spectrum of an M-type star is seen through a diffraction grating with 1000 lines/mm.

a) At what two angles are the first-order fringes of these two wavelengths seen?

b) These two wavelengths are to be resolved (see Problem 17 for the definition of wavelength resolution). Find the number of lines of the grating that will have to be illuminated to resolve the two wavelengths.

23.3 * 19. When light of wavelength 440 nm passes through a single slit, the width of the central maximum (the angle between the two zeros on each side of it) is found to be 73.74°. What is the width of the slit?

* 20. Light of wavelength 600 nm passes through a single slit of width 12 μm and hits a screen 80 cm behind the slit.

a) At what angles will the first three zeros of the pattern be found?

b) What will be the width of the central maximum in cm?

* 21. Light of wavelength 440 nm passes through a single slit of width 80 μm and forms a central maximum that is 4 cm wide.

a) How wide would the central maximum be if the slit were narrowed to 40 μm?

b) How wide would the central maximum be if the slit were widened to 160 μm?

** 22. Equation 23.6 determines the angular width of the central maximum in a single-slit diffraction pattern. But the angle in the formula depends on the wavelength of the light passing through the slit. This means that the central maximum will be wider for some wavelengths than for others, leading to a faint coloring of the edges of the central maximum. Consider a single slit of width 10 mm. What will be the difference between the width of 400-nm light and 750-nm light, and what color would you expect to see on the outside of the central maximum?

** 23. Finally, a practical problem! An annoying high school oboist in a straw hat stands in the apartment across the alley and pokes his oboe out his window to practice playing his concert A (440 Hz). The speed of sound is 1100 ft/s. The window in your apartment is 5 ft wide, and your room is covered in acoustic tile, so no sound bounces off the walls. At what angle to the incoming waves should you place your armchair so that you do not hear the annoying oboe? [Note: It is not really bad manners for us to insult the oboe because Ron is an oboist.] [Hint: This problem involves sound waves, not light waves, but the mathematics of sound waves passing through a slit is the same as that for light waves. You may use the wave results of Section 23.3. And, yes, there really will be a place where you can sit and hear almost no sound.]

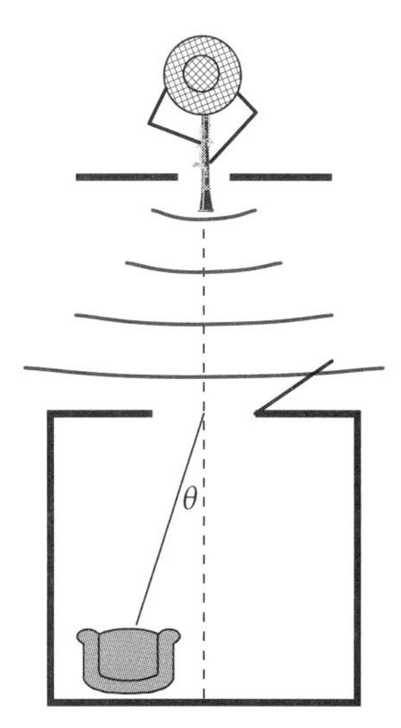

** 24. The central maximum of a single slit pattern is much brighter than the small higher-order maxima further out (see Figure 23.12). As a result, many higher-order fringes from a 2-slit pattern will be too dim to see if they lie outside the central maximum for light spreading out from each of the slits. If 500-nm light passes through two slits, each of width 2 μm, and if the separation between the 2 slits is 24 μm, how many bright fringes of the two-slit pattern will be seen within the central maximum of the light from each slit? [Hint: Remember that there are fringes on *both* sides of the central bright fringe, and don't forget to count the central bright fringe.]

** 25. Light of wavelength 400 nm hits a diffraction grating that has 100 slits, each of width 16 μm, separated from each other by distances of 48 μm.

a) At what angle will the third-order maximum of the grating be located?

b) At what angle will the first zero of the single-slit pattern be found?

c) Will there be a bright fringe at the angle found in part (a)? Explain.

*** **26.** Light passing through a grating will produce fringes by interfering constructively at the angles given by Equation 23.3, *unless* there is no light available to interfere constructively at that angle. When light of wavelength 400 nm passes through multiple slits, each 4 μm wide and separated by 10 μm, what are the orders of the fringes (the values of n in Equation 23.3) that will be completely missing from the resulting interference pattern?

23.4 * **27.** What is the angular resolution of Galileo's 1609 telescope, which had a primary lens of focal length 98 cm and diameter 3.7 cm? What is the angular resolution of the Hubble Space Telescope with its primary mirror having a 2.4-meter diameter and a 27-meter focal length?

** **28.** What is the angular resolution of a 64-meter circular radio antenna receiving signals of wavelength 2.3 cm from an interplanetary spacecraft?

** **29.** The primary lenses in two different pair of binoculars have the same 3-cm diameter, but one pair has a magnification of 10X, while the other has a magnification of 20X. Will the higher-power pair have a better resolution or will they both be the same? Explain your answer. If you are looking at small details in the two pair, what will the picture look like through the two instruments?

** **30.** The two rails on a standard railroad track are about 1.7 m apart. If a pilot has a pupil in her eyeball that is 3 mm in diameter, at what altitude above the ground will she no longer be able to tell that there are two rails on the track? Assume that her eyes are best able to resolve with light of wavelength 500 nm.

** **31.** Amateur astronomers know the importance of finding a telescope with a large aperture. A large aperture not only decreases the size of the central maximum into which the light passing into the telescope must spread the light, but it also collects more light with its larger collecting area. How much brighter will be the image of a particular star seen through a 20-cm-radius telescope, compared to the same star seen through a 10-cm-radius telescope? Explain.

** **32.** When a person enters a dark room from a lighted room, the pupil expands almost immediately from an average 2-mm diameter to about 5 mm in diameter. (This is not the much slower million-fold increase in retina sensitivity that occurs over the next 20–30 minutes in a darkened room.) Based on the decrease of diffraction disk size and increased energy collecting through the larger pupil alone, find the increase in brightness of a diffraction spot on the retina due to the increase from 2-mm to 5-mm pupil diameter.

23.5 * **33.** The soapy water in a soap bubble has an index of refraction $n = 1.34$. As the bubble floats through the air, the water forming the bubble will slowly evaporate. Just before the bubble pops, its thickness is nearly zero and its color is black. (Yes, this can actually be observed.) Explain why the color is black.

** **34.** A soap bubble of thickness 239 nm has an index of refraction of $n = 1.33$. Look at the colors of light that are not reflected from the film of the soap bubble and at those that are brightly reflected to explain why the bubble will appear red when illuminated with white light perpendicular to the surface.

** **35.** A mirror is to be designed that will reflect infrared laser light of wavelength 1064 nm. The designer chooses a thin layer of magnesium fluoride to coat a plane slab of glass. The index of refraction of the glass is 1.50 and that of the magnesium fluoride is 1.38. What is the thinnest MgF_2 coating that the designer can apply that will provide high reflection at this wavelength? Explain your answer.

** **36.** A wedge of air is formed by placing a human hair at one edge between two glass plates that are in contact at the other edge. In reflected green light of wavelength 540 nm, a total of 80 bright fringes are observed as one looks directly downward along the plate, from edge to edge. What is the diameter of the hair?

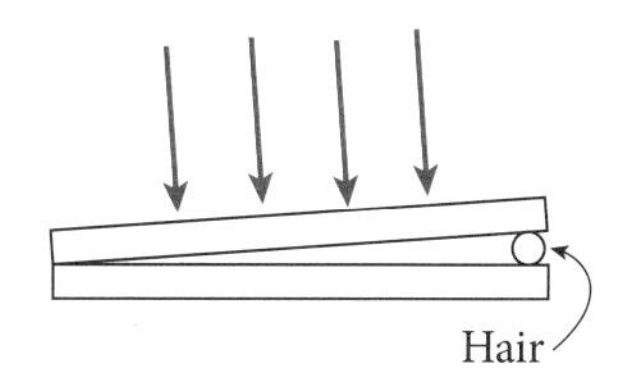

** 37. An oil slick (n = 1.45) covers the surface of the water with a film 750 nm thick. What wavelengths of visible light will be brightly reflected from the oil slick?

** **38.** Two plastic plates (index of refraction n = 1.3) are separated by an air gap of 0.9 μm. Analyze the reflections of light of wavelength 600 nm and conclude whether it will be brightly reflected or not.

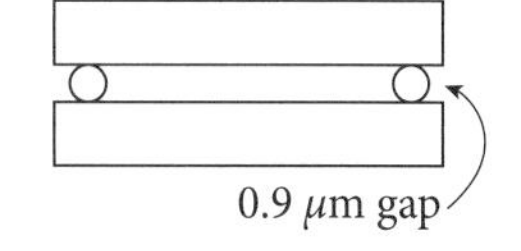

** **39.** The air gap in the previous problem is now filled with a mineral oil whose index of refraction is n = 1.5. Analyze the reflections of light of wavelength 600 nm and conclude if it will be brightly reflected or not.

*** **40.** You have probably been expecting us to ask you this ever since you first saw Figure 23.17. Yes, we want you to do the gory geometry of thin-film interference at angled incidence. So suppose light of wavelength 440 nm is incident at an angle θ = 30° from the normal toward a thin sliver of glass (n = 1.5), as shown in Figure 23.17. The thickness of the glass is 630 nm. Find the number of wavelengths by which the two reflected rays in the diagram will be out of phase when they finally combine far from the glass.

*** **41. Narrow-Band Filters.** Example 23.7 explains how a thin coating can provide high transmission for a particular wavelength by not reflecting that wavelength. However, other frequencies are only partly reflected by the coating and can still pass through the filter. In order to reject other frequencies more efficiently, successive filters are piled on top of each other. One common set of materials used for narrow-band filters is magnesium fluoride (MgF_2) and titanium dioxide (TiO_2), whose indices of refraction are 1.38 for MgF_2 and 2.4 for TiO_2. Example 23.7 finds that a layer of 96 nm of MgF_2 between air and glass provides complete transmission of 532-nm light. Find the minimum thicknesses of TiO_2 layers and of subsequent MgF_2 layers for perfect transmission at this same wavelength.

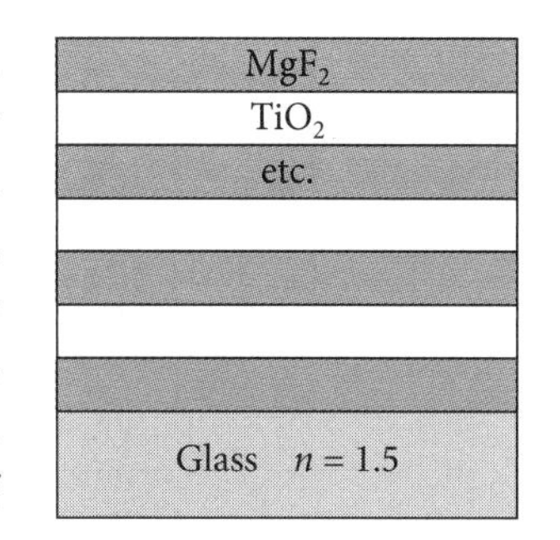

23.6 * **42.** Find the fraction of the electric field and the electrical power transmitted through a polarizer when polarized light with polarization 10° from the optical axis of the polarizer strikes it.

** **43.** Unpolarized light hits a first polarizer with a vertical axis of polarization. The light then passes through another polarizer. If the intensity after the first polarizer is I_0 and that after the second polarizer is $\frac{1}{4}I_0$, what is the angle between the polarization axes of the two polarizers?

** **44.** Two polarizers have their polarization axes offset by 60°, and unpolarized light falls on the two polarizers in sequence. If the intensity of the beam is 100 W/m^2 in the region between the polarizers, what will be the intensity to emerge from the second polarizer?

** 45. Suppose that the three polarizers in Example 23.10 are replaced by four polarizers, in which the second polarizer has its optic axis rotated by 30° relative to the first, the third polarizer is rotated by 60° relative to the first, and the fourth polarizer is rotated by 90°. Find the fraction of the light energy passing the first polarizer that emerges after the last polarizer, and compare with what was found in Example 23.10.

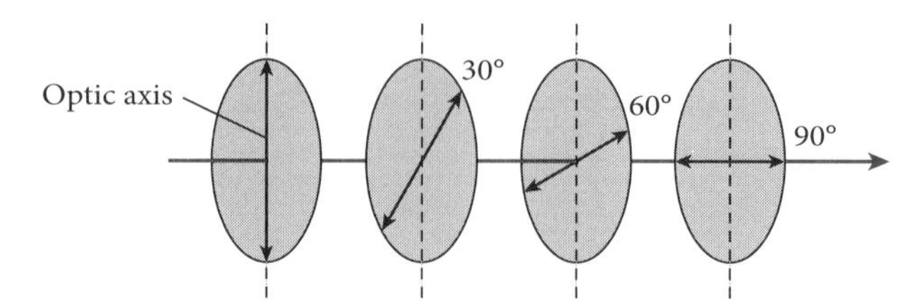

** 46. **How much unpolarized light gets through a polarizer?** Unpolarized light is as likely to have its polarization vector in one direction as another. Therefore, on average, the up-down component of the polarization vector for unpolarized light will be the same as the right-left component. These equal components in two perpendicular directions are in the same relationship as the E-field components when polarized light hits a polarizer with its polarization vector at 45° to the optic axis. We can use this connection to find out what fraction of the energy in an incident unpolarized beam makes it through the polarizer as polarized light.

a) If an electric field vector of a polarized light beam is 45° to the optic axis of the polarizer, find the ratio between the magnitude of the electric field vector that passes through the polarizer and the magnitude of an incident polarized electric field vector. This will be the same as the ratio of the magnitude of the electric field of the polarized light that passes a polarizer to that of the average unpolarized light before it hits the polarizer.

b) Find the ratio of the transmitted energy to the incident energy.

*** 47. Unpolarized light strikes a sequence of three polarizers, as shown below in the picture on the left. The first polarizer has its axis oriented vertically; the axis of the second is at an angle of 30° relative to the first; the third is at an angle of –30° to the vertical. Compare this situation with one in which the first and second polarizers are exchanged, as in the picture at right, below. What will be the relative intensities in the two cases and what will be the polarizations of the final light in each case?

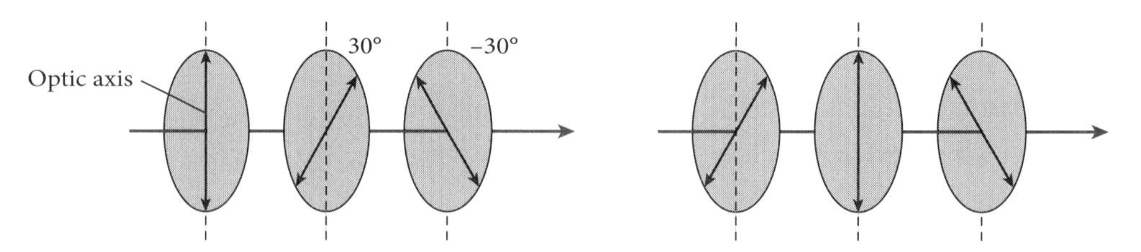

*** 48[f]. The condition for total polarization, at Brewster's angle, for a reflected beam from the interface between two media is that the reflected beam and the refracted beam are perpendicular to each other (see Figure 23.22). Use this information to determine a formula for Brewster's angle for an interface between vacuum and a medium of index of refraction n. [Answer: $\theta_B = \tan^{-1}(n)$.]

24

SPECIAL RELATIVITY

We would like to begin this chapter by saying, "Welcome to 1901." You have now read twenty-three chapters of physics, and everything you have learned was known by 1900 or earlier. It is all what is called "classical physics." We have not yet discussed any of the profound discoveries of the 20th or 21st centuries. Our reason for this is entirely honorable. The things that students in a general physics course are expected to know (i.e., things that show up on standardized tests) are almost all things from classical physics. Notice that we said *almost* all. There *are* a few topics from modern physics that you really should know about, and, frankly, that you should want to know about if you want to consider yourself well-educated in modern science. These are the subjects of our last three chapters.

24.1 The Two Principles of Relativity

Back in Chapter 5, we discussed the concept of inertial frames and explained that Newton's Laws did not work in accelerated reference frames, but that they work fine in any unaccelerated reference frame. In 1905, Albert Einstein incorporated this idea into one of the basic principles underlying his Special Theory of Relativity.

- First Principle: The laws of physics are the same in any inertial reference frame.

Maxwell's Equations specify that the speed of light is always $c = \sqrt{1/(\mu_0\varepsilon_0)}$, as we saw in Chapter 21. So, if the laws of physics include Maxwell's equations, and if they are to be the same in any inertial frame, then the next principle must also be true.

- Second Principle: The measured speed of light is the same in any inertial frame.

These two principles are the foundation of Special Relativity. However, if you think about the second principle, it will probably seem obvious to you that it is wrong.

In fact, let us begin our study of Einstein's theory by "proving" it wrong. Let's watch someone measure the speed of light from a railway "measurement car" moving at $3/5$ the speed of light (180,000 km/s). The car is 300 m long, and it has distances carefully measured out. It is equipped with synchronized clocks spaced out all along its length (take a look at Figure 24.1 on the next page). Each clock has a photodetector that rings a bell and punches out a timecard when it sees a flash of light. In Figure 24.1, we see a 600-m section of track with a light bulb at one end and a mirror at the 300-m mark, angled upward at 45° so that it will reflect light into the photodetectors. As the back of the car passes the bulb, the light flashes, setting off the detector at the zero-position on the train. We assume that all of the train's synchronized clocks were set so that they read $t = 0$ at that moment.

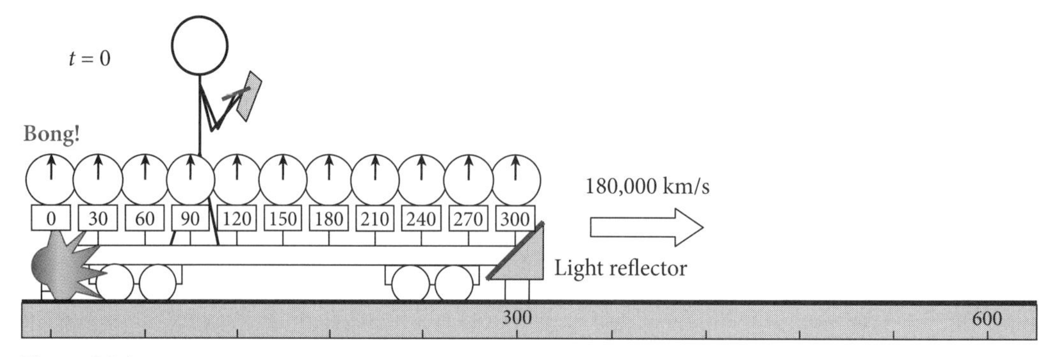

Figure 24.1 A measurement car passing a section of track at $t = 0$, as the red light flashes.

Now what should we see if we were to take a picture 1 μs later? As we show in Figure 24.2, the back of the moving train will be 180 m from the light bulb. The flash, moving at the speed of light, will be 300 m from the light bulb. So the train clock that will detect the flash will be the one at 120 m, as shown in the picture. The way to find the speed of light *relative to the train* is to measure the distance between the first flash and the second, *using rulers attached to the train*, and to divide that distance by the time between the two flashes, *using clocks on the train*. As the guy in Figure 24.2 says, that calculation gives 120,000 km/s for the velocity of light relative to the train. Let's be clear. This is exactly the right way to measure velocity relative to the train. It's what we discussed back in Section 4.6. And it is exactly the measurement that Einstein promised would give 300,000 km/s. But it doesn't, does it? So Einstein must be wrong, right?

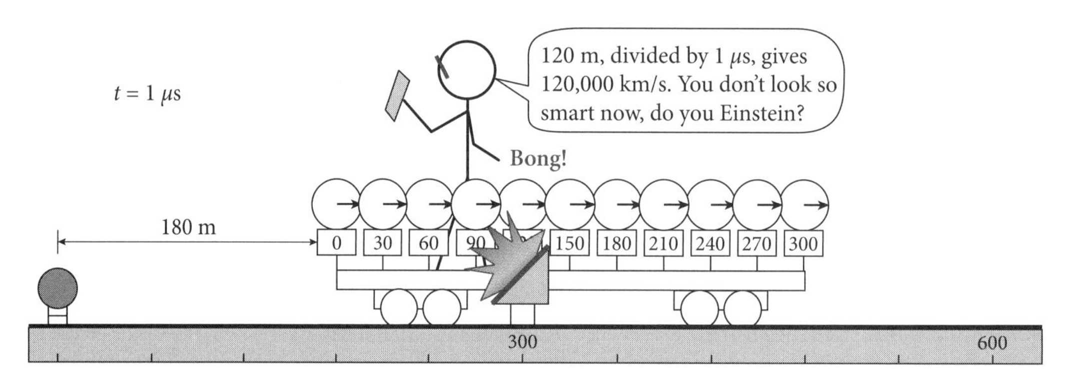

Figure 24.2 The car at $t = 1\ \mu$s, as the red light hits the mirror and reflects it upward.

Well, no. The problem with the calculation in Figure 24.2 is that the picture is badly drawn (and, by the way, so is Figure 24.1). Figure 24.3 shows what the train *really* looks like at $t = 1\ \mu$s. Three things probably look strange about the real picture. First, the train is contracted to a length of 240 m (and the wheels are oval). Second, the clocks on the train are not synchronized, with trailing clocks running ahead. Third, in the one microsecond that elapsed from Figure 24.1 to Figure 24.3, the zero-clock on the train has ticked off only 0.8 microseconds, because the moving clocks are all running slow. Put the three things together, and we see that the clock that records the second flash will be the 150-m clock, and the time on that clock will be 0.5 μs, as shown in Figure 24.3. In this chapter, we will see how these things come to be.

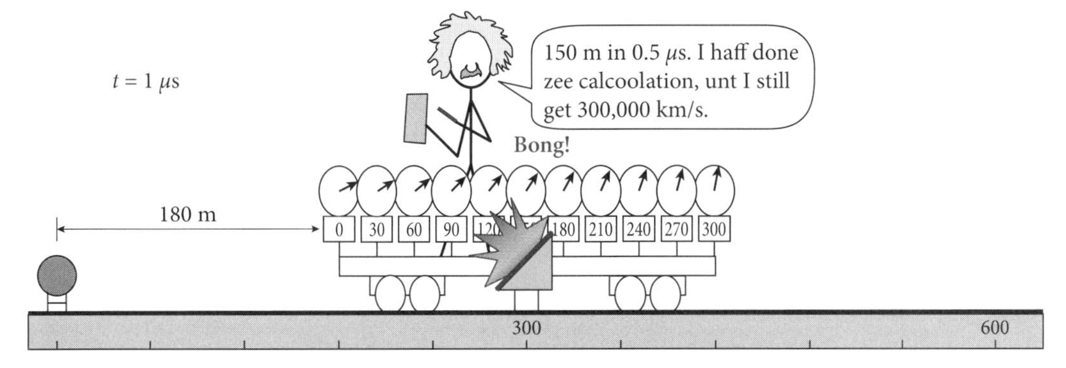

Figure 24.3 The true picture of a moving train according to Special Relativity.

But we want you to understand. The Newtonian ideas that produced Figures 24.1 and 24.2 are wrong. If you were ever to observe a "measurement car" moving at 3/5 the speed of light, it would never look like Figure 24.2. It would look like Figure 24.3. If you build a car that is 300 m long when it is sitting on the tracks, then every bona fide method you can devise to measure the length of the car when it is moving at $3/5\,c$ will tell you it is 240 m long. If a clock that you put on the train is identical to a clock in your pocket, every timing measurement you can make on the moving clock will show that it is running slow. And the best method you can suggest for synchronizing clocks in the moving frame will fail to produce synchronized clocks when observed from the ground. These effects are really observed in the laboratory and, especially, in space. And, even if you will never actually see a 300-m-long railway car moving at 3/5 the speed of light, a normal-length train car moving at a normal 60 mi/hr will never look *exactly* like Figure 24.2; it will always look a little bit like Figure 24.3.

As we discussed this hypothetical experiment, we loosely used the words "look," "see," or "observe." We need to be careful. When objects are moving close to the speed of light, the time it takes the light to go from the object to the observer is not negligible and will affect what is actually "seen." However, when we use these words from now on, we will assume that you are watching from a point perpendicular to the motion and far away, so that all distances that light must travel to you are the same. Since all signals will have the same delay, there will be no distortion of the images. The measurement car will really look just like the picture in Figure 24.3.

Now are these effects real or are they just optical illusions? Has the measurement car *really* shrunk in the direction of motion, or does it only appear that way? Well, how would you measure the length of a 300-meter measurement car moving at 3/5 the speed of light anyway? You might set out an array of synchronized clocks along the tracks and then measure the distance between a clock at the front of a car and one that is simultaneously at the back of the car. If you do that, the process will yield 240 m for the length of the car. If you were to take a picture of the train from a point perpendicular to the tracks and far away, the length of the car as seen in the photo would be 240 m. It would block out exactly 240 m of the buildings behind it in your picture. If you saw the speeding train from right alongside the tracks, and then made corrections for the light times from the front and the back of the car to you, you would calculate the length to be 240 m. If you measured the speed of the train and then timed how long it took the measurement car to pass you, you would get 240 m for its length. So let us say it this way. If the shrinking of a moving car in the direction of motion is an optical illusion, then that illusion is identically reproduced in any length measurement you can possibly make. There is no better definition of reality than that.

In the next section, we will use Einstein's second principle to prove that all three of the effects seen in Figure 24.3 must actually occur, and we will find the formulas that predict how large the effects will be. But the point of the little exercise we have done in this section, trying to disprove Einstein's second principle, is to show that the second principle is simply impossible if you insist that Figure 24.2 is a correct picture. The *only* way that the flash of light can have the same speed when that speed is measured relative to two different reference frames is if all three things shown in Figure 24.3 actually happen to moving rulers and clocks.

24.2 The Three Effects

Let us begin by explaining why moving clocks must run slow. First, we need to explain that this prediction applies to all clocks—your wrist watch, your heartbeat, the number of thoughts you can think, etc. All must slow by exactly the same amount when they are moving. Otherwise, two clocks that ticked the same in one frame would tick differently from each other in another frame, meaning that physics would not be the same in all frames, in contradiction to Einstein's First Principle.

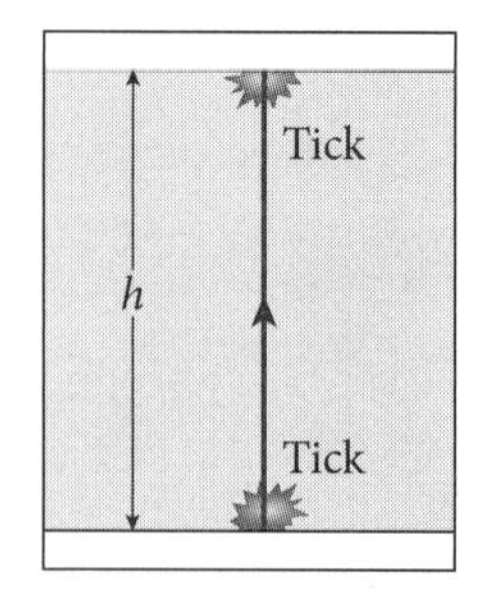

Figure 24.4 A clock based on light travel.

So let us invent a clock that Einstein guaranteed will work in any reference frame. In Figure 24.4, we have constructed a clock based on light travel time between two mirrors. Each time the light beam bounces off a mirror, the clock ticks. Since light always travels at the same speed in an inertial reference frame, the time between ticks of this clock will always be $\Delta t = h/c$, where h is the distance between the mirrors. If we make h = 30 cm, then the clock is guaranteed by Einstein's second principle to tick exactly once every nanosecond, since the time between ticks is $\Delta t = (0.3 \text{ m})/(3 \times 10^8 \text{ m/s}) = 10^{-9}$ s.

Before going on to analyze this clock when it is moving, it is important to prove one preliminary thing. If the clock moves perpendicular to the up-and-down direction of the light path, then that motion cannot affect the 30-cm height of the clock. We will prove this by first assuming the opposite—by supposing, say, that a moving height *shrinks* in a direction perpendicular to the motion. Let us then imagine attaching a paintbrush to the top of a moving stick one meter tall. If its motion makes the stick shrink, then, as it passes our meter stick, it will make a red stripe below the top of our stick, as shown in the top picture of Figure 24.5. But, from the reference frame of the brush, it is our meter stick that is moving and must have shrunk. When the meter stick passes the brush, it will be too short to even touch it, so there will be no stripe on the meter stick, as in the bottom picture of Figure 24.5. After the experiment is over, we can stop all motion and look for a stripe. It would be ridiculous to suggest that one observer will see a stripe and that the other will not. There are no logical contradictions in a correct theory. Therefore, the theory that a moving vertical stick shrinks in the direction perpendicular to its motion must be false. You can also think through for yourself the premise that a moving vertical stick *expands* in a direction perpendicular to its motion and see that this is also not allowed. Thus, in directions perpendicular to the motion, all lengths must be unaffected by the motion.

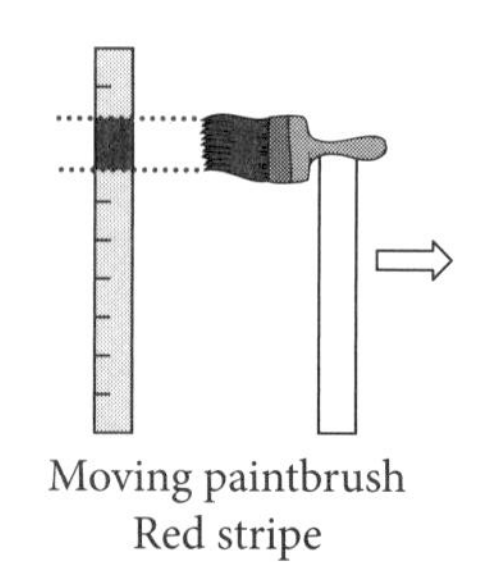

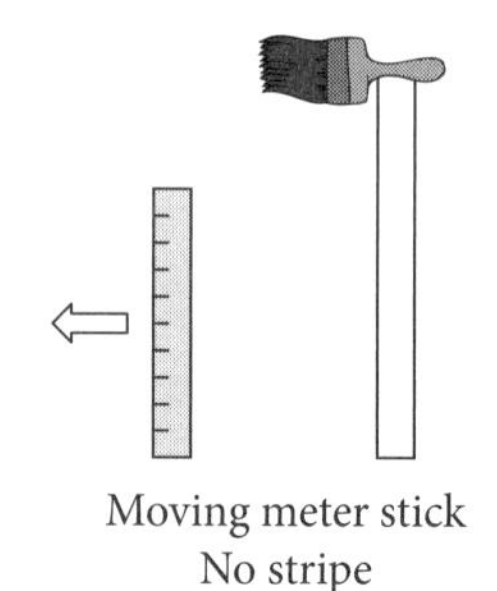

Figure 24.5 A moving height cannot shrink without creating a contradiction.

So let's go back to our clock. Think about it for a moment. If you build this clock in your laboratory, the light will go straight up a distance that all observers will agree is 30 cm, as in Figure 24.4. You can even insert a little apparatus that tests the speed of the light along its path, and your measurement is guaranteed by Einstein's Second Principle to give 3×10^8 m/s. If you compare this clock with a cesium atomic clock in your laboratory, you will see that your light-time clock ticks once every nanosecond. Whether your laboratory is on the earth orbiting the sun or in a space station orbiting the earth, this clock must work in your reference frame, because the 30 centimeters is 30 centimeters and the speed of light is the speed of light.

But now we will look at the clock differently. We will place one of these clocks in a laboratory that is moving horizontally as the light moves vertically along the 30-cm height, and we will compare it to a clock in our fixed laboratory. How long will it take the moving clock to tick? Figure 24.6 shows the light path. Due to the clock's motion, the distance the light will travel will be greater than h. The speed of the light is guaranteed by the second principle to be the same $c = 3 \times 10^8$ m/s in our frame as it is in the frame of Figure 24.4, so it will take the clock longer than $\Delta t = h/c$ to tick. We must conclude that the moving clock is running slow.[1]

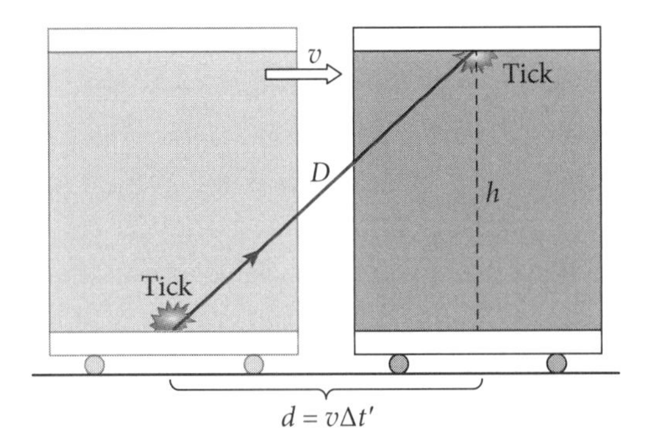

Figure 24.6 The light-travel clock when it is moving relative to the observer.

We can even figure out exactly how slow it will run. If the second tick occurs after an elapsed time $\Delta t'$, as measured by us, then our goal is to find what $\Delta t'$ corresponds to the moving clock's Δt. Figure 24.6 shows that the distance traveled by the light is

$$D = \sqrt{h^2 + d^2} = \sqrt{h^2 + (v\Delta t')^2}.$$

This distance will be covered at speed c in time $\Delta t'$, so we have

$$c\Delta t' = \sqrt{h^2 + (v\Delta t')^2}.$$

If we square both sides and then collect terms containing $\Delta t'$, we get

$$c^2\Delta t'^2 = h^2 + (v\Delta t')^2 \quad \text{and} \quad h^2 = (c^2 - v^2)\Delta t'^2.$$

We remember that the height is related to Δt by $h = c\Delta t$, so we can also write this as

$$c^2\Delta t^2 = (c^2 - v^2)\Delta t'^2 \quad \text{or} \quad \Delta t^2 = \left(1 - \frac{v^2}{c^2}\right)\Delta t'^2.$$

Taking the square root of each side of the equation and solving for $\Delta t'$ gives

$$\Delta t' = \frac{\Delta t}{\sqrt{1 - v^2/c^2}}. \tag{24.1}$$

As we will see in Section 24.5, no physical object can move faster than the speed of light, so the speed of a reference frame tied to any physical object cannot be greater than the speed of light. Since we will always have $v < c$, the quantity inside the square root sign in Equation 24.1 will always be positive, meaning that the square root will always be real and less than one. If we divide something by a quantity that is less than one, we will get an answer that is bigger than what we started with. If our clock was designed to have $\Delta t = 1$ ns, then $\Delta t'$ will be something more than 1 ns. We predict that the moving clock will take longer than 1 ns to tick. It is running slow.

1. This slowing of moving clocks is often referred to as "time dilation," but we don't really like this term since it is not very clear what "dilation" is supposed to mean.

Using the square root factor in the denominator of Equation 24.1 is a process that we will want to become familiar with. So let's pause and work an example.

EXAMPLE 24.1 Annie, the Aging Astronaut

Annie the Astronaut flies at 240,000 km/s toward a star 20 light years away. How long does it take her to make the trip as measured by mission control, and how old will she be when she arrives?

ANSWER First, let's be clear that the only non-intuitive predictions of relativity arise when we compare measurements of events in different reference frames. If we are working in a single frame, it is still true (from the *definition* of velocity) that $t = d/v$. A spaceship going 20 light years at $^4\!/_5$ the speed of light takes $(20\ c\cdot\text{yrs})/(^4\!/_5\, c) = 25$ yrs to make the trip. (Note that a light year can be written $c\cdot$yr, the product of c and a year, and that c can then be treated like a cancelable unit in calculations.)

But how much will Annie age during those 25 years? Her aging process is a kind of clock, and her clocks will run slow, so the answer must be less than 25 years. Let's see what the square root factor gives so we can see how much less she ages:

$$\sqrt{1-\frac{v^2}{c^2}} = \sqrt{1-\frac{(^4\!/_5 c)^2}{c^2}} = \sqrt{1-\left(\frac{4}{5}\right)^2} = \sqrt{1-\frac{16}{25}} = \sqrt{\frac{9}{25}} = \frac{3}{5}.$$

Please make sure you can follow each step in the calculation above. Since the quantity $(v/c)^2$ is always positive, we always take something positive away from one, so the square root will always be less than one, as it was here.

Now we strongly suggest that you do *not* do this kind of problem by looking at Equation 24.1 and trying to figure out which clock is Δt and which is $\Delta t'$. In fact, we make no effort in our chapter to be consistent as to which reference frame we choose to call primed. Rather, you should just go calculate the square-root factor first, as we have done, and then keep your wits about you, knowing whether you expect the calculation to give you a number that is bigger or smaller. If you need a bigger number, you divide by the factor; if you want a smaller number, you multiply by it. We know that the trip takes 25 years. We know that Annie's clocks are running slow, so the answer we need for how much time elapses on her spaceship must be less than 25 years. This means that we will have to *multiply* by $^3\!/_5$, giving $(^3\!/_5)(25\text{ yrs}) = 15$ yrs. Annie will be 15 years older when she arrives at the star than she was when she left earth.

Now we turn our attention to the shortening of lengths in the direction of motion. In Figure 24.7, a clock is moving to the left at speed v toward two flags planted in the ground a distance L apart. Someone on the ground would obviously measure a time $\Delta t = L/v$ for the clock to cover the distance between the two flags. However, the moving clock will be running slow. So, if it reads zero as it passes the red flag, it must read $\Delta t' = \Delta t\sqrt{1 - v^2/c^2}$ when it passes the checkered flag. (Note that we *multiplied* by the square-root factor, since we knew $\Delta t'$ would be less than Δt.) But physics must also make sense in the reference frame of the clock. The clock

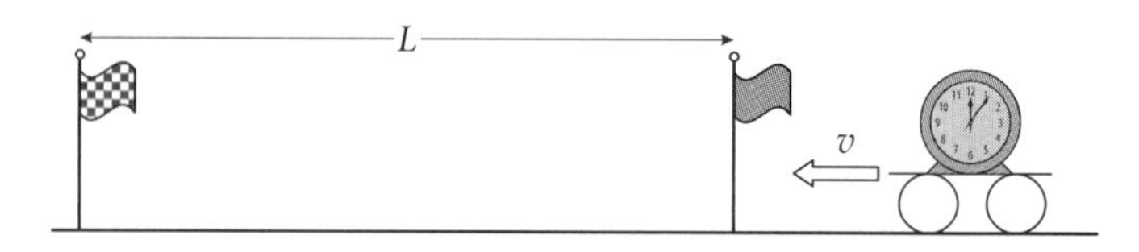

Figure 24.7 A moving clock passing two fixed flags.

sees two flags coming toward it at a speed v.[2] The only way that an observer riding with the clock could measure a time as short as $\Delta t'$ between the time when the red flag passes it and the time when the checkered flag passes it is if the distance between the two flags is shorter than L. In fact, this length would have to be

$$L' = v\Delta t' = v\Delta t\sqrt{1 - \frac{v^2}{c^2}} = L\sqrt{1 - \frac{v^2}{c^2}}. \tag{24.2}$$

Whenever you see a moving length, it will contract in the direction of its motion.[3] This formula uses the same square root factor, and, again, we encourage you to figure out the square root factor first and then, knowing that moving lengths contract, decide whether to multiply or divide by it, depending on whether you want to get a larger or smaller result from your calculation.

Before leaving the subject of moving lengths, let's point out that a "length" is the distance between two points that remain fixed in their own reference frame, like the two flags in the ground in Figure 24.7. It does not apply to the distance between two non-simultaneous events. If a boy throws a baseball to his sister in a moving train, then the distance between them when the ball is caught is indeed given by Equation 24.2. But this is not the distance traveled by the ball relative to the earth, since the boy has moved relative to the earth by the time his sister catches the ball. Thus, a clock measures the interval between two events that occur at different times at the same place, while a length is the distance between two events that occur in different places at the same time.

EXAMPLE 24.2 The Length of the *Enterprise*

We watch the U.S.S. *Enterprise* passing us at $\frac{3}{5}c$ and measure its length to be 236 m. What is the ship's length as measured in its own frame?

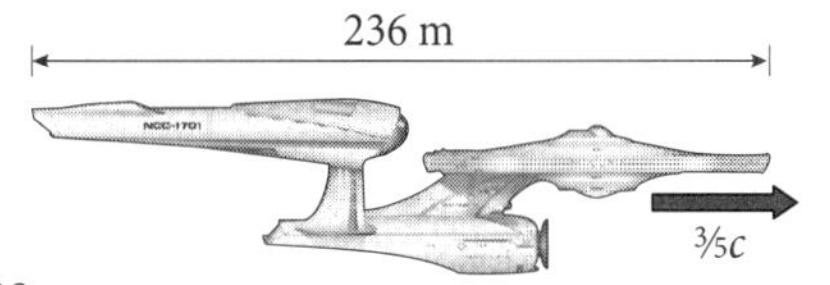

ANSWER *We* measure the length to be 236 m, but its motion means that it will have been contracted as we measured it. In its own reference frame, it must therefore be longer than 236 m. The square-root factor for a speed of $\frac{3}{5}c$ is

$$\sqrt{1 - \frac{v^2}{c^2}} = \sqrt{1 - \frac{(\frac{3}{5}c)^2}{c^2}} = \sqrt{1 - \frac{9}{25}} = \sqrt{\frac{16}{25}} = \frac{4}{5}.$$

Our answer will have to be more than 236 m, as we said, so we divide by $\frac{4}{5}$ to get

$$\frac{236 \text{ m}}{\frac{4}{5}} = 295 \text{ m}.$$

Note how we once again calculate the square root factor first and then decide what to do with it, based on our knowledge that moving lengths are shortened.

Finally, we turn to the question of how to synchronize clocks. One absolutely certain, Einstein-guaranteed way to do this is the method depicted in Figure 24.8. You fix the two

2. The idea of relative velocity means that the speed of the flag relative to the clock is the same as the speed of the clock relative to the flag.

3. This shortening of a moving length in the direction of its motion is often referred to as "Lorentz contraction."

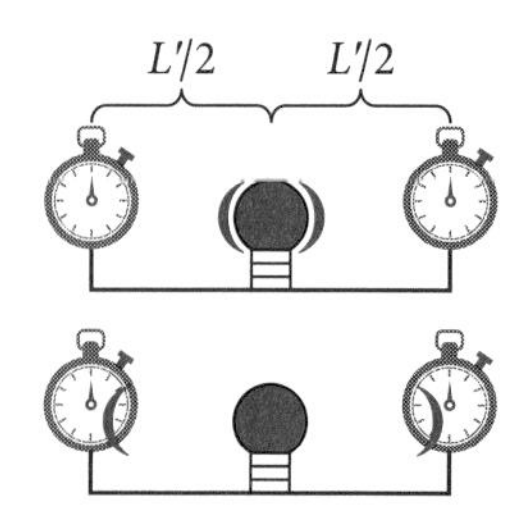

Figure 24.8 How to synchronize two clocks.

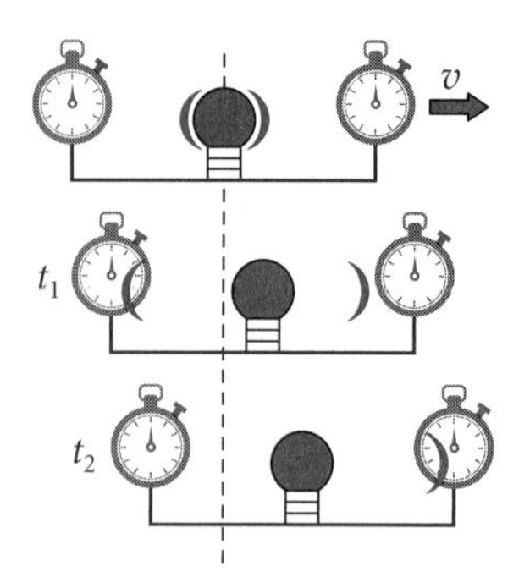

Figure 24.9 The same procedure seen from a frame in which the clocks are moving.

clocks you want to synchronize so that they will start when they detect a light. You then carefully go to a point equidistant between the two clocks and you set off a flash bulb. Both light waves are guaranteed by Einstein to travel at the same speed in both directions, and so both clocks *will* start simultaneously. It's perfect. There's no way this can fail.

Unless you're looking at it from a different reference frame. Let us watch the same synchronization procedure when the clocks are moving relative to us. In Figure 24.9, the light pulses travel in both directions at the same speed of light, but the motion of the leftmost clock toward the light means that it will start first and the clock on the right will start later. The clocks *were* synchronized in their own rest frame. (Look at Figure 24.8 again; it cannot fail.) But they cannot have been synchronized by this method, as seen from the rest frame of Figure 24.9. So, if two reference frames move relative to each other, clocks that are perfectly synchronized in the one frame will not be synchronized in the other, and things that are simultaneous in the one frame will not be simultaneous in the other.

In many ways, this is the most disturbing aspect of relativity. Certainly, we would argue, either two things happen at the same time or they don't. But we would be wrong. If two events happen at different places, then the question of whether the events occur at the same time or not has no answer. One must first ask, "As measured by whom?" We refer to this general phenomenon as "relativity of simultaneity."

A little further analysis will allow us to work out exactly how much the clock on the left will read ahead after the synchronization. We work in our reference frame and note that, if the length separating the two clocks is L' in its own reference frame, it will be a shorter $L = L'\sqrt{1 - v^2/c^2}$, as seen by us. Because the clock is moving toward the light as the light moves toward it, the left-half distance we measure to be $L/2$ will be covered at a rate $c + v$. The time at which the clock on the left sees the flash will be $t_1 = (L/2)/(c + v)$. Similarly, the rightward-moving signal covers its $L/2$ distance at a rate $c - v$, giving a $t_2 = (L/2)/(c - v)$ start time for the right clock. Thus, the elapsed time we measure between when the left clock starts and the right clock starts is

$$\Delta t = t_2 - t_1 = \frac{L/2}{c - v} - \frac{L/2}{c + v}.$$

Our goal, however, is to see by how much the reading on the clock on the left will be ahead of that on the clock on the right, when we look at it. For this, we are going to have to do a little algebra. If you really hate algebra, we invite you to rejoin us down at Equation 24.3. Otherwise, here we go. We start with the formula for Δt,

$$\Delta t = \frac{L/2}{c - v} - \frac{L/2}{c + v} = \frac{L}{2}\left[\frac{1}{c - v} - \frac{1}{c + v}\right] = \frac{L}{2}\left[\frac{c + v}{c^2 - v^2} - \frac{c - v}{c^2 - v^2}\right] = \frac{L}{2}\left[\frac{2v}{c^2 - v^2}\right].$$

This Δt is the elapsed time between the start times for the two clocks, *as measured by us.* As we said, what we want to know is the constant offset we will see that persists between the two moving clocks. This will be the same as the reading on the trailing clock when the leading clock starts. For this, we must take into account the fact that the trailing clock will be ticking slowly between the time it starts and the time when the leading clock finally starts. That elapsed time, as measured by us, is Δt, and so the fewer number of ticks for the moving clock during that time interval is found by *multiplying* Δt by the square-root factor to give $\Delta t' = \Delta t\sqrt{1 - v^2/c^2}$. The reading on the trailing clock when the leading clock starts is therefore

$$\Delta t' = \Delta t\sqrt{1 - \frac{v^2}{c^2}} = L\left[\frac{v}{c^2 - v^2}\right]\sqrt{1 - \frac{v^2}{c^2}}.$$

This expression is written in terms of the length L that *we* measure between the two clocks. The expression is actually simpler if we replace L by $L = L'\sqrt{1 - v^2/c^2}$, so that the final result will be written in terms of the distance L' between the two clocks as measured in their own reference frame. This substitution produces

$$\Delta t' = L'\sqrt{1 - \frac{v^2}{c^2}}\left[\frac{v}{c^2 - v^2}\right]\sqrt{1 - \frac{v^2}{c^2}}.$$

We can cancel the square roots with the $c^2 - v^2$ denominator, giving

$$\Delta t' = \frac{L'v}{c^2}. \tag{24.3}$$

If two clocks, moving with velocity v, are separated by a length L', *as measured in their own reference frame*, then, as seen by us, the reading on the trailing clock will always be ahead of the clock it is chasing by the $\Delta t'$ of Equation 24.3. We leave the prime on L' as a reminder that this formula uses the length as measured in the moving frame, and we leave the prime on $\Delta t'$ as a reminder that this is the amount by which the trailing clock will be ahead of the leading clock, as we look at it.

EXAMPLE 24.3 The Ultimate Story Problem

Gunslinger Greg had been on Rawhide Ron's trail for months. He finally caught up with him on the Einstein Express, a train known for its phenomenal top speed of 3⁄5 the speed of light. Gunslinger Greg walked forward from the rear of the train and Rawhide Ron walked back from the front. They met in the Pullman car, just as the train was speeding through the station. Both men drew, fired, and both men fell to the floor, badly wounded.

When the train was finally brought to rest, the sheriff took both men into custody and brought them back to the station to investigate.

"I saw it all," said Sweet Sue, the school marm. "I was standing on the platform at the time and I saw that no-good Gunslinger Greg shoot first."

"Did you correct for the time of flight of the light to your eyes?" the sheriff asked, wanting to make sure that justice was done.

"Of course I did," she answered. Sue was a little offended, but she patiently explained, "I was standing right next to Gunslinger Greg when he fired. He shot first and, after I corrected for the time of flight of the light from Rawhide Ron to me, I say that Rawhide Ron fired exactly 40 nanoseconds later than Gunslinger Greg."

"You seem pretty sure of yourself," the sheriff noted.

"I *have* read *College Physics: Putting It All Together,*" Sue explained, with more than a little pride.

The sheriff then stepped onto the train and measured the distance between the place where Gunslinger Greg had fallen and the place where Rawhide Ron had fallen. He found that it was exactly 30 meters.

Stepping once more onto the platform, the sheriff spoke to the crowd that had assembled. "You see," he explained, "the laws in this county say that I can only hang one of these two no-good bums. I can only hang the one who shot first in his own rest frame. The laws say the other bum was only acting in self-defense."

Turning to Sue, he said, "You say you have read *College Physics: Putting It All Together.* Can you tell me who shot first in the rest frame of the train, and by how much he shot first?"

ANSWER Sue was a little worried that the sheriff might not understand, so she first drew him a couple of pictures, showing the events as they transpired in her rest frame.

She then explained that she saw Gunslinger Greg shoot first at $t = 0$, as measured by her clock. After taking the light-time into account, she calculated that, in her frame, Rawhide Ron shot at $t = +40$ ns.

But she knew that Gunslinger Greg's clock would have been reading ahead (if he hadn't lost it in a poker game). The sheriff had measured the distance between the two shooters to be 30 m in the reference frame of the train. So, if the reading on Gunslinger Greg's clock had been $t' = 0$ at the moment he shot, and if his reading was ahead of Ron's, then Sue knew that Rawhide Ron's clock would have read

$$\Delta t' = -\frac{vL'}{c^2} = -\frac{(3/5c)(30\text{ m})}{c^2} = -\frac{18\text{ m}}{3 \times 10^8\text{ m/s}} = -60\text{ ns}$$

at that same instant (using Equation 24.3). Rawhide Ron's clock was a moving clock, so it was running slow, meaning that, during the 40 ns that Sweet Sue said elapsed between the two shots, Rawhide Ron's clock would have ticked off

$$t' = t\sqrt{1 - \left(\frac{v}{c}\right)^2} = (40\text{ ns})\sqrt{1 - \left(\frac{3}{5}\right)^2} = (40\text{ ns})\left(\frac{4}{5}\right) = 32\text{ ns}.$$

Thus, it would have read −60 ns + 32 ns = −28 ns when he shot. Now Gunslinger Greg's and Rawhide Ron's clocks would have been synchronized in their own frame, so, if Rawhide Ron shot at $t' = -28$ ns and Gunslinger Greg shot at $t' = 0$, then Sweet Sue could correctly report that that no-good Rawhide Ron shot first, as seen in the rest frame of the train.

The idea of relativity of simultaneity is probably the most difficult concept from Special Relativity and provides the explanation for most of the apparent paradoxes often associated with the theory. It is important to recognize what it means to give the readings on clocks that are synchronized in their reference frame, as we did in the last example. Greg's and Ron's clocks would have been synchronized in their own frame, so, if Greg's clock read −28 ns when he fired and Ron's clock read zero when *he* fired, then Greg really did shoot first, according to observers in the train's frame. But Sue's calculation is perfect for an observer in the station frame. If she takes the light time into account and assigns a time +40 ns to Greg's shot, then Ron really did shoot first, according to the synchronized clocks in her frame. And all of this is the same even if the clocks we refer to don't physically exist. It is what they *would* read that counts. The timing of the events depends on the definition in a particular reference frame of what is simultaneous with what, and simultaneity is relative.

24.3 Applications and Paradoxes

Special Relativity is a difficult concept because it contradicts the intuition we have developed over years of living in a world where physical things do not travel close to the speed of light and where we can't sense the tiny effects of relativity at the slow speeds we do live with. The key to understanding relativity is to suspend our intuition, to simply remember the three effects, and to apply them as needed to solve a problem. In this section, we will try to develop new intuition by working our way through several examples and seeing how the three effects are applied.

One of the common questions asked about relativity is at what speed you have to start using it. The answer is that there is no absolute answer to that question. The effects of relativity on moving lengths and clocks are always there at all speeds, but the effects are smaller at slower speeds. So the answer to the question must take into account the accuracy required in each application, as the next example illustrates.

EXAMPLE 24.4 When Do You Need to Use Relativity?

You and a lab assistant want to synchronize two laboratory clocks that are 300 m (about the length of three football fields) apart. You arrange for the clock where you are standing to send out a signal and for the clock your assistant is holding to start as soon as it receives the signal.

a) To make sure that the two clocks are synchronized, what should be the reading on the second clock when it starts?

b) In order to check the synchronization, your lab assistant hurries back with the synchronized clock, jogging along at 3 m/s. How much will the clock she carries now be offset from your own?

ANSWER

a) It will take the signal from your clock to your assistant's clock a time $t = d/v = 300\text{ m}/(3 \times 10^8\text{ m/s}) = 1\ \mu\text{s}$ to cover the 300 m. You start your clock at zero, as the signal is sent. So, if the assistant's clock is set to start with a reading of $+1\ \mu$s when it receives the signal, the two clocks will end up perfectly synchronized.

b) The second clock will be a moving clock during the 100 seconds it takes your assistant to run the 300 m at 3 m/s. The question we want to answer is how the slowing of the clock for that 1 minute and 40 seconds will affect the time it reads.

The square-root factor for a clock moving at 3 m/s is

$$\sqrt{1 - \left(\frac{v}{c}\right)^2} = \sqrt{1 - \left(\frac{3\text{ m/s}}{3 \times 10^8\text{ m/s}}\right)^2} = \sqrt{1 - (10^{-8})^2} = \sqrt{1 - 10^{-16}}.$$

Now it will take a special calculator just to recognize the number under the square root sign (sixteen 9s in a row), and certainly a special one to take the square root of this number. If you don't have such a calculator, this would be a good place to use the binomial approximation discussed in Appendix A.6. However you find it, the square root factor will turn out to be $1 - 5 \times 10^{-17}$. The result of the 100-second walk at 3 m/s will thus be a clock that has slowed by a total of $(100\text{ s})(5 \times 10^{-17}) = 5$ fs (where you should remember that fs stands for femtoseconds, or 10^{-15} seconds).

Now if you are using a quartz wristwatch (accurate to $\pm 5\ \mu s$ over an elapsed time of 100 s) or a laboratory cesium clock (accurate to ± 2 ps over 100 s), then you would never detect the loss of synchronization produced by moving the clock. However, the best strontium trapped-ion clocks achieve a stability of ± 3.4 fs over an averaging time of 100 seconds. If you carefully compared two of these clocks, the loss of synchronization would be detectable, and, if that matters, then you cannot ignore the tiny effects of relativity arising from running the second clock back to the first.

One of the apparent paradoxes that occurs to students of relativity, as soon as they realize that the choice of a reference frame is arbitrary, is the question of who is to say which clock is the moving clock and, therefore, which is running slow. Our next example addresses this question.

EXAMPLE 24.5 The Clock Paradox

We know that moving clocks will run slow. But, to someone in that clock's reference frame, *we* are the moving frame. So wouldn't it be our clocks that are running slow? Consider a 300-meter-long space barge passing our 300-meter-long space station at $^4\!/_5\,c$, or 2.4×10^8 m/s. We have clocks on the right and left of our station and the barge has clocks at the front and the back. Working entirely in our reference frame, explain why the driver on the barge will measure *our* clocks to be the ones that are running slow.

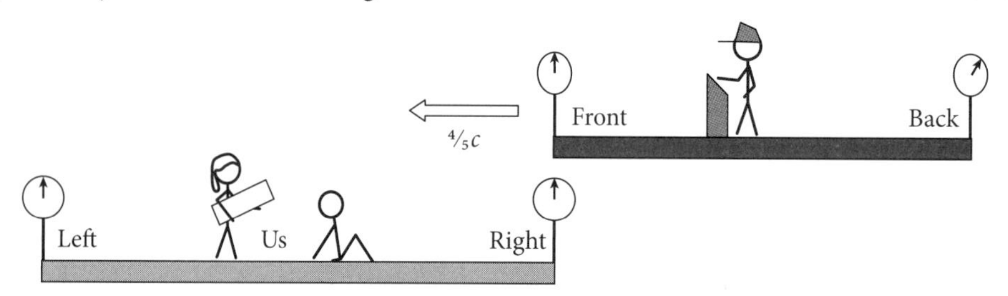

ANSWER Since the barge is moving at $^4\!/_5\,c$, it takes $t = 300\text{ m}/2.4 \times 10^8\text{ m/s} = 1.25\ \mu s$ for the front clock on the barge to move past our space station, moving from right to left. But the barge's clocks are all running slow. So, if that front clock reads zero when it passes our right clock, it will read $t' = (1.25\ \mu s)\sqrt{1 - (^4\!/_5)^2} = 0.75\ \mu s$ when it passes our synchronized left clock. Our left clock reads 1.25 μs when this occurs, and this is how we know that the barge's front clock is running slow.

Now the barge operator will want to compare our right clock with his synchronized front and back clocks. He will see that our right clock reads zero, just like his front clock, at the instant shown in the picture. Then he will compare it with his back clock when our right clock gets there, to see if our clock is running slowly or not.

Working in our reference frame, we know that his clocks are not synchronized. His back clock follows his front clock by the rest length $L' = 300$ m. So, at the instant when the front clock reads zero in the picture above, the back clock is ahead by

$$\Delta t' = \frac{vL'}{c^2} = \frac{(^4\!/_5 c)(300\text{ m})}{c^2} = \frac{(^4\!/_5)(300\text{ m})}{3 \times 10^8\text{ m/s}} = 0.8\ \mu s.$$

Since the barge is moving at $\frac{4}{5}c$, we will measure its length to be $L'\sqrt{1-(4/5)^2}$, or $L = (300\text{ m})(3/5) = 180$ m. The time it takes a 180-meter-long barge moving at speed 2.4×10^8 m/s to pass our right clock is $(180\text{ m})/(2.4 \times 10^8\text{ m/s}) = 0.75\ \mu\text{s}$, as measured by us. This means that our right clock will read 0.75 μs when the barge's back clock gets to it. But what will be the reading on that back clock?

Since that back clock is running slow, only $(0.75\ \mu\text{s})\sqrt{1-(4/5)^2} = 0.45\ \mu\text{s}$ of *its* time will have elapsed during that 75-μs time interval when our right clock is sweeping past the barge. But let us remember that, in the instant shown in the picture, when the front clock was passing our right clock, the back clock was already reading ahead by 0.8 μs. If the back clock then ticks off an additional 0.45 μs while it is coming to us, it will read 1.25 μs when it arrives. Of course, our clock reads 0.75 μs at this moment, as we said above. Therefore, if the barge pilot compares his back clock's 1.25 μs reading with our right clock's 0.75 μs reading, and notes that his synchronized front clock read zero when our same clock passed it, he will correctly conclude that our right clock has been running slow all this time.

The way we concluded that the barge's clocks were running slow was to compare the barge front clock to our right clock and then to our left clock. The comparison that led the barge operator to conclude that *our* clocks were running slow was to compare our right clock to his front clock and then to his back clock. Different things are being compared, and we all agree on what those comparisons will show. There is no paradox in saying that the barge clock is running slow compared to our clocks and that our clocks are running slow compared to his. And so, whatever reference frame you are in, if you see a moving clock, you may be sure it is running slow.

The last example we will take up in this section addresses another of the famous paradoxes of relativity, known as the twin paradox. Basically, it is this. If an astronaut travels to a far-away star at a speed close to the speed of light, and then returns, he will age more slowly due to his speed. Thus, he can return after, say, 20 years, having aged, perhaps, only 12 additional years during the trip. If he had a twin sister who remained at home, he will be aware that she has 8 years more of all the effects of aging, including 8 years more wrinkles, more knowledge, and more wisdom, than he, her astronaut brother.

But as seen by the brother, she is the one who was moving all that time. So why isn't she the one who has aged less? And here, we can't apply the method of Example 24.5, because, when the brother finally arrives home, they will see each other at the same time and at the same place. There is no way that he will see her older than he is while she sees him to be older than she is. So let us see how this paradox is resolved.

EXAMPLE 24.6 The Twin Paradox

Rocky, the rocketeer, travels at $\frac{4}{5}c$ to a star that is 8 $c\cdot$yrs away and returns to earth. His twin sister Terri stays on earth (so Terri stays on terra firma). How long will the trip take, as measured by Terri? How much will Rocky age during the trip? How do we explain Terri's aging from Rocky's point of view?

ANSWER The first two questions are easy. The first is just a $t = d/v$ problem. It will take the rocket $t = (8\ c\cdot\text{yrs})/(\frac{4}{5}c) = 10$ yrs each way. If clocks on earth and the star are synchronized in the earth frame, and if the clock on earth reads $t = 0$ when Rocky leaves, then the clock on the star will read 10 yrs when he turns around and the clock on earth will read 20 yrs when he returns. The second question is a slow-clock problem. On both the outbound and the inbound journey, Rocky's aging process is a moving clock that will tick off a time $(10 \text{ yrs})\sqrt{1-(^4/_5 c)^2}$, or 6 yrs, on each leg of the journey. This is a correct calculation. Rocky will return 12 years older than when he started, and 8 years younger than his sister in every way.

By the way, the reason he can cover the distance to the star and back in only 12 years is that, to him, the distance is not 8 $c \cdot$yrs, but $(8\ c \cdot \text{yrs})\sqrt{1-(^4/_5 c)^2} = 4.8\ c \cdot \text{yrs}$.

Now from Rocky's point of view, Terri is a moving clock on both 6-year legs of his journey. As measured by him, she will therefore age $(6 \text{ yrs})\sqrt{1-(^4/_5 c)^2} = 3.6$ yrs on each half of the trip, for a total of 7.2 yrs. So the question is: What happened to the missing 12.8 years between how much Rocky would expect Terri to age (7.2 yrs) and her actual aging (20 years)?

The answer lies in Rocky's acceleration. Let us say that Rocky turns around at the star by accelerating at 100 m/s^2 for exactly 4.8×10^6 seconds, about two months. At this rate, he will change his velocity by -4.8×10^8 m/s in those 4.8×10^6 seconds, going from $+2.4 \times 10^8$ m/s to -2.4×10^8 m/s. During this acceleration period, *Rocky is not in an inertial frame.* Instead, he is jumping from one inertial frame to another. He starts in a frame moving to the right at $^4/_5 c$ and ends in one moving to the left at $^4/_5 c$. His definitions of simultaneity will change as he changes reference frames.

So let's look at simultaneity from Rocky's point of view. As he leaves the earth at $t = 0$, he will see the earth–star distance flying past him at $^4/_5 c$. The clock on the star will be a trailing clock, so it will read ahead (see below left). The earth–star distance is 8 $c \cdot$yrs, measured in its own frame, so the *star clock* will be ahead of the earth clock by $\Delta t = (Lv)/c^2 = (8\ c \cdot \text{yrs})(^4/_5 c)/c^2 = 6.4$ yrs. Earth clocks and star clocks will run slow, ticking off only 3.6 yrs during Rocky's 6-year outbound journey. But, with its initial head start, the star clock will read 6.4 yrs + 3.6 yrs = 10 yrs when Rocky reaches it.

Outbound

Inbound

However, once he turns around, the earth will be coming toward him and the earth clock now trails the star. As he starts back, the *earth clock* will read ahead of the star clock by 6.4 yrs (see above right). The star clock still reads approximately 10 years (neglecting the two-month turn-around time), so earth clocks will read 10 yrs + 6.4 yrs = 16.4 yrs at that moment. As measured by Rocky, Terri will be 16.4 years older as he starts home, and she will then age 3.6 more years during his return trip. Rocky will see her actual 20 years of aging as partly due to what happened while he turned around.

Example 24.6 gives an explanation of what Rocky can observe during the times when he is in an inertial frame, where the laws of Special Relativity apply. As measured by Rocky, Terri will indeed age only 7.2 years during the 12 yrs in which he is in an inertial frame. And the reason that she will be 20 years older when he returns is the 12.8 years she ages during his acceleration, from 3.6 years to 16.4 years. We have seen that this is due to his changing his determination of what is simultaneous with what, as he accelerates. But we bet you are not very satisfied with this. How, you may still be asking, can Terri age 12.8 years during Rocky's 2 months of acceleration?

A deeper explanation requires finding laws of physics that apply in accelerating reference frames. The ones we will need are contained in Einstein's General Theory of Relativity.

General Relativity itself is beyond the scope of this text, but we will be able to pick up enough from the theory to see how it applies in explaining the slowing of clocks in an accelerating reference frame.

The first prediction from General Relativity we will need is the fact that clocks run slowly in a region of lower relative gravitational potential. (You may remember what a gravitational potential is from the discussion on page 333.) This prediction was first tested in an experiment that compared a clock at the top of Harvard University's Jefferson Laboratory with another clock in the basement. The clock in the basement was at a lower gravitational potential, so Einstein's theory predicted that it would tick more slowly. In General Relativity, the ratio of the elapsed time on a clock on the top floor to the time elapsed on a clock on the bottom floor is given by the formula

$$\frac{\Delta t_{\mathrm{TOP}}}{\Delta t_{\mathrm{BOTTOM}}} = 1 + \frac{\Delta U}{c^2}, \tag{24.4}$$

where $\Delta U = U_{\mathrm{TOP}} - U_{\mathrm{BOTTOM}}$ is the gravitational potential difference between the two clock locations. In a uniform gravitational field like that near the surface of the earth, the difference in potential is $\Delta U = gh$, where h is the difference in height of the two clock locations. The predicted difference in the clock rates on the two floors was only one part in 10^{15}, but it was still verified to within 1% in the Harvard experiment.

The second prediction we need from General Relativity is what is referred to as the *Equivalence Principle*. This states that physics in a gravitational field is the same as physics in an accelerated reference frame and vice versa. We may use this principle to predict that whatever happens in a region of gravitational field g will also happen in a reference frame accelerating at $a = g$. During Rocky's acceleration, Terri's clock is like the clock at the top of the building and Rocky's clock is like the clock at the bottom. The "gravitational field" in the equation is Rocky's acceleration (we assumed $a = 100\ \mathrm{m/s^2}$ in Example 24.6), and the height h is the $8\ c\cdot\mathrm{yr}$ distance between the star and the earth. Thus, according to Equation 24.4, if Rocky is accelerating for a time $\Delta t_{\mathrm{ROCKY}}$, then the time elapsed on Terri's clocks during that same period will be

$$\Delta t_{\mathrm{TERRI}} = \left(1 + \frac{ah}{c^2}\right)\Delta t_{\mathrm{ROCKY}} = \Delta t_{\mathrm{ROCKY}} + \frac{ah}{c^2}\Delta t_{\mathrm{ROCKY}}.$$

The first $\Delta t_{\mathrm{ROCKY}}$ in the expression is just the 2 months (4.8×10^6 s) that we decided to neglect in Example 24.6. The second term is the big one. It is

$$\frac{ah}{c^2}\Delta t_{\mathrm{ROCKY}} = \frac{(100\ \mathrm{m/s^2})(8\ c\cdot\mathrm{yrs})}{c^2}(4.8 \times 10^6\ \mathrm{s}) = \frac{3.84 \times 10^9\ \mathrm{m/s}}{3 \times 10^8\ \mathrm{m/s}}\ \mathrm{yrs} = 12.8\ \mathrm{yrs}.$$

This is exactly the missing additional time that elapses on Terri's clock during the time that Rocky accelerates. An accelerating clock ticks more slowly than a clock that is not accelerating, and the difference in the elapsed times will be proportional to how far away (h) the unaccelerated clock is located. This is an absolute statement on which all observers will agree. If I watch an accelerated clock from a distance h, it will run slow. If I am accelerating, I will observe non-accelerating clocks to be running fast.

24.4 Relative Velocity (Again)

We would like to take the unusual step of starting this section with an example.

EXAMPLE 24.7 Relative Velocity of Two Objects

An observer on a space station sees spaceship C, $1\frac{1}{2}\,c\cdot$yrs to her left, moving to the right at $\frac{3}{4}c$, and spaceship A, $1\frac{1}{2}\,c\cdot$yrs to her right, moving to the left at $\frac{3}{4}c$. At what rate are the two spaceships covering the initial $3\,c\cdot$yrs between them?

ANSWER This is not a case of comparing measurements in two different reference frames. All distances, times, and velocities are measured from a single reference frame, that of the space station. And so, there is nothing "relativistic" about this problem at all. If spaceship C is moving at $\frac{3}{4}c$, it is covering $\frac{3}{4}$ light years every year. Spaceship A is also covering $\frac{3}{4}$ light years every year. As they move together, they will be $1\frac{1}{2}$ light years closer every year, covering the distance between them at a rate of $1\frac{1}{2}$ times the speed of light. When we said, on page 515, that the speed of light is a speed limit, we meant that no single object will be able to move faster than the speed of light, as measured in any reference frame. But distance can still be covered at a higher rate when two objects are coming toward each other.

The purpose of this section is to show why we can say, despite the results of the last example, that no object will ever be measured to have a speed faster than the speed of light. Certainly, from the point of view of classical physics, Example 24.7 would indicate the opposite. It would indicate that observers on spaceship C would measure the speed of spaceship A to be $1\frac{1}{2}\,c$. But classical physics does not know about the fact that spaceship C would actually be doing its measurements with unsynchronized slow clocks and shortened measuring rods.

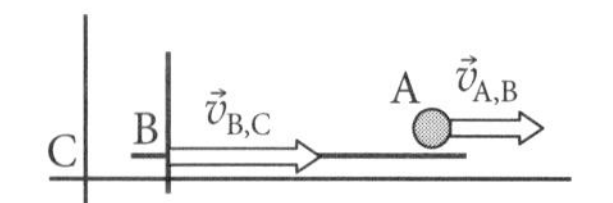

Figure 24.10 The velocity of origin B relative to origin C is $\vec{v}_{B,C}$ toward the right.

Because the three effects are really there when one compares a space or time measurement in one frame with a similar measurement in another frame, we may expect that the simple $\vec{v}_{A,C} = \vec{v}_{A,B} + \vec{v}_{B,C}$ velocity addition formula of Equation 4.9 will have to change. The new relativistically correct formula for velocity addition is the subject of this section. To help us compare the new result with the Chapter 4 result, we will use the same notation we used before, as shown in Figure 24.10.

Two of the three effects of relativity, length shortening and desynchronization of clocks, act differently in the direction of $\vec{v}_{B,C}$ than in a direction perpendicular to it, so the relativistic equations for velocity addition will be different for components parallel to $\vec{v}_{B,C}$ than for those perpendicular to $\vec{v}_{B,C}$. The formula for perpendicular components is actually quite complicated to derive or use since it mixes perpendicular and parallel velocity components, so we will restrict our discussion here to a velocity $\vec{v}_{A,B}$ that is parallel to the direction of the velocity $\vec{v}_{B,C}$ (as in Figure 24.10). Even with this restriction, we will be able to show how it is that the measured velocity of one object relative to another cannot give a result greater than the speed of light.

We have restricted ourselves to motion in a single dimension, parallel to the relative velocity of the two reference frames. but we will still need to do some serious algebra. We think it will be good for you to follow it, but you will only need the end result, Equation 24.7.

To help make this problem more concrete, let us choose frame B to be the inside of a railway car moving at a speed $v_{B,C}$ relative to the ground (the ground being frame C). And let's say that object A is a bullet that travels at velocity $v_{A,B}$, parallel to the direction of motion of the train, as shown in Figure 24.11. We will begin with a simple look at the motion in the train car only, so there is no relativity involved. If we use primed coordinates for frame-B values, we would say that the bullet is fired at time $t' = 0$ at the back of the train and that it hits a target a distance L' away at time $t' = \Delta t'$. These two events are shown in Figure 24.11. The velocity of bullet A in frame B is obviously the distance divided by the time, or $v_{A,B} = L'/\Delta t'$.

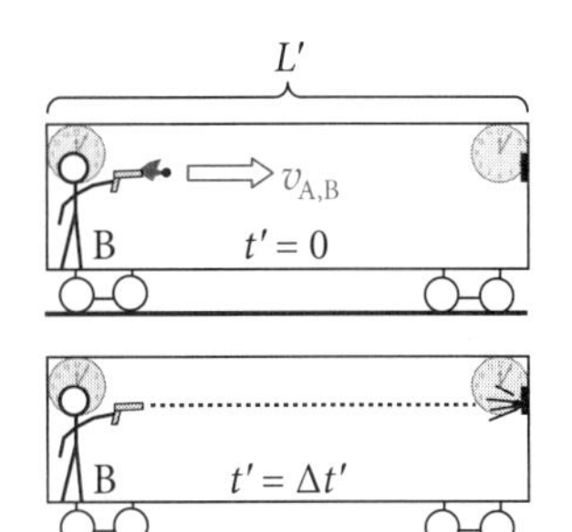

Figure 24.11 The events as seen in the train frame.

Now let's look at these same two events from the reference frame of the ground, frame C. In frame C, the distance between the event of firing the bullet and the event of the bullet hitting the target is shown in Figure 24.12 to be

$$\Delta x = v_{B,C}\Delta t + L, \tag{24.5}$$

where L is the shortened length of the car, as measured in frame C. The time of flight of the bullet, as it is measured in frame C, is a little harder to find.

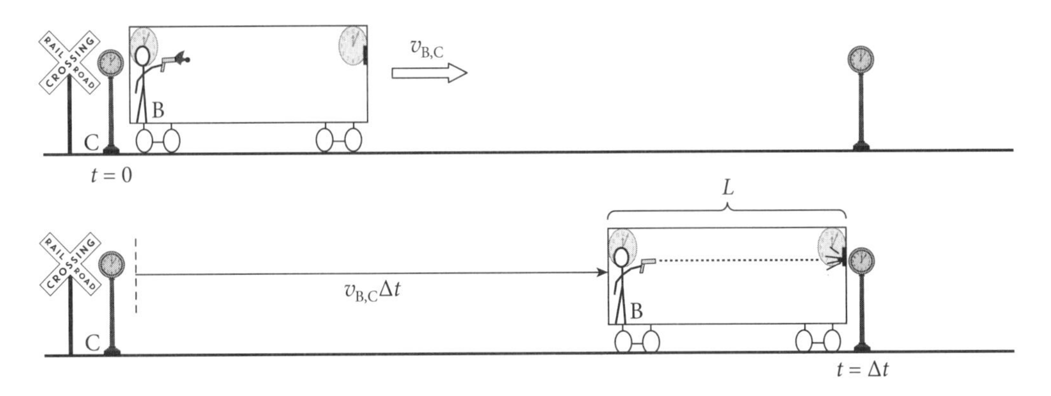

Figure 24.12 The events as seen in frame C, the reference frame of the ground.

The train clock at the back of the car in Figure 24.12 is a trailing clock, so it must run ahead, and so the train clock by the target must run behind. If the clock at the back reads $t' = 0$ as the bullet is shot, then the clock by the target reads $t' = -(L'v_{B,C})/c^2$ at the same instant. We know that the train clock by the target reads $t' = +\Delta t'$ when the bullet finally hits the target. So, between the two events, we see that the clock by the target must first tick forward up to zero, and then tick a bit more, up to $\Delta t'$, for a total elapsed time of $(L'v_{B,C})/c^2 + \Delta t'$. But that clock is a moving clock, so it would be running slow that whole time. Thus, the time interval measured in frame C would be

$$\Delta t = \frac{\Delta t' + (L'v_{B,C})/c^2}{\sqrt{1 - v_{B,C}^2/c^2}}. \tag{24.6}$$

The velocity of the bullet relative to the ground is $v_{A,C} = \Delta x/\Delta t$. To find $v_{A,C}$, we first substitute Δx from Equation 24.5 and then use Δt from Equation 24.6:

$$v_{A,C} = \frac{\Delta x}{\Delta t} = \frac{v_{B,C}\Delta t + L}{\Delta t} = v_{B,C} + \frac{L}{\Delta t} = v_{B,C} + L\frac{\sqrt{1 - v_{B,C}^2/c^2}}{\Delta t' + (L'v_{B,C})/c^2}.$$

We then replace L by L', using $L = L'\sqrt{1 - v_{B,C}^2/c^2}$. This gives

$$v_{A,C} = v_{B,C} + \frac{L'(1 - v_{B,C}^2/c^2)}{\Delta t' + (L'v_{B,C})/c^2}.$$

Finally, if we use the definition $v_{A,B} = L'/\Delta t'$ to write $L' = v_{A,B}\Delta t'$, this last equation can be rewritten as

$$v_{A,C} = v_{B,C} + \frac{v_{A,B}\Delta t'(1 - v_{B,C}^2/c^2)}{\Delta t' + (v^{\parallel}_{A,B}v_{B,C}\Delta t')/c^2}.$$

At this point, we can cancel the $\Delta t'$ in the big fraction and then put both terms on the right-hand side over a common denominator to get

$$v_{A,C} = \frac{v_{A,B} + v_{B,C}}{1 + (v_{A,B}v_{B,C})/c^2}. \tag{24.7}$$

So let's see where we stand. If some object A has a known velocity $\vec{v}_{A,B}$ with respect to an origin B, and if origin B has velocity $\vec{v}_{B,C}$ relative to origin C, and if $\vec{v}_{A,B}$ is parallel to $\vec{v}_{B,C}$, then $\vec{v}_{A,C}$, the velocity of A relative to C, may be found from Equation 24.7. This relativistic formula replaces the old $\vec{v}_{A,C} = \vec{v}_{A,B} + \vec{v}_{B,C}$ formula from Chapter 4. It is required if the velocities involved are significant fractions of the speed of light, or when the velocities are measured with fine enough accuracy for the effects of relativity to be important.

Note that Equation 24.7 was derived with reference to Figure 24.10, in which the velocities are all taken positive to the right. A case where origin B is moving to the left relative to origin C can also be accommodated by assigning a negative value to $v_{B,C}$ in Equation 24.7. Negative values are also used if the velocity of object A is to the left in either frame.

Let's do a couple of short examples to make sure we understand the process.

EXAMPLE 24.8 Velocity of One Object Relative to Another

An observer on a space station sees spaceship C, 1.5 $c \cdot$yrs to her left, moving to the right at $\frac{3}{4}c$, and spaceship A, 1.5 $c \cdot$yrs to her right, moving to the left at $\frac{3}{4}c$. What will be the velocity of spaceship A, as measured by spaceship C?

ANSWER This is a different question from that of Example 24.7. Now we are asking for the velocity of a single object, as measured in two different reference frames. This is a relativity problem. Let us define object A to be the spaceship A, frame B to be the space station, in which the velocity of object A is known, and frame C to be on spaceship C. If we again take positive to the right, we can write $v_{A,B} = -\frac{3}{4}c$. Then, since the spaceship on the left (frame C) is moving to the right, the velocity of the station (frame B), relative to C, is $v_{B,C} = -\frac{3}{4}c$. We may then use Equation 24.7 to find $v_{A,C}$. It gives

$$v_{A,C} = \frac{v_{A,B} + v_{B,C}}{1 + (v_{A,B}v_{B,C})/c^2} = \frac{-\frac{3}{4}c - \frac{3}{4}c}{1 + (-\frac{3}{4}c)(-\frac{3}{4}c)/c^2} = \frac{-1.5c}{1 + \frac{9}{16}} = -0.96c.$$

The measured velocity will be 96% the speed of light to the left. You should note that the numerator of this equation gives the usual $v_{A,B} + v_{B,C}$ one would expect from the Newtonian formula for the addition of velocities (Equation 4.9), but that the denominator quenches any addition results that would give something greater than the speed of light. Measured velocities will always be less than c.

EXAMPLE 24.9 The Relative Speed of a Light Signal

A space station sends a signal at the speed of light toward a star system in the Skyron Empire. A Skyron battle cruiser attempts to intercept the message by following after it at 0.9 times the speed of light. How fast will the signal be moving away from the battle cruiser, as measured in the cruiser's reference frame?

ANSWER Frame B, as usual, will be the frame in which the velocity is already known. In this case, it is the frame of the space station. Frame C would then be the Skyron battle cruiser. The signal is moving parallel to the cruiser, so we can use Equation 24.7. Let us picture the signal velocity positive to the right. The speed of the signal (object A) relative to the station is $v_{A,B} = c$. Since the battle cruiser is also moving to the right relative to the space station, the station (frame B) is moving to the left with respect to the cruiser, so $v_{B,C} = -0.9c$. If we plug in these values, we get

$$v_{A,C} = \frac{v_{A,B} + v_{B,C}}{1 + (v_{A,B}v_{B,C})/c^2} = \frac{c - 0.9c}{1 + (1c)(-0.9c)/c^2} = \frac{0.1}{0.1}c = c.$$

As we should have expected, the speed of light is the same to all observers.

24.5 Relativistic Mechanics

Our first four sections in this chapter have been about relativistic kinematics. They have addressed the questions about how time and distance measurements in different reference frames are related to each other. In this section, we will consider whether the postulates of relativity require changes in mechanics. In particular, we will see that the formulas for momentum and energy must change if those quantities are to continue to be conserved quantities in a world where Einstein's relativity holds.

Now we *could* work out what those new relativistically correct formulas are. We would start with the relativistic kinematics we have learned in the previous sections and consider collisions of slow-moving objects with fast-moving objects, comparing what is seen in two reference frames that are moving with respect to each other. We could thereby work out formulas for momentum and energy that would allow these quantities to be conserved in the collisions. You would be able to follow all we do, and we would love to do it for you. And we certainly haven't shied away from deriving difficult things for you in the past, have we? But, this time, the explanation and the algebra are just too long and involved to justify the effort. So we are going to take an unprecedented step (for us) and just *tell* you what these formulas are. If you were dying to see how these things are derived . . . we're sorry.

So here are the results for momentum. In Newtonian mechanics, we found that a momentum defined as $\vec{p} = m\vec{v}$ produced a total momentum that was conserved in isolated systems. In relativistic mechanics, we can still define momentum so that the total momentum of an isolated system will be conserved, but the new definition of momentum will have to be different. The relativistic momentum of a body of mass m moving at velocity $\vec{v}$ must be calculated with the formula

$$\vec{p} = \frac{m\vec{v}}{\sqrt{1 - (v/c)^2}}. \tag{24.8}$$

When the momenta of a system of particles are calculated using this formula, and vector-added together, the total momentum of the system will be exactly conserved. If one uses the old $\vec{p} = m\vec{v}$ formula, it will not be. But do note that, when the velocity of a particle is small compared to the speed of light, then the quantity v/c will be small compared to 1, and Equation 24.8 will come close to the simpler $\vec{p} = m\vec{v}$. Newtonian ideas are still approximately correct when the velocities are not too large (do Problem 36 to see what this means).

Then let us turn to energy. In Newtonian mechanics, we found that there was a kinetic energy that would be conserved in collisions where there was no way for the energy to escape, collisions that were termed "elastic." In relativistic mechanics, the kinetic energy is no longer an important quantity. It is replaced by the idea of the total energy of a particle. The *total energy* of a particle of mass m moving at velocity v, the quantity that is conserved in collisions, is

$$E = \frac{mc^2}{\sqrt{1 - (v/c)^2}}. \tag{24.9}$$

In the case of small velocity, Equation 24.9 can be approximated in a very interesting way. When v is much less than c, the quantity $\sqrt{1 - (v/c)^2}$ can be approximated using the binomial expansion (Appendix A.5) to give

$$E = \frac{mc^2}{\sqrt{1 - (v/c)^2}} = mc^2\left(1 - v^2/c^2\right)^{-1/2} \approx mc^2\left(1 + \frac{1}{2}\frac{v^2}{c^2}\right) = mc^2 + \frac{1}{2}mv^2.$$

In this form, we see that the total energy in the limit of low velocity is the sum of two terms. The second term is our old friend, the Newtonian kinetic energy ($\frac{1}{2}mv^2$); the first is mc^2. This first term has units of energy, but it does not require any velocity to exist. It is an energy that a particle possesses because it has mass. It is called the *rest energy* or the mass energy.

Let us be sure we are clear. A particle at rest has energy by virtue of its mass alone. The energy is $E_{rest} = mc^2$. If the particle is also moving, then its total energy is given by Equation 24.9. This is a new formula for total energy, not kinetic energy. If we want to define a quantity called kinetic energy to be the additional energy a moving object has, beyond its rest energy, then the relativistically correct formula for kinetic energy would be

$$\text{KE} = \frac{mc^2}{\sqrt{1 - (v/c)^2}} - mc^2. \tag{24.10}$$

This is awkward enough that you can see why the idea of kinetic energy is not often used in relativistic analysis of collisions. Only in the limit of low velocity may the kinetic energy be approximated as $\text{KE} \approx \frac{1}{2}mv^2$. In the analysis of collisions in relativity, we almost always use total energy, not kinetic energy.

There is an additional aspect of the relation $E_{rest} = mc^2$ that is important to recognize. Except for multiplying by c^2, *rest energy and mass are the same thing.* The inertia of a body at rest is $m = E_{rest}/c^2$, and Newton's second law for a body at rest can correctly be written $F = (E_{rest}/c^2)a$. This means, for example, that if you have two identical balls at different temperatures, then the hotter one will be harder to accelerate because it has the greater rest energy, because of its higher internal thermal energy. Or if a single body is formed by bringing together a positively charged particle and a negatively charged particle, the mass of that body will be less than the sum of the masses of the charged particles. This is because the two charged particles are attracted to each other, and so they have a lower potential energy when they are close than when they are far apart. The contribution to the mass from the temperature of an object is typically too small to measure directly, but the decrease in mass that arises when subparticles that make up a composite particle are strongly bound together by mutual attraction will be important in Chapter 26, where we discuss nuclear reactions.

Because of the fact that various forms of internal energy in objects are manifest in their masses, the term "elastic collision" basically loses its meaning. In Chapter 8, we saw that the way for a system of bodies to lose kinetic energy during a collision was typically to convert kinetic energy into internal energy inside the bodies, energy that we then ignored in our mechanical energy accounting. But, since we use *total* energy in relativistic mechanics, a quantity that includes the mass energy, internal energies are automatically included in the accounting, and energy conservation is the norm, rather than the exception. Of course, if we wanted you to keep track of this in solving any practical problem, we would have to tell you the initial and final masses of any objects in the collision, so you would automatically be including internal energies in the rest energy. We won't do any problems like this in this chapter, but you may expect to see these ideas again in Chapter 26.

One of the aspects of relativity we noted back on page 515 was the fact that all material objects must have speeds less than the speed of light. Now that we have Equation 24.9, we can see the reason for this. The energy of an object at rest ($v = 0$) is $E_{rest} = mc^2$. But, as v increases, the denominator of Equation 24.9 becomes smaller and smaller and the energy gets larger and larger. In the limit that $v \to c$, the energy goes to infinity. Thus, the only way for an object with mass to move at the speed of light is for all the energy in an infinite universe to be given to that single object. This cannot be done, and so no particle with mass can ever be accelerated to the speed of light.

The formulas of relativistic mechanics involve four quantities: velocity (the same old distance divided by time), mass (the rest energy), momentum (Equation 24.8), and total energy (Equation 24.9). But there is one more very useful formula that arises out of these relativistic definitions of momentum and energy. If we do a little algebra (all right, a lot of algebra) to eliminate v from these equations, we find

$$E^2 = (pc)^2 + (mc^2)^2. \tag{24.11}$$

In problems where we are more interested in conserved quantities than in velocities, Equation 24.11 will be very valuable.

In the following examples, and in a lot of the homework problems, we continue with the same kind of illustrative but unrealistic problems we saw in our last few sections (trains moving at $3/5$ the speed of light and so on). However, in Chapter 26, the equations of this section will be applied to some very realistic and practical problems in nuclear and particle physics. So do not forget Equations 24.8, 24.9, and 24.11 as you go on into the last chapter of our book.

EXAMPLE 24.10 The Rail Gun

An antimissile defense system uses a rail gun that can accelerate a 1-kg slug to a speed of $\frac{4}{5}c$. Assume 100% energy conversion efficiency, and calculate how many joules of energy are required to accelerate the slug to this speed.

ANSWER The speed here is clearly not a low speed, so we certainly will need to use the correct relativistic formula for energy. Equation 24.9 gives

$$E = \frac{mc^2}{\sqrt{1-(v/c)^2}} = \frac{(1\text{ kg})(3 \times 10^8\text{ m/s}^2)}{\sqrt{1-(4/5)^2}} = \frac{9 \times 10^{16}\text{ J}}{3/5} = 15 \times 10^{16}\text{ J}.$$

But this is the total energy. The slug started with some energy in its mass, even at rest. This rest energy was $E_{rest} = mc^2 = (1\text{ kg})(3 \times 10^8\text{ m/s})^2 = 9 \times 10^{16}$ J. So the additional (kinetic) energy required is $15 \times 10^{16}\text{ J} \quad 9 \times 10^{16}\text{ J} = 6 \times 10^{16}$ J.

As a reference, if you were to build up this much electrical energy by charging a capacitor at a rate of 10 kW (the power consumed by five 2-kW hairdryers operating simultaneously), it would take a time $t = (6 \times 10^{16}\text{ J})/(10^4\text{ J/s}) = 6 \times 10^{12}\text{ s} \approx 20{,}000$ yrs to charge the gun up.

EXAMPLE 24.11 A Zero-Momentum Relativistic Collision

A particle of mass m_1 is moving to the right at speed $\frac{3}{5}c$ while a particle of mass m_2 moves to the left at speed $\frac{4}{5}c$. When the two particles collide, they stick together to form a more massive particle at rest. What is the ratio m_1/m_2?

ANSWER Relativistic momentum is conserved in collisions in isolated systems. If the final particle is at rest, the total momentum of the system after the collision is zero. Momentum conservation then requires that the total momentum before the collision must also have been zero, meaning that the rightward momentum of the one particle must have the same magnitude as the leftward momentum of the other particle. Since we know the speeds, though not the masses, we can use Equation 24.8 to write

$$p_1 = \frac{m_1(3/5c)}{\sqrt{1-(3/5)^2}} = \frac{3/5\,m_1c}{4/5} = \frac{3}{4}m_1c \quad \text{and} \quad p_2 = \frac{m_2(4/5\,c)}{\sqrt{1-(4/5)^2}} = \frac{4/5\,m_2c}{3/5} = \frac{4}{3}m_2c.$$

Equating the magnitudes of the two momenta gives the ratio of the masses:

$$\frac{3}{4}m_1c = \frac{4}{3}m_2c \quad \Rightarrow \quad m_1 = \frac{16}{9}m_2 \quad \Rightarrow \quad \frac{m_1}{m_2} = \frac{16}{9}.$$

It may be instructive to note that, if we had incorrectly used the non-relativistic formula for momentum in this example, we would have arrived at a mass ratio by setting $\frac{3}{5}c\,m_1 = \frac{4}{5}c\,m_2$, a procedure that would have given the ratio $m_1/m_2 = 4/3$. That answer would have been wrong.

24.6 Summary

The concepts of Special Relativity are contrary to our intuition, but they are not difficult, if we simply apply what we know. The things to remember are these:

- Motion is relative, so there is no absolute definition of a "moving frame" or of a "fixed frame." If rods and clocks are moving relative to you, the things you will measure about them are summed up in the three effects.
- The three effects of relativity are 1) moving clocks run slow, 2) moving rods are shortened in the direction of their motion, and 3) trailing clocks read ahead.
- The slowing of clocks and shortening of lengths are governed by the square-root factor. If a problem involves moving clocks or lengths, you should go calculate the square root factor first. The result of the calculation will always be less than one. Then you should use what you know about moving clocks and moving rods to decide whether to multiply or divide by the factor.
- Whether or not two events are simultaneous depends on the reference frame. One may figure out the readings on moving clocks located at different places in a moving frame by using Equation 24.3, being careful to remember that L' is the distance between the clocks as measured in their own rest frame.
- If you have to use several of the effects together, as we had to do in many of the examples, remember that picking one rest frame and drawing pictures of the important events as seen in that rest frame will be helpful in keeping track of the three effects (think Sweet Sue on the platform).
- There are relativistically correct formulas for momentum and total energy that are conserved in collisions and interactions. The old Newtonian quantities are not exactly conserved.

CHAPTER FORMULAS

We will be true to our advice by not giving *formulas* for time slowing and length contraction, asking you instead to simply remember what happens to moving rods and clocks in order to decide how to use the factor, based on whether you need a bigger or smaller answer.

Kinematics:

Square root factor: $\sqrt{1-\left(\frac{v}{c}\right)^2}$ Trailing clocks read ahead by: $\Delta t' = \frac{vL'}{c^2}$

Gravitational slowing of clocks in a gravitational potential: $\frac{\Delta t_{\text{TOP}}}{\Delta t_{\text{BOTTOM}}} = 1 + \frac{\Delta U}{c^2}$

Velocity: Velocity addition for a parallel component of velocity: $v_{\text{A,C}} = \frac{v_{\text{A,B}} + v_{\text{B,C}}}{1 + (v_{\text{A,B}} v_{\text{B,C}})/c^2}$

Mechanics: $\vec{p} = \frac{m\vec{v}}{\sqrt{1-(v/c)^2}}$ $E = \frac{mc^2}{\sqrt{1-(v/c)^2}}$ $E^2 = (pc)^2 + (mc^2)^2$

PROBLEMS

24.2 * 1. It takes Rachel 10 years of earth time to complete an interstellar journey moving at a speed $^1/_{10}$ the speed of light. How much will she age during the trip?

* 2. A photon torpedo 2.4 m long, when it is at rest, misses your shuttle and whizzes by at 0.80 c. What will be the length of the torpedo, as measured by you, as it passes by?

** 3. The Federation has set up sensor buoys at intervals of 1 light second all along the edge of the neutral zone. The clocks on the buoys are at rest and synchronized in the earth's rest frame. If a Romulan ship traveling east passes the buoys at $^3/_5\,c$, by how much will successive clocks appear to be out of synchronization, as seen by the Romulans? Will the clocks that are running ahead be those toward the east or the west?

** 4. If Josette leaves earth when she is 20 years old and travels at half the speed of light to a nearby star, arriving when she is 30, how far is it from the earth to the star, as measured by earth observers? [Hint: If you answered 5 light years, you should look at this a little more carefully.]

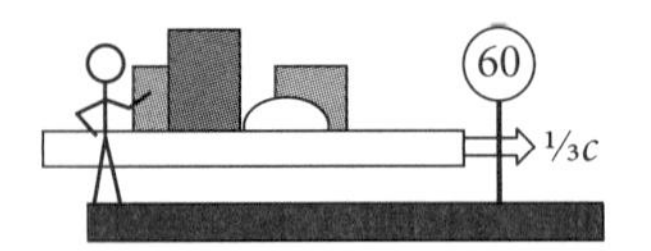

**5. A space barge passes your space station going $^1/_3\,c$ (see figure). At the instant the back of the barge passes you, you *see* the front of the barge at a point in your reference frame that you know is 60 m away. How long do you determine the barge to be? (This is what we call "making the light-time correction" in a length measurement.) [Hint: There is no relativity in this problem, just the simple kinematics of speed and distance.]

24.3 ** 6. We measure the length of a railway car passing us at $^3/_5\,c$ to be 100 m.

a) How long is the car, as measured in its own frame?

b) If the clock in the back of the car reads $t = 10\ \mu s$ as the back of the car passes us, what will be the reading of a clock in the front of the car, as seen by us (assuming that the clocks were synchronized in their own frame)?

** 7. A rocket ship is 50 m long when it is at rest. How long will it take the ship to pass a stationary observer who watches it go by at a speed $^1/_2\,c$?

** 8. An unstable K^+ particle has a lifetime of 12.4 ns when it is at rest. Certain K^+ particles are created in a particle interaction and are seen in the lab to be moving away from the place in the lab where the interaction took place at 0.96 times the speed of light, relative to the laboratory. How long will they survive according to laboratory clocks?

** 9. Consider the Section 24.1 example from the reference frame of the measurement car. A section of track moving to the left at $^3/_5\,c$ carries an oval light bulb and an angled mirror, 300 m apart, as measured in the earth frame. Working in the car's reference frame, explain which of the clocks will see the second flash and what the reading on that clock will be when the flash occurs.

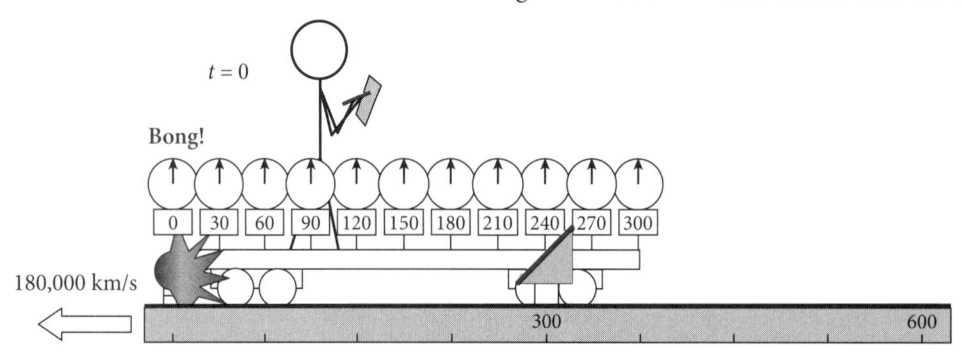

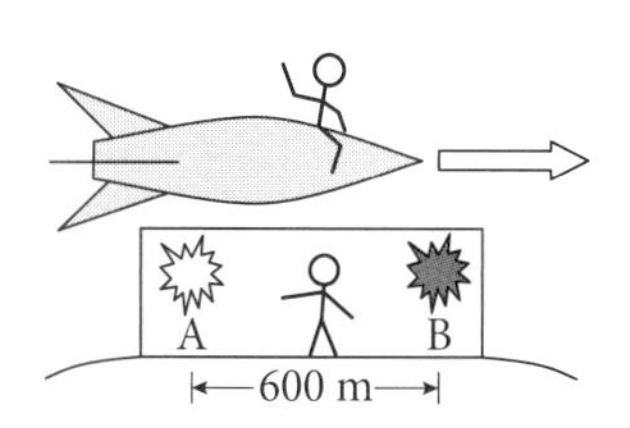

** 10. A laboratory observer notes that Event B occurs 800 ns later than Event A, and that the two events are separated in distance by 600 m, as shown. How fast must another observer be moving in order for those two events to be simultaneous, as measured by him? Is the rocket ship in the picture moving in the correct direction for this to occur? Explain your reasoning.

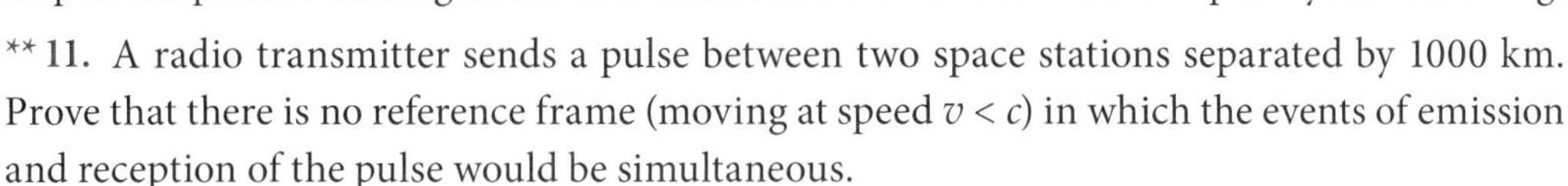

** 11. A radio transmitter sends a pulse between two space stations separated by 1000 km. Prove that there is no reference frame (moving at speed $v < c$) in which the events of emission and reception of the pulse would be simultaneous.

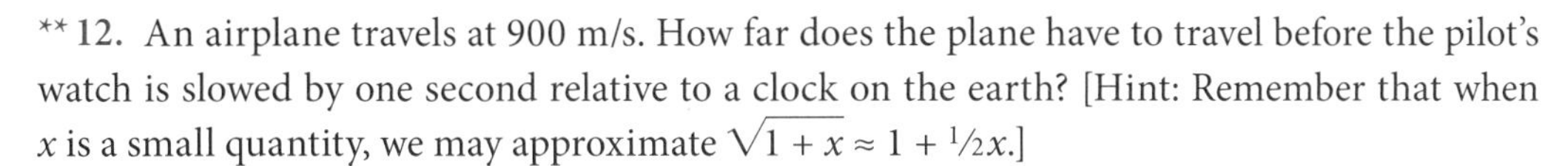

** 12. An airplane travels at 900 m/s. How far does the plane have to travel before the pilot's watch is slowed by one second relative to a clock on the earth? [Hint: Remember that when x is a small quantity, we may approximate $\sqrt{1+x} \approx 1 + \frac{1}{2}x$.]

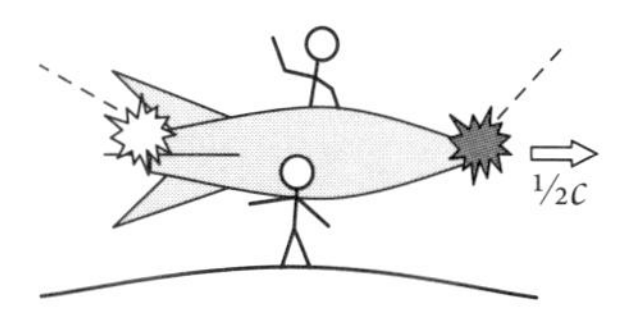

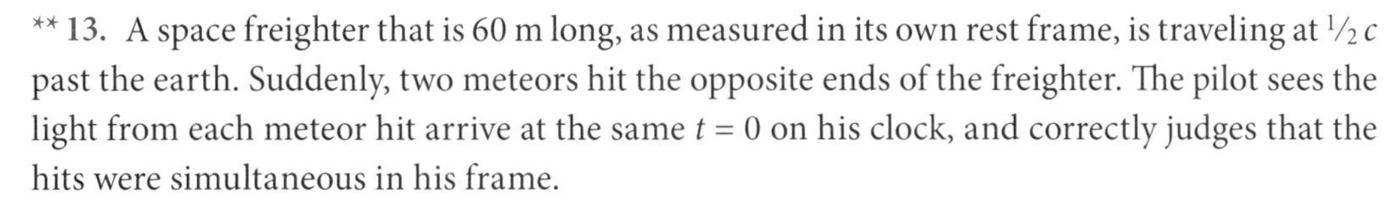

** 13. A space freighter that is 60 m long, as measured in its own rest frame, is traveling at $\frac{1}{2}c$ past the earth. Suddenly, two meteors hit the opposite ends of the freighter. The pilot sees the light from each meteor hit arrive at the same $t = 0$ on his clock, and correctly judges that the hits were simultaneous in his frame.

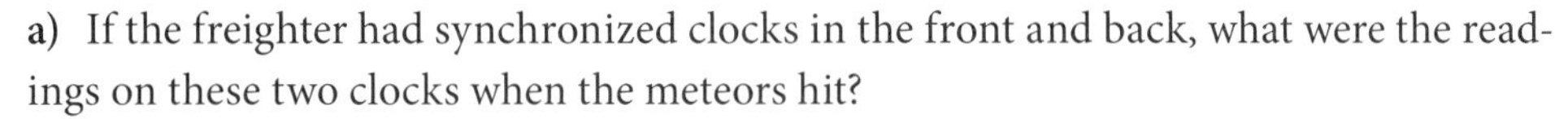

a) If the freighter had synchronized clocks in the front and back, what were the readings on these two clocks when the meteors hit?

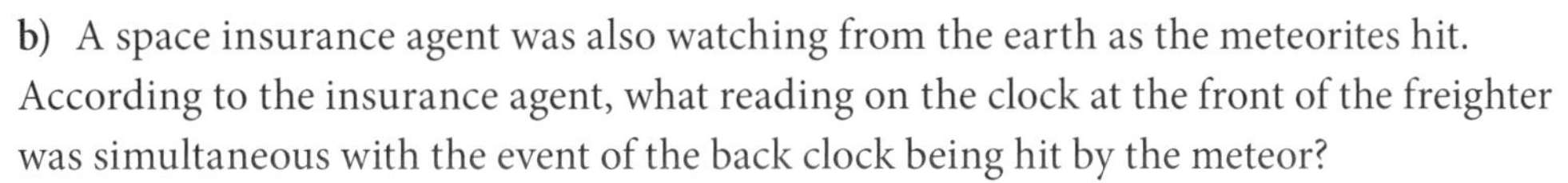

b) A space insurance agent was also watching from the earth as the meteorites hit. According to the insurance agent, what reading on the clock at the front of the freighter was simultaneous with the event of the back clock being hit by the meteor?

** 14. A spaceship of length 2 km, as measured in its own frame, moves past us at $v = \frac{4}{5}c$. There are two clocks on the ship, one in the nose and one at the tail. Both have been synchronized in their own frame. We, on the ground, have a number of clocks, synchronized with each other in our frame. Just as the nose of the rocket ship reaches us, both our clock and the clock in the nose read $t = 0$.

a) At this time, what does the clock in the tail of the ship read?

b) How long does it take the tail of the ship to reach us?

c) At this time, when the clock in the tail is beside us, what does it read?

** 15^{f}. **Proper Time.** An object in the laboratory moving at constant speed covers a laboratory distance Δx in elapsed time Δt. Find a formula for the time $\Delta t'$ that would elapse during this time interval, as measured by a clock on the object. Time measured by a clock at rest on a moving object measures what is called the object's "proper time."

*** 16. The red giant Canopus is 100 light years way. How fast would an astronaut have to travel in order to arrive there while aging only 10 years?

*** 17. A clock moves away from earth at a speed $v = \frac{4}{5}c$ and is watched by Aqir, the astronomer, through his telescope. The moving clock and the observatory clock both read $t = 0$ at the instant of departure. What reading does Aqir see on the moving clock in his telescope at the time his own clock reads 100 s?

** 18. Rocky, an astronaut, is sent off on a rocket at $\frac{4}{5}c$ toward a neighboring star system. After his clocks have ticked off 27 months, he receives a light-speed message from Terri, his wife, that she has just given birth to a little girl. Draw appropriate diagrams and answer the following:

a) According to Rocky's clocks, how long after his departure did the birth of his daughter occur? [Hint: Draw diagrams in Rocky's rest frame.]

b) How long was this as measured by Terri's clocks?

** 19. In VLBI tracking of a spacecraft (see Problem 23.10), measurement of the times of arrival of radio waves at two widely separated earth stations provides a very precise measurement of the angular position of the spacecraft. However, in order to get a correct measurement, the effects of relativity must be taken into account. The phase of the signal is simultaneously uniform on each wavefront, as seen in the solar system reference frame. But the earth moves relative to this reference frame, and so synchronization of the earth clocks at the tracking stations will not produce synchronized clocks in the solar system frame. The orbital speed of the earth is about 30 km/s. If the earth is moving along the baseline between two antennas separated by 10^7 m, in the direction shown in the figure below, by how much should the time tags on the data from station A be delayed relative to those from station B to account for the relativistic effects of the motion of the earth? [Note: These values are realistic and this correction must actually be done in order to get accurate VLBI tracking from the Deep Space Network.]

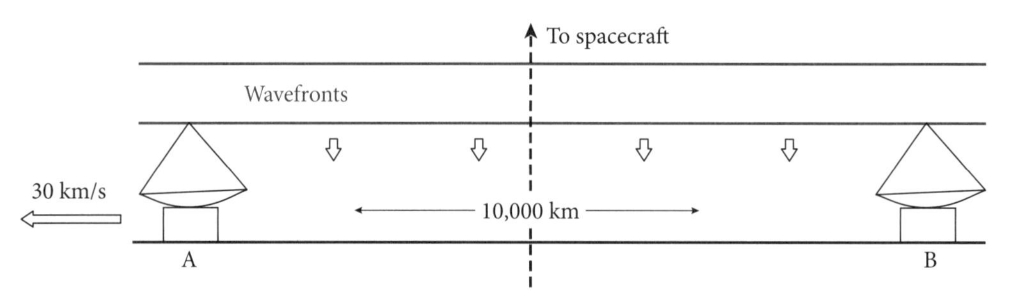

*** 20. Clocks on the *Millennium Falcon* read 4:00 am when its captain, Han Solo, detects a radar pulse. From the strength and frequency of the pulse, R2D2 is able to calculate that the pulse came from an Imperial cruiser that is approaching the *Millennium Falcon* at $v = {}^3\!/_5\, c$ and that the cruiser was 4 $c\cdot$hrs away (as measured in the *Millennium Falcon*'s rest frame) when the pulse was sent. Solo knows that when the reflected radar pulse is received by the Imperial cruiser, Darth Vader will know of the presence of the *Falcon* and will order an attack.

a) Draw three pictures showing (1) the time when the radar pulse left the Imperial cruiser, (2) the time when the pulse hit the *Falcon*, and (3) the time when the reflected pulse is finally received by the cruiser.

b) Use those pictures to calculate the time, according to the *Millennium Falcon*'s clocks, when the reflected radar pulse is received by the Imperial cruiser. [Remember, radar pulses travel at the speed of light.]

c) If the Imperial cruiser's clocks read midnight when the radar pulse was first sent, what time will their clocks read when the reflected pulse is received?

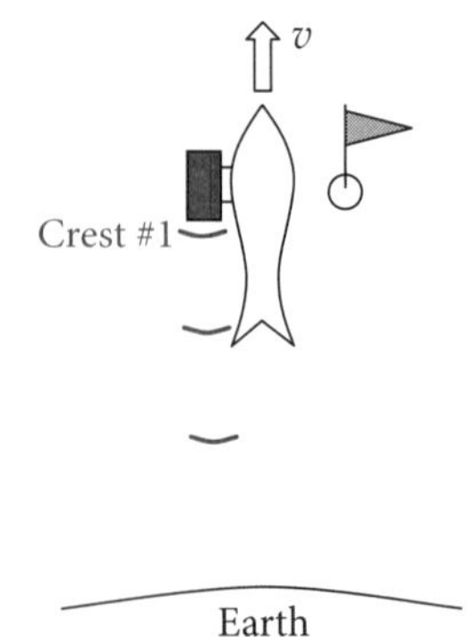

*** 21[f]. **Relativistic Doppler Shift.** A light source in the laboratory emits a wave crest every Δt seconds. Therefore, as the wave travels, its wavelength measured in the lab will be $\lambda = c\Delta t$. The source is now placed on a rocket ship moving away from earth at speed v. As the rocket passes a space marker at rest relative to the earth, it emits wave crest #1 back toward the earth.

a) How much earth time $\Delta t'$ will elapse before the wave crest #2 is emitted?

b) As measured in the earth frame, how far from the space marker will wave crest #1 be when crest #2 is emitted and how far from the marker will the source be when crest #2 is emitted? The difference between these two points is the wavelength of the wave as seen in the earth frame.

c) Find the wavelength λ' of the wave, as measured on earth, in terms of λ, v, and c. [Algebra hint: Remember that $c^2 - v^2 = (c + v)(c - v)$.]

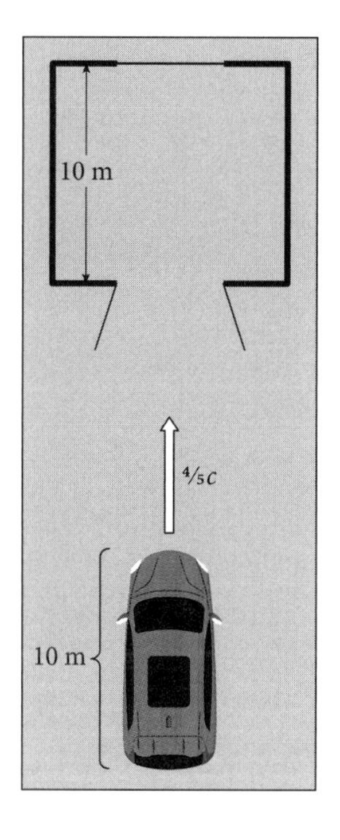

*** 22. **Car in the Garage Paradox.** A car 10 meters long is moving at a speed $v = {}^4\!/_5\, c$ toward a garage that is a little less than 10 m long. Because the car is moving, observers in the reference frame of the garage will see the car shortened to $L\sqrt{1 - v^2/c^2} = (10)\sqrt{1 - 16/25} = 6$ m. Clearly, such a short little car will easily fit inside the garage. But, in the car's rest frame, it is the garage that is moving. As seen from the car, the garage will be only 6 m long, and it will never be able to enclose a 10-meter-long car inside it. Of course the car cannot remain inside the garage, because, if it stopped, it would again be 10 meters long and would stick out in front and back. But, if the car keeps moving, and if the front door is closed first and the back door opened later, then we can clearly say the car was inside the garage for that short time interval. Assume that a clock on the front door reads $t_1 = 0$ and that a clock at the front of the car reads $t_1' = 0$ as the front of the car enters the garage. Then resolve the apparent paradox by calculating the following:

a) Find the reading on the front-door clock when the back of the car goes past the front door and it closes, and the reading on the back-door clock when the back door has to be opened to let the front of the car out.

b) By how much will the clock at the back of the car be reading ahead of the clock at the front of the car, as seen in the garage frame? Use this to find the reading on the clock at the back of the car when the front door closes and the reading on the clock at the front of the car when the back door opens to let it out.

c) Determine the order of the events of closing the front door and opening the back door, as seen in the two frames. Explain how this result resolves the paradox.

*** 23[f]. **Lorentz Transformation.** Two events occur in a rocket moving at speed v. We use x and t to label the events as measured in the rocket frame. A first event occurs at $x = 0$ and $t = 0$. The $x = 0$ end of the rocket passes the $x' = 0$ origin of ground coordinates at the first event, when both frames agree that $t = t' = 0$. A second event occurs at $x = \Delta x$ and at $t = \Delta t$. Note that Δx is also the length of the rocket, as measured in its own frame. Working in the ground frame, we will find formulas for $\Delta x'$ and $\Delta t'$, the difference in space and time coordinates of the two events as measured in the ground frame. The formulas we will derive, giving $\Delta x'$ and $\Delta t'$ in terms of Δx, Δt, and v, are known as the *Lorentz transformation*.

a) Find $\Delta t'$, the time between the two events as measured in the ground frame. [Hint: Remember that, between the two events, the rocket clock at $x = \Delta x$ starts out behind and ticks slowly till it reads $t = \Delta t$ at the second event.]

b) Find $\Delta x'$, the position of the second event, as measured in the ground frame.

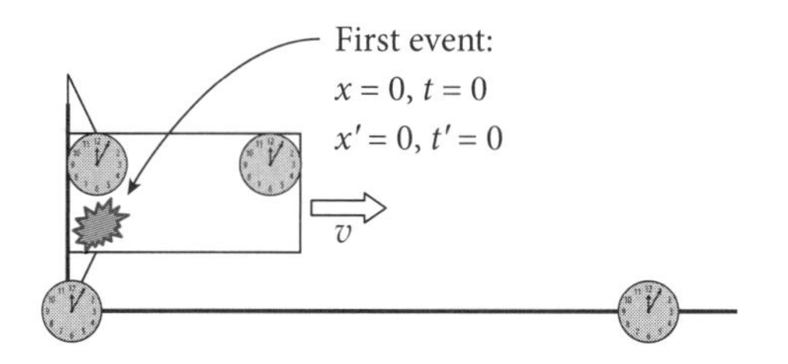

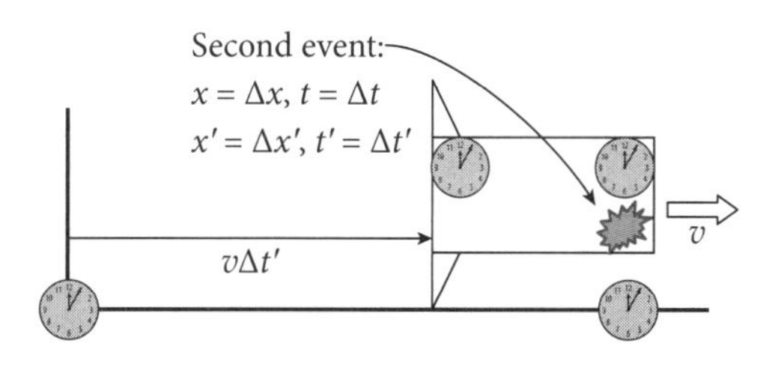

*** 24. **Another Big Story Problem.** At 1:00 am on April 1, in a year which, following advice of counsel, will not be named, John Dough was abducted by aliens from the planet Skyron and carried away to their home planet at ${}^4\!/_5\, c$. He immediately explained to his captors that they would have to return him to earth, because he had had income in the previous year and he had not yet filed his tax return. But the Skyronians informed him that they had all of his financial records in their computer memory banks and would be able to fill out his return for him. After exactly 5 days of computer calculations, they finally completed his return and sent it. In the rest frame of the Skyronian ship, the events would look like this:

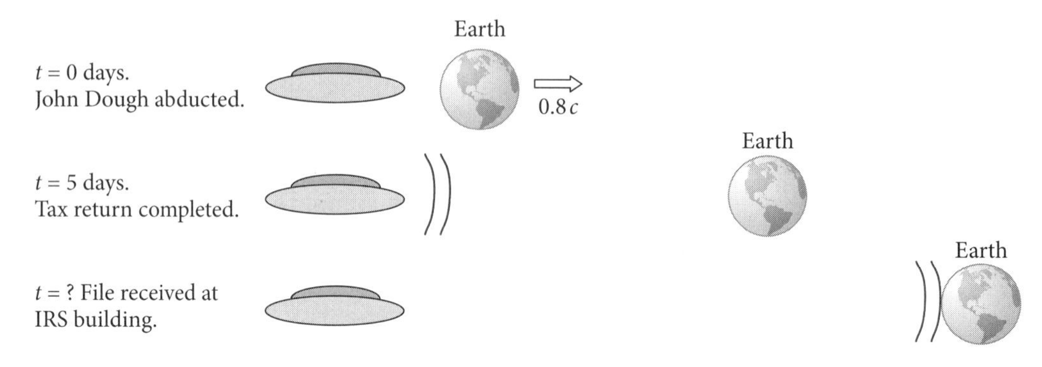

a) When did the return arrive on earth, as measured by the Skyronians?

b) Did the file arrive before midnight, April 15, as measured in the earth reference frame?

c) At what time, *as measured by clocks in the earth frame,* would John's tax return have been sent to earth? [Hint: The time stamp on the return would have read 1:00 am April 5, exactly 5 days of Skyronian time after the Skyronians left.]

d) Does the IRS determine lateness by when the return is received or by when it is posted? (All right. That last part isn't physics, so it shouldn't count.)

** 25. The latest trapped-ion clocks are able to produce a stability of a part in 10^{17}. (This means that they will be off by 10^{-17} seconds after 1 second has elapsed.) If one trapped-ion clock is placed on the ground and another is placed on a tabletop 1 meter high, will the gravitational redshift make a detectable difference in the rate of the clock ticking? Which clock will tick more slowly? Explain your answers.

** 26. A satellite carrying an atomic clock orbits the earth at an altitude of 550 km above the earth's surface at a speed of 8 km/s. The satellite stays in orbit for 10^7 seconds (about 4 months) before it is brought back down to earth.

a) Find the offset that will be seen between the satellite clock and an identical earth clock, due to the motion of the satellite. Which clock will be ahead?

b) Find the offset between the satellite clock and an identical earth clock due to the different gravitational potentials in each environment. Which clock will this effect cause to be advanced ahead of the other?

24.4 * 27. A rocket is moving to the right at $\frac{1}{2}c$. A gun in the rocket shoots a pellet at $\frac{1}{2}c$, as measured in the rocket frame. What is the speed of the pellet relative to the ground if

a) the pellet is fired to the right?

b) the pellet is fired to the left?

* 28. A boy throws a baseball forward at 60 mi/hr in a train that is itself moving forward at 50 mi/hr. Because of the effects of relativity, the speed of the ball relative to the ground will not be exactly 110 mi/hr. Will the actual velocity be more or less than 110 mi/hr? Explain how you arrived at your answer.

** 29. A photon torpedo is fired from the *Enterprise* at a speed of $\frac{4}{5}c$ toward a Borg vessel that is coming toward it at a speed of $\frac{3}{5}c$, both speeds as measured by the *Enterprise*. What will be the speed of the photon torpedo as measured by observers in the Borg vessel?

** 30. Two spaceships are approaching the earth at equal speeds and from opposite directions, as seen from the earth. The captain of one of the ships measures the speed of the other ship and sees that it is 0.90 times the speed of light. What is the speed of each ship as seen from the earth?

The next two problems involve velocity transformations when the velocity in the moving frame is completely perpendicular to the relative velocity of two reference frames. We know we mentioned on page 526 that perpendicular components are generally trickier to work out, but these simple cases may be solved by direct use of the three effects.

** 31[f]. Consider the following situations in which a car in a train, moving at horizontal speed v, is observed from the ground.

(a)

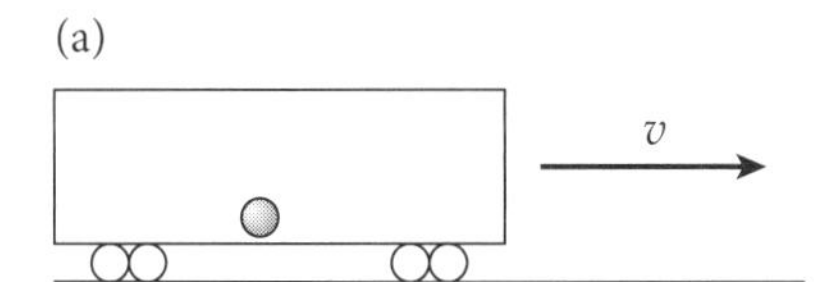

(b)

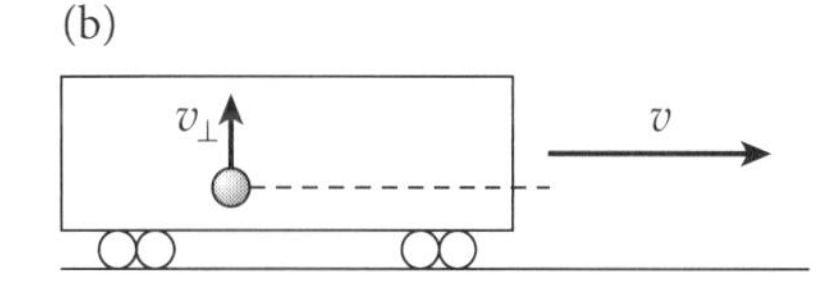

a) A ball is at rest in the car. What is the horizontal component of its velocity, measured from the ground? [Note: This is supposed to be easy.]

b) A ball is tossed directly upward relative to the car with velocity $v_\perp$. As measured from the car, the ball will rise a height h in time Δt. What will be the h' height moved and the $\Delta t'$ time elapsed as measured from the ground?

c) Find a formula for the vertical component of the ball's velocity, $v'_\perp$, as measured from the ground? Write this in terms of $v_\perp$, v, and c.

** 32. A 2-dimensional x–y coordinate system is centered on the sun, as shown, and a beam of starlight streams straight down along the negative y-axis, as shown. Its components in the sun frame are therefore $v_{||} = 0$ and $v_\perp = -c$. A flying saucer from Rigel is moving in the x-direction at a speed of $\frac{3}{5}c$. Note that, in the saucer reference frame, it is the sun that is moving. The sun's velocity relative to the saucer would be $v = -\frac{3}{5}c$.

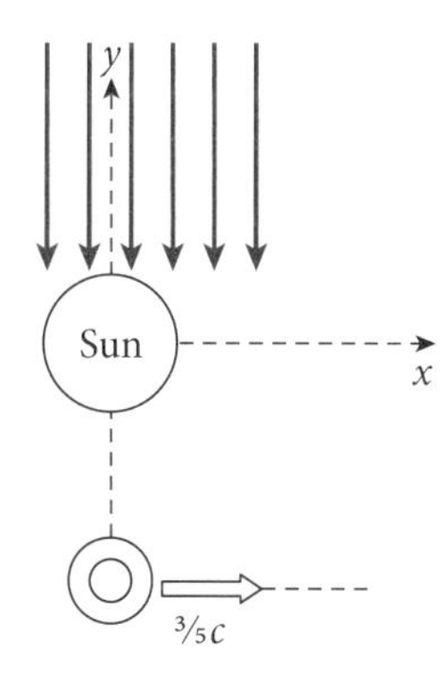

a) Find the two components of starlight velocity, $v'_{||}$ and $v'_\perp$, as measured by the Rigelians, relative to their saucer.

b) Find the magnitude of the starlight velocity, as measured by the Rigelians.

c) From what angle to their y-axis will the Rigelians see the starlight arrive?

[Hint: If you did not do Problem 31, you may find it useful to look at the answer to Problem 31 in the back of the book. There you will find formulas that are valid for a case where some object has a velocity, as measured in a moving frame, that is perpendicular to the direction of the motion of the moving frame relative to a fixed frame.]

24.5 * 33. The total power radiated from the sun is 3.9×10^{26} W. The source of this energy is the conversion of mass into energy in the nuclear reactions in the core of the sun. How many kilograms of mass does the sun lose each second? At this rate, how long would it take, in years, for the sun to completely consume its mass? (Conditions in the sun will radically change before such a thing occurs, of course.) [Hint: The mass of the sun is in Appendix B.]

* 34. A proton has a mass 1.67×10^{-27} kg and a total energy of 250 pJ. What is its momentum in kg·m/s? [Hint: See Equation 24.11.]

** 35. What is the rest energy in joules of a single atom of lead (mass 3.44×10^{-25} kg)? What is the total energy in joules of an atom of lead when it is moving at a speed of 0.999999 c?

** 36. Several 1-kg masses are moving at different speeds. Calculate the momentum of each mass using the (incorrect) Newtonian formula and the (correct) relativistic formula for each case. Then, in each case, determine the percent error made by using the Newtonian formula.

a) $v = \frac{1}{10}c$ b) $v = \frac{3}{5}c$ c) $v = \frac{99}{100}c$

** 37. In Example 24.10, we calculated the energy added to an object as it is accelerated to a speed v by subtracting the initial rest energy from the final total energy. As we discussed in Section 24.5, this energy could be called the kinetic energy, the relativistic formula for kinetic energy being given by Equation 24.10. Find the relativistic kinetic energy for a 1-kg object moving at $\frac{3}{5}c$ and compare it to the kinetic energy one would get by using the incorrect Newtonian KE = $\frac{1}{2}mv^2$.

** 38. A proton (mass 1.67×10^{-27} kg) is accelerated through a potential difference of 10^9 volts. Find its total energy in joules and its speed.

** 39. Electrons in a linear particle accelerator acquire a total energy exactly 4 times their rest energy. First work out the speed of these electrons, and then find:

a) How long it will take such an electron to travel a distance of 1 km after leaving the accelerator, as measured by clocks in the laboratory.

b) How far this electron will travel, as measured in a frame moving along with the electron.

** 40. A 10-gram bullet moves to the right at a speed of $\frac{3}{5}$ the speed of light, while another 10-gram bullet moves to the left at $\frac{3}{5}$ the speed of light.

a) What is the momentum of each bullet?

b) What is the total momentum of the system of the two bullets?

c) What is the total energy of the two bullets?

** 41. A star with total mass 4.0×10^{30} kg (twice the mass of the sun) runs out of nuclear fuel in its core and creates a supernova explosion, in which 9.0×10^{45} joules of energy are released. The star splits into two pieces in this explosion, and the energy released is carried away as the kinetic energy of the two pieces.

a) What is the total rest mass of the two remaining pieces?

b) One of the pieces has twice the mass of the other. What will be the approximate speeds of the two pieces as they leave the explosion site? [Note: The speeds will be large, but enough less than the speed of light that non-relativistic formulas for momentum can be used for an approximate answer.]

** 42. Cosmic rays consisting of alpha particles (rest mass 6.6×10^{-27} kg) moving at 0.90 c are absorbed by the test mass of an inertial sensor in a spacecraft. The test mass is a freely floating rectangular solid with dimensions 1 cm × 2 cm × 4 cm. It is composed of a platinum–gold alloy with density 20.0 g/cm^3. The purpose of the inertial sensor is to measure spacecraft accelerations from external forces. It is therefore important that other things have negligible effects on the test mass.

a) What is the momentum of a single alpha particle?

b) What is the speed of the test mass after one alpha particle has been absorbed?

** 43. The cosmic-ray alpha particles discussed in the previous problem maintain their rest masses when they are absorbed, so they simply add a little bit to the total mass of the test mass. But the kinetic energy they had is deposited into the test mass as they come to rest, slightly raising the temperature of the platinum and gold atoms in the test mass.

a) What is the total energy of one of the alpha particles and what is its kinetic energy?

b) The specific heat of the test mass is 0.13 J/g·K. If 200 alpha particles are absorbed, what will be the rise in the temperature of the test mass?

*** 44. An object of mass m moves at an initial speed $\frac{4}{5}c$ toward an initially stationary object of mass M. After a 1-dimensional collision, the mass m rebounds at speed $\frac{3}{5}c$, while mass M moves off to the right at speed $\frac{5}{13}c$.

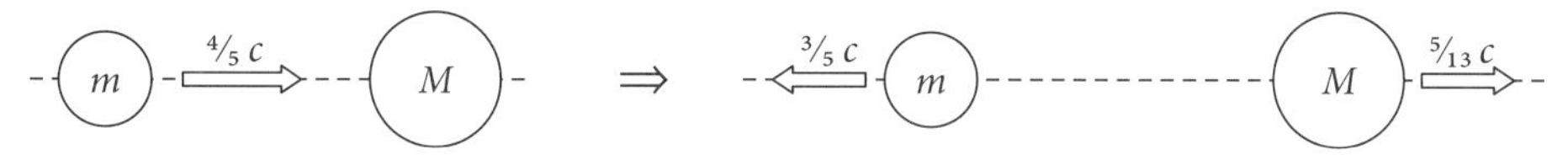

a) What is the ratio m/M?

b) If $m = 1$ kg, calculate the total energy in joules before and after the collision.

25

ATOMIC PHYSICS AND QUANTUM MECHANICS

The turn of the 20th century was probably one of the most self-satisfied moments in western culture, and physics was no exception. Scientists thought that they had the answers to everything. William Thompson, also known as Lord Kelvin (yes, this is the guy whose name is on the temperature unit), is reputed to have said, in the decade leading up to 1900, "There is nothing new to be discovered in physics now. All that remains is more and more precise measurement."[1] Still, physics recognized a few minor little problems that had not yet been completely resolved. One of these was what was called the *photoelectric effect.*

25.1 Photons

It was discovered, in the late 1880s, that when light fell on a clean piece of metal, electrons would be emitted from the surface, creating a current in a vacuum tube (see Figure 25.1). There was no surprise in this, of course. Light is an electromagnetic wave, with electric fields. And electric fields accelerate charges. It was even found, as expected, that a brighter light produced more current. But what was not expected, and, in fact, could not be readily explained, was that the effect could be produced with ultraviolet light, but not with visible light. No matter how bright the light, no electrons were emitted unless the frequency of the light was in the ultraviolet range or higher.

This was difficult to understand. It was expected that the electrons would be held inside the material by atomic forces, but the acceleration of the electrons should depend only on the magnitude of the electric field, and the magnitude of the electric field is what determines the brightness of the light. Nonetheless, no matter how bright the light beamed at the metal, the electrons responded, not to the magnitude of the electric field, but to its frequency. An experiment was also performed using light of even shorter wavelength, and electrons that were kicked out of the metal by ultraviolet light were subjected to a backward electric field, as shown in Figure 25.1. The result was that, when the voltage difference that the electrons saw was exactly right, the electrons would be stopped just before they reached the plate on the right, a condition that was noted by the current in the circuit dropping to zero. And it was found, again, that the required stopping voltage did not depend on the brightness of the light, but on its frequency. The higher the frequency, the higher would be the kinetic energy of the electrons as they left the metal.

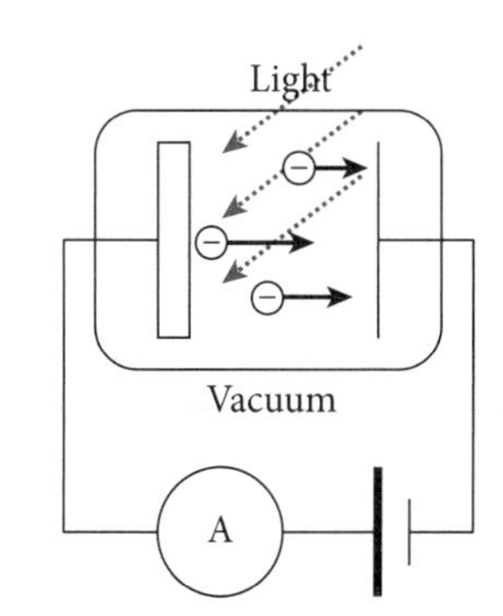

Figure 25.1 A circuit to test the stopping voltage for ejected photoelectrons.

1. Quoted in Paul Davies & Julian Brown, *Superstrings: A Theory of Everything?* (Cambridge University Press, 1992), although no first-hand source for the statement is known.

The solution to this mystery was suggested by Albert Einstein in 1905, the same year he published his paper on special relativity. In his solution, Einstein borrowed an idea that Max Planck had used in 1901 to explain the observed spectrum of the electromagnetic radiation from a blackbody (a concept we mentioned back on page 438). Einstein proposed that light, even though it exhibits all the properties of a wave, interacts with matter like a particle. The particle later came to be called a *photon*. Einstein's theory stated that light was emitted or absorbed by matter one photon at a time. The energy of each photon depended on the frequency of the light it represented, according to what is known as Planck's formula:

$$E = hf, \tag{25.1}$$

in which h, known as Planck's constant, has the value 6.626×10^{-34} J/Hz.

We are used to energy being carried by objects with mass, the relativistic total energy of the object being given by Equation 24.9, $E = mc^2/\sqrt{1 - (v/c)^2}$. However, since photons travel at the speed of light, the denominator in this formula would be zero. The only way to keep the fraction regular is for the numerator to go to zero also; the mass must be zero. So a photon's energy must not be thought of as the energy of a moving mass. There is no mass. It is just . . . energy.

But let us return to the problem of the photoelectric effect. The way Einstein's photon theory solves the problem should be more or less obvious. If light has a low frequency, it is composed of low-energy photons. A bright light will have many photons in the beam, but, since only one photon hits an electron at a time, the photon cannot kick the electron out of the metal unless it has sufficient energy to do so by itself. As light of some frequency falls on the metal plate in Figure 25.1, it should be thought of as a stream of particles, each with an energy given by Equation 25.1. The photons travel through space according to the wave theory for their wavelength, as discussed in Chapter 23, but they interact with matter like individual particles.

EXAMPLE 25.1 The Number of Photons in a Beam of Light

An experimenter collides two 1-mW laser beams in his laboratory. One beam contains green light of frequency $f = 5.64 \times 10^{14}$ Hz and the other contains red light of frequency $f = 4.62 \times 10^{14}$ Hz. How many photons per second are contained in each beam?

ANSWER The energy of each photon in the two beams will depend on the frequency of the light, according to Equation 25.1. The energy of each green-light photon is

$$E_{\text{green}} = hf = (6.626 \times 10^{-34}\ \text{J/Hz})(5.64 \times 10^{14}\ \text{Hz}) = 3.73 \times 10^{-19}\ \text{J}.$$

And the red-light photons in the problem each carry

$$E_{\text{red}} = hf = (6.626 \times 10^{-34}\ \text{J/Hz})(4.62 \times 10^{14}\ \text{Hz}) = 3.06 \times 10^{-19}\ \text{J}.$$

A milliwatt is 10^{-3} J/s, so the number of photons needed for 1 mW of power in each beam is

$$N_{\text{green}} = \frac{10^{-3}\ \text{J/s}}{3.73 \times 10^{-19}\ \text{J}} = 2.68 \times 10^{15}\ /\text{s} \quad \text{and} \quad N_{\text{red}} = \frac{10^{-3}\ \text{J/s}}{3.06 \times 10^{-19}\ \text{J}} = 3.27 \times 10^{15}\ /\text{s}.$$

It takes fewer of the green-light photons than it does the red-light photons to make up 1 mW of power in the beams.

It is probably obvious, after Example 25.1, that a joule is a very inconvenient energy unit for photons and atoms. Physicists almost always prefer a unit called the *electron-volt* (eV) for work in atomic or nuclear physics. One electron-volt is defined as the energy an electron acquires as it moves through a voltage difference of one volt. So one electron volt is $q\Delta V = (1.602 \times 10^{-19}\ \text{C})(1\ \text{V}) = 1.602 \times 10^{-19}\ \text{J}$. If we modify Equation 25.1 to give the energy of a photon in terms of wavelength, and if we convert units to eV and nm, we arrive at a most useful version of Planck's formula,

$$E = hf = \frac{hc}{\lambda} = \frac{(6.626 \times 10^{-34}\ \text{J/kg})(2.998 \times 10^{8}\ \text{m/s})}{\lambda}\left(\frac{1\ \text{nm}}{10^{-9}\ \text{m}}\right)\left(\frac{1\ \text{eV}}{1.602 \times 10^{-19}\ \text{J}}\right),$$

or

$$E = \frac{1240\ \text{eV}\cdot\text{nm}}{\lambda}. \tag{25.2}$$

This product hc has a very appropriate scale and units for working with atoms.

EXAMPLE 25.2 The Work Function of a Metal

The energy required to extract an electron from a metal is called the "work function" of the metal. For gold, it is found that photoelectrons can be ejected only if the wavelength of the light used is 244 nm or less. What is the work function of gold? What wavelength photon will give the ejected electron 0.90 eV of kinetic energy?

ANSWER The energy of each photon carried by 244-nm ultraviolet light can be found from Equation 25.2,

$$E = \frac{hc}{\lambda} = \frac{1240\ \text{eV}\cdot\text{nm}}{244\ \text{nm}} = 5.08\ \text{eV}.$$

When a photon of this energy hits an electron in the gold, it will give it just enough energy to be ejected from the metal; so this is the work function of gold.

If the wavelength of the light is shorter than 244 nm, the energy imparted to the electron will be more than 5.08 eV. In this case, the electron will give up 5.08 eV of energy as it leaves the surface of the gold, and the remainder will be its kinetic energy. Thus, the total energy given to the electron must be $E = 5.08\ \text{eV} + 0.90\ \text{eV} = 5.95\ \text{eV}$ and the photon must have wavelength $\lambda = 1240\ \text{eV}\cdot\text{nm}/5.95\ \text{eV} = 208\ \text{nm}$.

25.2 Atomic Physics

Einstein's characterization of the interaction of light with matter, as being more a question of the energy of a particle than of the amplitude and frequency of a wave, ultimately contributed to a new understanding of the nature of the atom.

Throughout the 1800s, scientists had discovered that, when various elements were placed in the flame of a Bunsen burner, they would emit a set of very precise wavelengths of light. These emission spectra, in fact, could be used to determine what elements were in an unknown sample of material. Once it was understood that light was electromagnetic radiation and that atoms contained electrons, a lot of work was done to try to explain the colors of light emitted as the natural frequencies of simple harmonic motion of vibrating electrons in the atom. But there was not much progress made in that direction. Then Einstein's photons opened up the idea that the solution to explaining the structure of atoms and the colors of light that they produce should focus on the energies of electrons in atoms, rather than their vibrational frequencies.

In the early 1900s, the British physicist Ernest Rutherford and a Danish physicist, Niels Bohr, were working on a model of the atom in which a single dense positively charged nucleus was orbited by a number of lighter electrons, looking like a tiny solar system of sun and planets. Niels Bohr, particularly, understood that he should *not* try to match the orbital frequencies to the frequencies of the light emitted, but that he should match the differences in the energies of the orbits to the energies of the photons produced. He had already realized that the atom must differ from a planetary system in that the orbits had to be what is called "quantized." This means that only orbits of certain radii and energies are allowed, unlike the planet orbits that can exist at any distance from the sun. But he had not been able to discover how to characterize the allowed orbits.

Then, in 1913, an old classmate of his, Hans Hansen, asked him if his model of the atom could explain the "Balmer series." When Bohr answered that he did not know anything about the Balmer series, Hansen explained to him that Johann Balmer, a Swiss mathematician, had found a mathematical formula that gave the wavelengths of the four observed colors of visible light emitted by hydrogen atoms. It was

$$\frac{1}{\lambda} = R\left[\frac{1}{4} - \frac{1}{n^2}\right] \quad \text{for } n = 3, 4, 5, \text{ or } 6, \tag{25.3}$$

where R is a proportionality constant known as *Rydberg's constant*. As soon as Bohr saw the formula, he knew he had the answer he was looking for. Now the answer may not be obvious to you, but that is because you have not spent the last five years thinking about energies of electrons orbiting protons. First, Bohr saw that Balmer's formula is an obvious difference between two terms, just like the energy lost by an atom when it emits a photon would be the difference between two electron energy levels in the atom. Second, Bohr knew (though you probably did not know off-hand) that one way of expressing the total mechanical energy of the electron (that is, no mass energy, but only kinetic energy plus electrical potential energy) is

$$E = -\frac{k^2 e^4 m}{2L^2},$$

where k is Coulomb's constant, e is the charge on the proton and on the electron, m is the mass of the electron, and L is the angular momentum of the electron as it circles the nucleus. The minus sign arises because the electrical potential energy is always more negative than the electron's kinetic energy is positive. Finally, Bohr knew that Planck's constant h actually had the same units as angular momentum.

So Bohr postulated that the electron orbits in a hydrogen atom were those in which the electron's angular momentum is quantized to be an integer multiple of Planck's constant. Actually, to get agreement with Balmer's formula, Bohr had to divide h by 2π, giving the allowed quantized angular momenta of the electrons as

$$L_n = \frac{nh}{2\pi} \quad \text{for } n = 1, 2, 3, \cdots. \tag{25.4}$$

With this requirement imposed, Bohr's formula for the quantized energies of the orbits of the electron in a hydrogen atom became

$$E_n = -\frac{2\pi^2 k^2 e^4 m}{h^2}\frac{1}{n^2} \quad \text{for } n = 1, 2, 3, \cdots.$$

All of the constants in the formula were known, allowing Bohr to calculate

$$E_n = -\frac{13.6 \text{ eV}}{n^2} \quad \text{for } n = 1, 2, 3, \cdots, \tag{25.5}$$

as the allowed energy levels for the hydrogen atom. The energy-level differences corresponded to photons whose wavelengths were those given in Balmer's formula. (If you love using your calculator it might be fun to check the calculation. You will conveniently find values for k, e, m, and h in Appendix B. Oh, and don't forget to convert the final energy units to electron-volts.)

As we said, Bohr pictured the electrons in an atom to be orbiting the nucleus like planets in a tiny solar system, although, as we will see in the next section, the reality is conceptually very different from this. Nevertheless, even with all the advances since Bohr's time, the energy levels given in Equation 25.5 remain correct.

Figure 25.2 shows a diagram of the energy levels in hydrogen. Note that all energy levels are negative, with E_n going to zero only in the limit that $n \to \infty$, and that the lowest energy the electron can have is –13.6 eV. If a hydrogen atom is excited to a higher state, say by collisions with other atoms or with passing electrons that are part of some current flowing through the gas, its electron can have energy $-13.6/2^2$ eV = –3.4 eV, or $-13.6/3^2$ eV = –1.51 eV, etc., as shown, but nothing in between these quantized values. Also shown in Figure 25.2 are six possible energy-level transitions, labeled a through f. The arrows are double-headed, reminding us that transitions can go either way. Each represents an amount of energy that a hydrogen atom can absorb as the electron is kicked up to a higher energy or the energy of the photon that will be emitted as the electron falls back down to a lower state. Finally, let us state explicitly that if light falls on hydrogen gas, but if its photon energies do not exactly correspond to one of the allowed electron transitions, the photon will pass right through the hydrogen atoms without being absorbed.

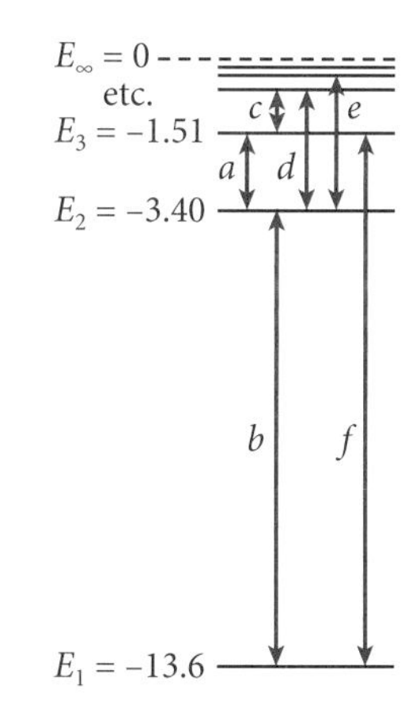

Figure 25.2 Energy levels of hydrogen, in eV.

EXAMPLE 25.3 Some Wavelengths of Light Emitted by Hydrogen Atoms

Find the wavelengths of light emitted by hydrogen corresponding to several of the labeled transitions in Figure 25.2.

ANSWER Let's start with transition b that goes from $n = 2$ to $n = 1$. As an electron goes from –3.4 eV to –13.6 eV, the energy lost will be

$$\Delta E = -3.4 \text{ eV} - (-13.6 \text{ eV}) = 10.2 \text{ eV}.$$

A photon will have to be created to carry off that energy. A photon with this energy must have a wavelength given by Planck's formula

$$\lambda = \frac{1240 \text{ eV}\cdot\text{nm}}{E} = \frac{1240 \text{ eV}\cdot\text{nm}}{10.2 \text{ eV}} = 121.6 \text{ nm}.$$

Table 21.1 tells us that this wavelength is in the ultraviolet. Any of the double-ended arrows in Figure 25.2 that are longer than the arrow for transition b correspond to a higher-energy photons and to correspondingly shorter wavelengths. So all such photons will be in the ultraviolet.

The transitions labeled a, d, and e all end on the $n = 2$ energy level. A look back at Balmer's formula (Equation 25.3), with its $2^2 = 4$ in the denominator of the first term, suggests that the wavelengths of the Balmer series are those that end on $n = 2$. Transition a drops from $E_3 = -1.51$ eV to $E_2 = -3.40$ eV, for a photon energy of 1.89 eV, corresponding to a wavelength

$$\lambda = \frac{1240 \text{ eV} \cdot \text{nm}}{1.89 \text{ eV}} = 656 \text{ nm}.$$

Similarly, transition d, from E_4 ($-13.6/4^2 = -0.85$ eV) to E_2 carries away 2.55 eV of energy at wavelength 486 nm, while transition e, from E_5 ($-13.6/5^2 = -0.544$ eV) to E_2, creates a photon of energy 2.86 eV, corresponding to $\lambda = 434$ nm. Table 21.1 shows that all of the Balmer wavelengths are in the visible portion of the spectrum.

Before we leave this section, let us be clear that the Bohr formula applies to hydrogen only. Similar procedures can be used for other atoms, but the energy levels will not be those given by Equation 25.5. So, if you are doing problems involving other elements, information about the energy levels will have to be given in the problem. However, once we know the energies of the photons produced, Equation 25.2 may always be used to find the wavelength of the light each photon carries.

25.3 Quantum Mechanics

In 1924, a French physics student at the Sorbonne, named Louis-Victor-Pierre-Raymond de Broglie [pronounced duh-broy′-uh], presented an interesting problem to his thesis committee. A doctoral thesis is supposed to be original, but de Broglie's was revolutionary. They could always reject his thesis and send him back to work on something else, but there were his family connections to consider. He was a member of French nobility, and the ducs de Broglie dated back to the time of Louis XIV. So, looking for advice, they sent the thesis to Einstein, who responded with his opinion that de Broglie had, in fact, unraveled one of the secrets of the universe. The PhD was awarded, and, in 1929, de Broglie received the Nobel Prize for his work, the only doctoral thesis ever to be so honored.

So what was so revolutionary? As Louis de Broglie returned to the university after World War I, he became interested in Einstein's idea that photons interacted with matter like particles, but traveled through space like waves. Young Louis wondered if all matter might not behave in the same way. He suggested that, when an electron collided with other matter, it did so as a particle with momentum p, as usual, but that when an electron traveled through space, it did so like a wave of wavelength

$$\lambda = \frac{h}{p}. \tag{25.6}$$

The immediate support for his idea was the way in which it could be used to explain the quantization of Bohr electron orbits in the hydrogen atom. Each of the allowed orbits in the Bohr atom had a particular radius and a particular angular momentum, allowing the electron's momentum to be determined. When its wavelength was calculated using Equation 25.6, it turned out to be a resonant wavelength of the Bohr orbit. Each wavelength was one in which an integer number of waves would just fit in the circumference of the circle corresponding to the Bohr orbit.

But the most important confirmation of de Broglie's theory came just a few years later when, in 1927, two American experimenters, Clinton Davisson and Lester Germer, performed an experiment that clearly demonstrated the existence of matter waves associated with electrons. What Davisson and Germer did was to aim a beam of electrons of known momentum (and, therefore, of known wavelength) at a crystal of nickel. As electrons reflected from the atoms in successive layers in the crystal, like light off a thin film of thickness d, a delay was introduced in electrons reflected at a particular angle (see Figure 25.3). It was found that, when the path length difference for the electrons reflecting off of successive layers into some direction was an integer multiple of their de Broglie wavelength, many electrons were detected, while, at other angles, the wave would interfere destructively and very few electrons would be found. Electrons were preferentially showing up at the places where constructive interference produced a wave of maximum amplitude. They were traveling through space like waves.

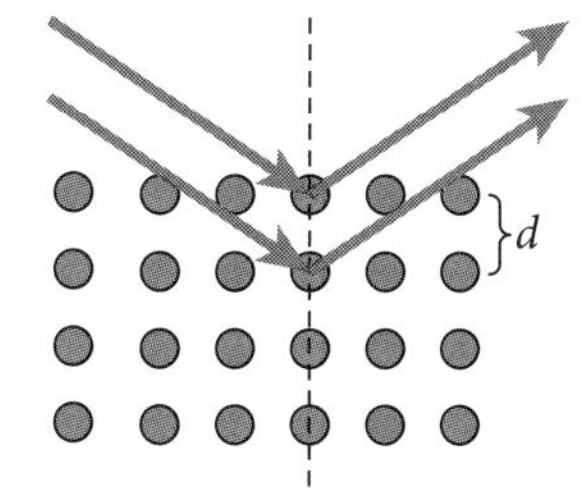

Figure 25.3 Diagram of the Davisson–Germer experiment showing interference in electron beams from de Broglie waves.

EXAMPLE 25.4 The Wavelength of an Electron Matter Wave

An electron is accelerated through a 10-kV voltage. What is its wavelength?

ANSWER To use Equation 25.6 for the wavelength, we will need to know the momentum of the electron. The energy added when the electron is accelerated is

$$\text{KE} = e\Delta V = (1.6 \times 10^{-19}\ \text{C})(10^4\ \text{V}) = 1.6 \times 10^{-15}\ \text{J}.$$

The voltage difference of 10 kV will not accelerate an electron to relativistic speeds, so we can simply use Newtonian formulas for kinetic energy and momentum. At low speed, we can write the momentum as $p = mv$ and the kinetic energy as $\text{KE} = \frac{1}{2}mv^2$, so we can also write kinetic energy as $\text{KE} = (p^2)/(2m)$. Solved for momentum, this equation gives $p = \sqrt{2m(\text{KE})}$. We know that the electron mass is 9.11×10^{-31} kg, so the momentum of this electron is

$$p = \sqrt{2m(\text{KE})} = \sqrt{2(9.11 \times 10^{-31}\ \text{kg})(1.6 \times 10^{-15}\ \text{J})} = 5.40 \times 10^{-23}\ \text{kg}\cdot\text{m/s}.$$

Knowing the momentum, we can find the wavelength from Equation 25.6. It is

$$\lambda = \frac{h}{p} = \frac{6.626 \times 10^{-34}\ \text{J/Hz}}{5.40 \times 10^{-22}\ \text{kg}\cdot\text{m/s}} = 1.23 \times 10^{-11}\ \text{m}.$$

This 12-picometer wavelength is much smaller than the hundreds of nanometers of wavelength typical for visible light. The shorter wavelength is the basic idea behind the electron microscope, used to see structures smaller than a wavelength of light.

If matter behaves like a wave, then it must be that a beam of particles passing through a slit will diffract and spread, just like a light beam. Let us consider a beam of electrons of momentum p and wavelength $\lambda = h/p$, as shown in Figure 25.4. We use coordinates x and y, with origin in

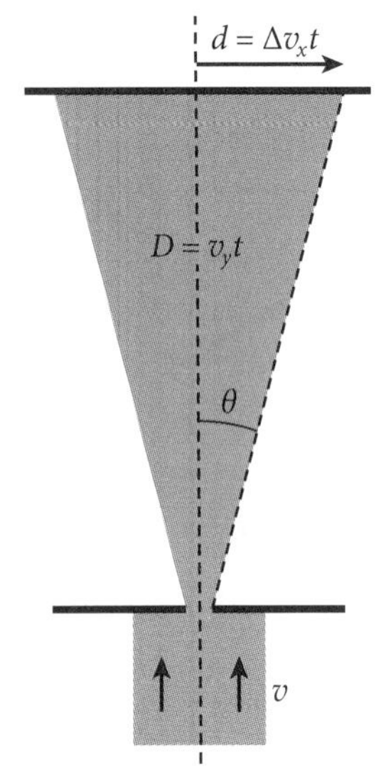

Figure 25.4 Matter–wave diffraction by a single slit.

the center of the slit, to locate the electron. Then, as we saw in Section 23.3, the diffraction of a beam of wavelength λ passing through a small opening of width w produces a first minimum at an angle satisfying $\sin\theta = \lambda/w$. Almost all the particles in the beam will hit within this central lobe of the diffraction pattern. Looked at another way, passing the electrons through a narrow slit is a way of localizing the electrons in the x-direction. We know that any electron passing into the region beyond the slit was once located somewhere inside a range of $\Delta x = \pm w/2$. On the other hand, the spreading of the beam means that, although the electrons in the beam may have started out with absolutely parallel velocity v_y, the passage of the beam through the slit gives each electron some unknown horizontal component of velocity, ranging between zero (for electrons that hit the center of the diffraction pattern) up to a maximum amount that we will call Δv_x (for electrons that end up at the edge of the lobe). The diffraction formula for a beam of wavelength λ relates the uncertainty in the horizontal position of an electron as it passes through the slit to the uncertainty in the horizontal velocity it acquires at the same time.

If θ is a small angle in Figure 25.4, then the formula $\lambda/w = \sin\theta$ becomes

$$\frac{\lambda}{w} \approx \frac{d}{D} \approx \frac{\Delta v_x t}{v_y t} \approx \frac{\Delta v_x}{v_y} \quad \Rightarrow \quad \Delta v_x w \approx v_y \lambda.$$

If we write w as $w = 2\Delta x$ and use de Broglie's formula for λ, we get

$$\Delta v_x \Delta x \approx \frac{v_y}{2}\frac{h}{p}.$$

If we now multiply both sides of this equation by the mass of the electron m, and remember that $p = mv_y$, the equation becomes

$$m\Delta v_x \Delta x \approx m v_y \frac{h}{2p} \quad \Rightarrow \quad \Delta p_x \Delta x \approx \frac{h}{2}. \tag{25.7}$$

In 1925, Werner Heisenberg suggested that this relation did not just apply to diffraction of matter waves through a slit, but represented a fundamental limitation on what can be known about matter. With a bit more care about the exact form of the wave and how the uncertainty is properly defined statistically, Equation 25.7 becomes what is now known as the *Heisenberg Uncertainty Principle*:

$$\Delta p_x \Delta x \geq \frac{h}{4\pi}. \tag{25.8}$$

According to this principle, no simultaneous measurement of the position and the momentum of an object can have a combined uncertainty lower than this limit. A similar uncertainty relation can be found for time and energy. It is

$$\Delta E \Delta t \geq \frac{h}{4\pi}. \tag{25.9}$$

The quantity ΔE in this last equation is obviously the accuracy with which the energy of a particle is measured, but the quantity Δt requires a little additional explanation.

While position, momentum, and energy are measured properties of a particle, the time is not. Time is measured by looking at a clock, not by looking at the particle. The accuracy of a time measurement can be whatever the laboratory clock can achieve, and it does not directly affect any other measurement of the particle. Thus, the Δt in the energy–time Heisenberg uncertainty relation of Equation 25.9 should not be thought of as the accuracy of the "time of a particle," whatever that might mean, but as the length of time that the energy must be measured before we can be sure that it has changed by an amount ΔE. This distinction will be important when we develop the ideas of virtual particles in Chapter 26. It also means that, if an energy is maintained at a single value for a time Δt, then the energy will be uncertain by an amount ΔE, as we will see below.

A couple of examples will help us understand these uncertainty relations.

EXAMPLE 25.5 Heisenberg Uncertainty in Target Shooting

A 25-gram rifle bullet is shot toward a target 1400 meters away (almost a mile) at a velocity of 700 m/s. If the position of the bullet is confined to a location inside the rifle barrel to within an uncertainty of ±10 μm, by how much might the bullet miss the target due to the uncertainty principle?

ANSWER If the position uncertainty is $\Delta x = 10^{-5}$ m, then the associated momentum uncertainty will be

$$\Delta p \geq \frac{h}{4\pi \Delta x} = \frac{6.626 \times 10^{-34}\ \text{J} \cdot \text{s}}{4\pi (10^{-5}\ \text{m})} = 5.3 \times 10^{-30}\ \text{kg} \cdot \text{m/s}.$$

In a 25-gram bullet, this will correspond to a velocity of

$$\Delta v = \frac{\Delta p}{m} = \frac{5.3 \times 10^{-30}\ \text{kg} \cdot \text{m/s}}{0.025\ \text{kg}} = 2.1 \times 10^{-28}\ \text{m/s}.$$

The bullet will require $t = d/v = 1400\ \text{m}/700\ \text{m/s} = 2$ s to reach the target, so this transverse velocity uncertainty can only move the bullet $\Delta vt = 4.2 \times 10^{-28}$ m. The Heisenberg uncertainty is actually much smaller than the position uncertainty inside the rifle barrel, though that uncertainty is itself negligibly small. If the shooter misses the target, the problem must lie in his aim, not in the laws of quantum mechanics.

EXAMPLE 25.6 The Natural Linewidth for the $n = 2$ State of Hydrogen

The lifetime of an electron in the first excited state ($n = 2$) of hydrogen is $\tau = 1.6$ ns before the electron drops down to the $n = 1$ state again. Because the energy is maintained for a short time only, the energy–time uncertainty relation can then be seen as requiring that the $n = 2$ state of hydrogen must have a small energy uncertainty. Thus, photons emitted in the transition between this excited state and the ground state do not have one precise wavelength, but are found with a small spread in wavelengths. This is referred to as the natural linewidth for the state. Calculate the natural linewidth for the first excited state of hydrogen.

ANSWER The two states of the hydrogen atom have an energy difference of 10.2 eV, corresponding to the creation of a photon of wavelength 121.6 nm, as we found in Example 25.3. But, since the excited state has a lifetime $\tau = 1.6 \times 10^{-9}$ s, there is an uncertainty in energy for this excited state, given by

$$\Delta E = \frac{h}{4\pi \Delta t} = \frac{h}{4\pi \tau} = \frac{6.626 \times 10^{-34}\ \text{J} \cdot \text{s}}{4\pi (1.6 \times 10^{-9}\ \text{s})} = 3.3 \times 10^{-26}\ \text{J}.$$

This energy range in joules converts to $\Delta E = 2.1 \times 10^{-7}$ eV. The wavelength of a photon with this additional energy would be

$$\lambda' = \frac{hc}{E + \Delta E} = \frac{hc}{E(1 + \Delta E/E)} = \frac{hc}{E}\left(1 + \frac{\Delta E}{E}\right)^{-1} \approx \lambda\left(1 - \frac{\Delta E}{E}\right),$$

where we have used the binomial expansion in the last step. The natural linewidth of this transition is the absolute value of the wavelength difference, or

$$\Delta\lambda = |\lambda' - \lambda| = \lambda\frac{\Delta E}{E} = (121.6 \text{ nm})\frac{2.1 \times 10^{-7} \text{ eV}}{10.2 \text{ eV}} = 2.5 \times 10^{-6} \text{ nm}.$$

Such a small difference in wavelength is actually very hard to measure.

We have talked about the wavelength of matter waves, but we have not yet said anything about their amplitude. The amplitudes of electromagnetic waves in Chapter 21 were the values of the electric field. But we saw that, when applied to light, this picture could not explain the photoelectric effect. And, for matter waves, there is not even an obvious field whose amplitude the waves might represent. Matter moves like a wave, but a wave of what?

Louis de Broglie thought of his matter waves as physical waves, made up of the energy of the particle they came from, so that the energy of the electron actually spread out to fill the interference pattern. In 1926, when Austrian physicist Erwin Schrödinger developed a partial differential equation whose solution gave the wave amplitude associated with a particle, he also thought of the wave as carrying the energy of the particle, spread out in space. But another interpretation soon came to dominate thinking about the physics of de Broglie's and Schrödinger's waves.

Because Heisenberg's principle implies that it is not possible to know where any particular electron passing through a narrow slit will end up—at the middle of the central lobe or off to the edge—it brings an element of chance into the mechanics of the very small. In the 1920s and 1930s, Neils Bohr, Werner Heisenberg, and Max Born came to explain Schrödinger's waves, not as waves of energy, but as waves of probability, where the square of the wave amplitude at any point gave the probability that the particle would be found there. This was a radical innovation. Physics, with its well-defined positions and velocities, was being replaced by statistics.

In one modern view, despite the wave behavior of the photons in Young's experiment or of the electrons in the Davisson-Germer experiment, one should think of both light and matter as fundamentally composed of particles. However, the paths that particles take and the places where they may be found are not determined by Newtonian-style kinematics and mechanics, but by the amplitude of a "probability wave" that is the solution of Schrödinger's equation for the particular system being considered. Particles will more often be found in places where the probability wave has a large amplitude.

In Figure 25.5, we show a computer simulation of the solutions to Schrödinger's equation for the hydrogen atom. Each tiny dot in the figure simulates the result of a measurement of the random position of the electron. Where there are many dots, the electron in the simulation showed up more often, so this indicates a region where the electron is more likely to be found. The different pictures represent solutions with different electron energies and angular momenta. The variables n and ℓ are what are called *quantum numbers.* Quantum number n labels the energy level from Bohr's formula (Equation 25.5), while ℓ labels the angular momentum of the electron.

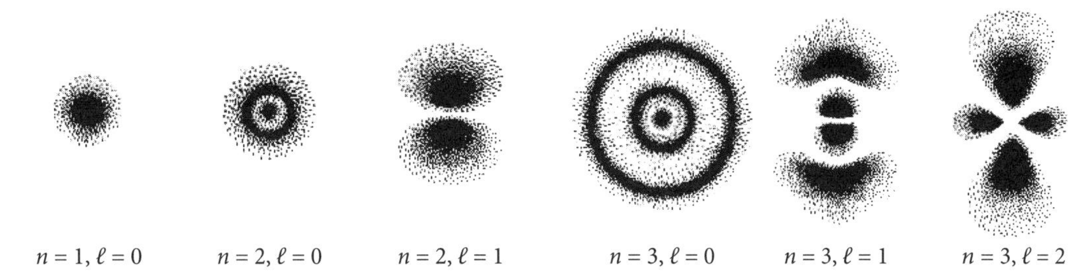

Figure 25.5 Simulation of electron positions for the $n = 1$, $n = 2$, and $n = 3$ states of hydrogen.

It should be noted how far this 1920s picture had come from Bohr's 1913 model of an electron orbiting the nucleus on a circle of well-defined radius, energy, and angular momentum. The radius is gone, though there are still indications of it in the high probability of finding an electron in spherical shells in the $\ell = 0$ patterns. But the angular momentum ideas have completely changed. Remember that it was the angular momentum L that Bohr quantized to develop his model, with a given L-value for each orbit. But we now see that the angular momentum can always have the value given by $\ell = 0$, regardless of the value of n, and that the angular momentum of an atom is actually independent of n, except for the restriction $\ell < n$. From Bohr's original atomic model, only the energy formula (Equation 25.5) remains, though it is still a reliable indicator of the principle energy levels in the hydrogen atom.

25.4 Spin

In the years that followed Niels Bohr's development of his model of the atom, he and other physicists quickly expanded on it. Different values of angular momentum were added (the ℓ values of Figure 25.5). But, even as Bohr was fitting his model to the Balmer formula for the wavelengths of hydrogen, it had long been known that most of the observed hydrogen wavelengths were what are called "doublets." Instead of a single wavelength of red light being emitted by an excited hydrogen atom, the atom actually emitted two wavelengths that were very close, but not exactly the same.

In 1925, two physicists from Holland, George Uhlenbeck and Samuel Goudsmit, realized that the doublet structure of the emitted wavelengths could be explained if the electrons in the atom each acted like a tiny magnet that interacted with the magnetic field produced by their own motion around the nucleus. When this little magnet was aligned with the orbital magnetic field, the energy would be slightly higher, and it would be slightly lower when it was oppositely aligned. It also became apparent that the electron had intrinsic angular momentum. Of course, an electron has mass and charge, so its angular momentum and its magnetic field could both be explained by assuming that the electron was spinning (a spinning charge carries charge in a circle, as if it were a current in an electromagnet). In fact, most of us *do* picture an electron as a tiny electrically charged grain of buckshot spinning around some axis. But, unfortunately, this picture quickly leads to conclusions that are completely wrong.

We need to remember that an electron is a very small object and that, for small objects, the laws of quantum mechanics are the rule. If an electron really were a small spinning sphere of some mass m, carrying a charge −1 e, with no quantum mechanics to it, we would expect that its angular momentum and its magnetic field could have any value and point in any direction. If we had an electron generator, as portrayed in Figure 25.6, the electrons in the beam could have their spin axes pointing in any direction, as indicated by the circle of random arrows in the figure. The question mark in the figure stands for two things. First, it tells us that the electrons in the beam have unknown spins until they are measured. But, second, it questions the very idea that the electrons *can* have spins as indicated by the array of arrows. Let's see what we mean by that last statement.

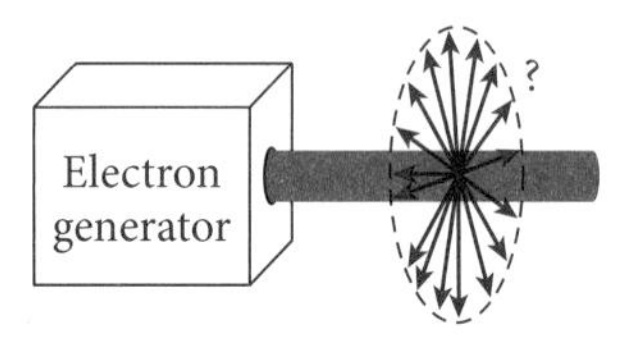

Figure 25.6 A generator of electrons with no spin alignment.

In Figure 25.7, we show a spin detector measuring the spins of the electrons in the beam. We turn the little double-arrow handle on the detector, setting it to measure spin components in the up–down direction. We saw in the Bohr atom that the electrons could only be in specific orbits and have particular energies, so it might not be surprising to us to learn that the angular momentum values we find in this experiment can only be $+\frac{1}{2}\hbar$ or $-\frac{1}{2}\hbar$, where $\hbar$ (pronounced "h-bar") is defined as $\hbar = h/2\pi$ (h being Planck's constant). But we wonder how that can be. The beam had no preferred orientation for the spins of the electrons it generated, so it would seem that there must be *some* spins that were pointing left or right, with little or no component in the up–down direction. But that doesn't ever happen. The angular momenta are $+\frac{1}{2}\hbar$ or $-\frac{1}{2}\hbar$, never anything in between. (We also define a term "spin," and say that the spins are $+\frac{1}{2}$ or $-\frac{1}{2}$.) It is as if the spin component we choose to measure affects the actual spins of the electrons in the beam.

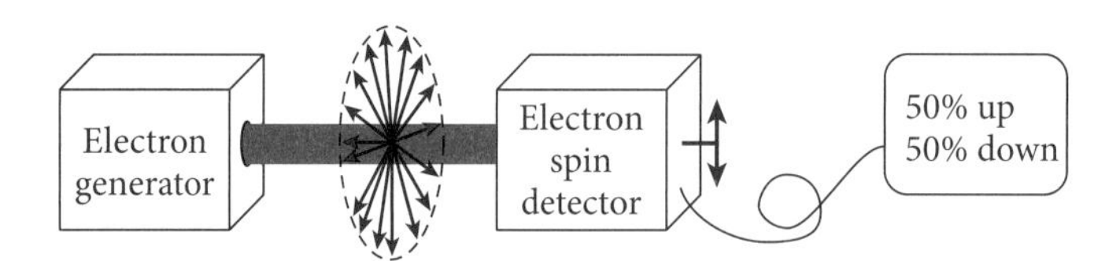

Figure 25.7 A spin detector aligned in the up–down axis.

But it is also possible to align the spins in the beam by using a magnetic field to put a spin-dependent force on the electrons and deflect all those with the wrong spins out of the beam. This way the beam will consist only of electrons with, say, spin up, as indicated by the arrow after the selector in Figure 25.8. In this case, as expected, 100% of the electrons detected will have spin up and none will have spin down.

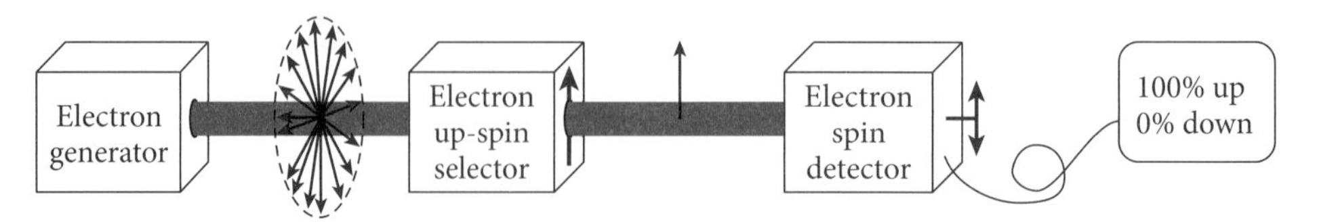

Figure 25.8 A spin selector is placed in the beam, passing only spin-up electrons.

However, if we chose instead to measure the component of the electron spin in the left–right direction, as shown in Figure 25.9, we would be in for another surprise. Even though all of the electrons in the selected beam had up–down spin components that were entirely in the up direction, and though all of them were exactly +½, the spin components in the left–right directions will not all be zero. In fact, none of them will be zero. Half will have spins to the left and half will have spins to the right.

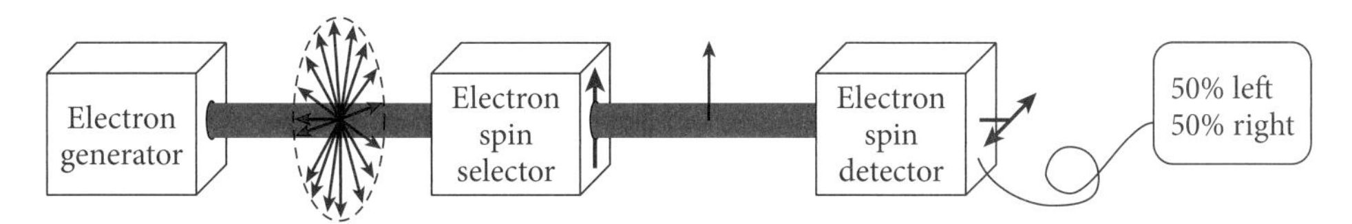

Figure 25.9 A spin detector oriented for left–right spins in a spin-up beam.

So we cannot think of an electron as a classical particle spinning about some axis. No matter how we choose to measure a component of an electron's spin angular momentum, there will only be two possible results, $+\frac{1}{2}\hbar$ or $-\frac{1}{2}\hbar$. Spin "up" is different than spin "down," and every electron will be one or the other. Spin is part of how we discriminate one electron from another. Let's see why this is important.

25.5 The Exclusion Principle

Chemistry arises in the electrons of an atom. If an atom has many electrons in it (equal to however many protons it has in its nucleus), then the number of electrons that are found in each possible energy state is determined by the descriptions of its electrons. An electron is described by its "quantum numbers." Each electron will have a value for its *principal quantum number n*, which gives its primary energy level. (The energies for hydrogen are given in Equation 25.5, but do remember that Equation 25.5 itself is valid only for hydrogen.) We also saw, in Figure 25.5, that the magnitude of the electron's angular momentum was set by its *angular quantum number* ℓ. The rule for ℓ is that it can take on positive integer values between 0 and $n - 1$. The wave function patterns shown in Figure 25.5 are determined by n and ℓ. And, to add to the confusion, the actual value of the angular momentum associated with the value of ℓ is not $\ell\hbar$, but turns out to be $L = \hbar\sqrt{\ell(\ell + 1)}$. So it is not like classical physics at all.

What's more, knowing the magnitude of the angular momentum vector is not enough. For $\ell = 1$ and higher, the patterns depicted in Figure 25.5 can be oriented in different spatial directions, corresponding to the orientation of the angular momentum vector. A third quantum number reflects that orientation. If we choose an axis, then there is a *magnetic quantum number, m*, that determines the component of angular momentum along that axis. The magnetic quantum number can take on positive and negative integer values between $m = -\ell$ and $m = +\ell$, the angular momentum component corresponding to each allowed value of m being $L_m = m\hbar$. Finally, the last quantum number describing an electron in an atom is its *spin quantum number, s*, whose values can be +½ or −½, as we have seen. The electron spin produces an additional angular momentum component $L_s = s\hbar$ along the chosen axis.

One of the reasons why these quantum numbers are important is because of a discovery made by the Austrian physicist, Wolfgang Pauli, in 1925. He noted that the number of electrons in each energy level of a complex atom could be predicted by requiring that each electron have a *different* set of values for these four quantum numbers. This discovery is known as the *Pauli*

Exclusion Principle, because, once there is one electron in an atom with a particular set of quantum numbers, other electrons with the same quantum numbers are *excluded* from the atom.

Both n and ℓ affect the energy of the electron, and electrons are added to a multi-electron atom in order of increasing energy. This is the basis for two rules that specify the order in which quantum numbers are assigned. The first is Janet's rule.[2] It states that the quantum numbers are taken in increasing order of the sum, $n + \ell$. The state $n = 4$, $\ell = 0$ ends up having lower energy than $n = 3$, $\ell = 2$, so the $n = 4$, $\ell = 0$ state would be taken before $n = 3$, $\ell = 2$. The second rule is Hund's rule. It says that, for a given set of n and ℓ values, all spins will be the same until all possible m options are used.

TABLE 25.1 Quantum Numbers for Oxygen's Eight Electrons

Number	n	ℓ	m	s
1	1	0	0	$-\frac{1}{2}$
2	1	0	0	$+\frac{1}{2}$
3	2	0	0	$-\frac{1}{2}$
4	2	0	0	$+\frac{1}{2}$
5	2	1	−1	$-\frac{1}{2}$
6	2	1	0	$-\frac{1}{2}$
7	2	1	+1	$-\frac{1}{2}$
8	2	1	−1*	$+\frac{1}{2}$

* The order of m values for an atom is arbitrary, so this could equally well have been 0 or +1.

The quantum numbers for the eight electrons in the oxygen atom are shown in Table 25.1. There are two electrons in the lowest energy state, one with spin up and one with spin down. In this $n = 1$ state, the only possible value for ℓ is 0 and the only possible value for m is 0. However, the 3rd electron in the oxygen atom cannot have $n = 1$, because all of the allowed $n = 1$ quantum numbers are taken by the first two electrons. The 3rd through the 8th electrons must have $n = 2$. With 2 as the principal quantum number, two values of ℓ are possible, $\ell = 0$ and $\ell = 1$. When $\ell = 0$, the only possible value for m is $m = 0$, but, when $\ell = 1$, there will be three possible values for m, −1, 0, and +1. Hund's Rule is expressed in the fact that all of the $n = 2$, $\ell = 1$ electrons have the same spin, which we have chosen to call "down," until the available m values are filled out. This accounts for electrons 5 through 7 coming in with spin down before the 8th electron comes in with spin up.

The rules for quantum numbers in an atom, and the Pauli Exclusion Principle itself, were invented in order to explain the experimental data for the ordering of the atomic electrons, but they succeeded far beyond that. They correctly gave the energies of electrons in atoms where not just the nucleus but the other electrons in the atom contributed to the energy of a single electron. And it was found that the Exclusion Principle applied not just to electrons, but to several other particles that were found to have spin ½, including the protons and neutrons in the nucleus. The energy levels in the nucleus are levels of the strong nuclear force rather than of the electrical force, but the same quantum rules apply and produce the so-called *shell model* of the nucleus.

The way in which energies are distributed among a set of spin-½ particles is called *Fermi–Dirac Statistics*, after two 20th-century physicists, Italian Enrico Fermi and the Englishman, Paul Dirac, who independently discovered the correct formula for the distribution. Spin-½ particles are referred to as *fermions* (Dirac generously chose the name), and all fermions obey the Exclusion Principle. But, as the knowledge of fundamental particles expanded during the 1950s and 1960s, it was found that there also existed particles with integer spin, rather than half-integer spin. We will mention some of these in our last chapter, in Section 26.4. These particles do not obey the Exclusion Principle. Particles with inte-

2. Named for French physicist Charles Janet (Ja-nay′). The rule is also called Madelung's Rule for German physicist Erwin Madelung, who discovered the same rule seven years later.

ger spin may share identical quantum numbers. Because of this fact, the way energies are distributed in a swarm of such particles is different from Fermi–Dirac statistics. Integer-spin particles obey what are called *Bose–Einstein Statistics*, named after Indian physicist Satyenda Bose and our old friend, Albert Einstein. Particles with spin 0 or 1 or higher integer values are called *bosons*.

25.6 Summary

Some of the concepts in this chapter are contrary to our experience, and often do not lead students to feel that they have really understood. Our experience is that you probably understand it better than you think. Anyway, here are some of the main concepts you should take away from the chapter.

- You should picture light as a collection of photons, which act like particles when they interact with atomic-level matter but whose motion is governed, not by Newtonian mechanics, but by the interference of their waves, the energy of the photons being related to the wavelength of the waves by Equation 25.2.
- You should understand that energy transitions in atoms are accompanied by the emission or absorption of photons, and that this explains why only certain wavelengths of light are emitted or absorbed by atoms. For the hydrogen atom, you should be able to relate the energy levels in the atom, as described by Equation 25.5, to the energies of the photons produced by hydrogen.
- You should understand that matter, like photons, behaves like a particle when it interacts with other matter, but that its motion is governed by its wave, with a wavelength given by Equation 25.6. However, for large objects, the wave nature may be ignored, as we saw in Example 25.5.
- You should also understand that the physics of photons and other particles is quantum mechanical and fundamentally probabilistic in nature, satisfying the Heisenberg Uncertainty Principle of Equations 25.8 and 25.9.

CHAPTER FORMULAS

Photons: $E = hf = \frac{hc}{\lambda} = \frac{1240\text{ eV}\cdot\text{nm}}{\lambda}$

The Bohr hydrogen atom: $E_n = -\frac{2\pi^2 k^2 e^4 m}{h^2}\frac{1}{n^2}$ for $n = 1, 2, 3, \cdots$ or $E_n = -\frac{13.6\text{ eV}}{n^2}$ for $n = 1, 2, 3, \cdots$

de Broglie wavelength: $\lambda = \frac{h}{p}$

Heisenberg Uncertainty Principle: $\Delta p \Delta x \geq \frac{h}{4\pi}$ and $\Delta E \Delta t \geq \frac{h}{4\pi}$

PROBLEMS

25.1 * 1. Ultraviolet light of wavelength 200 nm shines on a clean gold surface. What voltage is required to stop the electrons that are photoemitted by this light? [See Figure 25.1, and see Example 25.2 for the work function of gold.]

* 2. Light of wavelength 310 nm falls on a thin metal foil. The ejected photoelectrons may be stopped by a voltage of 0.6 volts. What is the work function of the metal?

** 3. As light shining on a metal surface is reduced in wavelength, photoelectrons begin to be detected when the wavelength reaches 400 nm. What will be the kinetic energy of photoelectrons ejected by light of wavelength 300 nm?

* 4. X-rays of wavelength 0.1 nm are detected coming from a star. What is the energy of the X-ray photons being produced by this star?

** 5. Estimate the number of photons emitted each second from a 60-watt light bulb. [Hint: You're going to have to just assume a wavelength for the light.]

** 6. A particular 1-watt laser produces light at 1064-nm wavelength. A typical cell phone emits 1 watt of power at a 900-MHz frequency.

a) What is the energy of each photon in the 1064-nm laser? How many photons per second make up the 1-watt laser beam?

b) Compare this result with the photon energy for signals from a cell phone and the number of photons per second emitted when the phone is radiating 1 W.

*** 7. Equation 24.8 gives the momentum of a massless photon as $p = E/c$. As a result of the photon momenta, a beam of light falling on a surface exerts a force on the surface. Find the force that a 1-watt beam of 500-nm light exerts on a perfectly reflecting surface by the following procedure:

a) Find the energy in joules of each photon in the beam and determine how many photons per second are required to produce the 1 watt of power.

b) Find the momentum of each photon in kg·m/s and determine the impulse required to reflect each photon, so it bounces back in the direction it came.

c) Find the total force exerted on the reflecting surface by the beam.

[Hint: You may want to review Section 8.1 on force and impulse and Section 14.1 on how to average multiple tiny collisions to find a macroscopic force.]

25.2 * 8. What is the energy of an electron in the $n = 1$ energy level of hydrogen? What is its energy in the $n = 5$ energy level?

** 9. A current passing through hydrogen gas kicks the electron in one atom up to the $n = 6$ energy level. Find the wavelength of the light emitted when this electron transitions down to the $n = 2$ level.

** 10. What are all the wavelengths of light that can be emitted by a hydrogen atom whose electron has been excited up to the $n = 4$ energy level?

** 11. An imaginary atom has energy levels $E_1 = -5$ eV, $E_2 = -2$ eV, and $E_3 = 0$, and no more.

a) Find the 3 wavelengths of photons that can be emitted by this atom.

b) Which of these wavelengths are in the visible? In what range are the others?

** 12. The principal energy levels in atoms with more than one electron are complicated by the interactions of the electrons with each other. However, an atom in which all but one electron have been removed will have a simple energy-level formula similar to the Bohr formula. The energy levels in ionized helium are given by $E_n = -54.4/n^2$. Find the energy of a photon emitted by ionized helium when

a) an electron drops from the $n = 2$ state to the $n = 1$ state;

b) an electron drops from the $n = 3$ state to the $n = 2$ state.

** 13. Show why you would expect an undisturbed single hydrogen atom at room temperature to be in its lowest $n = 1$ energy state. Show this with the following process:

a) Find the minimum energy required to kick an $n = 1$ electron up to the next highest energy level.

b) Convert this energy to joules and then to kelvins. Is this higher than room temperature?

*** 14. As stated on page 555, the values of angular momentum allowed by quantum mechanics are $L = \hbar\sqrt{\ell(\ell + 1)}$, for integer values of ℓ. The constraint to integer values of ℓ helps us understand the conclusion, stated without proof back on page 292, that rotational degrees of freedom of a molecule with small rotational inertia about some axis will not be excited at normal laboratory temperatures. To show the reason for this, we first need a new formula for the rotational kinetic energy of a spinning body. If we start with Equation 9.8, and substitute the connection $\omega = L/I$ coming from Equation 9.7, we find that the rotational kinetic energy of an object with angular momentum L and rotational inertia I is $\text{KE} = L^2/(2I)$. We may then explain the page 292 conclusion via the following steps:

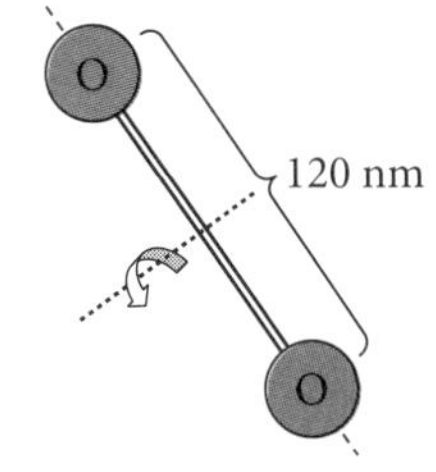

a) Consider the diatomic molecule O_2 as two tiny oxygen atoms, each of which has a mass 2.66×10^{-26} kg, separated by a distance 120 nm, and find the rotational inertia of this molecule for rotation about an axis through the center of mass of the molecule, perpendicular to the line joining the two oxygen atoms. This is shown as a rotation about the dotted-line axis in the picture at right. [Hint: See Figure 9.12 for rotational inertia formulas.]

b) The smallest non-zero angular momentum a molecule can have is given by $L = \hbar\sqrt{\ell(\ell + 1)}$, with $\ell = 1$. Calculate the smallest rotational kinetic energy the O_2 molecule can have, rotating about the dotted axis in the picture above right.

c) Use Boltzmann's relation between kinetic energy and temperature (Equation 14.2) to find the Kelvin temperature that corresponds to the energy found in part (b). Is this lower than common laboratory temperatures (so that you would expect this rotation to be excited by collisions in the oxygen gas)?

d) The formula for the rotational inertia of a sphere of mass m and radius r, rotating about its center, is $I = (0.4)mr^2$. For rotation about a line joining the two oxygen atoms (the double line in the picture), the rotational inertia of the molecule is the rotational inertia of two spheres of the radius of the oxygen nucleus ($\approx 2 \times 10^{-15}$ m), rotating about a common axis. Find the rotational inertia of an oxygen molecule about this axis.

e) What would be the rotational kinetic energy of an oxygen molecule rotating about this axis, when the angular momentum has its $\ell = 1$ value?

f) Use Boltzmann's relation to find the Kelvin temperature that corresponds to this rotational kinetic energy. Is this low enough, compared to common laboratory temperatures, that rotation about this axis should be excited?

25.3 ** 15. What is the wavelength of an electron whose total energy is 0.59 MeV? [Hint: First find the electron's relativistic momentum.]

* 16. The mass of a baseball is 0.145 kg. What is the de Broglie wavelength of a 90-mph (40-m/s) fastball?

** 17. A 0.5-kg tetherball is moving in a circular path of radius 2 m at a speed of 6 m/s. Suppose that the angular momentum of the tetherball is quantized according to Bohr's formula (Equation 25.4) and find the value of the quantum number n that gives the angular momentum of the tetherball. (Actually, the angular momentum of a tetherball *is* quantized, because quantum mechanics is how the universe works. It's just that n is so high, we don't notice the granularity.)

*** 18. The radii of the orbits in the Bohr model of the hydrogen atom are given by the formula $r_n = (0.0529 \text{ nm})n^2$.

a) Use this radius, along with Bohr's formula for quantizing angular momentum in the hydrogen atom (Equation 25.4), to find the linear momentum p of an electron in the $n = 3$ state. [Hint: Electrons have non-relativistic speeds in atoms, so Equation 9.6 can be used for the angular momentum.]

b) Find the de Broglie wavelength of this electron and calculate how many wavelengths would fit into the circumference of the $n = 3$ Bohr orbit.

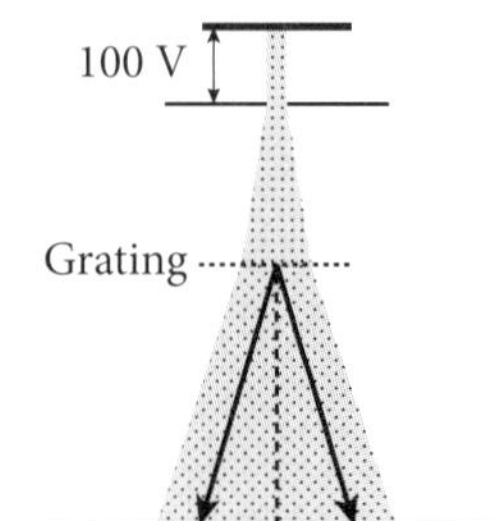

** 19. Electrons are accelerated through a potential difference of 100 volts and allowed to impinge on a grating with 3-nm slit separations.

a) Find the momentum of the electrons in the beam. [Hint: Remember that one way to write kinetic energy is $KE = \frac{1}{2}mv^2 = p^2/2m$.]

b) Find the de Broglie wavelength of the electrons.

c) What will be the angular location of the first-order bright fringe in the electron interference pattern?

** 20. A proton (1.67×10^{-27} kg) is accelerated through a potential difference of 2×10^9 V.

a) What is the rest energy of the proton?

b) What is its total energy after being accelerated?

c) What are the final momentum and wavelength of the proton?

25.5 * 21. How many electrons can be found in $n = 3$, $\ell = 2$ states in an atom?

* 22. How many electrons can be found in the $n = 3$ level of an atom?

** 23. Neon has 10 electrons in its electron cloud. When neon is in its lowest energy state, what will be the quantum numbers of each of its 10 electrons? Remember Janet's rule and Hund's rule.

** 24. Determine the lowest-energy electron configuration for calcium (20 electrons) and fill out a table like the one in Table 25.1. Remember Janet's rule and Hund's rule.

*** 25. One of the surprising manifestations of the Pauli Exclusion Principle is that it is important in the description of certain astronomical objects. One of these, a white dwarf, is a star in which an object with the mass of a typical star collapses down to a size slightly larger than the earth. At these high densities, the closeness of the electrons inside the star requires that the electrons occupy higher and higher energy states in order to satisfy the Exclusion

Principle. The pressure of these fast electrons is high enough that they will balance the otherwise overwhelming self-gravity of such a star. Estimate the effective temperature of the electrons in a white dwarf whose mass density is 10^9 kg/m^3 by the following procedure.

a) Knowing that white dwarfs are mostly composed of heavy nuclei, for which there are about as many neutrons as protons, determine the number of electrons per cubic meter of a white dwarf. [Hint: Remember that there are about 6.02×10^{23} nucleons in a gram of any material.]

b) Use this number to determine the average dimensions of a cube that contains a single electron. [Hint: A cube of side length d has volume d^3.]

c) If an electron is confined to a dimension of this size, the Heisenberg Uncertainty Principle insists that its momentum cannot be exactly zero, but that its average momentum must be on the order of the magnitude given by the uncertainty principle. Find the magnitude of this momentum.

d) Knowing the momentum, calculate the non-relativistic kinetic energy of the electron. [Hint: Remember KE = $\frac{1}{2}mv^2 = p^2/2m$.]

e) Convert this energy to units of kelvins to find the effective "temperature" of the electrons.

We have neglected a few factors of 2, 3, and π in various places, but, otherwise, this gives a good estimate of the size of the electron energies. This huge effective temperature provides the pressure that resists the enormous crush of gravity in these super-dense stars.

26

NUCLEAR AND PARTICLE PHYSICS

We mentioned in Section 16.1 that an atom consisted of a nucleus, composed of protons and neutrons, with electrons orbiting the nucleus. After reading Chapter 25, we are now a bit more sophisticated about the electrons, but the picture is otherwise sound. The number of positively charged protons in the nucleus sets the electric charge on the nucleus, and this determines how many electrons the atom will attract to itself. Since the chemistry (by which we mean what element an atom corresponds to) is completely determined by the electrons, any atom with the same number of protons in its nucleus will be an atom of the same element. We call the number of protons in a nucleus the *atomic number* of the atom. All atoms with the same atomic number are atoms of the same element. In addition to the protons, a nucleus will also include electrically neutral neutrons. These add nothing to the atomic number, but they add to the total mass of the nucleus. The combined number of protons and neutrons in the nucleus is called the *atomic weight* of the atom. Nuclei of the same atomic number (same element), but with different atomic weights, are called *isotopes* of that element.

26.1 Nuclear Forces

Students often wonder how protons can occupy the same nucleus, since they have like positive electrical charges and should repel each other. The answer is that they would indeed never form a nucleus if the electrical force was the only force in play. But it is not. It's time for us to learn about another fundamental force in nature.

The two fundamental forces we have seen so far—the gravitational force and the electromagnetic force—both have infinite extent. When things are far apart and electrically charged, the electric force dominates their motion. When things are far apart and electrically neutral, gravity dominates their motion. But when protons and neutrons are as close to each other as they are in the nucleus, a force stronger than the repulsive electric force, called the *strong force* (don't you *love* the creative name), takes over and dominates their motion, though you may wonder what "motion" means after our last chapter on quantum mechanics.

The theory of the strong force is still an area of active research, and so our understanding may change as time goes on. But, at present, we understand the strong force as arising out of the so-called *color force* that holds together the *quarks* that make up protons and neutrons. We will talk a little more about quarks and color later in this chapter, but, for now, let us just say that the strong force is the residual color force between color-neutral particles, like protons and neutrons.[1] It is

1. You may have learned about covalent bonding in your chemistry class. In covalent bonding, neutral atoms in a molecule are held together by sharing an electron. Similarly, color-neutral particles in a nucleus are held together by quarks, shared between them two at a time.

about one hundred times stronger than the electrical force between protons and its reach is limited to several femtometers (10^{-15} m), or about the size of the nucleus. The strong force is equally attractive between all *nucleons* (the generic term for the protons and neutrons), but there is a strong spin-dependence to the force, being greatly weakened when the particles are of opposite spin. Because the Pauli Exclusion Principle requires two neutrons or two protons in a nucleus to have opposite spins, the strong force is much weaker between two neutrons or two protons than it is between a proton and a neutron in a nucleus.

If we took two protons and two neutrons and placed them close enough that the strong force could take over and hold them tightly together, we would have formed them into a nucleus of helium (atomic number 2), with atomic weight 4. We choose to represent this nucleus with the notation ${}^4_2\mathrm{He}$, though the subscript 2 for the atomic number is redundant once we have written the chemical symbol He for helium, because all helium nuclei contain two protons. If we define the protons and neutrons to have zero "strong" potential energy when they are outside the range of the strong force, then they would have a very negative potential energy when they are in the nucleus. The magnitude of their negative total energy is called the "binding energy" of the nucleus, since that is the energy that would have to be supplied to break the nucleus apart into two protons and two neutrons again. It would be impossible to measure the work done on particles this small as they come together to form a nucleus, yet, surprisingly, the binding energy of a nucleus is easily found.

The masses of the proton, the neutron, and the helium nucleus are shown in Table 26.1. Let's just look at the second column for now; we will explain the third column in a minute.

TABLE 26.1 Particle Masses

Particle	Mass ($\times 10^{-27}$ kg)	Rest Energy (MeV)
proton	1.6726231	938.2723
neutron	1.6749286	939.5656
He nucleus	6.6446568	3727.3792

If we form a helium nucleus from two protons and two neutrons, the total mass we have available is $2m_p + 2m_n$, or

$$2(1.6726231 \times 10^{-27}\,\text{kg}) + 2(1.6749286 \times 10^{-27}\,\text{kg}) = 6.6951034 \times 10^{-27}\,\text{kg}.$$

But, after we form the helium nucleus, its total mass is only $6.6446568 \times 10^{-27}$ kg. So here's an interesting question. Where did the missing $0.0504466 \times 10^{-27}$ kg go?

You may remember that we discussed this possibility back in Section 24.5 (page 531). There we noted that the mass of a body is the same as its rest energy,

$$E_{\text{rest}} = mc^2. \tag{26.1}$$

Since the strong force that pulls two protons and two neutrons together to form a helium nucleus releases an amount of energy that is a discernible fraction of the mass energy of its four constituent particles, that energy change will show up as a lower mass for helium than for the four nucleons that comprise it. You may have learned in a chemistry class that mass cannot be created or destroyed, and for the chemistry that takes place out in the electron clouds around an atom, that is true enough. But at the energies available inside a nucleus, mass and energy really cannot be separated.

You have probably seen Equation 26.1 before and understood it as explaining that mass can be converted to energy. And so it can. But the modern way of looking at this equation is somewhat different. What it says is that, except for a factor of c^2, mass and rest energy are really the same thing. So Equation 26.1 can be seen as a kind of units conversion. You can measure a mass in kilograms or you can measure it in joules. In fact, it is common for those who work in high-energy physics to express the mass of a body in energy units, generally using the electron-volt units that are appropriate to atomic and nuclear scales. Thus, the third column in Table 24.1 is much more commonly used than the second column. We generally just say that the mass of the proton is 938 MeV (million electron-volts). In fact, as long as we can be certain we are talking about the energies of particles (and not about their inertia, as in $F = ma$), we can use the letter m to mean rest energy. Thus, we will often write the mass of the helium nucleus as $m_{\mathrm{He}} = 3727$ MeV.

26.2 Nuclear Reactions

As we have mentioned, chemists work out *chemical* reactions in which atoms and molecules react with each other to become other molecules, and they count on the principles of conservation of mass and of electrical charge to work out the reactions. But if mass can be converted into other forms of energy in a *nuclear* reaction, what then stays the same? What quantities can be counted on to be conserved?

Well, energy is still conserved, although some of the rest energy (mass) in a reaction can be converted to kinetic energy and vice versa. Electrical charge is still conserved, with the initial total charge and the final total charge of the reactants remaining the same. But there are additional conservation laws that nature appears to respect. One of these is the conservation of what is called baryon number. Protons and neutrons are classed as *baryons* (from the Greek *bary* for heavy), and there must be as many total baryons after a reaction as there were before. There are also conservation laws for *leptons* (from the Greek *lepto* for light). Electrons are leptons, but so is another particle, the neutrino. It turned up in nuclear reactions in such a way as to make total lepton number a conserved quantity.

Let us try to make the conservation laws clear with an example. One nuclear reaction that is a very important part of explaining why stars shine is the following:

$$ {}^{1}_{1}\mathrm{H} + {}^{1}_{1}\mathrm{H} \rightarrow {}^{2}_{1}\mathrm{H} + \bar{e}^{+} + \nu. \tag{26.2}$$

The symbols in this reaction are as follows: ${}^{1}_{1}\mathrm{H}$ is the hydrogen nucleus. ${}^{2}_{1}\mathrm{H}$ is an isotope of hydrogen, called "deuterium," in which a neutron accompanies the proton in the nucleus (for an atomic weight of 2). The ν stands for "neutrino," a particle with no significant mass and no charge. The symbol $\bar{e}^{+}$ requires a little more explanation.

Every particle we have mentioned so far—proton, neutron, electron, and neutrino—along with every particle we have not yet mentioned, has a counterpart in the antimatter world. The mass of an antiparticle is the same as that of its particle, and the spin is the same, but every other property of an antiparticle is the opposite of its particle counterpart.

Thus an antiproton has negative charge and negative baryon number, an antielectron (also called a positron) has positive charge and negative lepton number, and so on. Even for the photon, with no charge, no mass, and no lepton or baryon number, we say that it is its own antiparticle. In Equation 26.2, we have used the common notation of placing a bar over an antimatter particle. Thus $\bar{e}^+$ is the notation for a positron.[2] It has the mass of an electron, the opposite charge as the electron (it is positive), and it carries a –1 lepton number, because it is an antilepton.

Let us analyze Equation 26.2 to see if it satisfies the conservation laws. We start with two $^1_1$H particles for total charge +2. The final charge is +1 for the $^2_1$D plus the charge on the $\bar{e}^+$, for total charge +2. The initial baryon number is +2 from the two protons. The final baryon number is +2 for the $^2_1$D's single proton and single neutron. The initial lepton number is zero—no electrons or neutrinos—and the final lepton number is zero, the –1 for the anti-electron canceling the +1 for the neutrino.

Finally, let us look at the energy conservation. The initial kinetic energy of the two protons in Equation 26.2 has to be fairly high for them to overcome their mutual electrical repulsion enough to get to within a few femtometers of each other, where the strong force can take over. But this energy is still small compared to the huge mass-energies of the particles in the reaction. The initial energy available is the mass of the two protons, $E = 2m_p = 2(938.27\text{ MeV}) = 1876.54\text{ MeV}$. The deuterium mass is 1875.61 MeV, the electron is 0.511 MeV, and the neutrino mass is negligible.[3] Energy accounting shows that there is more mass in the parent particles, the original two protons, than in the daughter particles (the common term for the output of a reaction). As a result, there is an excess energy of

$$\Delta E = 2m_p - (m_D + m_e) = 2(938.27\text{ MeV}) - (1875.61\text{ MeV} + 0.511\text{ MeV}) = 0.42\text{ MeV}.$$

This excess energy is shared by the daughter particles as kinetic energy, as they fly away from the place where the reaction occurred.

We mentioned that every particle has its antiparticle, with identical positive rest energy (mass) and spin, but with opposite signs for all of its other properties. This means that one particular kind of reaction is always energetically permitted—the annihilation of a particle with its antiparticle. The positron in the reaction shown in Equation 26.2 is an antielectron, so the following reaction must be possible:[4]

$$e^- + \bar{e}^+ \rightarrow 2\gamma. \tag{26.3}$$

Both the total charge and the total lepton number of the initial particles are zero, so they can both completely disappear, leaving only their mass turned into energy. The energy will be carried away by the two photons (γ).

2. Actually, the notation $\bar{e}^+$ is redundant, since e^+ would be enough to identify the particle as an antielectron, but we have kept the $\bar{e}^+$ notation as a reminder. Some books do not do this.

3. The neutrino mass is presently not known, though it is known not to be zero. Best estimates put it less than 1 eV, so its mass is negligible in most nuclear reactions.

4. Since photons in nuclear reactions generally have energies in the gamma-ray range (see Table 21.1), the symbol γ (Greek letter gamma) is universally used for photons.

The reason that two photons are required, by the way, is that this is the only way to conserve momentum. Even though a photon has zero mass, it has momentum. The relationship between energy, momentum, and mass that we derived in Chapter 24 as Equation 24.8 is true for any particle, whether it has mass or not. This equation can be solved to give a formula for the momentum of a zero-rest-mass particle as

$$(cp)^2 = E^2 - (mc^2)^2 = E^2 - 0 = E^2 \qquad \text{or} \qquad p = \frac{E}{c}.$$

If we were to observe the reaction of Equation 26.3 in a reference frame in which the initial total momentum of the electron–positron pair is zero, the final momentum would also have to be zero. But a single energetic photon must carry non-zero momentum. So the only way to produce a zero final momentum is for the energy released in the reaction to be shared by two photons flying off in opposite directions.

Let us see how to determine the wavelengths of the photons in the reaction of Equation 26.3. If, for example, the electron and positron were initially at rest, or moving relatively slowly, then the total energy they would carry into the reaction would be their rest energy only, $2 \times (0.511\ \text{MeV}) = 1.022\ \text{MeV}$. Since the photons must likewise carry equal and opposite momenta, they would each have the same energy. It would be ½(1.022 MeV), or 0.511 MeV. Knowing the energy of each photon, we can calculate the wavelength from Equation 25.1, $\lambda = (1240\ \text{eV} \cdot \text{nm})/(0.511 \times 10^6\ \text{eV})$, giving $\lambda = 2.43 \times 10^{-3}\ \text{nm} = 2.43\ \text{pm}$. A glance back at Table 21.1 tells us that this photon will be a gamma ray. Astronomers search for gamma rays of wavelength around 2.4 pm in order to find evidence of free positrons in space.

EXAMPLE 26.1 Energy Released in a Hydrogen Fusion Reaction

The reaction shown in Equation 26.2 is one process in the chain of reactions that turn hydrogen into helium in the cores of stars, producing the energy that fuels the brightness of the star. The next step in the chain is the following reaction,

$${}^{2}_{1}\text{H} + {}^{1}_{1}\text{H} \rightarrow {}^{3}_{2}\text{He} + \gamma.$$

The mass of the helium-3 nucleus is 2809.40 MeV. How much energy is carried away by the photon in this reaction?

ANSWER First, just as a check, we verify that charge (+2) and baryon number (+3) are conserved. All that remains is the energy accounting. The mass lost in creating the ${}^{3}_{2}\text{He}$ nucleus from the deuterium and the hydrogen nuclei is

$$\Delta E = m_p + m_D - m_{{}^{3}_{2}\text{He}} = 938.27\ \text{MeV} + 1875.61\ \text{MeV} - 2809.40\ \text{MeV} = 4.48\ \text{MeV}.$$

This will be the energy of the single gamma-ray photon emitted in the reaction.

The reaction in Equation 26.2 is an example of a *fusion reaction*, in which two or more smaller nuclei "fuse" together to form a larger nucleus. Other kinds of nuclear reactions include those where a large nucleus spontaneously splits into two or more smaller nuclei, a process known as *nuclear fission*. There are also reactions in which a nucleus simply emits a small particle, a process known as *nuclear decay*. Nuclei that fission or decay are referred to as *radioactive*. An example of a very famous nuclear decay reaction is

$$^{14}_{6}\text{C} \rightarrow {}^{14}_{7}\text{N} + e^{-} + \bar{\nu}. \tag{26.4}$$

Here, an isotope of carbon, one with 6 protons (because all carbon has 6 protons) and 8 neutrons in the nucleus (for a total of 14 nucleons), spontaneously emits an electron and decays into nitrogen. A quick check will verify that charge and baryon number are conserved. Because lepton number must also be conserved, we know that there must also be an antineutrino produced in the reaction. We leave it as a chapter problem to work out the energy released by this reaction. But let us now go on to focus on another aspect of this and other nuclear reactions.

26.3 Reaction Rates and Half-Lives

The rate at which fusion reactions take place depends on the types of nuclei involved, on the number of nuclei per cubic meter, and on the ambient temperature. For a given choice of nuclei, higher density and a higher temperature lead to a higher reaction rate, as you might guess. But, for nuclear decay or nuclear fission, the reaction rate is simple. Within a wide temperature range, if the reaction can occur at all, the rate will be constant.

If we had a spoonful of carbon-14 atoms ($^{14}_{6}$C), every nucleus in the spoonful would be energetically able to undergo the reaction described by Equation 26.4, but they would not all react at once. One of the quantum mechanical aspects of nuclear reactions is that there is only a well-defined *probability* that a given nucleus will decay by way of a permitted nuclear reaction. The laws of physics do not specify when any particular nucleus will decay. As a result, nuclear decay proceeds a lot like popcorn popping in the microwave oven—some will go early and some will go late. For a $^{14}_{6}$C nucleus like the one described in Equation 26.4, the chances are 50:50 that the nucleus will decay in 5730 years. This 5730 years is called the *half-life* of carbon-14. What this means is that, if you start with 100 nuclei of $^{14}_{6}$C, you may expect there to be approximately 50 $^{14}_{6}$C nuclei left at the end of 5730 years. But you then need to remember that the remaining $^{14}_{6}$C nuclei are all still just $^{14}_{6}$C nuclei, and so each one of the 50 still has a 50% probability of decaying in 5730 years. This means that it would take another 5730 more years for there to be only 25 carbon-14 nuclei left.

Please note that this correct result is different from the incorrect result you would get by thinking that if 50 nuclei decay in one half-life (5730 years) then the remaining 50 would decay in the next half-life. It would also be incorrect to decide that 25 will decay in 2865 years (half of a half-life). If each nucleus in a sample has the same probability of decaying in a given time, then the number of decays seen will be higher when the number of nuclei is greater and lower when the number of nuclei is less.

Let us be clear that, with what we know so far, we can only say how many nuclei will remain after one half-life, after two half-lives, after three half-lives, etc. We have not yet explained what to expect if the time is not an integer number of half-lives.

EXAMPLE 26.2 Using the Radiocarbon-Dating Half-Life

Cosmic rays that hit the earth's atmosphere create ${}^{14}_{6}C$ from the normal ${}^{14}_{7}N$ that composes 80% of the atmosphere. Once formed, the ${}^{14}_{6}C$ decays back to ${}^{14}_{7}N$ via the reaction shown in Equation 26.4. If the rate of cosmic-ray bombardment of the earth's atmosphere remains constant at its current value, then these two reactions balance out at an atmospheric density of 1.5 atoms of ${}^{14}_{6}C$ for every 10^{12} atoms of ${}^{12}_{6}C$, the normal, stable carbon isotope. As long as a plant or animal is absorbing air from the atmosphere, it will maintain this same concentration of ${}^{14}_{6}C$ relative to ${}^{12}_{6}C$ in its cells. However, once the organism dies, it no longer replenishes its ${}^{14}_{6}C$ content from the atmosphere. The concentration in the cells is slowly reduced as ${}^{14}_{6}C$ becomes nitrogen via the electron decay shown in Equation 26.4.

Suppose that a sample from a piece of wood from an ancient stone-age camp site contains 8×10^{20} atoms of carbon and that an electron counter notes that the sample is emitting electrons at a rate consistent with there being a total of 3×10^8 atoms of ${}^{14}_{6}C$ present in the sample. How old is the wood?

ANSWER The concentration of ${}^{14}_{6}C$ in the atmosphere is 1.5 atoms of ${}^{14}_{6}C$ per 10^{12} atoms of ${}^{12}_{6}C$, a relative concentration of 1.5×10^{-12}. The concentration of ${}^{14}_{6}C$ in the wood sample is 3×10^8 atoms of ${}^{14}_{6}C$ in 8×10^{20} total atoms of carbon, giving a ratio $3 \times 10^8/8 \times 10^{20} = 0.375 \times 10^{-12}$. Fortunately (because otherwise we do not yet know how to solve the problem), this concentration is exactly ¼ of the initial atmospheric concentration of 1.5×10^{-12}, and this ¼ is ½ × ½. An original concentration of ${}^{14}_{6}C$ in a piece of dead wood drops by half in the first 5730 years and by another factor of ½ (for a total factor of ¼) in the next 5730 years. Therefore, we conclude that the sample was chopped down and put in the fire two half-lives, or 11,460, years ago.

We have seen that the number of some radioactive species of nucleus with some half-life will decrease in a known way. After one half-life, there will be ½ as many of the species as there were at first. After two half-lives, there will be ¼, or $1/2^2$ as many; after three half-lives, there will be ⅛, or $1/2^3$ as many. You may see a pattern here. After M half-lives, there are $1/2^M$ as many nuclei as there were to begin with. This obviously works for all integer values of M, but if we think about how the decay happens, being proportional to how many of the radioactive species are left in the sample, we can see that this formula works just as well if M is not an integer. If we start with a number N_0 of some radioactive species of nucleus, and if the half-life for the decay is $t_{1/2}$, then, after time t, the number of half-lives that have passed is $t/t_{1/2}$. In this case, the number of that species that is left will be

$$N(t) = \frac{N_0}{2^M} \quad \text{with } M = \frac{t}{t_{1/2}}. \tag{26.5}$$

In Example 26.2 we found the age of a sample when that age was an integer number of half-lives. Using Equation 26.5, we see that we can work out the age of any carbon-dated piece of wood, even if it is not an integer number of half-lives.

The purposes of our last example are, first, to show how to work with radioactive decay when the elapsed time is not an integer number of half-lives and, second, to serve as a reminder that time depends on the reference frame of the observer.

EXAMPLE 26.3 The Decay of Relativistic Neutrons from the Sun

Free neutrons are unstable and decay by the pathway $n^0 \rightarrow p^+ + e^- + \bar{\nu}$ with a half-life of 881.5 seconds. Suppose 1000 free neutrons are created at the surface of the sun and head toward the earth at a speed 1.69×10^8 m/s (a little more than ½ the speed of light). How many of those neutrons will be in the neutron cloud by the time it hits the earth? [Note: The sun is 1.49×10^{11} m from the earth.]

ANSWER First, let's figure how much time elapses while the neutrons are speeding toward the earth from the solar flare. At a speed 1.69×10^8 m/s, the 1.49×10^{11} m can be covered in a time $t = d/v = (1.49 \times 10^{11}\text{ m})/(1.69 \times 10^8\text{ m/s}) = 882$ s. Now this is so close to the half-life of 881.5 s that we might think that the problem was *set up* to give us exactly one half-life for the time of flight, allowing us to estimate the number of nuclei remaining as 500, ½ of the original 1000. But this would be wrong.

You see, radioactive decay is a kind of a clock, and the neutrons that constitute the clock are moving at high speed. And you remember what moving clocks do, don't you? Right, they tick slowly. So the time that elapses in the neutron cloud would be less than the 882 seconds measured from earth. At a speed 1.69×10^8 m/s, the square-root factor is $\sqrt{1 - (1.69 \times 10^8/3 \times 10^8)^2} = 0.826$, and so the time elapsed in the neutron rest frame is (0.826)(882 s) = 729 s. This gives $M = 729/881.5 = 0.827$. In 0.827 half-lives, the number of nuclei left in the cloud will be more than 500.

The calculation of the actual number, using Equation 26.5, will involve finding the value of $2^{0.827}$. If you have a scientific calculator with the normal entry method (i.e., a calculator with an equals sign as one of the buttons), here is how you proceed. Locate a button that says y^x on it. First enter a "2." Then hit the "y^x" button, then enter 0.827 and hit "=". Your display should read "1.77." The number of remaining neutrons in the cloud that reaches the earth is therefore

$$N = \frac{N_0}{2^{0.827}} = \frac{1000}{1.77} = 565.$$

An analysis of data from a neutron detector flying in a balloon over India in 1962 concluded that some of the neutrons they saw originated in the sun.* The observed neutron energies were consistent with the velocity chosen for this example.

* "Evidence for the possible emission of high-energy neutrons from the Sun," M. V. Krishna Apparao, R. R. Daniel, B. Vijayalakshmi, V. L. Bhatt, *Journal of Geophysical Research* 71, 1781–1785 (April 1966).

As we now leave nuclear physics to go on to elementary particle physics, let us just summarize by noting that, in the last two sections on nuclear reactions, we have learned how to do two things—to calculate the energies released or absorbed in nuclear reactions (Example 26.1) and to use the half-life of a reaction to determine how the number of a species of nucleus drops with elapsed time during nuclear decay (Examples 26.2 and 26.3).

26.4 Elementary Particles

Before 1930, the universe was simple. It consisted of atoms made up of protons, neutrons, and electrons, with photons carrying away the electromagnetic energy from atomic and nuclear reactions. But, in 1930, Wolfgang Pauli suggested that there needed to be a particle like the neutrino to explain how spin angular momentum could be conserved in reactions that emitted an electron or a positron, like the reactions we showed in Equations 26.2 and 26.4. Then, in 1935, a Japanese physicist named Hideki Yukawa postulated the existence of a particle with mass around 100 MeV that would help explain the strong nuclear force. In 1936, as that particle was being looked for in cosmic ray tracks, a particle of about the right mass was found, but it was later discovered that it did not participate in the strong nuclear force at all, so it could not be Yukawa's particle. This new particle was named the muon. It was not until 1950 that Yukawa's particles were found. They were the charged and neutral pions. But, by 1950, Pandora's box had been opened, and the number of known elementary particles had exploded, finally producing the complicated zoo of particles displayed in Table 26.2.

Let's look at this table for a moment. Leptons all have spin ½, while baryons can have spin ½ or 3⁄2. The things in the table called *mesons* all have spin 0 or 1. So leptons and baryons are fermions while mesons are all bosons. When we discussed the conservation laws on page 565, we stated that the number of leptons in a reaction was the same before and after the reaction, but it actually turns out that electron-lepton, muon-lepton, and tau-lepton numbers are all separately conserved in reactions. It is also true that baryon number is always a conserved quantity, but meson number is not necessarily conserved.

By 1960, the number of known elementary particles had grown so large that many began to wonder what the term "elementary" really meant. Physicists began to organize the properties of these particles in several "periodic tables" of sorts, and began to wonder if the orderliness of the elementary particle properties did not suggest that these particles might themselves be made up of more fundamental building blocks. Then, in 1964, American physicists Murray Gell-Mann and George Zweig came up with just such a scheme. It explained the properties of the mesons and baryons as arising from their being made up of more fundamental particles that they called *quarks*. In this prescription, a quark and an antiquark pair would produce a meson, while baryons were all composed of three quarks.[5]

The particles that were needed in the simple 1935 view of the nucleus, with protons and neutrons in the nucleus and Yukawa's pions holding them together, could all be created from just two kinds of quarks and their antiquarks. The quarks required were the "up" quark and the "down" quark, though we do need to be careful to explain that

TABLE 26.2 The Elementary Particles

Particle	Symbol	Mass (MeV)
Leptons		
electron	e^-	0.511
e-neutrino	ν_e	< 1 eV
muon	μ^-	105.7
μ-neutrino	ν_μ	< 0.3 eV
tau	τ^-	1784
τ-neutrino	ν_τ	< 30 eV
Mesons		
pion	π^+	139.6
	π^0	135.0
kaon	K^+	493.7
	K^0	497.7
	K^0_S	497.7
	K^0_L	497.7
eta	η	548.8
	η'	958.0
Baryons		
proton	p^+	938.3
neutron	n^0	939.6
lambda	Λ^0	1115
sigma	Σ^+	1189
	Σ^0	1192
	Σ^-	1197
delta	Δ^{++}	1230
	Δ^+	1231
	Δ^0	1232
	Δ^-	1234
xi	Ξ^0	1315
	Ξ^-	1321
omega	Ω^-	1672

5. The name "quark" arose from the baryon requirement of three, and derived from the phrase "Three quarks for Muster Mark," found in James Joyce's *Finnegan's Wake*.

up and down are just names and that the quarks do not point up or down along any direction. Each quark has baryon number $1/3$ and each antiquark has baryon number $-1/3$. All quarks are fermions, with spin $1/2$. These two quarks and their antiparticles are described in Table 26.3.

TABLE 26.3 Properties of the Up and Down Quarks

Quark	Up (u)	Antiup ($\bar{u}$)	Down (d)	Antidown ($\bar{d}$)
Charge	$2/3$	$-2/3$	$-1/3$	$1/3$
Baryon number	$1/3$	$-1/3$	$1/3$	$-1/3$

Combinations of these particles create protons, neutrons, and pions, as depicted in Figure 26.1. You should verify how each combination gives the correct charge and baryon number for each composite particle.

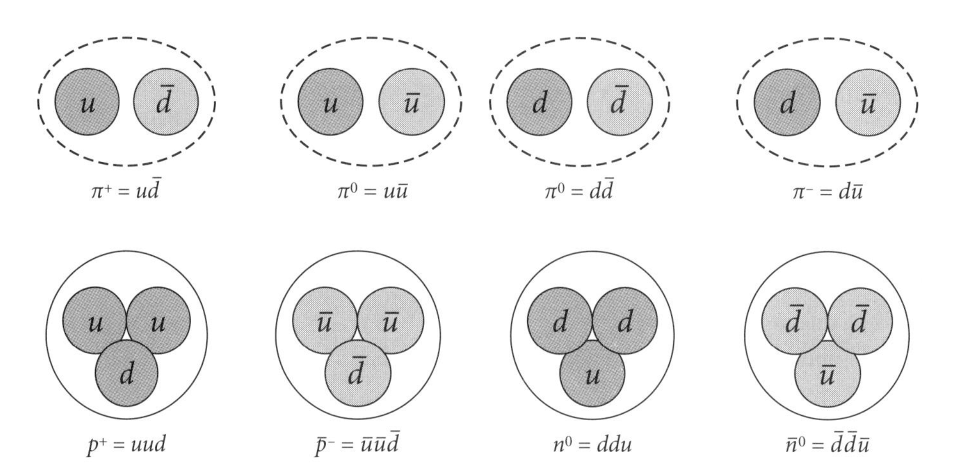

Figure 26.1 The quark content of pions, protons, and neutrons. Up quarks are u, down quarks are d. Quarks are in red, anti-quarks in gray.

This scheme explains three of the elementary particles from Table 26.2, the protons, neutrons, and pions. But what of the rest? Well, it turns out that additional quarks are needed to explain the rest of the baryons and mesons. During the years when the Table 26.2 particles were being discovered, it was found that the ways they could be created and the rates at which they decayed into other particles could be understood only if there were additional conserved quantities that the particles possessed, and additional quarks were invented to carry the prescribed amount of these new quantum numbers. The first new number discovered was what physicists came to call *strangeness*, requiring the existence of a *strange quark*. Subsequently discovered particles and reactions indicated that a property had to exist that they called *charm*. Before this process was over, particle theorists also had to add *top* and *bottom* quarks to the list.[6] There are

6. Actually there was a movement (with which we agreed) that wanted to call these properties *truth* and *beauty*, but it didn't stick. Maybe some physicists were worried about looking too ridiculous. Maybe they were right.

thus six quarks known (up, down, strange, charmed, top, and bottom) with similar relationships between them as between the six leptons in Table 26.2.

TABLE 26.4 Quark Properties

Quark	Symbol	B	Q	S	C	T	B'
Down	d	$\frac{1}{3}$	$-\frac{1}{3}$	0	0	0	0
Up	u	$\frac{1}{3}$	$+\frac{2}{3}$	0	0	0	0
Strange	s	$\frac{1}{3}$	$-\frac{1}{3}$	–1	0	0	0
Charmed	c	$\frac{1}{3}$	$+\frac{2}{3}$	0	1	0	0
Top	b	$\frac{1}{3}$	$\frac{2}{3}$	0	0	1	0
Bottom	t	$\frac{1}{3}$	$-\frac{1}{3}$	0	0	0	1

The columns show baryon number (B), charge (Q), strangeness (S), charm (C), and topness (T). The symbol for bottomness is (B') to distinguish it from baryon number. Antiquarks have the opposite values for B, Q, S, C, T, and B'. All quarks and antiquarks can have spin $\pm\frac{1}{2}$.

We will not go more deeply into this subject, since it is a little esoteric for our purposes. But, just to give you a sense of how these quarks work, let us write down the quark content of the Λ^0 (see Table 26.2). The Λ^0 was first discovered in cosmic rays in 1950 and its unexpectedly long lifetime[7] led to the discovery of strangeness. It is composed of an up quark u, a down quark d, and a strange quark s. By checking values from Table 26.4, we are able to see how the neutral lambda's charge 0, baryon number 1, and strangeness –1 derive from the properties of the quarks that make it up.

$$\Lambda^0 = uds.$$

26.5 Fundamental Forces

Let us now return to Table 26.2 and point out, as you may have noticed, that there is one particle that is missing from the table. This is the photon. The photon is different from the rest of the particles from Table 26.2 for a couple of reasons. First, the particles in the table all have mass (rest energy), but the photon does not. The photon carries energy, as given by Einstein's formula (Equation 25.1), but it has no mass, and indeed cannot ever come to rest. But the second reason for not including the photon in Table 26.2 has to do with another property of photons that we now want to discuss.

Back on page 550, we discussed how there was a limit on how accurately (ΔE) an energy could be measured if that energy was to be localized to a particular time with accuracy Δt. The limitation was given by $\Delta E \Delta t \geq h/4\pi$, the Heisenberg Uncertainty Relation of Equation 25.9. Looked at another way, this equation tells us the amount by which any energy, including rest energy, can be non-conserved, as long as it fails to conserve the energy for a short enough time. This idea is the

7. Well. Long is a relative term. Its lifetime was about 100 picoseconds, but that was 10^{13} times longer than it was expected to live.

basis for what are called *virtual particles*. In this scenario, an electron can spontaneously emit a particle of rest energy ΔE, with no loss to its own rest energy, as long as the particle is reabsorbed somewhere within time Δt, the two quantities being connected by Equation 25.9.

This phenomenon of virtual particles means that any electron is surrounded by a cloud of photons that it constantly emits and re-absorbs. However, if there is another electron nearby, the second electron can be the thing that absorbs one of the virtual photons from the first electron, as depicted in Figure 26.2. This procedure produces a change in the state of motion of both electrons. Electron 1 emits a photon without losing mass, but with a change in its momentum, and electron 2 absorbs the photon, likewise with no change in mass, but with a change in its momentum. The repulsive force between two electrons is thus seen to arise through their virtual photons, and the photon itself is recognized as a different kind of particle—a *field particle*—whose virtual emission and absorption *is* the electromagnetic field.

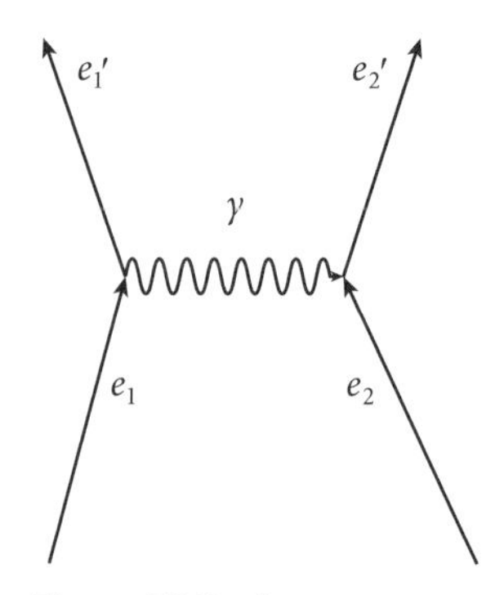

Figure 26.2 The emission and absorption of a virtual photon.

The picture in Figure 26.2, with an electron emitting a particle and recoiling in the opposite direction, seems a reasonably intuitive explanation of how exchanging a virtual particle can produce a repulsive force, but it is badly oversimplified. If the second electron were a positron (an anti-electron) instead of an electron, the two particles should be *attracted* to each other, and both the emission and the absorption events would need to bend the trajectories of the particles *toward* each other. In the end, this theory is not intuitively obvious. Nevertheless, the theory that fundamental forces arise through the exchange of field particles does provide a clear explanation of the origin of forces, when the quantum mechanics is done correctly (i.e., far beyond what we are going to explain here).

In modern quantum field theory, the nature of *all* fundamental forces is the same as that of the electromagnetic force, in that they all arise from the exchange of field particles. Table 26.5 lists the four fundamental forces in nature and the name of the field particle that produces the force. It also lists the strength of the field, with gravity being the weakest and the color force the strongest. We have already seen that gravity arises between particles with mass, while electromagnetism depends on a particle's electrical charge. The weak force is a force that acts on all fermions (particles with half-integer spin). The fourth force, the color force, is a little more complicated.

TABLE 26.5 The Four Fundamental Forces

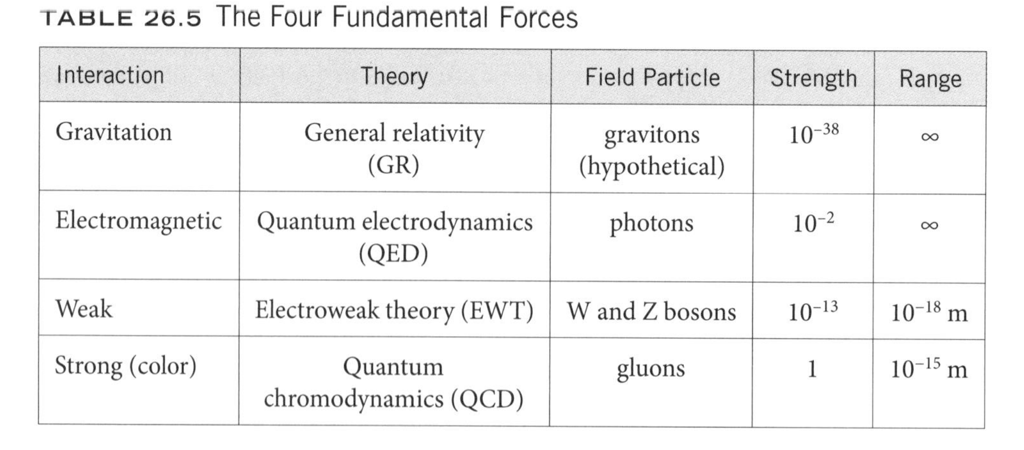

Interaction	Theory	Field Particle	Strength	Range
Gravitation	General relativity (GR)	gravitons (hypothetical)	10^{-38}	∞
Electromagnetic	Quantum electrodynamics (QED)	photons	10^{-2}	∞
Weak	Electroweak theory (EWT)	W and Z bosons	10^{-13}	10^{-18} m
Strong (color)	Quantum chromodynamics (QCD)	gluons	1	10^{-15} m

We will end this section by talking a bit about the color force. But, before we do that, let us work through an example that shows how the range of a force determines what the mass of its field particle can be.

EXAMPLE 26.4 The Finite Range of the Pion Exchange Force

Show how the mass of Yukawa's hypothetical pion was determined from the uncertainty principle and from the fact that the force had to be confined to the effective 10^{-15}-meter scale of the nucleus.

ANSWER The uncertainty principle gives the relationship between the mass–energy carried away by a field and the amount of time it can be gone from the particle that it left. The relationship is

$$\Delta E \Delta t = mc^2 \Delta t = \frac{h}{4\pi}.$$

The nuclear energies involved correspond to speeds close to the speed of light. If the field particle moves at the speed of light, the time it will take it to reach the edge of the nucleus is $\Delta t = (10^{-15}\text{ m})/(3 \times 10^8\text{ m/s}) = 3.3 \times 10^{-24}$ s. The mass that may be missing from the nucleons during such a time is

$$mc^2 = \frac{h}{4\pi\Delta t} = \frac{6.626 \times 10^{-34}\text{ J}\cdot\text{s}}{4\pi(3.3 \times 10^{-24}\text{ s})} = 1.6 \times 10^{-11}\text{ J}.$$

If we convert this energy to eV, we get

$$mc^2 = 1.6 \times 10^{-11}\text{ J}\left(\frac{1\text{ eV}}{1.6 \times 10^{-19}\text{ eV}}\right) = 10^8\text{ eV} = 100\text{ MeV}.$$

This was the mass of the meson that Hideki Yukawa proposed as the field particle for the strong nuclear force.

As we said, our last task in this chapter is to explain a little bit about the color force.

The particles affected by the color force are the quarks. As electrons have charge and interact by exchange of photons, so the quarks have properties that enable them to interact by exchange of gluons. Now physicists could have called this property "letter charge," and named the charges A, B, and C, or they could have called it "party charge," and named the charges republican, democrat, and independent. But they didn't. They decided to call this property *color charge* and named the charges *red*, *green*, and *blue*. We are sure you understand that these are just names and have nothing to do with the wavelengths of light emitted by the quarks. There are thus 18 kinds of quark, with each of the six basic kinds of quark (up, down, strange, charmed, top, and bottom) coming in three possible colors (red, green, and blue). There are also antiquarks for each quark, each carrying an anticolor, as we explain below.

When something contains a negative charge and a positive charge, as an atom does, then the particle has zero total charge, and we say it is neutral. Similarly, when a particle is composed of a red quark, a green quark, and a blue quark, the particle has zero color charge. We say that it is neutral, or, because of the color connection, we also call a color-neutral particle "white." Antiquarks have color charges antired, antigreen, or antiblue. A combination of a red, green, and blue quark make a white particle, as does a combination of a quark and its antiquark. White particles do not interact with other color charges, because they have no net color charge.

If an electron is removed from an atom, say by absorbing a high-energy photon, then the resulting positive ion will attract any nearby electrons, and the atom will become electrically neutral again. Similarly, if some combination of quarks were not color-neutral, it would attract nearby quarks until it became color neutral. So all stable baryons—protons, neutrons, pions, etc.—are white.

Let us complete our discussion of the color force by explaining the unusual way in which the color force depends on the distance the virtual gluons travel between quarks. Both the electromagnetic force and the gravitational force fall off with distance as $1/r^2$. The weak force weakens exponentially, due to the mass of the W and Z bosons. But the color force is different. It is very weak at short distance, meaning that the quarks inside a baryon or a meson act like free particles. This behavior is referred to as *asymptotic freedom* (the term "asymptote" refers to a limit, in this case the limit of high energy and zero distance). But, as the quarks get further and further away from each other, the gluon exchange force actually *increases* in strength. In this way, it is a little like the Hooke's law force that grows as the distance increases. The result of this strong force holding the quarks together is what is called *confinement* and is the reason for the non-existence of free quarks. Quarks do not occur outside the baryons and mesons which they make up, and the gluons never reach beyond the 10^{-15}-meter size of these particles. So where does the strong nuclear force come from? How do we get a force to hold all those positively charged protons together with the neutrons in a nucleus that is several times 10^{-15} meters in size?

Even though a nucleus is composed of protons and neutrons—combinations of quarks whose overall whiteness cancels their gluon exchange forces—it is possible for some white combinations of quarks to leak away from the protons and neutrons and act as field particles between them. The most common of these combinations are those that make up the neutral and charged pions. So we have, in effect, come back to Yukawa's view of the origin of nuclear forces in the exchange of pions and other mesons. However, we are now able to understand the nuclear forces as deriving from the underlying color forces that are mediated by the gluons. The strong nuclear force that results is weaker than the direct color force on confined quarks and it has a strength that falls off quickly with distance instead of increasing, as the direct color force does. Nevertheless, it remains much stronger than the electromagnetic force over nuclear distances of several femtometers and so it is able to strongly bind the nucleus together. The strength of the force we chose to enter in Table 26.5 is not the inherent strength of the color force, but the strength of the strong nuclear force that arises from the residual color force in pion exchange and that binds protons and neutrons within the nucleus.

26.6 Epilogue

At this point, after these last three chapters, we suspect that many of you are asking yourselves this question:

> Whatever happened to common sense? As we read the first twenty-three chapters of this book, some of it was easy and some of it was hard, but it all made sense. But then along came Chapter 24 and special relativity, teaching us that our reasonable answers were actually wrong. And then, in Chapter 25, it appeared that even our questions were wrong. Apparently, you can't ask, "Where is the electron?" The only question that has an answer is something like, "What is the probability of finding the electron here?" Do we really have to live in the world that 20th-century physics has given us?

The answer is that we do. We do live in a world that is fundamentally counter to our intuition and experience. Now we are not suggesting that you remove the first 23 chapters from your textbook. The things you learned there are reliable, just not fundamental. We can actually ignore these last three chapters if we are only trying to predict the behavior of things that are not too small and not too fast. If you want to calculate Tarzan's speed as he hits the surface of the pond, the energy method of Chapter 7 will give you the right answer, unless you want an answer that is correct to 20 significant figures, a requirement that we would never have and could never even measure. And, before we had to take relativity or quantum uncertainty into account, there would be things like air resistance and non-uniformity of the earth's gravitational field that would be far more important. Classical physics methods are completely appropriate for solving the kinds of problems you solved in the first twenty-three chapters. But it is true that the basic nature of things is different from what you probably expected.

As we pointed out at the beginning of Chapter 24, the topics we have now covered in our twenty-six chapters are the normal topics that everyone expects students in your class to have learned. This includes the topics we have selected from modern physics in these last three chapters. However, if you keep up with science news, you must be aware that there are many subjects of current scientific importance that we have not even mentioned. We discussed special relativity in Chapter 24, but only touched on the fact that Einstein also produced a theory of gravity that he called general relativity. It is general relativity that predicts and explains black holes and gravitational waves. We have touched on elementary particles, but have not really mentioned the recently discovered Higgs particle. We have also not said a word about cosmology, where the laws of general relativity and quantum mechanics are applied to explain the universe taken as a whole, from its big bang origin, through a period of accelerated expansion known as inflation, to its present state in a new era of acceleration that begs for an explanation and finds none. Presently, there is evidence from cosmology that the matter whose behavior we understand represents less than 5% of the content of the universe and that the rest of the universe is made up of *dark matter* (27%) and *dark energy* (68%). We don't know what dark matter is or how it behaves, but we know a lot of normal things that it cannot be. And for dark energy, we don't even know where to start to generate a good theory. Its only observed effect is gravitational, seen through the way it affects the expansion of the universe.

While most physicists in 1900 thought that they just about had the physical universe figured out, no one has come close to believably suggesting such a thing since. We live in a period of constant revolution that began in 1900 and has never stopped. As soon as it appears we are getting close to a final answer, nature throws us an unexpected observation that overturns our complacency and replaces it with baffled curiosity. It remains an exciting thing to live and study physics in times like ours.

Alice laughed. "There's no use trying," she said. "One can't believe impossible things."

"I dare say you haven't had much practice," said the queen. "When I was your age, I always did it for half an hour a day. Why, sometimes I've believed as many as six impossible things before breakfast."

26.7 Summary

As mentioned above, there are important things you should take from this chapter, just like from all the others. These include:

- You should understand that the mass of an object is its rest energy, which can be measured either in kilograms or in joules (or electron-volts), the units conversion being given by Equation 26.1.
- You should be able to check the conservation laws in a nuclear reaction and do the mass accounting to calculate the energy released or required in the reaction.
- You should understand the concept of the radioactive half-life and be able to use it to find the elapsed time from the observed concentration and vice-versa.

> **CHAPTER FORMULAS**
>
> Not many new formulas here. Just the not-really-new $E_{\text{rest}} = mc^2$
>
> and the radioactive decay formula: $N(t) = \dfrac{N_0}{2^M}$ with $M = \dfrac{t}{t_{1/2}}$

PROBLEMS

26.1 * 1. A 15-gram ice cube absorbs 5000 J of heat as it melts. By how many grams does its mass increase?

* 2. An electron has a momentum given by $pc = 1$ MeV. What is its total energy?

** 3. One proton is accelerated through a potential difference of 10^6 volts and another is accelerated through a potential energy of 10^{10} volts. For each proton, find the total energy, the momentum (actually find pc in MeV), and the wavelength.

26.2 * 4. Use the conservation laws to fill in the missing numbers or particle symbols indicated by the dots in the following reactions:

a) ${}^{238}_{92}\text{U} \rightarrow {}^{\cdots}_{90}\text{Th} + {}^{4}_{2}\text{He}$

b) ${}^{23}_{12}\text{Mg} \rightarrow {}^{23}_{11}\text{Na} + \cdots + \nu_e$

c) ${}^{239}_{94}\text{Pu} + {}^{1}_{0}n \rightarrow {}^{204}_{79}\text{Au} + {}^{\cdots}_{15}\text{P} + 5{}^{1}_{0}n$

d) ${}^{59}_{28}\text{Ni} + e^- \rightarrow {}^{59}_{27}\text{Co} + \cdots$

** 5. The bare mass of a carbon-12 nucleus (no atomic electrons included) is 11,174.9 MeV. Find the amount by which the rest energy of a carbon-12 nucleus is less than the sum of the rest energies of the individual nucleons that make it up (this is called "the binding energy" of carbon-12).

** 6. The bare mass of the oxygen-16 nucleus (no atomic electrons included) is 14,895.08 MeV. Find the amount by which the rest energy of oxygen-16 is less than the sum of the rest energies of all its nucleon components (this is called "the binding energy" of oxygen-16).

** 7. Uranium-235 can absorb a slow neutron to become uranium-236, which quickly decays into other lighter nuclei. One of the reactions that occurs through this process is $U^{235} + n = Ba^{141} + Kr^{92} + 3n$. The masses of the bare nuclei (no atomic electrons included with the nucleus) and the neutron are

Particle	U-235	Ba-141	Kr-92	Neutron
Mass (MeV)	218942.0	131261.0	85628.7	939.6

Find the energy in MeV released in the reaction. (Most of this energy will be carried away as kinetic energy of the neutrons.)

** 8. The rest energies of the bare nuclei (no atomic electrons included) and particles for the reaction displayed in Equation 25.13 ($^{14}_{6}C \rightarrow {}^{14}_{7}N + e^- + \bar{\nu}$) are

Particle	$^{14}_{6}C$	$^{14}_{7}N$	e	$\bar{\nu}$
Mass (MeV)	13040.870	13040.208	0.511	~0

Find the energy in MeV released in the decay.

** 9. A free neutron at rest decays into a proton, an electron, and an antineutrino (see Example 26.3). The masses of the particles involved are: neutron, 939.565 MeV; proton, 938.272 MeV; electron, 0.511 MeV; and neutrino, ~0. Find the total final kinetic energy that the three decay products will share.

** 10. The two reactions given in Equation 26.2 and in Example 26.1 are two of the important reactions that constitute the hydrogen fusion that uses up hydrogen to produce helium in the cores of massive stars. The third step is the reaction

$$^{3}_{2}He + {}^{3}_{2}He \rightarrow {}^{4}_{2}He + 2\,{}^{1}_{1}H.$$

The mass of the He-3 nucleus is found in Example 26.1 and the mass of the He-4 nucleus is in Table 26.1. Find the energy released in this reaction.

26.3 * 11. Radium-226 decays by emitting an alpha particle (a bare nucleus of $^{4}_{2}He$). A 0.5-gram sample of pure radium-226 emits 1.83×10^{10} decays per second. The half-life of the decay is 1600 years. What will be the decay rate of this sample after 4800 years?

* 12. Suppose a radioactive element has a half-life of 1200 years. If there are presently 3.4×10^{16} of the atoms in a particular sample and if the present rate of decay is 800 counts per second, what was the decay rate of the sample 3600 years ago and how many atoms of the radioactive element were in the sample at that time?

** 13. Radiometric dating is able to measure the ages of rocks over time scales of billions of years. It is assumed that a mineral begins as a relatively pure sample of some radioactive element, so that, by measuring the relative concentrations of the radioactive element and its decay products, one can determine the age of the material. Zircon ($ZrSiO_4$) easily incorporates uranium atoms into its crystalline structure, but strongly rejects lead; so freshly cooled zircon will have a significant concentration of uranium and almost no lead inside it. Uranium-235 decays to lead-207 with a half-life of 704 million years. A particular rock containing zircon is found to include three times as much lead-207 as uranium-235. How old is the rock?

** 14. Electrons with kinetic energy 0.016 MeV emitted in the decay reaction

$$^{14}_{6}\mathrm{C} \rightarrow {}^{14}_{7}\mathrm{N} + e^- + \overline{\nu}$$

are observed coming from a sample of papyrus taken from an Egyptian tomb. The rate at which the electrons are detected in live papyrus plants is 15 decays per minute per gram of carbon. If the papyrus is 4200 years old, what radioactivity rate per gram of carbon is to be expected? [Reminder: The half-life of C-14 is on page 568 as 5730 years.]

26.4 ** 15. Each of the reactions below would violate either conservation of charge, baryon number, lepton numbers, or energy–momentum. Identify the violation for each.

a) $\mu^- \rightarrow e^- + \overline{\nu}_e + \overline{\nu}_\mu$

b) $p^+ \rightarrow \pi^+ + \pi^+ + \overline{\pi}^-$

c) $\overline{\pi}^- + p^+ \rightarrow \Sigma^+ + \pi^0$

d) $\mu^- \rightarrow \overline{\pi}^- + \nu_\mu$

** 16. What particle does the X stand for in the antimuon decay $\overline{\mu}^+ \rightarrow \nu_e + \overline{\nu}_\mu + X$? Explain how you know.

** 17. The Σ^+ particle has charge +1, baryon number 1, and strangeness –1. It can be created in the reaction $\pi^+ + p^+ \rightarrow \Sigma^+ + X$, where X stands for an unknown particle. Use the conservation laws to determine the charge, baryon number, and strangeness of the X. Which particle from Table 26.1 do you think this is?

** 18. Use the fact that the Σ^+ has charge +1, baryon number 1, and strangeness –1 to find its quark content. [Note: The strange quark is described in Table 26.4.]

** 19. **Charmed Particles.** There are elementary particles that have the same properties as particles in Table 26.2, except that they have charmed quarks in place of other quarks. What will be the charge, strangeness, and charm of the Ω particle whose quark content is scc? [Hint: See Table 26.4.]

*** 20. A negatively charged muon decays into an electron and two neutrinos via

$$\mu^- \rightarrow e^- + \nu_\mu + \overline{\nu}_e.$$

(Note that both muon lepton number and electron lepton number are separately conserved.) The half-life for this decay process is measured to be 1.56 μs for muons at rest in the laboratory. A muon detector at an altitude of 2300 m detects 1000 muons per second from a cosmic-ray shower of particles. The energy of the muons corresponds to a direct downward velocity of 98% the speed of light, or 2.94×10^8 m/s. How many muons per second will be detected in the shower at another detector located at sea level? [Hint: Radioactive decay is a kind of a clock, and the muons in the shower are moving relative to the detectors.]

26.5 ** 21. There is no direct measure of the mass of the graviton, since it has never been directly observed. However, the fact that clusters of galaxies are gravitationally bound together means that gravity must extend at least to the 10^{23}-meter size of these astronomical objects. Assuming this as a range for the gravitational force, use the uncertainty principle to find a maximum mass for the graviton in eV.

** 22. In the late 1980s, it was proposed that there existed a fifth fundamental force that acted over a distance of around 100 m. Use the uncertainty principle to determine the mass of the field particle whose exchange would mediate this new force. (Later analysis and experiment, by the way, showed that this force did not actually exist.).

*** 23[f]. Use the equations for relativistic momentum (Equation 24.8) and relativistic total energy (Equation 24.9) to produce a formula for the velocity of any object in terms of its momentum and total energy (and the speed of light, but remember that that is just a conversion factor in these calculations).

The last few problems involve elementary particle reactions where conservation laws for energy and momentum must be used to answer the questions. You must remember that the energies and velocities in particle reactions are typically high enough that the correct relativistic formulas will need to be used.

** 24. An antineutrino of momentum 1.8043 MeV (as usual, this is actually the product pc) strikes a proton initially at rest. Given these conditions, is the reaction

$$\bar{\nu} + p^+ \rightarrow n^0 + \bar{e}^+$$

an allowed reaction, conserving both energy and momentum? Use Table 26.1 for the proton and neutron masses and note that the electron mass is 0.5110 MeV. [Hint: Do not solve for the final energies and momenta. Just test the mass-energy and momentum values to see if the reaction is possible.]

** 25. A π^+, initially at rest, decays by way of the reaction $\pi^+ = \bar{\mu}^+ + x$, where x is an unknown particle. From a bubble chamber (a way to measure the velocity of charged particles), it is found that the momentum of the $\bar{\mu}^+$ is 29.792 MeV (this is momentum in energy units, that is, it is the product pc). For this problem, you will need more accurate masses than given in Table 26.2. More accurate values are $m_{\pi^+} = 139.570$ MeV and $m_{\mu^+} = 105.658$ MeV. [Hint: Keep all calculations accurate to the nearest 0.001 MeV.]

a) Find the total energy of the μ^+ after the decay.

b) Use energy conservation to find the energy of the unknown particle.

c) Use momentum conservation to find the momentum (pc) of the unknown particle.

d) Find the mass of the unknown particle.

e) Based on your mass result, and on other conservation laws, explain what you think the particle might be.

*** 26. A neutral pion (mass 135 MeV) moving to the right with total energy 225 MeV decays into two photons that move right and left, as shown. Find the energy of each photon. [Hint: remember that both relativistic energy and momentum must be conserved in this decay.]

π 225 MeV $\longrightarrow$ $\Longrightarrow$ $\longleftarrow E_1$ γ $\bullet$ $\quad$ γ $\bullet$ $E_2 \longrightarrow$

Appendix A

ALGEBRA REVIEW

This appendix is for those whose algebra is a little rusty for one reason or another. It is not a complete review of all the algebra you learned in high school, but it does explain many of the things that will be needed for the problems you will solve in this textbook. We start with some pretty basic stuff.

1 Formulas and Arithmetic Expressions

Operations. The operations of arithmetic are:

- addition and subtraction, denoted by + and −.
- multiplication, denoted with ×, or by ∗, or by simply placing the numbers side-by-side (so ab means to multiply a and b). Note that, for the side-by-side notation, if we were using numbers instead of letters, it would not be possible to decide if 23 meant twenty-three or two times three. So if numbers to be multiplied are side-by-side, we must put them in parentheses. Thus (2)(3) means two times three.
- division, denoted by a horizontal or slanted bar, like $\frac{3}{2}$ or 3/2, which both mean three divided by two.
- powers, denoted with a superscript power, so 3^2 means three squared, or nine.
- roots, denoted with a root sign like $\sqrt{4}$ or as a fractional power, so $4^{1/2} = \sqrt{4} = 2$.

Order of Operations. In order to specify a set of operations that are to be performed sequentially, we use a *formula* or an *arithmetic expression.* Mathematicians have decided, for very good reasons, to set up the following rules for the order in which operations are to be performed:

1. **Powers and roots.** Before any other operations are performed, all powers and roots must be completed. If you need to do both, the order does not matter.
2. **Multiplication and division.** Any products or quotients must be done before adding or subtracting. If you need to do both, again the order does not matter.[1]
3. **Addition and subtraction.** Add and subtract last. Again, the order does not matter.

---

1. There is an exception to the general rule that order does not matter in products or quotients, and this is that triple quotients like $a/b/c$ are ambiguous. The notation 12/3/2, for example, must be clarified with parentheses, since (12/3)/2 is 4/2 = 2, while 12/(3/2) is 12/1.5 = 8.

As an example, let us look at the arithmetic expression $1 + 2 \times 3^2$.

- First we do powers and roots, $3^2 = 9$, so we end with $1 + 2 \times 9$.
- Next we multiply and divide, $2 \times 9 = 18$, so we end with $1 + 18$.
- Last we add and subtract, ending with $1 + 18 = 19$. The answer is 19.

Parentheses and Bars. Of course, there will be times where we want to prescribe a set of operations that differs from this standard order, say where you want to tell someone to add two things together first. In this case, *we specify things that need to be done first by putting parentheses around the thing we want done first.* The parentheses go around a sub-expression of the operations that need to be completed before other things are done. A similar grouping is defined in the case of a division or a square root by using a horizontal bar. Thus we have

- $7(5 - 3)$ means first subtract 3 from 5 to get 2, and then multiply by 7 to get 14.
- $(2 \times 3)^2$ means first multiply 2 times 3 to get 6, and then square it to get 36.
- $\sqrt{1 + 2 * 4} + 2$ (note the square root bar) means first multiply 2 times 4 to get 8 (we observe the usual order of operations within parentheses or under a bar). We then add 1 to get 9, take the square root of everything under the bar to get 3, then add 2 to get 5.

When needed, we use square brackets [] and even curly brackets {} for nested parentheses.

Formulas. A formula is a way of specifying the arithmetic operations that are to be done without specifying the particular numbers that are to be used. Thus, we might write down the formula $a + bc^n$, meaning to first take a number c, raise it to the power n, multiply the result by another number b and then add a number a to it. You should see that the special case with $a = 1$, $b = 2$, $c = 3$, and $n = 2$ gives the same $1 + 2 \times 3^2$ operations we did before, but the formula now applies for any values of a, b, c, and n. When the value of a letter is fixed ahead of time, it is called a *constant*. When it is unknown, or can take multiple values, it is called a *variable*.

To see why formulas are so useful, look at an example of how to use them. If an object starts from rest and accelerates at some rate for some time, the distance from where the object started to where it ends up can be found for any acceleration and any time by doing the operations prescribed by the following formula:

$$x = \frac{1}{2}at^2,$$

where a is the acceleration, t is the elapsed time, and x is the distance. What the formula tells you is that, whatever the time and the acceleration are, if you take the time, square it, multiply the result by the acceleration, and then multiply that by ½, you will get the distance covered in that time. This is a useful set of instructions to communicate. This is why we use formulas.

Note that the same order of operations, and the same meaning for parentheses and bars, applies for formulas as for arithmetic expressions. Thus,

- $a(b - c)$ means first subtract c from b, and then multiply the result by a.
- $(xy)^2$ means first multiply x times y, and then square the result.
- $\sqrt{a + xy} + c$ means first multiply x times y, then add a to the result. Take the square root of your result. Finally add c to the result of the square root operation.

Try It Out. To make sure you understand formulas and arithmetic expressions, you might want to verify the following calculations.

✓ $(4^2 - 7)^{1/2} = 3$.

✓ $(1 + x)^2 + b = 0$, if $x = 2$ and $b = -9$.

✓ $\frac{2+4}{3 \times 8} = \frac{1}{4}$

✓ Verify that order does not matter in most multiplications and divisions by calculating $(2 \times 3)/6 = 1$ and $2 \times (3/6) = 1$.

✓ Verify that order does not matter in roots and powers by calculating $\sqrt{4^3} = 8$ and then by calculating $(\sqrt{4})^3 = 8$.

2 Equations and Solving

The Equation as a Balance. When we write an equation, like $x = 7$, we mean that the things on each side of the equals sign are numerically the same. It is like a laboratory balance (Figure A.1a) where, when both sides weigh the same, the scales balance. If we had a 4-kg mass on one side of the scales and two 2-kg masses on the other, the scales would balance. They are the same with regard to weight. Then, from that point on, any time we change what is on both sides of the scales in the same way, the scales will continue to balance. If we added ¼ kilogram to each side (Figure A.1b), the scales would still balance (at 4¼ kg on each side). If we divided the weights in two (we would have to split the 4-kg weight, as in Figure A.1c), the scales would still balance. Similarly, if we do the same arithmetic operation to each side of an equation, the equation will continue to balance. The things on the two sides will remain numerically the same.

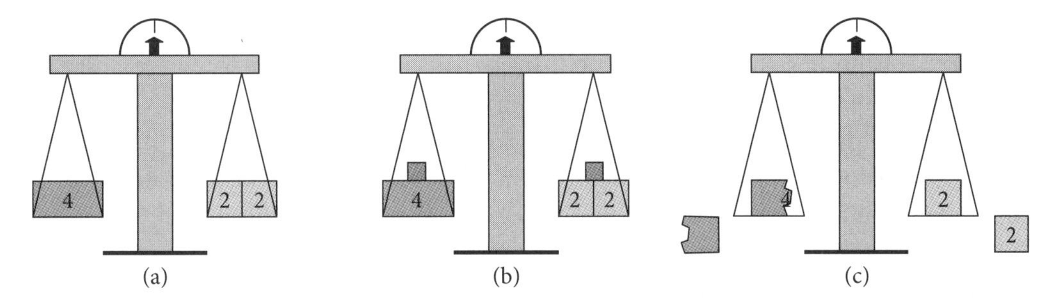

Figure A.1 (a) A 4-kg mass balances two 2-kg masses. (b) A ¼-kg mass added to each side still balances the scales. (c) Dividing both sides by 2 still balances the scales.

Suppose we apply this principle to an equation. If we start with $x = 7$, then we can multiply both sides of the equation by 2 to discover that $2x = 14$, or we can add 3 to each side to give $x + 3 = 10$. If we started with $\frac{1}{2}x = 3$, then we could multiply both sides by 2 to see that $x = 6$.

Solving for the Unknown. The last example in the last paragraph is typical of the process of doing algebra, of solving an equation to find the value of some unknown quantity. We begin with an equation involving the quantity whose value we wish to find, but in which the equation has it tied up in some other arithmetic operations, like x being multiplied by ½ in $\frac{1}{2}x = 3$. Then we find a way to do some operation to each side of the equation (thereby keeping the balance) that will undo the complication, leaving the quantity we want by itself. The resulting equation will then tell us what the unknown variable (like x)

is equal to. In the example above, we undid the multiplication by ½ by multiplying both sides by 2. This turned $\frac{1}{2}x$ into x, leaving the x alone on one side of the equation and telling us that x equals 6.

The key to isolating the quantity we wish to find lies in undoing the operations in which it is involved in the original equation. Here, we must remember the order of operations so that we can undo them *in the opposite order.* This way we leave the unknown standing by itself.

For example, suppose we knew that $1 + 2x^2 = 19$, and that we wanted to find the value of x that would satisfy this equation—the value that would make the equation true. We would need to do the following steps:

1. We know that, in calculating something with a formula, the *last* thing that we would do is to add and subtract. So addition and subtraction is the *first* thing we do to solve for x. We first subtract 1 from each side, producing the equation

$$1 + 2x^2 - 1 = 19 - 1 \quad \text{or} \quad 2x^2 = 18.$$

2. The next thing is to undo the multiplication by 2 by dividing both sides of the equation by 2, giving

$$\frac{2x^2}{2} = \frac{18}{2} \quad \text{or} \quad x^2 = 9.$$

3. Finally, x is complicated by being squared in the last equation. We undo a square by taking the square root of each side, giving

$$\sqrt{x^2} = \sqrt{9} \quad \text{or} \quad x = 3.$$

Actually, the last step could have been satisfied by either $x = +3$ or by $x = -3$, but we will talk about that a little later.

Note the order of the solution. As we solve for some variable, we reverse the order from the order of operations given in Section A.1. The exception to such a procedure, of course, is where there are parentheses in an equation. In this case, the order of operations tells us that we must first finish up everything inside the parentheses before going on to other operations. So we again do the opposite. When the variable we are solving for is part of an expression contained inside parentheses, we keep that expression intact while undoing everything else. In a way, we treat the expression in the parentheses as if *it* were the unknown and we only undo what is inside the parentheses at the very last, or when the parentheses are no longer needed.

Sometimes the unknown variable will occur more than once in an equation. In this case, the process of solution will involve operations that eliminate the unknown on one side of the equation and bring all instances of the unknown to the same side of the equation. As an example, suppose we started with $4x - 5 = x + 7$ and wanted to solve for x.

1. First, we can add 5 to each side, giving $4x = x + 12$.
2. Next, we can subtract x from each side, giving $3x = 12$.
3. Finally, we divide both sides by 3, giving $x = 4$.

Or suppose we wanted to find y in the equation $\frac{6y}{1 + y} = 3$.

1. Since $1 + y$ is under a bar, it is like an expression in parentheses and must be kept together until the bars or parentheses are no longer necessary. There is no other

addition or subtraction to do, so we turn to multiplication and division as our first step in disentangling y from the operations it is involved in. We always want to know what an unknown is, not what its inverse is, so our first step is to get $1 + y$ out of the denominator of the fraction. To do this, we multiply both sides of the equation by $(1 + y)$ to give

$$\frac{6y}{1+y}(1 + y) = 3(1 + y) \quad \text{or} \quad 6y = 3(1 + y).$$

Note how, on the left, we have used the fact that the operations of multiplying by $(1 + y)$ and dividing by $(1 + y)$ cancel each other out, leaving the $6y$ by itself.

2. Still keeping $(1 + y)$ by itself, we divide the equation by 3 to get $2y = 1 + y$.
3. At this point, the parentheses are no longer necessary to keep $1 + y$ together as a unit. So we proceed to get everything with y in it on the left side of the equation by subtracting y from each side. This gives $2y - y = 1 + y - y$, producing the final result, $y = 1$.

Try It Out. Follow along in the following solutions for the variable z.

- ✓ $1/z = 4$. Multiply both sides by z to get $4z = 1$. Divide by 4 to get $z = 1/4$.
- ✓ $7z + 2 = 23$. Subtract 2 from each side to give $7z = 21$. Divide by 7 to get $z = 3$.
- ✓ $(1 + z)/(1 - z) = 5$. Here we have two sets of parentheses to keep together. First, we multiply by $(1 - z)$ to get $1 + z = 5(1 - z)$. We then apply the distributive rule (if you don't remember what that is, don't worry; we will get to it first thing in the next section). This gives $1 + z = 5 - 5z$. Parentheses are now no longer needed to give the correct operations. We add $5z$ to both sides, giving $1 + 6z = 5$. We subtract 1 from each side for $6z = 4$, and, finally, we divide by 6 to find $y = 2/3$.

3 Expanding and Factoring

Distributive Rule. Although the operations that are performed for the expression $a(b + c)$ are perfectly clear and simple—you add b to c and then multiply the result by a—there are times when we want to come to the same final result by way of a different calculation. An example of this, and of a reason for it, was apparent in the last *Try It Out* in the last section. The expression $5(1 - z)$ was perfectly clear, but we wanted to separate the terms with z in them. To do this, we used the Distributive Rule.

The Distributive Rule states that $a(b + c) = ab + ac$. You know that it must be right, since it's a rule, but let's test it anyway with a couple of examples.

- $3(2 + 4) = 3 \times 6 = 18$. But it is also true that $3(2 + 4) = 3 \times 2 + 3 \times 4 = 6 + 12 = 18$.
- $(1 + 6)(2 + 3) = (7)(5) = 35$. But is it also true that we can come up with the same result using $(1 + 6)(2 + 3) = (1 + 6)2 + (1 + 6)3$, and another use of the Distributive Rule to make it $1 \times 2 + 6 \times 2 + 1 \times 3 + 6 \times 3$? Yes, this gives $2 + 12 + 3 + 18 = 35$.

Factoring. The opposite procedure from using the Distributive Rule to expand a product, as we did in the last paragraph, is to collect common multiplicative terms from a series of additions and subtractions, pulling them out of the terms they are in so that we can write the final expression as the product of the common term times a sum in parentheses. This is called factoring. It is the process of writing $ax + bx = (a + b)x$. We needn't test this procedure, since it's just the Distributive Rule in reverse. So let's just cite a couple of examples.

- $3x + ax = x(3 + a)$.
- $ax + bx^2 + cx^3 = x(a + bx + cx^2)$.

Binomials. In several of our examples so far, we have seen expressions like $1 + y$ or $1 - z$. A two-element expression like this is called a binomial. Products of binomials come up often in practical applications of algebra, so it is good to know how to handle them. Let us begin by expanding the product of two binomials using the Distributive Rule.

- The generic case is $(x + a)(y + b) = (x + a)y + (x + a)b = xy + ay + xb + ab$.
- A very common case has the same variable in both binomials, such as $(x + a)(x + b)$. This gives $x^2 + ax + xb + ab$. If we then factor the x out of the middle two terms, we end up with $x^2 + (a + b)x + ab$. Note that this final expanded version for $(x + a)$ times $(x + b)$ has an x^2 by itself, an x that is multiplied by a coefficient[2] that is the sum of a and b, and a term with no x in it that is the product of a and b. This is useful to remember if you ever want to undo the expansion and get back to the original factors. If we see an expression with x^2 by itself, plus some coefficient times x, plus some constant term, and if we can find two numbers that multiply to give the constant term and add to give the coefficient of the x term, then we can factor the expression into the product of two binomials.

A series of terms with various powers of some variable in them (including a constant term that can be thought of as the variable to the zeroth power) is referred to as a *polynomial.* A polynomial in which the highest power of the variable is 2 is called a *quadratic.* Let us look at some examples of expanding products of binomials or of factoring quadratic polynomials.

- Let us expand $(x + a)^2 = (x + a)(x + a) = x^2 + 2ax + a^2$.
- Let us expand $(x + a)(x - a) = x^2 + (a - a)x + (a)(-a) = x^2 - a^2$.
- Let us factor $x^2 - 2x - 15$. We note that -15 is $(-5)(3)$ or $(-3)(5)$, but that the -2 coefficient of x is $3 + (-5)$. So we can write $x^2 - 2x - 15 = (x - 5)(x + 3)$.

Try It Out. Here are some worked examples that you can verify.

✓ Expand $x(x + y + a)$ to $x^2 + xy + ax$.

✓ Expand $(a + y)(x^2 - y)$ to $ax^2 - ay + x^2y - y^2$.

✓ Factor $q^2 + 14q + 49$ to get $(q + 7)^2$.

✓ Factor $z^2 - 25$ to get $(z - 5)(z + 5)$.

4 The Quadratic Formula

Suppose we begin with an equation $x^2 = 9$, and we want to solve for x. We take the square root of each side to get $x = 3$. But, as we have already noted, there is another value for x that would satisfy the original equation. This is $x = -3$, because $(-3)(-3) = +9$. But how can algebra give us two different answers to the same question? Well, it can. If all you know is that some number squared equals nine, then there are two numbers that will work, and that's the right answer.

2. A constant or an expression that multiplies a variable is called the *coefficient* of the variable.

In fact, any time you have an unknown variable squared in an equation, you are taking away from that squared term any information about the sign of the thing being squared, and so you may expect the solution to the equation to have two answers to it, both correct.

The Quadratic Formula. There are several common places in physics where a quadratic equation occurs naturally and needs to be solved to find the answer to a problem. A quadratic equation has the general form

$$ax^2 + bx + c = 0, \tag{A.1}$$

where x is the variable we need to find and a, b, and c are known constants. The solutions to this equation (remember that we must expect two solutions to a quadratic equation) are given by[3]

$$x_+ = \frac{-b + \sqrt{b^2 - 4ac}}{2a} \quad \text{and} \quad x_- = \frac{-b - \sqrt{b^2 - 4ac}}{2a}. \tag{A.2}$$

An example is probably more useful than a long explanation of how to use this equation. Let us solve the equation

$$2x^2 - 7x - 4 = 0.$$

Comparing this equation with Equation A.1, we can identify $a = 2$, $b = -7$, and $c = -4$. Using these values for the constants in Equation A.2, we find

$$x_\pm = \frac{-b \pm \sqrt{b^2 - 4ac}}{2a} = \frac{-(-7) \pm \sqrt{(-7)^2 - 4(2)(-4)}}{2(2)} = \frac{7 \pm \sqrt{49 + 32}}{4} = \frac{7 \pm \sqrt{81}}{4} = \frac{7 \pm 9}{4}.$$

The two solutions are therefore $x_+ = (7 + 9)/4 = 4$ and $x_- = (7 - 9)/4 = -\frac{1}{2}$.

Let us also look at a physics example. Suppose a woman throws a ball directly upward with an initial velocity $v_0 = 5$ m/s. If the acceleration due to gravity is -10 m/s^2, the formula for the ball's height above the ground is $y = -5t^2 + 10t$. We want to find the time it will take the ball to rise to its maximum height above the ground and fall back down to the ground ($y = 0$). Since we want the time when $y = 0$, we need to specify $y = 0$ and solve the resulting quadratic equation for t. The equation to be solved is thus $-5t^2 + 10t = 0$. Comparing with Equation A.1 we see $a = -5$, $b = 10$, and $c = 0$. In Equation A.2 this gives

$$x_\pm = \frac{-b \pm \sqrt{b^2 - 4ac}}{2a} = \frac{-(10) \pm \sqrt{(10)^2 - 4(-5)(0)}}{2(-5)} = \frac{-10 \pm \sqrt{100}}{-10} = \frac{-10 \pm 10}{-10}.$$

As we expect, we have two solutions, $x_+ = (-10 + 10)/-10 = 0$ and $x_- = (-10 - 10)/-10 = 2$. But the physics question was well posed, and we know that there can only be one answer to it. What are we to do with two answers? You may already see the explanation. For the physics problem, we wanted the time at the end of the ball's flight, but the equation didn't know that. It only knows that $y = 0$, and, if you think about it, there are two times when the height is zero—one before it is thrown into the air and one when it comes back down. Algebra didn't lie; it gave the two times when $y = 0$. It is up to us to realize that only the $t = 2$ solution was the answer to the question we really wanted.

3. The $x_\pm$ were found by finding numbers that multiply to give c and add to give b, factoring the quadratic.

5 The Binomial Theorem

There is a lot that could be said about the binomial theorem, but, since we only use it in a narrow application in this book, we will concentrate on that application. In fact, we would go so far as to say that, in physics, this is by far the most useful application of the theorem.

The Binomial Expansion. When the value of x is less than 1, the binomial $(1 + x)$, raised to any power p, can be written as an infinite sum. The first few terms in the sum are

$$(1 + x)^p = 1 + \frac{px}{1} + \frac{p(p - 1)}{1 \cdot 2}x^2 + \frac{p(p - 1)(p - 2)}{1 \cdot 2 \cdot 3}x^3 + \cdots \tag{A.3}$$

from which you should be able to guess the form of the further terms in the series. When p is a positive integer, this process will eventually reach the point where one of the $p - n$ factors in the numerators of the coefficients in the series will be zero, and, since that same factor will be in every subsequent term, that term and all following terms will be zero. The series will therefore cut off after a finite set of terms. On the other hand, when p is *not* a positive integer, being either a negative integer or a non-integer of either sign, the series will be infinite.

The Small-Number Approximation. The real value of the binomial expansion for physics comes when the variable x is much less than 1. Let's look at a concrete numerical example to see how this works. Let us consider $\sqrt{1.03}$, which we can also write as $(1 + 0.03)^{1/2}$. According to Equation A.3, this can also be written as an infinite series

$$(1 + 0.03)^{1/2} = 1 + \frac{1}{2}(0.03) + \frac{\frac{1}{2}(-\frac{1}{2})}{1 \cdot 2}(0.03)^2 + \frac{\frac{1}{2}(-\frac{1}{2})(-\frac{3}{2})}{1 \cdot 2 \cdot 3}(0.03)^3 + \cdots$$

If we multiply out the coefficients, this becomes

$$(1 + 0.03)^{1/2} = 1 + \frac{1}{2}(0.03) - \frac{1}{8}(0.03)^2 + \frac{1}{16}(0.03)^3 + \cdots$$

Then let us then multiply out the 0.03 factors to get

$$(1 + 0.03)^{1/2} = 1 + \frac{1}{2}(0.03) - \frac{1}{8}(0.0009) + \frac{1}{16}(0.000027) + \cdots$$

At this point we would argue that we could find the square root of 1.03 to a good approximation by considering only the first two terms. This would give us $\sqrt{1.03} \approx 1 + 0.3/2 = 1.015$. The error we would make by neglecting the third term in the full expansion will be a small error out in the 4th decimal place, and the error in neglecting the fourth term only affects the 6th decimal place. Clearly, the error made in neglecting further terms would be smaller still, since higher powers of 0.03 are each smaller than the term before them. What this means is that, in a case where $x \ll 1$, the formula

$$(1 + x)^p \approx 1 + px$$

is a very good approximation to the correct result, getting better and better for smaller and smaller values of x.

Try It Out. Let's look at a few places where this binomial approximation makes calculations easy.

✓ $\sqrt{1 - 3 \times 10^{-16}} = (1 - 3 \times 10^{-16})^{1/2} \approx 1 - \frac{1}{2}(3 \times 10^{-16}) = 1 - 1.5 \times 10^{-16}$. And note that the only error made by using this approximation would be out in the 32nd decimal place.

✓ $1/(1 + z) = (1 + z)^{-1} \approx 1 - z$.

✓ $(1 - y)^{1/3} \approx 1 - \frac{1}{3}y$.

But do remember that these approximations are only valid if we are assured that the quantity in the expansion (like z or y) is actually small compared to 1.

6 Simultaneous Equations

Simultaneous Equations. When there are more unknowns than one in a problem, there must be multiple equations, each one specifying additional information about the unknowns. Often these equations will relate the unknowns to each other. When this happens, one cannot just concentrate on a single equation and find an answer for one of the variables all by itself. One must find a way to address the entire set of equations simultaneously.

Solving Simultaneous Equations. The most general way to solve simultaneous equations is to take one of the equations and to isolate one of the variables on one side of the equation with other variables and constants all forming an equivalent expression on the other side. The expression is then substituted for each occurrence of the variable in each of the remaining equations. The result is a smaller set of equations with one fewer unknown. As usual, a few examples are probably worth more than any more general explanation.

- Solve $2x + y = 6$ and $2y + x^2 = 9$ for the values of x and y.

  We could do this several ways. We can solve either equation for either variable and then substitute the result in the other equation to eliminate that variable. Let us choose to solve the first equation for y. Then $y = 6 - 2x$. If we make this substitution for y in the second equation, we get $2(6 - 2x) + x^2 = 9$. This is the quadratic equation $x^2 - 4x + 3 = 0$. Its solutions are found from Equation A.2 as $x = 1$ and $x = 3$.

- Solve $x + y + z = 17$, $x - 2y = 4$, and $x - y + 3z = 35$ for x, y, and z.

  Again, many options are available, but the second equation only has two of the unknowns in it, so that is a good place to start. We choose to eliminate x from the set of equations by writing the second equation as $x = 2y + 4$ and putting this in the remaining equations:

  $$(2y + 4) + y + z = 17 \quad \text{and} \quad (2y + 4) - y + 3z = 35,$$

  or $3y + z = 13$ and $y + 3z = 31$. So we now have two equations and two unknowns. We then run the program again. We solve the first of these for z, getting $z = 13 - 3y$ and put this in the second equation to produce $y + 3(13 - 3y) = 31$, or $-8y = -8$. So we find $y = 1$. To get z we note that we just had an equation that read $z = 13 - 3y$, so $z = 10$. And, to get x, we remember that we started with $x = 2y + 4$, so $x = 6$. That was long, but it was easy.

- Solve $\sqrt{x^2 - 2y^2} - 3 = 3x$ and $y - x = 1$ for x and y.

The second equation is easy to solve for either of the unknowns in terms of the other, so we start there, choosing to eliminate y using $y = x + 1$. When this is substituted in the first equation, we end with $\sqrt{x^2 - 2(x + 1)^2} - 3 = 3x$.

Square roots are poison in solutions. We don't have a technique for solving square-root equations, so we always end up having to get the square root by itself and then squaring both sides of the equation, eliminating the square root and living with whatever is left on the other side of the equation. We add 3 to each side of our last square-root equation

$$\sqrt{x^2 - 2(x + 1)^2} = 3x + 3.$$

We are now prepared to square each side, giving

$$x^2 - 2(x + 1)^2 = (3x + 3)^2.$$

We expand the two squares of the binomials to get $x^2 - 2(x^2 + 2x + 1) = (9x^2 + 18x + 9)$, or, collecting terms, $10x^2 + 22x + 11 = 0$. The solutions of the quadratic are

$$x_{\pm} = \frac{-22 \pm \sqrt{22^2 - 4(10)(11)}}{2(10)} = -1.10 \pm 0.33.$$

There are two possible values for y as well, one for each possible value of x. They are $y_{\pm} = x_{\pm} + 1 = -0.10 \pm 0.33$. (By the way, the necessary step where we squared both sides of the equation can sometimes create a situation where there are more solutions to the squared equation than for the original equation. Solutions that follow from this procedure should always be checked in the original equation to be sure they are solutions to the original equation. Both of the answers we found work in the original equation.)

Try It Out. Find x and y (and z, if needed) in the following sets of equations.

✓ $x + y = 7$ and $x - y = 5$ yield $x = 6$ and $y = 1$.

✓ $x^2 + 3y = 7$ and $x + y = 1$ yield $x_{\pm} = 1.5 \pm 2.5$ and $y_{\pm} = 1 - x_{\pm} = -0.5 \mp 2.5$, in which the symbol $\mp$ was used instead of $\pm$ because $y_+ = -3$ is the value that goes with $x_+ = 4$.

✓ $x = y + z$, $4 - 4y - x = 0$, and $3z - 4 + x + 2z + 9 = 0$. The first equation is already solved for x, so we eliminate x in the last two equations, giving $4 - 4y - (y + z) = 0$, or $5y + z = 4$ and $3z - 4 + (y + z) + 2z + 9 = 0$, or $6z + y = -5$. Let us use the first of these two to give $z = 4 - 5y$ and put that in the second to give $6(4 - 5y) + y = -5$, or $-29y = -29$. Thus $y = 1$, $z = 4 - 5y = 4 - 5 = -1$, and $x = y + z = 0$.

7 Powers of Ten Arithmetic

We end this algebra review appendix by returning to some arithmetic. One useful common notation for expressing numbers is to write them as a number between 1 and 10 (a number called the "coefficient"), times some power of 10. Arithmetic with numbers expressed this way is easy to learn and, perhaps, easy to forget. So let us review it here.

Expressing Numbers in Scientific Notation. This notation was invented in order to simplify the expression of very large or very small numbers. If we have the number 6740, we recognize that it is the same as 6.74 × 1000, or 6.74×10^3. This last version is in scientific notation. There is one digit to the left of the decimal point and the power of 10 gives the multiplier. An easy way to find the power of 10 needed is to put the decimal at the end of the original number and then see how many places it must be moved to the left to get it to where it should be in scientific notation. Thus

$$6740 = \underset{3\ 2\ 1}{6\,7\,4\,0.} = 6.74 \times 10^3.$$

For small numbers, we use negative powers. The number 0.0000674 is the same as 6.74/100000, or $6.74/10^5$, which is the same as 6.74×10^{-5}. To find the correct power of 10, we again start where the decimal point is found in the original number and count how many places it must be moved to the right to get it to where it should be in scientific notation. Thus

$$0.0000674 = \underset{1\ 2\ 3\ 4\ 5}{0.0\,0\,0\,0\,6}\,7\,4 = 6.74 \times 10^{-5}.$$

Adding Numbers in Scientific Notation. When adding two numbers, the powers of 10 must be the same. Thus

$$6.74 \times 10^3 + 6.74 \times 10^{-5} = 6.74 \times 10^3 + 0.0000000674 \times 10^3 = 6.7400000674 \times 10^3.$$

So, no short cuts involved, not easy, not pretty, but correct.

Multiplying Numbers in Scientific Notation. Here is where the power of scientific notation is apparent. An example is the best way to start. The product of 200 and 3120 is

$$200 \times 3120 = 624{,}000 \quad \text{or} \quad (2 \times 10^2)(3.12 \times 10^3) = (2)(3.12) \times (10^2)(10^3) = 6.24 \times 10^5.$$

Note how the coefficients are multiplied together and the powers are multiplied together. Note also that the product of 10^2 and 10^3 is 10^5. When multiplying powers of 10, the powers add. Let us look at another example.

$$300 \times 6740 = (3 \times 10^2)(6.74 \times 10^3) = 20.22 \times 10^5 = 2.022 \times 10^6.$$

Here, the multiplication of the two coefficients gave a number bigger than 10. The method of adding the powers still worked, but the final power of 10 had to be adjusted by hand to keep the scientific notation convention. As a final example, let us multiply

$$0.02 \times 3120 = (2 \times 10^{-2})(3.12 \times 10^3) = 6.24 \times 10^1,$$

the final power coming from adding the powers, as usual, with $-2 + 3 = 1$.

Dividing Numbers in Scientific Notation. We can see how to divide numbers in scientific notation by remembering that dividing by a number is the same as multiplying by its inverse. We also remember that, by definition of a negative exponent, $1/10^n = 10^{-n}$, for any power n. So we have

$$\frac{3120}{200} = \frac{3.12 \times 10^3}{2 \times 10^2} = 1.56 \times 10^{(3-2)} = 1.56 \times 10^1.$$

We also remember that subtracting a negative number is the same as adding a positive number.

$$\frac{3120}{0.02} = \frac{3.12 \times 10^3}{2 \times 10^{-2}} = 1.56 \times 10^{(3+2)} = 1.56 \times 10^5.$$

Try It Out. See that you can get the right answers for the following (remembering that, after the powers are added or subtracted, there might still be some adjustments needed by hand to get the final answer into scientific notation).

✓ $492 * 1500 = 7.38 \times 10^5$.

✓ $6.2 \times 10^{-11} \times 8 \times 10^2 = 4.96 \times 10^{-8}$.

✓ $0.042/840 = 5 \times 10^{-5}$.

✓ $6.67 \times 10^{-11}/5 \times 10^{-20} = 1.334 \times 10^9$.

✓ $2.42 \times 10^3 + 4.8 \times 10^2 = 2.9 \times 10^3$.

FUNDAMENTAL AND PHYSICAL CONSTANTS

Speed of light	$c \equiv 299792458$ m/s $\approx 3 \times 10^8$ m/s
Earth's gravitational field	$g = 9.8$ m/s^2 ≈ 10 m/s^2
Newton's gravitational constant	$G = 6.67 \times 10^{-11}$ N·m^2/kg^2
Boltzmann's constant	$k = 1.38 \times 10^{-23}$ J/K
Avogadro's number	$N_0 = 6.02 \times 10^{23}$/mole
Gas constant	$R = 8.31$ J/(mole·K)
Coulomb's constant	$k = 9.0 \times 10^9$ N·m^2/C^2
Permittivity of free space	$\varepsilon_0 = 8.85 \times 10^{-12}$ C^2/(N·m^2)
Permeability of free space	$\mu_0 = 4\pi \times 10^{-7}$ T·m/A
Planck's constant	$h = 6.626 \times 10^{-34}$ J·s $hc = 1240$ eV·nm
Elementary charge	$e = 1.602 \times 10^{-19}$ C
Electron mass	$m_e = 9.11 \times 10^{-31}$ kg $m_e c^2 = 0.511$ MeV
Proton mass	$m_p = 1.6726 \times 10^{-27}$ kg $m_p c^2 = 938.27$ MeV
Neutron mass	$m_n = 1.6749 \times 10^{-27}$ kg $m_n c^2 = 939.57$ MeV
Mass of earth	$M_\oplus = 5.974 \times 10^{24}$ kg
Mass of sun	$M_\odot = 1.989 \times 10^{30}$ kg
Mass of moon	$M_{☾} = 7.342 \times 10^{22}$ kg
Radius of earth	$R_\oplus = 6.37 \times 10^6$ m
Radius of moon	$R_{☾} = 1.74 \times 10^6$ m

Earth average atmosphere pressure at sea level	$p_A = 101{,}325$ Pa $\approx 1 \times 10^5$ Pa
Speed of sound in dry air at 20° C	$v_{air} \approx 343$ m/s
Density of water	$\rho_W = 1$ g/cm^3 $= 1000$ kg/m^3
Density of air at sea level at 15° C	$\rho_{AIR} = 1.225$ kg/m^3

The symbol ≡ is used when a number is set by definition. This is why we have given nine figures for *c*. The symbol ≈ is used for an approximate value that is acceptable in most problems in this book.

And a few trig values …

Angle	Sine	Cosine	Tangent
30°	0.500	0.866	0.577
37°	0.600	0.800	0.750
45°	0.717	0.717	1.000
53°	0.800	0.600	1.333
60°	0.866	0.500	1.732

Appendix C

UNITS CONVERSIONS

Length

1 in ≡ 2.54 cm

1 ft = 0.3048 m

1 mi ≡ 5280 ft

1 mi = 1.609344 km

1 $c \cdot$yr = 9.4607×10^{15} m

Area

1 $m^2 = 10^4$ cm^2

Volume

1 $m^3 = 10^6$ cm^3

1 liter = 10^{-3} m^3

Time

1 day = 86400 s

1 year = 3.15576×10^7 s

Velocity

1 ft/s = 0.3048m/s

1 mi/hr = 0.4470 m/s

1 mi/hr = 1.609 km/hr

60 mi/hr = 88 ft/s

Angle

1° ≈ $\pi/180$ rads

1 rad = 57.8°

360° = 6.283185307 rads

Force

1 N = 0.2248 lbs

1 lb = 4.448 N

Pressure

1 atmosphere = 1.013×10^5 Pa

1 atmosphere = 760 torr

1 lb/in^2 (psi) = 6890 Pa

Energy

1 Calorie = 4.186 J

1 KW·hr = 3.6×10^6 J

1 eV = 1.602×10^{-19} J

1 J = 6.242×10^{18} eV

Mass

1 lb (≡) 0.45359237 kg

The mass whose English weight is 1 lb is defined by agreement to be 0.45359237 kg. Inversely ...

1 kg (≈) 2.2 lb

1 u = 1.66054×10^{-27} kg

Mass: Energy

1 u = 931.5 MeV

Charge

1 e = 1.602×10^{-19} C

1 C = 6.242×10^{18} e

Units Abbreviations

ampere (A)
atomic mass unit (u)
Celsius (C)
coulomb (C)
degree (°)
diopter (D)
electron-volt (eV)
elementary charge (*e*)
Fahrenheit (F)
farad (F)
foot (ft)
gram (g)
henry (H)
hertz (Hz)
hour (hr)
joule (J)
kelvin (K)
kilogram (kg)
light year ($c \cdot$yr)
meter (m)
mile (mi)
newton (N)
ohm (Ω)
pascal (Pa)
pound (lb)
pound per square inch (PSI)
second (s)
tesla (T)
volt (V)
watt (W)
weber (Wb)
year (yr)

ANSWERS TO ODD-NUMBERED PROBLEMS

Chapter 1

5. 10^6 **7.** 24 m/s **9.** degree·ft^2·s/BTU **11.** 86400 s **13.** 1.524 m **15. a)** 10.76 ft^2; **b)** 452 ft^2 **17.** 1.893×10^{-2} m^3

Chapter 2

1. 88 ft **3.** Direct route: 120 s, alternate route: 126 s **5. a)** 40 mi/hr; **b)** 35 mi/hr; **c)** 11.67 mi; **d)** 20 mi **7.** 28 mi/hr **9. a)** 8 mi/hr; **b)** Less (the speed during the first second is all less than average); **c)** 6 mi/hr/s; **d)** 1 mi/hr/s **11. a)** 10 mi/hr/s; **b)** 0.4 s **13. a)** speeding up in the negative direction, the slope is negative and getting steeper; **b)** acceleration is negative; **c)** zero; **d)** A is passing B; **e)** equal; **f)** ≈–25 m/s **17. a)** B, F; **b)** C, D, E; **c)** A, G, H; **d)** A, D, H; **e)** C; **f)** E, G **19.** 20 m/s **21.** Yes. It requires only 40 m. **23.** 50 m/s **25. a)** 20 m/s down; **b)** 60 m above him **27.** 21.6 m/s **29.** 20 m, 6⅔ stories **31.** 5×10^5 m/s^2 **33. a)** 20 m/s; **b)** 30 m/s; **c)** 25 m/s; **d)** 25 m **35.** 2.76 s, the first solution is when she first catches up to the train, while the second (7.24 s) is when the train catches up with her if she keeps running. **37.** 10 s

Chapter 3

3. {8.9 cm, 27° rightward from forward} **7. a)** {57 m, 52°}; **b)** {48 m, 54°}; **c)** {13 mm, 67°} **9. a)** (18 m/s, 24 m/s); **b)** (0.51 cm/s, –1.09 cm/s); **c)** (–4.64 mi/hr, –1.87 mi/hr) **11. a)** {8 ft, east}; **b)** {5.8 ft, 31° north of east}; **c)** {2 ft, east} **13.** {14.0 lbs, 327°} = (11.76 lbs, –7.84 lbs) **15.** {1000 m, 37° north of east} **17.** {7.07 km, 45° south of east} **19.** 33.6° **21.** 1.56 mi **23.** 14 **25.** $x = R - \sqrt{R^2 - h^2}$

Chapter 4

1. {6 ft/s, 50° above horizontal} **3. a)** {3.38 mi/hr, 37°}; **b)** (2.68 mi/hr, 2.05 mi/hr) **5.** (–0.88 m/s^2, 2.1 m/s^2) **7.** {9.1 mi/hr/min, 50° below horizontal} **13. a)** (80 m/s, –60 m/s); **b)** 160 m; **c)** 140 m; **d)** 113 m/s **15. a)** 16 m/s, 12 m/s; **b)** No. After 0.5 seconds, the banana hits the canyon wall 0.25 m below the top on its way up (bad aim). **17. a)** 8 m/s, 6 m/s; **b)** By the time he crosses the canyon, he is –21 m below the point where he started, 9 m below the top of the canyon on the far side **19. a)** (4.8 m/s, –3.6 m/s); **b)** 7.3 m; **c)** 13.5 m/s, down at 69°; **d)** 2.88 m **21.** 0 m **23. a)** 0.4 s; **b)** 0.24 m **25.** $R = (v_0^2/g)\cos(2\theta)$ **27.** 8000 m/s **29.** 14.4 m, much more than 4.05 m **31.** $a_{cent} = (v_0 + at)^2/r$ **33.** 25 mi/hr, 37° north of west **35. a)** The shortest time will be when the cross-river component of the velocity is the greatest; **b)** 56.3°; **c)** 150 ft **37.** {1307 mi/hr, 2.2° north of east}

Chapter 5

1. a) Yes. A spaceship moving at constant velocity is unaccelerated, and any unaccelerated reference frame is inertial; **b)** No. The velocity does not matter. An accelerated reference frame is not inertial. **5.** 12 m/s^2 **7.** 6 s **9.** (80 N, 60 N) **13. a)** 4 m/s^2; **b)** 2 m/s^2
17. a) sliding: $W_{\text{earth,eraser}}$, $N_{\text{table,eraser}}$, $f_{\text{table,eraser}}$, falling: $W_{\text{earth,eraser}}$, rest: $W_{\text{earth,eraser}}$, $N_{\text{floor,eraser}}$; **b)** sliding: $W_{\text{eraser,earth}}$, $N_{\text{eraser,table}}$, $f_{\text{eraser, table}}$, falling: $W_{\text{eraser,earth}}$, rest: $W_{\text{eraser,earth}}$, $N_{\text{eraser,floor}}$
19. 1120 N **21.** 20 N in the vertical rope, 3 N in the horizontal rope **23.** 12 N
25. a) The accelerations are the same since the string will not let A move faster than B; **c)** The accelerations are the same and $m_B > m_A$, so $F_{\text{net,B}} > F_{\text{net,A}}$; **d)** $W_A = 3$ N, $W_B = 5$ N, $T_{A,B} = 7$ N, $T_{B,A} = 7$ N, $F_{A,\text{net}} = 12$ N; **e)** $T_{A,B} = 7$ N, $W_B = 5$ N, the tension is greater.
27. 49,000 N **29. a)** 2 m/s^2; **b)** 8 m/s; **c)** 3.2 m **31.** 180 N **33. a)** 5 m/s^2 down; **b)** 170 N down **35.** $T_1 = 28.9$ N, $T_2 = 57.7$ N **37. a)** 75 N; **b)** force comes from tension forces in his hip

Chapter 6

1. 7×10^{-7} N **3. a)** 2×10^{20} N; **b)** 2.7×10^{-3} m/s^2; **c)** 3.3×10^{-5} m/s^2 **5. a)** 3.56×10^{22} N, 5.9×10^{-3} m/s^2; **b)** 4.2×10^{17} N, 7×10^{-8} m/s^2 **7.** 80 N **9.** 12 N **13.** 0.8 **15. b)** 800 N; **c)** 7.5 m/s^2 right **19. a)** 10^4 N; **b)** 7520 N; **c)** 4.1 m/s^2 up the hill; **d)** 110 m **21.** 5 N
23. 450 N **25. a)** 15 kg; **b)** 180 N **27.** 4 m/s^2 **29. a)** The acceleration of B is toward the rotation axis; **c)** $N_{B,A} = -N_{A,B}, f_{B,A} = -f_{A,B}$; **d)** $N_{\text{wall,B}} > N_{A,B}$ for inward acceleration; **e)** $f_{B,A} = W_A = 40$ N, $W_B = 30$ N, $f_{\text{wall,B}} = 70$ N **31. b)** 660 N **33.** 2 N [Hint: Concentrate on the dolly as a free body.] **35.** 188 N **37. a)** 9.2 m/s^2 (0.94)g; **b)** 0.94
39. 42° (using $g = 10$ m/s^2)

Chapter 7

1. 2.7×10^{33} J, 3.75×10^{13} yrs **3.** 3.6 J **5. a)** 200 m/s; **b)** 1×10^4 J **7.** 0.8 m **9.** 6.32 m/s **11.** 1 m **13.** –104 J **15.** 2.8 mv **17. a)** W = 0 since the normal force is perpendicular to the motion; **b)** 2.5×10^4 N/m; **c)** 25 m/s^2 up; **d)** 2.8×10^4 N; **e)** 54 m/s **19.** 80 J
21. a) There is no friction and the normal force is perpendicular to the motion; **b)** 2.4×10^4 J; **c)** 10 m/s; **d)** 4 m/s^2; **e)** 480 N **23.** 300 W **25.** 1.8×10^7 J **27.** 1500 W **29. a)** 0.174 m/s; **b)** 1.44 m/s **31.** 40 J **33. a)** 1.54×10^3 J; **b)** 15.7 m/s **35.** 160 m **37. a)** 7.5 J; **b)** 17.3 m/s; **c)** 75 N **39.** 11 m **41. a)** 39.2 N; **b)** 148 N; **c)** 816 N/m **43. a)** $v = \sqrt{2gr(1 - \cos\theta)}$; **b)** $2mg(1 - \cos\theta)$; **c)** $N = mg(3\cos\theta - 2)$; **d)** 48.2°

Chapter 8

1. 3 N·s **3.** 5.5 N·s **5.** 240 kg·m/s north **7. a)** 5 kg·m/s; **b)** 250 N **9. a)** same times, since $\Delta t = I/\Delta p$; **b)** A stops in a shorter distance. **11.** 4 s **13.** 249 m/s **15.** 4.4×10^{-23} kg·m/s upward **17.** 2 m/s **19.** 67 kg **21.** 1×10^5 m/s **23.** 15 m/s south 63.4° above the horizontal, 167 J **25. a)** 288 N/m; **b)** –2 m/s; **c)** 36 J; **d)** 36 J; **e)** elastic. No change in KE; **f)** 1.6s **27.** 5.4 kg·m/s 42° above the horizontal **29. a)** 6×10^4 kg·m/s north; **b)** 6×10^4 kg·m/s north. The system is isolated, so momentum is conserved; **c)** 10 m/s north; **d)** 3×10^5 J; **e)** 25 m **31. a)** 4 m/s; **b)** +2 m/s; **c)** elastic collision **33. a)** 5 m/s; **b)** 1 m/s right; **c)** 25 J; **d)** 9 J; **e)** inelastic, but not maximally inelastic (the two blocks do not have the same final velocity) **35.** $v_2 = -2$ m/s, $v_4 = 4$ m/s

Chapter 9

1. 1.75 × 10^{-3} rad/s, into the wall **3. a)** 10 s; **b)** 37.7 rad/s out of the page **5.** 239 rev/min, to the rider's left **7. a)** 10 s; **b)** 25 rad/s; **c)** 12.5 m/s **9. a)** 127 revs; **b)** 100 rad/s^2 **11.** $\omega = \sqrt{(Mg)/(mr)}$ **13.** 33.6 cm from the end of the racket handle **15.** $x = 1, y = ½$ **17.** 3.33 cm **19.** 150 N **21. a)** 40 cm, 30 N; **b)** 40 cm, 30 N **23.** 200 N, 100 N **25.** 500 N, 600 N **27. b)** 500 N; **c)** 900 N down **29.** 9000 N **31. a)** 8 cm from the ball (32 cm from the bottom); **b)** 7.46 × 10^{-3} kg·m^2 **33. a)** 2.36 rad/s; **b)** 3.77 m; **c)** The pulley is still turning clockwise so ω is into the page; **d)** ω is getting less, so α is in the opposite direction, out of the page; **e)** The torque by the string is into the page. The net torque is out of the page. So the hand torque must be out of the page and greater than the string torque. **35. a)** −0.34 N and +9.3 × 10^{-3} N·m; **b)** −2 m/s^2 and +174 rad/s^2; **c)** 0.43 s; **d)** 2.1 m/s **37. a)** 2.12 × 10^{29} J; **b)** 3 million years **39.** 2.58 m/s and 272 rad/s

Chapter 10

1. 2 × 10^5 N, 2 × 10^5 N **3.** 13.0 m **5.** Air is much less dense than water, so letting air out of your lungs reduces the buoyant force more than the loss of the air reduces the force of gravity. **7.** 180 m **9.** 3000 N **11. a)** 0.5 N; **b)** 50 cm^3 and 16 g/cm^3; **c)** yes **13. a)** ρ_{green} = (600 g)/(800 cm^3) = 0.375 g/cm^3. The green object floats. ρ_{yellow} = (600 g)/(400 cm^3) = 1.5 g/cm^3. The yellow object sinks; **b)** B_{green} = 6 N, B_{yellow} = 4 N; **c)** a_{green} = 0, a_{yellow} = 3.33 m/s^2; **d)** 540 kg **15.** 2.5 cm **17. a)** The blue object floats in water and sinks in Z, so Z must be less dense than water; **b)** i: 25 g, ii: 30 cm^3, iii: 0.833 g/cm^3; **c)** the buoyant force is greater on the red object than on the blue; **d)** greater in water than in Z **19. a)** 30 m/s^2; **b)** 4800 cm^3; **c)** 12 N; **d)** ¾ **21. a)** 15 N; **b)** 9 N; **c)** 8.75 m/s^2; **d)** 233 cm^3; **e)** 5.5 N **23.** 133 N and 117 N

Chapter 11

1. c is in stable equilibrium, a is in unstable equilibrium, b is in neutral equilibrium, d is not in equilibrium. **3.** b, the same. The period depends only on k and m, neither of which are different on Mars. **5. a)** 200 N/m; **b)** 50 N/m **7. a)** ⅓ Hz; **b)** 26.3 N/m **9. a)** 5 s; **b)** 0.2 Hz; **c)** 12 m; **d)** 118 N/m **11.** 1.16 s **13. a)** 1.49 s; **b)** 445 N/m **15. b)** 3 m/s^2; **c)** 1.28 J **17. a)** 2 × 10^5 J; **b)** 20 m; **c)** 30 m/s; **d)** 50 m/s^2; **e)** There is no loss of energy from ejecting the rider and the energy A is always $½kA^2$, so the amplitude is the same. The period is $2\pi\sqrt{m/k}$, so it is less with the loss of mass. The energy is the same, but it is equal to $½mv^2$ at the equilibrium point, so the velocity must increase to compensate for the loss of mass. **19.** 0.3 m **21. a)** 80 J; **b)** 0.4 m **23.** 2.45 s **25.** 8 s **27. a)** 3.5 J; **b)** 2.3 rad/s **29.** 2.1 s **31. a)** $\alpha = 2(k/m)\theta$; **b)** $t = 2\pi$, $T = 2\pi\sqrt{m/(2k)}$

Chapter 12

1. 25.8 m/s **3.** 8 ms **5.** ½ **9. a)** The left side is steel (narrower pulse) and the right side is brass; **b)** The return pulse and the transmitted pulse are inverted, so the bounce was off of the slower segment (the steel) and the pulse began on the right; **c)** [your picture should show a left-moving down-pulse in the brass segment of the spring] **11. a)** The stretched string will put forces on the molecules both left and right of the pulse; **b)** 0.4 cm high, 10 cm wide **13.** 1.7 cm, 17 m **15.** 2¼ m/s **17.** (a). As the wave moves right, the new height of each segment of the stretched spring will be the height of the segment to its left. Thus B is going down, A is going up. **19.** 476 m/s **21.** 222 Hz. The choices are 218 Hz and 222 Hz, but tightening the string raises the frequency. **23.** The choices are 246 Hz and 254 Hz for the first case and 246 Hz and 264 Hz for the second case. This leaves only 246 Hz. **25.** 0.57 m **27.** Bill's. Elaine running away lowers her beeper frequency. Since this increases the beat rate, Elaine's beeper was already lower in frequency. **29.** 6.8 cm/s **31.** 30 m/s, 67 mi/hr **33.** $\Delta\lambda/\lambda = \sqrt{1 - 2(u/v)\cos\theta + u^2/v^2} - 1$ or $\Delta\lambda/\lambda = -(u/v)\cos\theta$ in the limit $u \ll v$

Chapter 13

1. 16 cm **3.** 16 m/s **5.** between 385 Hz and 400 Hz **7.** 204 Hz **9. a)** a node; **b)** 50 m; **c)** 100 m/s; **d)** 6000 N **11.** No. In air $\lambda = v/f$. If f is known, you still need to know the speed of sound in air to find λ. This v_{AIR} = is not the same as the speed of the wave along the string, so the wavelength of the wave on the string is not the wavelength of the sound wave in air. **13.** 25 cm **15. a)** 43 Hz; **b)** 2 m **17.** 1.14 m, open-closed pipe **19.** 584 Hz **21. a)** 436 Hz; **b)** 4 beats **23.** 283 Hz, 567 Hz **25. a)** parallel; **b)** in-between **27.** 12.5 cm, 37.5 cm, 62.5 cm **29.** 13.3 Hz

Chapter 14

1. 104° F **3.** 1447 K **5.** 8.3×10^{-12} J, 4×10^{11} K **7.** 190 m/s (not that different from 300 m/s) **9.** 11 mm **11.** 635° C **13.** 2.1 cm^3 **15.** 4.186 J **17.** 500 J/(kg·K) **19.** 32.6 K **21.** 13° C **23.** 0° C **25.** 0.925 kg **27.** 5.3×10^4 N

Chapter 15

1. 1.5×10^{23} atoms **3.** 12.42 J/mole **5.** The 5 gallons of water have 4×10^7 J/gal, while the 100,000-gallon pool has 5×10^6 J/gal. The number of molecules per gallon are the same for both quantities of water, so the 5 gallons has a higher internal energy per molecule, and is at the higher T. **7.** 4.8×10^{25} molecules **9.** 0.79 atm or 7.9×10^4 Pa **11.** 3.7×10^5 Pa (~69 lb/in^2) **13.** 90 g remain and 10 g leaked out **15.** 2.5×10^{22} $molecules/m^3$ **17.** 60.5 N **19.** 3.8 cm^3 **21.** 280 K **23.** 56% **25.** 33.2 J/mole **27.** 20.8 J/(mole·K), 5.9 mins **29. a)** 12.5 J/(mole·K); **b)** 3120 J; **c)** 0; **d)** 3120 J **31. a)** 20.8 J/(mole·K), 29.1 J/(mole·K); **b)** 1455 J lost; **c)** −1040 J; **d)** 415 J; **e)** 0.415 m^3 **33. a)** $T_A = 361$ K, $T_B = 722$ K, $T_C = 240$ K, $T_D = 120$ K; **b)** A→B: 3000 J, B→C: 0, C→D: −1000 J, D→A: 0; **c)** A→B: 7500 J, B→C: −6000 J, C→D: 2500 J, D→A: 3000 J; **d)** 2000 J; **c)** 10500 J; **f)** 19%

Chapter 16

1. 6.25×10^6 C **3.** Both coins end up with equal net positive charges. Charge depends on the net number of elementary charges on the coins, not on the ratio of net charge to the total number of elementary charges in the material. **5. (d).** The electric force is subject to Newton's third law. The forces will be equal and opposite **7.** 9 mN **9.** 3.33×10^{-10} C **11.** 6.25 N **13.** 9.2×10^{-8} N **15.** Negative charge on the rod repels electrons in the can to the far side of the can and leaves a positive charge on the near side of the can. The attractive force on the nearby positive charge is greater than the repulsive force on the far-away electrons. **17.** 1.17 μN, 60° above the horizontal **19.** 2.4×10^{-6} C **21.** 9.375×10^3 N/C south **23.** 2×10^{-9} N **25.** 1.2×10^3 N/C away from B **27.** 1.7×10^5 V/m **29.** 3.88 m **31. a)** zero; **b)** 600 N/C, left; **c)** +0.67 nC **33.** The final vector will point exactly to the right. **35.** Point A: 12.9×10^3 N/C along the line from B to A, Point B: 13.5×10^3 N/C along the line from B to A. **37. a)** a vector toward the left; **c)** The field due to −Q is not changed. An induced negative charge near the field point adds more field to the left, while the induced equal positive charge produces field to the right. The induced negative charge is closer to the field point, so the net field is increased to the left. **41. b)** negative. The particle accelerates in the opposite direction as the field. **45.** d. There is initially a uniform field to the right, due to both plates. When the rightmost plate is removed, the remaining field is still uniform to the right, but it is only that produced by the leftmost plate alone. **47. a)** The field at A is a uniform $\sigma/2\varepsilon_0$ upward, plus a uniform σ/ε_0 downward, for a net field $\sigma/2\varepsilon_0$ downward; **b)** The field at B is a uniform $\sigma/2\varepsilon_0$ downward, plus an additional σ/ε_0 down. The fields here combine to give $3\sigma/2\varepsilon_0$, a larger field than that at A; **c)** The field at A due to the $+\sigma$ sheet alone is $\sigma/2\varepsilon_0$ upwards. So the direction changes but the magnitude remains the same. **49. a)** The field on the outside cap is uniform outward, since the points on the top surface are all close to the surface. The field on the inside cap is zero, because that cap is inside a conductor. The field along the sides is zero inside the conductor and parallel to the surface for points outside and near the conductor; **b)** $E = \sigma/\varepsilon_0$; **c)** The field due to a sheet of uniform surface charge density σ is $\sigma/2\varepsilon_0$. This sheet produces an outward field of this size immediately to the right of the surface and an inward field of the same size just to the left of the surface. So, if the field must be zero on the inner cap, it must be that the field from all the rest of the conductor surface is equal and opposite (rightward) so that it cancels the field from the nearby charge. The location of the outer cap is very close to that of the inner cap, so the rightward field from the nearby surface and that from the rest of the whole surface will be acting in the same rightward direction. **51.** Symmetry requires that the electric field be uniform over the Gaussian sphere and directed everywhere radially outward. Gauss's law gives $E = \dfrac{Qr}{4\pi\varepsilon_0 R^3}$.

Chapter 17

1. 1.6×10^{-19} J **3.** 2.0×10^{-5} J **5.** Point B is 200 V higher than Point A. A positive charge going from A to B will gain potential energy, so an electron going from A to B will lose potential energy. It will therefore gain kinetic energy and its speed will increase. **7.** 5.36×10^5 m/s **9.** 4×10^{-16} J **11. a)** 5×10^4 V/m; **b)** 4.43×10^{-7} C/m^2 **13.** 2000 V/m **15. a)** Field lines to the left between the plates, none elsewhere; **b)** 6 V; **c)** 12 V/m; **d)** -9.6×10^{-19} J **17. a)** 375 V; **b)** the positive plate is at the higher voltage; **c)** 6×10^{-17} J; **d)** No. The proton KE is only 3.34×10^{-17} J, not enough to reach the positive plate. **19. a)** field lines to the right between the plates, none elsewhere; **b)** 7500 V/m; **c)** 3.2×10^7 m/s; **d)** No. If the electron only gets halfway before stopping, its initial KE must be 3000 eV. As long as the plates remain connected to the 6000-V power supply, the potential difference between the plates will be 6000 V and the 3000-eV electron will still only get halfway. **21.** 4800 kg **23. b)** point A is at the higher potential **25.** 60 μF **27.** 17.7 pF **29.** 310 pF **31.** 53 μm **33.** 330 nF **35 a)** 18.6 pF; **b)** 2×10^4 V/m **37.** 0.67 J **39. a)** 42 V; **b)** 3.78×10^{-3} J

Chapter 18

1. 3.2 mA to the left **3.** 150 mA **5.** Parallel, 12 V **7.** 6 **9.** 360 Ω **11.** 5.6×10^{21} electrons **13.** 818 C, 2.45×10^3 J **15.** 0.758 **17.** 15 Ω **19. a)** 200 V; **b)** 10^4 Ω **21.** 40 V across the 5 Ω, 9 V across the 3 Ω, 5 V across the 1 Ω **23.** When the switch is closed, the overall resistance decreases, so the current from the battery increases and bulb A brightens. All of the current now goes through the wire rather than through B, so bulb B is dark.

25. The brightness depends on the current flowing through each bulb, so $I_A < I_B < I_C$. More current will flow through the branch with the lesser resistance, so $R_1 < R_2 < R_3$. In arrangement 2, $R_{1\&3} < R_1$ and we already know that $R_1 < R_2$, so $R_{1\&3} < R_2$. Thus $I_{1\&3} > I_2$, so A is brighter than B. In arrangement 3, $R_{1\&3} > R_3 > R_2$, so $I_{1\&3} < I_2$, and B is brighter than A. **27.** $I_d = I_c > I_b = I_a$ **29. a)** B = C < D < A = E; **b)** All of the current from the battery goes through A and E. This current must divide to go through B, C, and D, so they will be dimmer; **c)** A gets dimmer. Opening the switch removes a path, increasing the total resistance of the circuit. The current through the battery decreases and so does the current through A; **d)** The current is proportional to the voltage drop across any bulb. Since A gets dimmer, the current through it is less, so the voltage drop across A is smaller; **e)** D gets brighter. In both cases $V_A + V_D + V_E$ must equal the same battery voltage. If V_A and V_E decrease, V_D must increase and the current through D must increase. **31.** 18 Ω, 9 Ω, 4 Ω, 2 Ω **33. a)** 18 Ω; **b)** 4 Ω **35. a)** 6 Ω; **b)** 3 A; **c)** 1 A **37. a)** The current from the battery divides into the 1-Ω branch and the 3-Ω branch. The current through the 1-Ω resistor divides at A. The current in the top 2-Ω resistor divides to return to the battery and go through the 4-Ω. The currents at B combine to produce the current through the 3-Ω; **b** $I_1 = I_{2,mid} + I_{2,top}$, $I_{2,mid} + I_4 = I_3$, $-2I_{2,top} - 4I_4 + 2I_{2,mid} = 0$, $-1I_1 - 2I_{2,mid} - 3I_3 = 0$, $46 - 1I_1 - 2I_{2,top} = 0$; **c)** $I_{2,mid} = 1$ A **39. a)** When the switch is closed, current starts to flow, but it has not yet built up any charge on the capacitor, so the voltage drop across the capacitor is zero; **b)** The ΔV across the capacitor is zero, so the current flow will favor the path with one bulb (less resistance). B is brighter; **c)** When the capacitor is fully charged, no more current will flow through it. All current will flow through A, C, and D. Bulb B will go out; **d)** Bulbs A, C, and D are identical and the voltage drop across all of them will equal the battery voltage. So $V_A + V_C + V_D =$ 12 V means that $V_A = 4$ V. No current flows through B, so $V_B = 0$ V. The voltage across the capacitor equals the sum of that across C and D, or 8 V; **e)** 4 C

Chapter 19

1. horizontal east **3.** 1.7×10^{-15} N, vertically downward **5.** B is into the page, so F_B is to the left. **7.** 3.6×10^{-11} N **9. a)** 8×10^{-16} N downward; **b)** 8.78×10^{14} m/s² down; **c)** 3 ns; **d)** 4 mm **11. a)** 3.2×10^{-12} N; **b)** 4.8×10^{14} m/s²; **c)** 0.83 m **13. a)** 31 nm; **b)** 65 μs; **c)** 260 m; **d)** 8.4 **15.** $v = E/B$ **17.** 0.03 N **19.** 2×10^{-3} N into the page **21.** C. Put your right thumb into the page and see the direction of your fingers at the field point. **23.** up **25.** d. The leftward field at A due to the $2I$ current is bigger than the downward field due to I. **29.** No net force. The leftward forces on A below B (toward the bottom of the page from B) are each canceled by rightward forces on A above B. But these forces will produce a net torque, rotating wire A clockwise in the plane of the page. **31. b)** counterclockwise **33.** 4.56×10^{-4} T into the tabletop **35.** 12 T **37.** 6×10^{-8} T, out of the page **39.** 5×10^{-4} T, to the right and rising at 37° out of the page. **41. a)** $C_B = 2\pi RB$; **b)** $I_{CUTTING} = NI$; **c)** $B = (\mu_0 NI)/(2\pi R)$; **d)** An Ampère's-law circle outside the toroid would have as many wires carrying current out of the page as there are wires carrying current into the page. The total $I_{CUTTING}$ would be zero, making $B = 0$.

Chapter 20

1. 8×10^{-7} Wb **3. a)** From $F = qvB$, we get units N = (c)(m/s)(T), or T = (N·C)/(m·s). A newton is N = (kg·m)/s², so T = Kg/(C·s). From $\Phi = BA$, we get Wb = T·m², or Wb = (kg·m²)/(C·s); **b)** Wb/s = (kg·m²)/(C·s²) = J/C = V. **5. a)** 1.5 mA; **b)** 7.5 μN **7.** 0.15 V

9. Counterclockwise. As current begins to flow, the B-field inside the loop is into the page and increasing. A counterclockwise current in the loop would produce a B-field out of the page, countering the increasing flux into the page. **11.** Clockwise. The B-field from the magnet is downward through the loop. As the wire loop falls, it gets further from the magnet, so the B-field weakens. A clockwise current in the loop would create a downward B-field, counteracting the weakening downward field from the magnet. **13. a)** 0.5 V, in the direction of the arrow; **b)** 0.5 V, still in the direction of the arrow. **15. c)** The current is clockwise in the loop. Since the B-field from the solenoid through the loop is out of the page, current will flow in the loop so as to counter the increasing flux by creating a B-field into the page. **17. a)** The north pole of the magnet goes into the solenoid first; **b)** The needle deflects to the left. [The current I creates a B-field up. The solenoid will respond by creating a B-field down $\Rightarrow$ clockwise current $\Rightarrow$ needle to the left.]; **c)** Needle deflects right. [The initial flux from the primary is up. As it is turned off, a current will flow in the secondary to keep an upward B-field. $\Rightarrow$ Counterclockwise current in the secondary. $\Rightarrow$ Needle to the right.] **19.** 7.3 km **21.** 2400 turns **23.** $L = (\mu_0 N^2 r^2)/(2R)$ **25.** 344 V. A is at a higher voltage than B. **27. a)** 4.77×10^8 V/m; **b)** Clockwise when looking along the current, from left to right; **c)** 2×10^{-5} T **29. a)** $\varepsilon = BDv \cos\theta$; **b)** $I = BDv \cos\theta / R$; **c)** $F = B^2 D^2 v \cos\theta / R$; **d)** $v = (mgR \sin\theta)/(B^2 D^2 \cos^2\theta)$ **31.** 4.2×10^{-11} T **33. a)** As the loop turns, the right-hand segment of the loop moves into the page. There will be a downward magnetic force on positive charges in the wire. The left-hand segment is coming out of the page, and so its positive charges experience an upward force. This produces a clockwise *emf* in the loop at this initial time. As the loop turns, it goes from no flux through the loop (in the picture) to positive flux toward the right. The induced *emf* creates a magnetic flux to the left, in accordance with Lenz's law; **b)** Consider the tiny segment of the loop on the extreme right. The magnetic field at that point has a component upward and a component to the right. As the loop moves down, the upward component will produce no force on charges in the wire, but the rightward component will create a force out of the page on positive charges. Symmetry around the circle means that an *emf* around the loop arises to create a current that creates a magnetic field downward. This counters the increasing upward magnetic flux created as the loop falls, in accordance with Lenz's law.

Chapter 21

1. 2.97 m **3.** The mirror should be 3 feet high, hung with the top of the mirror half-way between the man's eyes and the top of his head. The bottom will then be half-way between his eyes and his feet. **5.** 35° **7.** 2.17×10^8 m/s **9.** 229 nm, 5.4×10^{14} Hz **11. a)** 0° (the beam is unbent by the glass); **b)** 20.7° **13. a)** 50 ps; **b)** 0.5 cm = 5 mm **15.** 49.4° **17.** 2.26 m **19.** 41.8° **21.** 19.5° **23.** 62.4°, the ray must start in the glass **25.** The critical angle for glass in air is 41.8°. This means that any ray that enters the flat end, even one coming in 90° away from the normal to the flat face, will enter the glass 41.8° or less away from the cylinder axis. If it eventually hits the curved side of the cylinder, it will hit at an angle 90° – 41.8° = 48.2° or greater to the normal to this side. This is greater than the critical angle, so it will be completely internally reflected. If the reflection then takes it to the other side of |the curved cylinder wall, it will still make this angle of 48.2° or more to the normal to that side. Thus, the angle of incidence to the normal to the curved side of the cylinder is always greater than the critical angle. Only when it reaches the far flat end will it again make an angle 41.8° or less with the normal to this surface and finally be able to emerge from the glass.

27. 400 nm: 50.26°, 800 nm: 49.38°, the 400-nm violet light is bent the most. **29. a)** 5360 W; **b)** 2037 W; **c)** 710 V/m **31.** $f = R/(n - 1)$

Chapter 22

1. A baseball. Curvature is defined as $K = 1/R$, so the baseball, with its smaller radius, has the greater curvature **3.** Each point of the wave in the pinhole serves as a source for a forward spherical wave. The center of the wave should be close to flat and the edges should be curved. **5. a)** $+10\ m^{-1}$; **b)** ∞; **c)** $-2.5\ m^{-1}$ **7.** +22 D **9.** a virtual image 20 cm to the left of the lens **11.** +1.67 D, a converging lens **13. a)** $5\ m^{-1}$; **b)** 10 cm; **c)** 2 cm **15. a)** 5 cm; **b)** inverted; **c)** 1.6 cm **17. a)** erect; **b)** inverted; **c)** 9 cm **19.** +5 D **21. a)** inverted, 15 cm from the mirror; **b)** erect at ∞; **c)** erect, 90 cm behind the mirror **23. a)** virtual image 10 cm behind the mirror; **b)** 1.6 mm, erect **25.** 80 cm **27.** The final image is virtual an infinite distance to the left. **29.** 20 cm to the right of the 10-cm lens, real, erect, and 12 mm high **31.** 8.57 cm, real, inverted **33.** 0.025 rads, 0.075 rads **35.** 200X **37.** 12 cm **39. a)** the 2 D lens; **b)** 55 cm, **c)** 10X **41. a)** a final image at ∞; **b)** erect; **c)** 5X **43.** 17 cm

Chapter 23

1. 627.2 nm **3.** 300 nm **5. a)** 0.11459°; **b)** 0.11460°; **c)** $\approx$0.01% **7.** 0.029° **9.** 2.86° **11.** 3.44° **13.** Each bright fringe will remain at the same position, but will become half as wide. **15.** 3rd order red, 4th order green **17.** 0.5 nm **19.** 733 nm **21. a)** 8 cm; **b)** 2 cm **23.** 30° **25. a)** 1.43°; **b)** 1.43°; **c)** No. Each slit produces no light at 1.43°, since this is a null of the single-slit diffraction. Constructive interference from the different slits can therefore add nothing, since the amplitude from each slit is already zero. **27.** Galileo: 0.0010°, Hubble: $1.6 \times 10^{-5\circ}$ **29.** Same. Resolution depends only on the size of the aperture. The image in the higher-power binoculars will be bigger, but the details will be fuzzier. **31.** 4 times brighter. The amount of energy entering the telescope is proportional to the area of the opening, which is proportional to the square of the radius of the opening. All light from the star will be focused to a single point, so that point will be 4 times brighter in the bigger telescope. **33.** As the bubble thickness shrinks to zero, there is no phase delay between the reflections from the front and back surfaces. However, there will be a ½-cycle change at the first surface and not at the second, so the two reflected beams will be out of phase and interfere destructively. **35.** 386 nm. There is a ½-cycle phase change at both surfaces, so the path length must be a whole wavelength to keep the two reflected beams in phase. A coating of zero thickness is not possible. **37.** Visible wavelengths are 621 nm. 483 nm, and 395 nm. **39.** Light from the first surface has a ½-cycle phase change. Light from the second surface has no phase change at the reflection, but has a 9/2-cycle phase change from the path length. Both beams will thus be in phase, producing a bright reflection. **41.** 111 nm of TiO_2, 193 nm of MgF_2. **43.** 60° **45.** 42% **47.** The intensity in the second case will be 3 times brighter than the first case. The final polarizations will be the same in the two cases.

Chapter 24

1. 9.95 yrs **3.** 0.6 $c \cdot$s. Since the clocks are moving to the west, as seen by the Romulans, clocks to the east will be trailing, each one ahead of the one in front of it. **5.** At the instant the back of the barge passes me, I see the front of the barge at 60 m. The light from that point took $(60\ m)/(3 \times 10^8\ m/s) = 0.2\ \mu s$ to reach me. Between the time when the light left the

front of the barge and now, the front of the barge has moved 20 m. The length of the barge is 80 m. **7.** 289 ns **9.** The 300-m length between the bulb and the mirror will be contracted to 240 m. The light moves to the right at 3×10^8 m/s and the mirror moves left at 1.8×10^8 m/s, so the distance between the light and the mirror will decrease at 4.8×10^8 m/s. Thus, the time required to cover the 240 m to the mirror is 0.5 μs. After 0.5 μs, the light bulb will be $(1.8 \times 10^8 \text{ m/s})(0.5 \times 10^{-6} \text{ s}) = 90$ m to the left of the front of the barge, so the mirror that is 240 m to the right of it will be 150 m to the right of the front of the barge. **11.** [Hint: Solve for the velocity at which the time offset between the emission and reception events can be seen from a moving frame as being due to simultaneous events measured by offset clocks.]
13. a) $-0.1\ \mu$s; **b)** $-0.2\ \mu$s **15.** $\Delta t' = \sqrt{(\Delta t)^2 - (\Delta x/c)^2}$ **17.** 33.3 s **19.** 3.33 μs
21. a) $\Delta t' = \lambda/\sqrt{c^2 - v^2}$; **b)** $x_1 = c\lambda/\sqrt{c^2 - v^2}$, $x_2 = v\lambda/\sqrt{c^2 - v^2}$; **c)** $\lambda' = \lambda\sqrt{(c + v)/(c - v)}$
23. a) $\Delta t' = [\Delta t + (v/c)\Delta x]/\sqrt{1 - v^2/c^2}$; **b)** $\Delta x' = [\Delta x + v\Delta t]/\sqrt{1 - v^2/c^2}$ **25.** Yes. The difference in 1 s is 1.1×10^{-16} s. $\Delta t_{TOP} > \Delta t_{BOTTOM}$. The top clock records more time, meaning the clock on the floor is running slow. **27. a)** $4/5c$; **b)** 0 **29.** $0.946c$, or 2.84×10^8 m/s
31. a) $v_{\|}' = v$; **b)** $h' = h$, $\Delta t' = \Delta t/\sqrt{1 - v^2/c^2}$; **c)** $v_{\perp}' = v_{\perp}\sqrt{1 - v^2/c^2}$ **33.** 4.33×10^9 kg/s and 1.5×10^{13} yrs **35.** $E_{rest} = 3.10 \times 10^{-8}$ J, $E = 2.19 \times 10^{-5}$ J **37.** Relativistic: 2.25×10^{16} J, Newtonian: 1.62×10^{16} J **39. a)** 3.44 μs; **b)** 250 m **41. a)** 3.9×10^{30} kg; **b)** 9.6×10^7 m/s and 4.8×10^7 m/s **43. a)** $E = 1.36 \times 10^{-9}$ J, KE $= 0.77 \times 10^{-9}$ J; **b)** 7.4×10^{-9} K

Chapter 25

1. 1.1 V **3.** 1.03 eV **5.** Assuming $\lambda = 500$ nm, $N = 1.5 \times 10^{20}$ **7. a)** 3.97×10^{-19} J/photon, 2.52×10^{18} photons/s; **b)** $p = 1.32 \times 10^{-27}$ kg·m/s, impulse $= 2.65 \times 10^{-27}$ kg·m/s; **c)** 6.67×10^{-9} N **9.** 410.6 nm **11. a)** 248 nm, 413 nm, 620 nm; **b)** 413 and 620 are visible, 248 is UV **13. a)** 10.2 eV; **b)** 79,000 K **15.** 4.2 pm **17.** $n = 5.7 \times 10^{34}$
19. a) 5.40×10^{-24} kg·m/s; **b)** 123 pm; **c)** 2.35° **21.** 10 **23.** (1, 0, 0, –), (1, 0, 0, +), (2, 0, 0, –), (2, 0, 0, +), (2, 1, –1, –), (2, 1, 0, –), (2, 1, 1, –), (2, 1, 1, +), (2, 1, 0, +), (2, 1, 1, +)
25. a) 3.0×10^{35} electrons/m^3; **b)** 1.5 pm; **c)** 3.5×10^{-23} kg·m/s; **d)** 6.8×10^{-16} J; **e)** 3.3×10^7 K

Chapter 26

1. 5.6×10^{-11} g **3.** 10^6 V proton: $E = 939$ MeV, $pc = 43.3$ MeV, $\lambda = 28.6$ fm. 10^{10} V proton: $E = 10.9$ GeV, $pc = 10.9$ GeV, $\lambda = 0.114$ fm **5.** 92.1 MeV **7.** 173.1 MeV **9.** 0.782 MeV **11.** 2.29×10^9 decays per second **13.** 1.4 billion years **15. a)** μ-lepton number (+1 before, –1 after); **b)** baryon number (+1 before, 0 after); **c)** charge (–1 before, +2 after); **d)** energy (105.7 MeV before, >139.6 MeV after) **17.** $Q = +1$, $B = 0$, $S = +1$. It is a K^+. **19.** $Q = +1$, $S = -1$, $C = +2$ **21.** ~10^{-30} eV **23.** $v = (pc)^2/E$ **25. a)** 109.778 MeV; **b)** 29.792 MeV; **c)** 29.792 MeV; **d)** 0; **e)** a muon neutrino

PHOTO AND FIGURE CREDITS

Front Cover. Charlie Chaplin in *Modern Times* (© Roy Export SAS), used with permission, all rights reserved.

Chapter 1. *Figure 1.1,* Photograph courtesy of the BIPM; *Figure 1.2,* Used with permission from AdobeStock, all rights reserved.

Chapter 2. *Problem 2.21,* Used with permission from AdobeStock, all rights reserved.

Chapter 9. *Example 9.9,* Courtesy of Joseph Shaw.

Chapter 12. *Figure 12.4,* Courtesy of Educational Innovations, www.teachersource.com.

Chapter 13. *Figure 13.1,* Courtesy of Joseph Shaw; *Figure 13.11,* Used with permission from AdobeStock, all rights reserved.

Chapter 20. *Figure 20.3,* Used with permission from AdobeStock, all rights reserved; *Problem 20.8,* Used with permission from NASA.

Chapter 23. *Figure 23.1,* Redrawn from Thomas Young's diagram of double slit interference presented to the Royal Society (1803).

Chapter 25. *Figure 25.5,* Reproduced with permission from *Quantum Chemistry,* 2nd Edition, by Donald McQuarrie, published by University Science Books, ©2008, all rights reserved.

INDEX

D

E

L

M

X

Y

Z